NANOPARTICLE TECHNOLOGY HANDBOOK

NANOPARTICLE TECHNOLOGY HANDBOOK

Edited by

Masuo Hosokawa
Hosokawa Micron Corporation
Shoudai Tajika, Hirakata
Osaka, Japan

Kiyoshi Nogi
Osaka University, Joining and
Welding Research Institute
Ibraki, Osaka, Japan

Makio Naito
Osaka University, Joining and
Welding Research Institute
Ibraki, Osaka, Japan

Toyokazu Yokoyama
Hosokawa Micron Corporation
Shoudai Tajika, Hirakata
Osaka, Japan

ELSEVIER

Amsterdam • Boston • Heidelberg • London • New York • Oxford • Paris
San Diego • San Francisco • Singapore • Sydney • Tokyo

Elsevier
Linacre House, Jordan Hill, Oxford OX2 8DP, UK
Radarweg 29, PO Box 211, 1000 AE Amsterdam, The Netherlands

First edition 2007

Library of Congress Cataloging in Publication Data
A catalog record is available from the Library of Congress

British Library Cataloguing in Publication Data
A catalogue record is available from the British Library

ISBN: 978-0-444-53122-3

Printed and bound in The Netherlands

07 08 09 10 11 10 9 8 7 6 5 4 3 2 1

Preface

During the last few years the term "Nanotechnology" is increasingly employed to describe the process technologies and analytical techniques for material in the ultrafine range of the order of a millionth of a millimeter. Because they are sure to take an important part in shaping the 21st century, great attention is being paid to these technologies, with many countries actively involved in R&D. As the link between these new technologies and the established particle and powder technology, "Nanoparticle technology" includes the concepts and know-how to create, process and apply the ultrafine particles in the nanometer range, and is one of the key technologies for new material developments.

The technologies that are used to treat powders arrived with mankind, and countless inventions and improvements have been made during history. These particles and powders have very different properties from the bulk materials from which they are derived. There are applications to be seen in all industrial areas.

The history of the academic study of particle and powder technology is not so old. The first related society, Chubu Association of Powder Technology, was founded in Japan in 1956. It later became the Society of Powder Technology, Japan, and celebrated its 50th anniversary in 2006. Correspondingly, the Hosokawa Micromeritics Laboratory was established in 1956 and published its 50th anniversary issue of the annual technical journal *Funsai* (*The Micromeritics*) also in 2006.

Throughout this period a key issue has been to reduce the size of particles to maximise their functional properties and thus find new applications and create new products with superior performance. Great interest has been shown in submicron and even finer particles. Research and development has advanced at a rapid rate due to the cooperation of academia and industry in many areas, starting with particle creation and particle size analysis, and expanding to encompass particle design and processing in the micron- and nanometer-size ranges. Japan has been at the forefront in the conception and development of these technologies.

Due to this interest, the second World Congress in Particle Technology (WCPT) was held in Kyoto in 1990. Eight years later at the 3rd WCPT in Brighton, the author highlighted the importance of these ultrafine particles to an audience of about 700 researchers and engineers during the opening speech. Hosokawa Micron Corp., which celebrated its 90th anniversary in 2006, has been engaged in R&D on particle creation by the build-up (*bottom up*) method in both gaseous and liquid phases for more than 20 years. The result of this research, as combined with that on conventional grinding (*top-down*) processes, has led to the establishment of a mass production system for nanoparticles and to the foundation of a business based on application of these nanoparticles to functional materials.

Founded 15 years ago, the Hosokawa Powder Technology Foundation holds an annual symposium on powder technology for the exchange of information on particle engineering and powder technology. Since 2001, the main topics of the symposium have, in response to the requirements of industry, been related to nanoparticles and nanostructure control. The number of grant proposals received by our Foundation for research into nanoparticles continues to increase, and currently 40% of some 120 proposals relate to nanoparticles.

As a result of this trend, we published 3 years ago, the book *Nanoparticle Technology* to promote nanoparticle-related engineering by documentating the technologies constituting in this field. That book was very well received, and to continue contributing to the common welfare through the promotion of powder technology, we decided to systematically update *Nanoparticle Technology*, adding further developments and many examples of applications. The results of that effort were published in the form of a handbook, first in Japanese in the memorable year 2006, and with the present volume, in English this year. Although R&D in nanoparticle technology advances rapidly, and the contents of the future editions are sure to change, we hope the present collation of state-of-the-art knowledge and information will be of assistance to the researchers, engineers and others interested in this vitally important field.

In closing, I express my sincere sense of gratitude to the authors, the editing committee and the publishing staff for their great efforts in spite of their busy schedules.

Masuo Hosokawa
President, Hosokawa Micron Corporation
Chairman, Hosokawa Powder Technology Research Institute
President, Hosokawa Powder Technology Foundation

List of Contributors

Hiroya Abe, Dr.
Joining & Welding Research Institute, Osaka University

Tadafumi Adschiri, Dr.
Institute of Multidisciplinary Research for Advanced Materials, Tohoku University

Takashi Akatsu, Dr.
Materials and Structures Laboratory,
Tokyo Institute of Technology

Jun Akedo, Dr.
National Institute of Advanced Industrial Science and Technology (AIST)

Masanori Ando, Dr.
National Institute of Advanced Industrial Science and Technology (AIST)

Yoshinori Ando, Dr.
Department of Materials Science and Engineering,
Meijo University

Masanobu Awano, Dr.
National Institute of Advanced Industrial Science and Technology (AIST)

Akira Azushima, Dr.
Graduate School of Engineering, Yokohama National University

Tetsuya Baba, Dr.
National Institute of Advanced Industrial Science and Technology (AIST)

Kensei Ehara, Dr.
National Institute of Advanced Industrial Science and Technology (AIST)

Hitoshi Emi, Dr.
Association of Powder Process Industry and Engineering (APPIE)

Hiroshi Fudouzi, Dr.
Optronic Materials Center,
National Institute for Materials Science

Hiroshi Fukui, Dr.
Frontier Science Business Division, Shiseido Co., Ltd.

Takehisa Fukui, Dr.
Hosokawa Powder Technology Research Institute

Yoshinobu Fukumori, Dr.
Faculty of Pharmaceutical Sciences,
Kobe Gakuin University

Masayoshi Fuji, Dr.
Ceramics Research Laboratory, Nagoya Institute of Technology

Hidetoshi Fujii, Dr.
Joining & Welding Research Institute, Osaka University

Hideki Goda Mr.
R&D Department, Photo-electronic Materials Division,
Arakawa Chemical Industries, Ltd.

Kuniaki Gotoh, Dr.
The Graduate School of Natural Science and Technology,
Okayama University

Yukiya Hakuta, Dr.
National Institute of Advanced Industrial Science and Technology (AIST)

Kaori Hara Ms.
Hosokawa Powder Technology Research Institute

Kazuyuki Hayashi, Dr.
R&D Division, Toda Kogyo Corp.

Ko Higashitani, Dr.
Department of Chemical Engineering,
Kyoto University

Kazuyuki Hirao, Dr.
Department of Material Chemistry, Kyoto Univerisity

Masuo Hosokawa Mr.
Hosokawa Micron Corp.

Yuji Hotta, Dr.
National Institute of Advanced Industrial Science and Technology (AIST)

Hideki Ichikawa, Dr.
Faculty of Pharmaceutical Sciences
Kobe Gakuin University

Takashi Ida, Dr.
Ceramics Research Laboratory, Nagoya Institute of Technology

Manabu Ihara, Dr.
Research Center for Carbon Recycling Energy, Tokyo Institute of Technology

Shinji Inagaki, Dr.
Toyota Central R&D Labs., Inc.

Mitsuteru Inoue, Dr.
Toyohashi University of Technology

Eiji Iritani, Dr.
Department of Chemical Engineering,
Nagoya University

Naoyuki Ishida, Dr.
National Institute of Advanced Industrial Science and Technology (AIST)

Mikimasa Iwata, Dr.
Central Research Institute of Electric Power Industry

Kotaro Kajikawa, Dr.
Tokyo Institute of Technology

Toshio Kakui, Dr.
Chemicals Division, Lion Corp.

Hidehiro Kamiya, Dr.
Institute of Symbiotic Science and Technology,
Tokyo University of Agriculture & Technology

Kenji Kaneko, Dr.
Department of Material Science and Engineering,
Kyushu University

Kiyoshi Kanie, Dr.
Institute of Multidisciplinary Research for Advanced Materials, Tohoku University

Hitoshi Kasai, Dr.
Institute of Multidisciplinary Research for Advanced Materials, Tohoku University

Tomoko Kasuga, Dr.
Electrotechnology Applications R&D Center,
Chubu Electric Power Co. Inc.

Tsutomu Katamoto Mr.
Creative R&D Center, Toda Kogyo Corp.

Shinji Katsura, Dr.
Faculty of Engineering,
Gunma University

Masayoshi Kawahara Mr.
Hosokawa Powder Technology Research Institute

Yoshiaki Kinemuchi, Dr.
National Institute of Advanced Industrial Science and Technology (AIST)

Soshu Kirihara, Dr.
Joining & Welding Research Institute,
Osaka University

Akihiko Kondoh, Dr.
Department of Chemical Science and Engineering,
Kobe University

Katsuyoshi Kondou, Dr.
Joining & Welding Research Institute,
Osaka University

Yasuo Kousaka, Dr.
Hosokawa Powder Technology Research Institute

Kazue Kurihara, Dr.
Institute of Multidisciplinary Research for Advanced Materials, Tohoku University

Shun'ichi Kuroda, Dr.
The Institute of Scientific and Industrial Research,
Osaka University

Ken-ichi Kurumada, Dr.
Graduate School of Environment & Information Sciences,
Yokohama National University

Chunliang Li, Dr.
National Institute of Advanced Industrial Science and Technology (AIST)

Hisao Makino, Dr.
Energy Engineering Research Laboratory,
Central Research Institute of Electric Power Industry

Hiroaki Masuda, Dr.
Department of Chemical Engineering, Kyoto University

Yoshitake Masuda, Dr.
National Institute of Advanced Industrial Science and Technology (AIST)

Motohide Matsuda, Dr.
Graduate School of Environmental Science,
Okayama University

Tatsushi Matsuyama, Dr.
Faculty of Engineering, Soka University

Shuji Matsusaka, Dr.
Department of Chemical Engineering,
Kyoto University

Reiji Mezaki, Dr.
Nanomateria Center, Institute of Innovation,
The University of Tokyo

Takeshi Mikayama, Dr.
Kohno Patent Office

Kiyotomi Miyajima Mr.
Central Research Institute of Electric Power Industry

Minoru Miyahara, Dr.
Department of Chemical Engineering,
Kyoto University

Yoshinari Miyamoto, Dr.
Joining & Welding Research Institute,
Osaka University

Masaru Miyayama, Dr.
Research Center for Advanced Science and Technology,
The University of Tokyo

Hideki T. Miyazaki, Dr.
National Institute for Materials Science

Hidetoshi Mori, Dr.
School of Engineering, Aichi University of Technology

Tsutomu Morimoto, Dr.
Japan Chemical Innovation Institute

Atsushi Muramatsu, Dr.
Institute of Multidisciplinary Research for Advanced Materials, Tohoku University

Norio Murase, Dr.
National Institute of Advanced Industrial Science and Technology (AIST)

Toshihiko Myojo, Dr.
Institute of Industrial Ecological Sciences
University of Occupational and Environmental Health

Makio Naito, Dr.
Joining & Welding Research Institute, Osaka University

Noriyuki Nakajima Mr.
Institute of Nanotechnology, Kurimoto, Ltd.

Hachiro Nakanishi Mr.
Institute of Multidisciplinary Research for Advanced Materials, Tohoku University

Masami Nakamoto, Dr.
Osaka Municipal Technical Research Institute

Norikazu Namiki, Dr.
Kyoritsu Gokin Co., Ltd.

Kiyoshi Nogi, Dr.
Joining & Welding Research Institute, Osaka University

Yuji Noguchi, Dr.
The University of Tokyo

Toshiyuki Nomura, Dr.
Department of Chemical Engineering, Osaka Prefecture University

Satoshi Ohara, Dr.
Institute of Multidisciplinary Research for Advanced Materials, Tohoku University

Akira Ohtomo, Dr.
Institute for Materials Research, Tohoku University

Hidetoshi Oikawa, Dr.
Institute of Multidisciplinary Research for Advanced Materials, Tohoku University

Tomoichiro Okamoto, Dr.
Nagaoka University of Technology

Tatsuya Okubo, Dr.
The University of Tokyo

Kikuo Okuyama, Dr.
Graduate School of Engineering, Hiroshima University

Yoshio Otani, Dr.
Graduate School of Natural Science and Techonology, Kanazawa University

Yasufumi Otsubo, Dr.
Graduate School of Engineering, Chiba University

Fumio Saito, Dr.
Institute of Multidisciplinary Research for Advanced Materials (*IMRAM*), Tohoku University

Shuji Sakaguchi, Dr.
National Institute of Advanced Industrial Science and Technology (AIST)

Yoshio Sakka, Dr.
Nano Ceramics Center, National Institute for Materials Science

Takafumi Sasaki, Dr.
Institute of Multidisciplinary Research for Advanced Materials, Tohoku University

Norifusa Satoh, Dr.
Department of Chemistry, Keio University

Haruhide Shikano Mr.
Ibiden Co., Ltd.

Manabu Shimada, Dr.
Graduate School of Engineering, Hiroshima University

Tetsuya Senda, Dr.
National Maritime Research Institute

Yuichi Setsuhara, Dr.
Joining and Welding Research Institute, Osaka University

Akihiko Suda, Dr.
Toyota Central R&D Labs., Inc.

Hisao Suzuki, Dr.
Graduate School of Science and Technology, Shizuoka University

Michitaka Suzuki, Dr.
Department of Mechanical and System Engineering, University of Hyogo

Takahiro Takada, Dr.
Murata Manufacturing Co., Ltd.

Seiichi Takami, Dr.
Advanced Electronic Materials Center, National Institute for Materials Science

Hirofumi Takase, Dr.
R&D Division, Takiron Co., Ltd.

Hirofumi Takeuchi, Dr.
Laboratory of Phamaceutical Engineering, Gifu Pharmaceutical University

Junichi Tatami, Dr.
Graduate School of Environment & Information Sciences, Yokohama National University

Kenji Toda, Dr.
Graduate School of Science and Technology,
Niigata University

Hiroyuki Tsujimoto, Dr.
Hosokawa Powder Technology Research Institute

Tetsuo Uchikoshi, Dr.
Nano Ceramics Center,
National Institute for Materials Science

Keizo Uematsu, Dr.
Nagaoka University of Technology

Mitsuo Umetsu, Dr.
Graduate School of Engineering,
Tohoku University

Arimitsu Usuki, Dr.
Toyota Central R&D Labs., Inc.

Fumihiro Wakai, Dr.
Materials & Structures Laboratory,
Tokyo Institute of Technology

Akimasa Yamaguchi, Mr.
Energy Engineering Research Laboratory,
Central Research Institute of Electric Power Industry

Yukio Yamaguchi, Dr.
Department of Chemical System Engineering,
The University of Tokyo

Atsushi Yamamoto, Mr.
National Institute of Advanced Industrial Science and Technology (AIST)

Hiromitsu Yamamoto, Dr.
University School of Pharmacy,
Aichi Gakuin University

Kenji Yamamoto, Dr.
International Clinical Research Center,
International Medical Center of Japan

Kimihisa Yamamoto, Dr.
Department of Chemistry, Keio University

Masatomo Yashima, Dr.
Department of Materials Science and Engineering,
Tokyo Institute of Technology

Toyokazu Yokoyama, Dr.
Hosokawa Micron Corp.

Susumu Yonezawa, Dr.
Faculty of Engineering, University of Fukui

Table of Contents

FUNDAMENTALS

Chapter 1 Basic properties and measuring methods of nanoparticles

Chapter 2 Structural control of nanoparticles

Chapter 3 Characteristics and behavior of nanoparticles and its dispersion systems

Chapter 4 Control of nanostructure of materials

Chapter 5 Characterization methods for nanostructure of materials

Chapter 6 Evaluation methods for properties of nanostructured body

Chapter 7 Environmental and safety issues with nanoparticles

APPLICATIONS

From the Editors

As the size of a solid particle decreases in the order of one millionth of a millimeter, the number of atoms constructing the particle becomes small and in the order of several hundreds or thousands. At this state, the fundamental physical property such as the melting point can change drastically and ceramic materials may be sintered at a lower temperature. Also, as particles get smaller than the wavelength of visible light, they not only become transparent but also emit special light by plasma absorption. They show completely different electromagnetic or physicochemical properties from their bulk counterparts, although they are made of the same materials.

The authors published a book *Nanoparticle Technology* in Japanese in November 2003, which focused on the technology of handling nanoparticles that have unique properties and enormous potential usefulness. This book has drawn great attention from the readers and a growing demand to publish a handbook has developed, which systematically collects the basic information on nanoparticle technology with recent industrial applications.

Nanoparticle Technology to prepare, process, and apply nanoparticles plays a very important role in the development of nanotechnology. It also pays attention to various applications like life sciences, energy, environment, information technology, new materials, etc. However, there has been no handbook or manual on this technology so far. This is the first handbook written in English for handling nanoparticles and surveying their related processing technologies. It has been long awaited by researchers and engineers interested in nanoparticles or their use in the R&D of advanced materials.

This handbook systematically summarizes the fundamentals and state-of-the-art information in various industrial applications related to nanoparticles. However, since the advancement in the fields of concern is so rapid, not only the application developments but also the new physical properties and measuring methods from fundamental research become available as time goes by. Therefore, we plan to revise the contents of the handbook according to new technology developments in the future.

This handbook consists of fundamental and application sections including processing, evaluation, and application in a way different from other similar conventional handbooks. In the fundamental section, the basic properties, structural control of nanoparticles, nanostructural control, and property characterization with the measuring methods in the dispersed particle system are elucidated in detail mainly from the aspects of material processing and property evaluation. At the end of Fundamental Section, a chapter discussing the environmental and safety impact of nanoparticles is also included.

In the Application Section, various nanoparticle applications in the fields of life sciences, environment, energy, information technology, new materials, and production methods are listed according to their future market potential with focus on the new functionalities of nanoparticles.

To publish this handbook, we invited manuscripts from leading researchers and engineers specialized in a broad range of applications of concern as shown in the list of contributors. We would like to thank all the authors who contributed manuscripts despite their busy schedules and our colleagues in Hosokawa overseas operations as well as the staffs of the publisher for their generous supports. We are also deeply indebted to Dr. Y. Tsuji, Managing Director of Hosokawa Powder Technology Foundation and Dr. C. C. Huang of Hosokawa Micron Powder Systems, who gave us many useful comments on the English manuscripts, and to Ms. S. Nakai for her assistance in the preparation of the manuscript and proof.

Dr. Kiyoshi Nogi
Professor and Director, Joining and Welding Research Institute, Osaka University
Director, Hosokawa Micron Corporation
Director, Hosokawa Powder Technology Research Institute

FUNDAMENTALS

FUNDAMENTALS

CHAPTER 1

BASIC PROPERTIES AND MEASURING METHODS OF NANOPARTICLES

CONTRIBUTORS

Toyokazu Yokoyama (1.1)
Hiroaki Masuda (1.2)
Michitaka Suzuki (1.3)
Kensei Ehara (1.4)
Kiyoshi Nogi (1.5)
Masayoshi Fuji (1.6), (1.9)
Takehisa Fukui (1.7)
Hisao Suzuki (1.8), (1.11)
Junichi Tatami (1.10)
Kazuyuki Hayashi (1.12)
Kenji Toda (1.13)

1.1 Size effect and properties of nanoparticles

1.1.1 Definition of nanoparticles

The nanoparticles are ultrafine particles in the size of nanometer order. "Nano" is a prefix denoting the minus 9th power of ten, namely one billionth. Here it means nanometer (nm) applied for the length. One nm is extremely small length corresponding to one billionth of 1 m, one millionth of 1 mm, or one thousandth of 1 μm.

The definition of nanoparticles differs depending upon the materials, fields and applications concerned. In the narrower sense, they are regarded as the particles smaller than 10–20 nm, where the physical properties of solid materials themselves would drastically change. On the other hand, the particles in the three digit range of nanometer from 1 nm to 1 μm could be called as nanoparticles. In many cases, the particles from 1 to 100 nm are generally called as nanoparticles, but here they will be regarded as the particles smaller than those called conventionally "submicron particles", and concretely less than the wavelength of visible light (its lower limit is about 400 nm) as a measure, which need to be treated differently from the submicron particles. Figs 1.1.1 and 1.1.2 show examples of nanoparticles and the related phenomena and an example of electron microscope picture of gold colloid nanoparticles made by liquid phase chemical method [1], respectively.

1.1.2 Features of nanoparticles

(1) Activation of particle surface

All the solid particles consist of the atoms or the molecules. As they are micronized, they tend to be affected by the behavior of atoms or the molecules themselves and to show different properties from those of the bulk solid of the same material. It is attributable to the change of the bonding state of the atoms or the molecules constructing the particles. For example, as shown in Fig.1.1.3, if a cube with a side length of 1 cm is divided into a cube of 1 μm, the particle number increases to 10^{12} and being divided into the one of 10 nm, then it amounts to 10^{18}, where the fraction of the atoms or the molecules located at the surface on the particles plays a great role, since they are more active than those inside the solid particles because of the free hand, which leads to easy bonding with the contacting materials and causes various changes in particle properties.

The diameter of the smallest hydrogen atom is 0.074 nm, and that of the relatively large lead atom (atomic number is 82) is 0.35 nm. From these sizes, it is estimated that the particle with a size of 2 nm consists of only several tens to thousands atoms. When the particle is constructed by larger molecules, the number decreases furthermore. Table 1.1.1 shows the relationship between the particle size and the fraction of number of atoms located at the particle surface with an assumption of atomic distance of 0.2 nm [2]. It is indicated that the fraction of surface atoms of a 20 μm cubic particle is only 0.006%, but it increases to 0.6% for a 200 nm particle and then it is estimated almost half of the atoms are situated at the surface of a 2 nm particle.

(2) Increase of surface area

On the other hand, as the micronization of solid particles, the specific surface area increases generally in reversal proportion to the particle size. In the above-mentioned case, when the particle of 1 cm is micronized to 1 μm and 10 nm, the specific surface area becomes ten thousand times and million times, respectively. As the increase in the specific surface area directly influences such properties like the solution and reaction rates of the particles, it is one of major reasons for the unique properties of the nanoparticles different from the bulk material together with the change in the surface properties of the particles itself.

1.1.3 Evaluation of size of nanoparticles

In order to elucidate the change in properties and characteristics of nanoparticles with the particle size, it is essential first of all to measure the size of the nanoparticles accurately. The most basic method to measure the size of nanoparticles is the size analysis from the picture image using the transmission electron microscope, which could also give the particle size distribution. For this analysis, preparation of the well-dispersed particles on the sample mount is the key issue. The grain size of the particles can be obtained from peak width at half height in the X-ray diffraction analysis and it is regarded as an average primary particle size of particles.

Meanwhile, the laser diffraction and scattering method, which is popular for the size analysis of micron-sized particles, would hardly measure the particle size of individual nanoparticles but that of the agglomerated particles. The photon correlation method often used for the particle analysis in the nanosized range might not give accurate results in many cases, when the particle size distribution is wide.

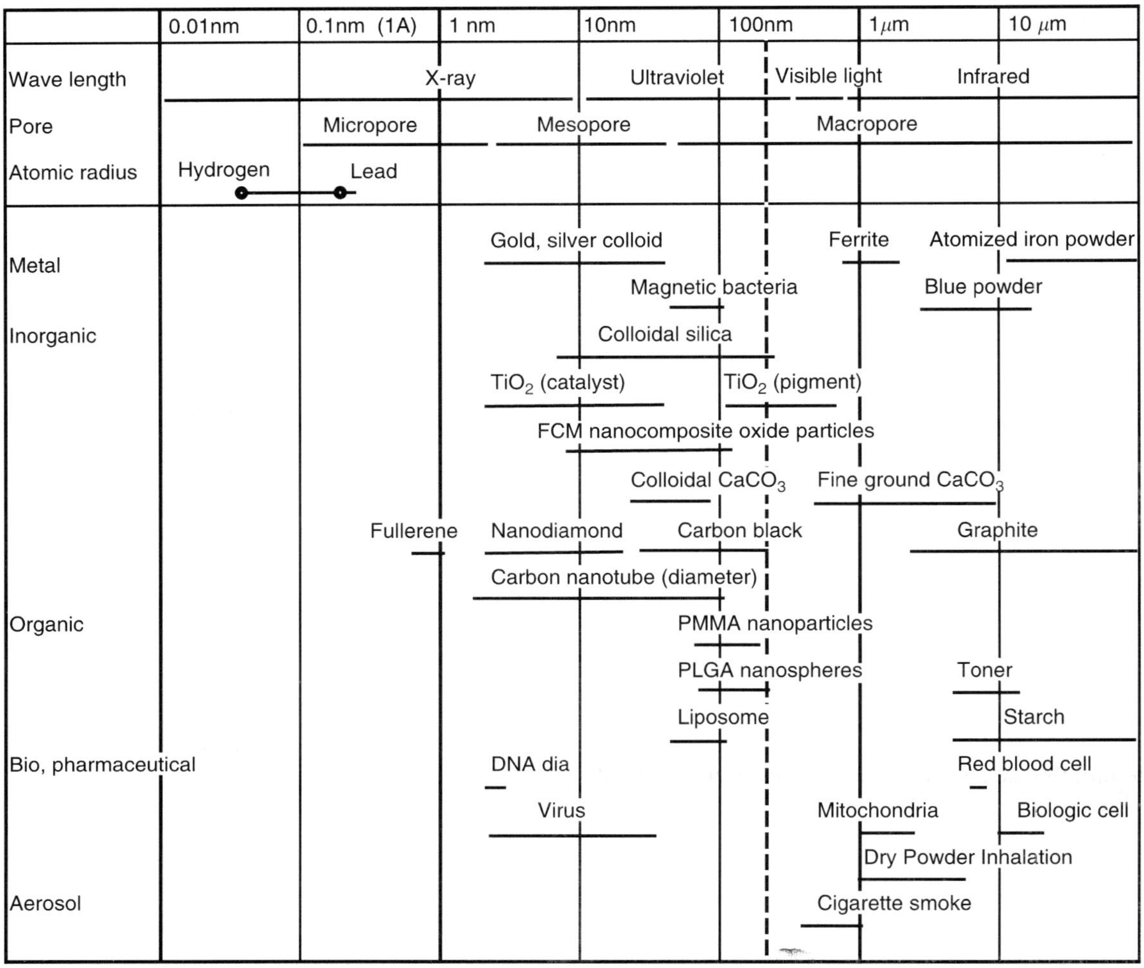

FCM : Flash Creation Method
PLGA : Poly-lactic-glycolic acid (Spherical crystallization method)
PMMA : Polymethylmethacrylate

Figure 1.1.1
Examples of nanoparticles and related phenomena.

Then the BET (Brunauer-Emmett-Teller) specific surface measurement based on the gas adsorption is often applied as a simple method to evaluate the size of nanosized primary particles.

By this method, it is possible to estimate the particle size from the specific surface area under the assumption of spherical particle shape. This equivalent particle size based on the specific surface area is useful for the evaluation of nanoparticle size, though it may differ from the particle size observed by the electron microscope depending upon the surface state and the inner structure of the particles.

1.1.4 Properties of nanoparticle and size effect

As mentioned above, with the decreasing particle size, the solid particles generally tend to show different properties from the bulk material and even the physical properties like melting point and dielectric constant themselves which have been considered as specific properties may change, when the particles become in several nanometer size. These changes in the fundamental properties with the particle size is called "size effect" in a narrower sense.

On the contrary, in a broader sense, it could also include the change in the various characteristics and

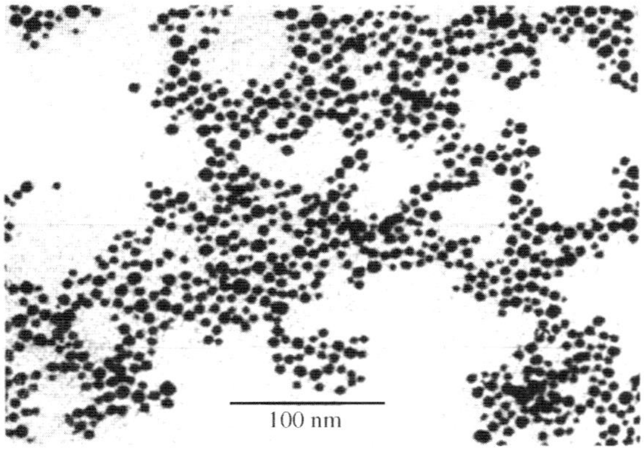

Figure 1.1.2
Transmission electron microscope picture of gold nanoparticles.

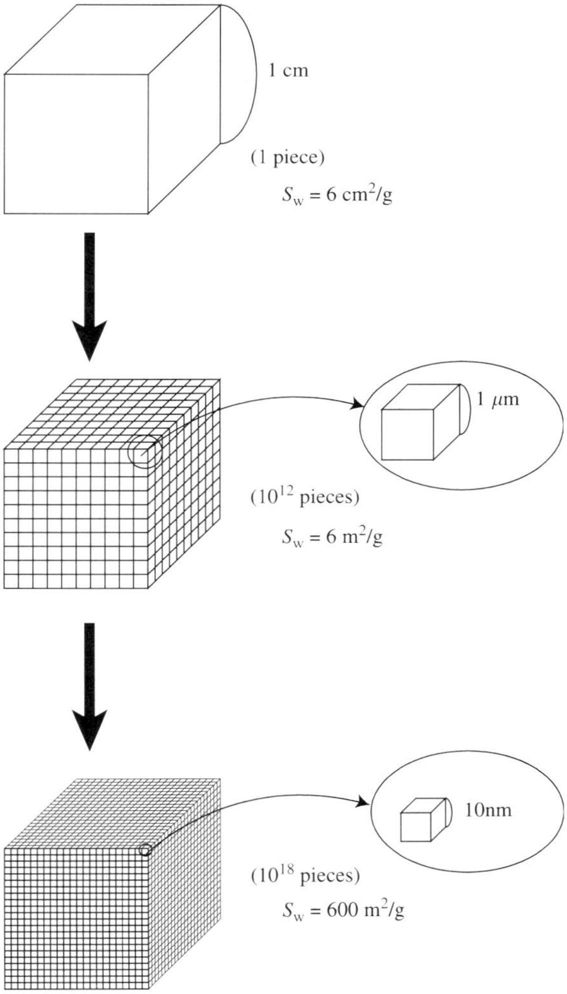

Figure 1.1.3
Change of specific surface area by miniaturization of a solid cube assuming the solid density of 1 g/cm^3.

behaviors of particles and powders with the particle size. The nanoparticles have various unique features in the morphological/structural properties, thermal properties, electromagnetic properties, optical properties, mechanical properties as described briefly in the following:

(1) Morphological/structural properties

The ultrafine size of the nanoparticles itself is one of useful functions. For example, the finer particles are apt to be absorbed more easily through the biological membrane. It is known as the enhanced permeation and retention (EPR) effect [3] that the particles having a particle size from about 50 to 100 nm, which would not be transferred to the normal cells through the vascular wall could be delivered selectively to a certain affected cells because of the enlarged cell gap of this part.

As mentioned above, the large specific surface area of the nanoparticles is an important property to the reactivity, solubility, sintering performance etc. related with the mass and heat transfer between the particles and their surroundings from the morphological viewpoint apart from the control of the surface and inner structures of the nanoparticles. Furthermore, the crystal structure of the particles may change with the particle size in the nanosized range in some cases.

Uchino et al. [4] reported that from the X-ray diffraction analysis of the lattice constant of $BaTiO_3$ powder prepared by hydrothermal synthesis method, the *c*/*a* axis length ratio showing the tetragonal characteristics decreased to indicate the increasing symmetric property with the decreasing particle size from about 200 nm as shown in Fig.1.1.4. This is considered to be attributable to the compressive force exerted on the particles as a result of the surface tension of the particle itself. For $PbTiO_3$, it is reported that the tetragonal crystals decreased and the cubical crystal increased in the particles from the particle size of about 18 nm [5]. In this way, the critical particle size for the crystal structure and the size effect differ with the materials concerned.

(2) Thermal properties

As the atoms and molecules located at the particle surface become influential in the nanometer order, the melting point of the material decreases from that of the bulk material because they tend to be able to move easier at the lower temperature.

For example, the melting point of gold is 1336 K as a bulk but starts to decrease remarkably below the particle size of about 20 nm and drastically below 10 nm and then becomes more than 500 degrees lower

Table 1.1.1
Solid particle size and the fraction of atoms located at the particle surface.

Number of atoms in a side	Number of atoms at the surface	Total number of atoms	Number ratio of surface atoms to the total (%)	Examples of particle size and powder
2	8	8	100	
3	26	27	97	
4	56	64	87.5	
5	98	125	78.5	
10	488	1,000	48.8	2 nm
100	58,800	1×10^{6}	5.9	20 nm (colloidal silica)
1,000	6×10^{6}	1×10^{9}	0.6	200 nm (titanium dioxide)
10,000	6×10^{8}	1×10^{12}	0.06	2 μm (light calcium carbonate)
100,000	6×10^{10}	1×10^{15}	0.006	20 μm (green tea powder, chalk)

1 m = 1×10^{6} μm = 1×10^{9} nm

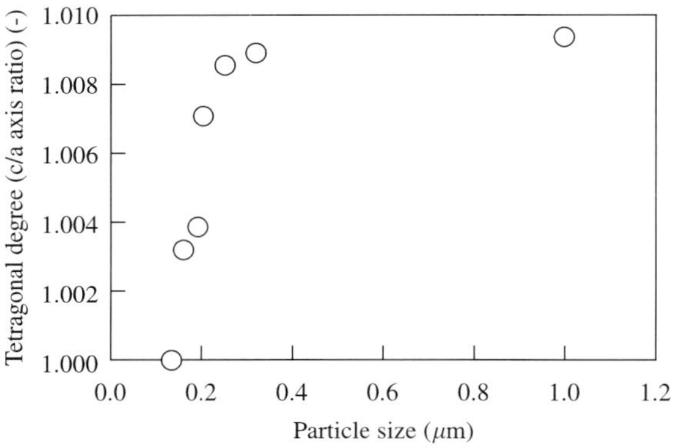

Figure 1.1.4
Relationship between particle size and tetragonal degree (c/a axis ratio) of $BaTiO_3$ powder.

than that of the gold bulk around 2 nm. The reduction of the melting point of ultrafine particles is regarded as one of the unique features of the nanoparticles related with aggregation and grain growth of the nanoparticles or improvement of sintering performance of ceramic materials [6].

(3) Electromagnetic properties

The nanoparticles are used as the raw material for a number of electronic devices. The electric properties and particle size of these nanoparticles play a great role for the improvement of the product performance [7]. As an example, there is a strong demand for the materials with a high dielectric constant to develop small and thin electronic devices. For this purpose, it has been confirmed by the X-ray diffraction analysis for instance that the dielectric constant of $PbTiO_3$ tends to increase considerably as the particles become smaller than about 20 nm. Meanwhile, it is also known that when the dielectric constant is measured with a pellet prepared by pressing these nanoparticles, it shows a peak with the raw material around 100 nm and decreases with the decreasing particle size, which is attributable to the influence of the grain boundary and void in the pellet [8].

On the other hand, the minimum particles size to keep the ferroelectric property (critical size) differs depending upon the kind and composition of the materials. According to the literature [9] summarizing the data of various kinds of materials, it varies from 7 nm for $PbTiO_3$ to 317 nm for Ba–Pb–Ti compounds. The Curie point defined as the point changing from the ferroelectric material to the paraelectric phase of $PbTiO_3$ reduces drastically with the decreasing particle size below 20–30 nm as shown in Fig. 1.1.5. As for the Curie point, some equations have been proposed for its estimation [10].

As for the magnetic property, ferromagnetic fine particles have a single magnetic domain structure as they become very small as in the order less than about 1 μm and show super-paramagnetic property, when they get further finer. In this case, although the individual particles are ferromagnetic with the single magnetic domain structure, the particles collectively behave as a paramagnetic. It is magnetized as a whole in the same direction of the external magnetic field but the magnetization disappears by the thermal fluctuation, when the external magnetic field is taken away. The time for disappearing of magnetization depends upon the particle size, namely the magnetization of the material responds promptly with the external magnetic field as a paramagnetic when the particles are small enough but it decreases gradually as the particle size becomes larger. As a result of such change in the electromagnetic properties of nanoparticles, it is known for instance that the gold which is a stable substance as a bulk shows unique catalytic characteristics as nanoparticles [11].

(4) Optical properties

As the size of particles becomes in the several nanometers range, they absorb the light with a specific wavelength as the plasmon absorption [12] caused by the plasma oscillation of the electrons and the transmitted light with different color depending upon the kind of metal and particle size is obtained [13].

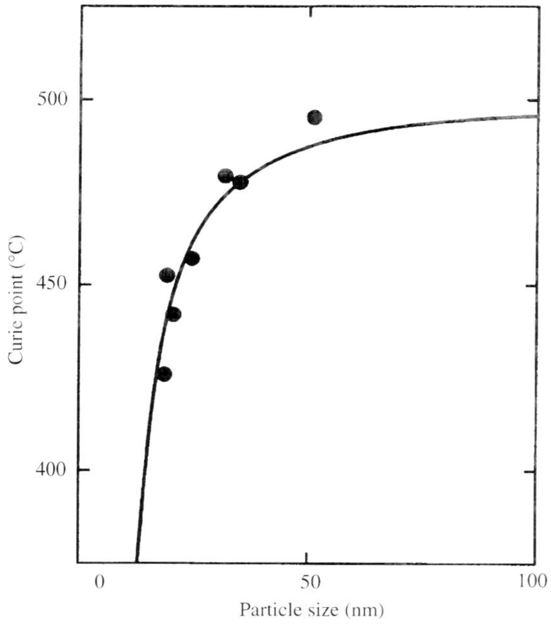

Figure 1.1.5
Change of Curie point $PbTiO_3$ with its particle size.

Fig. 1.1.6 shows the plasmon absorption of silver nanoparticles, where the spectral absorption intensity differs depending upon the particle size which is determined by the concentration of the surfactant used for their preparation [14]. In case of gold nanoparticles, it is reported that the maximum light absorption wavelength is 525 nm for the particles of 15 nm but it is enlarged by about 50 nm for 45 nm particles. In this way, these gold and silver nanoparticles show the color phenomena with splendid tinting strength, color saturation and transparency compared with the conventional pigments for the paint in the submicron size and the tinting strength per unit volume of silver nanoparticles becomes about 100 times higher than that of organic pigments. Furthermore, since the nanoparticles are smaller than the wavelength of visible light and the light scattering by the particles becomes negligible, higher transparency can be obtained with the nanoparticles than the conventional pigment.

On the other hand, concerning the light emitting performance, the indirect transition type substances like silicon and germanium, which do not emit the light as bulk material, give high light emitting efficiency as the direct transition type substances as a result of quantum effect, when the particle size is reduced down to several nanometers.

(5) Mechanical properties

It is known that the hardness of the crystalline materials generally increases with the decreasing crystalline size, and that the mechanical strength of the materials considerably increases by micronizing the structure of the metal and ceramic material or composing them in the nano range [15, 16]. Furthermore, with the ceramic

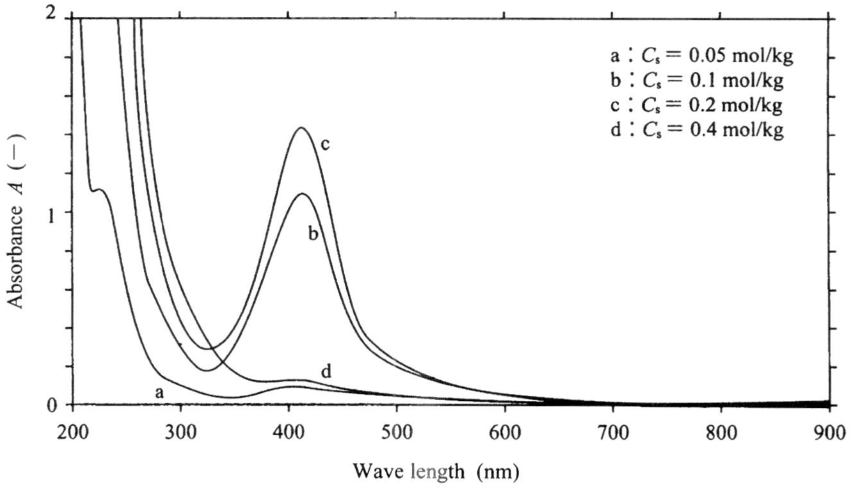

Figure 1.1.6
Effect of concentration of nonionic surfactants C_s on the plasmon absorption of silver nanoparticles. C_s = (a) 0.05, (b) 0.1, (c) 0.2, (d) 0.4 mol/kg.

material having crystalline size less than several hundred nanometers, the unique super-plastic phenomenon is seen that it is extended several to several thousand times from the original size at the elevated temperature over 50 % of the melting point [17], which may provide the possibility of forming and processing of ceramics like metallic materials.

1.1.5 Existing conditions of particles and their properties

The nanoparticles usually exhibit collective functions. Therefore, the dispersing state and the surrounding conditions in addition to the physical properties of the particles themselves are important. In many cases, the nanoparticles exist as aggregates of the primary particles by the adhesion and bonding during the production process because of their high adhesiveness.

The existing state of the nanoparticles is greatly influenced by the surrounding conditions if they are in gas, liquid, solid or in a vacuum and what sort of interaction they have with the surrounding materials. The nanoparticles are rarely used by themselves but dispersed in other materials or combined with them. The dispersing process of the nanoparticles is a key for the nanoparticle technology as well as their preparation methods, since the performance of the final products are affected by their dispersing conditions [18].

In this way, it is expected with great possibility to develop various new materials and applications by the nanoparticle technology producing and processing the nanoparticles, which have different properties from the bulk material by the size effects as mentioned above and in the following sections.

References

[1] M. Arakawa: *J. Soc. Powder Technol., Jpn.*, **42**, 582–585 (2005).
[2] M. Arakawa: *Funsai (The Micrometrics)*, No.27, 54–64 (1983).
[3] H. Maeda: *J. Control. Release*, **19**, 315–324 (1992).
[4] K. Uchino, E. Sadanaga and T. Hirose: *J. Am. Ceram. Soc.*, **72**(8), 1555–1558 (1989).
[5] H. Suzuki, T. Ohno: *J. Soc. Powder Technol., Jpn*, **39**, 877–884 (2002).
[6] N. Wada: Chem. Eng., **9**, 17–21 (1984).
[7] I. Matsui: *J. Chem. Eng., Jpn*, **38** (8), 535–546 (2005).
[8] M. Takashige, T. Nakamura: *Jpn. J. Appl. Phys.*, **20**, 43–46 (1981).
[9] K. Ishikawa: *J. Soc. Powder Technol., Jpn*, **38**, 731–740 (2001).
[10] K. Ishikawa, K. Yoshikawa and N. Okada: *Phys. Rev. B*, **37**, 5852–5855 (1988).
[11] M. Haruta: *Catalysts*, **36** (6) 310–318 (1994).
[12] Y. Kurokawa, Y. Hosoya: *Surface*, **34** (2) 100–106 (1996).
[13] K. Kobayashi: *J. Soc. Powder Technol., Jpn*, **41**, 473–478 (2004).
[14] S. Sato, N. Asai and M. Yonese: *Colloid Polym. Sci.*, **274**, 889–893 (1996).
[15] K. Niihara: *J. Ceram. Soc. Jpn*, **99** (10), 974–982 (1991).
[16] T. Sekino: *Mater. Integr.*, **13** (11) 50–54 (2000).
[17] F. Wakai, Y. Kodama, S. Sakaguchi, N. Murayama, K. Izaki and K. Niihara: *Nature*, **344**, 6265, 421–423 (29 March 1990).
[18] T. Yokoyama: *Sokeizai*, **3**, 6–11 (2005).

1.2 Particle size

Particle size is the most important information in practical applications of powder-particles. Usually, powder is constituted by particles of various sizes and, therefore, it is necessary to obtain not only the mean particle size but also the size distribution for the characterization. Recently, the methods for particle size analysis have been greatly developed. Especially, the analyzers with prominent characteristics such as rapid response, high repeatability and covering wide range of particle size are developed as in the case of laser scattering and diffraction method.

1.2.1 Definition of particle size

A particle is usually three-dimensional and it may take various shapes. "Particle size" is a term to represent the three-dimensional particle in one-dimensional scalar value. The size of any spherical particle can be represented by its diameter with no ambiguity. For a particle with irregular shape, the size is represented by a geometrically obtained one-dimensional scalar value, geometric size, or an equivalent size in relation to practical methods of particle size measurements.

The geometric size is obtained through three-dimensional measurements of a particle to get its width, thickness and length, and then calculating one-dimensional value such as arithmetic mean. In practice, however, one-dimensional value obtained based on the two-dimensional-projected silhouette is utilized such as a diameter of a circle having the same area as the projected area. Statistical diameter based on one-dimensional measurement is also well applied in practice such as a Feret diameter, which is determined as the distance between pairs of parallel tangents to the particle silhouette in some fixed direction.

As for the equivalent size in relation to practical methods of particle size measurements, there are many different definitions such as sieve diameter based on sieving, equivalent light-scattering diameter, Stokes diameter based on particle motion in fluid, and the equivalent diameter based on the Brownian motion. These equivalent diameters give, usually, different values depending on the measurement principles unless the particles are spherical. Specific surface area of powder or equivalent-specific surface diameter is well applied to the evaluation of nanopowders. However, in this case, the particles should not be porous. Besides, the method cannot be applied to get particle size distribution [1, 2].

1.2.2 Measuring methods

As discussed in the preceding section, geometric size is obtained in the microscopic measurements, and equivalent size is applied in the measurements by practically available instruments such as those based on laser diffraction and scattering, dynamic light scattering, differential mobility analysis, and so on. The equivalent size obtained by these instruments is practical but does not give a value directly connected to the unit of length.

The items discussed in the ISO relating to the particle size measurements are as follows:

1. ISO 13317: Determination of particle size distribution by gravitational liquid sedimentation methods (This ISO corresponds to JIS Z 8820).
2. ISO 13318: Determination of particle size distribution by centrifugal liquid sedimentation methods (JIS Z 8823).
3. ISO 13319: Determination of particle size distribution – Electrical-sensing zone method.
4. ISO 13320: Particle size analysis–Laser diffraction methods (JIS Z 8825).
5. ISO 13321: Particle size analysis – Photon correlation spectroscopy (JIS Z 8826).
6. ISO 13322: Particle size analysis–Image analysis methods.
7. ISO 13323: Determination of particle size distribution – Single particle light interaction methods.
8. Particle size analysis – Dynamic light scattering (DLS).
9. Determination of particle size distribution – Differential electrical mobility analysis for aerosol particles (DMA).
10. Particle characterization by acoustic methods – Ultrasonic attenuation spectroscopy.
11. Focused beam techniques.

Some of these methods can be applied to the measurements of nanoparticles. Japanese industrial standard will also cover the methods in keeping pace with ISO.

1.2.3 Key points in the measurements – Reference particles for calibration

Important point is that these particle sizes take different values depending on the principles of the measurements. The particles measured based on these different principles should, at least, show the same diameter as far as they are spherical particles. Therefore, we are now preparing spherical reference particles [3, 4]. Various mono-disperse spherical particles are available from The Association of Powder Process Industry and Engineering (APPIE, Japan) [5, 6]. However, they are not suitable for the calibration of such equipments for the measurement of particle size distribution for the following reasons [7]:

1. Even if the principle is the same, different equipments have different feeding system, different transportation system and different dispersion system. Therefore, particle fraction detected in a certain size interval may differ depending on the equipment through size-dependent deposition, imperfect dispersion and size-dependent segregation.
2. There may be various calculation software for the transformation of the measured data to particle size distribution.
3. Number of particles sampled, sample size, in the measurement may differ for each measuring system. If the sample size is not large enough, both the mean diameter and the size distribution cannot be correctly obtained [8, 9].

In the ISO 13320 (1999), laser diffraction, it is recommended that equipment should be tested using poly-disperse reference particles whose particle size is distributed over one decade. Reference particles for covering nano-range is not yet available. However, MBP 1–10 for 1–10 μm and MBP 10–100 for 10–100 μm, spherical transparent particles, are available from APPIE. Particle size analyzers should at least be calibrated by use of these spherical reference particles.

Through the calibration using the reference particles, equivalent size obtained by each instrument is, for the first time, connected to the unit of length. In other words, measured value obtained by the measurement other than microscopy can get traceability only through the calibration with reference particles. It is very tedious to measure a great number of particles by electron microscopy. Therefore, in practice, sedimentation method, standard sieves and electrical sensing zone method are admitted as definitive methods to the certification of reference particles [5]. However, this is only for the sake of convenience.

The purpose for using the reference particles is listed as follows [10]:

1. Calibration of equipments for the particle size measurement,
2. Verification of measurement procedure,

3. Establishing traceability of measurement,
4. Determination of the uncertainty of the obtained results.

For representing the particle size distribution in mathematical equation, log-normal distribution and Rosin–Rammler distribution can be well applied and special graph papers are available for them.

References

[1] K. Okuyama, H. Masuda and S. Morooka: *Biryuushi Kougaku – Fine particle technology*, Ohmsha, Tokyo, p. 3 (1992).

[2] Society of Powder Technology, Japan (ed.): *Terminology Dictionary of Powder Technology*, 2nd ed., Nikkan Kogyo Co., p. 223 (2000).

[3] Standard Powders Committee, APPIE: *Powder Sci. Eng.*, **35** (6), 75–79 (2003).

[4] H. Yoshida, H. Masuda, K. Fukui and Y. Tokunaga: *Adv. Powder Technol.*, **12**, 79–94 (2001).

[5] H. Masuda: *Powder Sci. Eng.*, **34** (5), 65–72 (2002).

[6] A. Itoh: *Powder Sci. Eng.*, **37** (7), 72–77 (2005).

[7] Y. Mori, H. Yoshida and H. Masuda: *Particulate Systems Analysis 2005*, Stratford-upon-Avon, UK (2005).

[8] H. Masuda, K. Iinoya: *J. Chem. Eng., Jpn.*, **4**, 60–66 (1971).

[9] H. Masuda, K. Gotoh: *Adv. Powder Technol.*, **10**, 159–173 (1999).

[10] ISO Guide35, Certification of reference materials – General and statistical principles, p. 25 (1989).

1.3 Particle shape

The fundamental particle properties such as particle diameter, particle shape of nano-size or fine particles influence the character of the particle-packed bed. In these particle properties, the particle diameter measurement equipment based on various principles can be marketed, and it is easy to measure particle diameter distribution. But a particle shape analyzer for nanoparticle cannot be easily found and the shape index of nanoparticles can be calculated from particle images observed using various types of microscopes.

1.3.1 Two-dimensional particle projection image

In order to measure the particle shape, the outline of particle shape of two-dimensional projection images captured from microscopic photograph is analyzed. Since the diameter of a nanoparticle is smaller than the wavelength of visible light, a nano-size particle cannot be observed by an optical microscope. Usually, the two-dimensional projection image of nanoparticles is captured by scanning electron microscope (SEM) or transmission electron microscope (TEM), and the particle shape indices can be calculated from the captured images by image analysis software. For the shape analysis of fine particle over micrometer order, automatic particle shape analyzers using two-dimensional image of particles in a sheath flow are available. This analyzer captures particle images automatically by an optical microscope with CCD camera under stroboscope flush lighting.

1.3.2 Three-dimensional particle image

In the particle shape measurement of flaky particle or porous particle including hole or space inside the particle, the shape analysis of a two-dimensional particle projection picture is inadequate, and three-dimensional shape analysis is necessary [1].

Although it is difficult to measure the thickness of a particle by the ordinary electron microscope, the thickness or surface roughness can be measured by three-dimensional scanning electron microscope (3D-SEM). 3D-SEM takes two microscopic pictures from slightly different angles and obtains the three-dimensional information including thickness and surface roughness geometrically. However, if one of the pictures has the hidden area in the shadow of a particle, the three-dimensional information of the area is hidden. When well dispersed nanoparticles adhere on a flat substrate, the height difference between the particle surface and the flat substrate can be measured accurately, and the thickness of nanoparticle is obtained.

Using the TEM, 120 transmission images are taken when a sample is rotated 1 degree interval from -60 to $+60$ degrees. This three-dimensional imaging technique called TEM-CT [2], which is similar to the computer aided tomography [1], is expected to be applied for detailed three-dimensional shape measurement of nanoparticles.

Using the scanning probe microscope (SPM) including atomic force microscope (AFM), the surface roughness can be measured by tiny probe with the high resolution under nano-meter order. SPM are very effective for thickness measurement of nanoparticles. The vertical direction length such as particle thickness can be measured accurately by SPM, but the measured horizontal length becomes bigger about the diameter of the probe. It means that the particle diameter in horizontal direction measured by SPM is bigger than real particle size. Moreover, for a soft particle or an adhesive particle, a particle position changes by contact of probe or a particle adheres to probe, and an accurate image is not obtained.

1.3.3 Particle shape index using particle diameter ratio

Once the microscopic particle image is captured, quantification of particle shape can be performed by the same method as a coarse particle even if it is a nanoparticle. Quantification of particle shape is classified roughly into the shape index which is the ratio of two kinds of different definition particle diameter and the other expression such as fractal dimension or Fourier analysis of particle perimeter. In the case of the ratio of two different definition particle diameter, there are many similar kinds of shape indices such as degree of elongation (aspect ratio) = major axis/minor axis, degree of flatness = minor axis/thickness, degree of circular = equivalent area diameter (Heywood diameter)/equivalent perimeter diameter, degree of sphericity = equivalent volume diameter/equivalent surface area diameter [3]. In addition, unidirectional maximum particle (Feret diameter)/equivalent perimeter diameter are also used as a particle shape index, and the value of the shape index of a particle perimeter without concave shows near unity. Each of these is a ratio of diameters of a particle, so these shape indices are non-dimensional values and should not be influenced by the particle size. However, the perimeter and the projection area of a fine particle have the tendency to decrease with the decrease in particle size, because the resolution of a particle image is getting worse. Therefore, strictly speaking, the shape indices are influenced by the resolution of microscopic particle images.

In these indices, the degree of elongation or aspect ratio is an index with which a particle expresses long and slender, and the index shows a large value for a slim fiber particle. The high degree of flatness value means flakey shape. The degree of circularity and the degree of sphericity are set to 1 of maximum when a particle has a spherical shape, and the index shows a small value for a non-spherical particle.

The shape indices, expressed with ratios of two particle diameters, are divided into two categories. The degree of elongation, the degree of flatness, etc. in the one category show the overall shape of a particle, and the degree of circularity and the degree of sphericity in the other category show the surface roughness of a particle. Using the relation between two shape indices in different categories, difference in particle shape can be expressed more clearly in the two-dimensional figure than the case using only one diameter ratio. As an example of such a relation, the unidirectional maximum particle diameter (Feret diameter)/diameter of equivalent particle perimeter is plotted on the horizontal axis and the degree of circular is plotted on the vertical axis [4].

1.3.4 Particle shape expression by fractal dimension

The fractal dimension is a dimension, taking a real numerical value, proposed by Mandelbrot [5], and is also used for particle shape expression. The divider method is used to measure the fractal dimension of particle projection image perimeters. In this method, the number of the polygonal lines $N(r)$ necessary to go around a particle perimeter is measured r as shown in Fig. 1.3.1. The line of length r is changed and the same operation is repeated. If the relation between r and $N(r)$ shows a straight line on log–log paper as shown Fig.1.3.1, the value corresponding to the inclination of this straight line is defined as the fractal dimension D [6].

$$N(r) \propto r^{-D} \qquad (1.3.1)$$

A high value of fractal dimension D means a rough rugged surface of a particle, and a value of fractal dimension close to 1 means a smooth surface like spherical beads. The fractal dimension is also obtained from the number of adsorbed gas molecules with different adsorption area instead of a line segment. In this method, a powder with larger surface area has higher measurement accuracy, and the average value showing the three-dimensional shape of many particles can be obtained. Thus, this method is effective in shape analysis of a nanoparticle.

In the covering method, a particle projection image is covered with the square of r instead of a line segment, and the relation between the number of squares containing a projection image or an outline of a particle $N(r)$ and size r, is also used for the particle shape analysis. Moreover, the turning-radius method is well used to obtain the fractal dimension of an agglomerate particle. In this method, the circle of a radius R is drawn from the center of an agglomerate particle, and the number of the primary particles contained in the circle $N(R)$ is counted. In this method, the fractal dimension D is obtained from the log–log plot of R and $N(R)$ [7]. From the log–log plot of the perimeter P and the projection area A of each particle, the average fractal dimension of many particle perimeters

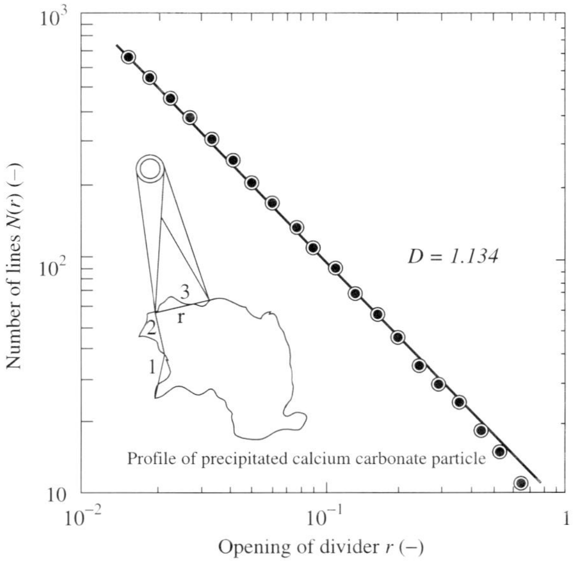

Figure 1.3.1
Fractal dimension measurement using divider method.

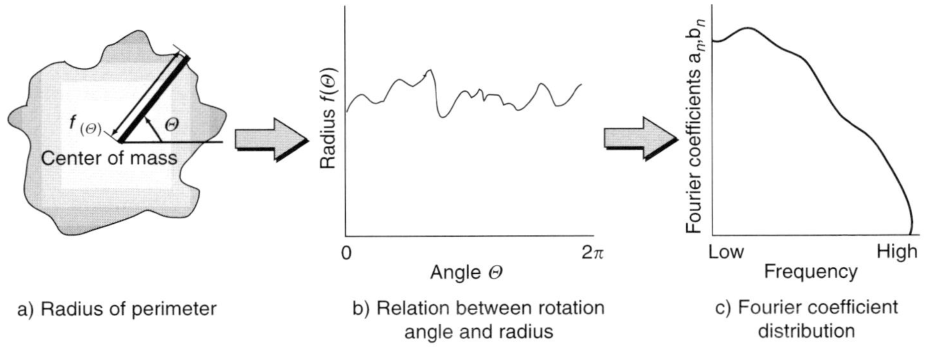

Figure 1.3.2
Fourier analysis of particle projective perimeter.

is obtained from inclination of the straight line using the following equation [8]:

$$P \propto A^{D/2} \tag{1.3.2}$$

The three-dimensional particle shape expressed by the fractal dimension is also possible from the log–log plot of the surface area and volume of each particle, instead of the perimeter and the projection area.

1.3.5 Particle shape analysis by Fourier analysis

In the Fourier analysis method, the particle shape is given as a function of the radius $f(\theta)$ from the center of mass to perimeter as shown in Fig. 1.3.2a. The Fourier analysis is carried out by the following equation [9]:

$$f(\theta) = a_0 + \sum_{n=1}^{\infty}\left(a_n \cos\frac{2n\pi}{T}\theta + b_n \sin\frac{2n\pi}{T}\theta\right) \tag{1.3.3}$$

The Fourier coefficients a_n and b_n represent the particle shape.

a_0 is the average radius of a particle image and T the cycle of trigonometric functions. In the series of Fourier coefficients a_n and b_n, the low-order coefficients with small n value express the large scale surface roughness and the high order coefficients with large n value express the small scale surface roughness. This Fourier-analysis method has the merit to synthesize the original particle perimeter from the Fourier coefficients. Synthesizing the particle shape is impossible by the other method. In order to rebuild the original outline completely, the infinite number of Fourier coefficients would be required.

1.3.6 Particle shape analysis of nanoparticle

A particle shape analyzer is not available for nanoparticle, and so a method based on the microscopic particle image is used usually. In order to obtain the average result about many particles, the ratio of two particle diameters obtained by different kinds of method such as a particle size by dynamic-light-scattering method and the specific surface diameter by the gas adsorption is used. These particle sizes include the influence of particle shape and, so the ratio of these diameters can be used for a simple particle shape expression.

References

[1] M. Suzuki, K. Kawabata, K. Iimura and M. Hitota: *J. Soc. Powder Technol.*, **41**, 156–161 (2004).
[2] H. Furukawa, M. Shimizu, Y. Suzuki and H. Nishioka: *JOEL News*, **36E**, 50 (2001).
[3] J. Tsubaki, M. Suzuki and Y. Kanda: Nyumon Ryushi·Funtaikougaku, Nikkan Kougyo Shinbunsha, 8 (2002).
[4] J. Tsubaki, G. Jimbo: *Powder Technol.*, **22**, 161–169 (1979).
[5] B.B. Mandelbrot: *Fractal Kikagaku*, Nikkei Sci. (1984).
[6] M. Suzuki, Y. Muguruma, M. Hirota and T. Oshima: *J. Soc. Powder Technol.*, **25**, 287–291 (1988).
[7] M. Matsushita, K. Itoh, M. Ohnishi, T. Ogawa, M. Kohno, M. Tanemura, H. Honda, K. Miyamoto, K. Miyazaki and N. Yoshimoto: *Katachi no kagaku*, Asakura shoten (1987).
[8] M. Suzuki, S. Yamada, H. Kada, M. Hirota and T. Oshima: *J. Soc. Powder Technol.*, **34**, 4–9 (1997).
[9] K. Otani, H. Minoshima, T. Uchiyama, K. Shinohara, K. Takayashiki and T. Ura: *J. Soc. Powder Technol.*, **32**, 151–157 (1995).

1.4 Particle density

1.4.1 Density measurement of powders composed of nanoparticles

(1) Definitions of powder density

Powder is an inhomogeneous material in that there are gaps between constituent particles, and there may be voids within particles themselves. Consequently, the density of a powder can be defined in several ways. Important definitions that are practical and useful are as follows.

(a) Material density

Material density of a powder is defined as the mass of the powder divided by the volume occupied by solid matter of the powder. It can be regarded as the density of a homogeneous material composed of the same material as the constituent particles. It is also referred to as true density. Material density might be measured after removing internal voids by breaking the particles.

(b) Particle density

Particle density is defined as the mass of the powder divided by the volume occupied by particles including internal closed voids. This volume is equivalent to the sum of volumes enclosed by the external surfaces of particles. When there are no closed voids within particles, the particle density is the same as the material density. In actual measurements, open voids that have only small openings at the surfaces of particles might behave as closed voids. Therefore, the particle density value obtained can depend on the details of the measurement procedure.

(c) Bulk density

Bulk density is defined as the mass of the powder divided by the volume of the space below the upper surface of the powder when it is placed in a container. The bulk density refers to the volume including both the gaps between constituent particles and the voids within particles. Two types of bulk density are often used: initial bulk density is the bulk density when well-dispersed particles are put gently into a container, while tap density is that obtained after a tapping procedure in which the powder container is repeatedly lifted and dropped onto a solid surface. The ratio of the tap density to the initial bulk density is called the Hausner ratio, and is used as a measure of powder flowability or compressibility.

(2) Measurement methods of powder density

In conventional methods of measuring the density of a powder, the mass of the powder is measured with a balance, and its volume measured either by the liquid immersion method or the gas replacement method. These methods are, of course, applicable also to powders composed of nanoparticles. For details of the methods see references [1, 2].

It has been reported that bulk density of a powder tends to remain unchanged by the tapping procedure when the powder is composed of nanoparticles, which suggests that the Hausner ratio might not be a good measure of the flowability or compressibility of nanoparticles [2]. Further study will be required to establish a method for characterizing nanoparticles in terms of bulk density.

When the density of individual particles is relatively homogeneous, the density reference liquid method [3] may be applicable: particles are put into density reference liquids that have been prepared to have several known densities, and the particle density is determined by judging whether the particles float to the liquid surface, settle at the bottom, or remain suspended in the liquid, after a certain holding time.

1.4.2 Density measurement of individual particles

(1) Effective density of individual particles

When particles are not in a powder state, but are dispersed in a gas or a liquid, the density of individual particles is often of concern. For particles dispersed in a gas, there are three major methods of measuring the density of individual particles: (A) the combined measurement of mobility and mass, (B) the combined measurement of mass and volume, and (C) the combined measurement of mobility and aerodynamic diameter.

Let m and d_b denote the mass and the mobility equivalent diameter (diameter of a sphere with the same mechanical mobility as the particle of concern) of a particle, respectively. The density obtained by methods (A) and (C) is the effective density defined by

$$\rho_e = \frac{m}{\pi d_b^{\,3}/6}. \tag{1.4.1}$$

For a spherical particle, d_b coincides with its geometrical diameter, and hence ρ_e is equivalent to the particle density defined in 1.4.1.[1] For a nonspherical particle, d_b depends on the particle shape.[2] Although the effective density in such cases is not an intrinsic material property, it is still useful for particle characterization. The density obtained by method (B), on the other hand, is the particle density, and leads to the material density, if there are no closed voids within the particles.

These methods are also applicable to particles suspended in a liquid, if the liquid suspension can be converted into an aerosol by atomizing the suspension and drying the droplets thus generated.

(2) Mass classification of aerosol particles

The method of classifying aerosol particles according to their mass, which is employed in methods (A) and (B), is briefly described here. Fig. 1.4.1 shows the

[1]Note, however, that the particle density in 1.4.1 is an average property of particles contained in a powder, whereas the density here is a property of an individual particle.

[2]If we denote the volume equivalent diameter of a particle by d_v, the shape dependence of d_b can be expressed by $d_b/C(d_b) = \kappa\, d_v/C(d_v)$, where κ is the dynamic shape factor and $C(d)$ the Cunningham slip correction.

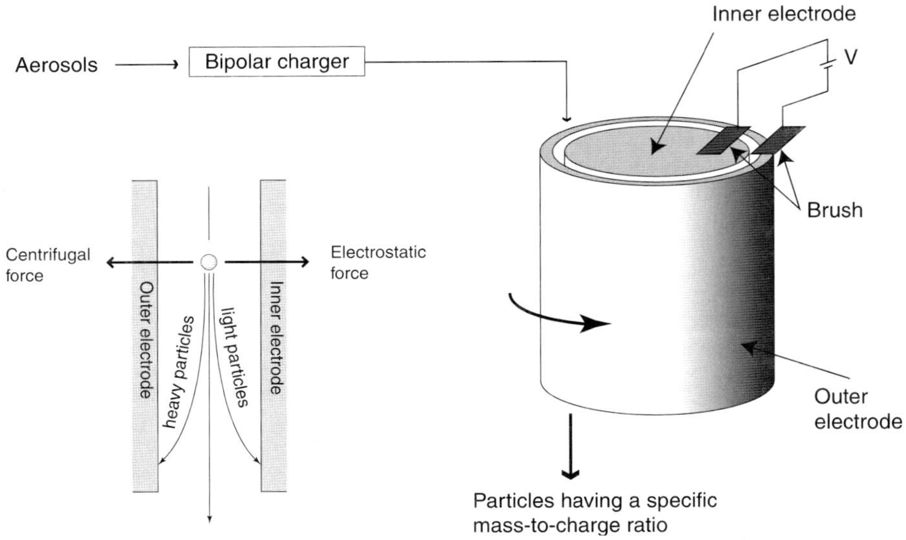

Figure 1.4.1
Principle of the aerosol particle mass analyzer (APM).

principle of the aerosol particle mass analyzer (APM) used for this purpose [4]. Aerosol particles are first passed through a bipolar charger in which bipolar ions generated by an ion source such as ^{241}Am, ^{85}Kr, or ^{210}Po bring the particles to an equilibrium charge state, and then introduced into a thin annular gap between coaxial cylindrical electrodes that rotate at the same angular velocity. In the gap, particles migrate in the radial direction under the influence of centrifugal and electrostatic forces, and only those particles for which the two forces balance can exit the electrodes. The particle mass that can be classified by this principle ranges roughly from 0.01 to 500 fg (1 fg $= 10^{-15}$ g). This range corresponds approximately to particle diameter of 30 nm to 1 μm for spherical particles of 1 g/cm^3 in density.

Fig. 1.4.2 shows a mass distribution spectrum obtained by classifying monodisperse polystyrene latex particles (PSL) with a known mass (208 nm diameter, 1.054 g/cm^3 density and 4.97 fg mass) with an APM and then by counting the number of exiting particles with a laser particle counter (LPC). The main peak structure coincides almost exactly with the theoretically predicted structure, and some other structures corresponding to doubly and triply charged singlet particles and singly charged doublet particles are also observed at theoretically predicted locations.

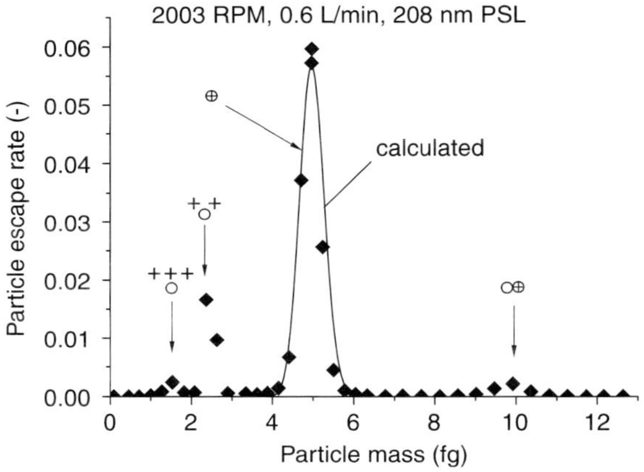

Figure 1.4.2
Mass distribution spectrum for PSL particles of 4.97 fg in mass.

(3) Combined measurement of mobility and mass

The effective density of aerosol particles can be determined by classifying them according to their mobility and mass using a differential mobility analyzer (DMA) and an APM connected in series, and then detecting the exiting particles using a particle counter such as a condensation nuclei counter (CNC) or an LPC. Fig. 1.4.3 shows a mass distribution spectrum of sodium chloride (NaCl) particles with a mobility equivalent diameter of 50 nm, obtained by using a DMA–APM–CNC system. The particles were generated by atomizing and drying an NaCl solution. The peak observed in the mass distribution is located at approximately 0.113 fg, indicating that the effective density of the particles is about 1.7 g/cm^3. This value

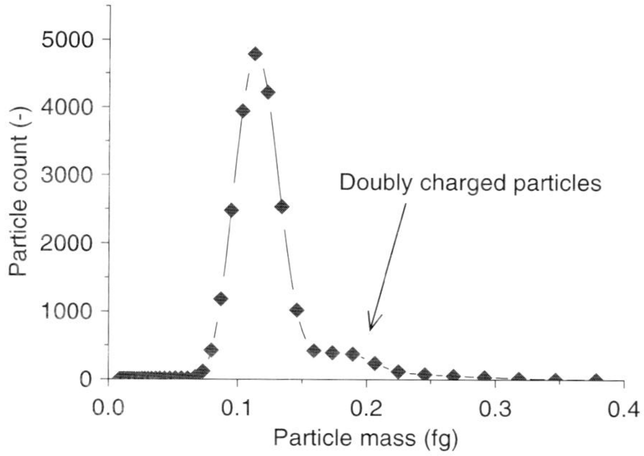

Figure 1.4.3
Mass distribution spectrum for NaCl particles having mobility equivalent diameter of 50 nm.

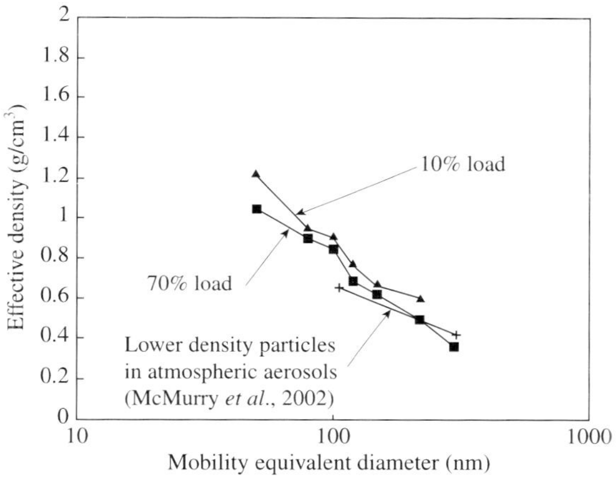

Figure 1.4.4
Size dependence of the effective density of diesel exhaust particles, reconstructed from Park, K. et al., *Environ. Sci. Technol.*, **37** 577-583 (2003).

is smaller than the density of NaCl bulk crystal (2.2 g/cm^3). The exact reason for this difference in density is not yet known, but it might be due to the dynamic shape factor of non-spherical NaCl particles.

By varying the mobility of particles classified by the DMA, d_b-dependence of ρ_e can also be determined. Fig. 1.4.4 shows the effective density of diesel exhaust particles (DEPs) with mobility equivalent diameter ranging from 50 to 300 nm, measured with this method [5]. The influence of the engine load on the effective density of DEPs is observed in Fig. 1.4.4. The same method has also been applied to atmospheric aerosols [6] and carbon nanotubes [7].

(4) Combined measurement of mass and volume

If the volume of a particle selected by an APM is measured directly with transmission electron microscopy (TEM), the density of the particle can be determined. The material density of DEPs ranging from 50 to 220 nm in mobility equivalent diameter was measured with this method [8]. It is reported that, after removing volatile components by heating particles to 300°C, the density was 1.77±0.07 g/cm^3 irrespective of their mobility equivalent diameter.

(5) Combined measurement of mobility and aerodynamic diameter

The effective density of an individual particle can also be determined if both the mobility and the aerodynamic diameter are known. This can be achieved by measuring the aerodynamic diameter of particles selected with a DMA. Traditionally, impactors have often been used to measure the aerodynamic diameter with this method [9, 10]. Recently, a time-of-flight type device has sometimes been employed instead of an impactor [11, 12]. The accuracy of the effective density obtained with this method is generally much higher than the accuracy obtained with the combined DMA and impactor method.

References

[1] R. Utsumi, in *Funtai kougaku binran (Powder engineering handbook)*, 2nd ed., Nikkan Kougyo Shinbun, pp. 42–47 (1998) (in Japanese).
[2] H. Yanagida, supervised: *Engineering System for Fine Particles*, Vol. 1, Fujitec Co. (2001) (in Japanese).
[3] JIS Z 8901: *Test Powders and Test Particles*, Japanese Standards Association (2006).
[4] K. Ehara, K.J. Coakley and R.C. Hagwood: *J. Aerosol Sci.*, **27**, 217–234 (1996).
[5] K. Park, F. Cao, D.B. Kittelson and P.H. McMurry: *Environ. Sci. Technol.*, **37**, 577–583 (2003).
[6] P.H. McMurry, X. Wang, K. Park and K. Ehara: *Aerosol Sci. Technol.*, **36**, 227–238 (2002).
[7] A.D. Maynard, B.K. Ku, M. Emery, M. Stolzenburg and P.H. McMurry: *J. Nanoparticle Res.*, **9**, 85–92 (2007).
[8] K. Park, D.B. Kittelson, M.R. Zachariah and P.H. McMurry: *J. Nanoparticle Res.*, **6**, 267-272 (2004).
[9] W.P. Kelly, P.H. McMurry: *Aerosol Sci. Technol.*, **17**, 199–212 (1992).
[10] S.V. Hering, M.R. Stolzenburg: *Aerosol Sci. Technol.*, **23**, 155–173 (1995).
[11] P. DeCarlo, J.G. Slowik, D.R. Worsnop, P. Davidovits and J.L. Jimenez: *J. Aerosol Sci.*, **38**, 1185–1205 (2004).
[12] A. Zelenyuk, Y. Cai, L. Chieffo and D. Imre: *Aerosol Sci. Technol.*, **39**, 972–986 (2005).

1.5 Melting point, surface tension, wettability

It is well known that melting point and surface tension, which are the physical properties of materials, and wettability (contact angle), which is the intrinsic value between liquid and solid, depend on particle size when the size is nano-meter order.

1.5.1 Melting point

A proportion of number of surface atoms to that of bulk atoms increases with decrease in particle size. Accordingly, the melting point of nanoparticle falls as much as corresponding surface energy.

Many researchers have reported melting points of nanoparticles [1–21] and some examples are shown in Figs. 1.5.1 and 1.5.2. A melting point of Au declines suddenly when a particle size is less than 15 nm and a decrease of almost 200 K can be observed at about 6 nm. As shown in Fig. 1.5.2, it has been reported that there is a linear relationship between a ratio of a melting point of nanoparticle (T_m) to that of bulk material (T_o) and a reciprocal number of a particle diameter.

1.5.2 Surface tension

Morokhov [22] calculated a change in liquid drop diameter with an evaporation of a liquid drop at a constant temperature based on a molecular dynamics and obtained an equation (1.5.1).

$$\ln(dr/dt) = \ln A + B(\sigma/r) \qquad (1.5.1)$$

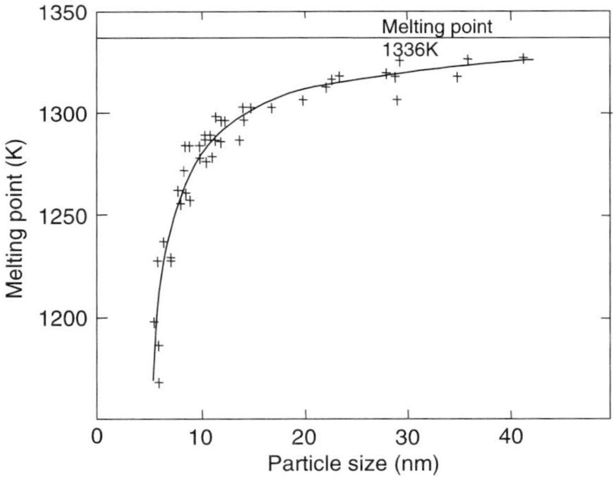

Figure 1.5.1
Relationship between melting point of gold and its particle size [8].

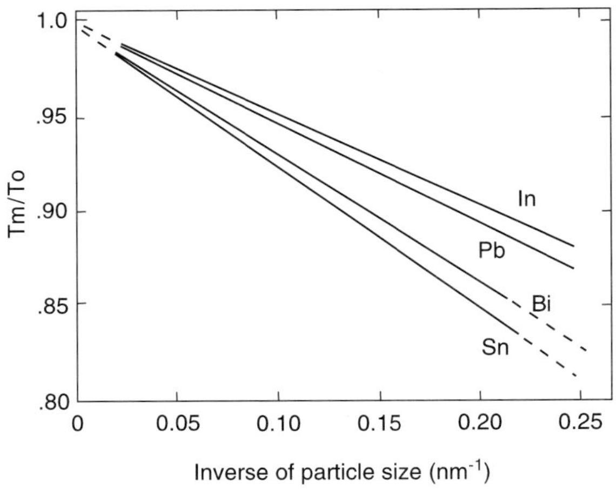

Figure 1.5.2
Relationship between inverse of particle size of various metal and T_m/T_o [19]. T_m, measured melting point; T_o, melting point of bulk material.

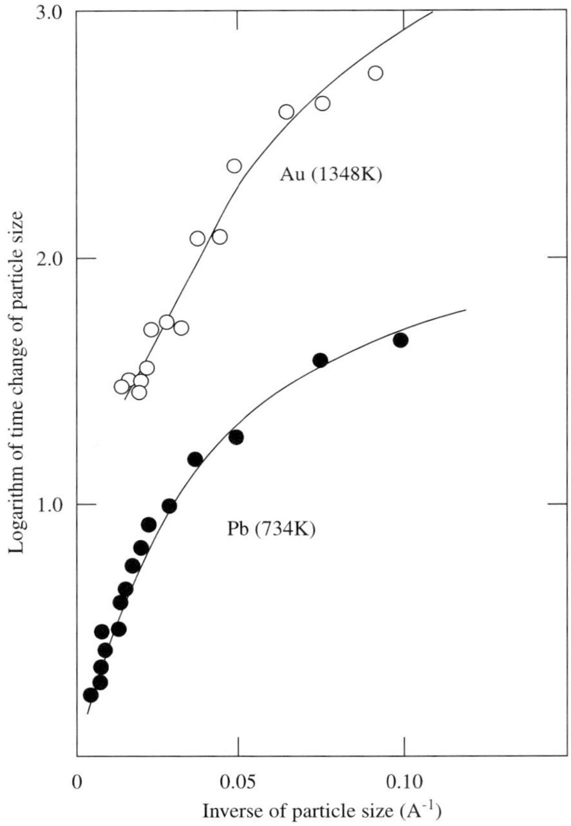

Figure 1.5.3
Relationship between particle size change of evaporating metal drop and particle size.

where r is the radius of liquid drop, σ the surface tension, t the time, and A and B are constants.

As can be seen from equation (1.5.1), when the surface tension value of liquid drop remains constant, Ln (dr/dt) should be proportional to ($1/r$). Fig. 1.5.3 shows changes of particle sizes of Au and Pb with time by evaporation. It can be found that Ln (dr/dt) and ($1/r$) does not show linear relationship when particle sizes of Au and Pb drops are less than 2 and 5 nm, respectively.

Gladkikh et al. [21] performed theoretical and experimental approach on surface energies of nanoparticles of metals and reported the results of Au and Pb particles. As can be found from Fig. 1.5.4, obtained result of Pb is fairly in good agreement with that by Morokov [22].

1.5.3 Wettability

As shown in Fig. 1.5.5, a contact angle value is used as a criterion for wettability of solid by liquid. In general, when a contact angle is larger than 90 degree (cf. Fig. 1.5.5a), it is defined as "non-wetting system" and when a contact angle is smaller than 90 degree(cf. Fig. 1.5.5b), it is defined as "wetting system".

Over 200 years ago, equation (1.5.2) was proposed by treating the contact angle of a liquid drop as the result of the equilibrium of mechanical energy between the drop and a solid surface under the action of three interfacial free energies.

$$\sigma_L \cos\theta = \sigma_S - \sigma_{SL} \tag{1.5.2}$$

where θ is the contact angle, σ_L the surface free energy of liquid, σ_S the surface free energy of solid and σ_{SL} the interface free energy between solid and liquid.

Since the surface free energy (surface tension) of liquid depends on a liquid drop size as mentioned above, the contact angle also depends on the liquid drop size. Fig. 1.5.6 shows particle size dependence of contact angles of various liquid metals on carbon

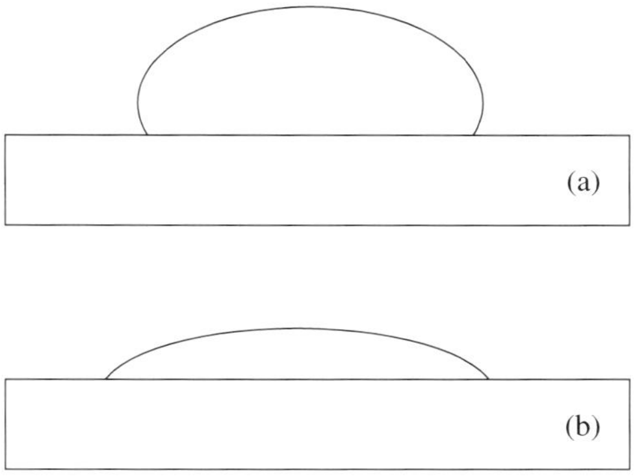

Figure 1.5.5
Liquid drop shape on solid.

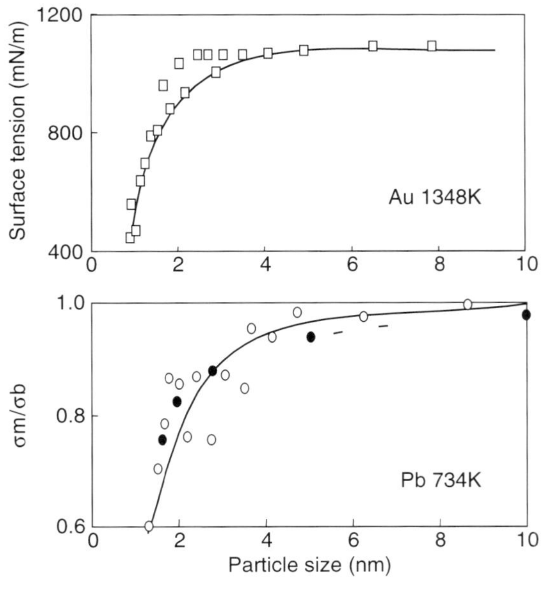

Figure 1.5.4
Effect of particle size of surface tension of Au and Pb. σm, surface tension of nanoparticle; σb, surface tension of bulk metal.

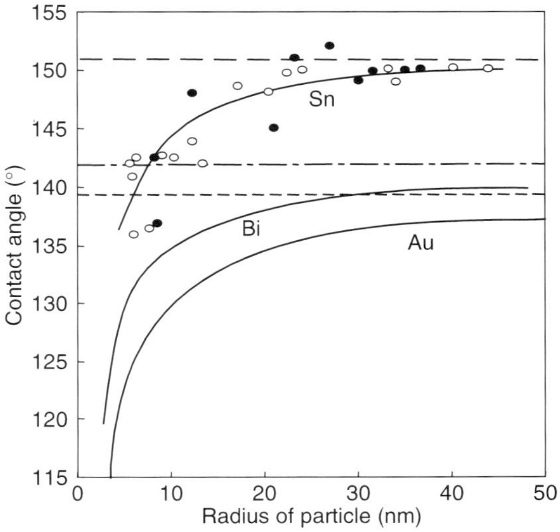

Figure 1.5.6
Relationship between contact angles of liquid metals on graphite and particle sizes.

substrate [23, 24]. It can be found that the contact angles of all liquid metals decrease at less than 40 nm and a remarkable decrease of the contact angles is observed when particle radius is less than 10 nm.

References

[1] M. Takagi: *J. Phys. Soc. Jpn.*, **9**, 359–363 (1954).
[2] F.O. Jones, K.O. Wood: *Brit J. Appl. Phys.*, **15**, 185–187 (1964).
[3] N.T. Gladkich, R. Niedermayer and K. Spiegel: *Phys. Stat. Sol.*, **15**, 181–192 (1966).
[4] C.R.M. Wronski: *Brit. J. Appl. Phys.*, **18**, 1731–1737 (1957).
[5] B.T. Boiko, A.T. Pugachev and V.M. Bratsykhin: *Sov. Phys. Solid State*, **10**, 2832–2834 (1969).
[6] J.F. Pocza, A. Barna and P.B. Barna: *J. Vac. Sci. Tech.*, **6**, 472–475 (1969).
[7] M. Blackman, J.R. Sambles: *Nature*, **226**, 938 (2970).
[8] M.J. Stowell, T.J. Law and J. Smart: *Proc. Roy. Soc. London A*, **318**, 231–241 (1970).
[9] J.R. Sambles: *Proc. Roy. Soc. London A*, **324**, 339–351 (1970).
[10] C.J. Coombes: *J. Phys. F: Metal Phys.*, **2**, 441–449 (1972).
[11] M. Bkackman, S.J. Peppiatt and J.R. Sambles: *Nature Phys. Sci.*, **239**, 61–62 (1972).
[12] R.P. Berman, A.E. Curzon: *Can. J. Phys.*, **52**, 923–929 (1974).
[13] S.J. Peppiat, J.R. Sambles: *Proc. R. Soc. Lond. A*, **345**, 387–399 (1975).
[14] S.J. Peppiatt: *Proc. R. Soc. London A*, **345**, 401–412 (1975).
[15] P. Buggst, J.P. Boreld: *Phys. Rev. A*, **13**, 2287–2298 (1976).
[16] G.L. Allen, W.W. Gile and W.A. Jesser: *Acta Metall.*, **28**, 1695–1701 (1980).
[17] V.P. Skipov, V.P. Koverda and V.N. Skokov: *Phys. State. Sol.*, **66**, 109–118 (1981).
[18] M.S. Rahman: *Micron*, **13**, 273–274 (1982).
[19] G.L. Allen, R.A. Bayles, W.W. Gile and W.A. Jesser: *Thin Solid Films*, **144**, 297–308 (1986).
[20] H. Saka. Y. Nishikawa and T. Imura: *Phil, Mag. A*, **57**, 895–906 (1988).
[21] N.T. Gladkikh, L.K. Grigoreva, S.V. Dukarov, V.E. Zilbervarg, V.I. Larin, E.L. Nagaev and S.P. Chizhik: *Sov. Phys. Solid State*, **31**, 728–732 (1989).
[22] I.D. Morokhov, S.P. Chizhik, N.T. Gladkikh, L.K. Grigoreva, S.V. Stepanova: *Izv. Akad. Nauk SSSR Metall.*, **6**, 159–161 (1979).
[23] N.T. Gladkikh: *Fiz. Khim. Obrabot. Mater.*, **2**, 96–102 (1979).
[24] S.P. Chizhik: *Izv. Akad. Nauk SSSR Metall.*, **4**, 73–79 (1981).

1.6 Specific surface area and pore

Nanoparticles have a large specific surface area, and hence their properties are dominated by surfaces rather than bulk. The specific surface area is often used as a basic unit for the particle properties of nanoparticles. In this section, methodologies and things to be considered for analyzing particle size from specific surface area will be discussed. The first half of this section will describe requirements for measuring particle size distribution from specific surface area. In the second half, pore size distribution will be reviewed as an application of the measurement of specific surface area.

It is important to access thoroughly the influence of particle shapes and particle size distribution for measuring particle size from specific surface area. For this reason, using electron microscopic observations or other particle size measurements are preferable in combination with the measurement of specific surface area. Comprehensive analysis for these measurements would enable to estimate reasonable and meaningful particle size.

The relationship between specific surface area and particle size is described here with an ideal particle model. As shown in Fig. 1.6.1, ideal size reduction where a dense cube, 1 cm on a side, is divided into cubes, l cm on a side, is supposed. The surface area of each divided cube is given as $6 \cdot l^2$ and the number of divided cubes is given as $1/l^3$. Therefore, the total surface area of all the divided cubes can be expressed as $(1/l^3) \cdot (6l^2)$, thus $6/l$. Supposing the true density of a particle ρ, specific surface area S is given by

$$S = 6/(\rho \cdot l) \qquad (1.6.1)$$

In contrast, supposing that the powder is composed of uniform cubic particles, the particle size can be given

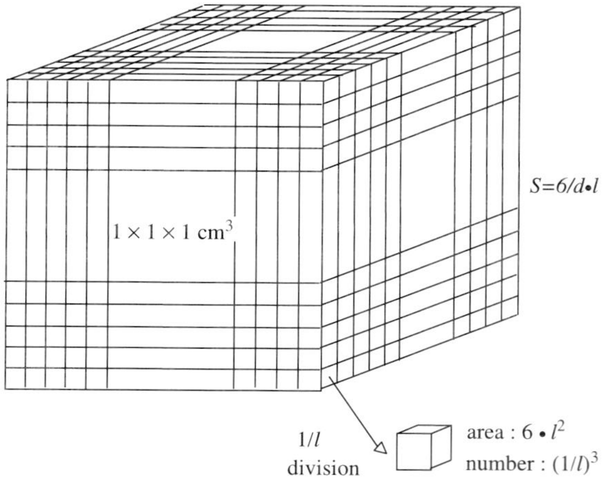

Figure 1.6.1
Relationship between particle size and specific surface area.

by the specific surface area and the true density of the particle. Although the particle shape was supposed to be cubic in the derivation above, the relationship between specific surface area and particle diameter for a perfect sphere can also be given by equation (1.6.1). Thus, the coefficient 6 in the equation is the shape factor for a particle with cubic or spherical shapes. For particles with other shapes, similar relationships can be obtained by using an appropriate shape factor. There are a number of definitions for shape factors [1, 2], whereas the details cannot be described here due to limitation of space. The particle size calculated from equation (1.6.1) is normally used as an equivalent specific surface diameter for practical convenience.

In general, it is very rare to deal with the ideal particle shown above. Actual particle has usually irregular shapes and a size distribution. In the following part, how particle size distribution and particle shape give effects for analyzing particle size from specific surface area will be discussed, and then points to be considered will be given.

If particles have ideal spherical or cubic shape with a mono-dispersed size distribution, the specific surface area can be related to the particle size as shown in equation (1.6.1). When a sample powder has a particle size distribution, relationship between specific surface area and the size distribution can be addressed as follows. Supposing that the spherical particles with a true density ρ and a diameter l_i exist n_i per unit mass, the specific surface area can be given by the following:

$$S = 6/\rho \cdot \sum [(n_i \cdot l_i^2)/[(n_i \cdot l_i^3)] \qquad (1.6.2)$$

For taking account of correspondence to particle size distribution, n_i is expressed here with a ratio v_i to the total number of particles N.

$$n_i = v_i N \qquad (1.6.3)$$

The particle diameter l_i is also expressed with a ratio λ_i to the mode particle size L.

$$l_i = \lambda_i L \qquad (1.6.4)$$

Substituting equations 1.6.3 and 1.6.4 into equation 1.6.2 gives the following:

$$S = f \cdot 6/(\rho L) \qquad (1.6.5)$$

Where

$$f \equiv \sum [(v_i \cdot \lambda_i^2)/(v_i \cdot \lambda_i^3)] \qquad (1.6.6)$$

v_i and λ_i can be obtained from data of particle size distribution, and f is a coefficient depending on shape of the size distribution. If the particle size distribution

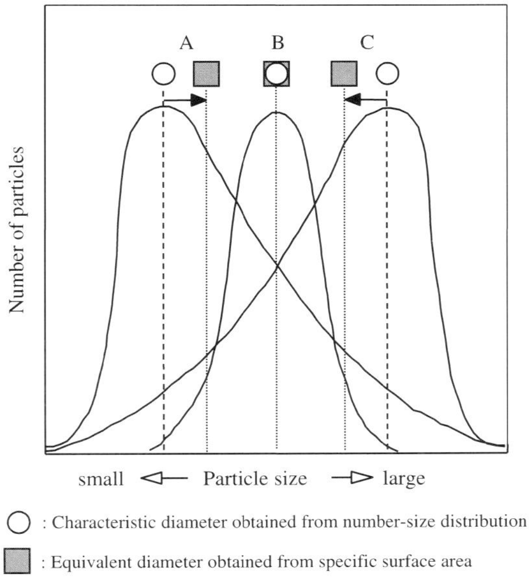

Figure 1.6.2
Relationship between particle size distribution and specific surface area.

has a good symmetry as shown in Fig. 1.6.2, the coefficient f is 1 and hence an equivalent specific surface diameter obtained from equation 1.6.1 is equal to a characteristic diameter obtained from number-size distribution. On the other hand, if the size distribution is not symmetric, the coefficient f is not equal to 1. For the distribution A shown in Fig. 1.6.2, the coefficient $f < 1$ and hence the equivalent specific surface diameter becomes smaller than the characteristic diameter obtained from number-size distribution. In contrast, for the distribution C shown in Fig. 1.6.2, the coefficient $f > 1$ and hence the equivalent diameter becomes larger than the characteristic diameter.

As described above, for a good symmetric size distribution, an equivalent diameter estimated by the measurement of specific surface area can be a characteristic diameter of a particle size distribution, but for an asymmetric size distribution the equivalent diameter is not always equal to the characteristic diameter. These points have to be considered thoroughly for estimating the equivalent diameter from specific surface area. As the measurement of specific surface area cannot give particle size distribution, concomitant use of an electron microscopic observation or other particle size measurements is necessary for more detailed discussions.

The particle shape has been supposed to be spherical or cubic in the discussions so far. However, it is very rare in practice to deal with such ideal shaped particles. In general, if primary particle has pores, specific surface area increases and a resultant equivalent diameter decreases apparently. Therefore, it is necessary to take account of voids or pores exiting in the particle for estimating particle size from the measurement of specific surface area.

Taking into account size of nanoparticles, pore size relating strongly to these particles would be less than approximately 5 nm (micropore) and 5–50 nm (mesopore) in radius.

From the scientific viewpoint of adsorption, micropore can be categorized and measured as follows. The surface on which multilayer of molecular adsorption develops is termed as "external surface", and the surface on which multilayer adsorption cannot occur due to steric hindrance of pore structure is termed as "internal surface". There is a distinct difference between the both surfaces in terms of adsorption mechanism.

In the region where the adsorption is not followed by the BET theory and multilayer adsorption does not occur, pore filling takes place in general and hence internal and external surface areas can be obtained by analysis of adsorption isotherm such as *V–t* plot or α_s-plot [3, 4]. *V–t* plot is the plot of *V* (adsorbed amount) against *t* (thickness of adsorbed layer), where *t* is derived from *t*-curve [relative pressure, P/P_0 versus *t*] obtained from the standard isotherm.

In general, the *V–t* plot forms a straight line passing through the origin for non-porous materials, and shows a folding point for materials with micropores. In the plot, slope of the straight line gives total surface area and external surface area, the intercept gives pore volume, and the *t* value at the folding point gives a radius of micropore. The α_s plot is normalized by $\alpha_s = V/V_{0.4}$ ($V_{0.4}$: adsorbed amount at $P/P_0 = 0.4$) without use of the *t*-curve in order to extend applicable range of the *V–t* plot. The specific surface areas can be obtained from a ratio of adsorbed amounts between a sample and a standard material at relative pressure 0.4.

For the mesopore, applying the theory of capillary condensation to isotherms of gas adsorption or desorption would give pore size distribution. Whereas the theory of capillary condensation is not described here in detail, vapor pressure *P* on meniscus of the liquid condensed in a pore is lower than vapor pressure P_s for a flat liquid surface (that is usual vapor pressure), and this vapor pressure drop is given by the following Kelvin equation:

$$\ln\frac{P}{P_s} = -\frac{2\gamma V_l}{rRT} \tag{1.6.7}$$

Where *r* is a radius of the meniscus of the liquid located in a pore, and is equal to the pore radius if contact angle is zero. V_l is the molecular volume of the liquid, γ the surface tension, *R* the gas constant, and *T* the temperature. As P_s, V_l, γ and *T* are constant at a certain temperature, smaller *r* gives smaller *P*.

This implies that under a vapor pressure, at which condensation does not occur within larger pores or for a flat surface, condensation may occur within smaller pores as the local pressure may attain the saturation level. In contrast, supposing that the gas pressure decreases in sequence from relative pressure 1, evaporation occurs at first from flat surfaces, then from larger pores, and eventually from smaller pores.

If all the gas adsorption takes place solely due to capillary condensation, the pore size distribution can be obtained from equation 1.6.7, which is a relatively easy and simple procedure. However, it is highly unlikely in practice that absorbed gas under a certain pressure is all in state of capillary condensation.

Wheeler [5] has taken a concept that pores having radius larger than r_p are covered with adsorption layer with thickness *t*, and pores smaller than r_p are filled with the liquid formed due to the capillary condensation. Therefore, as shown in Fig. 1.6.3, the Kelvin equation applies to r_k ($=r_p - t$) rather than r_p. For nitrogen absorbed at 77.35 K, we can now use the following equation:

$$r_k = r_p - t = \frac{2\gamma V_l}{RT\ln(P_s/P)} = \frac{4.14}{\log(P_s/P)} \tag{1.6.8}$$

While *t* is in general function of relative pressure P/P_s, Wheeler has suggested estimating *t* by using the following Halsey equation when P/P_s is larger than 0.5.

$$t = 4.3\left[\frac{5}{2.303\log(P_s/P)}\right]^{\frac{1}{3}} \tag{1.6.9}$$

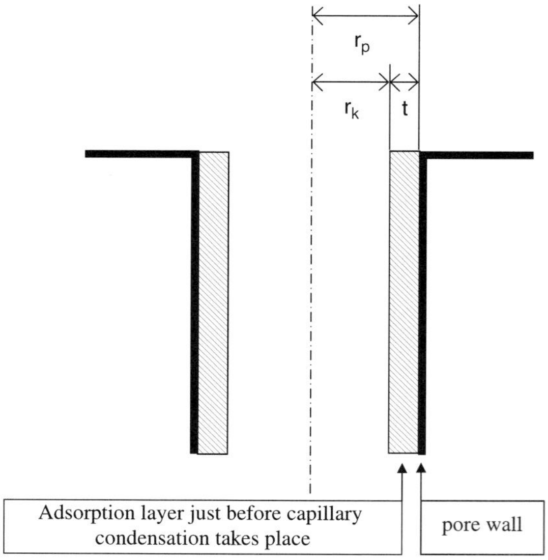

Figure 1.6.3
Relationship among r_p, r_k and *t* on a cylindrical pore model.

The values of t, r_k and r_p as a function of P/P_s can be calculated by using a combination of equations 1.6.8 and 1.6.9.

Supposing total pore volume V_p, which is expressed as a liquid volume equivalent to adsorbed amount at P_s, and adsorbed amount V at a pressure P (also as an equivalent liquid volume), $V_p - V$ is considered as a total empty pore volume that is generated by evaporation taking place at pores having radius less than r_p while the pressure is reduced from P_s to P. This is given by the following equation:

$$V_p - V = \int_{r_p}^{\infty} \pi(r-t)^2 L(r)dr \qquad (1.6.10)$$

where $L(r)$ is a total length of pore with radius r. V_p and V are obtained from adsorption data, and r_p and t can be determined with respect to P/P_s. However, $L(r)$ has to be obtained for each r.

For this reason, Wheeler has analyzed $L(r)$ with the assumption that its distribution follows Maxwell's or Gaussian distribution.

Thereafter, Barrett–Joyner–Halenda [6] have suggested a method that uses numerical integration without assuming any specific distribution, which is called BJH method. Since then this method has been improved by Pierece [7], Orr-Dalla Vallue [8], and Crauston-Inkley [9].

The measurement of pore size distribution using the gas adsorption method would have a problem in application of the Kelvin equation to pore radius less than 1 nm. Also, because measurement of relative pressure around saturation level is difficult and making some assumptions cannot be avoided for analyzing data, pore size distribution only less than 30 nm can be usually measured. Therefore, use of mercury intrusion porosimetry in combination with the gas adsorption method is preferable for measuring a wide range of pore size distribution.

References

[1] The Society of Powder Technology, Japan: *Funtai Kogaku Binran*, 2nd ed., Nikkan Kogyo Shimbun, p. 355 (1998) (in Japanese).

[2] G. Jimbo: *Funtai – Sono Kino to Oyo*, Nihon Kikaku Kyokai, p. 111 (1988) (in Japanese).

[3] S.J. Gregg, K.S. Sing: *Adsorption Surface Area and Porosity*, Academic Press, Inc., New York (1982).

[4] S. Kondo, T. Ishikawa and I. Abe: *Kyuchaku no Kagaku*, Maruzen, p. 52 (1991) (in Japanese).

[5] A. Wheeler: "*Catalysis Vol. II*", Reinhold Inc., p. 116 (1953).

[6] E.P. Barrett, L.G. Joyner and P.P. Halenda: *J. Am. Chem. Soc.*, **73**, 373 (2002).

[7] C. Pierce: *J. Phys. Chem.*, **57**, 149 (1953).

[8] C. Orr Jr., J.M. DallaValle: *Fine Particle Measurement*, Macmillan Inc., p. 271 (1959).

[9] R.W. Craust, F.F.A. Inkley: *Advance in Catalysis 9*, Academic Press Inc., New York, p. 143 (1957).

1.7 Composite structure

1.7.1 Composite structure of nanoparticle

A peculiar characteristic of the nanoparticle appears by the effect of size that is the general structure factor of particle. For example, basic physical properties such as melting point and boiling point drop by the superminiaturization. Various functions such as activity of catalyst are also improved by the nanosize effect. However, the cohesion of the nanoparticle remarkably increases with increase in the surface energy of particle by the nanosize effect, and the strong cohesion decreases the handling character of the nanoparticle. One useful method that improves handling character of nanoparticles is to apply the composite structure such as surface modification using nanoparticles. The composite structural control such as surface modification will reduce the cohesive property of nanoparticels, and the function of nanoparticles will appear smoothly. In addition, the composite structure that is formed by some kinds of nanoparticles will be able to combine some kinds of function, and it expects that new function will appear by the effect combined some kinds of function of nanoparticle. The composite structure can be roughly classified as follows: (1) the composite structure using nanoparticles, (2) the composite structure formed by agglomeration of nanoparticles, and (3) the composite body fabricated with nanoparticles.

The size of the nanoparticle is defined from single nano (less than 10 nm) to about 100 nm in the narrower sense, and also to a few 100 nm in the wider sense. In this chapter, the composite structure is assumed to be classified by the wide-range definition (from single nano to few 100 nm). Table 1.7.1 and Fig.1.7.1 show

Table 1.7.1
Classification of composite structure of nanoparticles.

Classification	Type of composite structure
1) The composite structure using nanoparticles.	(a) core–shell, (b) internal dispersion, (e) hollow, (f) porous.
2) The composite structure formed from agglomeration of nanoparticles.	(a) core–shell, (b) internal dispersion, (c) agglomeration, (d) coating (surface modification), (e) hollow (f) porous.
3) The composite structure body fabricated from nanoparticles.	(g) nano dens body, (h) nano porous body, (i) nano thin film.

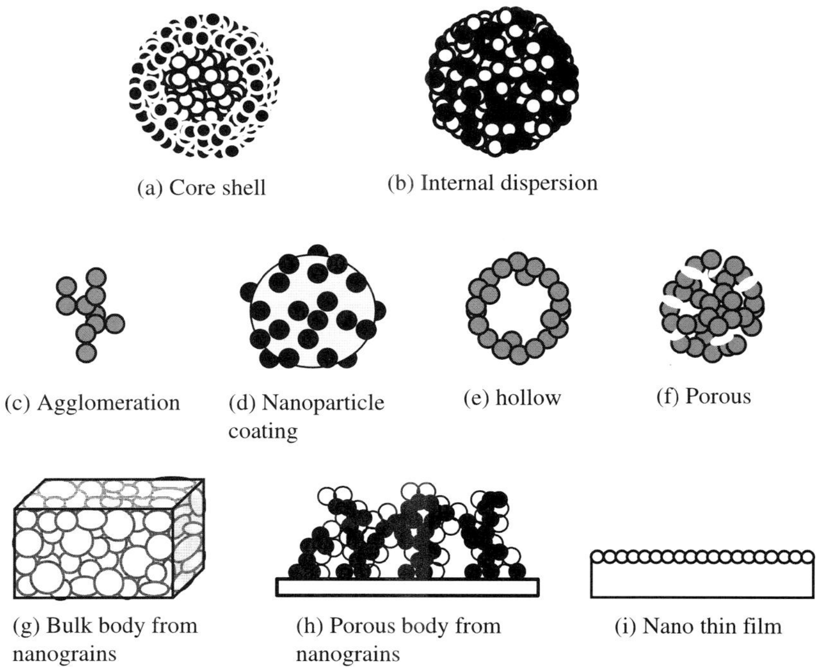

Figure 1.7.1
Images of composite structure fabricated from nanoparticles.

classification of composite structures of nanoparticles and their images. These composite structures are classified not only by the situation of agglomeration of nanoparticles but also by the pore structure. It can be thought that dispersibility, cohesive property, porosity and compact density are basic physical properties of the composite structures. In addition, these composite structures can be divided into the particle composite structure formed with more than two different kinds of particles, the element composite structure with various different elements, and the composite structure with various different composition grains. In short, the element and composition of the grains as well as their morphology become basic physical properties to form the composite construction.

1.7.2 Evaluation method of composite structure using electron microscopy

It is important for evaluating the composite structure of the nanoparticles to observe its morphology and microstructure that are the most basic physical properties. Scanning electron microscopy (SEM) and transmission electron microscopy (TEM) generally are used for observation of the morphology and microstructure, and can capture the composite structure in the sight directly. In addition, the composite structure becomes clearer by combining the electron microscopic observation with the elemental analysis. Specific surface area measurement, pore size distribution measurement in micro to nano area by gas adsorption method, and that in mm to μm area by mercury porosimetry will quantitatively supplement the evaluation by the electron microscopy observation for evaluating surface and internal pore structure. The specific surface area and pore size distribution measurements are already introduced in the previous chapter, and then, the present chapter explains the electron microscope observation which is an effective means to evaluate the composite structure.

In electron microscope observation, sample preparation can play a key role to the success. Especially, for evaluating the composite structure of the nanoparticles, it is necessary to evaluate its internal microstructure, and accurate observation is impossible even using high-performance electron microscope if the sample preparation is inapposite. The sample preparation methods for the electron microscope observation are as follows : (1) Dispersion method in which the sample is directly on a sample holder after dispersing in a solution by ultrasonic vibration, (2) Crushing method where the sample is set on a holder after being crushed with mortar etc., (3) Embedding method that forms observation side of the sample embedded in epoxy resin etc. by polishing, (4) Ultramicrotomy that forms observation ultrathin flake of the sample embedded in epoxy resin etc. by cutting, (5) Ion milling method that forms observation side of the sample or the sample embedded in resin using ion beam, (6) Etching method with which observation side is formed by etching of bulk sample. Embedding, ultramicrotomy, ion milling and etching are effective methods for observing the internal microstructure. In-depth knowledge of the sample and skill are required to prepare a proper sample for the observation.

SEM is a method of irradiating the electron beam to the sample, and obtaining the observation image based on the generated second electron beam. For SEM observation, conductivity is necessary for the sample, and the conductive thin film of gold etc. is formed by sputtering on the surface of the insulation material. There is little shape restriction in the sample, and an easy and effective evaluation is possible. Field emission type SEM can observe at a high magnification of about 300,000 times, and can observe nanosize microstructure. On the other hand, TEM is a method of obtaining observation images through the sample using the transmission electron beam, and then the thickness of the sample greatly influences the observation. It is necessary for the TEM observation that the sample thickness is less than 100 nm, and the sample preparation plays a very important role. Both SEM and TEM can set up energy dispersive spectrometer (EDS), which enable the elemental analysis at the observation point in addition to the shape and microstructure observation. The EDS is a device for analyzing the elements of the sample quantitatively or qualitatively, and can measure the elemental distribution of the surface of sample.

In the internal microstructure analysis of the nanoparticles, TEM–EDS with sample preparation by ultramicrotomy is most effective. An ultramicrotome is used for the ultramicrotomy, and can prepare an ultrathin flake of a thickness less than 100 nm. The sample such as powder is embedded in epoxy resin etc. The sample embedded in a resin is driven with a predefined thickness in each step continuously, and cut repeatedly with a knife to obtain the ultrathin flake. The thickness of ultrathin flake can be controlled by either thermal expansion or accurate mechanical drive. For obtaining a proper ultrathin flake in which the embedded particles are not separated from the resin or destroyed, it is important to match hardness of both the particle sample and the resin, and to select material of the knife used for cutting. There are glass, sapphire, and diamond as a knife material. The glass is used for the knife to cut the soft material such as biomechanical materials, and the sapphire and diamond are used to cut hard materials such as ceramics and metals.

1.7.3 Microstructure evaluation of several types of nano composite particles

Some examples of the composite structure evaluation of typical nanoparticles by TEM and SEM are introduced below.

(1) Nano particle coating (Core-partial shell) composite structure [1, 2, 3]

Fig. 1.7.2 shows a TEM image of platinum-supported carbon. The platinum-supported carbon is used as electrodes of polymer-electrolyte fuel cells (PEFC). In this figure, black points are platinum particles, and it is confirmed that the surface of carbon particle is

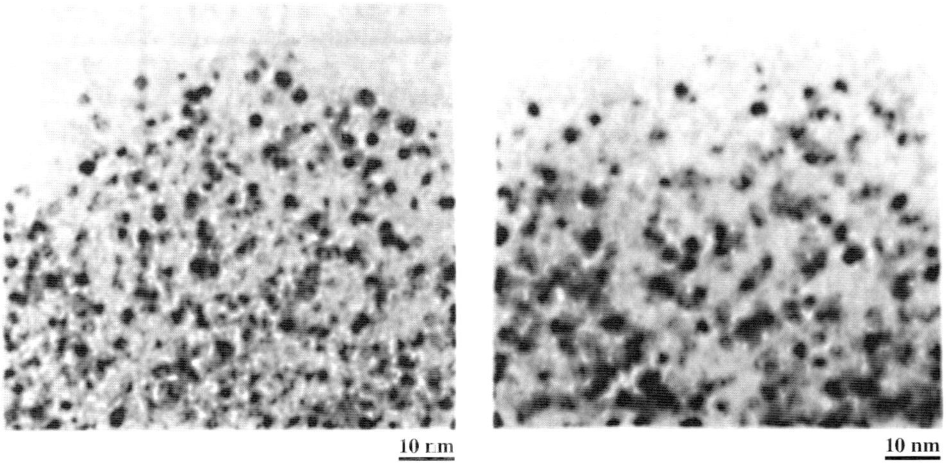

Figure 1.7.2
TEM image of Pt nanoparticles coated on the surface of carbon (electrode of PEM fuel cell) [1].

Figure 1.7.3
SEM image of NiO-YSZ composite particle [2].

coated with the nano platinum particles. For improving performance of PEFC electrode, it is essential to highly disperse finer platinum particles. The TEM observation can evaluate particle size and dispersion state of the nanoparticles.

Fig. 1.7.3 shows a TEM image of NiO-YSZ(Y_2O_3 stabilized ZrO_2) composite particles fabricated by mechanochemical technique [2] . This technique is defined as a novel method to create chemical bonds between particles in dry process without any binder. In this figure, surface fine particles with a size of about 100 nm are YSZ, and core particles are NiO. It is confirmed from this figure that fine YSZ particles partially cover the surface of NiO particles. In short, the core–shell structure, actually core-partial shell structure, has been made.

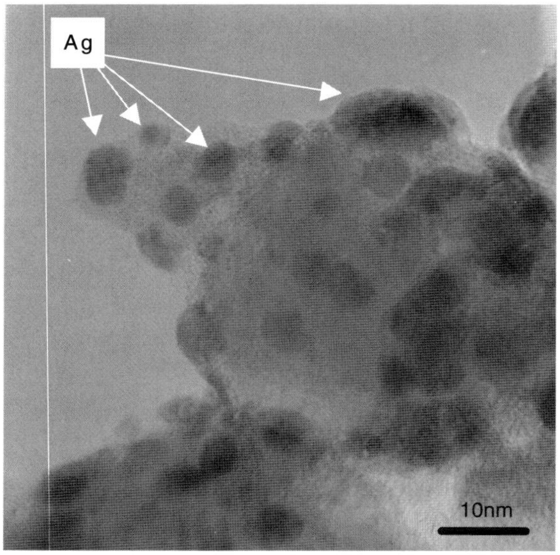

Figure 1.7.4
TEM image of TiO_2-Ag composite particle [3].

Fig. 1.7.4 shows a TEM image of TiO_2-Ag composite particles fabricated by spray-pyrolysis [3]. Element structure of this composite particle was also evaluated by EDS at the same time as TEM observation. The composite structure where the Ag nanoparticles of less than 10 nm are supported on the surface of TiO_2 can be confirmed from the TEM–EDS observation.

(2) Internal dispersion composite structure [4]

Fig. 1.7.5a shows a SEM image of La(Sr)MnO_3 (LSM)-YSZ composite particle fabricated by spray-pyrolysis, and Fig. 1.7.5b shows a TEM image of the cross section of the composite particle [4]. An ultrathin flake prepared by ultramicrotomy was applied for this TEM observation. It can be confirmed from these figures that the composite particles have been composed of dispersed nano LSM and YSZ particles.

(3) Core–shell composite structure [5]

Fig. 1.7.6a shows a SEM image of NiO-YSZ composite particle fabricated by spray-pyrolysis, and Fig. 1.7.6b shows a TEM image of the cross section of the composite particle [5]. An ultrathin flake was also applied for this TEM observation. Moreover, it was found by EDS and X-ray diffraction pattern that the surface shell nanoparticles were YSZ, and core particles were NiO. It can be confirmed from these observations that the composite structure consists of NiO core particles and YSZ shell particles. The above-mentioned results show that the TEM–EDS observation with sample preparation by ultramicrotomy is effective for evaluation of an internal microstructure. Moreover, it is understood that a composite structural becomes clearer by elemental analysis and X-ray diffraction measurement etc.

(4) Porous body composite structure [6]

Fig. 1.7.7 shows a SEM image of cross section of $LiCoO_2$-NiO cathode developed for molten carbonate fuel cells (MCFC). A sample prepared by embedding method was applied for this SEM–EDS observation. It is confirmed from this figure that the MCFC cathode has core–shell structure. Moreover, it is clear by line element analysis using EDS that Co exists only in the shell. The porous body with core–shell composite structure can be specified by these observations.

Figure 1.7.5
(a) SEM image of LSM-YSZ composite particles, (b) TEM image of the cross section of LSM-YSZ composite particle [4].

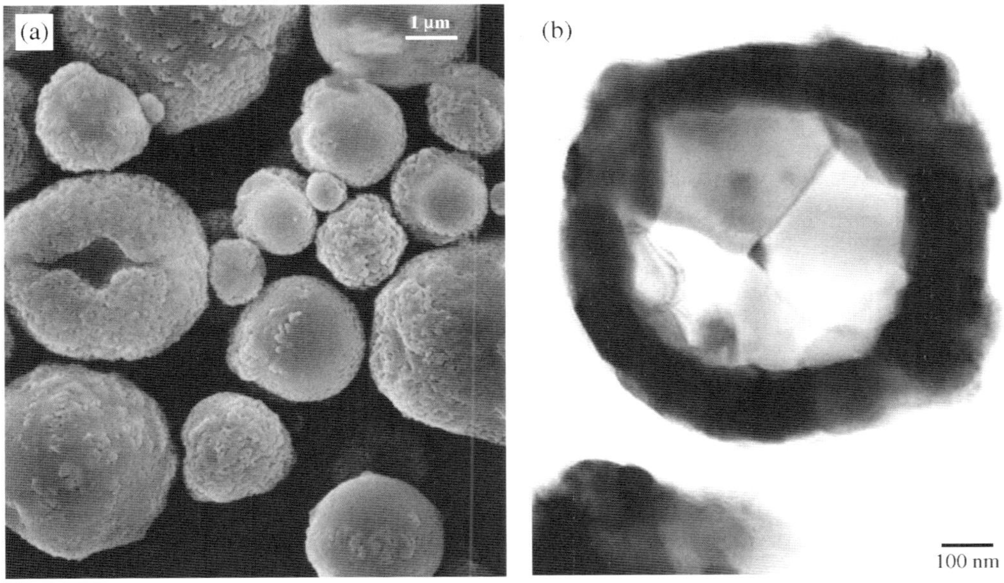

Figure 1.7.6
(a) SEM image of NiO-YSZ composite particles, (b) TEM image of the cross section of NiO-YSZ composite particle [5].

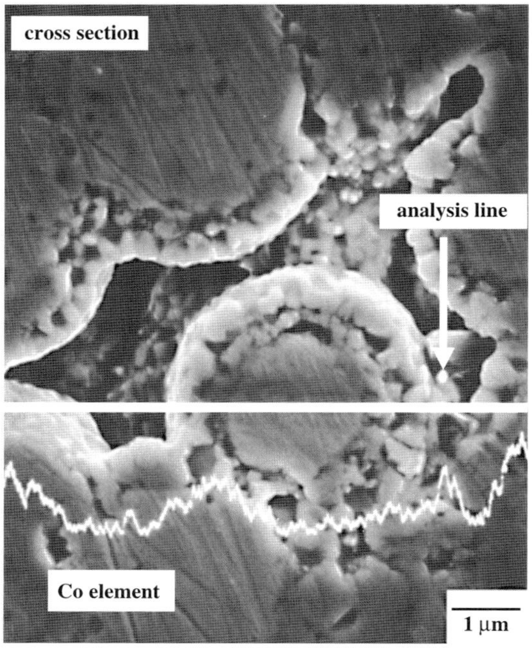

Figure 1.7.7
SEM image of the cross section of $LiCoNiO_2$ coated NiO grains [6].

References

[1] K. Yasuda, T. Ioroi: *Surface*, **76**, 563–567 (2000).
[2] K. Murata, T. Fukui, C.C. Huang, M. Naito, H. Abe and K. Nogi: *JCEJ*, **37**, 568–571 (2004).
[3] M. Matsumoto, K. Kaneko, Y. Yasutomi, S. Ohara and T. Fukui: *J. Ceram. Soc. Jpn.*, **110**, 60–62 (2002).
[4] T. Fukui, T. Oobuchi, Y. Ikuhara, S. Ohara and K. Kodera: *J. Am. Ceram. Soc.*, **80**, 261–263 (1997).
[5] T. Fukui, S. Ohara and K. Mukai: *Electrochem. Solid-State Lett.*, **1**, 120–122 (1998).
[6] T. Fukui, H. Okawa, T. Hotta, M. Naito and T. Yokoyama: *J. Am. Ceram. Soc.*, **84**, 233–235 (2001).

1.8 Crystal structure

The crystal structure of particles can be described by the same crystal symmetry as a bulk material, if the particle size is large enough. Therefore, we should take into account for the crystal chemistry of the bulk materials because the crystal symmetry of the materials is uniquely determined by the ambient atmosphere, independent of the morphology. Namely, crystal symmetry is determined only by the physical arrangement of the constituent elements, and the chemical bonding of the constituent elements makes no difference on the crystal symmetry. Crystal structure can be classified into the 7 crystal symmetries, 14 Bravais lattices and 32 crystal classes. These classifications are listed in Table 1.8.1. Typical crystal symmetries are expressed by two types of close packing structures and their stuffed structures of atoms and/or atom groups into the interstitial spaces of the close packing structures [1].

On the other hand, crystal structures of the nanoparticles for advanced materials depend on the particle size and the thermodynamic stability of the materials. Namely, we have to take into account such parameters for the crystal symmetries of the materials and these are recognized as so-called quantum effect or quantum size effect (hereafter, called size effect). The size effect of nanoparticles is observed if the particle diameter is small enough and then the surface energy becomes considerably large compared to the lattice energy of the nanocrystal. Therefore, the critical particle diameter, below which the size effect will be expected, can be changed by the intrinsic and/or the extrinsic conditions such as the kind of compound and the atmosphere, the preparation method and so on. Such variable factors make the interpretation of the size effect complicated. However, the development of recent analytical methods shows the possibility to provide the means to determine the critical size for advanced materials. In this section, all of these cannot be shown but some case studies are listed, which contain the crystal structure and the size effect of zirconium oxide (zirconia) and the ferroelectric nanoparticles to consider the crystal symmetry of the nanoparticle.

1.8.1 Particle size dependence of crystalline phases of zirconia

The phase transition of zirconia, which exhibits high fracture toughness and strength to apply as structural ceramics and high ionic conductivity at relatively low temperature to apply as oxygen sensor for automobile and solid electrolyte for fuel cell, is the typical example to show the crystal structure change by the thermodynamic reason. Zirconia undergoes unique phase transition behavior, so-called martensitic transformation. Crystal symmetry of zirconia changes from cubic to tetragonal at around 1000°C and tetragonal to monoclinic at lower temperature and these transition temperatures can be controlled by dissolving the yttria (Y_2O_3) and magnesia (MgO) to stabilize the tetragonal or cubic phase at room temperature (the so-called partially stabilized and stabilized zirconia, respectively). The application of zirconia ceramics to the structural materials or the functional materials by utilizing unique phase transition mechanism and its stabilizing technique are well-known and very useful [2, 3].

Table 1.8.1
Crystal symmetries and Bravais lattices.

Crystal symmetry	Bravais lattices			
Triclinic	$\alpha, \beta, \gamma \neq 90°$ 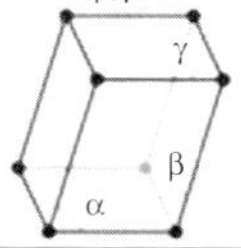			
	Primitive	Base Centered		
Monoclinic	$\alpha \neq 90°$ $\beta, \gamma = 90°$ 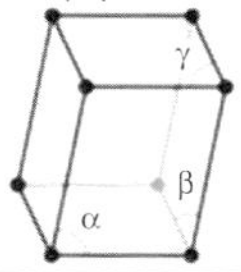	$\alpha \neq 90°$ $\beta, \gamma = 90°$ 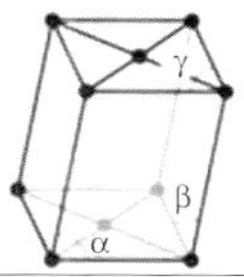		
	Primitive	Base Centered	Body Centered	Face Centered
Orthorhombic	$a \neq b \neq c$ 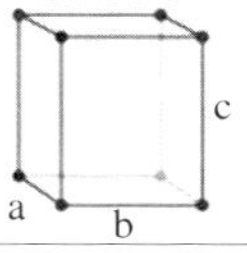	$a \neq b \neq c$ 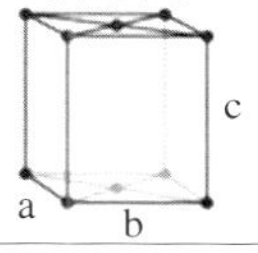	$a \neq b \neq c$ 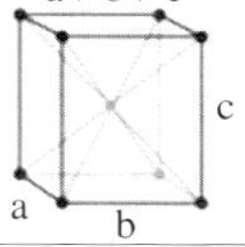	$a \neq b \neq c$ 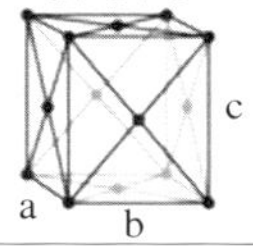
Hexagonal	$a \neq c$ 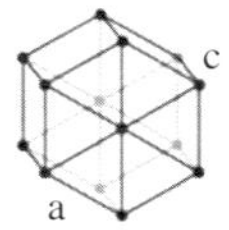			
Rhombohedral	$\alpha, \beta, \gamma \neq 90°$ 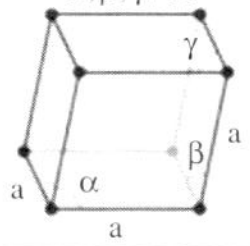			
	Primitive	Body Centered		
Tetragonal	$a \neq c$ 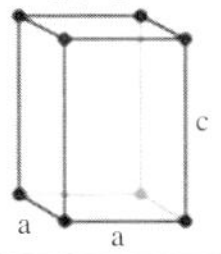	$a \neq c$ 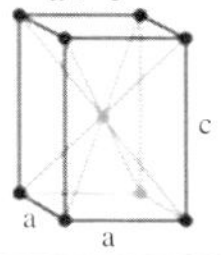		
	Primitive	Body Centered	Face Centered	
Cubic	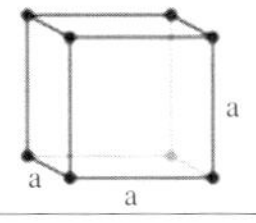	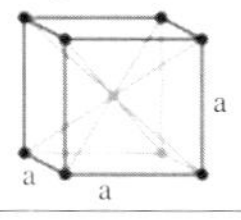	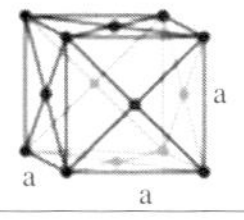	

There are two controversial theories for the formation of metastable tetragonal zirconia. One theory claims that a tetragonal phase forms thermodynamically [4, 5] and the other insists that it forms kinetically [6]. In the former theory a tetragonal phase is stabilized by the strain energy originated from impurity incorporation, surface energy of nanoparticles and constrained force by the matrix. And in the latter theory a tetragonal phase forms preferentially from a cubic phase and exists metastably at around room temperature by some reasons such as particle diameter and atmosphere. The latter is explained schematically in Fig. 1.8.1 [6]. From this figure, it is easily understood that higher potential or symmetric phase preferentially appears before the crystallization of the stable phase when the nanocrystal forms from homogeneous amorphous state. Considering the crystallization behavior of nanoparticles, first, very tiny instable preparticles (precursors) gather with each other to form embryo with a critical size to make a nuclei, followed by the stabilization and crystallization.

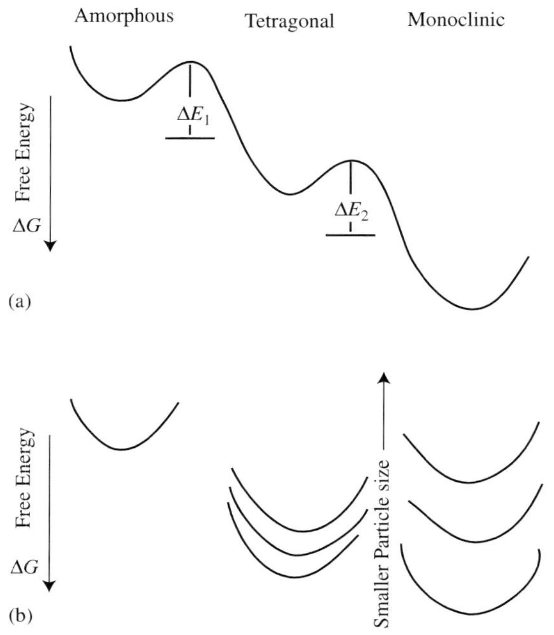

Figure 1.8.1
Free energy change for formation of metastable tetragonal zirconia. (a) Kinetic formation theory, (b) thermodynamic formation theory.

Therefore, preparation conditions such as matrix, atmosphere including temperature and so on significantly affect the crystallization behavior or the resultant crystal phase of the nanoparticles to be a metastable phase or stable phase. For the case of zirconia, it has been reported that the critical size which permits the metastable tetragonal phase is less than 30 nm [7]. On the other hand, tetragonal zirconia with a particle size larger than 30 nm can exist in some cases if the sophisticated preparation method is used and the matrix exerts a strong force of constraint on the zirconia particle crystallized in a composite [8].

As described above, crystal structure of the nanoparticles can change depending on the preparation condition and technique as well as the atmosphere including temperature and matrix and so on. Furthermore, strong force such as the milling can also change the crystal structure of the nanoparticles. Calcium carbonate and zirconia are the typical examples.

1.8.2 Size effect and crystalline phases of ferroelectric materials

The ferrodielectric materials are the promising materials as the core materials in the 21st century because ferroelectric thin films can make the development of high-performance memories and the micro-actuators on silicon and so on. For example, thinner the thickness of the dielectric layer in Multi-Layer Ceramic Capacitor (MLCC), larger the capacitance of the resultant MLCC. Deposition of the high-performance capacitors and the memories of ferroelectric thin films on a silicon wafer can lead to the development of various type nanosensors and Micro-Electro-Mechanical Systems (MEMS).

Functionalities such as piezoelectricity and ferroelectricity of the ferroelectric materials are originated from domain structure of dipole which forms by the slight asymmetry of the crystal structure. Therefore, the crystal structures or the electrical properties of the ferroelectric thin films are significantly affected by the substrate and deposition technique which lead to the difference of the residual stress in films. The control of the residual stress in the ferroelectric thin films is the current topic and many researchers are actively conducting the related studies on the residual stress in ferroelectric thin films.

The most famous ferroelectric materials are the barium titanate ($BaTiO_3$) family as dielectric materials and the lead zirconate titanate ($Pb(Zr, Ti)O_3$) family as piezoelectric materials. In this section, the size effects of these two materials are focused to discuss the crystal structure of the nanoparticles.

Electrical properties such as dielectric and ferroelectric properties are considered to be intrinsic independent of the shape and size of the materials. However, these physical properties of the materials can be variable if the particle size is below several tens of nanometers by the size effect. Therefore, recent semiconductor industry including information technology cannot ignore the size effect because the thickness of the thin films and the grain size in films used in the semiconductor industry ranges from several nm to several tens of nanometers. First, quantum size effect was reported by Känzig et al. in 1953 [9, 10]. Since then, many experimental [11, 12] and theoretical [13–16] approaches have been done to understand this quantum effect. Especially, many studies on the size effect of ferroelectric materials such as barium titanate [17] and lead titanate [18] have been carried out, whereas the origin and the mechanisms of the size effect are poorly understood. Ishikawa et al. proposed the surface relaxation model to explain the size effect, as shown in Fig. 1.8.2, where D and $l(D)$ are thickness and the particle diameter of the cubic phase, respectively. They also determined the critical particle size of barium titanate, strontium titanate and lead titanate from the precise X-ray diffraction analysis [18]. The ferroelectric materials with a particle size larger than the critical size exhibit ferroelectricity and the crystal symmetry of the ferroelectrics changes from tetragonal to cubic at this critical size at room temperature. So far, it has been reported that the critical size of barium titanate (BT) was in the range from 20 to 200 nm [17, 18] and that of lead titanate (PT) was in the range from 7 to 11 nm [19, 20].

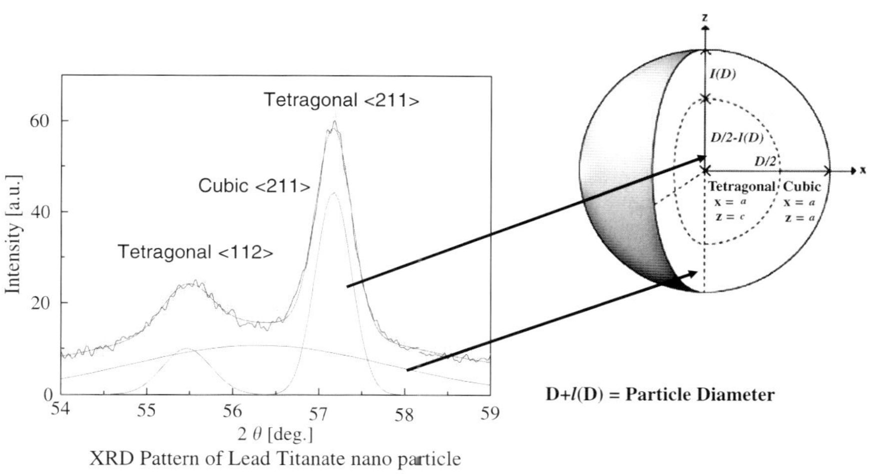

Figure 1.8.2
Surface relaxation model.

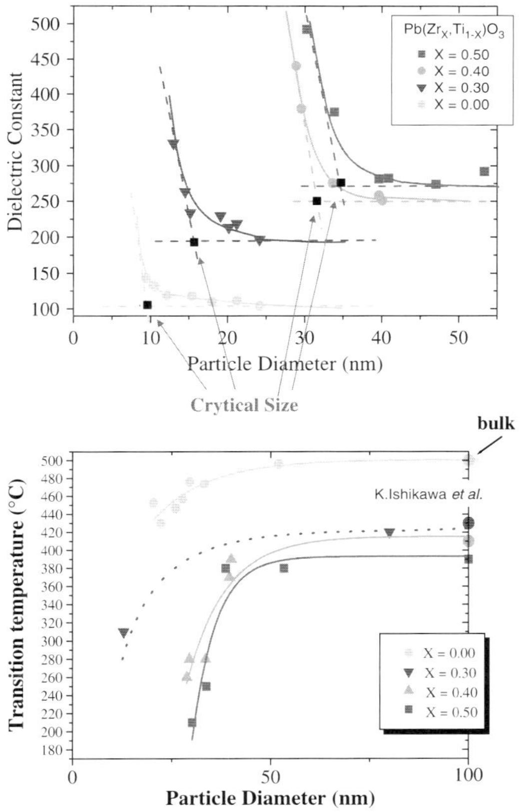

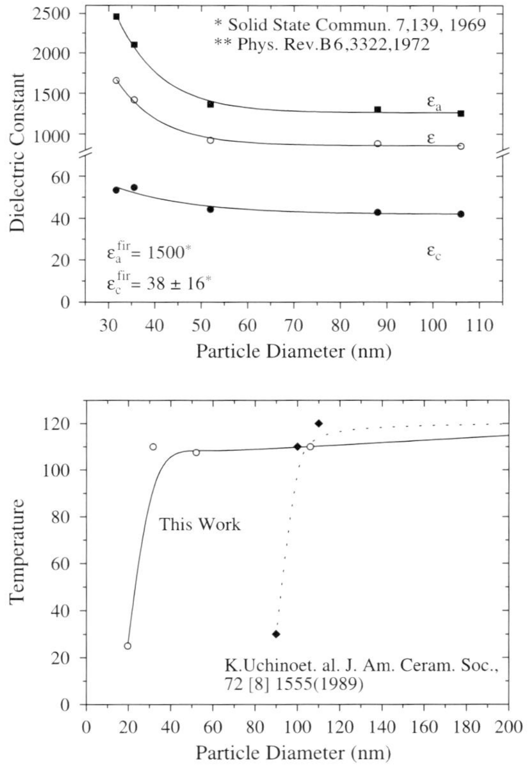

Figure 1.8.3
Size effect for lead zirconate titanate nanoparticles.
(a) Particle size dependence for dielectric constant of PZT nanoparticles calculated by LST relation. (b) Particle size dependence for Curie temperature of PZT nanoparticles.

Figure 1.8.4
Size effect for barium titanate nanoparticles. (a) Particle size dependence for dielectric constant of BT nanoparticles calculated by LST relation. (b) Particle size dependence for Curie temperature of BT nanoparticles.

On the other hand, few studies on the critical size of the attractive and useful materials of lead zirconate titanate (PZT) family have been reported. In addition, fabrication of bulk PZT single crystals for characterization is very hard, especially for the case of PZT with a morphotropic phase boundary (MPB; molar ratio of Zr to Ti is 53:47) composition which exhibits many excellent electrical properties such as ferroelectric and piezoelectric properties. Therefore, many intrinsic data for the PZT materials with a composition near MPB are basically insufficient from the theoretical prospect. Furthermore, characterization techniques have not been established for the evaluation of the intrinsic properties of the solid solution system of PZT family. Recently, analytical method to characterize the critical size and the dielectric property for the ferroelectric nanoparticles has been developed using Raman spectroscopy [21]. The critical size with which crystal symmetry of the particles changes can be measured by this method. The critical sizes of BT and PZT nanoparticles are shown in Figs 1.8.3 and 1.8.4, respectively, together with the particle size dependence of the Curie temperature at which crystal symmetry transforms from tetragonal to cubic. These figures clearly show that the chemical composition, particle size and the residual stress in the nanoparticles can change the crystal structure of the nanoparticles.

References

[1] F.S. Galasso (ed.), M. Kato, N. Mizutani and K Uematsu: *Fain-seramikkusu no Kessyou-kagaku, Agune-gijutusenta*, ISBN;4-900041-01-7.
[2] N. Claussen: *J. Am. Ceram. Soc.*, **61** (1-2), 85–86 (1978).
[3] S. Soumiya (ed.): *Zirukonia-Seramikkusu*, Uchidarou-kakuho, pp.109–125 (1983).
[4] R.C Garvie: *J. Phys. Chem.*, **69** (4), 1238–1243 (1965).
[5] J.E. Bailey, D. Lewis, Z.S.M. Librant and L.J. Porter: *Trans. J. Br. Ceram. Soc.*, **71**, 25–30 (1972).
[6] E. Tani, M. Yoshimura and S. Somiya: *J. Am. Ceram. Soc.*, **66** (1), 11–14 (1983).
[7] R.C. Gsrvie, R.H. Hannink and R.T. Pascoe: *Nature*, **258**, 703 (1975).
[8] T. Mitsuhashi, M. Ichihara and U. Tatsuke: *J. Am. Ceram. Soc.*, **57**, 97 (1974).
[9] C. Jaccard, W. Känzig and W. Peter: *Helv. Phys. Acta*, **26**, 521 (1953).
[10] W. Känzig, R. Sommerhalder: *Helv. Phys. Acta*, **26**, 603 (1953).
[11] V. Kuleshov, M.G. Radchenko, V.D. Dudkevich and Eu.G. Fesenko: *Cryst. Res. Technol.*, **18**, K56 (1983).
[12] T. Kanata, T. Yoshikawa and K. Kubota: *Solid State Commun.*, **62**, 765 (1987).
[13] I.P. Batra, P. Würfel and B.D. Silverman: *Phys. Rev. Lett.* **30**, 384 (1973).
[14] I.P. Batra, P. Würfel and B.D. Silverman: *Phys. Rev. B*, **8**, 3257 (1973).
[15] R. Kretschmer, K. Binder: *Phys. Rev. B*, **20**, 1065 (1979).
[16] A.J. Bell, A.J. Moulson: *Ferroelectrics*, **54**, 147 (1984).
[17] B.D. Begg, E.R. Vance and J. Nowotny: *J. Am. Ceram. Soc.*, 77, 3186 (1994).
[18] K. Ishikawa, T. Uemori: *Phys. Rev.B*, **60**, 11841 (1999).
[19] K. Ishikawa, K. Yoshikawa and N. Okada: *Phys. Rev. B*, **37**, 5852 (1988).
[20] S. Chattopadhyay, P. Ayyub, V.R. Palkar and M. Multani: *Phys. Rev. B*, **52**, 13177 (1995).
[21] H. Suzuki, T. Ohno: *J. Soc. Powder Technol., Jpn.*, **39** (12), 877–884 (2002).

1.9 Surface characteristics

In microscopic view of an atomic level, a solid surface is in an unsaturated state that the continuity of bonds is broken up. Atoms, ions or molecules constructing a solid surface cannot diffuse, even if potential energy differs between adjacent ones, because the activation energy required for surface diffusion is in general quite high, and hence homogenization of the surface cannot be achieved as taking place on liquid surface. Due to these two features, the solid surface shows unique characteristics that cannot be expected from bulk structure of the solid. Especially for nanoparticles, the number of atoms, ions or molecules making up the particle becomes larger on the surface rather than within the particle. For this reason, a number of unique characteristics specific to nanoparticles appear [1, 2].

These characteristics give strong effects not only to quantum size effect specific to nanoparticles but also to phenomena relating to powder handling such as adhesion and coagulation [3]. Therefore, to understand and control the characteristics of the particle surface are often key technology for success in researches and developments with regard to nanoparticles.

As described above, a solid surface newly formed is in a state that the bonds are broken up. For this reason, surface relaxations in various forms occur in order to stabilize the surface. The relaxations depend on the type of chemical bond such as covalent, ionic or metallic bond, and also on difference of materials. For convenience, the relaxation involving chemical changes is termed here as the chemical relaxation, and one without the changes are termed as the physical relaxation, whereas the both relaxations take place simultaneously in most cases. It is also desirable that nanoparticles are dealt with under atmospheric pressure in most industrial cases. In this case, surface relaxation due to adsorption of atmospheric components occurs simultaneously as well. For this reason, stabilization due to the adsorption will also be described in this section.

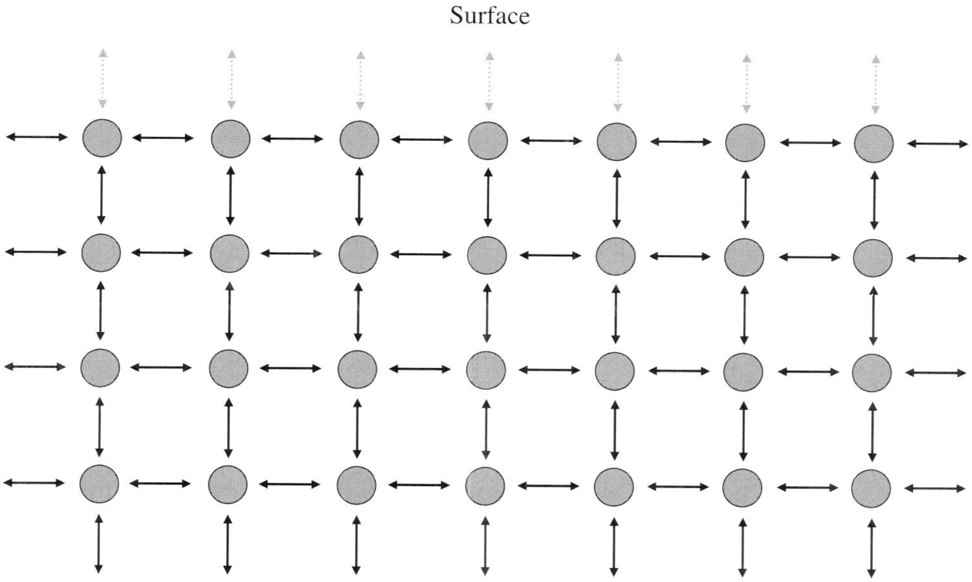

Figure 1.9.1
Surface and inside of solid.

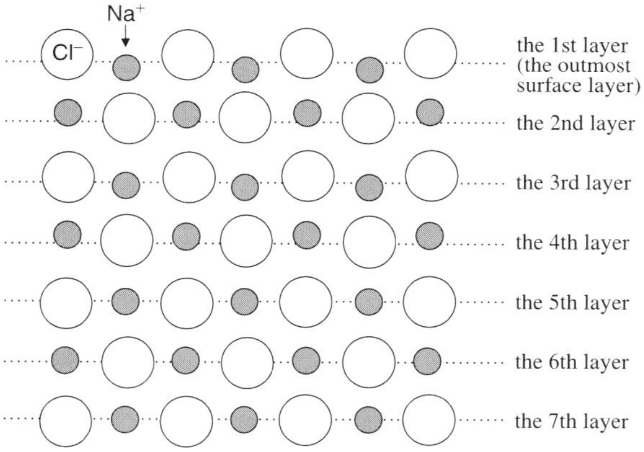

Figure 1.9.2
Physical relaxation model on NaCl (100) face: dotted lines indicate position of ions for inner crystal lattice.

As the bonds are broken up on the solid surface, the atoms, ions and molecules on the surface are subject to asymmetric interaction forces, as shown in Fig. 1.9.1. In the case of crystal materials, it is known that the position of atoms (or ions) located on the surface varies from that expected by periodicity of atoms (or ions) inside the crystal. This is the physical surface relaxation.

For example as shown in Fig. 1.9.2 [4], on the surface of halogenated alkali which is an ionic crystal material, vertical position of cations and anions near the surface varies above and below the lattice points expected as bulk. On the outmost surface layer of crystal (100) face, anions Cl^- polarized by adjacent cations Na^+ move toward outer side, in contrast the cations move toward inner side: this forms an electric double layer and consequently make the ions stabilized. Thus, the central positions of the both ions are not in the same plane. The gap between the 1st and 2nd surface layers, of which position is deduced from the average position of the both ions forming surface asperity, is compressed and shorter than the gap between layers inside the crystal. For the NaCl crystal, it is said that such variation in lattice points would reach to even the 5th layer.

Supposing a nanocrystal particle consisting of 10 layers, all the lattice gaps vary from the usual lattice gap. This is nothing but the change in the shape of electron cloud. Therefore, the value of the band gap may vary for semiconductive particles. Also, abnormity in melting point or solubility of nanoparticles may be understood by similar reasons above. These are a part of surface scientific interpretations to quantum size effect for nanoparticles.

Data relating to physicochemical relaxation phenomena are shown in Table 1.9.1 [5]. For the crystal face (001) of MgO which has a higher hardness and a larger bonding energy as compared to other ionic crystals of halogenated alkali, the distance between ions of the 1st and 2nd surface layer is compressed to around 85 % of the distance for the bulk. Manner of surface relaxation changes presumably due to differences in ion species, coordination number, crystal

Table 1.9.1
Physical surface relaxation.

	Crystal structure	(110)	(001)	(111)
Cu	Face-centered cubic	0.804	0.871	0.944
Al	Face-centered cubic	0.900	1.00	1.15
Fe	Body-centered cubic	1.00	0.986	0.85
NaF	Face-centered cubic	–	0.972	–
NaCl	Face-centered cubic	–	0.970	–
NaBr	Face-centered cubic	–	0.972	–
MgO	Face-centered cubic	–	0.85	–

Note: Values in this table indicate lattice gap between the 1st and 2nd layer divided by lattice gap for bulk crystal.

faces and bonding energy. Similar relaxation occurs on surface of metal crystals, and the surface relaxation is stronger on a crystal face having a lower atomic density. In this case, the region in which the relaxation occurs would be up to around two layers.

It has been already described that solid surfaces are highly reactive since they are unsaturated in terms of bonding. It is known for a solid left in the atmosphere that the highly reactive surfaces react chemically with gases in the atmosphere such as oxygen, carbon dioxide, water vapor, and then form a surface composition different from chemical composition of the bulk.

Oxidization and erosion of metal surface due to oxygen and water vapor in the atmosphere is usual phenomena taking place everywhere. Also on the surface of nitrides and fluorides, atmospheric oxygen adsorbs chemically and then oxides are formed. Chemical adsorption of water vapor takes place on the surface of the oxides, and then hydroxyl groups are eventually formed. This is the chemical surface relaxation.

Examples of the chemical adsorption of oxygen on the surface of metals, nitrides and fluorides are given as follows:

$$2Cu + O_2 \rightarrow 2CuO \tag{1.9.1}$$

$$4AlN + 3O_2 \rightarrow 2Al_2O_3 + N_2 \tag{1.9.2}$$

$$Si_3N_4 + 3O_2 \rightarrow 3SiO_2 + 4N_2 \tag{1.9.3}$$

In general, chemical adsorption of water vapor takes place easily on the surface of oxides, which is eventually covered with hydroxyl groups. An example for silica is as follows:

$$\equiv Si-O-Si \equiv + H_2O \rightarrow 2 \equiv Si-OH \tag{1.9.4}$$

For chemical adsorption of water vapor, there is a case to adsorb only on the surface layer and another case to adsorb on layers extending to the inside of the solid. It is known that for MgO the chemical adsorption takes place not only on the surface but also on the inside and hence layers of magnesium hydroxide are formed.

On the surface of nitrides and fluorides, the chemical adsorption of water vapor takes place and then oxides are formed as shown in the following equations. Eventually, the surface of the oxides is hydrolyzed due to further chemical adsorption of water vapor.

$$\begin{aligned} &Si_3N_4 + 6H_2O \rightarrow 3SiO_2 + 4NH_3, \\ &\equiv \text{Si-O-Si} \equiv + H_2O \rightarrow 2 \equiv \text{Si-OH} \end{aligned} \tag{1.9.5}$$

$$\begin{aligned} &2TiB_2 + 6H_2O \rightarrow 2TiO_2 + 2B_2H_6 + O_2, \\ &\equiv \text{Ti-O-Ti} \equiv + H_2O \rightarrow 2 \equiv \text{Ti-OH} \end{aligned} \tag{1.9.6}$$

The surface modified by chemical and physical relaxations as described above makes surface energy lower in the atmosphere, and hence a variety of physical adsorption phenomena takes place. On a polar surface, physical adsorption takes place and surface free energy decreases. On a hydrophilic surface, adsorption layer of around two molecules are formed at a relative humidity less than about 60 %.

The adsorbed water makes solid surface stabilized, and gives a strong effect on chemical, mechanical and electrical properties of the material. It also makes a change in the interparticle adhesion force as a particle property and gives a strong effect to phenomena of coagulation and consolidation [6–8].

The decrease in the surface free energy can be obtained from an adsorption isotherm [9]. The adsorption isotherms of water and *n*-octane with respect to nano-sized silica particles are shown in Fig. 1.9.3. The decrease in the surface free energy due to the formation of adsorption layers is called as surface pressure π, and can be related to the Gibbs' adsorption equation as follows:

$$\pi(\Gamma) = RT \int_{\Gamma=0}^{\Gamma} \Gamma d \ln(p/p_0) \tag{1.9.7}$$

where π is the surface pressure, R the gas constant, T the temperature, Γ the absorbed amount, p the equilibrium pressure and p_0 the saturated vapor pressure.

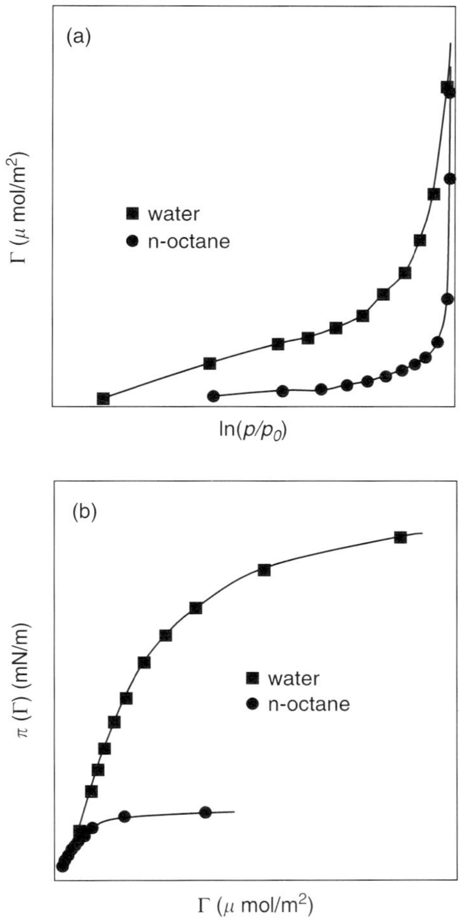

Figure 1.9.3
(a) Adsorption isotherm of water and *n*-octane for non-porous nano silica particle, (b) relationship between surface pressure and adsorbed amounts of water and *n*-octane.

The surface pressure can be given by using the Gibbs' equation, if the relationship between adsorbed amount and equilibrium pressure is obtained for a solid/gas interface. The surface pressure calculated from data shown in Fig. 1.9.3a is shown in Fig. 1.9.3b.

The surface hydroxyl groups formed by the chemical relaxation receive or release protons in water, which makes the surface charged positively or negatively, as shown in the following equations:

$$\text{M-OH} + \text{H}^+ + \text{OH}^- \rightarrow \text{M-OH}_2^+ + \text{OH}^- \quad (1.9.8)$$

$$\text{M-OH} + \text{H}^+ + \text{OH}^- \rightarrow \text{M-O}^- + \text{H}_2\text{O} + \text{H}^+ \quad (1.9.9)$$

The surface charging strongly relates to dispersibility of particles in a liquid. Since the ability of the hydroxyl group to receive protons depends on acidity and basicity, polarity and amount of charges differ from one oxide to another. Table 1.9.2 shows isoelectric point (pH at which interface charge is zero) of major oxides [10].

Table 1.9.2
Isoelectric point of various powders.

Material	Isoelectric point
WO_3	>0.5
SiO_2	1.8–2.5
SiC	3–4
Au	4.3
$Al(OH)_2$	5.0–5.2
SnO_2	6.6
α-FeO(OH)	6.7
TiO_2	6.7
CeO_2	6.75
Cr_2O_3	7.0
γ-FeO(OH)	7.4
$Zn(OH)_2$	7.8
γ-Al_2O_3	7.4–8.6
Y_2O_3	9.0
α-Fe_2O_3	9.04
α-Al_2O_3	9.1
ZnO	9.3
CuO	9.4
BeO	10.2
La_2O_3	10.4
ZrO_2	10–11
$Ni(OH)_2$	11.1
$Co(OH)_2$	11.4
MgO	12.4±0.3

The features of the receptors of H^+ and OH^- influence greatly by which type of the equations above the oxide particles dissociate when they are dispersed in a solvent. Thus, the configuration of interactions on an interface is different according to relative magnitude of acidity/basicity between the particle and the solvent. For instance, if the acidity of the particle surface is strong and the solvent is basic, the surface is charged negatively since the particle gives H^+ to the solvent. In contrast, if the basicity of the particle surface is strong and the solvent is acidic, the surface is charged positively since the particle receives H^+ from the solvent. If there is no or less difference in acidity/basicity between the particle surface and the solvent, the particle would not be charged.

References

[1] R. Kubo: *Phys. Lett.*, **1**, 49 (1962).
[2] M. Chikazawa, T. Takei: *J. Soc. Powder Technol., Jpn.*, **31**, 342 (1994) (in Japanese).
[3] M. Chikazawa, T. Takei: *Gypsum Lime*, **228**, 255 (1990) (in Japanese).

[4] G.C. Benson, T.A. Claxton: *J. Chem. Phys.*, **48**, 1356 (1968).
[5] A . Taroni, D. Haneman: *Surf. Sci.*, **10**, 215 (1968).
[6] M. Chikazawa, T. Takei: *J. Soc. Powder Technol., Jpn.*, **14**, 18 (1977).
[7] N. Harnb, A.E. Hawkins, I. Opalinski: *Trans. IChemE*, **74**, 605 (1996).
[8] M. Chikazawa, T. Kanazawa and T. Yamaguchi: KONA (Powder & particle), **2**, 54 (1984).
[9] L.J.M. Schlangen, L.K. Koopal, M.A. CohenStuart, J. Lyklema: *Colloid Surface A*, **89**, 157 (1994).
[10] R.H. Yoon, T. Salman, G. Donnay: *J. Colloid Interf. Sci.*, **70**, 483 (1979).

1.10 Mechanical property

Mechanical properties vary with the decrease in particle size as well as optical, electrical and magnetic properties. It is well known that the strength of metal and ceramic material improves by decreasing grain size to nano size or by making compositie in nano-scale. In metallic materials, yield stress increases with decrease in grain size (Hall–Petch relation) and then decreases with further decrease in grain size (Inverse Hall–Petch relation). Not only the mechanical property of such a bulk composed of nanoparticles but also that of a particle itself depend on the particle size. Fig. 1.10.1 shows the dependence of the strength of an alumina particle on its particle size. The strength increases with decrease in particle size, which results from decrease in flaw size in the particle [1]. Although such phenomena have been reported in the strength of a whisker, the mechanical properties of a nanoparticle was limited because of difficulty of measurement unlike in the case of the optical, electrical and magnetic properties which can be evaluated as nanoparticle ensemble.

However, because of advancement in technology on high-precision determination of very small load and displacement, the mechanical properties of a nanoparticle also has been measured in recent years. It has been reported that elastic modulus of Au cluster is as small as 2/3 of that of Au bulk as a result of nano indentation and scanning probe microscopy [2]. This result was caused by high internal compressive stress of a nanoparticle. Such inherent compressive stress suggests the possibility of super-hard material. Mechanical properties of a Si nanoparticle of 20–50 nm also have been directly measured. While the hardness of a bulk is ~12 GPa, that of a nanoparticle increases with decrease in particle size, which varied from 20 to 50 GPa (Fig. 1.10.2) [3].

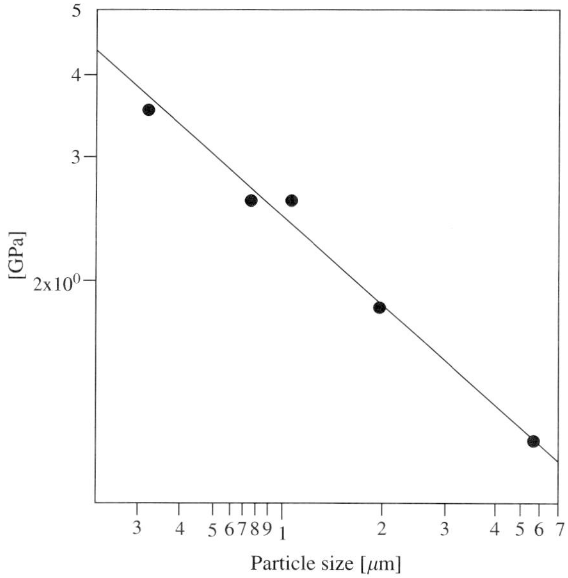

Figure 1.10.1
Dependence of the strength of an alumina particle on its particle size.

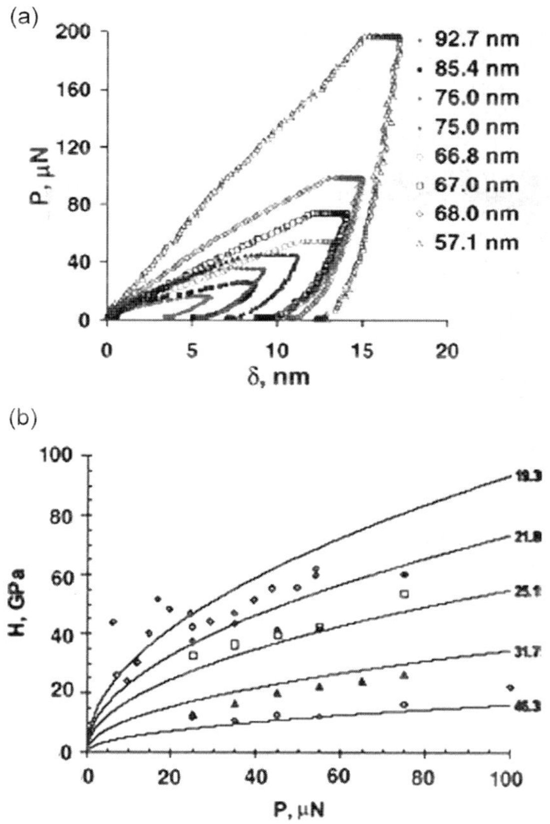

Figure 1.10.2
Mechanical properties of a Si nanoparticle evaluated by nano indentation technique. (a) Load–displacement curve, (b) load dependence of the hardness of a Si nano particle with different particle size.

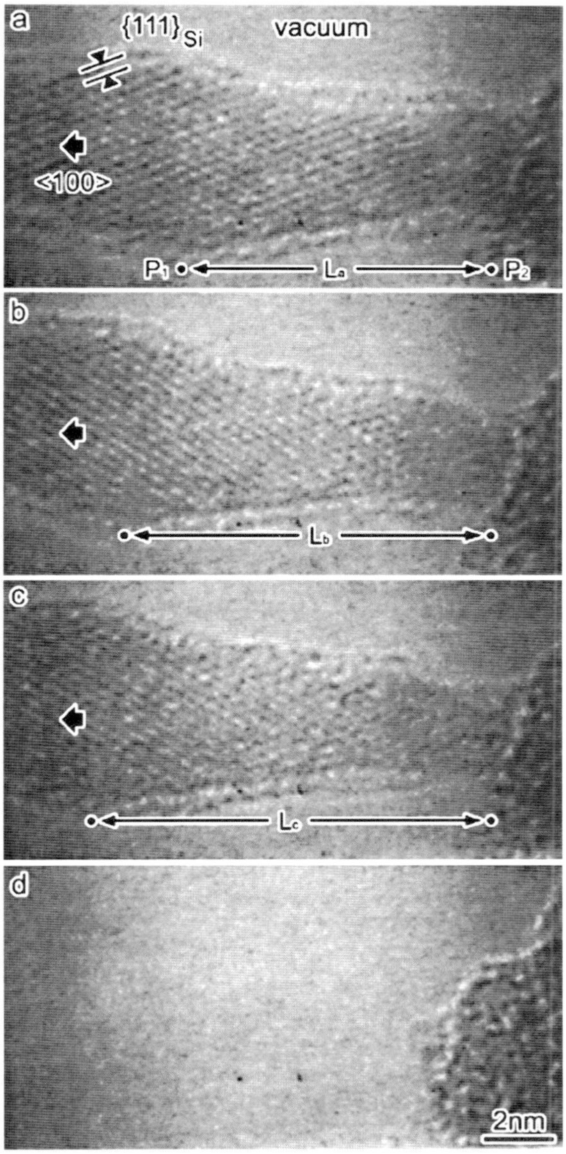

Figure 1.10.3
High resolution image of deformation process of single crystal Si nano wire in in-situ tensile test in transmission electron microscope. (Order of time: a→d.) Tensile direction is <100>

Recent development of transmission electron microscope (TEM) makes in-situ measurement of mechanical properties of nanoparticle possible. Fig. 1.10.3 shows high-resolution and continuous TEM photographs of deformation behavior of a single crystal Si nanowire under tensile stress [4]. Tensile direction is parallel to <110>. Si nanowire extends in

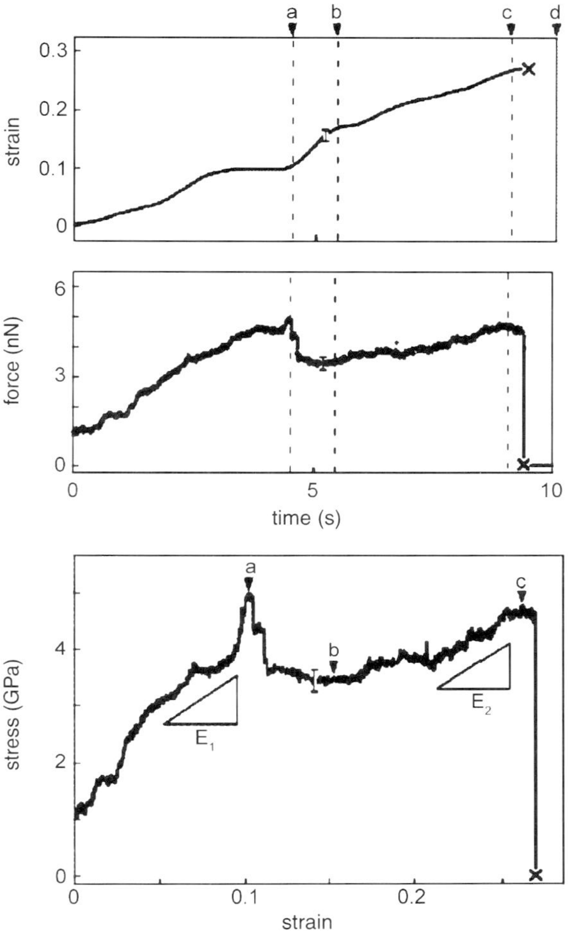

Figure 1.10.4
Stress strain curve of Si nano wire.

order corresponding to La, Lb, Lc (see distance between P1 and P2 in the figure). Finally, nanowire broke in Fig. 1.10.3d. Fig. 1.10.4 indicates a stress-strain curve in the tensile test of Si nanowire. Ultimate tensile strength of the Si nanowire is ~5 GPa. Furthermore, the slopes of the stress-strain curve from 0 to 0.1 and from 0.2 to 0.25 are almost the same. Young's modulus calculated from these regions was 18 GPa. Young's modulus of <110> of a Si bulk with few defects was 18 GPa. It has been also reported that Young modulus and bending strength of a minute Si specimen with 20–30 μm in thickness and 100μm in width were 140–170 GPa and 1–2 GPa, respectively. Consequently, it was shown that single crystal Si nanowire has higher strength and lower young modulus than a micrometer-size specimen. Although the strength of a crystalline material is generally explained by the formation and growth of defects, any dislocations or cracks in the Si nanowire are not observed. As a result, the Si nanowire seems to indicate high strength. Furthermore, Young's modulus of the nanowire was as small as 1/6 of a micrometer-size specimen. Elastic properties are originated from an atomic bonding. The reason why Young's modulus of Si nanowire is smaller is probably that the atomic bonding state was changed by decrease in particle size below 10 nm.

Carbon nanotubes have a cylindrical structure of graphene sheets, which show excellent mechanical properties because of their integrity. Theoretical calculation suggests over thousands GPa of elastic modulus [5]. In the experimental study, mechanical properties of CNTs have been evaluated. In bending test, it is observed that CNT deformed to undulate [6]. Mechanical properties of a CNT are evaluated using an atomic force microscope. As a result, over 1000 GPa is estimated as Young's modulus [7]. Furthermore, tensile test in SEM results in 11–63 GPa and 270–950 GPa for tensile strength and Young's modulus, respectively [8]. Such excellent mechanical properties are used to develop CNT dispersed composites.

References

[1] M. Yoshida, H. Ogiso, S. Nakano and J. Akedo: *Rev. Sci. Instrum.*, **76**, 093905 (2005).

[2] D.M. Schaefer, A. Patil, R.P. Andres and R. Reifenberger: *Phys. Rev. B*, **51**, 5322–5332 (1995).

[3] W.W. Gerberich, W.M. Mook, C.R. Perrey, C.B. Carter, M.I. Baskes, R. Mukherjee, A. Gidwani, J. Heberlein, P.H. McMurry and S.L. Girshick: *J. Mech. Phys. Solid.*, **51**, 979–992 (2003).

[4] T. Kizuka, Y. Takatani, K. Asaka and R. Yoshizaki: *Phys. Rev. B*, **72**, 035333 (2005).

[5] G. Overney, W. Zhong and D. Tomanek: *Z. Phys. D*, **27**, 93–96 (1993).

[6] Y. Ishida, T. Hayashi, H. Ichinose, T. Kuzumaki and K. Ito: *Proc. ICEM*, **13**, 9–10 (1994).

[7] E.W. Wong, P.E. Sheehan and C.M. Lieber: *Science*, **277**, 1971–1975 (1997).

[8] M.F. Yu, O. Lourie, M.J. Dyer, K. Moloni, T.F. Kelly and R.S. Ruoff: *Science*, **287**, 637–640 (2000).

1.11 Electrical properties

1.11.1 Introduction

In general, it is recognized that the electrical properties of the large particles are same as that of the bulk materials, as well as the crystal symmetry. Namely, the electrical properties such as dielectric and ferroelectric properties are the intrinsic independent of the shape and size. However, it is very difficult to estimate the electrical properties of the nanoparticles. For example, the dielectric property of the materials is usually measured by the following process. First, the capacitance of the sample is estimated by impedance analyzer, followed by the measurement of the sample thickness and the surface area of the electrode. Lastly, the dielectric constant of the sample is calculated from these data. However, if we use this method to evaluate the dielectric property, we cannot measure the intrinsic dielectric constant of the nanoparticles, because the measurement using the impedance analyzer requires the bottom and the top electrodes on the nanoparticles. Deposition of the electrodes on the nanoparticles seems to be very difficult.

To overcome this problem, some analysis methods have been attempted so far by many researchers. For example, nanoparticles were molded into the pellet and the capacitance of this pellet was measured by impedance analyzer, same as the bulk materials measurement [1]. Nakamura et al. reported the size effect on the dielectric property for lead titanate which was measured by the above-mentioned method. The result is shown in Fig. 1.11.1. From this figure, the highest dielectric constant was obtained for the lead titanate nanoparticle of 100 nm in diameter. However, in this method, the dielectric constant is affected by the void, the grain boundary and so on. Therefore, the intrinsic dielectric constant of the nanoparticles cannot be determined by this method. In fact, many reports have indicated that the critical size for lead titanate was below 9 nm [2], showing a big difference from the result in Fig. 1.11.1. Namely, it is concluded that the result in Fig. 1.11.1 was affected not only by the size effect but also by the other extrinsic effect such as the low dielectric phase of pores.

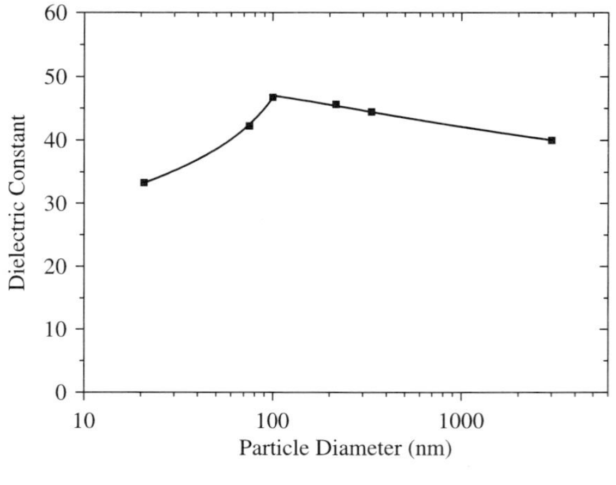

Figure 1.11.1
Particle size dependence for dielectric constant of lead titanate.

1.11.2 Novel characterization method for the dielectric property

The dielectric constant of nanoparticles can be estimated by analyzing the phonon modes of Raman spectra. It seems that we can estimate the intrinsic dielectric constant of nanoparticles by this method [3]. In this method, the intrinsic dielectric constant is calculated using Lyddane–Sachs–Teller relation (LST relation). LST relation is described in the next section [4, 5].

1.11.3 LST relation

First, frequency dependence of the dielectric constant for the ionic crystal (ε_ω) is discussed in the long wavelength limit region (wave number $k = 2\pi/\lambda$). The equation of motion is described as follows when the electric field $E(t)$ was applied to the crystal having a couple of a positive and a negative ion in the structure.

$$M^+\ddot{u}^+ = -f(u^+ - u^-) + qE(t)$$

$$M^-\ddot{u}^- = -f(u^+ - u^-) - qE(t) \tag{1.11.1}$$

where M^+ and M^- are the mass of a positive and a negative ion, respectively, u^+ and u^- the displacement from the balanced position of the positive and the negative ion, f the spring constant between the positive and negative ion, $\pm q$ the electric charge for the ion. At $1/M^* = 1/M^+ + 1/M^-$, $u = u^+ - u^-$, the equation can be arranged as follows:

$$M^*\ddot{u} = -fu + qE(t) \tag{1.11.2}$$

If $E(t)$ can be assumed as the sine wave of the angular frequency, the displacement (u) is calculated as follows:

$$u = \frac{qE(t)}{M^*(\omega_0^2 - \omega^2)} \tag{1.11.3}$$

where ω_0 is the angular frequency in the system. The dipole from the ion displacement is described as $\mu = qu$, therefore, the ionic polarizability (α_{ion}) is described as follows:

$$\alpha_{\text{ion}} = \frac{\partial u}{\partial E} = \frac{q^2}{M^*(\omega_0^2 - \omega^2)} \tag{1.11.4}$$

Then, the polarizability (α) is described as follows:

$$\alpha = \alpha_{\text{elec}} + \frac{q^2}{M^*(\omega_0^2 - \omega^2)} \tag{1.11.5}$$

where α_{elec} is the electronic polarizability.

If the Lorenz relation ($E_{\text{loc}} = E + P/3\varepsilon_0$) is used as the local electric field for the microscopic electric polarization ($P = n\alpha E_{\text{loc}}$),

$$P = n\alpha\left(E + \frac{P}{3\varepsilon_0}\right) \tag{1.11.6}$$

When equation (1.11.6) is solved for P, the susceptibility $\varepsilon_0\chi = P/E$ is obtained as below:

$$\chi = \frac{1}{\varepsilon_0}\frac{P}{E} = \frac{n\alpha/\varepsilon_0}{(1 - n\alpha/3\varepsilon_0)} \tag{1.11.7}$$

and, insert it into $\varepsilon = 1 + \chi$,

$$\frac{\varepsilon - 1}{\varepsilon + 2} = \frac{n\alpha}{3\varepsilon_0} \tag{1.11.8}$$

The above equation is well known as Clausius–Mosotti equation. This relates the dielectric constant to the electronic polarizability, but only for crystal structures for which the Lorentz local field is obtained. And, when the polarizability α is replaced by the above equation, the Clausius–Mosotti equation should be changed as follows:

$$\frac{\varepsilon(\omega) - 1}{\varepsilon(\omega) + 2} = \frac{n}{3\varepsilon_0}\left\{\alpha_{\text{elec}} + \frac{q^2}{M^*(\omega_0^2 - \omega^2)}\right\} \tag{1.11.9}$$

In addition, in the case of low frequency ($\omega = 0$),

$$\frac{\varepsilon(\omega) - 1}{\varepsilon(\omega) + 2} = \frac{n}{3\varepsilon_0}\left\{\alpha_{\text{elec}} + \frac{q^2}{M^*\omega_0^2}\right\} \tag{1.11.10}$$

In contrast, in the case of high frequency which the polarization cannot follow, the second part of the right–hand side of the equation can be ignored, therefore,

$$\frac{\varepsilon(\omega) - 1}{\varepsilon(\omega) + 2} = \frac{n}{3\varepsilon_0}\alpha_{\text{elec}} \tag{1.11.11}$$

Next, the Clausius–Mosotti equation was solved for $\varepsilon(\omega)$ using the above equations to give the following equation:

$$\varepsilon(\omega) = \varepsilon(\infty) + \frac{\varepsilon(0) - \varepsilon(\infty)}{1 - (\omega/\omega_T)^2} \tag{1.11.12}$$

and

$$\omega_T^2 = \omega_0^2\frac{\varepsilon(\infty) + 2}{\varepsilon(0) + 2} \tag{1.11.13}$$

When the frequency for $\varepsilon(\omega)=0$ is described as ω_L, the following equation.

$$\frac{\omega_L^2}{\omega_T^2}=\frac{\varepsilon(0)}{\varepsilon(\infty)} \tag{1.11.14}$$

This equation is well known as Lyddane–Sachs–Teller (LST) relation. In the case of lead titanate, the phonon mode is not the only one. Therefore, the LST relation should be rewritten as follows:

$$\frac{\varepsilon(0)}{\varepsilon(\infty)}=\frac{\omega_{1\mathrm{LO}}^2\,\omega_{2\mathrm{LO}}^2\,\omega_{3\mathrm{LO}}^2}{\omega_{1\mathrm{TO}}^2\,\omega_{2\mathrm{TO}}^2\,\omega_{3\mathrm{TO}}^2} \tag{1.11.15}$$

The subscripts indicate the order of the phonon mode from the lower frequency.

In LST relation, the intrinsic dielectric constant $\varepsilon(0)$ should be calculated using the phonon mode frequency which is determined by Raman analysis. However, the phonon mode using Raman scattering is complicatedly overlapped. Therefore, all the phonon modes need to be divided from the Raman spectrum to determine the exact phonon mode frequencies. The observed Raman spectra were fitted by the damped harmonic oscillators to determine the accurate phonon mode positions. Then the Rayleigh scattering was divided by the Debye relaxation equation. The equation for fitting is as follows:

$$I(\omega)=\left(\frac{1}{e^{\hbar\omega/kT}-1}+1\right)\times\left(\frac{F_r\omega\gamma_r}{\omega^2+\gamma_r^2}+\sum_i\frac{2\Gamma_iF_i\omega\omega_i^3}{(\omega_i^2-\omega^2)^2+4\Gamma_i^2\omega_i^2\omega^2}\right) \tag{1.11.16}$$

where the former term is the Bose–Einstein factor and, ω_i, Γ_i and F_i are the mode frequency, damping factor and oscillator strength, respectively. Fig. 1.11.2 shows the Raman spectra for the PT nanoparticles with different sizes and the fitting results by the above equation, together with the assignment for all the phonon modes. The Raman spectra were measured using triple monochrometer system in the range from 10 to 900 cm^{-1} at room temperature. The 488 nm line of an argon ion laser with 100 mW power was used as the excitation source. In addition, all the slit width was adjusted at 100 μm.

At room temperature, the crystal symmetry of lead titanate is reported to be tetragonal. In this case, the double-degenerated E mode ([100] direction) and the non-degenerated A1 mode ([001] direction) are observed, since the space group of the tetragonal perovskite structure is C_{4v}^1. In addition, the longitudinal

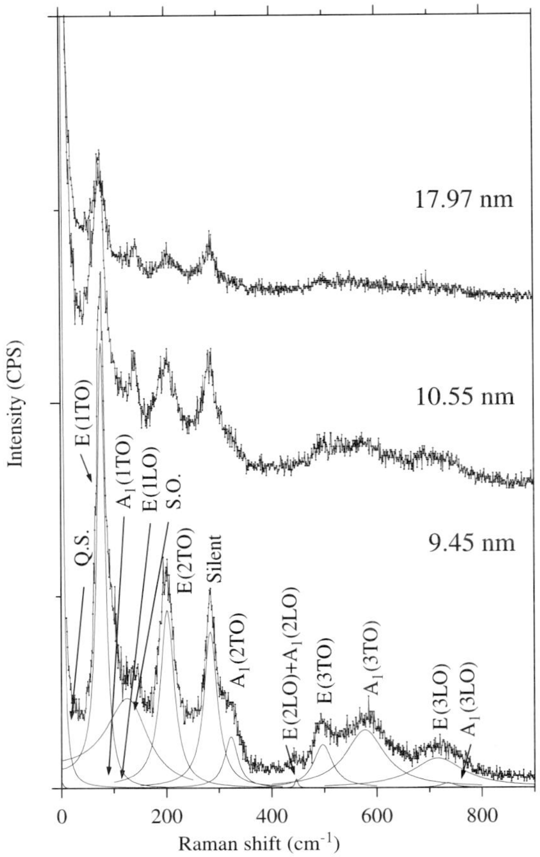

Figure 1.11.2
Change in Raman spectra of lead titanate with particle size and assignment of the phonon mode.

optical phonon mode (LO mode) and the transverse optical phonon mode (TO mode) are observed in E and A mode. In the case of lead titanate nanoparticles, 12 modes ($3(A_1+E)(LO)+3(A_1+E)(TO)$) generated from three T_{1u} mode and B + E mode (silent mode) generated from T_{2u} mode are observed by Raman scattering. In general, TO mode frequencies are lower than that of LO mode, and E mode frequencies are lower than that of A mode. The separation and the assignment of the phonon mode from Raman spectra have been done according to the above procedure.

1.11.4 Measurement of the dielectric constant of nanoparticles

Fig. 1.11.3 shows the particle size dependence of the dielectric constant of PT nanoparticles, which were calculated from the mode frequencies, determined

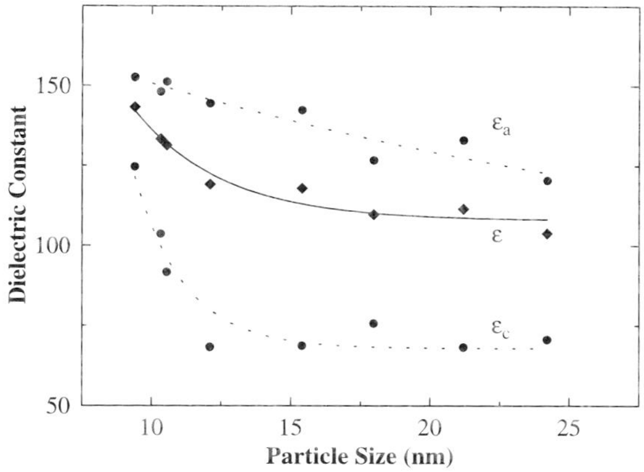

Figure 1.11.3
Particle size dependence for dielectric constant of PT nanoparticles.

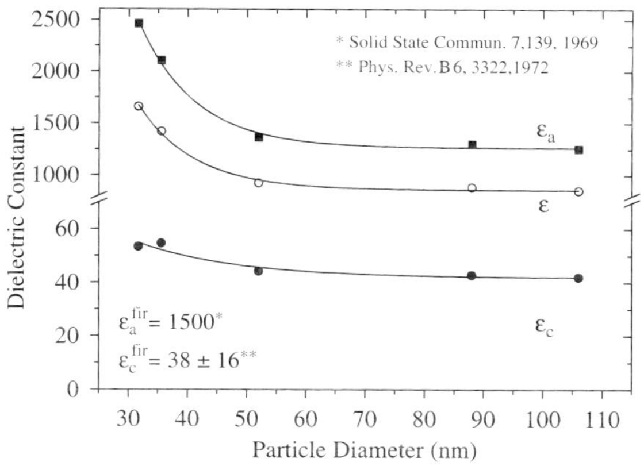

Figure 1.11.5
Particle size dependence for dielectric constant of BT nanoparticles.

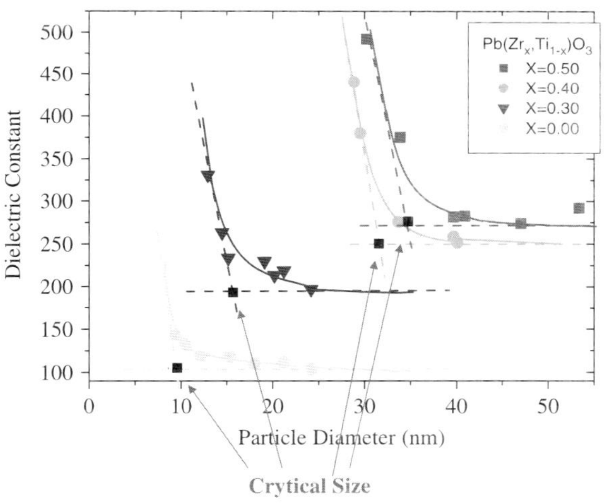

Figure 1.11.4
Particle size dependence for dielectric constant of PZT nanoparticles.

using LST relation in Fig. 1.11.2. From this figure, the dielectric constant started to increase at around 15–20 nm of the particle size. This result is in good agreement with the tendency in the *c/a* ratio change estimated by XRD analysis. In general, the critical size of lead titanate is reported to be up to 9 nm, whereas the electrical properties started to change as the particle size reached at around 15–20 nm. The dielectric constants of lead titanate single crystal are reported as $\varepsilon_a = 111$, $\varepsilon_c = 55$, respectively [6]. In contrast, the dielectric constant of lead titanate estimated by LST relation exhibited almost the same value as that of the normal one for the larger particles without size effect. Therefore, it is concluded that the dielectric constant calculated using LST relation is reasonable. The intrinsic dielectric constant of lead titanate increased with decreasing particle size, since the size effect lead to the decrease in Curie temperature [2]. It is well known that the dielectric constant is abruptly increased at around Curie temperature. Therefore, the size region to increase the dielectric constant also decreases with decreasing particle size, corresponding to the Curie temperature shift.

Furthermore, particle size dependences of the dielectric constant of lead zirconate titanate and barium titanate are shown in Figs 1.11.4 and 1.11.5, respectively. These results indicated that the intrinsic dielectric constant increases with decreasing particle size by the size effect. Other electrical properties should be affected by size effect. It is expected to develop other novel characterization methods to estimate the electrical properties of nanoparticles.

References

[1] T. Nakamura: *Jpn. J. Appl. Phys.*, **23**, 1265 (1984).

[2] K. Ishikawa, K. Yoshikawa and N. Okada: *Phys. Rev. B*, **37,** 5852 (1988).

[3] H. Suzuki, T. Ohno: *J. Soc. Powder Technol. Jpn.*, **39** (12), 877–884 (2002).

[4] T. Sakudo (ed.): *Kotai-Butsuri-Koushi-shindo/Yuudentai*, p. 72, Syoukabou (2003).

[5] T. Nakamura: Kyou-yuudentai to Kouzou-souteni, Syoukabou (1988).

[6] R.J. Pressley: *Handbook of Lasers with Selected Data on Optical Technology*, The Chemical Rubber Co., Cleveland (1971).

1.12 Magnetic properties

Nanoparticles have small volume in the similar range to molecule one. Magnetism of not only nanoparticles but also magnetic molecules appears by interactive effect between magnetic moments [1].

Electron motion on the atomic nucleus orbit is regulated by energy, angular momentum and spin (atomic magnetic moment). Energy is described by orbit radius around the nucleus. If it has larger energy, the orbit radius becomes larger. Angular momentum is a counterpart of integrated intensity. Spin corresponds to electron rotation. Magnetism of molecule depends on both orbital motion (energy and angular momentum) and spin.

1.12.1 Classification of magnetism

Many kinds of magnetism appear due to the characteristics of interaction between atomic magnetic moments. Magnetic substances are defined as the substance which has magnetic moment in a magnetic field. Magnetic substances are classified into several groups as follows [2]:

1. Feeble magnetism
   -Diamagnetism
   - Paramagnetism
   - Super-magnetism
2. Antiferromagnetism
   - Antiferromagnetism
   - Herimagnetism
   - Metamagnetism
3. Ferromagnetism
   - Parasitic ferromagnetism
   - Ferrimagnetism
   - Ferromagnetism

(1) Feeble magnetism

Feeble magnetism means the substances that are magnetized only a little by a magnetic field. In many cases, its magnetization I [Wb/m^2] is proportional to magnetic field H [A/m] and described as $I = \chi H$. χ is magnetic susceptibility. If it is described as a ratio to vacant magnetic permeability μ_0, it is called as specific magnetic susceptibility χ[3, 4].

Diamagnetism arises when I is caused in the inverse direction of H, and magnetic susceptibility is $\chi < 0$. Substances that do not have spin such as copper and organic compounds have diamagnetism.

Substances whose χ is negative and has an amount of $10^{-2} \sim 10^{-5}$ show paramagnetism. The reason of weak magnetism is that magnetic substance has a spin but the spin moves with random vibration when there are magnetic molecular with magnetic moment in it. If magnetic field charges on paramagnetic substances, their magnetization would be very little.

Super-magnetism often arises on nano magnetic substance. It appears when substance is composed of nano magnetic nanoparticles or magnetic fine particle exists in metal. Super-magnetic substances would not show magnetism without external magnetic field because the magnetic moment easily changes the direction with thermal vibration. It behaves as paramagnetic substance with external magnetic field.

(2) Antiferromagnetism

Each spin of magnetic molecule cannot be lined by the interaction between spins in antiferromagnetic substance. Two magnetic ions pinching oxygen ion are influenced by super-exchange interaction through p-orbit of oxygen atom. Antiferromagnetic substances are hardly magnetized because all the spins are cancelled. Then their magnetic susceptibilities become much smaller.

Herimagnetism arises when atomic spins line in not only regular direction but also spiral direction. It is observed in compounds such as MnO_2, metals such as $MnAu_2$ and rare earth metal such as Tb, Dy, Ho, Er and Tm. In the case of metal, conduction electron is magnetized by spin. The magnetization of conduction electron would decrease with turning about electric sign when the distance from the spin becomes large. This spin interaction with conduction electron is called RKKY-interaction. Spiral structure [5] that appears in rare earth metal is caused by RKKY-interaction [6–9].

Metamagnetism is a phenomenon that magnetism of substance changes from diamagnetic to ferromagnetic by external magnetic field. For instance, magnetization of several compounds such as $FeCl_2$ and $MnAu_2$ is proportional to a weak magnetic field. But their spins are reversed to line in the same direction and then the magnetization suddenly increases at a critical point of magnetic field.

(3) Ferromagnetism

Generally ferromagnetic substances show spontaneous magnetization. Spontaneous magnetization is formed by spin distribution in the ferromagnetic substance. When external magnetic field acts on the ferromagnetic substance, this spontaneous magnetization change the direction and reach saturated state. When the external magnetic field is removed, the saturated ferromagnetic substance has residual magnetization. Both saturated and residual magnetizations are very important parameters which indicate the characteristics of magnetic substance.

When spins of magnetic molecules line up with a slight angle against antiferromagnetic distribution, parasitic ferromagnetism raises as imperceptible spontaneous magnetization [10, 11].

Ferrimagnetism shows spontaneous magnetization when spin distribution becomes irregular because amount and kind of magnetic ion are different against antiferromagnetic distribution. Ferrite is a typical ferrimagnetic substance, and has formation of $M^{2+}O{\cdot}Fe_2O_3$ (M^{2+} is magnetic ion) generally.

Ferromagnetism arises when all the spins of magnetic substance line in parallel and show their spontaneous magnetization. Ferromagnetism is observed in 3d-transition metals such as Fe, Co and Ni, rare earth metals such as Gd, Tb, Dy and their alloy. Also it is observed in compounds such as CrO_2 and EuO.

1.12.2 Magnetism of metal materials

There are free conduction electrons in metal substances, and then their magnetic property is much different from that of insulator.

Transition metals and rare earth metals, which have d-unpaired electron, show complex magnetic characteristics. Magnetism of metals is classified as follows:

(1) Transition metals	(I)	αFe, Co, Ni Ferromagnetism
(2) Transition metals	(II)	Cr, Mn, γFe Antiferromagnetism
(3) Transition metals	(III)	Others Paramagnetism
(4) Rare earth metals		Ferromagnetism or Antiferromagnetism
(5) Other metals		Paramagnetism or Antiferromagnetism

Many kinds of alloys are suggested by using magnetic characteristics of transition metals. Useful industrial materials have been produced for many years.

1.12.3 Magnetism of oxide material

Conduction electron plays an important role in the magnetism of metals and alloys. However, effect of conduction electron can be almost neglected in the case of oxides. Many kinds of metal oxide substance can be regarded as ionic crystal. Metal atoms in ionic crystal are almost in ionic conditions. They are enclosed by positive and negative ions, and influenced by the crystal electric field.

There must be exchange interaction between magnetic ions in order that ionic crystal has ferromagnetism. Metal ions and oxygen ions are located alternately in oxide compounds. Iron ions are likely buried in the dense structure of oxygen ions in iron oxide compounds. In oxide compounds, it is impossible to apply the model of metal type exchange interaction. Negative ions affect on exchange interaction in oxide compound. P.W. Anderson suggested super-exchange interaction between positive ion spins carried by negative ions, and described the magnetism of oxide compounds [12].

The interaction between positive ion spins is super-exchange type, not directly act on itself. If super-exchange interaction is positive, its oxide compound shows ferromagnetism. In case it is negative, it would appear antiferromagnetism or ferrimagnetism.

(1) Monoxide

Typical examples of monoxide compounds are MnO, FeO, CoO and NiO. CoO belongs to tetragonal system, and MnO and FeO belong to trigonal system. Both systems are warped slightly, and then magnetic anisotropy would appear.

(2) Spinel oxide

Generally spinel oxide has ferrimagnetism. Oxide compounds with a spinel structure especially ferrites have been studied for many years. Magnetite, which has been known as a ferromagnetic substance, has the spinel structure. Industrial ferrite materials are often spinel structure compounds. Ferrite formula is defined as MFe_2O_4 (M = Mn, Fe, Co, Ni, Cu, Zn etc.). There are two kinds of ion distribution. One is normal spinel ($M[Fe_2O_4]$) and the other is inverse spinel ($Fe[MFe]O_4$). Ferrite formula is generally described as $M_{1-x}Fe_x[M_xFe_{2-x}]O_4$. In the case of $x = 0$ and $x = 1$, it is normal-spinel and inverse-spinel, respectively. It is known that x value of Cu-ferrite and Mg-ferrite can be varied by several heat treatments [13–15].

Ion distribution of spinel ferrite at high temperature can be partly controlled, and then various saturation magnetizations can be derived by suitable heat treatment.

(3) Perovskite oxide

Oxides with MTO_3 structure have perovskite structure when M is a cation with a large radius and T is a metal such as Cr, Mn or Fe. When T is trivalent metal such as Cr or Fe, M are often trivalent metal such as La or rare metal. If T is tetravalent Mn, M is alkali earth metal such as Ca and Sr.

Typical materials are antiferromagnetic substance such as $LaFeO_3$, $LaCoO_3$ and $LaNiO_3$. $LaMnO_4$ is usually antiferromagnetic substance. When La is replaced by Ca, Mn^{3+} changes to Mn^{4+}. It was reported that ferromagnetism appeared in these composite perovskite oxide [16].

(4) Garnet structure oxide

Magnetic substance ($3M_2O_3{\cdot}5Fe_2O_3$) which has the same structure as garnet ($Mg_3Al_2(SiO_4)_3$) has been studied [17,18]. First garnet structure oxides whose M is Y was studied. Next new magnetic materials whose M was replaced by rare earth metal have been studied.

When M is replaced by rare earth metals such as Gd, Tb and Dy, these garnet oxides are able to derive large magnetic moment by super-exchange interaction at low temperature. It is known that Yttrium–Iron–Garnet (YIG) remarkably reduces frequency loss on applications for high frequency device.

(5) Corundum structure oxide

Many kinds of corundum structure oxide such as Cr_2O_3 and Fe_2O_3 have existed widely. Cr_2O_3 and α-Fe_2O_3 have the same chemical lattice as rhombohedral system, but Cr_2O_3 has diamagnetic distribution due to different spin structure. Especially α-Fe_2O_3 has small spontaneous magnetization because of the declination of spin interaction.

The magnetization of these substances such α-Fe_2O_3 is so small, and then it is called as parasitic ferromagnetism. As the studies of magnetics have been advancing, it has been considered that this is intrinsic magnetism called "weak ferromagnetism" or D-M ferromagnetism [19, 20]. However, Cr_2O_3 do not have weak ferromagnetism.

(6) Rutile structure oxide

Rutile of natural resource has a composition of TiO_2 and body-centered cubic structure. Typical magnetic compounds are MnO_2 and CrO_2. MnO_2 seems to be antiferromagnetic substance, however, MnO_2 have a particular spin structure called "screw structure" [21, 22]. Several rare earth metal oxides have the same screw structure. CrO_2 is a peculiar ferromagnetic substance whose Cr^{4+} ions are lined as ferromagnetic distribution.

(7) Magnetoprumbite structure oxide

Magnetic oxides with a hexagonal structure which contains Fe^{3+} ion, divalent ions M^{2+} (M = Mn, Fe, Co, Ni, Cu, Zn, Mg etc.) and other divalent ion such as Ba^{2+}, Sr^{2+}, Ca^{2+} and Pb^{2+} are called as magnetoprumbite structure magnetic substance. Typical magnetic substance is barium ferrite ($BaFe_{12}O_{19}$ or $BaO{\cdot}6Fe_2O_3$). Barium ferrites have large magnetic anisotropy, and can be often applied to permanent magnets and magnetic recording materials.

1.12.4 Magnetic characteristics of nanosized materials

The characteristics of nanosized magnetic substance are deviated from the bulk ones. Specific surface area becomes larger with decreasing particle size, and then the influence of surface property becomes serious not to be neglected. Generally ferromagnetic substance would be metals or unstable oxides. Magnetic characteristics such as saturation magnetization decrease proportionally with size reduction of magnetic substance.

The relationship between particle size of spherical magnetite (Fe_3O_4 or $Fe^{2+}O{\cdot}Fe_3{}^{2+}O_3$) and saturation magnetization is described in Fig. 1.12.1. The abscissa shows particle diameter (nm), and ordinate shows the saturation magnetization (Am^2/kg) of spherical magnetite. It became clear that saturation magnetization suddenly decreased in proportion to particle size in the nanoregion.

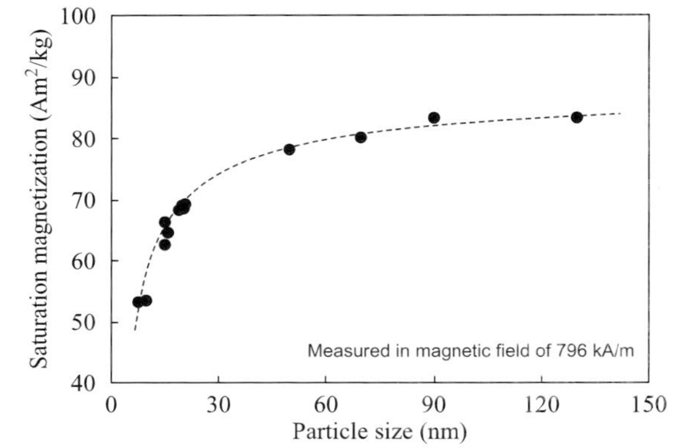

Figure 1.12.1
The relationship between particle size of spherical magnetite and saturation magnetization.

The decrease in saturation magnetization is not preferable, and the higher characteristics should be maintained in spite of nano-sizing. A lot of studies have been conducted by the modification of composition or the surface coating onto magnetic particles in order to find out excellent magnetic substance which maintains higher magnetic properties [23, 24].

For example, the oxidation resistance of iron oxide for magnetic recording media could be improved by the introduction of surface coating.

Iron-based acicular magnetic metal particles have been studied to achieve higher performance of magnetic recording. Reducing the size of iron-based metal particles is much required to obtain high signal and low noise of media such as magnetic tapes. The particle size of latest metal particles is about several decades nanometer, and the size reduction of metal particles would be further continued. Iron-based acicular metal particles have high magnetic performance such as saturation magnetization and coercive force. However, it was a problem that they were chemically unstable and consequently the magnetic properties might be deteriorated by surface oxidation for instance. The surface coating with carbon was investigated in order to improve the chemical stability such as oxidation resistance [25].

The timely change in saturation magnetization of magnetic metal particles with the lapse of time was measured at temperature of 60°C and relative humidity of 90% in air atmosphere. The oxidation stability of both 10 wt% carbon coated and uncoated metal particles is described in Fig. 1.12.2. The abscissa and the ordinate show time progress and the change of saturation magnetization, respectively. The magnetization of the uncoated metal particles reduced remarkably with time progress. However, reduction in the magnetization of the carbon-coated metal

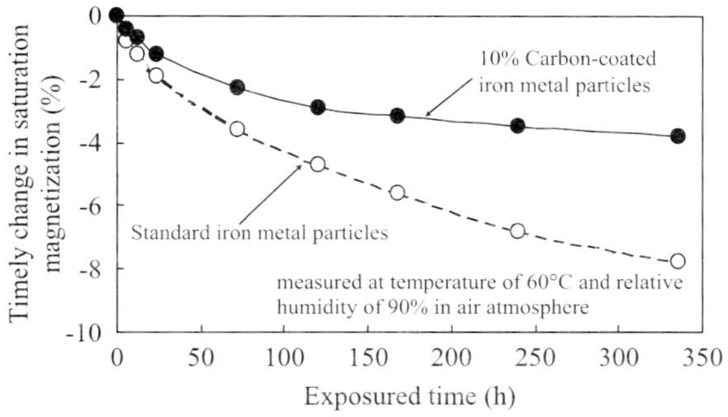

Figure 1.12.2
Timely change in saturation magnetization on powder at temperature of 60°C and relative humidity of 90% in air atmosphere.

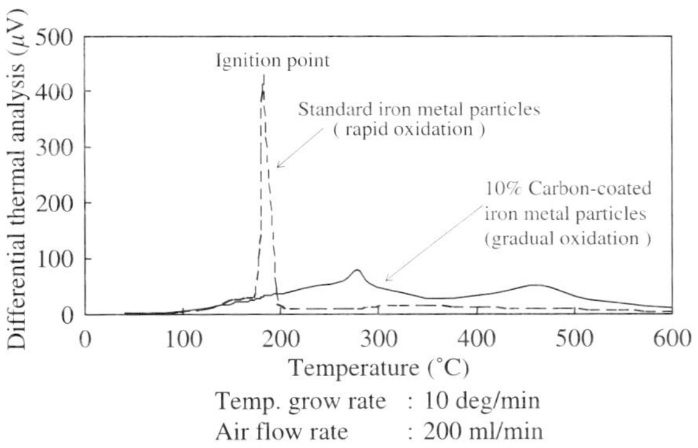

Figure 1.12.3
Disappearance of ignition point by carbon coating measured by DTA.

particles was very small. The oxidation of magnetism was suppressed by the surface carbon coating.

Fig. 1.12.3 describes the differential thermal analysis (DTA) result of iron-based metal particles. Uncoated metal particles had an ignition point of about 180°C. For 10 wt% carbon-coated metal particle, the oxidation gradually occurred and there was no ignition point. And then it became clear that carbon-coated metal particles were chemically stable enough compared to the uncoated metal particles.

Serious problems caused by reducing the size of metal particles would be chemical and thermal stabilities. They will be cleared by both modification of composition and introduction of surface coatings such as carbon coating to commercialize nanosized magnetic substances.

References

[1] S. Chikazumi: *Jikken Butsurigaku Kouza 17, Jiki*, Kyoritsu Shuppan (1968).
[2] S. Chikazumi, S. Miyahara, Y. Ishikawa, M. Asanuma, N. Wakiyama, Y. Gondo and K. Ohta: *Zairyou Kagaku Kouza 5, Busshitsu no Jikitekiseishitsu*, Asakura Shoten (1968).
[3] K. Souda, S. Chikazumi: *Daigaku Enshu, Denjikigaku*, Shoukabou (1956).
[4] S. Chikanobu: *Kyoujiseitai no Butsuri*, Shoukabou (1959).
[5] S. Chikanobu: *Kotai Butsuri 1*, **3**, 3 (1966).
[6] H. Miwa, K. Yoshida: *Prog. Theoret. Phys.*, **26**, 693 (1961).
[7] T.A. Kaplan: *Phys. Rev.*, **124**, 329 (1961).
[8] R.J. Elliot: *Phys. Rev.*, **124**, 340 (1961).
[9] K. Yoshida, A. Watanabe: *Prog. Theoret. Phys.*, **28**, 361 (1962).
[10] I. Dzialoshinski: *J. Phys. Chem. Solids*, **4**, 241 (1958).
[11] T. Moriya: *Phys. Rev.*, **120**, 91 (1960).
[12] P. W. Anderson: *Phys. Rev.*, **79**, 350 (1950).
[13] F. Bertaut: *CR Acad. Sci.*, **230**, 213 (1950).
[14] S. Miyahara, H. Ohnishi: *J. Phys. Soc. Jpn.*, **11**, 1296 (1956).
[15] H. Ohnishi, T. Teranishi and S. Miyahara: *J. Phys. Soc. Jpn.*, **14**, 106 (1959).
[16] G. H. Jonker, J. H. van Santen: *Physica*, **16**, 337 (1950).
[17] F. Bertaut, F. Forrat: *CR Acad. Sci.*, **242**, 382 (1956).
[18] S. Geller, M. A. Gilleo: *Acta Crystallogr.*, **10**, 239 (1957).
[19] I. Dzialoshinski: *J. Phys. Chem. Solids*, **4**, 241 (1958).
[20] T. Moriya: *Phys. Rev.*, **117**, 635 (1960); **120**, 91 (1960).
[21] A. Yoshimori: *J. Phys. Soc. Jpn.*, **14**, 807 (1959).
[22] T. Nagamiya: *J. Phys. Radium*, **20**, 70 (1959).
[23] H. Nishio, Y. Ota, M. Maekawa and H. Yamamoto: *J. Magn. Magn. Mater.*, **287**, 234–238 (2005).
[24] K. Hayashi, K. Iwasaki, Y. Tanaka and H. Morii: *Kagaku Kougaku Ronbunshu*, **25** (3), 361–366 (1999).
[25] K. Hayashi, M. Ohsugi, M. Kamigaki, B. Xia and K. Okuyama: *Electrochem. Solid-State Lett.*, **5** (7), J9–J12 (2002).

1.13 Optical property of nanoparticle

1.13.1 Band structure of nanoparticles

Optical property of nanoparticle is greatly different according to whether the nanoparticle is insulator, semiconductor or metal. Fig. 1.13.1 shows typical band structure model [1, 2]. In the insulator or the semiconductor, the valence band is completely occupied with the electron and the conduction band is emptied. Excitation of the electron is classified into

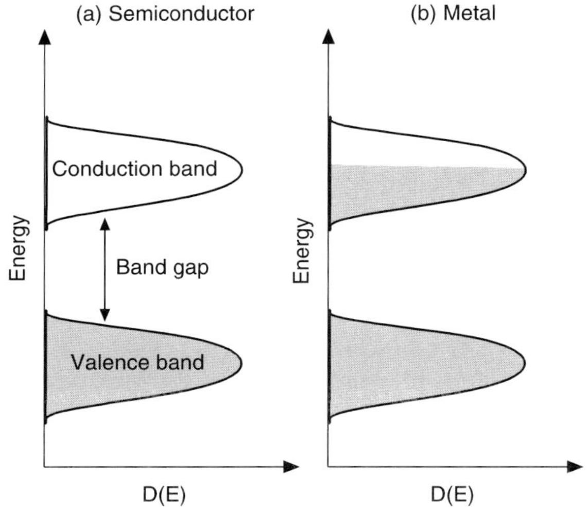

Figure 1.13.1
Band structures of (a) semiconductor and (b) metal.

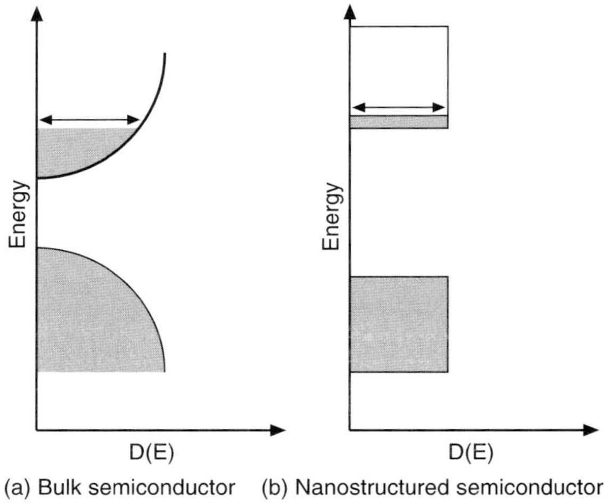

Figure 1.13.2
Density of states for the (a) bulk and (b) nanostructured semiconductor particle.

direct bandgap conserving momentum and indirect bandgap. In most cases, the minimum of the conduction band locates above the maximum of the valence band in momentum space. Therefore, the absorption spectra of photon can be understood in view of the direct bandgap transition. For the interband transition, generation of the electron and hole is an elementary excitation and an optical spectrum mainly depends on the density of states.

When an electron is confined in a nanostructured semiconductor, its band structure changes markedly. Since an electron is quantized in the thickness direction in a nanostructured sheet, the electronic density of states concentrates at particular energy levels, as depicted in Fig. 1.13.2. For that reason, the condition of photoexcitation differs. Assume a nanostructured semiconductor particle of several nanometer scale for which travel of an electron is limited in all three dimensions. The electron is therefore confined in the region; its density of states is discretized. For such nanostructured semiconductors, semiconductor nanoparticles like CdSe and ZnS, are anticipated for use as novel luminescent materials.

Strong Coulomb interaction between an electron and a hole establishes a mutually bound state, which reduces excitation energy by their binding energy. Such a couple in the bound state is referred to as an exciton. Light absorption by an exciton requires less energy than a band gap. The contribution to optical absorption of interband transition and an exciton depends strongly on the combined states of atoms in crystals. The traveling behavior of an electron and a hole that are mutually attracted by the Coulomb interaction is explained as similar to the relationship between an electron and a proton in a hydrogen atom. The Coulomb interaction is shielded by the effect of the large dielectric constant of a semiconductor. For that reason, the bound energy of an exciton becomes much smaller than that of a hydrogen atom. That is, the activity of an exciton extends over multiple atoms. The larger band gap that a semiconductor has, the more difficult it is to polarize, the smaller its dielectric constant gets and the more stabilized is an exciton. Namely, an ionic semiconductor generally has a larger band gap and is harder to polarize than a covalent semiconductor. Accordingly, in the former, an electron and a hole tend to approach one another and, optical transition readily occurs; therefore, an exciton is dominant.

The electronic density of states is discretized in nanostructured semiconductor particles because of the spatial confinement. In that case, the condition of an exciton is determined by the competition of the spatial confinement and the Coulomb interaction. Since the semiconductor becomes increasingly smaller, the magnitude of the Coulomb interaction extends over a bulk crystal, but the increase of the confinement energy prevails. Accordingly, the smaller it gets, the less the relative contribution of the Coulomb interaction becomes and the spatial confinement becomes dominant. In a region with a high obstacle potential and strong confinement, electrons and holes are no longer freely mobile. It is therefore impossible to approximate an exciton as a particle that is similar to a hydrogen atom.

On the other hand, conduction bands of a conductor (metal) are partially filled with electrons; no gap exists in electronic excitation. Conducted electrons in metal form a kind of plasma state, and the intraband transition of conduction electrons is described as the collective motion of free electrons. The oscillation of electrons by this collective motion, in other words, the repetitive oscillations of electrons to shield polarization caused by external fields such as light and electron rays, is referred to as plasma oscillation. This oscillatory wave is expressed as a particle and referred to as a plasmon. The contribution of a conduction electron is interpreted using the classical Drude model, which regards electrons as traveling within a stationary field with an ionic core resembling that of a gas, but it assumes their interaction with other electrons and ions as a mere damping term. An electric field, generated when an electronic gas deviates from a stationary ionic field as a whole, acts as the restoring force. This is the principle by which plasma oscillations occur. Interband transition also has a certain contribution in real particles. Therefore, optical response is described as their mixed state.

It is not often the case that real metal nanoparticles are completely isolated one by one in a vacuum: they are surrounded by a medium with different dielectric constants as a practically important morphology. Light incidence onto transparent insulators such as metal nanoparticle-dispersed glass brings about a mode of vibration called a surface plasmon that is localized on the metal nanoparticle surface. A dielectric function varies at the interface of different substances to try to be continuous. Consequently, a mode of vibration with its resonance frequency that is deviated from the ordinary plasma frequency occurs, which creates a surface plasmon. Beautiful colors appear when incident light and surface plasmons resonate. This phenomenon has been known and used commonly for many years, as in optical glasses such as stained glass in European cathedrals since the Middle Ages. The effect of surface plasmons is manifested brilliantly in ultrafine particles with a large surface fraction. Recently, metal nanoparticle systems have been attracting attention [3, 4] because of their enhanced nonlinear optics properties by single electronic electrical conduction phenomenon because of Coulomb blockade and surface plasmon mode. The nonlinear optical effect originates from nonlinear polarization that is proportional to the power of order. Therefore, a nonlinear optical effect can be implemented at high efficiency by reinforcing a local field remarkably by surface plasmons.

1.13.2 Measurement method of optical properties of nanoparticles

The optical properties of these nanoparticles are measured with spectrometry systems that are generally used for ordinary powder particles. The wavelength of light is far greater than the nanoparticle size. Consequently, it is likely that many particles are measured as a group. Luminescence measurement is a particularly effective measurement method for the optical measurement of nanoparticles. The basic procedures of the measurement are to excite a sample with excitation sources like light or electron rays, to carry out spectral analysis of the luminescence of the sample with appropriate means, and to measure light intensity with a detector. Especially for electron rays, their irradiation focused into a space of the order of tens of nanometers at low voltage has been realized, corresponding to the progress of the development of modern field emission display panels. Therefore, their significance as an excitation source in the emission spectrometry of nanoparticles has been increasing [5]. Since nanoparticles and nanostructured materials emit very little light as a result of their small size, the analytical system should have sufficient capability in emission collection, spectral efficiency and detector sensitivity. In emission measurements, an appropriate spectral method enables us to eliminate excitation light with deviated energy from the emissions, so that weak emission can be detected without being influenced by scattered light. Nevertheless, it is difficult to control the size, shape and composition of each nanoparticle precisely. Statistical measurement of many particles would therefore necessarily measure the spread of emission based on their inhomogeneity. Although the spread enables us to analyze characteristics such as their size distribution, it is difficult to measure the optical properties of a single nanoparticle with light. For example, the application of useful spectroscopic analytical methods, such as the fluorescence analysis, UV-Vis-IR spectroscopy, and Raman spectroscopy, to an aggregate of nanoparticles merely provides the average spectrum of nanoparticles with a distribution of size and composition. The optical measurement of a nanoparticle requires mapping of spatially limited spectroscopic information in combination with spatial information obtained with an optical microscope. However, the diffraction limit ($\lambda/2\sin\theta$) of an optical microscope based on the wave nature of light prevents imaging of light far smaller than its wavelength. Although light of a shorter wavelength can improve spatial resolution, the wavelength is typically dependent upon the measurement method and cannot be shortened easily. An alternative to shorten the wavelength without varying energy is to let the light pass through transparent matter of a refractive index n to convert the wavelength into λ/n. The oil immersion lens method using oil follows this principle, and is therefore adopted in biological observation with an optical microscope. A solid immersion lens (SIL) replaces oil with a transparent solid lens [6, 7]. However, the refractive index of transparent materials between visible radiation is about $n = 2$; there are also other restrictions, i.e., the necessity of evenness of the sample.

Near-field scanning optical microscopy (NSOM) is attracting attention [8, 9]. It enables formation of a fine spot below the diffraction limit of ordinary light using near-field light. Its fundamental concept was proposed by Synge in 1928 [10].

The diffraction limit of ordinary light is derived by solving Maxwell's equation under the assumption that light intensity follows a Gaussian distribution in the radial direction, which reflects the oscillatory nature of an electromagnetic wave that propagates in space.

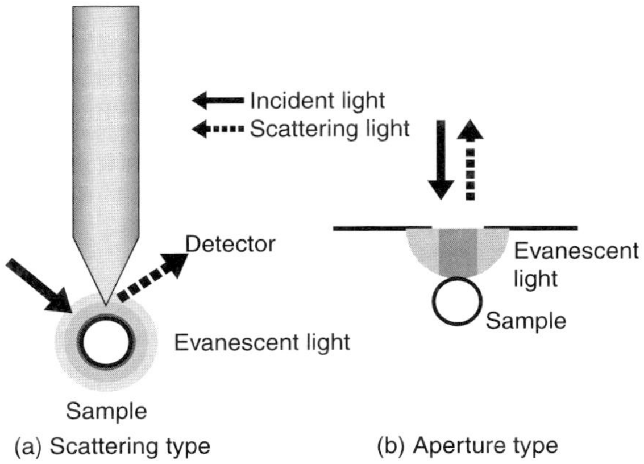

Figure 1.13.3
(a) Scattering type and (b) Aperture type NSOM.

On the other hand, a near-field light component representing the oscillation of an electromagnetic wave that attenuates without proceeding spatially is also derived from Maxwell's equation. The diffraction theory of near-field light was proposed by Bethe in 1944 [11]. It states that the near-field component of an electromagnetic wave can leak out at an aperture smaller than the wavelength of light up to about the aperture radius. This theory has encouraged experiments to apply light onto a region smaller than its wavelength through an aperture smaller than its wavelength. Although it had been difficult to obtain an image using a fine aperture on the wavelength of light, recent progress of microfabrication allows the use of NSOM for the analysis of nanoparticles or nanostructured materials as a common measuring apparatus. The NSOM is roughly categorized into two types, as illustrated in Fig. 1.13.3: the dispersion type and the aperture type. The dispersion type irradiates light at the sharpened tip of a dielectric or metal probe. An STM probe and a metal-coated AFM probe are useful as they are. Therefore, STM or AFM measurement can be carried out with the same instrument. The aperture type collects near-field light from a very small aperture. It is common today to fabricate a very small aperture at the tip of an optical fiber as a probe: an example is shown in Fig. 1.13.4. The sharpened tip of an optical fiber is wrapped by a shielding metal film, so that no scattered light is sensed. This probe is brought close to a sample. Then light is irradiated thereon through the aperture, and lights such as fluorescence and Raman scattering light are collected from the sample through the aperture again. Spectral analysis with respect to energy, if necessary, enables acquisition of optical properties from a spatial region of the order of tens of nanometers. The performance of the aperture type NSOM depends on the probe properties. The spatial resolution is determined by the aperture size and shape. The aperture face should be as flat as possible, and be parallel

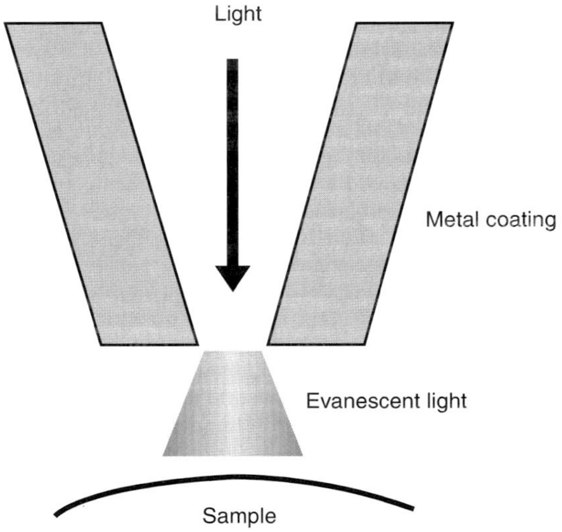

Figure 1.13.4
Metal-coated fiber probe of NSOM.

to the sample. Moreover, because the sensitivity depends strongly on the distance between the aperture and the sample, the distance should be controlled stably. The spectrometric performance is affected by the probe material. Although a common optical fiber passes from visible light to a part of infrared radiation efficiently, it tends to be degraded by ultraviolet light. Accordingly, the measurement of short wavelength ultraviolet rays should be carried out with caution.

References

[1] S.M. Sze: *Physics of Semiconductor Devices*, Wiley-Interscience (1981).

[2] C. Kittel, P. McEuen: *Introduction to Solid State Physics*, Wiley (2004).

[3] T.W. Ebbesen, H.J. Lezec, H. Hiura, J.W. Bennet, H.F. Ghaemi and T. Thio: *Nature*, **382**, 54 (1996).

[4] K. Kajikawa: *Jpn. J. Opt.*, **33**, 159 (2004).

[5] T. Mitsui, T. Sekiguchi and Y. Sakuma: *Comparison of Cathodoluminescence and Scanning near Field Optical Microscopy for the Study of Semiconductor Quantum Dots*, in 2nd International Workshop on Nano-scale Spectroscopy and Nanotechnology (2002).

[6] M. Yoshida, K. Koyama, M. Baba and H. Akiyama: *Butsuri*, **55**, 772 (2000).

[7] H. Akiyama, M. Yoshida and M. Baba: *Oyo Butsuri*, **71**, 716 (2002).

[8] M. Ohtsu, H. Hori: *Near-field Nano-optics*, Kluwer Academic/Plenum Publishers (1999).

[9] M. Ohtsu: *JSPE J.*, **66**, 661 (2000).

[10] E. Synge: *Philos. Mag.*, **6**, 356 (1928).

[11] H.A. Bethe: *Phys. Rev.*, **66**, 163 (1944).

FUNDAMENTALS

CHAPTER 2

STRUCTURAL CONTROL OF NANOPARTICLES

CONTRIBUTORS

Yoshinobu Fukumori (2.1)
Toshiyuki Nomura (2.2.1), (2.2.2), (2.3.2)
Tadafumi Adschiri (2.2.3), (2.4.3)
Satoshi Ohara (2.2.3)
Fumio Saito (2.2.4)
Makio Naito (2.2.5), (2.4.4)
Kikuo Okuyama (2.3.1), (2.5.1)
Masayoshi Kawahara (2.4.1)
Hisao Suzuki (2.4.2)
Takafumi Sasaki (2.4.3)
Masayoshi Fuji (2.5.2.1)
Shinji Inagaki (2.5.2.2)
Hirofumi Takeuchi (2.6)
Yoshinori Ando (2.7)

2.1 Structure construction and function adaptation of nanoparticles

Nanoparticles are designed and prepared with molecule and particle structures that can exhibit functions desired for applications. The size and shape of the prepared particles are primarily important in the function adaptation. Further, the surface and internal structures are commonly composed of multi-layers and/or multi-phases containing multi-components to provide the desired functions.

2.1.1 Structures of nanoparticles

Many case-studies have been reported on nanostructure construction so far. Figure 2.1.1 shows typical structures of nanocomposite particles.

First, the particle has a macrostructure consisting of the surface and internal part. The molecules and atoms in the surface layer are in the active state different from those in the bulk, even if the particle would be composed only of a single compound. Although this surface property itself is useful in applications of nanoparticles, the surface has often been modified with different compounds to provide special functions. The internal structure can be simply classified into monolithic matrix and core–shell construction. When the cores of particles would be empty or contain only medium, they can be called as the "hollow particles". Second, each layer of surface, shell and core has single phase, random multi-phases, or ordered multi-phases. When a certain phase in the multi-phases would be empty, the structure can be called "porous" (porous structure). A certain phase dispersed in the layer is not necessarily isolated, but sometimes spanned through the layer. Such a percolated structure has much effect on the properties of the layer. Third, each phase can be crystalline or non-crystalline (amorphous) in the molecular arrangement. Finally, the layers and phases are assembled by utilizing covalent bonding, van der Waals force, electrostatic force, hydrogen-bonding, and so on.

The structures of nanoparticles are required to be held permanently in some cases of application, but, in other cases, to be disassembled after playing a role at the beginning stage of application. In the latter case,

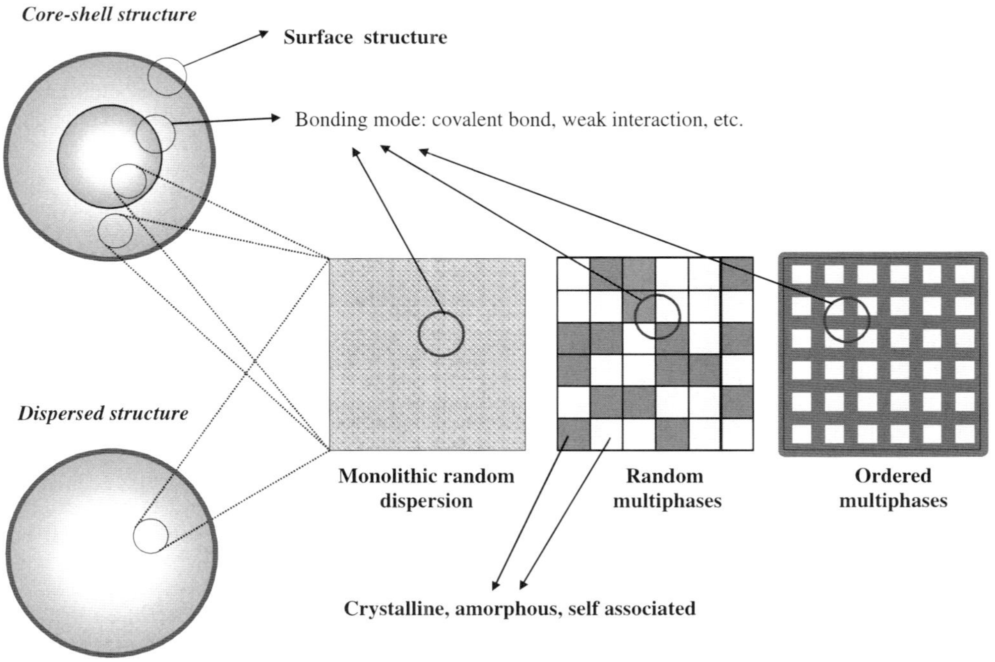

Figure 2.1.1
Typical structures of composite nanoparticles.

the new structures are sometimes required to be reassembled by a self-association mechanism in order to play a subsequent role. The timings of disassembling and reassembling are also critical to exhibit the functions desired. For medical and pharmaceutical uses, the prepared nanoparticles have to be finally disassembled; then, the components have to be utilized by the body or eliminated from the body. To efficiently and safely complete the whole process that the nanoparticles will go through, the components, the structures of particles, layers and phases, and the bonding modes have to be designed and prepared optimally.

Many kinds of fabrication process have been proposed to construct nanocomposite particles for providing desired functions. The phases and layers in Fig. 2.1.1 are constructed sequentially or simultaneously by depositing atoms, compounds, and particles physically or chemically in the gas or liquid phase: the intrinsic association forces and external mechanical forces can be the driving force in the structure-making process. In the present stage of nanotechnology, it is not so easy to complete the fabrication process in the controlled manner.

In this section, typical examples of nanoparticles designed and prepared for medical and pharmaceutical applications have been discussed (Table 2.1.1) [1].

2.1.2. Hollow particles

Liposomes are typical hollow particles (Table 2.1.1 (a–c)) [2–4]. Liposomes are small vesicles with phospholipid-bilayer shells. It is common that the cores contain just an aqueous phase, sometimes dissolving water-soluble drugs. Hydrophobic agents can be incorporated into the bilayer. Targeting agents that specifically associate the particles with target cells are put on the surface by binding phospholipids to them for anchoring to the bilayer. Liposomes enclosing magnetite nanoparticles are reported for applying them to hyperthermia treatment and diagnosis of cancer (Table 2.1.1(a)) [2].

2.1.3. Core–shell particles

Lipid microemulsions are aqueous dispersions of nanoparticles with liquid cores of lipid and shells of phospholipid monolayer. When the cores are solid lipids, they are called "solid lipid nanoparticles" (Table 2.1.1(w)). Therein, hydrophobic and amphiphilic agents are embedded in the cores and shells, respectively. The surface-modifying compounds are anchored in the same manner as liposomes.

A typical lipid nanoparticle is shown in Table 2.1.1(d) [5–7]. They were designed for delivering gadolinium in cancer neutron capture therapy [8]. The gadolinium is chelated with diethylenetriamine pentaacetic acid (DTPA) to make distearylamide (Gd-DTPA-SA). Gd-DTPA-SA is not dissolved in water and the soybean oil used as a core, embedded in the phospholipids shell. Cosurfactants with polyethylene glycol (PEG) chain are also embedded in the shells to hydrate the particle surface. When the nanoparticles thus prepared were intravenously injected into the blood of hamsters, the surface hydration contributed to the long-circulation of the particles in the blood. Since the size of the particles was smaller than 100 nm, these particles exhibited the enhanced penetration and retention effect [6, 7].

The so-called quantum dot (qdot) is shown in Table 2.1.1(f) [9]. The feasibility of in vivo targeting by using semiconductor quantum dots (qdots) was explored. Qdots are small (<10 nm) inorganic nanocrystals that possess unique luminescent properties; their fluorescence emission is stable and is tuned by varying the particle size or composition. ZnS-capped CdSe qdots coated with a lung-targeting peptide accumulated in the lungs of mice after intravenous (i.v.) injection, whereas two other peptides specifically directed qdots to blood vessels or lymphatic vessels in tumors. They also showed that adding polyethylene glycol (PEG) to the qdot coating prevented qdots from non-selective accumulation in reticuloendothelial system (RES).

The pharmaceutical application of nanoparticles can be found in latex systems that have been used as coating materials in the spray-coating processes such as fluidized bed, spouted bed, and tumbling fluidized bed process [10]. Commercially available latexes are formed as monolithic structures from random *co-* or *ter*polymers that are designed to be used chiefly for coating such coarse particles as granules and tablets. For further and broader application of this technique, the novel terpolymer and core–shell latexes have been proposed through development of fine particle coating technology and highly functional microcapsules, such as thermosensitive drug-releasing microcapsules using sophisticated latexes with temperature-dependent swelling properties.

An aqueous latex exhibiting a low degree of agglomeration, low membrane permeation, and high coating efficiency was developed using terpoly(ethyl acrylate (EA)/methyl methacrylate (MMA)/2-hydroxyethyl methacrylate (HEMA)), whose monomer molar ratio was 6:12:8 or 12:6:4 (Table 2.1.1(g)) [11]. Different from blend latexes, the composite latexes composed of the low-permeable 12:6:4 polymer core and the non-adhesive 6:12:8 polymer shell with a 6:4 core–shell weight ratio formed a low-permeable membrane by heat treatment. The composite latexes exhibited a very low degree of agglomeration, with the polymer yield remaining very high. These properties were still effective even in the coating of cornstarch as fine as 12 μm: at a 50% level of coating, the mass median diameter of the product was 16 μm and it contained only 3% agglomerates.

Another aqueous composite latex that suppressed the electrostatic particle-adhesion in the coating process

Table 2.1.1
Typical example of nanoparticles for medical or pharmaceutical applications.

		Therapeutic purpose	Delivery agent	Structures of core and shell	Surface property and component	Bonding to core
Hollow	a	Hyperthermia, cancer diagnosis	Magnetite	Phospholipid bilayer (liposome)	Antibody	Disulfide bonding
	b	Passive and active targeting	Amphotericin B/ doxorubicin	Egg PC liposome	DSPE-PEG/DSPE-PEG-Monoclonal antibody/ DSPE-PEG-Transferrin	DSPE-anchored
	c	Target radiotherapy	^{90}Y	Liposome	Integrin alphavbeta 3-antagonist, DTPA	DTPA-chelation, DSPE-anchored
Core–shell	d	Cancer neutron capture therapy	Hydrophobic chelating agent of gadolinium	Phospholipid monolayer on lipid	PEG	Hydrophobic anchor
	e	Intracellular diagnostics, therapeutic delivery	Gold	Gold (core) − BSA (shell)	Five peptides (nuclear localization signal, etc.)	Conjugated to BSA
	f	Disease-sensing, drug delivery	CdSe–ZnS	CdSe (core) + ZnS (shell)	Peptide/PEG	Covalent bonding
	g	Controlled release from microcapsules	Orally administered drugs	Core-shell of acrylic terpolymers	Hard polymer shell	Covalent bonding
	h	Thermosensitive release from microcapsules	Various drugs	Acrylic terpolymer (core)-poly (*N*-isopropylacrylamide (shell))	Thermosensitive swelling and shrinking	Covalent bonding
Inorganic simple core	i	MRI, hyperfusion region visualization, angiogenesis	Clariscan (superpara-magnetite)	Magnetite	Oxidized starch	Coated
	j	Cell-based therapy, cell labeling	Dextran-coated CLIO	CLIO	Tat peptide-FITC	Disulfide bonding
	k	MRI of apoptosis (phosphatidyl serine recognition)	CLIO	CLIO	Annexin V (protein)	Disulfide bonding
	l	MRI, hyperthermia, magnetic field-assisted radionuclide therapy	Magnetite	Magnetite	PEG, folate	Covalent bonding
	m	MRI, detection of gene expression	Dextran-coated CLIO, MION	CLIO, MION	Trasferrin (Tf-S-S-CLIO)	*N*-succinimidyl 3-(2-pyridyldithio) propionate linker
Organic simple core	n	Passive drug-targeting	Irinotecan	Matrix of PLA or PEG-PPG-PEG	PEG-PPG-PEG	Embedded
	o	Passive and estrogen-receptor targeting	Tamoxifen	Matrix of poly (e-caprolactone) (MW: 14.8 kDa)	Pluronic™ F-68	Adsorption
	p	Passive and active targeting	Doxorubicin/DNA	Matrix of poly(aspartic acid), PLA or poly(lysine)	PEG	Covalent bonding
	q	Anti-angiogenic gene therapy	Plasmid DNA (APTmu—Raf)	Matrix of polymerized cationic lipid	Integrin alphavbeta3-antagonist	Lipid anchor
	r	Gene therapy	DNA	Matrix of cyclodextrin polycation	Transferrin-PEG-adamantane conjugate	Adamantane-cyclodextrin complex formation
	s	Gene therapy	*Mycobacterium phlei* DNA	Matrix of chitosan (MW: 500,000, 84% deacetylated)	Cationic	Covalent bonding
	t	Cancer neutron capture therapy	Gadolinium–DTPA chelate	Matrix of chitosan	Cationic	Covalent bonding
	u	Neutron capture therapy	Gadolinium hexanedione	Wax matrix	Folate/Thiamine-PEG-DSPE, PEG-DSPE	DSPE-anchored
	v	Gene therapy	Plasmid DNA lipophilized with cationic lipid	Wax matrix with, Brij 78, Tween20, Tween80	Cholesterol-pullulan™	Lipid anchor
	w	Photodynamic therapy	Hemato-porphyrin	SLN	Folate, PEG spacer	Cholesterol-anchored

Note: BSA, Bovine serum albumin; Chol-pullulan, {*N*-[2-(cholesterylcarboxyamino)ethyl]carbamorylmethyl}-pullulan; CLIO, Crosslinked ion oxide; DSPE, Distearoylphosphatidylethanolamine MION, Monocrystalline ion oxide nanoparticle; MRI, Magnetic resonance imaging; PEG, Polyethylene glycol; PEG-PPG-PEG, Poly(ethylene glycol-block-poly(propylene glycol)-block-poly(ethylene glycol); PLA, Poly(lactic acid); SLN, Solid lipid nanoparticle.

and had a self-film-formability in water was developed as a coating material (Table 2.1.1(h)) [12]. This composite latex consisted of a terpoly(EA/MMA/HEMA) core and a non-crosslinked, thermosensitive poly(*N*-isopropylacrylamide (NIPAAm)) shell. When compared to homogeneous latexes with no poly(NIPAAm) shell, the composite latexes reduced the production of poorly coated particles and the particle size dependence of polymer yield when the coating operation was done at a temperature where poly(NIPAAm) shells were able to swell, i.e., below the lower critical solution temperature (LCST: 32°C) of poly(NIPAAm). Further, the surface-layer constructed with the composite latex particles also exhibited a self-film-formability in water at 37°C, resulting from shrinkage of the poly(NIPAAm) shells in the membrane at temperatures above the LCST. Thus, the coated particles exhibited a rapid release of drug at low temperatures, and a suppressed release at high temperatures (negative thermosensitivity).

The composite latexes composed of a hydrophobic poly(EA/MMA) core and a thermosensitive poly (NIPAAm) shell (Table 2.1.1(h)) [13] were synthesized by a semi-continuous two-stage emulsion polymerization technique. The poly(NIPAAm) shell was cross-linked with methylene bisacrylamide. The microcapsules with thermosensitive coats were prepared using a spouted bed coater assisted with a draft tube and bottom-spray, known as the Wurster process. The key structure of the microcapsules designed here was its composite coat, consisting of nanosized thermosensitive hydrogels dispersed in a thermo-insensitive polymeric matrix (ethylcellulose). At high temperatures, therefore, the poly(NIPAAm)-gel domains in the microcapsule membranes shrank, probably leading to the creation of many voids in the membranes. Consequently, the water-permeability increased as if a molecular valve had been opened. Owing to the voids thus formed, the drug release rate at high temperatures became higher than that at low temperatures (positive thermo-sensitivity).

2.1.4 Simple inorganic nanoparticles

Recently, inorganic nanoparticles that interact with biological systems have attracted widespread interest in biology and medicine. Such nanoparticles are thought to have potential as novel intravascular probes for both diagnostic (e.g., imaging) and therapeutic purposes (e.g., drug delivery). Critical issues for successful nanoparticle delivery include the ability to target specific tissues and cell types and escape from the biological particulate filter (RES).

Among inorganic materials, magnetite has been investigated most widely for cancer therapy and diagnosis (Table 2.1.1(i–m)) [14–18]. Magnetite nanoparticles were used directly or dispersed in the polymeric matrix. They have been used chiefly in the hyperthermia treatment of cancer and, in some cases, for magnetic field-assisted targeting of nanoparticles. For diagnostic purposes, they were used in magnetic resonance imaging (MRI) as contrast enhancing agents for the purpose of cancer diagnosis, targeted molecular imaging, hyperfusion region visualization, cell labeling in T-cell-based therapy, and for detection of angiogenesis, apoptosis, and gene expression. PEG or oxidized starch (Table 2.1.1(i)) [14] as a hydrophilic surface-modifier, antibodies, FITC-labeled Tat peptide (Table 2.1.1(j)) [15] or the Annexin V protein (Table 2.1.1(k)) [16] as a specific targeting agent, and folic acid (Table 2.1.1(l)) [17] or transferrin (Table 2.1.1(m)) [18] as a ligand of the receptors overexpressed in tumor cells have been used as surface molecules.

2.1.5 Simple organic nanoparticles

The active agents are often incorporated or dispersed in the matrix to control or suppress their release. Polymers [19–27] and waxes or lipids [28–34] have been widely used as matrix materials. Table 2.1.1(n–w) shows various matrix nanoparticles that have recently been investigated for delivery of active agents; they are often for doxorubicin, an anticancer drug, whose liposomes are on the market as Doxil™ (Alza Corp.). In these nanoparticles, biodegradable polymers, such as PLA [19], PLGA, poly(ε-caprolactone) [20], chitosan, PACA, poly(lysine), and poly(aspartic acid) [21], are used as the polymeric matrix. The matrixes, consisting of artificial polymers or naturally occurring polymers such as albumin and gelatin have also been used in nanoparticles for gene delivery. The polymers used therein are often cationic to enhance association with anionic cell surfaces and/or DNAs.

The Gd-DTPA-loaded chitosan nanoparticles (Gd-nanoCPs) for gadolinium neutron capture therapy (Gd-NCT) were reported (Table 2.1.1(t)) [25–27]. Gd-nanoCPs were prepared by a novel emulsion-droplet coalescence technique. Gd-nanoCPs with the highest Gd content, which were obtained using 100% deacetylated chitosan in 15% Gd-DTPA aqueous solution, were 452 nm in diameter and had a Gd-DTPA content of 45%. Gd-DTPA loaded onto Gd-nanoCPs was barely released in PBS (1.8%) over 7 days despite the high water solubility of Gd-DTPA. In contrast, 91% of Gd-DTPA was released in plasma over 24 h. When Gd-nanoCPs were intratumorally injected, 92% of Gd-DTPA injected efficiently without outflow was retained in the tumor tissue for 24 h, which was different from the case of Gd-DTPA solution injection (only 1.2%). Thus, Gd-nanoCPs with a high content of water-soluble Gd-DTPA were successfully prepared by the emulsion-droplet coalescence technique. Their high Gd content, releasing properties and ability for long-term retention of

Gd-DTPA in the tumor led to significant tumor growth suppression in Gd-NCT trials in vivo.

2.1.6 Summary

The structures of nanoparticles were briefly explained. Their structures should be as simple as possible for manufacturing purpose. However, functional adaptation in specific applications very often requires some complicated structures, and uses of very specific molecular interactions. Although the nanostructure construction to achieve desired functions is not easy in the present stage of nanotechnology, nanoparticles will be beneficial in wide applications due to their behaviors intrinsic in the nano scale.

Nanoparticles designed and prepared for wider applications will be introduced in the following section.

References

[1] Y. Fukumori, H. Ichikawa: *Ad. Powder Technol.*, **17**, 1–28 (2006).
[2] A. Ito, H. Honda and T. Kobayashi: *Chem. Eng. Jpn.*, **67**(12), 692–695 (2003).
[3] K. Maruyama, O. Ishida, T. Takizawa and K. Moribe: *Ad. Drug Delivery Rev.*, **40**, 89–102 (1999).
[4] L. Li, C.A. Wartchow, S.N. Danthi, Z. Shen, N. Dechene, J. Pease, H.S. Choi, T. Doede, P. Chu, S. Ning, D.Y. Lee, M.D. Bednarski and S.J. Knox: *Int. J. Radiat. Oncol. Biol. Phys.*, **58**, 1215–1227 (2004).
[5] M. Miyamoto, H. Hirano, H. Ichikawa, Y. Fukumori, Y. Akine, K. Tokuuye: *Chem. Pharm. Bull.*, **47**, 203–208 (1999).
[6] M. Miyamoto, K. Hirano, H. Ichikawa, Y. Fukumori, Y. Akine, K. Tokuuye: *Biol. Pharm. Bull.*, **22**, 1331–1340 (1999).
[7] T. Watanabe, H. Ichikawa, Y. Fukumori: *Eur. J. Pharm. Biopharm.*, **54**, 119–124 (2002).
[8] H. Ichikawa, H. Tokumitsu, M. Miyamoto and Y. Fukumori: Nanoparticles for neutron capture therapy of cancer, in *Nanotechnologies for the Life Sciences*, Vol. 6, Wiley-VCH, Berlin, pp. 87–121 (2006).
[9] M.E. Akerman, W.C. Chan, P. Laakkonen, S.N. Bhatia and E. Ruoslahti: *Proc. Natl. Acad. Sci. USA*, **99**, 12617–12621 (2002).
[10] Y. Fukumori, K. Takada and H. Takeuchi: *Nanotechnologies for the Life Sciences*, Vol. 10, Wiley-VCH, Berlin, pp. 255–308 (2006).
[11] H. Ichikawa, H. Tokumitsu, K. Jono, T. Fukuda, Y. Osako, Y. Fukumori: *Chem. Pharm. Bull.*, **42**, 1308–1314 (1994).
[12] H. Ichikawa, S. Kaneko and Y. Fukumori, *Chem. Pharm. Bull.*, **44**, 383–391 (1996).
[13] H. Ichikawa, Y. Fukumori: *J. Controlled Release*, **63**, 107–119 (2000).
[14] A. Bjornerud, L.O. Johansson and H.K. Ahlstrom: *MAGMA*, **12**, 99–103 (2001).
[15] M.F. Kircher, J.R. Allport, E.E. Graves, V. Love, L. Josephson, A.H. Lichtman and R. Weissleder: *Cancer Res.*, **63**, 6838–6346 (2003).
[16] E.A. Schellenberger, A. Bogdanov Jr., D. Hogemann, J. Tait, R. Weissleder and L. Josephson: *Mol. Imaging*, **1**, 102–107 (2002).
[17] Y. Zhang, N. Kohler and M. Zhang: *Biomaterials*, **23**, 1553–1561 (2002).
[18] D. Hogemann, L. Josephson, R. Weissleder and J.P. Basilion: *Bioconjugate Chem.*, **11**, 941–946 (2000).
[19] H. Onishi, Y. Machida and Y. Machida: *Biol. Pharm. Bull.*, **26**, 116–119 (2003).
[20] J.S. Chawla, M.M. Amiji: *Int. J. Pharm.*, **249**, 127–138 (2002).
[21] K. Kataoka, A. Harada and Y. Nagasaki: *Adv. Drug Delivery Rev.*, **47**, 113–131 (2001).
[22] J.D. Hood, M. Bednarski, R. Frausto, S. Guccione, R.A. Reisfeld, R. Xiang and D.A. Cheresh: *Science*, **296**, 2404–2407 (2002).
[23] N.C. Bellocq, S.H. Pun, G.S. Jensen and M.E. Davis: *Bioconjugate Chem.*, **14**, 1122–1132 (2003).
[24] M. Kabbaj, N.C. Phillips: *J. Drug Target.*, **9**, 317–328 (2001).
[25] H. Tokumitsu, H. Ichikawa and Y. Fukumori: *Pharm. Res.*, **16**, 1830–1835 (1999).
[26] F. Shikata, H. Tokumitsu, H. Ichikawa and Y. Fukumori: *Eur. J. Pharm. Biopharm.*, **53**, 57–63 (2002).
[27] H. Tokumitsu, J. Hiratsuka, Y. Sakurai, T. Kobayashi, H. Ichikawa and Y. Fukumori: *Cancer Lett.*, **150**, 177–182 (1999).
[28] M.O. Oyewumi, R.A. Yokel, M. Jay, T. Coakley and R.J. Mumper: *J. Control. Release*, **95**, 613–626 (2004).
[29] M.O. Oyewumi, S. Liu, J.A. Moscow and R.J. Mumper: *Bioconjugate Chem.*, **14**, 404–411 (2003).
[30] M.O. Oyewumi, R.J. Mumper: *Int. J. Pharm.*, **251**, 85–97 (2003).
[31] M.O. Oyewumi, R.J. Mumper: *Bioconjugate Chem.*, **13**, 1328–1335 (2002).
[32] M.O. Oyewumi, R.J. Mumper: *Drug Dev. Ind. Pharm.*, **28**, 317–328 (2002).
[33] Z. Cui and R.J. Mumper: *Bioconjugate Chem.*, **13**, 1319–1327 (2002).
[34] P.J. Stevens, M. Sekido and R.J. Lee: *Anticancer Res.*, **24**(1), 161–165 (2004).

2.2 Particle size

2.2.1 Gas-phase method

Homogeneous nucleation without seeds and heterogeneous nucleation with seeds have attracted much interest in the wide field of particle formation from gases and liquids by phase changes. Although many nucleation theories have been proposed, the complexity of nucleation theory prevents its practical use in applied fields such as particle processing.

In this section, particle formation by gas-phase and liquid-phase methods is described systematically. A simple model of liquid-phase nucleation and its expanded gas-phase nucleation model are introduced [1–3]. These models enable prediction of the effects of the operating conditions of particle processing on the steady-state concentration and size of the nucleated particle.

(a) Nucleation without seeds (Homogeneous nucleation)

In a gas-phase system, we have to take into account the fact that the size of the incipient nuclei is of the same order of magnitude as the size of gas molecules and that, consequently, we consider a limiting sphere around the critical nucleus. Outside the limiting sphere, monomers move according to the continuum diffusion equation. The monomers moving inside the limiting sphere travel as in a vacuum without colliding with air molecules, following free-molecular trajectories. Considering the above facts, the relation between the number concentration of nucleated particles n_0^* and the operating conditions is expressed as follows:

$$G_0^* = 4\pi DC^* n_0^* \langle r^* F(r^*) \rangle, \tag{2.2.1}$$

The expression of Fig. 2.2.1 leads to the relational expression of the liquid-phase method given in equation (2.2.13) which is corrected by using the Fuchs interpolation factor.

where

$$F(r^*) = \frac{1 + \dfrac{\lambda}{r^*}}{1 + \dfrac{4D}{r^* \bar{c}}\left(1 + \dfrac{\lambda}{r^*}\right)} \tag{2.2.2}$$

G_0^* is the critical monomer generation rate, D the monomer diffusion coefficient, C^* the critical monomer concentration, r^* the radius of the nuclei, λ the molecular mean free path, and $\bar{c}$ the monomer mean thermal speed, and these last two are given by:

$$\lambda = \frac{32D}{\pi \bar{c}} \frac{M_A}{M + M_A}, \tag{2.2.3}$$

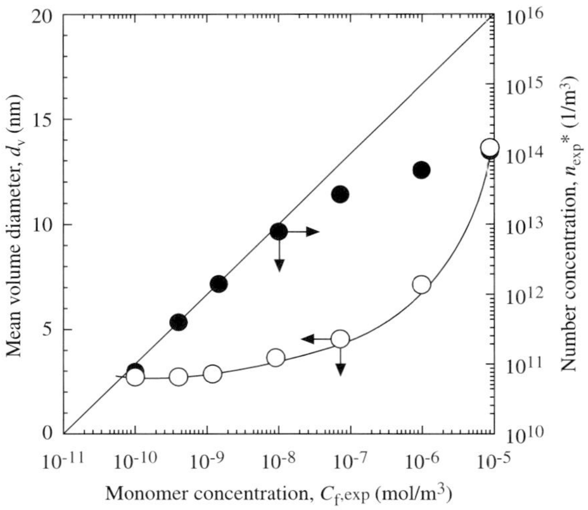

Figure 2.2.1
Number concentration and mean volume diameter of TiO_2 nanoparticle.

$$\bar{c} = \sqrt{\frac{8k_B T}{\pi m_1}} \tag{2.2.4}$$

where M_A is the molecular weight of gas medium, M the molecular weight of monomer, k_B the Boltzmann constant, T the absolute temperature, and m_1 the monomer mass. The equation (2.2.1) is the expression that is obtained from the equation (2.2.13) multiplied by the Fuchs interpolation factor.

Since the diffusion coefficient of monomer in the gas phase is three orders of magnitude larger than that in the liquid phase, the rate of growth by condensation is very rapid. The size of nucleated particles gradually increases with time, while new particles are being formed in the medium by homogeneous nucleation. We now present a simple way to account for the effect of particle growth on homogeneous nucleation, using the time-averaged value of $r^* F(r^*)$ appearing in equation (2.2.1).

The above model was applied to the case of a specific gas-phase reaction: the hydrolysis of titanium tetraisopropoxide (TTIP) to produce TiO_2 particles. Figure 2.2.1 shows the experimental results of particle number concentration $n_{0,exp}^*$ and mean volume diameter d_v as a function of precursor monomer concentration $C_{f,exp}$. The subscript "exp" refers to the experimental values. Figure 2.2.2 shows the TiO_2 particle size distributions measured at different initial TTIP concentrations. If entire TTIP vapor present in the system yields TiO_2, then the final monomer concentration without nucleation C_f is equal to the initial TTIP concentration. Since this reaction rate is proportional to the TTIP concentration, the monomer generation rate becomes $G_0^* \propto C_f$. If the reaction temperature

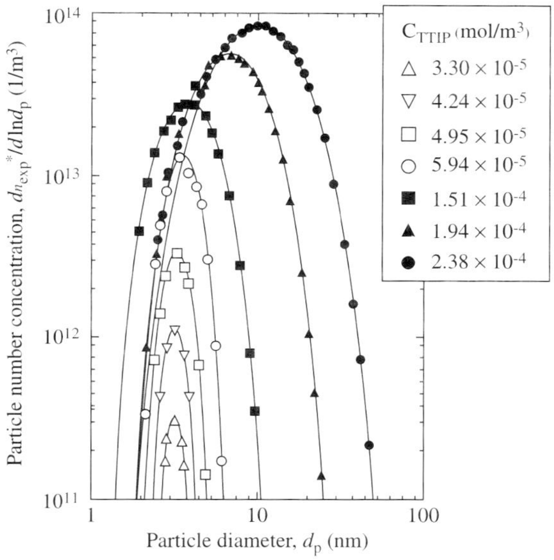

Figure 2.2.2
Particle size distribution of TiO_2 nanoparticle.

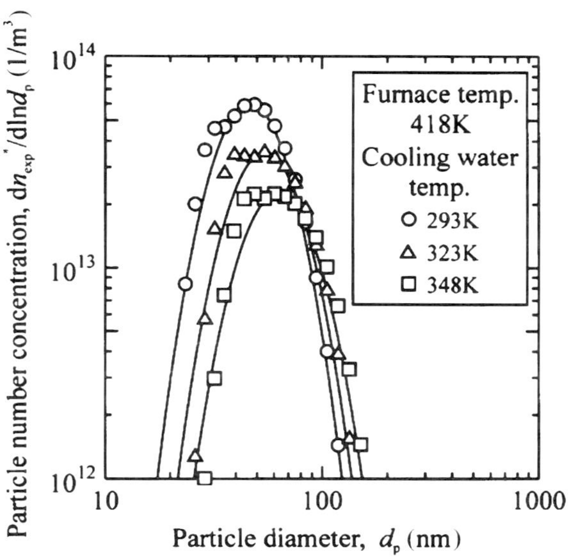

Figure 2.2.3
Shape and size control of DOS particles by condensation temperature.

is constant, the number concentration of nucleated particles is found to be proportional to the monomer concentration: $n_0^* \propto C_f$, because the other parameters appearing in equation (2.2.1) are constant. When diffusion losses are taken into account (penetration $P = n_{exp}^*/n_0^* = C_{f,exp}/C_f$), the above expression can be rewritten as $n_{exp}^* \propto C_{f,exp}$, d_v = constant. As seen, except in the region of very high monomer concentration, where coagulation effects cannot be neglected, the relationship between the number concentration of self-nucleated particles and the concentration of precursor monomers is linear.

Figure 2.2.3 shows an example of controlling the particle size of the dioctyl sebacate (DOS) particle generated by the evaporation–condensation method. It is proved that the size of the generated particle decreases with decreasing cooling temperature. This is because the generation rate of monomer becomes faster and the number concentration of generated particles is increased when the cooling temperature is lowered. As mentioned above, the size of the generated particles can be controlled if the difference between the gas-phase and the liquid-phase methods is considered.

(b) Nucleation with seeds (Heterogeneous nucleation)

When seed particles are present in the system, they are distributed randomly in the medium. As a result, new nuclei can be formed with a fraction v^*. The value 0.16 of v^* gives good agreement with experiments (refer to Section 2.2.2(b)). Since the diffusion coefficient of monomers in the gas phase is large, we have to consider the seed particles (number concentration; n_p and radius; r_p) and self-nucleated particles present, while new particles are being formed in the medium (τ is the mean aerosol residence time in the system, up to the particle measuring unit). With these two modifications, the total number concentration of particles n_T can then be estimated as [3]

$$n_T = \begin{cases} n_0^* + (1 - R')n_p & (n_p \leq n'_{pc}) \\ n_0^* v^* + n_p & (n_p \geq n'_{pc}) \end{cases} \tag{2.2.5}$$

with $n_{pc'} = n_0^*(1 - v^*)/R'$ and where, by definition

$$R' = \left\langle \frac{r_p F(r_p)}{r^* F(r^*)} \right\rangle = \frac{1}{\tau}\int_0^\tau \frac{r_p F(r_p)}{r^* F(r^*)} dt \tag{2.2.6}$$

Consider a particle (self-nucleated or seed) of radius r surrounded by vapor molecules (monomers). The equation of particle size growth rate is given by

$$\frac{dr}{dt} = Dv_1 N_{av} \frac{F(r)}{r}(C - C_s), \tag{2.2.7}$$

where v_1 is the volume of a monomer and N_{av} is Avogadro's number. Integration of equation (2.2.7) leads to

$$\left(\frac{4D}{\bar{c}} - \lambda\right)(r - r_0) + \frac{1}{2}(r^2 - r_0^2) + \lambda^2 \ln\left(\frac{r + \lambda}{r_0 + \lambda}\right) \tag{2.2.8}$$
$$= Dv_1 N_{av}(C - C_s)t,$$

where r_0 is the initial particle radius. The value of the particle radius at any time can be obtained from equation (2.2.8) by trial and error. In the above equations for particle size growth by condensation, the Kelvin effect and the latent heat have been neglected. Given the initial sizes of critical nuclei (taken as $r_0^* = 1$ nm) and seeds, the procedure to obtain n_T consists of calculating the sizes r_p and r^* at increasing times t by means of equation (2.2.8), and performing numerical integration of equation (2.2.6).

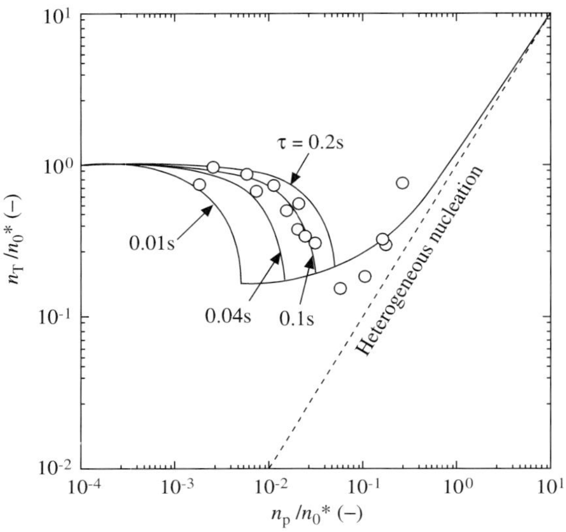

Figure 2.2.4
Relationship between total number concentration and seed number concentration.

DOS particles were generated by evaporation–condensation in a system containing $ZnCl_2$ seed particles. Figure 2.2.4 shows the measured total number concentration of particles against the number concentration of seeds, along with the theoretical curves calculated using equation (2.2.5). The line that best fits the experimental results is that calculated for the mean aerosol residence time of 0.1 s. The calculated results are in good agreement with the results obtained in gas-phase nucleation experiments. Figure 2.2.5 shows the typical size distributions obtained for DOS nucleation in the presence of $ZnCl_2$ seed particles. In the case shown in Fig. 2.2.5(a), the seed concentration is so low that most of the condensing vapor is spent for the formation of new particles (homogeneous nucleation predominant). Case (b) corresponds to the intermediate situation where both nucleation mechanisms, homogeneous and heterogeneous, coexist, resulting in a bimodal distribution for the total population of particles. Finally, Fig. 2.2.5(c) shows the situation where the monomer concentration is relatively low (in relation to that of the seeds). In this situation, most of the condensing vapor is used for seed growth and practically no new particles are formed by self-nucleation (heterogeneous nucleation predominant). As described above, the relationship between the number concentration and the mean volume diameter of growth of aerosol particles and the operating conditions can be controlled by considering the mean aerosol residence time.

2.2.2 Liquid-phase method

(a) Nucleation without seeds (Homogeneous nucleation)
According to the classical theory, the first stage of the nucleation process consists of the formation of nuclei

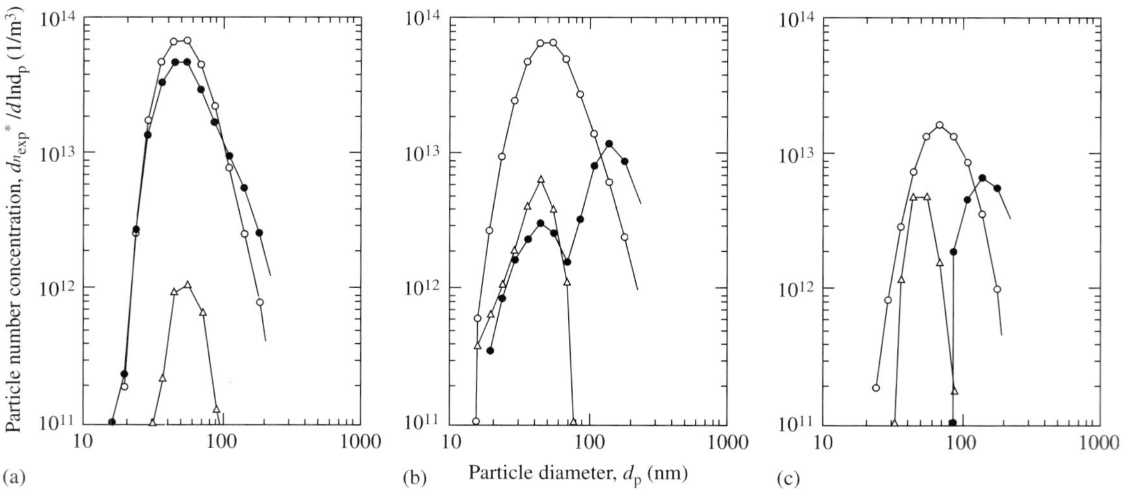

Figure 2.2.5
Typical particle size distribution (○ for generated particle without seed; △ for seed; ● for total particle).

(radius; r^*) when the precursor monomers (atoms or molecules) reach a certain critical concentration C^*. When coagulation of nuclei does not take place, each nucleus then grows in size as other monomers in its vicinity diffuse toward its surface and nucleate onto it. The attainable number concentration of nucleated particles n_0^*, is determined from the balance between the monomer generation rate and the rate of monomer depletion by diffusion toward previously formed nuclei.

In regions where the nuclei concentration is so low that the monomer generation rate is larger than the diffusion depletion rate, the monomer concentration C increases with time until the critical concentration is reached ($C = C^*$) and new nuclei can be formed. As the nuclei concentration increases, the depletion rate of monomers, resulting from diffusion toward their surfaces, also increases, until a time is reached at which the depletion rate balances the generation rate. From this time onward, the monomer concentration decreases gradually and no new nuclei can be formed. This local process occurs everywhere in the system and, hence, the concentration of nucleated particles attains a uniform value n_0^*. Considering the system overall, it can be stated that nucleation stops at a time when the concentration of nucleated particles attains its maximum possible value n_0^*, beyond which the monomer concentration decreases ($\partial C/\partial t < 0$).

The process of monomer diffusion toward the surface of a nucleus is modeled assuming that each nucleus is in the center of a spherical cell of constant radius b. The so-called cell model has been extensively used in the description of other phenomena, such as evaporation, condensation, and melting. The diffusion of monomers toward the nucleus can be expressed as:

$$\frac{\partial C}{\partial t} = G(t) + D\nabla^2 C \tag{2.2.9}$$

where $G(t)$ is the monomer generation rate and D its diffusion coefficient.

With a constant monomer generation rate, taking $t = 0$ as the time at which $C = C^*$, assuming $r^* =$ constant $\ll b$, and using the initial and boundary conditions: $C(r,0) = C^*$, $C(r^*,t) = 0$, $(\partial C/\partial t)_{r=b} = 0$, the approximate solution of equation (2.2.9) can be written as:

$$C(r,t) \approx \left(\frac{1}{r^*} - \frac{1}{r}\right)\left[e^{-\theta}\left(C^* r^* - \frac{Gb^3}{3D}\right) + \frac{Gb^3}{3D}\right], \tag{2.2.10}$$

where: $\theta = 3r^* Dt/b^3$.

Differentiating this equation with respect to time yields:

$$\frac{\partial C(r,t)}{\partial t} = -\left(\frac{1}{r^*} - \frac{1}{r}\right)\frac{3r^* D}{b^3} e^{-\theta}\left(C^* r^* - \frac{Gb^3}{3D}\right) \tag{2.2.11}$$

As stated above, homogeneous nucleation is suppressed whenever $\partial C(b,t)/\partial t < 0$, i.e., whenever the last term within brackets in equation (2.2.11) is positive. This gives us the necessary condition for homogeneous nucleation to occur:

$$G < \frac{3DC^* r^*}{b^3} \tag{2.2.12}$$

The constant cell radius b can be expressed as a function of the number concentration of nucleated particles n_0^* as $4\pi b^3 n^*/3 = 1$, so that the critical monomer generation rate G_0^*, i.e., the generation rate above which homogeneous nucleation occurs and below which nucleation is suppressed, can be written as:

$$G_0^* = 4\pi r^* DC^* n_0^* \tag{2.2.13}$$

The number concentration of nucleated particles can be predicted using equation (2.2.13), the mean volume diameter of nucleated particles d_v can be estimated from a mass balance:

$$d_v = \left(\frac{6MC_f}{\pi \rho_p n_0^*}\right)^{1/3} \tag{2.2.14}$$

where C_f is the final monomer concentration without nucleation, ρ_p the particle density, and M the molecular weight.

The critical nucleus radius r^* can, in principle, be estimated by the Gibbs–Thompson equation. However, this equation contains parameters whose actual values are only known approximately. The critical radius is usually taken as 1 nm. The monomer diffusion coefficient D can be estimated using the Wilke–Chang equation or the King equation. The monomer generation rate G_0^* can be estimated as $G_0^* = (\partial C/\partial t)_{t=t^*}$, where t^* is the nucleation time. The generation rate of monomer G_0^* can be approximated as $G_0^* = C^*/t^*$. Therefore, equation (2.2.13) can be written using the attainable number concentration of nucleated particles n_0^* as:

$$n_0^* = \frac{1}{4\pi r^* Dt^*} \tag{2.2.15}$$

This simple relation can be used to predict the number concentration of nucleated particles even if the reaction rate expression and the critical supersaturation concentration are unknown.

Figure 2.2.6 shows a comparison between theoretical (equation (2.2.15)) and experimental particle number concentrations. We used two types of particle formation methods. One is that of silver particles generated by chemical reduction, and the other is that of zinc sulfide particles generated by homogeneous precipitation. As seen, the calculated values are in good agreement with the experimental data.

Table 2.2.1 shows an example of controlling the particle size of silver particles generated by chemical reduction. It can be seen that the size of the generated particle decreases with increasing solution viscosity. The diffusion coefficient of monomers is inversely proportional to the solution viscosity. The nucleation time was constant regardless of the solution viscosity. The other parameters included in equation (2.2.13) can be considered to be constant. Therefore, since the number concentration of generated particles increases in proportion to the solution viscosity, the final size of the generated particles decreases with equation (2.2.14). In addition, the mean diameter of the generated particles can be estimated using these relational expressions.

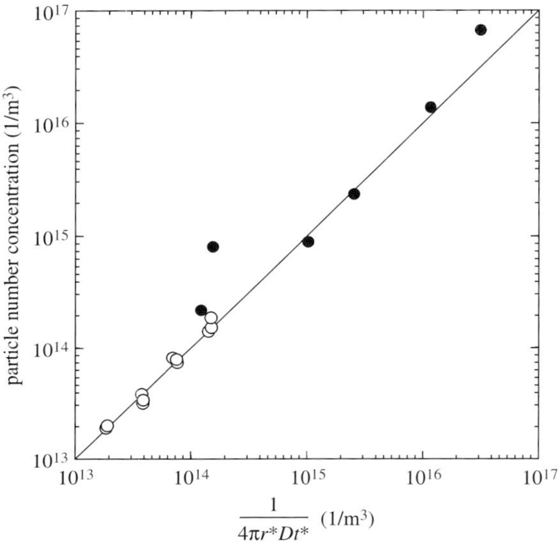

Figure 2.2.6
Prediction of number concentration of generated particles synthesized by liquid-phase method (○ for $ZnCl_2$ particle (homogeneous precipitation); ● for Ag particle (chemical reduction)).

(b) Nucleation with seeds (Heterogeneous nucleation)
When particles are present in the system, we must consider two types of condensation processes: condensation of monomers onto the seed surfaces (heterogeneous nucleation) and self-nucleation of monomers (homogeneous nucleation). The model for both homogeneous and heterogeneous nucleation is based on the development presented in the former section for homogeneous nucleation, using a cell model. To reproduce the experimental results of nucleation in the presence of seeds, modification of the cell model is necessary. The constant size cell model assumes that all the particles are equally spaced. Actually, the particles are distributed randomly in the medium, so that the distances between neighboring particles follow a certain distribution law. Monomers generated in regions where the seeds are separated by very small distances are soon depleted through condensation onto the seed surfaces. Conversely, new nuclei can only be formed in those places where the distance between particles is larger. We assume now that new nuclei can be formed with a fraction v^* of the total number of cells. Hence, the number concentration of self-nucleated particles is given by $v^* n_0^*$, because the number concentration of homogeneously nucleated particles without seeds is n_0^*. When the system contains seeds of radius r_p and number concentration n_p, the total particle number concentration after nucleation in a system containing seed particles n_T is given by [2]

$$n_T = \begin{cases} n_0^* + (1-R)n_p & (n_p \le n_{pc}) \\ n_0^* v^* + n_p & (n_p \ge n_{pc}) \end{cases} \tag{2.2.16}$$

where $R = r_p/r^*$, $n_{pc} = n_0^*(1-v^*)/R$.

Silver particles were generated by chemical reduction in a system containing silver seed particles. Figures 2.2.7 and 2.2.8 show the experimental and calculated results for the initial concentration of silver ions and the seed-to-nucleus size ratio, respectively. The solid lines in these figures were calculated from equation (2.2.16). The dotted lines correspond to the

Table 2.2.1
Shape and size control of Ag particle by solution viscosity.

Initial concentration of Ag^+ (mol/m³)	Solution viscosity (mPa·s)	Mean volume diameter d_v (μm)	
		Experimental	Calculation
1.47×10^{-1}	1.0	0.53	0.52
	9.9	0.25	0.24
	53.1	0.15	0.14
1.47×10^{1}	1.0	0.79	0.82
	9.9	0.42	0.38
	53.1	0.19	0.22

extreme case in which homogeneous nucleation does not occur at all, so that the total number concentration of particles in the system is simply equal to the number concentration of seeds. As seen in these figures, the model predictions are in good agreement with the results obtained in liquid-phase nucleation experiments. When the seed number concentration n_p is very low, the total number concentration of particles in the system is practically constant, independent of n_p. The reason for this is that when n_p is very low, the generation rate of monomers G is much larger than the monomer depletion rate by diffusion toward seeds, and homogeneous nucleation is predominant. On the other extreme, when the seed number concentration is very high, homogeneous nucleation is practically suppressed. At intermediate levels of seed concentration, both mechanisms, self-nucleation, and particle growth, coexist.

The model allows estimation of the operating conditions (concentration of seed particles and monomers, and seed-to-nucleus size ratio) under which (1) homogeneous nucleation is predominant, (2) particle growth is predominant, and (3) both phenomena, self-nucleation, and particle growth, coexist.

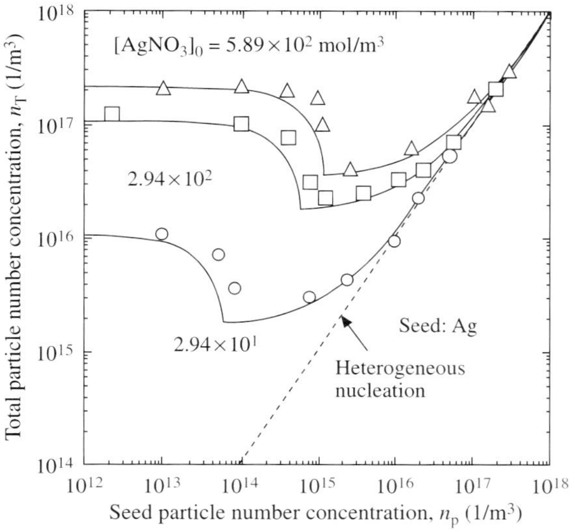

Figure 2.2.7
Effect of raw material concentration and seed number concentration on total particle number concentration ($R = 150$; ○ for $[Ag^+]_0 = 29.4\,mol\,m^{-3}$; □ for $[Ag^+]_0 = 294\,mol\,m^{-3}$; △ for $[Ag^+]_0 = 589\,mol\,m^{-3}$).

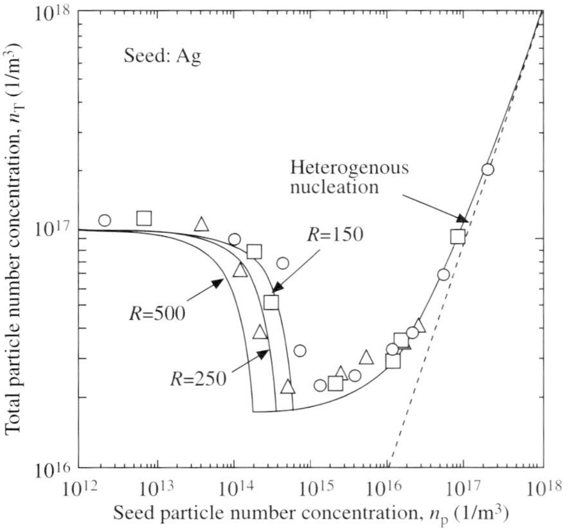

Figure 2.2.8
Effect of diameter and number concentration of seed on total particle number concentration ($[Ag^+]_0 = 294\,mol\,m^{-3}$; ○ for $R = 150$; □ for $R = 250$; △ for $R = 500$).

References

[1] Y. Kousaka, T. Nomura: *Kagaku Kogaku Ronbunsyu*, **23**, 666 (1997).
[2] T. Nomura, M. Alonso, Y. Kousaka, and K. Tanaka: *J. Colloid Interface Sci.*, **203**, 170 (1998).
[3] Y. Kousaka, T. Nomura, M. Alonso, M. Nishio, and E. Tenjiku: *J. Aerosol Sci.*, **31**, 519 (2000).

2.2.3 Supercritical hydrothermal method

The properties of water change drastically around the critical point and therefore the equilibrium and speed of reactions in water also change greatly. Consequently, the phase state also changes remarkably with the property changes and it forms a uniform phase with the gas. The authors have proceeded in the development with the idea that the nanoparticles with unique properties, which had not been obtained conventionally, can be synthesized using this supercritical field as a reaction zone of hydrothermal synthesis. Here the principle and features are introduced, above all the mechanism of nanoparticle generation.

The phase state is determined by the balance between the formation of agglomerated structure in liquid phase by the potential energy between molecules and the disordering in the gas phase by the heat energy. The critical point is an original value for each substance. The supercritical temperature and pressure of water are 374°C and 22.1 MPa, respectively. The change of density of water with temperature and pressure is shown in Fig. 2.2.9.

Since the molecular force and the kinetic energy are competitively antagonizing near the balancing point, the fluid density changes greatly with little change in temperature or pressure. The properties of fluid, namely the molecular distance also change drastically when the water density changes, because they are

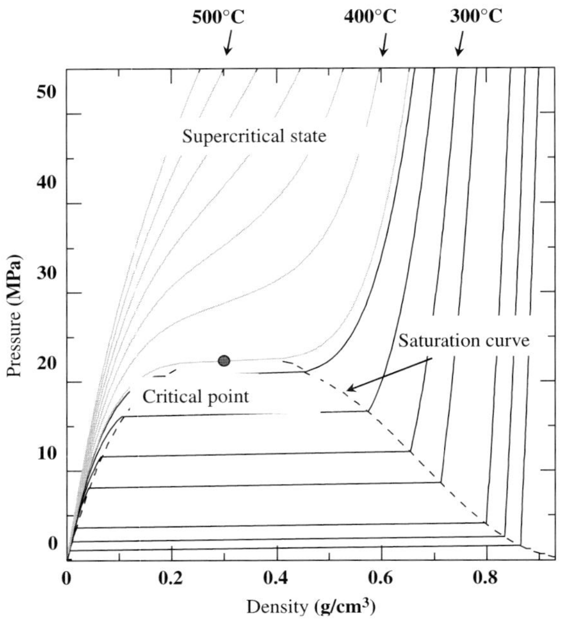

Figure 2.2.9
Dependence of water density on temperature and pressure.

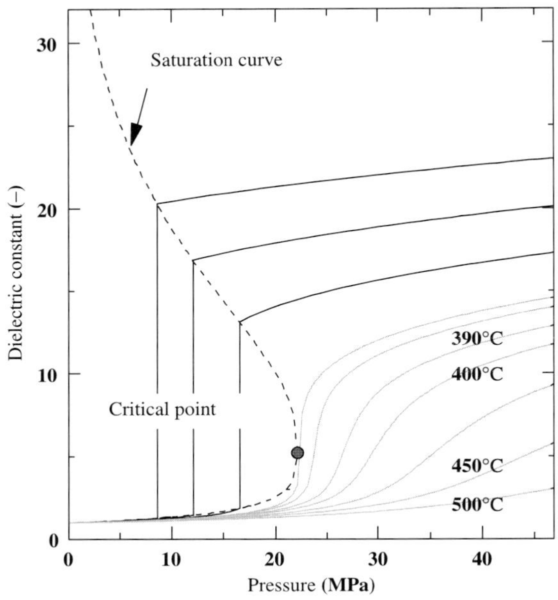

Figure 2.2.10
Dependence of dielectric constant of water on temperature and pressure.

determined by the molecular interactions. As an example, Fig. 2.2.10 shows the dielectric constant [1], which is one of the determinant factors of the equilibrium and speed of reactions.

The dielectric constant of water at a room temperature has such a high value as 78, which is an important factor as a polar solvent that can dissolve electrolytes stably. However, it decreases down to 2–10 with the increasing temperature, which is similar to that of polar organic solvent, near the critical point. The phase behavior of a water–gas two component system and that of water–organic substances near the critical point of water are shown in Figs. 2.2.11 and 2.2.12, respectively.

The curves in these figures show the gas–liquid critical boundary. It is indicated that in the higher temperature region on the right side of the curves, the two phases are mixed uniformly in an arbitrary ratio. The injection of air into water results in the generation of bubbles. However, as the supercritical water is in the state of high-density steam, it is mixed with the gas completely. Though the phrase "water and oil" is used as a phrase representing insoluble things, water and oil can produce a uniform phase at an elevated temperature and high pressure. It is caused by the reduction of the dielectric constant of water down to that of polar organic solvents and it loses its water-like properties.

Since the equilibrium shifts to the side of hydroxide or oxide when metallic salts are heated up, thermal hydrolysis reaction is regarded as the combination of hydrolysis and dehydration reactions. The method of

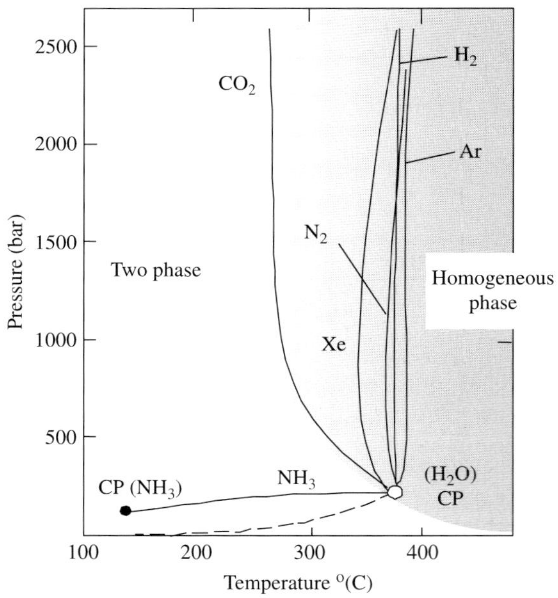

Figure 2.2.11
Phase behavior of water–gas two-component system (gas–liquid critical boundary).

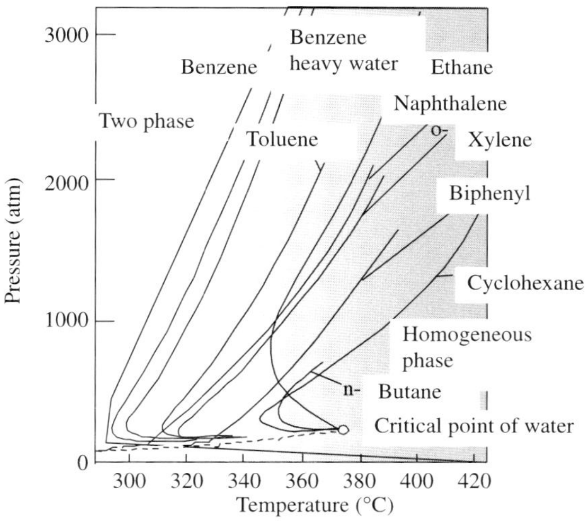

Figure 2.2.12
Phase behavior of water–hydrocarbon two component system.

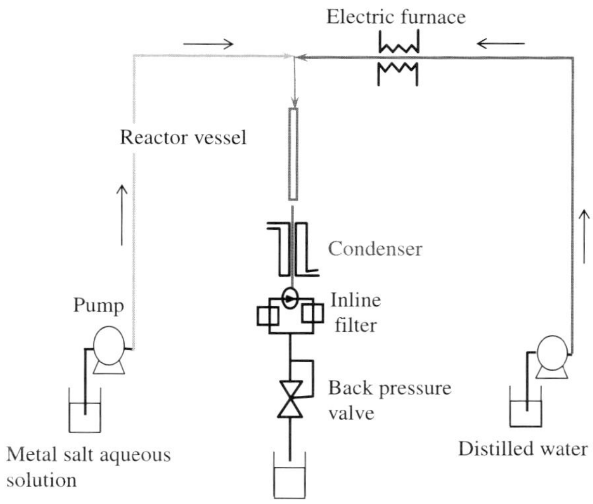

Figure 2.2.13
Rapid temperature rise type reactor for hydrothermal synthesis in supercritical fluid.

Table 2.2.2
Nanoparticles synthesized by the supercritical hydrothermal method.

Starting materials	Products	Particle size (nm)	Morphology
$Al(NO_3)_3$	AlOOH	80~1,000	Hexagonal Rhombus Needle-like
$Fe(NO_3)_3$	α-Fe_2O_3	~50	Spherical
$Fe_2(SO_4)_3$	α-Fe_2O_3	~50	Spherical
$FeCl_2$	α-Fe_2O_3	~50	Spherical
$Fe(NH_4)2H(C_6H_5O_7)_2$	Fe_3O_4	~50	Spherical
$Co(NO_3)_2$	Co_3O_4	~100	Octahedral
$Ni(NO_3)_2$	NiO	~200	Octahedral
$ZrOCl_2$	ZrO_2 (cubic)	~20	Spherical
$Ti(SO_4)_2$	TiO_2	~20	Spherical
$TiCl_4$	TiO_2 (anatase)	~20	Spherical
$Ce(NO_3)_3$	CeO_2	20~300	Octahedral
$Fe(NO_3)_3$, $Ba(NO_3)_2$	$BaO \cdot 6Fe_2O_3$	50~1,000	Hexagonal
$Al(NO_3)_3$, $Y(NO_3)_3$, $TbCl_3$	$Al_5(Y + Tb)_3O_{12}$	20~600	Dodecahedronal

synthesizing hydroxides or oxides making use of this shift of equilibrium is called the hydrothermal synthesis method. Supercritical hydrothermal synthesis applies supercritical water as the reaction zone for the hydrothermal synthesis. In most cases, the hydrothermal synthesis tests are conducted using a batch autoclave. After charging the feed material, the hydrothermal synthesis is carried out by heating the reaction vessel. However, since the hydrothermal synthesis takes place as the temperature increases by this method, the final product includes the compounds generated during the temperature rise, even when the process finally reaches the critical point.

To eliminate the effect of reaction during the temperature rise, a flow-type rapid temperature rise reactor [3] as shown in Fig. 2.2.13 has been developed. The metal salt solution is supplied by a high-pressure liquid pump and contact is made with the supercritical water fed through a high-temperature furnace in the other separated line.

In this method, the feed liquid is heated up rapidly to the supercritical state and the hydrothermal synthesis reaction takes place in the supercritical water. The results obtained with this flow-type rapid temperature rise reactor are shown in Table 2.2.2. It suggests wide application of this method because of the capability of synthesis of not only the fine metal oxide particles of a single component but also the composite oxides. The nanoparticles obtained were single crystalline in most cases and had smaller particle size compared with those synthesized under the subcritical conditions as seen in Fig. 2.2.14.

Next the mechanism of synthesis of nanoparticles in supercritical water will be explained. The experiment in subcritical water at the same concentration and flow rate using the same apparatus as that in Fig. 2.2.13 produced larger particles than in the supercritical water as shown in Fig. 2.2.15 [4]. On the other hand, when the volume of the tubular reactor was enlarged and the processing time was increased, the particle growth was confirmed with a processing time of less than the subcritical condition, while the particle size did not change by the supercritical hydrothermal synthesis. It was caused by the Ostwald-ripening [5, 6], which is the phenomena of solution of the finer particles and their recrystalization on the larger particles, in the subcritical water having higher solubility. The reaction speed was examined from the relationship between the hydrothermal synthesis and reaction time by further reducing the tubular reactor volume.

Figure 2.2.16 [4] shows the Arrhenius plot of the resulted data. Though the Arrhenius plot showed straight lines under subcritical conditions, the reaction speed increased abruptly in the supercritical water.

The hydrothermal synthesis proceeds easier in a fluid with a lower dielectric constant at the same temperature. When the temperature is changed at the same pressure, the dielectric constant decreases abruptly and the hydrothermal synthesis is accelerated above the supercritical region. While the solution rapidly heated up to the supercritical point in the mixing zone reacts quickly, the solubility of the generated product is extremely low in the supercritical state, which leads to a high degree of supersaturation, namely high speed of core generation. It is demonstrated in Fig. 2.2.17 that the supercritical hydrothermal synthesis by rapid temperature rise provides a suitable reaction zone for the synthesis of nanoparticles.

Finally, there is an important point to be considered in the development of nanomaterials. Even if nanoparticles are actually produced, when they would require special technology for the following processes like collection, handling, dispersion into resin, arrangement, or coating, innovation in nanotechnology by the industries would be limited. Therefore, the important aspect is regarded to exist in the preparation and processing of nanoparticles taking the final products into consideration.

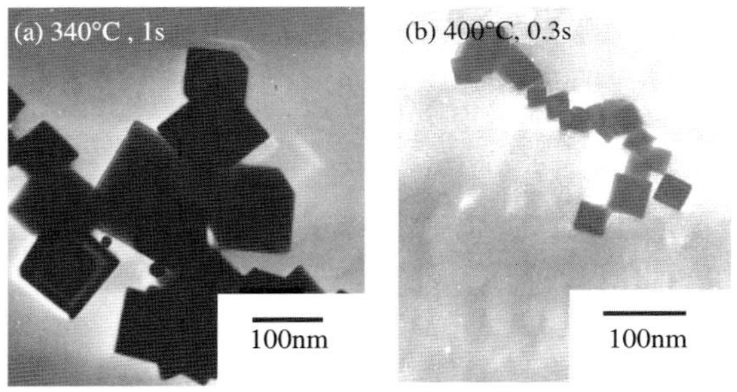

Figure 2.2.14
Ceria particles prepared in (a) subcritical water and (b) supercritical water.

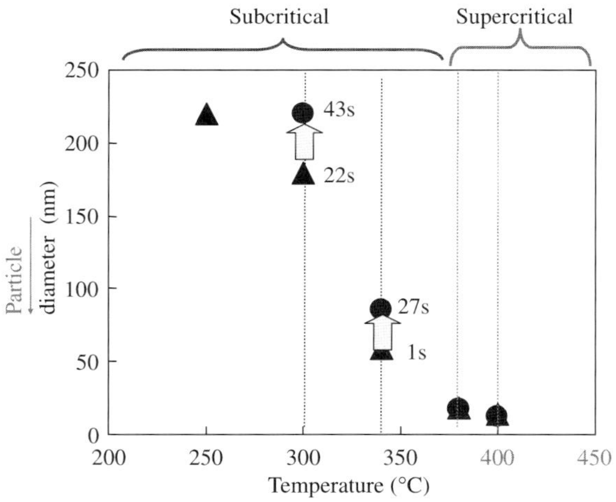

Figure 2.2.15
Dependence of particle size on temperature and reaction time.

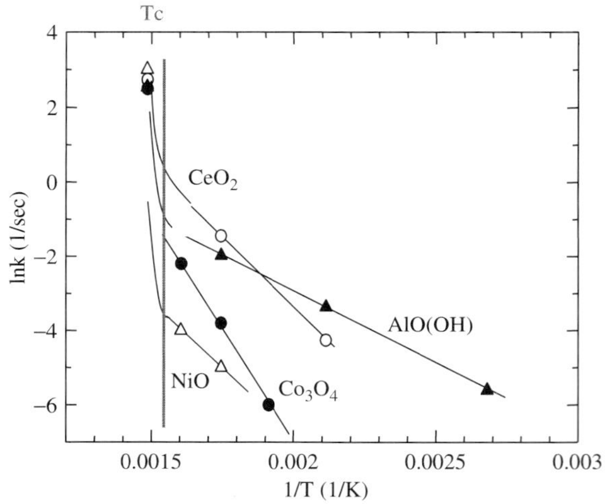

Figure 2.2.16
Arrhenius plot of the reaction speed of hydrothermal synthesis.

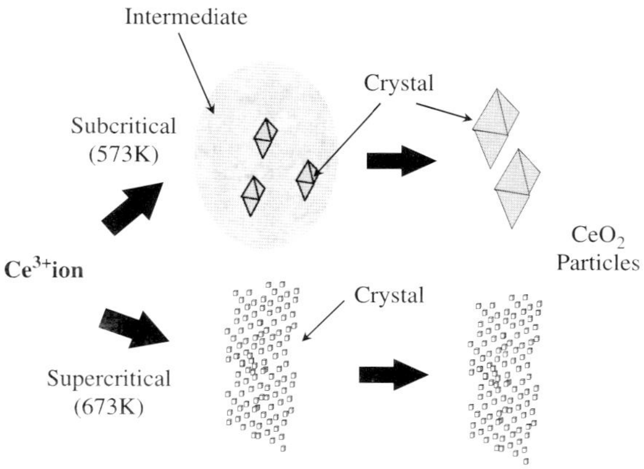

Figure 2.2.17
Mechanism of particle generation in supercritical water and subcritical water.

In this section, an overview of the preparation of nanoparticles by the supercritical hydrothermal synthesis has been explained. The authors are presently advancing the development of a method including in-situ organic surface modification of nanoparticles. Namely, the supercritical hydrothermal synthesis of the next-generation is being developed to create ideal hybrid-nanoparticles, which would realize the desired structure of the final product and have the modified surface designed for the optimized handling and processing for their production.

References

[1] M. Uematsu and E.U. Franck: *J. Phys. Chem.*, **81**, 1822 (1980).
[2] E.U. Franck: *Pure Appl.Chem.*, **53**, 1401 (1981).
[3] T. Adschiri, K. Kanazawa and K. Arai: *J. Am. Ceram. Soc.*, **75**, 1019–1022 (1922).
[4] T. Adschiri, Y. Hakuta, K. Sue and K. Arai: *J. Nanoparticle Res.*, **3**, 227–235 (2001).
[5] Y. Hakuta, S. Onai, H. Terayama, T. Adschiri and K. Arai: *J. Mater. Sci. Lett.*, **17**, 1211–1213 (1998).
[6] T. Aschiri, Y. Hakuta and K. Arai: *Ind. Eng. Chem. Res.*, **39**, 4901–4907 (2000).

2.2.4 Solid-phase method

Recently, nanoparticles can be produced by a grinding method; however, it has still lots of barriers to overcome in growth amounts of nanoparticles. Normally, nanoparticles can be synthesized by conventional methods such as PVD, chemical vapor deposition (CVD), crystallization and supercritical processes, and the products obtained are in aggregated states due to their high surface energy. This has to be disaggregated by a mechanical method such as an agitation mill with small beads.

As McCormick et al. [1, 2] have already reported, a mechanochemical (MC) method gives us an alternative for producing nanoparticles, and there has been a growing interest in these days. The nanoparticles produced by this method have unique characteristics such as narrow size distribution of particles besides well-dispersible powder. Of course, it is very important to choose the suitable condition in the milling operation. The MC method can produce nanoparticles of metals, metal oxides, complex oxides, and others.

Synthesis of nanoparticles of different materials by solid-phase method has been introduced as below:

2.2.4.1 Mechanochemistry and solid-state reaction

When different kinds of solids are subjected to grinding, their sizes are reduced; accordingly, their surface energy increases. The surface activated at the contact area and boundary induces solid-state reaction, and the reaction product formed near the boundary is scratched off from the solid surface for a fresh surface to appear. This fresh surface is the starting part of the next solid-state reaction. This kind of reaction takes place in molecular level, so that the particle sizes of the product are in nanometer range. The surface of the product synthesized by grinding operation has been stressed, and the structure may be disordered in the depth direction, and this may be amorphous one with small grain crystal size in the prolonged grinding. Therefore, the ground product is in a sub-stable or amorphous state, and this affects subsequent operations such as heating and leaching operations remarkably. For example, heating of the ground product causes easy sintering and dissociation, and leaching induces easy dissolution of selective elements or chemical species in the ground product. These phenomena are explained by difference in potential energy which is released to be stable, leading to mechanochemistry and MC effect of materials induced by the grinding operation.

The macro and microchanges of material caused by milling are shown in Fig. 2.2.18 [3].

2.2.4.2 MC synthesis of nanoparticles

a) Metal nanoparticles [4]

Metal nanoparticles such as Cu, Ni, Co, Fe can be synthesized from metal chlorides as starting materials by their grinding operation. Equation (2.2.17) is an example of the reaction, and the sizes of the product are ranged from 10 to 20 nm.

$$FeCl_3 + 3Na \rightarrow Fe + 3NaCl \qquad (2.2.17)$$

The data obtained by Ding [4] are shown in Table 2.2.3. Though most of them are the nanosize products synthesized by reductive reaction, some of them are classified by exothermic reaction. Therefore,

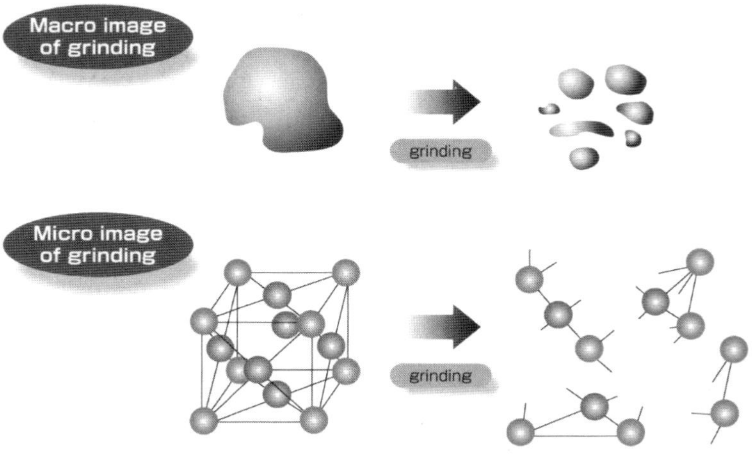

Figure 2.2.18
Macro and micro changes of solid during its grinding.

Table 2.2.3
Examples of metal nanoparticles synthesized by MC reaction.

Reaction	Metal volume ratio (%)	Size (nm)
$FeCl_3 + 3Na \rightarrow Fe + 3NaCl$	8.1	10–20
$CuCl_2 + 2Na \rightarrow Cu + 2NaCl$	11.6	20–50
$CoCl_2 + 2Na + 1.5NaCl \rightarrow Co + 3.5NaCl$	6.6	10–20
$NiCl_2 + 2Na + 1.5NaCl \rightarrow Ni + 3.5NaCl$	6.6	5–10

it is necessary to depress the thermal elevation, and for this purpose, it can be made by mixing a dilute agent in the starting materials, followed by grinding operation. NaCl plays a significant role as a dilute agent in the case of the formation of Co nanoparticles obtained from a mixture of $CoCl_2$ and Na, as given by the following equation (2.2.18):

$$CoCl_2 + 2Na + 1.5NaCl \rightarrow Co + 3.5NaCl \quad (2.2.18)$$

The dilute agent plays a role of dispersion to prevent from aggregation caused by grinding besides the depression of thermal elevation. As a result, well-dispersive nanoparticles can be separated by washing chloride out from the product with solvent. A TEM photo of the obtained nickel nanoparticles is shown in Fig. 2.2.19 [4]. From this figure, it is seen that fine and well-dispersive particles of 5–10 nm in sizes are recovered.

b) Metal oxide nanoparticles

It is possible to synthesize metal oxides nanoparticles through the MC reaction, and some of them are tabulated in Table 2.2.4 [5–11]. In this case, it is very important to add a dilute agent to the starting mixture.

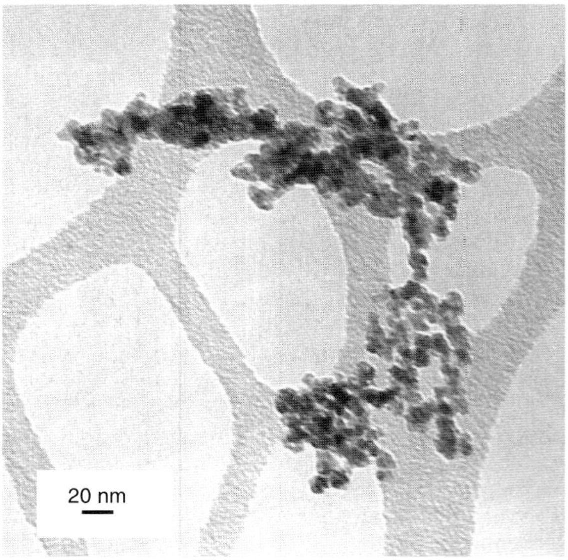

Figure 2.2.19
TEM photo of Ni-nanoparticles.

Table 2.2.4
Examples of metal oxide nanoparticles synthesized by MC reaction.

Starting mixtures		Reaction products		
$2AlCl_3$	$+ CaO$	$\rightarrow Al_2O_3$	$+ 3CaCl_2$	
$ZrCl_4$	$+ 2CaO$	$\rightarrow ZrO_2$	$+ 2CaCl_2$	
$GdCl_3$	$+ 3NaOH$	$\rightarrow Gd_2O_3$	$+ 3NaCl$	$+ 1.5H_2O$
$CeCl_3$	$+ NaOH$	$\rightarrow CeO_2$	$+ 3NaCl$	$+ H_2O$
$Na_2Cr_2O_7$	$+ S$	$\rightarrow Cr_2O_3$	$+ Na_2SO_4$	
$2NbCl_5$	$+ 5Na_2CO_3$	$\rightarrow Nb_2O_5$	$+ 10NaCl$	$+ 5CO_2$
$SnCl_2$	$+ Na_2CO_3 + O_2$	$\rightarrow SnO_2$	$+ 2NaCl$	$+ CO_2$
$2FeCl_3$	$+ 3Ca(OH)_2$	$\rightarrow Fe_2O_3$	$+ 3CaCl_2$	$+ 3H_2O$
$ZnCl_2$	$+ Na_2CO_3$	$\rightarrow ZnO$	$+ 2NaCl$	$+ CO_2$

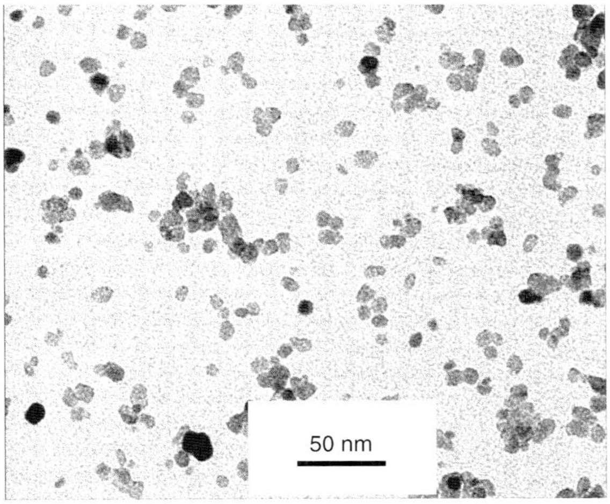

Figure 2.2.20
TEM photo of CeO_2 nanoparticles.

The addition of the dilute agent has a drawback of difficulty in improving the volume ratio of nanoparticles in the product. However, it is very important because it plays a significant role to prevent from both crystalization of material and sintering during heating. As an example, in the solid-phase reaction shown by equation (2.2.19), only relatively large particles of SnO_2 have been produced by heating the ground product, due to sintering of fine SnO_2 synthesized by the MC reaction. On the contrary, it is possible to form SnO_2 fine particles of 40 nm in size when the mixture is subjected to grinding with $CaCl_2$, followed by its heating.

$$SnCl_2 + Ca(OH)_2 + 0.5O_2 \rightarrow SnO_2 + CaCl_2 + H_2O \tag{2.2.19}$$

Figure 2.2.20 [12] is a TEM photo of CeO_2 nanoparticles obtained by washing the heated product at 500°C, after grinding the mixture of ($CeCl_3$ + NaOH + NaCl).

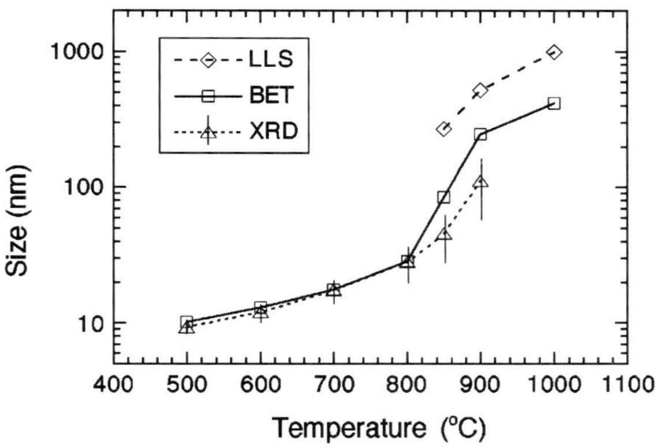

Figure 2.2.21
Size of CeO_2 nanoparticles as a function of temperature in heating operation.

$$CeCl_3 + 3NaOH + 12NaCl \rightarrow Ce(OH)_3 + 15NaCl \tag{2.2.20}$$

The role of heating is to form CeO_2 from $Ce(OH)_3$, and it is found from this photo that the mean particle size of CeO_2 is around 10 nm and its size distribution is quite narrow. Effect of heating temperature on the particle size of CeO_2 is shown in Fig. 2.2.21 [13], it is noted that the size of CeO_2 can be maintained in the nanometer range when the temperature is not elevated.

c) Complex metal oxide nanoparticles

Ito et al. [14] have synthesized fine complex metal oxide having nanometer size range by a MC method. The method is to grind a mixture of metal chloride and NaOH to produce metal hydroxide and NaCl. The product ground is subjected to heating to form complex oxide in which fine particles of NaCl are dispersed in the oxide phase. Subsequently, the heated sample is washed with water to remove NaCl from the sample. This washing operation implies purification of the metal complex oxide. The temperature in the heating operation must be reduced more than that for the mixture without grinding. The MC reaction equation is shown by Eq. (2.2.21), as below:

$$LaCl_3 + CoCl_2 + 5NaOH \rightarrow La(OH)_3 + Co(OH)_2 + 5NaCl \tag{2.2.21}$$

The ground product consists of ($La(OH)_3 + Co(OH)_2$ + NaCl), so that heating the product at 600°C makes it possible to form ($LaCoO_3$ + NaCl) in the heated sample. Consequently, nanoparticles of $LaCoO_3$ are obtained at high yield by washing the heated sample with water. The final product, $LaCoO_3$ is very fine and well-dispersed in comparison with the product synthesized by the MC treatment of a mixture of ($La_2O_3 + Co_2O_3$). Especially, the specific surface area of the former final product is about three times higher than that of the latter one. The reason why such fine and well-dispersed particles can be synthesized from the mixture of ($La(OH)_3 + Co(OH)_2 + NaCl$) by the MC method is attributed to the presence of NaCl in the heating, and NaCl plays a big role to prevent from agglomeration and sintering during heating, while the aggregated particles can be obtained by the MC method of ($La_2O_3 + Co_2O_3$).

d) Metal sulfide nanoparticles [11, 15]

The MC method is able to synthesize metal sulfide nanoparticles such as ZnS, CdS. For example the synthesis of ZnS, the reaction equations are given as below.

$$CaS + ZnCl_2 + CaCl_2 \rightarrow ZnS + 2CaCl_2 \quad (2.2.22)$$

$$Na_2S + CdCl_2 + 15.6NaCl \rightarrow CdS + 17.6NaCl \quad (2.2.23)$$

Subsequently, followings are the important issues to reduce the particle size of the final product:

- Use fine powder as starting samples.
- Use small size of media in ball milling.
- Decrease the volume ratio of metal sulfide such as ZnS.

According to the data obtained, it is possible to have fine product ranged from 4 to 8 nm in size by milling for 4 h when the media with 4.8 mm in diameter are used in wet media milling, as shown in Fig. 2.2.22 [15] which has shown size distribution of final product of CdS.

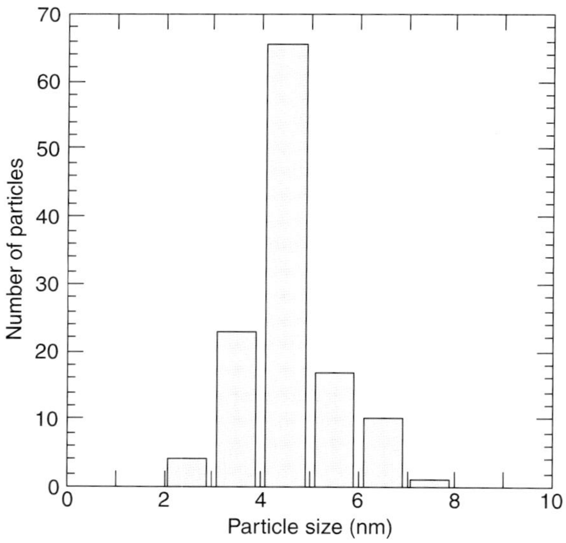

Figure 2.2.22
Size distribution of CdS nanoparticles obtained from TEM photos.

We have introduced several examples of metal, metal oxide, and sulfide nanoparticles, each synthesized by MC solid-state reaction based on dry grinding starting mixtures.

As is known, grinding and mechanochemistry are traditional operation and subject, but they have intrinsic properties including novel technology for synthesizing nanoparticles. This technology has a lot of potential to be able to synthesize nanoparticles of different kinds of materials by the combination of starting mixture and subsequent operations such as leaching and heating. It is known that, in fact, the energy efficiency of grinding is relatively low, however, it is not excessively low for the MC operation when we compare its total costs with that using other operations such as high temperature and solution devices. This technology of grinding operation combined with subsequent operations will be developed further in the near future, contributing to a new material processing. This is a great expectation of this kind of research work, leading to futuristic creative and profitable advanced material science and technologies.

References

[1] P.G. McCormick: *Mater. Trans. JIM*, **36**, 161 (1995).
[2] P.G. McCormick: *Handbook on the Physics and Chemistry of Rare Earths,* Gschneidner, Jr. K.A., Eying, L. (eds), Vol. 24, pp. 47–82 (1997).
[3] F. Saito: *Materia Japan*, **42**(8), 559–564 (2003).
[4] J. Ding, T. Tsuzuki and P.G. McCormick: *J. Mater. Sci.*, **34**, 1 (1999).
[5] J. Ding, T. Tsuzuki and P.G. McCormick: *J. Am. Ceram.*, **79**, 2956 (1996).
[6] J. Ding, T. Tsuzuki and P.G. McCormick: *Nanostr. Mater.*, **8**, 75 (1997).
[7] A.C. Dodd, T. Tsuzuki and P.G. McCormick: *Mater. Sci. Eng.*, **A301**, 54 (2001).
[8] T. Tsuzuki, P.G. McCormick: *J. Alloys Comp.*, **281**, 146 (1998).
[9] T. Tsuzuki, E. Pirault and P.G. McCormick: *Nanostr. Mater.*, **11**, 125 (1999).
[10] T. Tsuzuki, P.G. McCormick: *Proc. INCOME-2003*, p. 39, Braunschweig (2003).
[11] P.G. McCormick, T. Tsuzuki: *Proc. NEPTIS-13*, pp. 7–16 (2004).
[12] P.G. McCormick, T. Tsuzuki, J.S. Robinson and J. Ding: *Adv. Mater.*, **13**, 1008 (2001).
[13] T. Tsuzuki, P.G. McCormick: *J. Am. Ceram. Soc.*, **84**, 1453 (2001).
[14] T. Ito, Q. Zhang and F. Saito: *Powder Technol.*, **143–144**,170 (2004).
[15] T. Tsuzuki, P.G. McCormick: *Appl. Phys., A*, **65**, 607 (1997).

2.2.5 Grinding method

The grinding is a fine particle production method by applying mechanical energy on the solid materials to break the bonding between atoms or molecules. It is one of the oldest unit operations in the history of human beings. It can be applied not only to the existing nature solids but also those artificially synthesized organic and inorganic materials. Although grinding is a common technology, the grinding process is still a black box due to the complication of the fracture phenomena of solid particles as well as the relationship between the particle fracture mechanism and the particle size of ground particles. It requires thorough statistical investigation.

Nowadays, the particle size requirement for grinding is getting finer and finer. For example, about 20 years ago, grinding limit ("Grinding Barrier") was said to be at 3 μm in the continuous grinding of brittle materials. The grinding limit has long passed the submicrons and reaches the nanosized range in recent days because of the development of advanced technologies and industries.

It was said impossible to produce the nanoparticles less than about 100 nm by grinding; but the gas- or liquid-phase method accompanying the phase change had potentials for this purpose. However, examples of producing nanoparticles by grinding can already be found in some cases these days. Furthermore, the synthesized nanoparticles often form strongly aggregated agglomerates in the production process, and in many cases, they take apparent bulk form from sintering and the like. Therefore, the dispersion or disintegration of these agglomerates into the primary particles with high-specific surface area is regarded as grinding in a broader sense. Lately, in the production of nanoparticles, the grinding is often applied as the subsequent dispersion or disintegration operation.

From coarse to fine grinding, various types of mills are used depending on the particle size of the feed materials and the desired products. In the fine grinding, the mills based on the principle of surface grinding are usually used. The surface grinding is a grinding principle where ultrafine particles are generated from the particle surface by strongly rubbing them against each other as realized by the same material friction method [1]. Examples of this type of mills are impact mill, roller mill, attrition mill, ball mill, media agitation mill, jet mill, and so on. The grinding results using these mills are explained in the followings.

For example, Fig. 2.2.23 [2] shows the particle structure of calcium carbonate ground by an attrition mill observed with TEM. The particle size distribution of the ground product was at $D_{50} = 0.7\,\mu m$ in volume-basis measured by the liquid-sedimentation method. However, as seen from the TEM picture, the ground powder consists of nanosized particles. It suggests that a number of nanosized particles are generated from the solid surface during grinding, and they tend to form agglomerate instantaneously in the gas and become stable, coarser particles.

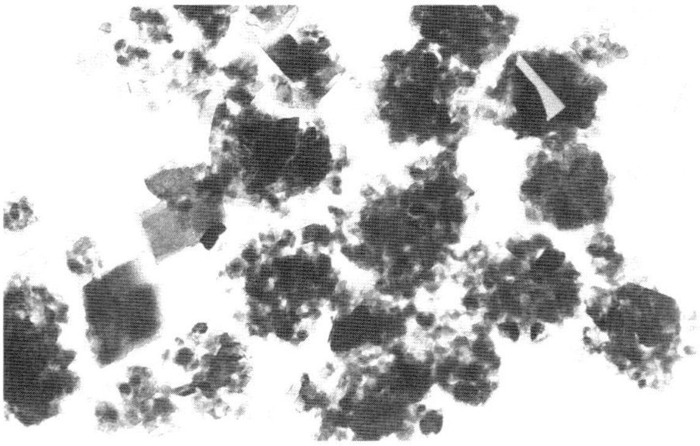

Figure 2.2.23
TEM picture of the particle structure of calcium carbonate ground by an attrition type mill.

In this way, it is very difficult to produce stable nanosized particles by the dry grinding method, since active nanoparticles form fresh agglomerates as soon as they contact each other. Therefore, the grinding has to be carried out in a liquid (wet grinding).

In the wet grinding, as the surface of the generated particles is surrounded by a solvent instantaneously, it is possible to control the agglomeration of particles much better than in the dry grinding. For the purpose of producing the nanoparticles, media agitation mills has drawn much attention recently.

The media agitation mill is a grinding mill which agitates ball media in the mill with agitation blades which exerts stronger force on the ball media, made by ceramics and the like, than a typical tumbling ball mill. Besides the agitation mills, some other type of mills, like the planetary mills, apply centrifugal acceleration to the ball media by revolution as well as rotation of the container for nanogrinding.

Figure 2.2.24 shows the relationship between the product average size and the grinding time (residence time) of the wet grinding of piezoceramic material as a function of ball diameters [3]. As seen in the figure, the ground product with an average size of less than 0.5 μm can be obtained easily by the continuous wet grinding, and it is interesting to find that the product particle size gets finer with the decrease of ball diameters.

The grinding performance is determined by the frequency and the intensity of ball collisions affecting the fracture of the particles. As the particle size decreases, the force required to break the particle itself reduces, although the fracture strength of the particle increases. Therefore, the grinding capability is kept even when the ball diameter decreases. Additionally, with the decreasing ball diameter, the frequency of ball collisions on the solid particles increases, which presumably results in the effective grinding due to higher grinding speed.

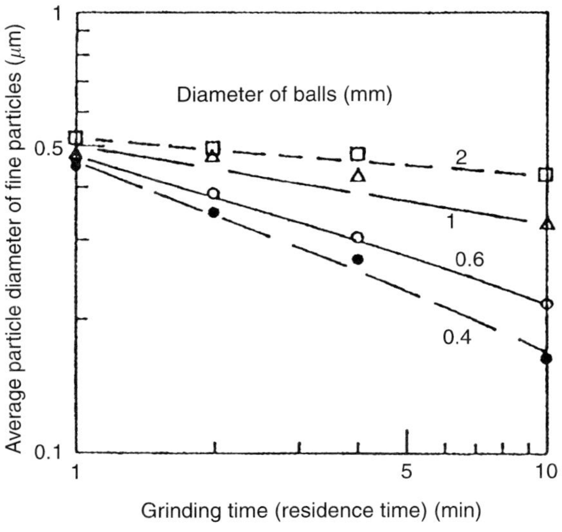

Figure 2.2.24
Results of wet grinding of piezoceramic material with a continuous media agitation mill [3].

The attempt to produce nanoparticle by reducing media size has drawn much attention recently and known as the disintegrating and dispersion technology [4–6]. For example, Fig. 2.2.25 shows a dispersing result of titanium dioxide by a wet media agitation mill [6].

The abscissa of the figure indicates input power per unit mass of the feed material. As shown in Fig. 2.2.25, the particles are well-dispersed to nearly the primary particles after the processing, while the feed materials are agglomerates with an average size of 3 μm. Additionally, this effect is more notable with the reducing ball size. In this way, the grinding operation with proper ball media sizes can be an effective technique to disperse nanoparticles.

However, the excessive energy and force may adversely lead to the deterioration of particle properties such as crystallinity and cause particle growth such as the agglomeration of ground products. For example, also shown in Fig. 2.2.25, the crystallite structure is disturbed and consequently the catalytic properties of TiO_2 are deteriorated, when the excessive energy is exerted onto the particles.

On the other hand, Fig. 2.2.26 shows the influence of forces of media balls on the product particle size during the grinding operation [7], where a planetary ball mill was used and the force applied on the particles was represented by the maximum centrifugal force working on a single ball. The product particle size decreased with the grinding time up to some point, after that it tended to increase with it. This phenomenon is called "reversed grinding" and the minimum particles size at the turning point is defined as the minimum grinding equilibrium size, which is indicated on the ordinate in the figure.

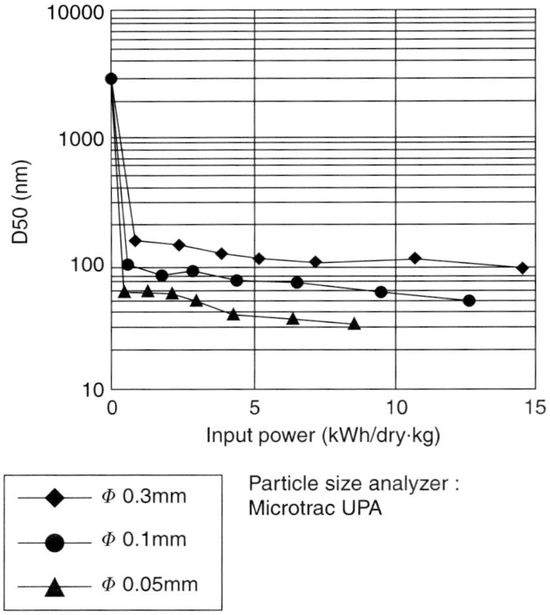

Figure 2.2.25
Result of dispersion of titanium dioxide by a media agitation mill using different ball sizes [6].

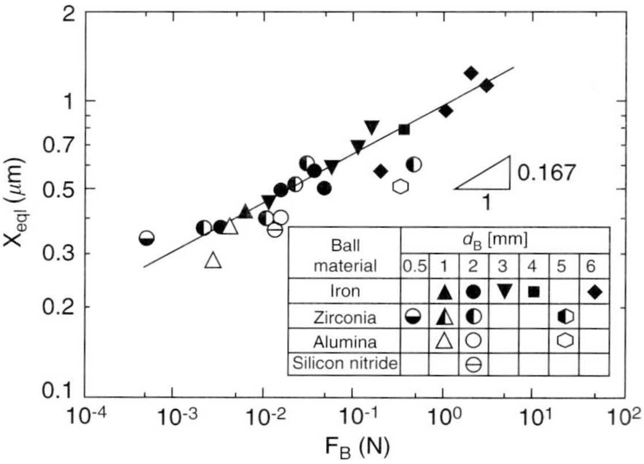

Figure 2.2.26
Relationship between the grinding equilibrium particle size and the maximum centrifugal force given to a media ball [7].

From Fig. 2.2.26, it is seen that the minimum equilibrium particle size defined as above increases with the maximum centrifugal force given to the media ball, which implies that the excessive force on the particles results in the growth of the finally attainable particle size of the ground product. Therefore, in order to disperse powder to its primary nanoparticles, it is necessary to select the optimum operational conditions and avoid applying excessive force on the particles.

The grinding operation is a very effective technique not only to break solid particles but also to disperse nanoparticle agglomerates. For example, a useful mill can disperse nanoparticle agglomerates generated in gas or liquid phase in the dry state. As to dispersing nanoparticles in slurry, liquid jet mills are also as effective as the media agitation mills and planetary ball mills. Grinding plays an important role both in the conventional particle production and nanoparticle production systems.

References

[1] S. Morohashi, Y. Sawahara: *J. Soc. Mater. Sci., Jpn.*, **22**, 689–692 (1973).
[2] Hosokawa Micron Corporation Micromeritics Laboratory: *Mechanofusion*, p. 46 (1989).
[3] K. Kugimiya: *The Micromeritics*, **36**, 177–180 (1992).
[4] H. Yoden, N. Ito: *J. Soc. Powder Technol., Jpn.*, **41**, 457–464 (2004).
[5] M. Inkyo, T. Tahara: *J. Soc. Powder Technol., Jpn.*, **41**, 578–585 (2004).
[6] T. Ishii: *Powder Sci. Eng.*, **37**(8), 51–57 (2005).
[7] T. Yokoyama, Y. Taniyama, G. Jinbo and Q. Zhao: *J. Soc. Powder Technol., Jpn.*, **28**, 751–758 (1991).

2.3 Particle shape

2.3.1. Gas-phase process

Evaluation and control of the morphology of the particles produced by chemical vapor deposition (CVD) and spray methods in the gas phase are more difficult than those prepared by liquid-phase methods. Aerosol particles synthesized via aerosol routes, including CVD and spray methods, are either droplets (liquid particles) or solid particles. In the case of the droplets, the morphology is typically spherical. In contrast, in the case of solid particles, the morphology is strongly dependent on the deposition process of crystalline particles from gas and liquid phases.

(1) Morphology of particles synthesized by the CVD method

When the particles are produced by the CVD method, as shown in Fig. 2.3.1, the morphology depends on

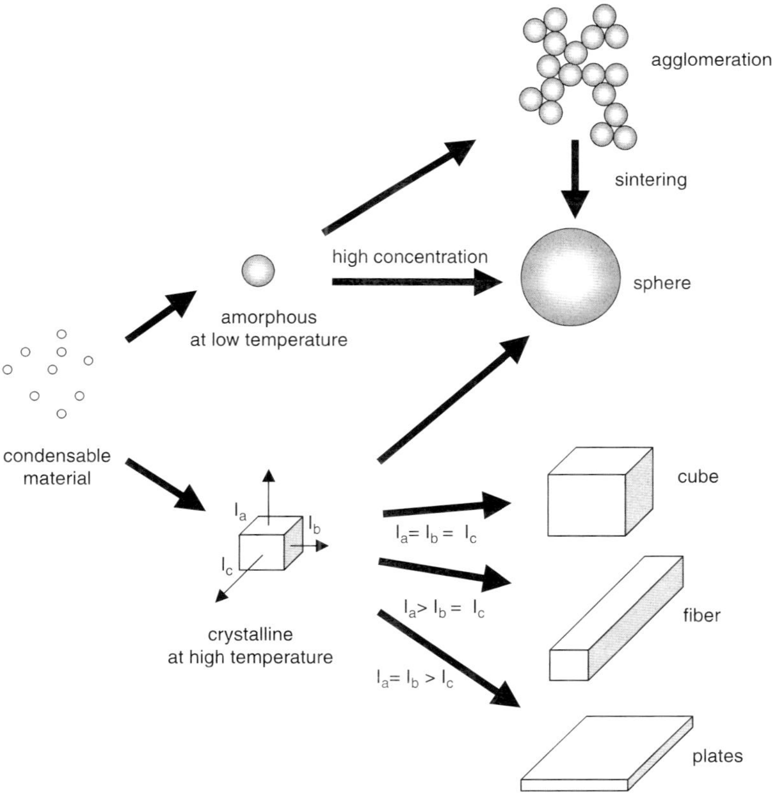

Figure 2.3.1
Variation of crystal structure.

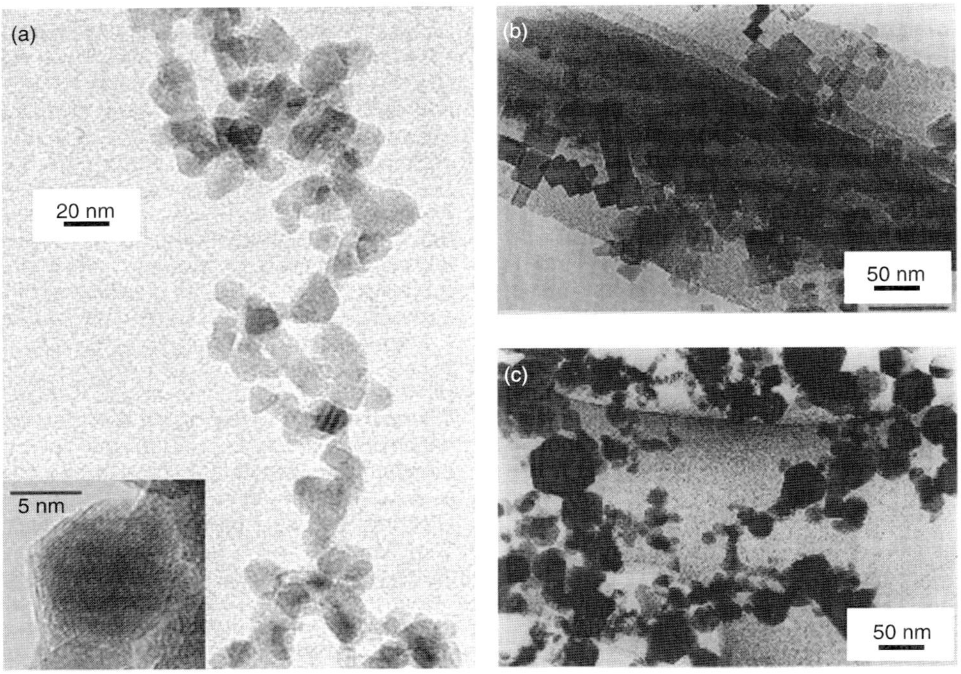

Figure 2.3.2
SEM image of particle with various crystal structure.

their crystallinity, the angle dependence of growth velocity, and the concentration of condensable species around the particles. A major factor that affects morphology is whether the structure of the particles is amorphous or crystalline. Amorphous particles that are produced at a low temperature, typically have a spherical structure because their growth velocity in the radial direction is the same at all positions of the surface. On the other hand, in the case of crystalline particle synthesized at high temperature, two-dimensional nuclei are formed and grow on the particle surface. When the two-dimensional nuclei are not influenced by crystal direction, the particles are also spherical. However, in the case of crystalline particles that grow in the crystal direction (epitaxial growth), the characteristic morphology, also called crystal habit, is expressed by a difference in growth velocity on the crystal facet. As a result by a relative velocity difference in crystal growth in each direction, the particles have various shapes: rods, cubes, agglomerates, polyhedra, and even fibers.

Figure 2.3.2 shows SEM images of particles with various crystal structures produced by a CVD method [1, 2]. The particles have a specific structure due to their crystalline characteristics. In the case of the oxidation of zinc vapor, as can be seen in Fig. 2.3.3, the zinc oxide particles are in the form

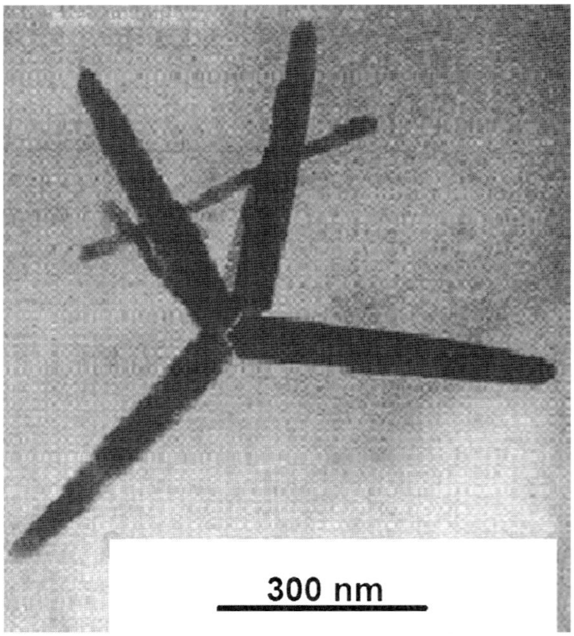

Figure 2.3.3
TEM image of zinc oxide particle with tetrapot-type structure.

of a tetrapot-type structure, also called four-rings, consisting of four needles [3]. It is generally thought that this structure is due to the larger growth velocity in the *C* axis than in the other axes. Therefore, when the zinc vapor content decreases, the particles do not grow to a needle shape and become an aggregate consisting of ultrafine spherical particles.

Depending on coagulation and sintering behavior, the morphology of particles synthesized by a CVD method can be spherical or agglomerates [4]. When the characteristic time for coagulation (τ_c), i.e. the velocity of coagulation, is sufficiently shorter than the characteristic time for sintering (τ_f), i.e. the velocity of sintering, the particles first become nonspherical agglomerates. These nonspherical aggregates gradually become spherical due to the sintering after τ_f. However, when another particle agglutinates before τ_f passes, the particles are agglomerated. That is to say, at $\tau_f < \tau_c$, the particles are spherical, at $\tau_c < \tau_f$, the particles are nonspherical aggregates. The equation for expressing τ_f has the functions of sintering mechanism, particle properties (melting point, diffusion coefficient, and so on), temperature, and size. Consequently, it is important to evaluate coagulation and the sintering phenomena in the characterization of the size of product particles. Both phenomena have an effect on the temperature profile during the particle production process.

Figure 2.3.4 shows one example of qualitative changes in thermal profile in a thermal CVD reactor, the characteristic time for coagulation (τ_c) and sintering (τ_f). Clusters are generated near the reactor inlet, and generally, when the temperature increases in

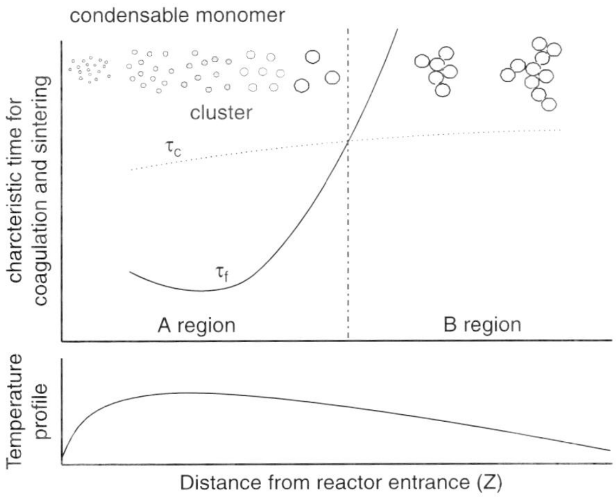

Figure 2.3.4
Qualitative changes of thermal profile in a thermal CVD reactor, characteristic time for coagulation (τ_c), and characteristic time for sintering (τ_f).

the reactor, τ_c is longer than τ_f (A region). In this case, the particles grow by coagulation, and are spherical because the sintering ends instantaneously. Therefore, the particle diameter depends on the frequency of the collisions between the particles greatly. On the other hand, when τ_f is longer than τ_c (B region), as mentioned above, the particles are agglomerated. We can conclude that the morphology of agglomerated particles is dependent on both the coagulation and sintering behavior.

Figure 2.3.5 shows the change in morphology of titanium dioxide particles synthesized by a CVD method at various temperatures [5]. The titanium dioxide particles were synthesized by the thermal decomposition of tetraisopropoxide (TTIP). In all cases, the particles are agglomerations of the nanometer order of primary particles, formed at a low reaction temperature. When the reaction temperature increases above 800°C, where titanium dioxide particles undergo sintering, the primary particles are larger because they are growing up by sintering.

In general, nanoparticles generated by CVD processes are produced in the form of aggregates due to coagulation. An electrospray-assisted CVD (ES-CVD) method has recently been reported for generating nonagglomerated nanoparticles because unipolarly charged particles undergo mutual electrostatic repulsions, and particle collisions and growth are decreased [6].

(2) Morphology of particles prepared by spray methods

Figure 2.3.6 shows the morphology of particles preparing by a spray method. To prepare final particles by spray pyrolysis, a starting solution is typically prepared by dissolving, the metal salt of the product in the solvent. The droplets, which are atomized from the starting solution, are introduced into the furnace. Evaporation of the solvent, diffusion of solute, drying, precipitation, reaction between precursor and surrounding gas, pyrolysis, or sintering may occur inside the furnace to form the final product. The size or morphology of the final particles produced can also be determined by the concentration and velocity of the droplet generated by the atomizers. The relation between the characteristic solvent evaporation time (τ_{sv}) and the characteristic solute diffusion time (τ_{sl}) determine the final morphology of the particle, whether a porous particle or a hollow particle, in the solvent evaporation and solvent diffusion process.

In the case of a lower heating rate or small droplet size, a polycrystalline particle having a somewhat large crystalline size or a nearly single crystal particle can be produced. Temperatures near or over the melting point are necessary to achieve a satisfactorily dense particle and to achieve a higher density with a more spherical-shaped particle. However, "plane" particle morphology also can be obtained whenever there is an orientation in the crystal growth, during the solidification process.

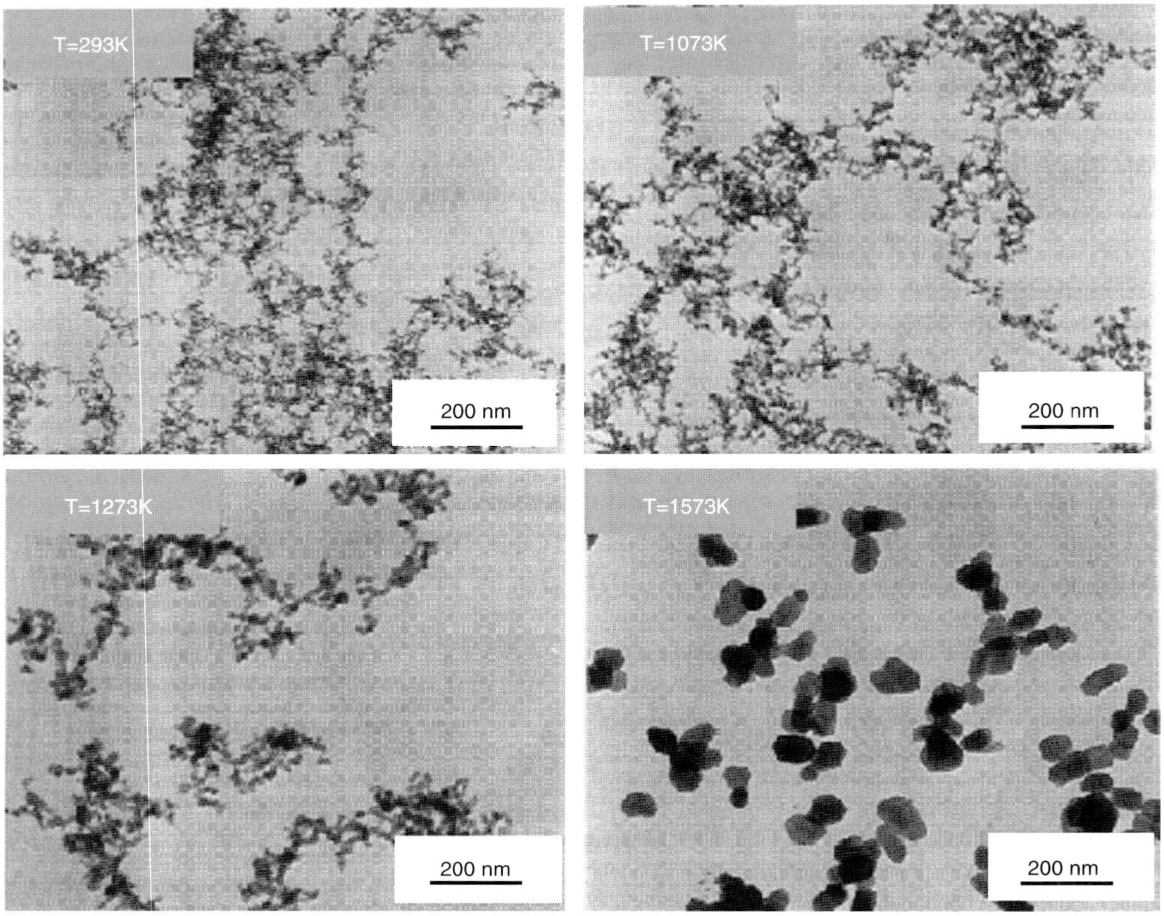

Figure 2.3.5
Morphology change of titanium dioxide particles prepared by a CVD method with various temperatures.

In the case of a high heating rate or large droplet size, a porous particle is produced. These morphological conditions are undesirable for most applications. Single-crystal or polycrystal particles are also formed, depending on the operating conditions.

Figure 2.3.7 shows an SEM image of ZrO_2 particles prepared by spray pyrolysis. In the case of low heating, evaporation of the solution and the crystallization progress become slow. As a result, a solid (dense) particle will be produced, as shown in Fig. 2.3.7(a). In the case of a high heating rate, as the first stage in the spray pyrolysis, solvent evaporation can occur rapidly near the surface of the droplet. As a result, the solute concentration in the droplet will not be uniform. Furthermore, when the inside or center of the particle is still in the liquid phase, the generation of gas from the reaction allows the particle to be fragmented or to acquire a hollow, porous morphology (Fig. 2.3.7(b)).

Figure 2.3.8 shows an SEM image of $Bi_2CaSr_2Cu_2O$ particles prepared by spray pyrolysis[7]. When the heating temperature is set at 800°C, the crystallization of particles occurs rapidly. As a result, a particle comprised of small crystals is prepared. However, after the precursor solution is heated to over the melting point (950°C), and followed by holding the temperature at 800°C, a flake-morphology and a dense particle is produced.

The morphology of particles prepared by a CVD process and spray pyrolysis is affected by particle crystal characterization and the process used to generate it. Furthermore, in the case of generation involving CVD process, agglomeration and sintering phenomena are important determinants of particle morphology. On the other hand, in the case of spray pyrolysis, the evaporation of droplets and the diffusion of solute are important factors.

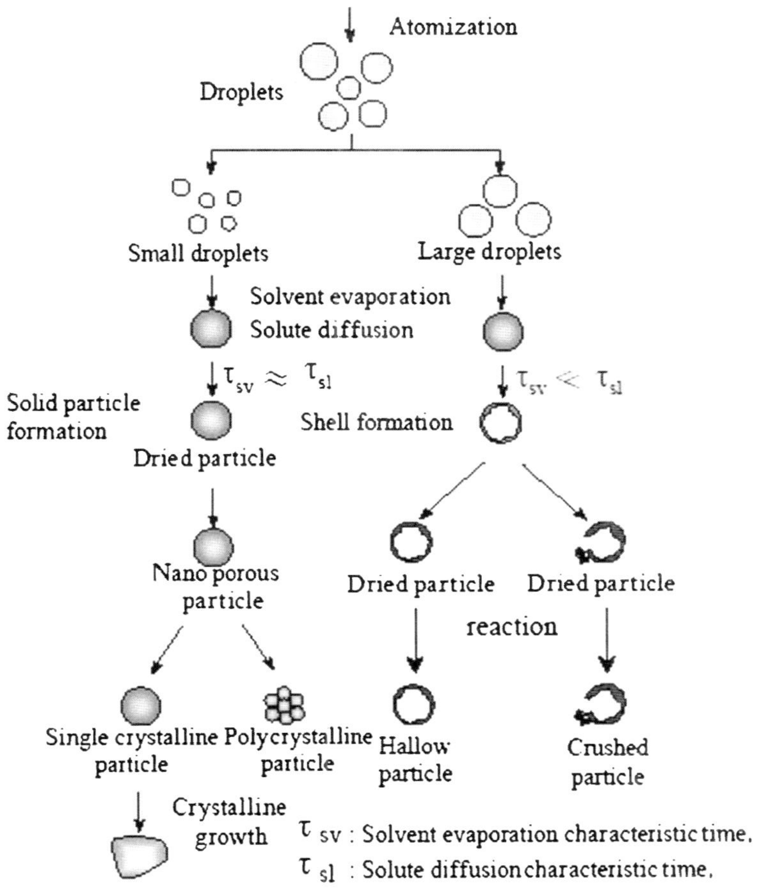

Figure 2.3.6
Morphology of particle prepared by spray pyrolysis.

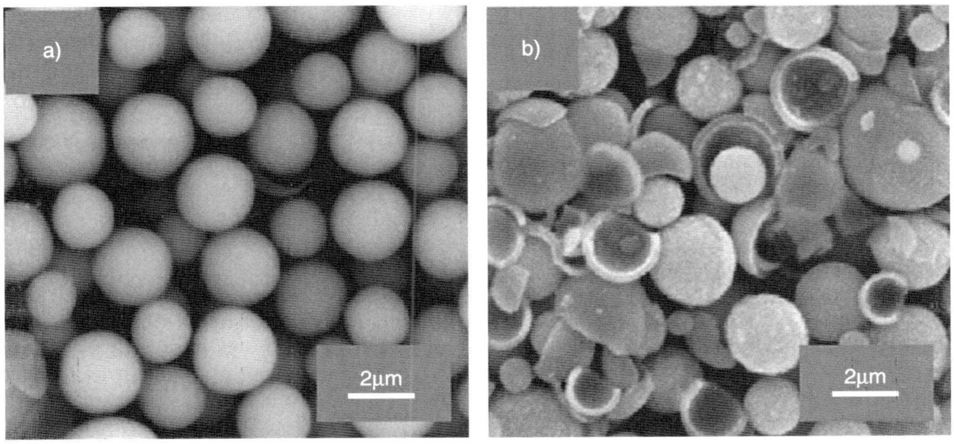

Figure 2.3.7
Morphology of ZrO_2 particles. (a) Spherical particle. (b) Hollow particle.

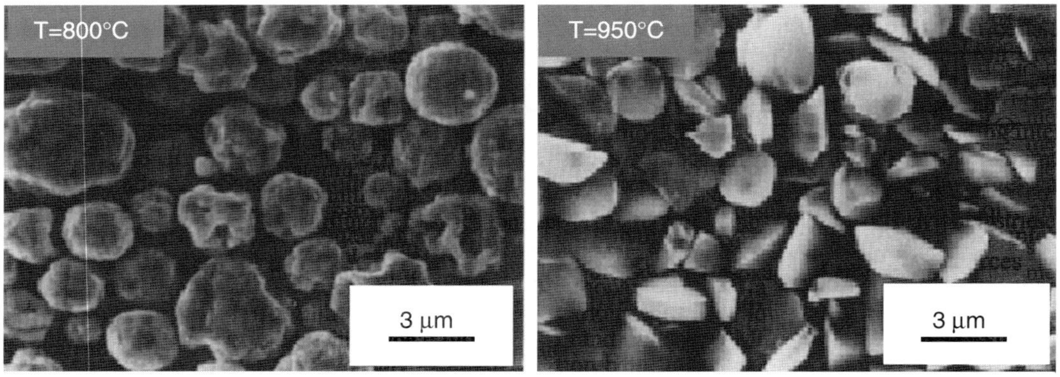

Figure 2.3.8
SEM images of $Bi_2CaSr_2Cu_2O$ particle prepared by spray pyrolysis.

References

[1] Y. Azuma, M. Shimada and K. Okuyama: *Chem. Vapor Deposition*, **10**(1), 11–13 (2004).
[2] T.T. Kodas, H. Smith: *Aerosol Processing of Materials*, WILEY-VCH, p. 33 (1999).
[3] A. Seto, N. Aya, K. Okuyama and K. Hata: *J. Aerosol Res.*, **13**(4), 337–342 (1998).
[4] F. Ogino: *Kagaku Kogaku Handbook*, Asakura Syoten, pp. 313–335 (2004).
[5] T. Seto, M. Shimada and K. Okuyama: *Aerosol Sci. Technol.*, **23**(2), 183–200 (1995).
[6] K. Nakaso, B. Han, K.H. Ahn, M. Choi and K. Okuyama: *J. Aerosol Sci.*, **34**(7), 859–881 (2003).
[7] N. Tohge, M. Tatsumisago, T. Minami, M. Adachi, Y. Kousaka and K. Okuyama: *J. Am. Ceram. Soc.*, **74**(9), 2117–2122 (1991).

2.3.2 Liquid-phase method

Several methods have been proposed for controlling the shape and size of nanoparticles generated by the liquid-phase method: template synthesis using a surfactant [1], the hot soap method which suppresses growth of specific crystal faces by injection of precursor molecules into a hot surfactant [2], the solvothermal method which uses high temperature and pressure [3], and the microemulsion method which uses interfacial activities of reverse micelles and microemulsions [4]. It is important to understand the mechanism of nanoparticle formation and its controlling factors in order to control the shape and size of generated nanoparticles.

(1) Factors controlling the shape and size of nanoparticles

It is necessary to consider the balance between the supply and the consumption rates of monomer in order to control the shape and size of nanoparticles generated by the liquid-phase method. In open systems such as a semibatch reactor, a monomer is continuously fed from outside the system using a constant rate pump, etc. In a closed system such as a batch reactor, the monomer is fed from inside the system, using a chemical reaction such as hydrolysis of alkoxide and urea.

The particle formation process is classified into the prenucleation, nucleation, and growth stages. The monomer concentration increases with time (prenucleation stage). When the precursor monomer concentration reaches a critical value, generation of nucleus begins (nucleation stage).

The smallest diameter of stable nucleus is determined by the free energy of the aggregation of nuclei and the surface area. Then, monomers supplied are consumed for the growth of the stable nuclei (growth stage). To synthesize a monodispersed particle, it is desirable to separate the nucleation and the growth stages. However, in general, it is difficult to separate these stages. Consequently, the supplied monomers are consumed competitively by the nucleation and the growth of the generated nuclei.

Therefore, to synthesize monodispersed particles, we have to increase the nucleation rate or decrease the growth rate as much as possible. It is necessary to know the change in the supersaturation ratio and the aggregation of nuclei with time. It is possible to obtain the situation in which the growth of seed only occurs if the monomer concentration is kept constant between the saturation and the supersaturation monomer concentrations by adding seeds to the system. The crystallinity (crystalline or amorphous) and the type of particle growth (diffusion control or surface reaction control) are also important. Furthermore, the reaction temperature, raw material concentration, coexistence of ion species, and solution pH, etc., which are directly related to the controlling factors, are very important. In this section, the key points for controlling the shape and size of nanoparticles are described.

(2) Shape and size control of nanoparticles

When generated particles are amorphous, two-dimensional nucleation does not occur on the particle surface. Thus, the generated nuclei can grow in any direction at the same growth rate, and form spherical particles. The typical example of this type of particle synthesis is the sol-gel method. In this case, the template method controls the shape and size of spherical amorphous particles satisfactorily. For example, the surfactant in the solvent forms a hexagonal structure and a lamellar structure, etc., depending on the concentration. Using these structures as templates, rod-like particles, plate particles, etc., can be synthesized.

When the generated particles are crystalline, it is important to keep a balance between the nucleation and the growth rates of nuclei. If the nucleation rate is faster than the growth rate, the generated particles form polycrystalline spherical particles. In the case of sulfide and oxide particles, the critical supersaturation concentration is low, and the growth rate of nuclei is fast since these solubilities are quite small. Consequently, the generated nuclei concentration becomes high, and polycrystalline particles are formed.

The methods for controlling the shape and size of polycrystalline particles are as follows: controlling the nucleation rate to increase the growth rate of nuclei, or suppression of the aggregation of generated nuclei. The former can be controlled, for example, by slowing the supply rate of raw material or by causing a drastic rise in the supersaturation concentration by adding a chelating reagent, and the latter can be controlled by the addition of a disperser or thickener, or by controlling the solution pH so that is moves away from the isoelectric point of the generated particles.

The shape and size of generated particles can be controlled by the above-described methods and by using anisotropic growth of the crystalline particles as described below. It is also effective to use the template method as described above.

When the supplied monomers are consumed at a nucleus growth rate that is faster than the nucleation rate, the crystalline particles become anisotropic. The nucleus growth is classified into diffusion and surface reaction controlled-growth by the balance between the diffusion coefficient of monomers and the reaction rate on the crystal face. The former shape is determined by the differences between the surface energies of the crystal faces, and the latter shape is determined by the reaction rate at each crystal face.

In this case, the shape and size of nanoparticles can be controlled if the growth of the crystal face is suppressed by an additive that is selectively adsorbed to a specific crystal face. Additives used for suppressing the growth are anions, amines, carboxylic acids, surfactants, etc.

As a specific example in which the shape and size of nanocrystalline particles are controlled, the microemulsion method using the interfacial activities of reverse micelles and microemulsions was introduced [5]. Barium chromate was used as the model particle, and AOT was used as the surfactant.

Figure 2.3.9 shows TEM photographs of the product material obtained when the reverse micelle solution ($Ba(AOT)_2$/isooctane) and the W/O microemulsion solution (Na_2CrO_4aq/NaAOT/isooctane) were mixed at equal initial concentrations of barium and chromate ions (standard condition; $[Ba^{2+}]:[CrO_4^{2-}] = 1:1$). The product material was a bundle in which the nanowire was self-organized, and the space between the generated particles was about 2 nm. This interval corresponds to that of the adsorption layer of the surfactant molecules on the product. Electron diffraction (SAED) showed that the nanowire was elongated along the *a* axis (Fig. 2.3.9 inset).

Formation of the bundle of nanowires is considered to occur as follows. When coalescence of the microemulsion and the reverse micelle occurs, the

(a) low magnification (b) large magnification

Figure 2.3.9
TEM photographs of $BaCrO_4$ nanostructure ($[Ba^{2+}]:[CrO_4^{2-}] = 1:1$).

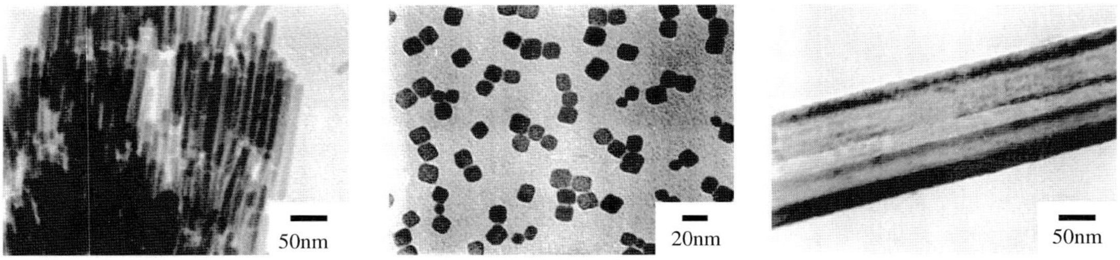

(a) $[Ba(AOT)_2] = 0.01$ mol/l $[Na_2CrO_4] = 0.08$ mol/l $[Ba^{2+}] : [CrO_4^{2-}] = 1 : 4$

(b) $[Ba(AOT)_2] = 0.01$ mol/l $[Na_2CrO_4] = 0.2$ mol/l $[Ba^{2+}] : [CrO_4^{2-}] = 1 : 10$

(c) $[Ba(AOT)_2] = 0.1$ mol/l $[Na_2CrO_4] = 0.02$ mol/l $[Ba^{2+}] : [CrO_4^{2-}] = 10 : 1$

Figure 2.3.10
Effect of raw material concentration on product morphology.

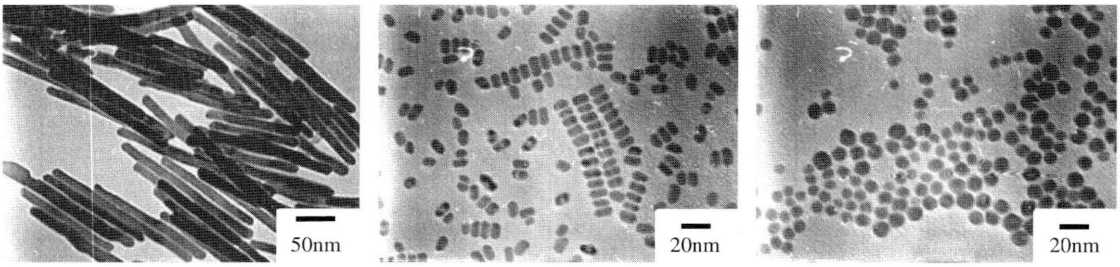

(a) [EtOH] = 25vol% (b) [EtOH] = 50vol% (c) [EtOH] = 75vol%

Figure 2.3.11
Effect of poor solvent on product morphology.

AOT molecules selectively adsorb on specific crystal faces along the *a* axis of the generated particles and suppress the growth of the adsorbed AOT. Thus, the generated particle grows only in the *a* axis direction and forms a nanowire. Then, the nanowires are self-organized by the hydrophobic interaction between the surface-anchored AOT molecule chains, and a bundle of nanowires is formed.

Figure 2.3.10 shows TEM photographs of the product material obtained when the concentration of barium or chromate ions is increased from that for the standard conditions (experimental conditions: (a) $[Ba^{2+}]:[CrO_4^{2-}] = 1:4$; (b) $[Ba^{2+}]:[CrO_4^{2-}] = 1:10$; (c) $[Ba^{2+}]:[CrO_4^{2-}] = 10:1$). With an increase in the relative concentration of the chromate ion, the morphology of the product material changed from a bundle of nanowires (Fig. 2.3.9) to a chain of nanorods (Fig. 2.3.10(a)) and then to dispersed nanodots (Fig. 2.3.10(b)). The aspect ratio of the generated particles decreased with the increase in the relative concentration of chromate ion.

In contrast, regardless of any relative increases in the barium ion concentration, the morphology of the product remained that of a bundle of nanowires (Fig. 2.3.10(c)). Briefly, it is assumed that the generated particle is monodisperse, the total volume of generated particles is V, the number of generated particles is N, and the (100) face area of a particles is A. The length L of a particle along the *a* axis can be expressed as $L = V/NA$. Since the growth of the crystal face with absorbed surfactant AOT is suppressed, the (100) face area A is almost constant. The particle number N is a function of the supersaturation ratio S defined in the following equation:

$$S = ([Ba^{2+}][CrO_4^{2-}]/K_{sp})^{1/2}, \qquad (2.3.1)$$

where K_{sp} is the solubility product of $BaCrO_4$. Since $Ba(AOT)_2$ has a poor solubility in water and NaAOT exists in excess at the interface of the emulsion, the AOT ion concentration always remains saturated and constant within the water-filled pools as nanometer-sized reaction fields, the free barium ion concentration in equation (2.3.1) always seems to be constant at a low concentration. Thus, the number of generated particles N exponentially increases with increasing chromate ion concentration regardless of the barium ion concentration. Since the total volume of generated particles V is also constant from the stoichiometry, the aspect ratio of the generated particle decreases with an increase in the chromate ion concentration.

Figure 2.3.11 shows TEM photographs of the product material obtained when a poor solvent (ethanol) was added in the standard condition. When the amount of additional poor solvent increases, the supersaturation ratio S increases with a decrease in the solubility product of $BaCrO_4$. Thus the number concentration of the generated particle N increases exponentially, and the aspect ratio of the generated particle is decreased. As mentioned above, it is possible to control the shape and size of nanoparticles by controlling the supersaturation ratio, which depends on the raw material concentration and the amount of poor solvent, etc., and by suppressing the growth of specific crystal faces with a selectively adsorbing surfactant.

References

[1] Y. Li, J. Wan and Z. Gu: *Mat. Sci. Eng. A*, **286**, 106 (2000).

[2] X. Peng, L. Manna, W. Yang: *Nature*, **404**, 59 (2000).

[3] S. Yu, M. Yoshimura: *Adv. Mater.*, **14**, 296 (2002).

[4] M. Li, H. Schnablegger and S. Mann: *Nature*, **402**, 393 (1999).

[5] T. Nomura, H. Arima, T. Mori and Y. Konishi: *Trans. Mater. Res. Soc. Jpn.*, **29**, 2387 (2004).

2.4 Composite structure

2.4.1 Gas-phase method

There are specific functional requirements for making fine particles as ceramics materials. Figure 2.4.1 shows these applications and material names. Ceramic powders are not always in nanosizes. However, in the practical applications and the development of new materials, the properties of nanoparticles, such as chemical, mechanical, optical, thermal, electric, and magnetic characteristics, are investigated. Depending on applications, the specifications of nanoparticles can be their particle diameters, shapes, surface properties, purities, and manufacturing histories.

Recently, in addition to single metal oxide, the nanoparticles with complex metal oxides become popular. The near-term potential industrial applications of nanoparticles are in the fields of electronic device, energy, environment, polymer, film, and biomedical. In these areas, research and development are geared to explore the functional effects of various nanomaterials. On the other hand, the production method for producing nanoparticles with highly advanced characteristics in commercial quantity has been studied. This section describes a plasma enhanced gas-phase method to produce nanoparticles with complex compounds and structures.

Ceramics		
Chemical	Catalysts	: ZrO_2, CeO_2, ZnO, SnO_2, Al_2O_3, MnO_2, La_2O_3, V_2O_5, MgO, CuO
	Battery electrodes	: $LiCoO_2$, MnO_2
	Photocatalyst	: TiO_2
	Explosives	: KNO_3
	Fire retardants	: ZrO_2, Sb_2O_5
Mechanical	Solid lubricants	: BN, MoS_2
	Mold	: $CaSO_4 \cdot 1/2H_2O$
	Fillers	: MgO, Fe_2O_3, ZrO_2
	Abrasives	: ZrO_2, Al_2O_2, CeO_2, SnO_2, SiC, Cr_2O_3
Optical	Cosmetics	: TiO_2, ZnO, SiO_2, $CaCO_3$
	Paints, Printing in	: TiO_2, Fe_2O_3, Fe_3O_4, ZnO, SiO_2, Cr_2O_3, $BaSO_4$, $PbCrO_4$, MnO_2
	UV stabilizers	: TiO_2, ZnO, CeO_2
	Fluorescents	: Y_2O_3:Eu, $BaMgAl_{14}O_{23}$:Eu, Zn_2SiO_4:Mn, ZnS (Ag)
Thermal	Catalytic supports	: Al_2O_3, SiO_2
	Fillers	: Al_2O_3, SiO_2, SiC, AlN
Electric	Semiconductors	: ITO, SnO_2, $LaTiO_3$
Magnetic	Magnets, Disks	: Ferrite, $3CoFe_2O_4 \cdot Fe_3O_4$, YIG

Figure 2.4.1
Ceramic powders used for various purposes.

The production method is shown in Fig. 2.4.2.

High-temperature plasma is irradiated onto raw materials. The raw materials including various metal molecules are vaporized by heat and subsequently quenched instantaneously by cooling gas to form nanoparticles. For example, nanoparticles containing AB oxide can be made by processing raw materials containing metals A and B at the same time. Applying this method, nanoparticles with multi-component oxide as well as single-component oxide can be made. The composition ratio of the multi-component oxides is controllable by using various raw materials; and the nanoparticle production can be carried out in a continuous process with a production rate of several kilograms per hour. Table 2.4.1 shows the examples of nanoparticles with multi-component oxides produced by this nanoparticle production system. In this table, 38.9 wt% $(CeO_2)_{0.92}$ $(Sm_2O_3)_{0.08}$ - 61.1wt% NiO is not a compound, but a mixture of these elements dispersed in one nanoparticle, which is a unique feature of this production method. Various particles with BET equivalent spherical diameter from 10 to 100 nm are produced at a production capacity more than 1 ton per month depending on compositions.

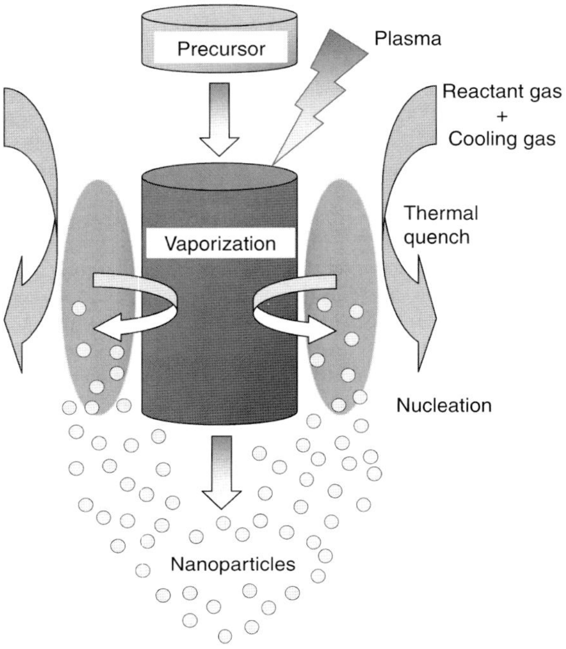

Figure 2.4.2
Nanoparticle production system.

In view of particle structure, nanoparticle composites produced by this method can be classified as shown in Fig. 2.4.3. The compound and solid solution particles in Fig. 2.4.3 represent AB-type oxides as shown in Figs. 2.4.4 and 2.4.5. Figure 2.4.6 shows the changes of the X-ray diffraction pattern as the composition ratio of La and Sr in the $(La_xSr_{1-x})MnO_3$ varies. It is understood that even if the composition ratio of La and Sr changes, the basic structure does not change. The particle in Fig. 2.4.3 has core–shell structure with the core of oxide B covered by the shell of oxide A. Figure 2.4.8 shows the elementary analyses of Spot 1 and Spot 2 on the particle in Fig. 2.4.7 by EDX. The core–shell particle structure is further proven by the fact that the major component at Spot 1 is Si, while that at Spot 2 being Zr (Fig. 2.4.8). Since nanoparticles with core–shell

Table 2.4.1
Example products of multi-component oxide nanoparticles.

Products	Chemical formula (% means percent by mass)	Equivalent spherical diameter (nm)
Yttria stabilized zirconia	YSZ:$(Y_2O_3)_{0.08}(ZrO_2)_{0.92}$	30
Titanium silicon oxide	94%TiO_2-6%SiO_2	60
	67%TiO_2-33%SiO_2	70
	25%TiO_2-75%SiO_2	100
Barium titanate	BTO:$BaTiO_3$	115
Samarium doped ceria – Nickel oxide	NiO–SDC:x%$[(CeO_2)_y(Sm_2O_3)_{1-y}]$-$(100-x)$%NiO	60
Lanthanum strontium manganese oxide	LSM:$(La_xSr_{1-x})MnO_3$	65
YSZ – Nickel oxide	x%YSZ-$(100-x)$%NiO	45
Ferrite	MO-nFe_2O_3	70
Cerium zirconium oxide	$(CeO_2)_x/(ZrO_2)_{1-x}$	15
Aluminum silicon oxide	x%Al_2O_3-$(100-x)$%SiO_2	110
Zirconium silicon oxide	50%ZrO_2-50%SiO_2	25
Calcium silicon oxide	50%CaO-50%SiO_2	93
Manganese doped zinc silicon oxide	Zn_2SiO_4:Mn^{2+}	50
Indium tin oxide	ITO:95%In_2O_3/5%SnO_2	20

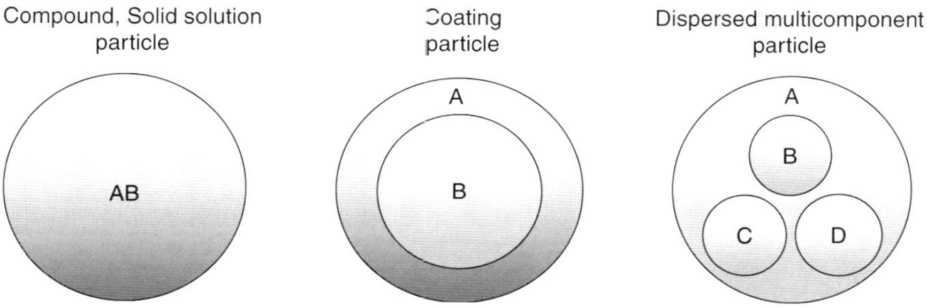

Figure 2.4.3
Composite structure of the nanoparticle.

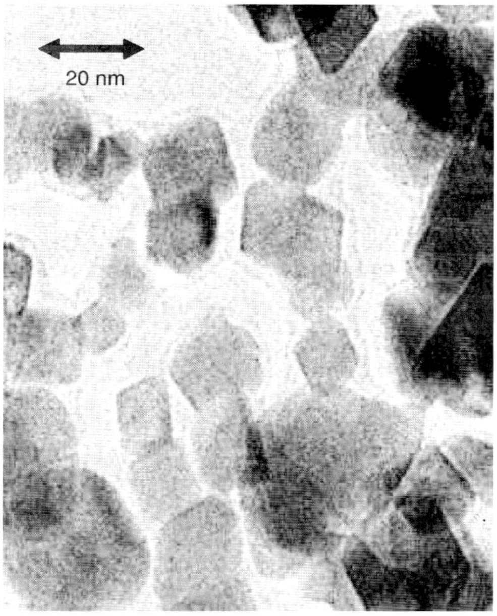

Figure 2.4.4
$(CeO_2)_x/(ZrO_2)_{1-x}$ nanoparticles (solid solution).

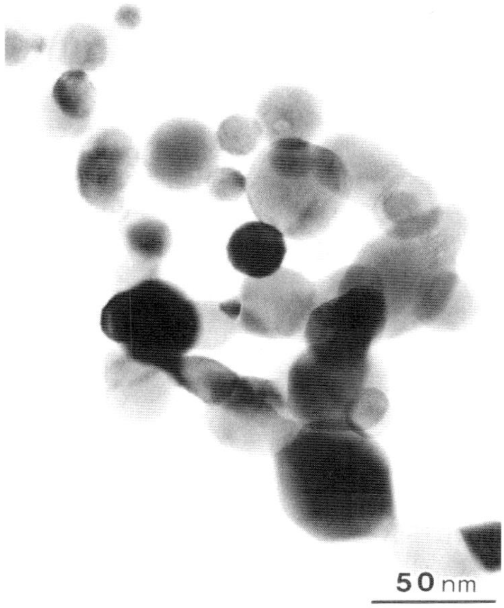

Figure 2.4.5
$(La_xSr_{1-x})MnO_3$ nanoparticles.

structure can have the characteristics of each composing metal oxide, it can potentially be used to increase the affinity of the particle (in this case, the part of the oxide A of the coated particle in Fig. 2.4.3) to the resin and restraint its deterioration due to the contact between the particle and the resin (in this case, the part of the oxide B of the coated particle in Fig. 2.4.3). The dispersed multi-component nanoparticle in Fig. 2.4.3 represents the oxide A encapsulating the oxides B, C, and D as shown in Fig. 2.4.9. Figure 2.4.10 shows the elements analyzed by EDX at two locations (Spot u and Spot v) on the nanoparticle consisting of the oxides of Al, Ca, Si, Ti, and Zr in Fig. 2.4.9. It is found that the particle has approximately homogeneous dispersion of each component in its structure. In addition, Fig. 2.4.11 shows nanoparticles of Al_2O_3–SiO_2 with the structure of the dispersed multi-components. Minute crystalline Al_2O_3 can be found in the amorphous SiO_2 by X-ray diffraction in this case, as shown in Fig. 2.4.12.

Multi-component oxide consisting of more than two kinds of metal elements can often have unique functionality, such as semiconduction, ferromagnetism, ferroelectricity, and piezoelectric. Combining different kind of metal elements through oxygen-binding and

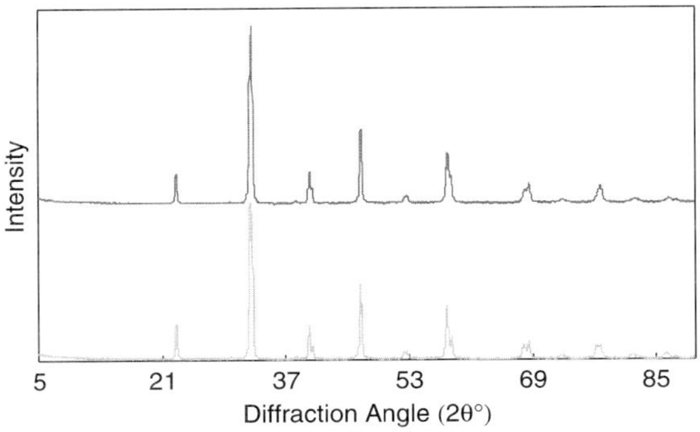

Figure 2.4.6
X-ray diffraction pattern of $(La_xSr_{1-x})MnO_3$ nanoparticles.

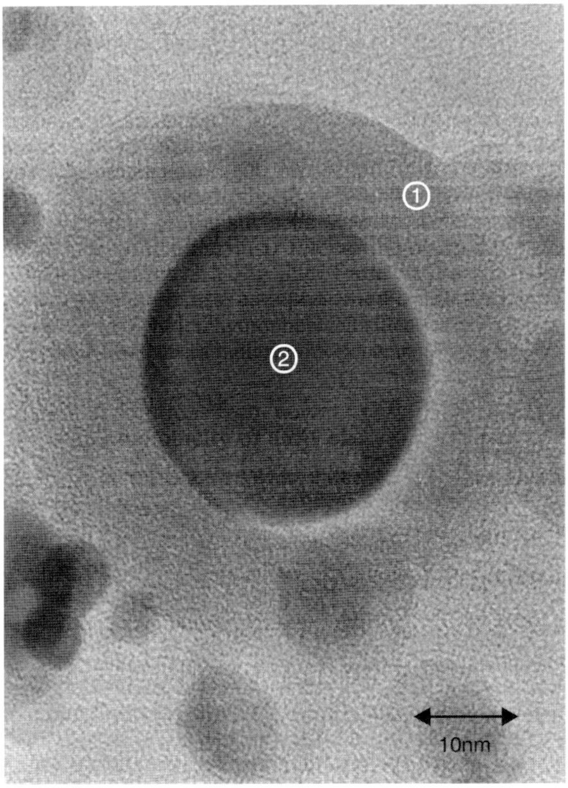

Figure 2.4.7
50% ZrO_2-50% SiO_2 nanoparticles.

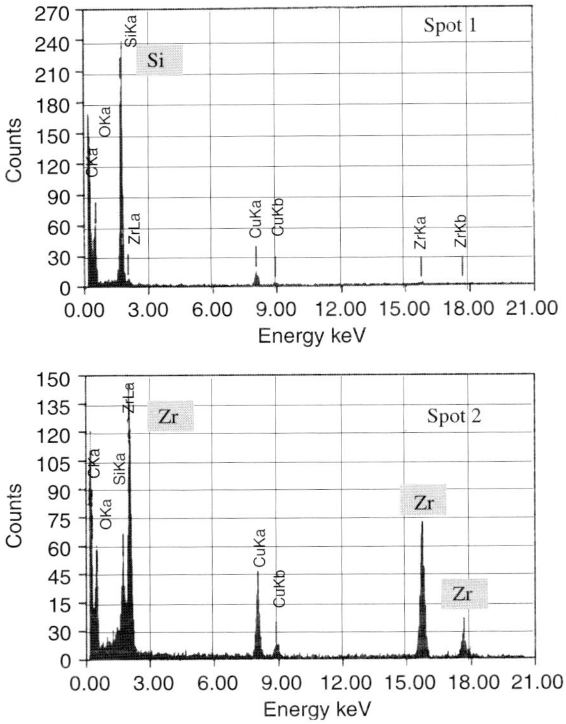

Figure 2.4.8
Analytical EDX of 50% ZrO_2-50% SiO_2 nanoparticle.

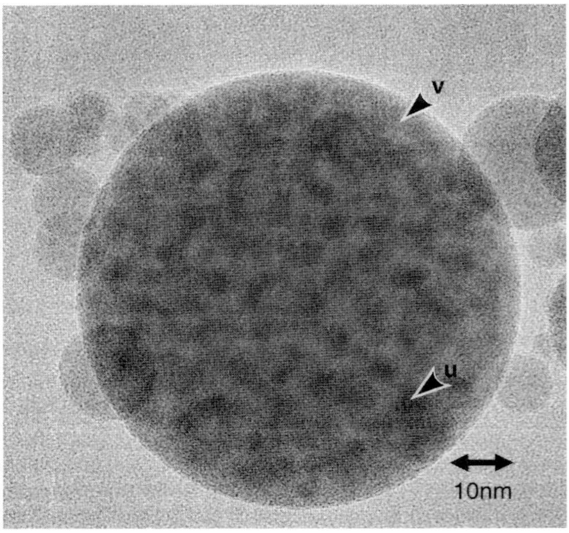

Figure 2.4.9
Al, Ca, Si, Ti, and Zr oxide nanoparticles.

independent oxides by additive and doping, materials with new characteristics can be generated. The materials for making sensors and catalysts with superior characteristics and mechanical property are created

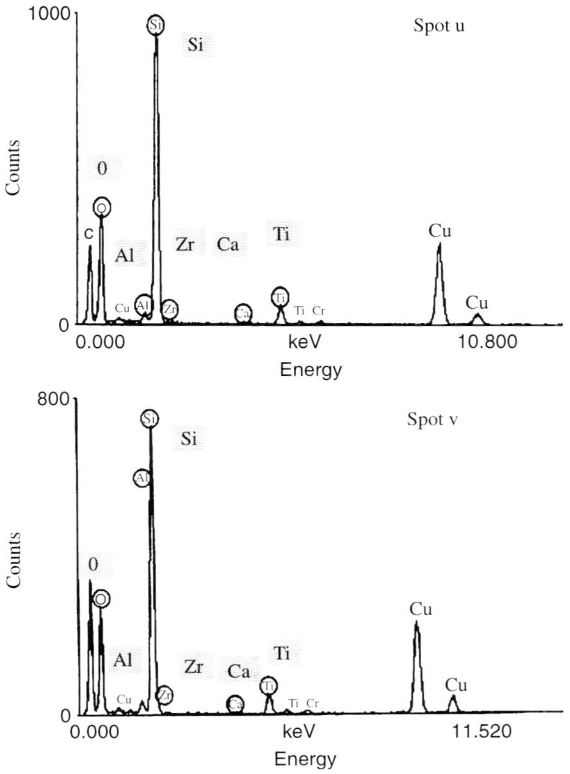

Figure 2.4.10
Analytical EDX of Al, Ca, Si, Ti, and Zr oxide nanoparticle.

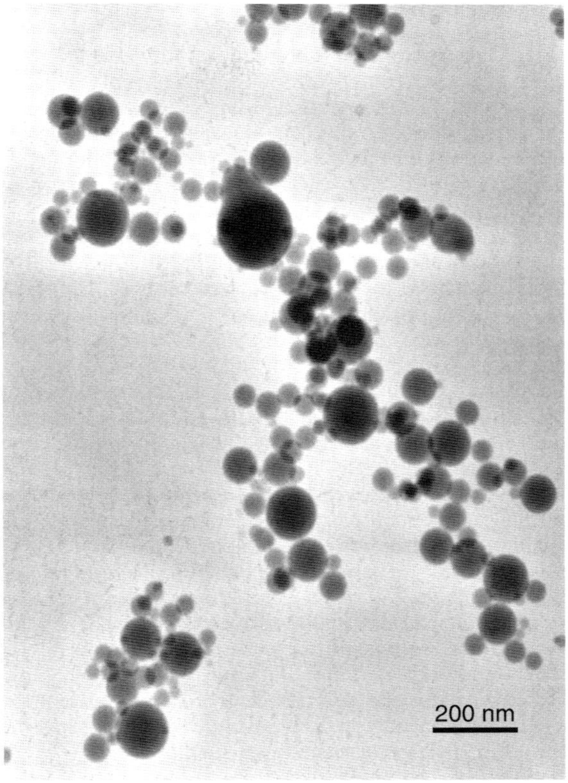

Figure 2.4.11
79% Al_2O_3-21% SiO_2 nanoparticles.

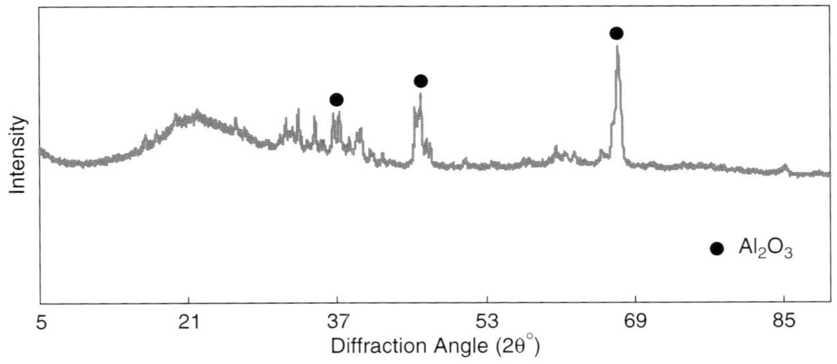

Figure 2.4.12
X-ray diffraction pattern of 79%Al_2O_3-21%SiO_2 nanoparticles.

from the multi-component oxide with both specific stoichiometric and non-stoichiometric compositions. When the uniformity of the multi-component oxide is high, the chemical bond becomes heterogeneous because of the symmetry of the metal element that causes the missing of oxygen bonds. Hereby, there is a possibility to generate new function and characteristics for the materials.

2.4.2 Solution method

Advanced materials including nanomaterials have been attracting much attention because of the NNI project that was proposed by the former U.S. President Mr. Clinton. In many cases, film is a typical form of nanomaterials, while the other is the nanoparticles or nanopowders. Recently, the advancement of the nanotechnology in the semiconductor industry is remarkable and its convenience can be easily recognized in the daily life. For example, the thickness of the dielectric layer in multi-layer ceramic capacitor (MLCC), more than 100 pieces of which are used for a mobile phone, has been decreasing day by day and is going to be in the order of nanometer [1]. Therefore, the grain-size in the dielectric layer of MLCC should be several tens of nanometers and the size effect of ferroelectric materials is going to be necessarily taken into account in the case of MLCC [2]. In addition, the remarkable improvement of the MLCC performance has been attained by controlling the nanostructure of the dielectric nanoparticles in MLCC to be a core–shell structure. This type of MLCC has already been commercialized [3]. In other words, the nanoparticles with controlled nanostructure exhibit the possibility to be the core materials of the nanomaterials in the 21st century.

Here, it is useful to consider the essential demand for the nanoparticles as the advanced materials. As for the nanoparticles, the specific surface area is larger than that of the conventional fine particles, and so, the nanoparticles are easy to coagulate. Therefore, the handling is often difficult, compared with the conventional fine particles. To overcome this problem, it is necessary to have a narrow size distribution and a controlled particle size as well as the well-controlled surface properties. Furthermore, excellent functionality is also required for the nanoparticles. The nanoparticles that meet these requests include composite nanoparticles or hybrid nanoparticles.

There are three solution-based methods to prepare the composite or hybrid nanoparticles with a particle diameter less than 100 nm. The first method is a nanocoating on the nanoparticle in a microemulsion as a reaction vessel. The second one is the heterocoagulation of the nanoparticles with different particle size, and the last one is the nanocoating on the nanoparticle by the sol-gel method. However, there is no established method for nanocoating on a nanoparticle. Therefore, a novel method to prepare the hybrid nanoparticles has increasingly been demanded.

(1) Nanocoating on nanoparticles by sol-gel method

One of the excellent methods to prepare hybrid nanoparticles is the alkoxide method, so-called sol-gel method, in which the hydrolysis and the following condensation reaction are well-controlled. The sol-gel method is famous as a method to deposit the functional thin films and the nanoparticles in recent years, but about 30 years before, it attracted attention as the low-temperature synthesis method of glasses. At present, the sol-gel method is well-known as a deposition method of the functional thin films and widely called as chemical solution deposition (CSD) to distinguish it from chemical vapor deposition (CVD) and physical vapor deposition (PVD) [4]. In this method, nanoparticles and thin films are deposited from the molecular-designed precursor solutions or sols. The most outstanding advantage over the other methods is the easy deposition of the thin films and the nanoparticles with excellent properties and well-controlled structures at relatively low cost. To utilize this advantage efficiently, it is necessary to know as much about the science (the chemistry and the physics as well as the engineering) in the preparation of the precursor solution and the deposition process of thin films or the nanoparticles. To prepare the precursor solution or sol with high quality in the sol-gel method, metal alkoxides are the common raw materials, because metal alkoxides are Lewis base and easy to hydrolyze at room temperature followed by the further polycondensation to form metal oxides [5]. If the hydrolysis and condensation reactions of the metal alkoxide could be controlled precisely, it is possible to prepare the nanohybrid particles consisting of organic compounds, inorganic compounds, and/or metals, leading to a novel nanoparticles and thin films which are important not only in the chemistry but also in the applied physics.

However, nanocoating on the nanoparticles has hardly been achieved so far. This is because that it is difficult to prepare the monodispersed nanoparticles with narrow size distribution less than 100 nm, and moreover, the selective nanocoating on the surface of the nanoparticles is very difficult from the technical viewpoint. In this section, two typical case studies of hybrid nanoparticles by sol-gel method are given, in which the silica (SiO_2) monodispersed nanoparticles were prepared by the Stöber method [6] and the functional nanofilms of several or several tens of nanometers were nanocoated on the surface of the monodispersed silica nanospheres. The first example is the hybrid nanoparticle of VO_2–SiO_2 which can be used as the smart window by the hybridization with the plastic film. The second one is the hybrid nanoparticle of TiO_2–SiO_2 that exhibits photocatalytic property, and so on.

(2) Smart window by hybridization of hybrid nanoparticles of VO_2-SiO_2 and plastic film

A nanocoating technique to prepare hybrid nanoparticles is basically different from the coating technique onto the surface of metals or glasses, because the reaction control of the selective deposition on the surface of the nanoparticles is essential to prepare the hybrid nanoparticles by nanocoating through wet-chemical processing. It is well-known that monodispersed silica

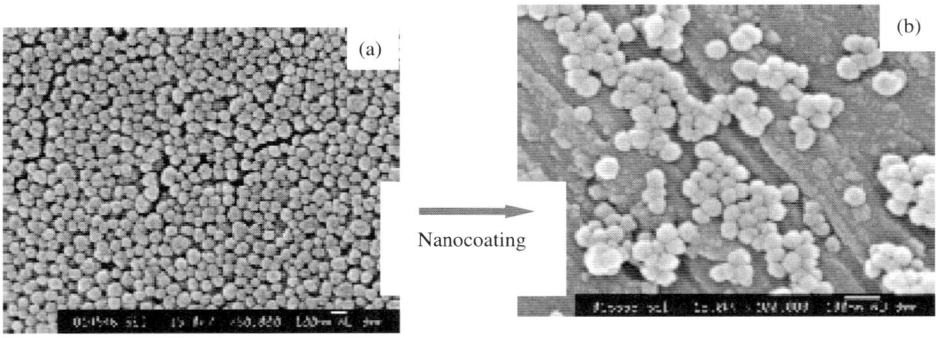

Figure 2.4.13
SEM images for VO_2–SiO_2 hybrid nanoparticles.

nanoparticles with well-controlled particle size can be prepared by the Stöber method. However, it was difficult to prepare the monodispersed silica nanospheres with smaller size less than 100 nm by the Stöber method. Then, the two stages for the nucleation and the growth of the particles were effectively separated to prepare the monodispersed particles of 10 nm in diameter [6].

In this section, a nanocoating method on the surface of monodispersed silica nanoparticles was described including the applications to cost-effective smart windows by dispersing the VO_2–SiO_2 hybrid nanoparticles into the transparent biodegradable polylactic acid (PLA) [7–9].

The actual preparation procedure is as follows; For the water removal, the monodispersed SiO_2 nanoparticle sol (solvent was ethanol including small amount of residual water) was substituted for solvent by 1-butanol. Vanadium precursor solution was prepared by dissolving $VO(O\text{-}i\text{-}Pr)_3$ in the mixed solvent of 2-propanol and 2-methoxy ethanol or 2-propanol solvent. Acetic acid was added to the vanadium precursor solutions to form the chemically modified vanadium alkoxide by the chelation that controls the following hydrolysis and condensation on the surface of the monodispersed silica nanospheres (nanocoating). The resulting hybrid nanoparticle sol was filtrated by suction to obtain the VO_2–SiO_2 hybrid precursor nanoparticles. This precursor nanoparticles were oxidized for the crystallization and then reduced or were crystallized by the in-situ combustion method of the chemically modified hybrid precursor to form the VO_2–SiO_2 hybrid nanoparticles. Crystallinity of the as-prepared VO_2–SiO_2 hybrid nanoparticles by the in-situ combustion method was not so good that the further annealing at 600°C for 1 h was conducted in N_2 atmosphere to increase the crystallinity. The smart film consisting of the VO_2–SiO_2 hybrid nanoparticles and PLA was prepared by dissolving PLA and dispersing the VO_2–SiO_2 hybrid nanoparticles into the

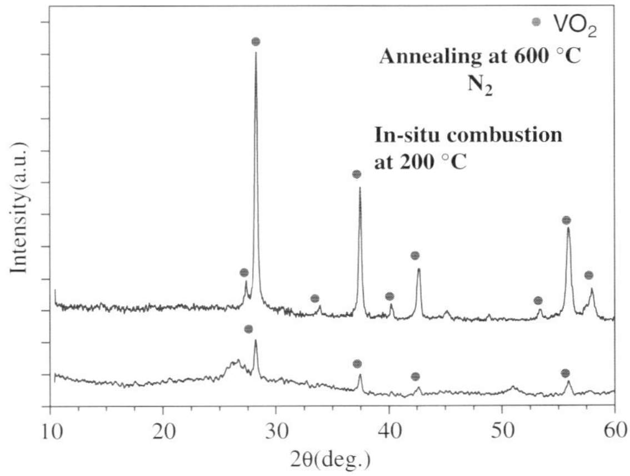

Figure 2.4.14
X-ray diffraction patterns for VO_2–SiO_2 hybrid nanoparticles.

dissolved PLA followed by casting it on a glass plate at 110°C to solidify.

The scanning electron microscope (SEM) image and XRD patterns of the resultant VO_2–SiO_2 hybrid nanoparticles are shown in Figs. 2.4.13 and 2.4.14, respectively to exhibit the particle size distribution and crystalline phase. The well-crystallized VO_2–SiO_2 hybrid nanoparticles were successfully prepared by the reaction control of the precursor alkoxide using controlled chemical modification method. As shown in the SEM photograph, the hybrid nanoparticles maintain the shape of the original monodispersed silica nanospheres to show the achievement of the nanocoating (Fig. 2.4.15).

For the molecular designing of alkoxide-derived precursors, deep knowledge on the organic chemistry is essential. The controlled chemical modification method enables the partial hydrolysis of the precursor

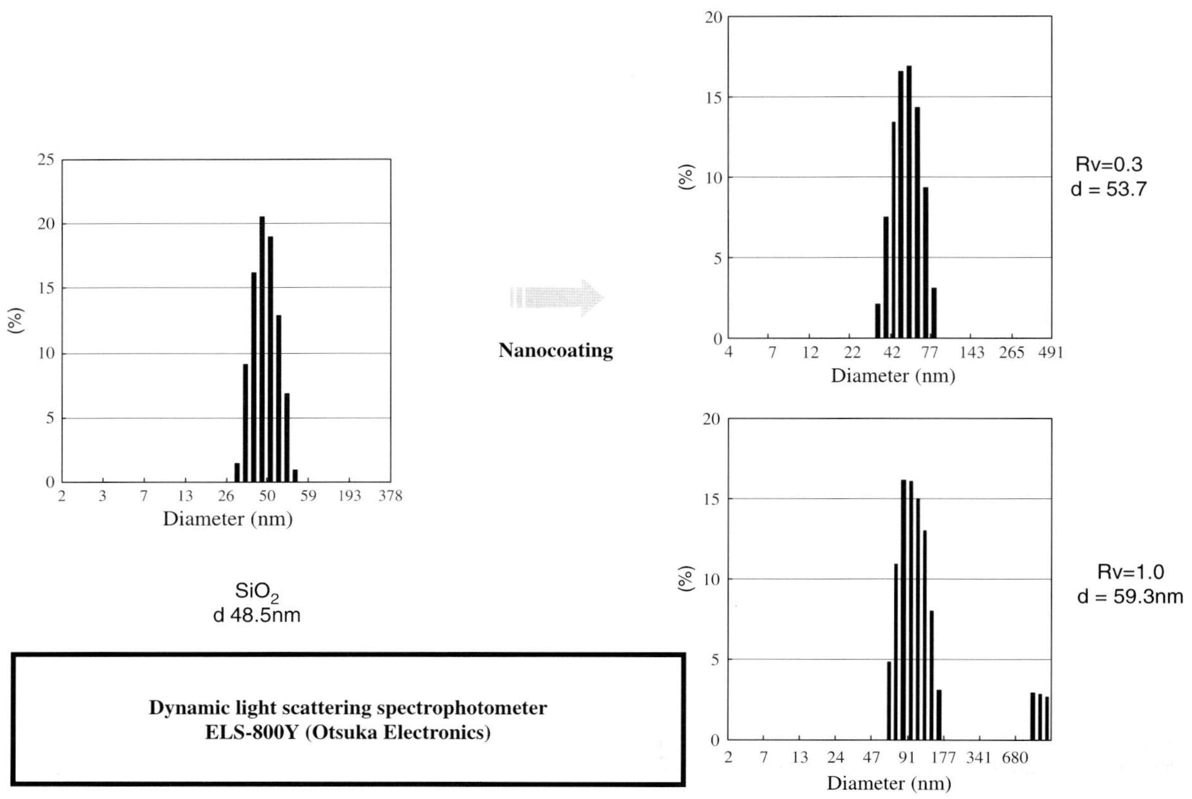

Figure 2.4.15
Particle size distribution for VO_2–SiO_2 hybrid nanoparticles.

alkoxide to suppress the following polycondensation of the metal alkoxide, even for the case of the metal alkoxide with rapid hydrolysis and condensation rate, to form the hydroxide precipitates. The deduced molecular structure of the vanadium alkoxide precursor prepared by the chemical modification in the vanadium precursor solution was shown in Fig. 2.4.16. The nanometer-level uniform coating could be successfully attained by the preferential condensation on the silica nanosphere of the partially hydrolyzed alkoxide precursor due to steric hindrance and introduction of the preferential active side chain group in an alkoxide molecule.

Thermochromic property of the resultant smart film is shown in Fig. 2.4.17. In this figure, infra-red (IR) spectra of the smart film at a room temperature and a temperature higher than the phase transition temperature were shown to investigate the performance of the resultant smart film, together with the same IR spectra for the film composed of the commercial VO_2 particles and PLA. As a result, the transmittance change of the film consisting of the commercial VO_2 particles and PLA between the

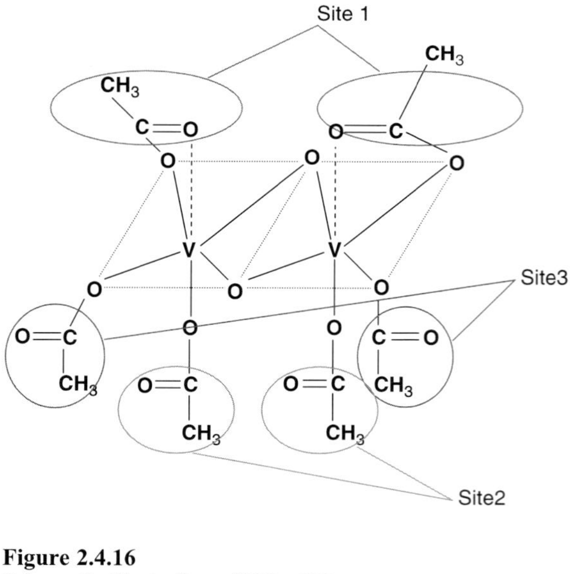

Figure 2.4.16
Schematic illustration of VO_2–SiO_2 precursor.

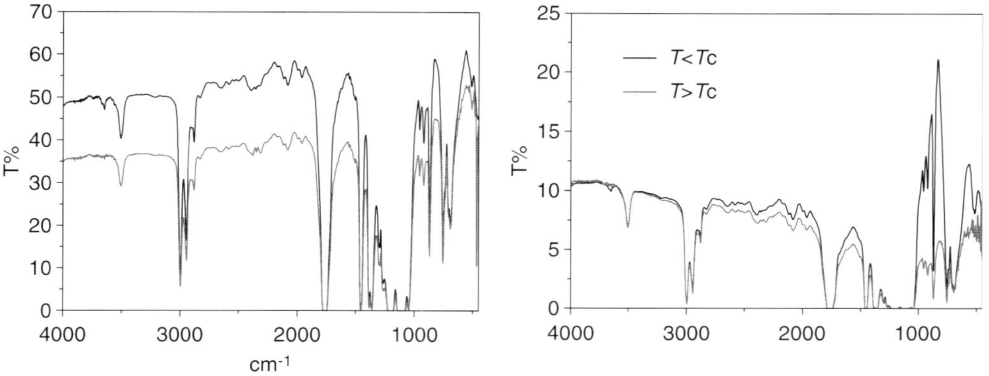

Smart film of VO_2-SiO_2 hybrid nanoparticles and PLA

ΔI=12.8%(at 4000cm^{-1})

Smart film of commercial VO_2 particles and PLA

Figure 2.4.17
Thermochromic property for smart film of VO_2–SiO_2 hybrid nanoparticles and PLA.

higher temperature and the room temperature was too low to show the thermochromic property. On the other hand, the transmittance of the smart film using VO_2–SiO_2 hybrid nanoparticles was about 50% which was lower by 20% compared with that of PLA film, confirming enough thermochromic property. The transmittance change ($\Delta I = I - I_0$) in this case was about 13% (4000 cm^{-1}). These results show the potentiality for the excellent smart window, if the dispersion state of the hybrid nanoparticles in a plastic film will be further improved.

The hybrid nanoparticles by the nanocoating onto the monodispersed silica nanoparticle surface exhibit excellent thermochromic property and good handling characteristics to be a key material for the nanotechnology in the 21st century. As described above, chemical processing from chemically designed precursor solutions is very promising for the excellent nanomaterials. The smart window was shown as an example of application of the hybrid nanoparticle system of VO_2–SiO_2. The nanocoating process (the sol-gel method with a controlled chemical modification) from the alkoxide precursor solution was considered to be very useful to prepare hybrid nanoparticles, not only for the case of the sophisticated VO_2–SiO_2 nanoparticles but also the other types of hybrid nanoparticles. Moreover, chemical modification of the monodispersed silica surface by the metal oxide and the organic groups exhibits the possibility for the various type of applications such as a deoxyribonucleic acid analysis, higher information technology, or the fuel cell material to support ubiquitous society. That is, hybrid nanoparticles are expected as the key materials to support a core technology in the 21st century. The further advancement in this field is expected in the near future.

References

[1] N. Wada: *J. Soc. Powder Technol., Jpn.*, **41**(1), 35–43 (2004).
[2] H. Suzuki: *J. Soc. Powder Technol., Jpn.*, **39**(12), 877–884 (2004).
[3] H. Chazono, H. Kishi: *J. Am. Ceram. Soc.*, **83**, 101 (2000).
[4] H. Suzuki: *Ceramics*, **37**(3), 152–156 (2002).
[5] S. Sakka: *Zoru-geru-hou no Kagaku*, Agune-Shofu, (1988).
[6] H. Kamiya, H. Suzuki, D. Kato and G. Jimbo: *J. Am. Ceram. Soc.*, **76**, 54–64 (1993).
[7] S.M. Babulanam, T.S. Eriksson, G.A. Niklasson and C.G. Granqvist: *Solar Energy Mater.*, **16**, 347–363 (1987).
[8] P. Jin, S. Tanemura: *Thin Solid Films*, **281–282**, 239–242 (1996).
[9] I. Takahashi, M. Hibino and T. Kudo: *Jpn. J. Appl. Phys.*, **35**, L438–L440 (1996).

2.4.3 Supercritical approach

Lately, the composite materials including nanomaterials draw an increasing attention and the R&D on the composite materials is conducted with a focus on the nanoparticles. There have been various synthetic methods invented to provide the specific fields for the synthesis of nanoparticles till now.

The supercritical fluid (SCF) has been applied in wide areas like medicine, foods, and electronic devises. In this section, some examples of trials to prepare the composite materials making use of SCF are introduced.

First, the synthesis method of nanoparticles using the supercritical fluid is briefly summarized, and then some examples of applications for the structure control of composite particles based on it are introduced, finally referring to the future development for creation of the composite materials.

2.4.3.1 Synthesis of nanoparticles using the supercritical fluid

The major synthesis methods for nanoparticles by using the supercritical fluid proposed so far are listed up in Fig. 2.4.18 [1]. They are classified into the physical method using it as a good or poor solvent and the chemical one where it is utilized as a reaction field in itself. All these methods are making use of the properties of the supercritical fluid, whose density and dielectric constant change depending upon the temperature and pressure and which consequently can control the solubility of solute and then the degree of supersaturation.

Among them, the supercritical hydrothermal synthesis, RESS and SAS are already in practical use for the production of composite materials (Table 2.4.2).

2.4.3.2 Nanocomposite particles

(a) Inorganic/inorganic nanocomposite particles

Gupta et al. [2] reported the application of supercritical hydrothermal synthesis to prepare nanocomposite particles of ZnO–TiO_2. This synthesis was carried out using an apparatus having two reactors connected in series. In the first reactor, ZnO nanoparticles were generated and then they were transferred to the second one to synthesize TiO_2, which produced ZnO–TiO_2 nanocomposite particles. These nanocomposite particles did not have core–shell structure but agglomerates with two kinds of randomly dispersed oxides.

Arai et al. [3] reported the synthesis of Ni/Fe_3O_4 composite particles in the subcritical water as reaction field using a flow-type synthesis apparatus. This composite particle was formed in two stages. First, the Fe_3O_4 particles were synthesized hydrothermally as a core and then Ni was deposited onto the Fe_3O_4 particles. Although the particle size increased with the temperature rise in case of synthesis of Fe_3O_4 particles only, it was kept constant when the formation of Ni took place at the same time. The authors consider that it is because Ni deposited so rapidly to suppress the growth of Fe_3O_4 particles.

(b) Carbon nanotube/inorganic composite particles

Using supercritical CO_2, the carbon nanotube (CNT)/metal composites have been synthesized [4]. The mixture of supercritical CO_2 and H_2 was pumped into the stainless-steel reactor with CNT and metal salt of like Ni, Pd, and Cu, etc., beforehand. After the metal salts were dissolved in the supercritical fluid in this way, the metals were deposited by the reduction process, with H_2 increasing the temperature. Figure 2.4.19(a) and (b) show the TEM images of Pd/CNT composite nanoparticles synthesized.

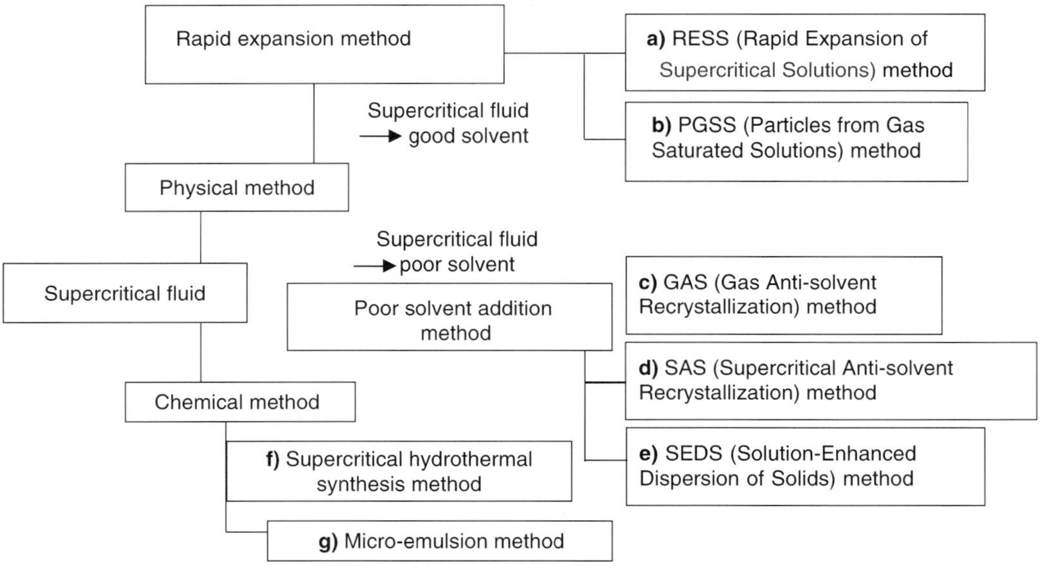

Figure 2.4.18
Technology for the synthesis of fine particles making use of supercritical fluid.

Table 2.4.2
Overview of approaches to produce fine particles by the use of supercritical fluid.

a) RESS method	Method of producing particles by rapid expansion and reduction of density of supercritical fluid with dissolved substance to the atmospheric pressure through nozzle
b) PGSS method	Using freezing point depression due to the presence of CO_2, solid material that is normally impossible to be sprayed can be sprayed as a liquid. Method of depositing solid by cooling effect and decompression of CO_2 caused by rapid expansion
c) GAS method	Method of depositing crystal by introducing supercritical fluid as a poor solvent into the solution
d) SAS method	Method of causing crystallization by injecting solution into the supercritical fluid as a poor solvent through nozzle
e) SEDS method	Method of depositing particles by raising dispersion efficiency of supercritical fluid and solvent by mixing at the nozzle tip, introducing solvent, supercritical fluid, and entrainer through two or three fluid coaxial nozzle
f) Supercritical hydrothermal synthesis method	Nanoparticle synthesis method for metallic oxide made by hydrolysis and dehydration reaction of metal salt solution in supercritical water. By this method, it is possible to synthesize nanoparticles with high crystallinity and monodispersibility, and size and shape of particles can be controlled by adjusting temperature and pressure
g) Micro-emulsion method	Method of synthesizing particles in the droplet of water in CO_2-type microemulsion formed in the supercritical CO_2

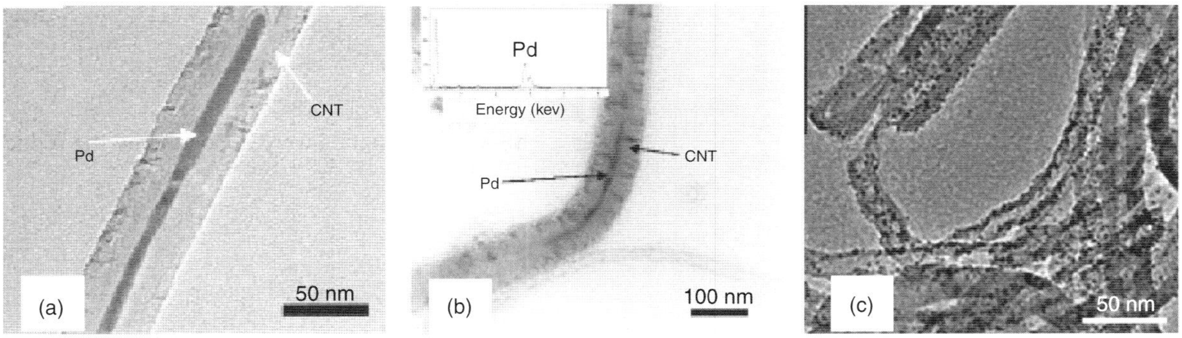

Figure 2.4.19
TEM images of CNT/metal composite. (a) Pd/CNT composite; (b) Pd/CNT composite and its EDS spectrum: (c) Ru/CNT composite.

It was confirmed by energy dispersion spectrometry (EDS) that the nanowire inside the CNT was Pd. The diameter of this Pd-nanowire was 7–9 nm corresponding to the inner diameter of CNT and its length was over 200 nm. It was also found that metal nanowire took the shape of straight line or curved structure in accordance to the CNT shape.

Ye et al. reported to have deposited the metals selectively onto the surface of CNT introducing the functions of –COOH or –OH by the treatment with the HNO_3/H_2SO_4 solution [5, 6]. Figure 2.4.19(c) shows a TEM image of CNT with Ru nanoparticles deposited on it.

Besides using metals, a method to form polystyrene inside the CNT was reported by Han et al. [7] In place for CNT, the formation of metal on SiO_2 nanowire was also reported [8].

(c) Polymer/inorganic composite nanoparticles

Wang et al. [9] synthesized nanocapsules consisting of SiO_2 nanoparticle and Eudragit (polymer) using SAS method. After solving the Eudragit into acetone, SiO_2 nanoparticles were dispersed in it. By injecting this solution into supercritical CO_2 as a poor solvent, nucleation of polymer took place and the particles grew on the surface of SiO_2 nanoparticles to form SiO_2/polymer nanocapsules. The average particle size of SiO_2 was from about 16 to 30 nm and that of nanocapsules was controllable from about 50 to 100 nm. However, because of inhomogeneous formation of polymer layer on the SiO_2 particles and agglomeration between the nanocapsules, the composites had irregular shape. These synthesized nanocapsules were observed by TEM and their components were analyzed by electron energy loss spectrometry (EELS).

Figure 2.4.20 shows TEM and TEM–EELS images. As a result of silicon mapping (Fig. 2.4.20(b)) and carbon mapping (Fig. 2.4.20(c)), it was made clear that the carbon signal derived from the polymer are concentrated on the surface of SiO_2 nanoparticles and that the SiO_2 nanoparticles were encapsulated by the polymer.

Han et al. [10] reported the synthesis of Ag/polystyrene composite nanoparticles using the supercritical CO_2. The Ag nanoparticles were synthesized in the surfactant/water/cyclohexane reverse micelles and then polystyrene was dissolved in its solution. By introducing the supercritical CO_2 into this solution, the solvent and the surfactant were removed and polystyrene was deposited on the Ag particles. With the increasing pressure and temperature of the reaction, the agglomeration was suppressed and the particle size of the composite nanoparticles was reduced. Han et al. synthesized also the composite nanoparticles of CdS/polymethyl methacrylate, and so on [11, 12].

(d) Biomolecule/inorganic composite nanoparticles

Finally, the synthesis of the nanoparticles conjugated with protein is introduced. Sun et al. made the composites consisting Ni, Co, Fe nanoparticles and polymers using Rapid Expansion of a Supercritical Solution into a Liquid SOLVent (RESOLV), the improved RESS method [13]. The metal salt is dissolved into the supercritical ethanol and then injected into the receiving solution of polyvinyl pyrrolidone (PVP), which resulted in the generation of nanoparticles and the coating with PVP at the same time.

By the similar method, Meziani et al. synthesized Ag composite nanoparticles using the supercritical ammonia [14]. For the solution to receive the supercritical fluid, the PVP or bovine serum albumin (BSA) was dissolved as the coating material. The size of the composite particles was controllable in the range between 4.8 and 43 nm depending upon the kind of coating material and the composition of the receiving solution [15]. $AgNO_3$ was dissolved in the supercritical ammonia and injected into the water solution including BAS and then BSA-conjugated AgS nanoparticles were synthesized.

Figure 2.4.21 shows a TEM image and AFM image of the BSA-conjugated AgS nanoparticles. From the TEM observation, the particle size was confirmed to be 6.3 nm and from the AFM image it was found that the AgS nanoparticles coated with the protein were well-dispersed and the particle size was from about 20 to 30 nm. The authors presume from these results that the particles are covered uniformly by the protein, which prevents the agglomeration of the nanoparticles.

In the above section, several synthetic method of composite nanoparticles using the supercritical fluid were demonstrated. All these methods make use of the feature of the supercritical fluid that the solvent properties can be widely controlled by the pressure and temperature.

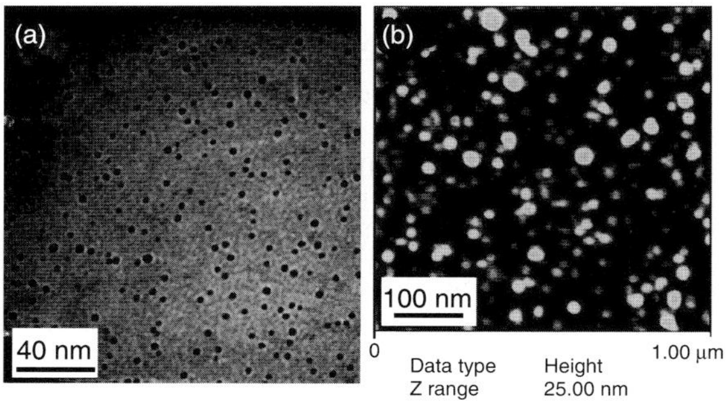

Figure 2.4.21
(a) TEM image; (b) AFM image of BSA-immobilizing AgS nanoparticles.

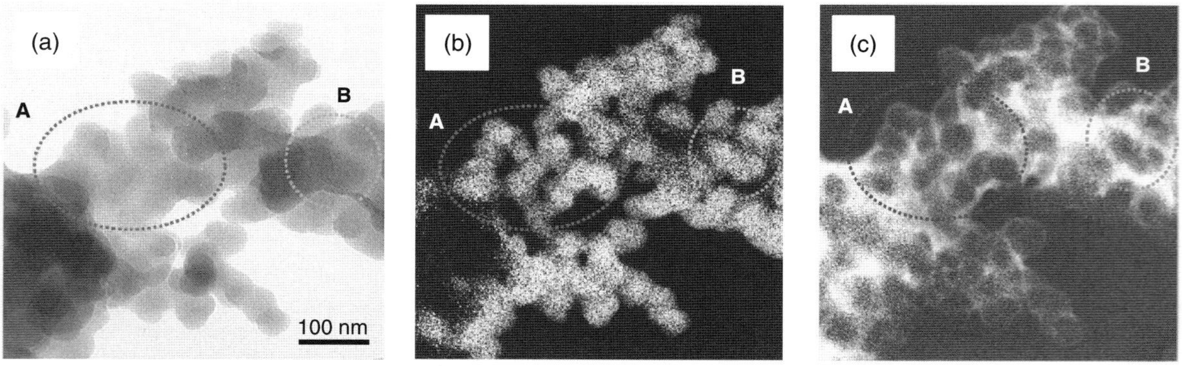

Figure 2.4.20
TEM image of SiO_2/polymer nanocapsule. (a) Standard observation; (b) Silicon mapping; (c) Carbon mapping.

The composite materials are required widely in the industrial and medical fields and expected to become a key technology for the development of nanotechnology. The application of the supercritical fluid is expected to advance as a technology indispensable for the creation of such composite materials in the future.

References

[1] H. Uchida: *NanoTechnology with Supercritical Fluids*, p. 64, CMC Publishing Co., Ltd., Tokyo (2004).

[2] R. Viswanathan, G.D. Lilly, W.F. Gale and R.B. Gupta: *Ind. Eng. Chem. Res.*, **42**, 5535–5540 (2003).

[3] K. Sue, N. Kakinuma, T. Adschiri and K. Arai: *Ind. Eng. Chem. Res.*, **43**, 2073–2078 (2004).

[4] X.-R. Ye, Y. Lin, C. Wang and C.M. Wai: *Adv. Mater.*, **15**, 316–319 (2003).

[5] X.-R. Ye, Y. Lin and C.M. Wai, *Chem. Commun.*, 642–643 (2003).

[6] X.-R. Ye, Y. Lin, C. Wang, M.H. Engelh, Y. Wang and C.M. Wai: *J. Mater. Chem.*, **14**, 908–913 (2004).

[7] Z. Liu, X. Dai, J. Xu, B. Han, J. Zhang, Y. Wang, Y. Huang and G. Yang: *Carbon*, **42**, 458–460 (2004).

[8] X.-R. Ye, H.-F. Zhang, Y. Lin, L.-S. Wang and C.M. Wai: *J. Nanosci. Nanotechnol.*, **4**, 82–85 (2004).

[9] Y. Wang, R.N. Dave and R. Pfeffer: *J. Supercrit. Fluids*, **28**, 85–99 (2004).

[10] J. Zhang, Z. Liu, B. Han, D. Liu, J. Chen, J. He and T. Jiang: *Chem. Eur. J.*, **10**, 3531–3536 (2004).

[11] Z. Li, J. Zhang, J. Du, T. Mu, Z. Liu, J. Chen and B. Han: *J. Appl. Polym. Sci.*, **94**, 1643–1648 (2004).

[12] Z. Li, J. Zhang, T. Mu, J. Du, Z. Liu, B. Han and J. Chen: *Collids. Surf. A*, **243**, 63–66 (2004).

[13] Y.-P. Sun, H.W. Rollins and R. Guduru: *Chem. Mater.*, **11**, 7–9 (1999).

[14] M.J. Meziani, H.W. Rollins, L.F. Allard and Y.-P. Sun: *J. Phys. Chem. B*, **106**, 11178–11182 (2002).

[15] M.J. Meziani, Y.-P. Sun: *J. Am. Chem. Soc.*, **125**, 8015–8018 (2003).

2.4.4 Mechanical processes

Among the methods to customize the structure of particles, a technique can produce composite particles by direct bonding the nanoparticles onto the surface of fine particles using mechanical treatment.

Figure 2.4.22 schematically shows two typical structures of composite particles obtainable by this method [1]. Figure 2.4.22(a) illustrates the core–shell type of composite particle with nanoparticles on the surface of the core particle. This particle structure is widely used in the particle surface modification and the like.

Using the mechanical processing, it is possible to design various particle surface structures from dense to porous ones. It can also make multi-layer structure by bonding different components or gradients onto the particle surface.

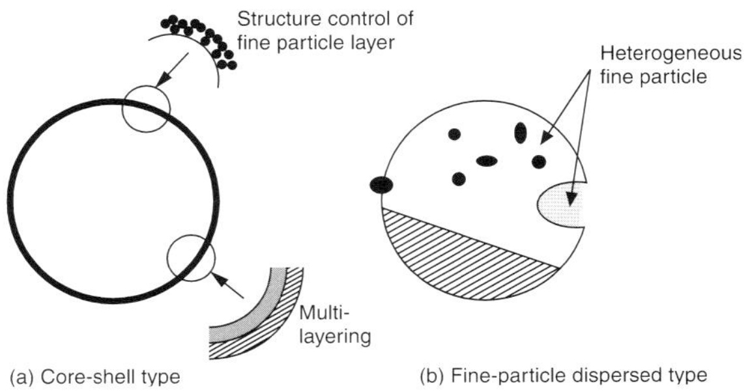

Figure 2.4.22
The structures of composite particles obtainable by the mechanical process.

On the other hand, Fig. 2.4.22(b) shows the composite structure (fine-particle dispersed type) made by burying or dispersing nanoparticles in the coarser particles. The mechanical alloying applied in the field of powder metallurgy is one of the technologies for producing this type of composite particles.

The major principles used for producing composite particles by the mechanical methods are summarized in Table 2.4.3 [2], which classifies the particle composing methods into three categories. The first category is the composing machines conventionally used as grinding mills. Although grinding is the mechanical operation to break solid particles, most of their size reduction principles have effective functions to make composite particles as seen in the table. It becomes obvious to explore the possibility of applying mechanical effects for making particle composites. As a result, some particle composing machines have been developed based on the basic grinding mechanisms, as shown in the right column of the table. On the other hand, machines taking new mechanical principles and equipment design are also being developed and listed as others in the table.

Since these mechanical processes are normally carried out in a dry state, there is no solvent contamination on the composite particles and its production process is simple and environment friendly in comparison to the wet processing. Consequently, most of the particle-composing machines are used in a dry state.

In the following, the mechanical particle-composing processes and their critical operating parameters for controlling the structure of particle composites as shown in Fig. 2.4.22 will be explained.

2.4.4.1 Particle-composing process

To make composite particles, the mechanical force needs to be exerted on the powder. In this case, the powder is a mixture of core particles and the guest nanoparticles. In general, the size of the core particles

Table 2.4.3
Major principles used to produce particle composites by the mechanical process [2].

	Principle	Example of products developed as a composing machine
Application of pulverizer	High-speed impact type pulverizer	
	Pin mill	Hybridization system (Nara machinery Co., Ltd.)
	Disc mill	
	Centrifugal classification type	Cosmos (Kawasaki Heavy Industries, Ltd.)
	Attrition-type mill	Mechanofusion system (Hosokawa Micron Corp.)
	Ball mill	
	Tumbling type	
	Vibration type	
	Planetary type	
	Centrifugal fluidized type	
	Media agitating-type mill	
	Agitation vessel type	
	Jet mill	
	Fluidized-bed type	
Others	Motar and pestle:	
	Bottom plate of cylindrical vessel containing particles rapidly rotates	Mechanomill (Okada Seiko Co., Ltd.)
	Elliptic cylindrical vessel rotor rapidly rotates in a slowly rotating elliptic cylindrical vessel	Thetacomposer (Tokuju Co., Ltd.)

should be sufficiently larger than that of the guest nanoparticles. Most of the particle-composing machines shown in Table 2.4.3, the mechanical force is given onto the particles by the rotational motion of the rotor. The nanoparticles are composed onto the surface of core particle depending upon the contact number between the particles and various effects, including the mechanical and thermal ones, at the contact points caused by the rotor movement.

Figure 2.4.23 shows the modeling of composing nanoparticles onto the particle surface [2]. As shown in the figure, it proceeds in two steps. In the first step, the surfaces of the core particles are activated and the different kinds of nanoparticles adhere to it one after another. As a result, the adhesion ratio R of the nanoparticles to the core particles increases, while the specific surface area S_w of the powder mixture decreases. In the second step, after sufficient nanoparticles being built-up, the nanoparticle layer itself is strongly pressed and bonded onto the core particle surface. The boundary surface between the composite particles made by this process becomes very strong.

For example, in case of composing silica-based glass beads with titanium dioxide using the MechanoFusion system in Table 2.4.3, strong bonding was observed at the interface between the two components due to the electron transfer. Therefore, it is possible to control the structure of coating layer and its bonding condition to some extent by changing the processing time and other factors.

2.4.4.2 Factors to control the particle composing

The operating parameters of the machine affect the particle composing, including mechanical intensity, processing temperature, ambient conditions, and the type of the machine, and so on. These parameters are often difficult to control independently. Furthermore, from the viewpoint of particles, the particle size and combination ratio of core particles to the guest nanoparticles and the way to add nanoparticles also influence the composing behavior considerably. As an example, the remarkable effects of mechanical conditions and the processing temperatures on the particle composing are explained below.

First regarding the mechanical effect, it is considered that the higher the revolution of the machine is, the stronger is the force on the powder and on the particles between the interfaces. In addition, with the increasing revolution, the frequency of giving the mechanical force on the particles also increases. Therefore, the adhesion ratio in Fig. 2.4.23 increases generally with the increase of machine revolution, which in turn reduces the specific surface area of the powder.

However, the composing effect of the mechanical force may differ depending on the combination of the core particles and the nanoparticles. For example,

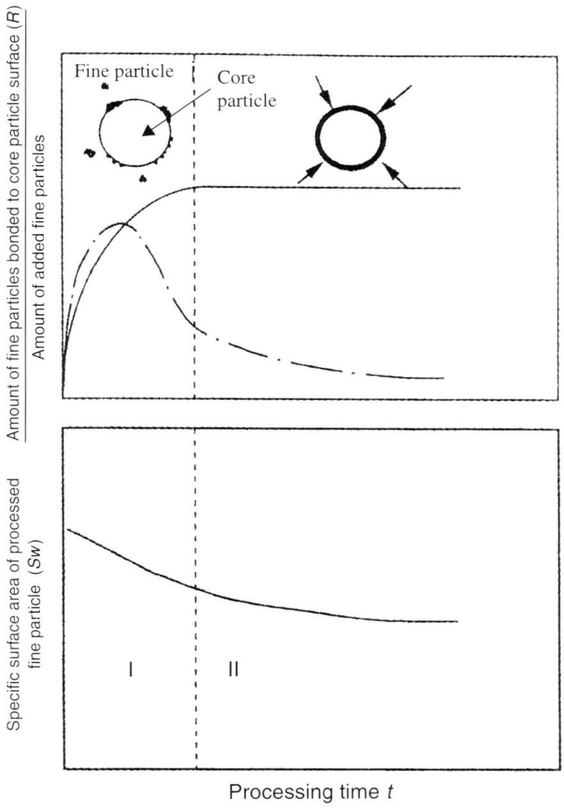

Figure 2.4.23
The process of making core–shell type composite particles by the mechanical process.

in case of composing titanium dioxide nanoparticles on the surface of glass beads, the nanoparticles can be layered on the core particles under certain processing time with proper machine revolution. However, the nanoparticle-coating layer can be ripped off from the surface of core particles as seen in Fig. 2.4.23, when the machine revolution exceeds a critical value. The critical revolution tends to reduce with the increasing size of core particles [2]. This is attributed to the increase in the mechanical energy per unit area due to the reduction in specific surface area of the larger core particles.

Furthermore, there is an apparent difference in the structure of particle composites depending on how the nanoparticles are added, namely either charging all the feed mixture at once or feeding the nanoparticles little by little repeatedly with a certain time interval. By feeding the nanoparticles to the core particles little by little repeatedly, smooth and uniform coating layer on the core particles can be obtained [3].

The effect of processing temperature on the composing process is the next to discuss. Since the mechanical processing generates heat in the equipment, it is important to control the processing temperature to prevent melting or deformation of the particles. For example, when resin is used as the feed material, the processing temperature needs to be kept below its glass-transition temperature. On the other hand, it is reported that the higher the temperature is, the denser becomes the nanoparticle layer on the core particles when composing metal and ceramic particles [2].

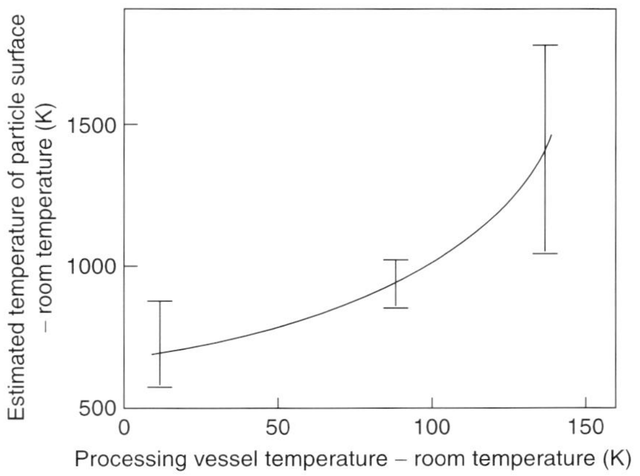

Figure 2.4.24
The relationship between the vessel temperature and the particle surface temperature during the particle composing operation [4].

To investigate the processing temperature effect on the properties of particle composites, it is necessary to know the actual temperature at the particle surface during the composing process. Figure 2.4.24 shows an example of the relationship between the vessel temperature and the particle surface temperature during the particle composing with the MechanoFusion System as explained in Table 2.4.3 [4].

The particle surface temperature was estimated thermodynamically from the reaction of oxide layer on the surface of metal particle and the added carbon. It is believed that the temperature at the particle surface reaches about ten times higher than that of the composing vessel. This temperature effect can initiate highly specific phenomena at the particle surface and enhance the bonding strength between the particles to a great extent during the particle-composing process.

2.4.4.3 Examples of particle composing

Figure 2.4.25 shows an example of particle composing of nanoparticles on the surface of carbon nanofiber [5]. This technology enables the composing of nanoparticles not only on the surface of spherical particles but also on plate-like and other shaped particles. As seen in the figure, the surface of nanofiber is coated densely with nanoparticles by

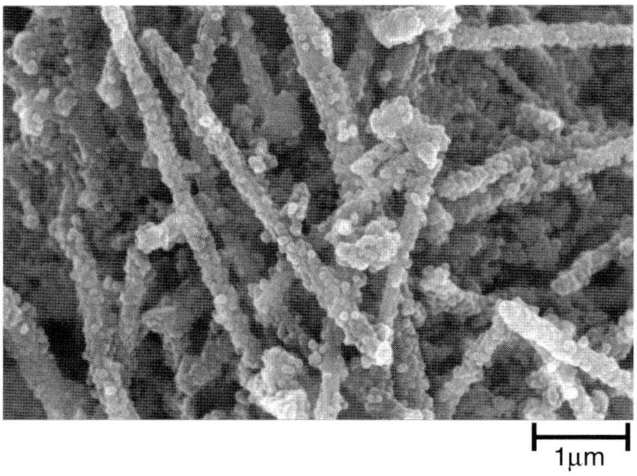

Figure 2.4.25
The structure of carbon nanofiber and nanoparticles composite [5].

using this method. On the other hand, it is also possible to make the coating layer on the core particle with nanopores by changing the processing conditions. In fact, nanoparticle layers with nanopores have been made on the surface of glass fiber or ceramics fiber in commercial applications [6]. The application examples of nanoparticle composites made by this method are introduced in the Section 4.3.2.

The reaction temperature between the particles can be reduced using the composite particles made by this method as a precursor. For instance, the magnesium–boron mixture layer can be formed on the surface of magnesium particle by strongly pressing the fine boron particles on the magnesium particle. The composite particles with the surface layer of fine magnesium–boron mixture can develop super-electroconductivity on their surfaces by heat treatment at 500°C, which is far below the typical temperature for producing the super-electroconductive phase of MgB_2 [7].

Furthermore, plasma discharging is investigated as additional field energy to enhance this mechanical particle-composing process. For example, nitrogen can be doped onto the surface of TiO_2 nanoparticles by processing the nanoparticles under the glow discharging with a gas mixture of ammonia and argon using MechanoFusion System as shown in Table 2.4.3. As a result, the photocatalytic performance of the TiO_2 nanoparticles can be significantly improved as confirmed by visible light tests [8]. Taking the same approach, the mechanical processing can be combined with other potential working mechanisms to develop new processes for further application development of nanoparticles in the future.

References

[1] M. Naito: Kagaku Kougaku no Shimpo 30 (Progress of Chemical Engineering), in *Biryushi Seigyo (Fine Particle Control)*, Maki-Shoten, p. 69 (1996).

[2] M. Naito, A. Kondo and T. Yokoyama: *ISIJ Int.*, **33**, 915–924 (1993).

[3] M. Naito, T. Hatta, S. Asahi, T. Tanimoto and S. Endo: *Kagaku Kogaku Ronbunshu*, **24**, 99–103 (1998).

[4] K. Nogi, M. Naito, A. Kondo, A. Nakamura, K. Niihara and T. Yokoyama: *Powder and Powder Metal.*, **43**, 396–401 (1996).

[5] M. Naito: *Nanoparticle Technol.*, The Nikkan Shimbun, Ltd., p. 159 (2003).

[6] H. Abe, I. Abe, K. Sato and M. Naito: *J. Am. Ceram. Soc.*, **88**, 1359–1361 (2005).

[7] H. Abe, M. Naito, K. Nogi, M. Matsuda, M. Miyake, S. Ohara, A. Kondo and T. Fukui: *Phisica C*, **391**, 211–216 (2003).

[8] H. Abe, T. Kimitani, K. Sato, M. Naito and K. Nogi: *J. Soc. Powder Technol., Jpn.*, **42**, 409–412 (2005).

2.5 Pore structure

2.5.1. Gas-phase method

Particles having porous structures that are synthesized from gas-phase reaction methods can be obtained if the particles consist of blocky aggregate structures. In the case of the particles synthesized by a chemical vapor deposition (CVD) method, it typically consists of chained aggregated/agglomerated structures. In contrast, particles prepared by spray pyrolysis or spray-drying methods are occasionally formed with porous structures. In this section, the preparation of particles having porous structures using gas phase (i.e. CVD), spray pyrolysis, and spray-drying methods are described.

2.5.1.1 Preparation of porous structured particles by CVD method

Many particles prepared by CVD methods are typically in the form of chained agglomeration structures, and only a few of them have pores if they are in the form of blocky aggregation structure. An example of porous structured particles prepared by the CVD method includes the formation of spherical-shaped zinc oxide particles. As the preparation method, zinc vapor is first formed, followed by a cooling treatment, during which, zinc particles are formed. In the next process, only the surface of the zinc particles are oxidized to form a zinc oxide layer, and finally the elemental zinc is removed to obtain zinc oxide cage particles, as shown in Fig. 2.5.1 [1].

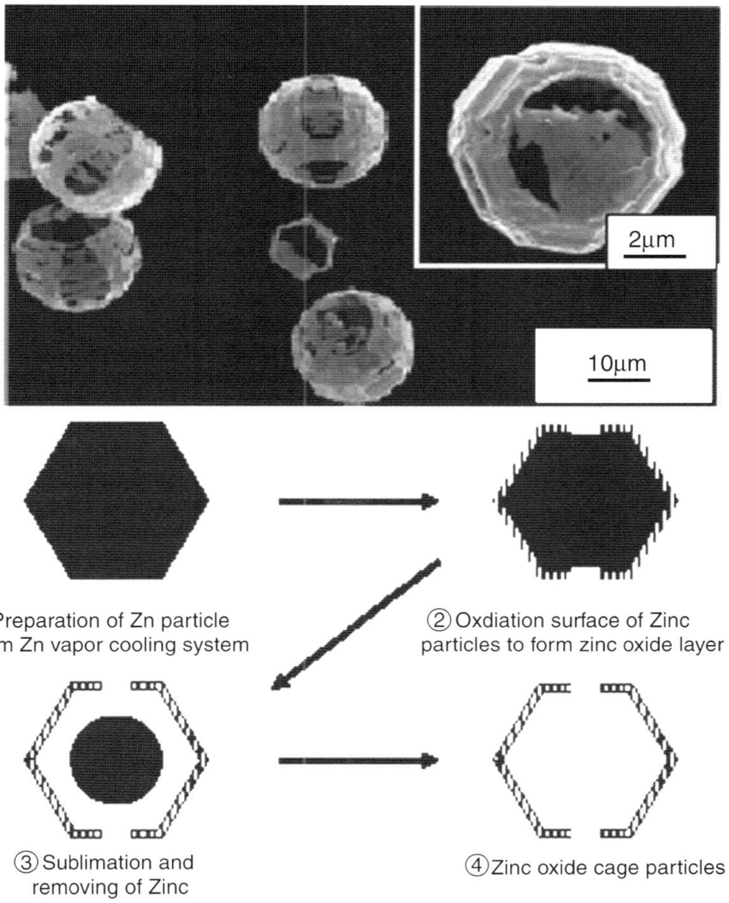

Figure 2.5.1
SEM images of zinc oxide hollow particles and the synthesis mechanism.

2.5.1.2 Preparation of porous structure particles by spray-pyrolysis method

As discussed in Section 2.3.1, in the spray-pyrolysis method, a starting solution (or precursor), typically consisting of a dissolved metal salt, nitrate salt, acetate salt, etc., are sprayed to form droplets. The droplets are then directly introduced into a high-temperature reactor furnace. Inside the reactor furnace, the solvent in the droplets evaporates and followed by the reaction of the reactants inside the droplets, finally, producing the solid particles. As shown in Fig. 2.3.6 in Section 2.3.1, as the crystallization process is undertaken during the evaporation of solvent from droplet, particles having a porous structure can be produced if the sintering temperature is sufficiently high.

Figure 2.5.2(a) shows a TEM picture of cross-section of ZnO particles prepared from an aqueous solution of a metal nitrate salt [2]. In the case of the large size of the ZnO particles, the particles consist of a large pore inside. The other examples, zinc sulfate (ZnS) and cadmium sulfate (CdS) particles, prepared from thiourea and metal nitrate acid mixed solution, are shown in Fig. 2.5.2(b) and (c), respectively [3, 4]. From the figures, the ZnS particles have a dense morphology since the particles consist of that are several nanometers in diameter and in other samples, the CdS particles have a porous morphology since the prepared particles consist of primary particles with a diameter of around 20 nm.

Furthermore, using a template method, particles having a porous structure could be produced after the removal of the template. A spray solution consisting mixed polystyrene (PSL) particles and the reactant solution is used in the spray pyrolysis. Figure 2.5.3 shows scanning electron micrographs (SEM) of various

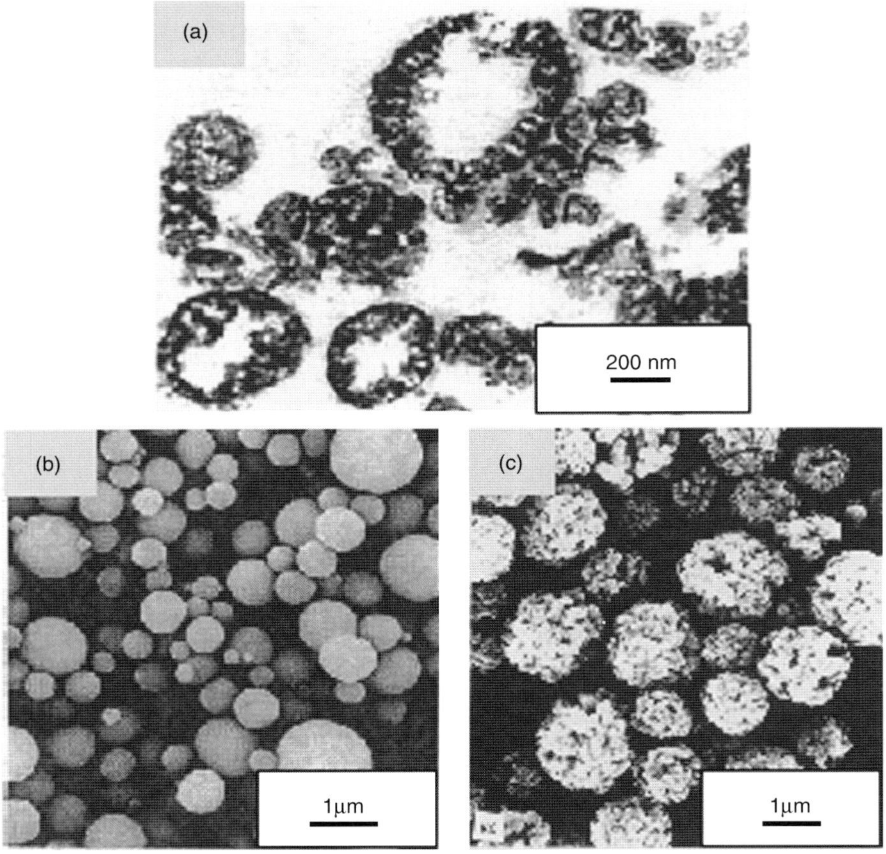

Figure 2.5.2
SEM images of porous structure particles. (a) ZnO, (b) ZnS, (c) CdS.

porous particles produced by the spray-pyrolysis method [5]. Within this method, the sprayed droplets are first heated at a low temperature, i.e. less than 400°C, to evaporate the solvent and to obtain particles that contain both PSL particles and reactant. The particles are then heated at a relatively high temperature to burn out the PSL particles, thus producing for obtaining porous particles. Various types of oxide porous particles can be successfully produced by the spray-pyrolysis method [5].

2.5.1.3 Preparation of porous structure particles by spray-drying method

Spray-drying involves spraying a suspension of droplets followed by a drying process, and results in the production of solid particles.

As shown in Fig. 2.5.4, spherical SiO_2 particles can be produced by spray-drying a sol of SiO_2 nanoparticles under conditions of small droplet size, high concentration, and very large silica particle size. On the other hand, doughnut-type SiO_2 particles can be produced an under conditions of high evaporation rate [6].

Figure 2.5.5 shows SEM images of particles prepared from different sizes of silica nanoparticles [7]. Figure 2.5.6 shows variation in effective refractive index and porosity as a function of 8 and 100 nm primary particles content used as the precursor [8].

In the case of a suspension containing two different sizes of nanoparticles, the resulting particles may have a composition in which larger particles are encapsulated by smaller particles, since the smaller particles tend to move faster to surface of the droplet during evaporation. Figure 2.5.7 shows variation in the effective refractive index for spray-drying microencapsulated particles derived from small Al_2O_3 or ZrO_2 particles mixed with large SiO_2 particles, as a function of their content [8]. It is known that by using a spray-drying process involving two different sizes of particles and/or nanoparticles having two different refractive indexes; the refractive index of the prepared

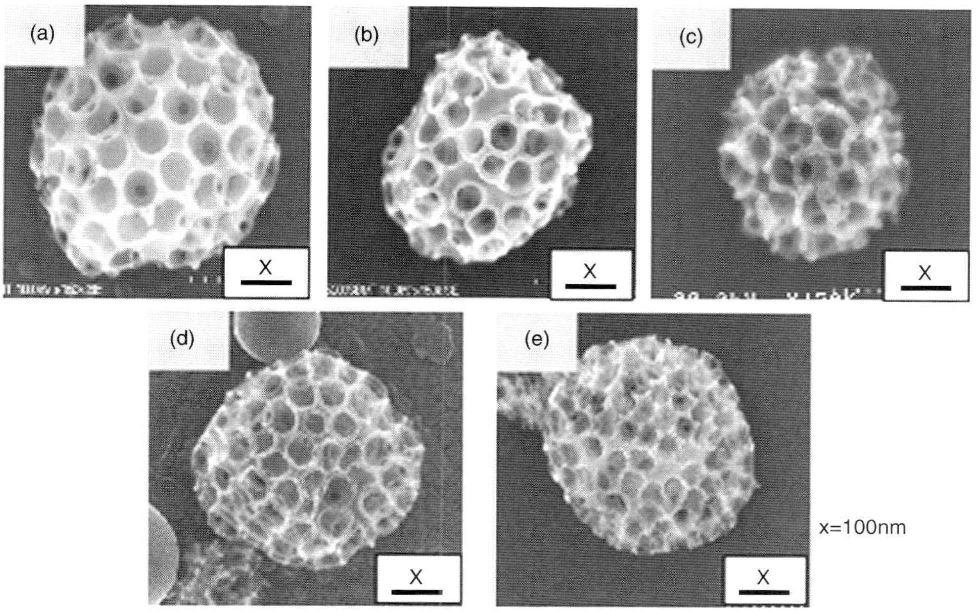

Figure 2.5.3
SEM images of porous particles prepared by spray-pyrolysis method. (a) SiO_2, (b) TiO_2, (c) ZrO_2, (d) Al_2O_3, (e) Y_2O_3.

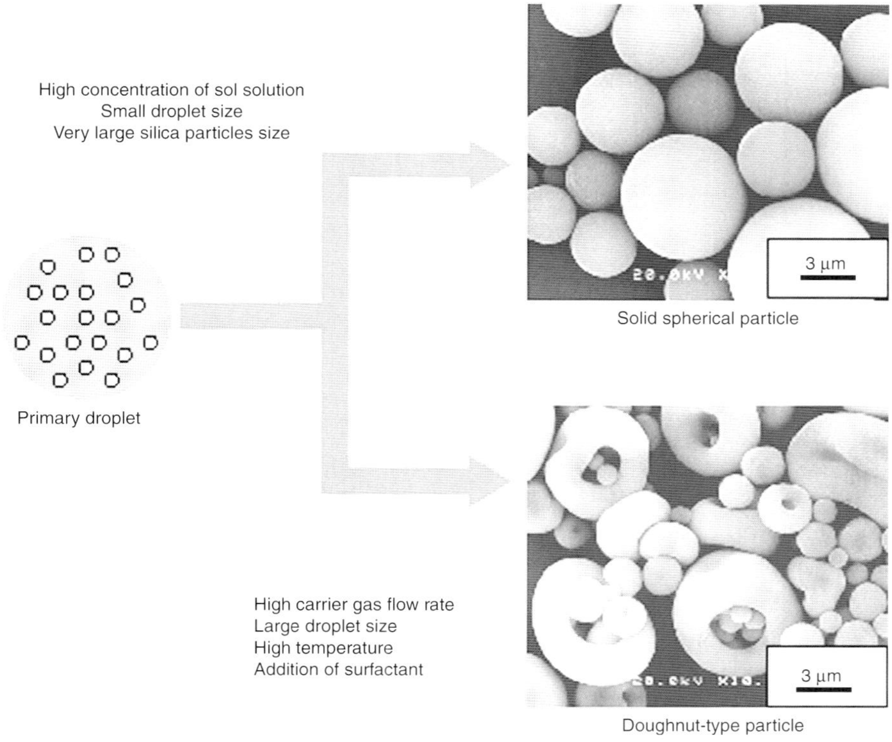

Figure 2.5.4
Spherical and doughnut-type SiO_2 particles prepared by spray-drying method.

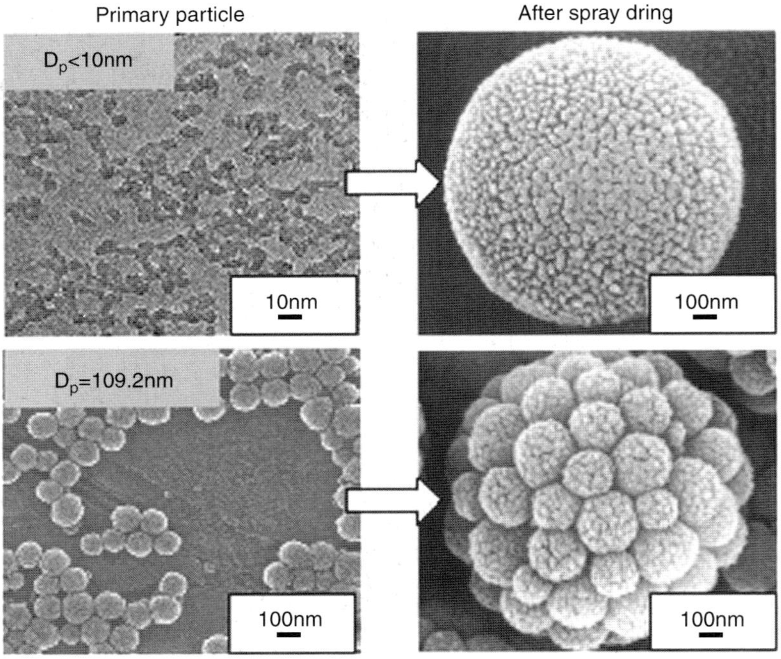

Figure 2.5.5
Morphology change of particles prepared by spray-drying method derived from different primary particles size.

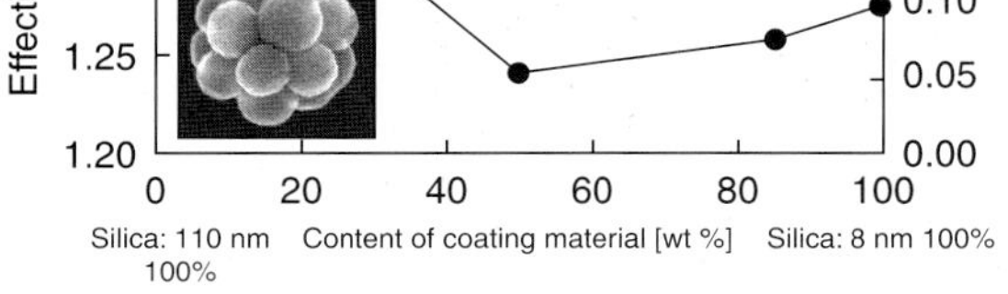

Figure 2.5.6
Effective refractive index and porosity of SiO_2 particles prepared with different primary particles size.

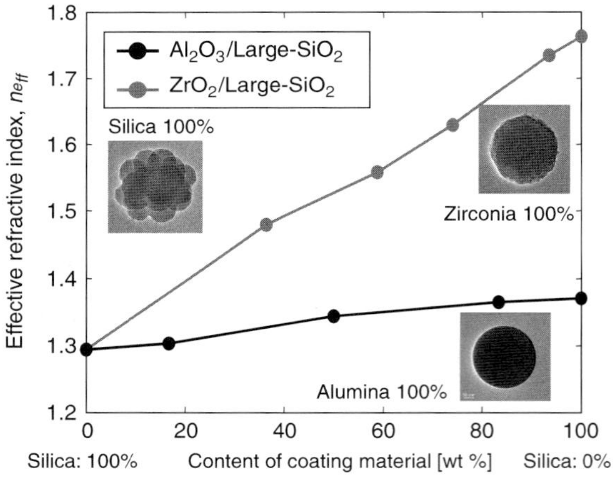

Figure 2.5.7
Variation of effective index of SiO_2/ZrO_2 or Al_2O_3 multi-component particles prepared from the different content.

particles can be easily controlled by changing their content in the precursor.

On the other hand, porous SiO_2 particles can be also produced by spray-drying, as shown in Fig. 2.5.8 [9]. The process is as follows: first, a mixed solution of SiO_2 nanoparticles and polystyrene latex (PSL) particles are sprayed to form droplets, and the solvent is then evaporated at low temperature, leading to the self-organization of particles in the droplet, and finally the PSL particles are removed at a high temperature. Pore size can be controlled by changing the PSL size, as shown in Fig. 2.5.9. For successfully

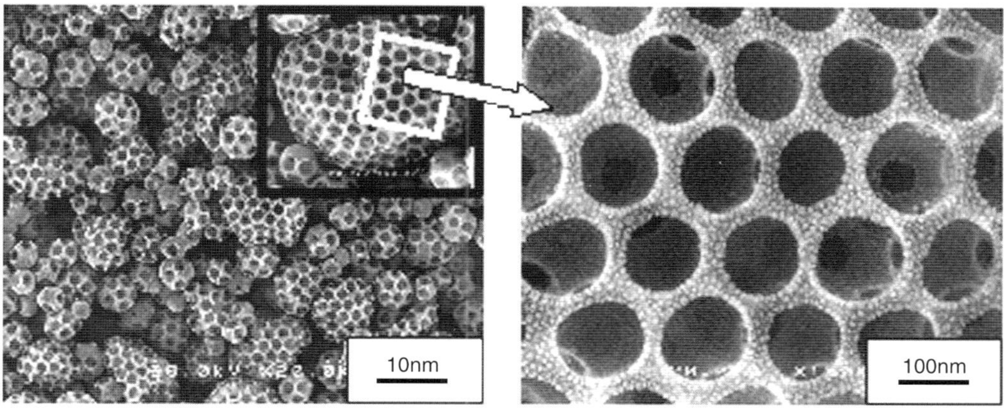

Figure 2.5.8
Porous SiO_2 particles prepared by removing PSL particles after it arranged via self-organization process.

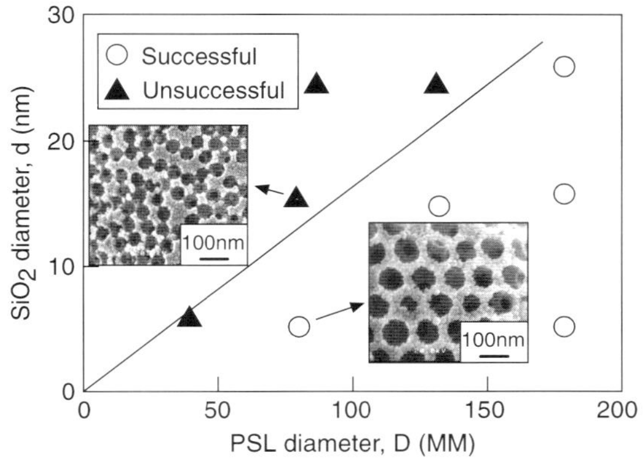

Figure 2.5.9
Preparation conditions for obtaining successful ordered porous SiO_2 particles.

organized pores, the ratio between the size of the silica nanoparticles size, d, and the PSL size, D, must be known to be lower than $d/D = 0.155$.

Figure 2.5.10 shows an SEM image of porous titania particles prepared from titania nanoparticles and PSL particles mixed precursor. Since the titania nanoparticles used are not monodispersed and some are in an agglomerated condition, the porous arrangement in the particles is not as good as the result obtained for the porous silica particles. However, if the size of the PSL particles used is relatively larger, well-organized porous particles can be easily prepared.

The phenomena of the porous arrangement in particles prepared by the present method can be explained quantitatively by the topology theory [10].

Particles having a porous structure are of interest because of their potential applications such as in optical functional particles, oral inhalation drug particles, etc. To control the porous structure, a spray-drying

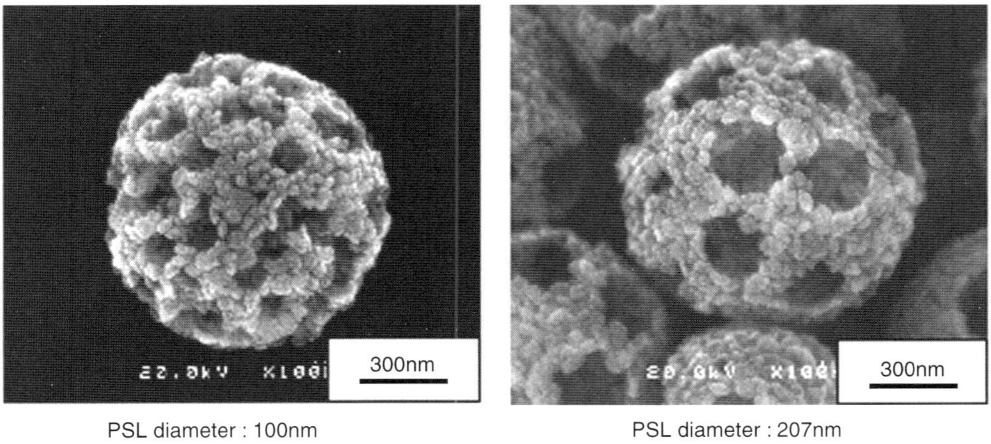

Figure 2.5.10
Porous titania particles prepared from the mixed solution of titania nanoparticles and PSL particles.

method using two different types of sol precursors is currently the most effective method.

References

[1] H.J. Fan, R. Scholz, F.M. Kolb, M. Zacharias and U. Goesele: *Sol. State Commun.*, **130**, 517—521 (2004).

[2] K. Oshima, K. Tsuto, K. Okuyama and N. Tohge: *Kagaku Kogaku Ronbunsyu*, **18**(3), 288–295 (1992).

[3] N. Tohge, S. Tamaki and K. Okuyama: *Jpn. J. Appl. Phys.*, **34**, L207–L209 (1995).

[4] S. Tamaki, N. Tohge and K. Okuyama: *J. Mater. Sci. Lett.*, **14**, 1388–1390 (1995).

[5] M. Abdullah, F. Iskandar, S. Shibamoto, T. Ogi and K. Okuyama: *Acta Materialia*, **52**(17), 5151–5156 (2004).

[6] F. Iskandar, L. Gradon and K. Okuyama: *J. Colloid Interface Sci.*, **265**, 296–303 (2003).

[7] F. Iskandar, I.W.Lenggoro, B. Xia and K. Okuyama: *J. Nanoparticle Res.*, **3**, 263–270 (2001).

[8] F. Iskandar, H. Chang and K. Okuyama: *Adv. Powder Technol.*, **14**(3), 349–367 (2003).

[9] F. Iskandar Mikrajuddin, K. Okuyama: *Nano Lett.*, **2**(4), 389–392 (2002).

[10] L. Gradoń, S. Janeczko, M. Abdullah, F. Iskandar and K. Okuyama: *AIChE J.*, **50**(10), 2583–2593 (2004).

2.5.2 Liquid-phase synthesis

2.5.2.1 Hollow particles

Synthesis and application of hollow particles with their diameter ranging from nano to micrometer are attracting much attention these days. Because hollow particles have unique properties which are different from solid particles, for example low density, high-specific surface area, and ability to incorporate a material.

Owing to these properties, hollow particles are applied for various materials such as lightweight filler, an insulator [1], composite materials [2], and color materials [3]. Meso- and macroporous hollow particles with controlled shell thickness are used as analytical isolation and encapsulating materials [4]. Especially, encapsulating materials are used for drug delivery and inhibition of hypersensitiveness of enzymes or proteins [5]. Optical properties due to the large difference in refractive index of core and shell materials are also unique. Hollow particles are applied to photoelectric devices and coating materials because of their optical properties [6, 7].

As described above, hollow particles have more potentialities of application based on their unique properties compared with solid particles that have been widely used so far. In the following, organic bead template method will be described as one of the most famous ways to prepare hollow nanoparticles.

The feature of this organic bead template method is usage of organic bead as a template. Poly(stylene) (PS) is most frequently used in this method, but basically, any template can be applied to this method if it can be removed.

Preparation of hollow particles of silica [4, 5, 8], titania, zinc sulfide [6], and cadmium sulfide [9] by using the organic beads template method have been reported. These hollow particles are prepared via core–shell (coated) particles that are formed by coating a shell material around the core organic beads by using selective precipitation on the charged surface of the core particles [10]. The hollow particles are prepared by removing the organic core particles after filtration and drying. The thermal (pyrolytic) and the chemical decomposition are known as the way to remove the organic core. A schematic illustration of the organic bead template method is given in Fig. 2.5.11.

The hollow particles that were prepared by using this organic beads template method are introduced below.

Caruso et al. [5] have reported about the hollow silica and silica-polymer core–shell particles synthesized by this method. At first, PS latex particles are coated with polymer in order to control the surface charge of the core. This coating layer provides positively charged surface that promotes adsorption of negatively charged SiO_2 as well as it provides smooth surface.

A multi-layer of SiO_2/PDADMAC (poly(diallyldimethylammonium)chloride) is formed by alternative adsorption of SiO_2 and PDADMAC. The template PS latex polymer is removed by calcination at 500°C. Then, hollow particles with the shape of internal cavity templated by core particles were obtained.

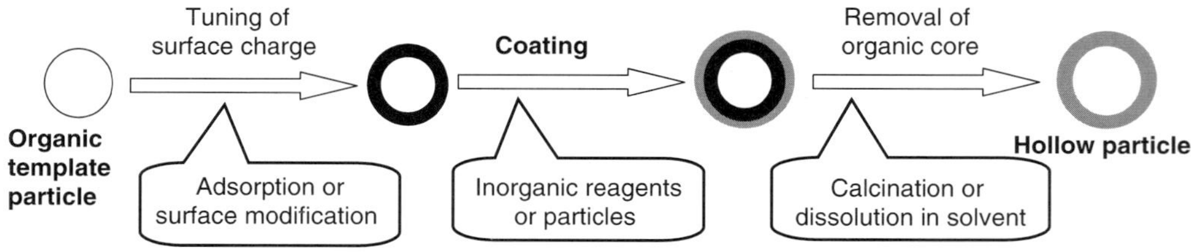

Figure 2.5.11
Organic beads template method.

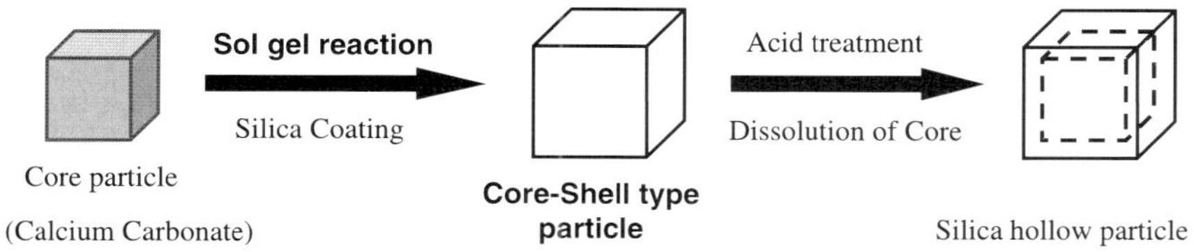

Figure 2.5.12
Inorganic template method.

The preparation of sulfide hollow particles have been attempted by this method. Song and co-workers [9] have prepared PSA (poly(styrene-acrylic) acid) latex core/CdS shell particles, and silica/TiO_2 composite hollow spheres. Yin et al. [6] prepared ZnS hollow shell by the similar way.

Walsh et al. [11] synthesized crystalline calcium carbonate porous hollow particles. They found that thin aragonite shell was formed in O/W type microemulsion saturated with calcium carbonate. Formation of the aragonite shell proceeded in the presence of PS beads that were added to make the shell incorporate the template. In other words, calcium carbonate precipitated and coated around the PS particles which were incorporated in the organic phase of the O/W emulsion. They reported that removal of the core PS by calcination of dried core–shell particles resulted in the formation of honey-comb-structured hollow spheres.

Organic materials have been utilized in the template method so far, although, in general, removal process of the organic core causes environmental pollution. In the cases of the PS beads core, organic solvents, e.g. toluene, are employed for removal of the core, and release organic waste. Thermal decomposition of the polymer core generates a large amount of CO_2 gas. Inorganic template methods are proposed as alternative less harmful preparation process for hollow particles [12]. Schematic illustration of the concept of inorganic particle template method is shown in Fig. 2.5.12.

When nanosized calcium carbonate particles are used as a core, inorganic acids, e.g. hydrochloric acid, can be used to remove them. Calcium chloride waste solution and generated CO_2 are reusable to synthesize calcium carbonate particles.

As described above, the inorganic template method can be applied to plan a less environmental load process compared to the organic template process. In addition, various shapes of hollow particles can be achieved by using the intrinsic morphology of the inorganic particles, and expected for further applications.

References

[1] T. Tani: *R & D Review of Toyota CRDI*, **34**(3) (1999).
[2] H. Toda, S. Kaga, K. Hosoi, T. Kobayashi, Y. Ito, T. Higashihara and T. Aida: *Zairyo*, **50**(5), 474–481 (2001).
[3] J. Park: *J. Colloid Interface Sci.*, **266**, 107–114 (2003).
[4] D. Walsh, B. Lebeau and S. Mann: *Adv. Mater.*, **11**(4), 324–328 (1999).
[5] F. Caruso, R.A. Caruso and H. Mohwald: *Science*, **282**, 1111–1114 (1998).
[6] J. Yin, X. Qian, J. Yin, M. Shi and G. Zhou: *Mater. Lett.*, **57**, 3859–3863 (2003).
[7] T. Enomae: Proceedings of the 5th Asia textile conference, Vols. **1464–1467** (1999).
[8] O.D. Velev, T.A. Jade: *Nature*, **389**, 447–448 (1997).
[9] C. Song, G. Gu, Y. Lin and H. Wang: *Mater. Res. Bull.*, **38**, 917–924 (2003).
[10] G. Decher: *Science*, **277**, 1232–1234 (1997).
[11] D. Walsh, S. Mann: *Nature*, **377**, 320–323, (1995).
[12] M. Fuji, M. Chikazawa, T. Takei, K. Tanabe, K. Mitsuhashi and N. Tagami, *Japan Kokai Tokkyo Koho*, JP2005-263550A (2005).

2.5.2.2. Mesoporous structures

(1) Structure

Ordered mesoporous materials such as MCM-41 [1] and FSM-16 [2] are relatively new materials that were first synthesized between 1990 and 1992 [3]. A mesopore is defined as a pore with a diameter of 2–50 nm (IUPAC), and is intermediate between micropores ($\leq$2 nm) typical of zeolites, and macropores ($\geq$50 nm) typical of porous glass. Silica gel and some activated charcoals are mesoporous, but such materials do not possess an ordered mesopore structure and hence generally exhibit a wide distribution of pore diameter. Mesoporous materials having mesopores of a uniform size in a regularly ordered structure (Fig. 2.5.13) are reported here.

Mesoporous materials can be synthesized in various forms as particles, thin film, or as monolithic blocks (Table 2.5.1). Particles are synthesized by precipitation in a solvent, and the particle shape and diameter (10 nm to 1 mm) can largely be controlled by the synthesis conditions. Spherical monodisperse particles of hundreds of nanometers in size can be synthesized using the Stöber method as discussed below.

The pore structure can be controlled using surfactants with various liquid crystal structures as a template to form one-dimensional tunnel structures [1, 2], three-dimensional channel structures [1], or three-dimensional cage structures [4] (Fig. 2.5.14).

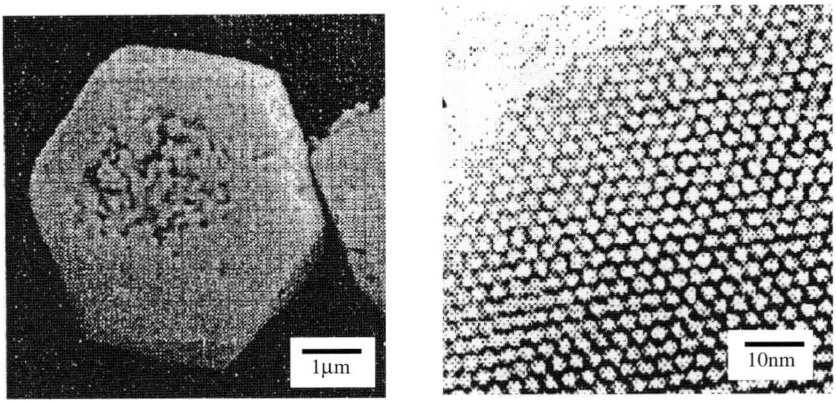

(a) SEM image of MCM-41 (b) TEM image of MCM-41

Figure 2.5.13
Particle morphology (a) and pore-structure (b) of mesoporous silica.

Table 2.5.1
Diversity of mesoporous structure.

Morphology	Powder, supported film, self-standing film, transparent film, oriented film, sphere, fiber, micropatterning, nanoparticle with defined shape	
Pore-structure and material name	2D-hexagonal	MCM-41, FSM-16, SBA-15
	3D-hexagonal	SBA-2
	Cubic *Ia3d*	MCM-48
	Cubic *Pm-3n*	SBA-1, SBA-16
	Cubic $I4_1/a$	CMK-1
	Orthorombic *C2mm*	KSW-2
	Disordered	HMS, MSU, KIT-1
	Lamellar	MCM-50
Pore diameter	1.5~10 nm, 3~7 m, 5~30 nm, 0.4~1.5 nm	
Framework composition	SiO_2, SiO_2-$MO_{n/2}$ (M = Al, Ti, V, B, Mn, Fe, Ga, Zr) Al_2O_3, TiO_2, ZrO_2, Ta_2O_5, Nb_2O_5, SnO_2, HfO_2, $AlPO_4$ Pt, Au, Carbon, Organic/silica hybird	

Synthesis of a one-dimensional channel structure with orthorhombic symmetry has also been reported [5]. The various types of mesoporous materials shown in Table 2.5.1 have been synthesized, including structures with different space groups. Pore diameter can be finely adjusted by altering the chain length of the surfactant [1, 2]. Pore diameter can be controlled in a wide range (2–30 nm) by the addition of swelling agents [1] and polymeric surfactants [6], as well as temperature control [7], and the addition of short-chain amines [8]. Originally, the framework of mesoporous materials was made of silica or silica alumina. However, mesoporous materials have more recently been synthesized with frameworks composed of various transition metal oxides [9], metals (Pt, Au) [10], carbon [11], and organic/silica hybrid materials [12]. The pore wall of most mesoporous materials is amorphous in structure, exhibiting just a halo X-ray diffraction pattern observed only in the intermediate-angle region ($10^\circ < 2\theta < 60^\circ$). On the other hand, the X-ray diffraction in the small-angle region ($\theta < 10^\circ$) showed one or more peaks due to the ordered pore arrangement structure with a periodicity of several nanometers. While mesoporous silicate materials have relatively high thermal stability (800–1,000°C), the mesoporous structure in transition metal oxides usually collapses during the crystallization process (400–600°C).

(2) Formation mechanism

In aqueous solution, surfactants are known to form a lyotropic liquid crystal phase. For example, an aqueous solution of hexadecyltrimethyl ammonium bromide ($C_{16}TMABr$), one type of surfactant, forms a liquid crystal phase that is hexagonal, cubic, or lamellar in structure, depending on the solution conditions. A periodic pore-structure of mesoporous materials is obtained by transferring the structure of the liquid crystal onto an inorganic material. Figure 2.5.15 shows a scheme for the synthesis of a mesoporous material in which a liquid crystal template method is used. A self-organized mesostructure is formed by adding an inorganic precursor (tetraethoxysilane, for example) to an aqueous solution of a surfactant under appropriate conditions. A mesoporous material with uniform pores can be synthesized by removing the surfactant from a mesostructured substance composed of a complex of the inorganic framework and the surfactant. Generally, a very low concentration of the surfactant is used for the synthesis of mesoporous material, in which the surfactant is present in the form of molecules or free micelles. The mesoporous structure is believed to be formed by a co-operation of the surfactant and an inorganic molecules because a self-assembly of ordered mesostructure is induced by the addition of an inorganic precursor to a dilute surfactant solution [13]. Alkyltrimethylammonium chloride/bromide and tri-block copolymers are commonly used as a surfactant template. A mesoporous silica (FSM-16) with higher stability can be synthesized using kanemite ($NaHSi_2O_5 \cdot 3H_2O$), a layered polysilicate, as the inorganic precursor [2].

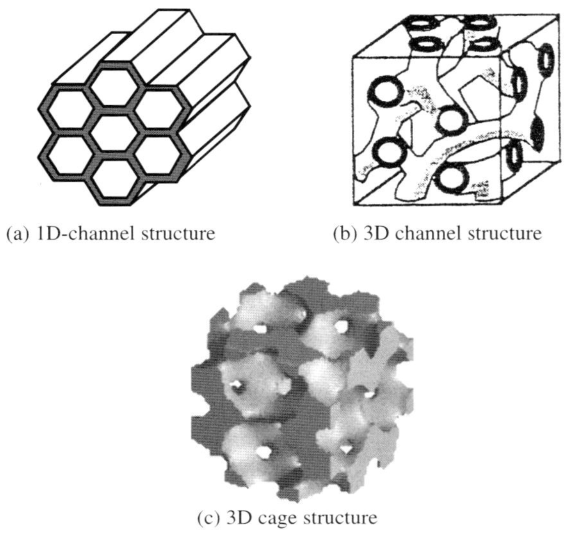

Figure 2.5.14
Pore-structures of ordered mesoporous materials.

(3) Synthesis of monodisperse particles

There are various methods for synthesizing spherical monodisperse mesoporous particles. The Stöber method is typically used, in which precipitation is induced under alkaline conditions in a mixed solvent of water and alcohol [14–17]. A surfactant ($C_nH_{2n+1}N(CH_3)_3Br$) and sodium hydroxide are dissolved in a mixed solvent of water and methanol,

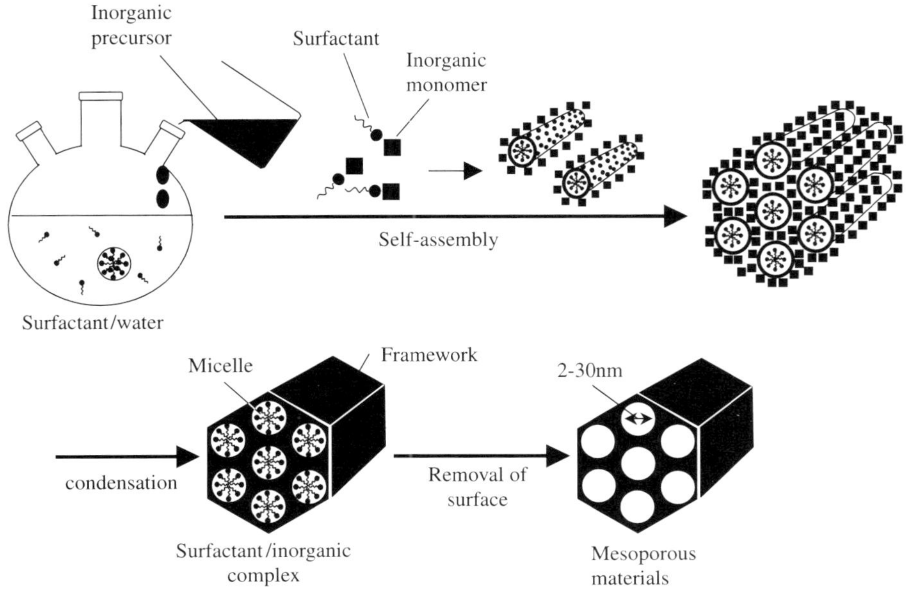

Figure 2.5.15
Synthesis of ordered mesoporous material using surfactant as a template.

then tetramethoxysilane is added while stirring vigorously. After several hours, precipitation occurs, and after ageing overnight, the solution is filtered and washed. A mesoporous material is obtained by removing the template by calcination at 550°C [15, 16]. Figure 2.5.16 shows a scanning electron microscopy (SEM) image of mesoporous silica synthesized by this method. The mesoporous silica shows very uniform spherical morphology and monodisperse particles. The mean diameter of these particles is 655 nm, with very little distribution in particle size (standard deviation: 3.7%). Moreover, with a pore diameter of 2 nm, the particles have an extremely large surface area: approximately 1000 m^2/g. Mesoporous particles from 100 nm to 3 μm in size are currently being synthesized using this method.

In addition to the Stöber method, other techniques for synthesizing mesoporous particles include emulsion [18], and gas-phase methods [19]. Particles of around 0.5 mm in size can be synthesized using emulsion methods. There is also a method for converting silica particles (diameter: 10 μm) into a mesoporous material while maintaining the particle shape of the base material [20].

(4) Drug delivery system applications

Mesoporous materials have a large adsorption capacity due to the large void fraction of approximately 50%. Because the pore size is neither too large nor too small, adsorbed molecules can be released easily. In this respect, the application of mesoporous materials in drug delivery systems (DDS) is being investigated.

Ibuprofen in solution was adsorbed onto mesoporous silica powder (MCM-41) with a pore diameter of 1.8–2.5 nm at amount of 30 wt% [21], and the release of ibuprofen from MCM-41 in simulated body fluid continued for 70 h. Moreover, the release rate of the drug can be controlled by organic modification of the mesoporous silica surface. Mesoporous silica modified by aminopropylsilane [$NH_2(CH_2)_3Si(OCH_2CH_3)_3$] selectively adsorbs anionic proteins, and its release can also be controlled by adjusting the pH of the solution [22]. Mesoporous systems that actively respond to external stimuli are also being formulated, making it possible to release the drug when it reaches the affected area. A method in which an coumarin group dimerized by irradiation is bonded to the pore opening has been proposed, where light is used to control whether the pore is closed or open [23]. Other attempts have been made to seal the pores with CdS particles, which are chemically removed when the drug has reached the affected area [24].

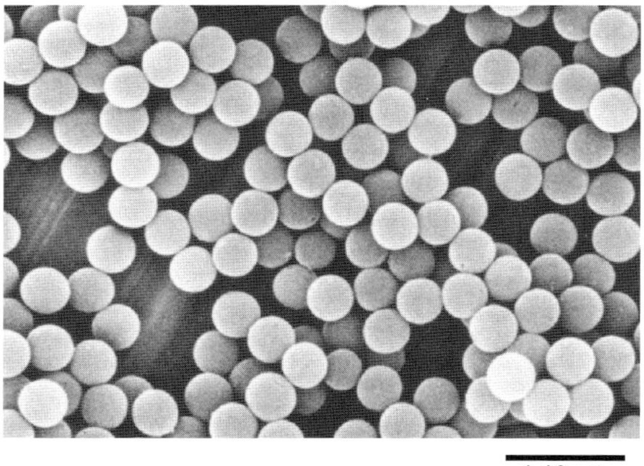

Figure 2.5.16
SEM image of mono-dispersed mesoporous silica particles.

References

[1] C.T. Kresge, M.E. Leonowicz, W.J. Roth, J.C. Vartuli and J.S. Beck: *Nature*, **359**, 710–712 (1992). J.S. Beck, J.C. Vartuli, W.J. Roth, M.E. Leonowicz, C.T. Kresge, K.D. Schmitt, C.T.W. Chu, D.H. Olson, E.W. Sheppard, S.B. McCullen, J.B. Higgins and J.L. Schlenker: *J. Am. Chem. Soc.*, **114**, 10834–10843 (1992).

[2] S. Inagaki, Y. Fukushima and K. Kuroda: *J. Chem. Soc., Chem. Commun.*, Vol. 8, 680–682 (1993). S. Inagaki, A. Koiwai, N. Suzuki, Y. Fukushima and K. Kuroda: *Bull. Chem. Soc. Jpn.*, **69**, 1449–1457 (1996).

[3] S. Inagaki: *J. Soc. Powder Technol., Jpn.*, **33**, 868–874 (1996).

[4] Q. Huo, R. Leon, P.M. Petroff and G.D. Stucky: *Science*, **268**, 1324–1327 (1995).

[5] T. Kimura, T. Kamata, M. Fuziwara, Y. Takano, M. Kaneda, Y. Sakamoto, O. Terasaki, Y. Sugahara and K. Kuroda: *Angew. Chem. Int. Ed.*, **39**, 3855–3859 (2000).

[6] D. Zhao, Q. Huo, J. Feng, B.F. Chmelka and G.D. Stucky: *J. Am. Chem. Soc.*, **120**, 6024–6036 (1998).

[7] A. Sayari, P. Liu, M. Kruk, M. Jaroniec: *Chem. Mater.*, **9**, 2499–2506 (1997).

[8] T. Sun, J. Ying: *Nature*, **389**, 704–706 (1997).

[9] P. Yang, D. Zhao, D.I. Margolese, B.F. Chmelka and G.D. Stucky: *Nature*, **396**, 152–155 (1998).

[10] G.S. Attard, P.N. Bartlett, N.R.B. Coleman, J.M. Elliott, J.R. Owen and J.H. Wang: *Science*, **278**, 838–840 (1997).

[11] R. Ryoo, S.H. Joo, M. Kruk and M. Jaroniec: *Adv. Mater.*, **13**, 677–681 (2001).

[12] S. Inagaki: *J. Soc. Powder Technol., Jpn.*, **39**, 518–526 (2002).

[13] A. Monnier, F. Schuth, Q. Huo, D. Kumar, D. Margolese, R.S. Maxwell, G.D. Stucky, M. Krishnamurty, P. Petroff, A. Firouzi, M. Janicke and B.F. Chmelka: *Science*, **261**, 1299–1303 (1993).

[14] M. Grun, I. Lauer and K.K. Unger: *Adv. Mater.*, **9**, 254–257 (1997).

[15] K. Yano, Y. Fukushima: *J. Mater. Chem.*, **13**, 2577–2581 (2003).

[16] K. Yano, Y. Fukushima: *J. Mater. Chem.*, **14**, 1579–1584 (2004).
[17] N. Shimizu, M. Ogawa: *Bull. Chem. Soc., Jpn.*, **78**, 1154–1159 (2005).
[18] Q. Huo, J. Feng, F. Schuth and G.D. Stucky: *Chem. Mater.*, **9**, 14–17 (1997).
[19] P.J. Bruinsma, A.Y. Kim, J. Liu and S. Baskaran: *Chem. Mater.*, **9**, 2507–2512 (1997).
[20] T. Martin, A. Galarneau, F. D. Renzo, F. Fajura and D. Plee: *Angew. Chem. Int. Ed.*, **41**, 2590–2592 (2002).
[21] M. Vallet-Regi, A. Ramila, R.P. del Real and J. Perez-Pariente: *Chem. Mater.*, **13**, 308–311 (2001).
[22] Y.-J. Han, G.D. Stucky and A. Butler: *J. Am. Chem. Soc.*, **121**, 9897–9898 (1999).
[23] N.K. Mal, M. Fujiwara and Y. Tanaka: *Nature*, **421**, 350–353 (2003).
[24] C.-Y. Lai, B.G. Trewyn, D.M. Jeftinija, K. Jeftinija, S. Xu, S. Jeftinija and V.S.-Y. Lin: *J. Am. Chem. Soc.*, **125**, 4451–4459 (2003).

2.6 Nanoparticle design for DDS

2.6.1 Drug delivery with nanoparticle

Drug Delivery System (DDS) means a system by which the suitable amount of drug is delivered to the suitable regions of body when it is needed. Strictly, it is the targeting of drug, taken into body, to the specific region. There are two methods of targeting: one is the active targeting that utilizes the specific affinity (e.g. antibody) in human bodies, the other is passive targeting that is completed by extending the blood-circulating time of drug intravenously administered and increasing the efficiency of drug accumulation in the inflammatory region or tumor cells.

On the other hand, in considering the concept of DDS as an ideal formulation of drug (DDS in a broad view), it is necessary to focus on delivery of drug into bodies by selecting the most suitable administration route. As a lot of water insoluble drugs have been developed in recent years, it is an important task to find the new administration methods in order to get the certain efficacy of drugs. Although some drugs are administered restrictedly as injections, it is preferable for the patient to select the lower invasive routes such as an oral administration. It will lead to the good patient compliance. The demand on development for this type of dosage forms has been increased. Figure 2.6.1 shows such a concept of DDS as a diagram. In any aspect of drug delivering, nanoparticles are expected to play an important role to complete it.

To design the dosage form by using nanoparticles, it is needed to understand its roles and behaviors precisely. For example, when drugs are enclosed in the fine particles and administered into vein, the particle diameter must be in the submicron size, around 100 nm. The particulate drug carriers might plug the extremely narrow part of blood tube. The particles that are administered externally are recognized as nonselves (foreign material, extraneous substance) in

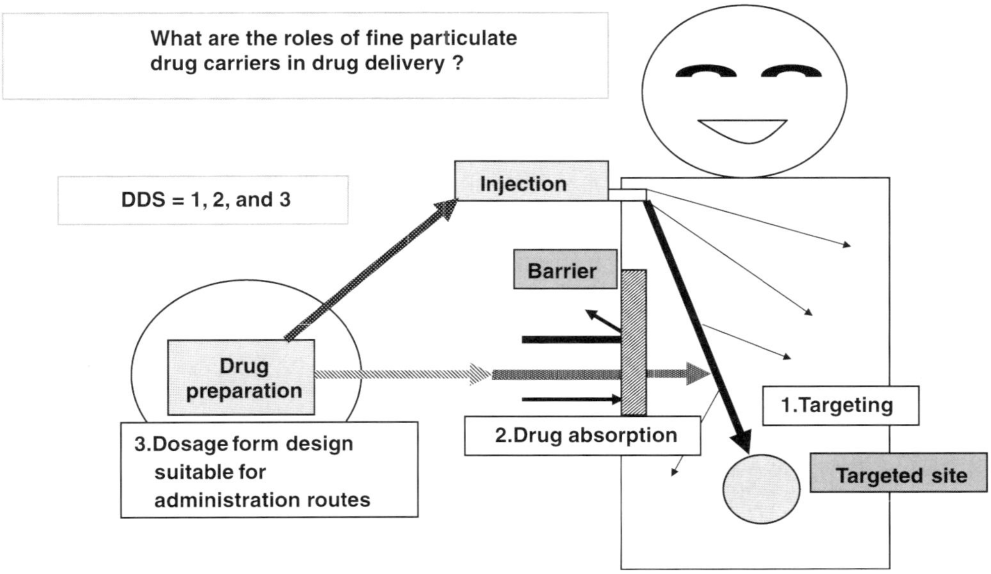

Figure 2.6.1
Design concept for drug delivery systems.

the body, whenever the particles are fine enough. They are easily captured by the reticuloendothelial system (RES) in liver and spleen. So, it is very difficult to increase their circulating time in the blood and it is one of the most serious subjects for the research to escape out (avoid, bypass, circumbent) RES uptake. When liposome, which is one of the typical fine particles, is prepared with combination of polyethylene glycol-conjugated phospholipids and the usual phospholipids, its blood circulation time is prolonged by decreasing RES uptake. This liposome is called a stealth liposome after a stealth fighter plane flying away from the radar [1].

The drug administration method via absorption through a mucous membrane such as gastrointestinal mucosa, nasal mucosa, and lung mucosa, has received a great attention as an alternative method of injection, as well as trans-dermal systems. In comparison with the absorption trans-dermal system, the absorption from the mucosa is very efficient, because the surface area of mucosa is about two hundred times larger than those of epidermia. However, some drugs cannot be absorbed from mucosa owing to their own characteristics: the high molecular weight and hydrophilic or hydrophobic property. Peptide drugs are the typical ones, which have these inconvenient characteristics and require the new dosage forms instead of injection. These are also susceptible to enzymatic degradation on the mucosa. Fine particles are expected as a carrier to accomplish the drug delivery. For effective drug delivery, the choice of administration methods and the control of the drug releasing properties with these particles are very important. The suitable designs and selection on their particle size, the type of material, and particle structure are required for the effective drug delivery.

There are many aspects for designing optimum formulation and preparation with nanoparticle systems. In this section, the preparation methods and efficacy of fine particles in drug delivery are shown, including surface modification of nanoparticles for effective drug delivery.

2.6.2 Design of nano drug carrier

A drug carrier in submicron size is called a colloidal drug delivery system [2]. Typical colloidal drug particles are categorized into the phosphor-lipid types (liposome and liquid emulsion) and the polymeric types (polymer nanosphere, nanocapsule, and polymer micelle). Figure 2.6.2 shows schematic diagrams of the typical nano drug carriers.

(1) Nanoparticles with phospholipids

Phosphor-lipid is one of cell component, and liposome is a finer particle with a double-membrane structure of phospholipids. It is usually prepared by the thin lipid layer hydration method. In this method, phospholipids' thin layer is formed at the bottom of flask by removing organic solvent from the solvent dissolving phosphor-lipid, and then the layer is hydrated with a suitable buffer solution, in order to obtain phospholipids vesicles. There are some other methods: the heating method to change dispersed phospholipids into liposome and the reverse layer solvent-evaporating method (REV method). The details are not here, but they are available in the special technical books [3].

Liposome particle prepared in the thin lipid layer hydration method has a wide particle size distribution and its average particle size is usually a few micrometers. This particle is relatively soft, so during preparation it can be sized down into the submicron range by the ultrasonic wave method or the extruder method. The probe-type ultrasonic apparatus is usually used for this purpose. In the extruder method, liposome suspension is passed through a polycarbonate filter with objective-sized holes with a high pressure. It is possible to get particles with intended size, like 100, 200, 400 nm, and so on.

There are some commercial products of liposomal systems such as Doxil (adriamycin), DaunoXome (daunorubicin), and AmBisome (amphoterycin). It has been reported that these show a good accumulation onto the disease regions, one of the targeting formulations. Among them Doxil is the stealth liposome with

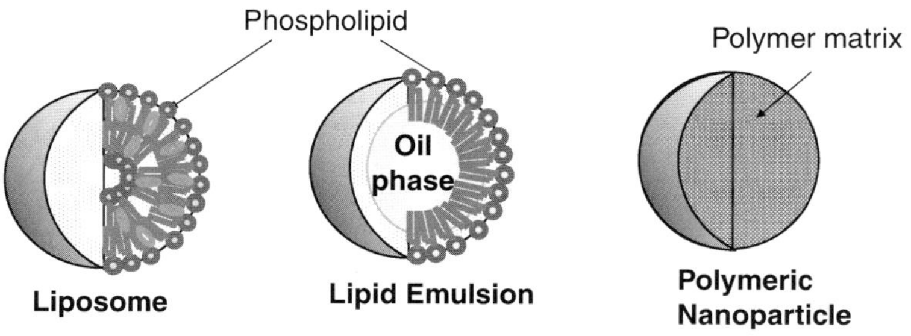

Figure 2.6.2
Various types of nanosized drug carriers.

a surface modification with polyethylene glycol. So far they are approved only in the Western countries.

One of other nanoparticles of phospholipid-type is a lipid emulsion. This is also a colloidal dispersed system, with about 200 nm particles of soyabean oil emulsified with phospholipid, shown in Figure 2.6.2. To prepare the colloidal-dispersed system, a high-pressure homogenizer is usually used. The submicron-sized particles having narrow particle size distribution are prepared with controlling the pressure in the system, as a result of collisions between particle/particle and between particle/wall. Mantongohlin homogenizer is a typical apparatus used for a long time and Microfluidizer and Nanomizer are also available.

As mentioned above, this lipid-emulsion formulation has been used as an injection satisfactorily to supply nutrition. Recognizing the safety of this formulation, DDS formulation, which delivers drug to the aimed region by enclosing it in the oily area of core, has been developed in Japan (lipo-formulation). As the drug is enclosed in the oily part of particle, it is limited to the hydrophobic one. These formulations used currently are prostaglandin (PGE_1), dexamethasone, and flurbiprophen. The latter two formulations are derived into dexamethasone palmitate and then into flurbiprophen akycetyl to increase its hydrophobicity. It has been reported that these preparations show good collectiveness to the disease regions.

(2) Polymeric nanoparticles

Polymeric particles are more rigid than the phospholipid particles. Thus, size control of polymeric nanoparticles is not easier than in the case of phospholipid nanoparticles. The preparation method of polymeric nanoparticles is classified into two types by the starting material: monomers or polymers, and there are the individual preparation methods in each case. In common with other particles, it is important to enclose a drug to form the capsule of drug efficiently.

Polyalkylcyanoacrylate nanoparticle is one of the typical polymeric nanoparticles, prepared from monomers with simultaneous polymerization. It uses the high reactivity of alkylcyanoacrylate, a monomer, in the existence of water. Since Couvreur et al. [4] first reported this method, this type of polymeric nanoparticles has been investigated and developed mainly in Europe. In the typical procedure of this method, alkylcyanoacrylate monomers are added into hydrochloric acid solution containing a surface-active agent (surfactant) under mixing. On the other hand, Al Khouri-Fallouh et al. [5] showed that nanocapsules, enclosing drugs perfectly, were produced from monomers (isobutylcyanoacrylate) by applying the interfacial polymerization technique. It was found that water-soluble substances can be enclosed by applying the same interfacial polymerization and by using aqueous drug solution as the dispersed phase in the system [6].

The technique to prepare nanospheres using polymers as a starting material is also important. As those polymers are usually water-insoluble, it is needed to dissolve them in an organic solvent to make emulsion and then to evaporate the solvent off to form nanospheres in the system. These methods are characterized by their evaporation techniques of solvent (Table 2.6.1). One of the nanosphere preparation methods in which polylacticacid-glycolicacid copolymer (PLGA), a typical biodegradable polymer, is used has been investigated in the author's laboratory. Two techniques were developed: the emulsion solvent diffusion method and the phase separation in oil method.

The two types of emulsion solvent diffusion methods: in water [7] and in oil [8] have been developed. In the former method, PLGA is dissolved into the admixture of acetone, which is miscible with water, and dichloromethane, which is immiscible with water. When it is added to the Polyvinylalcohol (PVA) solution, acetone can diffuse into the water phase rapidly and form the submicron-sized polymer droplet in the water phase. The following evaporation of the organic solvent in the polymer droplet at reduced pressure leads to formation of suspension of polymeric nanospheres. In the latter one, oil was used as the continuous phase of the emulsion based on the same principle. The rate of enclosing water-soluble materials in this method improves 10 times more than that in water system. In addition, to improve the enclosing capacity, the phase separation in oil method [9] and the modified phase separation method with combination of phase change in the emulsion system [10] have been developed. The latter method can improve the encapsulation rate of water-soluble material of insulin and also can decrease the burst release phenomenon at the initial stage of release test.

To prepare the suitable polymeric nanoparticles for drug delivery, the suitable choice of the polymeric materials of the particle and techniques by considering the properties of the drug entrapped is required.

Table 2.6.1
Preparation methods of polymeric nanosphere.

a)	Evaporation method: Evaporation of the good solvent for the polymer under a reduced pressure
b)	Good solvent extraction method: Extraction of good solvent for the polymer by adding the good solvent to the continuous phase
c)	Poor solvent adding method: Adding a poor solvent, which is miscible with the good solvent, for precipitation of polymer
d)	Solvent diffusion method (in water and oil): Spontaneous diffusion of the good solvent to the continuous phase
e)	Phase separation method in oil: Phase separation in the good solvent diffusion into the continuous phase

2.6.3 Design of nanoparticle surface and application for DDS

Surface properties of fine particles affect its stability and pharmaceutical functions. When relatively soft particle like a liposome is mixed with water-soluble polymer solution having a hydrophobic moiety just like cholesterol, its surface can be modified with the polymer by penetrating the hydrophobic part of polymer molecule into the lipid double layer of the liposome (anchoring) [11]. Anchoring method is available only for layers with high-fluid particle surface like phospholipid layer.

For designing the novel dosage form, surface modification of liposomes was investigated in the author's laboratory. Polyvinylalcohol (PVA), which is popular as a protective colloid, was chosen as the coating polymer; long alkyl chain was chemically bonded at the end of the polymer molecule (PVA-R). The surface of liposomes was coated with PVA and PVA-R. As the passive targeting of drug entrapped in the polymer-coated liposomes was expected, the blood circulation was evaluated in rats. As a result, much longer circulation was observed for PVA-R-coated liposomes, while PVA-coated liposomes by physical adsorption could cause little effect. It confirmed that the thick coating layer with PVA-R is effective in prolonging the circulation time in blood after injection. The longer circulation of PVA-R-coated liposomes than non-coated and PVA-coated liposomes was attributed to the lower RES uptake [12, 13]. In comparison to PEGylated liposome, PVA-R-coated liposome could give the same effect even with less amount of coating because of its high molecular weight [14].

To identify the effectiveness of liposome coating with polymers in drug delivery, the liposome enclosing doxorubicin was applied with injection for tumor-bearing rats. The circulation time in the blood of those animals also increased significantly with the polymer-coated liposome, and it lead to the increase in the amount of drug accumulated in the tumor tissue (Fig. 2.6.3) [15, 16]. In addition, the passive targeting effect of liposome coated with polymers was also confirmed by significant inhibition in tumor cell growth on the tumor-bearing mice [17].

As another DDS dosage form utilizing surface modification, mucoadhesive liposomes were developed in the author's laboratory [18, 19]. It was expected that the adhesive properties of polymer-coated liposomes containing drug lead to prolonging the residence time in the intestinal tract for effective drug absorption, and so insulin was encapsulated into the liposome coated with chitosan, which is the most popular adhesive polymer to mucous membrane; it was administered orally into rats. The pharmacological test (the measurement of blood sugar) showed the continuous decrease of blood sugar value for 12 h. Liposome without coating could show no pharmacological effect for hours. These results confirmed us the effectiveness of mucoadhesive property of chitosan-coated liposomes in vivo. Liposome used was multi-lamellar vehicles (MLV), and its average particle diameter was a few micrometers. So, in this case, it was estimated that the liposome coated with chitosan released the drug during the residence time on or within the mucous layer in the intestinal tract. It leads to increase in drug concentration there and following absorption of drug. When calcitonin was used as another model peptide drug to be enclosed, liposomes coated with chitosan showed much higher pharmacological effect (the decrease of calcium concentration in blood) than non-coated liposomes [20, 21]. Decreasing diameter of liposome coated with chitosan into submicron size, the pharmacological effect could be prolonged much more [22, 23]. After the administration of liposome coated with chitosan in submicron size to rats, their intestinal tubes were observed with scanning cofocal microscopy. It could be observed that liposome particles penetrated into the enteric mucous layer and then into the mucous layer.

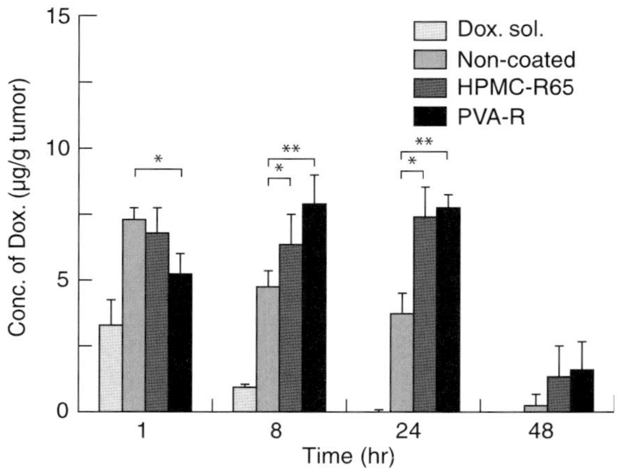

Figure 2.6.3
Drug concentration in tumor after i.v. administration of liposomal doxorubicin in the tumor bearing rats (Walker rat carcinoma); (EPC:Chol = 5:5, pore size = 100 nm), $^{*}p < 0.05$, $^{**}p < 0.01$: significantly different.

Modifying the surface of fine particle with chitosan is coincident with the formation of ion complex on the particle surface. So, it was applied to the polymeric nanoparticles. Using PLGA nanosphere coated with chitosan, its effectiveness was confirmed in calcitonin absorption in oral administration and the drug absorption through lung mucous membrane [24, 25]. It was also observed that PLGA nanospheres coated with chitosan have longer residence time in lung after inhalation than PLGA nanosphere without surface modifying. Moreover, chitosan adsorbing on the surface of nanosphere opened the lung epithelial cell gap (tight junction), and accelerated the absorption of drug released from PLGA.

2.6.4 Pharmaceutical nanotechnology

Generally, the definition of nanotechnology means the optional control of molecules and atoms, finer than nanometer size, and the control of their conformation in nanoorder. So far, the pharmaceutical industry has oriented to the manufacture in molecule level. Historically, we can say that the base of pharmaceutical research is nanotechnology. In a dosage-form design of pharmaceutical molecules, the importance of viewpoint in nanolevel has been recognized in plenty, and many researches are continued broadly. One of them is the investigation about the drug-carrier particles in submicron size, mentioned in this section.

In historical point of view, the researchers have studied on liposomes and biodegradable polymeric nanoparticle before the term of "nanotechnology" began to be used frequently. However, since nanotechnology age has come, the observation of particles and the measurement of particle size have become much easier by using atomic force microscope, dynamic light-scattering method, and so on. Evolution of nanotechnology may accelerate the pharmaceutical design of dosage form and DDS investigation, and they will keep the position at the center of the "nanotechnology".

References

[1] D. Papahadjopoulos: *Stealth Liposomes*, Lasic D., Martin, F. (eds), CRC Press, pp. 1–7 (1995).

[2] K. Kreuter: *Colloidal Drug Delivery Systems* K. Kreuter (ed.), Marcell Dekker, (1994).

[3] S. Nojima et al. (eds): *Liposome,* Nankodo, (1988).

[4] P. Couvreur, B. Kante, M. Roland, P. Guiot, P. Bauduin and P. Speiser: *J. Pharm. Pharmacol.*, **31**, 331–332 (1979).

[5] N. Al Khouri Falloug , L. Roblot-Treupel, H. Fessi, J. Ph Devissaguet and F. Puisieux: *Int. J. Pharm.*, **28**, 125–132 (1986).

[6] M.S. EL-Samaligy, P. Rohdewald and H.A. Mahmoud: *J. Pharm. Pharmacol.*, **38**, 216–218 (1986).

[7] T. Niwa, H. Takeuchi, T. Hino, N. Kunou and Y. Kawashima: *J. Controlled Release*, **25**, 89–98 (1993).

[8] Y. Kawashima, H. Yamamoto, H. Takeuchi, T. Hino and T. Niwa: *Eur. J. Pharm.*, **45**, 41–48 (1998).

[9] T. Niwa, H. Takeuchi, T. Hino, T. Nohara and Y. Kawashima: *Int. J. Pharm.*, **121**, 45–54 (1995).

[10] H. Hanada, Y. Kawashima, H. Takeuchi, T. Hino and H. Yamamoto: Proceedings of 15th Symposium on Particle Design and Preparation, 119–124 (1998).

[11] H. Takeuchi, H. Yamamoto, A. Toyoda, H. Toyobuku, T. Hino and Y. Kawashima: *Int. J. Phrarm.*, **164**, 103–110 (1998).

[12] H. Takeuchi, H. Kojima, A. Toyoda, H. Yamamoto, T. Hino and Y. Kawashima: *European J. Pharm. Baiopharm.*, **48**, 123–129 (1999).

[13] H. Takeuchi, H. Kojima, H. Yamamoto and Y. Kawashima: *J. Control. Rel.*, **68**, 195–205 (2000).

[14] H. Takeuchi, H. Kojima, H. Yamamoto and Y. Kawashima: *J. Control. Rel.*, **75**, 83–91 (2001).

[15] H. Takeuchi, H. Kojima, H. Yamamoto and Y. Kawashima: *Pharm. Pharmacol. Commun.*, **5**, 445–448 (1999).

[16] H. Takeuchi, H. Kojima, Yamamoto and Y. Kawashima: *Biol. Pharm. Bull.*, **24**, 795–799 (2001).

[17] H. Takeuchi, H. Kojima, H. Yamamoto, Y. Kawashima: *S.T.P. Pharm. Sci.*, **11**, 271–274 (2001).

[18] H. Takeuchi, H. Yamamoto, T. Niwa, T. Hino and Y. Kawashima: *Chem. Pharm. Bull.*, **42**, 1954–1957 (1994).

[19] H. Takeuchi, H. Yamamoto, T. Niwa, T. Hino, Y. Kawashima: *Pharm. Res.*, **13**, 896–901 (1996).

[20] H. Takeuchi, Y. Matsui, H. Yamamoto and Y. Kawashima: *J. Control. Rel.*, **86**, 235–242 (2003).

[21] H. Takeuchi, H. Yamamoto and Y. Kawashima: *Adv. Drug Delivery Rev.*, **47**, 39–54 (2001).

[22] H. Takeuchi, Y. Matsui, H. Sugihara H. Yamamoto and Y. Kawashima: *Int. J. Pharm.*, **303**, 160–170 (2005).

[23] H. Takeuchi: *Pharm. Tech. Jpn.*, **19**, 1227–1239 (2003).

[24] Y. Kawashima, H. Yamamoto, H. Takeuchi and Y. Kuno: *Pharm. Dev. Technol.*, **5**, 77–85 (2000).

[25] H. Yamamoto, Y. Kuno, S. Sugimoto, H. Takeuchi, and Y. Kawashima: submitted to *J. Control. Rel.*, **102**, 373–381 (2005).

2.7 Nanotubes (CNT)

Nanotube means an ultrathin tube of diameter in nanometer size, whose constituent elements are carbon, boron, silicon, etc. Here, carbon nanotubes (CNT), found first, are described. CNT was found by S. Iijima in 1991 [1], which is closely related to the main title of this handbook "Nanoparticles".

The original sample of the discovery of CNT was obtained as cathode deposit of dc arc discharge evaporation of graphite rod in inert gas [2]. The evaporation method itself is an improvement of mass production method [3] for fullerenes [4] represented by C_{60}. Originally, the evaporation apparatus was developed by the present author to produce ultrafine particles [5] of SiC as a kind of gas evaporation method [6]. As Si is hard to evaporate by resistive heating, arc discharge between Si block and graphite rod was employed. Later, by replacing the Si block with another graphite rod, dc arc discharge between two graphite rods was applied. This resulted in mass production of fullerenes in the carbon soot generated in the chamber. In this process, the anode graphite was fully evaporated leaving a deposit on the cathode graphite. CNT was discovered as a byproduct in that cathode deposit [1, 2].

Table 2.7.1
Main production methods of CNT.

Kind of CNT	Production method	Use of catalyst
SWNT	Arc	Yes
	Laser	Yes
	CVD	Yes
DWNT	Arc	Yes
	CVD	Yes
MWNT	Arc	No
	CVD	Yes
	Oxidation [15] of SiC	No

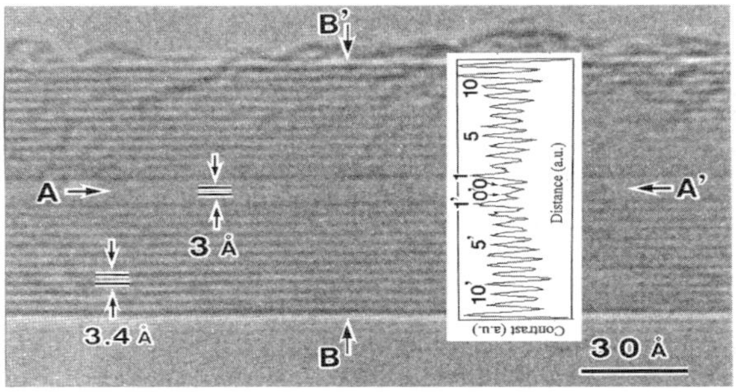

Figure 2.7.1
High resolution TEM image [24] of MWNT including 3 Å tube at the center. Insert is intensity profile of lattice fringes.

Two years after the discovery of CNT, Iijima et al. [7] and IBM's group [8] independently found single-wall carbon nanotubes (SWNT), which is a single roll of graphene sheet with honey-comb lattice. On the other hand, first-found CNT having a Russian dole structure is called multi-walled carbon nanotubes (MWNT). Recently, double-wall carbon nanotubes [9] (DWNT) composed of two graphene sheets are noticed in the field of their application, because DWNT has intermediate characteristics of SWNT and MWNT.

As described above, the first CNT was produced by arc discharge method. Later on, other evaporation sources were also used for CNT production, such as laser source [10] and chemical vapor deposition [11] (CVD). In the case of MWNT production by arc discharge method, catalyst is not necessary, while for SWNT production by arc discharge method the use of catalyst is essential. However, in the case of CVD, CNT cannot be produced without catalyst. In the case of laser evaporation, SWNT can be produced by using appropriate catalyst metal, but no MWNT by any catalyst. Including the case of DWNT, production methods of CNT are summarized in Table 2.7.1. Many books [12] and reviews [13, 14] are published on CNT.

In this section, CNT production by arc discharge method is mainly described. Especially, the case of MWNT and SWNT production is focused.

2.7.1 Production of MWNT by arc discharge method

After a vacuum chamber is filled with an appropriate gas under a pressure of about 10 kPa, arc voltage of dc 20–30 V is applied between the two graphite electrodes 1–2 mm apart. Then, the anode is evaporated by the heat of arc discharge and carbon soot generates in the chamber. In the soot, about 10% fullerenes are included [16]. At the same time, some part of evaporated carbon deposits on the cathode and MWNT is included within and at the surface of the cathode deposit [2, 17].

The atmospheric gas due to arc evaporation, usually inert gas such as He or Ar, is used for the production of fullerenes. However, in the case of CNT production, such as CH_4 gas [18, 19] including hydrogen atom is better than inert gas to prepare high-crystalline MWNT. This is a big difference in the case of fullerene, which cannot be produced in a gas including hydrogen atom. The best gas condition to prepare high-crystalline and densely packed MWNT is found to be pure hydrogen gas [20, 21] at a pressure about 10 kPa. Purification of these MWNTs is also easily done by heat treatment at 500^{0}C in air, because co-existing carbon nanoparticles are very little in hydrogen-arc produced MWNTs [22].

Usually, the smallest diameter [23] of the innermost tube of MWNT produced by hydrogen-arc discharge is only 4 Å. Exceptionally, thinner innermost tube of diameter 3 Å is also found [24], where the tube diameter is smaller than the spacing of graphene layer in graphite. Figure 2.7.1 shows a high-resolution transmission electron microscope (HRTEM) image of such thin innermost tube, 12 parallel symmetric lines of 3.4 Å spacing are seen. At the center, 3-Å diameter tube of weak contrast is seen between A and A′; the distance AA' is about 120 Å. The insert of Fig. 2.7.1 is a densitometer curve along the line BB′, in which a pair of weak contrast peaks 00′ are observed.

2.7.2 Production of SWNT by arc discharge method

The first found SWNT was also produced by an arc discharge method [7, 8], but the quantity was limited. After then, SWNT was produced by laser evaporation [10] and CVD method [25] as well. In arc discharge method, by using two kinds of catalyst metals such as Ni and Y, a large amount of SWNT was produced by Journet et al. [26]. The present authors also improved the method by setting the two electrodes at a sharp angle of 30°, and produced soot [27] including SWNT at a rate 1 g/min. However, the soot included a large quantity of impurities, not only catalyst metals but also amorphous carbon.

Now, a kind of CVD called HiPCo method [28] is used for mass production of SWNTs, by using high-pressure CO gas and Fe catalyst. However, the crystallinity of produced SWNTs is not so high. Recently, Hata et al. [29] made film-like SWNT by CVD method, using a control of moisture in the process. By this method, SWNT of the diameter 2–3 nm and specific surface of 1,000 m^2/g was produced.

On the contrary, SWNTs produced by arc method have high crystallinity, because growth temperature is higher than CVD method. By utilizing the feature, we achieve [30] mass production of SWNTs by arc discharge in hydrogen mixed gas using only Fe catalyst. We call it FH-arc method. When we conduct evaporation by FH-arc method, SWNTs are produced as macroscopic net similar to the production of cotton cake. The produced SWNT nets are just put in the 1-L bottle, and the mass [31] measured is to be only 1 g as shown in Fig. 2.7.2. The green density is only 1 mg/cm^3, but the specific surface area measured by BET method is several 100 m^2/g.

Low-magnification TEM image of as-grown SWNT is shown [30] in Fig. 2.7.3(a). HRTEM of the same specimen is shown in Fig. 2.7.3(b). Catalyst iron particles are observed as dark spots uniformly dispersed on SWNT net in Fig. 2.7.3(a). The size of each iron particle is less than 10 nm, and covered by thin amorphous carbon (2–3 nm thick) as seen in Fig. 2.7.3(b). Purification of this SWNT can be easily done by first thermal removing of amorphous carbon,

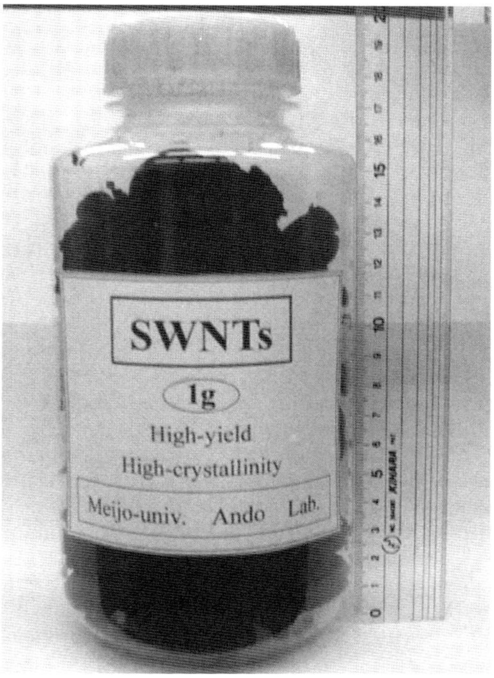

Figure 2.7.2
SWNT produced by FH-arc method bottled in 1-L vessel [31]. Mass is only 1 g.

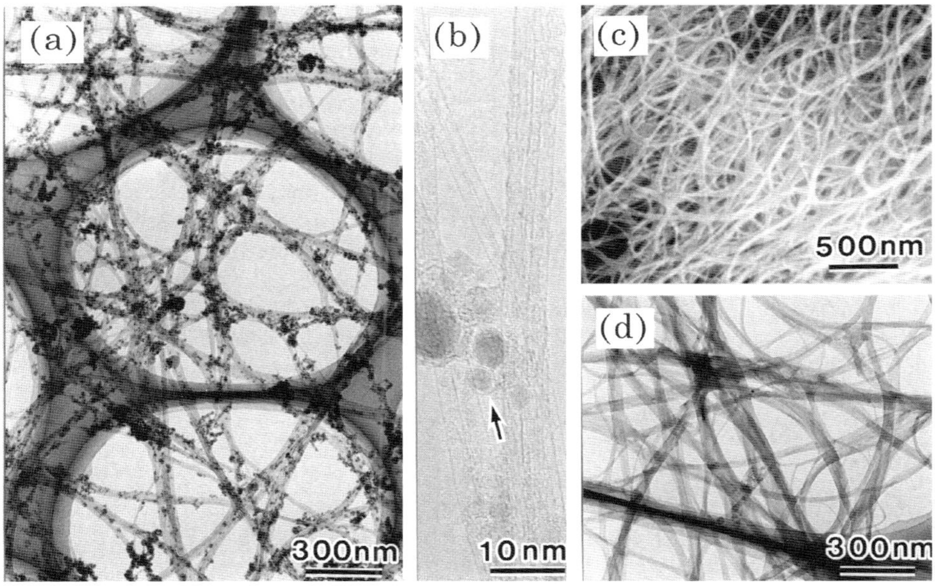

Figure 2.7.3
Electron microscope images of SWNTs produced by FH-arc method [30]. (a) Low-magnification TEM image of as-grown SWNT; (b) High-resolution TEM image of as-grown SWNT; (c) Low-magnification SEM image of purified SWNT; (d) Low-magnification TEM image of purified SWNT.

and then removing of iron by hydrochloric acid [30]. The damages of SWNT themselves are not so serious under these purification process. Scanning electron microscope (SEM) and TEM image of these purified SWNTs are shown in Fig. 2.7.3(c) and (d). It can be seen in their images of catalyst iron particles, they are almost removed. By the analysis of energy dispersive X-ray (EDX), the content of iron reduced from 11 to 0.5% by this purification [31].

Among the three kinds of production methods of CNT, here the arc discharge method is described to prepare MWNT and SWNT. In the case of arc discharge in pure hydrogen gas we can produce high-crystalline MWNTs with very thin innermost tube. In this case, the diameter of MWNT is 10–30 nm, but the length is larger than 100 μm. So-called aspect ratio is larger than 1,000. It is far from "nanoparticles".

On the other hand, in order to produce SWNT, addition of catalyst metal particles is necessary. Diameter of each SWNT is 1–2 nm and the length is longer than 1 μm, but they usually form bundles of a number of SWNTs. In appearance, they seem like powder similar to nanoparticles. By FH-arc method using Fe catalyst and H_2 gas mixture, macroscopic SWNT net of low-green density, 1 mg/cm^3, and high-specific surface area, several 100 m^2/g, can be obtained.

References

[1] S. Iijima: *Nature*, **354**, 56–58 (1991).

[2] Y. Ando, S. Iijima: *Jpn. J. Appl. Phys.*, **32**, L107–L109 (1993).

[3] W. Krätschmer, L.D. Lamb, K. Fostiropoulos and D.R. Huffman: *Nature*, **347**, 354–358 (1990).

[4] H.W. Kroto, J.R. Hearth, S.C.O'Brien, R.F. Curl and R.E. Smalley: *Nature*, **318**, 162–163 (1985).

[5] Y. Ando, M. Ohkohchi: *J. Crystal Growth*, **60**, 147–149 (1982).

[6] R. Uyeda: *J. Crystal Growth*, **45**, 485–489 (1978).

[7] S. Iijima, T. Ichihashi: *Nature*, **363**, 603–605 (1993).

[8] D. S. Bethune, C.H. Kiang, M.S. de Vries, G. Gorman, R. Savoy, J. Vazquez and R. Beyers: *Nature*, **363**, 605–606 (1993).

[9] T. Sugai, H. Yoshida, T. Shimada, T. Okazaki, H. Shinohara and S. Bandow: *Nano Lett.*, **3**, 769–773 (2003).

[10] A. Thess, R. Lee, P. Nikolaev, H. Dai, P. Petit, J. Robert, C. Xu, Y.H. Lee, S.G. Kim, A.G. Rinzler, D.T. Colbert, G.E. Scuseria, D. Tomànek, J.E. Fisher and R.E. Smalley: *Science*, **273**, 483–487 (1996).

[11] M. José-Yacamán, M. Yoshida, L. Rendon and J.G. Santiesteban: *Appl. Phys. Lett.*, **62**, 657–659 (1993).

[12] Y. Saito and S. Bandow: *Basic of Carbon Nanotubes* (In Japanese), Corona Campany (1998), R. Saito, G. Dresselhaus and M.S. Dresselhaus: *Physical Properties of Carbon Nanotubes* Imperial College Press (1998).

[13] T. W. Ebbesen: *Physics Today*, **49**, 26–32 (1996); M. Terrones: *Int. Mater. Rev.*, **49**, 325–377 (2004); Y. Ando, X. Zhao, T. Sugai and M. Kumar: *Mater. Today*, 7, 22–29 (2004).

[14] Y. Ando, X. Zhao: *New Diamond Frontier Carbon Technol.*, **16**, 123–137 (2006).

[15] M. Kusunoki, M. Rokkaku and T. Suzuki: *Appl. Phys. Lett.*, **71**, 2620–2622 (1997).

[16] Y. Saito, M. Inagaki, H. Shinohara, H. Nagashima, M. Ohkohchi and Y. Ando: *Chem. Phys. Lett.*, **200**, 643–648 (1992).

[17] Y. Ando: *Jpn. J. Appl. Phys.*, **32**, L1342–L1345 (1993).

[18] Y. Ando: *Fullerene Sci. & Technol.*, **2**, 173–180 (1994).

[19] M. Wang, X. Zhao, M. Ohkohchi and Y. Ando: *Fullerene Sci. & Technol.*, **4**, 1027–1039 (1996).

[20] Y. Ando, X. Zhao and M. Ohkohchi: *Carbon*, **35**, 153–158 (1997).

[21] X. Zhao, M. Ohkohchi, M. Wang, S. Iijima, T. Ichihashi and Y. Ando: *Carbon*, **35**, 775–781 (1997).

[22] Y. Ando, X. Zhao and M. Ohkohchi: *Jpn. J. Appl. Phys.*, **37**, L61–L63 (1998).

[23] L.C. Qin, X. Zhao, K. Hirahara, Y. Miyamoto, Y. Ando and S. Iijima: *Nature*, **408**, 50 (2000).

[24] X. Zhao, Y. Liu, S. Inoue, T. Suzuki, R.O. Jones and Y. Ando: *Phys. Rev. Lett.*, **92**, 125502/1–4 (2004).

[25] H. Dai, A.G. Rinzler, P. Nikolaev, A. Tess, D.T. Colbert and R.E. Smalley: *Chem. Phys. Lett.*, **260**, 471–475 (1996).

[26] C. Journet, W.K. Maser, P. Bernier, A. Loiseau, M. Lamy de la Chapelle, S. Lefrant, P. Deniard, R. Lee and J.E. Fisher: *Nature* **388**, 756–758 (1997).

[27] Y. Ando, X. Zhao, K. Hirahara, K. Suenaga, S.R. Bandow and S. Iijima: *Chem. Phys. Lett.*, **323**, 580–585 (2000).

[28] P. Nikolaev, M.J. Bronikowski, R.K. Bradley, F. Rohmund, D.T. Colbert, K.A. Smith and R.E. Smalley: *Chem. Phys. Lett.*, **313**, 91–97 (1999).

[29] K. Hata, D.N. Futaba, K. Mizuno, T. Namai, M. Yumura and S. Iijima: *Science*, **306**, 1362–1364 (2004).

[30] X. Zhao, S. Inoue, M. Jinno, T. Suzuki and Y. Ando: *Chem. Phys. Lett.*, **373**, 266–271 (2003).

[31] Y. Ando, X. Zhao, S. Inoue, T. Suzuki, M. Ohkohchi and T. Kadoya: *Trans. Mater. Res. Soc. Jpn.*, **30**, 1193–1198 (2005).

FUNDAMENTALS

CHAPTER 3

CHARACTERISTICS AND BEHAVIOR OF NANOPARTICLES AND ITS DISPERSION SYSTEMS

CONTRIBUTORS

Hidehiro Kamiya (3.1), (3.6.1), (3.6.3), (3.7.2)
Kuniaki Gotoh (3.2.1)
Manabu Shimada (3.2.2.1)
Tetsuo Uchikoshi (3.2.2.2)
Yoshio Otani (3.3)
Masayoshi Fuji (3.4)
Shuji Matsusaka (3.5.1.1)
Tatsushi Matsuyama (3.5.1.2)
Junichi Tatami (3.5.1.3)
Ko Higashitani (3.5.2)
Kazue Kurihara (3.5.3.1)
Naoyuki Ishida (3.5.3.2)
Michitaka Suzuki (3.5.3.3)
Hiroya Abe (3.6.2)
Yasufumi Otsubo (3.7.1)
Minoru Miyahara (3.8)

CHAPTER 3 CHARACTERISTICS AND BEHAVIOR OF NANOPARTICLES AND ITS DISPERSION SYSTEMS

3.1 Introduction of nanoparticle dispersion and aggregation behavior

Recently, the technological innovation of low-cost and large-scale synthesis process of nanoparticles with less than 100 nm in diameter is developing, and various kinds of nanoparticles are applied as the raw materials in the different fields, for example, cosmetic, medical supplies, catalysts, pigments, toner and ink. Since the commercial products in such fields are colloidal suspension with relatively low solid content or powdery condition, it is rather easy to apply nanoparticles for commercial product. On the contrary, for the application of functional properties of nanoparticles to new materials, it is difficult to produce the final commercial products. Nanoparticles are expected to be raw materials for many kinds of new functional materials, for example nanodevice with high-density circuit and wiring in use of nanoparticles, two-dimensional high-density nanodot memory element, and hybrid composite materials with high-density packed nanoparticles. For the commercial implementation of such materials, it needs to develop many kinds of nanoparticle processing. For nanometer-scaled devises, it is necessary to develop nanoparticle dispersion in suspension without aggregation, nanoparticles assemble on the circuit, drying and sintering technology to cause neither disconnection nor deformation.

The aggregation/dispersion behavior control, which is the first process for the preparation of new functional materials using nanoparticles, is very difficult for nanoparticles with less than 100 nm in diameter. In this chapter, based on the introduction of the basic reason why the aggregation and dispersion behavior control of nanoparticles is very difficult, the basic concept and method for the aggregation and dispersion control will be reviewed.

3.1.1. Surface interaction between nanoparticles

Since dispersed nanoparticles form aggregates in gas phase with the progress of the time, the dispersion and aggregation behavior of nanoparticles is generally controlled in liquid phase. The main mechanisms of surface interaction between particles in liquid phase are summarized in Table 3.1.1. The first and second items in this table, van der Waals force and electrostatic force by the overlap of electric double layer had been systematized in famous DLVO theory [1]. If it is possible to control the aggregation and dispersion behavior of nanoparticles in liquid by the adjustment of electrostatic interaction, we can design the dispersion behavior by using DLVO theory. However, non-DLVO type interaction, steric, bridging and depletion force, is generally important to control dispersion behavior of nanoparticles. The fundamentals of dispersion control in liquid are introduced based on DLVO theory.

3.1.2. Difficulty in nanoparticle dispersion control based on DLVO theory

DLVO theory is based on the van der Waals and electrostatic interaction by the overlap of electric double layer which is generated by the counter ions concentrated at the surface of particles by the surface charge. By using this theory, it is possible to discuss the dispersion and aggregation behavior of particles in aqueous suspension.

One example of potential curve of surface interaction calculated by this theory is shown in Fig 3.1.1. Surface charge on particles and counter ion concentration in

Table 3.1.1
Examples of surface interaction between particles in liquid phase.

Surface interaction	Generation mechanism
van der Waals interaction	Short-ranged electromagnetic force between molecule and/or atoms, which has neutral charge only.
Overlap of electric double layer	Electrical interaction by the overlap of electric double layer around particle in solution
Steric interaction of adsorbed polymer	Short-ranged interaction by the overlap of adsorbed polymer layer on particles
Bridge force	Formation of the bridge of polymer binder and/or surfactant between particles
Hydration force	Overlap of hydrogen-bonded water molecule on hydrophilic surface on particle
Depletion	Negative adsorption of solute and polymer by having less affinity for the surface than the solvent

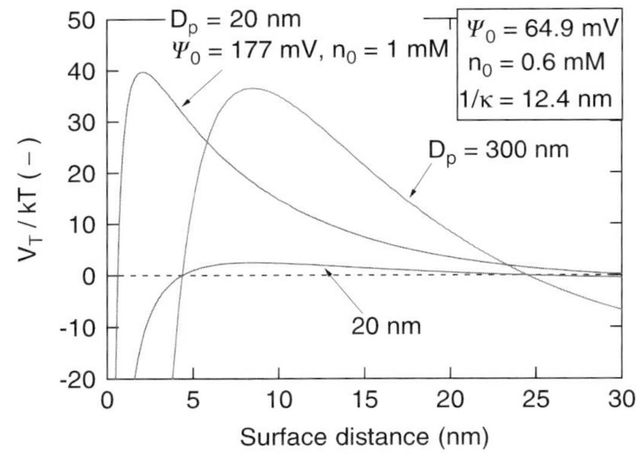

Figure 3.1.1
Examples of calculated potential curve based on DLVO theory, effect of particle diameter and surface charge.

suspension were assumed to be 64.9 mV and 0.6 mM for the calculation, respectively. Since the thickness of double layer, $1/\kappa$, was estimated as 12.4 nm from the counter ion concentration and it needs that the particle diameter should be larger than the thickness of double layer, the minimum particle diameter was selected to be 20 nm. One important conclusion from this calculation is to know if the particle can be dispersed or not. If the peak value of calculated potential curve, V_{Tmax}, is 10–20 times larger than the product of absolute temperature (T) and Boltzmann constant (k), particles can be dispersed in suspension. Low maximum value of surface potential ($V_{Tmax} < 10$ kT) promotes the particle aggregation.

If particle size is larger than 100 nm in diameter, particles are able to disperse in suspension at the surface charge and counter ion concentration of 64.9 mV and 0.6 mM, respectively. For nanoparticles less than 100 nm, the maximum value at the same surface charge and ion concentration was much lower than 10 kT. In order to increase the maximum potential up to 15 kT and disperse nanoparticles, it is necessary to increase surface charge up to one hundred and several tens mV as shown in this figure. If anionic or cationic surfactant is used, since the density of adsorbed surfactant has a limit value, it is difficult to obtain such high surface charge. For such high surface charge it is necessary to develop special method and technology, for example, electron hole presentation by addition of small amount of Al on silica surface.

3.1.3. Difficulty in nanoparticle dispersion, discussion based on the effect of particle diameter and solid fraction on distance between particle surface

On the basis of the DLVO theory, the aggregation and dispersion behavior control of nanoparticles in suspension was able to be discussed by the potential curve as shown in Fig. 3.1.1. It also shows another important information concerning nanoparticle dispersion. The maximum repulsive potential appears at the distance of about several nanometers between the surfaces. If mean surface distance between particles is less than several nanometers, the aggregation of nanoparticles cannot be prevented by surface repulsive potential of electrostatic double layer. The mean surface distance, h_{susp}, between particles is given in the function of solid fraction, F, and particle diameter, D_p, and estimated by the following Woodcock's equation [2]:

$$h_{susp} = D_p[\{1/(3\pi F) + 5/6\}^{0.5}] \tag{3.1.1}$$

If the central position of the particle is placed at the hexagonal closest packing structure, the relationship between surface distance, solid fraction, and particle diameter is estimated by the following equation:

$$h_{susp} = D_p[\{\pi/(3 \times 2^{0.5})\}^{1/3} - 1] \tag{3.1.2}$$

By using the above equations, the estimated surface distance was calculated and shown in Fig. 3.1.2. For submicron particle with 300 nm in diameter, at the solid fraction higher than 40–50 vol%, the estimated surface distance reached to several nonometers where the maximum repulsive potential appeared in Fig. 3.1.1. For the dispersion of particles at such high solid fraction, it is difficult to disperse by only the electrostatic repulsive force, and it needs to generate other repulsive interaction, for example, steric force. On the contrary, for 20 nm in diameter, the surface distance reached to several nanometers at relatively

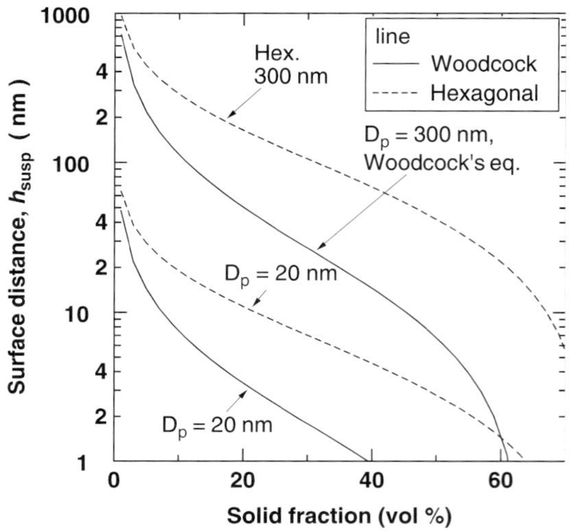

Figure 3.1.2
Effect of solid fraction and particle diameter on estimated surface distance between particles.

low solid fraction about 20–30 vol%. If solid fraction is not so high, less than 10%, it is possible to control aggregation and dispersion behavior by using DLVO type interaction, however, with increase of solid fraction in suspension, the additional repulsive interaction is needed to disperse nanoparticles.

In order to produce the steric repulsive force for the dispersion of particles at relatively high solid fraction condition, surface modification by the adsorption of surfactant or reaction with silane coupling agent is generally used. Special consideration on such technique is necessary for the nanoparticles. For example, the molecular weight of polymer dispersant is generally recommended at about 10,000 g/mol for the dispersion of fine powder with submicronmeter in diameter. Since the size of dispersants with such molecular weight is ranging from several angstroms to several nanometers in liquid, it is difficult for them to enter between the particle surfaces as estimated in Fig. 3.1.2. Furthermore, since such large dispersants have almost equivalent size to the nanoparticles, polymer dispersants sometimes form bridge between particles, and promote the aggregation of particles and consequently suspension viscosity increases. The detail examples will be introduced in Section 3.7.2.

3.1.4. Surface molecular-level structure of nanoparticles [3]

Surface molecular-level structure also changes with decrease of particle size down to nanometer scale. For example, Stöber's process is a very popular method to prepare uniform spherical alkoxide-derived silica particles. The particle diameter of the silica was controlled by the concentration of the catalyst. Fig. 3.1.3 shows the influence of silica particle diameter on FT–IR spectra in the fundamental absorption region from wavenumber 2,000 to 4,000 cm^{-1}. Absorption bands of free and hydrogen-bonded silanol groups are present. The free silanol band at wavenumber 3,750 cm^{-1} decreased with increasing particle diameter, almost disappearing for the 260-nm particle. In contrast, the ratio of adsorption of the hydrogen-bonded silanol band at wavenumber 3,660 cm^{-1} to that of the free group increased with increasing particle diameter. This tendency was recognized also in the silica particles prepared by other methods.

Changes of the silanol group structure with decreasing particle diameter seem to relate to the distance between neighboring silanol groups. Increasing the distance between neighboring silanol groups promoted the formation of free silanols. The calculated effect of curvature on the distance between neighboring silanol groups is shown in Fig. 3.1.4. The bonding angle of silanol groups was fixed at 104 degrees, and the distance between neighboring Si atoms was calculated from curvature and the mean density of amorphous silica (2.2 g/cm^3). The calculated distance

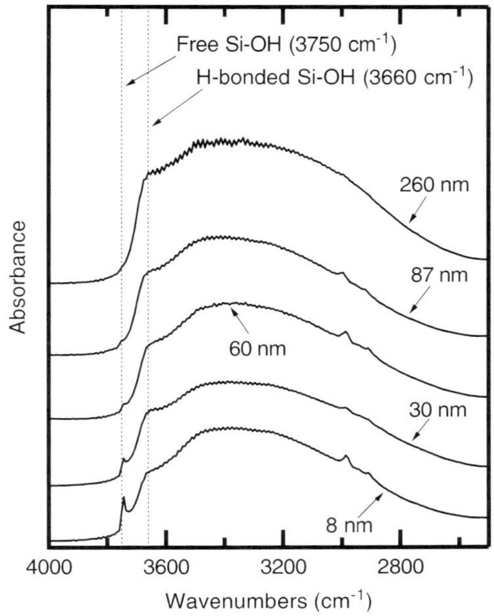

Figure 3.1.3
Effect of particle diameter on FT-IR spectrum (prepared by Stöber method).

between O and H atoms in neighboring Si–OH group increased from 3.0 to 3.17 Å as particle size decreased from several hundreds nanometer to 8 nm. If the distance between O and H atoms changed by 0.2 Å, the potential energy would have decreased about 20 %, as calculated from an ab initio study [4] of the interaction of molecular hydrogen with isolated silica. Furthermore, in the classical Lennard–Jones potential between molecules, the attractive van der Waals potential varies with inverse-sixth power of the distance. The effect of the distance between O and H on the attractive potential was calculated from the change of mean distance with particle diameter by using Lennard–Jones potential and shown in Fig. 3.1.4. The attractive potential of each particle, E, has been made dimensionless in the energy in case of the plane, E_0. For example, the attractive potential between O and H atoms decreased about 28% as a result of calculation of $(3.17/3.0)^{-6} = 0.714$ based on the L.-J. potential as particle size decreased from several hundreds nanometers to 8 nm. Such a decrease in potential energy, however, is not sufficient to explain the increase in isolated silanol structures. The additional mechanism was the formation of a relatively loose structure on the surface of the ultrafine silica powder. The density of amorphous silica is about 2.2 g /cm^3, however, the density of ultra-fine silica powders had a lower value [3].

On the basis of the above research, for relatively large particles, since the hydrogen bond between O–H in the silanol promoted polarization in the other O–H group,

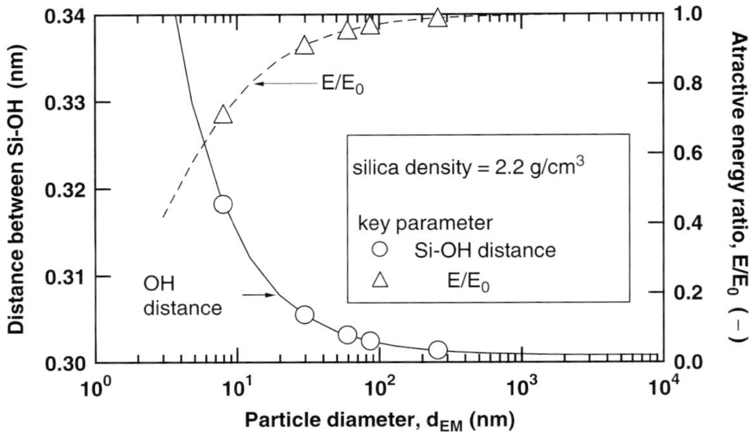

Figure 3.1.4
Analysis of the change of silanol structure due to particle diameter. Effect of curvature on the mean distance between nearest oxygen and hydrogen in silanol group and the attractive potential energy. The attractive potential energy varied with inverse-sixth power of the mean distance between nearest oxygen and hydrogen.

the amount of hydrogen-bonded molecular water on the silica surface increased. The surface layer of hydrogen-bonded molecular water on the silica particles caused additional hydration repulsion force between particles. Apparently, the stability of dispersions of silica particles larger than 60 nm in diameter can be attributed to the additional hydration force associated with the hydrogen-bonded molecular water layer. For ultrafine powders with a particle size of about 10 nm, disappearance of the bonded water layer on the silica surface weakened any additional hydration force between particles. Since polarization of the isolated silanol group was less than that of the hydrogen-bonded silanol group, the amount of hydrogen-bonded molecular water decreased, and dispersion stability by the additional hydration force disappeared. It was reported that such increase in the distance between O and H atoms of neighboring silanol groups decreased the reaction rate of silane coupling agent per unit surface area of silica particles [5]. It is estimated that the surfactant and dispersant adsorption behavior is changed by such surface molecular structure change on nanoparticles.

3.1.5. Basic approach to control nanoparticle dispersion behavior

For the control of aggregation and dispersion behavior of fine particles in suspension, physical and mechanical methods such as ultrasonic and milling method, and chemical surface modification are generally used. Since the fluid drag on nanoparticles is not so strong, general physical and mechanical methods are not useful for dispersion of nanoparticles. High energy ultrasonic homogenizer, subsonic slurry collision or special milling method by using small beads with several micrometer in diameter is applied for dispersion of nanoparticles. Ultrahigh-pressure pressing higher than 1 GPa under dry condition is also useful for the collapse of aggregate structure and dense packing of nanoparticles up to hexagonal close packing density [6]. Gleiter's group. [7] developed new preparation process of nanocrystalline with almost full densification at room temperature in a vacuum vessel by using ultrahigh-pressure uniaxial pressing at 5 GPa. In this process, nanoparticles were condensed and grown on the frozen metal surface by PVD process. Prepared nanoparticles were packed into a mold and given ultrahigh-pressure in a vacuum vessel. It is necessary for dispersion and uniform dense packing of nanoparticles by physical and/or mechanical process to use special high-energy ultrasonic wave, pressure, and milling with small size beads.

On the contrary, for colloidal process for nanoparticle dispersion, there are two types of general surface modification methods. One is adsorbing or grafting organic groups on nanoparticles by using various surfactants or silane alkoxides, after the nanoparticle synthesis. In this case, since a multi-step procedure which involves particle synthesis, drying and redispersing is required, nanoparticles tend to form aggregates during the surface modification procedure. It is difficult to disperse down to the primary particles after preparation and aggregation of particles. Another is the in-situ modification process, which introduces organic groups on the nanoparticle surface during the particle synthesis procedure. Since the organic ligand in the synthesis solution controls the particle growth and attaches to the particle surface at the same procedure, redispersible nanoparticles can be obtained.

Several different methods have been reported to obtain organic-capped nanoparticles by the in-situ modification process including reactions in reversed micelles [8], sol-gel reactions of metal alkoxides with existence of capping agents [9], non-hydrolytic sol-gel reactions of metal halides with capping agents [10], and thermal decomposition of metal complexes [11]. Since these processes prepared nanoparticle in a nanoscaled liquid poor using organic compound the organic surfactant groups remained on particle surface and well dispersed particles were obtained. However, since most studies of the above process such as modified reverse micelle method have been made under extremely dilute conditions, it is difficult to apply the prepared nanoparticles for new material and composite except expensive products such as quantum dot.

Recently, new preparation processes [12, 13], where the complex of surfactant and metal ions were used for raw materials in aqueous and oil phase, have been developed. Since nucleation and growth of nanoparticles occur from the complex and it does not require the special nanopool structure such as reverse micelle, it is possible to prepare dispersed nanoparticles with relatively high solid concentration.

Large-scale and low-cost synthesis method of nanoparticles is developing recently, however, the aggregation, dispersion design and uniform packing and arrangement of nanoparticles based on the surface interaction control are at the initial stage for development. In this chapter, based on the above background, the basic theory, characterization and control as well as analytical simulation method will be introduced to elucidate various aspects of nanoparticles including motion, surface structure and properties, interaction, dispersion/aggregation.

References

[1] E. Verwey, J.Th.G. Overbeek: *Theory of the Stability of Lyophobic Colloids*, Elsevier, Amsterdam, Netherlands, (1948).

[2] L.V. Woodcock, Proceedings of a Workshop held at Zentrum für Interdisziplinäre Forschung University Bielefield, November 11–13, 1985, Edited by Th. Dorfmüller and G. Williams (Lect. Notes Phys., 277, 113–124 (1987)).

[3] H. Kamiya, M. Mitsui, S. Miyazawa and H. Takano, *J. Am. Ceram. Soc.*, **83**(2), 287–293 (2000).

[4] E. Garrone, V.B. Kaznsky, L.M. Kustov, J. Sauer, I.N. Senchenya and P. Ugliengo, *J. Phys. Chem.*, **96**, 1040–1045 (1992).

[5] M. Iijima, M. Tsukada and H. Kamiya, *J. Colloid Interf. Sci.*, **301** (2), 418–424 (2007).

[6] H. Kamiya, H. Suzuki, D. Kato and G. Jimbo, *J. Am. Ceram. Soc.*, **76**(1), 54–64 (1993).

[7] J. Karch, R. Birringer and H. Gleiter., *Nature*, **330**, 556–558 (1987).

[8] M.A. Lopez-Quintela, *Curr. Opin. Colloid Interf. Sci.* **8**, 137 (2003).

[9] S.O'Brien, L. Brus and C.B. Murray, *J. Am. Chem. Soc.*, **123**, 12085 (2001).

[10] T.J. Trentler, T.E. Denler, J.F. Bertone, A. Agrawal and V.K. Colvin, *J. Am. Chem. Soc.*, **121**, 1613 (1999).

[11] J. Rockenberger, E.C. Scher and A.P. Alivisatos, *J. Am. Chem. Soc.*, **121**, 11595 (1999).

[12] H. Kamiya, K. Gomi, Y. Iida, K. Tanaka, T. Yoshiyasu and T. Kakiuchi, *J. Am. Ceram. Soc.*, 86(12), 2011–2018 (2003).

[13] J. Park, K. Au, Y. Hawang, J.-G. Park, H.-J. Noh, J.-Y. Kim, J.-H. Park, N.-M. Hwang, T. Hypon, *Nat. Mat.*, 3, 891–895 (2005).

3.2 Single nanoparticle motion in fluid

3.2.1 Single particle motion

(1) Dynamic equation

The motion of a single particle is classified into two categories; one is a motion determined by the inertia, the other is a random motion induced by the brown motion of fluid or medium. In this section, the deterministic motion is described.

The motion of spherical particle having a diameter of D_p is described by the motion equation [1]

$$m_p \frac{dv}{dt} = -C_D A\left(\frac{\rho_f v_r^2}{2}\right) + \frac{\pi}{6}\rho_f D_p^3 \frac{du}{dt} + \frac{\pi}{12}\rho_f D_p^3\left(\frac{dv}{dt} - \frac{du}{dt}\right) + \frac{3}{2}D_p^2\sqrt{\pi\mu\rho_f}\int_0^t\left(\frac{dv}{d\xi} - \frac{du}{d\xi}\right)\frac{d\xi}{\sqrt{t-\xi}} + F_e \quad (3.2.1)$$

v, velocity vector of particle;
u, velocity vector of fluid;
v_r, relative velocity ($=v-u$);
ρ_p, particle density;
ρ_f, fluid density;
A, project area of particle ($=\pi D_p^2/4$).

The first term on the right-hand side of equation (3.2.1) is the fluid resistance. The second term is the force induced by the pressure gradient of fluid surrounding the particle. The third term is the force for accelerating the virtual added mass of the particles relative to the fluid. The fourth term is called as the Basset term. This term describes the change of total fluid momentum due to the change of particle velocity during the time $\chi=0$ to t [2]. The last term F_e is an external force.

The second, third and fourth terms on the right-side of equation (3.2.1) can be neglected, except for an extremely unsteady particle motion induced by extremely strong external force. In most cases of particle motion, the equation of motion can be simplified as follows [2]:

$$m_p \frac{dv}{dt} = -C_D A\left(\frac{\rho_f v_r^2}{2}\right) + F_e \quad (3.2.2)$$

Coefficient C_D in equations (3.2.1) and (3.2.2) is called as the resistance coefficient. The coefficient C_D is expressed as the function of the particle Reynolds number defined by the relative velocity v_r as a representative velocity and the particle diameter D_p as a representative length.

$$C_D = \frac{24}{Re_p} : Re_p < 2 \quad \text{(Stokes regime)} \quad (3.2.3)$$

$$C_D = \frac{10}{\sqrt{Re_p}} \; 2 < Re_p < 500 \quad \text{(Allen regime)} \quad (3.2.4)$$

$$C_D = 0.44 : Re_p > 500 \quad \text{(Newton regime)} \quad (3.2.5)$$

Here,

$$Re_p = \frac{\rho_f v_r D_p}{\mu} \quad (3.2.6)$$

Substituting these equations into the first term on the right-hand side of equations (3.2.1) and (3.2.2) gives the equation of fluid resistance force F_r. For example, the force F_r in Stokes regime is as follows:

$$F_r = 3\pi\mu D_p v_r \quad (3.2.7)$$

Equations (3.2.3) and (3.2.7) are derived analytically from the Navier–Stakes equation under the assumption of without inertia. Therefore, the equation can be applied to the range of small Reynolds number. As for a nano-size particle, most cases are in the Stokes regime because of the small particle size.

Several equations describing the resistance coefficient C_D have been proposed besides of equations (3.2.3–3.2.5). For Reynolds number Re_p larger than the Stokes regime, Oseen proposed the following equation with considering the inertia effect of fluid surrounding the particle [2],

$$C_D = \frac{24}{Re_p}\left(1 + \frac{3}{16} Re_p\right) : Re_p < 5 \quad (3.2.8)$$

Here, it should be noted that the resistance coefficient C_D should be corrected by the Cunningham's correction factor or slip correction factor when a particle diameter is less than 1 μm. For example, equation (3.2.7) describing the resistance F_r in the Stokes regime becomes the following equation:

$$F_r = \frac{3\pi\mu D_p v_r}{C_c} \quad (3.2.9)$$

(2) Cunningham's correction factor

When particle diameter D_p is in the same order of the mean free path λ of gas molecule (λ is about 0.0645 μm at 1 atm, 20°C) and/or less than that, the gas cannot be considered as a continuous medium. Because of the molecular motion, the gas velocity at the surface of the particle cannot be assumed to be zero. In other words, the fluid slips at the surface. The decrease of the fluid resistance due to the slip is corrected by the Cunningham's correction factor or slip corrections factor.

On the basis of the theoretical analysis or experimental data, several equations giving the Cunningham's correction factor have been proposed. One of the equations is as follows [3]:

$$C_c = 1 + kK_n \quad (3.2.10)$$

$$k = 1.257 + 0.400\exp(-1.10/K_n) \quad (3.2.11)$$

Here, K_n is Knudsen number defined by the ratio of mean free path λ to particle diameter.

$$K_n = 2\lambda/D_p \quad (3.2.12)$$

The mean free path λ of the air is expressed by the following equation:

$$\lambda = \frac{\mu}{0.499P}\left(\frac{\pi M}{RT}\right)^{1/2} \quad (3.2.13)$$

μ, viscosity;
P, pressure;
M, molecular weight;
R, gas constant;
T, temperature.

(3) Characteristic value of particle motion

As one of the simplest case of particle motion, we consider the case where a spherical particle having diameter D_p is injected into a stationary fluid with the initial velocity v_0. When no external force F_e and the Stokes law are assumed, the equation of motion of the particle can be written as follows:

$$\frac{\pi}{6}\rho_p D_p^3 \frac{dv}{dt} = -\frac{3\pi\mu D_p v}{C_c} \quad (3.2.14)$$

The distance S of particle movement at time t is as follows:

$$S = \tau v_0 \{1 - \exp(-t/\tau)\} \tag{3.2.15}$$

$$\tau \equiv \frac{\rho_p D_p^2 C_c}{18\mu} \tag{3.2.16}$$

Index τ defined by equation (3.2.16) has the dimension of time and it is called as "particle relaxation time". The distance S_∞ at time $t=\infty$ is called as the "stopping distance". The distance S_∞ represents the inertia effect on the particle movement.

If the gravity mg (m: mass $=(\rho_p - \rho_f)\pi D_p^3/6$, g: acceleration of gravity) is acting on the particle as an external force, the motion of equation and the particle velocity at time t can be written as follows:

$$\frac{\pi}{6}\rho_p D_p^3 \frac{dv}{dt} = -\frac{3\pi\mu D_p v}{C_c} + m_p g \tag{3.2.17}$$

$$v = \tau g\{1 - \exp(-t/\tau)\} \tag{3.2.18}$$

From equation (3.2.18), it can be found that the velocity will become terminal velocity $v_t(=\tau g)$ at $t=\infty$. The particle relaxation time τ is equal to the time when the particle velocity v reaches $(1-\exp^{-})v_t$ (= about $0.632v_t$). It characterizes how fast the particle reaches the steady state.

By using the characteristic length D and the characteristic velocity U specifying the system in which the particle is moving, equation (3.2.14) can be rearranged into non-dimensional form,

$$\Psi\frac{d\bar{v}}{d\bar{t}} = -\bar{v} \tag{3.2.19}$$

$$\Psi \equiv \frac{C_c \rho_p D_p^2 U}{18\mu D} = \frac{Sk}{2} = \frac{\tau U}{D} \tag{3.2.20}$$

The coefficient Ψ of the inertia term (=left-hand side) of equation (3.2.19) is "inertia parameter". If the inertia parameter Ψ and Reynolds number Re characterizing the fluid flow are the same in two different systems, the particle motions in these systems are similar. The coefficient of the inertia term obtained by the characteristic length $D/2$ is called as Stokes number $Sk = 2\Psi$. The inertia parameter Ψ and Stokes number Sk are the ratio of the stopping distance τU to the characteristic length D.

As it can be found in the definition equation (3.2.20) of the inertia parameter, the particle shows the similar motion with the larger particle when the Cunningham correction factor Cc is increased, for example, by increasing the mean free path λ by means of the low-pressure operation. This means that performance of the inertia classifier such as the impactor can be improved by means of the operation in low pressure.

References

[1] C. Crowe, M. Sommerfeld and Y. Tsuji: *Multiphase Flows with Droplets and Particles*, CRC Press, Boca Raton, Florida, USA, 86 (1997).

[2] Y. Otani, H. Emi: *J. Soc. Powder Technol., Jpn.*, **23**, 31–43 (1986).

[3] P.C. Reist: *Introduction to Aerosol Science*, Macmillan Pub., New York, USA, 53 (1984).

3.2.2 Phoretic phenomena

3.2.2.1 Phoretic phenomena in gas phase

The phoretic phenomenon is defined as the particle motion caused by a force acting non-uniformly on a particle surface or by the motion of a medium in the very vicinity of the surface because of interactions between the particle surface and the medium. In general, a phoretic phenomenon is distinguished from the particle motion caused by forces acting on the entire part of each particle, such as external forces.

The phoretic motion of particles in the gas phase is induced, as shown in Fig. 3.2.1, when the momentum transferred from the gas molecules to the particle surface is not uniform over the surface. Particles suspended in a gas show Brownian motion due to collision with the surrounding gas molecules that show thermal motion. For uniform temperature and gas composition, the motion of gas molecules is isotropic and random. In such a case, the momentum transferred to the particles is uniform over their entire surface after the collisions of many molecules, resulting in isotropic particle movement.

However, in a field with non-uniform temperature and composition, the momentum transferred to a particle surface is dependent on the position on the surface, resulting in a biased motion and the net transport of particles, i.e., phoretic motion. Typical phoretic phenomena in the gas phase include thermophoresis originating from a temperature gradient around a particle and diffusiophoresis originating from a non-uniform distribution of gas composition. The evaluation of phoretic velocity caused by such phoretic phenomena is described below.

(1) Thermophoresis

Thermophoresis occurs owing to the difference in the momentum transfer to a particle between gas molecules with a high thermal velocity and those with a low thermal velocity. Because this phenomenon enhances

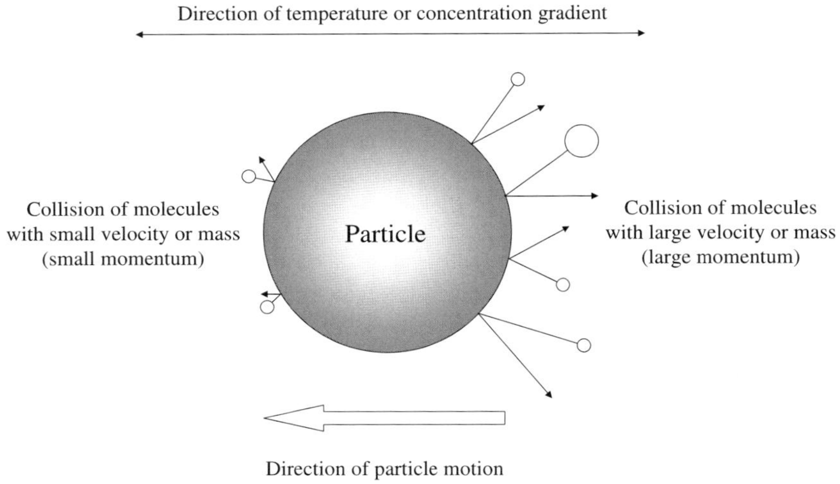

Figure3.2.1
Mechanisms of phoretic motion of particles in gas phase.

particle deposition onto cooled surfaces, its importance has been pronounced in the fields involving particle collectors equipped with cooled plates, scaling phenomena in heat exchangers, electrical and optical material manufacturing process by particle deposition, particulate contamination control and so on.

When particle size (d_p) is much smaller than the mean free path of gas molecules (λ), in which case the particle–gas system is called the free molecular regime (Knudsen number: $Kn = 2\lambda/d_p \gg 1$), and heat transfer is governed by conduction and convection and not by radiation, the particle velocity vector by thermophoresis, $\boldsymbol{v}_T$, is given as [1, 2]

$$\boldsymbol{v}_T = -0.55\nu\frac{\nabla T}{T}, \quad (3.2.21)$$

where ν and T are the kinetic viscosity and temperature of the gas, respectively. Equation (3.2.21) indicates that particles move from a high-temperature region to a low-temperature region and that thermophoretic velocity depends on temperature gradient, not on particle size.

On the other hand, for a wide range of conditions including the continuum regime characterized by $Kn \ll 1$, the next equation was proposed considering the temperature gradient inside a single particle and a non-zero gas velocity on the particle surface [3, 4].

$$\boldsymbol{v}_T = -\frac{2.34(\alpha + 2.18Kn)C_C}{(1 + 3.42Kn)(1 + 2\alpha + 4.36Kn)}\nu\frac{\nabla T}{T} \quad (3.2.22)$$

Here, α is the ratio of thermal conductivity of the gas to that of the particle, and C_C the Cunningham's correction factor.

Fig. 3.2.2 shows the dependence on particle size d_p of the thermophoretic velocity v_T derived from equation (3.2.22). For particles in air under normal conditions and smaller than about 0.5 μm in diameter, v_T is almost independent of d_p and is expressed by equation (3.2.21). However, for larger particles, a temperature distribution can be formed inside the particle, and v_T exhibits a dependence on the d_p and thermal conductivity of the particles. Furthermore, as shown in the figure, the v_T of particles less than 0.1 μm in diameter is higher than the gravitational settling velocity in the presence of the temperature gradient dT/dx = 100 K/m.

The phenomenon by which particles irradiated by light move in the direction of the light source is called photophoresis. Light absorption by a particle leads to the formation of a temperature distribution in the particle. Photophoresis is induced by a temperature gradient of the gas surrounding the particle having such a temperature distribution. Thus, one can consider photophoresis as a special case of thermophoresis. Equations similar to equation (3.2.22) have been proposed for photophoretic velocity. However, the expression of photophoretic velocity is very complicated and depends on the refractive index of particles and the wavelength of light as well as Knudsen number because the increase in particle temperature is influenced by the light absorption property of particles [5]. Moreover, it is also difficult to determine the association between the temperature distributions inside and the surroundings of the particle.

Generally, photophoretic velocity reaches its maximum if the particle diameter and light wavelength are comparable. For particles with a high light absorbance, particles move in the direction away from the light source because the temperature of the particle surface on the irradiated side is higher than that on

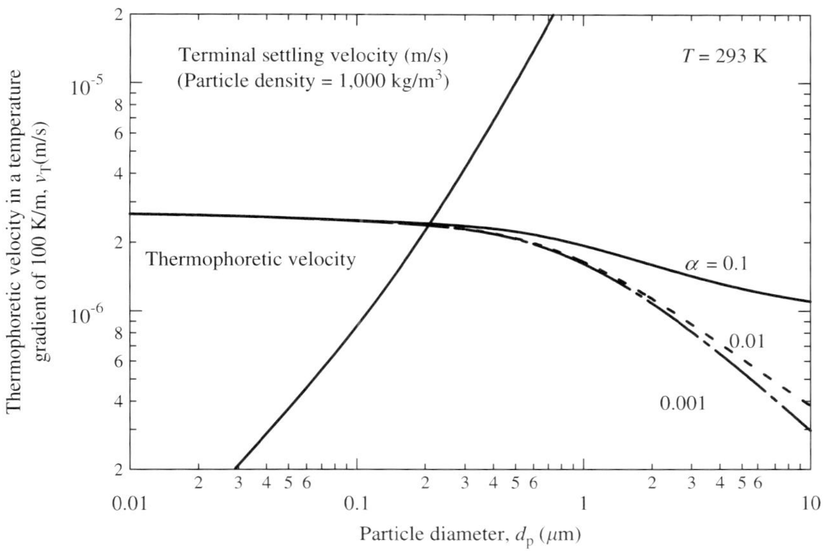

Figure 3.2.2
Dependence of thermophoretic velocity on particle size (α=(thermal conductivity of gas)/(thermal conductivity of particle)).

the opposite side. In contrast to this, particles with a low light absorbance move toward the light source depending on their size and light absorbance. This occurs when the particle behaves like a lens focusing the light backward and thus inducing a temperature increase there.

(2) Diffusiophoresis

Consider a particle suspended in a gas mixture with a spatially non-uniform composition. In such a multi-component mixed gas system, each component has its own concentration gradient and diffuses accordingly along the gradient. The particle suspended in such a gas experiences a net force in the direction to which the gas component with a larger molecular weight is diffusing. In other words, the side of the particle surface facing a higher concentration of heavier molecules receives a larger momentum due to molecular collisions. The force thus induced is called the diffusiophoretic force, and the diffusiophoresis of particles in the gas phase is defined as the particle motion caused by the diffusiophoretic force.

To express the diffusiophoretic velocity v_D in a binary gas mixture system, the following equation has been used [1]:

$$\boldsymbol{v}_D = -\frac{\sqrt{m_1} - \sqrt{m_2}}{\gamma_1\sqrt{m_1} + \gamma_2\sqrt{m_2}} D_{12}\, \nabla \gamma_1 \qquad (3.2.23)$$

Here, m is the mass of a molecule, γ the molar fraction, and subscripts 1 and 2 are the two components. D_{12} is the counter diffusion coefficient between components 1 and 2. It is shown by the above equation that diffusiophoretic velocity is dependent on the gradient of the field, not on particle size, similarly to that of thermophoretic velocity. Also note that particles move in the direction of heavy-molecules diffusion; however, phoretic motion ceases if the masses of the molecules of the two components are the same. A more rigorous expression of v_D is discussed in the literature [6, 7]. The particle velocity in the mixed gas consisting of three or more components is cited from a published book, for example [8].

Aside from diffusiophoretic forces due to concentration gradients, a flow called the Stephan flow arises near a surface on which evaporation or condensation takes place, and imposes fluid resistance on particles suspended there. This corresponds, for instance, to the case in which water vapor condenses on a wall placed in a humid atmosphere or, conversely, to the case in which water vapor evaporates from a wall.

In a water vapor–air system over a water surface, air diffuses toward the surface where evaporation takes place to maintain the total pressure of the vapor–air mixture constant, but is not absorbed on the surface. This results in a net fluid flow in the direction away from the surface. These mechanisms cause the formation of a thin, particle-free layer over an evaporation surface under an isothermal condition. Conversely, fluid flow toward a condensation surface arises.

Diffusiophoresis and Stephan flow usually act simultaneously on particles above an evaporation or condensation surface [1, 9]. For water vapor in air under the normal conditions, the following equation in which the gradient of the partial pressure of water vapor (dp/dx) is expressed in kPa/m and velocity v_{Dsf} in m/s was reported to agree well with the experimental values [9].

$$v_{Dsf} = -1.9 \times 10^{-7} \frac{dp}{dx} \quad (3.2.24)$$

References

[1] L. Waldmann, K.H. Schmitt: Chapter 6 in *Aerosol Science*, C.N. Davies (ed.), Academic Press, New York (1966).
[2] D.E. Rosner, D.W. Mackowski, M. Tassopoulos, J. Castillo and P. Garcia-Ybarra: *Ind. Eng. Chem. Res.*, **31**, 760–769 (1992).
[3] J.R. Brock: *J. Colloid Sci.*, **17**, 768–780 (1962).
[4] L. Talbot, R.K. Cheng, R.W. Scheffer and D.R. Willis: *J. Fluid Mech.*, **101**, 737–758 (1980).
[5] D.W. Mackowski: *Int. J. Heat Mass Transf.*, **32**, 843–854 (1989).
[6] P.J. Whitmore: *J. Aerosol Sci.*, **12**, 1–9 (1981).
[7] S.P. Bakanov, V.I. Roldughin: *Aerosol Sci. Technol.*, **7**, 249–255 (1987).
[8] M.M.R. Williams, S.K. Loyalka: Chapter 7 in *Aerosol Science Theory and Practice*, Pergamon (1991).
[9] P. Goldsmith, F.G. May: Chapter 7 in *Aerosol Science*, C.N. Davies (ed.), Academic Press, New York (1966).

3.2.2.2 Electrokinetic effects in liquid

(1) Electrophoresis

Electrophoresis is the movement of an electrically charged particle toward an oppositely charged electrode under the influence of an electric field. Suppose the shape of a particle is spherical and the electrical force acting on the double layer of the particle is ignored. The forces acting on the particle are the direct electric force, $f_e = qE$, and the viscous drag of the liquid given by the Stokes equation, $f_v = 6\pi\eta av$. When the particle settles down to a constant velocity v, these two forces are balanced:

$$qE = 6\pi\eta av \quad (3.2.25)$$

where q is the net charge on the particle, E the electric field strength, η the viscosity of the liquid and a the radius of the particle.

The electrophoretic mobility u is defined as follows:

$$u = \frac{v}{E} = \frac{q}{6\pi\eta a} \quad (3.2.26)$$

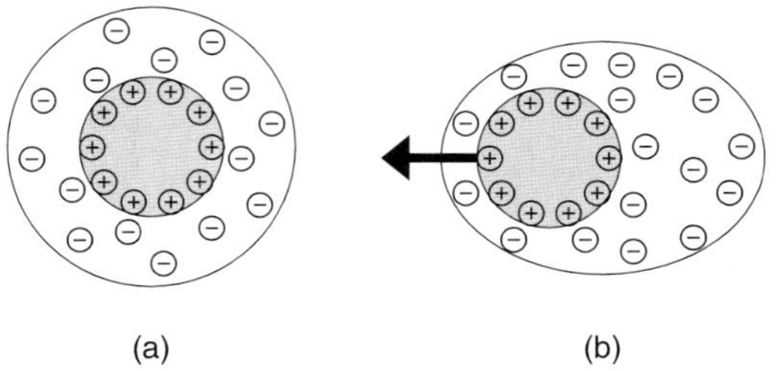

Figure 3.2.3
The electric double layer around the charged particles: (a) stationary state; (b) during electrophoresis.

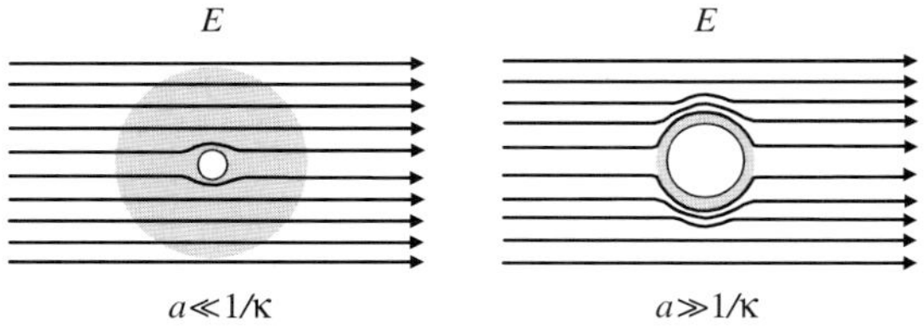

Figure 3.2.4
The double layer thickness ($1/\kappa$) and the electric field lines around the particles.

The ζ-potential on a charged sphere is related to the charge by the expression $\zeta = q/4\pi\varepsilon_r\varepsilon_0 a$; therefore, the electrophoretic mobility is given by

$$u = \frac{2\varepsilon_r\varepsilon_0\zeta}{3\eta} \quad (3.2.27)$$

As an actual colloidal particle is accompanied by the double layer as shown in Fig. 3.2.3., the influence of the double layer should be considered resting on its thickness as shown in Fig. 3.2.4.

(a) Hückel equation

When the double layer is very thick compared with the particle radius, $a \ll 1/\kappa$, a charged particle may be treated as a point charge in an unperturbed electric field. Therefore, equation (3.2.27) is applicable without amendment.

$$u = \frac{2\varepsilon_r\varepsilon_0\zeta}{3\eta} \quad (3.2.28)$$

Hückel equation is not likely to be applicable to particle electrophoresis in aqueous media, however, it has possible applicability in non-aqueous media of low conductance.

(b) Smoluchowski equation

When the double layer is very thin compared with the particle radius, $a \ll 1/\kappa$, the double layer is effectively flat and may be treated as such. Under those conditions, the forces imparted to the liquid by the applied electric field are transmitted to the particle as the liquid flows along its surface. The electrophoresis is given by

$$u = \frac{\varepsilon_r \varepsilon_0 \zeta}{\eta} \tag{3.2.29}$$

Smoluchowski equation is applicable in aqueous media.

(c) Henry equation

Henry showed that the two equations could be reconciled by taking account of the effect of the particle shape and size on the electric field. His equation is

$$u = \frac{2\varepsilon_r \varepsilon_0 \zeta}{3\eta} f(\kappa a) \tag{3.2.30}$$

where the function $f(\kappa a)$ varies as $1 < f(\kappa a) < 3/2\kappa$ for $0 < \kappa a < \infty$.

The equations of electrophoresis are very important since they are connected to the electrophoretic mobility with the ζ-potential. The conventional method for measuring zeta potential is to observe the migration of particles in a liquid with a microscope and measure their velocity using a stopwatch. The technique of laser Doppler velocimetry (LDV) has mostly replaced this laborious measurement procedure.

(2) Electro-osmosis

Electro-osmosis is the motion of liquid which is adjacent to a flat, charged surface under the influence of an electric field applied parallel to the surface. If the surface is negatively charged, there will be a net excess of positive ions in the adjacent liquid and as they move under the influence of the applied field they draw the liquid along them.

The electro-osmotic velocity v_{eo} under an electric field strength E is defined as follows:

$$v_{eo} = u_{eo} E \tag{3.2.31}$$

where u_{eo} is the electro-osmotic mobility which is given by

$$u_{eo} = \frac{\varepsilon_r \varepsilon_0 \zeta}{\eta} \tag{3.2.32}$$

In a fused silica capillary whose inside walls is negatively charged, considerably fast flow toward cathode is observed in the wide range of pH. The magnitude of the electro-osmotic mobility is usually much larger than the mobility of other species. As a result, the electro-osmotic flow carries any species toward cathode through the capillary regardless of the polarity of surface charge of the species. This phenomenon has been utilized to separate ionic species by their charge and frictional forces. This technique is well-known as capillary electrophoresis. More recently, the electro-osmosis has been applied to the transportation, manipulation and patterning of nano particles [1].

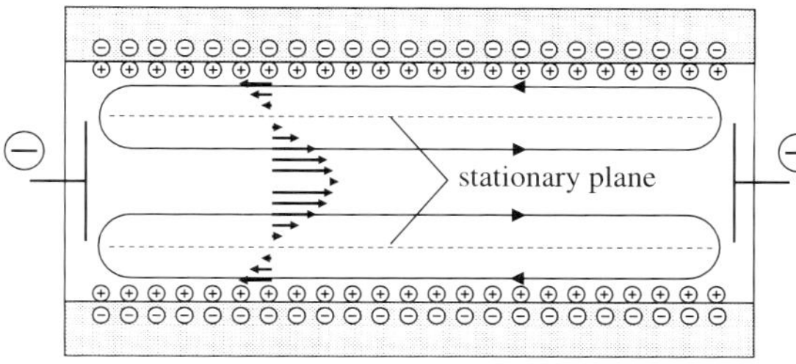

Figure 3.2.5
The electro-osmosis in a quartz cell for zeta potential measurement.

Figure 3.2.5 shows the schematic structure of an electrophoresis cell for the zeta potential measurement. It is composed of a quartz glass capillary with electrodes at both ends. The capillary is sealed by valves to form a closed tube. The inside wall of the cell has negative charge and this leads to a liquid flow in the cell due to electro-osmosis. In a closed capillary, the fluid flow along the cell wall is cancelled at a special position called the stationary layer or stationary plane because of the return flow in the center of the cell. Measurements of the particle's motion at this position give the true electrophoretic mobility. The position of the stationary plane is dependent on the cell geometry and can be calculated for symmetrical shapes of capillary.

(3) Dielectrophoresis

The motion of a particle, with a dielectric permittivity different from that of a surrounding liquid medium, is termed "dielectrophoresis". This force does not require the particle to be charged and all particles exhibit dielectrophoretic activity in the presence of electric fields. However, the strength of the force depends strongly on the dielectric properties of the particles and the medium, on the particle shape and size, as well as on the frequency of the electric field. Consequently, fields of a particular frequency can manipulate particles with great selectivity. This has allowed, for example, the separation of cells or the orientation and manipulation of nanoparticles and nanowires. The dielectrophoretic behavior of a single particle has been characterized using a hyperbolic quadrupole microelectrode [2].

References

[1] M. Fuji: *Ceram. Soc. Jpn. Symp. Abst.*, 107 (2004).
[2] S. Tsukahara, T. Sakamoto and H. Watarai: *Langmuir*, **16** (8), 3866–3872 (2000).

3.3 Brownian diffusion

Particles with a diameter smaller than 1 μm exhibit irregular and random motion because their masses are small enough to render fluctuation by the bombardments of gas molecules. As a result of random motion, particles as a whole move toward to a low concentration region from a high concentration region. This phenomenon, which is similar to gas molecules, is referred to as Brownian diffusion of particles. The diffusion coefficient of particles both in liquid and air is given by the following Stokes–Einstein equation:

$$D = \frac{kT}{f} = \frac{C_c kT}{3\pi\mu D_p} \tag{3.3.1}$$

where f is the Stokes' drag coefficient given by equation (3.3.2), k the Boltzmann constant, T the temperature, μ the viscosity of fluid, and D_p the particle diameter.

$$f = \frac{3\pi\mu D_p}{C_c} \tag{3.3.2}$$

C_c is the Cunningham's slip correction factor, which is equal to unity for particles in a liquid and given by equation (3.3.3) for particles in a gas.

$$C_c = 1 + Kn\left(1.257 + 0.4\exp\left(-\frac{1.1}{Kn}\right)\right) \tag{3.3.3}$$

In the above equation, Kn is the Knudsen number which is the ratio of mean free path of gas molecules to particle radius, $D_p/2$.

$$Kn = \frac{2\lambda}{D_p} \tag{3.3.4}$$

The mean free path of gas molecules is 0.065 μm for air at the normal temperature and pressure. As seen form equation (3.3.1), Brownian diffusion is more significant for smaller particles and therefore the motion of nanoparitcles are governed by it.

When a thin film is formed on a substrate by the deposition of nanoparticles synthesized in gas phase, Brownian diffusion mainly determines the structure and growth rate of the film. In order to predict the deposition flux of particles, particle–gas system is regarded as a continuum and the following convective diffusion equation is solved:

$$\frac{\partial C}{\partial t} + u\nabla C = D\nabla^2 C - \nabla\frac{F_{ext}}{f}C \tag{3.3.5}$$

where C is the particle concentration, u the gas velocity, F_{ext} the external force acting on particles.

Equation (3.3.3) cannot be used to predict the microscopic structure of film formed by the particle deposition. For obtaining the structure of particle-accumulated layer, discrete or stochastic model is employed with the aid of computer simulation. Rosner et al. [1] introduced Diffusion Limited Aggregation model to simulate growth process of particle layer and studied the effects of Peclet number and the mean free path of particles on the porosity, thickness and surface roughness of the particle layer.

When an external force, F_{ext}, acts on a particle, the equation of motion of a particle is given by the following equation:

$$m\frac{dv}{dt} = -fv + F_D(t) + F_{ext}(t) \tag{3.3.6}$$

where m is the mass of a particle, v the particle velocity, t the time, F_D is the fluctuating force acting on a particle by the bombardment of gas molecules. Equation (3.3.6) is integrated step by step with small time increments to obtain particle trajectories starting at arbitrary positions. The direction of fluctuating force acting on a particle at each time step is given by generating random numbers which follow Gaussian distribution with zero mean and the standard deviation equal to $2D\delta t$ for one-dimensional particle diffusion.

Equation (3.3.6) is based on the following two assumptions:

1. F_D is independent of particle velocity and F_D averaged over many particles is equal to zero.
2. Since F_D fluctuates in a very short time period, the time scale of particle motion is considerably large compared to the time scale of particle fluctuation. Consequently, the particle velocity is constant over a short time period of δt and there is no correlation between F_D's at the start and end of the time interval.

These assumptions do not seem to hold for practical problems, but is verified through molecular kinetic theory [2]. These assumptions hold when δt is significantly large compared to the relaxation time of particle, τ $(=C_c\rho_p D_p^2/18\mu)$ and the particle's displacement during δt is smaller than the mean free path of particles.

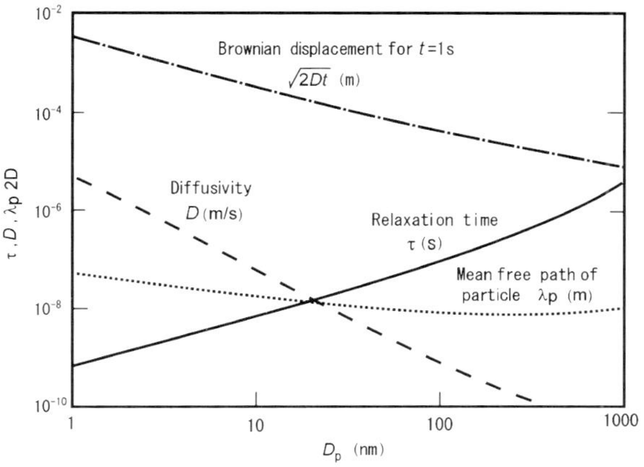

Figure 3.3.1
Particle properties related to Brownian motion at the normal temperature and pressure.

The mean square displacement of particles, $\sqrt{x^2}$, for one-dimensional diffusion, and the mean free path of particles are given by the following equations:

$$\sqrt{x^2} = \sqrt{2Dt} \tag{3.3.7}$$

$$\lambda_p = \frac{(mkT)^{1/2}}{f} \tag{3.3.8}$$

Fig. 3.3.1 shows, as a function of particle size, the diffusion coefficient, the relaxation time of a particle, the mean free path of particle, and the mean square displacement of particles due to Brownian diffusion at the normal temperature and pressure. From this figure, the relaxation time of 5-nm particle is in the order of 10^{-9} s and the mean square displacement is in 10^{-3} s and the mean free path of the particle is in 10^{-8} m. Consequently, in order for the time increment δt to satisfy the aforementioned assumption, δt should be in the order of $10^{-9}-10^{-5}$ s because $\delta t > \tau = 10^{-9}$ and $\delta t < \lambda_p/\sqrt{2Dt} = 10^{-8}/10^{-3} = 10^{-5}$.

References

[1] M. Tassapoulos, J.A. O'Brien and D.E. Rosner: *AIChE J.*, **35**, 969–979 (1989).
[2] S.K. Friedlander: *Smoke, Dust, and Haze* (2nd ed.), Oxford University Press, p. 31 (2000).

3.4 Adsorption properties and wettability of nanoparticle surface

To deal with a suspention of nanoparticles, polymer dispersant is added or modification of particle surface is made to maintain the dispersion stability. In terms of physical chemistry, the former is physical adsorption and the latter is chemical adsorption.

The interaction between each particle is usually regarded as sum of the forces due to van der Waals forces, electrostatic forces etc. The van der Waals force is a universal force that always exists. The electrostatic force is an electric double-layer force that is determined by zeta potential. Hence, without any secondary interaction described below and when the influence of gravity is not so strong, dispersion stability in aqueous suspension is determined by the balance of these two forces. The concept of interparticle potential is illustrated in Fig. 3.4.1. When the working distance of repulsive force of electric double layer is longer than that of the van der Waals force, no aggregation occurs and dispersion is stable.

The van der Waals force cannot be controlled because it is intrinsic to each material. Consequently, the distance to which a repulsive force of electric double layer reaches and the interparticle distance determine whether the system is dispersed or aggregated. It is because interparticle distance in the slurry becomes smaller when the slurry is prepared with small particles such as nanoparticles that even for the same solid volume fraction, the dispersion of the slurry of the smaller particles is not stable and the solid loading cannot be increased. For the so-called nanoparticles, which have smaller diameter than submicron scale, when volume fraction of the particles exceeds 40–50%, the average interparticle diameter becomes smaller than a few nanometers. And almost the same as the distance of maximum of potential barrier. Then a stable dispersion cannot be maintained by repulsion force of electric double layer.

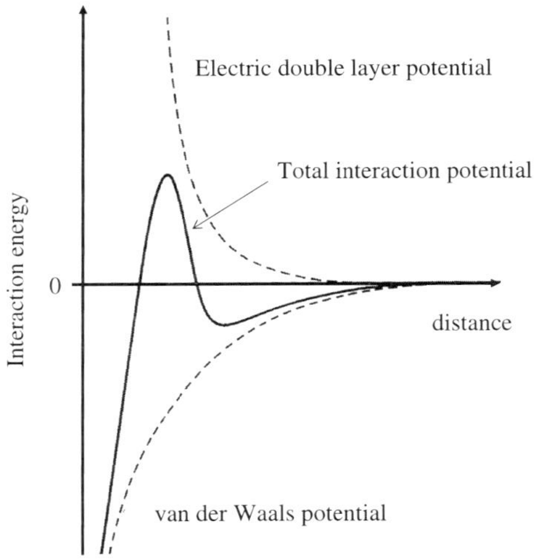

Figure 3.4.1
Interparticle interaction potential curve.

Therefore, the particle distance in the slurry at a high solid loading of nanoparticles is small and the particles aggregate easily. Physical adsorption of polymeric dispersant or surface modification using chemical adsorption is effective in order to disperse the particles at such high concentration. Adsorbed polymeric dispersants increase the interparticle distance by steric repulsion and the dissociation of modifier molecules increases electrostatic repulsion force.

From these reasons above, dispersants or modifiers with steric adsorption structure that keeps the interparticle distance and dissociates to increase electrostatic repulsion should be selected.

The selection of polymeric dispersant considering the properties and structure of adsorbing material is introduced below. It is said that polymeric dispersant of the molecular weight about 10,000 effectively stabilize the dispersion. Dispersants of excess molecular weight sometimes promote aggregation by interparticle bridging effect.

Less amount of dispersant than the optimum concentration in which the dispersant perfectly covers the surface of the particles, causes collision of the particles and bridging with the polymer, and for excess amount, bridging causes aggregation of the particles. When more dispersant is added, non-adsorbed dispersant molecules increase in the solvent, and depletion effect that promotes collision of the particles will appear.

Since the addition of polymeric dispersant is basically a physical adsorption process, changes in temperature or pH of the slurry causes adsorption or desorption of the dispersant. For example, if the temperature of the slurry decreases and desorption occurs, even though an optimum amount of dispersant was added, the slurry changes to dispersant poor state as shown in Fig. 3.4.2 and the state of dispersion will change from dispersed state (b) to pearl-chain like aggregation (a). Inversely if additional adsorption occurred, the structure of the aggregate will be massive as shown in Fig. 3.4.2(c) [1, 2]. For example, difference between the temperature of slurry preparation and that of molding process sometimes induces such phenomena. In usage of nanoparticles, which have very large specific surface area, these affects are considered to cause considerable change of dispersion state.

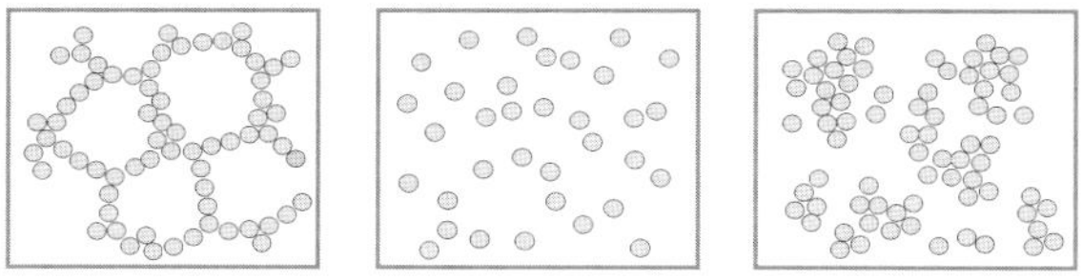

Figure 3.4.2
Relationship between amount of dispersant and state of the particle aggregation.

Here, taking polymeric dispersant as an example, relationship between interacting particles and adsorbing behavior of dispersant to nanoparticles in aqueous system has been described. It is important to notice that the effects arise from adsorption of dispersant depend on the surface area or weight of the particles. Proper usage and selection of dispersant can be derived from aforementioned concepts in the cases of non-aqueous slurry or surface modification by chemical reaction.

It is ideal that dispersion is achieved during particle synthesis process in liquid phase to prepare nanoparticle suspension. However, in many cases, raw materials for chemical industry are supplied as dry powder. To prepare stable nanoparticle dispersion from these dry materials, affinity between a solvent and a surface of particle (wettability) is important.

Contact angle measurement is one of the methods to evaluate the wettability of a particle. For large particles, contact angle can be measured directly using microscope [3]. The contact angle is also obtained from force curve of a particle attached on the probe of atomic force microscopy [4], although the contact angle of nanoparticles cannot be measured directly.

Therefore, a contact angle of nanoparticles is calculated taking an assumption below. In a pillar filled with powder as shown in Fig. 3.4.3, small gap between particles are assumed to be composed of uniform cylindrical capillary. From the penetration rate of a liquid into the powder bed, the contact angle is calculated. Taking the radius of internal

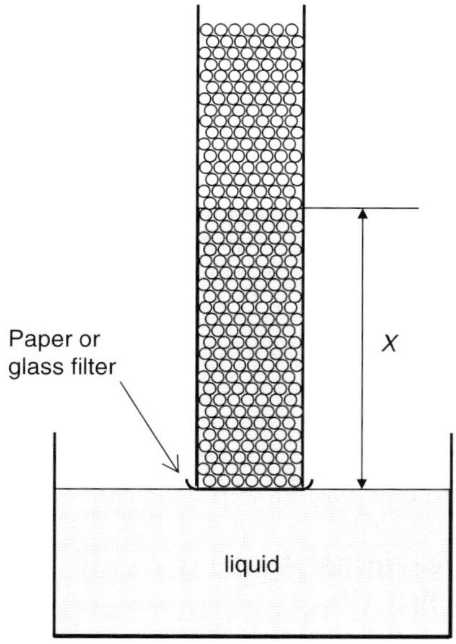

Figure 3.4.3
Illustration of instrument for contact angle measurement using penetration rate.

cylindrical capillary as r, penetration rate can be expressed by the next equation.

$$\frac{dx}{dt} = \frac{r\gamma\cos\theta}{4\eta x} - \frac{r^2\rho g}{8\eta} \tag{3.4.1}$$

where x is the height of liquid column, t the time, γ the surface tension of the liquid, η the viscosity of the liquid, ρ the density of the liquid, g the acceleration of gravity and θ the contact angle. The first term of right-hand side of equation (3.4.1) is a contribution of capillary suction force, and the second term is a contribution of gravity. When the gravity is negligible, integration of equation (3.4.1) gives the following equation:

$$x^2 = \frac{r\gamma\cos\theta}{2\eta}t \tag{3.4.2}$$

The capillary radius r is determined in advance by the slope of x^2–t plot of powder and well-spreading (contact angle $\theta = 0$) liquid. Next, by plotting x^2 vs. t for a particular liquid, contact angle of the powder and the liquid is calculated.

Evaluation of wettability is also done by dispersion test of powder into various liquids or mixed solvent. Besides aforementioned methods, wettability is also evaluated by analysis of adsorption isotherm, heat of adsorption, or heat of immersion [5]. These methods are effective to evaluate wettability of nanoparticles whose contact angle cannot be measured directly.

References

[1] M. Takahashi, M. Oya, M. Fuji: *Adv. Powder Tech.*, **15**(1), 97–107, (2004).
[2] M. Takahashi, M. Oya and M. Fuji: *J. Soc. Powder Technol. Jpn.*, **40**, 410–417 (2003).
[3] M. Fuji, H. Fujimori, T. Takei, T. Watanabe and M. Chikazawa: *J. Phys. Chem. B*, **102**, 10498–10504, (1998).
[4] M. Preiss, H.-J. Butt: *J. Colloid Interf. Sci.*, **208**, 468–477, (1998).
[5] M. Fuji, T. Takei, T. Watanabe and M. Chikazawa: *Colloid Surf. A*, **154**, 13–24, (1999).

3.5 Interactions between particles

3.5.1 Interactions between particles in gases and control of adhesion

3.5.1.1 Van der Waals force and liquid bridge force [1, 2]

Van der Waals force that is caused by the charge fluctuation of atoms or molecules acts between all surfaces in contact. The interaction force between two spherical particles of diameter D_{p1} and D_{p2} is represented by the following equation:

$$F_v = -\frac{Ad}{12z^2} \tag{3.5.1}$$

where

$$d = \frac{D_{p1}D_{p2}}{D_{p1} + D_{p2}} \tag{3.5.2}$$

The negative sign in equation (3.5.1) indicates that the force is attractive. A is the Hamaker constant [3], z the separation distance between particles, and d the reduced particle diameter defined by equation (3.5.2).

The Hamaker constant depends on the materials (see Table 3.5.1 [4, 5]). For a rough estimation, the values of the constant are in the range of (4–10) $\times 10^{-20}$ J for hydrocarbons, (10–15) $\times 10^{-20}$ J for metal oxides and (15–50) $\times 10^{-20}$ J for metals. For the interaction between two different materials 1 and 2 in gases is approximately given by:

$$A_{12} = \sqrt{A_{11}A_{22}} \tag{3.5.3}$$

If a medium 3 such as water exists between the surfaces of material 1, the Hamaker constant is approximated by:

$$A_{131} = \left(\sqrt{A_{11}} - \sqrt{A_{33}}\right)^2 \tag{3.5.4}$$

In the presence of medium 3 between materials 1 and 2, the Hamaker constant is given by:

$$A_{132} = \left(\sqrt{A_{11}} - \sqrt{A_{33}}\right)\left(\sqrt{A_{22}} - \sqrt{A_{33}}\right) \tag{3.5.5}$$

Table 3.5.1
Hamaker constants [4, 5].

Material	Hamaker constant, A (10^{-20} J)
Methanol	3.1
Ethanol	3.9
Water	4.4
Polystyrene	6.2–6.6
MgO	10.6
Al_2O_3	15.5
Mica	9.5
Diamond	28.4
Graphite	47.0
Cu	28.4
Ag	40.0
Au	45.5

If the Hamaker constant A_{33} is between A_{11} and A_{22}, the resulting Hamaker constant A_{132} is negative. Thus, the two bodies of materials 1 and 2 experience repulsive force. The Hamaker constants in air are almost the same as those in vacuum. However, when water caused by capillary condensation exists between the surfaces, the effect of the medium 3 should be taken into consideration [6].

The attractive force increases as the separation distance decreases as in equation (3.5.1). However, when the distance is very small, the electron clouds of atoms on the surfaces overlap, and a strong repulsive force acts on the surfaces, which is known as Born repulsion [1]. Therefore, there is a stable separation distance z. In general, $z = 0.4$ nm is used for smooth surfaces in gases [7].

On the other hand, when the surfaces are an appreciable distance apart, the van der Waals force becomes smaller than the value calculated by equation (3.5.1). This is because the finite speed of light causes a phase lag in the charge fluctuation interaction between atoms or molecules. This is referred to as the retardation effect [1, 8, 9]. At distances beyond about 5 nm, the van der Waals force begins to decrease more rapidly, and at 100 nm separation, the retarded van der Waals force is about one order smaller than the non-retarded one. Therefore, over 100 nm separation, the van der Waals force is negligibly small compared to other forces exerted on the surfaces.

Surface roughness also affects the van der Waals force (see Fig. 3.5.1). The force decreases with increasing roughness, as represented by the following equation [10, 11]:

$$F_{vb} = -\frac{Ad}{12(z+b)^2} \tag{3.5.6}$$

where b is the mean value calculated by the thicknesses of surface roughness layers b_1 and b_2.

$$b = \frac{b_1 + b_2}{2} \tag{3.5.7}$$

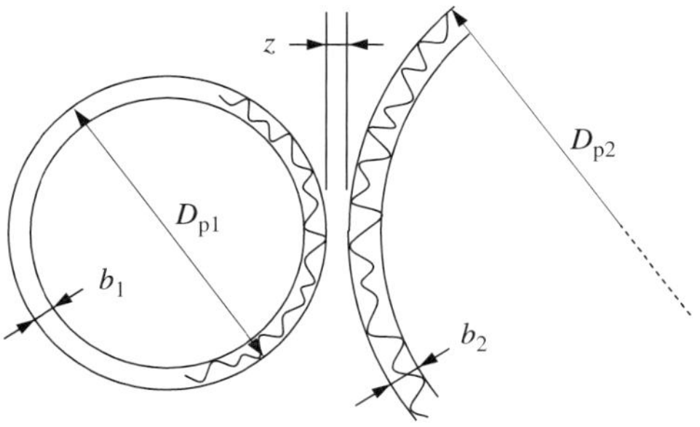

Figure 3.5.1
Nano-roughness on particle surfaces.

For more quantitative estimation, a method using a root mean square roughness is proposed [12, 13]. To reduce the van der Waals force through the use of the roughness, nanoparticles are added on the surface of primary particles. This method is widely used in various fields because of the simplicity and effectiveness.

For large surface roughness, the radius of curvature of asperity summits should also be taken into account for the estimation of the van der Waals force. [14]. In addition, particles of irregular shape or complicated surface shape contact at different points on the surface, and therefore the total force of the adhesion should be calculated by summing the force components in the direction of the adhesion.

In general, particles elastically deform under an applied force if the load is small. The deformation of a spherical particle can be analyzed by the Hertz theory [15]. In order to clarify the relationship between the adhesive force and the deformation, several models based on the Hertz theory were proposed. The JKR theory, developed by Johnson, Kendall and Roberts [16], gives the following equation relating the external compressive force F and the van der Waals force to the radius of the contact area a.

$$a^3 = \frac{3kd}{8}\left[F + \frac{Ad}{8z^2} + \sqrt{\frac{Ad}{4z^2}F + \left(\frac{Ad}{8z^2}\right)^2}\right] \tag{3.5.8}$$

where k is the reduced elastic constant for two contacting bodies of different materials with Young's moduli (E_1,E_2) and Poisson's ratios (ν_1,ν_2), i.e.

$$k = \frac{1-\nu_1^2}{E_1} + \frac{1-\nu_2^2}{E_2} \tag{3.5.9}$$

The values of Young's modulus and Poisson's ratio are listed in Table 3.5.2 [17]. To obtain a real solution of equation (3.5.8), the following relationship should be satisfied:

$$\frac{Ad}{4z^2}F + \left(\frac{Ad}{8z^2}\right)^2 \geq 0 \tag{3.5.10}$$

Table 3.5.2
Young's moduli and Poisson's ratios [17].

Material	Young's modulus E (Gpa)	Poisson's ratio ν (−)
Fe	206	0.28
Cu	123	0.35
Al	68.5	0.34
Quartz glass	75.0	0.17
PMMA	2.33	0.34
Polystyrene	1.39	0.35

Therefore, the limiting value of the external force F_s is given by:

$$F_s = -\frac{Ad}{16z^2} \qquad (3.5.11)$$

As an external force F (< 0) is gradually applied to separate the particle from the other particle or flat surface, the radius of the contact area a decreases, and the particle is separated at $F = F_s$. The separation force F_s gives the adhesive force, which is also called 'pull-off force'. The magnitude of F_s in equation (3.5.11) is 3/4 of the magnitude of F_v in equation (3.5.1). Substituting equation (3.5.11) into equation (3.5.8) gives the radius of the contact area a_s on the separation, i.e.:

$$a_s = \sqrt[3]{\frac{3kAd^2}{128z^2}} \qquad (3.5.12)$$

One difficulty with the JKR theory is that the stress at the edge of the contact area has an infinite value. However, it is widely used for the analysis of the force because of the reasonable agreement with the actual phenomena [18].

On the other hand, DMT theory [19], developed by Derjaguin, Muller and Toporov, has no such an unphysical situation, i.e. the attractive force does not act in a line but a band. The contact area by DMT theory is reduced to zero as the separation force increases. Therefore the resulting pull-off force is the same as that calculated by equation (3.5.1).

Muller et al. [20, 21] introduced a Lennard–Jones potential (6–12 potential) into a model and showed that the JKR theory is applicable to soft materials and the DMT theory is applicable to hard materials, and also referred to the intermediate materials. Table 3.5.3 summarizes the difference of these theories [18].

In the case that a small amount of liquid is held between particles or between a particle and flat surface, an attractive force arising from the liquid bridge acts between these bodies. The force is known as "liquid bridge force" or "capillary force". The liquid bridge force is predominant in adhesive forces between surfaces in gases. The formation of liquid bridge depends on the wettability of the surfaces.

Table 3.5.3
Theories for elastic deformation and adhesion between spherical particles [18].

	Hertz	JKR	DMT	Muller et al.
Force between surfaces				
Stress under compressive load				
Shape under compressive load				
Shape under zero load				

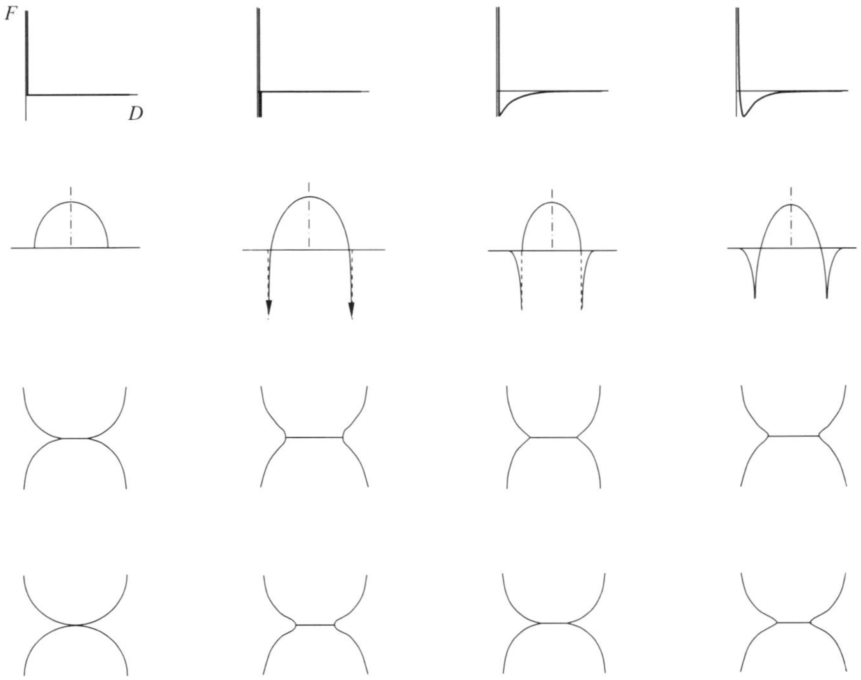

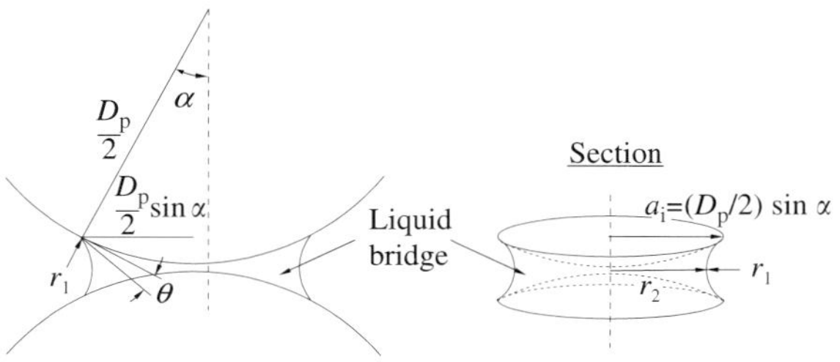

Figure 3.5.2
Liquid bridge between spherical particles.

For hydrophobic materials, the liquid bridge is hardly formed between the surfaces [22]. Here the liquid bridge caused by capillary condensation is discussed for hydrophilic surfaces.

Fig. 3.5.2 schematically shows a liquid bridge formed between two spherical particles of the same diameter D_p. The two principle radii of curvature of the liquid bridge are represented by r_1 and r_2. The pressure of the liquid bridge is smaller than that of the surrounding gas, and the pressure difference Δp is given by the following Young–Laplace equation.

$$\Delta p = \gamma\left(\frac{1}{r_2} - \frac{1}{r_1}\right) \tag{3.5.13}$$

where γ is the surface tension of the liquid. Since $r_2 \gg r_1$, equation (3.5.13) is rewritten as:

$$\Delta p \approx -\frac{\gamma}{r_1} \tag{3.5.14}$$

The attractive force F_c by the capillary pressure is given by the product of the projected area πa_i^2 of the liquid bridge shown in Fig. 3.5.2 and the pressure difference Δp. When the liquid bridge is much smaller than the particle, the projected area is approximated as $\pi D_p r_1 \cos\theta$ where θ is the contact angle between the liquid and the particle. Therefore the force F_c is approximated by the following equation:

$$F_c \approx -\pi D_p \gamma \cos\theta \tag{3.5.15}$$

Equation [3.5.15] can be applied both for the contact between particles of different diameters and for the contact between a particle and flat surface through introducing the reduced particle diameter d. i.e.

$$F_c \approx -2\pi d \gamma \cos\theta \tag{3.5.16}$$

Attractive force arising from the surface tension around the circumference of the liquid bridge is much smaller than that by the capillary pressure, and is usually neglected.

The relationship between the relative humidity and the capillary condensation is represented by the following Kelvin equation:

$$\ln\frac{p}{p_s} = \frac{M\gamma}{\rho_L RT}\cdot\left(\frac{1}{r_2} - \frac{1}{r_1}\right) \tag{3.5.17}$$

where p is the vapor pressure of the capillary condensed liquid, p_s the saturated vapor pressure of the same liquid, M the molecular weight of the liquid, ρ_L the density of the liquid, R the gas constant and T the absolute temperature. Substituting the relative humidity into p/p_s in equation (3.5.17), the meniscus curvature $(1/r_2 - 1/r_1)$ of the capillary condensed liquid can be obtained. In order to estimate the attractive force by the capillary condensation, the projected area πa_i^2 should be obtained through the geometric analysis of the liquid bridge. The attractive force is obtained as the product of the projected area and the pressure difference Δp. In addition, the liquid bridge force is influenced by many factors, such as bridge shape [23, 24], surface roughness [25], adsorbed water [26], and impurities in the condensed liquid [27], and some of them have been studied in detail.

References

[1] J.N. Israelachvili: *Intermolecular and Surface Forces* (2nd ed.), Academic press, London (1992).
[2] K. Okuyama, H. Masuda, K. Higashitani, M. Chikazawa and T. Kanazawa: *J. Soc. Powder Technol., Jpn.*, **22**, 451–475 (1985).
[3] H.C. Hamaker: *Physica*, **4**, 1058–1072 (1937).
[4] J. Visser: *Adv. Colloid Interf. Sci.*, **3**, 331–363 (1972).
[5] Chemical Society of Japan (eds): *Chemical Handbook (Kagaku-Binran Kisohen)* (4th ed.), pp. II. 99–100, Maruzen, Tokyo (1993).
[6] X. Xiao, L. Qian: *Langmuir*, **16**, 8153–8158 (2000).
[7] H. Krupp: *Adv. Colloid Interf. Sci.*, **1**, 111–239 (1967).
[8] E.J. Clayfield, E.C. Lumb and P.H. Mackey: *J. Colloid Interf. Sci.* **37**, 382–389 (1971).
[9] J. Gregory: *J. Colloid Interf. Sci.*, **83**, 138–145 (1981).
[10] J. Czarnecki, T. Dąbroś: *J. Colloid Interf. Sci.*, **78**, 25–30 (1980).
[11] J. Czarnecki, V. Itschenskij: *J. Colloid Interf. Sci.*, **98**, 590–591 (1984).
[12] Y.I. Rabinovich, J.J. Adler, A. Ata, R.K. Singh and B.M. Moudgil: *J. Colloid Interf. Sci.*, **232**, 10–16 (2000).
[13] Y.I. Rabinovich, J.J. Adler, A. Ata, R.K. Singh and B.M. Moudgil: *J. Colloid Interf. Sci.*, **232**, 17–24 (2000).
[14] H. Rumpf: *Chemie-Ingenieur-Technik,* **46**, 1–11 (1974).

[15] S.P. Timoshenko, J.N. Goodier: *Theory of Elasticity* (3rd ed., chapter 12), McGraw-Hill, New York (1970).
[16] K.L. Johnson, K. Kendall and A.D. Roberts: *Proc. R. Soc. Lond., A*, **324**, 301–313 (1971).
[17] Chemical Society of Japan (eds.): *Chemical Handbook* (*Kagaku-Binran Kisohen*) (4th ed.), pp. II.26, Maruzen, Tokyo (1993).
[18] R.G. Horn, J.N. Israelachvili and F. Pribac: *J. Colloid. Interf Sci.*, **115**, 480–492 (1987).
[19] B.V. Derjaguin, V.M. Muller and Yu.P. Toporov: *J. Colloid. Interf. Sci.*, **53**, 314–326 (1975).
[20] V.M. Muller, V.S. Yushchenko and B.V. Derjaguin: *J. Colloid. Interf. Sci.*, **77**, 91–101 (1980).
[21] V.M. Muller, V.S. Yushchenko and B.V. Derjaguin: *J. Colloid. Interf. Sci.*, **92**, 92–101 (1983).
[22] M. Fuji, K. Machida, T. Takei, T. Watanabe and M. Chikazawa: *Langmuir*, **15**, 4584–4589 (1999).
[23] Y. Kousaka, Y. Endo and Y. Nishie: *Kagaku Kogaku Ronbunshu*, **18**, 942–949 (1992).
[24] A. Fukunishi, Y. Mori: *J. Soc. Powder Technol., Jpn.*, **41**, 162–168 (2004).
[25] Y.I. Rabinovich, J.J. Adler, M.S. Esayanur, A. Ata, R.K. Singh and B.M. Moudgil: *Adv. Colloid. Interf. Sci.*, **96**, 213–230 (2002).
[26] M. Fuji: *J. Soc. Powder Technol., Jpn.*, **40**, 355–363 (2003).
[27] Y. Endo, Y. Kousaka and Y. Nishie: *Kagaku Kogaku Ronbunshu*, **18**, 950–955 (1992).

3.5.1.2 Electrostatic interaction

(a) Coulomb's law

The force $\boldsymbol{f}$ [N] working between two particles with charges q [C] and Q [C] is given by Coulomb's law as

$$\boldsymbol{f} = \frac{1}{4\pi\varepsilon}\frac{qQ}{r^3}\boldsymbol{r} = \frac{1}{4\pi\varepsilon}\frac{qQ}{r^2}\hat{\boldsymbol{r}} \qquad (3.5.18)$$

where ε [F/m] is the permittivity of the medium between the particles, and $\hat{\boldsymbol{r}}$ the unit vector co-directional with $\boldsymbol{r}$. In air (or any gas phase, in general), the permittivity of free space, $\varepsilon_0 = 8.854 \times 10^{-12}$ F/m, can be used to approximate the permittivity of the medium.

The superposition principle holds for Coulomb's law. Therefore, when there are multiple point charges q_i ($i = 1 \ldots N$) around a particle with charge Q, and an external electric field $\boldsymbol{E}_0$ [V/m] due to boundary conditions, the electric field generated at the position of the particle, excluding the charge Q itself, is given as,

$$\boldsymbol{E} = \sum_{i=1}^{N}\frac{1}{4\pi\varepsilon_0}\frac{q_i}{r_i^3}\boldsymbol{r}_i + \boldsymbol{E}_0 \qquad (3.5.19)$$

Thus, the force working on the particle of the charge Q is given as,

$$\boldsymbol{f} = Q\boldsymbol{E} \qquad (3.5.20)$$

Even if a particle is neutral, a so-called gradient force works on the particle when an applied electric field is non-uniform on the size scale of the particle. This is a force due to the non-uniform electric field working on the polarization charge induced by the field. The gradient force, $\boldsymbol{f}_g$, working on a spherical particle is given by

$$\boldsymbol{f}_g = 2\pi a^3\varepsilon_0\frac{\varepsilon_p - \varepsilon_0}{\varepsilon_p + 2\varepsilon_0}\nabla|\boldsymbol{E}|^2, \qquad (3.5.21)$$

where a is the radius of the particle and ε_p the relative dielectric constant of the particle.

Note that it is rarely the case that an external electric field applied to a nanoscaled particle will be non-uniform on that size scale. However, it can also be noted that the local electric field becomes non-uniform when charged two particles of similar size approach each other. In this case, the electric field generated by each charge on the particles is indeed non-uniform in the vicinity of the particle, and the non-uniform field generated by one particle results in a gradient force on the other one. When the permittivity of the particle is high, and if the distance between two particles is less than the diameter of the particle, this effect is not negligible compared to the Coulomb interaction due to the true charge on the particle. The detailed calculation of this effect for two same-sized spherical particles with opposite charges of the same absolute value is available in reference [1].

For a spherical particle with uniform charge on its surface, the electric field outside the particle is equivalent to that produced by a point charge located on the center of the particle and with the same charge as the total surface charge. This equivalence is only a first order approximation if the charge is distributed unequally on the particle surface. The actual surface charge distribution on the surface of a nanoparticle is not understood at this moment.

(b) Amount of charge on a particle in the gas phase

A particle suspended in the gas phase gains charge when an ion attaches to its surface. It is assumed that all the ions reaching the surface are captured by the particle. Impact charging (field charging) and diffusion charging are two mechanisms to bring ions to the particle surface. The former is dominant for bigger particles than 1 μm, and vice versa.

(i) Impact (field) charging

Suppose a spherical particle with a diameter a and with a charge Q is situated in a uniform electric

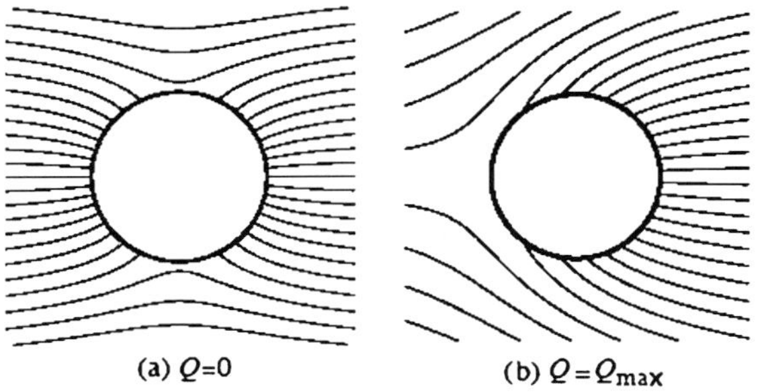

Figure 3.5.3
Schematic illustration of electric flux lines around a particle.

field, $\boldsymbol{E}_0$. The electric field around the particle, including the effect of polarization of the particle, is calculated as,

$$\boldsymbol{E} = \boldsymbol{E}_0 + \frac{\varepsilon_p - \varepsilon_0}{\varepsilon_p + 2\varepsilon_0}\left(\frac{a}{r}\right)^3 (3\boldsymbol{E}_0 \cdot \hat{\boldsymbol{r}}\hat{\boldsymbol{r}} - \boldsymbol{E}_0) + \frac{Q}{4\pi\varepsilon_0 r^2}\hat{\boldsymbol{r}}, \tag{3.5.22}$$

where $\hat{\boldsymbol{r}}$ is the unit radius vector. Fig. 3.5.3 shows a schematic illustration of the electric flux lines. Note that the number of the lines in the schematic illustration does not quantitatively correspond to the strength of the electric field. Because ions in the air migrate along the electric field lines, the ion can reach the particle surface and attach itself if the flux line is terminated by the particle. The component in the radial direction, E_r, of the electric field, $\boldsymbol{E}$, at the particle surface ($r=a$) is obtained from equation (3.5.22) as,

$$E_r\Big|_{r=a} = |\boldsymbol{E}_0|\cos\theta\left\{2\frac{\varepsilon_p - \varepsilon_0}{\varepsilon_p + 2\varepsilon_0} + 1\right\} + \frac{Q}{4\pi\varepsilon_0 a^2} \tag{3.5.23}$$

With the condition of $E_r \geq 0$ for all θ (the angle between E and r), no electric field can reach the particle surface, thus no more ions can impact the particle via this mechanism. Therefore, this condition gives the maximum charge $Q_{\max}$ for the impact (field) charging:

$$Q_{\max} = 4\pi\varepsilon_0 a^2 E_0 \frac{3\varepsilon_p}{\varepsilon_p + 2\varepsilon_0} \tag{3.5.24}$$

The time dependence of this charging is given as,

$$Q(t) = Q_{\max} \cdot \frac{t}{t + 4\varepsilon_0/eBN} \tag{3.5.25}$$

where e ($=1.60 \times 10^{-19}$ C) is elementary charge, B [$\mathrm{m^2s^{-1}V^{-1}}$] and N[$\mathrm{m^{-3}}$] are the electric mobility of the ion and the ion number density apart from the particles, respectively. This is called Pauthenier theory [2].

Obviously in the above theory the charge on the particle is assumed to distribute uniformly on the surface. Such an assumption can be satisfied only when a particle has either low electrical resistance or rotational motion. For a particle with high resistance and no rotation, especially when the permittivity of the particle is low, the particle can capture only a small amount of charge, less than half of the Pauthenier limit [3].

(ii) Diffusion charging [4]

Ions in the gas phase reach the particle surface by thermal diffusion. For diffusion charging by unipolar ions, the evolution of the number concentration of particles n_p[$\mathrm{m^{-3}}$] with p elementary charges is governed by the following birth and death equation, assuming that the ion concentration N apart from the particles is sufficiently high compared to the concentration of the particles:

$$\frac{dn_0}{dt} = -\beta_0 n_0 N, \tag{3.5.26}$$

$$\frac{dn_p}{dt} = (\beta_{p-1}n_{p-1} - \beta_p n_p)N \qquad p > 0, \tag{3.5.27}$$

where n_0 is the number concentration of neutral particles.

Letting n_T represent the total number concentration of all the particles, the solutions of the above equations are given as,

$$\frac{n_0}{n_T} = \exp(-\beta_0 Nt) \tag{3.5.28}$$

$$\frac{n_p}{n_T} = \left(\prod_{k=0}^{p-1}\beta_k N\right)\sum_{j=0}^{p}\frac{\exp(-\beta_j Nt)}{\prod_{\substack{k=0\\ j\neq i}}^{p-1}(\beta_i - \beta_j)N} \tag{3.5.29}$$

The rate constant β_p[$\mathrm{m^3s^{-1}}$] is called the combination coefficient, which represents the probability of impact between a single ion and a single particle with p elementary charges per unit time and unit ion concentration. In order to determine the combination coefficient, a lot of theoretical and experimental effort has been expended. Fuchs theory [5], demonstrated to have the widest applicability in terms of particle size (including the nanoscale), gives the combination coefficient as

$$\beta_p = \frac{\pi\gamma c_i\delta^2\exp\left(-\dfrac{\phi(\delta)}{kT}\right)}{1 + \exp\left(-\dfrac{\phi(\delta)}{kT}\right)\dfrac{\gamma c_i\delta^2}{4D_i}\displaystyle\int_\delta^\infty \frac{1}{r^2}\exp\left(\frac{\phi(r)}{kT}\right)dr}, \tag{3.5.30}$$

where c_i[m/s] and D_i[m^2/s] are the mean thermal velocity and diffusion coefficient of the ions, respectively, T[K] is the absolute temperature, and $k = 1.38\times10^{-23}$ [J/K] is the Boltzmann constant. The radius of the limiting-sphere in Fuchs theory δ[m] is given as,

$$\delta = \frac{a^3}{\lambda^2}\left\{\frac{\left(1+\frac{\lambda}{a}\right)^5}{5} - \frac{\left(1+\frac{\lambda^2}{a^2}\right)\left(1+\frac{\lambda}{a}\right)^3}{3} + \frac{2}{15}\left(1+\frac{\lambda^2}{a^2}\right)^{5/2}\right\}, \quad (3.5.31)$$

where a is the radius of the particle and λ [m] is the mean free path of the ions.

Including the effect of the image force, the potential of an ion $\phi(r)$ [J] is calculated as

$$\phi(r) = \frac{e^2}{4\pi\varepsilon_0}\left\{\frac{p}{r} - \frac{\varepsilon_p - \varepsilon_0}{\varepsilon_p + \varepsilon_0}\cdot\frac{a^3}{2r^2(r^2-a^2)}\right\} \quad (3.5.32)$$

γ is a correction coefficient given as $\gamma = b_m^2/\delta^2$ [6], and b_m^2 is a parameter given as the minimum of the following function [7]:

$$b^2 = r^2\left\{1 + \frac{2}{3kT}(\phi(\delta) - \phi(r))\right\} \quad (3.5.33)$$

Bipolar diffusion charging is also called charge neutralization. When the concentrations of both positive and negative ions are sufficiently higher than that of particles, the birth and death equations are

$$\frac{dn_0}{dt} = \beta_{+1}^- n_{+1} N^- - \beta_0^- n_0 N^- + \beta_{-1}^+ n_{-1} N^+ - \beta_0^+ n_0 N^+, \quad (3.5.34)$$

$$\frac{dn_p}{dt} = \beta_{p+1}^- n_{p+1} N^- - \beta_p^- n_p N^- + \beta_{p-1}^+ n_{p-1} N^+ - \beta_p^+ n_p N^+ \qquad |p| \geq 1 \quad (3.5.35)$$

Under the condition of $N^- = N^+$, the equilibrium charge distribution corresponding to the steady state ($dn_0/dt = 0$, $dn_p/dt = 0$) can be calculated as

$$\left.\begin{aligned} \frac{n_0}{n_T} &= \frac{1}{\Sigma} \\ \frac{n_p}{n_T} &= \prod_{k=+1}^{p} \frac{\beta_{k-1}^+/\beta_k^-}{\Sigma} \quad p \geq +1 \\ \frac{n_p}{n_T} &= \prod_{k=-1}^{p} \frac{\beta_{k+1}^-/\beta_k^+}{\Sigma} \quad p \leq -1 \end{aligned}\right\} \quad (3.5.36)$$

$$\Sigma = \sum_{p=+1}^{+\infty}\left(\prod_{k=+1}^{p}\frac{\beta_{k-1}^+}{\beta_k^-}\right) + \sum_{p=-1}^{-\infty}\left(\prod_{k=-1}^{p}\frac{\beta_{k-1}^-}{\beta_k^+}\right) + 1. \quad (3.5.37)$$

References

[1] T. Matsuyama, H. Yamamoto: *J. Soc. Powder Tech., Jpn.*, **34**, 154–159 (1977).
[2] M.M. Pauthenier, M.M. Moreau-Hanct: *J. Phys. Radium*, **3**, 590–613 (1932).
[3] S. Masuda, M. Washizu: *J. Inst. Electros., Jpn.*, **3**, 153–159 (1979).
[4] Editorial committee for basis of powder technology (eds): *Funtai-kougaku no kiso*, Nikkankogyo, p. 86 (1992).
[5] N.A. Fuchs: *Pure Appl. Geophys.*, **56**, 185–193 (1963).
[6] W.A. Hoppel, G.M. Frick: *Aerosol Sci. Tech.*, **51**, 1–21 (1986).
[7] G.L. Natanson: *Soviet Phys. Tech. Phys.*, **5**, 538–551 (1960).

3.5.1.3 Solid bridging (solution and precipitation, sintering)

When fine particles are set at high temperature below the melting point, atoms diffuse to reduce their total surface energy. As a result, strong bonding between particles forms to be a sintered body. Depending on diffusion path, pores are excreted to be densified. Since specific surface area of particles increases with decrease in particle size, nanoparticles have large driving force for sintering, namely, their sinterability is extremely high.

In the contact region between two particles (neck region), stress generates due to surface energy (Fig. 3.5.4). The stress results in increase in excess vacancy concentration and decrease in vapor pressure. These enhance mass transfer to advance sintering. In solid state sintering, mass transfer occurs by evaporation–condensation, surface diffusion, boundary diffusion and bulk diffusion (Fig. 3.5.5). Representative time dependences of neck size x assuming each mass transfer are listed in Table 3.5.4. In any case, neck size, x, is proportional to time and the mth powers of grain size. The exponent, m, depends on the mass transfer route. In surface diffusion and evaporation–condensation route, the distance between two particles does not change only causing neck growth but not being densified [1]. On the other hand, in case of bulk diffusion [2] and boundary diffusion [3] route, the distance between particles shrink during the neck growth. In these cases, shrinkage rates are obtained as shown in Table 3.5.4. Furthermore, sintering theory in viscous flow [4] is also listed in Table 3.5.4. Time-dependence of shrinkage ratio is different in sintering mechanism. Although grain-size effect also depends on sintering mechanism, smaller grain size has higher shrinkage rate in any case. Furthermore, two step sintering has been studied in which a green body is fired at a higher temperature to obtain higher relative density, followed by firing at a lower temperature for a long time to restrict grain growth (Fig. 3.5.6) [5]. As a result, a dense sintered body composed of nano-size grains has been obtained.

Carbides and nitrides have high surface energy but low self-diffusion constant. Therefore, it is difficult for them to be densified by themselves. Such sintering resistant materials are densified by liquid phase sintering. The liquid phase sintering technique has been applied not only to such sintering resistant materials but also to porcelain or refractory. TiC-Ni, ZrC-Ni, WC-Co, ZnO-Bi_2O_3-CoO, AlN-Y_2O_3, Si_3N_4-Y_2O_3-Al_2O_3, SiC-Y_2O_3-Al_2O_3, are taken for instance of densifiable system by liquid phase sintering. Necessary conditions for densification by liquid phase sintering are: (1) enough wettability of liquid to solid, (2) some solubility of solid to liquid, (3) low viscosity of liquid for fast diffusion of solid through liquid and (4) existence of liquid as much as is sufficient. In the initial stage of liquid phase sintering, grain rearrangement occurs due to capillary force between particles (Fig. 3.5.7). Sintering rate is experimentally given by the following equation [6]:

$$\Delta l/l \propto t^{1+y} \tag{3.5.38}$$

where $\Delta l/l$ is the shrinkage ratio, $1+y$ is 1.1–1.3. In the middle stage, dissolution–precipitation process is dominant sintering mechanism to result in densification by smoothing of contact interface and dissolution of finer particles. There are two rate-controlling processes in liquid phase sintering, diffusion controlling process and interfacial reaction controlling process and they are formulated as the following equations, respectively [7]:

$$\left(\frac{\Delta L}{L_0}\right)^3 = \frac{12\delta\Omega\gamma DC}{r^4 kT}t \tag{3.5.39}$$

$$\left(\frac{\Delta L}{L_0}\right)^2 = \frac{4K_r\Omega\gamma C}{r^2 kT}t \tag{3.5.40}$$

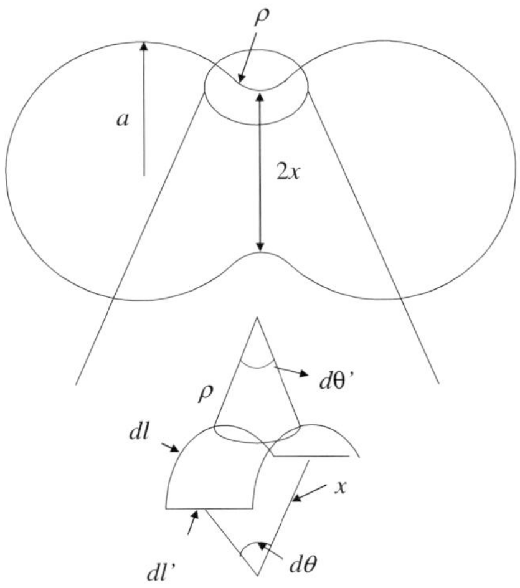

Figure 3.5.4
Stress at the neck between two particles.

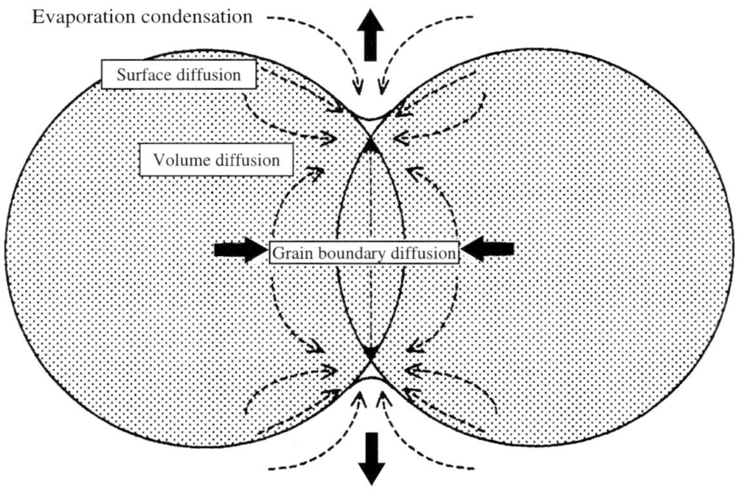

Figure 3.5.5
Mass transform in solid state sintering.

Table 3.5.4
Typical equations of sintering rate for each sintering mechanism.

Sintering mechanism	x/r vs. t	$\Delta L/L_0$ vs. t
Surface diffusion (Kuczynski [1])	$\left(\frac{x}{r}\right)^7 = \left(\frac{56\gamma\Omega\delta D_s}{kTr^4}\right)t$	
Evaporating condensation (Kuczynski [1])	$\left(\frac{x}{r}\right)^3 = \left(\frac{3\sqrt{\pi}\alpha\gamma M^{3/2}p_0}{\sqrt{2}R^{3/2}T^{3/2}d^2r^2}\right)t$	
Volume diffusion (Kingery [2])	$\left(\frac{x}{r}\right)^5 = \left(\frac{80\gamma\Omega D_V}{kTr^3}\right)t$	$\left(\frac{\Delta L}{L_0}\right)^{5/2} = \left(\frac{20\gamma\Omega D_V}{\sqrt{2}kTr^3}\right)t$
Grain boundary diffusion (Coble [3])	$\left(\frac{x}{r}\right)^6 = \left(\frac{192\gamma\Omega\delta D_b}{kTr^4}\right)t$	$\left(\frac{\Delta L}{L_0}\right)^3 = \left(\frac{3\gamma\Omega\delta D_b}{kTr^4}\right)t$
Viscous flow (Frenkel [4])	$\left(\frac{x}{r}\right)^2 = \left(\frac{3\gamma}{2\pi\eta r}\right)t$	$\left(\frac{\Delta L}{L_0}\right) = \left(\frac{9\gamma}{4\eta r}\right)t$

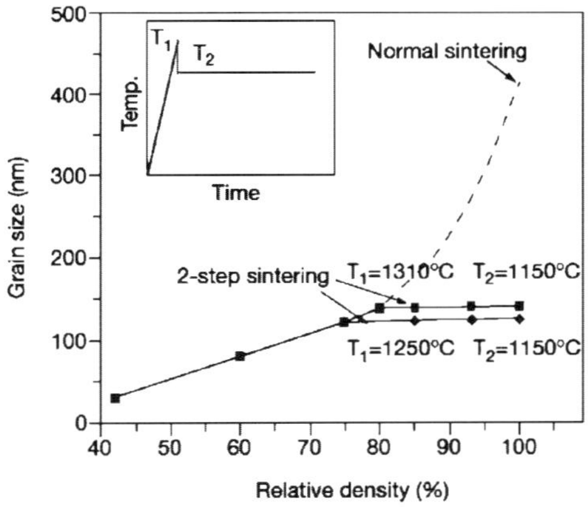

Figure 3.5.6
Densification behavior of Y_2O_3 nano particles and microstructure of the sintered body fired by adding 1 wt% MgO at 1,080°C to 76% and at 1,000°C for 20 h.

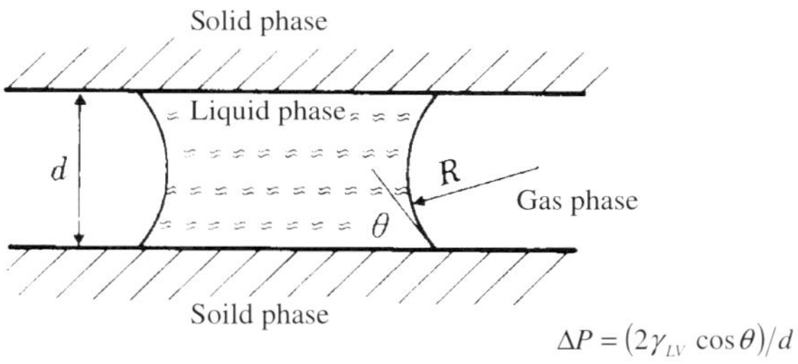

Figure 3.5.7
Capillary force ΔP between two particles.

where δ is the thickness of liquid phase between particles, Ω the atomic volume, γ the vapor/liquid interfacial energy, D the diffusion constant of solid in liquid, C the solid content in liquid, t the sintering time, k the Boltzmann constant, T the temperature, r the particle size and K_r the reaction constant. The effect of time and grain size on shrinkage rate is the same formula as solid phase sintering. The exponent depends on the rate controlling process; shrinkage ratio is proportional to 1/3 power and 1/2 power of time in diffusion controlling process and interfacial reaction controlling process, respectively. In the final stage, grain growth and shape relaxation occur. Microstructure of sintered body by liquid phase sintering includes secondary phase. Some of the secondary phase is glass, which influences the mechanical, chemical and thermal properties.

Since diffusion for densification is enhanced under applied pressure, the pressure sintering, such as hot-pressing or hot isostatic pressing, is carried out in order to densify sintering resistant material.

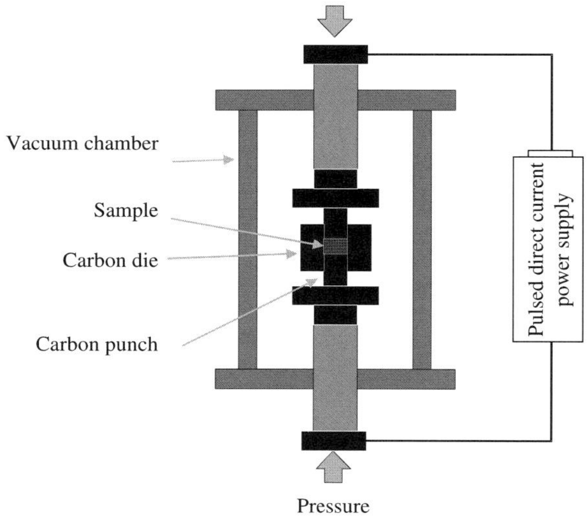

Figure 3.5.8
Apparatus of spark plasma sintering.

Furthermore, spark plasma sintering technique, in which an electrical conductive mold is filled with powder and then heated up by turning on electricity into the mold and/or powder, has been proposed (Fig. 3.5.8) [8]. By this technique, it is possible for many kinds of sintering resistant materials to be densified at lower temperature and shorter time. Ceramic materials composed of nanoparticles have also been fabricated by SPS (Fig. 3.5.9) [9].

References

[1] G.C. Kuczynski: *Trans. AIME*, **185**, 169–177 (1949).
[2] W.D. Kinkery, M. Berg: *J. Appl. Phys.*, **26,** 1205–1212, (1955).
[3] E.L. Coble: *J. Am. Ceram. Soc.*, **41**, 55–62 (1958).
[4] J. Frenkel: *J. Phys.*, **IX**, 385–391(1945).
[5] I.-W. Chen, X.-H. Wang: *Nature*, **404**, 168–171 (2000).
[6] W.D. Kingery: *J. Appl. Phys.*, **30**, 301–306(1959).
[7] W.D. Kingery, M.D. Naraslman: *J. Appl. Phys.*, **30**, 307–310 (1959).
[8] M. Omori, T. Hirai: *Mater. Sci. Technol.*, **40**, 138–142 (2003).
[9] Q. Li, K. Komeya, J. Tatami, T. Meguro and L. Gao: *Key Eng. Mater.*, **247**, 75–78 (2003).

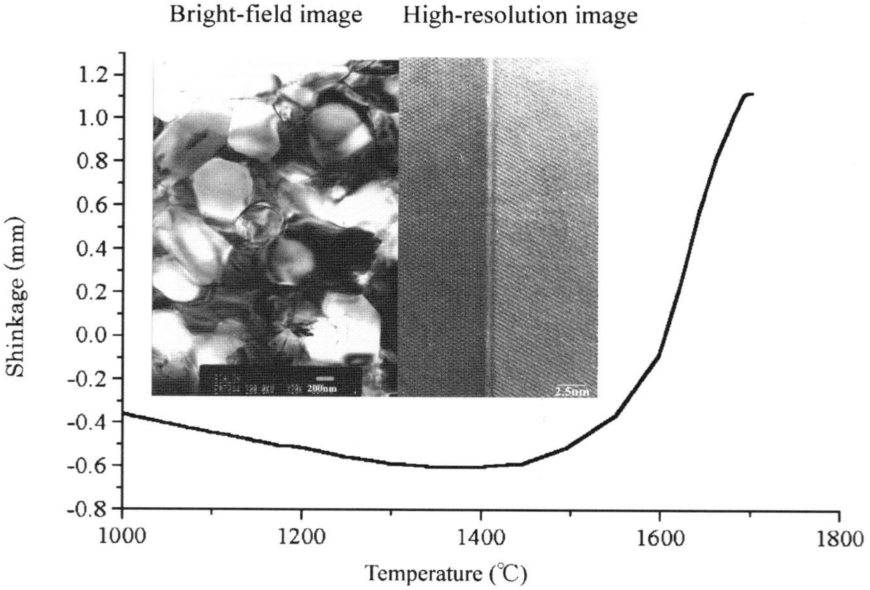

Figure 3.5.9
Sintering behavior of β-SiAlON nano particles and microstructure of the sintered body.

3.5.2 Control of interactions between particles in liquids

Many colloidal particles in industrial processes are in flow fields, where particles are flowing, colliding with each other. The stability of these particles is decided by the combination of hydrodynamic and static interactions between particles. In other words, it depends on the balance between their kinetic energy and interaction potential whether the collision of particles results into their coagulation or the dispersion by their rebound. The relative motion of i particle against j particle are determined by the equation of motion.

$$m_i(d\boldsymbol{v}_{ij}/dt) = \boldsymbol{F}_{\mathrm{I}ij} + \boldsymbol{F}_{\mathrm{B}ij} + \boldsymbol{F}_{\mathrm{F}ij} \quad (3.5.41)$$

where m_i is the mass of colliding particle i, $\boldsymbol{v}_{ij}$ the relative velocity of i and j particles, t the time and $\boldsymbol{F}_{\mathrm{I}ij}$, $\boldsymbol{F}_{\mathrm{B}ij}$, $\boldsymbol{F}_{\mathrm{F}ij}$ are the vectors which represent the static interactions between particles, the time-averaged force due to the Brownian motion and the hydrodynamic interactions, respectively. $\boldsymbol{F}_{\mathrm{I}ij}$ is called the static interaction force independent of the fluid motion, such as the electrostatic and the van der Waals forces. On the other hand, $\boldsymbol{F}_{\mathrm{B}ij}$ and $\boldsymbol{F}_{\mathrm{F}ij}$ are called the dynamic interaction force generated by the motions of particles and fluids. There exist two concepts on the stability, that is, coagulation and dispersion, of colloidal particles.

(1) One is that the stability of dispersions is determined qualitatively only by the absolute magnitude of $\boldsymbol{F}_{\mathrm{I}ij}$. When the repulsive force of $\boldsymbol{F}_{\mathrm{I}ij}$ is large enough, the collisions of particles do not result into their coagulations, so that the dispersion is regarded as stable. In this way, the stability of dispersions is able to be determined without knowing the detailed relative motion of particles. This kind of decision on the stability of dispersions is very useful when either the absolutely stable dispersions or the absolutely unstable dispersions are needed. Studies of these kinds have been carried out extensively in the field of the science of colloid chemistry [1, 2].

(2) The other concept is that the stability of dispersions should be determined quantitatively by solving equation (3.5.41), including the dynamic interaction forces between particles. The dynamic behavior of particles is especially important in the particle processing, where the collision of particles always occurs. In this case, it is important to know the degree of each contribution of the kinetic energy and interaction potential to the dispersion stability [3].

The most fundamental concept (1) described above is explained below. The details of the concept (2) are given elsewhere [3].

3.5.2.1 Charging mechanism of colloidal particles

When particles are dispersed in solutions, their surfaces are more or less charged. The charging mechanism is classified as follows: the charging by functional groups on the particle surface, the charging by adsorbed ions and the charging by isomorphic substitution.

(1) Charging by functional groups

When the particle surface has the functional group, such as –OH, –COOH and $-NH_3$, the surface will be charged by the dissociation in an aqueous solution. For oxide particles, such as SiO_2, TiO_2 and Al_2O_3, water molecules adsorb and form –OH groups on the surface, as shown below.

$$\mathrm{M-OH_2^+} \overset{\mathrm{H^+}}{\rightleftarrows} \mathrm{M-OH} \overset{\mathrm{OH^-}}{\rightleftarrows} \mathrm{M-O^-} + \mathrm{H_2O}$$

where M indicates a metal atom. In the cases of oxide particles, as well as zwitterionic particles, the surface potential is always positive at low pH and becomes negative at high pH. Hence there exists the point of zero charge (pzc) (pH_0) in-between as shown in Table 3.5.5.

(2) Charging by ionic adsorption

For AgI crystal particle, the surface charge is determined by the excess amount of either Ag^+ or I^- ions adsorbed on the particle surface. This surface potential Ψ_0 is able to be calculated by the so-called Nernst equation:

$$\Psi_0 = -\left(\frac{2.3kT}{e}\right)(\mathrm{pAg} - \mathrm{pAg_0}) \quad (3.5.42)$$

where k is the Boltzmann constant, T the temperature, e the elementary charge. A few other crystal particles, such as AgBr, AgCl, Ag_2S, AgCNS, and $BaSO_4$, are charged by the same mechanism. Values of the point of zero charge, pAg_0, are given in Table 3.5.5.

(3) Charging by isomorphic substitution

This charging mechanism is found particularly in clay minerals. When there are defects in the crystal lattice of particles in which Si^{4+} is substituted by Al^{3+}, the deficit of charge results in charging particles. This charge is not affected by the solution pH.

Table 3.5.5
Point of zero charge (pH_0, pAg_0).

Materials	pH_0	Measuring method	Materials	pH_0	Measuring method
α-Al_2O_3	9.1~9.2	sp	NiO	10.3	mep
γ-Al_2O_3	7.4~8.6	sp	$Ni(OH)_2$	11.1	mep
α-AlOOH	7.7	sp	PuO_2	9.0	mep
	9.4	mep	Sb_2O_5	<0.4	eo
γ-AlOOH	5.5~7.5	mep	SiO_2 (Quartz)	1.8~2.5	mep
α-$Al(OH)_3$	5.0~5.2	mep		2.2~2.8	so
γ-$Al(OH)_3$	9.3	mep	(Sol)	1~1.5	mep
BeO	10.2	eo	ThO_2	9.0~9.3	mep
CdO	10.4	mep	SnO (Hydrate)	6.6	mep
$Cd(OH)_3$	>10.5	mep	SiO_2	6.6~7.3	mep
CeO_2 (Hydrate)	6.8	mep	TiO_2		
			(Synthesized rutile)	6.7	sp
$Co(OH)_2$	11.4	mep	(Natural rutile)	5.5	sp
$Cu(OH)_2$ (Hydrate)	7.7	mep		4.8	mep
CuO	9.5	mep	(Synthesized anatase)	6.0	mep
Cr_2O_3 (Hydrate)	6.5~7.4	mep	WO_3 (Hydrate)	0.5	mep
$Fe(OH)_2$	12.0	eo	V_3O_8	4	sp
Fe_3O_4	6.5	eo	V_2O_3 (Hydrate)	9.3	mep
α-Fe_2O_3	8.3	mep	ZnO (Hydrate)	9.3	mep
γ-Fe_2O_3	6.7~8.0	mep	ZnO_2	4	eo
α-FeOOH	6.1~6.7	sp	La_2O_3	10.5	mep
γ-FeOOH	7.4	mep	Y_2O_3(Hydrate)	9.0	mep
$Pb(OH)_2$	9.8	mep	HgO	7.3	mep
MgO	12.4	sp	AgI	(pAg_0) 5.63~5.65	titr
$Mg(OH)_2$	12.4	eo	AgBr	(pAg_0) 5.4	titr
$Mn(OH)_2$	12.0	mep	AgCl	(pAg_0) 4.6	titr
MgO	7.3	mep	Ag_2S,pH 4.7	(pAg_0) 10.2	titr

Note: sp, streaming potential; eo, electroosmosis; mep, electrophoresis; titr, titration in water. [C.A. Parks: *Chem. Rev.*, **65**, 177 (1965); R.H. Yoon, T. Salman and G. Donnay: *J. CotioidInterface Sci.*, **70**, 483 (1979); J. Lyklema: *Fundamentals of Interface and Colloid Science*, Academic Press (1995)]

3.5.2.2 DLVO theory based on static interactions [1, 2]

The stability of colloidal particles in solutions is essentially determined by both contributions of the electrostatic repulsive and van der Waals forces. This mechanism was established by the research group of Derjagui and Landau and the group of Verwey and Overbeek almost simultaneously, so that the theory has been called as DLVO theory.

As described above, particles in solutions are generally charged. Because of the charge neutralization principle, the counter ions of the particle charge were attracted electrostatically toward the particle surface. A part of these ions are adsorbed firmly on the surface and the rest of ions are distributed like an ionic cloud around the particle, as shown in Fig. 3.5.10, balancing their electrostatic attractive force with their thermal diffusion force. The above-mentioned firmly adsorbed layer and the ionic cloud are called as Stern layer and the electrostatic double layer, respectively. The potential decreases linearly from the surface potential Ψ_0 down to the Stern potential Ψ_d within the Stern layer. Because the value of Ψ_0 is not easily determinable, either Ψ_d or the so-called zeta potential Ψ_ζ is regarded as the surface potential.

The potential within the diffusion layer is given by the Poisson–Boltzmann equation, as follows:

$$\nabla^2\Psi = -\sum(n_{0i}z_ie/\varepsilon)\exp(-z_ie\Psi/kT) \qquad (3.5.43)$$

where n_0 is the ionic concentration of bulk solution, z is the ionic valency, $\varepsilon(=\varepsilon_r\varepsilon_0)$ the permitivity of the medium, ε_r and ε_0 are the relative and vacuum permitivities, respectively.

When the solution is symmetric aqueous solution ($z_+=-z_-=z$, $n_{0+}=n_{0-}=n_0$) and the surface potential is so low that the Debye–Hückel approximation holds, i.e., $ze\Psi<kT$ [$\Psi<25.7/z$ [mV] at 25°C], equation (3.5.43) can be solved analytically. For flat plate, Ψ is given as follows:

$$\Psi = \Psi_0\exp(-\kappa x) \qquad (3.5.44)$$

$$\kappa = (2n_0z^2e^2/\varepsilon kT)^{0.5} = (2\times10^3C_eN_{av}z^2e^2/\varepsilon kT)^{0.5} \qquad (3.5.45)$$

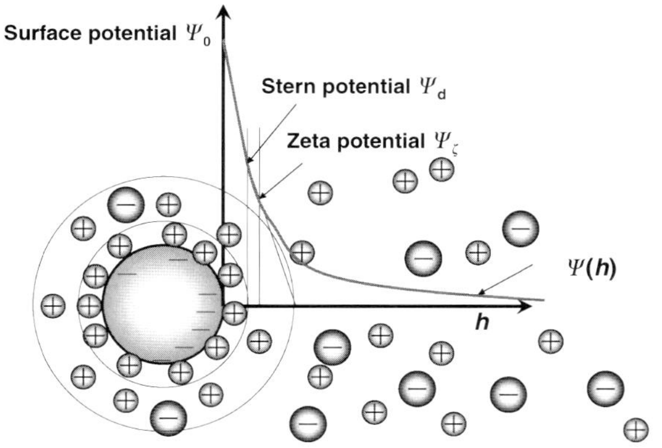

Figure 3.5.10
Schematic drawing of the ionic cloud and potential variation around a charged particle in a solution.

where Ψ_0 may be replaced by either Ψ_d or Ψ_ζ. x is the distance from the surface, C_e the electrolyte concentration [mol/l], and N_{av} the Avogadro number. The distance at the intersection of the tangential line at $x=0$ with the x-axis is equal to $1/\kappa$, which is called the thickness of double layer. This value is often used as a measure of the thickness of double layer. For the aqueous solution at 25°C, $1/\kappa = 3\times10^{-10}/(z\sqrt{C_e})$, where C_e is the electrolyte molar concentration [mol/dm^3]. Because the surface charge density σ_0 is equal to the sum of the charge ρ of counter ions per unit area within the double layer, the following equation holds.

$$\sigma_0 = \int_0^\infty \rho dx = -\varepsilon(d\Psi/dx)_{x=0} = (4n_0 ze/\kappa)\sinh(ze\Psi_0/2kT) \quad (3.5.46)$$

The surface potential Ψ for a spherical particle of radius a and the total charge Q are given respectively, as follows:

$$\Psi = (\Psi_0 a/r)\exp\{-\kappa(r-a)\} \quad (3.5.47)$$

$$Q = -\int_a^\infty 4\pi r^2 \rho dr = 4\pi\varepsilon a(1+\kappa a)\Psi_0 \quad (3.5.48)$$

where r is the distance from the center of the particle. When two surfaces approach within the distance where their double layers overlap each other, the repulsive interaction force arises between the surfaces. This interaction is normally discussed in terms of either the inter-surface potential V or force F. The correlation between V and F is given by,

$$F = -dV/dr \quad (3.5.49)$$

The repulsive potential V_R between similar particles is given explicitly by the following equations, when the magnitude of Ψ_0 is sufficiently low.

$$V_R = \pm 2\pi\varepsilon a\Psi_0^2 \ln\left[1 \pm \exp\{-\kappa(r-2a)\}\right] (\kappa a > 10) \quad (3.5.50)$$

$$= 4\pi\varepsilon a\Psi_0^2 \exp\{-\kappa(r-2a)\}/r (\kappa a < 5) \quad (3.5.51)$$

where the + sign in equation (3.5.50) indicates the constant surface potential, and the – sign indicates the constant surface charge. At $\kappa a < 5$, V_R is given by equation (3.5.51), independently of the surfaces of constant charge and constant potential. The above equations are widely used as V_R, but there exist the other expressions which may be used under the different conditions. The details are given elsewhere [4].

It is well know that the van der Waals attractive force acts between particles of any kind. The potential between the similar spherical particles is given by the following equation:

$$V_A = -(A/6)\left[2a^2/(r^2-4a^2) + 2a^2/r^2 + \ln\{(r^2-4a^2)/r^2\}\right] \quad (3.5.52)$$

A is the Hamaker constant which depends only on the properties of the particle and medium. The experimental values are investigated extensively but the theoretical prediction is also possible [4]. The magnitude of A is in the order of 10^{-20} J in gas, and 10^{-21} J in liquids.

When the separation distance of two surfaces is extremely small, the Born potential V_B, which is generated from the overlapping of electron clouds, appears as an infinitely repulsive potential at the separation distance of 4 Å. Hence the total potential V_T is given by the following equation:

$$V_T = V_R + V_A + V_B \quad (3.5.53)$$

Typical potentials are shown in Fig. 3.5.11. It depends on the balance between the energy of particles given by the Brownian motion and/or the fluid motion and the potentials at the peak and minima, whether particles are coagulated at the primary or secondary minima or they are dispersed. However, one can know only by equation (3.5.53) without knowing the contribution of fluid motion in equation (3.5.41) that the dispersion is completely unstable when the maximum peak does not exist, and that the dispersion is stable when the maximum peak is large enough, for example, larger than 20 kT.

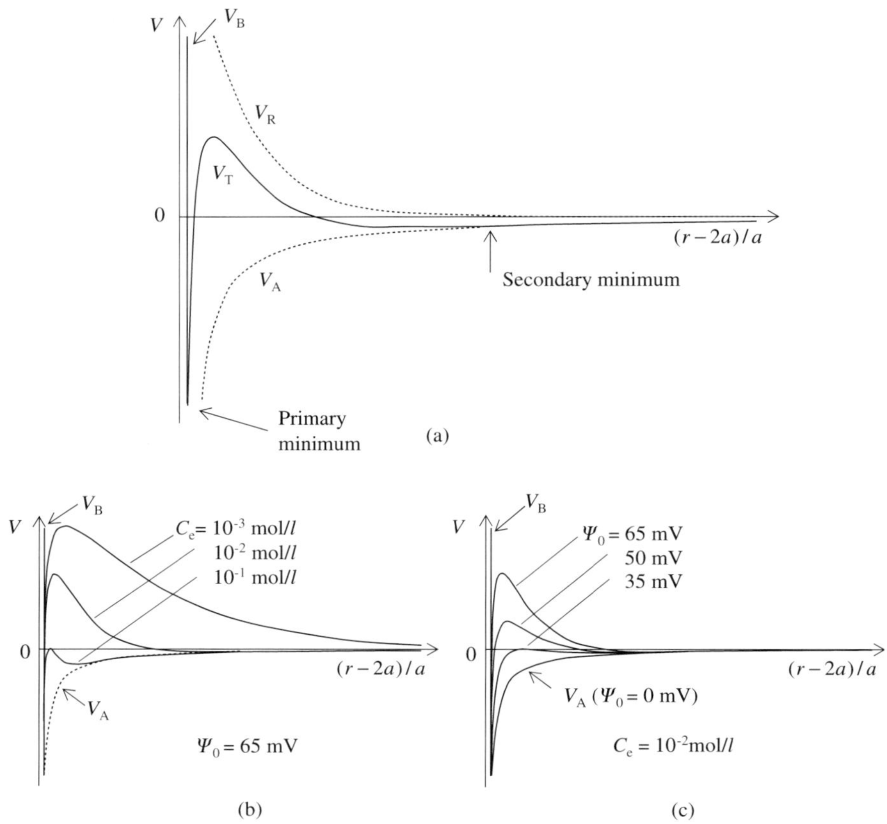

Figure 3.5.11
Typical interaction potentials V between charged particles in solutions given by DLVO theory (a), the dependence on the electrolyte concentration (b), and the dependence on the surface potential (c).

3.5.2.3 Control of dispersion stability using DLVO theory

As known from equations (3.5.50) and (3.5.52), the variable parameters in these equations are Ψ_0 and C_e (or κ) only, because the others belong to the specific properties of dispersions. The value of C_e may be altered by adding salts, and that of Ψ_0 by the solution pH. As illustrated in Fig. 2(b), (c), the dispersion becomes unstable, as C_e increases or Ψ_0 decreases.

When the value of C_e increases, the electrostatic repulsive potential disappears and the potential nearly coincides with the van der Waals potential at the so-called critical coagulation concentration (*CCC*). At $C_e > CCC$, the coagulation rate becomes constant. This region is called the rapid coagulation region, and that of $C_e < CCC$ is called the slow coagulation region. It is known that the value of *CCC* depends on z. According to the theoretical prediction, *CCC*(mol/l) for particles of high surface potential is expressed by the following equation:

$$CCC = 7.2 \times 10^{-57} \varepsilon_r^3 T^5 / z^6 A^2 \qquad (3.5.54)$$

where the temperature and Hamaker constant are given by T(K) and A(J), respectively. This indicates that the value of *CCC* for trivalent cations is 1/729 of *CCC* value for mono-valent cations. This dependence of *CCC* on z is confirmed experimentally, and the relation is known as the Schulze–Hardy law. This coincidence between the prediction by the DLVO theory and experimental results is one of the reasons why the DLVO theory has been widely accepted as the fundamental theory for hydrophobic colloids. In the many real colloids, the following relation may be used.

$$CCC \propto z^{-n} \ (n = 2 \sim 6) \qquad (3.5.55)$$

This rule indicates that stable dispersions are obtainable by reducing the concentration of multi-valent ions, while the addition of multi-valent ions is very effective to promote the coagulation and the solid–liquid separation. The stability of dispersions is controllable also by the surface potential Ψ_0. At the small value of Ψ_0, the repulsive force is reduced and particles will be coagulated. The control of Ψ_0 depends on the charging mechanism. For example, the value of Ψ_0 is changed by the solution pH for oxide particles. It is known that the dispersion is unstable at $\Psi_0 < 20$ mV.

3.5.2.4 Non-DLVO interactions

The non-DLVO interactions include all the interactions which cannot be explained by the DLVO theory. The non-DLVO interaction appears for the surfaces with adsorbed layers of water molecules, ions, hydrated ions, surfactants, polymers and nanobubbles.

(i) Non-DLVO interactions in solutions and the relation with the stability of nanoparticles

Not only polarized water molecules but also ions and hydrated ions, more or less, adsorb on the charged surface of particles in solutions. The thickness of these adsorbed layers depends on the properties of particles and the medium. For silica and mica surfaces, there exists the layer of ca. 1 nm thickness, that is, the thickness of two or three layers of water molecules [5]. Fig. 3.5.12 shows the data of interaction force F_T in a 10^{-3}M LiCl solution measured by the Atomic Force Microscope (AFM). The data at the separation $h > 10$ nm agree well with the prediction by the DLVO theory. However, it is clear that the interaction at $h < 10$ nm is always repulsive in spite of the van der Waals attraction predicted by the DLVO theory. This is caused by the structured force generated by the adsorbed layers of hydrated Li^+, which is sometimes called as the solvation force. The thickness of adsorbed layer follows the order of hydration enthalpy of ions, $Cs^+ < K^+ < Na^+ < Li^+$ and $Ca^{2+} < Mg^{2+}$, and the degree of coagulation of particles also follows this order.

These experimental results are very important because the salvation forces do influence the stability of nanoparticles.

1. When the particle size becomes smaller than 100 nm, the thickness of adsorbed layer is of the same order with the distance where the van der Waals attraction influences. Hence, the probability of coagulation by the collision of particles reduces exponentially, as shown in Fig. 3.5.13. This implies that the dispersion becomes more stable as the particle size decreases, if all the surface properties are the same [6].
2. Because the strength of the van der Waals attraction changes greatly at the small separation distance, the slight difference of the thickness of adsorbed layers affects extensively the adhesive force between particles [7, 8]. Fig. 3.5.14 shows that the adhesive force F_{ad} is influenced greatly by the contact time of two surfaces, t_c, and the kind of ions. This effect becomes significant with decreasing particle size, because the adsorbed layer becomes more influential with increasing relative thickness of adsorbed layer to the particle size.

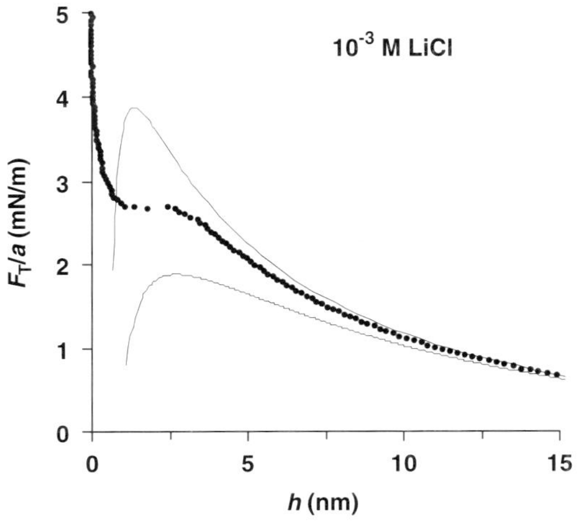

Figure 3.5.12
Comparison of force curves between silica surfaces in a 10^{-3}M LiCl solution between the measurement by AFM and the prediction by DLVO theory. (The upper and lower thin lines represent the predictions under the conditions of constant charge and constant potential of particle surface, respectively.)

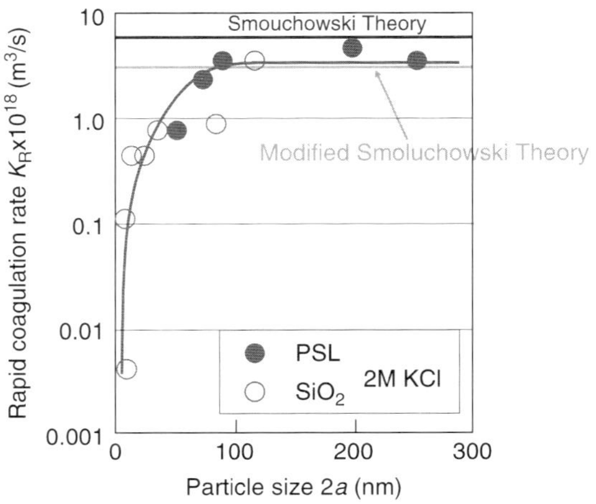

Figure 3.5.13
Dependence of rapid coagulation rate K_R on particle size.

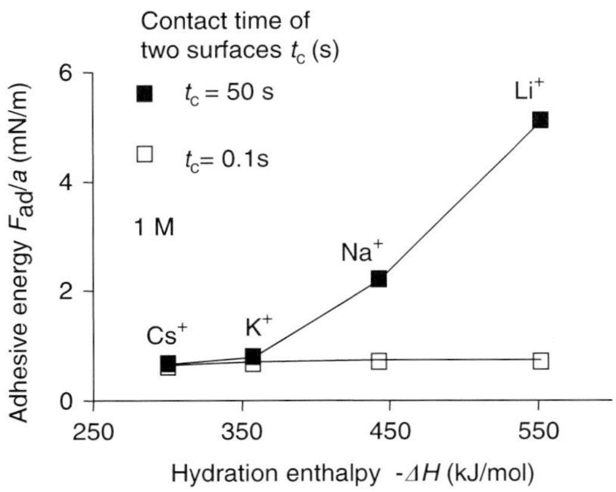

Figure 3.5.14
Dependence of the adhesive force F_{ad} on the contact time t_c between two surfaces and the hydration enthalpy ΔH of monovalent cations.

(ii) Non-DLVO interactions by surfactants

The stability of particles in surfactant solutions depends on the many factors of adsorbed surfactants, such as their charge, hydrophobicity, structure, and so on. If the particle surface is hydrophobic, surfactants adsorb on the surface such that the aliphatic tail is adsorbed on the surface and the polar head is directed toward the solution. Then particles are dispersed and become stable because of the inter-particle repulsive force due to the charge of the adsorbed surfactants.

When surfactants with the charge opposite to the particle surface are dosed, the head groups are adsorbed on the surface by the electrostatic attraction. If the particle surface is adsorbed completely in this case, the surface becomes hydrophobic and particles will be unstable by the hydrophobic attractive force. When the surfactant is dosed further, the secondary adsorbed layer will be formed such that the polar heads are directed toward the solution. Then the surface charge is reversed and the suspension becomes stable again.

(iii) Non-DLVO interactions by polymers

As shown in Fig. 3.5.15, polymers may become either the flocculant or the dispersant, depending on the dosage. When the dosage is sufficiently small compared with the amount of saturated adsorption, polymers act as the flocculants bridging between particle surfaces. At the saturated adsorption, particles are structurally repulsive for good solvent, while particles are attractive for poor. This depends on how affinitive the polymers are either to polymers on the other particle surface or the medium. However, when polymers are much more concentrated, such that the concentration of polymers in the gap between the neighboring particles becomes smaller than that of the bulk, the attractive force acts between surfaces, because of the osmotic pressure. This flocculation is called the depletion flocculation [2].

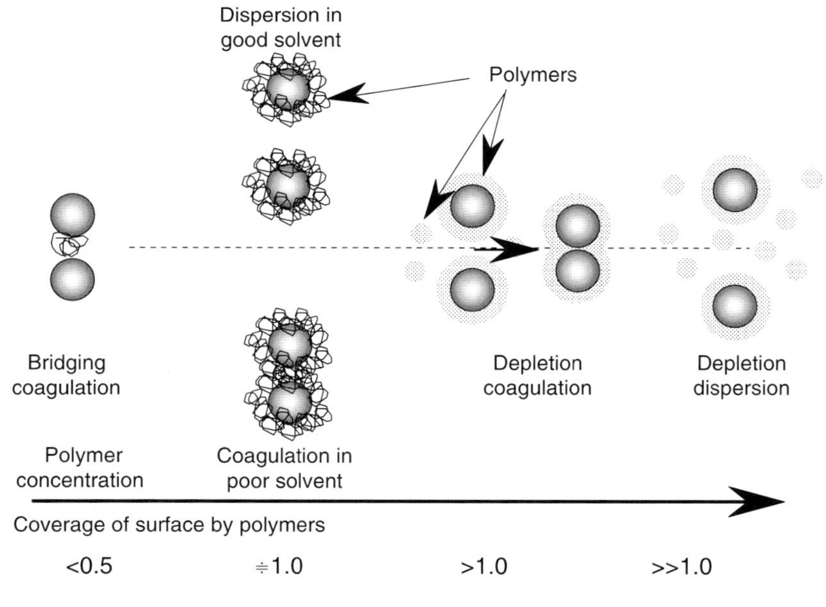

Figure 3.5.15
Schematic drawing of the dependence of dispersion stability on the polymer concentration in solutions.

The roles of polymer flocculants have been estimated by the results of macroscopic experiments, but sometimes the estimated mechanism differs from the real one. Fig. 3.5.16 shows the in-situ AFM images of cationic high polymers transferred from the solution onto the mica surfaces on 20 days after the solution was prepared. It is found that the polymers are coiled at the beginning, and 10 days are needed for them to relax completely. Then it is examined which state is better as polymer flocculants; the initial stage of the island-like adsorption coiled polymers or the final stage of the adsorption of fully extended polymers. The former stage is found to be better, because the strong attractive force acts between the highly-charged coiled sites and the bare sites of opposite charge [9, 10]. This indicates that the molecular-scale information is directly related with the macroscopic behavior of particles, so that the understanding of phenomena on the molecular scale is extremely important.

(iv) Non-DLVO interactions by hydrophobic interaction and nanobubbles

It has been well known especially in the field of flotation that the strong attractive force acts between hydrophobic surfaces in solutions. This is explained as follows: water molecules between hydrophobic surfaces of the small separation distance, say 10 nm, become extremely unstable because of the poor affinity of water molecules to the surface. This phenomenon was confirmed experimentally [5].

However, according to the recent measurements with AFM, many data have been reported as shown in Fig. 3.5.17, in which the hydrophobized surfaces attract strongly with each other from the long separation distance, such as a few hundreds nanometers. The origin of this long attractive force F has been a mystery for a long time. The author clarified that the attractive force is attributable to the nano-size bubbles attached on the hydrophobized surfaces. When the hydrophobic surfaces are immersed into solution, it is extremely difficult to be completely free from the attachment of nanobubbles on the surface, and also to get rid of the nano bubbles completely afterwards. Hence the long-range strong attraction is due to surface tension of the bridging gas between surfaces, which is attributable to the coalescence of bubbles on particle surfaces. The details are given elsewhere [11, 12].

It is also found recently that the long-range attraction appears even between hydrophobized surfaces by the adsorption of surfactants [13]. The origin of this force is considered to be due also to the nanobubbles which are introduced into solution in the preparation

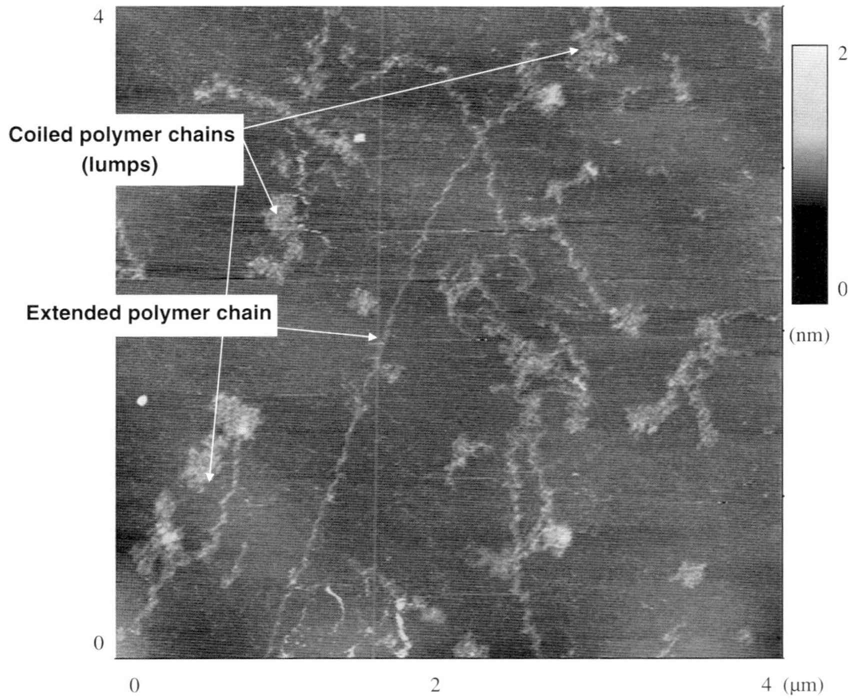

Figure 3.5.16
In-situ AFM image of polymers which are dissolved in water for 20 days and then transferred on the mica surface.

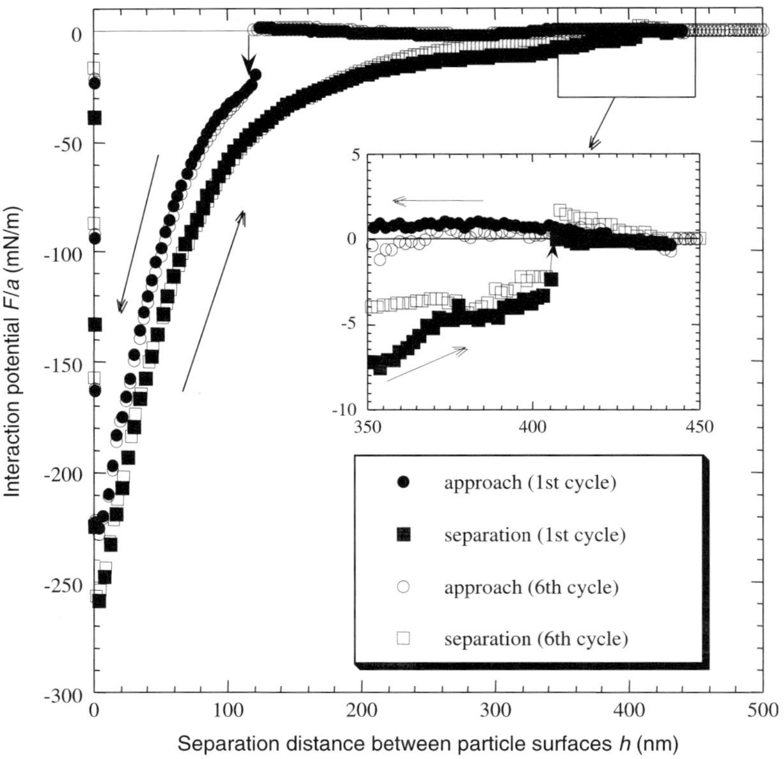

Figure 3.5.17
Force curves between hydrophobized silica surfaces in water measured repeatedly by AFM. The jumps of the approaching and separating curves indicate the existence of bubbles on the surfaces.

process of solutions from powder surfactants, because of the hydrophobicity of alkyl chains of surfactants.

References

[1] H.R. Kruyt: *Colloid Science*, Elsevier (1952).
[2] R.J. Hunter: *Foundations of Colloid Science*, Vol. 1, Clarendon Press (1987).
[3] K. Higashitani: *Biryushi Kogaku* (Japanese), Asakura (1994).
[4] W.B Russel, D.A. Saville and W.R. Schowalter: *Colloidal Dispersions*, Cambridge University Press, Cambridge (1989).
[5] N.J. Israelachvili: *Intermolecular and Surface Force*, (2nd ed), Academic Press (1992).
[6] K. Higashitani, M. Kondo and S. Hatade: *J. Colloid Interf. Sci.*, **142**, 204 (1991).
[7] I.U. Vakarelski, K. Higashitani: *J. Colloid Interf. Sci.*, **242**, 110 (2001).
[8] I.U. Vakarelski, K. Ishimura and K. Higashitani: *J. Colloid Interf. Sci.*, **227**, 111 (2000).
[9] T. Arita, Y. Kanda, H. Hamabe, T. Ueno, Y. Watanabe and K. Higashitani: *Langmuir*, **19**, 6723 (2003).
[10] T. Arita, Y. Kanda and K. Higashitani: *J. Colloid Interf. Sci.*, **273**, 102 (2004).
[11] N. Ishida, M. Sakamoto, M. Miyahara and K. Higashitani: *Langmuir*, **16**, 5681 (2000).
[12] N. Ishida, T. Inoue, M. Miyahara and K. Higashitani: *Langmuir*, **16**, 6377 (2000).
[13] M. Sakamoto, Y. Kanda, M. Miyahara and K. Higashitani: *Langmuir*, **18**, 5713 (2002).

3.5.3 Characterization techniques for interactions between particles

3.5.3.1 Surface forces measurement

Regulation of dispersion and aggregation properties of colloidal particles, including nanoparticles, is one of the key issues of colloid science. Their dispersion and aggregation states can be controlled by utilizing the surface charge or polymer adsorption layer of the particles. The interaction is attractive when the particles aggregate, while it is repulsive when the particles are dispersed. The forces acting between macroscopic surfaces are called the surface forces, which we should be able to thoroughly understand only after measuring them directly and precisely.

(1) Surface forces measurement

(a) Surface forces

The surface forces are proportional to the surface area involved. Frictional and adhesion forces are also a part of the surface forces. Typical surface forces are van der Waals force, electrical double-layer force between charged surfaces, steric forces due to the size and shape of molecules, and the solvation force. The first two forces, i.e. van der Waals force and electrical double-layer force, are called the DLVO force, and other forces are often classified as the non-DLVO force (see Sections 3.5.1 and 3.5.2).

(b) Surface forces measurement

This method directly measures the distance dependence of interaction (interaction potential) between surfaces using a spring balance. It provides the knowledge about the distance dependence and magnitude of the surface forces. It is also possible to calculate the force, adhesive force, necessary for separating the surfaces in the adhesive contact [1].

Surface force measurement was first developed to realize the dispersion forces. The initial attempt of this measurement dates back in the 1930s [2]. The precise measurement of distance, which is necessary for the present nanometer scale measurement, became possible by introducing fringes of equal chromatic order (FECO) interferometry developed by Tabor et al. [3]. The DLVO theory for discussing the stability of colloidal dispersions was proved in the late 1970s, when a prototype of the present surface forces apparatus (SFA) was completed for the measurements of surface interactions in liquids [4].

The surface force apparatus as well as the colloidal probe atomic force microscopy [5] are employed for surface force measurements. When we use the former instrument, the resolutions of distance and force are 0.1 nm and 1 nN, respectively. The geometry of the sample surfaces is in two crossed cylinders for the former, and a sphere and a flat plate for the latter. From the early stage of this research, these geometries have been adopted to avoid the complicated adjustment such as keeping two flat surfaces parallel. It is known that the measured surface force (F) normalized by the radius of the surface curvature (R) is proportional to the interaction energy between flat surfaces (G_f) ($F/R = 2\pi G_f$: Derjagnin approximation), therefore, this geometry is also convenient for comparing the experimental results with the theory.

(c) Measurement method [6]

The conventional surface force measurement is based on the spring balance (Fig. 3.5.18). The interaction

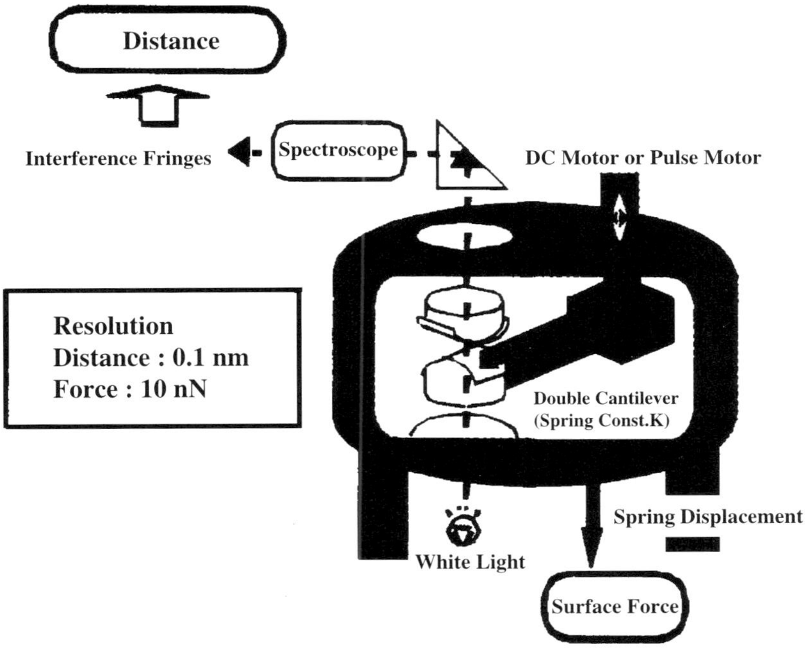

Figure 3.5.18
Surface forces apparatus.

forces between surfaces can be obtained from the displacement of the spring that is connected to one of two opposing surfaces. Sequentially, the separation distance between surfaces can be varied from submicrometers to the adhesive contact. The surface jumps-in to the contact when the gradient of the attractive force, dF/dD, exceeds the spring constant (K), $dF/dD \geq K$ (Fig. 3.5.19). It is possible to investigate the distance dependence of an attractive force by varying the spring constant and finding the distance where jump-in occurs [7]. It is also possible to estimate the adhesive force, which is the force necessary to separate the surfaces in adhesive contact.

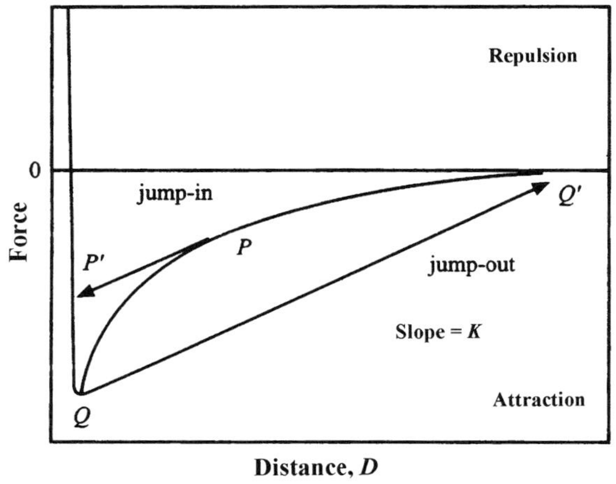

Figure 3.5.19
The surface force (F/R) – distance (D) profiles. Force calculation: (a) $F=K\Delta D$ using change in the separation distance; (b) $dF/dD \geq K$ from jump-in; (c) F (pull-off force) $=K \cdot D$ (Q–Q′) from jump-out.

The force measurement system could be divided into two parts. One is a main chamber (Fig. 3.5.18) that consists of a surface driver (distance controller) and a spring. The driver has a high resolution for varying the separation between two surfaces. Another part is an optical interferometry system (Fig. 3.5.20) to precisely measure the separation distance between surfaces.

1. Main device: In a narrow sense, this unit is called as the surface forces apparatus. This instrument is commercially available. Mark 4 is marketed from Australian National University and SFA2000 can be purchased from Surface LLC (USA). These apparatus are originated from the prototype developed by Tabor, Winterton, and Israelachvili. Construction of a machine by oneself is also possible.
2. Optical system and distance determination: The optical system needs to be set up by the operator. White light source (halogen lamp) is used. The light focused by a lens passes through a quartz lens window at the bottom of the main chamber and the sample surfaces, leading to a spectroscope through a microscope and a prism

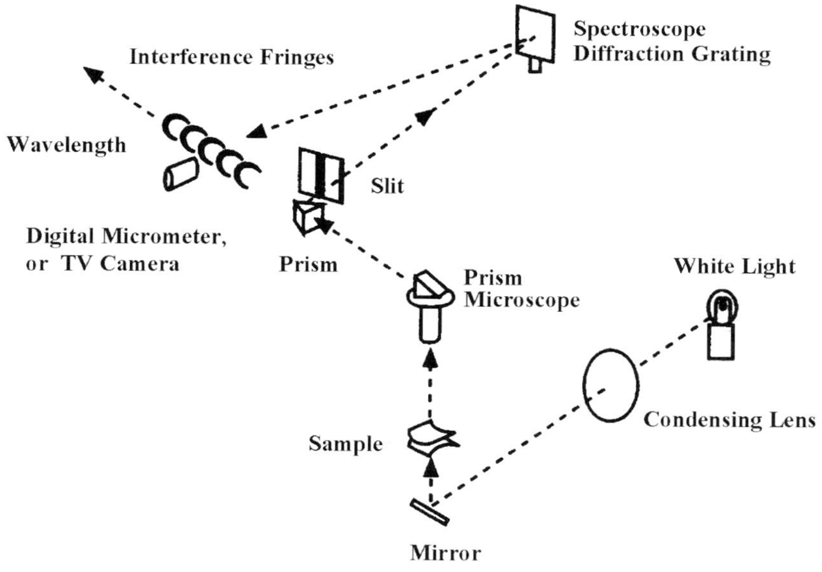

Figure 3.5.20
Schematic diagram of the optical system for surface force measurement.

(See Fig. 3.5.20). The spectroscope (the resolution is 3.2 nm/mm) is the core part of this optical system and used to read the wavelength of the interference fringe (FECO). The wavelength of FECO is determined using micrometer equipped with an encorder and/or a CCD camera with a video gauge, at the exit side of a spectroscope.

Generally, mica, which is molecularly smooth and transparent, is used as a substrate surface. A thin film mica sheet (2~3 μm thick, 10 mm^2) is glued on the cylindrical quartz lens (20 mm radius, 10 mm diameter). These lenses are then mounted in an apparatus in the crossed cylindrical geometry. Interference occurs by the reflected light between the backside of mica surfaces coated with 50-nm silver films, which enhance the interference. Since the surfaces are in the crossed cylindrical geometry, the interference fringes show a parabolic shape. When two surfaces approach, the parabolic curves move toward the short wavelength region, and the spaces of adjacent fringes increase. The separation distance between surfaces can be calculated from the wavelength of these interference fringes [6].

(2) Experimental example

Israelachvili et al. measured the interaction forces between mica surfaces in aqueous solutions in the distance range of 0–100 nm using SFA. Figure 3.5.21 compares the force curves measured by SFA and the DLVO theory for mica surfaces in KNO_3 solution (10^{-4}~1 M) [4]. The repulsive force decays exponentially with distance, and turns to the short-range attractive force. The decay length of the long-ranged repulsive force is in agreement with that of the electrical double layer (Debye length) that changes depending on the salt concentration, i.e. the Debye length decreases with the increased salt concentration. This is a direct experimental demonstration of the DLVO theory.

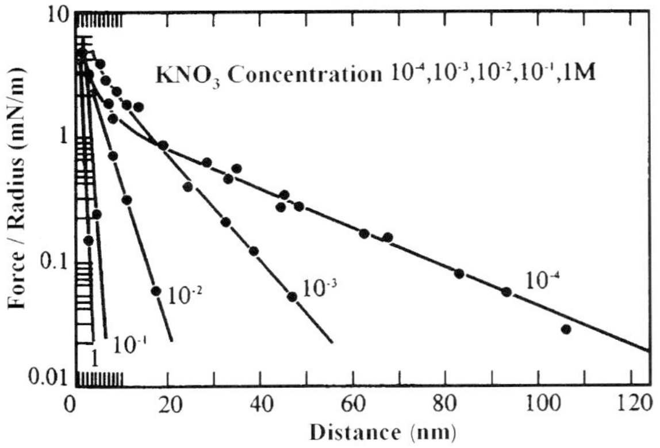

Figure 3.5.21
Surface forces profile between mica surfaces in an aqueous solution.

The surface forces measurements enable us to investigate the electrical double-layer force, which plays an important role for the stability of colloidal dispersions, and the various interactions due to adsorption of polymers or polyelectrolyte [8–10]. Recently, extending research area, the authors have investigated the adsorption in binary liquids [11], and also developed a resonance shear measurement [12] to characterize nano-rheological and/or nano-tribological properties of liquid in a confined nanospaces. Surface force measurement becomes one of essential techniques in nanomaterial science and technology.

References

[1] J.N. Israelachvili: *Intermolecular and Surface Forces* (2nd ed.), Academic Press, London (1991).

[2] V.S.J. Craig: *Colloids Surfaces A*, **129/130**, 75 (1997).

[3] D. Tabor, H.S. Winterton: *Nature*, **219**, 1120 (1968); D. Tabor, H.S. Winterton: *Proc. R. Soc. London Ser. A*, **312**, 435 (1969).

[4] J.N. Israelachvili, G.E. Adams: *J. Chem. Soc. Faraday Trans. I*, **74**, 975 (1978).

[5] W.A. Ducker, T.J. Senden and R.M. Pashley: *Langmuir*, **8**, 1831 (1992).

[6] CSJ (ed): *Colloid Science IV. Experimental Method in Colloid Science*, Kagakudojin Press, Tokyo (1996).

[7] K. Kurihara, T. Kunitake: *J. Am. Chem. Soc.*, **115**, 10927 (1992).

[8] S.S. Patel, M. Tirrell: *Annu. Rev. Phys. Chem.*, **40**, 597 (1989).

[9] K. Kurihara, Y. Murase:*Handbook of Polyelectrolytes and their Applications*, Vol. 1, American Science Publisher, p. 207 (2002).

[10] P. Berndt, K. Kurihara and T. Kunitake: *Langmuir*, **8**, 2486 (1992).

[11] M. Mizukami, K. Kurihara: *Chem. Lett.*, 1005 (1999); 256 (2000).

[12] K. Kurihara: *Ekisyo*, **6**, 34 (2002).

3.5.3.2 AFM colloid probe method

The invention of the surface force apparatus (SFA) has helped create important and fundamental knowledge about the interaction forces between solid surfaces. However, applying the SFA to the measurement of a variety of surfaces is sometimes difficult because of the following: (1) the test surface material must be atomically smooth and transparent, which is practically restricted to mica; (2) it usually takes a long time to obtain data; and (3) the instruments are rather less versatile for handling and operation. Due to these

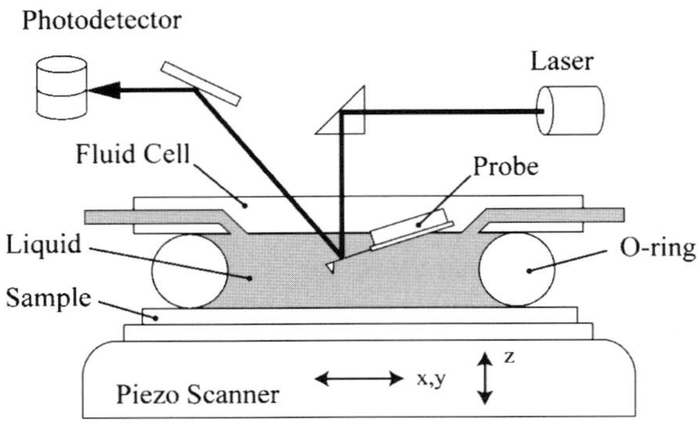

Figure 3.5.22
Schematic drawing of the atomic force microscope.

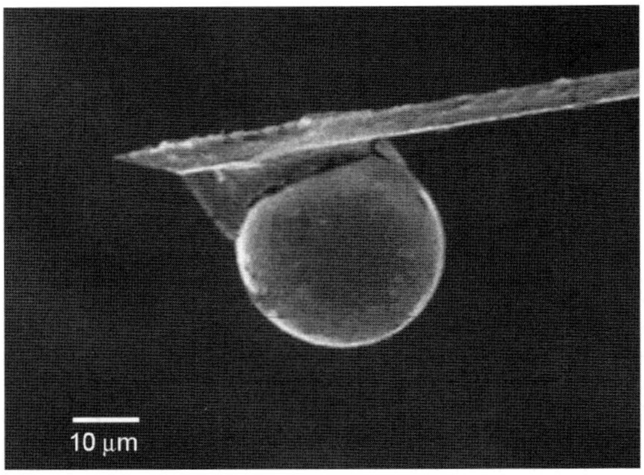

Figure 3.5.23
Scanning electron microscope image of a colloid probe.

reasons, several other methods have been developed to measure the interaction forces between the solid surfaces of a wider variety of materials. Among them, the colloid probe method [1] using an atomic force microscope (AFM) has recently become one of the most important methods in various research fields and has seemed to become as popular as the SFA measurement or more.

The AFM was originally designed to provide high-resolution topographic images of surfaces with a molecular (in the ideal case, atomic) order. Its principle is shown in Fig. 3.5.22: a probe with a cantilever spring and a pyramidal-shaped sharp tip attached to the end of the cantilever is brought in contact with a sample surface mounted on a piezo scanner. The tip scans over the surface by the movement of the piezo scanner on which the sample is mounted. During scanning, the displacement of the cantilever along the undulation of the surface is measured with the laser reflected onto the back of the cantilever. By mapping the displacement, a surface image can be obtained.

In order to apply the AFM to force measurements, the sample surface is moved vertically (z direction) by the expanding–shrinking of the piezo scanner to alter the distance between the probe and the surface. The cantilever deflection by the interaction forces acting between the surfaces can be measured with the laser as is the case in imaging (most commercial AFMs have an operation mode to conduct these procedures automatically). The interaction forces are calculated by Hooke's law, multiplying the spring constant of the cantilever by the cantilever deflection. The interaction forces between macroscopic surfaces, which are a particle and a flat surface, can also be measured by attaching a small particle onto the top of the cantilever (Fig. 3.5.23), which is often called the "colloid probe". For the colloid probe, a wide variety of materials can be used and the surface can also be modified with a variety of molecules by adsorption from a solution or chemical reaction. Due to its versatility and easiness to operate, the colloid probe method has recently been applied to a wider variety of research fields.

The details of the practical procedure of the AFM colloid probe method are as follows. For the colloid probe, spherical particles typically in the diameter range of 1–30 μm are used. A particle is attached to the probe with epoxy glue or hot-melt epoxy resin, using a micropositioning device such as a micromanipulator by observing with an optical microscope or a CCD camera. The probes for imaging are applicable to force measurements and a large variety of probes are now available commercially. Thus, a probe can be chosen based on their spring constant, depending on the strength of the interaction force to be measured. Typically, probes with a soft, triangular cantilever are used to measure repulsive forces, while those with a stiff, rectangular cantilever are used to measure strong attractive forces and/or adhension forces. Although the nominal spring constant is usually provided by the manufacturers, it is highly desirable to determine the spring constant precisely to obtain the accurate value of the interaction force. To determine the spring constant, various methods have been proposed thus far. The most popular and reliable method among them would be the one developed by Cleveland et al. [2]. In this method, the resonant frequency of the oscillating probe is measured both by attaching and without attaching particles of known mass to the cantilever, and the spring constant is calculated from the shift in frequency.

Figure 3.5.24(a) shows a schematic representation of the typical force data obtained by a force measurement between symmetrically charged solid surfaces in an aqueous solution. This measurement cycle entails the following procedure: approaching the probe to the flat sample, making contact with both surfaces, and retracting the probe from the flat sample. During the measurement, the photodetector output voltage V that indicates the deflection of the cantilever by the position

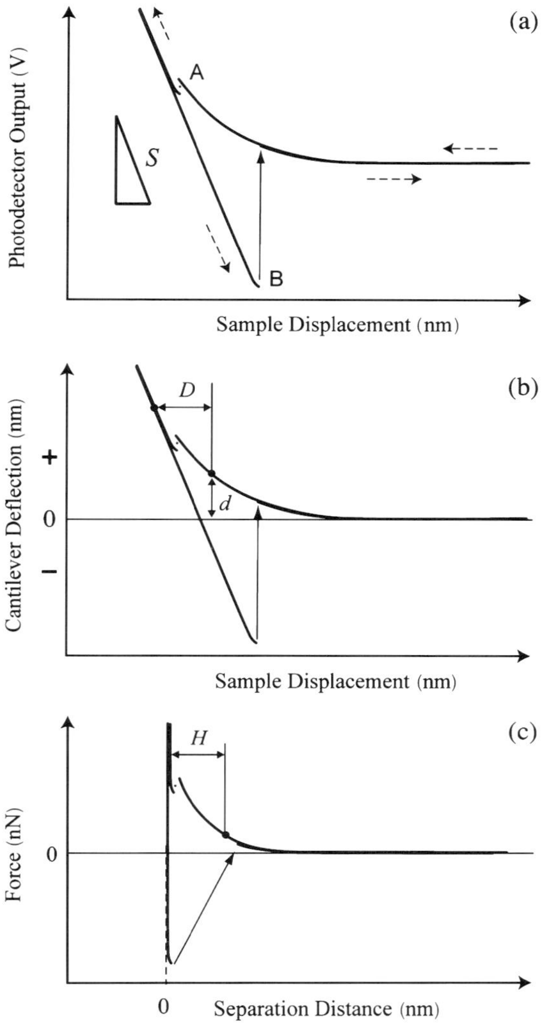

Figure 3.5.24
Typical force data measured in an aqueous electrolyte. (a) The raw data graph; (b) Cantilever deflection-sample displacement data obtained from (a); and (c) The force-separation distance curve converted from (b).

of the laser spot on the detector is monitored as a function of the relative displacement of the sample by the motion of the piezo scanner. As the surfaces approach, the electrostatic repulsion between the surfaces is detected as an upward deflection of the cantilever. Then, the surfaces are brought into contact, which is shown as point A in the figure, due to the van der Waals force acting at a shorter range. If the attraction is strong enough, the particle often "jumps-in" the flat surface due to the instability of the cantilever. When the surfaces are apparently brought into contact, the curve becomes linear because the displacement of the sample should be equal to the cantilever deflection. This linear region is often called the "constant compliance region". As the surfaces are separated, the cantilever gets displaced downward by the adhesion force between the surfaces, and they are separated suddenly (jump-out) at point B in the figure.

Unlike the SFA, the AFM contains no device or system to measure the separation distance between the surfaces directly. Therefore, the force vs. distance relationship has to be calculated from the photodiode voltage vs. piezo position curve, which is obtained directly from the AFM [3]. The cantilever deflection d is calculated from the slope S of the constant compliance region shown in Fig. 3.5.24(a) as

$$d = V/S. \tag{3.5.56}$$

If the distance is assumed to be zero at the constant compliance region, the separation distance H at each point of the data is

$$H = D + d, \tag{3.5.57}$$

where D is the difference in the displacement between each data point and the constant compliance region. Then, one can obtain the force–distance relationship by multiplying d with the spring constant of the cantilever.

It should be noted, however, that the method explained above will not give an accurate zero distance, whereas it is the only method to estimate the separation distance in AFM measurements. In particular, if there is an adsorbed layer such as a polymer layer or adsorbed surfactants and they are not excluded from the surfaces, a certain offset between the obtained and real distance should be expected. Therefore, the zero distance obtained by this method should be regarded as the point of closest approach and the distance H is considered as the relative distance from the point. This is a critical problem encountered in the AFM force measurement so far.

The colloid probe method was first reported by Ducker et al. [1], and they have shown that the accuracy of this method is similar to that of the SFA at a short distance down to several nanometers. After this study, numerous reports have been published for a number of systems. Using different materials as the surfaces, which is one of the advantages of the AFM, the interaction forces between dissimilar metal oxides [4], inorganic mineral oxides [5, 6], and polymer latexes [7] in aqueous solutions have been obtained thus indicating the accuracy of the DVLO theory for dissimilar systems (heterocoagulation theory). As a good example of the fast operation of the AFM, Hu and Bard [8] investigated the adsorption kinetics of thiol on gold by analyzing the dynamic variation of the interaction forces between gold surfaces in thiol solution. In addition, it is also possible to conduct measurements between a colloid probe and a metal surface, whose potential is altered [9].

Not only solid–solid but also solid–fluid interaction forces can be measured with the AFM, including the interactions between an air bubble and a particle [10, 11], and an oil droplet and a particle [12]. In addition to the tangential force, the friction force can be measured by scanning the colloid probe laterally over sample surfaces [13]. By approaching the particle to the surface with high velocity, the effect of the hydrodynamic force can also be estimated [14]. An interesting example is that specifically reacting complimented molecules are attached to the free end of the polymers on the probe and flat surface, and they are hooked in a manner similar to "fly-fishing" to estimate the bonding energy between the molecules [15]. This method has particularly become an important and useful method for the investigation of biological systems. These studies have suitably shown the versatility of the AFM.

On the other hand, one of the disadvantages of the colloid probe method might be that very small particles such as nanoparticles that are not observed optically with a microscope or CCD cannot be used. Typically, the size of the smallest particles used in this method is around 1 μm. As the behavior of the particles is affected by the size and/or the shape of the particle, it is unclear whether the force data for spherical particles with a diameter of several micrometers could directly be applied to analyze the behavior of real nanoparticles. Therefore, some methods to use small materials such as nanoparticles [16] and carbon nanotubes [17] have been proposed so far, which have not been, however, popular yet. In order to overcome this problem fundamentally, it would be also useful to use other approaches such as computer simulation together with experiments.

One of the other important methods to measure the interaction force is total internal reflection microscopy [18]. This method measures the separation distance between a prism and free particles by the interaction with the evanescent wave formed upon the total internal reflection of light emitted in the prism. As this method has an advantage in that the interaction energy of the free particle in a liquid can be obtained, while it is not suitable to measure short-range forces, this method is expected to be developed further.

References

[1] W.A. Ducker, T.J. Senden and R.M. Pashley: *Nature*, **353,** 239–241 (1991).

[2] J.P. Cleveland, S. Manne, D. Bocek and P.K. Hansma: *Rev. Sci. Instrum.*, **64**, 403–405 (1993).

[3] W.A. Ducker, T.J. Senden and R.M. Pashley: *Langmuir*, **8**, 1831–1836 (1992).

[4] I. Larson, C.J. Drummond, D.Y.C. Chan and F. Grieser: *J. Phys. Chem.*, **99**, 2114–2118 (1995).

[5] S. Veeramasuneni, M.R. Yalamanchili and J.D. Miller: *J. Colloid Interface Sci.*, **184**, 594–600 (1996).

[6] G. Toikka, R. Hayes and J. Ralston: *J. Chem. Soc. Faraday Trans.*, **93**, 3523–3528 (1997).

[7] Y.Q. Li, N.J. Tao, J. Pan, A.A. Garcia and S.M. Lindsey: *Langmuir*, **9**, 637–641 (1993).

[8] K. Hu, A.J. Bard: *Langmuir*, **14**, 4790–4794 (1998).

[9] R. Ratieri, M. Grattarora and H.-J. Butt: *J. Phys. Chem.*, **100**, 16700–16705 (1996).

[10] M.L. Fielden, R.A. Hayes and J. Ralston: *Langmuir*, **12**, 3721–3727 (1996).

[11] M. Preuss, H.J. Butt: *Langmuir*, **14**, 3164 (1998).

[12] P. Mulvaney, J.M. Perera, S. Biggs, F. Grieser and G.W. Stevens: *J. Colloid Interface Sci.*, **183**, 184–186 (1996).

[13] S. Biggs, R. Cain, N.W. Page: *J. Colloid Interface Sci.*, **232**, 133–140 (2000).

[14] V.S.J. Craig, C. Neto and D.R.M. Williams: *Phys. Rev. Lett.*, **80**, 054504 (2001).

[15] M. Rief, F. Oesterhelt, B. Heymann and H.E. Gaub: *Science*, **275**, 1295–1297 (1997).

[16] I.U. Vakarelski, K. Higashitani: *Langmuir*, **22**, 2931–2934 (2006).

[17] J.-M. Cho, W.M. Sigmund: *J. Colloid Interface Sci.*, **245**, 405–407 (2002).

[18] S.G. Flicker, J.L. Tipa and S.G. Bike: *J. Colloid Interface Sci.*, **158**, 317–325 (1993).

3.5.3.3 Flowability of powder bed

It is difficult to handle nanoparticles individually, and so usually an assemblage of particles such as a powder bed is handled for nanoparticles. In this case, it is necessary to know the adhesion and flowability of nanoparticle bed. In addition, the average value of adhesion and flowability can be easily obtained from the investigation of powder bed. The measuring method of adhesion and flowability, and several quantitative evaluation method of the powder bed including the nanoparticles are outlined below.

a) Packing method

Packing characteristics of the fine powder have the strong influence on adhesion and flowability of powder in comparison with the coarse particle. In the packing method, adhesion and flowability are presumed from the packing characteristic that can be measured easily. The dispersed fine particle packed in the suitable container through a screen, void fraction or porosity ε is calculated from measured apparent density ρ_b and the particle material density ρ_p, or particle packing volume V and mass M by following equation [1].

$$\varepsilon = 1 - \frac{\rho_b}{\rho_p} = 1 - \frac{M}{\rho_p V} \tag{3.5.58}$$

For adhesive fine powder with low flowability, the porosity shows a high value and, so adhesion and flowability of a powder can be judged from the absolute value of the measured porosity. In addition, from the tap apparent density ρ_t and loose apparent density ρ_b of sieve packing, the compressibility C_c can be calculated by the following equation.

$$C_c = \frac{\rho_t - \rho_b}{\rho_t} \qquad (3.5.59)$$

The compressibility C shows the degree of flowability of a powder. The relationship between particle diameter D_p and void fraction ε is shown in Fig. 3.5.25 which has been obtained in a compression packing test. The figure shows that the void fraction is increasing with the decreasing of particle size. The effect of the particle size on the void fraction can be expressed by Roller's equation (3.5.60) [2]. The high exponent value n in the equation means the strong effect of particle size on packing and this result can be usually found in the adhesive fine powder

$$\varepsilon = \frac{1}{1 + \left(\frac{1}{\varepsilon_c} - 1\right)\left(\frac{Dp}{Dp_c}\right)^n} \qquad (3.5.60)$$

where Dp_c is the critical particle size beyond which the effect of particle size on the porosity disappears, and ε_c is the constant value of void fraction above critical particle size Dp_c.

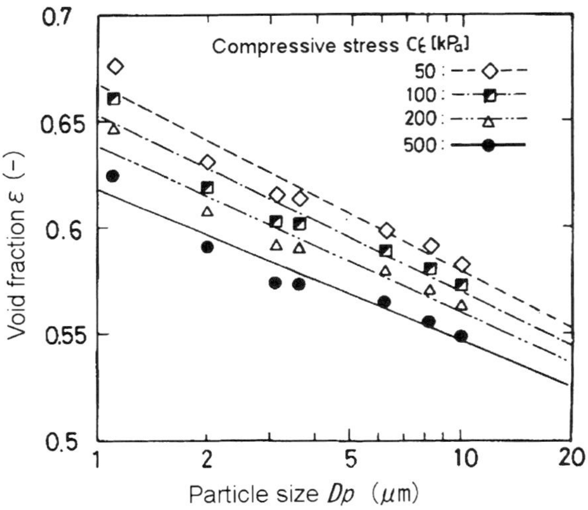

Figure 3.5.25
Effect of particle size on void fraction.

b) Angle of repose

The angle of repose is the angle between the horizontal plane and the free surface of the powder bed. The angle of repose is widely used practically because adhesion and flowability of the powder bed can be appraised simply based on it. As shown in Fig. 3.5.26, the measuring method of the angle of repose is classified into three kinds as the feeding method where the powder feeds from the funnel, the discharge method where the powder discharge from a hole on bottom of the bed, and the gradient method where the powder bed in tabling bin collapse by rotation. There is a report [3] of the correlation between the angle of repose and the internal friction angle by powder shear tests. Carr's flowability index [4] can be calculated from the angle of repose, compressibility, the angle of spatula, homogeneity, and the degree of agglomeration. Using the Carr's flowability index, powder can be classified into seven types such as excellent flow: 90–100, good flow: 80–90, fair flow: 70–79, average: 60–69, poor flow: 40∼59, very poor: 20∼39, and sticky: 0–19.

c) Tensile test

The concept of powder bed tensile test is shown in Fig. 3.5.27. The procedure of the test is as follows. A powder bed is packed in the divided cell, and a half of the cell is fixed and another cell is moving with tensile force. The tensile strength of powder bed σ_T can be calculated from the tensile force at the breaking point divided by the breaking area. The tensile tester is roughly classified into two types such as horizontal-type tensile and vertical-type tensile by the tensile force direction [5]. In each of the tester, the powder is packed in the cell, and pre-consolidated in vertical direction before the tensile test. The measurement result of the vertical type tensile tester which pulls the cell in the same direction with pre-consolidation is larger than the horizontal type tester which pulls the cell in a direction perpendicular to pre-consolidation. For the vertical-type tensile tester, the packed powder slide on the container wall surface and is easy to fall down. The dense packing of nanoparticle is very difficult, in general, and therefore the tensile strength of nanoparticles becomes small. Therefore, it is necessary to make the friction of the device small. In order to achieve small friction, the horizontal tensile tester uses the suspended type rather than the ball bearing system as in the figure. Tensile strength σ_T decreases linearly with void fraction on the semi-logarithmic paper with σ_T on the vertical axis of the logarithmic scale. Thus, the following equation can be derived experimentally

$$\sigma_T = k_1 \exp(-\varepsilon / k_2) \qquad (3.5.61)$$

where k_1 and k_2 are the constants which show the adhesion of powder and the void fraction dependency of tensile strength, respectively. The following

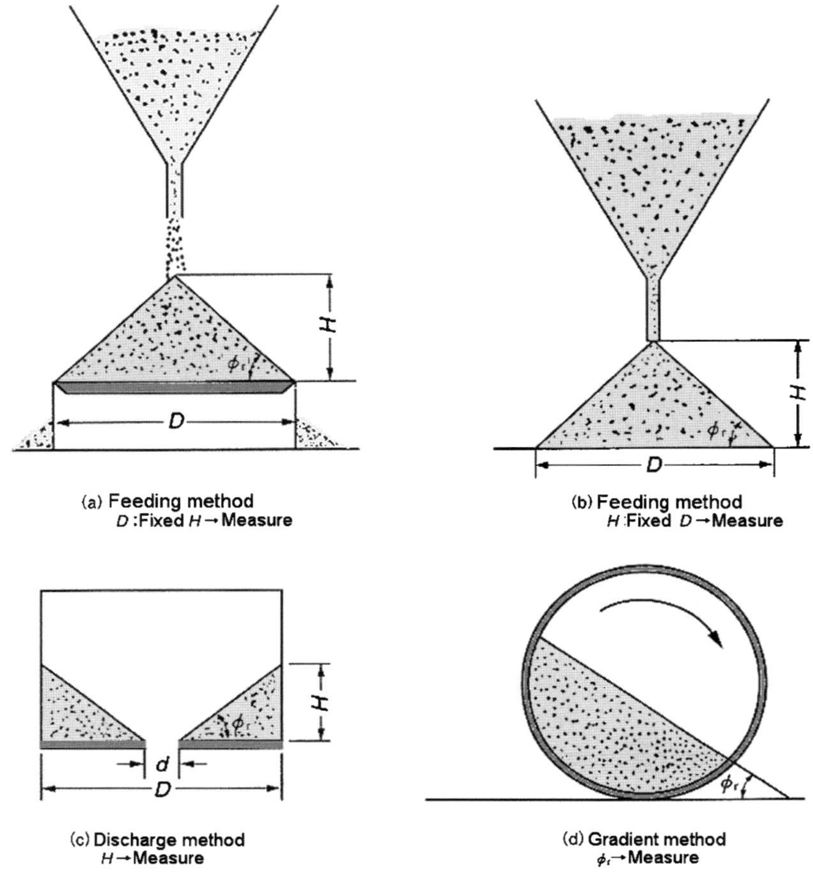

Figure 3.5.26
Measuring method of angle of repose.

Rumpf's model equation shows the relation between adhesion force F_T of the particles and tensile strength σ_T of the powder bed [6]

$$\sigma_T = \frac{1-\varepsilon}{\pi} N_c \frac{F_T}{D_p^2} = \frac{1-\varepsilon}{\varepsilon} \frac{F_T}{D_p^2} \tag{3.5.62}$$

where N_c is the number of contact points on a particle called coordination number and Dp is the particle size. Using Rumpf's equation, the adhesion force F_T at a contact point in the powder bed can be calculated from the tensile strength σ_T of the powder bed.

d) Shear test

As shown in Fig. 3.5.28, the procedure of shear test [7] is as follows; (1) pack the powder in the shear cell, (2) pre-consolidate and pre-shear the packed bed, (3) acting the shear force on the upper cell under normal load, (4) measure the shear force at the collapse point. The shearing strength of the powder bed τ can be calculated as the value which divides the shear force by the fracture surface. Fig. 3.5.29 shows the schematic diagram of Jenike's shear cell which is the ISO standard [8]. Hirota et al. [9] invented the parallel plate-type shear tester which test powder shears between two parallel plates. The shearing strength for various normal stresses concerning the same pre-consolidation pressure is calculated, and the data are plotted on the stress plane like Fig. 3.5.29. The curved line which ties these plotting is called Powder Yield Locus (or PYL). The powder showing the linear relationship is called the Coulomb powder. The tilt angle and the slope of the straight line are called the angle of internal friction and the internal friction factor, respectively. These values are used as flowability index of powder.

$$\tau = \mu_i \sigma + \tau_C = \sigma \tan\phi_i + \tau_C \tag{3.5.63}$$

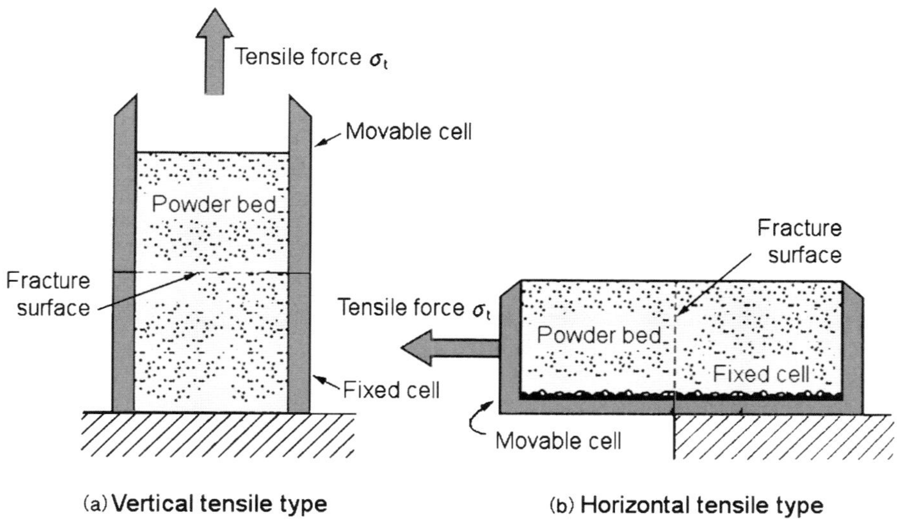

Figure 3.5.27
Concept of two types of tensile tester.

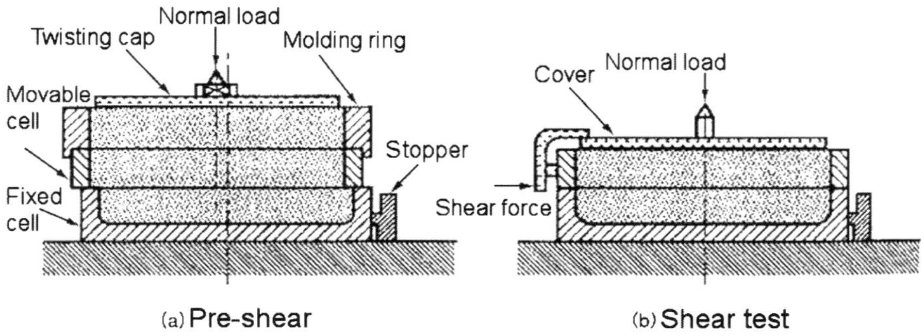

Figure 3.5.28
Concept of shear test using Jenike's cell.

τ_C in the equation means τ axial intercept of the line, it is called the adhesion which shows the shearing strength of adhesion power at $\sigma = 0$. Generally, the powder yield locus of fine powders such as nanoparticles does not show the straight line, shows a convex curve. In this case, the following Warren spring equation [10] can be used for the approximation of PYL

$$\left(\frac{\tau}{\tau_c}\right)^n = \frac{\sigma_T + \sigma}{\sigma_T} \quad (3.5.64)$$

where τ_c and σ_T are the adhesion and the tensile strength of the powder, respectively, and n is the shear index. For shear index $n = 1$, the equation agrees with Coulomb's equation, and the flowability of powder is going bad with increment of n.

e) Flow factor

The powder yield locus changes with the pre-consolidation pressure or void fraction of the powder bed, but as is indicated the powder flowability that does not depend on packing condition of the flow factor was proposed by Jenike [11]. The flow factor means the maximum height of a self-sustaining powder without collapse by the weight, and calculated from the shear test result. As shown in Fig. 3.5.30, when the Mohr's stress circle is drawn, the flow factor F_f can be calculated from the maximum principal stress of the Mohr's stress circle or called the unconfined

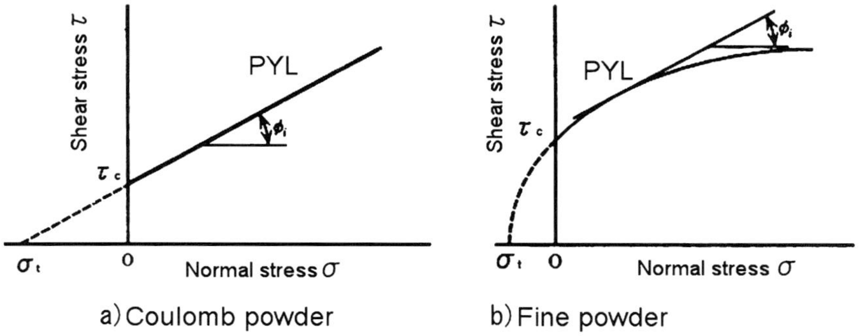

Figure 3.5.29
Powder yield locus (PYL) and internal friction angle ϕ_i.

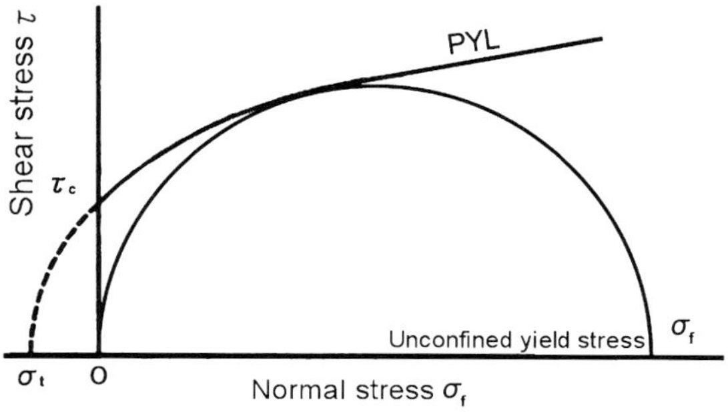

Figure 3.5.30
Powder yield locus and unconfined yield stress σ_f.

yield stress σ_f. The Mohr's stress circle passes the origin of stress plane and is tangential to PYL. The flow factor F_f is given by

$$F_f = \frac{\sigma_f}{\rho_b g} = \frac{\sigma_f}{(1-\varepsilon)\rho_p g} \qquad (3.5.65)$$

where ρ_b and ρ_p are the apparent and particle densities, respectively, and g the gravitational acceleration. The powder with high value of F_f can stack high, and the high F_f value means high adhesive and low-flowability powder.

f) Packing characteristics and flowability of nanoparticles

The nanoparticle is very fine powder and has unique packing characteristics and flowability. Generally, the packing characteristics depend on the force balance of packing force such as powder weight and external force and impeditive force such as the wall friction and cohesion of powder. When particle size becomes small, the relative adhesion force increases because the particle weight is proportionate to cubic of particle diameter, but adhesion force between particles is proportionate to 1 powers or square of particle size. Nanoparticles make the dendrite-like structure by the larger adhesion force than particle weight and so nanoparticles show the quite larger porosity than 0.95 or more, and very difficult to achieve high-packing density. If this phenomenon is utilized, the characterization of the nanoparticle can be quantized with simple packing test.

On the other hand, particles can penetrate into the space which is 7–10 times larger than the particle size, but difficult to flow into the opening less than 7 because of the powder bridge formation. The nanoparticle is easy to leak from the tiny space, and to entrain with micro airflow from the microscopic opening, because particle size of nanoparticle is very small. The nanoparticles show specific floodability when they are packed in the cell or container such as in shearing test, caution during operation is necessary.

References

[1] J. Tsubaki, M. Suzuki and Y. Kanda: *Nyuumon Ryushi Funtaikougaku*, Nikkankogyo Shinbunsya, p. 38 (2002).
[2] P.S. Roller: *Ind. Eng. Chem.*, **22**, 1206–1208 (1930).
[3] S. Toyama: *J. Soc. Powder Technol.*, **7**, 57–62 (1970).
[4] R.L. Carr : *Chem. Eng.*, **72**, 163–168 (1965).
[5] H. Yanagida et al.: *Biryushi Kogaku Taikei*, Vol. 1, Kihongijyutyu, Fuji-Techno-System, p. 526 (2001).
[6] H. Rumpf: *Chem. Ing. Tech.*, **42**, 538–544 (1970).
[7] A.W. Jenike, P.J. Elsey and R.H. Wooley: *Proc. Am. Soc. Test. Mat.*, 1168–1190 (1960).
[8] International Standard Organization: 11697(1994)
[9] M. Hirota, T. Kobayashi and T. Oshima: *J. Soc. Powder Technol.*, **20**, 493–499 (1983).
[10] M.D. Ashton, D.C.H. Cheng, R. Farley, F.H.H. Valentin: *Reologica. Acta*, **4**, 206–212 (1965).
[11] A.W. Jenike: *Bull. Univ. Utah*, **53**, 1–198 (1964).

3.6 Aggregation and dispersion, characterization and control

3.6.1 Aggregation and dispersion in gas phase

Aggregation phenomena of nanoparticles in gas phase are generated by the collision and adhesion between particles, and controlled by the Brownian motion, drag force, surface interaction such as van der Waals and electrostatic forces, etc. Based on the fundamentals of aggregation of nanoparticles in gas phase, the characterization and control methods of nanoparticles will be introduced.

(1) Aggregation and dispersion mechanisms in gas phase

The decreasing rate, dn/dt, of number base density of uniform spherical particle with a diameter, D_p, dispersed to the primary particle at the initial stage, is described by the following equation;

$$dn/dt = -K_0 n^2 \qquad (3.6.1)$$

where K_0 is an aggregation rate constant which is a function of particle diameter and aggregation mechanism. If the aggregation of primary particles is controlled only by the Brownian motion and collision of particles without interaction between particles, the estimated value of K_0 is shown in Fig. 3.6.1. In three different regions of Kn number ($Kn = 2l/D_p$, l: mean free path of molecule), i.e. continuum regime, transition regime, and free molecule regime ($Kn > 10$), K_0 is calculated by three different theories. In this figure, the rate of particle aggregation has the maximum value ranging from 10 to 50 nm in diameter, and increases with increasing of particle density. The aggregation rate in gas phase is much higher than that in liquid since it increases with decreasing solvent viscosity and density. Because of high aggregation rate, it is difficult to control the aggregation and dispersion of nanoparticles in gas phase.

The aggregation phenomena of particles depend on various parameters, for example, surface interaction such as electrostatic and van der Waals forces [2], velocity gradient and turbulent flow of gas [3], acoustic wave and ultrasonic wave, and wave motion such as microwave [4]. By using several types of external force and interaction, the control of aggregation and dispersion of nanoparticles has been studied.

(2) Characterization of aggregation and dispersion

Many kinds of the characterization method of particle aggregation behavior in gas phase have been investigated. One traditional indirect and macroscopic method is Carr's dispersibility determined by the equipment as shown in Fig. 3.6.2 [5]. Powder with a mass, m_0 charged from the upper side of a cylindrical tube is collected on a disk with 4 inches in diameter, which is placed at 24 inches lower from the cylinder bottom. By measuring the weight m_f of the collected powder on the disk, the dispersion behavior of powder is characterized. For dispersed powder, since the sedimentation speed of dispersed particles is slow and the particles diffuse by the Brownian motion and adhere onto the inside surface of the cylindrical tube, mass of the collected powder on the disk is reduced. Based on this mechanism, the dispersibility was defined by the ratio of m_f/m_0. This method has been

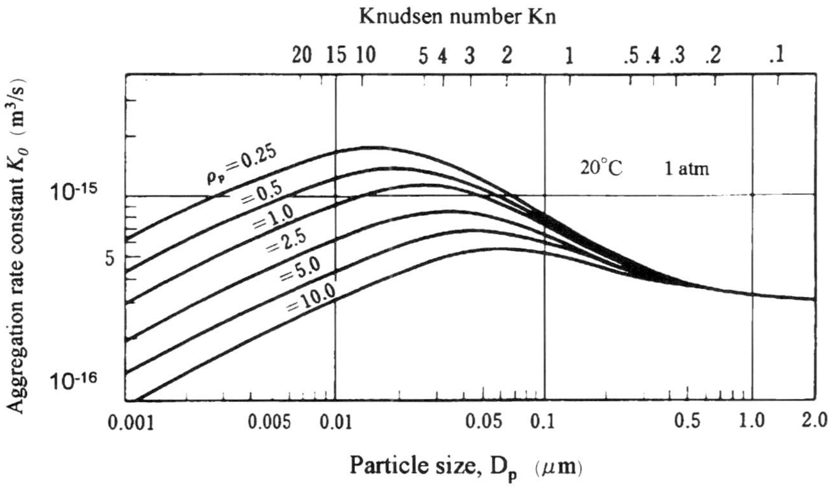

Figure 3.6.1
Calculated result of aggregation rate constant with the particle size and K_n number as variables in air.

used to characterize the dispersion behavior of particles larger than submicronmeter in diameter. Few papers have been published for nanoparticles using this method.

For the characterization of dispersion behavior of nanoparticles in gas phase, aggregate size distribution was measured by nanoparticle measurement method in gas phase such as low-pressure cascade impactor or differential mobility analyzer DMA. The ratio of mean aggregate size to the mean primary particle size determined from the distribution by the image processing of TEM observation as shown in Fig. 3.6.3 was calculated and defined as aggregation index. The overlapped area of both distributions in this figure was defined as the amount of primary particles without aggregation. The ratio of this amount to the total particles was defined as dispersibility [6]. If all particles are dispersed to the primary particles, since both size distributions perfectly agree, the dispersibility is 1. If the aggregation of particles in aerosol is promoted, the overlapped area and the value of dispersibility are reduced.

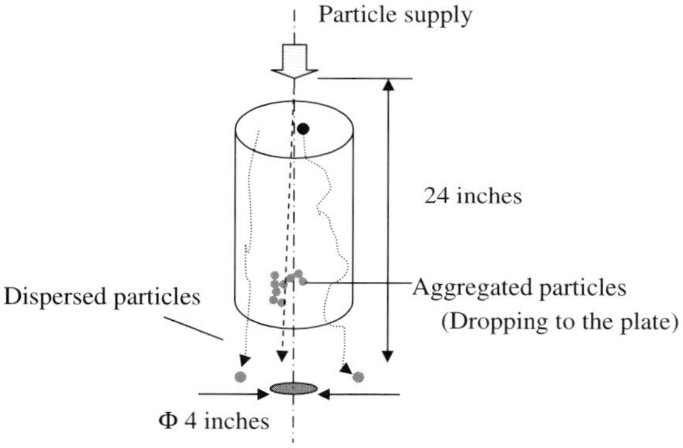

Figure 3.6.2
Principle of measurement of dispersibility by Carr.

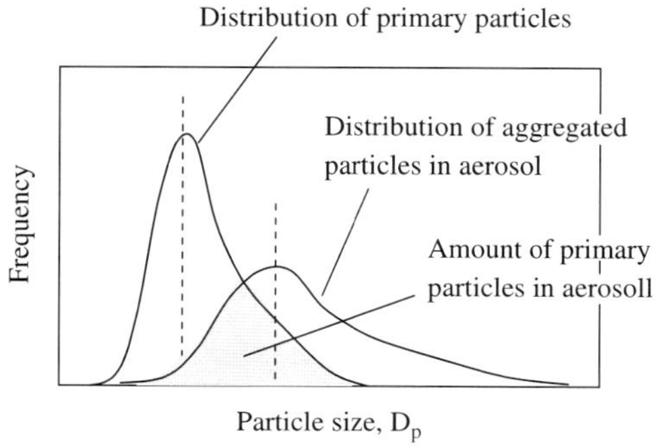

Figure 3.6.3
Characterization method for dispersibility by aggregated particle size distribution.

(3) Control of aggregation and dispersion

For the control of dispersion and aggregation of nanoparticles in gas phase, various methods have been investigated. One method is mechanical or hydrodynamic dispersion of aggregate structure. This method is applied for particles with larger than submicronmeter in diameter in the field of aerosol science and technology. This method is classified into some types according to the application method of mechanical and/or hydrodynamic interaction between particles, for example, acceleration by airflow, collision to the obstructed plate, and mechanical dispersion by the moving medium. The representative dispersing principles by each method are shown in Fig. 3.6.4. Airflow acceleration method gives a fluid drag to particles by using ejector or orifice with rapid reduction in the cross section. Collision method disperses and collapses aggregate structure by the collision of aerosol to the plate. Fluidized bed or vibration is a popular method that uses fluidized medium. By the movement of medium, the aggregate structure is dispersed. However, it is difficult to disperse the aggregates of nanoparticles completely down to primary particles by such mechanical and hydrodynamic methods.

Another new approach for nanoparticle dispersion in gas phase is the control during synthesis process of nanoparticles. Since it is difficult to keep the dispersion of nanoparticles in gas phase, synthesized nanoparticles in gas phase are dispersed in liquid. However, this method has to control the bubble formation during the aerosol injection into liquid, and the dispersion control in liquid phase needs to further develop for the maintenance of dispersion stability. The salt-assisted method has been developed to prevent the aggregation by the deposition of the water-soluble salt between nanoparticles during chemical

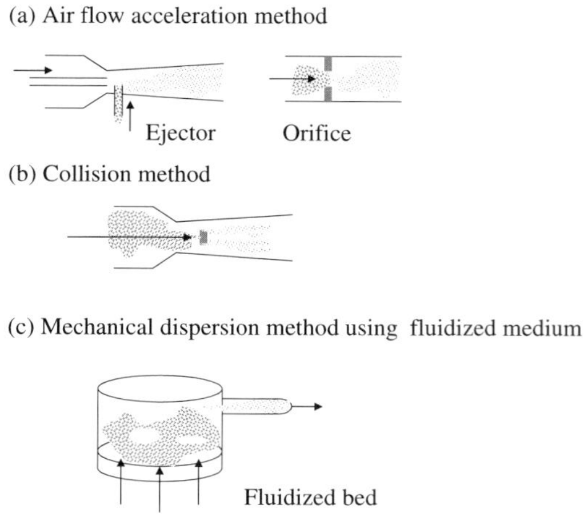

Figure 3.6.4
Basic structure of dispersing machines.

vapor deposition process for nanoparticle synthesis [7]. The deposited salt was dissolved during dispersion of nanoparticles into liquid with surfactant or polymer dispersant. By the adsorption of surfactant, nanoparticles are dispersed in suspension.

The packing and assembling structure control of synthesized nanoparticles in the film or bulk material are being developed by the use of special technology. By Akedo et al. [8], aerosol gas containing fine particles with several hundreds of nanometers was jetted and collided on the surface of a substrate. By the collision of particles on the substrate, the particles were dispersed to nanoparticles and formed dense powder layer on the substrate. Gleiter's group [9] prepared almost fully densed nanopolycrystalline by the following process. Nanometer-sized inorganic particles were prepared by PVD method in high-vacuum chamber. Prepared nanoparticles were packed into small metal mold, and consolidated by one axial pressing with ultra high pressure up to 5 GPa in vacuum chamber. Based on this work, the aggregated structure of oxide nanoparticles in atmosphere was able to be collapsed and packed to almost hexagonal closed-packing structure by cold isostatic pressing up to 1 GPa [10]. These closely packed green bodies were able to be sintered to almost fully dense ceramic at several hundred degrees lower than usual sintering temperature [11].

Based on the above research, if nanoparticles were prepared in gas phase, it is necessary to develop new process engineering such as coating, packing, and forming process for the preparation of new materials with nanoparticles.

References

[1] N.A. Fuchs: *The Mechanism of Aerosol*, p. 288 (1989).

[2] R.L. Drake: *Topics in Current Aerosol Research, Part 2*, Pergamon Press, p. 201 (1972).

[3] K. Okuyama, Y. Kousaka and T. Yoshida: *J. Aerosol Sci.*, 399 (1978).

[4] D.T. Shaw: *Recent Development in Aerosol Science*, Wiley, p. 279 (1978).

[5] R.L.Carr: *Chem. Eng. Jan.*, 163 (1965).

[6] H. Masuda, K. Goto: *J. Powder Technol., Jpn*, 703 (1993).

[7] C. Panatarani, I.W. Lenggoro and K. Okuyama: *J. Nanoparticle. Res.*, **5**(1), 47–54 (2003).

[8] J. Akedo, M. Lebedev: *Jpn. J. Appl.Phys.*, **38** (P1-9B), 5397 (1999).

[9] J. Karch, R. Birringer and H. Gleiter: *Nature*, **330**, 556 (1987).

[10] H. Kamiya, H. Suzuki, D. Kato and G. Jimbo: *J. Am. Ceram. Soc.* **76**(1), 54 (1993).

[11] H. Kamiya, H. Suzuki, T. Ichikawa, Y.I. Cho and M. Horio: *J. Am. Ceram. Soc.*81, 173 (1998).

3.6.2 Liquid phase

Fine particles including nanoparticles are generally supplied in the aggregated state, because they are easily aggregated during the drying process of production and transportation due to pressure caused by stacking. Dispersion of these aggregates in the liquid phase requires the input of mechanical energy to break them.

Breaking the aggregates increases the surface area of the solid–liquid interface, which causes re-aggregation to impede the dispersion process if the interface is unstable. Usually, the solid–liquid interface is stabilized in the liquid phase by forming an electric double layer or adsorption of the polymer around fine particles. For the dispersion of fine particles in liquid, how to mechanically disperse fine particles in liquid and control their stability are very important together with the evaluation technique.

(1) Mechanical dispersion

The mechanical dispersion methods are classified into stirring, high-speed revolution shearing, milling, high-pressure jet, and ultrasonic waves. In these methods, a bead mill is mainly evaluated for dispersing nanoparticle aggregates. A schematic depiction of this apparatus is shown in Fig. 3.6.5 [1, 2]. The mechanism of this milling to break the aggregated particles is as follows [2]. In the milling, the beads stirred by

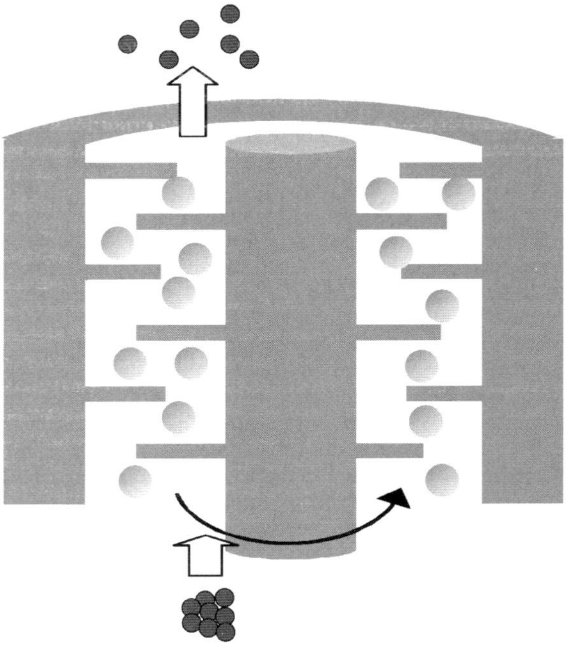

Figure 3.6.5
Schematic illustration of a bead mill.

the rotating pins impinge against the feed slurry introduced from the inlet located at the bottom of the mill. The impingement breaks aggregates in the slurry coming up through the mill. The slurry and the beads are separated by centrifugation in the upper part of the mill and only the slurry is discharged through the hollow shaft to the outside of the mill.

Usually, the separated slurry is circulated to a feed vessel for repetition of the process. The beads less than 100 μm are effective for breaking and dispersing the agglomerated nanoparticles [1–3]. Successful dispersion of nano-sized titanium oxide in liquid with beads of 30-μm diameter has been reported [2].

(2) Colloidal method

Fine particles repeatedly collide with each other by their thermal motion and van der Waals force arising among themselves. As a result, they agglomerate unless sufficient repulsion force arises. For stable interaction, interparticle potential should be sufficiently large, while the agglomerating speed should be extremely slow. The concept of stabilization by electrostatic repulsion potential is well-known as DLVO theory [4].

The combination of the attractive van der Waals interaction and the repulsive double-layer repulsion is the foundation of the well-known DLVO theory [4], which provides an overall net interaction energy. This theory is successful in describing the interactions between charged particles immersed in water. At a low surface potential or a high ionic strength, the repulsive barrier vanishes, allowing particles to flocculate. For highly charged surfaces, there is a strong long-range repulsion, and the energy barrier is too high for the particles to overcome during any reasonable time period. When dispersing metal oxide particles in water, the net charge is controlled by pH. The built-up of a charge at the solid–liquid interface is resulted from proton-transfer reaction with the surface hydroxyl groups.

The relation between zeta potential and pH of alumina fine particles (average particle size is 200 nm) is shown in Fig. 3.6.6. The isoelectric point is at around pH 9. This point is a boundary, that is, the particle surface has a positive or negative potential caused by adsorption or desorption of a proton in the acid or basic side of the isoelectric point, respectively. Good dispersion can be maintained by adjusting the pH of the slurry to a value as far as possible from the isoelectric point to increase surface electric potential. In this example, high electrostatic repulsion potential can be generated in the acid region by a positive charge. Actually, high-concentration alumina water slurry is obtained at pH 4, which is used for the fabrication of ceramics [5].

In the preparation of aqueous slurry of silicon nitride fine particles by ball milling, the reaction between the surface of the fine particles and water ($Si_3N_4 + 6H_2O \rightarrow 3SiO_2 + 4NH_3$) forms a nanoscale SiO_2 layer on the particle surface and NH_3, which increases pH of the slurry up to about 10.5. Since the isoelectric point of SiO_2 is around pH 2, high surface electric potential is generated on the surface of the fine particles. This self-dispersion allows the formation of silicon nitride water slurry at about 35 vol% concentration [6].

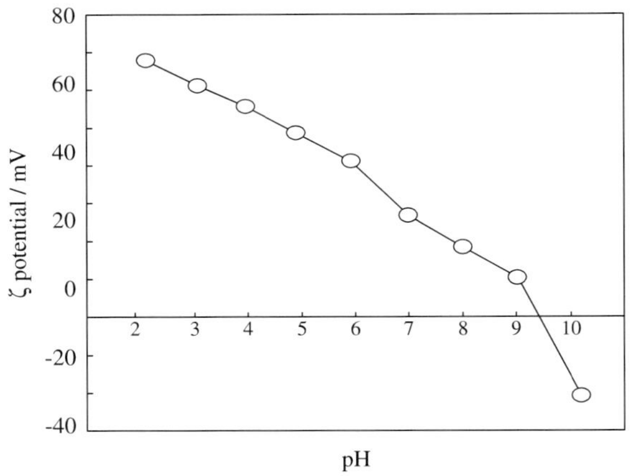

Figure 3.6.6
Relation between pH and zeta potential (alumina).

(3) Evaluation of surface electric potential

Measurement of zeta potential is one of the methods to evaluate surface electric potential in liquid, like electrophoresis, streaming potential, and ultrasonic methods [7]. Today, excellent devices based on these methods are commercially available domestically and abroad. Zeta potential is a little lower than surface potential, however, is often used instead of the latter technically.

Care must be taken in evaluating zeta potential, since the measurements vary depending on the sample preparation method and titration procedure as well as the measurement method. For example, the hysteresis of pH–zeta potential characteristic may be observed as shown in Fig. 3.6.7, if appropriate stirring is not performed during titration [8]. It is noted that zeta potential is not related to the colloidal stability, when non-DLVO factors such as steric hindrance contribute to the stabilization.

(4) Control with dispersants

In the dispersion of nanoparticles, it is often difficult to create sufficient surface electric potential by only adjusting pH. In this case, it is necessary to add an appropriate amount of a polymer electrolyte with dissociable functional groups, for utilizing the electrostatic and the steric repulsion effects. Anionic surface-active agents such as an ammonium salt of polyacrylic acid are used for the dispersion of basic

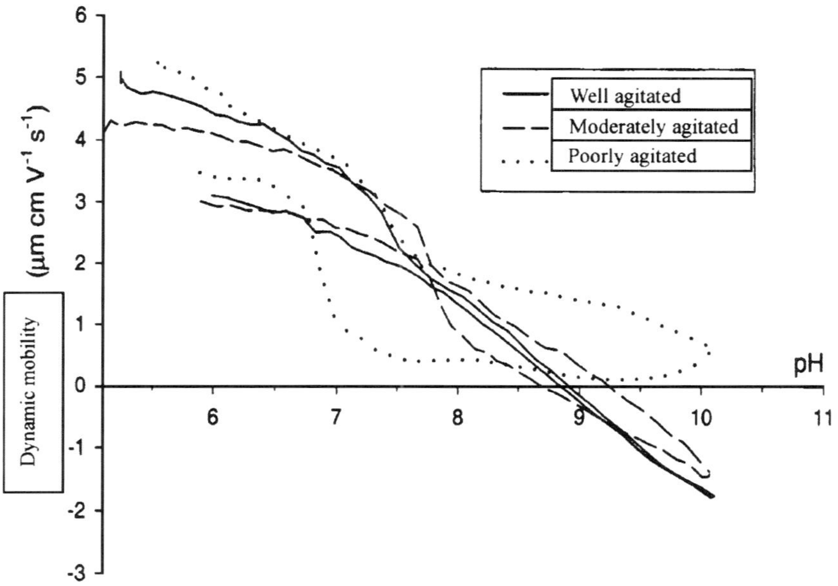

Figure 3.6.7
Effect of agitation conditions on ESA mobility in 10 vol% alumina slurry [8].

metal oxide particles such as alumina. This agent is dissociated in water to form COO-and adsorbs on the surface of alumina, dispersing the particles by the electric repulsion and steric hindrance. The effect of polyacrylic acid on dispersion remarkably depends on the pH of the system, that is, better in an alkaline pH region. In the acid region, the dispersion effect is poor because the carbolic acid group is not dissociated well.

The surface of fine particles with strong solid acidity such as SiO_2 is highly negatively charged in the alkaline environment, which inhibits adsorption of this dispersant that has the same charge as the particle surface.

Isoelectric points of functional fine particles such as titania and zirconia are acidic [9]. Cationic polymers ($-[CH_2-CH_2-NH]_n-$) such as polyethylenimine (PEI) are effective for water dispersion of these fine particles. In water with a pH less than 10, –NH– basic groups is positively charged, resulting in $-NH_{2+}-$. Some NH_{2+} in PEI is adsorbed on surface acid sites (negative sites, Ti or $Zr-O^-$), giving the particles sufficient positive charge under neutral pH conditions. Stable slurry of zirconia nanoparticles has been obtained using PEI [10].

The effect of steric repulsion depends on factors such as the structure and molecular weight of the dispersants [11]. It is very important to select a dispersant and a solvent such that their advantages can be fully utilized to maintain the best dispersion of each system.

(5) Various factors related to dispersion stability

The amount of polymer dispersant used is one of the most important control factors for stabilization. Figure 3.6.8 shows the relation between the amount of dispersant and apparent viscosity at 30°C (solid concentration of the slurry is 35 vol%). An ammonium salt of polyacrylic acid (Chukyo Yushi Co., Ltd., Cerna D305, molecular weight – 10,000) was used as the dispersant. The addition of 0.2 mass% dispersant gave high viscosity. Since the adsorbed amount of polymer to the particles is smaller than that of saturation adsorption, particles cross-linked with the polymer agglomerate. The addition of 0.4 mass% gave the lowest viscosity and reached almost saturation adsorption status. Further addition (2 mass%) results in an increase of the apparent viscosity due to depletion aggregation.

Figure 3.6.9 shows flow curves (viscosity η-shear rate D) of 35 vol% slurry for an appropriate and deficient amount of a dispersant. The measurement temperature was kept constant at 25°C, and the shear rate was changed at a constant speed in the range from 1 to 1,000 s^{-1}. Use of an appropriate amount of the dispersant (0.4 mass%) gave the low viscosity and it hardly depends on the shear rate.

This slurry is almost a Newtonian fluid, indicating good particle dispersion. The slurry with a deficient amount of the dispersant (0.2 mass%) has high apparent viscosity and its η–D plot indicates a decrease in the apparent viscosity with increasing shear rate

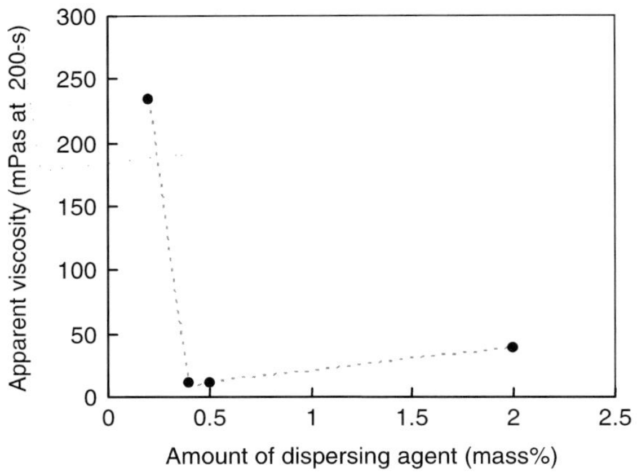

Figure 3.6.8
Relation between amount of dispersant added and apparent viscosity.

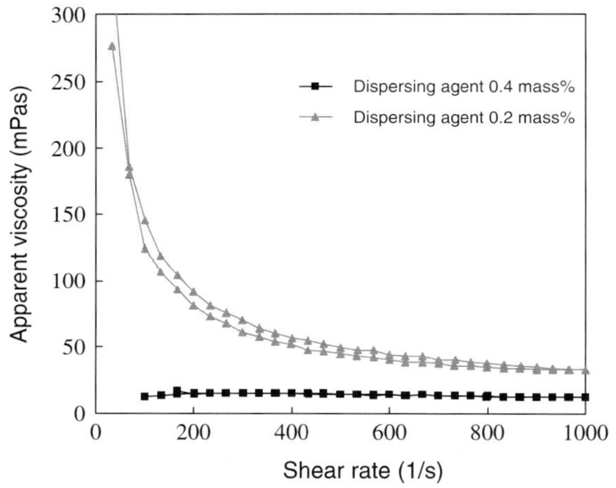

Figure 3.6.9
Flow curves of alumina slurries.

(pseudoplastic flow). This phenomenon is a result of breaking the aggregate structure of the particles in the liquid by increasing the shear rate [9]. There is a hysteresis loop observed, that is, two different slurry flow curves with the increasing and the decreasing shear rates. This is a thixotropic behavior caused by imbalance between destruction of the aggregate structure in the course of increasing shear rate and reaggregation in the course of decreasing shear rate.

Slurry temperature is also an important factor that influences the adsorption of dispersants. The adsorption affinity of polymer dispersants varies greatly in the temperature range of around 0–40°C [12]. For example, desorption of the dispersant at a higher temperature brings about the condition where the dispersant is insufficient in aggregating the particles, even if an appropriate quantity of the dispersant is added to the slurry. Accordingly, the temperature must be carefully controlled in the fabrication processes of ceramics.

The rheological properties shown in Fig. 3.6.9 are widely used for evaluating the dispersion/aggregation behavior of high concentrated slurry systems. In addition, other properties are available for the evaluation. This includes shear stress change with time to keep the shear rate constant and sedimentation, deformation, and compaction properties under mechanical fields [13]. The AMF colloid probe technique is used in liquid for evaluating microscopic interaction between the particles (Refer to Section 3.5.3). It is also attempted to observe the dispersed or aggregated structure in liquid directly or indirectly.

In the past, dispersed/aggregate structures were observed using freeze-dried samples that maintained the slurry structures. Recently, a new technique has been developed to observe the particles fixed by gelling with a solvent [14]. Another technique is direct observation of slurry in a transparent slit under microscope CCD. This technique allows observation of large inhomogeneous structures in slurry up to a concentration of about 30%, although the resolution is around 1 μm [15].

References

[1] M. Inkyo, T. Tahara and K. Okuyama: Preprint of 40th Particle Technology Meeting. The Society of Powder Technology, Japan. The Association of Powder Process Industry and Engineering, Japan, pp.5–6 (2005).
[2] M. Inkyo, T. Tahara: *J. Soc. Powder Technol., Jpn.*, **40**, 578–585 (2004).
[3] H. Yoden, N. Ito: *J. Soc. Powder Technol., Jpn.*, **41**, 457–464 (2004).
[4] N. J. Israelachvili: *Intermolecular and Surface Force*, Academic Press (1985).
[5] A. Krell, P. Blank, H. Ma, T. Hutzler and M. Nebelung: *J. Am. Ceram. Soc.*, **86**, 546–553 (2003).
[6] H. Abe, T. Hotta, M. Naito, N. Shinohara, M. Okumiya, H. Kamiya and K. Uematsu: *Powder Technol.*, **199**, 194–200 (2001).
[7] F. Kitahara, K. Furusawa, M. Ozaki and H. Ohshima: *Zeta-potential*, Scientist Inc. (1995).
[8] V.A. Hackley, J. Patton, L.H. Lum, R. Waeshe, M. Naito, H. Abe, Y. Hotta and H. Pendse: *J. Dispersion Sci. Tech.*, **23**,601–617 (2002).
[9] J.S. Reed: *Ceramic Processing* (2nd ed.), Wiley (1995).
[10] C. Duran, Y. Jia, Y. Hotta, K. Sato and K. Watari: *J. Mater. Res.*, **20**, 1348–1355 (2005).
[11] T. Kakui: *The Micromeritics, No. 48*, 46–55 (2004).

[12] Y. Tomita, L. Guo, Y. Zhang, N. Uchida and K. Uematsu: *J. Am. Ceram. Soc.*, **78**, 2153–2156 (1995).
[13] Japan Society for the Promotion of Science 124th Committee on Preparation and Application of Advanced Ceramics. The Nikkan Kogyo Shimbun, Ltd. (2005).
[14] M. Takahashi, M. Oya, M. Fuji: *J. Soc. Powder Technol., Jpn.*, **40**, 410–417 (2004).
[15] H. Abe, M. Naito, K. Okamoto, T. Hotta, S. Ohara, T. Fukui: *J. Soc. Powder Technol., Jpn.*, **41**, 10–14 (2004).

3.6.3 Dispersion in organic solvent and polymer resin

(1) Function and application of nanoparticle composite material

Various types of nanoparticle-dispersed composite materials, whose matrix material is polymer, metal, glass, or ceramics, have been developed aiming at the improvement of various functions. The examples are, thermal properties such as heat-resistance and thermal conductivity, mechanical properties such as strength, fracture toughness, and abrasion resistance, electromagnetic properties such as dielectric property and electromagnetic wave shielding, optical properties such as refractive index. For fine inorganic particles larger than submicronmeter, various polymer composites dispersed with fine powders have been already developed and utilized in various applications.

If the primary size of dispersed particles in polymer is less than several nanometers, the additional function and behavior will be given to the composite materials. In Table 3.6.1, the material of nanoparticles, function, and application field of composites are summarized. For example, if titanium oxide nanoparticles with high refractive index larger than 2.5 are uniformly dispersed at high concentration into polymer lens material with relatively low refractive index ~1.4–1.8, they are expected to apply to the new optical lens and optical fiber cable with high refractive index. To obtain a high transparency and prevent scattering of the visible light, it is needed to disperse the nanoparticles to the primary particles without any aggregation in the polymer. For the application of shield material of electromagnetic wave with high frequency than several gigahertz for mobile phones and precise medical instrument, etc., it is necessary to develop highly concentrated dispersion of magnetic nanoparticles less than several 10 nm into polymer resin.

Table 3.6.1
Examples of nanoparticle dispersed polymer composite.

Material of nanoparticle	Function	Application
TiO_2	High refractive index	Lens, optical fiber
$BaTiO_3$	High dielectric constant	Capacitor
Silica-coated Fe, Co	High permeability	Electromagnetic absorber

(2) Dispersion method and process of nanoparticles into resin

The general process to prepare powder-dispersed polymer composites is the kneading of particles into resin with fluid. However, for nanoparticles, the particle aggregation and the rapid increase of polymer viscosity by the addition of nanoparticles inhibits the highly concentrated filling up to several tens of volume percent by kneading process. For the preparation of nanoparticle/polymer composites, it is necessary to develop new preparation processes. The outline of two kinds of approach to prepare nanocomposites will be introduced. One is the dispersion of synthesized nanoparticles into organic solvent and resin, and the other is in-situ synthesis of nanoparticles during polymerization process of monomer [1, 2].

1) In-situ synthesis of nanoparticles during polymerization process of monomer

In this popular process, raw monomer materials and organic metal compound such as metal alkoxide are mixed uniformly, and during the polymerization process, nanoparticles are generated by nucleation and growth in the polymer matrix material. If the initial mixing condition of monomer and alkoxide is good, the uniform distribution of nanometer-sized particulate matter in polymer will be obtained in comparison with other methods. Some composite polymers dispersed with amorphous silica nanoparticles of high fracture strength and heat resistance has already been developed and brought out to the market. Lü et al. reported to have succeeded in dispersion of nanometer-sized titanium oxide in polymer with a high concentration of more than 60 vol% and increase of reflective index with the increasing addition of inorganic particles [3]. However, by using this process, if the concentration of inorganic nanoparticles increases over a certain critical value, the coalescence and growth of dispersed nanoparticles occur during the in-site synthesis process. In order to keep the uniform dispersion of nanoparticles in the composite, it is necessary to control the polymerization reaction process precisely.

The inorganic phase generated by this process is generally amorphous. In some cases, the interface between the inorganic and organic compounds is not clear. It is rather difficult to control the crystal phase of nanoparticles, because the heat resistance of polymer compounds is not so high. If the expected function of composite does not depend on the crystal phase of inorganic component such as heat resistance, etc., it is not necessary to control the crystal phase. However, when the function of composite materials depends on the crystal phase, for example, optical properties such as refractive index and electromagnetic ones like dielectric constant and magnetic susceptibility, the crystal phase control is important. For optical materials, the refractive indexes of anatase phase (2.5) and rutile phase (2.7) are much higher than that of amorphous titanium oxide. If the crystal phase can be controlled during the synthesis process of composite, it is needed to prevent the crystal growth of inorganic particles less than several tens of nanometers. For these reasons, commercial-level products of the polymer composite have been developed only for high heat-resistance materials of amorphous silica composite.

2) Dispersion of synthesized nanoparticles into organic solvent and polymer resin

In order to disperse inorganic particles into organic solvent and polymer resin, hydrophobic group and hydrocarbon chain were formed on the surface of particles by various surface treatment methods. Since it is difficult to modify hydrocarbon chain on the surface of metal or metal oxide directly, silane-coupling agents and thiols have been used to modify the particle surface of metal and metal oxide. Thiols were used for surface modification of metal compounds and semiconductors such as Au, Ag, CdS, and CdSe, etc. The fundamental structure of thiol is shown in Fig. 3.6.10(a). The sulfur atom in thiol combines with metal atom on the particle surface and hydrocarbon chain or functional group such as NH, COOH, formed on the metal surface and promotes an increase in reactivity between the metal particle and polymer or organic solvent. By using hydrocarbon chain with double bond, grafting polymers from the double-bonded structure in the organic chain of chemisorbed thiols will be an effective modification process.

For metal oxides that cannot be chemisorbed with thiols, silane-coupling agent was used and its fundamental structure is shown in Fig. 3.6.10(b). The silane-coupling agent has a long history, and applied to various fields, for example, silica filler dispersed in the insulator resin for a semiconductor and water repellency processing of the glass. The molecular structure and weight of hydrocarbon chain in both surface modification agents are designed for the purpose of treatment.

The reaction process of both agents is shown in Fig. 3.6.10. The SH group in thiols reacts with metal surface directly. The $-OCH_3$ or $-OC_2H_5$ group in silane-coupling agents is firstly hydrolyzed with water molecule and condensed with OH groups at the surface of oxide particles. If the ideal reaction control is achieved, OH groups on the oxide particles are replaced by hydrocarbon chain of the coupling agent.

Nanometer-scale surface coating is a different approach for surface modification of nanoparticles. The surface of metal/inorganic compound except metal oxide nanoparticles experiences oxidation or hydrolysis reaction in air or water. Since it is difficult to prevent such oxidation only by the surface modification at the molecular level, nanometer-scale surface coating is carried out on the surface of nanoparticles. One example is silica layer coating by the hydration and condensation of silicon alkoxide such as *tetra*-ethyl-ortho-silicate (TEOS). Silica surface layer prevents the surface oxidation of metal particles in air and water. By the surface modification with silica using a silane-coupling agent with diverse functional groups, various surface functional properties have been improved.

Recently, a new particle synthesis process, which is a simultaneous process of particle synthesis and surface modification, has attracted wide attention. The first example is the particle synthesis in the nanopool structure made by surfactants such as

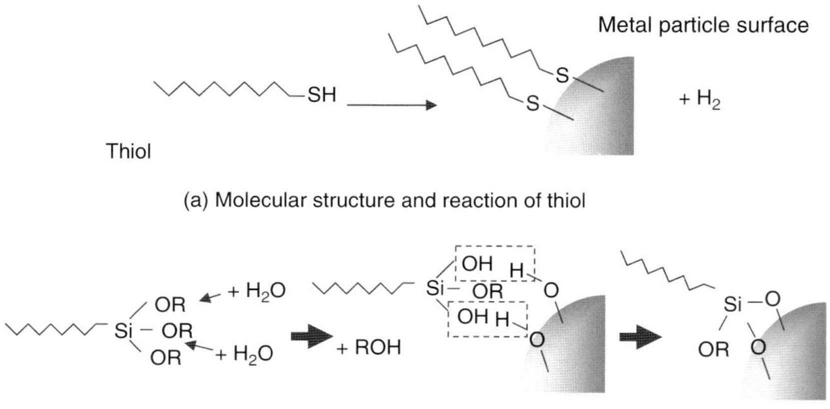

(a) Molecular structure and reaction of thiol

(b) Molecular structure and reaction process of silane coupling agent

Figure 3.6.10
Molecular structure and reaction process of thiol and silane-coupling agent.

reverse micelle and hot soap methods. Various semiconductor nanoparticles and quantum dots can be prepared by this process. However, since it is not possible to increase the particle concentration, the yield is relatively low. Thus the prepared particles can only be applied for valuable products. Recently, new processes for nanoparticle synthesis using a metal–surfactant complex as a reagent have been reported [4]. Since these methods can be operated under relatively high reagent concentrations, further development in this research area is expected.

Since the viscosity of organic solvent with monomer or polymer and high solid content of nanoparticles is relatively high, the design of mechanical mixing tool such as kneader and mixer is also important to prepare uniform nanoparticle-dispersed polymer composites.

References

[1] Z. Sun, J.S. Gutmann: *Physica A*, **39**(1–2), 80–85 (2004).

[2] C. Lü, C. Fuan, Y. Lju, Y. Cheng, B. Yang: *Chem. Mater.*, **17**, 2448–2454 (2005).

[3] C. Lü, Z. Lui, C. Fuan, J. Guan, B. Yang, J. Shen: *Macromol. Mater. Eng.*, **288**, 717–723 (2003).

[4] J. Park, K. Am, Y. Hwang, J.-G. Park, H.-J. Noh, J.-Y. Kim, J.-H. Park, N.-M. Hwang, T. Hyeon: *Nat. Mater.*, **3**, 891–895 (2005).

3.7 Rheology of slurry

3.7.1 Fundamentals of suspension rheology

(1) Viscosity behavior of non-interacting suspensions

The most basic equation on the viscosity of suspension is the Einstein equation. For suspensions of monodisperse spherical particles without colloidal interactions in a liquid with viscosity η_0 at a volume fraction ϕ, the viscosity η is given by the following equation:

$$\eta = \eta_0(1 + 2.5\phi) \qquad (3.7.1)$$

This equation is theoretically derived for an isolated single particle. The viscosity increase in the presence of particles is qualitatively explained by the increase in hydrodynamic energy dissipation due to the distortion of the velocity field in the vicinity of each particle. The theory deals with one particle and the equation is valid only for very dilute suspensions in which the hydrodynamic interactions such as collisions in shear fields can be ignored. Since the distortion of streamline is responsible for the increase in energy dissipation, the Einstein theory leads to a very important conclusion that the viscosity of emulsion containing droplets of immiscible liquid, the viscosity of which is lower than that of medium, also increases at low shear rates.

Equation. (3.7.1) indicates that the viscosity of dilute suspensions proportionally increases only as a function of volume fraction of particles, because the region subjected to distortion of velocity field is constant, relative to the particle diameter. However, at high concentrations the particles cannot behave independently in shear fields, but the interparticle interactions including collisions are hydrodynamically induced. The flow field outside a particle is influenced by the neighboring particles and the overlapping of distortion region takes place. The interference between particles in shear fields leads to additional energy dissipation. Therefore, when the actual viscosity data are compared with the Einstein prediction, the former becomes much higher and exponentially increases in concentrated suspensions, whereas the same values are obtained at low concentrations.

Because of no colloidal interaction between particles, the viscosity behavior of non-interacting suspensions is governed by the balance between Brownian motion and hydrodynamic forces. The hydrodynamic forces exerted on a single particle in shear fields are given as $6\pi\,\eta_0 a^2\dot{\gamma}$ and forces due to thermal motion as kT/a. Here, η_0 is the viscosity of medium, a the particle radius, $\dot{\gamma}$ the shear rate, and kT the thermal energy. The relative importance of shearing forces to Brownian diffusion can be represented by the Peclet number $\eta_0 a^3\dot{\gamma}/kT$. Referring to the previous work, the relative viscosity η/η_0 measured using different particle radii, fluid viscosity, and temperatures are superimposed when plotted against the Peclet number[1]. The viscosity behavior for suspensions of non-interacting particles can be scaled on the Peclet number and this can be treated as a master curve. The important feature for concentrated suspensions in which the particles are completely dispersed in a Newtonian liquid without flocculation is that the hydrodynamic effects give rise to shear-thinning flow (often called pseudo-plastic flow), defined as decreasing viscosity with increasing shear rate.

When the particle concentration is increased beyond 47 vol%, another unique viscosity behavior appears at high shear rates. Figure 3.7.1 shows the effect of particle concentration on the viscosity for concentrated suspensions of 45 vol% and above [2]. At 45 vol% the suspension is slightly shear-thinning over the wide range of shear rates, above 50 vol% the flow becomes shear-thickening flow (often called dilatant flow), defined as increasing viscosity with increasing shear rate, and beyond 53 vol% a discontinuous jump of viscosity is induced [3]. For suspensions at 53 and 55 vol%, only the data before viscosity jump are plotted, because the reproducibility is considerably poor at higher shear rates due to flow instability.

The viscosity jump is an extreme manifestation of shear-thickening flow and peculiar characteristic of

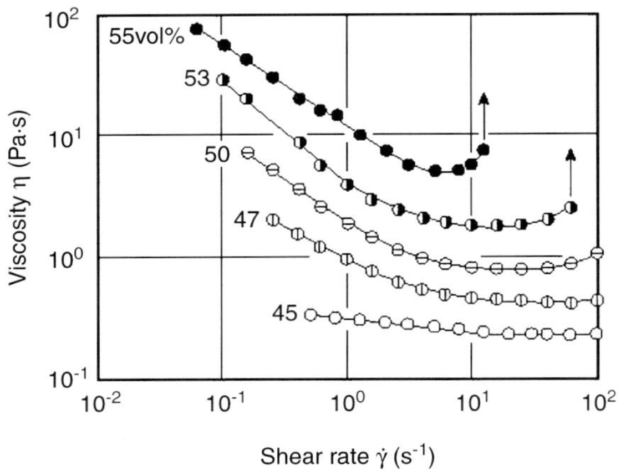

Figure 3.7.1
Effect of particle concentration on the viscosity behavior for suspensions of non-interacting particles.

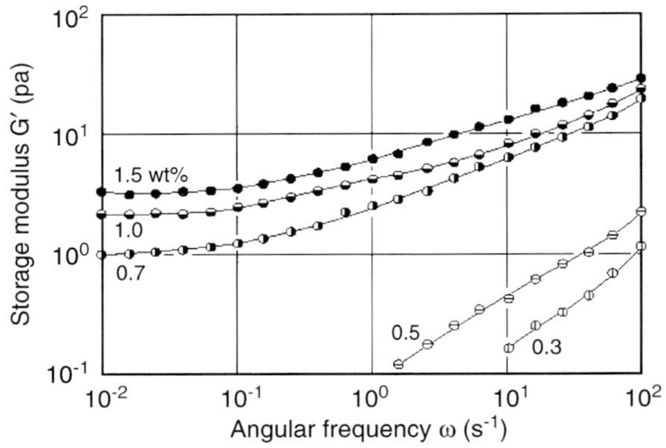

Figure 3.7.2
Effect of flocculant concentration on the elastic behavior of suspensions.

concentrated suspensions of non-interacting particles. To explain the viscosity jump, two types of mechanisms are proposed: rupture of two-dimensional layering and hydrodynamic clustering.

(2) Rheology of flocculated suspensions

(a) Dynamic viscoelasticity

In many cases, particles dispersed in a liquid are bound together to form flocs due to interparticle attraction. The interparticle forces can be of three types [1]: London-van der Waals attractive forces, electrostatic repulsive forces, and those connected with adsorbed polymer. At low particle concentrations, the flocculated suspensions consist of a collection of discrete flocs. As the concentration is increased, the formation of floc–floc bonds progresses and above some critical volume fraction three-dimensional network is developed through the system. When the network structure is sufficiently strong to transmit the forces through floc–floc bonds, the flocculated suspensions behave as solids, that is, the suspensions respond elastically to small strains. Although the elasticity arises only from the attractive forces between two particles, the appearance of elasticity can be associated with the structural changes from discrete flocs to unbounded network. The extreme changes in material transport properties can be discussed in relation to the long-range interactions which are cooperatively generated in the infinite network made up of contiguous sequences of primary bonds between the most neighboring units [4].

Figure 3.7.2 shows the effect of polymer concentration on the frequency (ω) dependence of storage modulus (G') for 30 vol% suspensions of latex particles with a diameter of 240 nm. The polymer is polyvinyl alcohol (PVA) with a molecular weight of 1×10^5 and acts as a flocculant in the suspensions. Since the flocculation is induced by a bridging mechanism in which one polymer chain adsorbs onto two or more particles to bind them together, the flocculation level is considered to increase with PVA concentration. In the absence of PVA, the suspension is electrostatically stabilized and the elastic responses are not detected. The additions of PVA cause the increase in storage modulus. Below 0.5 wt%, the storage modulus linearly decreases with decreasing angular frequency and hence the huge flocculated structures are not expected. As the particle concentration is increased to 0.7 wt%, the storage modulus clearly indicates the existence of plateau, showing that the relaxation does not occur even after long times. Because the suspensions begin to respond elastically, one can accept that the unbounded flocs are developed at this point. The flocs with finite sizes are regarded as independently distributed and have mobility to some extent, while the network structure of unbounded flocs hold a certain shape. The former systems are characterized as liquids and the latter as solids. Therefore, it is considered that the phase transition from liquids to solids takes place, at a point where the frequency-dependent curve of storage modulus shows a plateau at low frequencies. In many flocculated suspensions, the interparticle bonds generally have solid-like properties that are not broken down by thermal energy and the viscoelastic behavior can be basically understood through the same concept.

(b) Viscosity behavior

The floc structures in ordinary suspensions are not broken down in a quiescent state. But they are easily ruptured in shear fields, since the interparticle bonds are not very strong. For suspensions that show solid-like responses due to the unbounded network developed three-dimensionally, at high stresses the systems can flow as the result of structural ruptures.

The critical stress at which the suspension undergoes a transition from solids to liquids is referred to as yield stress. The requirement for the development of yield stress for flocculated suspensions is the formation of unbounded network over the system and this behavior may be grounded on the same mechanism for the appearance of plateau in dynamic viscoelastic functions at low frequencies.

Figure 3.7.3 shows the creep and recovery behavior at different constant stresses for suspension with 1.0 wt% PVA in Fig. 3.7.2 [5].

The strain was measured as a function of time under the condition where the stress was instantaneously applied at $t = 0$, maintained constant for 60 s, and suddenly removed. The system responds like a viscoelastic body at low stresses, because the time-dependent curve shows instantaneous and retarded elasticity on the application of stress and striking elastic recovery after the removal of stress. When the stress is increased above 2.0 Pa, the strain is increased by a factor of about 1,000, the time-dependent profile of which is approximated by a straight line with very small recovery after the removal of stress. The behavior is categorized as a response of purely viscous liquid. Therefore, at high stresses the unbounded floc structures developed over the system is considered to be broken down to a great extent. Such non-linear viscoelasticity measurements at various different stresses serve to evaluate the yielding behavior due to the ruptures of floc structures.

Figure 3.7.4 shows the shear rate dependence of viscosity for suspensions previously described in Fig. 3.7.2. The suspension in the absence of PVA is a Newtonian fluid with low viscosity. The additions of PVA cause an increase in viscosity in the entire range of shear rates and the flow becomes markedly shear-thinning. The formation of flocs brings about immobilization of a fraction of the suspending medium trapped in the flocs. Because of larger effective volume of dispersed phase, the flocculated suspensions usually have higher viscosity than the suspensions completely dispersed to primary particles. However, the floc structures are easily broken down in high-shear fields and reversibly rebuilt in low-shear fields. As a result, the flow becomes shear-thinning even for suspensions without development of three-dimensional network of flocs. The general characteristic accepted for ordinary flocculated suspensions is that the flocculation of particles causes a viscosity increase, the increment of which is larger at lower shear rates, and thus the shear-thinning tendency becomes striking in a wide range of shear rates. When the time-scales of measuring programs are comparable to those of structural formation and rupture in shear fields, the time-dependent behavior such as thixotropy can be observed.

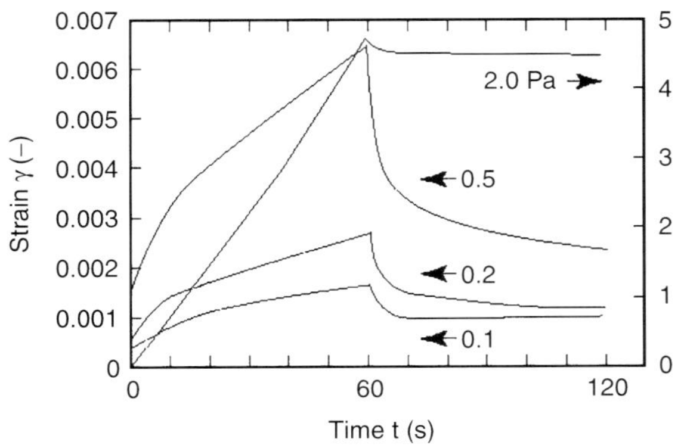

Figure 3.7.3
Creep behavior of flocculated suspension under different stresses.

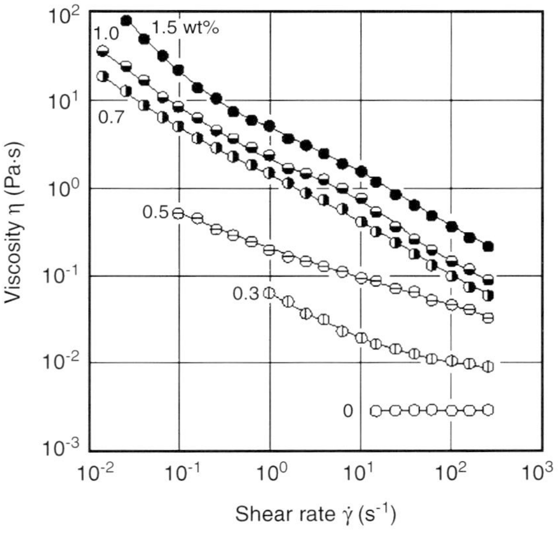

Figure 3.7.4
Effect of flocculant concentration on the viscosity behavior of suspensions.

As mentioned in Figs. 3.7.2 and 3.7.3, since the three-dimensional network structures are formed in suspensions with PVA at concentrations above 0.7 wt%, the dynamic viscoelastic curves show the plateau at low frequencies and the yield of plastic is observed at the onset of structural ruptures. However, it seems difficult to verify the drastic changes in structures from the viscosity curves. With increasing degree of flocculation the slope of viscosity curve become steeper and the formation of three-dimensional network structures causes the curve to be approximated by a straight line with a slope of −1 at low shear rates. This behavior indicates that independent of shear rate, a constant stress is generated, once the flow starts after the yielding. Therefore, the constant stress in flow curve is often used as a measure of yield stress. But the yield stress is defined as a critical stress, below which the substance shows a solid-like response and above the viscous flow occurs. Even though the constant stress is observed in flow curve, the value determined through the measurements in which the substances are subjected to steady flow does not necessarily show the critical stress corresponding to the transition from solid to liquid. To understand the flocs structures, the yield stress as the minimum stress at which a solid-like substance starts to flow is important and for the determination of the value the creep measurements shown in Fig. 3.7.3 are required. The values determined by creep experiments and extrapolation of flow curve to zero shear rate can be called static- and dynamic-yield stresses, respectively. In industries, the dynamic-yield stress is mainly used for rheological evaluation and control of flocculated suspensions.

References

[1] I.M. Krieger: *Trans. Soc. Rheol.*, 7, 101–109 (1963).
[2] Y. Otsubo: *J. Soc. Rheol. Jpn.*, **22**, 75–79 (1994).
[3] R.L. Hoffman: *Trans. Soc. Rheol.*, **16**, 155–173 (1972).
[4] Y. Otsubo: *Langmuir*, **6**, 114–118 (1990).
[5] Y. Otsubo: *Langmuir*, **11**, 1893–1898 (1995).

3.7.2 Rheological property of nanoparticle dispersed suspension

Since large and irregular aggregates are formed in highly concentrated nanoparticle-dispersed suspension, non-linear rheological property is often observed. The aggregation of nanoparticles is promoted with increase in solid fraction of suspension, because the attractive interaction is much stronger than the repulsive interaction with the reduction in the distance between nanoparticles less than several nanometers. In order to prevent the aggregates formation, many kinds of surface treatment such as adsorption of surfactant and surface modification by silane-coupling agent or thiols have been applied. Such surface treatment is useful to disperse fine particles whose size is larger than 100 nm, however it is necessary to consider different factors for the control of rheorlogical behavior of nanoparticle suspension.

For example, the effect of molecular weight of polyethyleneimine, PEI [1], and particle size of alumina particles ranging from 7 to 300 nm on ethanol suspension viscosity is shown in Fig. 3.7.5. The solid fraction of alumina in each suspension was controlled such that the suspension viscosity without dispersant was almost of the same value as shown in Table 3.7.1. To obtain the minimum suspension viscosity, the optimum molecular weight of the submicron alumina powder was about 10,000 g/mol. For nanoparticles, the optimum molecular weight to obtain the minimum viscosity was 1,200 g/mol. With decrease of particle diameter, the optimum molecular weight to obtain the minimum viscosity was decreased.

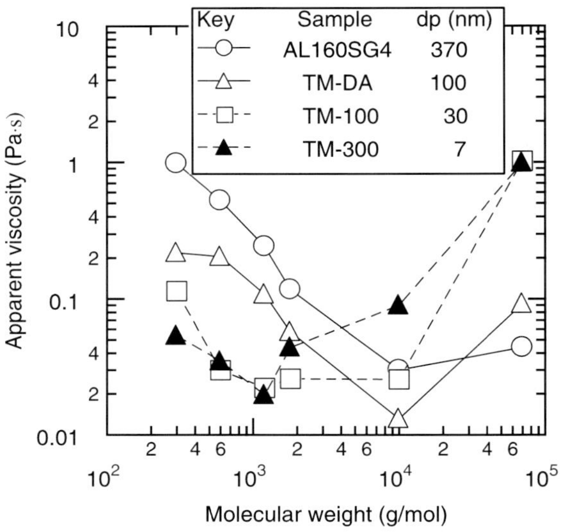

Figure 3.7.5
Effect of molecular weight of polymer dispersant and particle size on apparent viscosity.

Table 3.7.1
Solid fraction to obtain same suspension viscosity with different particle diameter.

Particle diameter [nm]	370	95	30	7
Solid fraction [vol%]	36.0	23.1	12.5	9.79

In order to analyze the relationship between molecular weight and suspension viscosity with different particle size, the surface interaction between nanometer-scaled alumina surfaces adsorbing polymer dispersant with different molecular weight was measured by colloid probe AFM method and shown in Fig. 3.7.6 [2]. When polymer dispersant with relatively low molecular weights, 300 and 1,200 g/mol, was adsorbed, the adhesion force was disappeared. However, for 10,000 g/mol adsorption, non-linear and long-range attractive interaction up to several 10 nm was observed. For submicron alumina particles, such long-rang attractive force cannot be observed. It seems that this long-range attractive force was generated by the bridge formation of free polymer in the solution and adsorbing polymers.

The rheological behavior of nanoparticle-dispersed suspension depended on not only the particle diameter but also solid fraction. The effect of solid fraction and molecular weight of polymer dispersant on the apparent aqueous suspension viscosity of titanium oxide with 50 nm in diameter is shown in Fig. 3.7.7 [3]. Two kinds of polymer dispersant, polyacrylic acid (P100) and copolymer of acrylic acid and methyl acrylate

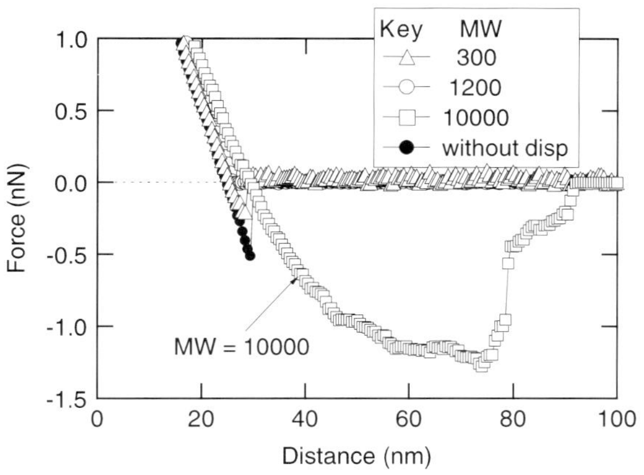

Figure 3.7.6
Influence of molecular weight of PEI on surface interaction characterized by colloid probe AFM method.

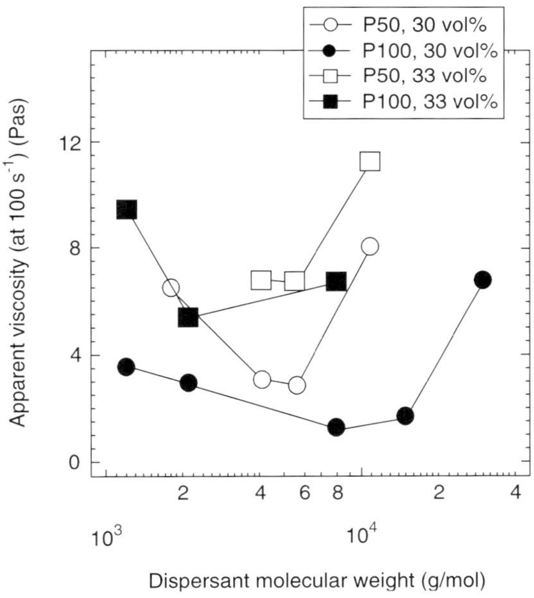

Figure 3.7.7
Effect of molecular structure and weight and solid fraction on apparent suspension viscosity.

(P50) were used. For P100, at the volume fraction of 30 vol%, the optimum molecular weight to obtain the minimum viscosity was 10,000 g/mol. However, when the volume fraction increased to 33 vol%, the optimum molecular weight decreased to 2,000 g/mol. The suspension viscosity adsorbing 100% hydrophilic dispersant (P100) is lower than that of 50% hydrophilic dispersant.

To explain the relationship between molecular weight and suspension viscosity with different particle size and solid fraction, the analysis of estimated surface distance in Section 3.1 is useful. The surface distance decreased with increase in solid fraction and was same as the size of polymer dispersant less than 1 nm. Large molecular weight polymer dispersant cannot invade between particle surfaces of formed bridge between the particles. Such bridging phenomena of large molecular weight dispersant between particles promoted particle aggregation and increase in suspension viscosity.

References

[1] T. Kakui, T. Miyauchi and H. Kamiya: *J. Euro. Ceram. Soc.*, **25**, 655–661 (2005).
[2] H. Kamiya: 54th Proceedings of Theory Applied Mechanics Lecture in Japan, pp. 45–48 (2005).
[3] S. Kondo, M. Tsukada, H. Kamiya, T. Kakui and T. Ooguro: Proceedings of 18th Fall Meeting of Ceramic Society, Japan, p. 349 (2005).

3.8 Simulation of colloidal dispersion system

As described so far in this chapter, various kinds of interaction forces act between solid surfaces, which will then affect the behavior of colloidal systems resulting in, e.g., stabilized dispersion, coagulation, and structure evolution. In general, it is quite difficult to predict analytically such behaviors of masses of particles, which is especially the case for non-equilibrium or dynamic features. It is thus desirable to apply a computational approach, combining with experimental characterization technique for particles, to analyze or to understand the behavior of nanoparticles and evolved nanostructures.

Numerical simulation methods for the molecular scale, such as the molecular dynamics (MD) and Monte Carlo (MC) simulations, are utilized recently for predicting phase equilibria and dynamic properties for engineering purposes, with the advancement of computer technology. Those for nanoparticle systems, however, are still under development because of their scale and highly heterogeneous nature: A unit cell for molecular simulations cannot handle important phenomena of nanoparticle systems that have a far larger size compared with molecules, and that must involve liquid–solid interface across which densities change drastically. Usual methods for computational fluid dynamics, on the other hand, must struggle to accommodate numerous particles and interaction forces acting between them.

Nonetheless, what are expected to support and develop the high-tech society in the near future would be the functional materials that cannot be produced from homogeneous systems, and the engineering tools for predicting the behavior in heterogeneous systems would be highly desired because they would speed up precise development, without which one must undergo carpet bombing or rely only upon empirical knowledge.

This section describes a possible scheme for predicting the behavior of mesoscale particulate systems by appropriate combination of simulations for molecules and colloidal dispersions, based on the microscopic molecular properties such as intermolecular potential functions. The scheme must take multi-scale structure to handle far different sizes of simulation elements such as simple molecules, complex molecules like surfactants and polymers, mesoscale particles, and macroscopic flow-fields. What is important in this multi-scale structure is to connect each unit with appropriate information that contains molecular-level properties, with which microscopic properties can be correctly reflected in a meso/macroscopic behavior: The interaction forces between the simulation elements are to bridge them as described later. Standing on this "bridging scheme", notable simulation techniques are reviewed and a possible structure composed of them will be discussed.

3.8.1 Space–time mapping of simulation methods

(i) A gap lying in the mesoscale

There have been many examples of application of molecular simulation methods for engineering purposes in the microscale, while in the macroscopic methods such as the discrete element method (DEM) and the Stokesian Dynamics (SD) [1] also have shown their application for micron-sized or larger particles. The only lack is the simulation method from several nanometers to submicron sizes, or the mesoscale simulation techniques, for engineering applications.

This gap makes it impossible to connect molecular properties to macroscopic phenomena. The quantum mechanics, for example, can provide intermolecular interaction energies that can be applied to molecular simulations, which can bridge the atomic scale to the molecular scale. However the SD, for example, is not supported by a simulation technique just under itself. Thus one would have to conduct a molecular simulation with the size of micronscale, which is excessively huge for molecules, if one would try to obtain surface forces based upon molecular properties. For the moment, the interparticle forces can only be obtained through the direct measurement by Atomic Force Microscopy (AFM), or through the theoretical prediction standing on continuum assumption, e.g., by the DLVO theory for electrostatic systems. In all cases the pathway from the molecular-level properties are closed.

(ii) A possible pathway from microscale properties to macroscopic dynamics

a. Brownian dynamics

Figure 3.8.1 schematically shows a structure in the space–time mapping that can bridge the gap in the mesoscale. (Note that the figure shows a typical time and space scale in which each technique is applied, and does not show the limit for the simulation technique.)

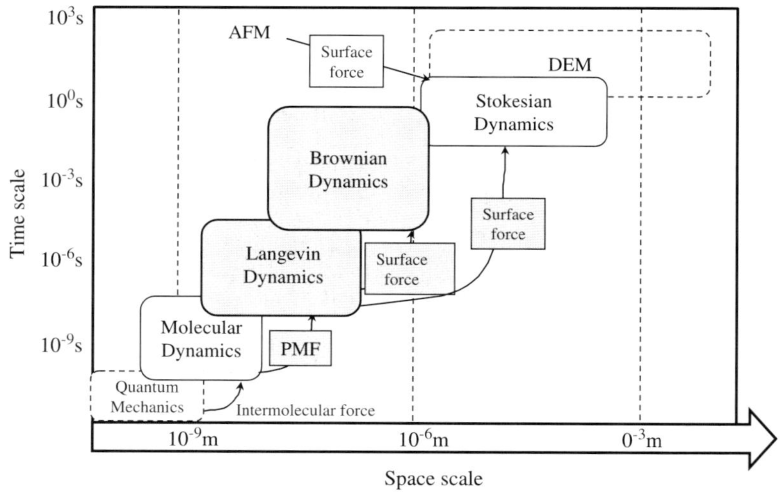

Figure 3.8.1
Space–time mapping of simulation methods for possible micro-macro connection.

Firstly desired would be a simulation method for nanoparticles in the broad sense, or those including up to ca. 100 nm, which would provide various functional materials by controlling the dispersed state or their array structure. The Brownian Dynamics (BD) should be appropriate for this purpose because it expresses the thermal Brownian motion, which become prevailing for particles smaller than submicron scale. The BD has been proposed by Ermak [2], and followed by certain number of researches after 80s mainly in the field of physics, but its application to engineering aspects, especially for concentrated dispersion systems, is insufficient at present. As a result, one cannot know to what extent the method is effective in solving problems for engineering/technological purposes, and what point should be improved in the technique. While the MD faithfully traces the thermal motion of the computational elements, or molecules, the BD inevitably approximates the thermal motion of nanoparticles for simulating them without surrounding fluid molecules. Thus the comparison of the simulation results with corresponding experimental ones must show its applicability before it stands as a "predictive" method. In this sense the authors have recently started a study with the BD method applied to the adsorption and order formation by colloidal nanoparticles adsorbing onto a substrate with counter charge [3, 4]. Figure 3.8.2 illustrates some examples: Adsorption process and

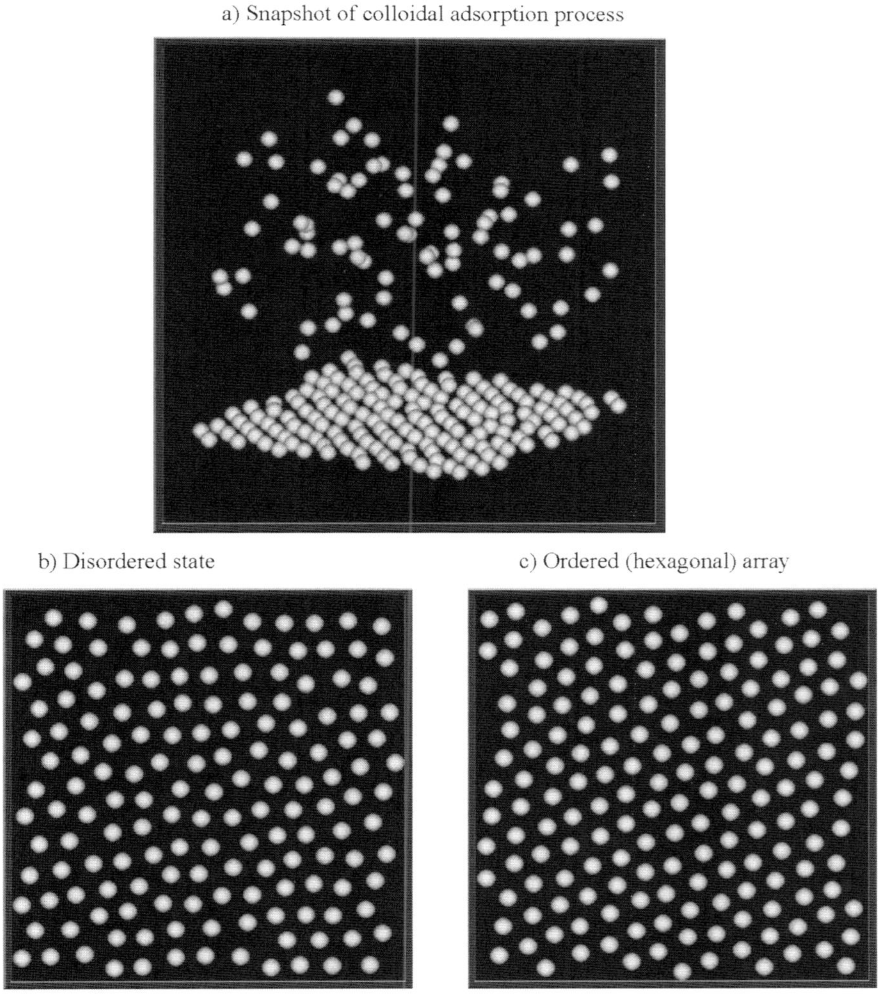

Figure 3.8.2
Order formation of mesoscale particles by adsorption onto substrate.

order formation of electrostatically stabilized colloidal particles with radius of 50 nm onto a planar surface with counter charge, as shown in (a), are examined, and the adsorbed particles show ordered or disordered arrays, as shown in (b) and (c), depending strongly on the operating conditions such as the concentration in the bulk phase, electrolyte concentration, and surface charge densities. The simulation results in various systems are to be compared, quantitatively, with experimental findings to brush up the BD method applicable to various engineering problems.

b. Langevin dynamics

The BD needs interparticle interactions as a bridging property from a lower layer of the multi-scale simulation structure. The molecular simulations, as stated earlier, suffer from the huge difference in the scale to provide it, and further they are not appropriate to predict the effects of surfactants, polymers, and additives that often drastically change the nature of the interaction. On the other hand, one may employ a direct measurement by AFM, using a "colloid probe" which has a particle glued on the top of the cantilever apex, but the lower limit for the measurement is a few microns in general. Thus we need a lower layer just beneath the BD to obtain the surface forces, which should be the Langevin dynamics (LD).

The essence of the method is the elimination of the solvent molecules, which is in principle same as the BD. Because both the BD and LD share their basic equation as the Langevin equation, they sometimes classify the two methods into the same category, calling either Brownian or stochastic dynamics. The LD, similar to the BD, has rather a long history in the physics field, but has only limited studies for engineering aspects. Utilization of the LD would benefit the multi-scale simulation structure by capability of providing, e.g., interaction forces between surfaces adsorbed by surfactants, to the BD or SD. The LD, on the other hand, needs the molecular-level information called the potential of mean force between (coarse-grained) elements as shown in Fig. 3.8.1, which will be explained in the next section.

The above is the possible multi-scale simulation structure, and the next section describes the basic concept and the feature of each simulation unit, reversely from molecular scale to meso/macroscale.

3.8.2 Simulation methods in nano/mesoscale

(i) Evaluation of interaction forces between nanoparticles by large-scale molecular simulations

The interparticle forces directly affect the behavior of the particulate system, which is especially of crucial importance for nanoparticles because they prevail more for smaller particles over other forces like gravity or inertia. The colloidal-probe AFM, as described earlier, can only give those for particles larger than a few microns. The molecular simulation cells, on the other hand, cannot accommodate micron or submicron particles, but it becomes possible if the particle size is extremely small.

Shinto et al. [5, 6] succeeded in obtaining interaction forces between nanoparticles with a few nanometers in diameter immersed in simple fluids by large-scale MD simulations, and discussed the effect of solvophobicity or solvophilicity onto the interparticle interaction. Another study showed interaction forces between nanoparticles and a substrate exposed to vapor with various pressures lower than the saturated one by the grand canonical MC simulations, discussing the relation between the condition of capillary bridge and the interaction forces [7]. Note here that the particle diameter has its UPPER limit around several nanometers.

The simulation methods explained in the following are summarized in Fig. 3.8.3. The smallest one, MD, has its typical scale of a few nanometers as described above. The basic idea and features of other methods are to be contrasted with this smallest size.

(ii) Langevin dynamics

For a larger size of particles, say, above 10 nm, the MD or MC cannot be used. Also difficult to be applied are the effect of surfactants or polymers onto surface forces because of their large-scale and slow behavior for relaxation and/or structure change. The LD should fill this portion in the multi-scale simulations.

Solvent molecules are not explicitly included in this simulation, and the solute molecules such as surfactants, polymers, and additives are the elements in the simulation that explicitly appear in a cell. Instead, the basic equation should include a random force to express thermal motion by solvent molecules, and a friction term that mimics viscosity attenuation proportional to the solute's velocity, resulting in the so-called Langevin equation.

Another important point in this method is to use, not the direct interaction forces as employed in MD, but the one including the effect of solvent molecules. For example, ions and sites in a surfactant molecule with hydrophilic and hydrophobic parts exchange far different forces from those of direct interactions in vacua, because of the effect of water molecules existing between them. This kind of interaction forces or potential energies can be determined by conducting a MD simulation at a fixed distance between the elements surrounded by water molecules that explicitly appear in the simulation. Conducting it with various distances and integrating against the distance, one can obtain the interaction potential for the LD, or the potential of mean force (PMF, also called as solvent averaged force). Setting this potential in LD without solvents, the elements (ions, sites in complex molecules, etc.) would feel a force as if the solvent molecules exist and surround them. This PMF is the one

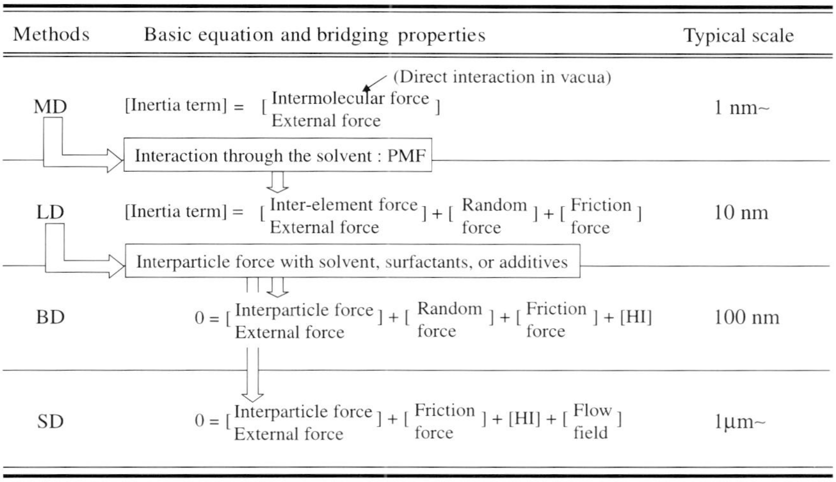

Figure 3.8.3
Basic idea and features of mesoscale simulations.

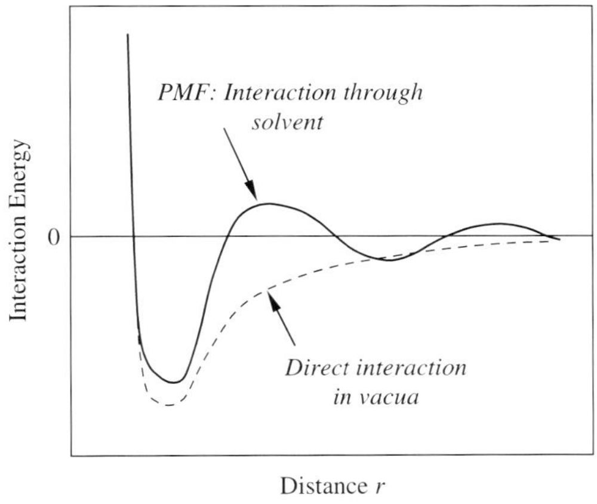

Figure 3.8.4
Schematic comparison of PMF and direct interaction.

that the MD should provide to the LD as a bridging property. As schematically shown in Fig. 3.8.4, the direct interaction used in MD is in general a smooth one, while the PMF between a pair of sites mediated by solvent would exhibit oscillations in the molecular scale periodicity, which is a clear reflection of discreteness of solvent molecules. Recent example of thus determined PMFs can be seen in Ref. [8].

The random force and friction attenuation must be coupled in the LD: The actual sites receive random force as the result of the collision from solvent molecules, and the motion by the random force will be attenuated by the friction with the solvent. Series of this process can mimic the random motion by thermal effect. Details are not shown here, but the strength of the random force and friction attenuation acting on a site can be determined from the temperature and the diffusivity in infinite dilution because they are connected by the fluctuation–dissipation theorem. In other words, the random force is, upon coupled with friction attenuation, playing the role of thermostat to maintain the system at a constant temperature. Even removing solvent molecules from molecular simulations, the LD can be thermodynamically consistent because of use of the PMF and because of the coupled fluctuation–dissipation relation. Further details of the theory can be found in literature [9, 10].

The scheme of finite difference equation against time and the algorithms of numerical integration can be given by, e.g., Gunsteren & Berendsen [11] and Allen [12]. For example, Hatton et al. [13, 14] conducted the LD simulation for micelle formation by surfactant molecules. The problem would be, however, that they employed unrealistic PMF for LD, or those without molecular-scale oscillations. A smooth PMF without realistic oscillation is reported to be incapable of representing coordination structure of solutes [15]. To clarify the behavior of complex molecules on and between solid surfaces and to make the prediction of mesoscale interparticle forces, realistic PMFs that honestly express oscillations in molecular scale through determination by molecular simulations should be used in the LD, as seen in Ref. [16].

If one needs to simulate smaller solutes or short-time phenomena, it is necessary to introduce the so-called memory function into the friction term and a random force with time correlation. This scheme is termed as generalized Langevin dynamics, and applied for studying behavior of ions in aqueous solution [17]. For the purpose to determine the surface forces with complex molecules, the above extent of preciseness seems not necessary. On the other hand, this literature gives an important and useful finding: By definition the PMF must have concentration dependence, but use of that from infinite dilution, which is usually the case, can give fair degree of prediction even for concentrated systems.

(iii) Brownian dynamics (Overdumped Langevin)

With the surface force obtained from the LD, one would be able to conduct the BD. The inertia term can be neglected for particles with diameters ranging from tens to several hundreds of nanometers, resulting in the simple basic equation shown in Fig. 3.8.3, which is sometimes called as the overdumped Langevin equation. The random force and friction term are the same as those for the LD. In this case the velocity can be directly determined by the force balance, and the algorithm becomes quite simple. One should, however, note that time increment for integration Δt must be set as an intermediate value that compromises both of the following two conditions: long enough to smear out momentary thermal motion of particles; short enough to allow only a small displacement for particle movement so as to treat surface force as a constant.

This method is applicable to relatively dilute systems such as the electrostatically stabilized colloidal dispersion. An example is seen in literature, which studied adsorption of colloidal nanoparticles onto a substrate [18].

(iv) Brownian dynamics with hydrodynamic interaction (BD with HI)

In a concentrated dispersion, the motion of a particle will affect those of other particles mediated by solvent, which is termed as the hydrodynamic interaction (HI). Ermak et al. [2] quite beautifully established its basic equations, but the long-ranged and multi-body nature of the HI, coupled with the existence of the random force, makes its equations and the algorithm far more complicated than those for the BD and SD [2, 10]. The HI must be expressed as a matrix of $3N$ x $3N$ for a three-dimensional system with N-particles, which needs large memory area, and whose computational time must be proportional to N^2. Further the existence of the HI brings correlation between random forces of particles, resulting in a huge computational cost for generating special type of random numbers, each of which relates to all other particles' random displacements.

The difficulty, therefore, rises in the case where one has to conduct a large-scale simulation for concentrated dispersion system of fine particles with non-negligible degree of Brownian motion. New development of methodology for the above difficulty would be desperately desired, some of which will be briefly reviewed in the next section.

3.8.3 Recent simulation methods including hydrodynamic interaction

Much difficulty exits, as explained above, in simulating concentrated dispersion of submicron particles. Further difficulty would add if the system is subjected to a flow field. This kind of system must be, however, one of the most important dispersion operations applied for producing functional materials by nanoparticles. Thus a new approach beyond conventional ones would be desired. A typical method of particulate dynamics that includes HI is firstly the Stokesian dynamics [1], as mentioned in the beginning of this chapter, which does not include fluid explicitly, based upon the idea of Ermak, but expresses its effect as a resistance tensor including relative velocities of all the pair of particles. This method suffers from complexity in coding, instability in computation and high computational cost, which would limits its applicability to the large-scale of concentrated dispersion. In the following some of recent approaches will be briefly described.

(i) Dissipative particle dynamics

One of the recent methods attracting attention would be the dissipative particle dynamics (DPD) proposed by Hoogerbrugge and Koelman [19]. The details are given by literature [20, 21], but this approach does not remove solvent contrary to the case with the BD and LD. Fluid is treated as composed of many coarse-grained mesoscopic "particles" each of which means a mass of fluid molecules. Both of the fluid and solid "particles" are subjected to the random force and the friction attenuation. The existence of the fluid makes it possible to treat the HI honestly with its multi-body nature, which is in principle impossible in the case of BD. Also, the computational load is linear with the N number of particles

Since the principle comes from the fluctuation–dissipation theorem, the basic equation for the DPD is the same as the LD. This point is of much interest, because much similarity in computational scheme must exists between the two methods, which leads to a possibility to establish a new systematized computational platform that can handle broad range of scales from nano to micron, which may be called as a "generalized mesoscale simulation". At present, however, the DPD seeks general guiding principles for, e.g., the size of the coarse-grained fluid, time increment for integration, and boundary condition for surface of colloidal particles. If these points are clarified, the

DPD would become useful as a powerful simulation method for colloidal dispersion, which would be desperately desired at an early date.

(ii) Fluid particle dynamics (FPD)

If any of the computational fluid dynamics (CFD) simulations can be adapted to a particulate dispersion system, it automatically is capable of expressing the HI. A direct introduction of solid particles into the CFD simulations, however, would bring a significant difficulty to handle moving boundaries in a fluid. Tanaka et al. [22], who had been conducting researches on the phase separations in complex fluids, noticed that the difficulty of moving boundaries would be able to be eliminated if a particle is treated as phase-separated component with extremely high viscosity, and proposed this method naming as the fluid particle dynamics (FPD). A particle is expressed as a concentration field of the high-viscosity component, and thus the whole system can be treated as a continuum, which can then be expressed by the Navier–Stokes equation. The value 50 for the ratio of the viscosities of the phase-separated component, which stands for solid particles, was reported to be enough to simulate colloidal systems. This method was applied to coagulation process, clarifying interestingly that the coagulated structure differs significantly depending on whether the HI exists or not in the system.

(iii) Lattice Boltzmann method (LBM)

Similar to the case of FPD, the Lattice Boltzmann method (LBM) would be able to express colloidal dispersion system if it can accommodate particles in its scheme: The LBM is a method for fluid dynamics, based, in principle, on the Boltzmann equation: The fluid is expressed by fictitious particles with some of finite velocities that come from grid spacing of a lattice for the flow field, and the time evolution of the flow field can be followed by the process of translational motion and collision between the fictitious particles employing velocity distribution functions for all of the velocity components. This method is reported to be capable of fast and large-scale calculations, compared with the finite-element method, for example, because of its simplicity in algorithm and suitability in parallelization. Typical example of its potential is given by Ladd [23]: By working out with the boundary conditions and treatment of velocity distribution functions at the particle surface, he succeeded in expressing a particle with five units of the grid and conducted simulations expressing behavior of tens of thousands of colloidal particles. The method for accommodating particles in the LBM scheme, however, has not yet been well-established, and the method needs further study on how to adopt surface forces and Brownian motion. Further development on the above points is expected to make the LBM stand as a powerful tool for colloidal simulations.

3.8.4 Closing remark

Many of mesoscale simulation methods have been under research mainly in the field of physics. They have shown, however, not so many examples of application to realistic engineering problems, and have not been organized to a level at which engineers can use them efficiently without difficulty. Meanwhile the dynamics in the mesoscale should be the most important phenomena to be pursued for the development and production of functional materials. We may have to wait until far future if we would expect the contribution only from the detailed research in science area. It is, instead, highly desired that many engineers and researchers in the engineering field have awareness of this issue in common, to thrust the development of the meso-simulations by tightly connecting them with experimental knowledge.

References

[1] J.F. Brady, G. Bossis: *J. Fluid Mech.*, **155**, 105–129 (1985); G. Bossis, J.F. Brady: *J. Chem. Phys.*, **80**, 5141–5154 (1984).

[2] D.L. Ermak, J.A. McCammon: *J. Chem. Phys.*, **69**, 1352–1360 (1978).

[3] M. Miyahara, S. Watanabe, Y. Gotoh and K. Higashitani: *J. Chem. Phys.*, **120**, 1524–1534 (2004).

[4] S. Watanabe, M. Miyahara and K. Higashitani: *J. Chem. Phys.*, **122** (104704), 1–10 (2005).

[5] H. Shinto, M. Miyahara and K. Higashitani: *J. Colloid Interface Sci.*, **209**, 79–85 (1999).

[6] H. Shinto, M. Miyahara and K. Higashitani: *Langmuir*, **16**, 3361–3371 (2000).

[7] H. Shinto, K. Uranishi, M. Miyahara and K. Higashitani: *J. Chem. Phys.*, **116**, 9500–9509 (2002).

[8] H. Shinto, S. Morisada and K. Higashitani: *J. Chem. Eng. Jpn.*, **37**, 1345–1356 (2004).

[9] M.P. Allen, D.J. Tildesley: *Computer Simulation of Liquids*, Clarendon Press, pp. 257–269 (Chapter 9) (1987).

[10] S. Kamiyama, A. Sato: *Ryutai Micro Simulation*, Asakura, pp. 52–77 (Chapter 7) (1997).

[11] W.F. van Gunsteren, and H.J.C. Berendsen: *Mol. Simul.*, **1**, 173–185 (1988).

[12] M.P. Allen: *Mol. Phys.*, **40**, 1073–1087 (1980).

[13] F.K. von Gottberg, K.A. Smith and T.A. Hatton: *J. Chem. Phys*, **106**, 9850–9857 (1997).

[14] F.K. von Gottberg, K.A. Smith and T.A. Hatton: *J. Chem. Phys*, **108**, 2232–2244 (1998).
[15] J. Trullas, A. Giro and J.A. Padrp: *J. Chem. Phys*, **91**, 539–545 (1989).
[16] H. Shinto, S. Morisada, M. Miyahara and K. Higashitani: *Langmuir*, **20**, 2017–2025 (2004).
[17] M. Canales, G. Sese: *J. Chem. Phys*, **109**, 6004–6011 (1998).
[18] M.R. Oberholzer, N.J. Wagner and A.M. Lenhoff: *J. Chem. Phys*, **107**, 9157–9167 (1997).
[19] P.J. Hoogerbrugge, J.M.V. Koelman: *Europhys. Lett.*, **19**, 155–160 (1992); J.M.V. Koelman, P.J. Hoogerbrugge: *Europhys. Lett.*, **21**, 363–368 (1993).
[20] H. Ohashi: *Nagare*, **18**, 5–10 (1999).
[21] P.B. Warren: *Curr. Opin. Colloid Interface Sci.*, **3**, 620–624 (1998).
[22] H. Tanaka, T. Araki: *Phys. Rev. Lett.*, **85**, 1338 (2000).
[23] A.J.C. Ladd: *Phys. Rev. Lett.*, **76**, 1392 (1996); A.J.C. Ladd: *Phys. Rev. Lett.*, **88**, 048301 (2002).

FUNDAMENTALS

CHAPTER 4

CONTROL OF NANOSTRUCTURE OF MATERIALS

CONTRIBUTORS

Hiroya Abe (4.1)
Yoshinari Miyamoto (4.2.1)
Mitsuo Umetsu (4.2.2)
Tetsuo Uchikoshi (4.2.3)
Tatsuya Okubo (4.3.1)
Makio Naito (4.3.2)
Yuji Hotta (4.3.3)
Tomoko Kasuga (4.3.4)
Akihiko Suda (4.4.1)
Hidetoshi Mori (4.4.2)
Reiji Mezaki (4.4.3)
Tsutomu Morimoto (4.4.4)
Akira Azushima (4.4.5)
Katsuyoshi Kondou (4.4.6)
Keizo Uematsu (4.5.1)
Takahiro Takada (4.5.2)
Kiyoshi Nogi (4.5.3)
Hidetoshi Fujii (4.5.4)
Jun Akedo (4.5.5)
Yoshiaki Kinemuchi (4.5.6)
Yoshio Sakka (4.5.7)
Yukio Yamaguchi (4.6.1)
Yoshitake Masuda (4.6.2)
Shinji Inagaki (4.6.3)

4.1 Assembly of nanoparticles and functionalization

Implementation of excellent functional materials or devices using nanoparticles requires to not only identify nanostructures with desirable functions, but also assemble those nanostructures up to a practical size maintaining the nanostructure arrangement. This process is similar to the biological technique of assembling small structural blocks of the atomic or molecular level up to a large structure, or the bottom-up technology. The bottom-up assembling and compounding of plural different materials and nanospace will enable the creation of nanoparticle structures with more versatile functionality.

In the field of the wet process of fine particles, for example, many reports have been published recently on self-organization and self-assembling technologies to construct structures by utilizing the laws of nature. In these technologies, the phenomenon of convective assembling is used, which accompanies meniscus (liquid surface) movement. In this process, a field is formed changing from a dispersed to an assembled state of particles so that the free energy of the system is reduced. In the future, understanding and systematization will be increasingly required of these new process technologies from the perspective of both basic research and application.

In this chapter, process technologies are summarized for nanostructural controls using mainly fine particles including nanoparticles as a starting material. In Section 4.2, the assembly structures of nanoparticles are discussed, introducing nanobiotechnologies and colloid processes. In addition, there is an explanation of fractal structures, rather than periodic or random structures, and optical properties. Section 4.3 looks at nanoporous structures and their control technologies, including zeolite, creation technologies of nanoporous structures by dry processes, control technologies of nanoporous structures, and the control of tubular porous structures.

In Section 4.4, the relation between nanocomposite structures used in catalysts and fuel cell electrodes and their functions is explained together with polymer nanocomposite technologies. In addition, plastic deformation technologies are discussed for controlling the nanostructures of metal and alloy. The distinctive process technologies of sintering and bonding of nanoparticle assembly and self-organization of nanoparticles are covered in Sections 4.5 and 4.6, respectively. The latest information is also introduced on various technologies useful for forming nanostructures, including sintering technologies of nanoparticles, low temperature sintering technologies of ceramics, aerosol deposition, colloid chemical processes, self-organization phenomena of nanoparticles in liquid phase, assembly patterning technologies and organic/inorganic mesoporous materials.

4.2 Nanoparticles arranged structures

4.2.1 Photonic fractal

A fractal is defined as a rough and irregular structure with self-similarity. In other words, the local configuration or substructure is similar to the whole configuration [1, 2]. The characteristic feature of fractal is found in complex forms of nature such as thunderhead, complex coastline, forest, and wrinkled wall of intestine. These forms have statistical self-similarity because the enlarged local configuration does not coincide with the whole, but resemble it.

The geometric dimension D is defined by equation (4.2.1), where N is the number of self-similar units newly created when the size of the initial unit decreases to $1/S$. It is called self-similar dimension or fractal dimension.

$$N = S^D \tag{4.2.1}$$

For example, when each edge of a cube is divided into three equivalent segments, the initial cube consists of 27 smaller identical cubes as seen in Fig. 4.2.1(a), and the equation (4.2.1) will be $27 = 3^3$. This result coincides with our recognition of three-dimensions. In case of Fig. 4.2.1(b), seven smaller cubes are extracted from the body-and face-centers. The equation is $20 = 3^D$. Therefore, the fractal dimension is about 2.73. When this division and extraction process is repeated three times, it is called the stage 3 Menger sponge as shown in Fig. 4.2.1(d). The fractal dimension is non-integer number. Menger sponge structure can be imaged as an intermediate structure between two- and three-dimensions when the stage number is increased.

A lot of research has been carried out [3, 4], because it is interesting to investigate what will happen when electromagnetic waves or light travels through a fractal structure. However, all the fractal structures investigated were one- or two-dimensional ones because of difficulty in fabrication of complex 3D structures. Japanese researchers have first fabricated cube fractals of Menger sponge structure with dielectric materials and found the localization of electromagnetic waves in 2003 [5–7]. They named such a fractal having localization function of electromagnetic waves or light as photonic fractal.

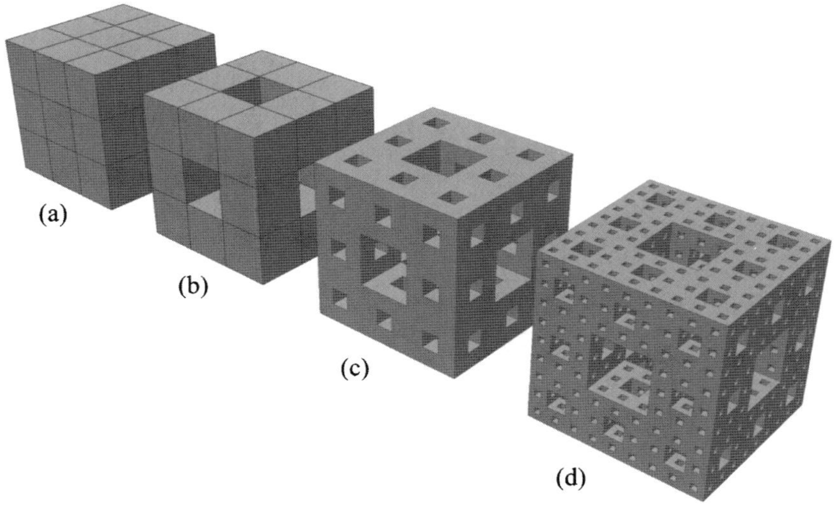

Figure 4.2.1
Model structures of Menger sponge (a) Cube initiator; (b) Stage 1 Menger sponge called generator; (c) Stage 2; (d) Stage 3.

3D-CAD is used to design of photonic fractals. It enables various designs such as modifying, reversing, and combining fractal structures as well as cutting at specific planes or joining connectors. The designed structure is converted into a rapid prototyping format (STL file), sliced into a set of thin sections, and transferred to a stereolithographic machine. It forms a 3D object layer-by-layer by scanning a UV laser over a liquid photopolymer resin. The dielectric constant of a formed object can be increased by incorporating ceramic particles into resin [8].

Figure 4.2.2 shows a stage 4 Menger sponge made of 10 vol% TiO_2–SiO_2/epoxy resin composite. The dielectric constant was increased to 8.8 comparing with 2.8 of resin. High dielectric contrast between dielectric material and air cavities can act to enhance the localization effect and reduce the structure size. Incorporation of nanosized ceramic particles is essential to fabricate microfractal structures.

Figure 4.2.3 shows the transmission and reflection spectra for the sample in the previous figure when electromagnetic waves with TE10 mode were irradiated. Both transmission and reflection amplitudes show large attenuations below 1/1,000 at the same 13.5 GHz suggesting the confinement of electromagnetic wave at this frequency. Figure 4.2.4 illustrates the electric field profile at the inner and outer space of Menger sponge structure. The relatively high electric field intensity is observed in cavities against very low intensity around the sponge.

The wavelengths of localization mode in a dielectric Menger sponge can be predicted using the empirical equation [9]:

$$\lambda = 2^l a\sqrt{\varepsilon_{\mathrm{eff}}}/3^{2l-1} \qquad (4.2.2)$$

where λ is the wavelength of the localized mode in air, a is the side length of Menger sponge cube, l is the order number of the localized modes, and $\varepsilon_{\mathrm{eff}}$ is the volume averaged effective dielectric constant of the Menger sponge which is calculated by the mixing rule of $\varepsilon_{\mathrm{eff}} = V_f\varepsilon_A + (1-V_f)\varepsilon_B$, where ε_A and ε_B are dielectric constants of air and object material, respectively. V_f is the volume fraction of material in a Menger sponge which is expressed by $V_f = (N/S^n)^m$, where n is the dimension number of the structure (0, 1 ,2 ,3...) and m is the stage number of Menger sponge.

Although the physical meaning of this empirical equation is not well-understood yet, the relation among the localized wavelength for the first mode, fractal size, and effective dielectric constant is given as a simple equation (4.2.3). It is surprising that only 1.5 times larger size Menger sponge against the wavelength is required to localize.

$$\lambda/\sqrt{\varepsilon_{\mathrm{eff}}} = 2/3a \qquad (4.2.3)$$

The theoretical analysis of localization is carried out by using finite differential time domain (FDTD) code and much more localization modes are obtained than those observed in experiment [10]. The localized wave is scattered to all directions though a part of wave energy is absorbed by component material during localization. Less scattering and absorption would increase the localization time.

Figure 4.2.5 shows a 90° scattering spectrum when a receiver monopole antenna was placed at right direction against the incident beam in order to avoid the influence from the incident beam. A peak appeared at the same frequency of 13.5 GHz as measured in both

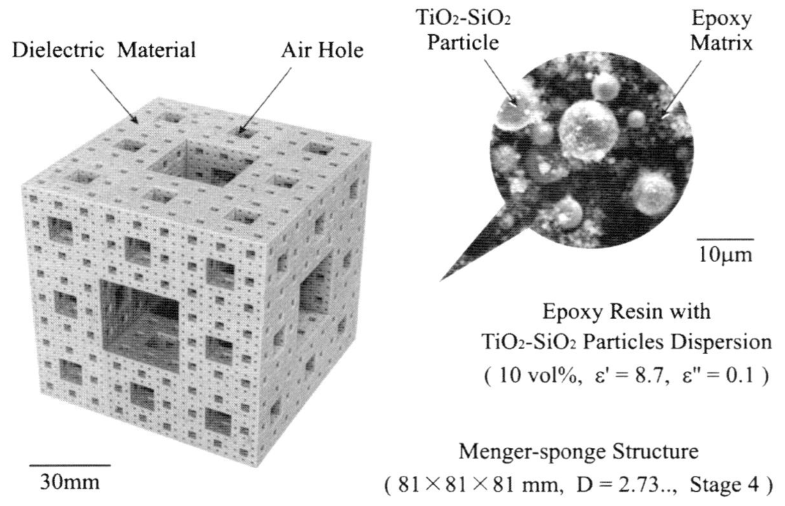

Figure 4.2.2
A stage 4 Menger sponge formed by stereolithography. The cube edge size is 81 mm. Four different size square holes with 27, 9, 3, 1 mm in edge size are opened against each face.

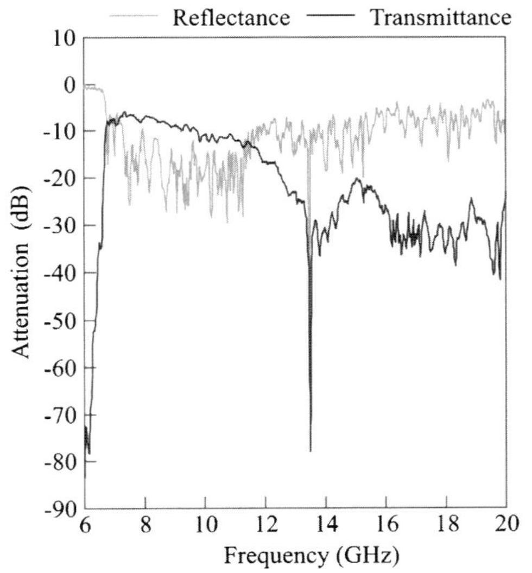

Figure 4.2.3
Transmission and reflection spectra of electromagnetic waves against a stage 4 Menger sponge structure. The second order localization peak appears at 13.5 GHz. Another deep transmission attenuation at around 6 GHz corresponds to the cutoff frequency by a horn antenna.

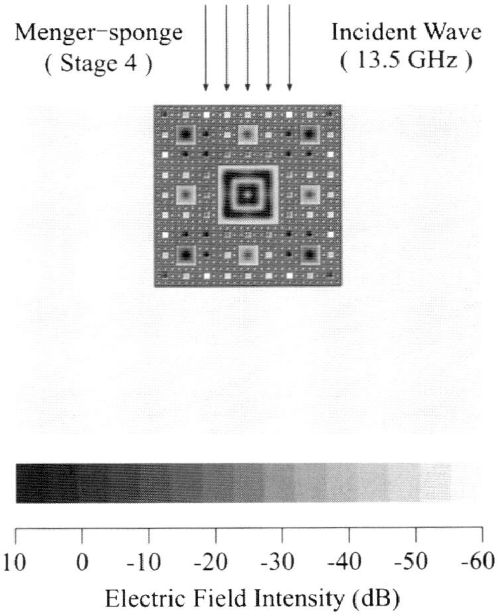

Figure 4.2.4
Electric field mapping image at the inner and outer space of a stage 4 Menger sponge on the mid-plane.

transmission and reflection spectra indicates that the scattered intensity of a localized mode increases than that of delocalized mode.

The quality factor Q was about 180, which was calculated from the half width of the scattered peak and peak frequency. The estimated confinement time from this value was about 2 ns. The Q value and the confinement time are very low at present. The survey of low loss ceramics and design of effective fractal structure to reduce scattering are investigated.

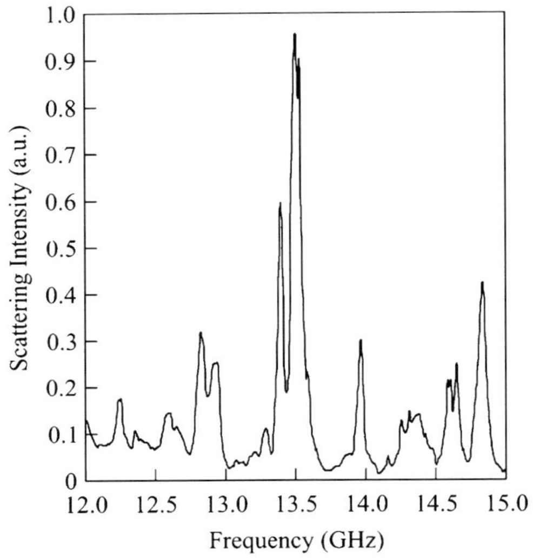

Figure 4.2.5
90° scattering spectrum for a stage 4 Menger sponge.

Figure 4.2.6
Photo of a stage 3 Menger sponge with sintered alumina.

Figure 4.2.6 shows an alumina photonic fractal. It was fabricated by dewaxing and successive sintering of the fractal made of nanosized alumina particles dispersed at 40 vol% in acrylic resin. Various applications in communication such as filter, cavity, attenuator, antenna are expected for such cube fractals like Menger sponge because they can confine electromagnetic waves at the specific frequency for a limited time. If it is possible to increase Q value to 10^5 or 10^6, the application will be expanded. When nano or micrometer size photonic fractals can be fabricated, applications to terahertz wave or optical devices would be expected. It may be possible to draw a dream to develop a photonic fractal battery to accumulate electromagnetic waves or light when the lossless confinement will be achieved in future.

References

[1] K. Honda: *Fractal*, Asakura Publishing Ltd., Tokyo (2002).
[2] M. Matsushita: *Physics of Fractal*, Shokabo Ltd., Tokyo (2002).
[3] W. Wen, Z. Yang, G. Xu, Y. Chen, L. Zhou, W. Ge, C.T. Chan and P. Sheng: *Appl. Phys. Lett.*, **83**, 2106–2108 (2003).
[4] D.P. Tsai, J. Kovacs, Z. Wang, M. Moskovits, V.M. Shalaef, J.S. Suh and R. Botet: *Phys. Rev. Lett.*, **72**, 4149–4154 (2004).
[5] M.W. Takeda, S. Kirihara, Y. Miyamoto, K. Sakoda and K. Honda: *Phys. Rev. Lett.*, **92**, 093902-1-4 (2004).
[6] Y. Miyamoto, S. Kirihara, S. Kanehira, M.W. Takeda, K. Honda and K. Sakoda: *Int. J. Appl. Ceram. Technol.*, **1**, 40–48 (2004).
[7] Y. Miyamoto, S. Kirihara, M.W. Takeda, K. Honda and K. Sakoda: *Mater. Integration*, **18**, 40–45 (2005).
[8] S. Kirihara, Y. Miyamoto: *Exp. Mech.*, **1**, 210–211 (2001).
[9] Y. Miyamoto, S. Kirihara and M.W. Takeda: *Chem. Lett.*, **35**, 342–347 (2006).
[10] K. Sakoda, S. Kirihara, Y. Miyamoto, M.W. Takeda and K. Honda: *Appl. Phys.*, **B81**, 321–324 (2005).

4.2.2 Nanoparticle patterning by nanobiotechnology: Peptide

Shellfishes and corals synthesize calcium carbonate from carbon dioxide in vivo, and mammals synthesize calcium phosphate for bone and teeth from calcium ion. Diatoms make exquisite silicate membranes on their own surfaces [1], and magnetic bacteria let magnetic particles form a line for the detection of earth magnetism [2]. These syntheses in vivo are called biomineralization, in which the crystal growth of synthesized minerals is strictly controlled in nanoscale on the basis of gene information. Nanotechnology starts from the expression of unique function by decreasing the size of materials, and currently, it is expected to express a novel function by the assembly of functional nanocrystals. Nanocluster assembly is observed in biomineralization and the mild synthesis condition, "at room temperature in neutral solution", attracts immense interest in the hybridization with heat-labile or pH-sensitive organic compounds. Here, we show you recent biomineralization studies by combinatorial procedure.

4.2.2.1 Functional biomolecules: Peptide and protein

In general, peptide and protein play an important role on biomineralization, as a framework and catalyst. Peptide/protein is a polymer produced by the polycondensation of amino acids, and its complicated steric structure and function depend solely on the order of amino acids (Fig. 4.2.7). Considering that there are 20 kinds of amino acids, the combination number of amino acids order is numerous. Living organisms can respond to diverse environmental changes by combinatorial evolution of peptides and proteins utilizing the numerous combination number of amino acids order. Typical functions of peptides and proteins are specific binding ability (e.g. antibody), catalysis (e.g. enzyme), and "forming" ability (e.g. collagen). These functions are considered to be necessary for biomineralization. Shimizu et al. focused on the silica synthesized by marine sponge, and identified the protein used for silica synthesis, called silicatein [3]. Silicatein can hydrolyze tetraethoxysilane in neutral solution to make silica in vitro. The silica synthesis by peptide/protein is called "silicon biotechnology" [4], which is expected for a novel silicon synthesis.

4.2.2.2 Phage display

In nature, there are peptides and proteins that bind and synthesize inorganic materials, such as silicatein. However, their identification from in vivo not only takes a long time and is inefficient, but also is logically hard to find the peptide/protein for non-biological materials. Recently, the peptides and proteins with affinity for non-biological inorganic materials have been discovered by means of a combinatorial library approach. Here, we show you a method using phage.

Phage display method is a combinatorial chemistry, in which a molecule with desired function is selected by screening from a great diversity of molecular population called library. The problems for the combinatorial library approach are how to make library, how to screen desired molecule, and how to analyze the selected low-concentrated molecule. In the case that peptide/protein is an element in library, the use of a kind of virus, phage, can solve the problems.

Bacteriophage M13 used in phage display method, is made of five kinds of coat proteins (gp III, gp VI, gp VII, gp VIII, and gp IX), and gp III functions to infect host *Escherichia (E.) coli* (Fig. 4.2.8). A peptide or protein is displayed at the *N*-terminus of gp III, and the

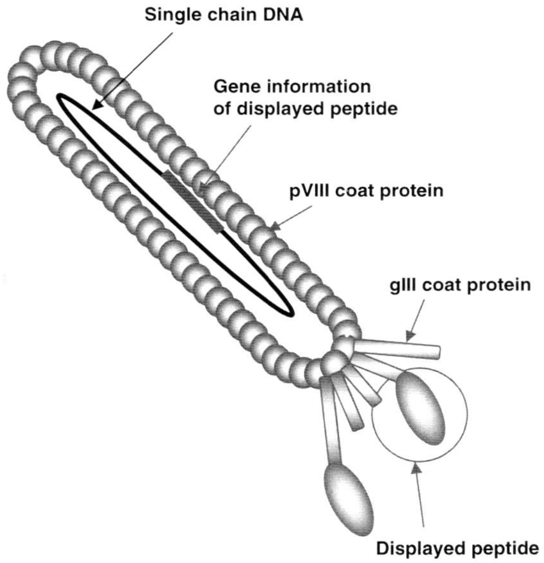

Figure 4.2.8
Bacteriophage M13.

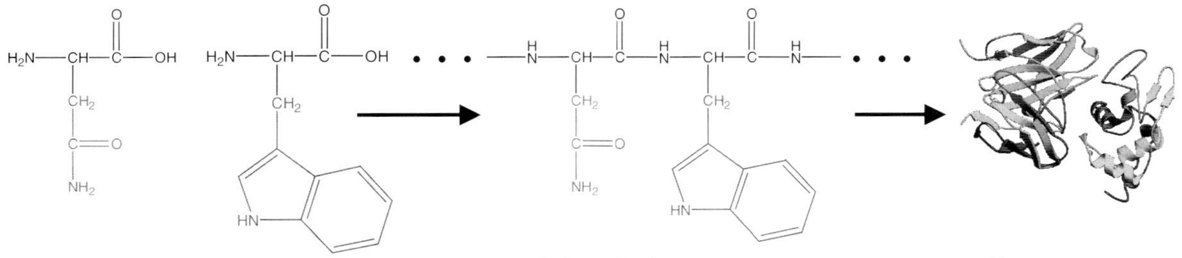

Figure 4.2.7
From amino acids to peptide/protein.

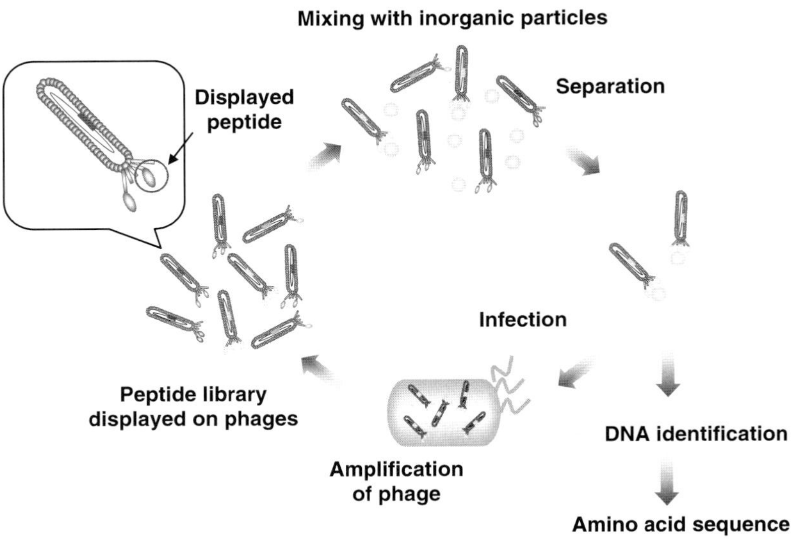

Figure 4.2.9
Phage display method.

amino acid sequence is randomly mutated by error-prone method, and so on (In the case of 12-amino acid sequence, we can prepare $10^7 \sim 10^9$ kinds of peptides in the library). The selected low-concentrated peptide/protein can be analyzed by identifying the DNA code of the selected phage after DNA amplification. Figure 4.2.9 shows a selection of the peptide with affinity for inorganic material using the phage display method. The phage viruses displaying peptides are mixed with inorganic particles. The bound phage viruses to added inorganic particles are separated from the surface of inorganic particles after the unbound phages are removed, and this screening procedure is repeated for several times to remove the phage viruses accidentally bound to added particles. The screening procedure is repeated after the eluted phage viruses are amplified in *E. coli*, so that the magnitude of library decreases as the number of phage viruses is kept. Finally, the amino acid sequences of the peptides with affinity for added inorganic material are analyzed by identifying the DNA code of the residual phage on added material.

Table 4.2.1 lists the amino acid sequences for the peptides with affinity for inorganic materials that have been selected by combinatorial library approaches. To our knowledge, the peptide motif that can bind to iron oxide (Fe_2O_3) is the first report for the peptides with affinity for inorganic materials [5]. The peptide was selected by cell surface display method. The first material-binding peptide selected by phage display method is the one with affinity for GaAs [6]. At present, the peptides with affinity for gold [7], chrome oxide [8], lead oxide [8], cobalt oxide [8], manganese oxide [8], zinc oxide [9, 10] and zeolite [11], and for calcium carbonate [12], silver [13], silica [14], zinc sulfide [15], titanium [16], carbon nanocompound [17, 18], platinum [19] and palladium [19] have been selected by cell surface display and phage display methods, respectively.

4.2.2.3 Bio-patterning and mineralization of inorganic materials by peptide

The combinatorial library approaches enable us not only to find the peptides with affinity for non-biological materials, but also to mineralize materials by the selected peptides; for example, the addition of the peptides with affinity for gold and silver into each metal ion solution results in reduction and unique crystalline of gold and silver, respectively [7, 13]. Further, Belcher et al. made a ZnS nanowire with the width of 10 nm by mineralizing ZnS on the phage with the ZnS-binding peptide in p VIII coat protein [20, 21]. Here, we show you our resent studies for bio-immobilization and biomineralization by the ZnO-binding peptide we selected [10].

To immobilize ZnO particles using the ZnO-1 peptide, we synthesized the ZnO-binding peptide with a cysteine residue at the *C*-terminus, and the peptide was immobilized on the surface of gold-coated polypropylene plate via the thiol group of cysteine residue. The plates with or without immobilized ZnO-binding peptide were soaked in an aqueous suspension containing fluorescent ZnO particles at room temperature (20°C), and the plates were then analyzed by fluorescence detection (Fig. 4.2.10). Although the plates looked identical under no ultraviolet radiation, an intense ZnO-specific fluorescence was clearly observed only on the peptide-bearing plate under ultraviolet radiation at 254 nm (Fig. 4.2.10). In order to estimate the specificity of the ZnO-binding peptide for ZnO, a plate with

Table 4.2.1
Reported peptides with affinity for inorganic materials [19].

Peptide selected by phage display method		Peptide selected by cell surface display method	
Material	Amino acid sequence	Material	Amino acid sequence
Gallium arsenic (GaAs)	AQNPSDNNTHTH	Ferric oxide (Fe_2O_3)	RRTVKHHVN
	RLELAIPLQGSG		
	TPPRPIQYNHTS	Gold (Au)	MHGKTQATSGTIQS
			LGQSGQSLQGSEKLTNG
Calcium carbonate ($CaCO_3$)	HTQNMRMYEPWF		EKLVRGMEGASLHPA
	DVFSSFNLKHMR		
		Chrome oxide (Cr_2O_3)	VVRPKAATN
Silver (Ag)	AYSSGAPPMPPF		RIRHRLVGQ
	NPSSLFRYLPSD		
	SLATQPPRTPPV	Palladium oxide (PbO_2)	YPPFHNNDHRS
			SKPLARSSGA
Silica (SiO_2)	MSPHPHPRHHHT		
	RGRRRRLSCRLL	Cobalt oxide (CoO)	GRMQRRVAH
	KPSHHHHHHTGAN		LGKDRPHFHRS
Zinc sulfide (ZnS)	NNPMHQN	Manganese oxide (MnO_2)	HHMLRRRNT
			HINASQRVA
Titanium (Ti)	RKLPDAPGMHTW		
		Zinc oxide (ZnO)	NTRMTARQHRSANHKSTQRA
Carbon nanotube	HWSAWWIRSNQS		TRRGTHNKD
Carbon nanohorn	DYFSSPYYEQLF	Zeolite	VKTQATSREEPPRLPSKHRPG
	WPGWHHVPPAVS		MDHGKYRQKQATPG
	GHWHHITKVSKQ		
Platinum (Pt)	DRTSTWR		
	QSVRSTK		
	SSSHLNK		
Palladium (Pd)	SVTQNKY		
	SPHPGPY		
	HAPTPML		
Zinc ion (Zn^{2+})	HFQAQMRHGHGH		
	HQSHHYGPRDHT		
Fullerene (C60)	NMSTVGR		

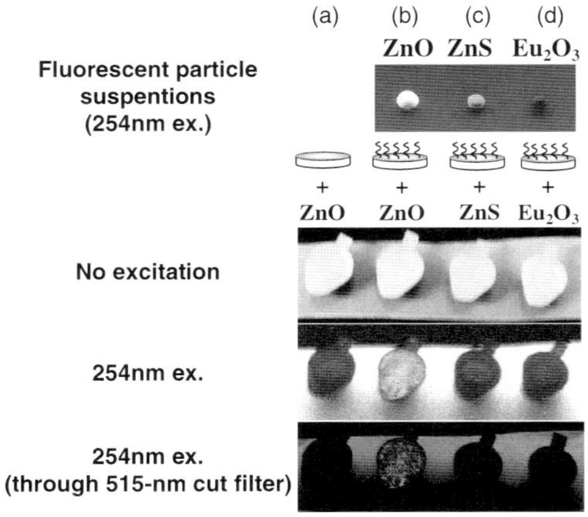

Figure 4.2.10
Immobilization of ZnO particles on a polypropylene–gold plate through ZnO-binding peptide (a) The plate without ZnO-binding peptide was soaked in a suspension containing fluorescent ZnO particles; (b–d) The plates with ZnO-binding peptide were soaked in a suspension containing fluorescent ZnO particles (b), ZnS particles (c), or Eu_2O_3 particles (d).

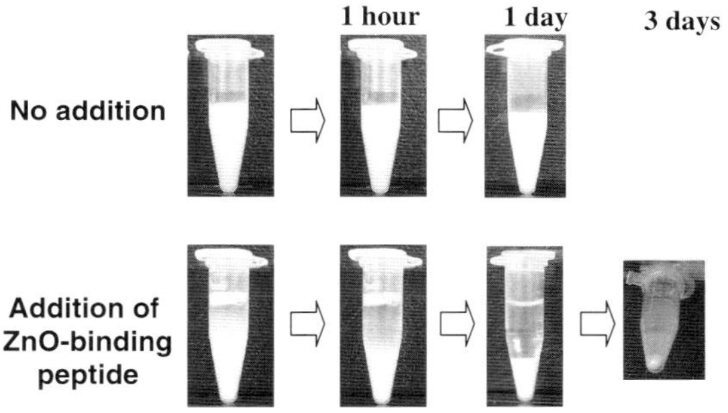

Figure 4.2.11
Biomineralization of ZnO by ZnO-binding peptide.

immobilized ZnO-binding peptide was soaked in a suspension of fluorescent ZnS or europium oxide (Eu_2O_3). However, these plates exhibited no fluorescence when excited at 254 nm, indicating that the ZnO-binding peptide binds specifically to the ZnO surface by recognizing both the zinc and oxygen atoms. This result implies that ZnO particles can be patterned along ZnO-binding peptide lines at room temperature in neutral solution, and the bio-patterning can be expected to be a

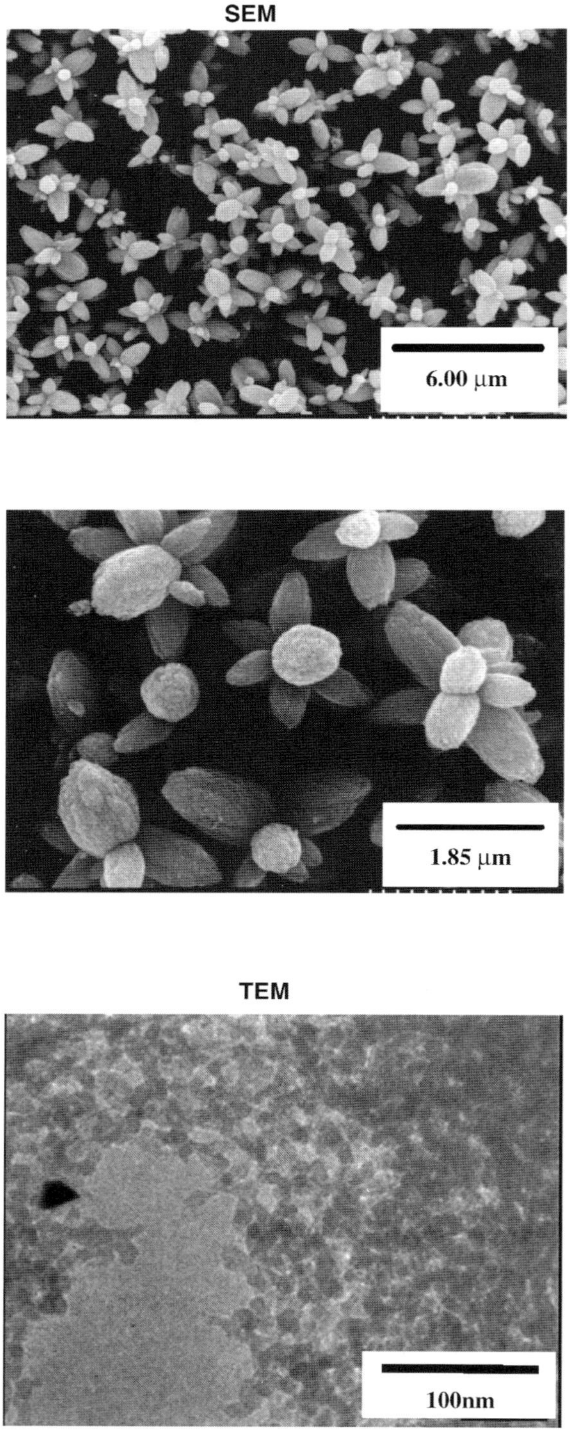

Figure 4.2.12
SEM and TEM images of the ZnO particles synthesized by ZnO-binding peptide.

patterning method on the substrate of heat-labile or pH-sensitive organic compounds.

For mineralization, the ZnO-binding peptide with a cysteine residue at the *C*-terminus was added into a sol solution of zinc hydrate ($Zn(OH)_2$). In the absence of peptide, the $Zn(OH)_2$ sol solution was stable during the time course of the experiment; in fact, no change was observed after a month. In contrast, ZnO-binding peptide appeared to immediately interact with the $Zn(OH)_2$ sol phase, so that the sol state became more condensed and finally ZnO particles were precipitated in the solution after 3 days (Fig. 4.2.11). The addition of a mixture of free amino acids found in the ZnO-binding peptide shows no condensation of $Zn(OH)_2$ nor deposition of ZnO. The mineralization function is expressed by making the peptide with amino acids ordered according to the ZnO-binding peptide. In addition, a scanning electron microscope (SEM) and transmission electron microscope (TEM) images for synthesized ZnO showed a flower-type anisotropic morphology composed of ZnO nanoparticles with a diameter of 10–20 nm (Fig. 4.2.12). Our results show the potential that the peptide with the capability to synthesize inorganic materials can assist in the formation of inorganic matter with unique structures.

In order to pattern inorganic nanoparticles on organic substrates, selection of appropriate adhesion between organic and inorganic materials is necessary. When various different inorganic nanoparticles are arranged on an organic substrate, specific organic–inorganic adhesions for each combination should be required; however, the number of possible covalent bonding between organic and inorganic materials is not sufficient for the requirement. Whereas, peptide can supply various organic–inorganic adhesions, utilizing a numerous combination number of amino acids order. In addition, biomineralization has the potential of the production of hybrid materials, because of the mineralization at room temperature in neutral solution. The advance in peptide/protein engineering will open the way to a novel inorganic nanoparticle assembly.

References

[1] http://www.ndu.ac.jp/~t-nagumo/
[2] R.P. Blakemore: *Annu. Rev. Microbiol.*, **36**, 217–238 (1982).
[3] K. Shimizu, J. Cha, G.D. Stucky: *Proc. Natl. Acad. Sci. USA*, **95**, 6234–6238 (1998).
[4] D.E. Morse: *Trend. Biotechnol.*, **17**, 230–232 (1999).
[5] S. Brown: *Proc. Natl. Acad. Sci. USA*, **89**, 8651–8655 (1992).
[6] S.R. Whaley, D.S. English, E.L. Hu, P.F. Barbara, A.M. Belcher: *Nature*, **405**, 665–668 (2000).
[7] S. Brown: *Nat. Biotechnol.*, **15**, 269–272 (1997).
[8] M.A. Schembri, K. Kjaergaard, P. Klemm: *FEMS Microbiol. Lett.*, **170**, 363–371 (1999).

[9] K. Kjaergaard, J.K. Sørensen, M.A. Schembri, P. Klemm: *Appl. Environ. Microbiol.*, **66**, 10–14 (2000).
[10] M. Umetsu, M. Mizuta, K. Tsumoto, S. Ohara, S. Takami, H. Watanabe, I. Kumagai, T. Adschiri: *Adv. Mater.*, **17**, 2571–2575 (2005).
[11] S. Nygaard, R. Wendelbo, S. Brown: *Adv. Mater.*, **14**, 1853–1856 (2002).
[12] D.J.H. Gaskin, K. Starck, E.N. Vulfson: *Biotech. Lett.*, **22**, 1211–1216 (2000).
[13] R.R. Naik, S.J. Stringer, G. Agarwal, S.E. Jones, M.O. Stone MO: *Nat. Mater.*, **1**, 169–172 (2002).
[14] R.R. Naik, L.L. Brott, S.J. Clarson, M.O. Stone: *J. Nanosci. Nanotechnol.*, **2**, 95–100 (2002).
[15] S.W. Lee, C.B. Mao, C.E. Flynn, A.M. Belcher: *Science*, **296**, 892–895 (2002).
[16] K. Sano, K. Shiba: *J. Am. Chem. Soc.*, **125**, 14234–14235 (2003).
[17] D. Kase, J.L. Kulp, M. Yudasaka, J.S. Evans, S. Iijima, K. Shiba: *Langmiur*, **20**, 8939–8941 (2004).
[18] S. Wang, E.S. Humphreys, S.Y. Chung, D.F. Delduco, S.R. Lustig, H. Wang, K.N. Parker, N.W. Rizzo, S. Subramoney, Y.M. Chiang, A. Jagota: *Nat. Mater.*, **2**, 196–200 (2003).
[19] M. Sarikaya, C. Tamerler, A.K. Jen, K. Schulten, F. Baneyx: *Nat. Mater.*, **2**, 577–585 (2003).
[20] C. Mao, C.E. Flynn, A. Hayhurst, R. Sweeney, J. Qi, G. Georgiou, B. Iverson, A.M. Belcher: *Proc. Natl. Acad. Sci. USA.*, **100**, 6946–6951 (2003).
[21] C. Mao, D.J. Solis, B.D. Reiss, S.T. Kottmann, R.Y. Sweeney, A. Hayhurst, G. Georgiou, B. Iverson, A.M. Belcher: *Science*, **303**, 213–217 (2004).

4.2.3 Preparation of ceramic films by liquid-phase processing: Electrophoresis

Colloidal processing, which is composed of the dispersion of ceramic powders in liquid media followed by consolidation, is superior to conventional dry processing in the control of density and microstructure of green and sintered bodies. Electrophoretic deposition (EPD) is one of the most promising colloidal processes wherein ceramic bodies are directly shaped from a stable colloidal suspension by a dc electric field. Electrically conductive metals or graphite are often used both as an electrode and a substrate, however a non-conductive porous material, which is placed in front of an electrode non-contactedly, is also possible to use as a substrate. EPD is a combination of two processes: electrophoresis of charged particles in a suspension and deposition on a substrate. The mechanism of deposition is explained based on the DLVO theory; the

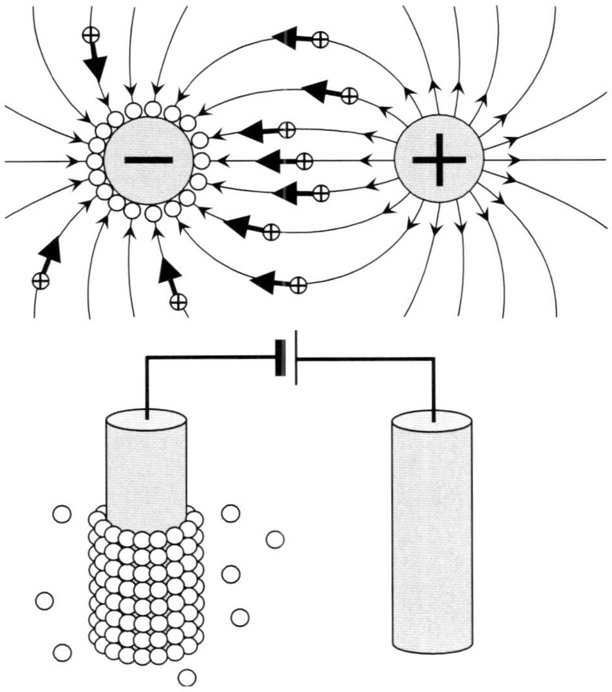

Figure 4.2.13
The electrophoresis of particles and electric field lines during electrophoretic deposition between a pair of cylindrical electrodes.

deposition is due to the coagulation by the van der Waals attraction accompanied by the drop of the repulsive potential at an electrode. Figure 4.2.13 shows the electrophoresis of particles and electric field lines during electrophoretic deposition between a pair of cylindrical electrodes. The surface-charged particles do not take the shortest straight route but move along the electric field lines to an oppositely charged electrode in a liquid. Therefore, the particles deposit even on the back of a substrate. This characteristic enables the EPD to make uniform surface coating on a substrate with uneven surface and complex shape.

1. Kinetics of EPD processing

Figure 4.2.14 shows the electrophoresis of the particles and ions in a suspension under applying dc field. The charged particles attract oppositely charged ions, i.e., counterions, around themselves; this "atmosphere" plus the particle is the "lyosphere". When the lyospheres reach the cathode, the double layer is thinned. If the particles are close enough to each other, attractive forces dominate and coagulation/deposition occurs. During the local lyosphere thinning and coagulation, the still-charged particles are thrust into the substrate due to the electric-field force. In the case that the drop of repulsive potential is insufficient, the attractive force

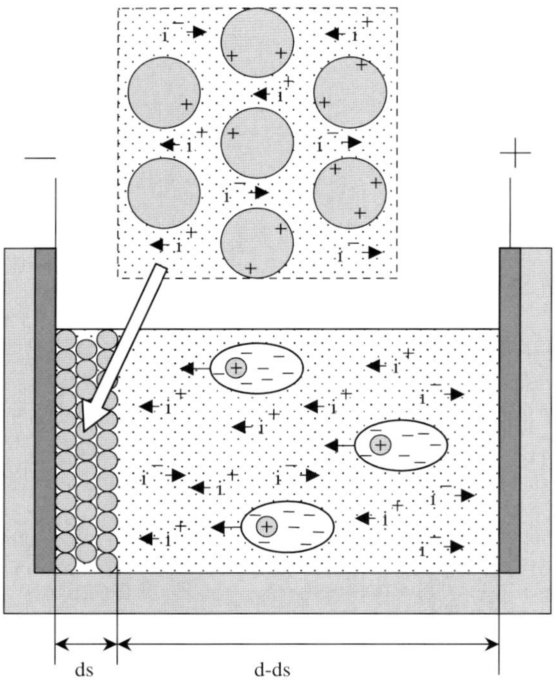

Figure 4.2.14
The electrophoresis of the particles and ions in a suspension under applying DC field.

between the particles is weak; it may cause slip-off of the deposits when the applied dc is turned off.

The weight of deposit, W(g), during electrophoretic deposition is estimated based on the mass balance law of Hamaker [1];

$$\frac{dW}{dt} = f\mu CES \tag{4.2.4}$$

where t is the deposition time (s), μ the electrophoretic mobility (m^2/Vs), E the electric field strength (V/m), C the solid loading (g/m^3), S the surface area of the electrode (m^2), and f the sticking probability of the particles that arrived at the substrate ($0 \leq f \leq 1$). The electric field strength E is given by the following equation with the current flowing through the circuit I(A) and the conductivity of the suspension Λ (S/m).

$$E = \frac{I}{S\Lambda} \tag{4.2.5}$$

The E changes during the deposition. The applied potential to the circuit V_a is consumed by the potential drops at both anode and cathode, and the ohmic loss over the deposit and suspension. The V_a is given as;

$$V_a = \Delta\phi_{\text{anode}} + IR_{\text{sus}}(d - d_s) + IR_s d_s + \Delta\phi_{\text{cathode}} \tag{4.2.6}$$

where $\Delta\phi_{\text{anode}}$ and $\Delta\phi_{\text{cathode}}$ are the potential drops (V) at the anode and cathode, respectively, R_{sus} and R_s are the specific resistance (Ω/m) of the suspension and deposit, respectively, and d and d_s the electrode distance and the deposit thickness (m), respectively. Under the deposition of constant voltage mode, the progress of deposition can be monitored indirectly by checking the decrease of current with time which is caused by the IR drop. Under the deposition of constant current mode, the progress can be monitored by checking the increase of voltage.

2. Preparation and evaluation of suspension

For the EPD processing, optimization of the colloidal suspension is very important. Aqueous suspensions are generally used for conventional colloidal processing since an aqueous system has the advantages of low-cost processing, lower electrical potential requirement, and lower environmental cost, however, non-aqueous suspensions are usually preferred for EPD to avoid the electrolysis of the solvent and obtain a bubble-free deposit. It is essential that the colloidal particles in a solvent should be electrostatically stabilized for the electrophoresis. The charge on a colloidal particle could originate from various sources. Dissociation or ionization of surface groups on the particles is commonly observed with adsorbed carboxylic acid, amine, and oxide surfaces. In these systems, the degree of charge development and its sign depend on the pH of the solution.

For example, surfaces of oxide particles tend to coordinate water molecules to form amphoteric hydroxyl groups that can charge positively or negatively depending on pH of liquid media:

$$(M - OH) + H^{+} \leftrightarrow (M - OH_2) \quad \text{(at low pH)} \tag{4.2.7}$$

$$(M - OH) + OH^{-} \leftrightarrow (M - O)^{-} + H_2O \quad \text{(at high pH)} \tag{4.2.8}$$

where H^+ and /or OH^- behaves as potential-determining ions. Addition of acid or base and/or polyelectrolytes helps to enhance the surface charge potentials of particles, however too much addition of them can be countereffective on the stability of a suspension since non-adsorbed free ions causes the double-layer compaction.

Though the measurement of zeta potential is important, the measurements of acidity and electric conductivity are simple and easy ways to evaluate the stability of a suspension. The acidity of aqueous solutions is indicated by $pH = -\log[H^+]$ measured using a pH meter. However, an ordinary pH meter is also applicable to evaluate the acidity of non-aqueous media which is apt to contain water such as lower alcohol and ketone. The acidity of non-aqueous media is given by the following equation, in which the concentration of proton $[H^+]$ is replaced to the activity of proton a_H.

$$pa_H = -\log a_H \tag{4.2.9}$$

$$\mu_H = \mu_H{}^0 + RT \ln a_H \tag{4.2.10}$$

When the acidity of non-aqueous solvent is measured by using a pH electrode (KCl electrolyte) calibrated for aqueous media, the "operational" pH, pH_{op}, differs from the real pa_H in a non-aqueous solvent; i.e.,

$$pH_{op} - pa_H = \frac{\Delta E_j}{\left(\frac{RT \ln 10}{F}\right)} \tag{4.2.11}$$

or

$$pH_{op} - pa_H = \frac{\Delta E_j}{0.05916} \quad \text{at } 25°C \tag{4.2.12}$$

where ΔE_j is the residual liquid-junction potential encountered in the standardization and testing step of a standard pH meter, R the gas constant, and F the Faraday constant. For the ethanol suspension, the following is calculated.

$$pa_H = pH_{op} + 1.23 \tag{4.2.13}$$

The pH_{op} measurement is simple but effective to evaluate the stability of a suspension even for non-aqueous system.

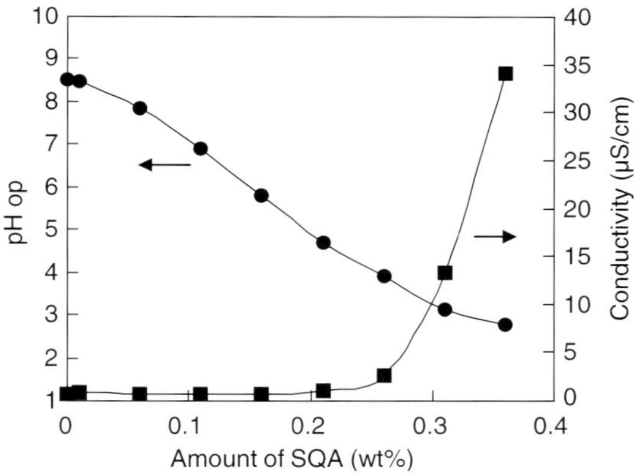

Figure 4.2.15
The operational pH and conductivity of an alumina–ethanol suspension as a function of amount of squaric acid (SQA).

The electric conductivity is an index of the concentration of free ions in a suspension. It suddenly increases when the addition of acid, base, or dispersant exceeds the adsorption saturation on the surface of particles. Figure 4.2.15 shows the operational pH and conductivity of an alumina–ethanol suspension as a function of amount of squaric acid (SQA). SQR dissociates in an organic solvent and liberates the protons from the hydroxyl groups. The protons adsorb on the powder and make its surface positively charged. As the amount of SQR is increased, the amount of adsorbed protons becomes gradually saturated and the electrical conductivity increases. Excessive free ions disturb the electrophoresis of the surface charged particles. Electrical conductivity should be checked when the particles does not deposit properly.

The addition of binders sometimes helps to increase the adherence and strength of the deposit and prevent cracking during the subsequent drying. For example, polyvinyl alcohol (PVA) and polyvinyl butyral (PVB) are often used for aqueous and non-aqueous suspensions, respectively.

3. Synthesis of novel materials via EPD

EPD is a powerful technique for the design and control of ceramic composites, particularly laminates. Figure 4.2.16 shows the procedure to fabricate a laminate composite. Layers are formed by the sequential deposition on a same substrate. The thickness of each layer is easily controlled by altering the deposition time. Recently, multi-layered, "monolithic" ceramic composites with textured layers have been fabricated by EPD in a strong magnetic field. EPD is applicable to fabricate ceramic products with irregular shapes or free-standing objects with complicated patterns by

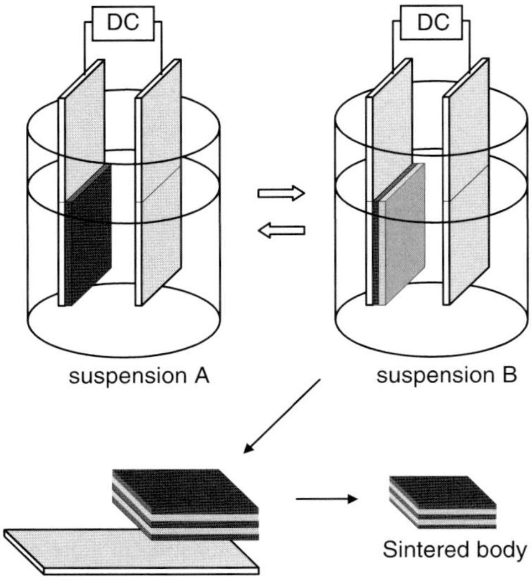

Figure 4.2.16
The procedure to fabricate a laminate composite.

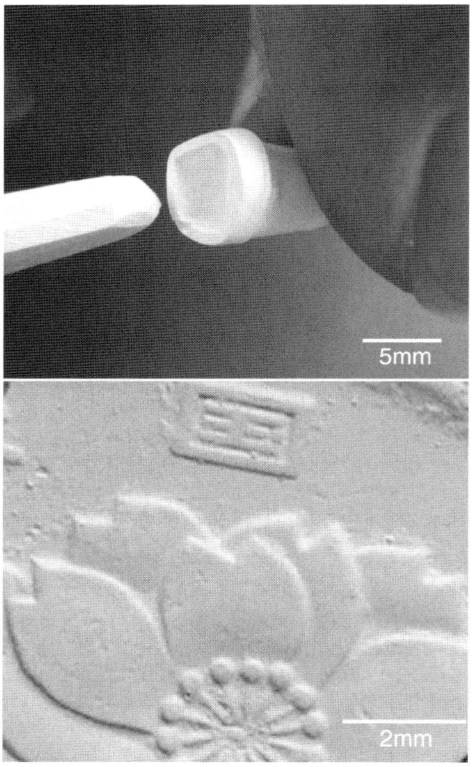

Figure 4.2.17
The examples of alumina ceramics fabricated by sintering the deposits together with conductive polymer-covered ceramic substrates in air.

using pre-shaped substrates. Figure 4.2.17 shows a few examples of alumina ceramics fabricated by sintering the deposits together with conductive polymer covered ceramic substrates in air. This technique can be utilized for fabricating various shape-controlled ceramics by EPD. Further fundamentals and applications of EPD are referenced in the following review papers [2–7].

References

[1] S. Put, J. Vleugels and O. Van der Biest: *Acta Mater.*, **51**, 6303–6317 (2003).
[2] P. Sarkar, P.S. Nicholson: *J. Am. Ceram. Soc.*, **79**, 1987–2002 (1996).
[3] O. Van der Biest, L.J. Vandeperre: *Annu. Rev. Mater. Sci.*, **29**, 327–352 (1999).
[4] I. Zhitomirsky: *Adv. Colloid Interface Sci.*, **97**, 279–317 (2002).
[5] A.R. Boccaccini, I. Zhitomirsky: *Curr. Opin. Sol. State Mater. Sci.*, **6**, 251–260 (2002).
[6] Y. Fukada, N. Nagarajan, W. Mekky, Y. Bao, H.-S. Kim and P.S. Nicholson: *J. Mater. Sci.*, **39**, 787–801 (2004).
[7] A.R. Boccaccini: *J. Ceram. Soc. Jpn.*, **114**, 1–14 (2006).

4.3 Nanopore structure

4.3.1 Microporous material: Zeolite

Zeolite is a class of microporous aluminosilicate crystals including molecular-sized, void spaces and ion-exchange sites within the structure. It is called as "molecular sieve" since zeolite recognized molecules by their sizes and shapes. Owing to the unique, structural features, zeolites have been already utilized as adsorbents, catalysts, and ion-exchangers widely in industrial and agricultural fields as well as in our daily necessities [1, 2].

Natural zeolites were first recognized in the mid-eighteenth century. After 200 years, R.M.Barrer [3] succeeded in the synthesis of some zeolites based on the formation condition of natural zeolites. Since the success, several synthesis methods have been developed, which resulted in formation of a variety of framework structures; now framework types more than 160 are recognized by International Zeolite Association, the Structure Commission [4]. However, there remain a lot of unexplained details on the synthesis routes. Still, most of novel zeolites have been found through trial-and-error approaches, and some of them have been industrially produced in the scale of 2 million tons per year worldwide.

(1) Structure of zeolite

Some of typical zeolite structures are shown in Fig. 4.3.1. T atoms (Si or Al) with four bonds are connected each other via O atoms, and zeolite

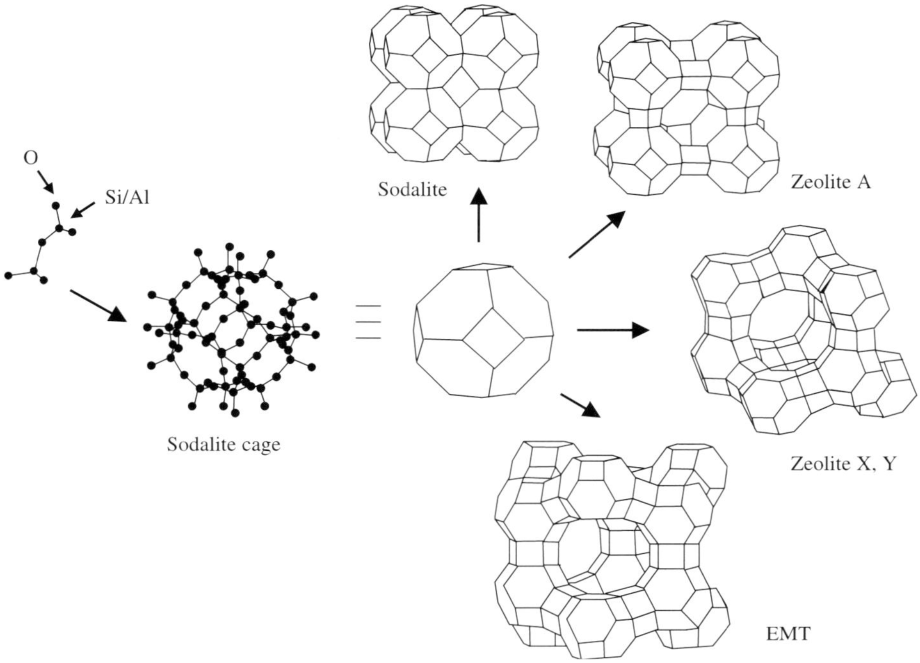

Figure 4.3.1
Typical zeolite structures.

frameworks are constructed. Since zeolite is crystal, each void has the same shape and size; it is a great feature of zeolite in porous materials. When all T atoms are Si(IV), the framework is electrically neutral, while the framework is non-charged (neutral), when a part of Si(IV) at T sites are replaced by Al(III). In order to compensate for the negative charge, metal cations can be stabilized in the framework by electrostatic interaction. Since such cations can be exchanged with other species, the location where the cation is held is called as ion-exchange site. Also metals other than Si or Al can be doped at T sites.

The structures of void spaces are rich in variety: from 1D to 3D and from pipes to cages. The sizes of the spaces are, however, all in microporous region (< 2 nm, refer chapter 5, section 5.4 for the classification). The size of zeolite pores is represented by the number of O atoms included in zeolite pores. Typical zeolites contain 8, 10, or 12 rings.

(2) Introduction to zeolite synthesis

Natural zeolites are formed under hydrothermal condition in geologically long-time scale. In order to synthesize zeolites, they have to be formed within a few hours to a few days under mild conditions. Thus, relatively low-temperature process (< 250°C) was developed, and a novel zeolite of which structure has never been found in natural one was synthesized [5]: Zeolite A in Fig. 4.3.1. In these conditions, zeolites are formed as meta-stable phases; therefore, formation routes must be controlled kinetically. The products have to be recovered before the continued conversion of the meta-stable products to stable phases.

Thereafter, a novel route was explored; organic molecules or ions are embedded in the crystalline products as "template", followed by the removal to open void space in the structures [6, 7]. Since these organic "template" is larger than alkali metal cations, the resulted framework contains less Al and becomes less charged. In other words, organic-template approach enabled us to synthesize zeolites with larger Si/Al and hydrophobic properties. After ion-exchange with H^+, such high-silica zeolites show strong acidity, which have been utilized for catalytic conversion. These organic "template" is designated as structure-directing agent (SDA), and bulky organic molecules or ions have been investigated to synthesize zeolites.

The starting materials of zeolites should contain silicon source (sodium silicate, colloidal silica, fumed silica, alkoxide, and so on), aluminum source (aluminum hydroxide, sodium aluminate, alkoxide, and so on), mineralizer (alkali metal hydroxide and fluoride), and water. If necessary, an organic SDA is added. Silicon and aluminum are the main components of the framework. The mineralizer lets the framework components dissolve in the water phase. At the same time, alkali metal cations act as the charge-compensating elements for negatively charged

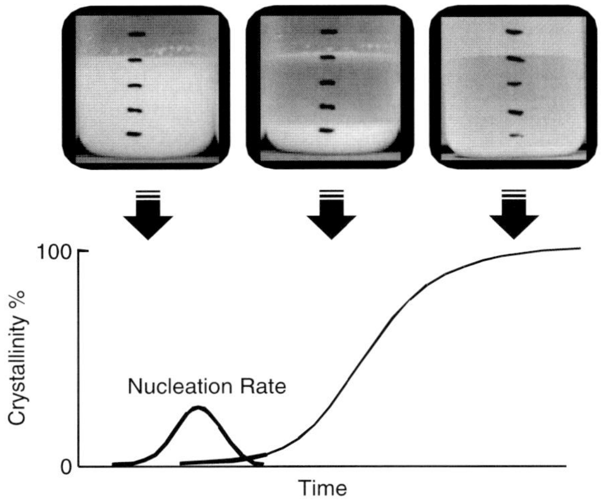

Figure 4.3.2
Outline of zeolite synthesis.

frameworks. Highly reactive, amorphous hydrogels are formed by mixing these reactants, which is encapsulated in an autoclave and heated at prescribed temperature ($< 250°C$) to result in the crystallization of zeolites. Apparent changes are shown in Fig. 4.3.2.

Several investigations have been carried out in order to understand the formation mechanism of framework structure with the void spaces. Since it is difficult to perform in-situ observation in dense hydrogel under high temperature, high pressure, and high pH conditions, the details of the formation have been only speculated based on the crystallization condition of the zeolites. Thanks to the recent development of analytical tools, formation mechanism of zeolites has been getting clearer, especially for high-silica zeolites.

(3) Formation of microporous structure of zeolite

Zeolite synthesis route consists of nucleation and crystal growth. As for both of the steps,

- solution-mediated transport mechanism,
- solid-hydrogel transformation crystallization mechanism

have been proposed. There have been several experimental results to support each of them [8]. In some cases, the same product can be obtained via different routes depending on the synthesis condition and resources. It should be noted that water is highly concerned in solid–hydrogel transformation since a lot of water is included in the gels.

The system that has been studied most intensively is the synthesis of high- or pure-silica zeolites with SDAs. Several tools have been employed to understand the nucleation and the crystal growth, and the schemes in molecular levels are getting clearer. Here, synthesis of Si-ZSM-5 (Zeolite Socony Mobil #5) is introduced since several groups provide a unified view [9].

^1H-^{29}Si CP Magic Angle Spinning Nuclear Magnetic Resonance (MAS NMR) was applied to understand the interaction between an SDA ($(C_3H_7)_4N^+$:tetrapropyl ammonium ion, TPA^+) and silicate at the early stage of the synthesis, and behavior for the SDA to organize silicate was investigated. H in the SDA and Si in the precursor of framework have weak intermolecular interaction of non-covalent bonding. Based on the analysis, it was revealed that the TPA^+–silicate complex is formed prior to the onset of the long-range ordering in – 10 nm scale that corresponds with crystal lattice of the zeolite as illustrated in Fig. 4.3.3(a–c). TPA^+ and silicate are soluble in water, but not very hydrophilic; therefore, water molecules are ordered around them by hydrophobic hydration. When both of them approach each other and the hydrated layers are overlapped, TPA^+ is covered with silicate and ordered water molecules are released. Thus, this complex formation is thermodynamically reasonable.

Continued nucleation and crystal growth processes were investigated with in situ combined small angle X-ray scattering (SAXS)–wide angle X-ray scattering (WAXS) [9] using a clear solution instead of hydrogel. The schemes proposed from the results are Fig. 4.3.3(d–f). Throughout the process, 2.8 nm clusters are observed, and their aggregation proceeds prior to the onset of the long-range ordering (crystallization). WAXS can detect the formation of crystalline nuclei and the particle size distribution can be determined from SAXS. From these analyses, it is proposed that the nuclei are converted from aggregates of the clusters, and the crystal growth proceeded through the transportation of the cluster decomposed from the aggregates. Also for hydrogel systems and on other pure-silica zeolites (Si-beta, Si-ZSM-12), experimental results to support a similar scheme have been reported.

As shown in the case of Si-TPA-ZSM-5, the role of SDA is to organize silicate or aluminosilicate around it to form precursor complex [8]. Hydrophobic hydration provides the driving force for the organization. Thus, SDA should

- be soluble in water, but less hydrophilic,
- be stable under hydrothermal synthesis condition, and
- have suitable size and shape.

Thus, some quaternary alkylammonium cations and organic amines have been employed as SDAs.

As mentioned above, the initial stages of zeolite synthesis with SDAs is getting clearer although this synthesis method was developed after inorganic synthesis

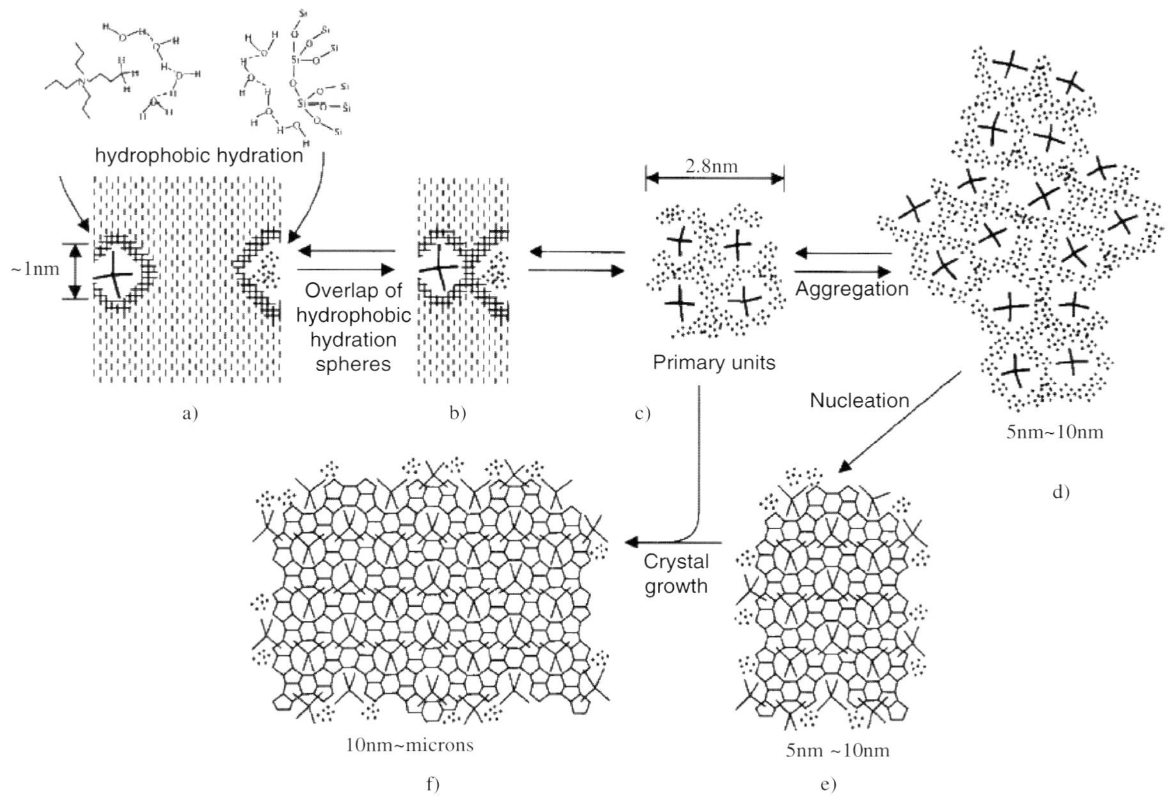

Figure 4.3.3
Nucleation and crystal growth model of Si-ZSM-5 zeolite synthesized with TPA^+ as SDA.

of zeolites without using SDAs. Still, most of zeolites are industrially produced without using SDAs although the details of the synthesis have not been well-understood yet. Thanks to the recent progress of analytical tools, the details of the processes will be clearer before long [10].

(4) Ion exchange

Porous structure of zeolites can be further tuned precisely by the post modification. Typical example of pore size control of zeolite A is illustrated in Fig. 4.3.4. When the cation is Na^+, 4 Å molecules can go into zeolite structure after adsorbed water is removed. Since the inner diameter of the 8-member ring of zeolite A is 5 Å, it is clear that the effective pore size is reduced by Na^+. After exchange of Na^+ with K^+, the effective pore size is further reduced to 3 Å. The effect of K^+ is larger than Na^+, since larger K^+ is located at the same site as Na^+. On the contrary, the effective pore size is 5 Å when it is exchanged with Ca^{2+}. When a cation is divalent, the total number of cations is reduced to one-half, and Ca^{2+} exists only in inner, stable sites selectively, resulting in the reduced effect for diffusing molecules.

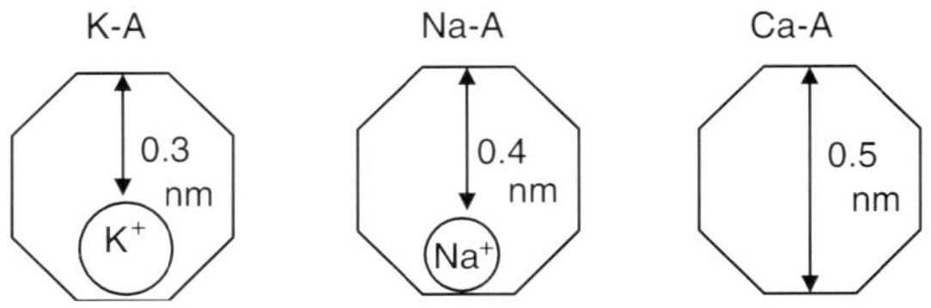

Figure 4.3.4
Pore size control by ion exchange.

Based on the precise control of pore size, separation of molecules with only slight differences in size has been realized.

References

[1] Y. Ono, T. Yashima (eds): *Science and Engineering of Zeolite*, Kodansha, Tokyo (2000).
[2] H. Tominaga (ed.): *Science and Application of Zeolite*, Kodansha, Tokyo (1987).
[3] R.M. Barrer: *Hydrothermal Chemistry of Zeolites*, Academic Press (1982).

[4] http://www.iza-structure.org/databases/
[5] D.W. Breck: *Zeolite Molecular Sieves*, Wiley (1974).
[6] R.M. Barrer, P.J. Denny: *J. Chem. Soc.*, 971–982 (1961).
[7] G.T.Kerr, G. Kokotailo: *J. Am. Chem. Soc.*, **83**, 4675–4675 (1960).
[8] M.E. Davis, R.F. Lobo: *Chem. Mater.*, **4**, 756–768 (1992).
[9] P.P.E.A. de Moor, T.P.M. Beelen, B.U. Komanschek, L.W. Beck, P. Wagner, M.E. Davis and R.A.V. Santen: *Chem. Eur. J.*, **5**, 2083–2088 (1999).
[10] M. Ogura, Y. Kawazu, H. Takahashi and T. Okubo: *Chem. Mater.*, **15**, 2661–2667 (2003).

4.3.2 Preparation of nanoporous material by dry processing

(1) Preparation methods for nanoporous materials

It is known that the nanoporous materials having a pore size below about 100 nm show completely different properties from the conventional materials with micron-sized pores. For example, the convective heat transfer becomes extremely difficult inside the porous materials with pore sizes less than about 70 nm, which is the mean free path of the molecules in the air. Also, various other properties, like catalytic or absorptive properties, etc., exhibit different characteristics in the nanosized space from those of their bulk counterparts.

Many methods have been proposed to prepare the porous materials with nanosized pores so far. The method introduced here is to create nanoporous materials with nanoparticles by using the dry mechanical processing. Since the whole processing, from the preparation of the starting material to its forming, is conducted in a dry state, neither solvent nor liquid binder is used. Consequently, the residual contamination inside the porous material can be avoided and the process becomes quite simple. Furthermore, it is an environment-friendly process because of no dying and no gas to exhaust.

Furthermore, since it is desirable to use the nanoparticles as close as possible to their produced condition, it is expected to be a low cost, environment-friendly process. As to the nanoparticles, practically there is no restriction on whether they are inorganic or organic materials. Moreover, this process seems to have relatively high possibility for commercialization from the industrial viewpoints, because the material-forming method can be scaled up by proper equipment design. The formed nanoporous components have advantages in lightness and good workability because of their stable strength and high porosity.

The key point of this method for producing the porous materials is to create nanosized pores by direct bonding of the nanoparticles. Figure 4.3.5 shows its schematic structure [1]. For example, it is possible to make nanosized pores below 100 nm inside the material by bonding ~10 nm nanoparticles together. As the particle size decreases, the number of atoms located at the particle surface approaches to the total amount of atoms constructing the particle. Therefore, nanoparticles are expected to be bonded together at lower temperatures than the bulk materials. By applying strong mechanical energy to nanoparticles, their surfaces can be activated and they can be bonded together in the dry state.

Using this method, bulk nanoporous components can be produced by bonding nanoparticles alone; but, they are often reinforced with fibrous materials like fiberglass to improve their strength and workability. Figure 4.3.6 shows the process to prepare composite particles for the production of nanoporous materials [2]. During the material processing, the nanoparticles are bonded onto the surface of the fibrous particles to make composite particles, which can form a layer with nanosized pores. These nanocomposite particles can be prepared by using the above-mentioned strong mechanical energy in a dry state.

The flowability of the composite particles consisting of the nanoparticles and fibrous particles can be improved by controlling their structures. As a result, the shaping die can be filled with the dry composite particles uniformly; and, the porous materials with high uniformity and stable strength can be produced by subsequent press-forming operation.

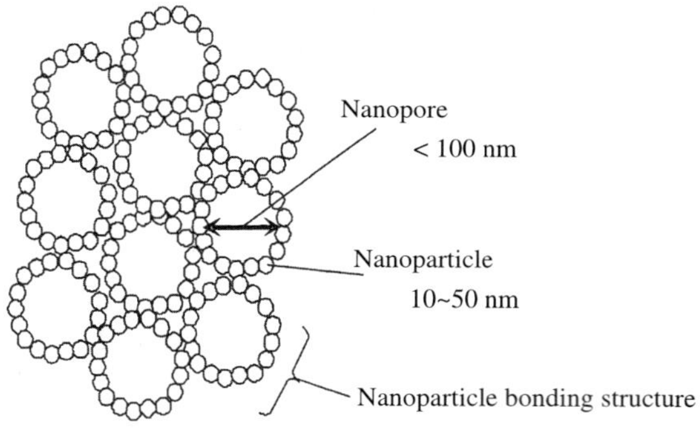

Figure 4.3.5
A schematic structure showing how to control the nanosized pores by the nanoparticle bonding [1].

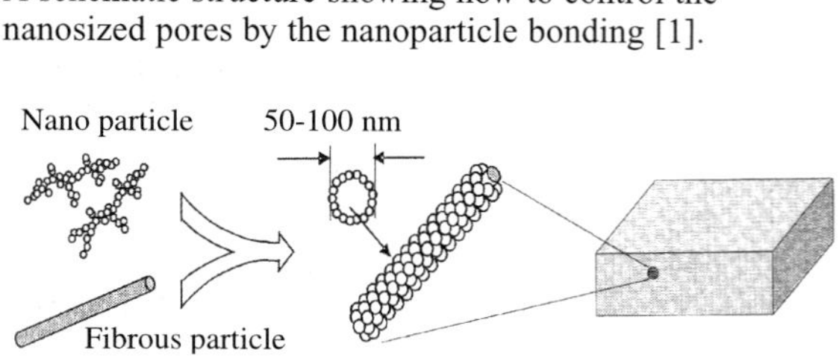

Figure 4.3.6
Production of nanoporous component by composing nanoparticles and fibrous particles [2].

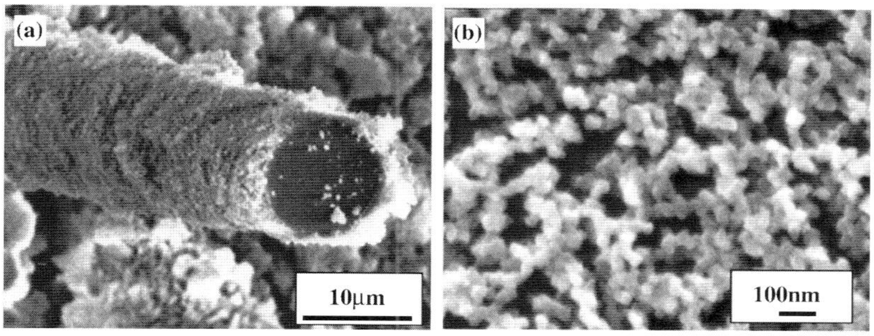

Figure 4.3.7
Fibrous composite particles made by a dry processing method [3]. (a) SEM picture of the surface of a fibrous particle; (b) Enlarged picture at the surface of the fibrous particle.

(2) Example of production of nanoporous materials

As an example, a porous material can be fabricated by using fumed silica (particle size about 10 nm) as nanoparticle and glass fiber as fibrous particle (fiber diameter about 10 μm, length about 3 mm). Figure 4.3.7 shows the SEM pictures of the surface of the composite particle made by the dry mechanical method [3]. As seen in the picture, fibrous composite particles, having highly porous nanoparticle layer on the surface of the fiber, can be formed by applying proper stress conditions that will not break the glass fibers in the particle composing process. Examining the bonded interface between the fiber and the nanoparticle with the SEM, it is interesting to find that the particle surface layer has relatively high porous structure, while the fiber surface and the nanoparticles are tightly bonded together. Making use of this gradient porous structure, composite particles having good adhesiveness between the fiber and the nanoparticles, and nano-sized pores can be produced.

Porous material components can be easily made by press-forming after filling these composite particles into dies. Furthermore, there is no unstable part, which may lead to ripped-off at the surface of finished components. The relationship between the fracture strength and apparent density of the porous component is plotted in Fig. 4.3.8 [3]. As seen in the figure, the fracture strength increases with the increase in apparent density; and, it can be up to 1 MPa even at a relative low density, which makes it possible to process the components without breakage.

These nanoporous materials are known to have various interesting properties. For example, Table 4.3.1 shows the relationship between the density, porosity, and thermal conductivity of a component. The thermal conductivity here was measured by periodical heating method. As shown in the table, the component has extremely low thermal conductivity while keeping its high porosity.

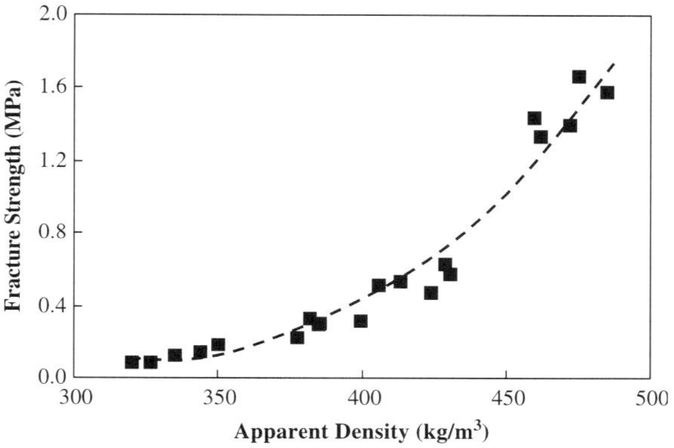

Figure 4.3.8
Relationship between the fracture strength and apparent density of a porous component.

Table 4.3.1
Relationship between the density, porosity, and thermal conductivity of a porous compact [3].

Specimen	Density (kg/m^3)	Porosity (%)	Thermal conductivity (W/m K) 100°C	400°C
#1	459	81.2	0.0266	0.0269
#2	485	80.1	0.0266	0.0282

References

[1] M. Naito, H. Abe: *Application of Porous Materials in New Age*, CMC Publishing Co. LTD., Japan, pp. 204–209 (2004).

[2] M. Naito, H. Abe: *Ceram. Transact.*, **57**, 69–76 (2004).

[3] H. Abe, I. Abe, K. Sato and M. Naito: *J. Am. Ceram. Soc.*, **88**, 1359–1361 (2005).

4.3.3 Ordered porous structures

Highly porous inorganic materials, particularly those exhibiting well-defined pore sizes, have been exploited in a wide range of applications, such as catalytic supports and surfaces, optical devices, bioreactor supports, and chemical filters [1–4]. Porous materials are classified according to their pore size, into micropores (below 2 nm), mesopores (2~50 nm), and macropores (above 50 nm). In order to get the maximum functionality from a porous material, it is important to control their size, distribution, and shape. Many simple methods, e.g. immature sintering [5], polymer foam replication [6], foam decomposition [7], and freeze-drying [8], have been utilized to produce inorganic porous materials. However, it is difficult to produce a well-ordered porous matrix with controlled pore size using these methods.

Recently, the fabrication of micro-, meso-, and macroporous materials utilizing surfactant and block copolymers have been reported. Porous materials with well-controlled pore structures can be produced by use of the self-assembly characteristics of organic molecules or particles.

This section will give an explanation about the technique to control pore size, pore arrangements, and pore distribution in inorganic materials prepared using organic templating materials, like micelle and colloid crystal.

(1) Micelle template method

Highly ordered porous materials prepared from micelle forming molecules have been fairly studied as a template method after Mobil reported the synthesis of MCM-41 from alkyltrimetylammonium (CnTMA) [9] and TOYOTA group demonstrated the synthesis of FSM-16 from kamemite [10] since the beginning of 1990. Either ionic or non-ionic type surfactants can be used to form an assembled morphology from micelles. In the former case, an inorganic source is assembled onto the micelle surface by electrostatic attraction to fabricate an organic–inorganic complex with ordered structures (Fig. 4.3.9). The ordered porous material that is reflected by the micelle structure with hexagonal and/or cubic shape is fabricated by removing the organic template molecules [11–13]. On the other hand, in the latter case, an organic–inorganic complex is formed by assembling an inorganic source on the micelle surface by hydrogen or coordinate bond formation. The R-$(EO)_n$ molecules, like the ethylene oxide, are commonly used in such bind of systems. The organic–inorganic complex approach using a non-ionic surfactant is used to fabricate the porous materials with controlled pore morphology which is reflected by the micelle shape, e.g. hexagonal, cubic, and so on, and moreover the wall thickness can be controlled [14, 15].

(2) Colloid crystal template method

Recently, the template techniques using colloid crystal structures have been demonstrated to result in ordered porous matrix with controlled pores (Fig. 4.3.10) [16]. It consists of the preparation of a biphasic composite comprising of a continuous matrix phase of inorganic precursors and a dispersed sacrificial organic phase that is homogeneously distributed throughout the matrix, and is then extracted to generate the controlled pores. In this technique, first a colloid solution is prepared from monosized latex or SiO_2 particles, and then dropped on a substrate and allowed to dry. It is the action of capillary forces during evaporation that leads to a colloid crystal formation. Moreover, the structure can also be prepared by sedimentation and centrifugation methods. After preparing the colloid crystal, raw material sources (sol solution) are immersed into the colloid crystal body. After that, an organic–inorganic complex

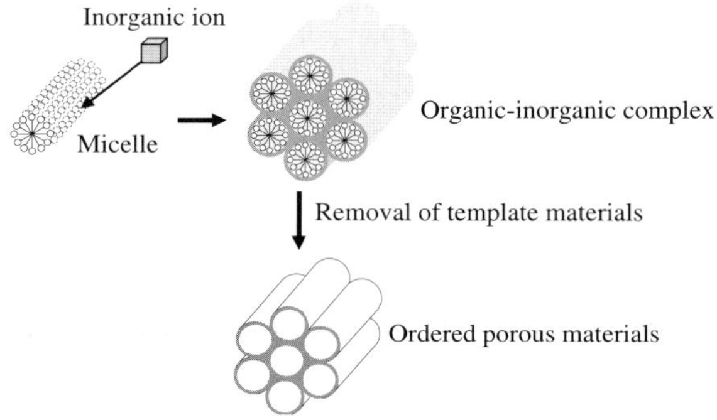

Figure 4.3.9
Scheme of ordered porous materials produced by micelle template method.

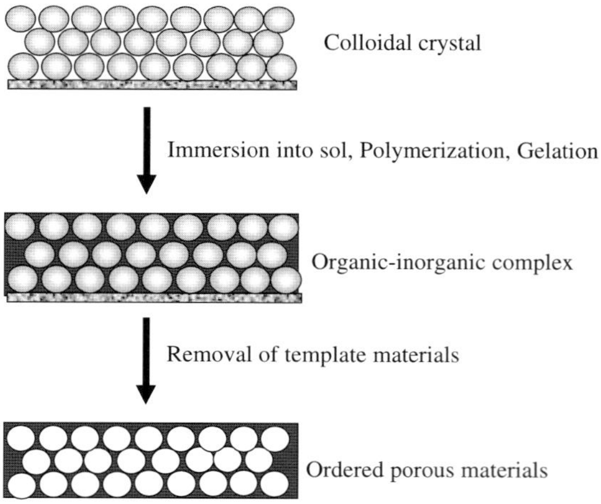

Figure 4.3.10
Scheme of ordered porous materials produced by colloidal crystal template method.

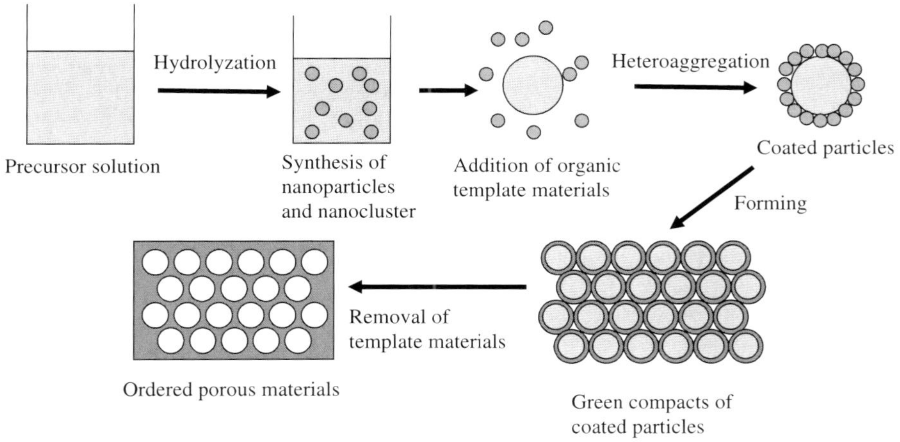

Figure 4.3.11
Scheme of ordered porous materials produced by coated organic particle template method.

formation proceeds via hydrolyzation, polymerization, and heat treatment reaction. Removal of colloid crystals by pyrolysis or chemical treatment results in a porous material with a well-ordered pore structure. This technique is quite flexible, so the processing method can be applied to many materials. Preparation of highly controlled porous structures has been demonstrated not only for simple oxides such as SiO_2, Al_2O_3, and TiO_2 [16–18], but also for complex metal compounds and carbon compounds [19]. The main advantages of this technique are its simplicity and utilization of readily available template particles of various morphologies and sizes. Moreover, the large-scale application of this technique in the field of catalyst, sensor, and so on is expected, because the colloid crystal template technique briefly leads to the highly controlled porous materials with a well-ordered pore structure.

(3) Coated organic particle template method

Fine control of pore wall thickness and structural integrity are of importance because these parameters determine the mechanical stability and the properties of porous materials, such as density, thermal conductivity, and dielectric permittivity.

Recently, it was demonstrated how silica-coated particles could act as building blocks for well-ordered porous materials. Nanosized and nanocluster particles can be fabricated by hydrolysis reaction of ceramic precursors. Adsorption and coating reaction take place between oppositely charged particles in solution (hetero-aggregation). Particles coated with nanosized powders are produced by the hetero-aggregation mechanism.

First nanosized silica particles are synthesized from controlled hydrolysis of tetraethylorthosilicate (TEOS), which is then adsorbed on the oppositely charged polystyrene (PS) particles via the hetero-aggregation reaction to make silica-coated PS particles [20]. These particles are used as the building block (template) for the

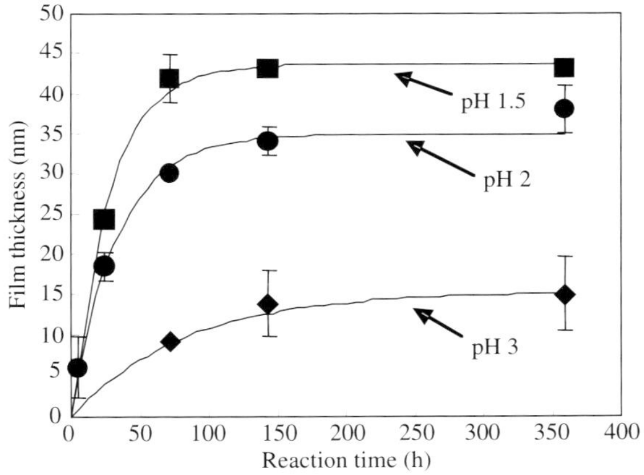

Figure 4.3.12
Relationship between reaction time and silica film thickness at pH 1.5, 2, and 3.

ordered porous structures. After consolidation of silica-coated submicron-sized PS particles, the ordered porous silica matrix was produced by removing the PS template materials from the green compacts (Fig. 4.3.11). Figure 4.3.12 shows the relation between the reaction time and the film thickness on PS particles. The thickness of the film grows parabolically with an initial rapid increase with time and an asymptotic approach to a maximum film thickness. The growth of the film thickness, δ, can be fitted by a simple first-order equation (4.3.1) describing the film growth kinetics.

$$\delta = \delta_M(1 - e^{-\kappa t}) \qquad (4.3.1)$$

where δ_M represents the maximum film thickness, t the reaction time, and κ the rate constant. Figure 4.3.13 shows the relationship between reaction pH and δ_M and

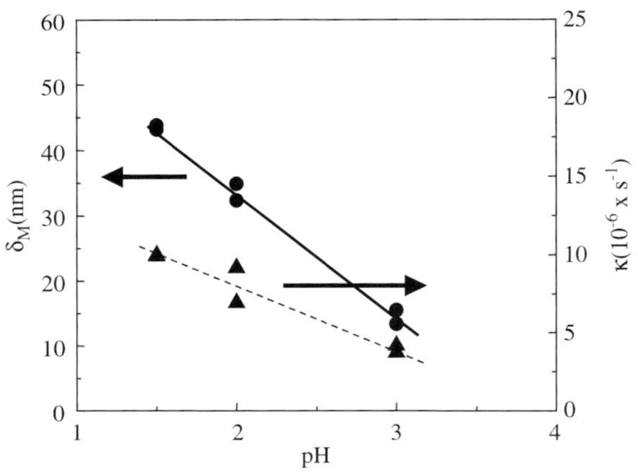

Figure 4.3.13
Relationship between reaction pH and maximum silica film thickness (δ_M; circle) and rate constant (κ; triangle).

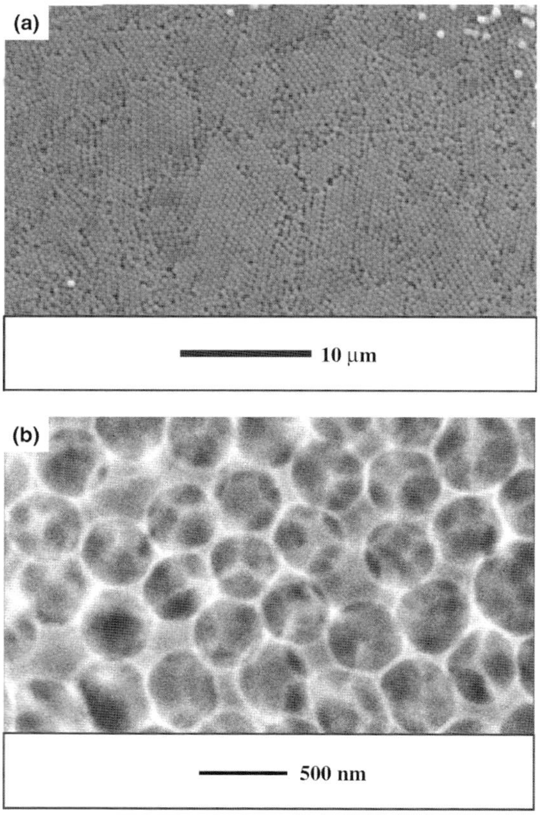

Figure 4.3.14
SEM micrographs of close-packed structures formed by centrifuging silica-coated PS particles (a) prior to sintering and (b) after sintering.

κ estimated from the fitting by equation (4.3.1) to the dependence in Fig. 4.3.12. The maximum film thickness strongly depends on the reaction pH. The hydrolysis reaction rate increases with decreasing pH, while the adsorption is related to the charge of the spices in solution and at the particle surface. At a pH below the isoelectric point of silica, there is an electrostatic attraction between the silica precursors and the negatively charged PS particles due to the opposite charges.

Figure 4.3.14(a) shows that the silica-coated PS particles can be close-packed by centrifuging. Porous

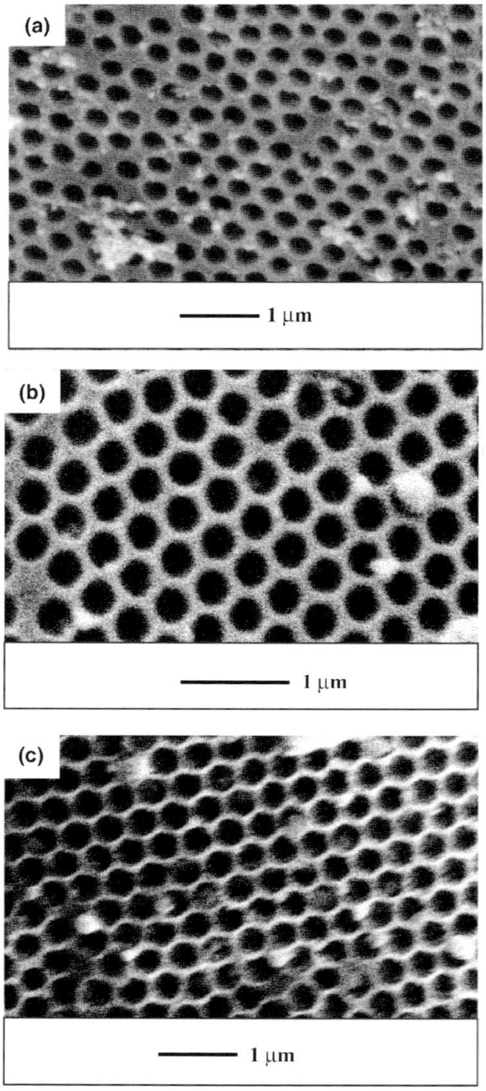

Figure 4.3.15
SEM micrographs of ordered porous silica materials produced by sintering silica-coated PS particles: (a) pH 1.5; (b) pH 2; (c) pH 3.

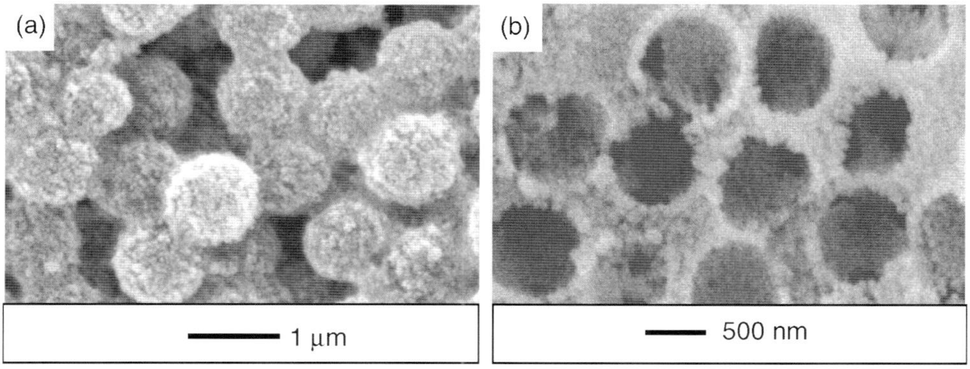

Figure 4.3.16
SEM micrographs of (a) nano-ZrO_2-coated PS particles and (b) sintered surface after sintering the coated particles.

silica structures characterized by a well-defined order are formed by removing PS particles (Fig. 4.3.14(b)). The structure consists of hexagonally close-packed uniform spherical pores and connecting channels. Figure 4.3.15 shows SEM micrographs of porous sintered samples synthesized under different reaction conditions (pH 1.5, 2, and 3). The pore structures are replicate of the hexagonal close-packed PS spheres. The silica wall thickness becomes thicker as the pH is decreased, i.e. consistent with the relation between the film thickness and pH shown in Figure 4.3.12. Thus, it is possible to control the wall thickness well by use of coated particles.

Ordered porous microstructures with controlled pores have been reported not only from silica precursors but also from zirconia precursors [21]. Well-dispersed nanosized ZrO_2 particles were coated on PS particles which are used as a building block to obtain porous microstructures [22]. The resulting ZrO_2 microstructures composed of both macropores and mesopores due to the PS template particles and assembly of nanoparticles, respectively (Fig. 4.3.16).

References

[1] N. Takeda, T. Torimoto, S. Sampath, S. Kuwabara and H. Yoneyama: *J. Phys. Chem.*, **99**, 9986–9991 (1995).

[2] I. Hattori: *Ceram. Jpn.*, **29**, 307 (1994).

[3] T. Takebayashi, T. Inada: *Ceram. Jpn.*, **32**, 356 (1997).

[4] A. Mori, N. Matsumoto and C. Imai: *Biotechnol. Lett.*, **11**, 183–188 (1989).

[5] M.E.V. Costa, P.Q. Mantas: *J. Eur. Ceram. Soc.*, **19**, 1077–1080 (1999).

[6] Y. Hotta, T. Banno and K. Oda: *J. Mater. Sci.*, **37**, 417–423 (2002).

[7] P. Colombo, M. Modesti: *J. Am. Ceram. Soc.*, **82**, 573–578 (1999).

[8] T. Fukasawa, M. Ando, T. Ohji and K. Kanzaki: *J. Am. Ceram. Soc.*, **84**, 230–232 (2001).

[9] C.T. Kresge, M.E. Leonowics, W.J. Roth, J.C. Vartuli and J.S. Beck: *Nature*, **359**, 710–712 (1992).

[10] S. Inagaki, Y. Fukushima and K. Kuroda: *J. Chem. Soc. Chem. Commun.*, 680–682 (1993).

[11] Q. Huo, D.I. Margolese, U. Ciesla, D.G. Demuth, P. Feng, T.E. Gier, P. Sieger, A. Firouzi, B.F. Chmelka, F. Schuth and G.D. Stucky: *Chem. Mater.*, **6**, 1176–1191 (1994).

[12] Y.M. Setoguchi, Y. Teraoka, I. Moriguchi, S. Kagawa, N. Tomonaga, A. Yatsutake and J. Izumi: *J. Porous Mater.*, **4**, 129–134 (1997).

[13] J. Xu, Z. Luan, H. He, W. Zgao and L. Kevan: *Chem. Mater.*, **10**, 3690–3698 (1998).

[14] D. Zhao, Q. Huo, J. Feng, B.F. Chmelka and G.D. Stucky: *J. Am. Chem. Soc.*, **120**, 6024–6036 (1998).

[15] S.A. Salty, T. Hanaoka: *Adv. Mater.*, **15**, 1893–1899 (2003).

[16] O.D. Velev, W.E. Kaler: *Adv. Mater.*, **12**, 531–534 (2000).

[17] B.T. Holland, C.F. Blanford, T. Do and A. Stein: *Chem. Mater.*, **11**, 795–805 (1999).

[18] S.H. Park, Y. Xia: *Adv. Mater.*, **10**, 1045–1048 (1998).

[19] B.H. Juarez, M. Ibistate, J.M. Palacios and C. Lopez: *Adv. Mater.*, **15**, 319–323 (2003).

[20] Y. Hotta, P.C.A. Alberius and L. Bergström: *J. Mater. Chem.*, **13**, 496–501 (2003).

[21] Y. Hotta, Y. Jia, M. Kawamura, K. Tsunekawa, K. Sato, K. Watari: *J. Mater. Sci.*, **40**, 1–7 (2005).

[22] Y. Jia, C. Duran, Y. Hotta, K. Sato and K. Watari: *J. Colloid. Inter. Sci.*, **291**, 292–295 (2005).

4.3.4 Nanoporous materials (Titania nanotubes)

Recently, much attention has been paid to research on the novel functionalities of materials by controlling their structures and morphologies in nanometer-size level. Carbon nanotubes, discovered by Iijima have been reported to possess unique electrical and mechanical properties different from those of conventional carbons such as graphite and fullerene [1]. The fundamental studies on the electrical conductivities and the formation mechanism of carbon nanotubes are in progress. Carbon nanotubes have realized a high performance as a cantilever (mechanical probe) for a scanning probe microscope, and field emission display. Carbon nanocoils have been reported to show excellent electromagnetic wave absorption properties.

The novel functionalities are expected to be generated by controlling the shape, size, and composition of materials: (1) shapes such as tube, layer, sheet, or wire; (2) sizes in nanometer or micrometer level; and (3) composition such as carbon, alumina, or titania are quite important.

Since the discovery of carbon nanotubes, there have been numerous works on the preparation of various non-oxide nanotubes such as, BN, B-C-N, WS_2, and oxide-based nanotubes such as SiO_2, Al_2O_3, and TiO_2.

In the present section, TiO_2-based nanotubes with photocatalytic activities [2], in various fields involving (1) the removal of NO_x and gases associated with aldehyde decomposition; (2) environmental purification processes such as water purification and decomposition of pollutants; (3) catalytic reactions for hydrogen generation; and (4) electrode materials for dye-sensitizing solar batteries, are briefly reviewed.

(1) Preparation methods of titania nanotubes

Two significant routes for the preparation of titania nanotubes have been reported: (a) a replica or template method, (b) a chemical treatment or hydrothermal method. Some of these works were summarized in Ref. [3] by Suzuki et al.

(a) The replica method

In 1996, titania nanotubes were prepared for the first time using a replica method by Hoyer [4, 5]. Anodically grown aluminum oxide was used as the starting material. A gold film was made on the sheet by evaporation for the electrical condition of the material, and then methyl methacrylate (PMMA) was polymerized inside of the pores. After the resulting films were treated with NaOH aqueous solution to remove the Al_2O_3, resulting in the formation of the PMMA replicated negatively. Subsequently, electrochemical deposition techniques in the PMMA mold were used to form TiO_2. After the chemical dissolution of the polymer, titania nanotubes with inner diameter of 70–100 nm were obtained. The porous alumina membrane prepared by an anodic oxidization method consists of highly oriented alumina pillars and, the shapes of the resulting titania crystals are reflected by the mold derived from the pillars.

After this report, some works on the preparation of the nanotubes using polycarbonate, carbon nanotubes, surfactants as replica or template materials were reported. Table 4.3.2 shows some reports on titania nanotubes prepared by replica methods.

Imai et al. reported that the direct preparation of titania nanotubes in nanochannels of porous alumina membranes using a deposition technique from TiF_4 solution. After dissolving the alumina membrane, titania nanotubes with an outer diameter of 200 nm, and an inner diameter of 100–150 nm were obtained [6].

Michailowski et al. reported that titania nanotubes with a diameter of 50–70 nm and a wall thickness of 3 nm were prepared by filling pores in aluminum oxide membrane with titanium isopropoxide and subsequent heating [8]. Varghese et al. prepared titania nanotubes by the anodic oxidation of a pure titaniumu sheet in an aqueous HF solution [10, 11].

(b) The soft-chemical method

In 1998, Kauga et al. reported for the first time that titania nanotubes were prepared by using a soft-chemical method [14]. Needle-shaped titanium oxide crystals with an inner diameter of approximately 5 nm and an outer diameter of 8 nm were obtained when titania nanopowders were treated chemically with

Table 4.3.2
Titania nanotubes prepared by a template or a replica method.

Author	Template mold	Raw solution
Hoyer[4, 5]	Anodic alumina, PMMA	$TiCl_3$
Imai[6]	Anodic alumina	TiF_4
Adachi[7]	Surfactant	Titanium alkoxide
Michailowski[8]	Anodic alumina	Titanium alkoxide
Caruso[9]	Fibrous polymers	Titanium alkoxide
Gong[10, 11]	Anodic oxidation of pure titanium	HF solution
Sun[12]	Carbon nanotubes	$TiCl_4$
Shin[13]	Polycarbonate filters	Titanium alkoxide

Table 4.3.3
Titania nanotubes prepared by soft-chemical methods.

Author	Raw material	Condition of alkali treatment	Product
Kasuga[14, 15]	A,R		
	A+R	5–17 M NaOH, 25–150°C	TiO_2
Wang[16]	A	10 M NaOH, 110°C, 20H	TiO_2
Ma[17]	A	10 M NaOH, 110–150°C, 12–72 H	Layered titania oxide $H_xTi_{2-/4}\square_{x-4}O_4$
Seo[18]	A	5 M NaOH, 100–200°C, 12H	TiO_2
Yao[19]	A	10 M NaOH, 150°C, 12H	TiO_2
Chen[20]	A	10 M NaOH, 130°C, 72H	
Du[21]	A	10 M NaOH, 130°C, 24–72H	$H_2Ti_3O_7$
Sun[22]	A	10 M NaOH, 100–180°C, >48H	$Na_xH_{2-x}Ti_3O_7$

Note: A, anatase; R, rutile; A+R, composite of anatase and rutile.

5–15M NaOH aqueous solution at 60–120°C for 20 h, and subsequent HCl aqueous solution. Table 4.3.3 shows some reports on titania nanotubes prepared by soft-chemical methods.

Figures 4.3.17 and 4.3.18 show scanning electron micrographs (SEM) and transmission electron micrograph (TEM) of TiO_2 powders treated with 10M NaOH aqueous solution at 110°C for 20 h.

Electron microscopy and Raman spectroscopy of the titania nanotubes revealed a soft-chemical reaction: when TiO_2 crystals are treated with aqueous NaOH, some of Ti–O–Ti bonds are broken and Ti–O–Na and Ti–OH bonds are formed [15]. And subsequently when they are treated with acidic solution/water, titania nanotubes form. Wang et al. reported that TiO_2 crystals treated with alkaline solution at 180°C for 30 h grow into a tube structure [16]. Ma et al. reported that, when anatase-type TiO_2 crystals are subjected to hydrothermal treatment in aqueous NaOH, high-purity nanotubes and nanobelts form depending on the treatment temperature and duration time [17]. Such nanotubes and nanobelts have layered titanate structures.

Formation of the several types of crystals, such as TiO_2 [18, 19], $H_2Ti_3O_7$ [20, 21], and $Na_xH_{2-x}Ti_3O_7$ [22] are influenced by the chemical treatment.

Kasuga et al. reported that it is possible to form either titania or titanate (residual Na^+ ions) nanotubes by controlling the amount of Na^+ ions remaining in the sample [23].

Long-size titania nanofibers or ribbons were successfully prepared by controlling the alkali treatment conditions, i.e., heating at temperature above 150°C and prolonging duration time.

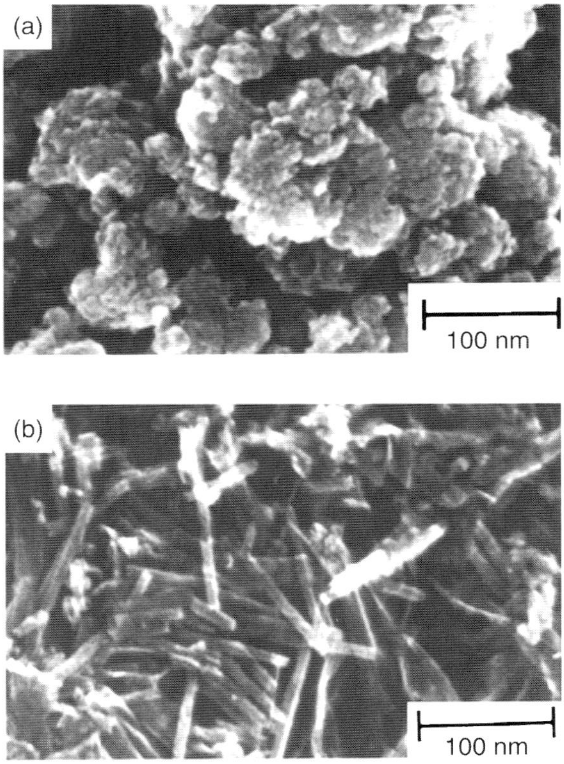

Figure 4.3.17
SEM image of TiO_2 powders before and after soft-chemical treatment. (a) Powders before chemical treatment; (b) Powders treated with 10 M aqueous NaOH at 110°C for 20 h.

(2) Functionality of titania nanotubes and their applications

Application of titania nanotubes to photocatalists, dye-sensitized solar cells, hydrogen sensors, lithium-ion secondary batteries and so on will be discussed.

Recently, some efforts have been carried out to improve the performance of the nanotubes.

Lin et al. suggested sulfated titania nanotubes as catalysts for esterification reaction. Mao et al. reported

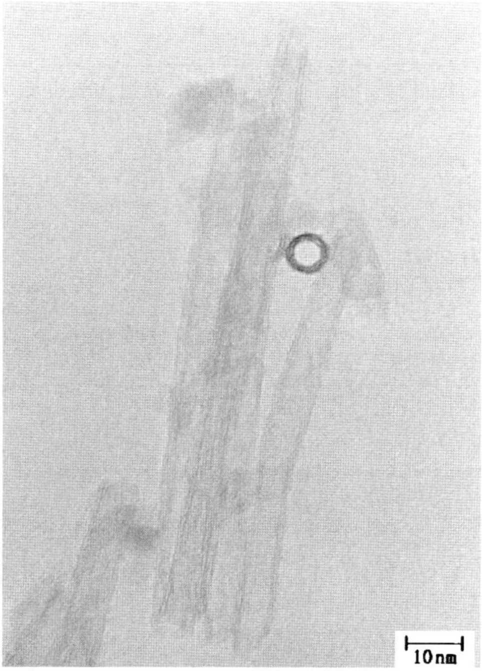

Figure 4.3.18
TEM image of TiO_2 powders treated with 10 M aqueous NaOH at 110°C for 20 hours.

that $BaTiO_3$ and/or $SrTiO_3$ were obtained by refluxing titania nanotubes in an inert atmosphere with Ba(OH) and/or $SrCl_2$.

Sekino et al. prepared metal (Nb, V, Cr, Mn, Co)-doped titania nanotubes for improving photocatalytic activities. Wang et al. reported the preparation of RuO_2/TiO_2 nanotube composites and the electrochemical properties.

Kasuga reported that titania nanotubes show excellent ion adsorption activities and contain a large amount of hydroxyl group in comparison with raw titania powders. These characteristic properties lead to new performance such as biocompatibility and proton conduction [23].

Ca^{2+} ion-doped titania nanotubes are expected to be used as biomaterials, and oxyacid-doped titania nanotubes are expected to be applied to solid electrolytes. Ca-titania nanotubes have good bioactivity for bone regeneration in living body [24]. The oxyacid-doped titania nanotubes exhibit high proton conductivities at temperatures of 100°C or higher. Therefore, they are expected to lead to the development of technologies for electrolytic films used in fuel cells.

These nanotubes are expected to be used as applications in the fields of medicine and/or energy-generation as well as photocatalytic applications. In future, new performances may be found by examining various ion-species as dopants.

Since the finding of titania nanotubes 10 years ago, numerous works for clarifying the formation mechanism have been carried out.

References

[1] S. Iijima: *Nature*, **354**, 56–59 (1991).
[2] T. Kasuga, M. Hiramatu, M. Hirano, A. Hoson and K. Oyamada: *J. Mater. Res.*, **12**, 607–609 (1997).
[3] Y. Suzuki, T. Sekino: *Mater. Integration*, **18**, 3–10 (2005).
[4] P. Hoyer: *Langmuir*, **12**, 1411–1413 (1996).
[5] P. Hoyer: *Adv. Mater*, **8**, 857–859 (1996).
[6] H. Imai, Y. Takai, K. Shimizu, M. Matsuda and H. Hirashima: *J. Mater. Chem.*, **9**, 2971–2972 (1999).
[7] M. Adachi, Y. Murata, M. Harada and S. Yoshikawa: *Chem. Lett.*, 942–943 (2000).
[8] A. Michailowski, D. Almawlawi, G.S. Chemg and M. Moskovits: *Chem. Phys. Lett.*, **349**, 1–5 (2001).
[9] R.A. Caruso, J.H. Schattka and A. Greiner: *Adv. Mater.*, **13**, 1577–1579 (2001).
[10] D. Gong, C.A. Grimes, O.K. Varghese, W. Hu, R.S. Singh, Z. Chen and E.C. Dickey: *J. Mater. Res.*, **16**, 3331–3334 (2001).
[11] O.K. Varghese, D. Gong, M. Paulose, K.G. Ong, E.C. Dickey and C.A. Grimes: *Adv. Mater.*, **15**, 624–627 (2003).
[12] J. Sun, L. Gao and Q.H. Zhamg: *J. Mater. Sci. Lett.*, **22**, 339 (2003); J. Sun, L. Gao, Q.H. Zhamg and A. Iwasa: *Carbon*, **42**, 895–899 (2004).
[13] H. Shin, D.K. Jeong, J.G. Lee, M.M Sung and J.Y. Kim: *Adv. Mater.*, **16**, 1197–1200 (2004).
[14] T. Kasuga, M. Hiramatu, A. Hoson, T. Sekino and K. Niihara: *Langmuir*, **14**, 3160–3163 (1998).
[15] T. Kasuga, M. Hiramatu, A. Hoson, T. Sekino and K. Niihara: *Adv. Mater.*, **11**, 1307–1311 (1999).
[16] W. Wang, O.K. Varghese, M. Paulose and C.A. Grimes: *J. Mater. Res.*, **19**, 417–422 (2004).
[17] R. Ma, K. Fukuda, T. Sasaki, M. Osada and Y. Bando: *J. Phys. Chem.*, **B.109**, 6210–6214 (2005).
[18] D.-S. Seo, J.-K. Lee and H. Kim: *J. Cryst. Growth*, **229**, 428 (2001).
[19] Q. Chen, G.H. Du, S. Zhang and L.-M. Peng: *Acta Cryst.*, **B 58**, 587–593 (2002).
[20] B.D. Yao, Y.F. Chen, X.Y. Zhang, W.F. Zhang, Z.Y. Yang and N. Wang: *Appl. Phys. Lett.*, **82**, 281–283 (2003).
[21] Q. Chen, L.-M. Peng, W.Z. Zhou and G. Du: *Adv. Mater.*, **14**, 1208–1211 (2002).

[22] X. Sun, Y. Li: *Chem. Eur. J.*, **9**, 2229–2238 (2003).
[23] T. Kasuga: *Thin Solid Films*, **496**, 141–145 (2006).
[24] S. Kubota, K. Johkura, K. Asanuma, Y. Okouchi, N. Ogiwara, K. Sasaki and T. Kasuga: *J. Mater. Sci. Mater.*, **15**, 1031–1035 (2004).

4.4 Nanocomposite structure

4.4.1 Catalyst microstructure

This chapter explains the catalyst microstructure. Though a broad range of contents are included in the term of "catalyst microstructure", this chapter covers the microstructure of the solid catalyst which is deeply linked to this handbook. The catalyst support is used with catalytic component for controlling its specific surface area or pore size distribution in a solid catalyst. In this case, the term, catalyst microstructure, includes that of the catalyst support. The solid catalyst is classified into the following nine kinds by the element that composes it, a metal, a metallic oxide, a metallic sulfide, a metallic sulfate, a metallic phosphate, a solid phosphoric acid, a cation-exchange resin, a zeolite, and a heteropoly acid. In them, a metal or a metallic oxide is the chief object of the technological development in the field of the nanoparticle technology. Materials used as catalyst supports are listed as follows, MgO, Al_2O_3, TiO_2, SiO_2, SiO_2-Al_2O_3, ZrO_2, zeolite, active carbon, diatomaceous earth, mesoporous materials made by organic templates [1]. Among them, zeolite and mesoporous materials made by organic templates are explained in detail in other paragraphs of this book. Microstructure of active carbon and diatomaceous earth came from naturally occurring one, so that, it is difficult to treat them in the same way as the structure made by the agglomerated nanoparticles. Therefore, catalyst microstructures of nanoparticle aggregate of metals or metallic oxide (which is only called "oxide" in the following) are involved in this section.

Pore volume, pore size distribution, and specific surface area are used generally as representative parameters to describe a catalyst microstructure. Cylindrical pore shape whose one side shut is assumed when those parameters are measured. For the assumption, such a cross-sectional drawing as Fig. 4.4.1 is often used. However, excluding zeolite and mesoporous materials made by organic templates, general oxide catalyst consists of an aggregation of nanolevel crystallites or amorphous particulates. So the pore space measured as a pore volume is composed of a gap between the nanolevel crystallites or amorphous particulates, and a gap between the secondary particles of the crystallites of the particulates. Pore size among the nanolevel ones is from several nanometers to several ten nanometers. And pore size among the secondary particles is from

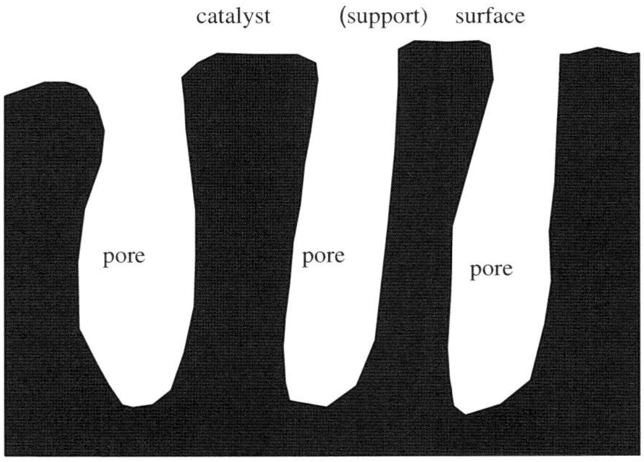

Figure 4.4.1
Schematic diagram of catalyst microstructure.

sub-micrometers to several micrometers [1]. Pore size distribution and pore volume of those size ranges are usually measured by mercury porosimetry. Mercury porosimetry will be explained in the Section 6.8 "Catalytic characterization". Size of a macropore is affected by milling or compression of the powder, so the macropore size could be controlled by pelletization process or milling and wash-coat process of a catalyst. However, mesopore is not affected by mechanical compression with usual crushing operation with mortar, metal mold, and hydrostatic pressure press. If you want to control mesopore, it is necessary to control the structure of mesopores, that is, the size of the crystallites, their size distribution, and their aggregation, when the catalyst powder is synthesized. Moreover, because the surface energy, that becomes driving force of sintering, is large in the crystallite at the nanolevel, some sort of sintering inhibition is necessary in a catalyst for high temperature use. In the case of noble metal loaded catalyst, catalyst supports not only have to have a sufficient specific surface area for keeping high-dispersivity of noble metal particles, but also have an adequate interaction between their surface and noble metal particles. The following four technologies are introduced in this chapter; pH swing method for controlling a size of mesoscopic pores of mesoporous alumina; sintering control of γ alumina by using additives; effect of specific surface area of ceria-zirconia solid solutions to the Pt dispersion of Pt loaded ceria–zirconia solid solution catalysts; and the effect of Au particle size to the catalytic activity in Au-loaded titania catalyst.

The pH swing method is a typical successful example to control pore size of porous γ alumina in the industrial scale [2, 3]. For instance, both of the aluminum nitrate (acidity) and the sodium aluminate (basicity) are prepared as an aluminum source. The pH inside the reactor is swung by alternately putting these two solutions into the reactor. To become a target pH swing value (for instance, the pH at both ends is from 2 to 10), the raw material solution of the obtained amount is added from the titration curve. The precipitation of pseudo-boehmite is formed in the neutral area of the pH. A small pseudo-boehmite crystallite dissolves and disappears because the pH at both ends is the dissolution areas of pseudo-boehmite. Relatively large pseudo-boehmite that remains without dissolving becomes a nucleus, and pseudo-boehmite grows up in the following pH swing operation. Only a limited number of particles from initial stage of synthesis can grow up by this operation because the nucleus newly generated disappears. The pH swing operation is repeated several times until pseudo-boehmite grows up and reaches a target pore size. As a result, the pseudo-boehmite crystallite diameter grows up with the pH swing operation frequency. Because a pore of pseudo-boehmite consists of the space of the crystallite, pore size increases with the crystallites growing up (Fig. 4.4.2). The sharp pore distribution is obtained in the uniform crystallite size powder, and the broad pore size distribution is obtained in the not uniform crystallite size distribution powder. The pH swing method is suitable for preparing the amphoteric element oxide with dissolubility on both an acid and basicity side. The amphoteric elements are Zn, Al, Pb, and Sn. Pseudo-boehmite is baked to transform into γ alumina after the adequate pore size distribution is obtained, and the pore size distribution of pseudo-boehmite come down to γ alumina. γ alumina maintains a high-specific surface area that exceeds 100 m^2/g at about 1,000°C. However, the specific surface area decrease by the sintering remarkably at around 1,100°C, and the transformation into α alumina also happens at the temperature.

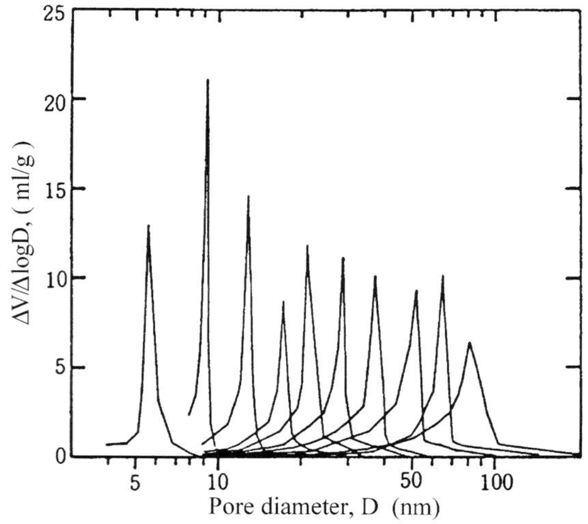

Figure 4.4.2
Pore size distribution of alumina [2].

The specific surface area of γ alumina decreases remarkably through the transformation into α alumina. It is thought as a main current before Schaper's paper [4] that the decrease of specific surface area at the high temperature of γ alumina is caused by the rearrangement from the cubic arrangement of the oxygen ion into the hexagonal one. Schaper et al. reported that sintering is a phenomenon of γ alumina itself, and the phase transition to α alumina is caused as a result. It is known that some γ alumina shows a large decrease of specific surface area at much lower temperature than the temperature of transformation of γ alumina to α alumina. It proves the correctness of this idea. Afterwards, Burtin et al. [5] proposed the model that the nucleus of α alumina is generated at the moment that the oxygen vacancy of γ alumina meets cation in the sintering process of γ alumina. The discussion is still divided now on the issue that which comes first, sintering of γ alumina or transformation into α alumina. When a γ alumina is used as a catalyst support, if the γ alumina sinters, loaded noble metal particles' sintering is also promoted by that, and moreover some noble metal particles have been buried into the support with sintering of the support. Then, the catalytic activity of the catalyst deteriorates [6]. Therefore, it is devised to control sintering of γ alumina. They are roughly classified into two ways, the one is with the additive element, and the other is by the microstructure control.

The alkaline earth metal, the rare earth element, Si, and Zr, etc., are proposed as an additive element. The most general one of these for γ alumina is La. The compound of alkaline earth elements easily dissolves into the acidic solution, so that there is a considerable difficulty in catalyst making with using alkaline earth elements. Moreover, the Si addition to alumina has the problem of causing caulking easily in rich atmosphere though the effect of making to high heat-proof is the most remarkable. The La_2O_3 layer is formed on the surface of alumina at 800°C or less, and the $LaAlO_3$ layer is formed at a temperature more than 800°C in the La-addition alumina. As a result, the action of driving force decreases of sintering works, such as by the stabilization of the surface of the γ alumina, by the obstruction of the surface diffusion of alumina, and by the obstruction of nucleation of α alumina. These mechanisms suppress sintering of γ alumina [5, 6]. Ba and Si show the similar effect as La for suppressing sintering of γ alumina [7–9]. The sintering suppression by controlling microstructure is basically done according to the following idea. The initiation of sintering or nucleation of α alumina usually begins at a contact point of alumina crystallites (grain boundary). So reducing the contact point is the only way to suppress sintering in microstructure control. As one of the typical example of a preparation method of a microstructure with few grain contact points, low-density agglomeration of needle-like γ alumina by the atomization pyrolysis of

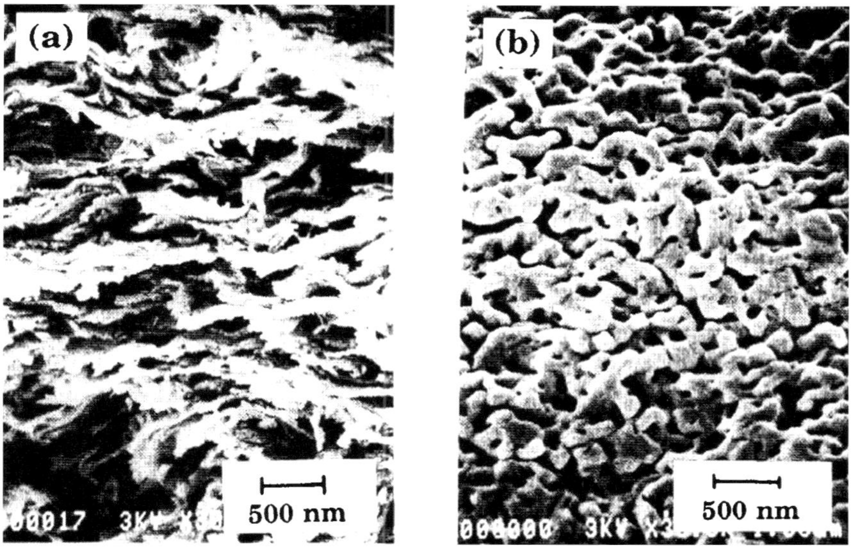

Figure 4.4.3
SEM photograph of α alumina thin films calcined at 1,000°C (a) prepared by using template; (b) prepared without template [10].

boehmite sol [8], or aero gel of alumina by supercritical drying [9] have been reported. In addition, Kunitake et al. [10] reported to have got crystallographically oriented mesoporous θ-alumina film whose specific surface area is 200 m^2/g (Fig. 4.4.3). This research is also one of the archetypal examples of microstructural control to prevent sintering. This mesoporous θ-alumina film is prepared by filtration of well-dispersed suspension of both bimolecular membrane of amphiphile molecule and boehmite sol by using the fluoroplastic membrane filter, and subsequent baking at 800°C. This alumina film keeps high-specific surface area of 100 m^2/g even after heat treatment at 1500°C and after transformation into α alumina. (Fig. 4.4.4).

Keeping high-specific surface area is important for noble metal-loaded catalyst as above mentioned. Then, minimum required level of that will be shown in the following. This is a platinum (Pt) loaded ceria–zirconia solid solution catalyst whose Pt concentration is 1 wt%. As for about 30 m^2/g or more, a dispersion level of Pt is constant. When specific surface area of ceria–zirconia solid solution becomes 30 m^2/g or less, dispersion level of Pt on it decreases with its specific surface area decrease (Fig. 4.4.5) [11]. Evaluation method of Pt dispersion will be explained in the Section 6.8 about catalyst characterization. The role of the catalyst support is not only to keep the high-specific surface area steady, and to keep high dispersed state of noble metals. But also, the role of the support must to provide adequate electronic circumstance for noble metal particles by adequate interaction between them. Gold (Au) shows no catalytic activity as a bulk state, which is as opposed to Pt. However, Au changes to show high catalytic activity as nanosize particles loaded on catalyst support. And at the case, the catalytic activity of Au varies depending upon the kinds of the support [12, 13]. The oxidation reaction of hydrogen with the Au catalyst is hardly influenced by the kind of the catalyst support. That is because the reaction occurs on the surface of the Au nanoparticles.

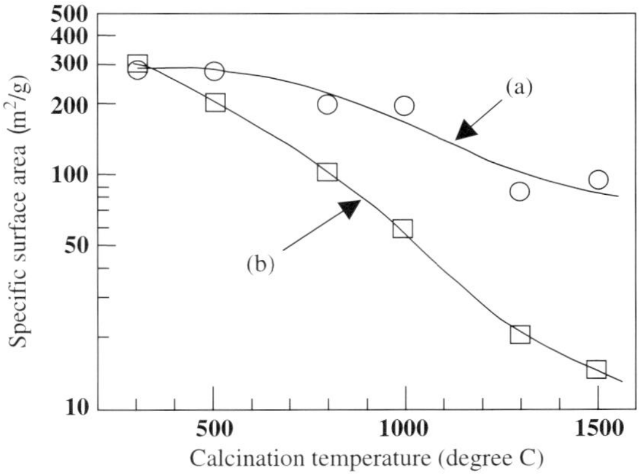

Fig.4.4.4
Temperature dependence of the specific surface area of alumina thin films prepared with or without templete (a) prepared by using template (b) prepared without templete [10].

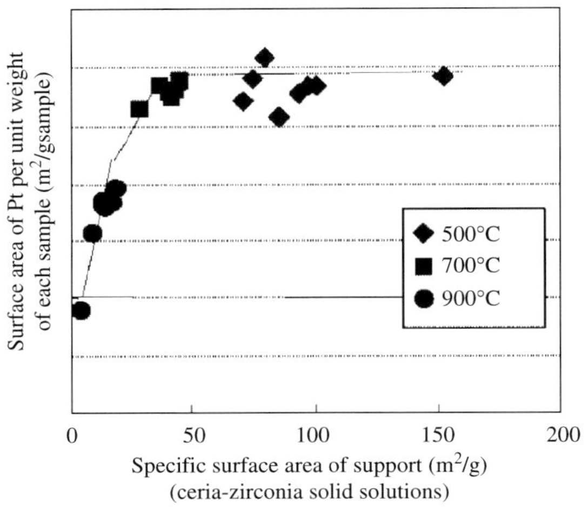

Figure 4.4.5
Effect of specific surface area of ceria–zirconia support on the surface area of Pt per unit weight of each sample [10].

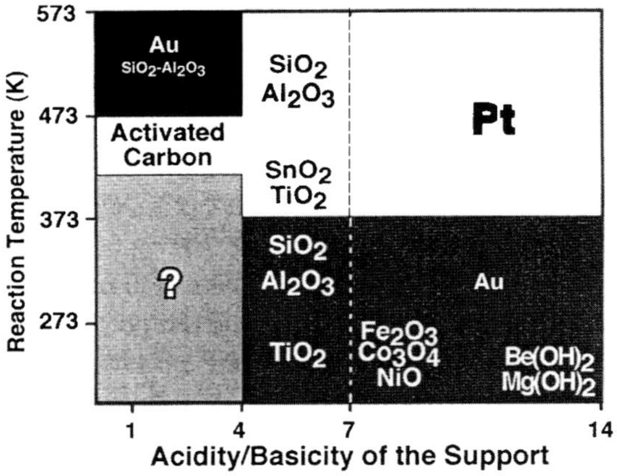

Figure 4.4.6
Catalytic activities of the supported Au and Pt catalysts for CO oxidation as a function of support acidity/basicity [11].

On the other hand, in the oxidation reaction of CO, with the Au catalyst, CO adsorbs on the edge of Au particle or at the boundary of Au and the catalyst support. It is thought that the reactive point is at its boundary. This is thought to be the reason why a CO oxidation reaction of Au catalyst is strongly affected by acid–base degree of catalyst support (Fig. 4.4.6). Because hydroxides of alkaline earth metals are stable in the vicinity of the room temperature, Au particles can be kept at 1 nm of high dispersed state. And, they show excellent CO oxidation activity at the low temperature of 200 K. When the transition element oxide is used as a catalyst support, Au shows good CO oxidation activity as well as that of alkali earth metal. CO oxidation activity of Au decreases when a catalyst support is alumina or silica. If high acidic silica-alumina is used as a support of Au, higher temperature than 500 K is necessary for CO oxidation. The CO oxidation activity of Au/alkali earth metal catalyst tends to decrease within several months, which of Au/transition element catalyst tends to decrease within several years. This is the reason why catalyst activity decreases, because of an accumulation of reaction inhibitor such as carbonate on the boundary of Au particles and the support. A lot of research reports have been published lofty apex for Au/TiO_2 catalyst among Au-loaded catalyst. Because it is easy to obtain high-specific surface area, easy to achieve high Au dispersion, and easy to refresh by heating, by the light irradiation, and the moisture addition, etc. The CO oxidation rate rises by decreasing the Au particle diameter, and the catalyst activity rises by increasing the contact boundary of the Au particle and titania.

References

[1] E. Kikuchi, K. Segawa, A. Tada, Y. Imizu and H. Hattori: *Atarashii Shokubai Kagaku*, Sankyo Syuppan (in Japanese) (1997).
[2] S. Asaoka: *Shokubai*, **28**(4), 256–261 (1986).
[3] S. Inoue, T. Takatsuka, Y. Wada, S. Nakata and T. Ono: *Catalysis Today*, **43**(3–4), 225–232 (1998).
[4] H. Schaper, E.B.M. Doesburg, P.H.M. Dekorte and L.L. Vanreijen: *Solid State Ionics*, **16** (1–4), 261–265 (1985).
[5] P. Burtin, J.P. Brunelle, M. Pijolat and M. Soustelle: *Appl. Catal.*, **34**(1–2), 239–254 (1987).
[6] M.A. Fraga, E. Soares de Souza, F. Villain and L.G. Appel: *Appl. Catal. A*, **259**(1–8), 57–63 (2004).
[7] H. Arai, M. Machida: *Shokubai*, **33**(5), 328–334 (1991).
[8] H. Arai, M. Machida: *Shokubai*, **35**(4), 231–236 (1993).
[9] T. Horiuchi, L. Chen, T. Osaki, T. Sugiyama, K. Suzuki and T. Moril: *Catal. Lett.*, **58**(2–3), 89–92 (1999).
[10] N. Tsutsumi, K. Sakata and T. Kunitake: *Chem. Lett.*, **8**, 1465–1468 (1992).
[11] A. Suda, K. Yamamura, Y. Ukyo, T. Sasaki, H. Sobukawa, T. Tanabe, Y. Nagai and M. Sugiura: *J. Ceram. Soc. Jpn.*, **112**(11), 581–585 (2004).
[12] M. Date, M. Haruta: *Shokubai*, **47**(1), 8–13 (2005).
[13] M. Haruta, M. Daté: *Appl. Catal. A: General*, **222**(1–2), 427–437 (2001).

4.4.2 Percolation structure

Solid oxide fuel cells generally have porous anode and cathode electrodes into which fuel and oxidant gases are introduced. The triple-phase-boundaries (TPB), the electronic conductor-ionic conductor-gas (three) phase boundaries, are the active site for electrochemical reactions. The phenomena that take place in the anode can be summarized as (1) transport of oxide ions from the electrolyte through the ionic conductors and transport of reactants from the anode surface to the active sites through the pores; (2) electrochemical reaction at the active sites; and (3) transport of electrons from the active sites to the current collector through the electronic conductors and transport of products from the active sites to the anode surface through the pores.

Since the electrode performance is highly dependent on its microstructure, the anode electrode must be designed as a homogeneous or graded structure consisting of three phases, providing percolation paths for electrons, oxide ions, and gaseous hydrogen and water, respectively. An analogous framework holds for the cathode electrode in which oxide ions are generated via reduction of oxygen molecules.

(1) Realization of percolation clusters

Figure 4.4.7 shows a scheme of anode microstructure. The particles of the same type (electronic conductor or ionic conductor) have to be kept in contact with each other so that a network of particle chains (i.e. percolation clusters) is formed through the electrode; this is the condition under which an adequate conductivities of electrons or ions can be reached. Adequate contact between the particles of different types is also important factor to ensure a large TPB active site. The pores after heat treatment are generally percolated through the electrode, suggesting that the optimal design of electrodes should lead to a discussion on a relation between percolation and particle coordination in binary particle mixtures.

The theory of particle coordination number in a random packing of binary spheres have been widely developed by numerical simulation. Kuo et al. [1] discussed the above theory in a wide range of particle size ratio α (r_{io}/r_{el}=0.154~6.464, r_{el}; electric conductor size and r_{io}; ionic conductor size) and claimed that the experimental data are satisfactorily fitted by setting the coordination number $z_{i/i}=1.764$ (i denotes electronic or ionic conductor) at the percolation threshold, below which no electronic or ionic conducting cluster connects to both ends of electrode. The percolation probability that an arbitrary particle belongs to a percolation cluster is given as

$$p_i = \left\{1 - \left(\frac{4.236 - z_{i/i}}{2.472}\right)^{2.5}\right\}^{0.4} \quad (4.4.1)$$

The coordination number of electronic to electronic conductors, $z_{el/el}$, and that of ionic to ionic conductors, $z_{io/io}$, are

$$z_{el/el} = \frac{zn_{el}}{n_{el} + (1 - n_{el})\alpha^2}$$
$$z_{io/io} = \frac{zn_{io}}{n_{io} + (1 - n_{io})\alpha^{-2}} \quad (4.4.2)$$

where z is the average total coordination number in a random packing of monosized spheres, which is equal to 6.

The theoretical predictions for the percolation thresholds are shown in Fig. 4.4.8, where the volume fraction of electronic conductor f_{el} at the percolation threshold $z_{ii}=1.764$ is plotted as a function of the particle size ratio. In the region bounded by the percolation thresholds, the electronic and ionic conductors can percolate through the electrode simultaneously. When $\alpha=1$, for instance, the both components form percolation clusters in the range of $0.294<f_{el}<0.706$. The percolation range becomes thinner and thinner if the particles become very different in size.

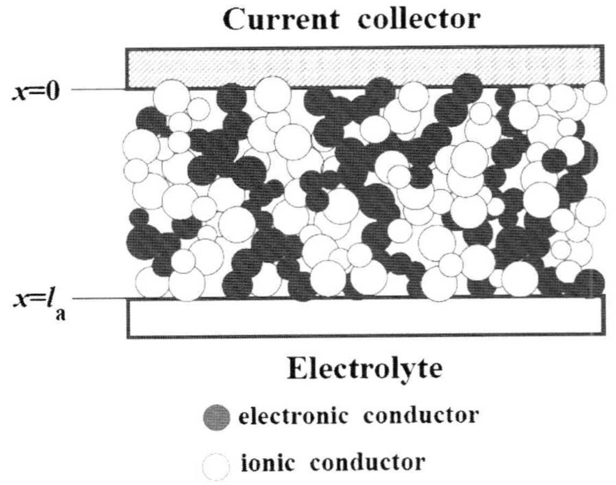

Figure 4.4.7
Microstructure of solid oxide fuel cell electrode (anode).

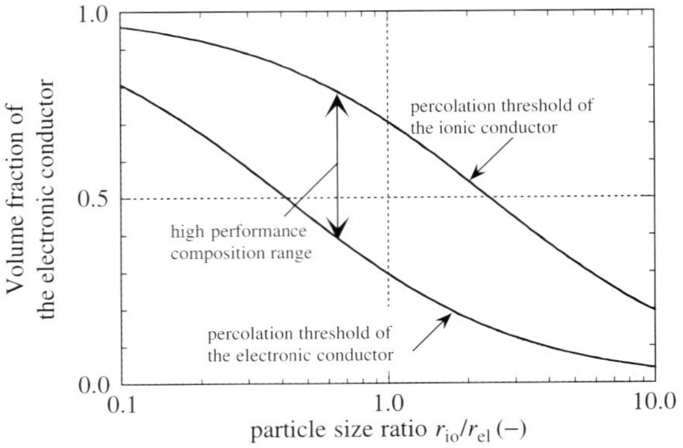

Figure 4.4.8
Relation of the volume fraction of electronic conductor to the particle size ratio at the percolation threshold.

(2) Active TPB area

Figure 4.4.9 shows a schematic illustration of the TPB formed after heat treatment at contacts between the particles of different types. If we assume that the active area derived from the formed TPB is the contact area of different types of particles and the latter is proportional both to the number of contact points per unit volume and to the probability of the whole range connection of the same type particles, the active area can be estimated.

$$
\begin{aligned}
s_t &= \pi r_{el}^2 \sin^2\theta \, n_t n_{el} z_{el/io} p_{el} p_{io} \\
&= \pi r_{el}^2 \sin^2\theta \, n_t n_{el} n_{el} \frac{z_{el} z_{io}}{z} p_{el} p_{io}
\end{aligned}
\tag{4.4.3}
$$

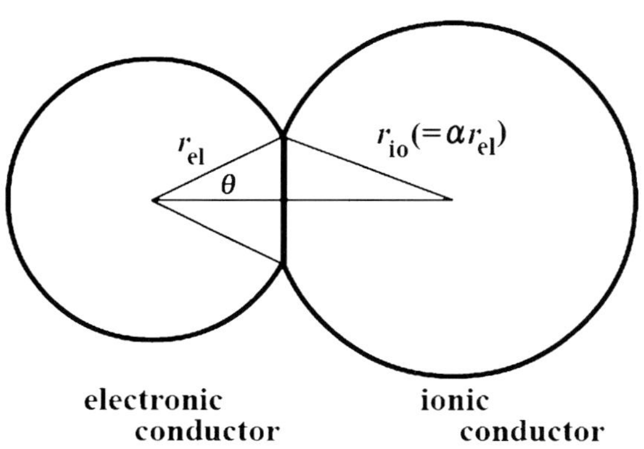

Figure 4.4.9
Scheme of particles contact after heat treatment.

where n_t is total number of particles per unit volume, n_{el} and n_{io} are number fraction of electronic and ionic particles, respectively, z_i the average coordination number of i-component, $z_{el/io}$ number of ionic particles in contact with electronic particles. Some results for $r_{el}=0.5$ μm are shown in Fig. 4.4.10, where porosity and contact angle are set at 0.4° and 15°, respectively. The active area is zero at the percolation thresholds, where the probability p_i is zero. For intermediate compositions a maximum is present: if $\alpha=1$ the maximum is reached for $f_{el}=0.5$.

The average coordination number z in a random packing of monosized spheres is 6. When the particle size decreases, the initial porosity is expected to increase because of the properties fine particles hold essentially. The influence of porosity on the active area is shown in Fig. 4.4.11, in which the calculation for an electrode with particle size $r_{el}=0.1$ μm ($\alpha=1$ and $f_{el}=0.5$) is performed in the range between the percolation limit ($z_{ii}=1.764$ and $\varepsilon=0.552$) and the closed packing ($z_{ii}=6$ and $\varepsilon=0.259$). The packing model proposed by Suzuki et al. [2] can be used to estimate the coordination number in a wide range of porosity. The result demonstrates that the active area decreases with an increase in the porosity. The change is conspicuous near the percolation limit, suggesting the control of porosity to be an important factor for the optimal active area.

(3) Effective conductivity

The percolation probability expressed by equation (4.4.1) and the conductivity, for instance, of a system are shown in Fig. 4.4.12, which clearly indicates

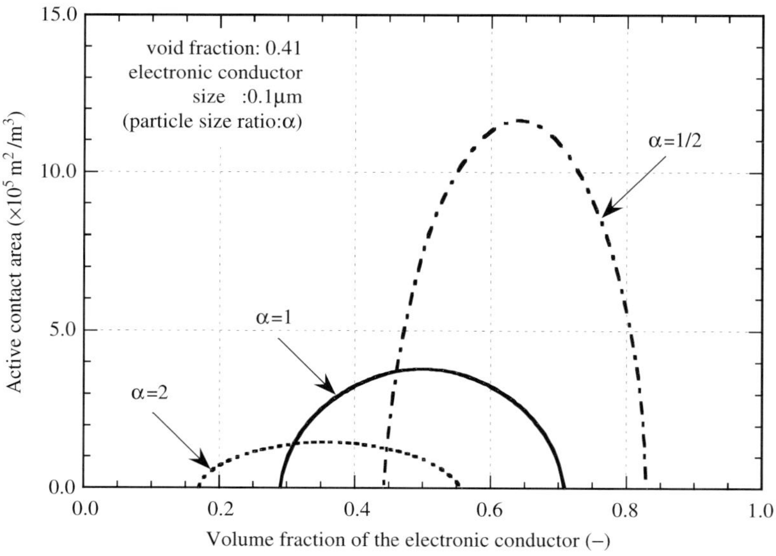

Figure 4.4.10
Effect of the particle size ratio on the active contact area.

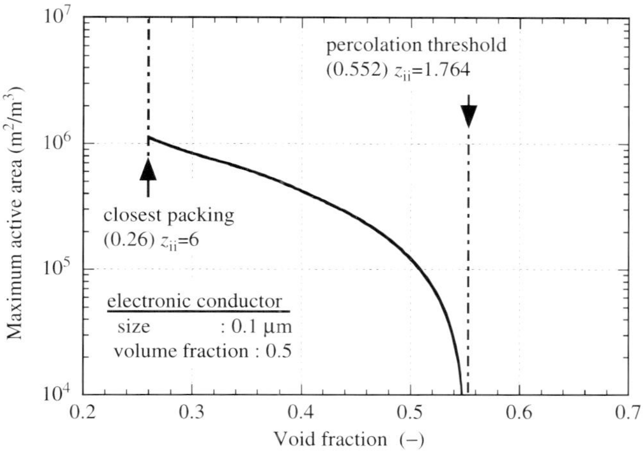

Figure 4.4.11
Effect of the void fraction on the maximum active contact area.

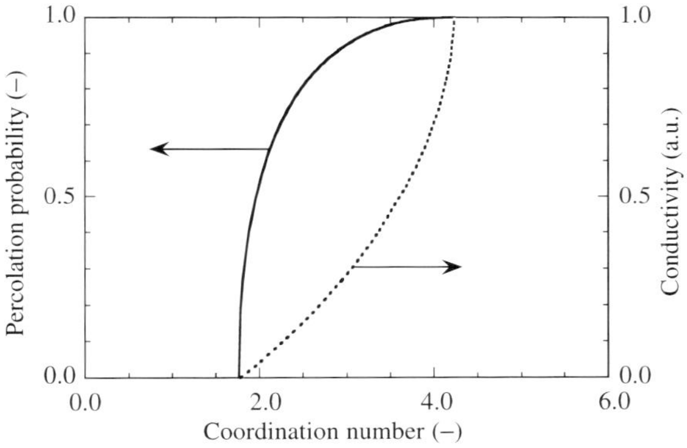

Figure 4.4.12
Percolation probability and conductivity of a system as a function of the coordination number.

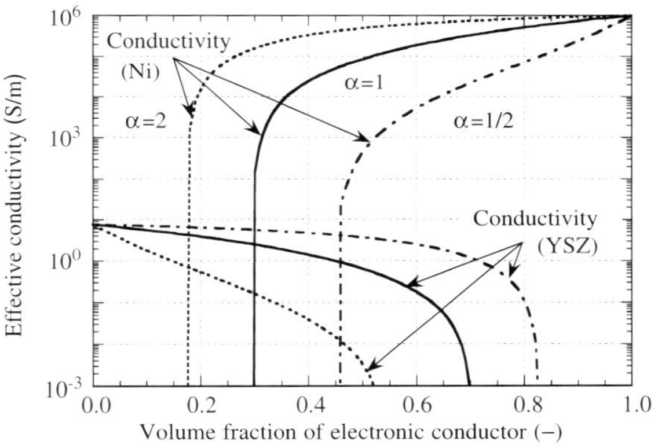

Figure 4.4.13
Estimation of the effective conductivity of Ni/YSZ electrode.

that the two quantities are not proportional. The conductivity curve seems to end at the threshold with zero slope although the percolation probability curve has infinite slope there, because most of the infinite network at the threshold belong to "dangling" or "dead ends", not to the "backbone". Thus most of the clusters make no contribution to the conductivity of a system, and therefore the critical exponent for conductivity differs from that for the percolation probability. This means we have to estimate the conductivity under the various conditions on the coordination number.

The effective conductivity can be calculated by [3]

$$\sigma_i^{\text{eff}} = \gamma \frac{\sigma_i (n - n_c)^2}{(1 - n_c)^2} \tag{4.4.4}$$

where σ_i is the conductivity of i-component, n the fraction number of particles of either the electronic or ionic conductor, and n_c the percolation threshold. The γ value is an adjustable parameter so that $\sigma_{ie}^{ff} = \gamma\sigma_i$ when $n = 1$, taking into account the effect of the necks between the particles on the conductivity. Some results for an electrode formed by a mixture of Ni ($\sigma_{el} = 2 \times 10^6$ S/m) and YSZ ($\sigma_{io} = 15$ S/m) at 1273 K are shown in Fig. 4.4.13, in which the adjustable parameter γ is 0.5. The two conductivities σ_{Ni} and σ_{YSZ} are almost equal for a composition near to the percolation threshold of the electronic conductor phase. However, the estimation of anode performance requires exact calculation of the conductivity because of sudden changes in the conductivity at the percolation threshold. It should also be noted that the effective conductivity is 1/10~1/20 of the Ni or YSZ original conductivity.

References

[1] C.H. Kuo, P. Gupta: *Acta. Metall. Mater.*, **43**, 397–403 (1995).
[2] M. Suzuki, T. Oshima: *Powder Technol.*, **35**, 159–166 (1983).
[3] S.H. Chan, Z.T. Xia: *J. Electrochem. Soc.*, **148**, A388–A394 (2001).

4.4.3 Structure of filler orientation in matrix

Recently polymer-nanofiller composites have been attracting considerable worldwide attention as cutting edge materials which benefit human welfare. The composites exhibit unique properties and functions, both of which are different from those of component materials. During the past 10 years a large number of research and development projects have been conducted to investigate relationships between the structures and the various properties of the composites. The properties cover a broad area including mechanical, electronic, electrical, optical, thermal, chemical, transport, biological, medical, and so forth. Obviously, the foregoing properties of polymer–nanofiller composites originate partly from the properties of materials used for the composites. However, the elicitation of the unique properties is mainly attributed to the following structural characteristics which nanocomposites usually possess. First, both filler-matrix contact area and filler-matrix interaction increase significantly when the nanoscale filler and matrix are intimately mixed. Second, many nanocomposites have an ultrafine layered structure that helps to improve some properties such as mechanical, thermal, barrier, and the like. There exist at least two other important factors to be considered especially for the improvement of mechanical strength of polymer-based nanocomposites. They are the degree of dispersion and the orientation of fillers. As one of the short-term research projects in the domain of nanotechnology research, the United States has selected a project to develop construction materials with substantially increased mechanical strength per unit weight as compared to existing construction materials. Presumably the construction materials will be used for vehicle parts in automobiles, aircrafts, and spaceships. In short, the project aims to invent light-weight and mechanically strong materials to be used in these vehicles. Development of polymer-based nanocomposites appears to be very promising toward attaining this goal. Many polymer-based nanocomposites contain less than 10% in weight of fillers. The amount is much less compared to the filler contents of conventional composite materials. To aim for a substantial increase of mechanical strength of polymer-based nanocomposites with such a small amount of nanofillers, both dispersion and orientation could be the most dominating parameters. In this review we intend to show the results of recent investigations related to the structure of polymer-based nanocomposites. Special emphasis is placed on the methods useful for obtaining satisfactory dispersion and orientation of nanofillers in polymer matrix.

In order to discuss the dispersion and orientation of fillers in host materials it is adequate to define the shapes and sizes of fillers. The aspect ratios of fillers were used for the definition. Three different ranges of aspect ratios were considered: (1) aspect ratio is closed to unity, (2) aspect ratio is approximately 100 (fillers with the aspect ratio are exemplified by clay minerals) (3) aspect ratio ranges from 10 to 10,000 (the aspect ratios of many rod-like and fiber-like fillers fall in this range).

(1) Aspect ratio is close to unity

For fillers with almost perfect sphericity (the aspect ratio is close to unity) much effort is placed on how to arrange the spherical particles rather than how to orient. One possible method is to prepare the one- or two-dimensional arrays of nanoparticle precursors on the surface of polymer. After the arrangement, the precursors are allowed to react with other chemicals to form the arrays of nanoparticles [1]. This method is expected to be used to fabricate electronic and optical devices. Another plausible method is to arrange various dendorimers in polymer, which possess abilities to bond to metal clusters. The addition of the metal clusters to the dendrimers leads to form polymeric composites with aligned nanoclusters [2]. By selecting various combinations of metals and dendric polymers, it is possible to synthesize nanocomposites with many varieties of characteristics and functions. Kumacheva and his coworkers [3] proposed a method to form a three-dimensional configuration of nanoparticles. First, they fabricated a three-dimensional structure of nanoscale latex particles composed of a hard core and a soft shell. Then, this structural material was heat-treated to melt the soft shell made usually of a polymer. The treatment yielded three-dimensionally assembled hard spheres in a continuous medium of the polymer. It is thus possible to synthesize various functionalized nanocomposites when materials with different kinds of functionalities are chosen for the hard-core components. The hard core materials include chromophores, fluorescent dyes, metal complexes, and the like.

(2) Aspect ratio is approximately 100

Layered silicates are representative clay minerals often used as the fillers of polymer-based nanocomposites. The silicates contain silicate platelets with a thickness of about 1 nm and a side length of about 100 nm. Thus, they are considered to be fillers with an aspect ratio of about 100. The clay minerals are quite attractive as fillers especially for polymer-based composites due to the following reasons. First, the minerals are natural products and are easily obtainable at low cost. Second, they have nanoscale layered structures by nature, which are ideally suited for fabricating nanostructured composites. Third, the clay minerals are inorganic substances with high chemical and heat resistance. For synthesis of the composites one often uses organic dispersing agents and surface modifiers to increase dispersibility of the clay minerals in matrix polymer. Sometimes, one employs organic swelling agents to enlarge the interlayer spacing of clays. Even if these additives are used, not much chemical change occurs for these additives and the clays. This indicates that the clay minerals are rather non-reactive with organic chemicals. Fourth, as Okamoto [4] reported, the clay minerals showed self-assembling abilities when a proper environment was provided. Moreover, the minerals were able to assist crystallization of matrix polymer, which was helpful to improve the mechanical properties of the resulting composites [5]. In general, both self-assembling and crystallization of polymer are processes conducive to improving mechanical properties of polymer-clay composites.

In the following part, discussions will be presented regarding the dispersion and orientation of clay minerals in polymer matrix. For the fabrications of polymer-clay mineral composites two methods are usually employed. One is to blend a clay mineral with a molten polymer using extruders or injection molding machines. The other is in-situ polymerization of monomer in the presence of clay. For these two processes efforts are mainly placed on finding efficient techniques to prepare polymers with uniformly dispersed fillers. At present, not much consideration is given to the orientation of clay in polymer matrix. Needless to say, the dispersion of clay has the most profound influence on the properties and functions of the composites in general and the mechanical properties in particular. Thus, it is quite reasonable to give the first priority to investigations, the objects of which are to clarify the effect of filler dispersion on the mechanical properties of nanocomposites. Note however that orientation is a factor which cannot be ignored when the anisotropic properties of composites are studied. Here a brief explanation is given on the mechanical blending of clays and polymers. In the blending, one frequently employs rotating screw-type equipment to allow clay minerals to disperse in polymers under the influence of high shear generated by the rotating action of screw. The equipment yields both uniformly dispersed and well-exfoliated clays in matrix polymer. In certain cases the equipment is also useful to allow fillers to orient toward a desired direction in polymer. As shown in Fig. 4.4.14 three different kinds of basic structures are formed by this method depending upon the physical and chemical properties of clays and polymers, physicochemical characteristics of clay surfaces, and operating conditions of the rotating screw blenders. In Fig. 4.4.14, we see two significantly different structures for polymer–clay composites. One is a structure in which matrix polymer is intercalated into the nanoscale layers of clay. The other is an exfoliated structure in which the clay layers are well-dispersed in polymer matrix [6]. Results of many studies have indicated that the latter gives more favorable mechanical properties as compared to the former especially in terms of tensile strength and impact strength. Masenelli-Variot and his coworkers [7] investigated the mechanical properties of a polyamide–clay composite and found that the tensile properties were not much different for both an exfoliated and an

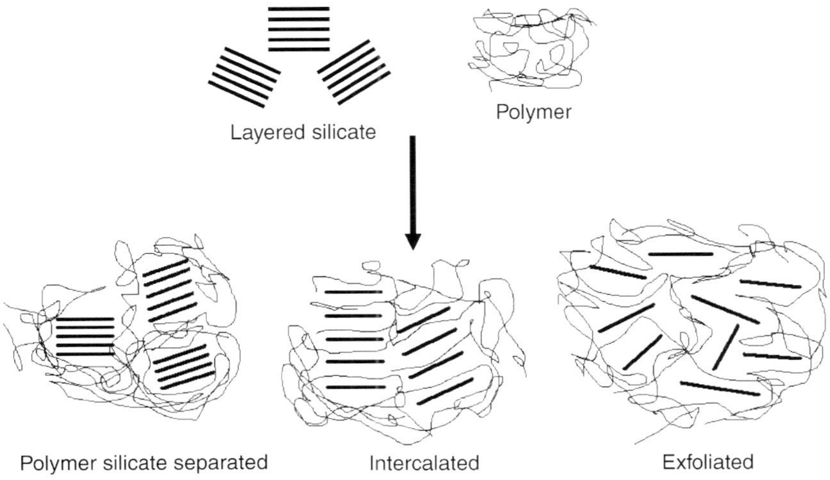

Figure 4.4.14
Structure of polymer-clay nanocomposite.

intercalated composite. This result seems to be different from those obtained by other investigations. For kneading of polymers and clays, Dennis and his coworkers [8] have made a comprehensive study on the relationships among the operating conditions of hardware, the degrees of both dispersions and exfoliations of fillers in the resulting composites and their mechanical properties. The results of this study provide a useful guide for predicting the mechanical properties of polymer–clay composites obtained by conventional kneading equipment. Besides the dispersed and exfoliated conditions of clays in polymer, there is another important factor influential on the mechanical properties of polymer–clay composites. That is an interaction between polymer chains and clay minerals on the interfaces of these two components.

Concerning the orientation, Kojima and his colleagues [9] investigated the effects of flow conditions of a montmorillonite-Nylon 6 composite and shear stress exerted upon molten composite on the orientations of these two components. They prepared the composite using in-situ polymerization for Nylon 6 in the presence of the layered montmorillonite, formed films by employing an extruder and examined specimens prepared by injection molding. The composite specimens thus prepared had a laminated structure composed of ultrafine crystallites of Nylon 6 and the montmorillonite layers comprised of a few or several monolayers of montmorillonite. In terms of the orientation of the montmorillonite and the Nylon 6 crystallites, each of specimens had three distinct layers in the direction of the thickness of specimen. They are surface, intermediate, and middle layers. The surface layer was located in the vicinity of the specimen surface, while the middle layer was situated in the central section of the specimen. The intermediate layer was formed between these two layers. It should be noted that the shear stress was very high in the surface layer. In this layer both the montmorillonite monolayers and the chain axes of Nylon 6 crystallites showed orientations parallel to the specimen surface and the directions of the axes were random within the layers of the Nylon 6 crystallites. In the intermediate layer where the shear stress was reduced, the montmorillonite layers aligned parallel to the specimen surface and the direction of the monolayers were random. In this zone the chain axes of the Nylon 6 crystallites were perpendicular to the montmorillonite layers. In the middle layer with the lowest shear stress as compared to two other layers, the directions of the montmorillonite layers were perpendicular to the flow direction of the composite and the directions of monolayers were random. The Nylon 6 crystallites were randomly oriented around the flow axis of the composite and the chain axes of Nylon 6 were perpendicular to the montmorillonite layers. The results indicated that the direction of clay orientation was influenced significantly by the flow conditions of polymeric material during the course of composite fabrication. Simply speaking, the direction tends to become parallel to the flow direction of polymer in a high-shear zone. In a lower-shear zone, interestingly enough, the chain axes of polymer crystallites aligned perpendicularly to the layers of montmorillonite. As an extension of studies related to the orientation of fillers for polymer-based composites, Okamoto and his coworkers [10] investigated the orientation of clay under an elongational flow of polypropylene–clay system and found that a card-house structure of clay was formed in the polymer. The formation was caused by the interaction of negatively charged surfaces of clay layers and positively charged end sections of the clays. Fong and his coworkers [11] studied the orientation of montmorillonite clay in Nylon 6 for the electrospinning process. In their study the clay layers aligned in the direction perpendicular to the axes of Nylon fibers. As shown above a number of investigations have been conducted on the orientations of fillers in polymer-based nanocomposites. Regardless of the fact that the filler orientation is an important factor for the design of advanced composite materials, not much data are available to find quantitative relationships between the orientation of fillers and physical properties of composites. For dispersion, the degree of dispersion can be controlled by the kneading temperature and distribution of shearing velocity. On the contrary, it is not easy to control the orientation of fillers during the course of composite fabrication. As already stated, the anisotropy in mechanical, transport, and optical properties are closely related to the orientations of polymer crystals, polymer chains, and laminated fillers. Therefore, it is urgent to establish technical methods by which the orientations of fillers can be precisely regulated.

(3) Aspect ratio ranges from 10 to10,000

Fibrous fillers such as glass fibers, various whiskers and tube-type fillers represented by carbon nanotubes are considered to fall into this category. Particularly single-walled carbon nanotubes (SWNTs) exhibit unique characteristics in mechanical, thermal, optical, electrical, electronic, and transport properties. For this reason, SWNTs attract a good deal of attention as a filler for advanced composite materials and the worldwide competition is quite keen in development of composites containing SWNTs. The followings show the recent techniques related to how to obtain well-dispersed SWNTs in polymer and how to align SWNTs in a desired direction. They are equally challenging problems to overcome for the fabrication of the polymer–SWNT composites. Because of the highly aggregative nature of SWNTs it is rather hard to obtain uniform and stable dispersion of SWNTs. To solve this problem many methods have been proposed. Chemical modification of SWNT surfaces is an effective way to weaken the aggregation force of SWNTs. In addition, synthetic methods [12, 13] have been developed to prepare unbundled or not highly entangled SWNTs. Another plausible method is to apply a high shear on the blend of SWNTs and polymer when they are kneaded. With the application of high shear, however, there exists a fair chance for the destruction of SWNT structure and the breakage of SWNT tubes. This may result in obtaining composites with property levels much lower than those expected originally especially for mechanical strength and electrical conductivity. Barrera [14] summarized various methods applicable to improve the orientation of SWNTs in polymer matrix. Here, we briefly describe three feasible methods selected from his article. First, a monomer is polymerized to the corresponding polymer in the presence of SWNTs, which are aligned beforehand in a desired direction. Second, SWNTs are oriented in a direction under a simultaneous application of high shear and magnetic field. Third, molten composites are strongly elongated under heating to allow fillers to align in a direction of elongation. Note that a Japanese patent [15] was requested for a method similar to the second method. Composite materials presented in Sections (2) and (3) are expected to be commercialized in the near future. In industrial-scale productions of composites one will face the same problems as those encountered in the laboratory-scale productions. They are the dispersion and orientation of filler. Simply speaking, desirable results will be obtained if the motion of the filler is not restricted in matrix. This may be attained by decreasing the viscosity of filler-polymer mixture. For this purpose an attempt [16] was made to inject supercritical carbon dioxide into the mixture during the course of kneading. Needless to say caution should be exercised in regard to the property changes of products, which may be caused by the injection of carbon dioxide. Yet, the attempt should be considered one of the promising techniques for obtaining composites with well-dispersed and designed orientation of filler in matrix.

References

[1] L.L. Brott, R.R. Naik, D.J. Pikas, S.M. Kikpatrick, D.W. Tomlin, P.W. Whitiock, S.J. Clarson and M.O. Stone: *Nature*, **413**, 291–293 (2001).

[2] F. Grohn, G. Kim, B.J. Bauer and E.J. Amis: *Macromolecules*, **34**, 2179–2185 (2001).

[3] E. Kumacheva, O. Kalinina and L. Lilgel: *Adv. Mater.*, **11**, 231–234 (1999).

[4] M. Okamoto: *Seikei Kakou (J. Jpn. Soc. Polymer Process.)*, **16**, 574–578 (2004).

[5] N. Ogata, G. Jimenez, H. Kawai and T. Ogihara: *J. Poly. Sci. Part B Poly. Phys.*, **35**, 389–396 (1997).

[6] M. Alexandre and P. Dubois: *Mater. Sci. Eng.*, **28**, 1–63 (2001).

[7] K. Masenelli-Varlot, E. Reynaud, G. Vigier and J. Varlet: *J. Polymer Sci. Part B Polymer Phys.*, **40**, 272–283 (2002).

[8] H.R. Dennis, D.L. Hunter, D. Chang, S. Kim, J.L. White, J.W. Chao and D.R. Paul: *Plastics Eng.*, **56**–60 (2001).

[9] Y. Kojima, A. Usuki, M. Kawasumi, A. Okada, T. Kurauchi, O. Kamigaito and K. Kai: *J. Polymer Sci. Part B Polymer Phys.*, **33**, 1039–1045 (1995).

[10] M. Okamoto, P.H. Nam, P. Maiti, T. Kotaka, N. Hasegawa and A. Usuki: *Nano Letts.*, **1**, 295–298 (2001).

[11] H. Fong, W. Liu, C.-S. Wang and R.A. Vaia: *Polymer*, **43**, 775–780 (2002).

[12] L. Huang, X. Cui, B. White and S.P.O'Brien: *J. Phys. Chem. B*, **108**, 16451–16456 (2004).

[13] K. Hata, D.N. Futaba, K. Mizuno, T. Namai, M. Yumura and S. Iijima: *Science*, **306**, 1362–1364 (2004).

[14] E.V. Barrera: *JOM-J. Mater. Metals & Mater. Soc.*, **52**, A38–A42 (2000).

[15] Japanese Publication of Unexamined Patent Applications: Tokukai 2002-273741.

[16] M.D. Elkovitch, L.J. Lee and D.L. Tomosako: *Polymer Eng. Sci.*, **41**, 2108–2125 (2001).

4.4.4 In situ particle polymerization

In this section, two techniques are described for preparing composite materials by polymerization in the presence of particles or precursors of particles. Also information on these techniques is presented.

(1) Polymerization in the presence of particles

This technique along with the mixing and kneading technique of particles with polymers by a 2-screw extruder is a typical method of obtaining composite materials. Many patents have been published using various polymers.

Well-known nanocomposites include nylon, polyester, and epoxy resin in which layered particles such as clay are dispersed. Many of them are produced by "in situ particle polymerization"; that is, layered particles such as clay are first dispersed in monomers or oligomers and then the mixture is polymerized. Typical layered particles are clay minerals of several micrometers such as montmorillonite or originally large kaolin. These particles are formed by many laminated layers several nanometers in size. How to delaminate and disperse them in the polymer is a key point. As an example, a schematic depiction of the structure of montmorillonite is shown in Fig. 4. 4. 15.

Generally, as pretreatment, interlayer metal cations in layered particles are replaced with cationic organic compounds or monomers modified with organic cations to widen the interlayer space and make it hydrophobic (intercalation). The modified particles

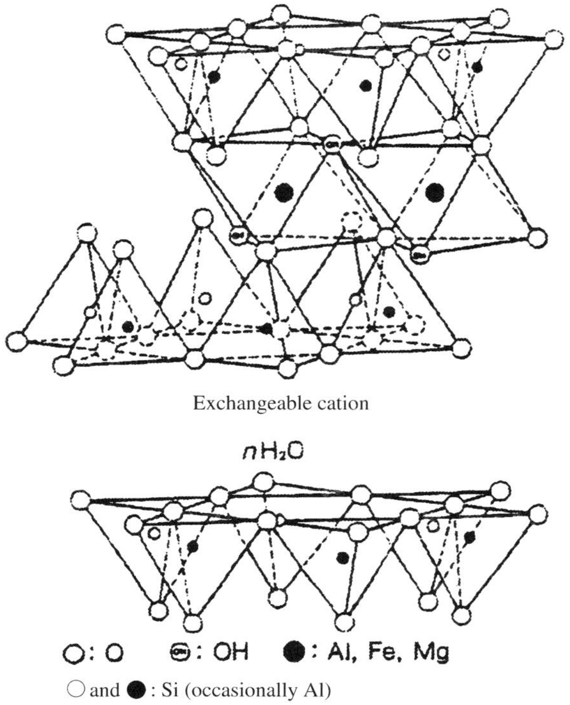

Figure 4. 4. 15
Structure of montmorillonite [1].

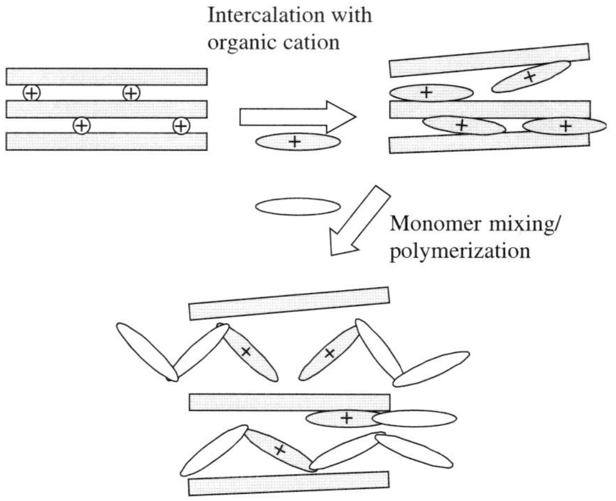

Figure 4. 4. 16
Schematic depiction of synthesis of layered particle nanocomposite by intercalation method.

are mixed with monomers and then the mixture is polymerized. Simultaneous delamination of the layered particles and polymerization produce composite materials with the delaminated particles dispersed in them (Fig. 4.4.16). This is a general method for preparing composite with delaminated layered particles.

These techniques are renowned throughout the world and many studies have been published on various combinations of resins and layered particles. Studies on the in situ polymerization are still being conducted, although reports have appeared recently in which nanocomposites are obtained by mixing and kneading intercalated particles with polymers without polymerization [2, 3].

The oldest study on preparing layered particle/polymer nanocomposite by polymerization is on nanocomposite made from layered particles and polyamide, in which nanocomposite is commercialized [4]. Amine compounds can easily be inserted between layers by ion exchange; probably this is the reason why this combination was chosen for basic and commercialization studies. Many reports have been published on the preparation of polyamide nanocomposite by intercalation polymerization. One of the methods Usuki et al. [1] reported is that montmorillonite is treated with 12-aminododecane acid and hydrochloric acid to produce organic clay, then molten ε-caprolactam is inserted into the interlayer, and finally the mixture is polymerized at 250°C.

Recently, studies have been published on polyester [5], polyacrylic acid, and polyolefin such as polystyrene in addition to polyamide. Ke et al. obtained PET nanocomposite by mixing molten polyethylene terephthalate (PET) oligomer with clay whose interlayer ion has been exchanged, and then polymerizing the mixture [6, 7]. Also, the preparation of polystyrene-layered silica by living polymerization has been reported. Since interlayer ion exchange with polyolefin monomer is difficult due to its almost non-polar property, monomers modified by organic amine compounds are often used. Weimer et al. inserted an organic ammonium cation, a polymerization initiator, into the interlayer, added styrene monomer, and polymerized the mixture using living polymerization to obtain a nanocomposite (Fig. 4. 4. 17) [8].

All of these layered particle nanocomposites have improved mechanical properties such as tensile strength and flexural modulus, thermal properties such as thermal deformation temperature and crystallization speed, and gas barrier.

Many reports have been published on nanoparticle in situ polymerization using particles other than layered particles. For example, Choi et al. reported synthesis of PET and nylon Ag nanoparticle composites [10]. In preparation of PET composite polymer, oligomer was first produced from dimethyl terephthalate and ethylene glycol (EG), and then to this oligomer Ag nanoparticle EG slurry was added, finally the mixture was polymerized at 280°C for 3 h under reduced pressure. For nylon, ε-caprolactam dispersed with Ag nanoparticle was reacted at 260°C for

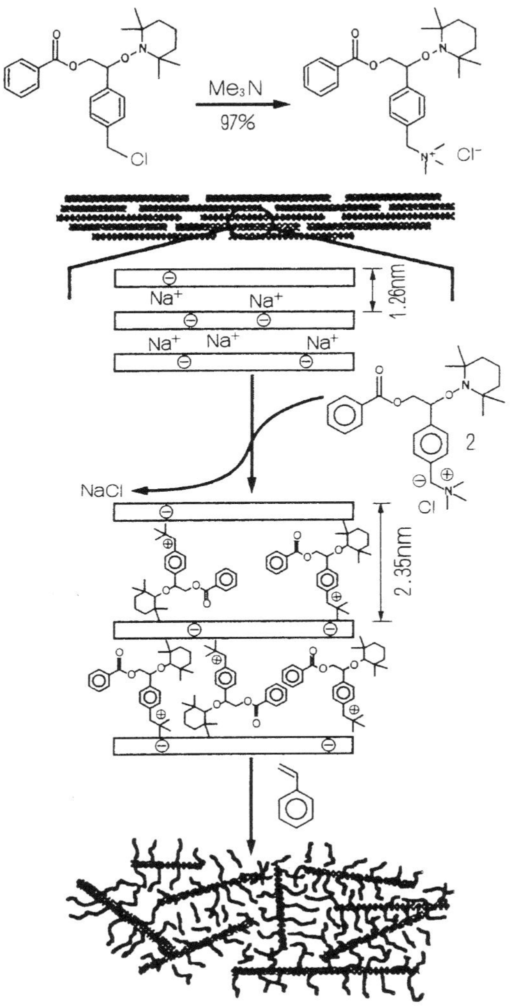

Figure 4. 4. 17
Schematic depiction of synthesis of PS nanocomposite by living polymerization [9].

1 h under a pressure of 15 kgf/mm^2 and then 1 h under normal pressure.

The surface of the Ag particle was treated with mercaptosuccinic acid that has both carboxylic acid and thiol radicals. The thiol radical adsorbs on the Ag metal surface to prevent aggregation of the nanoparticles. However, the Ag nanoparticle aggregates in PET composite since this dispersing agent does not work in PET, although it is confirmed that Ag particles are highly dispersed in nylon composite. This means that a dispersing agent does not always work in polymer, even if it has a carboxylic acid group that is capable of forming ester. This shows the difficulty of dispersing nanoparticles in polyester.

In this report, Choi et al. considers simply that the difference is due to the difference in affinity to water, that is, the main chain of nylon is hydrophilic while that of PET is hydrophobic. There is another consideration that the dispersing agent may be desorbed from the particles, but the metal particles may be stabilized by the unshared electron pair on a nitrogen atom of the amido group of the nylon main chain. That is, the polymer itself may work as a dispersing agent.

Acrylic monomers are used in many studies on the preparation of composite materials from metal particles and polymer precursors. For example, Fang et al. dispersed synthesized CoPt nanoparticles with a mixture of methyl methacrylate, ethylene glycol dimethacrylate, and 2-hydroxyethyl methacrylate monomers, then polymerized and cross-linked the mixture by UV radiation to obtain CoPt–PMMA nanocomposite [11]. Many reports have been published on metal nanoparticle composite using acrylic polymers.

(2) Simultaneous syntheses of particles and polymers

The Pechini process [12, 13] is a sol-gel method for obtaining particles from a homogeneous mixture of precursors of the particles, polycarboxylic acid, and polyol (a precursor of polymer) that are capable of forming complexes with the precursors of the particles. One example of syntheses is that citric acid, an excess amount of ethylene glycol and a metallic salt compound are mixed to obtain a homogeneous solution, and the mixture is esterified at 160–300°C. Further polymerization reaction to increase the polymerization degree and particle formation reaction such as hydrolysis were performed at the same time to obtain gel-type polymer/particle composite [14, 15].

Studies have been published on the use ethylenediamine tetraacetic acid instead of citric acid [13]. That is, a feature of the Pechini method is to obtain homogeneous particles using chemical compounds capable of forming easily stable complexes in the particle reaction field (in this sense, both the methods using PEG or PVA described in the former section may be defined as Pechini methods).

Since polycarboxylic acids that can be used for this method are limited and moldability of the product is poor, this composite has only limited applications as molding materials. Usually, the method is used to obtain only particles by incinerating the composite. Generally speaking, use as a composite material is unthinkable.

In the Pechini method, it is considered that reactions proceed gradually with the consumption of carboxylic acid groups by esterification; accordingly control of particle size is difficult in polymerization. It is mainly controlled in incineration after the polymerization [13].

Nakao et al. reported a method of in situ polymer/particle simultaneous synthesis other than the Pechini method. In this method, a mixture of methyl methacrylate monomer and palladium or platinum acetyl acetate is polymerized and reduced at the same time at 60°C to obtain metal nanoparticle/poly methyl methacrylate composite [16]. Since the polymerization of the monomer capable of forming the complex and the particle formation reaction proceed at the same time, this process can be defined as one of the Pechini methods, although the esterification reaction does not occur. In the simultaneous syntheses of polymers and particles (or particle syntheses in polymers), the stability is considered important for inorganic compounds as precursors of particles and complex compounds to be formed.

In this field, many reports concentrate on acrylic polymer and polyalkylene glycol, the reason being, probably, that the stability of the complex is a key issue. Other examples of polymer systems include a study on the preparation of nanocomposites in the PET synthesis reaction field [17]. They may be classified as organic/inorganic hybrid materials rather than polymer/particle nanocomposites.

References

[1] The Chemical Society of Japan: *Kikan Kagaku Sosetsu No.42, Organic–Inorganic Hybrid Material*, Japan Scientific Society Press, Tokyo, p. 35 (1999).
[2] L. Liu, Z. Qi and X. Zhu: *J. Appl. Polym. Sci.*, **71**, 1133–1138 (1999).
[3] J. Chang, D. Park: *J. Polym. Sci. Part B*, **39**, 2581–2588 (2001).
[4] K. Chujo: *Nanokonpojitto No Sekai (in Japanese, World of Nanocomposite)*, Kogyo Chosakai Publishing, Tokyo, p. 113 (2000).
[5] Z. Ke, B. Yongping: *Mater. Lett.*, **59** (27), p. 3348–3351 (2005).
[6] Y. Ke, C. Long and Z. Qui: *J. Appl. Polym. Sci.*, **71**, 1139–1146 (1999).
[7] Y. Ke, Z. Yang and C. Zhu: *J. Appl. Polym. Sci*, **85**, 2677–2691 (2001).
[8] M.W. Weimer, H. Chen, E.P. Giannelis and D.Y. Sogah: *J. Am. Chem. Soc.*, 1615–1616 (1999).
[9] K. Chujo: *Nanokonpojitto No Sekai (in Japanese, World of Nanocomposite*, Kogyo Chosakai Publishing, Tokyo, p. 161 (2000).
[10] S. Choi, K. Lee and S. Park: *Stud. Surf. Sci. Catal.*, **146**, 93 (2003).
[11] J. Fang, L.D. Tung, K.L. Stokes, J. He, D. Caruntu, W.L. Zhou and C.J. O'Commor: *J. Appl. Phys.*, **91**, 8816 (2002).
[12] M.S. Pechini: US Patent No. 3, 330, 697 (1967).
[13] B.L. Cushing, V.L. Kolesnichenko and C.J. O'Connor: *Chem. Rev.*, **104**(9), 3893 (2004).
[14] S. Yamamoto, M. Kakihana and S. Kato: *J. Alloys Compos.*, **297**, 81–86 (2000).
[15] P. Duran, F. Capel, D. Gutierrez, J. Tarataj and C. Moure: *J. Euro. Ceram. Soc.*, **22**, 1711–1721 (2002).
[16] Y. Nakao: *Kobunshi: High Polymers, Japan*, **43**, 852–855 (1994).
[17] T. Morimoto, M. Yoshida and K. Okuyama: *J. Soc. Mater. Sci., Jpn.*, **54**, 393–398 (2005).

4.4.5 ECAP

(1) ECAP Process

Figure 4.4.18 shows the schematic representation of side extrusion processes, which are a kind of the double-action extrusion. In this figure, (d) and (e) indicate the process in which pure shear deformation can be repeatedly imposed on materials so that an intense plastic strain is produced with the materials within any change in the cross-sectional dimensions of the workpiece. This processes are named as an equal channel angular extrusion (ECAE) or an equal channel angular press (ECAP).

Segal [1] proposed this process in 1977. In 1990s, many studies for the development of ultra-fine grained materials were carried out by Valiev et al. [2], Horita et al. [3, 4], and Azushima et al. [5]. Azushima et al. proposed the repetitive side extrusion process. It is a process in which a high back pressure is applied in the process of (d) type as shown in Fig. 4.4.18 in

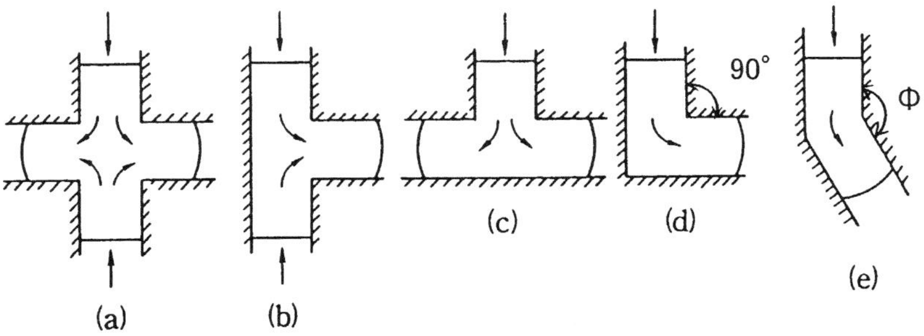

Figure 4.4.18
Schematic representation of side extrusion processes in double-acting extrusion.

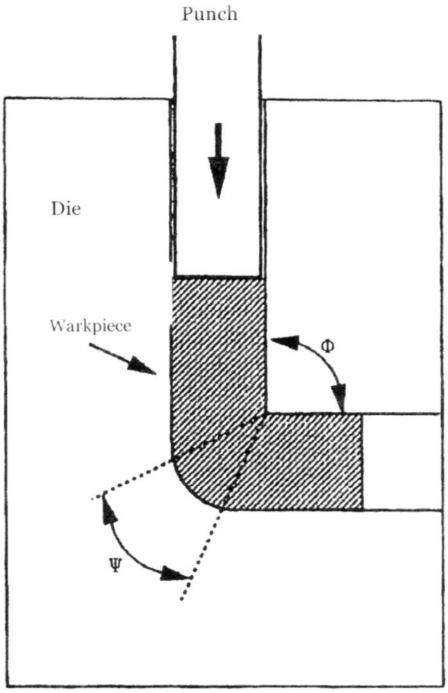

Figure 4.4.19
Schematic representation of side extrusion process of ECAP type.

order to produce the uniform shear deformation and present the defect of workpiece.

The schematic representation of this process is shown in Fig. 4.4.19. The specimen is side extruded through the shear deformation zone with and due to the dead zone in the outer corner of the channel. When the workpiece is side extruded through the channel, the total strain is

$$\varepsilon = \frac{1}{\sqrt{3}}\left\{2\cot\left(\frac{\Phi}{2}+\frac{\Psi}{2}\right)+4\operatorname{cosec}\left(\frac{\Phi}{2}+\frac{\Psi}{2}\right)\right\} \qquad (4.4.5)$$

In this figure, when $\Phi = 90$ and $\Psi = 0$, the total strain follows from above equation that $\varepsilon = 1.15$. After n passes, it becomes ε_n.

$$\varepsilon_n = n\varepsilon \qquad (4.4.6)$$

(2) Ultra-fine grained metal produced by repetitive side extrusion

(a) Microstructure

From the optical microstructure of metals over 5 passes of repetitive side extrusion, it has been found that the strong filamentary microstructure is developed with increasing number of pass. In these conditions, the observation of microstructure must be made using the TEM analyzer. From the TEM microstructure analysis, it has been confirmed that the many metals with an ultra-fine grain size under 1 μm are developed by ECAP processes. Valiev et al. [2] reported that the ultra-fine grain with a sub-micron was created by ECAP processes in the many metals and the grain size of the AL-4%Cu-0.5%Zr alloy became about 200 nm by ECAP with a plastic strain of 7 at 160°C. Horita et al. [3] reported that the aluminum and aluminum alloys with a submicron-scaled grain size were developed by ECAP processes.

Azushima et al. [6] reported that the ultra-fine grain size with a major axis length of 0.5 μm and a minor axis length of 0.2 μm was developed by 10 passes of repetitive side extrusion as shown in Fig. 4.4.20.

(b) Mechanical properties

From the TEM observation of microstructure, it is expected that the hardness and the tensile strength of metals with the ultra-fine grain become higher by the repetitive side extrusion. Figure 4.4.21 shows the relationship between the hardness, tensile strength, and pass number of ECAP for Armco steel [7]. The hardness and the tensile strength increase with increasing pass number. The Vickers hardness is increased from 100 to 250 and the tensile strength is increased from 300 to 750 MPa after one pass. The tensile strength is

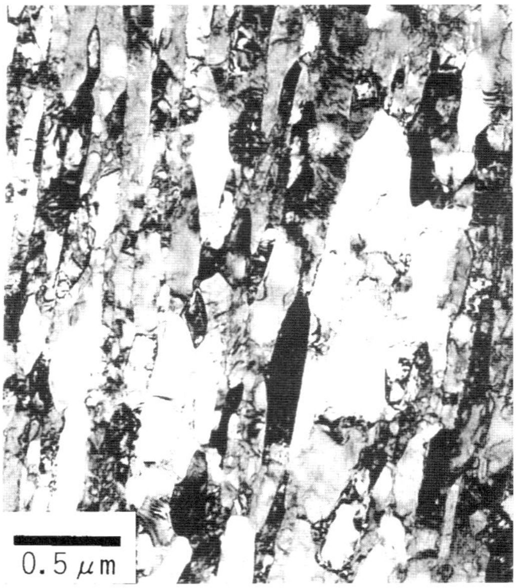

Figure 4.4.20
TEM micrograph of ultra low-carbon steel after 10 passes of side extrusion.

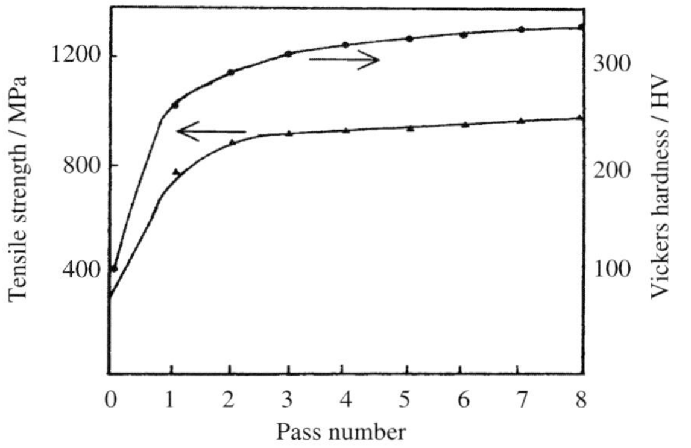

Figure 4.4.21
Relationship between tensile strength or Vickers hardness and pass number.

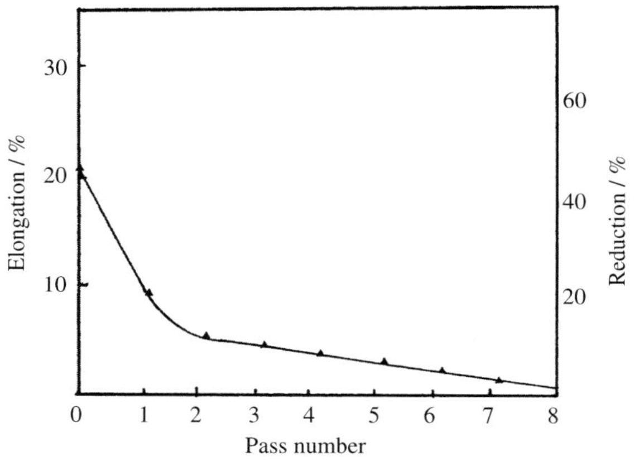

Figure 4.4.22
Relationship between elongation or reduction and pass number of Irmco steel.

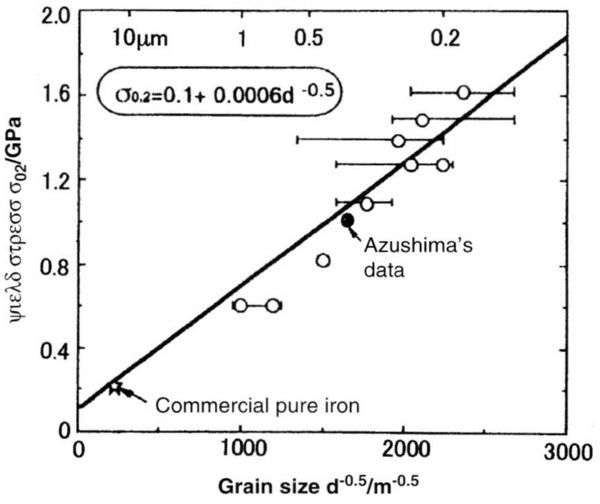

Figure 4.4.23
Relationship between yield stress and grain size of ultra low-carbon steel.

increased by a factor of 2.5 after one pass in comparison with the specimen before ECAP process and it increases with increasing pass number up to 8 passes. The tensile strength is over 800 MPa after 8 passes. On the other hand, the elongation shows a reduction from 20% for the specimen before ECAP process to several percent after 8 passes as shown in Fig. 4.4.22 [7].

Horita et al. and Azushima et al. reported the same results for the aluminum alloys and for the steels, respectively. In particular, Azushima et al. [5] reported that the tensile strength after 10 passes of repetitive side extrusion was over 1,000 MPa. It has increased by a factor of 3 in comparison with the as-received material. The experimental data of the specimen after 10 passes are plotted in the Hall–Petch relationship of the yield stress against the root grain size as shown in Fig. 4.4.23. In this figure, the results for these specimens show good agreement with the standard Hall–Petch relationship of iron obtained by Takagi [8].

(3) Advanced ECAP processes

Figure 4.4.24 shows the fundamental process of metal flow during ECAP. As shown in the figure, the channel is bent through an angle equal to 90°. The specimen is inserted within the channel and it can be pressed through the die using a punch. There are four basic processing routes in ECAP. In route A, the specimen is pressed without rotation, in route B_A the specimen is rotated by 90° in alternate direction between consecutive passes, in route B_C the specimen is rotated by 90° in the counterclockwise between each pass, and in route C the specimen is rotated by 180° between passes.

Figure 4.4.25 shows the macroscopic distortion introduced into a cubic element for up to a maximum of 98 passes through the die [9]. From these distortions, the influence of the processing route on the development of an ultra-fine grained microstructure can be considered. Iwahashi et al. [10] reported that the ultra-fine grained microstructure of pure aluminum after 10 passes in route A was the same that of after 4 passes in route B_C.

(4) Material development by ECAP

Although ECAP is generally used in the processing of solid metals, it may be used also for the consolidation of metallic powder. Kudo et al. [11] carried out the pressing of pure aluminum powder by the repetitive side extrusion applied to the back pressure at a room temperature and they reported that the relative density after pressing became over 99.6%.

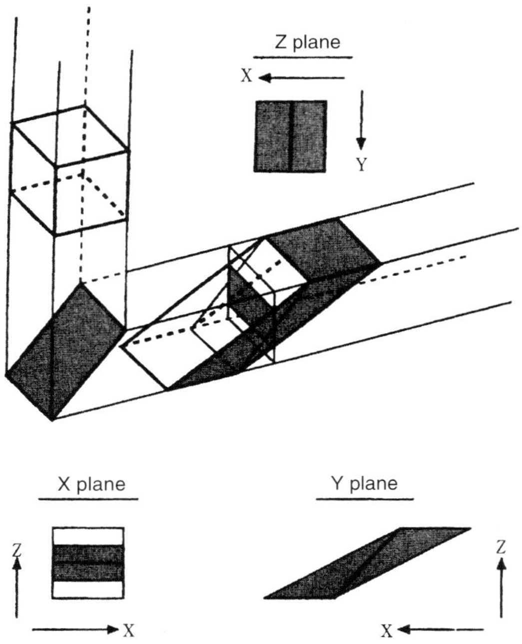

Figure 4.4.24
Geometry change of cubic element after one pass of side extrusion.

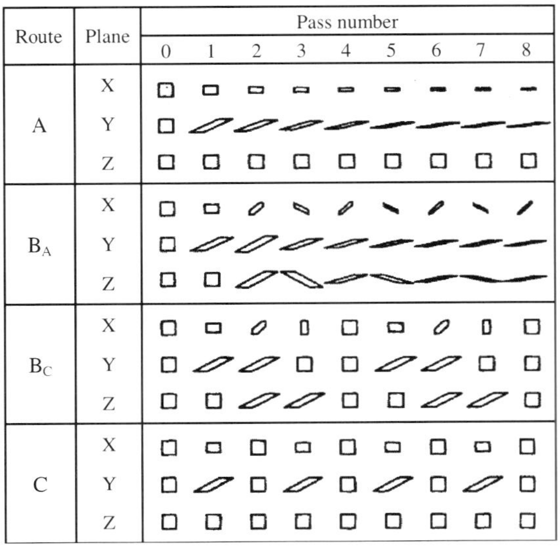

Figure 4.4.25
Geometry change of cubic elements when viewed on X, Y, and Z planes for processing routes A, B_A, B_C, and C after repetitive side extrusion.

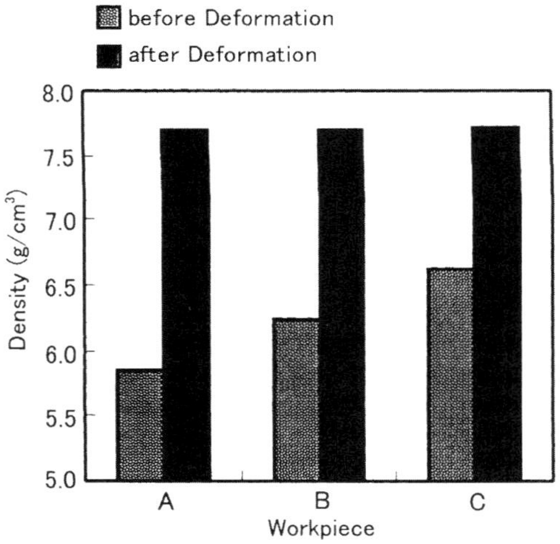

Figure 4.4.26
Density change of compacts before and after side extrusion.

Azushima et al. [12] carried out the pressing of pure iron powder compacts with several densities of 5.86 (A), 6.24 (B), and 6.63 g/cm3 (C) by repetitive side extrusion and they reported that the relative density became the same 98% as shown in Fig. 4.4.26.

References

[1] V.M. Segal: Patent of the USSR, No. 575892, (1977).
[2] R.Z. Valiev, A.V. Korznikov and R.R. Mulyukov: *Mater. Sci. Eng.*, **A168**, (1993), 141–148.
[3] K. Nakashima, Z. Horita, M. Nemoto and J.G. Langdon: *Acta Mater.*, **46**, 1589–1599 (1998).
[4] M. Furukawa, M. Iwahashi, Z. Horita, M. Nemoto, N.K. Tsenev, R.Z. Valiev and T.G. Langdon: *Acta Mater.*, **45**, 4751–4757 (1997).
[5] A. Azushima, K. Aoki, Y. Kataoka and T. Inoue: CAMP ISIJ, 12-3, 347–350 (1999).
[6] A. Azushima, K. Aoki and T. Inoue: Proceedings of the 1998 JSME Annual Meeting, Vol. 2, pp. 267–268 (1998).
[7] V.M. Segal: *Mater. Sci. Eng.*, **A197**, 157–164 (1995).
[8] H. Hidaka, Y. Kimura and S. Takaki: CAMP-ISIJ, 12-3, 357–360 (1999).
[9] M. Furukawa, Y. Iwahashi, Z. Horita, M. Nemoto and T.G. Langdon: *Mater. Sci. Eng.*, **A257** (1998), 328–332.
[10] Y. Iwahashi, Z. Horita, M. Nemoto and T.G. Langdon: *Acta Meter.*, 45–11, 4733–4741 (1997).
[11] Y. Mastumoto, M. Uda and H. Kudo: Proceedings of the 1988 Japan, Spring Conference Technology Plasticity, pp. 159–162 (1988).
[12] A. Azushima, K. Aoki and M. Hirama: *CAMP-ISIJ*, **15**(2), 379 (2002).

4.4.6 Nanostructure control of alloy

Plastic working with heat treatment is a useful process to prepare nanostructured and nanocomposite materials. This subject of the present section is the alloying process of high-performance materials with nanostructure by solid-state reaction in employing powder particles. Here bulky mechanically alloying process to produce Mg_2X (X; Si, Ge, Sn) via solid-state synthesis at low temperature based on the repeated plastic working is described in particular.

(1) Creation of nanostructured compact by repeated plastic working process

Mechanical alloying [1] (MA) is one of the conventionally effective processes to easily prepare metallic materials with nanostructure of amorphous phases in employing powder particles. It is also applied to the mass production process by using the large-scale equipment. The important point, however, is to prepare the bulky compacts consisting of MAed powder with refined structures. In general, it is difficult to consolidate MAed powder in powder metallurgy (PM) process because of the low flowability caused by their segregation or the large friction between very fine powder particles with micron or submicron order. Furthermore, the large specific surface area causes some pollution on the particle surface, and the poor compactability of MAed powders induced with large strain by severe plastic work hardening makes it difficult to consolidate them. The advanced process, repeated plastic working (RPW) [2], to prepare the bulky compact with nanostructure has been developed as schematically illustrated in Fig. 4.4.27. First of all, raw material powders are input into the die installed in the screw-driven high-speed press machine. After consolidating them by Punch I, the green compact with a relative density of 80~90% is obtained. Without knocking it out from the die, the backward extrusion is carried out by inserting the Punch II into the compact at high speed with 150~200 mm/s to serve a severe plastic deformation. The large strain is induced into the powder compact, and the refined microstructure is formed. Re-compression of the compact after backward extrusion by Punch I is effective to the uniform mixture. By repeating the above plastic working on the raw powder particles in the die with a suitable number of cycles, the columnar bulky green compact with nanostructure as shown in Fig. 4.4.27 is directly created and consolidated under a thermal history by sintering, hot extrusion, and hot forging processes.

When employing magnesium (Mg) and silicon (Si) powder mixture, the RPW process possibly supplies the Mg–Si mixture compact with nanostructure. Pure Mg powder with a mean particle size of 111.5 μm and Si powder with that of 22.3 μm are prepared as input raw materials. The mixtures of the chemical compositions with Mg:Si = 2:1 (at%)

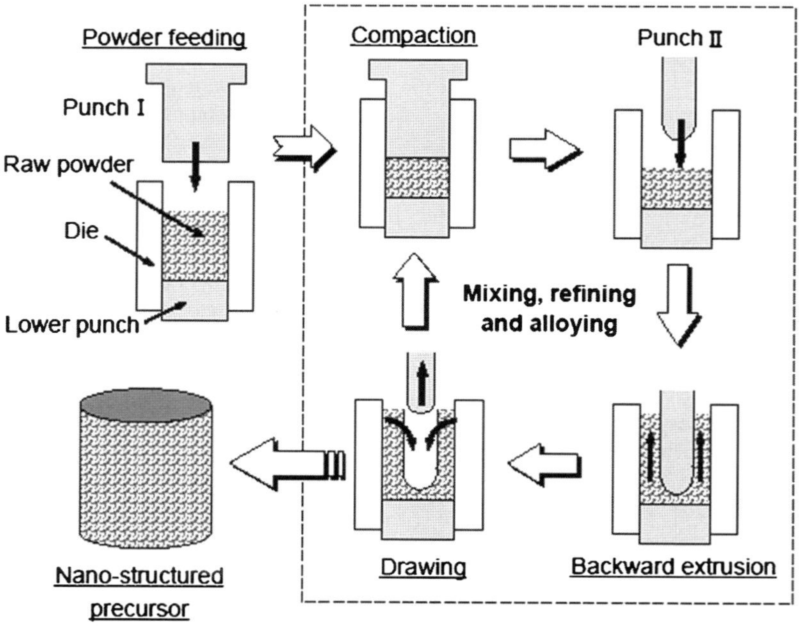

Figure 4.4.27
Schematic illustration of repeated plastic working process.

are supplied into the die with a inner diameter of 35 mm installed in the 1,000 kN screw-driven press machine. After consolidating them by applying 600 MPa, Punch II with an outer diameter of 11 mm is inserted into the compact for the backward extrusion. The maximum number of cycles is 600 in this study. X-ray diffraction (XRD) analysis and transmission electron microscope (TEM) observation on the green compact after RPW process is carried out to evaluate the microstructure. As shown in Fig. 4.4.28, the peak intensities of Mg and Si decrease, and Mg_2Si peak becomes larger with increasing the number of RPW cycles. This is similar to the mechanically alloying process, that is, Mg elements are super-saturated into Si, and Mg_2Si is synthesized by the solid-state reaction between Mg and Si particles by the severe plastic working [3]. Figure 4.4.29 reveals TEM observation results of the green compact after 300 (a) and 600 (b) cycles. In the former, both fine Si particles with about 500 nm and 10~30 nm Mg_2Si particles exist. The specimen of 600 cycles includes the refined Si particles and coarser Mg_2Si particles [4]. Furthermore, the oxygen and carbon contents of the compact after RPW process is almost same as those of raw powder, that is, this process obstructs the increase of impurities. In conclusion, the RPW process is effective to prepare the precursor bulky compacts with nanostructure in employing raw powder particles as the input materials.

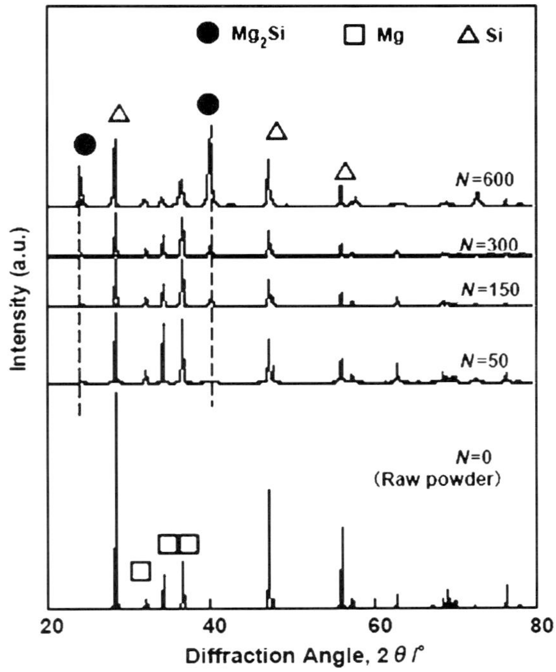

Figure 4.4.28
XRD patterns on Mg-Si pre-mixed powder precursors via RPW process.

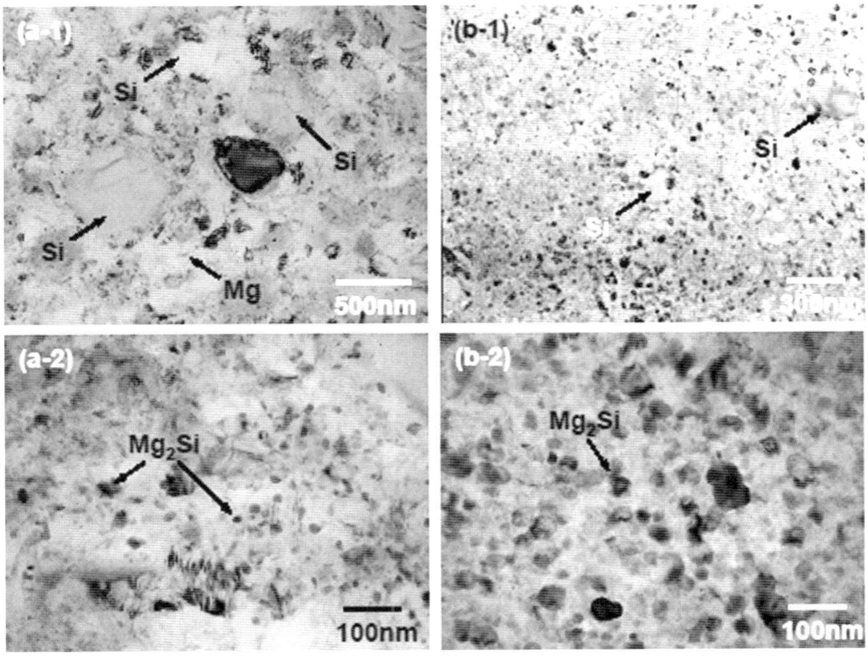

Figure 4.4.29
TEM observation results on Mg-Si precursors via RPW process with (a) $N = 300$ and (b) $N = 600$ cycles.

(2) Nanostructure control of alloy by low temperature synthesis

On the other hand, a suitable thermal history is necessary to produce the bulky Mg_2Si materials via the complete reaction between Mg and Si. Differential thermal analysis (DTA) evaluation is carried out to optimize the heating temperature of RPWed Mg-Si compacts because Mg-Si reaction is accompanied with an exothermic heat of 89.2 ± 11.4 kJ/mol [5]. As shown in Fig. 4.4.30, the originally raw Mg-Si mixture without RPW ($N = 0$) reveals the remarkably exothermic heat at 873 K and a small endothermic at 923 K due to melting of the Mg elements. With increasing the number of RPW cycles, the starting temperature of the exothermic reaction decreases. That is, the solid-state synthesis of Mg_2Si occurs at lower temperature; for example it starts at about 473 K in employing Mg-Si compact after $N = 100$ cycles. The dominant factors to accelerate the low temperature synthesis are considered as follows; (a) the formation of new-created active surface after breaking oxide films of raw powders, (b) the increased specific surface area of refined powder particles, and (c) the progress of Mg-Si solid-state diffusion due to the inert energy accumulated in powders during severe plastic deformation. Figure 4.4.31 shows a TEM observation result of Mg-Si compact via RPW with $N = 600$ cycles after heat treatment at 573 K at argon gas atmosphere. The Mg_2Si grain size is 50~100 nm. Coarsening and growth crystal grain are completely suppressed by heating lower temperature at 573 K. As the result, Mg_2Si sintered materials with nanostructures are prepared by the combination of RPW process and consolidation at lower temperature. This process is also useful to create nanostructured materials in employing Mg-Sn or Mg-Ge binary system, and AZ91 conventional magnesium alloy [6].

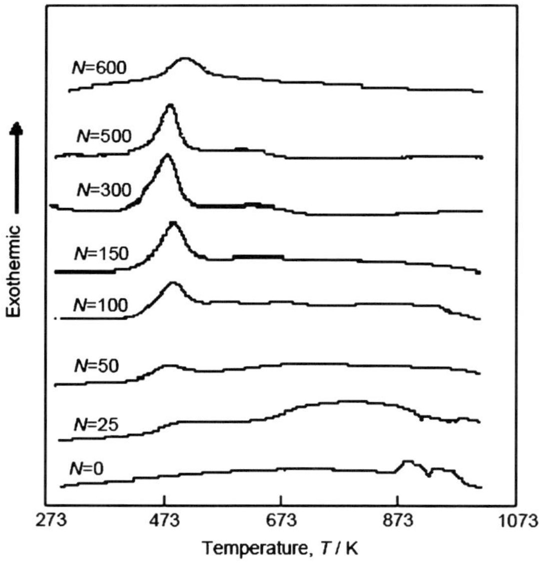

Figure 4.4.30
DTA curves of Mg-Si pre-mixed powder precursors after RPW.

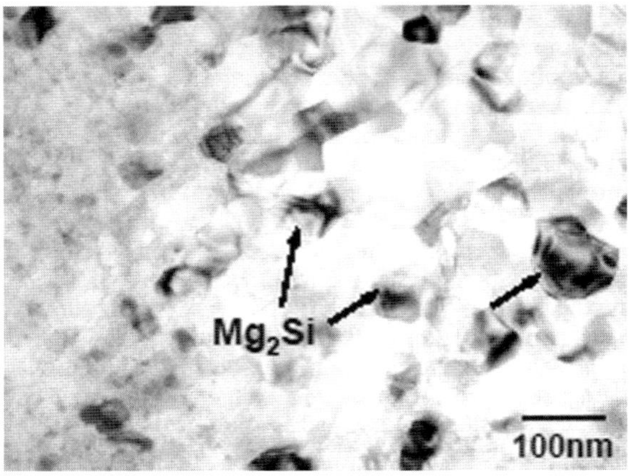

Figure 4.4.31
TEM observation on RPWed Mg-Si precursor after heat treatment ($N = 600$ cycles).

References

[1] R. M. German: *Powder Metallurgy Science* (2nd ed.), Metal Powder Industries Federation, Princeton, NJ (1994).
[2] K. Kondoh, T. Aizawa: *Mater. Trans.*, **44**(7), 1276–1283 (2003).
[3] K. Kondoh, H. Oginuma, E. Yuasa and T. Aizawa: *Mater. Trans.*, **42**(7), 1293–1300 (2001).
[4] K. Kondoh, H. Oginuma, A. Kimura, S. Matsukawa and T. Aizawa: *Mater. Trans.*, **44**(5), 981–985 (2001).
[5] R. Blachnik, D. Kunze and A. Schneider: *Metall. Trans.*, **25**(2), 119–121 (1971).
[6] K. Kondoh, T. Luangvaranunt and T. Aizawa: *Mater. Trans.*, **42**(7), 1254–1257 (2001).

4.5 Structure control of nanoparticle collectives by sintering and bonding

4.5.1 Sintering of nanoparticles

Sintering is the phenomenon in which a powder compact is transformed into a body with desired strength upon heating under the melting point of the main component through the formation of bonds between particles. Sintering is widely applied in industries such as ceramic and powder metallurgy where particles of submicron to micron levels are now used.

Nanoparticles may drastically reduce the sintering temperature as compared to the traditional powders. They also reduce the grain sizes in microstructure, and thus the properties of resultant products. However, a very high barrier is present before reaching these goals. Properly controlled microstructure is essential for excellent materials. Starting powder of excellent characteristics is only a single fulfillment among numerous requirements for the production of excellent materials. Nanoparticles have a difficult aspect other than its high potential. Their handling and treatment are extremely difficult. Improperly handled and treated, the properties of resultant product could be inferior to those made with conventional powders. In this section, fundamentals of sintering and microstructure development will be explained briefly [1, 2] and recent technology in sintering of nanoparticles will be discussed.

1. Microstructure

Two phenomena proceed simultaneously upon heating of powder compact as shown in Fig. 4.5.1. One is the formation of bonding between particles (neck formation) and the other the growth or annihilation of particles (Fig. 4.5.2). Particles in microstructure are largely single crystals. Grain boundary corresponds to the bonding region between them and the atomic arrangement is irregular there. The properties of grain boundary may differ significantly from those of crystal, and thus the net properties of polycrystalline material may differ completely from those of single crystal. Proper control of microstructure is often a key in designing properties of material.

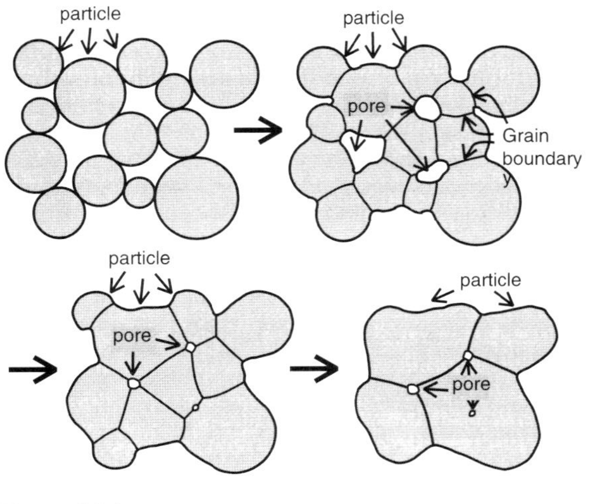

Figure 4.5.1
Formation of bonding between particles, grain growth, and microstructure development.

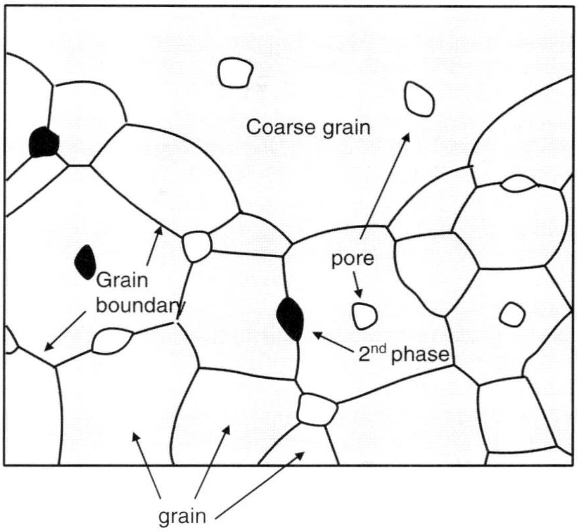

Figure 4.5.2
Schematic diagram of microstructure.

Desired features for microstructures depend widely on the intended application of material. Generally, high density, as well as fine and uniform microstructures are needed in the field of nanotechnology. Namely, the pores must be small and few in number. Size and shape of grains must be fine and uniform. The distribution of secondary phases must be uniform. Abnormally large grains, pores, and secondary phases are extremely undesirable, for they exert especially bad effects on properties. To fulfill these requirements in microstructure, precise control of sintering process is absolutely needed, in addition to the achievement of uniform green structure.

2. Sintering and mechanism of microstructure development

The driving force for sintering is the change of interfacial energy in the process. It is the difference between the total surface energy of particles in the powder compact and the sum of total grain boundary energy and surface energy of residual pores in the sintered body. The similar driving force works on grain growth; the reduction of grain boundary area and thus the total grain boundary energy. Tendency of grain growth is inevitable in sintering. The extreme goal of sintering is the single crystal without grain boundary or residual pore. The significance of the same driving force involved in densification and grain growth is that it is difficult to precede one of them preferentially. A special technique is needed to achieve densification with minimal grain growth. Controlled grain growth is used in the commercial production of ferrite single crystal. A seed crystal is used to convert polycrystalline feed ceramic to a large single crystal.

Figure 4.5.3 shows various mechanisms of solid-stage sintering and the route of mass transport. During sintering, substances move from the surfaces or grain boundaries of particles as the starting points to the neck regions formed between particles as terminal points, forming solid bonding. The starting point of substance governs shrinkage in sintering. If it is the grain boundary, net shrinkage occurs in powder compacts, since the centers of adjacent particles get closer as the substance is removed from the contact region between them. There is no shrinkage in powder compact, if the substance is removed from the surface of particles. The particles are thinned, but the centers of particles remain at the same position in this case. Explicit routes of mass transport include bulk diffusion in solid, grain boundary diffusion, surface diffusion on particle surface, and evaporation-diffusion in gas phase-condensation.

Liquid-phase sintering is widely used industrially and is the major sintering technique for commercial production of materials. It uses additives which form considerable amount of liquid phase between particles and/or at particle surface at the sintering temperature [3]. The liquid phase plays two roles. One is to pull particles together by the surface tension and the other the rapid mass transport. In general, mass transport is much faster in liquid than in solid. The bulk and/or film liquid phase dissolves substance, and allows rapid transport through them. Compared to the industrial success, the science of liquid-phase sintering has not been well-established. It involves too many complicated phenomena for detailed analysis.

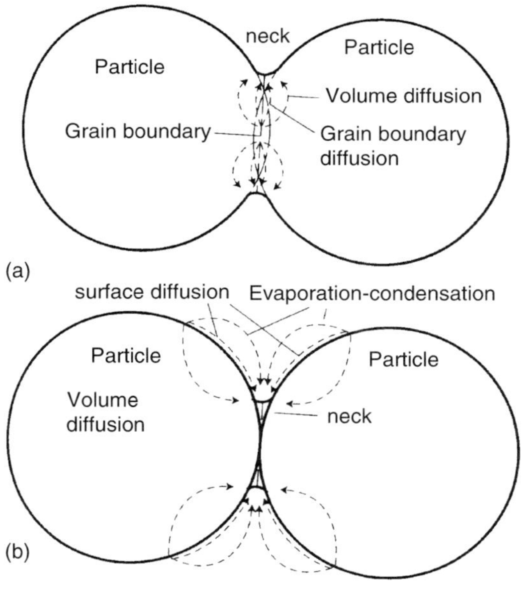

Figure 4.5.3
Solid-stage sintering and the route of mass transport
(a) Mass transport with contraction and (b) Mass transport without contraction.

Nanoparticles contribute to a very high sintering rate, due both to their high driving force and the short transport distance for substance. The driving force governed by the surface curvature of particle increases with decreasing particle size. The small particles in the system also reduce the distance between the starting and terminal points. In general, the sintering rate varies with the minus second to third powder of particle size.

Unfortunately, the grain growth rate also increases rapidly with decreasing particle size, i.e., with the minus second to third powder of particle size. Large grains are often noted after sintering of nanoparticles. It is difficult to form nanostructure after sintering. Unlike conventional technique, additive for suppressing grain growth are ineffective in most cases of nanosystems. The driving force for grain growth is too strong to suppress it by additives.

3. Sintering and control of microstructure in nanosystem

Special technique is needed to obtain homogeneous microstructure with fine grains in the sintering of nanoparticle systems. Conventional sintering does not lead to a desired microstructure. Rapid grain growth accompanying densification often results in microstructures similar to those formed by conventional systems.

A novel two-step technique has been reported recently by I.W. Chen to sinter nanoparticles into ceramics with nanograins [4].

Figure 4.5.4 shows the relationship between the relative density and the grain size for the yttria nanoparticle system that was sintered by the conventional technique with the time–temperature pattern shown in the figure. The grain size increased with increasing density, reaching about 100 nm, and cannot be categorized as nanoparticle at the relative density 70%. The grain growth continues in the subsequent heating and densification. This is the similar results often noted in the conventional systems. Addition of niobium did not help the formation of nanoceramics, although it suppressed the grain growth slightly.

Figure 4.5.5 shows the two-step sintering, in which the powder compact is heated for a very short period of time at a temperature normally selected for conventional sintering (T_1), and then sintered for a long period of time at a lower temperature (T_2) with the temperature–time profile shown in the figure. Densification was accompanied by the little grain growth at 1,310°C and 1,250°C in the first step. In the second step, only the densification proceeded with the grain size kept virtually constant. Both high density and fine grain with size under 100 nm have been achieved with this technique.

Figure 4.5.6 shows the grain size after the sintering of the first stage which is required for the achievement of densification in the sintering of the second stage.

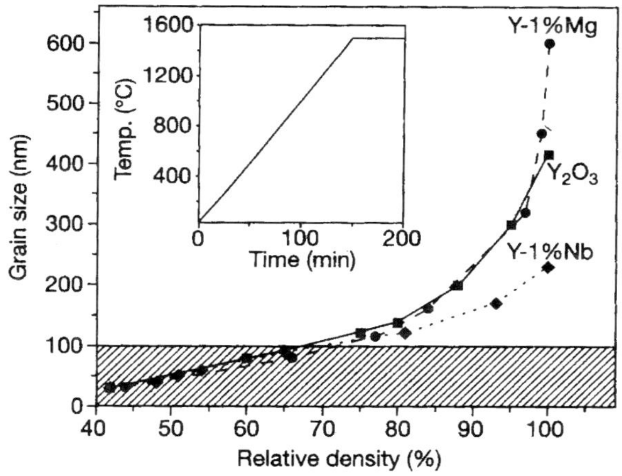

Figure 4.5.4
Relation between relative density and the grain size.

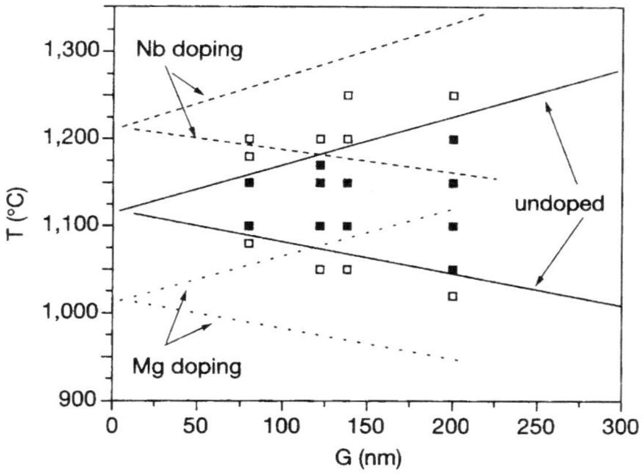

Figure 4.5.6
Relation between initial grain size and calcination temperature required for the achievement of densification in the sintering of the second stage.

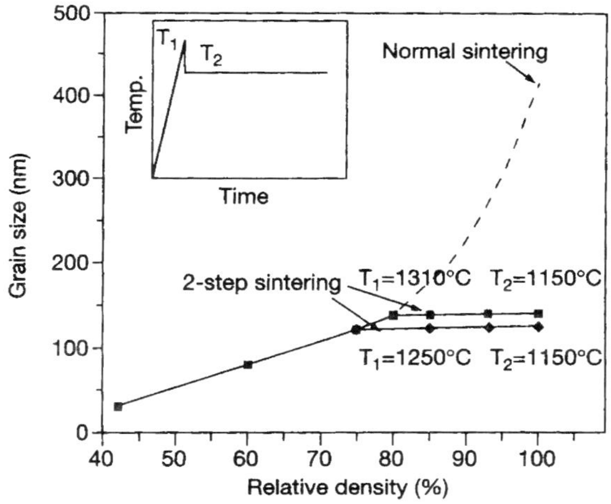

Figure 4.5.5
Relation between relative density and grain size with temperature control in two-stage sintering.

In the system without additives, high density with nanograins can be achieved when the sintered bodies with the grain sizes shown by the solid line are heated in the respective temperature ranges. High densification is not possible in the sintering of second stage for the sintered bodies with the density below the line. In contrast, significant grain growth occurs for the sintered bodies above the line, and ceramics with nanoparticles is not obtained. The conditions for successful densification, i.e., simultaneous achievement of high density and nanograins, become narrow as the particle size decreases. This two-step sintering technique is effective also for the system with additives. The necessary heating conditions are shown by the broken line and the dotted line for the systems with grain growth suppresser and accelerator, respectively.

The applicability of the present technique is noted in various systems, including alumina [5], ceria [6], and silicon carbide [7]. In the sintering of silicon carbide in particular, a drastic reduction of sintering temperature was possible in the two-step sintering technique compared to the conventional sintering method; 1,700–1,750°C for the first stage and 1,550–1,600°C for the second stage. Achievement of high relative density over 98% was suggested with this technique.

References

[1] JSPS124 Committee (eds): *Production and Application of Advanced Ceramics*, Nikkan Kogyo Sinbunsha, Tokyo, Japan (2005).
[2] Y. Moriyoshi, T. Sasamoto, K. Uematsu, Y. Ikuma, H. Monma, T. Ikegami and T. Maruyama: *Sintering of Ceramics*, Uchida Rokakuho, Tokyo, Japan (1995).
[3] R.M. German: *Liquid Phase Sintering*, Plenum, New York (1985).
[4] I.-W. Chen, X.-H. Wang: *Nature*, **404**, 168–171 (2000).
[5] A. Rosenflanz, M. Frey, B. Endres, T. Anderson, E. Richards and C. Schardt: *Nature*, **430**, 761–764 (2004).
[6] C. Kleinlogel, L.J. Gauckler: *Adv. Mater.*, **13**, 1081–1085 (2001).
[7] Y.-I. Lee, Y.-W. Kim and M. Mitomo: *J. Mater. Sci.*, **39**, 3801–3803 (2004).

4.5.2 Low temperature cofired ceramics (LTCC)

(1) Features of LTCC

Low temperature cofired ceramics (LTCC) is a ceramic material, including an inner conductive metal with low resistance and melting point, which can be sintered together simultaneously. Low-electroresisting metals, such as Ag, Au, and Cu, are mainly applied for the metal part of LTCC, while the cofiring temperature of these low-electroresisting metals, with a melting point of around 1,000°C, is set about 900°C below the melting point of the metal.

The major features required for the LTCC board are (1) thermal expansion coefficient, (2) low dielectricity, (3) low dielectric loss, and (4) low electroconductor loss during the cofiring with the low-electroresisting material. The production process for a multi-layer ceramic board widely using LTCC is shown in Fig. 4.5.7.

The LTCC was developed in the 1980s, targeting an increased number of layers of hybrid IC (HIC) board and applying the technology to handle the resistant materials accumulated with low-electroconductive materials and HIC. These have been widely used in industry to produce supercomputer mainframe boards and automobile module boards. Since the low-electroresistant material can be used as inner wiring, its potential for use in applications for light, high-density, high-frequency boards, and module parts with moving element functions has attracted considerable attention. A typical structure of the board is shown in Fig. 4.5.8.

(2) Classification of LTCC materials

LTCC materials are classified into three types; namely (1) "Glass composite system" made by mixing aggregates like Al_2O_3 and glass, (2) "Crystallized glass system" generated as crystals from the glass, and (3) "Non-glass system " firing only the starting material at a lower temperature without adding glass. The major materials previously developed and applied are shown in Table 4.5.1. The bending strength of most materials is about 200 MPa and their dielectric constant is less than 10. The average particle size of the glass and Al_2O_3 used for LTCC

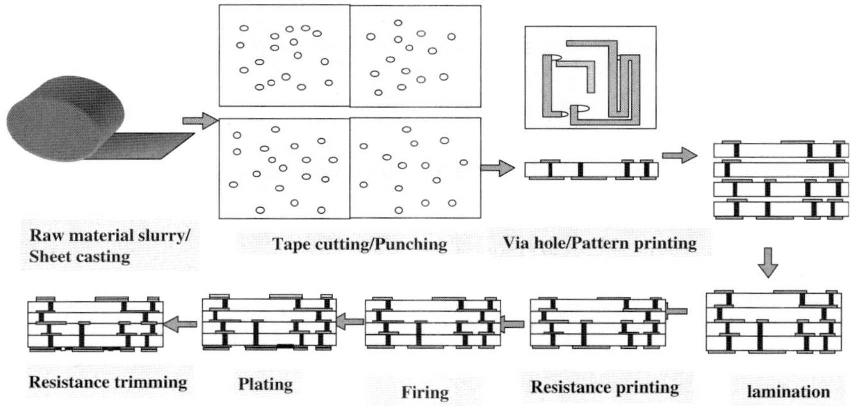

Figure 4.5.7
Process flow to manufacture the LTCC board.

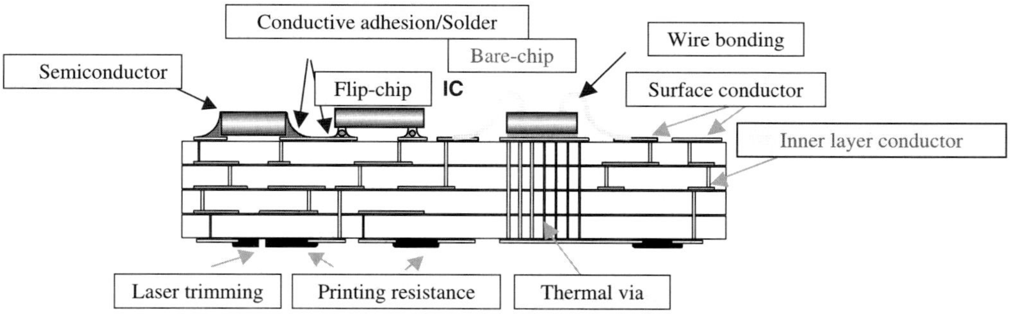

Figure 4.5.8
Example of a cross-sectional view of the LTCC board structure.

Table 4.5.1
Major materials for LTCC and their properties [1].

	Material system	Conductor[a]	Bending strength (MPa)	Thermal expansion coefficient (ppm/°C)	Thermal conductivity (W/m°C)	Dielectric constant at 1 MHz	Manufacturer
Glass composite system	Al_2O_3+ Forsterite+BSG	(3) (4)	235	5.9	4.2	7.4	Asahi Glass Co., Ltd.
	BSG+Al_2O_3	(2)	204	6.3	2.5	7.9	
	BSG+Al_2O_3+ SiO_2+Cordierite	(2)	190	4.0	2	5.0	Kyocera Corp.
	Crystallized Glass+Al_2O_3	(1)	210	4.2	3	6.2	
	Al_2O_3+ $CaZrO_3$+Glass	(1)(3)(4)	200	7.9	—	8.0	Dupont
	PbO-BSG+ Al_2O_3	(3)	300	4.2	4.2	7.8	NEC Corp.
	BSG+ Cordierite+SiO_2	(2)	140	3.5	3.5	4.7	
	BSG+Cordierite+SiO_2	(1)	85	3.2	—	3.4	
	Al_2O_3+Forsterite+Glass	(2)(3)	140	7.6	8.4	7.4	Noritake Co., Ltd.
	BaO-Al_2O_3- BSG+ Al_2O_3+$ZrSiO_2$	(1)(3)	200	5.5	1.7	7.0	Hitachi, Ltd.
	Al_2O_3+BSG	(2)	200	4.0	4	4.9	Fujitsu Limited
	PbO-BSG+ Al_2O_3	(2)	260	6.0	3	7.4	Matsushita Electric Industrial Co., Ltd.
	CaO-Al_2O_3-SiO_2- B_2O_3Glass+ Al_2O_3	(1)(2)(3)	200	5.5	2.5	7.7	Murata Manufacturing Co., Ltd.
	Li_2O-MgO-ZnO- B_2O_3-SiO_2+$MgAl_2O_4$	(2)	250	10.6	5.6	7.4	
	Cordierite	(2)	210	3.0	—	5.0	IBM Corp.
Crystallized glass system	ZnO-MgO-Al_2O_3-SiO_2	(4)	200	3.0	3	4.9~5.0	NGK Spark Plug Co., Ltd.
Non-glass system	Al_2O_3-CaO-SiO_2- MgO-B_2O_3	(2)	250	4.8	8.4	7	Taiyo Yuden Co., Ltd.
	$BaSnB_2O_6$	(3)(4)	200	5.4	5.4	8.5	Toshiba Corp.
	BaO-Al_2O_3-SiO_2	(2)	200	8.0	8.0	6.1	Murata Manufacturing Co., Ltd.

Note: BSG, boron silicate glass.
[a]Conductive metal: (1) Ag, (2) Cu, (3) Ag-Pd, (4) Au.

generally ranges from submicron to a few microns and no report using nanoparticles has been issued to date.

There is a unique LTCC of CaO-Al_2O_3-SiO_2-B_2O_3-system glass ceramics, which starts from the "Glass composite system" after the above classification of a mixture of aggregates like Al_2O_3 and glass as raw materials. However, this shows the properties of a "Crystallized glass system" generating crystals from the glass during the firing.

As shown in Fig. 4.5.9, the anorthite ($CaAl_2Si_2O_8$) comes out as crystals, via the reaction of the raw materials of CaO-Al_2O_3-SiO_2-B_2O_3 glass and Al_2O_3 with the firing time. The fired board consists of the three components of Al_2O_3, glass as aggregates and anorthite. The state of the interface between the glass and Al_2O_3 particles in the fine region is considered to influence the sintering behavior and microstructure of the material during the firing process.

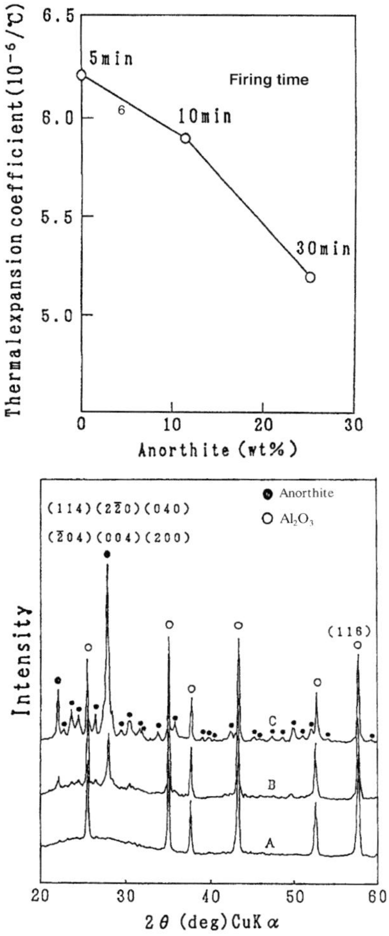

Figure 4.5.9
Powder X-ray diffraction analysis of Al_2O_3 and the glass system LTCC. (Sintered at 890°C, sintering time: A; 5 min, B; 10 min, C; 30 min) [2].

(3) Control of the microstructure

The major important factors influencing the sintering behavior and microstructure are the particles, forming and firing, respectively. The influencing factors of the particles are particle size, size distribution, surface condition, synthetic method, and agglomeration, etc. As for the forming, the density, microstructure, and wet or dry conditions will affect them. In particular, in order to apply LTCC for the layer structure of boards and module parts, sheet forming is generally used, where the properties of powder and slurry, after mixing with a binder, are closely related to the forming performance. Namely, the density of the wet sheet, particle shape, size distribution, and the microstructure of the sheet are important for the sintering behavior and post-firing microstructure.

The sintering is influenced by the temperature, processing time, environmental conditions, and presence or absence of restraint. Recently, in order to obtain a printed board integrated with a high-precision and high-density electric circuit, various kinds of constrained sintering methods have been proposed. These cause no contraction in the x-y plane but only that in the normal direction of the sheet during firing, via the partial layering of a ceramic sheet which does not sinter at the sintering temperature of LTCC.

Other methods may also be used, like firing under pressure. These are required as a necessary technology to achieve high-accuracy and high-density packaging.

As an example, the effect of the component ratio of anorthite and Al_2O_3 on board properties, like the thermal expansion coefficient, will be introduced here.

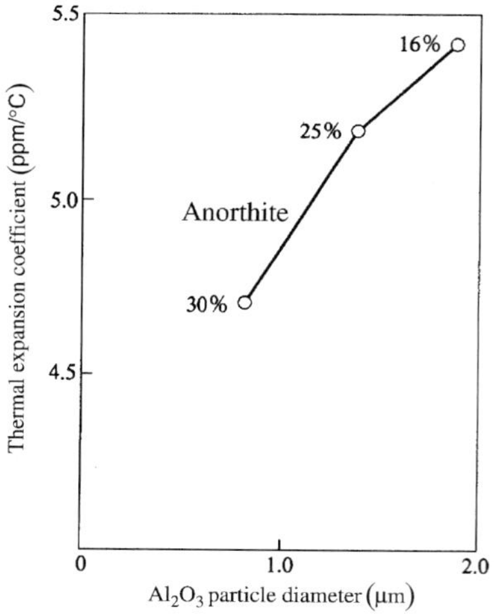

Figure 4.5.10
Factors affecting the thermal expansion coefficient [2].

As shown in Fig. 4.5.10, the amount of post-anorthite firing increases with the decreasing Al_2O_3 particle size, leading to a corresponding drop in the thermal expansion coefficient. The increase of anorthite due to the prolonged firing time also causes its reduction.

The anorthite is considered to be a product of the reaction between Al_2O_3 and glass and the microstructure is reportedly very important in obtaining the required board strength. Figure 4.5.11 shows the bending strength of the sample piece with a different mass fraction of anorthite by firing Al_2O_3 and glass at 900°C. Data marked with D in the figure were obtained by initially adding the anorthite and the others, marked with A, B, and C, by controlling Al_2O_3 particles to change the anorthite amount.

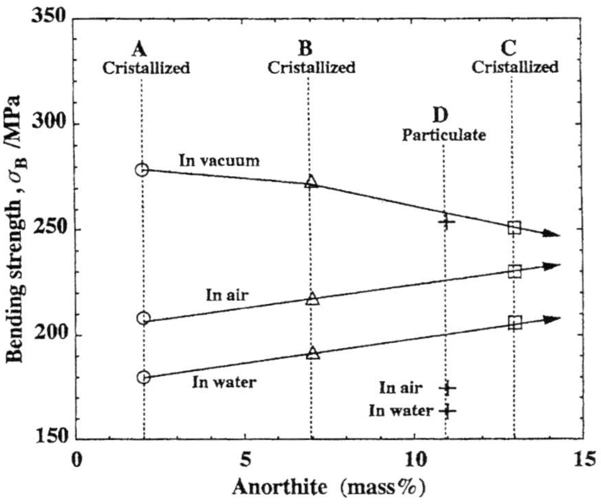

Figure 4.5.11
Relationship between the anorthite amount and bending strength (A, B: Pb-Al_2O_3-SiO_2-B_2O_3 glass, C, D: CaO-Al_2O_3-SiO_2-B_2O_3 glass) [3].

It is presumed that strength is influenced, not only by the anorthite amount but also the difference in the structure shown in Fig. 4.5.12. As seen in Fig. 4.5.11, there is a tendency for the bending strength in air and water to increase, with the increasing anorthite amount leading to an improvement in the depression of stress corrosion. However, under the condition of D, the strength decreases despite the rise in the anorthite amount, which suggests the considerable influence of many initial defects. As well as the anorthite amount, the microstructure with the anorthite between the Al_2O_3 particles and the glass is considered to contribute toward the improvement of the strength by the crack, as shown in Fig. 4.5.12.

Figure 4.5.11 indicates that the bending strength in a vacuum is higher than that in air and reduces with the increasing anorthite amount. It is suggested that the strength in vacuum is not affected by the stress corrosion but simply by the hindered densification due to the crystallization.

(4) Structural control in the submicron to nanosized region

Evaluation of the microstructure is conducted using an optical microscope, scanning electron microscope (SEM), transmission electron microscope (TEM), scanning probe microscope (SPM), and so on. These pieces of equipment provide information on the particle size, sintered particle shape, necking, surface state, and the generated phase. Figure 4.5.13 shows an SEM picture of the microstructure of LTCC, which is a composite material, with the anorthite generated from the Al_2O_3 particles and the glass.

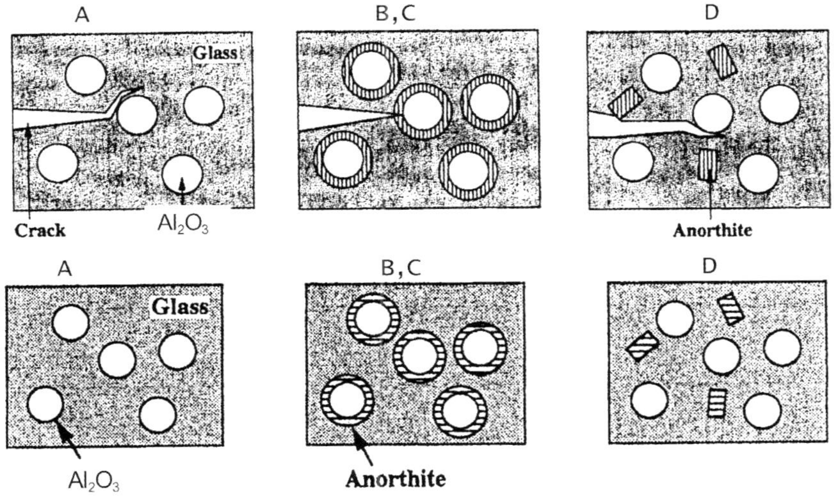

Figure 4.5.12
Schematic diagram of microstructure [3].

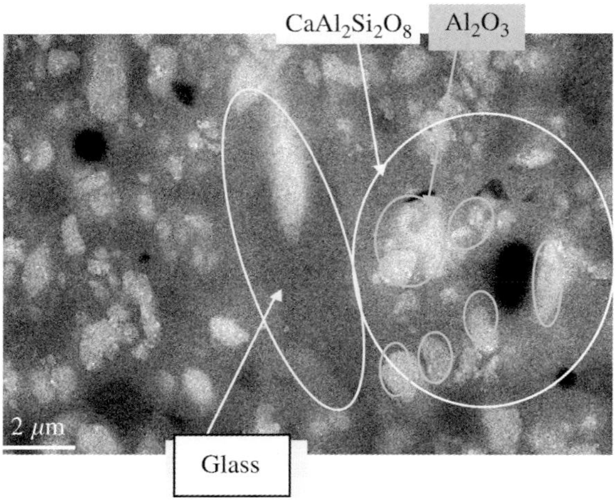

Figure 4.5.13
Microstructure of LTCC of the Al_2O_3-glass-anorthite system.

The existence of an anorthite layer of thickness 1–2 μm around the Al_2O_3 particles, as schematically shown in Fig. 4.5.12, can be confirmed by observation of the microstructure of the sintered material of Al_2O_3 particles and glass. Energy dispersive X-ray analysis (EDX) reveals that the white particles, dark gray parts, and light gray layer around the particles are regarded as Al_2O_3 particles, glass, and the anorthite, respectively. Presumably the anorthite crystal emerges after the reaction of the submicron Al_2O_3 particles with the glass. Subsequently the lack of difference in the reactive sintering performance within the nanosized region is considered to have a significant influence on the structure of the final board and sintering control in a further finer region is required in future.

As mentioned above, the LTCC of the glass phase system shows different properties, depending upon the microstructure of ceramics and glass, like an aggregate or generated crystal, even if the composition is the same. To create a dense structure, for which the sintering in glass liquid phase plays a significant role, the elucidation of the crystallization mechanism and the control of the microstructure under a firing condition suitable for the electrode materials are needed, as well as the densification. Above all, in order to acquire excellent electric properties, the microstructure and crystal need to be controlled more precisely, which requires technical advancement in the control of powder properties from submicron to nanosized range and the firing process in future.

References

[1] Electronic Materials Manufacturers Association of Japan: *Denshikairoyou koukinou seramikku kiban (in Japanese, Advanced Ceramic Board for Electric Circuit)*, Japan Industrial Publishing Co., Ltd., (1994).

[2] A. Inoue, J. Fukuta, Y. Matano and Y. Matsumoto: *J. Ceram. Soc. Jpn.*, **100**(2), 208–210 (1992).

[3] K. Kageyama, M. Enoki, T. Kishi, K. Ikuina and M. Kimura: *J. Jpn. Inst. Metals* (in Japanese), **57**(7), 756–760 (1993).

4.5.3 Nanostructure control of a joined interface

A joined interface plays an important role to reveal high functionalities of materials including composite materials and electronic devices.

It is a very important issue to clarify mechanism of revealing of high functionalities and to control high functionalities at the joined interface. Remarkable progresses of measurement, evaluation, and analytical technology at atomistic level have been performed by progresses of quantum beam technologies such as an electron beam, a laser beam, and a synchrotron radiation, a nanoscale analytical technology and a simulation method by a first principle molecular dynamics.

Economos et al. reported more than 40 years ago that 4 types of joined interfaces between oxides and metals, which are (1) a new phase is formed at the interface, (2) a ceramics is chemically attacked by metal, (3) a metal penetrates along with grain boundaries of ceramics and (4) a reaction zone is not observed at the interface, are observed by an optical microscope [1]. But, progress of analytical and characterization technologies of interfaces revealed that joined interfaces cannot be classified into simple categories and formation of interface is affected by many factors such as a joining time, an atmosphere, and a temperature. This has already been suggested by the study of wettability of Al_2O_3 by liquid Ni [2].

Many researchers have conducted direct observations and simulations of joined interface between metal and ceramics. Especially, the interface structure between Nb and Al_2O_3 have been studied as a model system, since thermal expansion coefficient of Nb is similar to that of Al_2O_3.The interface have been systematically investigated by using Al_2O_3 single crystal which has different crystal orientation and by using samples made by a solid phase joining, an internal oxidation and a molecular beam epitaxial (MBE) methods [2–4]. It has been revealed that Nb and Al_2O_3 is directly joined without a formation of reaction layer and interface structures depend on a crystal orientation of Al_2O_3 and joining method [5]. An information of bonding status between Nb and Al_2O_3 has also been obtained by X-ray photo-emission spectroscopy (XPS) [6, 7]. Electrons in Nb atoms migrate to oxygen atoms that are the components of Al_2O_3. The results obtained by XPS are in good agreement with simulation result by Yamamoto et al. [8, 9].

When joined interfaces between dissimilar materials such as metal–ceramics systems are in an epitaxial relationship, one of the important factors which affect interface characteristics is a lattice mismatching. An interface structure is considerably affected by

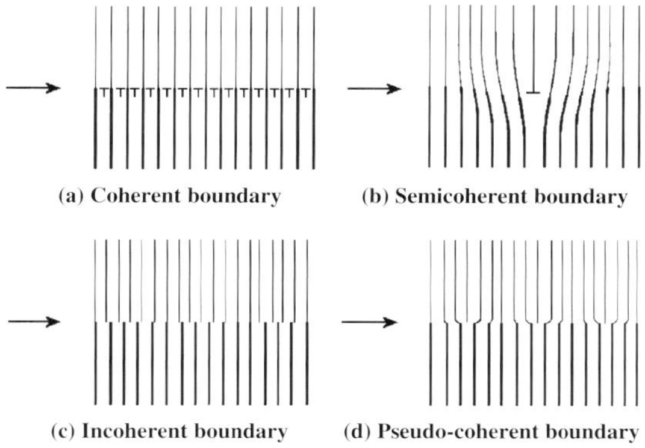

Figure 4.5.14
Types of dissimilar materials interfaces.

a degree of lattice mismatching. In general, it is called a large mismatching when the mismatching is larger than 4~5% and a small one when the mismatching is smaller than 4~5%. The Nb–Al_2O_3 system belongs to the latter one.

A dislocation, which is one of lattice defects, is introduced to relieve a lattice mismatching at the interface. When the lattice mismatching between metal and ceramics is small, a coherent boundary (see Fig. 4.5.14(a)) is formed by a uniform strain of crystal itself [10]. A condition to form such a coherent boundary depends on a degree of mismatching and a thickness of crystal. When the thickness of crystal exceeds a critical one, the dislocations (see Fig. 4.5.14(b)) are introduced to relieve an increased strain and a semi-coherent boundary is formed which is shown in Fig. 4.5.14(b). As shown in Fig. 4.5.14(c), an incoherent boundary is formed when an interaction does not exist at the interface. This state represents that two kinds of crystal contact only at the interface. When two crystals are joined at the interface, atomic interaction takes place at the interface and atomic displacements occur at near interfaces shown in Fig. 4.5.14(d). Accordingly, the coherent region takes the structure that is divided by geometric miss-fit dislocation. The miss-fit dislocation is formed in the crystal whose Young's modulus is smaller than the others'.

An oxide ceramics is superior in a thermodynamical stability to a carbide and a nitride ceramics in general. Accordingly, a steep interface is formed in metal/oxide ceramics systems as shown in Fig. 4.5.14. On the other hand, thick interface layers are formed between the carbide and the nitride ceramics and metals which are even non-reactive ones. As an example, result of Cu/SiC interface is shown in Fig. 4.5.15 [11]. It can be seen that about 100 μm of a reaction layer is observed at the interface between Cu and SiC. The interface layer is composed of metallic (Cu) layer in which carbon is not found and a mixed layer of carbon

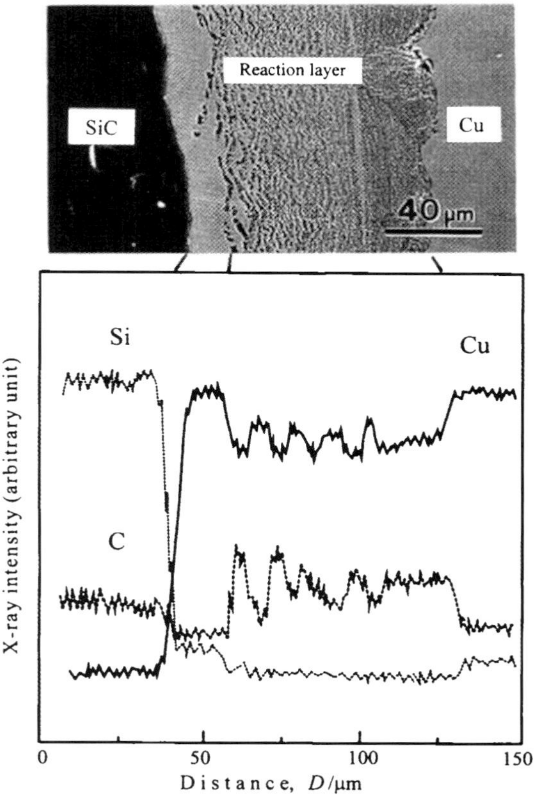

Figure 4.5.15
SEM observation at the interface between α-SiC and Cu.

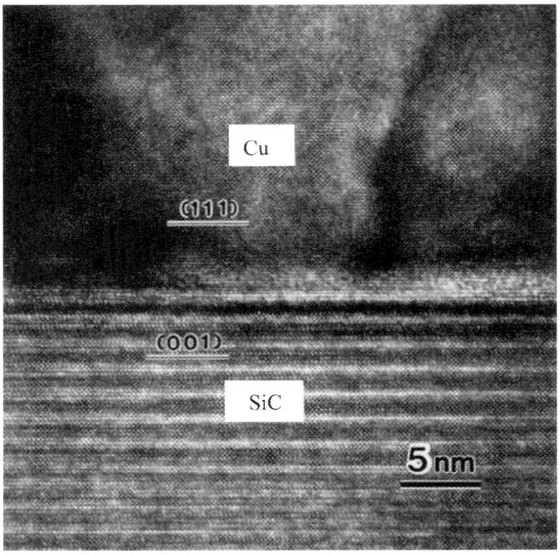

Figure 4.5.16
HREM observation at the interface between α-SiC and Cu.

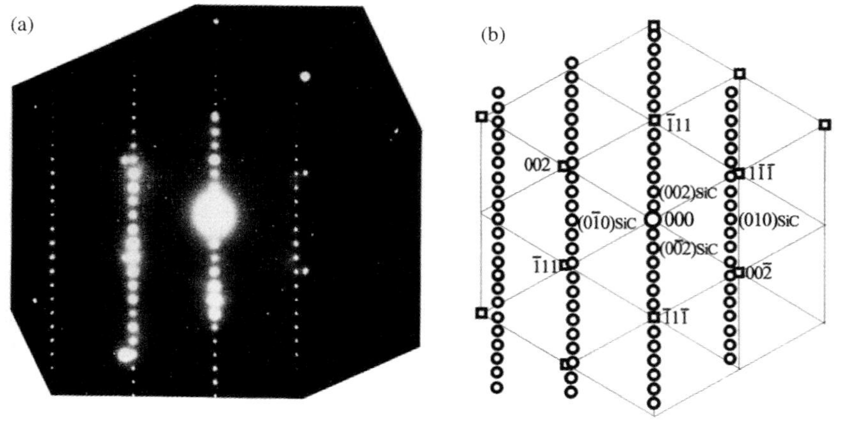

Figure 4.5.17
Electron diffraction at the interface between Cu and α-SiC.

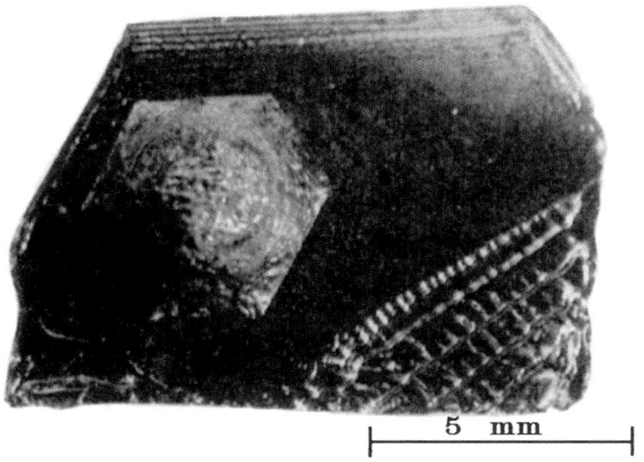

Figure 4.5.18
Appearance of solidified Cu drop on α-SiC.

and metallic (Cu) layer. Figures 4.5.16 and 4.5.17 are a result observed by high resolution electron microscope (HREM) and an electron diffraction pattern at the interface between Cu and SiC. It can be concluded that the interface between Cu and SiC is formed by (001) plane of SiC and (111) plane of Cu which is the most closed packed plane and a preferred orientation is {100}SiC//{111}Cu and <100>SiC//<110>Cu at the interface. This relationship depends on a thin oxide film on SiC surface, an atmosphere and a temperature. Especially, the result shown in Fig. 4.5.16 is not obtained when oxide film exists at the SiC surface. It is also well-known that a joined interface structure is significantly affected by a sintering agent when a ceramics is a sintered body [12]. In other words, it is suggested that the joined interface structure can be controlled by changing a kind of sintering agent, quantity, a temperature, and an atmosphere.

It has also been found in some combinations of metals and ceramics that nanostructure reflects macro- and micro-structure at the joined interface. As an example, a macrostructure of Cu/SiC system is shown in Fig. 4.5.18 [13]. The macrostructure of interface is reflected by the nanostructure of α-SiC and shows a hexagonal shape. This phenomenon has also been observed in In/Ge system [14].

References

[1] G. Economos, W.D. Kingery: *J. Am. Ceram. Soc.*, **36**, 403–409 (1963).

[2] K. Ogino, H. Taimatsu: *Trans. JIM*, **46**, 871–876 (1982).

[3] W. Mader, M. Ruhle: *Acta Metall.*, **37**, 853–866 (1989).

[4] J. Mayer, C.P. Flynn and M. Ruhle: *Ultramicroscopy*, **33**, 51–61 (1990).

[5] Y. Ishida, J. Wang and T. Suga: *ISIJ Int.*, **30**, 1041–1045 (1990).

[6] Y. Ikuhara: *Materia*, **34** (1995), 751–756.

[7] F.S. Ohuchi: *J. Mater. Sci. Lett.*, **8**, 1427–1429 (1989).

[8] M. Kohyama, R. Yamamoto: *Trans. JIM*, **29**, 893–901 (1990).

[9] R. Yamamoto, M. Kohyama, Y. Ebata and M. Kinoshita: *Proc. MRS Int. Meet. on Adv. Mater. (Mat. Res. Soc.)*, **8**, 183–188 (1989).

[10] G.B. Olson, M. Cohen: *Acta Metall.*, **27** (1979), 1907.

[11] K. Suganuma, K. Nogi: *Trans. JIM*, **59** (1995), 1292–1298.

[12] M.E. Brito, Y. Hirotsu: *Trans. JIM*, **29** (1990), 910–917.

[13] K. Nogi, K. Ogino: *Trans. JIM*, **29** (1988), 805–811.

[14] A.A. Berg: *J. Electrochem. Soc.*, **110**, 908–914 (1963).

4.5.4 Joining by FSW

4.5.4.1 Friction stir welding

Friction stir welding (FSW) has already entered the practical use stage in various industrial fields such as railroad vehicles, vessels, structures, and cars [1–5], after it was developed by TWI (British Welding Institute) in 1991 [6, 7]. Research and development in this area has significantly increased in every country around the world. In this joining method, as shown in Fig. 4.5.19(a), the columnar rotating tool at a high speed contacts the materials to produce frictional heat, and the materials can then be joined using the frictional heat. As shown in Fig. 4.5.19(b), the tool consists of a large shoulder part and a probe part at the tip.

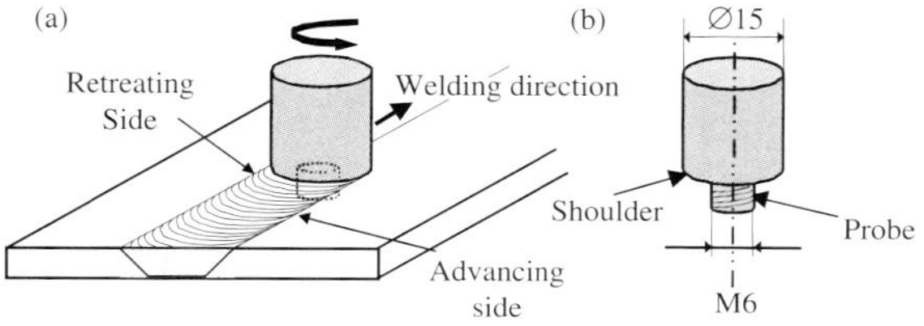

Figure 4.5.19
Friction stir welding and tool (example for 5-mm thick plate).

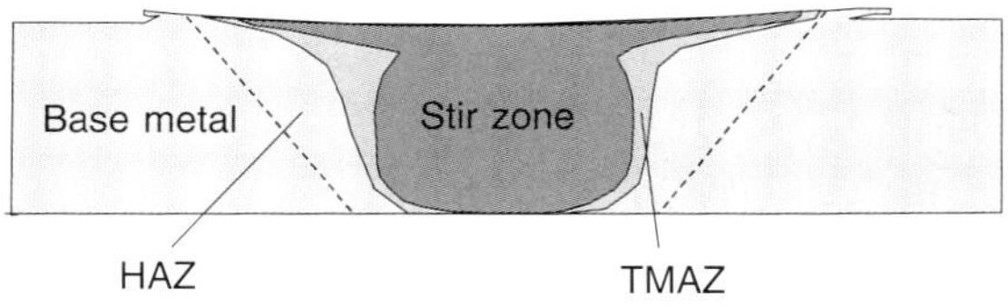

Figure 4.5.20
Schematic illustration of microstructure.

Only the probe is forced into the material during welding, and the tool is moved along the interface. The material is restricted to the back plate. The joining is performed by plastic flow caused by the tool in the solid state. Because the peak temperature does not reach the melting point and the material is in the solid state at this time, the grains are refined, the decrease in the strength by welding is much lower than that of fusion welding, and, in some cases, the strength of the joints is higher than that of the base metal. The details of the tool are to be referred to a textbook on FSW [8].

A schematic diagram of the cross-sectional microstructure of the welded part is shown in Fig. 4.5.20. The center, called the stir zone, consists of a recrystallized microstructure which consists of equiaxial grains of hundreds of nanometers to several micrometers in diameters. Figure 4.5.21 [9] shows TEM images of the stir zones of pure aluminum and pure Fe (IF steel). Outside the stir zone, there is the thermo-mechanically affected zone (TMAZ), in which the crystal grain was extended by plastic modification. Outside the TMAZ, the heat affected zone (HAZ) exists, which is affected by the heat, but not affected by the plastic deformation. As a reference, for the normal fusion welding, the grains grow from several tens of microns to several hundreds of microns.

According to Frigaard et al. [10], the heat input during FSW, $Q(W)$, can be expressed as follows:

$$Q = \frac{4}{3}\pi^2 \mu PNR^3 \tag{4.5.1}$$

where μ is the friction coefficient, P the pressure (N/m^3), N the tool rotation speed (/s), and R the shoulder diameter (m). For this equation, it is assumed that the heat is generated only between the shoulder of the tool and the material. P is the load per unit area of the shoulder part. Equation (4.5.1) can be used when the tool is not moved. When the tool is

Figure 4.5.21
Recrystallized microstructure in stir zone of pure Al and IF steel.

moved, the revolution pitch, which is the welding speed V divided by the rotation speed N, can be an index for the heat input.

$$\text{Rotation pitch} = V/N \qquad (4.5.2)$$

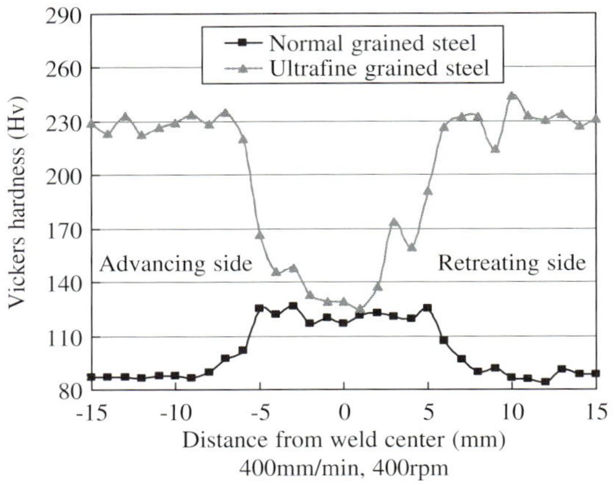

Figure 4.5.22
Hardness distribution on cross-section of FSWed IF steel.

The revolution pitch is the distance over which the tool moves while the tool rotates one time, and is equivalent to the interval of the lines formed on the surface of the welded part after FSW. A higher revolution pitch, namely, greater movement during one rotation, indicates a lower heat input. In FSW, the quantity of the heat input is one of the important parameters, which determines the success or failure of the welding or friction stir process [8].

4.5.4.2 Grain refinement and FSW of ultrafine grained steel

As mentioned above, since FSW can refine the grains or restrict the grain growth, it should be a good candidate for joining ultrafine grained materials. Figure 4.5.22 shows a horizontal hardness distribution in a cross-section perpendicular to the welding direction of FSWed IF steel (ultralow carbon steel), which is used for the steel plate in cars, etc. For steel with a normal grain diameter, the hardness of the stir zone, i.e., the center, increases over the hardness of the surrounding mother material. This is because the grains are refined due to the stirring by the tool, as shown in Fig. 4.5.23(a and d).

Figure 4.5.22 also shows the hardness distribution of an ultrafine grained material. The ultrafine grained

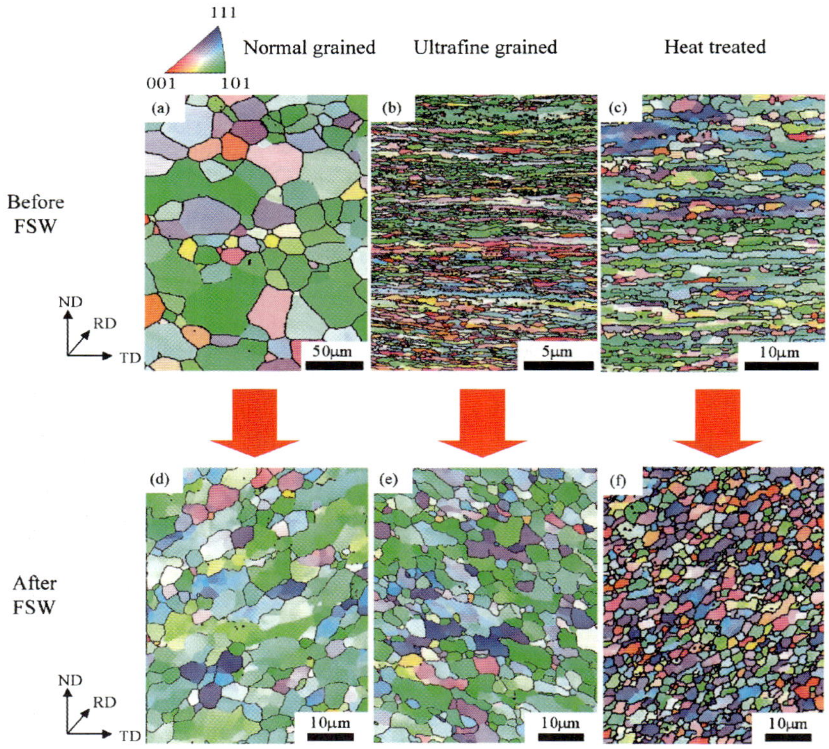

Figure 4.5.23
EBSPmapping of stir zone in FSWed IF steel.

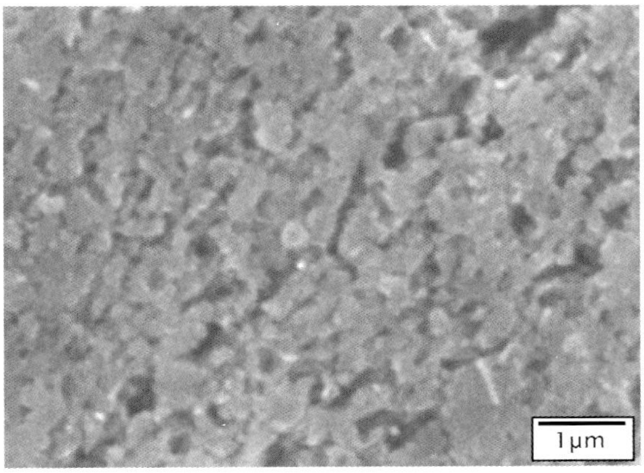

Figure 4.5.24
Carbon nanotube distributed AZ31 by FSP.

IF steel, whose grain size is 0.65 μm, was produced by accumulative roll bonding (ARB). If the crystal grain of the base material is too small for a pure material, such as IF steel, grains become relatively larger in the stir zone, and the hardness of the stir zone approaches that of the annealed material. For such a material, the grain growth of the stir zone can be prevented during FSW [11] by a heat treatment before FSW, to some extent, that makes the grains grow as shown in Fig. 4.5.23(c). In this case, the heat treatment conditions are 30 min. at 600°C. The grain sizes in the stir zone after FSW are not always in the same order of the grain sizes before FSW. The minimum grain size cannot be obtained after FSW when the grain size is too large or too small before the FSW. Thus, there is an optimal initial grain size for obtaining the final highest strength, indicating the requirement of a material design especially for FSW. The same result has also been obtained for an aluminum alloy.

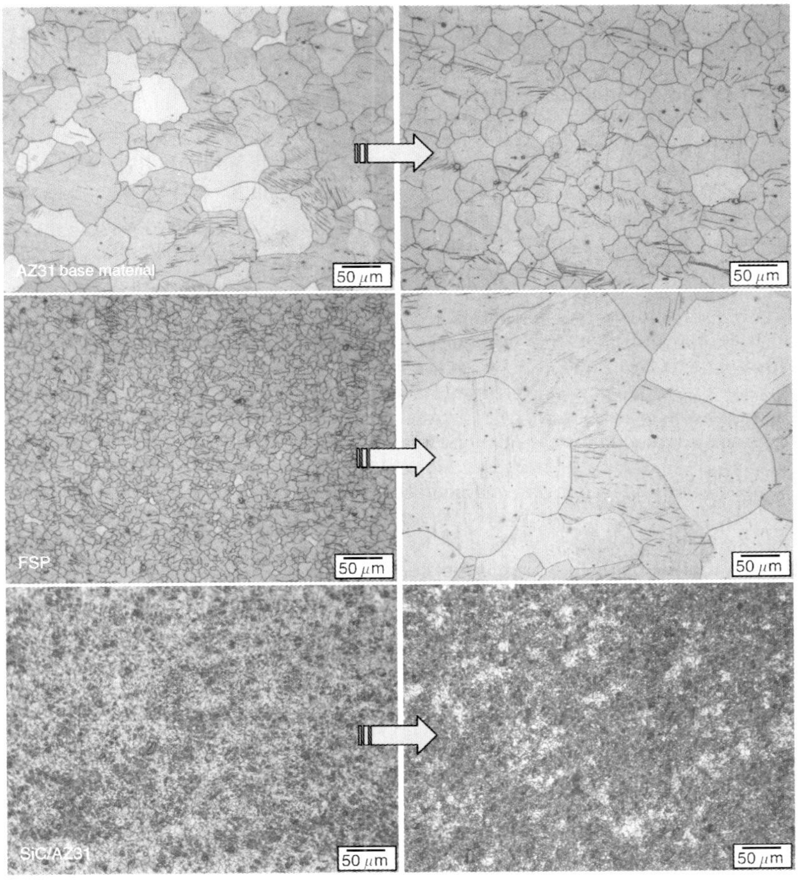

Figure 4.5.25
Thermal stability of SiC distributed AZ31by FSP.

4.5.4.3 FSP and formation of surface partial composite

Although FSW was initially a technology developed as a joining method of metals, a field called friction stir process (FSP) has spread, since it is possible to obtain a refined microstructure and to partially strengthen a material using this technique [12–17]. Furthermore, it has recently become clear that the grains are further refined by dispersing small particles such as ceramic particles [18, 19], carbon nanotubes [20], and C_{60} [21] into metal.

Morisada et al. [19–21] engraved the sample surface (AZ31) with a slot having a 1 mm width and 2 mm depth, filled it with carbon nanotubes, C_{60} or ceramic particles, and then performed FSP on the plate, as nanostructures, such as carbon nanotubes, cannot easily be dispersed in a metal by a liquid-phase method. Because the distribution of these particles can restrict the grain growth, the crystal grain is much smaller than that in the simply FSPed sample, as shown in Fig. 4.5.24, which can significantly improve the hardness from 41 Hv of the base material to 78 Hv. When only FSP is performed, the hardness increases only to 55 Hv.

Because the thermal stability of the FSPed part is lower due to the introduction of a distortion into the metal, the crystal grain can become larger than that of the base material due to remarkable grain growth when the temperature is raised after the FSP as shown in Fig. 4.5.25. However, when SiC particles are distributed, the grain growth is restricted and the thermal stability is significantly improved.

References

[1] H. Okamura, K. Aota and M. Ezumi: *J. Jpn. Inst. Light Met.*, **50**, 166–171 (2000).

[2] G. Campbell, T. Stotler: *Weld. J.*, **78**, 45–47 (1999).

[3] M.R. Johnsen: *Weld. J.*, **78**, 35–39 (1999).

[4] K.E. Knipstron, B. Pekkari: *Weld. J.*, **76**, 55–57 (1997).

[5] C.J. Dawes, W.M. Thomas: *Weld. J.*, **75**, 41–45 (1996).

[6] W.M. Thomas, E.D. Nicholas, J.C. Needhan, M.G. Murch, P. Temple-Smith and C.J. Dawes: *International Patent Application PCT/GB92/02203 and GB Patent Application 9125978.8*, UK Patent Office, London, December 6, 1991.

[7] C.J. Dawes, W.M. Thomas: *Weld. J.*, **53**, 41 (1996).

[8] Japan Welding Society: *Friction Stir Welding*, Sanpo Publication, Tokyo (2006).

[9] H. Fujii, R. Ueji, Y. Takada, H. Kitahara, N. Tsuji, K. Nakata and K. Nogi: *Mater. Trans.*, **47**, 239–242 (2006).

[10] O. Frigaard, O. Grong, B. Bjorneklett and O.T. Midling: Proceedings of the Ist International Symposium on FSW, Thousand Oaks, USA, 14–16 June, 1999, 8-2CD-ROM.

[11] H. Fujii, R. Ueji, Y. Takada, H. Kitahara, N. Tsuji, K. Nakata and K. Nogi: *Mater. Trans.*, **47**, 239–242 (2006).

[12] R.S. Mishra, M.W. Mahoney, S.X. Mcfadden, N.A. Mara and A.K. Mukherjee: *Scripta Mater.*, **42**, 163–168 (1999).

[13] H.J. Liu, H. Fujii and K. Nogi: *Mater. Sci. Tech.*, **20**, 399–402 (2004).

[14] Z.Y. Ma, R.S. Mishra and M.W. Mahoney: *Scripta Mater.*, **50**, 931–935 (2004).

[15] M.L. Santella, T. Engstron, D. Storjohann and T.Y. Pan: *Scripta Mater.*, **53**, 201–206 (2005).

[16] I. Charit, R.S. Mishra: *Acta Mater.*, **53**, 4211–4223 (2005).

[17] J.Q. Su, T.W. Nelson and C.J. Sterling: *Mater. Sci. Eng. A*, **405**, 277–286 (2005).

[18] R.S. Mishra, Z.Y. Ma and I. Charit: *Mater. Sci. Eng. A*, **341**, 307–310, (2003).

[19] Y. Morisada, H. Fujii, T. Nagaoka and M. Fukuzumi: *Mater. Sci. Eng. A.*, **433**, 50–54 (2006).

[20] Y. Morisada, H. Fujii, T. Nagaoka and M. Fukuzumi: *Mater. Sci. Eng. A.*, **419** 344–348 (2006).

[21] Y. Morisada, H. Fujii, T. Nagaoka and M. Fukuzumi: *Scripta*, **55**, 1067–1070 (2006).

4.5.5 Aerosol deposition method for nanostructuring of crystal layer and its applications

Within the field of nanotechnology research, the study of nanoparticles has become an active research topic. In addition, fabrication technologies to realize nanostructure materials and to use nanoparticles for thin-film coatings are also being actively studied. Although thermal spray-coating technology using powders as a source material has been used for a long time, its applications have been limited because the high temperatures needed do not permit the formation of films with acceptable electrical and mechanical properties. Many of the problems with existing coating techniques have been solved, however, by impacting dry powders with substrates to be coated at impact velocities of several hundreds of meters per second. It is thought that reaction zones with high temperature and high pressure are formed when particles release kinetic energy during high-energy impacts, permitting them to collide and bond firmly with a substrate. The advantage of such a process is that nanostructured layers can be formed at lower temperatures than other spray-coating techniques, producing films with superior electrical and mechanical properties.

This method is called the aerosol deposition method (AD). In this paper we explain the principle of the AD method and its application to ceramic film formation for information/communication and energy technology applications.

1. Aerosol deposition method

The AD method is based on shock-loading solidification due to the impact of ultrafine ceramic particles with a surface [1, 2]. First, particles are mixed with a gas to generate an aerosol. This aerosol is ejected through a nozzle at low pressure, and impacted onto a substrate to form a thin film. During impact with the substrate, part of the particle's kinetic energy is converted into thermal energy that raises the temperature of the particle and promotes bonding between the substrate and the particles and also between multiple particles. However, the detailed mechanism occurring during particle impingement with a surface in the AD method has still not been fully clarified.

Figure 4.5.26 shows the deposition apparatus and Table 4.5.2 gives the deposition conditions for the AD method. The AD apparatus is made up of two vacuum chambers connected by a tube. The first chamber is an aerosol-generation chamber, and the second chamber is a deposition chamber. The deposition chamber is used for the formation and patterning of films.

The aerosol chamber has a carrier gas system and a vibration system to mix the powder with the carrier gas. The aerosol generated in the aerosol-generation chamber is delivered to the deposition chamber by a pressure difference between the two chambers. The deposition chamber contains a nozzle, substrate with heating system, and a mask alignment system, used for making patterned films. A rotary vacuum pump coupled to a mechanical booster pump is used to evacuate this chamber to a pressure of about 1–50 kPa during deposition. The particles flow through a micro-orifice nozzle and deposit onto the substrate through a patterning mask.

Sintered, ceramic powders with a particle size range of about 0.08–2 μm are typically used as the deposition particles. After suspension in the carrier gas to form an aerosol, the aerosol is accelerated to several hundreds of meters per second through an orifice with diameter less than 1 mm. Not all of the particles deposit onto the substrate, and the ratio of deposited to non-deposited particles strongly depends on the size and degree of aggregation of the particles. To form films with acceptable density and material properties, particles with a particular size and morphology must be preferentially used. To generate a jet of particles with acceptable size and morphology, between the aerosol-generation and deposition chamber a deagglomeration device is used to break the particles apart as much as possible, and then an aerodynamic filter is used to select particles in a prescribed size range.

2. Room temperature impact consolidation

For the AD method, high-speed film formation of ceramic films at room temperature with high density and high transparency is possible by optimizing the particle diameter and deposition conditions. The result is a process that yields acceptable solidification

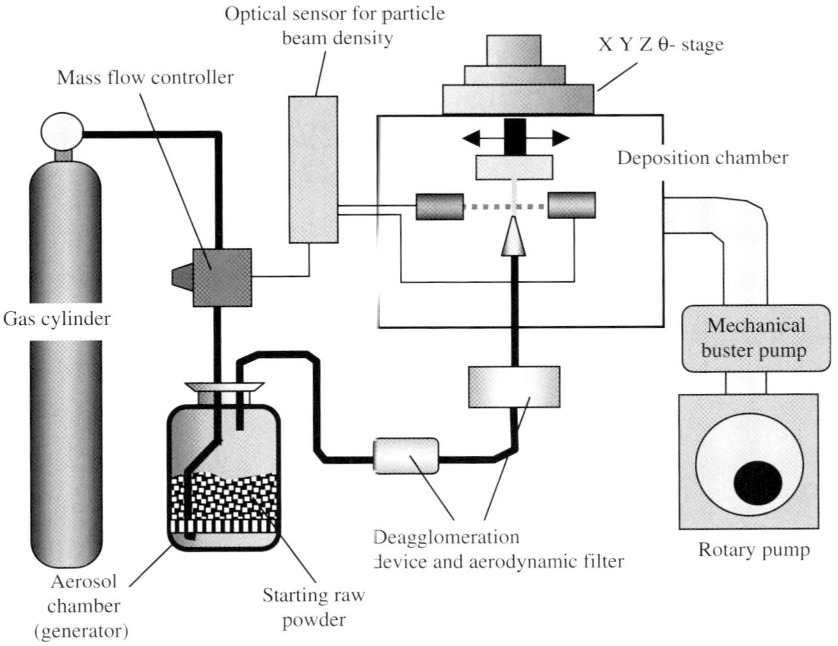

Figure 4.5.26
Fundamental setup of aerosol deposition apparatus.

Table 4.5.2
Typical deposition condition.

Pressure in deposition chamber (operation)	0.4~2 Torr
Pressure in aerosol chamber	80~600 Torr
Size of nozzle orifice	5 × 0.3 mm^2 10 × 0.4 mm^2
Accelerated carry gas	He, Air, N2
Consumption of accelerated carry gas	1~8 l /min
Substrate heating	Non
Relative scanning speed of nozzle motion along substrate	0.125~1.25 mm/sec
Distance between nozzle and substrate	1~20 mm

without the need for high-temperature heat treatment. We call this process room temperature impact consolidation (RTIC) [2, 3].

To show the characteristics of RTIC films, Fig. 4.5.27 shows X-ray diffraction (XRD) profiles, scanning electron microscopy (SEM) images, and transmission electron microscopy (TEM) images of an α-Al_2O_3 film deposited at room temperature. The film shown in Fig. 4.5.27 has high density and randomly oriented polycrystalline nanostructures with crystal size less than 20 nm. TEM and electron diffraction imaging did not show either amorphous layers or heterostructures at the boundary of crystal grains. XRD profiles confirmed that the spectral phases of the α-Al_2O_3 particles were retained in the deposited layer. However, broadening of the spectra and a slight shifting of the spectral angle were observed. The reason for the change between the spectra of the particles and the deposited layer may be due to reduction of the film crystal size or distortion during deposition. Clear lattice images in crystal grains less than 10 nm were observed, as well as uniform microstructures at the boundary between the substrate and the deposited layer. For α-Al_2O_3 layers deposited at room temperature, film density was over 95% of theoretical density and Vickers hardness was over 1,600 Hv. Such α-Al_2O_3 layers are acceptable for use as abrasion–resistant coatings [4]. The layer hardness increased with increasing particle-impact velocity, and sometimes was higher than that of the bulk material, which was sintered at a high temperature. Critical particle velocities for acceptable RTIC ranged from 100–500 m/s, and the velocity needed to create films with acceptable hardness tended to increase with increasing sintering temperature of a particular ceramic material.

Acceptable room temperature deposition was observed not only for oxide materials such as lead

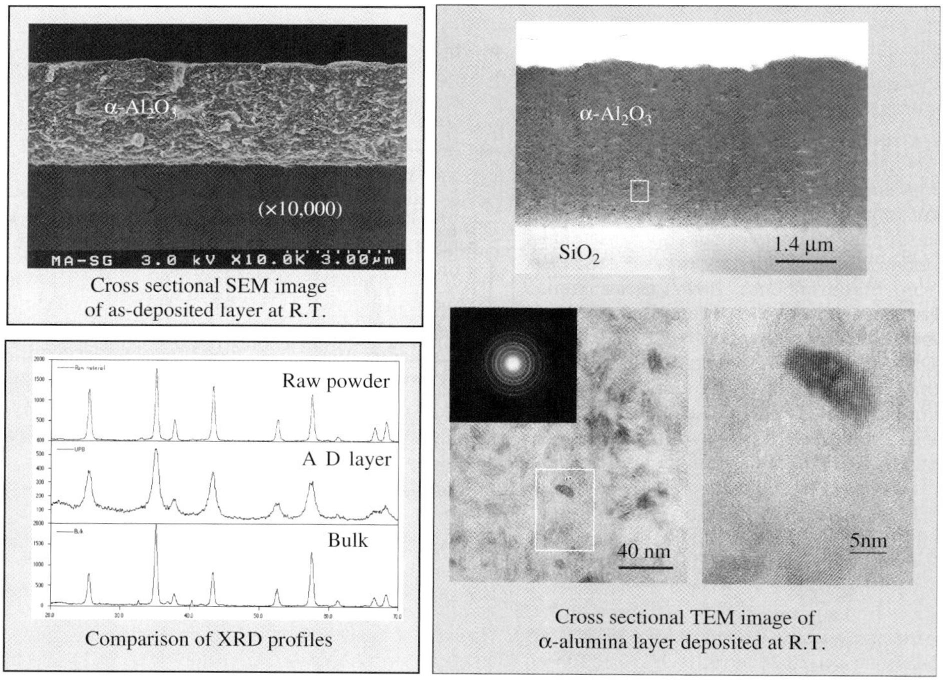

Figure 4.5.27
Microstructure of α-Al_2O_3 layer deposited at room temperature (RT) by aerosol deposition method.

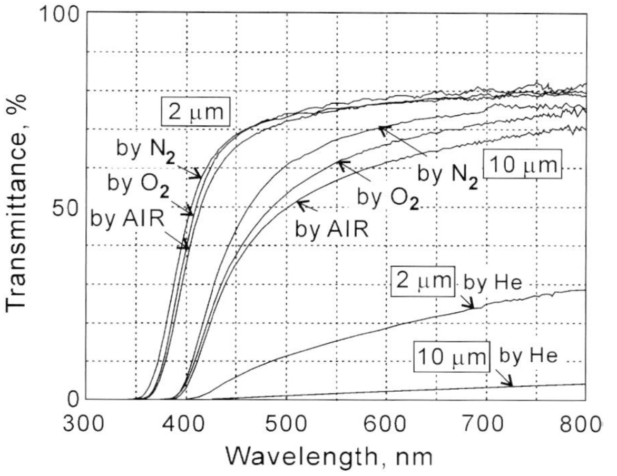

Figure 4.5.28
Influence of carry gas to transmittance of PZT layer deposited at RT.

Zirconate Titanate (PZT:$Pb(Zr_{52},Ti_{48})O_3$) and Ni-Zn-Fe_3O_4, but also for non-oxide materials such as AlN and MgB_2. In either case, particles with diameter greater than 80 nm and with single crystal structure are needed to make films with acceptable hardness. The crystal grain size of As-deposited films was less than 50 nm, which was smaller than that of the starting particles. The reason for this is apparently that the starting particles break down during collisions and then each particle bonds together to form a nanocrystal layer.

If the carrier gas and pressure in the deposition chamber are properly selected so as to reduce the electrical discharge during particle deposition, transparent PZT and alumina films with transmission efficiencies of 60–80% at wavelengths ranging from 450–800 nm could be formed at room temperature as shown in Fig. 4.5.28 [5]. The reason this can be achieved is probably because during deposition the size of the defects in the starting particles is reduced to crystal sizes that are smaller than those found in sintered, bulk materials, and that are also smaller than the wavelength of visible light. Although distortions were included inside the films, a rise of substrate temperature due to particle collisions was not observed, and ceramic powders could be solidified at room temperature without the use of binders. Although the formation of thin films using RTIC does not require sintering, it produces high-density films.

3. Deposition properties and film patterning

Deposition of PZT at film-deposition rates of 10–30 μm/min over an area 5 mm^2 was achieved. The particle velocity and concentration in the aerosol jet downstream of the nozzle increased with increasing carrier gas flow rate, which increased the deposition efficiency. The deposition efficiency was also strongly affected by properties of the starting particles, such as the average particle size and size distribution, and mechanical and surface properties [6]. The deposition efficiency of the AD method is about 30 times higher than conventional thin-film-formation methods such as sputtering, which have a deposition rate of only about 10 nm/min for an area of 5 cm^2. Therefore, the AD method is attractive for manufacturing processes because it permits high throughput.

In conventional thin-film and thermal spray-coating methods, surface cleaning before deposition is needed to achieve acceptable adhesion with a substrate. On the other hand, the AD method does not require such pre-cleaning, because during the initial stage of particle impaction with the surface, similar to sand-blasting processes, surface contaminants such as dirt and oils are removed by the initial particle collisions. Deposition automatically begins when the surface becomes sufficiently clean. The film adhesion strength to glass and metal substrates was in excess of 20 MPa. However, to realize maximum adhesive strength, suitable hardness and elasticity of the substrate are needed.

Ceramic film patterning can be achieved by using a mask deposition method, in which a particle jet is impacted onto the substrate through a pattern mask that contains many openings, each of which is in the order of 50 μm. In this case, it is important to consider the aerosol jet flow in the deposition chamber and through the mask orifices. If the pressure in the deposition chamber is not sufficiently low, the particle jet is scattered by the edge of the openings in the mask and the resulting mask pattern is not preserved on the substrate.

The effect of ceramic particles sprayed onto a substrate change from deposition to erosion [7], depending on the particle diameter, velocity, and angle of incidence of the particle jet to the substrate. These factors also influence the layer density and surface roughness. To achieve acceptable patterns through a mask with acceptable detail, however, the angle of incidence of the particle jet must be kept within a specified range. This is because the angle of incidence affects the flow patterns on the downstream side of the mask, which in turn affects the deposition efficiency and the degree to which the mask profile is distorted. Figure 4.5.29 shows a thick, patterned PZT layer deposited onto Si, SUS, and Pt/Si substrates [8] under optimum deposition conditions. A ceramic microstructure with a 50-μm line width and aspect ratio greater than 1 can be patterned by controlling the substrate heating temperature and strating particle properties. The AD method is useful for making piezoelectric films more than 10 μm thick, for applications such as ultrasonic devices.

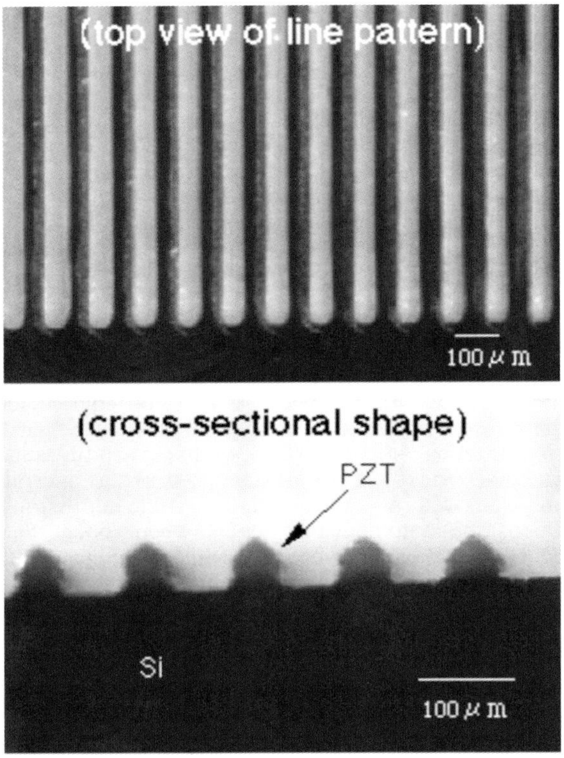

Figure 4.5.29
Micropatterning of PZT thick film by mask deposition method.

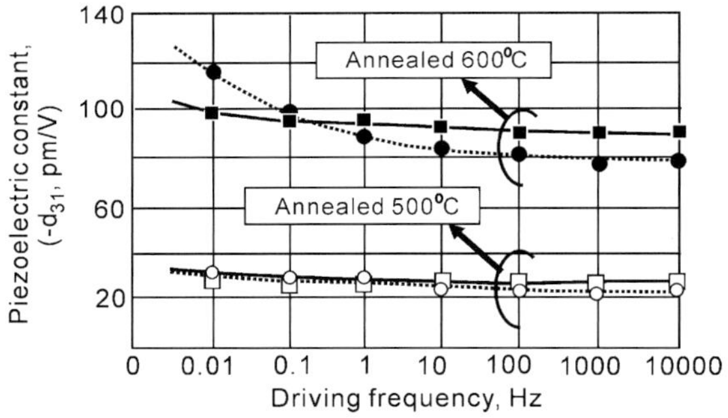

Figure 4.5.30
Piezoelectric constants of PZT thick film deposited by AD method.

4. Electrical properties and recovering properties by heat treatment

As-deposited AD layers deposited at room temperature generally have thermal insulation and electrical breakdown characteristics that exceed that of the bulk material. For example, the electrical breakdown of α-Al_2O_3 and PZT exceeds 3 MV/cm and 500 kV/cm, respectively. Such electrical characteristics can be useful for developing devices such as electrostatic chucks [9]. Although PZT layers deposited at room temperature exhibit piezoelectric and ferroelectric behavior, their properties are unacceptable for practical applications because of structural defects introduced during deposition. By post-annealing in air at temperatures ranging from 500–600°C, grain growth of fine crystals and defect recovery in AD layers has been observed, dramatically improving ferroelectric properties. The dielectric constant (ε) and piezoelectric constant (d_{31}) of post-annealed layers formed at 600°C were 800–1,200 and –100 pm/V as shown in Fig. 4.5.30 [10], respectively, which is comparable to that achieved with conventional thin-film formation methods. Moreover, the electrical breakdown (< 1 MV/cm) and Young's module (>80 GPa) of AD films exceeded those obtained with conventional thick-film formation technology. By post annealing at 850°C, remnant polarization (Pr) of 38 μC/cm^2 and coercive field strengths (E_c) of 30 kV/cm were obtained in the PZT films [11]. Even without adding special additives to the feed particles or using special procedures, compared to conventional screen-printing methods, the AD method permits a 300–400? reduction of the process temperature. For films deposited by using both conventional and AD processes, Fig. 4.5.31 shows the effect on electrical properties of heat treating films. For both conventional-bulk and thin-film processes, heat treatment at more than 600°C was required to obtain crystallization and densification of the films. The most important characteristic of the AD method is that compared to the properties of bulk materials at room temperature, the As-deposited film has a more dense, crystallized structure.

5. Application to MEMS devices

Because piezoelectric materials act both as sensors and as actuators, they have a wide range of potential application, such as ink-jet printers, high-speed actuators for nanopositioning, and micro, ultrasonic devices. To realize integrated microdevices with piezoelectric materials, thin-film deposition technologies and associated microfabrication methods are being actively studied in the research fields of micro electro mechanical systems (MEMS) and micro total analysis systems (μ-TAS). The required thickness of piezoelectric films for these applications is from 1 to about 100 μm (this thickness range is called "thick films").

Resonance type optical microscanners have been made as one practical application of the AD film-formation method [12, 13]. These types of optical scanners are expected to be a key component in future laser displays and retina projection-type displays. For such applications the requirements are scanning frequency greater than about 30 kHz, scanning angle

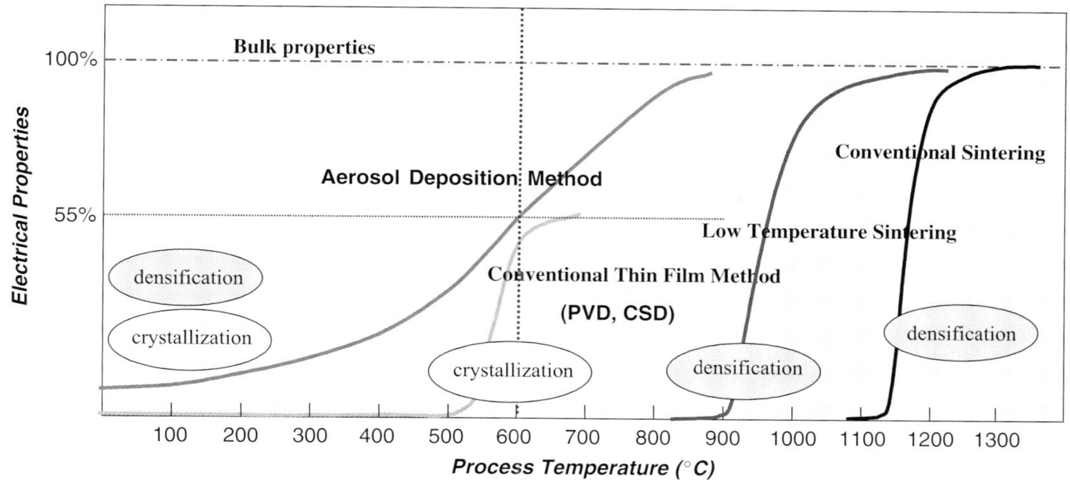

Figure 4.5.31
Recovering electrical properties by heat treatment.

greater than 20°, mirrors sufficiently large to prevent distortion, and reduction of drive voltage. It is thought that piezoelectric films will be desirable for such applications because of their simple structure and high actuation force. However, these devices are difficult to make using conventional thin-film deposition technologies combined with MEMS fabrication processes, because of low deposition rates and complicated etching processes. Figure 4.5.32 shows an optical microscanner and an enlarged image of a piezoelectric actuator. To make this devise, first the scanner structure was made by using conventional Si micromachining, and then the AD method was used to deposit a thick PZT layer. This layer was formed on an Si cantilever without any damage or deformation to the cantilever from the particle impingement associated with the AD method. Proper alignment of the mask prevented either adhesion or stacking of raw PZT particles on the mirror part. The mirror was not influenced by the deposited film. The maximum mirror scanning angle and resonance frequency in air were 26° and 33 kHz, respectively, representing high-speed performance. Neither mirror bending nor deformation of the scanning laser beam were observed by the thick structure of the microscanner device. This scanner performance was higher than that of scanners made with conventional methods.

In addition, diaphragm actuators [14] used for micropumps can be made from thin Si membranes deposited with the AD method and micromachined. These have flat frequency response and good symmetrical elastic deformation over a wide frequency range. During fatigue testing (stability of long-time performance) under high driving electrical fields, there was no de-polarization nor peeling of the AD deposited film from the substrate [11]. The amplitude

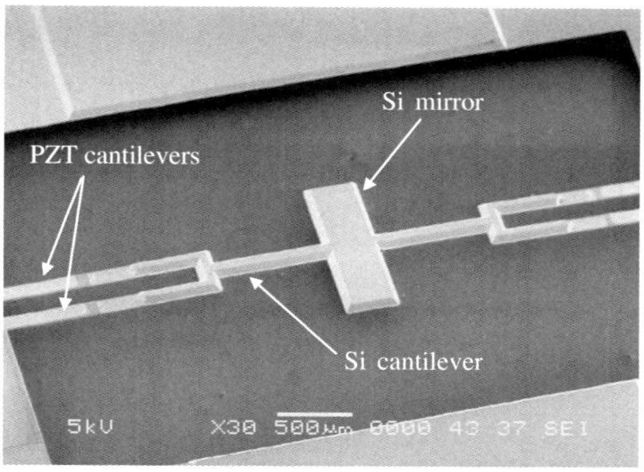

Figure 4.5.32
Resonance type optical microscanner driven by PZT thick layer fabricated with AD method.

of this actuator was 25 mm at a resonance frequency of 22.4 kHz and a driving voltage of 8 V. This performance is suitable for applications using micromixers and micropumps.

6. Future prospects for using AD methods in material integration technology

The AD method is a non-thermal equilibrium process, which solidifies feed particles at room temperature to form films. This differs from thermal spray coating methods, which use higher temperatures for film deposition. There are two important features of the AD method. One is that the AD deposition rate is higher than that of conventional thin-film processes, because the deposition material is particles, which carry substantially more mass than the molecules used in conventional methods such as sputter deposition. In addition, it is easy to form complicated oxide thin films with the AD method, because the crystal structure of the starting particles is preserved during deposition. Therefore, the AD method is useful for developing composite or integrated materials with various ceramics, metals, and polymers. However, because many defects are introduced during deposition, a process such as annealing is needed to achieve acceptable electrical properties of the deposited films. One of the origins for these defects may be that the crystals in the starting particles are crushed during collision with the substrate. Or it may be that surface defects of the starting particles are introduced into the deposited layer.

Because integration of ceramics with low melting-point materials is important for a broad range of applications, reduction of process temperature is a requirement of future fabrication techniques. To achieve these goals with the AD method, the deposition mechanism, formation of defects in the layer, and recovery of electrical and mechanical properties by post-deposition annealing will be investigated in detail. Establishment of methods for controlling the particle size and degree of aggregation of the starting particles are also important. We expect that AD films will be useful for making MEMS devices, high-frequency components, optical integration devices, and components for fuel cells, which require ceramic thick films greater than 1 μm thick.

References

[1] J. Akedo, M. Lebedev: *Materia*, **41**(7), 459–466 (2002) (in Japanese).

[2] J. Akedo, M. Lebedev: *Jpn. J. Appl. Phys.*, **38**(9B), 5397–5401 (1999).

[3] J. Akedo: *J. Am. Ceram. Soc.*, **89**(6), 1834–1839 (2006).

[4] M. Kiyohara, Y. Tsujimichi, K. Mori, H. Hatono, J. Migita, T. Kusunoki, N. Minami, M. Lebedev and J. Akedo: *Proc. 15*th *Ceram. Soc. Jpn. Autumn Symp.*, 228 (2002).

[5] J. Akedo, M. Lebedev: *Jpn. J. Appl. Phys.*, **40**, 5528–5532 (2001).

[6] J. Akedo, M. Lebedev: *Jpn. J. Appl. Phys.*, **41**, 6980–6984 (2002).

[7] M. Lebedev, J. Akedo: *IEEJ Trans.*, **120-E**(12), 600–601 (2000).

[8] J. Akedo: *Microsystem Technol.*, **6**(11), 205–209 (2000).

[9] T. Ide, T. Ito, H. Hatono, M. Kiyohara, M. Lebedev, J. Akedo: *Proc. 15th Ceram. Soc. Jpn. Autumn Symp.*, **229** (2002).

[10] J. Akedo, M. Lebedev: *Appl. Phys. Lett.*, **77**, 1710–1712 (2000).

[11] J. Akedo, M. Lebedev: *J. Cryst. Growth*, **235**, 397–402 (2002).

[12] N. Asai, R. Matsuda, M. Watanabe, H. Takayama, S. Yamada, A. Mase, M. Shikida, K. Sato, M. Lebedev and J. Akedo: *Proc. MEMS 2003, Kyoto, Japan*, pp. 247–250 (2003).

[13] J. Akedo, M. Lebedev, H. Sato and J-H. Park: *Jpn. J. Appl. Phys.*, **44**(9B), 7072–7077 (2005).

[14] M. Lebedev, J. Akedo and Y. Akiyama: *Jpn. J. Appl. Phys.*, **39**, 5600–5603 (2000).

4.5.6 Suppression of particle growth in sintering nanoparticles

Increasing the density of bulk products with the suppression of particle growth is important to fully utilize the benefits of nanoparticle functions. In middle and later periods of the sintering process, usually, densification and particle growth proceed simultaneously; accordingly, crystalline size of bulk products is of the order of micrometers, even if nanoparticles are used as the starting materials.

Two techniques are available to suppress the particle growth: one increases the contribution of surface and grain boundary diffusions in sintering by controlling sintering conditions and the other accelerates plastic flow by applying external pressure. The former method is called two-step sintering because of its specific temperature profile, while the latter is called pressure sintering.

(1) Two-step sintering

Two types of two-step sintering are available: one keeps nanoparticles at a relatively low temperature where surface diffusion predominates and then increases the temperature for densification [1, 2] (Pattern 1 in Fig. 4.5.33); and the other first brings particles to a high temperature and then, without maintaining that high temperature, decreases the temperature by about 100°C and finally maintains that temperature for a long period of time [3] (Pattern 2 in Fig. 4.5.33). In both of them, the objective of the first step is to induce pre-coarsening.

These are effective methods for homogeneous sintering of particles with a wide size distribution or with aggregates often observed in nanoparticles. For example, when using powder with a wide size distribution, first sintering of small-sized particles starts; the pore size distribution of the bulk products is wide reflecting the inhomogeneous particle size distribution. That is, both large and small pores coexist; the larger pores being stable even in the last period of the sintering process, where high sintering temperature is required for densification. A similar situation occurs in the presence of aggregates. That is, with the packing density of the aggregates different from that of the other parts, the pore structure becomes inhomogeneous. Particularly, the sintering of nanoparticles that easily aggregate depends largely on the sintering property of aggregates. As a result, a high final

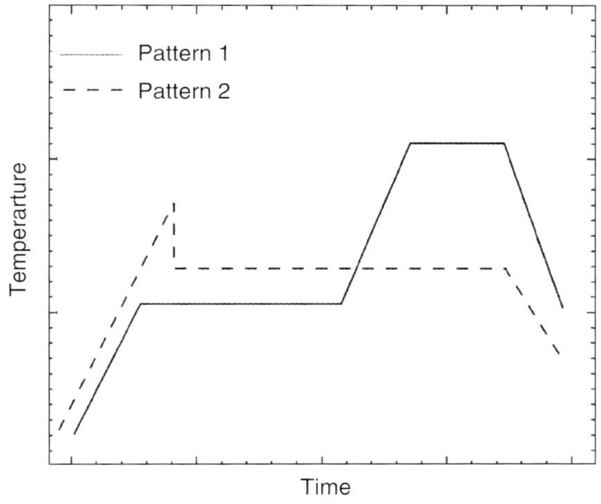

Figure 4.5.33
Temperature ramping profile for the two-step sintering.

sintering temperature is often necessary for densification, even if nanoparticles are used.

In the two-step sintering of Pattern 1, pre-coarsening can dissolve uneven inter-particle network structures in a mold, forming homogeneous microstructures. Since the pre-coarsening proceeds in a low temperature region where material transfer by surface diffusion predominates, microstructures grow without densification. In this process, salient features of the microstructures are (1) homogeneous particle size distribution with small-sized powder particles decreased, (2) round microstructures and grown neck structures. These structures are confirmed by methods such as direct observation of the microstructures, measurement of specific surface area, measurement of pore-size distribution by a mercury porosimeter and other methods. The homogenization mechanism of microstructures is related to strong dependence of the neck growth rate on the neck radius of curvature. The neck growth rate by surface diffusion is expressed by the following equation [4]:

$$\frac{dX}{dt} = \frac{32r^3\gamma\Omega^{\frac{4}{3}}D_s}{C^3X^6kT} \tag{4.5.3}$$

where, X is neck radius, r particle radius, γ. surface energy, Ω atomic volume, D_s surface diffusion coefficient, C shape factor, k Boltzman's constant, and T temperature. The neck radius of curvature, ρ, is expressed by the following equation:

$$\rho = \frac{CX^2}{4r} \tag{4.5.4}$$

where, the shape factor, C, is 2 for the material transfer accompanied by densification, and 1 for that without densification. Considering material transfer without densification ($C=1$), the neck growth rate is a function of neck radius of curvature as expressed by the following equation:

$$\frac{dX}{dt} = \frac{\gamma\Omega^{\frac{4}{3}}D_s}{2\rho^3kT} \tag{4.5.5}$$

The equation indicates that sharper the neck is with a smaller neck radius of curvature, larger the neck growth rate is; the rate depends on cube of the neck radius of curvature. In an early sintering step, distribution of neck radius of curvature is related to heterogeneity in size distribution of raw materials and packing density. Regarding the neck growth rate in an early sintering step (that is, neck growth by surface diffusion), the sharp neck grows rapidly, while the neck with large radius of curvature grows slowly, as indicated by equation (4.5.5). As a result, homogenization of neck size occurs.

By this mechanism in the pre-coarsening at a low temperature, round-shaped microstructure forms, neck grows, and neck size is homogenized, while relative density scarcely changes. Holding a sintering body at a high temperature after this treatment causes densification by volume diffusion. Sintering proceeds more homogeneously in the body that is pre-coarsened than that not pre-coarsened; this is because of the homogenized neck. As a result, particle size distribution of bulk products becomes narrow, large particles decrease, and average particle size becomes smaller than that of products sintered by the normal method. Since the neck grows at a low temperature, sintering speed is low in the middle and final sintering periods. With large pores scarcely remaining due to the homogeneous microstructure, the density of the final product becomes high.

In the two-step sintering of Pattern 2, the density increases, with particles hardly growing, during the period of constant temperature in the 2nd step, reaching the theoretical value. Figure 4.5.34 shows the relation between relative density and particle size for the ZnO nanoparticle (1st step temperature: 750°C, 2nd step temperature: 650°C). For the heat profile, the temperature of the 1st step is raised to a temperature such that 75–85% relative density is obtained. The 1st step temperature, without being held, is lowered by about 100°C and this temperature is held in the 2nd step. If the 2nd step temperature is too low, densification does not proceed. Since the 2nd sintering temperature is lower than that of normal sintering, the sintering time is, generally, more than several tens of hours – a long time profile. This method was first proved effective with

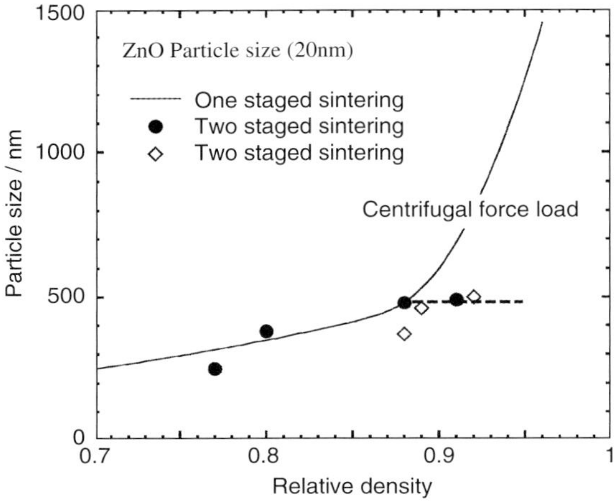

Figure 4.5.34
Particle growth in two-step sintering (Pattern 2).

solid-phase sintering of Y_2O_3; later it was shown that the method is also effective in liquid-phase sintering [5].

Although this phenomenon is very interesting, no reports have been published on details of this mechanism. Chen et al. suggest that the activation energy of the particle growth is higher than that of sintering, because the densification proceeds with suppression of the particle growth [3]. Particle growth is influenced by relative size among nearby particles and presence of pore (pinning effect). Therefore, it can be said that particle growth is difficult in the samples that have homogeneous particle size and pore distributions; these microstructures are probably formed in the 1st step of the two-step sintering process.

Figure 4.5.35 shows the fractured surfaces of ZnO sintered by a normal temperature profile and the two-step profile of Pattern 2. The relative density of bulk products is almost the same in these two examples. No significant difference is observed in the particle distribution, although the two-step sintering gives somewhat larger particle size.

Figure 4.5.36 shows electron microscopic images of polished surfaces of the same samples. Large voids originated from aggregates are observed in the sample prepared by the one-step sintering (a normal sintering pattern). In the one-step sintering, similar voids are observed very frequently, which clearly shows the tendency of easy aggregation of nanoparticles.

On the other hand, the chance of finding these voids decreases in samples prepared by two-step sintering; the sintering is more homogeneous in the two-step than in the one-step. Both the molds are prepared from nanoparticles by unidirectional

Figure 4.5.35
Comparison of microstructures.

molding at 30 MPa followed by CIP treatment at 100 MPa. Accordingly, the above difference is not caused by the difference in the molding method. The relative density is the same in both the samples, that is, the porosity is the same, suggesting that the pores are homogeneously distributed in the sample obtained by the two-step sintering process. From these observations, it is considered that one of the reasons why particle growth is suppressed in the two-step sintering is the homogeneous distribution of pores.

(2) Pressure sintering

The pressure sintering is another method to minimize particle size other than the method described in the former section, the heat profile controlling method. An advantage of pressure sintering is relatively low temperature with the suppression of particle growth. Generally, the effect of externally applied pressure is expressed by adding an external pressure term to the sintering stress (driving force of sintering such as surface tension) term. That is, the driving force for

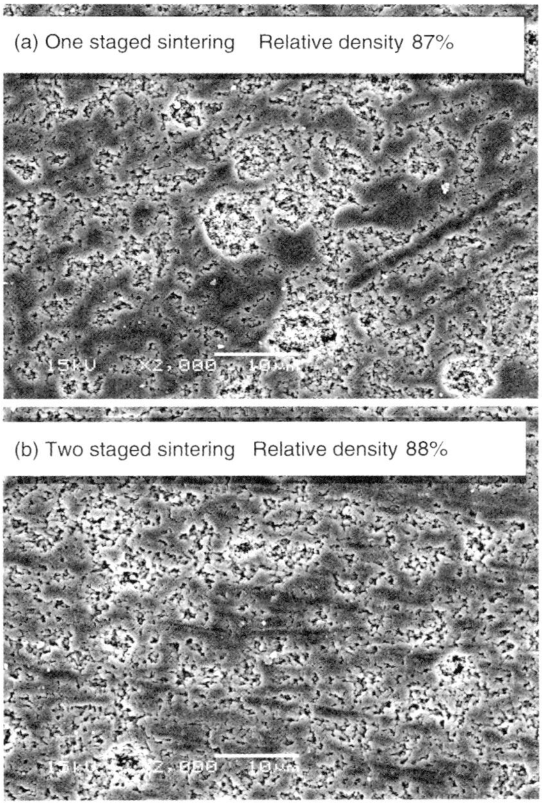

Figure 4.5.36
Distribution of aggregates.

pressure sintering, σ, is expressed by the following equation:

$$\sigma = \sigma_{\text{sinter}} + \frac{\sigma_{\text{appl}}}{3} \tag{4.5.6}$$

where, σ_{sinter}, is sintering stress, and σ_{appl} is unidirectional pressure stress applied. The sintering rate is often evaluated using the following equation:

$$\frac{dl}{ldt} = A\frac{\sigma^n}{a^p} f(d)\exp\left(\frac{-Q}{kT}\right) \tag{4.5.7}$$

where, dl/l is a coefficient of linear contraction, A a constant, a particle size, $f(d)$ a function of density d, and Q apparent activation energy.

For example, it is reported that n is 2.2–2.6 ($20 < \sigma$ MPa < 100, $550 <$ T°C <750) in the case of TiO_2 with a particle size of 12 nm; n is 3 ($5 < \sigma$ MPa < 50, T: 1,400°C) in the case of ZrO_2 with a particle size of 25 nm [6]. The value of n is 1 for the sintering by diffusion. Accordingly, the value of n larger than unity suggests that plastic flow of crystalline particles, such as sliding, occurs during the sintering process. For TiO_2, the value of n is large, although the sintering temperature is very low. This is a very interesting feature of nanoparticles.

Various techniques are available in the pressure sintering method; centrifugal sintering is introduced here, which has been developed recently [7]. Figure 4.5.37 shows the principle. The centrifugal sintering is a method to sinter materials under centrifugal force; the

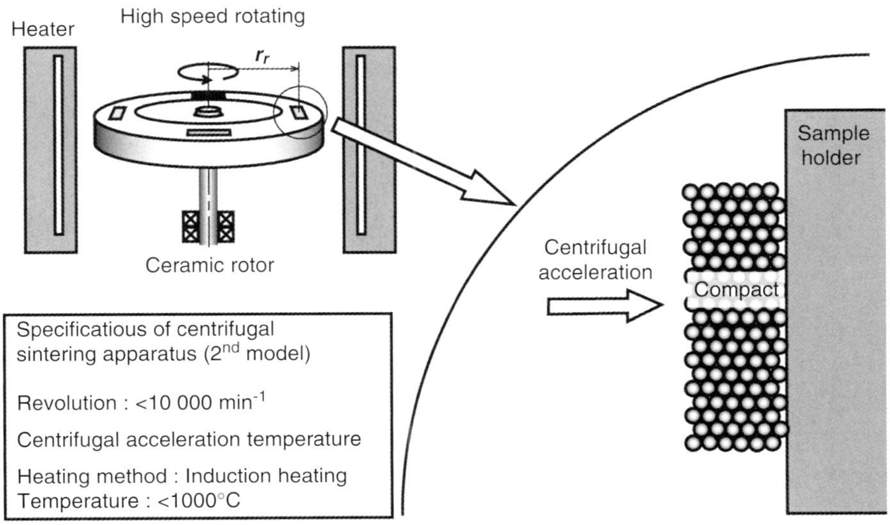

Figure 4.5.37
Principal of centrifugal sintering.

materials are pressured by their own weight. In this method, a sample is sintered under the acceleration of about 10,000 times that of gravity. Since the pressure is caused by own weight of materials, they can be pressured uniformly, in principle.

The relation between density and particle size is shown in Fig. 4.5.34 for samples prepared by the same two-step sintering (Pattern 2) [sintering pattern, 1st step temperature: 750°C, 2nd step holding temperature 650°C; centrifugal acceleration: 90,000 m / s^2 (kept constant during processing, pressure generated by the acceleration corresponds to several Mega Pascal)]. The relations between density and particle size are similar in both the samples prepared by centrifugal sintering and normal sintering. This is because the process temperature is the same.

Figure 4.5.38 shows increase of the density during the holding time in the 2nd step. The density of the sample is always higher than that of sample by the normal sintering; while the shrinkage rate of the centrifugal sintering is low. Judging from the fact that the particle sizes are almost the same, improvement in densification by the centrifugal process is considered to be attributable to plastic flow caused by sliding of particles during the initial sintering period.

To suppress particle growth in the sintering of nanoparticles, it is important to suppress uneven sintering caused by aggregate structures and to increase density. For this purpose, the two-step sintering is an easy and effective method. In the pressure sintering, rearrangement of particles and plastic flow can be expected which improve sintering property. Since the sintering temperature is low, this method is effective to suppress particle growth.

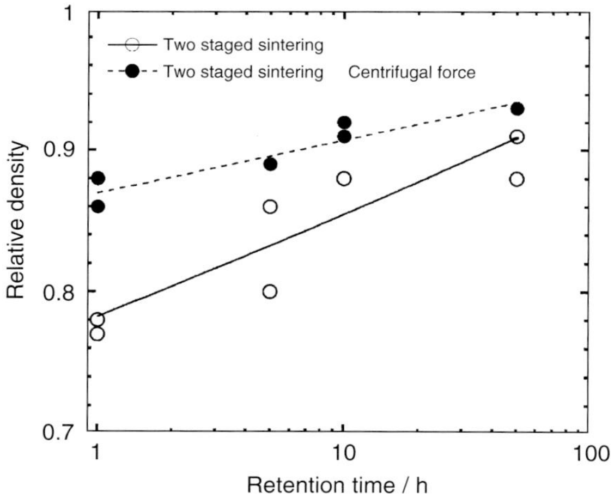

Figure 4.5.38
A relation between relative density and holding time in the two-step sintering (Pattern 2).

References

[1] M.Y. Chu, L.C. De Jonghe, M.F. Lin and F.J.T. Lin: *J. Am. Ceram. Soc.*, **74**, 2902–2911 (1991).
[2] F.J.T. Lin, L.C. De Jonghe and M.N. Rahaman: *J. Am. Ceram. Soc.*, **80**, 2269–2277 (1997).
[3] I.W. Chen, W.H. Wang: *Nature*, **404**, 168–171 (2000).
[4] G.C. Kuczynski: *J. Metals. Trans. AIME*, **185**(2), 169–178 (1949).
[5] Y.I. Lee, Y.W. Kim, M. Mitomo and D.Y. Kim: *J. Am. Ceram. Soc.*, **86**, 1803–1805 (2003).
[6] R.S. Averback, H.J. Höfler and R. Tao: *Mater. Sci. Eng.*, **A166**, 169–177 (1993).
[7] Y. Kinemuchi, S. Uchimura and K. Watari: *J. Euro. Ceram. Soc.*, **24**, 2061–2066 (2004).

4.5.7 Fabrication of nanoceramics by colloidal processing

Colloidal processing is a wet method for consolidating particles with a high density and homogeneous microstructure by controlling the interparticle interaction in a suspension. Table 4.5.3 shows typical colloidal processing accompanied by a drain step, in which the transport media and their driving forces are indicated [1].

Slip casting is a common method in which a suspension is poured into a porous mold such as plaster and a consolidated layer is obtained. As is shown in Fig. 4.5.39, the consolidation rate is controlled by the drain rate in which the solvent is removed through the filter formed by the consolidated layer consisting of packed particles. Therefore, during the drain process, the consolidation rate becomes slower when the particle size becomes smaller. Pressure or vacuum casting is an enhanced filtration method in which the suspension is filtered from the outside of the mold by pressure application or evacuation. Centrifugal casting is the method for the sedimentation and consolidation of particles by centrifugal force in which the solvent is removed at the top; therefore a porous mold is not necessary.

Tape casting is the method for forming a sheet film in which the suspension is poured onto a moving sheet *via* a knife edge so-called doctor blade. This process is widely used for making a laminated layer.

On the other hand, electrophoretic deposition (refer to Section 4.2.3) is a colloidal processing technique, in which charged particles are transported to the plate and consolidated on it by an external DC application. The consolidation rate is fast and independent of the particle size; therefore, this process is suitable for the consolidation of fine particles. Moreover, not only the layer thickness but also the assembling of particles can be controlled by adjusting the external electric field.

As another colloidal processing without accompanying dehydration, gel casting, and floc-casting have received increased attention because of isotropic

Table 4.5.3
Classification of collide process.

Method	Force	Actors	
		Watchers	Movers
Slip casting	Capillarity	Particles Both ions	Liquid
Pressure/vacuum casting	Capillarity and/or pressure and/or suction	Particles Both ions	Liquid
Centrifugal force casting	Centrifugal force	Both ions	Particles Liquid
Tape casting	Mechanical	Liquid Particles Both ions	
EPD	Electrohydrodynamics Electrochemical	Liquid	Particles Both ions

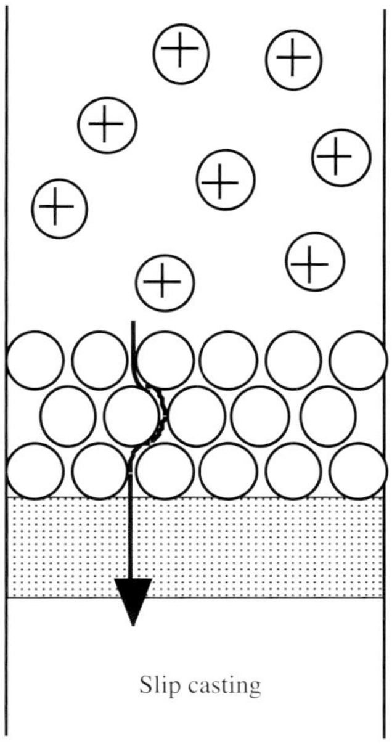

Figure 4.5.39
Schematic diagram of slip casting.

shrinkage and precise dimensional accuracy [2]. These methods are based on in-situ solidification by the polymerization of monomers or flocculation upon heating, etc., using a high-solid loading suspension. When using fine particles, such as a high-solid loading suspension cannot be prepared and a special technique is necessary for applying such a process.

The most important point of colloidal processing is how to control the stability of each particle in a suspension. Therefore, to understand the characteristic of each particle in the solvent is essential. A ceramic particle is charged in the solvent especially in an aqueous solution due to the interaction between the particle surface (or surface adsorbed species) and solvent. The magnitude of the surface potential is estimated by measuring the zeta potential.

During practical processing, the systems where particle dispersion can be controlled by pH are very limited. In some systems, particles do not have a high enough zeta potential or problems such as hydration or dissolution happen. Therefore, the adsorption of a polyelectrolyte with –COOH or $-NH_3$ on the powder surface is usually conducted. In this case, an electrosteric stabilization is expected due to the surface charge of the electrolyte and adsorption of the polymer [3].

According to the DLVO theory [4], the colloidal stability is governed by the total interparticle potential energy (V_T), which is the summation of the repulsive potential energy (V_R), and van der Waals interaction potential energy (V_A).

Figure 4.5.40 shows a typical interparticle potential energy curve. To increase the energy barrier (V_{max}) in the potential curve, V_R should be increased. Both values of V_A and V_R are dependent on the particle size, and V_{max} becomes smaller as the particle size decreases, which results in the difficult particle stabilization due to the smaller particle size.

A metastable phase diagram illustrates the dispersion characteristics of colloidal particles for studying the map of the surface potential and solid loading of the colloidal system. If a simple cubic model is used for the packing of a colloidal solid for relating the interparticle distance x to the solid loading S of a colloidal suspension consisting of particle size d, the following equation, $S=\pi/6(d/(x+d))^3$ is derived [5]. Figure 4.5.41 shows the calculated metastable phase diagram for one-component colloidal systems

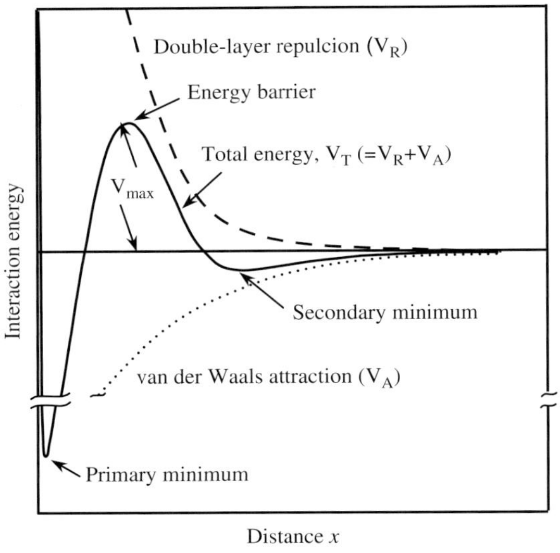

Figure 4.5.40
Interparticle potential energy curve.

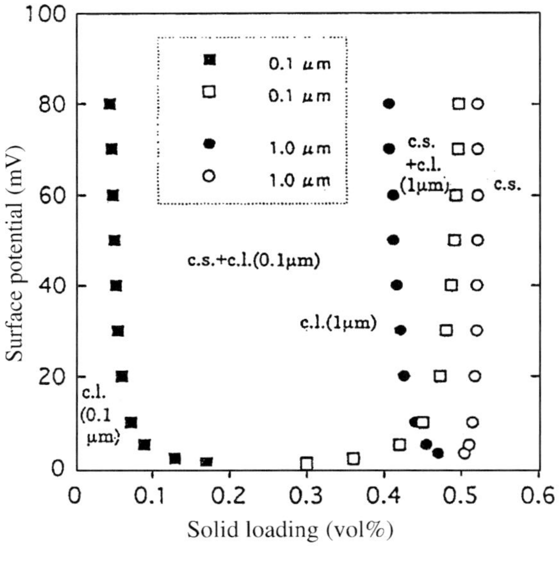

Figure 4.5.41
Metastable phase diagram for one-component colloidal systems of particles with the diameter of 1 and 0.1 μm based on a simple cubic model, in which c.l. and c.s. indicate the colloidal liquid and colloidal solid, respectively.

according to Hirata's method based on a simple cubic model [5]. The case where particle diameters are 1 and 0.1 μm is shown. Here, the interparticle potential energy is equivalent to the free energy describing the thermodynamic stability of the colloidal system consisting of a one-component system. The boundary range of the solid loading between the colloidal liquid (fluid suspension) and colloidal solid with a high viscosity is approximated by the range between the solid loading showing a secondary minimum and energy barrier as shown in Fig. 4.5.38. It is seen from Fig. 4.5.41, that even at the same surface potential energy, the range of the colloidal liquid becomes narrow as the particle size becomes smaller. In other words, if the particle size becomes smaller, the viscosity of the suspension becomes higher and loses its fluidity; therefore, each particle cannot rearrange into a dense packing, which indicates the difficulty in obtaining a dense green body using nanosized particles.

Figure 4.5.42 shows the relationship between the applied pressure and relative density of the green body, in which various types of suspensions of α-alumina particles with a 0.4 μm diameter and boehmite (aluminium hydroxide) particles with a 20–30 nm diameter were prepared by adjusting the pH and consolidating by pressure casting [I.A. Aksay: Private communication]. The pressure dependence of the green density is not observed for the well-dispersed suspension prepared at pH=3. However, a high pressure is necessary to obtain a high green density for the flocculated suspensions, but the density is lower than that of the well-dispersed one. Also, a higher pressure is necessary for the pressure casting of nanosized particles, but the green density is much lower than that for the sub-micron sized particles. For the nanosized particles, the results indicate that the range of c.l. is narrow and the dispersion by a pH adjustment is difficult.

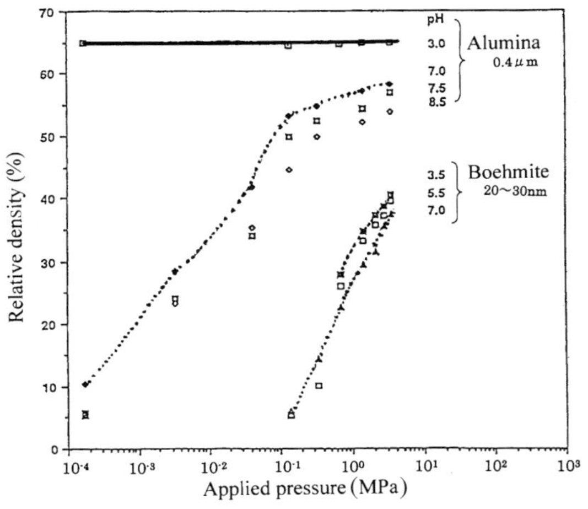

Figure 4.5.42
Relation between applied pressure of pressure casting and relative green density pressure-casted from the suspension of sub-micron sized alumina and nanosized boehmite prepared by changing the pH.

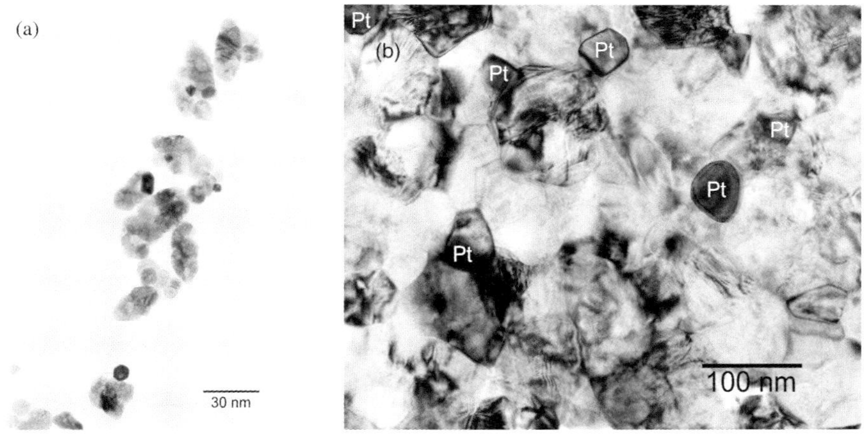

Figure 4.5.43
TEM photographs of 3 YTZ aggregate prepared by using optimum conditions (a) and 1.5 wt% Pt/3Y-TZP composite sintered at 1150°C for 30 h (b).

Therefore, another method is necessary for the dispersion of nanosized particles such as a special polyeletrolyte addition, etc.

Nanosize 3YTZ (3 mol%Y_2O_3 doped tetragonal zirconia) powder was prepared by hydrothermal coprecipitation from a sol of metal chlorides and urea followed by dispersion and calcinations [6]. The non-agglomerated 3YTZ powder consisting of uniform nanosized (about 45 nm) aggregates can be prepared using the optimum synthesis conditions, powder treatment, controlled calcinations, etc., [6] as shown in Fig. 4.5.43. Aqueous suspensions were prepared by adding the dispersant (ammonium polycarboxylate). Consolidations of the suspensions by slip casting and subsequent CIP at 400 MPa were applied. By using the uniform nanosized aggregates, the densification can be completed at the lower temperature of 1,150°C and a fine-grained microstructure with a grain size of less than 0.1 μm was obtained. The possibility of engineering the Pt/3Y-TZP composite nanoaggregates was also studied, in which Pt metal ions were sonochemically reduced from aqueous solutions onto the surface of the 3YTZ. An aqueous suspension of the Pt/3Y-TZP composite nano-aggregates was prepared by a method similar to that for 3Y-TZP. After slip casting and CIP treatment, the sample was sintered at 1,150°C in air. A dense Pt–zirconia nanocomposite was fabricated as shown in Fig. 4.5.43(b) [7].

For enhancement of the sintering, inhabitation of grain growth and improvement of the properties, multi-component systems are commonly used. In this case, multi-component colloidal processing is necessary [8].

Figure 4.5.44 schematically shows the relationship between the pH and zeta potential of the A and B components, and the dispersion state of both components.

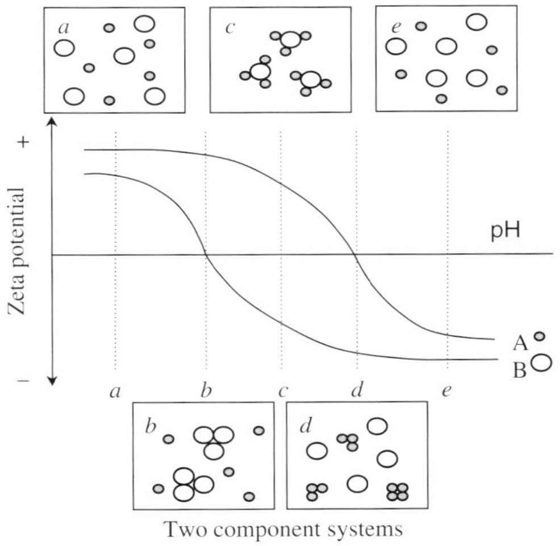

Figure 4.5.44
Schematic diagram of dispersion state of two-component systems from the standpoint of the relationship between the zeta potential and pH.

In regions *b* and *d* in the figure, flocculation of one component occurs due to low zeta potential, and dispersion state cannot be established as schematically shown. In the *a* and *e* regions, a well-dispersed suspension is obtained due to higher zeta potential of both components. In the *c* region, either component is dispersed, but a heterocoagulated suspension is obtained due to the opposite zeta potential. In the multi-component systems, either the well-dispersed suspension or hetero-coagulated suspension is used

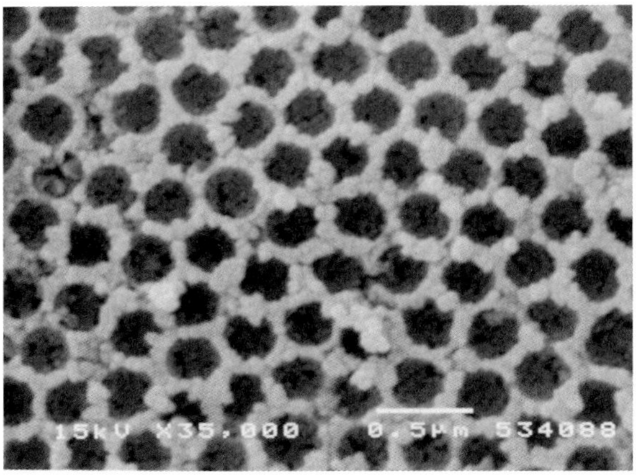

Figure 4.5.45
SEM photograph of porous γ-Al_2O_3.

for obtaining a homogeneous microstucture. For the dispersed suspension, segregation during colloidal filtration is a common problem due to differences in the sedimentation rate, but it can be minimized using a suspension with a high solid loading.

As an example of the multi-component system, some colloidal processing trials of SiO_2 added 3YTZ are described. The addition of SiO_2 is one method for enhancing the superplastic elongation by modification of the 3YTZ grain-boundary. Using commercial 3YTZ powder with an aggregate size of 90 nm, the following three types of suspensions were prepared [9]; SiO_2-coated 3YTZ particles used for the colloidal processing in which the composite particle is prepared by an alkoxide precipitation method, coating by SiO_2, a hetero-coagulated suspension is used, and a well-dispersed suspension with 30 vol% solids loading. In either case, a homogeneous microstructure with a fine grain size of 0.2 μm and high density is established by the low-temperature sintering [9].

The hetero-coagulation method is also applicable for the preparation of porous materials. This is based on the template-assisted approach of a core–shell composite [10]. Monodispersed polymer spheres are used as templates, and ceramic particles act as the target materials. By particle surface modification, well-dispersed suspensions of the polymer and ceramic particles with a high opposite charge could be obtained at the same pH. The flocculated particles are subsequently closely packed by vacuum filtration. The polymers are finally removed by calcination at round 500°C for 4 h, resulting in the porous structures. Figure 4.5.45 shows porous γ-Al_2O_3 with a particle size of about 30 nm from a 350-nm template after heating at 850°C. This method is applicable for obtaining various types of porous ceramics with a controlled pore size and porosity by changing the particle sizes of both the ceramic and polymer [10].

References

[1] Y. Sakka, T. Uchikoshi: *Mater. Sci. Technol.*, **41**, 205–210 (2004).
[2] K. Uematsu: *Ceram. Jpn.*, **40**, 686–702 (2005).
[3] J. Cesarano III, I.A. Aksay: *J. Am. Ceram. Soc.*, **71**, 1062–1067 (1988).
[4] F. Kitahara, K. Furusawa: *Bunsan-Nyuka-Kei no Kagaku*, Kogyo Shuppan, Tokyo, p. 104 (1991).
[5] Y. Hirata, S. Nakagama and Y. Ishihara: *J. Ceram. Soc. Jpn.*, **98**, 316–321 (1990).
[6] O. Vasylkiv, Y. Sakka and V.V. Skorokhod: *J. Am. Ceram. Soc.*, **86**, 299–304 (2003).
[7] O. Vasylkiv, Y. Sakka, Y. Maeda and V.V. Skorokhod: *J. Euro. Ceram. Soc.*, **24**, 469–473 (2004).
[8] Y. Sakka: *Ceram. Jpn.*, **30**, 1017–1020 (1995).
[9] Y. Sakka, K. Hiraga: *Nippon Kagaku Kaishi*, 497–508 (1999).
[10] F. Tang, F. Fudouzi, T. Uchikoshi and Y. Sakka: *J. Euro. Ceram. Soc.*, **24**, 341–344 (2004).

4.6 Self-assembly

4.6.1 Self-organization of nanoparticles

It is expected that various kinds of functions can be realized by thin films consisting of two- or three-dimensional structures of nanoparticles on a substrate. One of the most realistic methods of ordering nanoparticles is coating and drying process [1]. It is a typical bottom-up method based on self-organization of nanoparticles in solution. The self-organization method observed anywhere in nature is not well-understood. The nanomaterials design from an engineering point of view can be established by understanding the mechanisms of self-organization in nature. As a result, the innovation of nanoprocessing to fabricate nanomaterials can be established. The nanostructure control is essential to design sophisticated functional materials. In fact, nanostructures observed in functional materials are not well-designed but obtained by chance. It is the same as a recipe of cooking accumulated as personal skills.

Nanoparticles consisting of metals, ceramics, semiconductors, organics, carbons, and polymers are synthesized by various kinds of methods for realizing desired both physical properties and functions. The necessary conditions to explore new functions by employing nanoparticles are as follows.

1. The size of nanoparticles is preferable less than 10 nm.
2. The size distribution of nanoparticles is smaller than 10 % in CV value whose definition is the ratio of the standard deviation to the average value.

3. The nanoparticles are dispersed by preventing aggregation.
4. The surface of nanoparticles is modified to avoid aggregation.

There are a lot of trade-off relationships between keeping dispersion and desired function.

The applications of thin films of self-ordered nanoparticles are a light emitted diode [2] of semiconductor nanoparticles, a solar cell [3], and an optical switching [4]. A magnetic memory media [5] with high-density memory is also studied by using magnetic nanoparticles such as FePt. FED displays expected as next generation need phosphor nanoparticles such as Y_2O_3:Eu. Furthermore, there are also a lot of applications of nanocomposites consisting of nanoparticles dispersed in polymers. The anticipated characteristics of nanocomposites are; improving the thermal resistance, high optical reflection, high stiffness of mechanical strength [6], and improving tribology characteristics such as friction and wear. There are also bio-medical applications such as DDS and biosensors using antigen-antibody reactions with antibody surface modification on nanoparticles. The thin films of nanoparticles above mentioned can be fabricated by conventional coating–drying processes.

In general, there are two methods to fabricate thin films of nanoparticles. They are wet and dry processes summarized in Table 4.6.1. Here we focus on coating and drying.

1. *Self-organization process of nanoparticles in colloidal suspensions*

Coating and drying processes are employed to fabricate thin films of nanoparticles. Coating process is categorized in two methods [1] which are a continuous coating such as die coating and a discontinuous coating such as ink jet and microcontact stamp. The continuous coating is usually employed for nanocomposites, memory media, and photoluminescence device. The discrete coating is used for patterning of thin films such as electrode lines and organic LED.

Nanorheological properties for the high concentration of nanoparticle suspensions more than 10% strongly affect the final structure of thin films. The processes of dispersing, coating, and drying are responsible for the formation of highly ordered structure of nanoparticles. The surface modification of nanoparticles and the surface treatment of a substrate characterize nanointerfaces responsible for the formation of nanoparticles aggregation. The highly ordered structures of nanoparticles are due to the balance between field conditions determined by processes and nanointerfaces characterized by molecules modified on nanoparticle surface. In materials nanotechnology, the control of dispersion and aggregation of nanoparticles is a key technology. The geometrical shape of aggregates must be also controlled for the nanomaterials design. The relationships among colloidal structures, thin film structures of nanoparticles, and functions are schematically shown in Fig. 4.6.1.

The aggregation dynamics of fine particles in a suspension under steady state are well studied in colloidal science, but there are few studies about both the size distribution and the shape of aggregates. The aggregation of fine particles is a typical example of self-organization, and therefore the prediction of both the size and the shape of aggregates becomes a key issue. For instance, the structure of aggregates of nanoparticles in nanocomposites affects the physical properties such as optical transparency, mechanical strength, and electrical conductivity. In general, the balance between dispersion and aggregation must be controlled to obtain high quality functions. In another word, the control of isolation and interaction among nanoparticles is expected for the design of nanocomposites.

There are two methods of self-ordering. One is self-assembly realized in the vicinity of equilibrium, and the other is self-organization [7] operated at non-equilibrium far from equilibrium. The typical example of self-assembly is epitaxial growth in a single crystal growth. The self-organization is typically observed in the nucleation or the generation of nanoparticles in a bulk. The structure formation

Table 4.6.1
Comparison between wet and dry processes of thin films of nanoparticles.

Method	Process	Characteristics	Merit	Demerit	Applications
Wet	Coating Ink jet SAM	Surface modification of nanoparticles and a substrate	Control of dispersion	Purity	Magnetic tapes phosphor display solar cells color filters
Dry	Evaporation CVD Sputtering MBE	Charge control of nanoparticles	High purity	Aggregation	Hard disk quantum dot laser SWNT

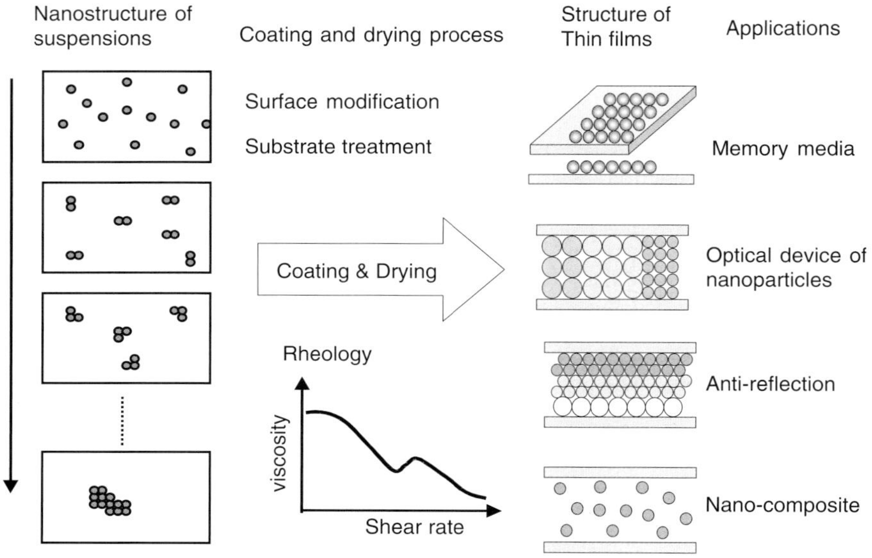

Figure 4.6.1
Thin film structures desired for various applications.

determined by a surface potential on substrates is self-assembly. For example, the pattern formations due to the combination of both hydrophilic and hydrophobic characteristics and also due to positive and negative electric potentials are classified as self-assembly. On the other hand, the pattern formation under uniform surface potentials is due to self-organization. The difference of self-ordering between self-assembly and self-organization on the same substrate is due to the dependency on surface potentials. As a result, the pattern of nanoparticles generated on the latent pattern of surface potentials of a substrate is due to self-assembly. On the other hand, the pattern formation due to self-organization is controlled by particle–particle interactions and transport properties such as the diffusion of nanoparticles and the viscosity of suspensions. Furthermore, the structure formation due to self-organization, which is a rate process under non-equilibrium, depends on both the process time and the relaxation time of nanoparticles structures. For instance, the complicated structure formation of nanoparticles in coating and drying process is mainly determined by the balance between drying time and particle diffusion time. The drying time is a sort of quenching time to fix the structure of nanoparticles. The fast quenching rate results in disorder structures of nanoparticles. The transition from disorder to order is a function of the degree of non-equilibrium. The relationship is schematically shown in Fig. 4.6.2. The structure formation due to self-organization is not well-understood because of high non-equilibrium and strong non-linearity.

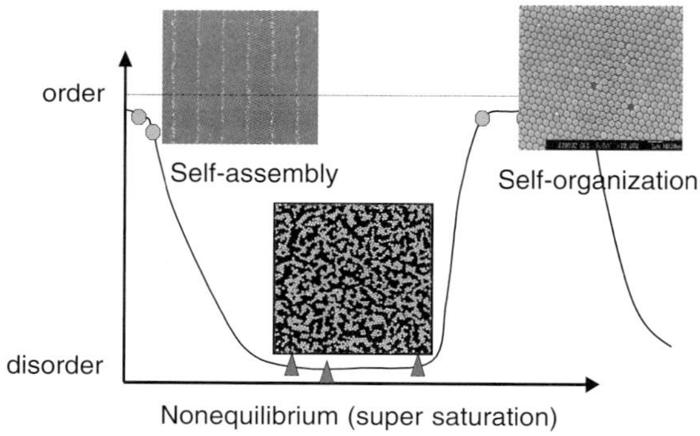

Figure 4.6.2
Schematic characteristics of self-assembly and self-organization.

2. Self-ordering modeling under drying process

The mechanism of self-ordering [8] in colloidal system can be explained by self-organization of nanoparticles during drying. The colloidal system during drying on a substrate is schematically shown in Fig. 4.6.3. The nanoparticles in solvent reach the substrate with the decreasing thickness of liquid film during drying. The ordering of nanoparticles takes place just before the dry up of solvent, because of the capillary force, the attractive force, due to the change of the free surface shown in the middle of Fig. 4.6.3. The fcc ordering shown in the bottom of Fig. 4.6.3 can be observed [9–14]. Here, we explain the

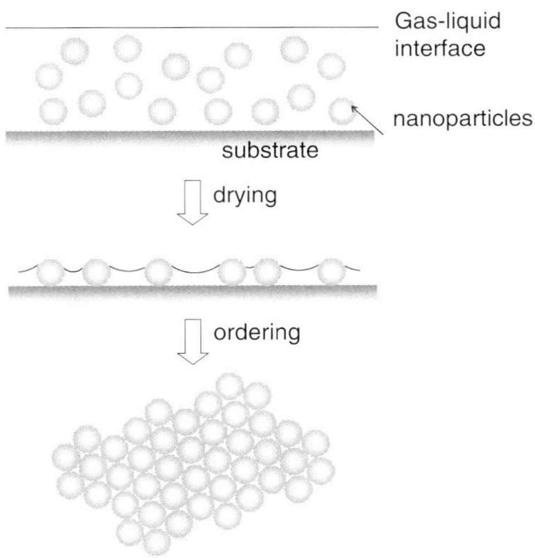

Figure 4.6.3
Schematic figures during drying.

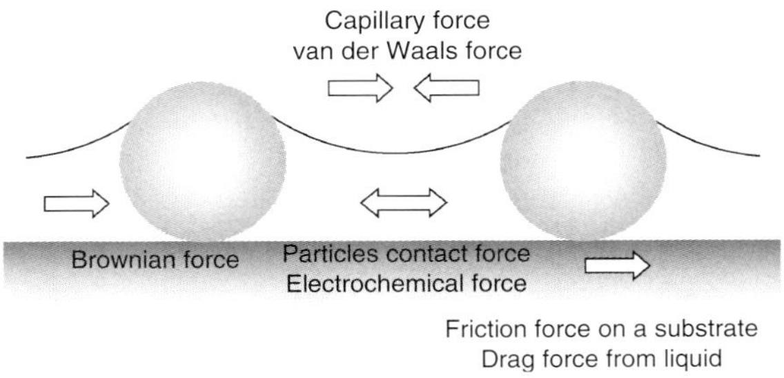

Figure 4.6.4
Forces exerted on nanoparticles during drying.

modeling [15, 16] of structure formation just before and after the end of drying. At the moment, the height of the liquid film becomes almost the same as the diameter of nanoparticles.

Nanoparticles move around in liquid based on Newton's law consisting of various kinds of forces. The horizontal capillary force [17] due to the shape deformation of free surface caused by interaction between nanoparticles and solvent, and the electrical force due to the surface charge of nanoparticles can be described by a particle–particle potential depending on the particle–particle distance. The Brownian force based on the molecular movement of solvent can be modeled as a random force with Gaussian distribution under the continuous media hypothesis of solvent. As a result, the Brownian dynamic model including the drag force exerted from solvent can be applied. Van der Waals force between nanoparticles can also be modeled as the volume integration of the molecular force. When nanoparticles contact together, a mechanical model with spring and dashpot as used in powder technology is employed. The interaction force between particles and a substrate is defined as friction force based on the Coulomb rule. The forces on nanoparticles considered here are shown in Fig. 4.6.4.

The force modeling on nanoparticles in liquid is a multi-scale and multi-interdisciplinary model such as molecular dynamics, Brownian dynamics, continuous fluid mechanics, and powder mechanics. Then, the translational and the rotational motions can be described by using the Newton's law and the conservation rule of angular momentum.

The result of two-dimensional simulation of nanoparticles on a substrate is shown for polystyrene particles of 50 nm diameter in aqueous solution. The surface coverage of particles, corresponding to the particle density after drying, is set to be 0.7. The liquid temperature is 20°C. The contact angle between the particle and water is 60°. The zeta potential, the friction coefficient between particles, the friction coefficient between particles and the substrate, and the drying rate are set to be −100 mV, 0.1, 0.025, and 0.01 m/s, respectively.

The liquid thickness as an initial condition is the same as the diameter of nanoparticles. The arrangement of nanoparticles at the beginning is random. The area assigned in calculation is the square of 2.7 μm and the periodical condition is imposed on the simulation boundary.

The snap shots of the structure of nanoparticles during drying are shown in Fig. 4.6.5 with the drying time t and the height of the liquid film h normalized by the particle radius. First, the amorphous state (Fig. 4.6.5(b)) appears because of the repulsive force with −100 mV after the initial arrangement in Fig. 4.6.5(a). As drying proceeds, the rearrangement of nanoparticles due to electrostatic repulsive force and Brownian motion takes place, and results in the relatively ordered structure shown in Fig. 4.6.5(c). The structure shown in Fig. 4.6.5(c) looks like colloidal crystals formed by the repulsive force with point and line defects. The capillary force, the attractive force, becomes stronger than the electrostatic force, the repulsive force, with the decreasing liquid thickness. When the liquid thickness reaches some height less than the diameter of particles, the abrupt change of structures takes place, because the attractive force exceeds the repulsive force. The onset of the structure transition from dispersion to aggregation takes place from a position where the balance between attractive and repulsive forces is broken. Then, the voids shown in Fig. 4.6.5(d) are formed, which is similar to the nucleation of vacancy. The domain of the voids shown in Fig. 4.6.5(e) grows with drying.

It is preferable to express the characteristics of the ordering of nanoparticles [15] as simple as possible. The isotropic ordering factor (IOF) which indicates the spatial isotropy of ordered nanoparticles

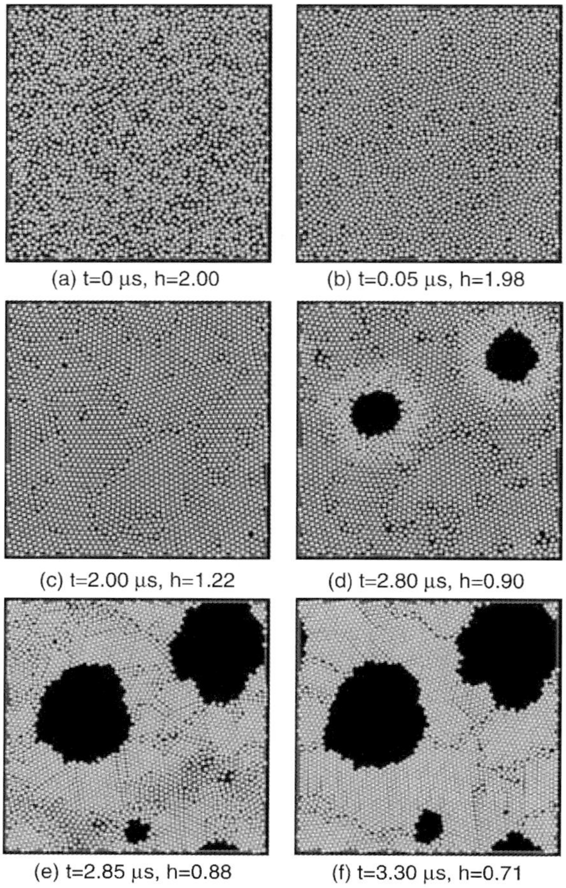

Figure 4.6.5
Evolutional change of structures during drying.

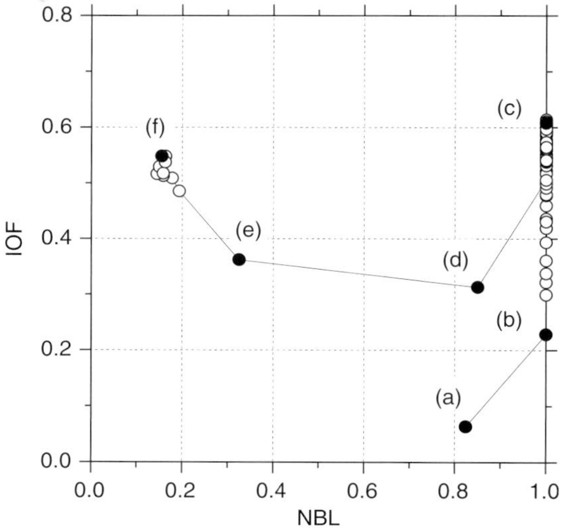

Figure 4.6.6
Structure changes of NBL and IOF during drying.

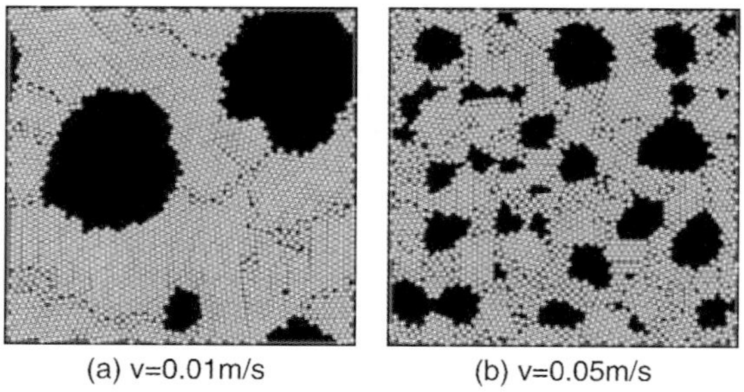

Figure 4.6.7
Structure change due to drying rate.

is defined as the first criterion using Delaunay triangulation. The second criterion, the non-dimensional boundary length (NBL) is defined to quantify the global structure of self-organized nanoparticles that cannot be classified using the IOF. The value of NBL is equal to one when every particle is dispersed, and it becomes smaller as aggregation proceeds.

The structure transforms during drying, already shown in Fig. 4.6.5 are plotted in Fig. 4.6.6. The figures from (a) to (f) in Fig. 4.6.5 correspond to the points from (a) to (f) in Fig. 4.6.6. Since NBL and IOF indicate the degree of dispersion and ordering, respectively, we can easily understand the structure changes during drying by tracking the points (a) to (f). The first structure (a) is just the starting state which is disorder with low IOF, and a little aggregation with high NBL.

The markers in Fig. 4.6.6 show the time evolution for every 0.05 μs. It is important to notice that the structure changes are due to drying which causes the change of dominant force because of the decrease of liquid thickness. The point (b) is totally dispersed because of large repulsive force due to high zeta potential. Next, the point (c) shows the progress of ordering with keeping dispersion. Then, the point (d) is just after the creation of voids and starts aggregation because of strong attractive force due to capillary force which becomes stronger as shown in the middle of Fig. 4.6.3 with the decreasing liquid thickness. In other words, the colloidal crystal, the point (c), breaks down because of the increase of attractive force, and results in the void formation. The point (e) is the growth of the voids with keeping the same IOF, ordering structure. Finally, the point (f) reaches the final ordered structure with the evolution of the voids, where is characterized with relatively high IOF and low NBL.

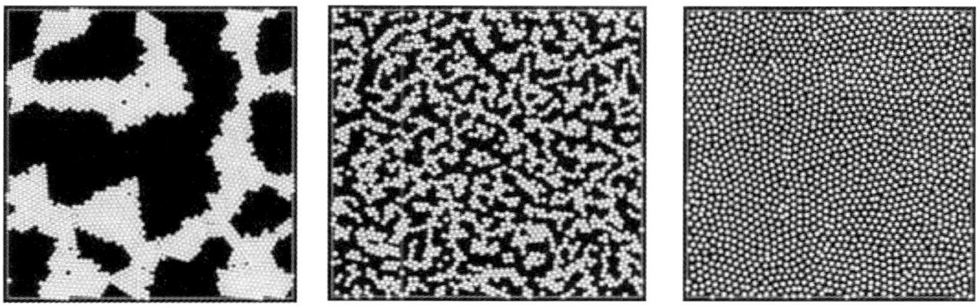

Figure 4.6.8
Capillary force and zeta potential effects (the coverage is 0.5).

The structure changes of nanoparticles during drying can be easily and visually understood from the simulated results on the IOF–NBL map, the structure map. There are a lot of factors to control the ordered structure of nanoparticles. It is concluded that the desired structure of nanoparticles can be obtained by drying.

3. Structure changes by process conditions

The structure of nanoparticles changes with the drying rate as shown in Fig. 4.6.7. The drying rate is increased by five times. Then, the number of voids increases as seen in Fig. 4.6.7(b). It is explained that the time for rearrangement is not enough to create larger domains which reach the equilibrium state. Since the rapid rate of drying results in disorder, it is necessary to decrease the drying rate when the ordered structure is desired.

The liquid thickness and the zeta potential are changed to study the effect of capillary, electrical, and Brownian forces. The nanoparticles employed here are polystyrene spheres with the diameter of 20 nm in aqueous solution. The crystal structure of the islands in Fig. 4.6.8(a) is fcc structure, because the attractive force or the capillary force becomes dominant. When the Brownian motion is dominant under the attractive force, the non-isotropic domains in Fig. 4.6.8(b) become smaller than the crystal domain of Fig. 4.6.8(a). When the electrostatic force, which corresponds to −100 mV, is dominant, the colloidal crystal as shown in Fig. 4.6.8(c) is obtained.

4. Conclusions

It was found that all forces considered here could become a primary factor in a coating-drying process, because the magnitude of each force changes considerably during the evaporation of solvent due to the change of the inter-particle distance and the contact height of the interface. Therefore, a multi-scale and multi-disciplinary modeling proposed here is indispensable for simulating the motion of colloidal nanoparticles during drying.

The ordered structure of nanoparticles, self-organized structure, is determined by the balance of various kinds of forces. They are attractive and repulsive forces on particles which are capillary, electrostatic, and van der Waals forces with Brownian force as a random force. It is concluded that the present study clarified two-dimensional self-organization of nanoparticles, which is indispensable to establish the structure design of nanomaterials.

References

[1] S. Maenosono, T. Okubo and Y. Yamaguchi: *J. Nanoparticle Res.*, **5**(1–2), 5–15 (2003).

[2] S. Coe, W.K. Woo, M. Bawendi and V. Bulovic: *Nature*, **420**, 800–803 (2002).

[3] W.U. Huynh et al.: *Adv. Funct. Mater.*, **13**(1), 73–79 (2003).

[4] M. Ohtsu et al.: *IEEE J. Selected Topics in Quantum Electronics*, **8**(4), 839–862 (2002).

[5] C.T. Black et al.: *Science*, **290**, 1131–1134 (2000).

[6] L. Monette, M.P. Anderson, H.D. Wagner, R.R. Mueller: *J. Appl. Phys.*, **75**, 1442–1455 (1994).

[7] Y. Yamaguchi: *Powder Sci. Eng.*, **33**, 63 (2001).

[8] M. Fujita, Y. Yamaguchi: *J. Surf. Sci. Soc. Jpn.*, **25**(10), 32–39 (2004).

[9] K. Nagayama: *Nature and Science of Self-assembly*, Maruzen ISBN: 9784621044117, (1997).

[10] M. Shisido: *Powder Sci. Eng.*, **34**(10), 62–68 (2002).

[11] M. Inoue, M. Yoshiga: *J. Jpn. Soc. Color Mater.*, **76**(1), 24–33 (2003).

[12] C. Wang et al.: *Chem. Mater.*, **12**(12), 3662–3666 (2000).

[13] M.P. Pileni: *J. Phys. Chem. B*, **105**(17), 3358–3371 (2001).

[14] E. Rabani et al.: *Nature*, **426**, 271–274 (2003).

[15] M. Fujita et al.: *Jpn. J. Appl. Phys.*, **43**(7A), 4434–4442 (2004).

[16] H. Nishikawa et al.: *J. Nanoparticle Res.*, **5**(1–2), 103–110 (2003).

[17] P.A. Kralchevsky, K. Nagayama: *Particles at Fluid Interfaces and Membranes*, Elsevier, Amsterdam (2001).

4.6.2 Assembling and patterning of particles

Particle collectives can be used for photonic crystals and the like by producing two- or three-dimensional particle structures on a micron- or nanosized scale based on preparation of bulky colloid crystals (opal structure), formation of a particle deposition layer on the board surface, and patterning of particle collectives in an arbitrary shape. Examples include two-dimensional pattering of a particle monolayer, two-dimensional pattering of particle multilayers with a controlled number of layers, and fine particle wiring with straight and curved particle alignments.

Since photonic properties with the band gap obtained from particle collectives in an fcc (hcp) structure are restricted, preparation of particle collectives with other crystal structures (periodical structure) is necessary. New technology for preparing nanomicro structures with full command of design in three-dimensional space is required for advanced photonic crystals and other new functions. For example, high-order structures have been prepared by making spherical particle collectives and its patterning and integration.

The following section describes the "two-solution method" for particle assembling and patterning, which uses a thin organic membrane called the self-assembled mono-layer as a template. It makes use of "liquid-phase pattering" in colloid solutions, the "drying patterning" process of drying the colloid solution and the interface between two kinds of solutions.

In describing this process, references are made to other studies: Liquid-phase pattering [1–6], drying patterning [1], preparation and patterning of a self-assembled mono-layer [7, 8], details of the two-solution method and the structure, defects, evaluation of the forming mechanism [9, 13], thermodynamic explanation of self-assembling phenomena (formation of a self-assembled mono-layer, formation of particle collectives) [9], survey of Japanese name of SAM [9], introduction of self-assembling pattering of particles using a functional template [11].

(1) Liquid-phase pattering

Liquid-phase pattering is capable of arrangement, alignment, and patterned integration of fine particles in only the specified region on the patterned self-assembled mono-layer by the chemical reaction and electrostatic interaction between the nano-/microparticles and the functional group at its surface after forming it on the board (Fig. 4.6.9) [14–16].

This method realizes particle monolayer patterning (Fig. 4.6.10(a)), straight/curved particle wiring (Fig. 4.6.10(b–d)), and precise arrangement of single particles [14]. In the particle monolayer patterning, a self-assembled mono-layer with an octadecyl group is prepared all over the surface of a silicon board. The specified area exposed to the emission of ultraviolet light by use of photo masking is converted into a silanol group and used as a template (Fig. 4.6.9).

The surface of the silica particles is also modified with a carboxyl group by converting the cyano group at the end of the self-assembled mono-layer into a carboxyl group and then the patterned self-assembled mono-layer is immerged in water dispersed with these particles for a few minutes. In water with a pH of about 7, the amino group has a positive zeta potential and the carboxyl group has a negative zeta potential. In this way, the particles are integrated in only the amino group region of the board for patterning using the electrostatic effect between the particles and the board (Fig. 4.6.10) [14].

(2) Drying patterning

The solution drying process has advantages in that the fcc (hcp) structure and closest packing structure are easily obtained by use of the contraction and meniscus force of colloid solution. Figure 4.6.11 shows the

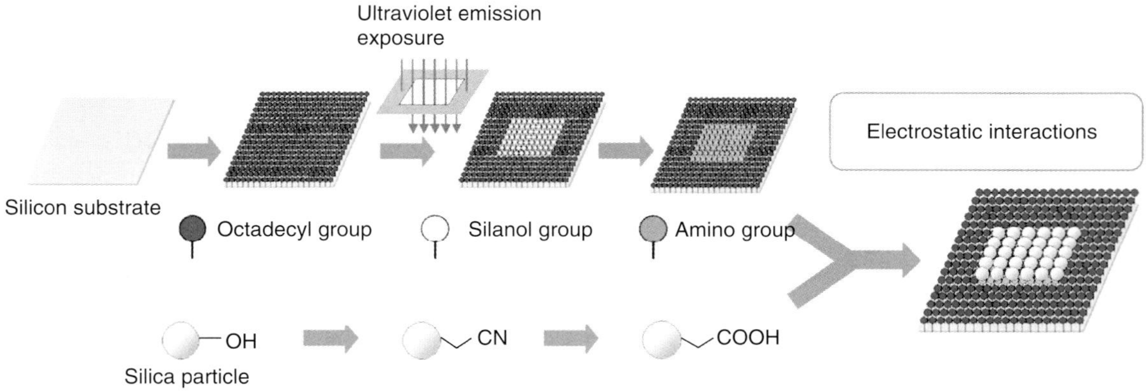

Figure 4.6.9
Liquid phase pattering using the self-assembled mono-layer.

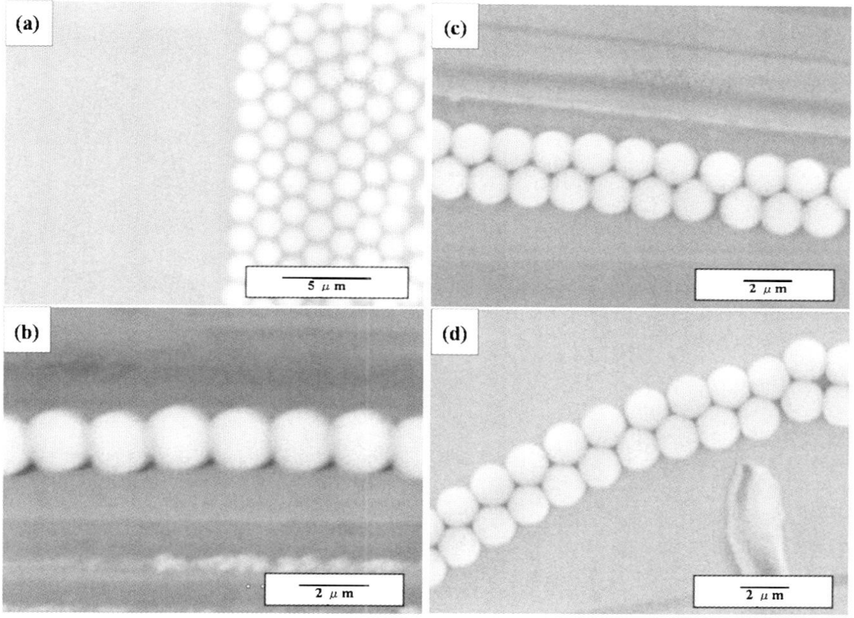

Figure 4.6.10
Particle patterns made by liquid patterning (a) Mono-layer pattern; (b) Particle wiring; (c) Fine double particle wiring; (d) Curved fine double particle wiring.

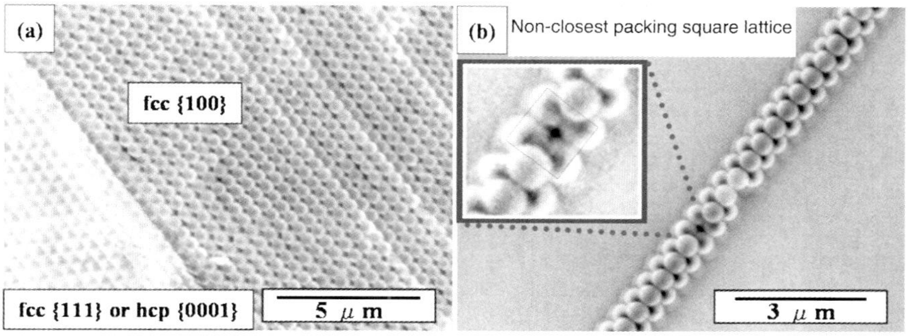

Figure 4.6.11
The colloid crystalline pattern prepared by drying patterning (a) Particle collectives with fcc (hcp) structure; (b) Fine particle wiring with square lattice of non-closest packing.

colloid crystalline with arbitrarily shaped patterning prepared by using a self-assembled mono-layer patterned with hydrophilic and hydrophobic regions as a template, dropping the solution to align the colloid solution along the hydrophilic region and drying out the solvent [17].

Fine wiring arrayed with particle lines at the same interval was prepared by drying the solvent from the surface of the board covered by the self-assembled mono-layer keeping it standing in a colloid solution. It is based on the phenomena that the colloid solution behaves discontinuously on the board surface (Fig. 4.6.12 upper (a~f), lower (A~E) [18, 19]). Furthermore, it is possible to prepare by this method particle fine wiring with a squared lattice of non-closest packing structure (Fig. 4.6.12(b) [17]) and particle fine wiring with a NaCl crystal structure (Fig. 4.6.12(F) [18]).

It was thought that only the fcc (hcp) structure of closest packing represented by opals could be prepared from a colloid solution. However, these processes show the possibility of preparing structures

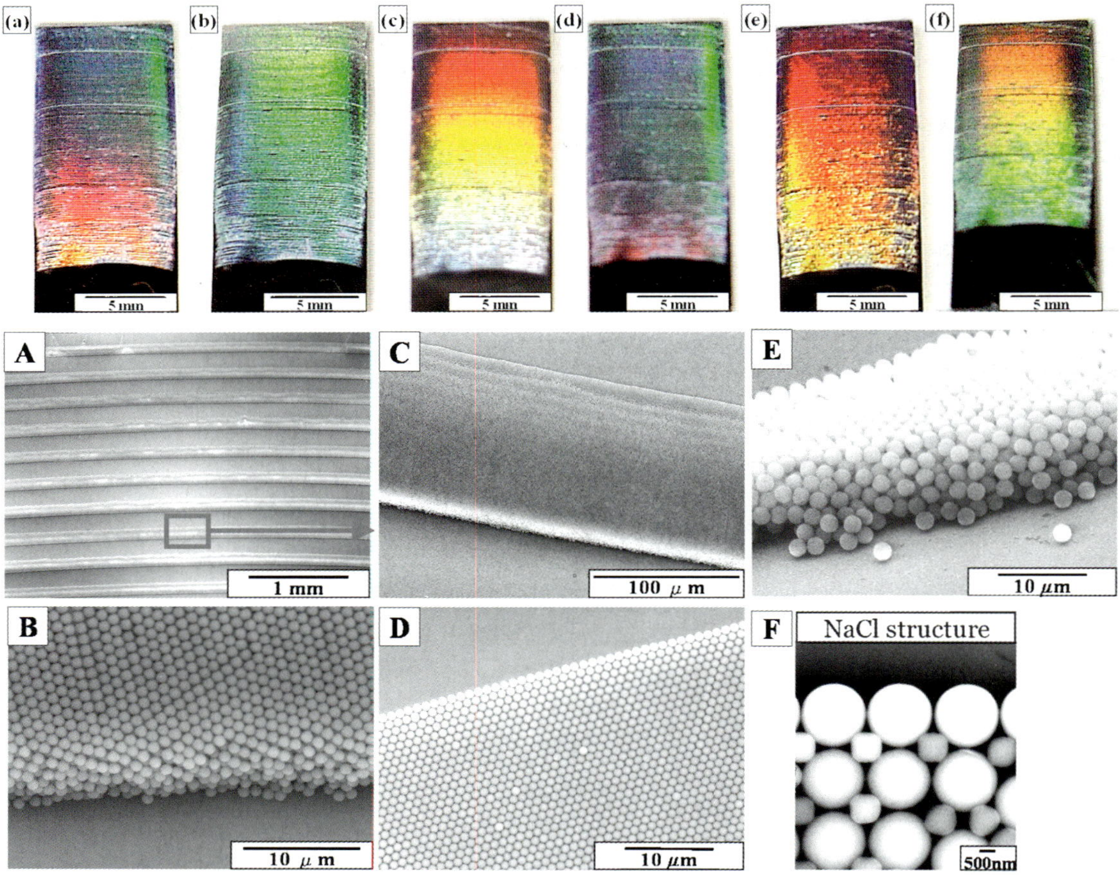

Figure 4.6.12
The fine particle-wiring array prepared by drying patterning (a~f): The same sample showing different rainbow structural colors depending upon the direction of observation. (A~E): Microstructure of the fine particle-wiring array.

other than fcc (hcp) by the particle self-assembling phenomena from the colloid solution, which indicates the great potential of the self-assembling process.

(a) Two-solution self-assembly method

Liquid-phase patterning and drying patterning have advantages and disadvantages. In the liquid phase, there is the advantage of utilization of the electrostatic interaction and reactions between the particles and the board against the disadvantage of less driving force for the closest packing of particles.

In the process of the drying solution, the meniscus force generated at the contraction of the colloid solution and drying of solvent between the particles can be used as the driving force for the closest packing of particles, while there is the disadvantage that chemical reactions are hardly made use of because of the dynamic process. For these reasons, the two-solution method having the advantages of both processes has been developed in order to realize the preparation of particle self-assembled agglomerate pattering [20] and the preparation and patterning of spherical particle agglomerates [21].

(b) Preparation of particle self-assembled collectives pattering by two-solution method

After making a self-assembled monolayer with the patterning of hydrophobic and hydrophilic regions on the silicon board, methanol solution with dispersed particles is dropped on it (Fig. 4.6.13). Then the solution is repelled from the hydrophobic region and patterned along with the hydrophilic region. After that, it is soaked in decalin solution, which is barely soluble to methanol. With gradual solution of methanol in decalin, the area of methanol including the particles shrinks and the particles become closer to each other.

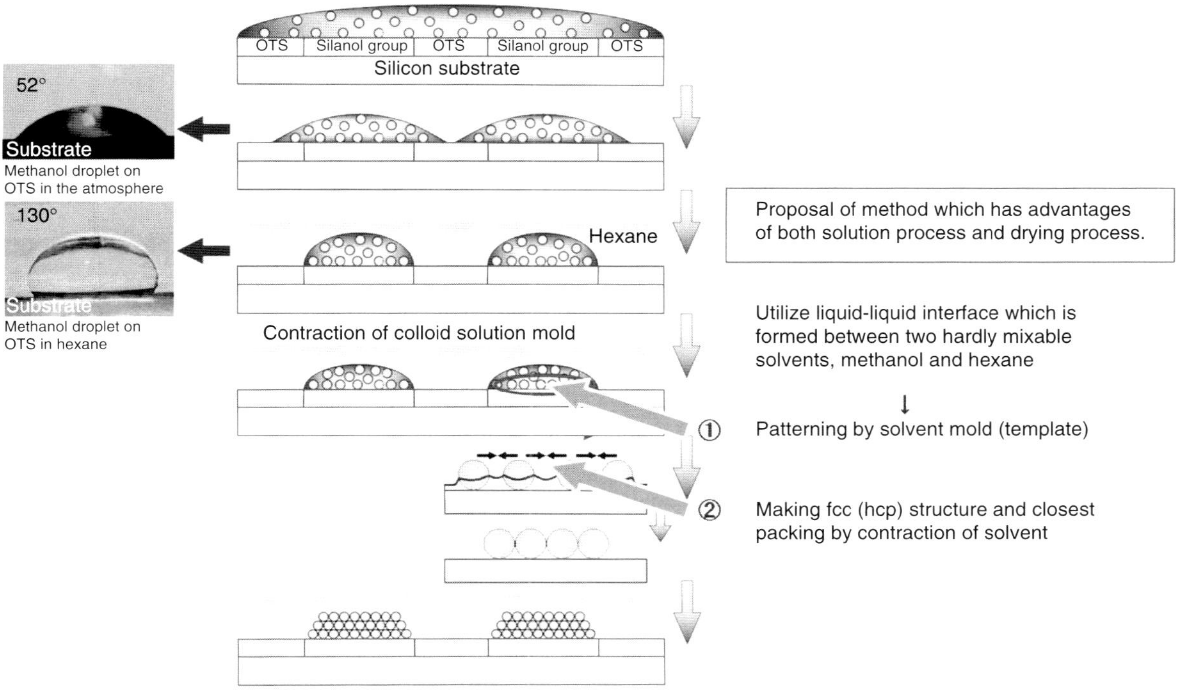

Figure 4.6.13
Particle collective patterning made by two-solution method.

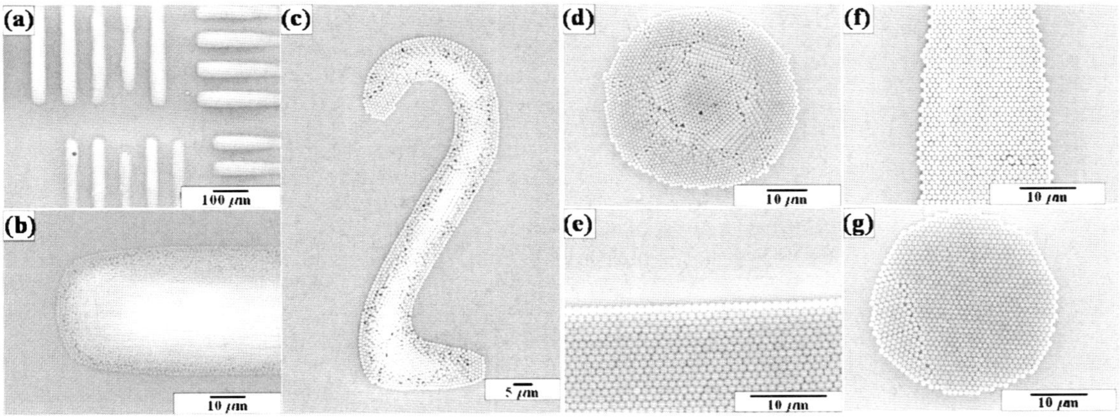

Figure 4.6.14
Particle collective patterns prepared by two-solution method. (a, b) Multi-layer particle collective patterns; (c, d) Double-layer particle patterns; (e–g) Mono-layer particle patterns.

In the stage of further solution of methanol, the methanol forms a bridge between adjacent particles at the interface between the methanol and the decalin, which causes a meniscus force attracting the particles to each other. The contraction of colloid solution occurring in these processes and the meniscus force at the last stage of drying accelerate the preparation of fcc (hcp) construction and closest packing of the particles and form the particle collectives along the hydrophilic patterning on the board.

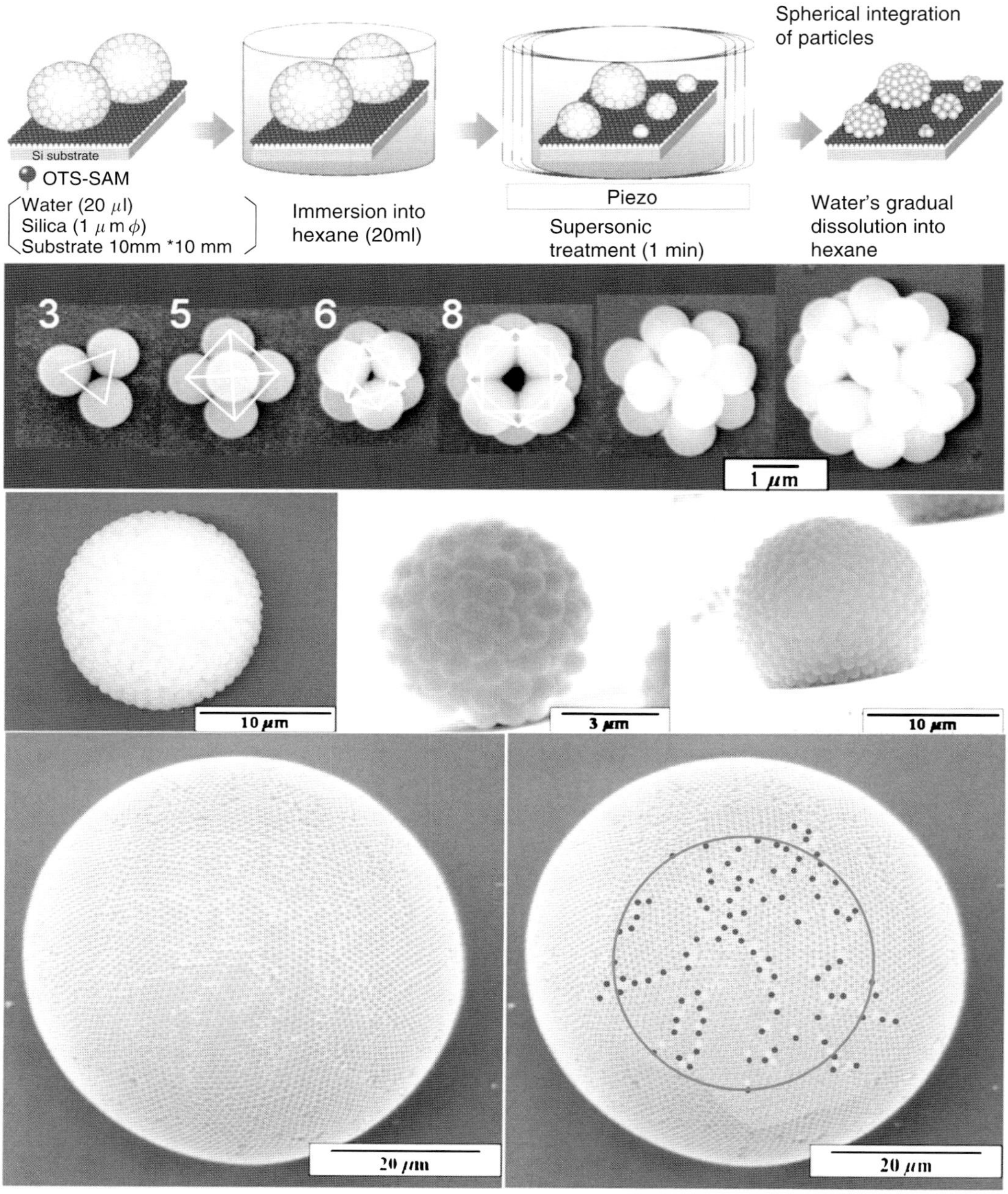

Figure 4.6.15
Preparation of spherical particle collectives made by two-solution method.

Additionally, since the whole process is conducted completely in a liquid phase, it is possible to add other chemical reactions easily to this reaction system. This process can realize multi-layer particle collective patterns in an arbitrary shape, double-layer particle pattern, and mono-layer particle pattern (Fig. 4.6.14) [20].

As seen in Fig. 4.6.14(e), the particles on the edge of the particle assembled layer pattern are arranged in a straight line with an accuracy less than a few hundreds of nanometers showing high patterning precision. The centers of the individual particles are positioned in a regular arrangement with an accuracy of less than a

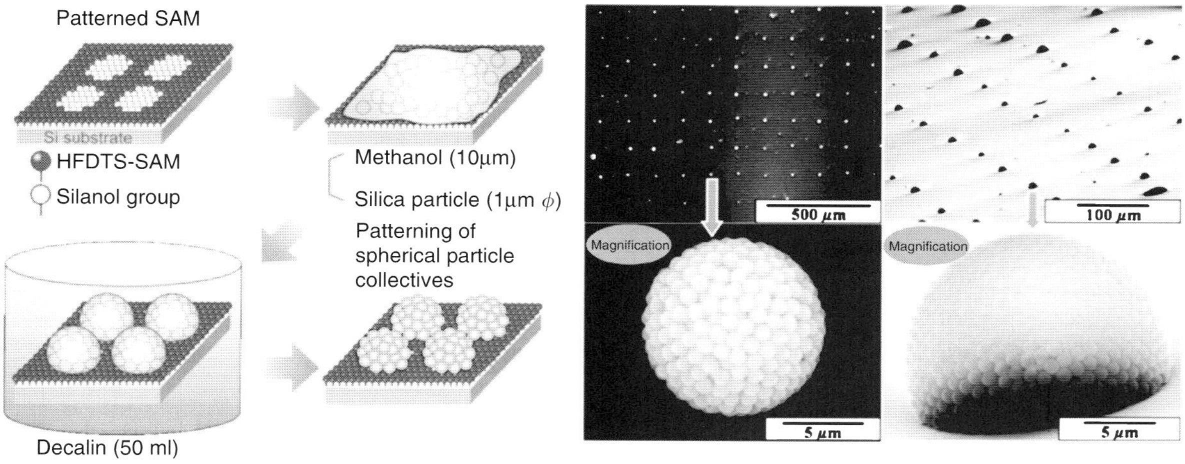

Figure 4.6.16
Preparation of spherical particle collective patters prepared by two-solution method.

few nanometers over the whole particle collective and a rainbow structure color is observed from it.

(c) Preparation and pattering of spherical particle collectives by two-liquid method

Based on this process, the preparation of spherical particle collectives made of an arbitrary number of particles (Fig. 4.6.15) and its patterning (Fig. 4.6.16) have been realized [21]. For the preparation of spherical particle collectives, particles-dispersed water is dropped on a hydrophobic self-assembled monolayer and soaked in hexane, which is barely soluble to water. After the size control divides the droplets in the colloid solution by ultrasonic treatment, the particle collectives are sphericalized due to the hexane/water solution.

By this method, particle collectives of any size can be prepared by changing the particle concentration and droplet size. In the patterning, the hydrophobic/ hydrophilic self-assembled monolayer is used as a template to arrange the collectives at the same interval.

For these structures, it is possible to control the shape from a sphere to a flat dome by changing the combination of self-assembled monolayers and solution as well as the density ratio of the two solutions and controlling the contact angle of the colloid solution and the buoyant force exerted on it. Concerning the process of preparing these particle collectives, the possibility of producing multi-dimensional periodical structures and controlling the photonic band gap is being investigated by making second-stage periodical structures by the use of spherical particle collectives.

By these particle assembling and patterning processes, the particle collective patterning of any shape can be created using a solution process that has low energy consumption and low environmental load. However, in order to materialize photonic devices, the manufacture of various kinds of structures with high accuracy is required to control the photonic band gap and to produce different types of functional integration. Further development in this field is expected.

References

The related papers of the author are found at the Internet site http://staff.aist.go.jp/masuda-y/index.html

[1] Y. Masuda: *Metamorphosis*, Murata Manufacturing Co., Ltd., Vol. 10, pp. 26–27 (2005).

[2] Y. Masuda, K. Koumoto: *Ceramics Data Book*, Vol. 28, pp. 47–49 (2000).

[3] Y. Masuda, N. Saito and K. Koumoto: *Bull. Ceram. Soc. Jpn.*, **37**, 615–620 (2002).

[4] Y. Masuda, K. Tachibana, M. Itoh and K. Koumoto: *Mater. Integration*, **14**, 37–44 (2001).

[5] Y. Masuda, K. Tachibana, M. Itoh and K. Koumoto: *Fabrication of Microstructure for Novel Property and their Application*, TIC, Tokyo, Japan, (2003).

[6] Y. Masuda, K. Koumoto: *J. Soc. Inorg. Mater. Jpn.*, **7**, 4–12 (2000).

[7] Y. Masuda: *Mater. Sci. Technol., Kinzoku*, **76**(3), 284–292 (32–40) (2006).

[8] Y. Masuda, K. Koumoto: *Nano/micro-patterning Process Technology in Next Generation*, Science & Technology Co., Ltd, Tokyo, Japan, pp. 180–188 (2006).
[9] Y. Masuda: *J. Soc. Powder Technol., Jpn.*, **43**, 362–371 (2006).
[10] Y. Masuda: *Annu. Report Murata Sci. Foundation*, **19**, 345–355 (2005).
[11] Y. Masuda: *Bull. Ceram. Soc. Jpn.* **41**(5), 346–351 (2006).
[12] Y. Masuda: *Bull. Nano Sci. Technol.*, **4**(2), 59–66 (2006).
[13] Y. Masuda, K. Koumoto: *Assembly Films of Functional Materials and their Applications*, CMC Publishing Co., Ltd., Tokyo, Japan, pp. 103–118 (2006).
[14] Y. Masuda, M. Itoh, T. Yonezawa, and K. Koumoto: *Langmuir*, **18**, 4155–4159 (2002).
[15] Y. Masuda, W.S. Seo and K. Koumoto: *Jpn. J. Appl. Phys.*, **39**, 4596–4600 (2000).
[16] Y. Masuda, W.S. Seo and K. Koumoto: *Thin Sol. Films*, **382**, 183–189 (2001).
[17] Y. Masuda, K. Tomimoto and K. Koumoto: *Langmuir*, **19**, 5179–5183 (2003).
[18] Y. Masuda, T. Itoh, M. Itoh and K. Koumoto: *Langmuir*, **20**, 5588–5592 (2004).
[19] Y. Masuda, M. Itoh and K. Koumoto: *Chem. Lett.*, **32**, 1016–1017 (2003).
[20] Y. Masuda, T. Itoh and K. Koumoto: *Langmuir*, **21**, 4478–4481 (2005).
[21] Y. Masuda, T. Itoh and K. Koumoto: *Adv. Mater.*, **17**, 841–845 (2005).

4.6.3 Fabrication of organic/inorganic mesoporous materials

Ordered mesoporous materials with uniform nano-sized pore structures are expected to be applied in a wide range of fields as catalysts, adsorbents, and in electronic devices [1]. Originally, the framework of mesoporous materials was composed of silica (SiO_2) or silica-alumina (SiO_2-Al_2O_3), but the compositions of the framework have now been extended to include metalo-silicate (SiO_2-$MO_{n/2}$), transition metal oxides ($MO_{n/2}$), sulfides, phosphates, metals, carbon, and organic materials. These frameworks themselves can provide a variety of functions, and expanded the range of application for mesoporous materials. However, most of these mesoporous materials have an amorphous pore wall structure, and it has been difficult for the high-level functions inherently possessed by the inorganic crystals to be expressed. On the other hand, attempts have been made to introduce organic groups into mesoporous materials, and use the various functions of the organic groups in the stable nanospace of the pores [2–5]. Organic/inorganic hybrid mesoporous materials include surface-bound types of materials in which an organic group is attached to the pore surface, and uniform framework types in which an organic group and silica moieties are uniformly distributed in the framework (Fig. 4.6.17). The uniform framework type is especially interesting because of the highly ordered mesostructure and the crystal-like periodic structure within the pore walls.

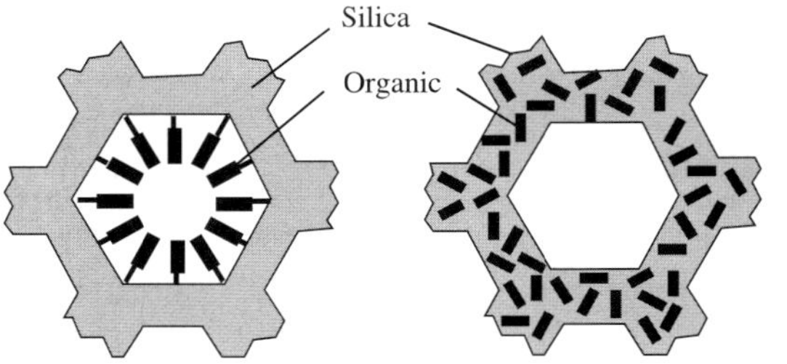

Figure 4.6.17
Framework structure of organic–inorganic hybrid mesoporous materials.

(2) Mesoporous ethane-silica

A hybrid mesoporous material with a uniform framework can be synthesized from organic silane in which there are 2 or more Si alkoxides ($Si(OR')_3$) bound to an organic group (R). In 1999, the author's group first reported on a uniform framework type of hybrid mesoporous material synthesized from an ethane-bridged organic silane ($(CH_3O)_3Si$-CH_2CH_2-$Si(OCH_3)_3$) using alkyltrimethyl ammonium (ATMA) as the template (Fig. 4.6.18) [6]. By controlling the synthesis temperature and the alkyl chain length of the ATMA, it was found possible to synthesize 3 types of mesoporous materials, not just two-dimensional (2D) hexagonal structures [6], which have a one-dimensional pore structure, but also cubic *Pm-3n* [7–9] and three-dimensional hexagonal [6] with a three-dimensional (3D) cage structure (Fig. 4.6.19). These mesoporous ethane-silica structures are highly ordered, and the particle morphology is well-defined in the form of a hexagonal column (2D hexagonal), sphere (3D hexagonal), or 18-facet polyhedron (cubic *Pm-3n*), reflecting the symmetrical characteristics of the pore structure [6–9]. Especially cubic mesoporous particles are particularly uniform, as all particles have the same size (5 μm) and shape (18-facet).

(3) Synthesis of various mesoporous organic silicas

Later, a family of various types of mesoporous organic silica were synthesized containing organic groups such as ethylene [10–12], phenylene [13–15], methane, and toluene, and highly functional organic groups such as phenyl groups [16],

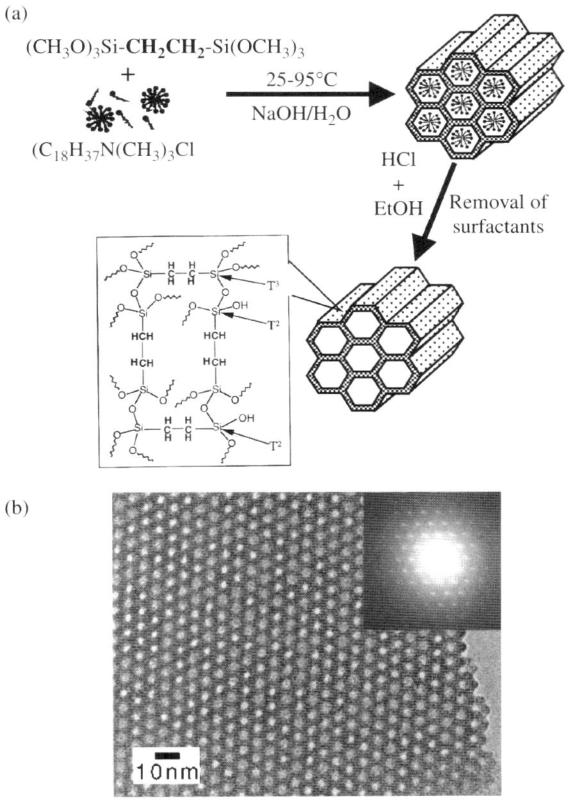

Figure 4.6.18
(a) Synthesis of mesoporous ethane-silica and (b) TEM image of pore structure.

Figure 4.6.19
SEM images of mesoporous ethane-silicas
(a) 2D-hexagonal; (b) 3D-hexagnal; (c, d) Cubic *Pm-3n*.

violegen [17], and cyclic amine compounds [18]. Adsorption analysis clearly indicates that the hydrophobic properties of the pore surface are increased by the introduction of an organic group such as ethane [19, 20]. Violegen is an electron acceptor that appears in various charge transfer complexes and during the course of electron transfer. It has been confirmed that radical cations are produced when a mesoporous material to which violegen has been introduced is irradiated with a 308 nm laser beam [17]. Mesoporous materials in which a cyclic amine has been introduced, are expected to have various applications in catalytic reactions, since the metal on the pore wall can be coordinated [18]. In addition, by co-condensation of two types of organic silane precursors, mesoporous materials can be synthesized in which different organic groups have been introduced both in the pore wall and on the wall surface [21].

It is difficult to directly synthesize mesoporous materials that contain bulky organic groups or highly reactive organic groups. Thus, methods have been proposed in which a mesoporous material with reactive organic group such as phenylene or ethylene group is chemically modified by a post-treatment process. By treating a mesoporous benzene silica with fuming sulfuric acid, it was found possible to sulfonize some of the phenylene groups while maintaining the ordered structure of the material [14]. Benzocyclobutene has also been successfully bonded to mesoporous ethylene-silica via a Diels–Alder reaction [22]. The benzene ring bonded to the surface was easily sulfonized. The solid acid catalytic properties of these sulfonized mesoporous organic silicas are currently being investigated.

(4) Crystal-like mesoporous organic silica

Mesoporous silica with a crystal-like periodic pore wall structure has been synthesized from a benzene-bridged organic silane ($(EtO)_3Si\text{-}C_6H_4\text{-}Si(OEt)_3$) (Fig. 4.6.20) [14]. The mesoporous benzene-silica has 2D hexagonal structure. The silica and benzene moieties are arranged in rings on the pore wall encircling the pore (Fig. 4.6.21). Moreover, an ordered surface structure is formed in which the silica and benzene layers are alternately arranged with a spacing of 7.6 Å in the direction of the pore. Silanol (Si-OH) covers the surface of the silica, and presents hydrophilic properties, while the benzene layer displays hydrophobic properties. This type of periodic surface, being able to align and adsorb molecules and clusters, is expected to dynamically increase catalytic selectivity and activity and enhance the properties of the molecules and clusters enclosed in the material. This material exhibits a high degree of thermal and hydrothermal stability. Thermogravimetry has confirmed that the benzene is maintained on the pore walls at temperatures up to 500°C in both air and nitrogen. Moreover, the periodicity is maintained on

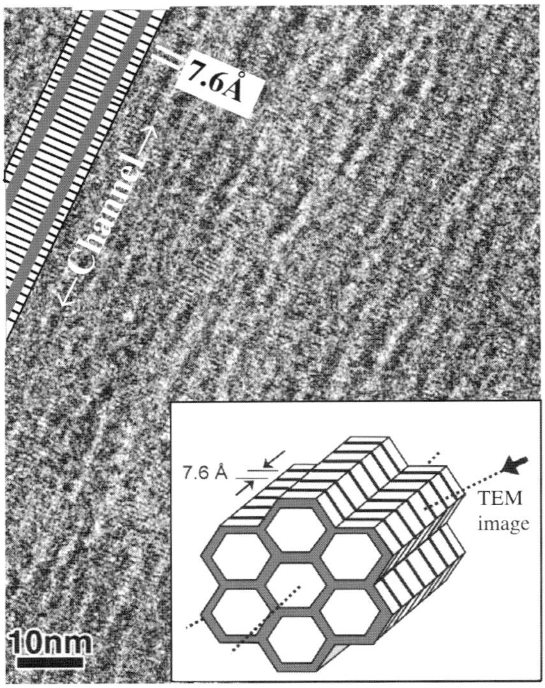

Figure 4.6.20
TEM image of crystal-like mesoporous phenylene-silica.

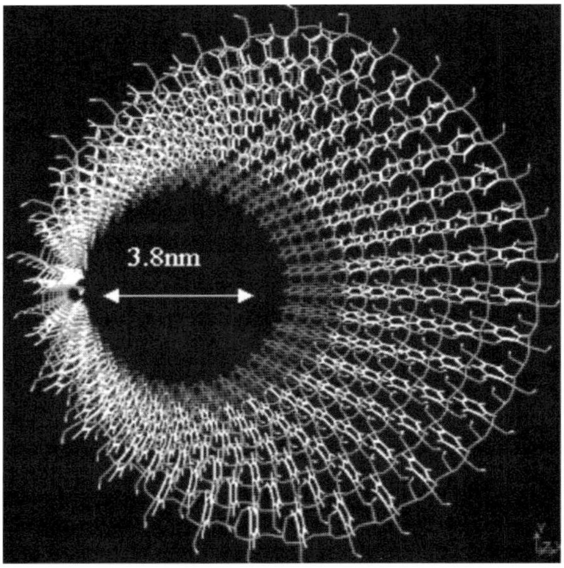

Figure 4.6.21
Simulation model of crystal-like mesoporous phenylene-silica.

both the mesoscale and molecular scale even when this material is boiled in water for 8 h.

Similar molecular-scale periodicity has been observed with biphenyl ($-C_6H_4$-C_6H_4-) [16], 1,3-benzene [23], and toluene. Although a highly ordered mesostructure is formed with ethane and methane, which undergo weak interaction, a periodic structure does not develop on the pore walls. This indicates that strong interaction with the organic silane molecules plays a crucial role in the formation of the molecular-scale periodicity. A benzene-silica material with a lamellar mesostructure and a molecular-scale periodicity was also synthesized by controlling the synthesis conditions [24]. Crystallization of amorphous pore walls of mesoporous benzene-silica has also been performed as a post-treatment process [25].

Silanol (Si-OH) or organic groups on the pore wall can be readily modified, and up to now, this method has been applied to introduce sulfone groups (-SO_3H) and to solid acid catalysts [26–28].

References

[1] K. Kuroda: *Ceramics*, **36**, 902–905 (2001).
[2] S. Inagaki: *J. Soc. Powder Technol., Jpn.*, **39**, 518–526 (2002).
[3] J.Y. Ying, C.P. Mehnet and M.S. Wong: *Angew. Chem. Int. Ed.*, **38**, 56–77 (1999).
[4] A. Stein, B.J. Meldw and R.C. Schroden: *Adv. Mater.*, **12**, 1403–1419 (2000).
[5] A. Sayari, S. Hamoudi: *Chem. Mater.*, **12**, 3151–3168 (2001).
[6] S. Inagaki, S. Guan, Y. Fukushima, T. Ohsuna and O. Terasaki: *J. Am. Chem. Soc.*, **121**, 9611–9614 (1999).
[7] S. Guan, S. Inagaki, T. Ohsuna and O. Terasaki: *J. Am. Chem. Soc.*, **122**, 5660–5661 (2000).
[8] S. Guan, S. Inagaki, T. Ohsuna and O. Terasaki: *Microporous Mesoporous Mater.*, **44–45**, 165–172 (2001).
[9] M.P. Kapoor, S. Inagaki: *Chem. Mater.*, **14**, 3509–3514 (2002).
[10] B.J. Melde, B.T. Holland, C.F. Blanford and A. Stein: *Chem. Mater.*, **11**, 3302–3308 (1999).
[11] T. Aefa, M.J. MacLachlan, N. Coombs and G.A. Ozin: *Nature*, **402**, 867–871 (1999).
[12] K. Nakajima, D.L. Lu, J.N. Kondo, I. Tomita, S. Inagaki, M. Hara, S. Hayashi, K. Domen: *Chem. Lett.*, **32**, 950–951 (2003).
[13] C.Y. Ishii, T. Asefa, N. Coombs, M.J. MacLachlan and G.A. Ozin: *Chem. Commun.*, 2539–2540 (1999).
[14] S. Inagaki, S. Guan, T. Ohsuan and O. Terasaki: *Nature*, **416**, 304–307 (2002).
[15] Y. Goto, S. Inagaki: *Chem. Commun.*[J], 2410–2411 (2002).

[16] M.P. Kapoor, Q. Yang and S. Inagaki: *J. Am. Chem. Soc.*, **124**, 15176–15177 (2002).
[17] M. Alvaro, B. Ferrer, V. Fornes and H. Garcia: *Chem. Commun.*, **24**, 2546–2547 (2001).
[18] R.J.P. Corriu, A. Hehdi, C. Reye and C. Thieuleux: *Chem. Commun.*, **13**, 1564–1565 (2003).
[19] M.P. Kapoor, N. Setoyama, Q. Yang, M. Ohashi and S. Inagaki: *Langmuir*, **21**, 443–449 (2005).
[20] M. Kruk, M. Jaroniec, S. Guan and S. Inagaki: *J. Phys. Chem. B*, **105**, 681–689 (2001).
[21] Q. Yang, M. Kapoor and S. Inagaki: *J. Am. Chem. Soc.*, **124**, 9694–9695 (2002).
[22] K. Nakajima, I. Tomita, M. Hara, S. Hayashi, K. Domen and J.N. Kondo: *Adv. Mater.*, **17**, 1839–1842 (2005).
[23] M.P. Kapoor, Q. Yang and S. Inagaki: *Chem. Mater.*, **16**, 1209–1213 (2004).
[24] K. Okamoto, M.P. Kapoor and S. Inagaki: *Chem. Commun.*, **11**, 1423–1425 (2005).
[25] Y. Goto, K. Okamoto and S. Inagaki: *Bull. Chem. Soc. Jpn.*, **78**, 932–936 (2005).
[26] Q. Yang, J. Liu, J. Yang, M.P. Kapoor, S. Inagaki and C. Li: *J. Catal.*, **228**, 265–272 (2004).
[27] Q. Yang, M.P. Kapoor, N. Shirokura, M. Ohashi, S. Inagaki, J.N. Kondo and K. Domen: *J. Mater. Chem.*, **15**, 666–673 (2005).
[28] Q. Yang, M.P. Kapoor, S. Inagaki, N. Shirokura, J.N. Kondo and K. Domen: *J. Mol. Catal. A: Chemical*, **230**, 85–89 (2005).

FUNDAMENTALS

CHAPTER 5

CHARACTERIZATION METHODS FOR NANOSTRUCTURE OF MATERIALS

CONTRIBUTORS

Satoshi Ohara (5.1)
Tadafumi Adschiri (5.1)
Takashi Ida (5.2.1), (5.2.2)
Masatomo Yashima (5.2.3), (5.2.4)
Takeshi Mikayama (5.3.1), (5.3.2)
Hiroya Abe (5.3.3)
Yuichi Setsuhara (5.3.4)
Kiyoshi Nogi (5.3.5)
Minoru Miyahara (5.4)
Kenji Kaneko (5.5)
Akira Ohtomo (5.6)

5.1 Nanostructure and function (characterization of local nanostructure)

The 20th century was the so-called "age of new material synthesis", as shown in Fig. 5.1.1 [1]. "Synthesis" of new substances had been essential to attain a new property for materials. However, in the recent R&D of material, keen attention is attracted on controlling material function by the material structure. Nanotechnology is a typical example of this stage of technology. In the control of material structure, "process" is regarded more important than "synthesis". For example, in the nanoparticle production, size and shape of particle significantly depends on the operating conditions of crystallization process. It was found that some semiconductor materials show strong photo-luminescence and the wavelength can be controlled with its particles size ("quantum size effect"). With a same wavelength of excitation, a wide range of colors can be obtained by changing the particle size of the nanoparticles of CdSe, CdTe, CdS, etc., the so called quantum dots (QD) [2]. Photonic crystal can be produced by self-assembly of nanomaterials (homo-assembly) and the control of self-assembly phenomena is a critical issue of process design and operation. In the fabrication of ultraviolet (UV) laser [3–5] and light-emitting diode (LED) [6] based on ZnO materials, laser molecular beam epitaxy is a powerful process to grow fine ZnO nanocrystals self-assembled and arrayed parallel on substrates. The next stage of the nanotechnology will be "programmed assembly" of nanomaterials, because this is a foreseeable future target for the industries of various fields. In the device production by assembling the structured materials, "fabrication" is the key technology. Thus, science and technology for materials is being shifted from "synthesis" toward "processing" and "fabrication".

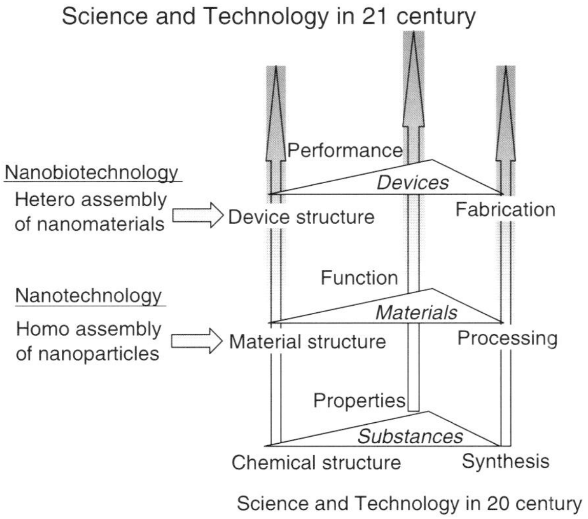

Figure 5.1.1
Science and technology in 20th and 21st century.

Nanobiotechnology has substantial potential for the programmed fabrication of structured materials, in which specific interactions between molecules of DNA or peptide are used to make "hetero-assemblies" of multi-component materials or nanoblocks (biomaterials or bioreactive nanomaterials). Control of assembly and disassembly of a variety of biomolecules, and the nanoblock, will be a key of fabrication of multi-component and multistructured materials. Furthermore, if the assembly and disassembly of the structured materials can be controlled by manipulating the atmosphere, the material will be designed to change its structure dynamically.

Not only the trend of academic researches mentioned above but also a significant shift has occurred for R&D in chemical industry, in which most of new resources are now devoted to the design and manufacture of specialty chemicals, structured materials, and chemical devices rather than traditionally dominant commodity chemicals. Application of the new products covers semiconductors, environmental, health and medical industries among others. Determining how to manipulate the properties of nanomaterials and their size variation has become a worldwide trend in nanotechnology. This has accelerated the shift in R&D of new products. Control of size and shape of nanomaterials and its assembly to generate structured materials will give rise to create new products with new functions that will dominate the new economies.

Nanoanalysis/measurement is very important to develop "processing" and "fabrication" for nanomaterials, because the functions of nanomaterials are influenced by their size, shape, and structures. Characterization methods for analysis and measurement for nanophenomena is essential in the development of nanobiotechnology. Especially, development of precise analysis with atomic level for local nanostructures such as chemical composition and bonding state, defects, and impurities, is a key to elucidate the mechanism of nanophenomena. Recently, improvements in analytical equipment have allowed researchers to study and to understand the local nanostructures in functional materials from atomic level. In this Chapter 5, the top-end characterization methods for nanomaterials are introduced.

References

[1] T. Adschiri: *Chem. Eng. Jpn.*, **66**, 554 (2002).

[2] I.L. Medintz, H.T. Uyeda, E.R. Goldman and H. Mattoussi: *Nat. Mater.*, **4**, 435–446 (2005).

[3] Z.K. Tang, G.K.L. Wong, P. Yu, M. Kawasaki, A. Ohtomo, H. Koinuma and Y. Segawa: *Appl. Phys. Lett.*, **72**, 3270–3272 (1998).

[4] M. Kawasaki, A. Ohtomo, I. Ohkubo, H. Koinuma, Z.K. Tang, P. Yu, G.K.L. Wong, B.P. Zhang and Y. Segawa: *Mater. Sci. Eng. B*, **56**, 239–245 (1998).

[5] M.H. Huang, S. Mao, H. Feick, H. Yan, Y. Wu, H. Kind, E. Weber, R. Russo and P. Yang: *Science*, **292**, 1897–1899 (2001).

[6] A. Tsukazaki, A. Ohtomo, T. Onuma, M. Ohtani, T. Makino, M. Sumiya, K. Ohtani, S.F. Chichibu, S. Fuke, Y. Segawa, H. Ohno, H. Koinuma and M. Kawasaki: *Nat. Mater.*, **4**, 42–46 (2005).

5.2 Crystal structure

5.2.1 X-ray diffraction method

X-ray diffraction (XRD) method is the most basic method for characterizing the crystal structures. Evaluation of crystal structure is generally important even for the nanoscale materials, the properties of which might be affected by the structures of several to several hundreds nanometer scale, while the crystal structures are specified by the arrangement of atoms separated by about 0.1 nm. In some cases, the crystal structure of each nanoparticle should be more appropriately characterized by electron beam diffraction (ED) method using a transmission electron microscope (TEM) rather than XRD. However, XRD method certainly has the following advantages: (i) measurements in the air or necessary atmosphere are allowed, (ii) the preparation of sample is easier, and (iii) the average crystal structures can be quantitatively evaluated.

The XRD and small-angle X-ray scattering (SAXS) methods are based on the measurements of X-ray intensities scattered by the statistically distributed electrons belonging to the atoms in the material. The arrangement of atoms or the population of electrons is determined by analysis of angular dependence of scattered X-ray in both the methods. The XRD method is suitable to determine the crystal structures by analyzing the positions and intensities of diffraction peaks typically observed for the well-crystallized material in the range of diffraction angle from 10 to 150°. The method is also used for evaluating the microstructures (crystallite size and microstrain) by analyzing the width and the shape of the peak profiles. The XRD method is sometimes called as wide-angle X-ray diffraction (WAXD) method, to be distinguished from the SAXS method.

Since the most stable structure of a pure material is crystal, where the atoms are periodically arranged, the pattern of the positions and intensities of XRD peaks can be uniquely assigned to the material. Therefore, XRD measurement is important to identify the main component of materials. The most commonly used database for the identification is called JCPDS-ICDD. JCPDS database consists of sets of data about interplanar distances and intensities of the significant diffraction peaks. Since the diffraction angle 2θ the interplanar distance d and X-ray wavelength λ are connected with each other by the Bragg's law:

$$\lambda = 2d\sin\theta, \qquad (5.2.1)$$

the observed diffraction angle 2θ can straightly be related to the interplanar distance d, applying the X-ray wavelength used for the measurement.

A conventional powder X-ray diffractometer commonly used in laboratories is enough for the purpose of evaluating the structures in most cases. Since the characteristics of the diffractometer with the conventional design (Bragg–Brentano geometry) is well known, it is often more reliable than diffractometers with more advanced designs. The CuK_α characteristic X-ray (mean wavelength: 0.15418 nm) is usually used as the X-ray source for the powder XRD measurements. Since the CuK_β characteristic X-ray (wavelength: 0.13922 nm) is also radiated from the X-ray tube with a Cu target, a Ni filter or curved graphite monochromator is used to eliminate the XRD peak caused by the K_β X-ray. Cu$K\alpha$ X-ray is a 2:1 mixture of $K\alpha_1$ (0.15405 nm) and $K\alpha_2$ (0.15443 nm) doublet. It is recommended to use the value of $K\alpha_1$ wavelength and the data obtained by eliminating the effect of $K\alpha_2$ subpeak to precisely determine the interplanar distance. The software for the Rachinger method [1], which is well known as a conventional method to eliminate the $K\alpha_2$ subpeak, is often attached as standard support software to a commercial powder X-ray diffractometer. It should be noted that one should be careful on interpretation of the data after the elimination of $K\alpha_2$ subpeaks, because the application of the method generally introduces additional errors into the data. In the Rietveld method, which will be described later, the raw data without applying elimination of $K\alpha_2$ subpeaks are usually used for the analysis.

Combination of a scintillation counter and a receiving slit has been conventionally used to detect the diffracted X-ray, but an advanced type of diffractometers attaching position-sensitive detectors (PSD) are currently commercially available, which can virtually improve the efficiency of detection. It is worth taking into consideration the use of those new-type detection systems, especially for such cases where rapid measurement or reduction of statistical uncertainty is strongly required.

Evaluation of crystallite size by peak profile analysis is one of the important applications of powder

XRD method to study on nanoparticles. The range of crystallite size that can be evaluated by powder XRD method depends on the accuracy of the instrument and analytical methods. Since the guaranteed precision of a commercial X-ray diffractometer is better than 0.01°, evaluation of line broadening larger than 0.1° will sufficiently be evaluated. The broadening of 0.1° roughly corresponds to the crystallite size for frequently used Cu$K\alpha$ of the wavelength 0.15 nm as follows, (180° × 0.15nm)/(0.10° × π) = 90nm. It means that crystallite size under 100 nm can quantitatively be evaluated with a usual powder diffractometer. On the other hand, it is often difficult to distinguish a diffraction peak from the neighbor peak or from background scattering, when the width of the peak is larger than several degrees, which corresponds to the crystallite size of several nm. Therefore, we can conclude that the powder XRD method is especially suitable for evaluating crystallite size ranging from several to several tens nanometers, while the methodology of XRD-based evaluation of crystallite size is now aiming at the size of several hundreds nanometers [2].

Since the experimentally observed width and shape of diffraction peaks are affected by the instrument and measurement conditions, the detailed experimental parameters should properly be taken into account, if accurate evaluation of crystallite size is needed. The basic resolution of the instrument is evaluated by measuring the powder diffraction profile of a well-crystallized sample as standard material. The effect of the instrument can be removed from the observed data by Fourier deconvolution of an empirically obtained instrumental function in the Stokes' method [3], where part of the sample is annealed at high temperatures and used as the standard sample to determine the instrumental resolution. More advanced methods, where the effects of instruments are predicted by *a priori* calculation based on the geometry of the optics and diffraction condition [4, 5], and a Fourier-based deconvolution method, applicable straight to the wide-angle diffraction data have been proposed [6].

There may be intrinsic difficulty in the evaluation of crystallite size by powder diffraction method, when the observed diffraction peak profile is strongly affected by strain or fault in the crystallites. However, the effects of smallness of the crystallites and the internal strain can be separated by using the method of Warren–Averbach [7] or Williamson–Hall [8]. In the Williamson–Hall method, the following dependence of the integral breadth β on the diffraction angle is assumed,

$$\beta = \frac{\beta_X}{\cos\theta} + \beta_Y \tan\theta \tag{5.2.2}$$

where the first term of the above equation means the broadening caused by finite size of crystallites, and the second term can be attributed to the broadening caused by internal strain in the crystallites and the spectral width of the source X-ray. Integral breadth is defined by the division of the integrated intensity by the peak intensity. The ratio of the integral breadth to the full width at half maximum (FWHM) of the Lorentzian (Cauchy-distribution) function: $f_{\mathrm{Lor}}(x) = \beta^{-1}\,[1 + (\pi x/\beta)^2]^{-1}$ is equal to ($\pi/2$). Since the following relation is derived from equation (5.2.2),

$$\beta\cos\theta = \beta_X + \beta_Y \sin\theta, \tag{5.2.3}$$

the parameters β_X and β_Y are estimated as the intersect and slope of a straight line fitted to the plot of $\beta\cos\theta$ versus $\sin\theta$ evaluated from the experimental data (Williamson–Hall plot). When the integral breadth is used to specify the diffraction peak width, the volume-weighted average size (the ratio of the average of the fourth power of the diameter D to the average of the third power of the diameter) for spherical crystallites can be calculated from the value of β_X by using the following equation,

$$\langle D\rangle_V = \frac{\langle D^4\rangle}{\langle D^3\rangle} = \frac{4\lambda}{3\beta_X} \tag{5.2.4}$$

where radian is used as the unit of the angle.

Recently, it has been reported that evaluation of the broadness of statistical distribution of crystallite size can be evaluated as well as the average crystallite size by fitting a theoretical peak profile calculated by assuming some distribution models to the experimental diffraction peak profile [2, 9, 10]. When the spherical or ellipsoidal shape of crystallites is assumed, the diffraction peak profile of a crystallite with the diameter D along the diffraction vector is given by the following equation,

$$p_{\mathrm{sphere}}(s;D) = \frac{\pi D^4}{6}\Psi(2\pi s D) \tag{5.2.5}$$

$$\Psi(x) = \frac{3}{x^2}\left[1 - \frac{2\sin x}{x} + \frac{4\sin^2(x/2)}{x^2}\right] \tag{5.2.6}$$

for the deviation of the diffraction vector from the peak value, $s = 2\,(\sin\theta - \sin\theta_0)/\lambda$. The theoretical peak profile of the spherical or ellipsoidal crystallites with the size distribution modeled by the log-normal distribution (median m and logarithmic standard deviation ω):

$$f_{\mathrm{LN}}(D;m,\omega) = \frac{1}{\sqrt{2\pi}D\omega}\exp\left[-\frac{(\ln D - \ln m)^2}{2\omega^2}\right] \tag{5.2.7}$$

is given by

$$P_{\mathrm{SLN}}(s;m,\omega) = \int_0^\infty p_{\mathrm{sphere}}(s;D) f_{\mathrm{LN}}(D;m,\omega)\,dD \tag{5.2.8}$$

The recent studies on the above theoretical peak profile have elucidated that the theoretical profile becomes close to the Lorentzian shape in the case where about 70% of the whole crystallites have the size within the range from half to twice of the median size, and "super-Lorentzian" shape with sharpened peak-top and long tails is predicted for broader distribution of the crystallite size [2]. Figs. 5.2.1 and 5.2.2 show the probability density function of the log-normal distribution function and the corresponding theoretical peak profiles.

Although the Rietveld method [11] is mainly aimed to refine the crystal structure from the powder diffraction data, it can also be applied to the evaluation of crystallite size. The application of the Rietveld method to evaluate the crystallite size will be less ambiguous, if sufficient size broadening is observed and the experimental line profile can be approximated by the Lorentzian peak shape. In such a case, log-normally distributed size and spherical shape of the crystallites can be assumed as described above. The dependence of the Lorentzian width on the diffraction angle is appropriately modeled by equation (5.2). Since the two parameters β_X and β_Y can be treated as adjustable parameters in the Rietveld method to fit the experimental data, the optimized values are automatically estimated by iterative calculations. The analytical method assuming the Voigtian profile, which is defined by the convolution of the Lorentzian and Gaussian (normal distribution) functions, has also been proposed [12].

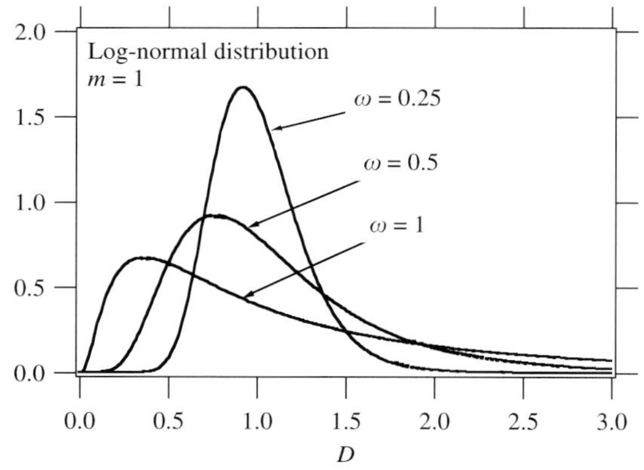

Figure 5.2.1
The density function of log-normal distribution for various logarithmic deviation.

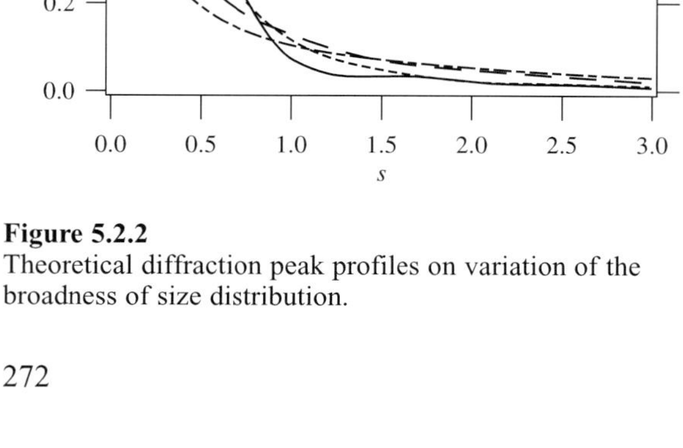

Figure 5.2.2
Theoretical diffraction peak profiles on variation of the broadness of size distribution.

References

[1] W.A. Rachinger: *J. Sci. Instrum.*, **25**, 254–255 (1948).
[2] T. Ida, S. Shimazaki, H. Hibino and H. Toraya: *J. Appl. Crystallogr.*, **36**, 1107–1115 (2003).
[3] A.R. Stokes: *Proc. Phys. Soc.*, **61**, 382–391 (1948).
[4] R.W. Cheary, A. Coelho: *J. Appl. Crystallogr.*, **31**, 862–868 (1998).
[5] T. Ida, K. Kimura: *J. Appl. Crystallogr.*, **32**, 982–991 (1999).
[6] T. Ida, H. Toraya: *J. Appl. Crystallogr.*, **35**, 58–68 (2002).
[7] B.E. Warren, B.L. Averbach: *J. Appl. Phys.*, **21**, 595–599 (1950).
[8] G.K. Williamson, W.H. Hall: *Acta Metall.*, **1**, 22–31 (1953).
[9] J.L. Langford, D. Lour and P. Scardi: *J. Appl. Crystallogr.*, **33**, 964–974 (2000).
[10] N.C. Popa, D. Balzar: *J. Appl. Crystallogr.*, **35**, 338–346 (2002).
[11] H.M. Rietveld: *J. Appl. Crystallogr.*, **2**, 65–71 (1969).
[12] K. Santra, P. Chatterjee and S.P. Sen Gupta: *Bull. Mater. Sci. (India)*, **25**, 251–257 (2002).

5.2.2 Small-angle X-ray scattering

The small-angle X-ray scattering method (SAXS) is the method to evaluate the size of small particles or modulation of the electronic density in the scale-range from several nm to 100 nm, by analyzing the scattered intensity of X-rays in small angular range from about 0.1 to 5°. In contrast to the WAXD method, the SAXS method can be applied not only to the crystalline materials but also amorphous materials or polymer molecules. The distribution of small-angle scattering intensity is caused by any particles, which may be defined by the region where the electron density is different from the media. The method can be applied to evaluation of the size and distribution of pores in solid materials [1], local inhomogeneity in amorphous material, and also colloidal particles and the coagulation. Analysis of SAXS data is sometimes aimed at evaluating long-range order or interparticle distance in collection of polymer molecules by applying some structure models [2]. Application of SAXS to evaluate particle size and distribution is described in this section.

Specially designed optics system should be used for collecting SAXS data, because accurate evaluation of small-angle scattering intensities strongly requires elimination of source X-ray and scattering from the incident beam path. When laboratory X-ray source is applied, collimation using curved crystal monochromator, thin slits or small pinholes and precise alignment of them are usually applied. Recently, efficient SAXS measurement systems with laboratory X-ray source and multilayered mirror are commercially available. Since the use of synchrotron radiation as the X-ray source is especially advantageous in SAXS measurements, SAXS beamlines are available in many synchrotron facilities (e.g., BL-10C and BL-15A in KEK-PF; BL40B2 in SPring-8). PSD, charge-coupling device (CCD), and imaging plate (IP) are also effectively applied in SAXS measurements.

The most basic application of SAXS method is evaluation of particle size in dilute dispersion system, where interference caused by scattering from different particles is considered to be negligible [3]. The scattering intensity curve for the electron density of the media ρ_0 and the particle ρ, wavelength of the X-ray λ and scattering angle 2θ is given by

$$I(s) = (\rho - \rho_0)^2 \left| \int_V \exp(2\pi i \vec{s} \cdot \vec{r})\, dv \right|^2 \tag{5.2.9}$$

where $\vec{s}$ is the scattering vector, the length of which is given by $s = |\vec{s}| = 2(\sin\theta)/\lambda$.

When we restrict our attention in the small-angle range, the length of the scattering vector is approximated by $s \sim 2\theta/\lambda$. The integral in the above equation means the three-dimensional (3D) integral within the interior range of the particle, and uniquely determined by the shape, size, and orientation of the particle. When the particle has spherical shape with the radius a, the scattering intensity curve versus the length of the diffraction vector s is given by

$$I(s) = (\rho - \rho_0)^2 V^2 \Phi^2 (2\pi s a) \tag{5.2.10}$$

$$\Phi(x) = \frac{3(\sin x - x\cos x)}{x^3} \tag{5.2.11}$$

Fig. 5.2.3 shows the SAXS intensity curve of the particles with the radius of 5, 10, 20 nm for the wavelength of the X-ray 0.15 nm.

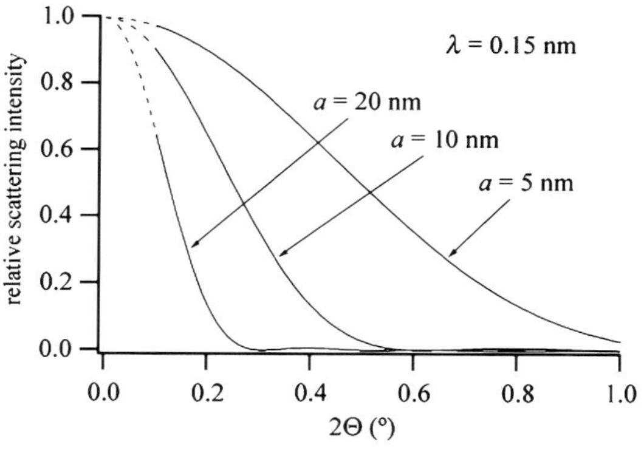

Figure 5.2.3
Theoretical scattering curves from a spherical particle.

When the scattering intensity curve near the origin is approximated by the Gaussian function, the following equation is derived,

$$I(s) = (\rho - \rho_0)^2 V^2 \exp\left(\frac{-4\pi^2 s^2 R_g^2}{3} \right) \tag{5.2.12}$$

which is called Guinier approximation. R_g is the radius of gyration defined by

$$R_g^2 = \frac{1}{V} \int_V r^2 dv \tag{5.2.13}$$

The radius of gyration for a spherical particle with the radius a is given by $R_g = \sqrt{3/5a}$. When the behavior of the small angle scattering near the origin is approximated by equation (5.2.12), linear dependence is expected for the plot of ln $I(s)$ versus s^2 (Guinier plot), and the radius of gyration R_g can be evaluated from the slope $-4\pi^2 R_g^2/3$ (Fig. 5.2.4). Practically, the data within the range $2\pi s R_g < 1.3$ are used for the analysis.

When the size of the particles are statistically distributed, the radius of gyration for the collection of particles with the radius a_j is nominally given by

$$\bar{R}_g = \sqrt{\frac{3}{5}} \left(\frac{\sum_j a_j^8}{\sum_j a_j^6} \right)^{1/2} \tag{5.2.14}$$

The above equation implies that the radius of gyration for the collection of particles is heavily affected by larger particles as compared with smaller particles.

In order to evaluate the statistical distribution of particle size from SAXS data, theoretical scattering curve

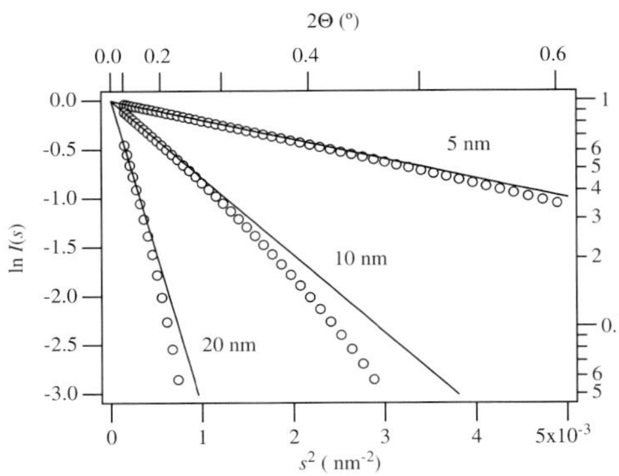

Figure 5.2.4
Guinier plot. Theoretical scattering intensity curves from spherical particles are plotted by markers, and the solid lines are calculated by Guinier approximation for $s^2 \to 0$.

assuming a model for statistical distribution is fitted to the experimental curve. Log-normal distribution [4] given by

$$f_{\mathrm{LN}}(a;m,\omega) = \frac{1}{\sqrt{2\pi}a\omega}\exp\left[-\frac{(\ln a - \ln m)^2}{2\omega^2}\right] \quad (5.2.15)$$

or the Gamma distribution [5] given by

$$f_{\Gamma}(a;p,\sigma) = \frac{1}{\sigma\Gamma(p)}\left(\frac{a}{\sigma}\right)^{p-1}\exp\left(-\frac{a}{\sigma}\right) \quad (5.2.16)$$

is usually used as the model size distribution. Here m is the median radius, ω the logarithmic standard deviation, $\Gamma(\mathrm{p}) = \int_0^\infty t^{p-1} e^{-t}\, dt$ is the Gamma function, and p and σ are the parameters to specify the Gamma distribution, which give the average $p\sigma$ and the variance $p\sigma^2$ for the distribution. When the probability density function of the radius is given by the function $f(a)$, the scattering intensity curve from the collection of the spherical particles is given by

$$I(s) = \frac{16\pi^2}{9}(\rho - \rho_0)^2\int_0^\infty a^6\Phi^2(2\pi sa)f(a)da \quad (5.2.17)$$

Porod has shown that the asymptotic behavior at $s \to \infty$ of the scattering intensity curve is given by

$$I(s) \to \frac{(\rho - \rho_0)^2 S}{8\pi^3 s^4} \quad (5.2.18)$$

where S is the total surface area of the particle, no matter what the shape or size distributions of the collection are, when the orientations of the particles are randomly distributed [6]. Therefore, the plot of ln $I(s)$ versus ln s generally approaches to the straight line with the slope of -4. Total surface area of the collection of particles can be evaluated from the scattering intensity curve, if the calibration using a standard sample with known composition and surface area is applied.

The plot of $s^2I(s)$ versus the scattering vector s (Kratky plot) is used to evaluate the shape of the polymer particle, where characteristic change is observed on transformation from chain-like to granular shape of particles.

References

[1] A.V. Semenyuk, D.I. Svergun, L. Yu, Mogilevsky, V.V. Berezkin, B.V. Mchedlishvili and A.B. Vasilev: *J. Appl. Crystallogr.*, **24**, 809–810 (1991).
[2] T. Veki: *The Fourth Series of Experimental Chemistry*, Diffraction, edited by the Chemical Society of Japan, Vol. 10, Chapter 7, Maruzen Publishing, Tokyo (1992) (Japanese).
[3] A. Guinier: *X-ray Diffraction in Crystals, Imperfect Crystals, and Amorphous Bodies*, Chapter 10, Dover Publications, New York (1994).
[4] R. Kranold, S. Kriesen, M. Haselhoff, H.J. Weber and G. Goerigk: *J. Appl. Crystallogr.*, **36**, 410–414 (2003).
[5] W. Ruland, B. Smarsly: *J. Appl. Crystallogr.*, **38**, 78–86 (2005).
[6] G. Porod: *Kolloidn Zh.*, **124**, 83–111 (1951); G. Prod: *Kolloidn Zh.*, **125**, 51–57 (1952); G. Porod: *Kolloidn Zh.*, **125**, 108–122 (1952).

5.2.3 Neutron diffraction

By neutron diffraction, we can obtain information on the crystal structure, crystallite size and strain as well as X-ray and electron diffraction. Neutron diffraction measurements need a neutron source such as a reactor (steady neutron source) or an accelerator (pulse neutron source) [1]. In case of using a monochromatized neutron beam with a fixed wavelength (e.g., 1.82 Å), it is irradiated to a sample and then the diffraction signals are collected by detectors. As of 2005, at the research reactor JRR-3M of the Japan Atomic Energy Agency (JAEA), two neutron diffraction instruments of the HERMES, installed by Tohoku University, and of the HRPD, installed by the JAEA, are working. Diffraction experiments utilizing the pulse neutron source are conducted by the time-of-flight (TOF) method. In the TOF method, the speed of the neutron and the wavelength are determined by measuring the time of the flight of white pulse neutrons over a constant distance. There were Vega and Sirius that were the TOF neutron powder diffraction instruments in neutron research facility KENS of KEK as of 2005. The KENS facility shut down in 2006, however the successor, whose resolution is much higher, the range of lattice spacing is much wider, and intensities are much higher, is under construction at the J-PARC in Tokai village [2]. Usual measurement requires a large amount of samples of several cc in volume. Since the intensity of the incident neutron beam might be about 100 times more intense in the J-PARC, it will be easier to collect the neutron diffraction data using much less quantity of samples. There exist only a few facilities for neutron diffraction experiments in Japan and the beam time is limited. Therefore, one should not utilize neutrons, but X-rays for the subjects one can study through the X-ray diffraction (XRD).

Next we explain what we know from a diffraction profile in order to understand the difference between the neutron and X-ray diffractometry. From the peak positions, the unit-cell parameters can be determined. From the peak positions and widths, we can estimate the lattice strains. From the peak width, the crystallite size can be estimated. Since the distribution of peak positions occurs due to the compositional distributions in compositionally inhomogeneous samples, the peak widths and anisotropy in peak profiles change. Details of the characterization of crystallite size and lattice strains are described in Section 5.2.1 (X-ray diffraction). Integrated peak intensities include the information on site occupancy g_j of jth atom, atomic displacement parameters and atomic positional parameters x_j, y_j, z_j. The integrated intensity is proportional to the square of the absolute value of the structure factor F. The structure factor is expressed as,

$$F = \sum_j g_j f_j T_j \cdot \exp[2\pi i(hx_j + ky_j + lz_j)] \qquad (5.2.19)$$

In the Rietveld method, the unit cell and structural parameters (site occupancy, atomic coordinates, atomic displacement parameters, etc.) are determined by a least square means so that the intensity data calculated using equation (5.2.19) and so on reproduce the measured intensity data. A noticeable difference in equation (5.2.19) between neutron and XRD is the scattering power of jth atom. The X-ray is scattered by electrons, thus the X-ray scattering power (atomic scattering factor) increases monotonously with the atomic number (Fig. 5.2.5). On the contrary, neutrons are scattered by the nucleus and its scattering power (neutron scattering length) does not change monotonously (Fig. 5.2.5). All the X-ray scattering powers are plus, while some of neutron scattering powers are minus (Fig. 5.2.5). The X-ray scattering power decreases with Bragg angle, while the neutron scattering power is independent of Bragg angle.

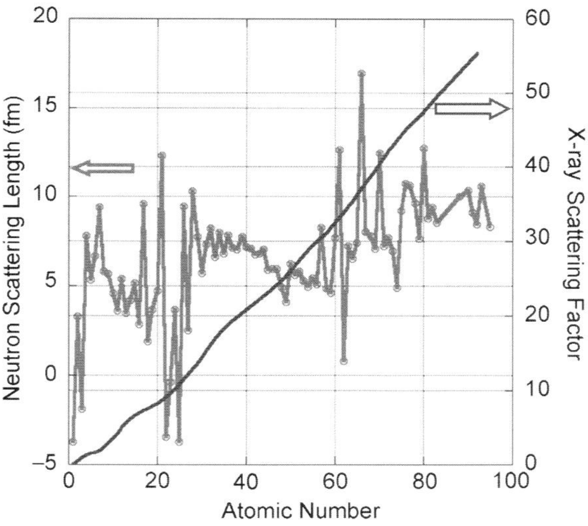

Figure 5.2.5
Dependence of neutron scattering length and X-ray scattering factor ($\sin\theta/\lambda = 0.5$ Å^{-1}) on atomic number.

Comparing with X-ray, neutron diffraction has the following traits [1, 3, 4]:

1. Structural parameters of light elements in a crystal consisting of heavy elements and light elements (H, Li, O, N, C, etc.) can be precise. For example, neutron diffraction is useful to investigate the structural parameters and spatial distributions of oxygen ions in ceria (CeO_2). In fact, the neutron scattering power of oxygen atoms is relatively much larger than X-ray scattering power where both neutron and X-ray scattering powers of Ce atoms are assumed to be equal (Fig. 5.2.6). Neutron diffraction is appropriate to study the structural parameters and spatial distribution of Li ions in the Li-doped lanthanum titanate. Moreover, neutron diffractiometry is very effective in determining the position of hydrogen atoms. Because hydrogen raises the background, it is preferable to carry out experiments by using the sample substituted for the deuterium as long as it is possible to do.
2. Neutron can distinguish atoms with similar atomic numbers from each other. For example, the occupancy factors of oxygen and nitrogen atoms can be determined precisely, which is difficult in the case of XRD. Neutron diffractometry is effective for determining the positions of nitrogen and oxygen atoms in oxynitrides.

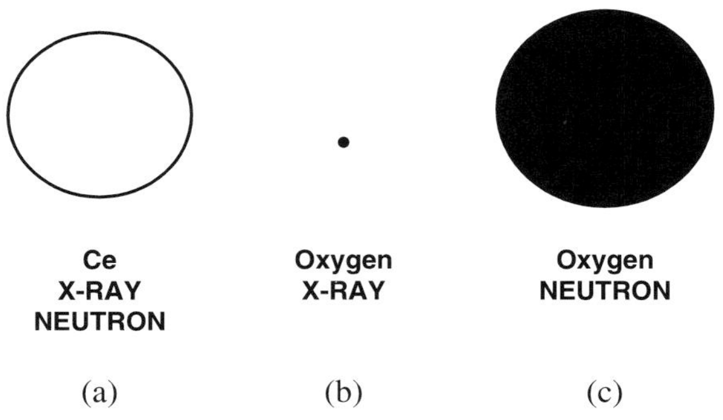

Figure 5.2.6
Comparison of scattering power of an oxygen atom when the neutron and X-ray scattering powers of Ce atom: (a) are assumed to be equal. (b) X-ray atomic scattering factor of oxygen atom ($\sin\theta/\lambda = 0.5\text{Å}^{-1}$). (c) Neutron scattering power of oxygen atom. The area of the circle corresponds to the square of the scattering power [3].

3. Neutron can distinguish the isotopes from each other. For example, it is able to distinguish hydrogen (neutron scattering length $b = -3.7423$ fm), deuterium ($b = 6.674$ fm) and tritium ($b = 5.10$ fm) from each other. It can distinguish ^{58}Ni ($b = 14.4$ fm) from ^{60}Ni ($b = 2.8$ fm). XRD cannot distinguish the isotopes from each other, because it examines the electrons.
4. The X-ray atomic scattering factor steeply decreases with an increase of Bragg angle, while the neutron scattering length is independent of Bragg angle. Contrary to XRD, neutron diffraction does not need to consider the polarization factor. Therefore, in the neutron diffraction analysis, precise atomic displacement parameters can be obtained.
5. Penetration of neutron into a material is much deeper than that of X-ray. Thus, the neutron diffraction experiments are usually performed in transmission geometry. It is easy to design and fabricate the sample environmental attachments such as high-temperature furnace [3] and high-pressure apparatus. It is easy to collect high-quality intensity data with higher precisions, because there are a large number of grains in the polycrystalline or powdered sample, which contribute to the diffraction intensity data, leading to less influence of preferred orientation and large grains. A part of the inside of a big specimen (e.g., practical industrial material) can be examined by neutron diffraction, thus we are able to examine the distribution of stress in a structural material by this technique. There exists less influence of the pollution and oxidation of a sample surface, thus, it is easy to obtain the information of bulk inside a specimen. It is recommended to examine the neutron absorption cross sections of the material before neutron studies, because some elements such as B and Cd have quite large neutron absorption coefficients. The neutron beam intensity I is expressed by the equation, $I = I_0 \exp(-N\sigma t)$ when the incident beam with intensity I_0 passes a planer specimen with the thickness of t. Here the N and σ are an atomic number in a unit volume and the total scattering cross section, respectively. The σ is calculated by adding the scattering cross section σ_s to absorption cross section σ_a. These cross sections σ_s and σ_a are known, thus it is easy to calculate the absorption.
6. Neutron has magnetic moment, thus can analyze the configuration of spin in a crystal, namely, the magnetic structure analysis is possible.
7. Using reciprocal vector $\boldsymbol{k}$, the structure factor can be described as,

$$F(\boldsymbol{k}) = \int \rho(\boldsymbol{x}) \cdot \exp(2\pi i \boldsymbol{k} \cdot \boldsymbol{x}) d\boldsymbol{x} \qquad (5.2.20)$$

In XRD the $\rho(\boldsymbol{x})$ is the electron density at the position $\boldsymbol{x}$, while it is nuclear density (strictly speaking, scattering amplitude distribution) in neutron diffraction. Application of the maximum-entropy method (MEM) for the structure factors and their errors obtained with a Rietveld analysis program enables determination of precise $\rho(\boldsymbol{x})$, compared with conventional Fourier method. The MEM technique is effective for investigating the spatial distribution of atom positions (positional disordering), ionic conduction (diffusion) path, and anisotropic thermal motions. It has been demonstrated that the mobile oxygen ions in a fast-ion conductor bismuth oxide with the fluorite-type structure have a large spatial distribution and anisotropic thermal motions. Anisotropic thermal motions and diffusion path of oxygen ions were studied for the lanthanum gallate perovskite (Fig. 5.2.7) [4].

In the present section, we have described mainly neutron powder diffractometry. Neutron studies have a wide range of applications such as single-crystal diffractometry, small-angle scattering method, inelastic

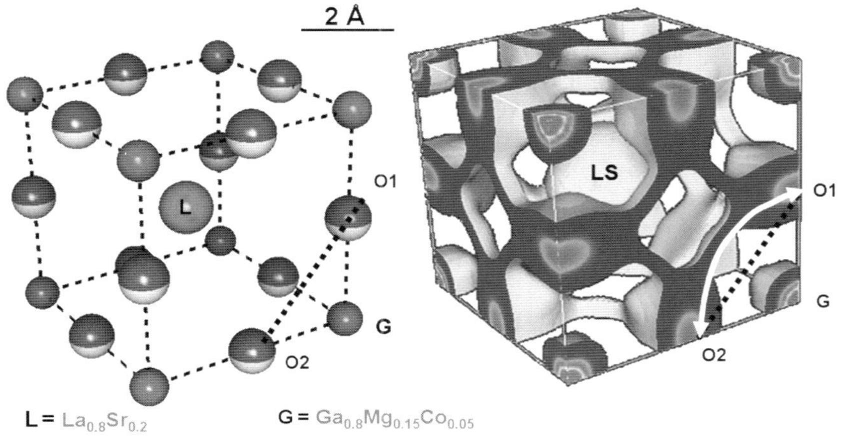

Figure 5.2.7
(a) Crystal structure and (b) equi-density surface of nuclear density of lanthanum gallate (1,392°C) determined by the Rietveld refinement and MEM analysis of neutron diffraction data. Diffusion path of oxygen ions (white arrows) is not a straight O1–O2 line but curved [4].

scattering (lattice vibration, phonon, motions of molecules, atomic motions in liquids, colloid and liquid crystals, spin wave) [1]. Neutron techniques can be applied not only on inorganic crystals but also a wide variety of materials such as organic crystals, biomaterials, polymers, magnetic materials, amorphous materials and liquids. The research of the nanoparticle by using the neutron techniques will become more active in the future.

References

[1] For example, N. Niimura: *Lectures for Experiments in Chemistry*, Vol. 10, *Diffraction*, Maruzen Publishing, Tokyo, pp. 481–548 (1992).
[2] M. Yashima: *J. Jpn. Soc. Neutron Sci. (Hamon)*, **14** (1), 17–21 (2004) and some papers in this number.
[3] M. Yashima: *J. Crystallogr. Soc. Jpn.*, **44**, 121–126 (2002).
[4] M. Yashima: *J. Crystallogr. Soc. Jpn.*, **46**, 232–237 (2004); M. Yashima, K. Nomura: *Electrochemistry*, **73**, 71–76 (2005).

5.2.4 Raman scattering [1–5]

Raman effect is a phenomenon in which the scattered light includes components whose frequencies shift by ν peculiar to the material from the incident light frequency ν_0 (frequency $\nu_0 \pm \nu$) when the incident light with frequency ν_0 is irradiated to the material. Raman scattering is a characterization technique for materials, which applies the Raman effect. This technique is applicable to not only crystals but also to a variety of materials such as amorphous materials, liquids, and gases. Raman scattering is a powerful, convenient, nondestructive, and noncontact technique not requiring pre-treatments of the specimen, which enables the investigation of intermediate-range order and formation phases in nanoparticles, where the degree of long-range order is low, and the study of molecular species absorbing on the surface of materials. In many cases, Raman data can be collected at room temperature at atmospheric conditions.

Raman scattering light with longer wavelength (frequency $\nu_0 - \nu$) is called Stokes light and that with shorter wavelength (frequency $\nu_0 + \nu$) anti-Stokes light. The anti-Stokes intensity is much lower than the Stokes intensity. The frequency (energy) shift ν is attributed to the transfer of energy due to the lattice vibrations (phonon), molecular vibrations, rotation of molecules and electronic transitions. Thus, one of the unique characteristics of the Raman scattering technique is the investigation of each vibration and each bonding. In a Raman spectrum, scattering light intensity is plotted against the shift of frequency (energy) (unit is cm^{-1} or eV). Raman spectrum is strongly dependent on crystalline phases. For example, as shown in Fig. 5.2.8, the Raman spectra of monoclinic, tetragonal, and cubic zirconia phases are very different from each other, thus, the Raman spectrum enables the phase identification [4, 5]. Note that in Raman scattering, the Raman signals must be distinguished from other scattering lights such as fluorescence. To prevent invalid interpretation (e.g., a fluorescence is regarded as a Raman signal), it is useful to measure the Raman spectra with a different excitation wavelength [5]. We should carefully examine the signals. It is useful to check the anti-Stokes Raman signal corresponding to the Stokes Raman peak. Raman spectra have a lot of information, but you should interpret them very carefully.

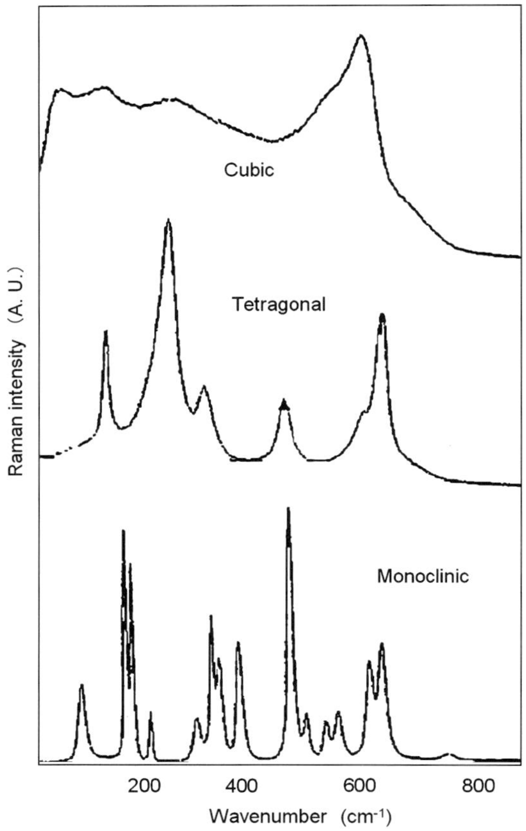

Figure 5.2.8
Raman spectra of monoclinic, tetragonal, and cubic zirconia phases.

Peak positions ν of the optical Raman bands are governed by the force constants of lattice vibrations, so to speak, by the spring constant. Vibrations of a simple molecule can be classified into stretching vibrations of bonds and bending vibrations. The interpretation of phonon vibrations of a little complicated molecules and of crystals is not very easy. A phonon vibration does not always contribute to the Raman scattering. A vibration is classified into Raman active mode and infrared (IR) active mode, which are determined by the selection rule governed by the symmetry of crystal and molecule. Therefore, the Raman and IR (Section 5.3.3) techniques are alternative characterization methods. The selection rule is strongly dependent on the crystal structure. For example, the fluorite-type structure has one Raman-active mode, while the tetragonal and monoclinic zirconia phases have six and eighteen Raman-active modes, respectively. Thus, the Raman spectrum is strongly dependent on the material and crystalline phase, which enables the phase identification of materials and crystalline phases from Raman spectrum. Comparing with the X-ray and neutron diffraction, the Raman scattering technique has the following features [1–5]:

1. Raman scattering is useful for the identification of crystalline phases with less long-range order, less long-range periodicity and intermediate periodicity. Thus, this technique is also effective for characterizing the nanoparticle samples with small crystallite sizes. For example, in the initial stage of crystallization, the Raman spectroscopy enables to investigate how ordering exists in the sample. From the Raman peak width, the crystallite size can be estimated.
2. In substitutional solid solutions such as (Zr, Yb)$O_{2-\delta}$, the Zr and Yb ions coexist at the cation site, and the oxygen ions and vacancies coexist at the anion site. Thus, strictly speaking, the periodicity is broken. In the system where the periodicity is broken, various vibration modes appear due to the breaking of the selection rule [4, 5]. For example, in the cubic fluorite-type structured (Zr,Yb)$O_{2-\delta}$ solid solution, the Zr and Yb ions coexist at the cation site, and the oxygen ions and vacancies coexist at the anion site. According to the factor group analysis, only one Raman band is expected for a defect-free fluorite-type structure, but the cubic fluorite-type structured (Zr,Yb)$O_{2-\delta}$ solid solution exhibits many bands and continuous spectrum due to the disordering and coexistence of some ions and vacancies (Fig. 5.2.8). In this way, the Raman scattering technique is useful for investigating the disordering.
3. Raman scattering is sensitive to the atoms with large polarizability. Since the anions such as oxygen and fluorine have large polarizability, the Raman scattering is often effective for studying the displacements and defects of oxygen atoms in oxides [4, 5].
4. The micro-Raman technique enables the phase identification from a small area with the spatial resolution of about 1 μm. Recently, near-field Raman scattering instruments have been developed and their spatial resolution can be several tens of nanometers. Measurements of the Raman spectra with these instruments enable the phase identification and the characterization of stresses in a very small amount of sample (less than several nanograms). Utilizing the high spatial resolutions, the Raman scattering technique has

been applied to various studies of the phase characterization near a crack, deformation of a fiber, self-diffusion coefficient of oxygen, and stresses of a local area in semiconductors. Mapping of micro-Raman data enables the visualization of the spatial distribution of stresses and existing phases. Average structure in any area can easily be obtained by changing the probe size from 1 μm to 1 cm with different lens and/or optical systems.

5. Applying the stresses to a material changes the spring constant, thus, we can characterize the stresses from the Raman peak position. The peak width of a phonon Raman band is governed by the lifetime of phonons. The Raman peak is broadened by the stresses without directions and compositional distributions.
6. It is possible to conduct in situ observations at high temperatures, at high pressures and in aqueous solutions. By installing an attachment such as furnace, high-pressure cell and aqueous solution cell around the sample, we can observe in situ and non-destructively the phase transition and kinetics of chemical reactions. In a conventional Raman technique with visible-light excitations, the Raman measurements from the materials at high temperatures above 1,000°C are difficult due to strong thermal radiation from the specimen, but through the Raman excitations by light with short wavelength such as ultraviolet light, the influence of the thermal radiation can be reduced considerably [6, 7]. There is an example of Raman measurements from the material at about 2,000 K, by ca. 364 nm excitations. An alternative effective technique for measuring the Raman spectra from the materials at high temperatures is the time-resolved Raman method. Raman intensity and scattering efficiency are strongly dependent on sample temperature. Sample temperature can be estimated from the intensity ratio between the Stokes and anti-Stokes signals.
7. The Raman scattering technique is effective for investigating the phase transition. In particular, the mechanism of phase transition can be studied by measuring the soft mode.
8. Raman spectra from a polycrystalline material consist of components, which depend on the crystal orientation and polarization. Each component can be obtained by the polarization Raman spectra from some single-crystal samples with different orientations. The orientation of a crystal can be determined by polarization Raman spectra.
9. The development of time-resolved Raman techniques enabled measurements of Raman spectra of a phenomenon occurring within picoseconds.
10. Raman excitations with different wavelengths enable the reduction of the influence of fluorescence and thermal radiation, and the resonance Raman studies.
11. High carrier density in a semiconductor induces an LO-phonon–plasmon-coupled mode. Carrier concentrations can be studied from the plasma frequency. If the surface recombination velocity is relatively low and the carriers induced by the incident laser remain, the velocity can be studied from the remaining carrier quantity.

The advance of Raman scattering techniques is a rapid progress. The range of the application will extend more and more because of the progress of lasers, detectors, and measurement methods.

References

[1] T. Katoda: *Characterization of Semiconductors by Laser Raman Spectroscopy*, University of Tokyo Press, Japan (1988).

[2] T. Kitagawa, A.T. Tu: *Introduction to Raman Spectroscopy*, Kagaku-Dojin Publishing Co., Kyoto, Japan (1988).

[3] H. Hamaguchi, A. Hirakawa: *Raman Spectroscopy*, Japan Scientific Societies Press, Japan (1988).

[4] M. Yashima, M. Kakihana and M. Yoshimura: *J. Tech. Assoc. Refractories Jpn.*, **46**, 150–156 (1994).

[5] M. Yashima, K. Ohtake, M. Kakihana, H. Arashi and M. Yoshimura, *J. Phys. Chem. Solids*, **57**, 17–24 (1996).

[6] M. Yashima, M. Kakihana, R. Shimidzu, H. Fujimori and M. Yoshimura, *Appl. Spectrosc.*, **51**, 1224–1228 (1997).

[7] H. Fujimori, M. Yashima, M. Kakihana, and M. Yoshimura, *J. Am. Ceram. Soc.*, **84**, 663–65 (2001).

5.3 Surface structure

5.3.1 AFM

AFM (atomic force microscopy) is one of SPM (scanning probe microscope) for imaging of surface profile. A typical horizontal distance resolution is about 0.2 nm. AFM can provide higher resolution than normal SEM (scanning electron microscope). It is comparable in resolution to TEM.

TEM provides a phase contrast imaging of a thin bulk. However, AFM provides real topographical image of sample surface. A typical vertical distance resolution is very high with 0.01 nm, which is much better than normal SEM and TEM. A scanning area of typical AFM is from several nanometers to several hundreds nanometers. The maximum height for imaging also decreases with a decrease in scan area.

Fig. 5.3.1 shows each imaging size of SPM and other typical microscopes. AFM can work perfectly well with atomic resolution in vacuum, ambient air, or even a liquid environment. AFM is used not only for surface observation but also for microfabrication or measurement of surface physicality. AFM is useful for imaging of both conducting sample and insulating one, e.g., polymer material, biological macromolecules, chemically modified nanoparticles. Comparison of AFM and other microscopes is shown in Table 5.3.1.

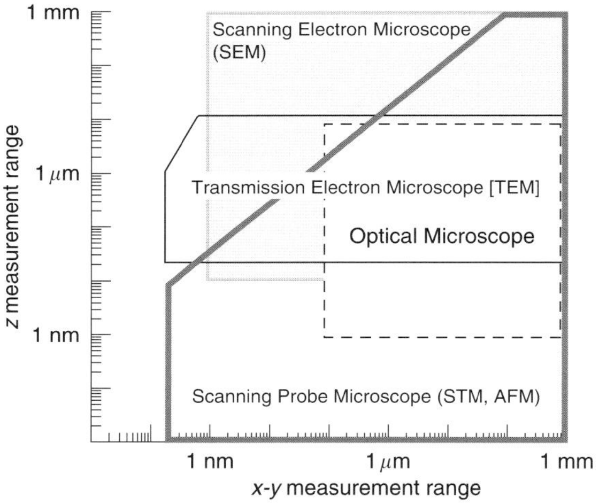

Figure 5.3.1
Each imaging size of SPM and other typical microscopes.

5.3.1.1 Interatomic force

Interatomic force between two electrically neuter atoms is described by Lennard–Jones equation [1].

$$U(r) = 4\varepsilon\left\{\left(\frac{\sigma}{r}\right)^{12} - \left(\frac{\sigma}{r}\right)^{6}\right\} \tag{5.3.1}$$

where, σ and ε are interatomic distance and energy in balance conditions, respectively. Both σ and ε are constant values determined by molecular species. r is interatomic distance. Fig. 5.3.2 shows the interatomic force versus interatomic distance. The interatomic force at large distance shows weak attraction due to induced dipole moment (dispersion interactions). However, the force at small distance shows repulsion due to exchange interaction from Pauli exclusion principle. The interatomic distance, r_1 at the lowest force shown in Fig. 5.3.2 corresponds to the interatomic distance in closed-packing.

5.3.1.2 Detection technique of force

The AFM consists of a sharp tip (probe) with a tip radius of curvature on the order of nanometers at the end of a leaf spring or "cantilever". The cantilever is microfabricated from silicon or silicon nitride. Several manufacturers produce many cantilevers with various geometries, various values of spring constant, and tip radius of curvature. The cantilever selection for each sample is one of important factors to obtain clear and reproducible images.

As the tip of the cantilever approaches the sample surface, the force between the tip and sample surface induces the vertical deflection of the cantilever. AFM obtains the force to detect the deflection of the cantilever. There are two optical methods to detect the deflection of the cantilever, optical lever method, and laser interferometry.

Fig. 5.3.3a shows the concept of the optical lever method used mostly. The sample is located on the tube piezo scanner, which moves the sample in the x–y–z directions (In some AFMs, the piezo scanner is mounted to the cantilever.). The four-segment

Table 5.3.1
Comparison of AFM and other microscopes.

	AFM	TEM	SEM	Optical microscope
Maximum resolution	Atomic resolution	Atomic resolution	Several nanometers	Several hundreds of nanometers
Observation environment	In air, in liquid, in vacuum, in gas	In vacuum	In vacuum	In air, in liquid
In situ observation	Possible	Impossible	Impossible	Possible
Preparation of sample	Easy	Difficult	Easy	Easy

photodetector detects the angle of reflection from a laser beam focused on the metal-coated top surface of cantilever.

Symbols A, B, C, and D in Fig. 5.3.3(a) denote the output from four segments of the photodetector, respectively. The vertical deflection, ΔZ can be represented as the signal of (A+C)−(B+D). Meanwhile, the signal of (A+B)−(C+D) is called lateral force microscope (LFM) signal, which represents horizontal deflection. Surface friction can be obtained from LFM signal. This is called LFM.

LFM is sometimes useful to obtain clear image of phase domain structure, which has little difference in roughness but large difference in surface friction between domains. The concept of another optical method, laser interferometry, is shown in Fig. 5.3.3b.

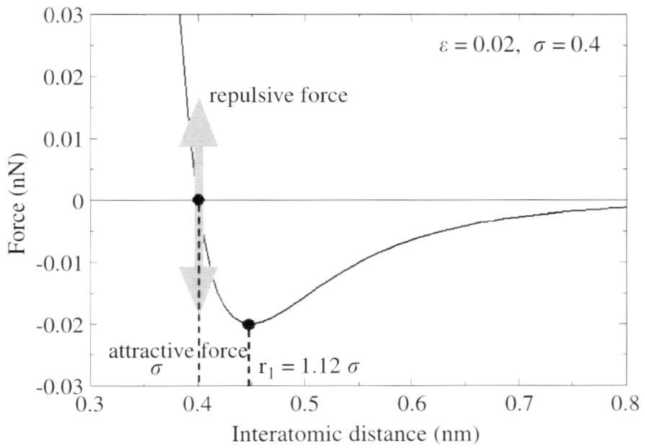

Figure 5.3.2
Force versus interatomic distance curve.

If the beam is focused on the cantilever through a single-mode optical fiber, a part of beam reflects on the top surface of the cantilever. At the same time, a part of the beam reflects on the edge surface of the optical fiber. Thus two laser beams return to the optical fiber and interfere with each other in it.

The optical path difference of the two beams is the distance, d, between the top surface of the cantilever and the edge surface of the optical fiber. AFM measures phase difference of interfered beams to obtain d and detect the deflection of the cantilever, ΔZ.

As self-detective methods, there are piezoelectric cantilever method and tuning fork-cantilever method. In the former method, the deflection of the cantilever is converted to voltage by the use of piezoelectric sensor combined with the cantilever. In the latter method, the deflection can be obtained to measure the change of resonance frequency of tuning fork electrically. Tuning fork is a kind of crystal resonator and it is combined with the cantilever. These methods do not need optical adjustment and it is easier to use commercial cantilevers for these methods in recent days. Thus these non-optical methods will be more commonly used.

5.3.1.3 Force curve

Force curve is a plot of the force obtained from the product of spring constant and cantilever deflection as a function of tip position along the z-axis. It is recorded as the tip of cantilever is brought close to and even indented into a sample surface and then pulled away.

Fig. 5.3.4 shows typical force curve observed in ambient air. Fz and z are the force on the cantilever and the tip position along the z-axis (the extension of the piezoelectric scanner), respectively. When the tip

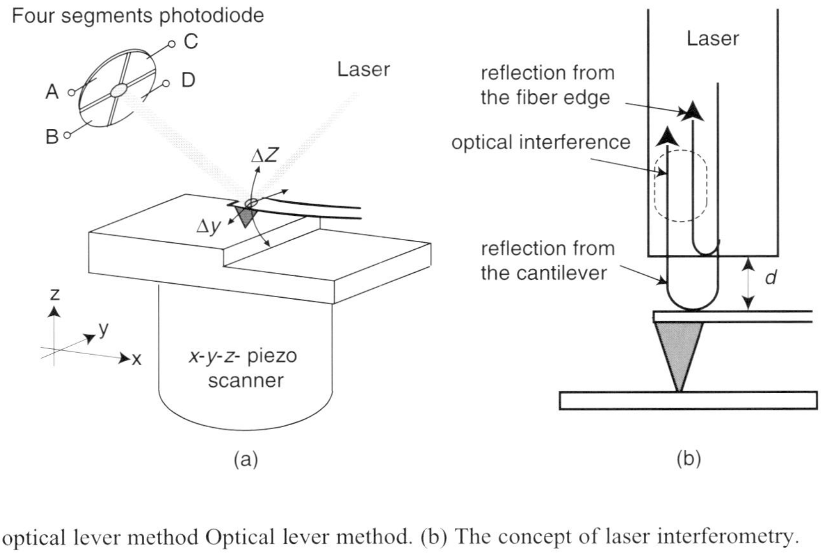

Figure 5.3.3
(a) Concept of the optical lever method Optical lever method. (b) The concept of laser interferometry.

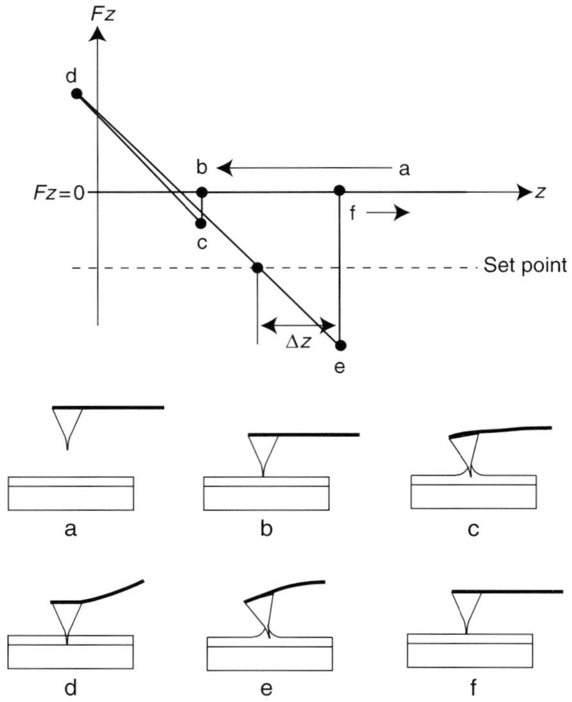

Figure 5.3.4
Force curve observed in ambient air.

is far from the sample surface, the force Fz is zero (a). As the scanner extends, the tip approaches to the sample surface. In air, a single or a few monolayers of water are present on the sample surface. When the tip contacts to the water layer (b), the capillary force due to this water layer pulls the tip into the surface strongly (c). This attractive force is called as jump-in force. As the scanner continues to extend, the force Fz becomes repulsive (d). Next, the cantilever is pulled away from the sample surface with the piezoelectric scanner. The water layer holds the tip in contact with the sample surface and the large attractive force due to adhesion is observed (e). When the adhesion is broken, the tip becomes free from the sample surface (f). When imaging is carried out, the cantilever approaches to sample surface until the force Fz reaches setpoint.

When the surface of the tip or the sample surface is contaminated, it is difficult to carry out reproducible measurement of force curve. In such situation, stable AFM measurement with small setpoint is not obtained because the force cannot be maintained to the setpoint. Meanwhile, large setpoint sometimes damages the sample surface during scanning of the cantilever. Thus the reproducible and clear image cannot be obtained. The force during imaging is calculated from ΔZ in the curve if the spring constant, k, of the cantilever is known. The force is defined by the equation:

$$Fz = k\,\Delta Z \qquad (5.3.2)$$

The force curve provides a clue to determine the optimum setpoint for getting clear image. The shape and reproducibility of the force curve is used for the detection of unstable contact between the tip and surface, called "false engage", which disturbs imaging. The jump-in force sometimes damages the surface of soft sample. Measurement in vacuum and liquid decreases jump-in force due to capillary force of water layer on the sample surface and enables imaging with small setpoint to reduce the possibility of damage to the sample surface.

5.3.1.4 Topographic imaging 1: contact mode

Two primary modes of imaging are contact mode and dynamic mode. In the contact mode, the tip scans while it is in mechanical contact with the sample surface.

An electronic feedback control of the Z voltage applied to piezoelectric scanner keeps the resulting deflection corresponding to desired setpoint shown in Fig. 5.3.5 by adjusting the z position of the cantilever. The tip scans along x-axis with recording of the x position, x and z displacement of the cantilever, Δz. During scanning, the deflection of the cantilever is constant ($\Delta z_1 = \Delta z_2$). The plot of x versus Δz shows topographic profile. This principle is same as the stylus surface profiler.

The Ruster scan of the tip provides the distribution of Δz in x–y plane. The distribution is shown in color-contrast scale and the topographical image of the sample surface is obtained. The number of scan line per second is called scan rate. The scan area (the size of image) and the characteristics of the sample surface influence and determine the optimum scan rate

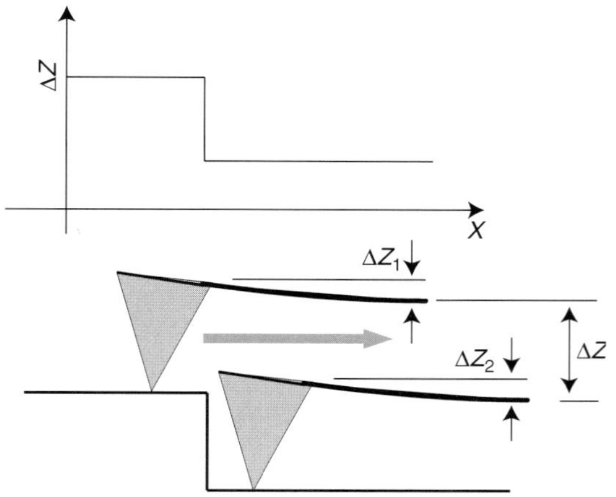

Figure 5.3.5
Contact mode.

Table 5.3.2
Setting of scan rate in several conditions.

	Scan rate
To prevent the distortion in image (reduce drift)	Higher
To clear the image (improve surface tracking)	Lower
To reduce scratching in image	Lower
To image with wide scan area	Lower

for getting good image. During scanning, the drift of tip sometimes occurs due to vibration and thermal expansion of AFM instrument. The drift depends on measurement environment and individual instrument. It also influences the optimum scan rate and it is determined by trial and error as shown in Table 5.3.2.

In the contact mode, AFM provides the atomic image of apical oxygen on the freshly cleaved mica surface easily. Thus cleaved mica is commonly used as a standard sample for a validation test of the AFM instrument. However, the image is not real atomic image because single point-like defects are not observed. In addition, the radius of curvature is not small enough to get atomic image in the contact mode.

Today, it is believed that cyclic friction during scanning induces pseudo-atomic image. However, it is still useful to obtain the lattice parameter and crystal orientation of the sample. This problem is not so important for observation of nanoparticles because the size of the nanoparticles is same as or larger than the radius of curvature of the tip of the cantilever. However it is often difficult to get good image of nanoparticles because the tip often grabs a nanoparticle during the scan and moves it. In such situation, dynamic mode is useful as described in next section.

5.3.1.5 Topographic imaging 2: dynamic mode

Dynamic mode is classified into tapping mode (tapping mode is a registered trademark of Veeco Instruments) and noncontact mode. Small piezoelectric device on the cantilever holder is oscillated by applying the AC voltage and the cantilever is externally oscillated at or close to its resonance frequency. The frequency characteristic of the cantilever called Q curve is derived from a plot of frequency versus amplitude (Fig. 5.3.6a).

The resonance frequency is found by the Q curve. When the probe tip interacts with a surface, the amplitude of oscillation generally shifts to a lower value.

In the tapping mode, a constant amplitude shift is maintained by the feedback control during the Ruster scanning and the topographical image is obtained. While scanning, the oscillating cantilever contacts the sample surface intermittently in repulsive force region and the lateral friction between the tip and the sample surface is very small. Thus it moderates the wear of the tip while scanning and it provides reproducible and good images of the sample surface that are otherwise easily damaged. In addition, this mode allows scanning without scratching and imaging the sample surface with large roughness, e.g., nanoparticles.

Fig. 5.3.7 shows tapping-mode AFM image of gold nanoparticles with 30 nm diameter immobilized onto Langmuir–Blodgett film by electrostatic adsorption [2].

In the noncontact mode, the tip hovers above the sample surface with small distance while scanning. A constant frequency shift is maintained by the feedback control during the Ruster scanning (Fig. 5.3.6b). The mechanical resonance frequency (Q) of the cantilever is low, several hundreds in air. However it reaches several tens of thousands in vacuum and it allows controlling very small force in attractive force region. The AFM in this mode is called non-contact AFM (NC-AFM).

The real atomic image is derived by NC-AFM without contact during scanning when atomically sharp tip is used. After the instruments to detect frequency shift precisely are commercially available, NC-AFM becomes widely used.

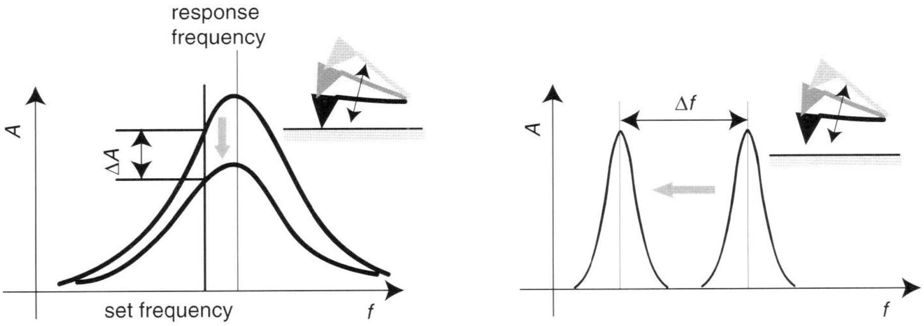

Figure 5.3.6
(a) Tapping mode. (b) Non-contact mode.

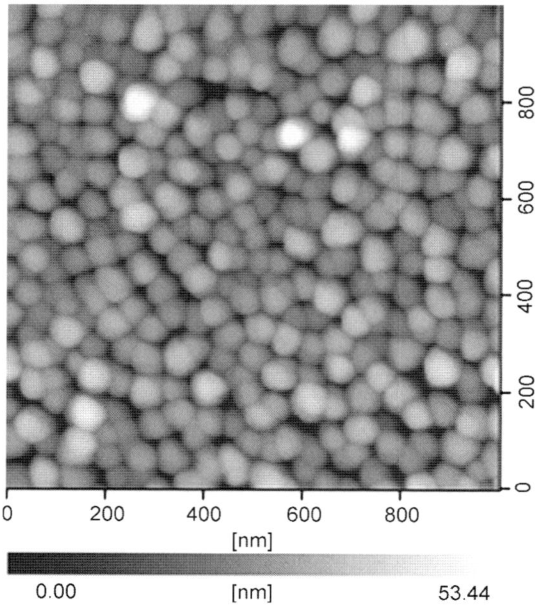

Figure 5.3.7
Tapping-mode AFM image of gold nanoparticles.

5.3.2 STM

STM (scanning tunneling microscope) is one of SPMs for imaging of surface state and profile like AFM (see Fig. 5.3.1 in the former section to make a comparison of STM with other microscopes). In STM, the voltage is applied between the metallic tip sharpened to a single atom point and a conductive sample. Then the metallic tip is very close to the sample surface (near 1 nm). STM scans the tip over the sample surface to get atomic-resolution topographical image by the detection of a tunneling current flowing between the tip and the sample. In fact, STM provides images of the electric state of sample surface. Thus in a precise sense, STM image is different from the real topographical data using AFM. The sample STM requires a certain level of electrical conductivity in samples. The imaging is carried out in air and vacuum. The special tip enables the imaging in liquid.

5.3.2.1 Tunneling current

There is a vacuum gap between a sharp metallic tip and metallic sample surface (Fig. 5.3.8a). When the tip approaches within about 1nm of the sample while applying bias voltage between the tip and sample surface, the tunneling current flows from the tip to the sample through the gap (see Fig. 5.3.8a). The energy diagram shows Fig. 5.3.8b. The bias voltage, V, is smaller than the work function of the sample, Φ_s and the work function of the tip, Φ_t ($eV<(\Phi_t+\Phi_s)/2$), the tunneling current, I, is described by the following equation

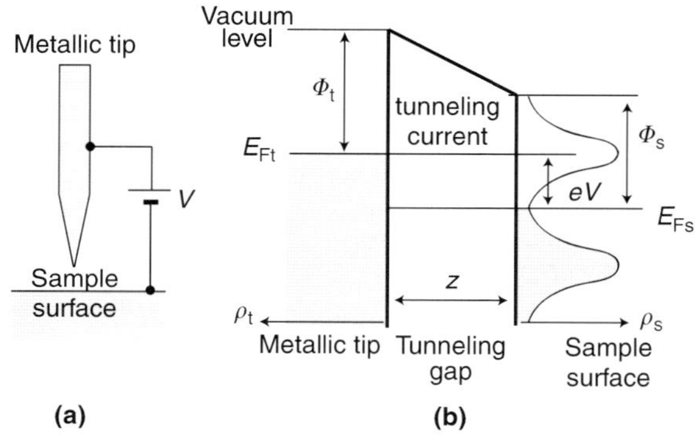

Figure 5.3.8
(a) Configuration of a metallic tip and a sample surface.
(b) Energy diagram.

which is based on the Wentzel-Kramers-Brillouin (WKB) approximation.

$$I(V) = \int_0^{eV} \rho_s(E)\rho_t(-eV+E)T(z,eV,E)\,\mathrm{d}E \qquad (5.3.3)$$

where, ρ_s and ρ_t are the density of state of the sample and the tip, respectively. E is energy relative to Fermi level. T is the probability of tunnel transition with bias voltage, V, and tunneling distance, z. T is derived from the following equation [3].

$$T(z,eV,E) = \exp\left(-\frac{2z\sqrt{2m}}{\hbar}\sqrt{\frac{\phi_s+\phi_t}{2}+\frac{eV}{2}-E}\right) \qquad (5.3.4)$$

where m is electron mass, h is Planck constant ($=h/2p$), T is a monotonously increasing function of bias voltage, V, but an exponentially decay function of the distance z. Thus the tunneling current is very sensitive to the distance z. It is sensitive enough to control the distance between the tip and the sample surface with atomic resolution.

5.3.2.2 Instrumental configuration

Two primary modes of STM imaging are constant-current mode and constant-height mode. Fig. 5.3.9 shows schematic diagram of STM. The sample is located on the tube piezo scanner, which moves the sample in the x–y–z directions with atomic resolution (In some STMs, the piezo scanner is mounted on the tip). Proportional-Integral-Derivative (PID) controller keeps the tunneling current constant with applied bias voltage between the tip and the sample surface by adjusting the distance between the tip and the surface using feedback control of z voltage of the piezoelectric scanner (constant-current mode). The Ruster scan of the tip provides the distribution of z voltage corresponding to the sample profile in x–y plane.

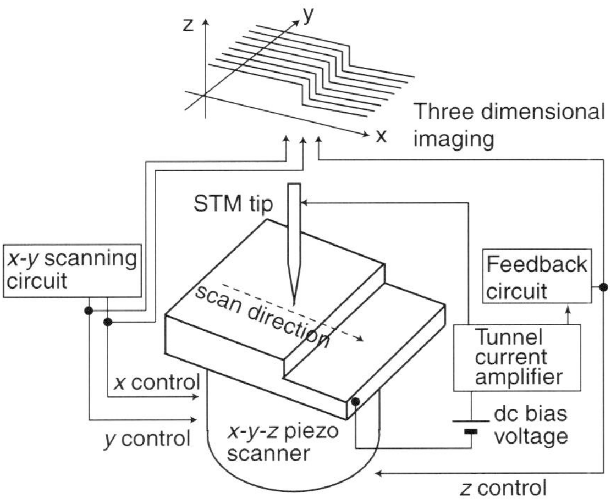

Figure 5.3.9
Schematic diagram of STM.

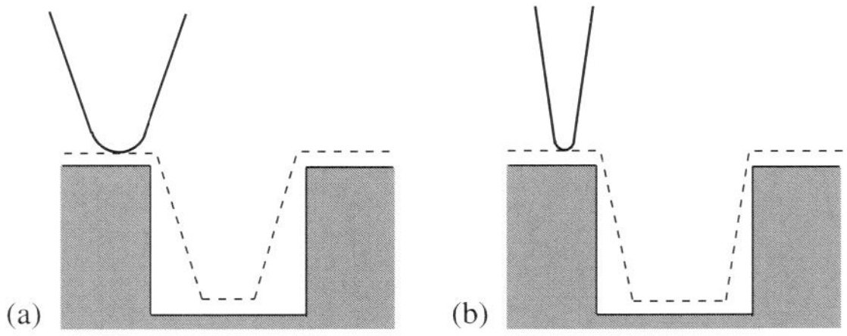

Figure 5.3.10
Relationship between surface tracking and tip radius curvature. A tip with (a) large radius curvature and (b) small radius curvature.

The distribution is shown in color-contrast scale and the topographical image of the sample surface is obtained. In the constant-height mode, STM keeps the z voltage almost constant during the Ruster scan and obtain the distribution of tunneling current, which provides image of the sample surface. The constant-height mode enables imaging with high scan rate and clearer image.

However, when the tip scans over the rough sample surface or wide area, it often contacts the sample surface. The coarse approach mechanism is used for bringing the tip and sample to close within a range of z movement of piezoelectric scanner. To prevent tip crash into the sample surface, this mechanism is realized by the combination of mechanical coarse approach with precision screw driven by stepping motor- and microapproach with piezoelectric device. It is absolutely necessary to isolate the system from external vibrations for getting high-resolution image. STM has suspension system or stacked plate system as simple and effective isolation. In suspension system, a slab of heavy stone or metal is suspended by rubber cords. The stacked plate system consists of a stack of steel plates with gel or rubbery material in between. For acoustical insulation, the STM unit is located in the box, whose inside wall is covered by sound-proof sheet. The STM units for observation in vacuum and at low temperature are large in size and heavy. Thus, these STM units have an eddy current damper or a hybrid damper with air damper and spring [3].

5.3.2.3 STM tip

The tip shape influences the image and the different image of the same sample is sometimes obtained when the tip shape is changed (Fig. 5.3.10). Thus the fabrication of the tip is very important for good quality of STM image. To achieve atomic resolution, single or a few atoms at the end of the tip must contribute to the flowing of tunneling current. The tip sometimes crashes into the sample surface during the coarse approach before detection of tunneling current or the optimization of the imaging parameters after starting of imaging in case of very rough sample, e.g., nanoparticles. This destroys the atomically sharpened tip easily and needs frequent exchange of the tip.

Some commercial tips have insufficient quality for atomic resolution and the problem of aging degradation due to contamination of tip. In addition, they would cost much if frequent exchange of the tip is needed. Using of homemade tips fabricated by simple instrument provides a solution of these problems.

(1) Materials of the tip

The major material of the tip for imaging in air is Pt/Ir alloy, which is the inoxidizable platinum alloyed with 10–20% iridium content to get higher strength [4–6]. Tungsten is commonly used for imaging in vacuum. The fabrication of nickel, gold, and cobalt tip has also been reported [7–9].

(2) Mechanical cutting

Mechanical cutting consists of cutting the end of the Pt–Ir wire with wire cutters. The cut should be made with an angle of approximately 45° at a stroke like snapping the wire. As the wire is being cut, the wire cutters should be pulled away (Fig. 5.3.11a). Some skills are required to get a sharp tip for atomic imaging with the yield ratio of almost 100%. When the surface of cutting tip is rough, the tunneling point is changed each time after approaching (Fig. 5.3.11b).

Thus, the reproducibility of image is poor. In addition, the asymmetric shape of the cutting sometimes induces a tip crash onto the sample surface or a pseudo image called artifact, which is different from real image when the rough sample surface,

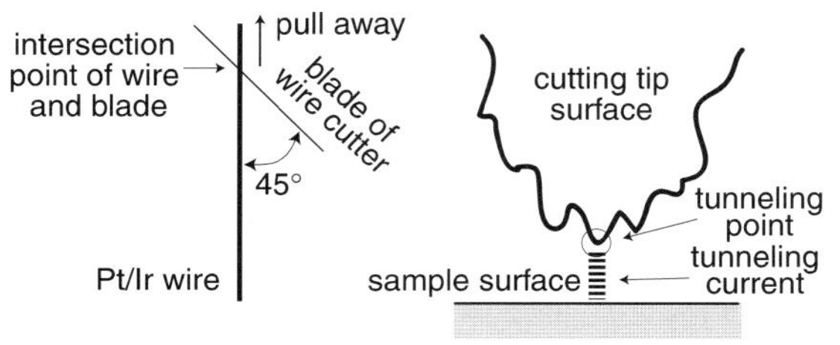

Figure 5.3.11
(a) Mechanical cutting and (b) electrochemical etching.

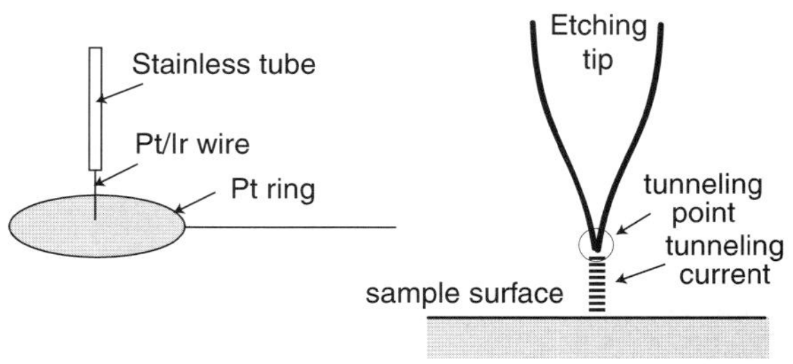

Figure 5.3.13
(a) Precision electrochemical etching for sharp STM tip.
(b) Tip shape of etching tip.

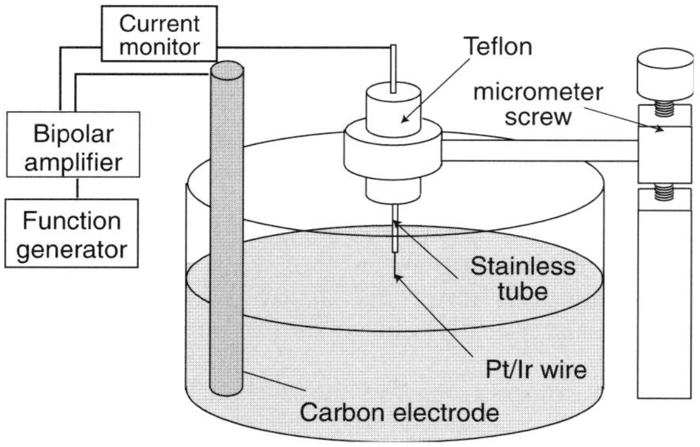

Figure 5.3.12
Configuration of electrochemical etching.

e.g., nanoparticle is observed. However, mechanical cutting is still useful when easy and rapid preparation of the tip is needed, e.g., for testing of STM.

(3) Electrochemical etching

Fig. 5.3.12 shows a typical instrument configuration of electrochemical etching. Etching solution for Pt/Ir wire is for instance the mixture solution of $CaCl_2$ aqueous solution and HCl [4]. The acetone is sometimes added to the etching solution [5].

Pt/Ir wire is submerged about 1 mm into the etching solution using a micrometer screw. The sinusoidal voltage of 25–50 Vac is applied between the Pt/Ir wire and the carbon electrode (DC offset voltage with several volt is sometimes applied). When a positive voltage is applied to the Pt/Ir wire, Pt/Ir is etched with the reaction. $Pt^{4+}+6Cl^{-}\rightarrow PtCl_6^{2-}$ Meanwhile when a negative voltage is applied to it, H_2 gas is produced, $2H^{+}+2e^{-}\rightarrow H_2$.

The bubble of H_2 gas moves upward from the tip apex to solution level and remove $PtCl_6^{2-}$ ion. The addition of acetone makes the bubbles escape from the Pt/Ir wire smoothly. Applied voltage is stopped when the etching current measured reaches zero. The tip shape depends on composition of the etching solution, amplitude of sinusoidal voltage, frequency (50–60 Hz) and DC offset voltage.

The optimization of these parameters has to be carried out in each experimental circumstance. The atomic image of cleaved Highly Oriented Pyrolytic Graphite (HOPG) and the molecular image of octadecane thiol SAM (self-assembled monolayer) are commonly used to test the tip using STM in air with the tunneling current of 100 pA–1 nA and several pA, respectively [10, 11].

The second etching after the first etching mentioned above can be used to sharpen the tips more precisely without the effect of external vibration for high resolution imaging using the experimental configuration shown in Fig. 5.3.13a [4, 7, 9]. The film of etching solution is suspended inside the Pt (or Au) ring by surface tension. Pt/Ir wire was clamped and threaded at the center through the Pt/Ir ring, which acts as the cathode. Sinusoidal voltage of about 2 Vac is applied between Pt/Ir wire and the ring for a certain period of time or a certain number of sine waves. Optimization of voltage and etching time also has to be carried out in each experimental circumstance.

When the shape of etching tip is symmetric, the tunneling point is localized at one or a few atoms at the end of the tip shown in Fig. 5.3.13b. The etching tip provides good image reproducibly without artifacts. The etching tip coated with an insulator film except tip apex should be used for STM in liquid to reduce Faraday current [12].

5.3.2.4 Surface imaging

Fig. 5.3.14 shows STM image of nanoparticles chemically modified with SAM and structure model of a nanoparticle [13]. STM image shows the clear structure of SAM on each nanoparticle with a diameter of 10 nm. As mentioned above, STM provides high-resolution image without contact during scanning. Thus STM is a powerful technique for imaging of nanoparticle surface, which has a certain level of electrical conductivity.

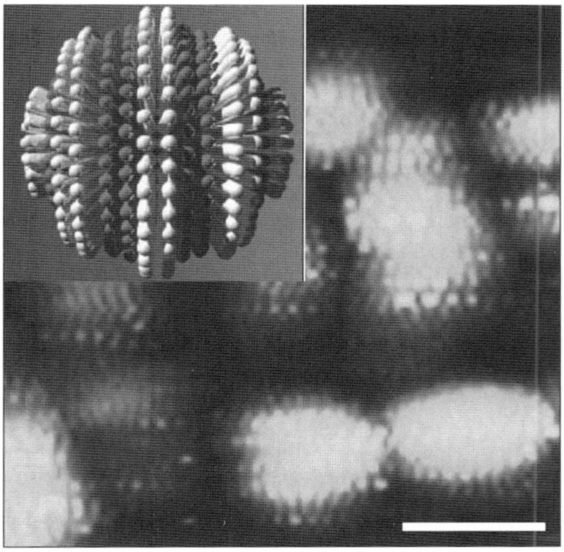

Figure 5.3.14
STM image of chemically surface-modified nanoparticles. A scale bar shows 10 nm [13].

References

[1] J. N. Israelachvili: *Intermolecular and Surface Forces*, 2nd edn., Academic Press, London (2005).
[2] H. Tanaka, M. Mitsuishi and T. Miyashita: *Langmuir*, **19**, 3103–3105 (2003).
[3] A. Koma (Ed.): *Hyoumen bussei sokutei*, Butsuri Zikken Kouza Vol. 10, Maruzen Publishing, Tokyo, p. 133 (2001).
[4] I.H. Musselman, P.A. Peterson and P.E. Russell: *Precision Eng.*, **12**, 3–6 (1990).
[5] A.H. Sorensen, U. Hvid, M.W. Mortensen and K.A. Morch: *Rev. Sci. Instrum.*, **70**, 3059–3067 (1999).
[6] H. Bourque, R.M. Leblanc: *Rev. Sci. Instrum.*, **66**, 2695–2697 (1995).
[7] M. Cavallini, F. Biscarini: *Rev. Sci. Instrum.*, **17**, 4457–4460 (2000).
[8] B. Ren, G. Picardi and B. Pettinger: *Rev. Sci. Instrum.*, **75**, 837–841 (2004).
[9] C. Albonetti, L. Bergenti, M. Cavallini, V. Dediu, M. Massi, J.F. Moulin and F. Biscarini: *Rev. Sci. Instrum.*, **73**, 4254–4256 (2002).
[10] M. Levlin, A. Laakso, H.E.-M. Niemi and P. Hautojarvi: *Appl. Surf. Sci.*, **115**, 31–38, (1997).
[11] D.W. Wang, F. Tian and J.G. Lu: *J. Vac. Sci. Technol. B*, **20**, 60–64 (2002).
[12] L.A. Nagahara, T. Thundat and S.M. Lindsay: *Rev. Sci. Instrum.*, **60**, 3128–3130 (1989).
[13] A.M. Jackson, J.W. Myerson and F. Stellacci: *Nat. Mater.*, **3**, 330–336 (2004).

5.3.3 FT-IR

IR spectroscopy and Raman spectroscopy, both designed to measure vibration spectra, yield information pertaining to chemical bonds. These analytical methods allow the measurement of substances in any state (gas, liquid, or solid), with only a small amount of sample required and minimal interference from coexisting substances. Moreover, the process is swift, involving simple manipulations. These advantages mean the techniques have been widely used in various fields of research related to materials, environments, organisms, and so on.

IR absorption takes place when bipolar molecular moments are altered by molecular vibration, while Raman scatter is seen when the polarization rate of molecules change (Raman activity). Generally, IR and Raman spectroscopies make up for the shortcomings of the other. This section will focus on IR spectroscopy and briefly detail its method as well as how to apply it to surface analysis. Details of the principles and features of IR spectroscopy have been described in numerous excellent monographs [1, 2].

5.3.3.1 IR spectroscopy

IR spectroscopy is based on molecular vibrations, accompanied by changes in the bipolar moment in the middle IR region (2.5–25 μm) and the wave number region between 4,000 and 400 cm^{-1}. This mode of vibration includes bending/stretching, rotating, pinching, twisting, and oscillating vibration and so on, which collectively cause intense IR absorption. IR activity is not seen during the bending/stretching vibration of CO_2 or diatomic molecules, which is also not accompanied by any change in the bipolar moment, although Raman activity is visible during the process.

The IR spectrum is often used for direct identification of certain specific functional groups constituting organic molecules. Among inorganic solids, the hydroxide group, bound water, oxyanions (carbonates, nitrates, and sulfates), and so on are known to cause intense IR peaks.

In the qualitative analysis of IR spectra, major absorption is divided into two regions, i.e., the wave number region specific to functional groups (4,000–1,400 cm^{-1}) and the fingerprint region, which presents vibration of the molecular frame (below 1,400 cm^{-1}). Analysis is performed in both regions and when IR spectra are used for quantitative analysis, the latter is based on Lambert–Beer's rule. If a proportion between the concentration of components and the absorbance is predetermined for a reference material, the concentration in a given sample can be measured.

The device often used for this method is the Fourier-transformation IR (FT-IR) spectroscope. With the FT-IR, coherent light is usually applied to the sample via a Michelson interferometer and the obtained interference signals are subjected to Fourier transformation.

5.3.3.2 Observation of the powder surface

If light is applied to a powder sample, complex optical paths will be generated by transmission, scattering, reflection, and so on. With this in mind, the method of measurement needs to be changed depending on which of these optical paths is to be utilized and addressed, and transmission and diffuse reflection methods are often used for this purpose.

With the transmission method, the light transmitted through a disk-shaped sample is measured with the latter usually prepared by pressure forming. If disk forming is difficult, alternative methods available are compressive forming with KBr and spraying of the sample onto the IR ray-transmitting material board.

With the diffuse reflection method, i.e., diffuse reflectance IR Fourier transform spectroscopy (DRIFT), spectra are obtained by measuring the diffuse reflective light, which returns after diffusion within the sample. If light is applied to the powder layer, it diffuses in various directions, while repeating reflection on the particle surface, as shown in Fig. 5.3.15, and is eventually released out of the sample from the uppermost layer.

During this process, light is repeatedly transmitted through the superficial layer of the particles. For this reason, the diffuse reflective light contains information about the IR absorptive property of the material in the vicinity of the particle surface. In research related to fine particles, this method is often used for identification of the surface and observation of adsorbed species. The Lambert–Beer's rule is inapplicable to the data collected by DRIFT, and instead of it, the Kubelka–Munk equation (K–M equation) is used for such data [1, 2].

The K–M function is proportional to the sample concentration if the absorptive power of the sample is low and the scattering coefficient is constant for all particles constituting the sample. Because the K–M function can also be affected by the particle size, shape, filling structure, and so on, these parameters need to be kept constant when this function is used for quantitative analysis. However, concerning the nanosized particles, it is difficult to keep these parameters constant in a highly reproducible manner, and particular care is needed when the peaks are compared. Other methods available include reflection absorption spectroscopy, photoacoustic spectroscopy, and so on [3].

Fig. 5.3.16 shows the DRIFT spectra for nanosized silica particles prepared by different methods. One type of silica was prepared in the liquid phase (precipitation method, 160 $m^2 g^{-1}$, Nipsil, Tosoh Corporation), while the other was synthesized in the gaseous phase (flame method, 300 $m^2 g^{-1}$, Aerosil, Nippon Aerosil Co., Ltd.).

The sharp peak in the highest wave number region near 3,750 cm^{-1} can be attributed to an isolated SiOH group, while the slightly broad peak at about 3,660 cm^{-1} is attributable to the hydrogen-bonded SiOH group. The extensive absorption around 3,400–3,450 cm^{-1} is attributable to –OH vibration associated with physically adsorbed water. Qualitatively speaking, a larger amount of the isolated SiOH group is formed by the flame method, which is explained by hydrolysis occurring at high temperatures (1,000°C or higher).

In the case of silica prepared by the precipitation method, the amounts of the hydrogen-bonded SiOH group and physically adsorbed water are large. In practice, the water content was about 3 mass% for silica prepared by the flame method and 6–9 mass% for that prepared by the precipitation method respectively. As far as nanosized particles of metal oxides

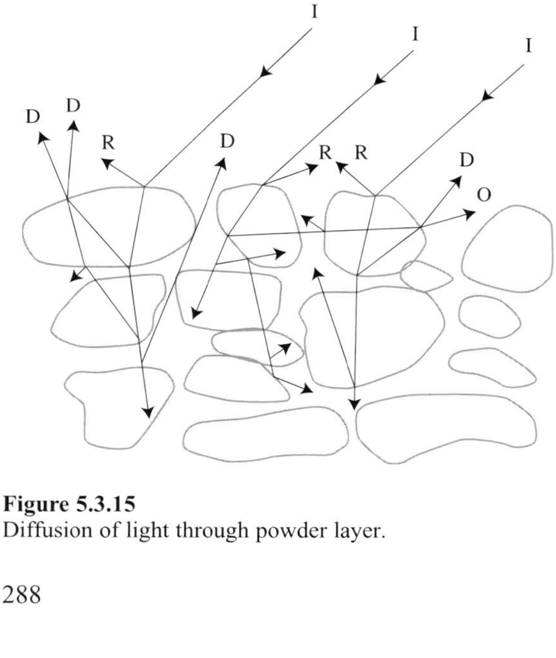

Figure 5.3.15
Diffusion of light through powder layer.

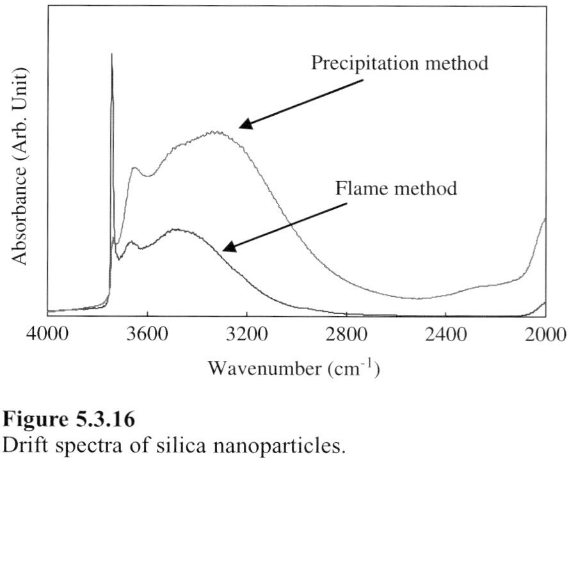

Figure 5.3.16
Drift spectra of silica nanoparticles.

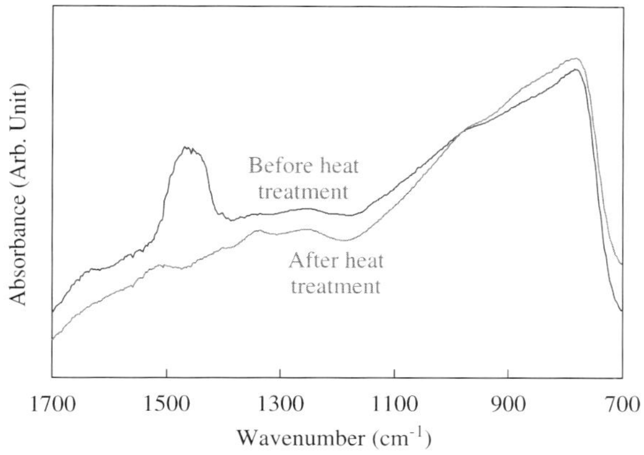

Figure 5.3.17
Drift spectra of barium titanate particles.

are concerned, precise evaluation of the properties and adsorption profiles of the superficial hydroxide group (a polar site) is essential for material and process development.

Fig. 5.3.17 shows a DRIFT spectrum of barium titanate particles (mean particle size: about 300 nm). If assayed immediately after purchase, the particles show an evident peak of $CO_3^{2!}$ at about 1,450 cm^{-1}, which disappears, however, after heating at 1,000°C. After about one week, a peak reappears at about 1,450 cm^{-1}. Changes in the crystal quality associated with this reaction cannot be detected by XRD.

In addition to these purposes, DRIFT has been used for various other applications, ranging from the evaluation of physico-chemical characteristics of surfaces to analysis of the relationship with process factors, e.g., identification of the solid acid on the fine particle surface [4], pressure-caused structural deformation of nanosized silica particles [5], and so on.

5.3.3.3 Observation of the circuit board surface

Attempts have been made to use a combination of IR spectroscopy and ATR (attenuated total reflection) for the quantitative analysis of organic contaminants adsorbed on the surface of silicon wafers. For example, contaminants of 0.5% ML (single molecule layer) or below have been detected by checking the C–H bending/stretching vibration of organic substances [6]. This technique has also been used to evaluate the influence of cleaning conditions and recontamination by organic substances within clean rooms [7, 8].

The ATR method uses a prism with a high refractive index, such as that made of ZnSe or Ge. If IR light is applied to a prism with an incident angle greater than the critical angle keeping the sample in close contact with the prism, the applied IR light permeates across the ultra-superficial layer and is then reflected. The absorption spectrum of the penetrating IR light (evanescent light) is evaluated. Fig. 5.3.18 illustrates the concept of the ATR method. The depth of penetration (d_p) is defined by the following equation:

$$d_p = \frac{10^4}{2\pi \nu n_1}\left\{\sin^2\theta - \left(\frac{n_2}{n_1}\right)^2\right\}^{-1/2} \tag{5.3.5}$$

In this equation, ν, n_1, n_2 and θ denote the wave number, refractive index of the ATR prism, that of the sample, and the incident angle, respectively.

Using equation (5.3.5), the relationship between d_p and the wave number was analyzed with a prism made of ZnSe ($n_1 = 2.4$) or Ge ($n_1 = 4$) and obtained results are shown in Fig. 5.3.19. A prism made of quartz glass was denoted as n_2. In the wave number region shown in this figure, the refractive index of the quartz glass ranged from 1.4 to about 1.45, but it was assumed to be 1.45 in this analysis, while θ was deemed to be 45°. As illustrated in this figure, d_p was smaller for Ge with a higher refractive index. It was also shown that the decrease in the wave number resulted in an increase in

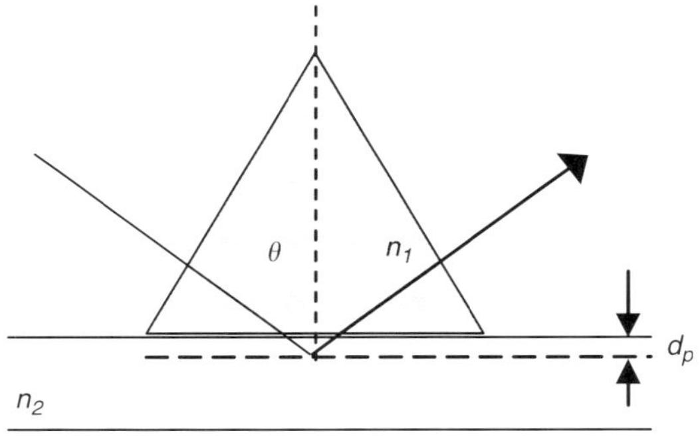

Figure 5.3.18
Schematic depiction of ATR method.

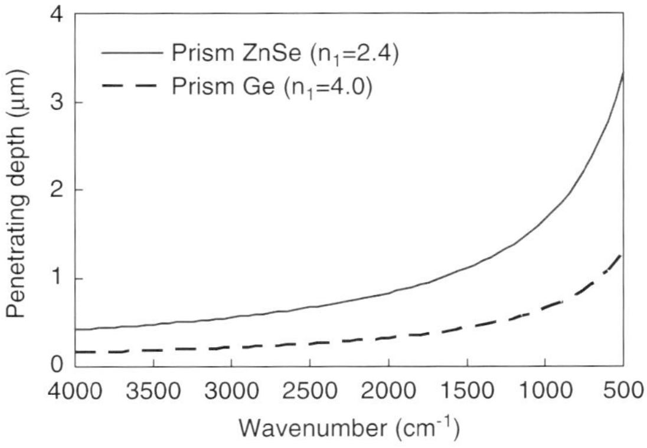

Figure 5.3.19
Dependence of penetration depth on wave number.

d_p (which was particularly sharp at a wave number of 1,500 cm^{-1} or less).

The conventional ATR method was based on multiple total reflection. This method involved the sample being kept in close contact with the surface of a prism made of a material with high refractive potential, and measured by light having undergone multiple reflections on the prism–sample interface. In recent years, the prism-fixed horizontal type ATR method, which involves compression of the sample placed horizontally on the prism from the upper direction (single-reflection ATR), has been developed and for reasons of convenience, this new ATR method has replaced the conventional method.

Fig. 5.3.20 shows the FT-IR-ATR spectra of the surface of a quartz-glass circuit board immediately after purchase (as-received sample) and after UV dry cleaning, respectively. The as-received sample generated evident peaks at 2,850 cm^{-1} and 2,920 cm^{-1}, both attributable to the symmetrical and reversely symmetrical bending/stretching motions of the CH_2 group.

A low peak in the vicinity of 2,960 cm^{-1} is probably attributable to $-CH_3-$. These results indicate that the board was contaminated with the organic substances adsorbed. With this in mind, it is known that organic contaminants in ambient air can be adsorbed onto the hydroxyl group of the quartz glass circuit board, and that their adsorption isotherm curve assumes a Langmuir pattern.

After the UV dry cleaning of the circuit board, none of these peaks was seen any more. The UV ray used for the dry cleaning of the circuit boards has a relatively short wavelength (UV-C: 100–280 nm) and UV rays in this range have high energy, approximately equivalent to the chemical bond energy of many organic substances. Furthermore, this UV ray induces the formation of ozone from oxygen in ambient air, resulting in the rapid formation of oxygen radicals and relatively effective oxidation of organic substances. In this way, the organic contaminants on the solid surface are oxidized by UV rays and eventually degrade into CO_2 and H_2O, which are eliminated from the surface by means of evaporation.

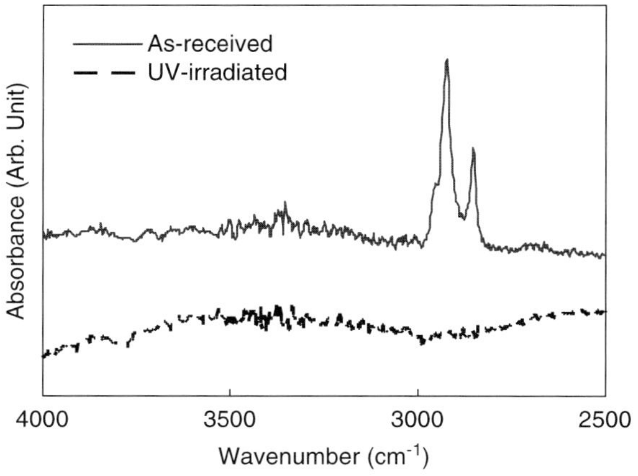

Figure 5.3.20
FTIR-ATR spectra of the surface of quartz glass substrates (as-received and after UV dry cleaning).

References

[1] J. Hiraishi, S. Minami, K. Masutani, S. Kondo, K. Ishitani: *Fourier Transform Infrared Spectrometry*, Japan Scientific Societies Press, Tokyo (1982).
[2] M. Tasumi: *Fundamentals and Practice of FT-IR*, Tokyo Kagaku-Dojin, Tokyo (1986).
[3] T. Yamashita: *Characterization of Solid Surface*, Kodansha Scientific Books, Tokyo (2005).
[4] A. Satsuma: *Appl. Catal. A*, **264**, 229 (2004).
[5] T. Uchino: *J. Phys. Condens. Matter.*, **14**, 1111–1114 (2002).
[6] V. A. Burrows: *Appl. Phys. Lett.*, **53**, 998 (1998).
[7] T. Ohmi: *Ultraclean Technology*, Chapter 3, Baifukan, Tokyo (1995).
[8] T. Ohmi: DISPLAY, No. 11, p. 1, Techno Times Co., Ltd. (2000).

5.3.4 XPS

X-ray photoelectron spectroscopy (XPS), which is also referred to as X-ray photoemission spectroscopy, is a surface analysis technique that is based on energy-spectrum measurements of photoelectrons emitted from a material surface under irradiation with a monochromatic soft X-ray radiation [1]. XPS is routinely used for a qualitative and/or quantitative analysis of surface elemental compositions and a chemical or an electronic state analysis of each element in the sample surface. A significant feature of the XPS analysis is that the technique is one of the ultimate surface-sensitive methods for analyzing the elements that are present within 1–10 nm of the sample surface. XPS is also known by the alternative acronym, ESCA, an abbreviation for electron spectroscopy for chemical analysis, as a technique that provides the surface elemental composition. Traditionally, XPS is one of the surface-analysis techniques that is categorized as photoelectron spectroscopy, which has been subdivided according to the source of exciting radiation into XPS using soft X-ray (200–2000 eV) radiation to examine core levels (inner-shell electrons) and UPS, an abbreviation for UV photoelectron spectroscopy, using vacuum UV (VUV; 10–45 eV) radiation to examine valence levels (outer-shell electrons).

First physical basics involved in the photoelectron emission can be briefly reviewed as follows. When a material surface is irradiated with a photon with an energy of $h\nu$, in which h is Planck constant and ν is frequency of the radiation, the photon energy is totally absorbed by an electron that is bound in either core levels or valence levels of an atom in a molecule or solid with a binding energy of E_b, leading to ionization and the emission of the bound electron out of the atom with an energy of $h\nu - E_b$ through the photoelectric effect. Hence, the photo-emitted electron is called a photoelectron.

Now let us consider a physical situation to measure an energy spectrum of photoelectrons emitted from a solid-material sample, which is irradiated with a monochromatic soft X-ray radiation beam, using an electron energy analyzer that is electrically connected to the solid–material sample, as schematically illustrated in Fig. 5.3.21. Main components of the XPS system for the photoelectron measurements include a source of monochromatic X-ray radiation beam, a sample stage with a set of stage manipulator and a sample introduction–transfer mechanism, an electron energy analyzer with an electron-collection lens and electron optics, an electron detector system, and mu-metal magnetic field shielding, which are all equipped in an ultra-high vacuum (UHV) chamber with UHV pumps. In the commonly employed X-ray sources, monochromatic X-rays are usually produced

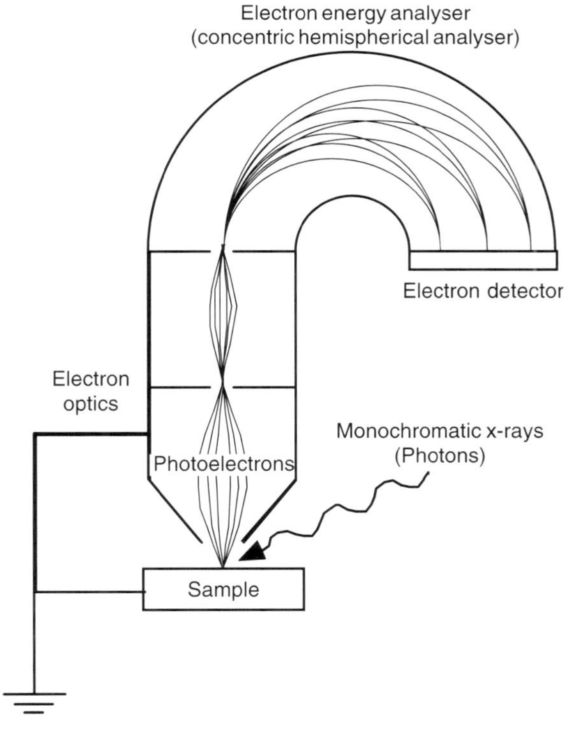

Figure 5.3.21
Basic components of XPS system.

by diffracting and focusing a non-monochromatic K_α X-ray radiation emitted from an electron-bombarded thin metal anode (usually Mg or Al), which gives rise to MgK_α radiation with a photon energy of 1253.6 eV and AlK_α radiation with a photon energy of 1486.6 eV. Photoelectrons, which are emitted from the sample surface through the photoelectric effect, are collected into the electron energy analyzer. There exist a variety of designs for the electron analyzer, however, the most preferred and the most commonly employed design for the photoelectron spectroscopy is a concentric hemispherical analyzer, in which a static electric field between two hemispherical surfaces disperses electrons according to their kinetic energy. The energy-analyzed electrons are normally detected with electron-multiplication devices such as a channeltron and a multichannel plate (MCP). The channeltron is employed for measurements of the electrons at a fixed point of the detector position by converting the flux of electron signals to a multiplied electron current, while the MCP is employed in the advanced systems for 2D measurements of the electrons reaching (to) the detector position according to the energy-dispersed spectrum versus the angular and/or the spatial distribution of the photoelectron collection.

The kinetic energy of the photoelectron measured with the energy analyzer is determined as follows. Here it is noted that the Fermi levels at the sample surface and the electron energy analyzer are kept at a common level since the sample surface and the electron energy analyzer are electrically connected to each other and thus both of the objects are in thermal equilibrium. Therefore, the kinetic energy of the photoelectrons measured with the electron analyzer (E_k) can be determined using an equation:

$$E_k = h\nu - E_b - \phi \quad (5.3.6)$$

where E_k is the kinetic energy of the photoelectron as measured with the electron energy analyzer, E_b the binding energy of the photoelectron emitted from one orbital within the atom as normalized to the common Fermi level, $h\nu$ is the photon energy of the X-ray radiation being used and ϕ is the work function of the electron energy analyzer as normalized to the common Fermi level. Thus we can measure the binding energy of the photoelectron that is normalized to the common Fermi level [1]. The energy spectrum of photoelectrons measured with an XPS system reflects the electronic structures of the atoms, since each element produces a characteristic set of XPS peaks at characteristic binding energy values. Therefore, the elements that are present in the sample surface can be directly identified by the characteristic set of XPS peaks. However, it is important to note that an XPS analysis of a non-conductive insulation material requires special care since the XPS measurements of such materials generally give rise to unwanted shifts of the energy spectrum because the electric potential of the sample surface tends to be hiked due to charge-up of the surface by the electron emissions from the sample and the determination of the Fermi level tends to become indefinite due to insufficient electric contact of the surface to the XPS system.

Dependence of binding energy for electrons in core levels (K, L, and M shells) is shown roughly in Fig. 5.3.22 [1, 2] as a function of the atomic number Z of elements. The binding energy generally increases with increasing atomic number Z of elements, being roughly proportional to the square of the atomic number. The characteristic combinations of the binding energies, which are unique to each element, provide significant advantage of XPS analysis for direct identification of elements, in which discrimination of elements can be easily performed even for neighbor elements in the periodic table.

Furthermore, the exact binding energy of the photoelectron depends not only upon the orbital level from which the photoelectron is emitted, but also upon the local chemical state of the atom, which gives rise to small shifts in the XPS peak positions in the spectrum, being so-called chemical shifts. The small changes in the binding energy occur due to Coulomb interactions between the photo-emitted electron and the ion core that is rearranged by the chemical bonding with the neighbor atoms and/or molecules in the solid. Therefore, information about the chemical states for the specific elements of interest can be analyzed by observation of the chemical shifts. For example, the peak position of Si 2p spectrum, which is observed around 99 eV for Si–Si bonding state, is known to shift to 103 eV when SiO_2 bonding is formed [1].

Next let us consider what kind of conditions are required for photoelectrons to escape from the solid surface into the vacuum. A photoelectron emitted from

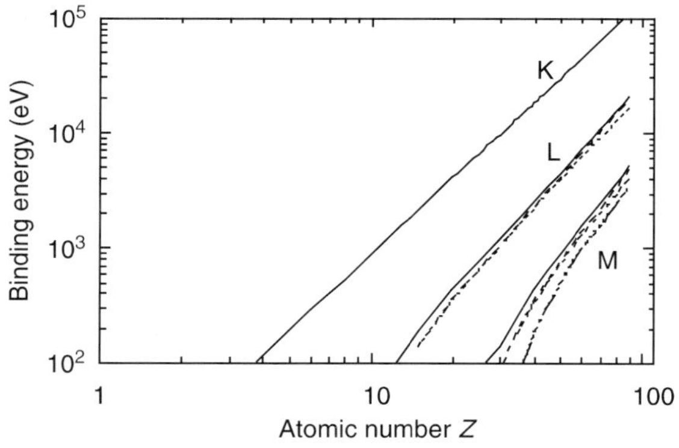

Figure 5.3.22
Binding energy of elements for K, L, and M shells as a function of atomic number.

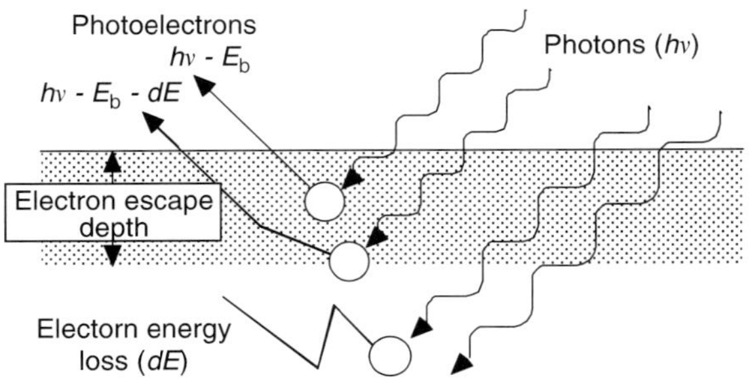

Figure 5.3.23
Behaviors of photoelectrons in solid.

an atom in the solid suffers energy loss due to elastic and non-elastic collisions with electrons in the solid. Therefore, one should note that the electrons that are detected in XPS measurements are only those originated in the vicinity of the solid surface and actually escaped into the vacuum of the instrument, as schematically illustrated in Fig. 5.3.23. The photoelectrons that are emitted in the deeper region by the penetrated X-rays are either captured in various excited states in the materials or lose significant energy even though they can escape from the solid surface.

The number of electrons that can escape from a region at a depth of x from the surface is known to decrease exponentially with increasing distance from the surface according to $\exp(-x/\lambda)$, where λ is the mean free path of electrons for elastic and inelastic collisions in the solid and the value gives rise to rough criterion of the electron escape depth. It is known that the mean free path of electrons in solid does not significantly depend upon elements; thus the dependence of the mean free path as a function of the electron energy is so-called "universal" curve [3] of the electron escape depth. The universal curve reveals that the mean free path takes a minimum value (0.5–1 nm) in the electron energy region of 50–100 eV, meanwhile the value at 1–2 keV region is 1.5–2 nm. Here it is noted that the penetration depth of soft X-ray radiation is much deeper than the value of this electron mean free path by several orders of magnitude. Thus the photoelectrons that can be measured in XPS analysis are limited to only those from the elements that are present within several nanometers of the sample surface and the X-ray penetration depth is substantially insignificant in terms of considering the sensitive depth region.

Furthermore, a non-destructive depth analysis in the region of subnanometer to several nanometers can be achieved by varying the escape angle of photoelectrons. Especially, when the XPS measurements are performed using a hard X-ray radiation beam (~10 keV) instead of the soft X-ray radiation in the conventional XPS instruments, a non-destructive depth analysis in the region that is much deeper than those with the conventional XPS system can be performed by properly selecting the X-ray energy and the electron escape angle.

In routine analysis by XPS, two kinds of spectrum-survey modes are usually taken in scanning the kinetic energy by properly setting the energy width and the resolution for measurements; (a) wide-scan survey spectrum (wide energy width and low energy resolution; mainly employed for qualitative analysis of elements) and (b) narrow-scan survey spectrum (narrow energy width and high energy resolution; mainly employed for quantitative analysis of elemental composition and for analysis of chemical state identification).

Examples are shown for XPS analysis of a boron carbon nitride (BCN) film, which was prepared with ion-beam-assisted deposition, in Fig. 5.3.24 showing a wide-scan survey spectrum, and in Fig. 5.3.25 showing a narrow-scan survey spectrum for C1s signal. First, the peaks found in the wide-scan survey spectrum correspond to photoelectrons that are considered to originate in the vicinity of the sample surface and to

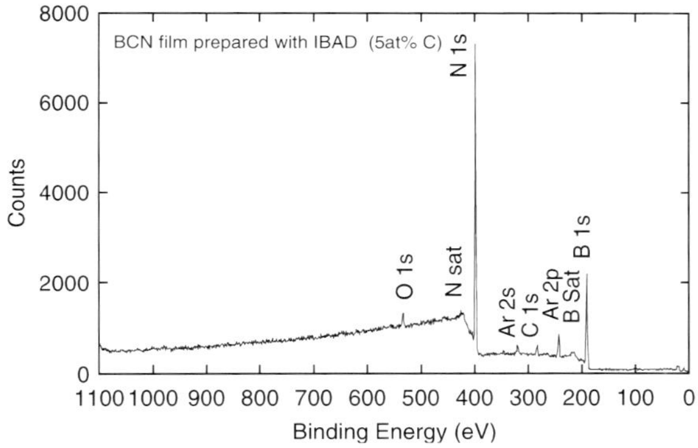

Figure 5.3.24
XPS wide-scan survey spectrum measured for a BCN film prepared with ion-beam-assisted deposition.

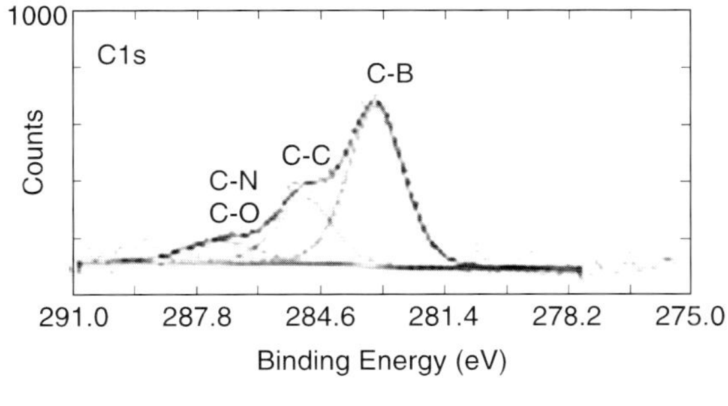

Figure 5.3.25
XPS narrow-scan survey spectrum for C1s signal measured for a BCN film prepared with ion-beam-assisted deposition.

escape into the vacuum without suffering significant energy loss, while the electrons that originated in relatively deeper regions are considered to form tails and background of the XPS spectrum in the lower kinetic energy region (i.e., higher binding-energy region in the spectrum). In addition, it is shown that the signals for boron, carbon, and nitrogen, which are the neighbor elements in the periodic table, are clearly discriminated in the spectrum.

In the narrow-scan survey spectrum shown in Fig. 5.3.25, which is measured for photoelectrons emitted from the C1s orbital, in this example, it is evaluated that the chemical bonding state of carbon atoms is dominated by their bonding with boron atoms. In addition, narrow-scan survey spectra are routinely used to perform quantitative analysis of elemental composition, since the signal background level can be defined with better precision. Furthermore, depth-profile measurements of the elemental composition can also be performed by alternating the narrow-scan survey and the surface removal with ion-beam sputtering.

Following are the points that require special attentions in XPS analysis. First it is important to note that an XPS analysis of non-conductive (insulation) materials requires special attentions since the XPS measurements of such materials generally give rise to unwanted shifts of the energy spectrum due to a charge-up by the electron emissions from the sample surface and indefinite determination of the Fermi level. The problem can be minimized by means of coating the sample surface with conductive films such as Au, which should be thinner than the electron mean free path and chemically inert to the sample, and taking the electrical contact of the coated surface to the XPS system using a carbon tape. In addition, it is somewhat effective to calibrate the energy shifts by measuring the known XPS peaks such as surface contaminated carbon atoms. In the XPS instruments equipped with an electron gun, the electron supply to the sample can also be performed to compensate for the charge-up. Second, a special attention must be taken to the surface damage of the samples when cleaning and/or sputtering of the sample surface are performed using ion beam sputtering in the XPS instruments. For example, it is important to note that the ion irradiation onto the oxide materials has been reported to lead sometimes to the deficiency of oxygen atoms by more than 10%.

In the state-of-the-art advanced systems, an ultra-high energy resolution of less than 1 meV has been achieved in the electron spectroscopy, which has enabled precise determination of fine electronic structures near the Fermi level and band structures in nanostructured materials such as carbon nanotubes and quantum dots (QD) [4]. Furthermore, advanced analysis techniques based on X-ray photoelectron diffraction, which was evaluated by measuring the patterns of photoelectron emission angles, has been developed to determine surface structures by analyzing diffraction and scattering of photoelectrons in the vicinity of the surface [5]. This new technique is expected to develop into a method for simultaneous determination of the electronic structures and the surface fine structures.

Finally, major features and limits of the XPS technique are summarized as follows.

1. XPS is routinely used to measure:
   - the elemental compositions (qualitative and quantitative measurements)
   - the chemical state identification of one or more elements of interest
2. Sensitivities and limitations of measurements:
   - Depth sensitivity of analysis: 0.3–3 nm (photoelectron escape depth, depth profiling via angular variation of electron collection)
   - Detectable elements: All elements with atomic number larger than or equal to 3
   - Detection limits: 0.1–1.0 atomic%
3. Analysis area and depth-analysis limitations:
   - Minimum analysis area: > 9 μm
   - Depth-profile capabilities: several tens of nanometers to several micrometers in depth (sputter-depth profiling)

References

[1] L.C. Feldman, J.W. Mayer: *Fundamentals of Surface and Thin Film Analysis*, Elsevier Science, North-Holland, NY and Amsterdam (1986).
[2] K. Siegbahn: *ESCA, Atomic, Molecular and Solid State Structures Studied by Means of Electron Spectroscopy*, Almquist and Wiksells, Uppsala (1967).
[3] G. Somerjai: *Chemistry in Two Dimensions: Surfaces*, Cornel University Press, Ithaca, NY (1981).
[4] T. Sato, H. Matsui and T. Takahashi: *Oyo Buturi*, **74**, 1305–1315 (2005).
[5] Y. Nihei, M. Nojimai: *Oyo Buturi*, **74**, 1341–1344 (2005).

5.3.5 Wettability

As we often observe wetting phenomena in our daily life, in this meaning, it can be considered that the wetting phenomena are macroscopic ones. From a view point of material science, many phenomena in manufacturing and processing of materials are influenced by wettability of solid by liquid. Accordingly, a lot of the wettability measurements have been conducted and reported in aqueous solution and solid systems at low temperature and in melt and solid systems at high temperature. As the factor determining wetting or non-wetting systems is an atomistic interaction between atoms composing liquid and those composing solid, it is important for us to understand wettability as microscopic and/or nanoscopic phenomena by approaching in atomistic theory.

Wettability of solid by liquid depends not only on a liquid drop size as mentioned in 1.5 but also on a crystal structure of solid surface and surface adsorption at a solid surface.

It is well known that the factors affecting the wettability are:

1. Atmosphere (especially, partial pressure of oxygen).
2. Chemical reaction at the interface between solid and liquid and formation of reaction layer.
3. Mutual solubility.
4. Surface morphology of solid (surface roughness, crystal orientation, adsorption, and strain induced by working.
5. Temperature.
6. Thermodynamical stability of solid and liquid.
7. Others (like impurities and additives).

The wettability of solid by liquid is determined by above factors, but it is difficult to conclude which factor is the most dominant, since the most effective factor depends on a combination of liquid and solid.

The wettability is generally evaluated by a contact angle of a liquid drop on a plane solid surface as shown in Fig. 5.3.26. It is called "wetting system" when the contact angle is smaller than 90° (see Fig. 5.3.26a) and "non-wetting system" when the contact angle is larger than 90° (see Fig. 5.3.26b). It is usually impossible to expect whether an objective

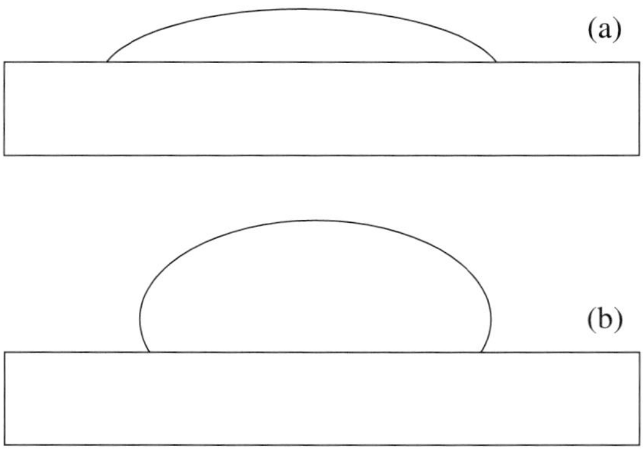

Figure 5.3.26
Liquid drop shape on solid.

system is a wetting or a non-wetting one since many factors affect the wettability.

Relation between contact angle and surface free energy of solid, surface free energy of liquid (surface tension), and interfacial free energy can be expressed by an equation (5.3.7), which is so-called Young's equation:

$$\gamma_s = \gamma_l \cdot \cos\theta + \gamma_{sl} \qquad (5.3.7)$$

where θ is the contact angle of liquid drop on a smooth surface of solid, γ_l is the surface free energy (surface tension) of liquid, γ_s is the surface free energy of solid, and γ_{sl} is the interfacial free energy between solid and liquid.

The work of adhesion, W_{ad}, which is defined as a reversible work to separate liquid from solid, is expressed by an equation (5.3.8).

$$W_{ad} = \gamma_l \cdot (1 + \cos\theta) \qquad (5.3.8)$$

Equation (5.3.9) has been proposed to a rough surface of solid [1].

$$\cos\theta_r = R\cos\theta \qquad (5.3.9)$$

where θ_r is the contact angle on a rough surface, θ the contact angle on smooth surface and the roughness factor, R, is defined as a proportion of a surface area of rough surface to that of smooth surface ($R \geq 1$).

As can be seen from the equation (5.3.9), a contact angle (θ_r), which is called as an apparent contact angle, on a rough surface becomes larger when a contact angle (θ), which is called as a real contact angle, is larger than 90°, and an apparent contact angle becomes smaller when a true contact angle is smaller than 90°.

Effects of a crystal structure of a solid surface and an adsorbed gas on a solid surface on the wettability have been systematically studied at high temperature [2–22].

As to the effects of the adsorbed gas and surface structure change on the wettability, it has been investigated in a liquid metal-diamond system [16, 17]. It is considered that hydrogen gas is adsorbed on a diamond surface and its desorption temperature depends on a crystal orientation of diamond [23–25]. Reported values of desorption temperature of hydrogen on diamond surfaces are summarized in Table 5.3.3. It is clear from Table 5.3.3 that the desorption temperature of hydrogen depends on a crystal orientation of diamond surface. Diamond is also a metastable structure and changes to a graphite structure at a high temperature under atmospheric pressure. Accordingly, as shown in Fig. 5.3.27, the surface structure of diamond changes with an increase of temperature.

Table 5.3.3
Hydrogen desorption temperature on diamond surface [18–21].

Surface orientation	Desorption temperature (K)
(1 0 0)	1,073
(1 1 0)	1,173, 1,223
(1 1 1)	1,273

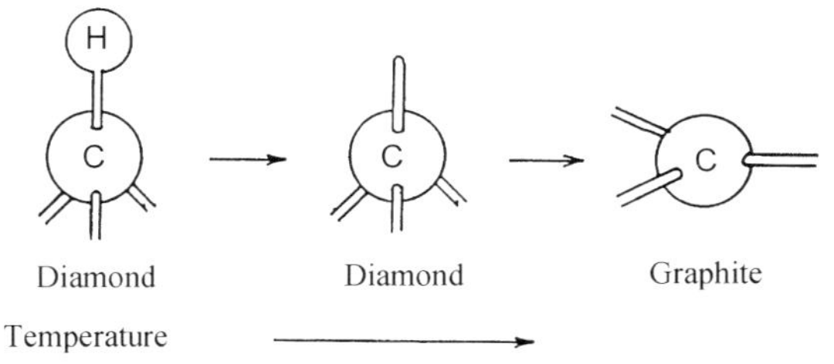

Figure 5.3.27
Surface structure change with temperature increasing.

As an example, a temperature dependence of a contact angle of liquid Sn on (1 1 0) plane of diamond is shown in Fig. 5.3.28 [16]. It can be seen from Fig. 5.3.28 that the contact angle dramatically changes at near a desorption temperature of hydrogen and also near a graphitization temperature. Similar results were obtained on (1 0 0) and (1 1 1) planes of diamond [16].

The effect of crystal orientation of solid surface on a contact angle has been investigated in detail in liquid metals (Sn, Pb, and Bi)/oxide single crystals (alumina and magnesia) system in which no reaction between liquid metals and oxides occurs and also liquid Al/oxide single crystals (alumina and magnesia) systems in which reaction between liquid Al and oxide occurs [2–22].

In no reaction systems, time dependence of contact angles is not observed. Values of work of adhesion between liquid metals and single crystal of magnesia are shown in Table 5.3.4 [15]. It can be seen from Table 5.3.4 that the work of adhesion considerably depends on a surface orientation of magnesia single crystal. Surface orientation dependence is the largest in (1 0 0) plane of magnesia, then (1 1 1) plane, and is the smallest in (1 1 0) plane. This tendency does not depend on the kind of metals. This fact suggests that wettability is affected not only by a thermodynamical stability of liquid metals, which described by McDonald et al. [26], but also a surface structure of solid.

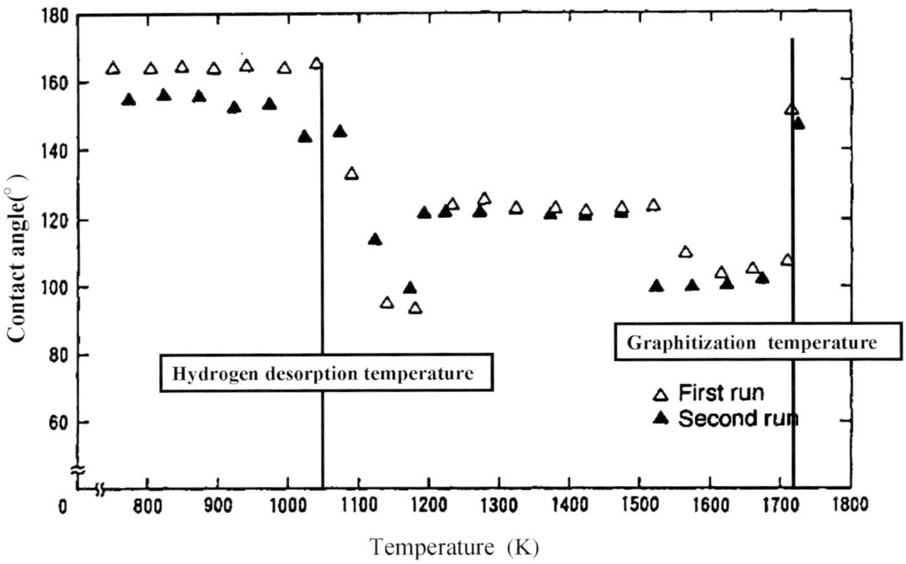

Figure 5.3.28
Temperature dependence of contact angle of liquid Sn on (1 1 0) plane of diamond.

Table 5.3.4
Work of adhesion between liquid metal and magnesia single crystal.

Metal	Work of adhesion (erg cm^{-2})		
	(1 0 0) plane	(1 1 1) plane	(1 1 0) plane
Sn	121	84	8
Pb	97	63	14
Bi	88	52	32

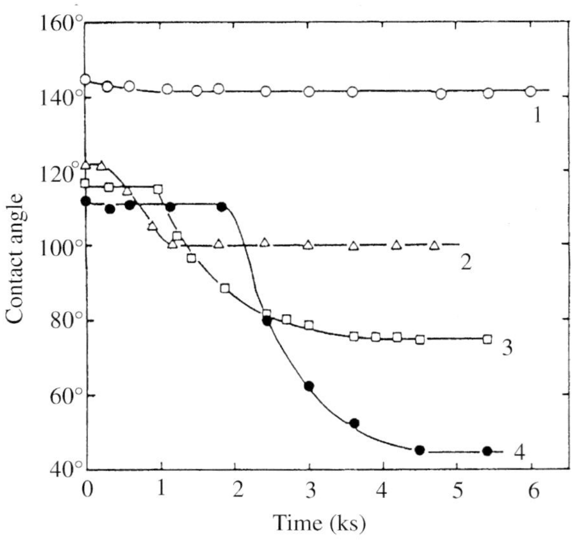

Figure 5.3.29
Contact angle of liquid Cu–Ti alloys on polycrystal of SiC.

On the other hand, in reaction systems, contact angles change with a time elapsing. As an example, a time dependence of contact angle of liquid Cu–Ti alloy on polycrystal SiC is shown in Fig. 5.3.29 [27]. It is clear from Fig. 5.3.29 that the contact angle decreased with time as Ti forms a reaction layer (TiC) at the interface between liquid CuTi alloy and SiC and Si dissolved in liquid Cu–Ti alloy. This tendency is also observed in single crystal [2–14].

A contact angle is also affected by surface reconstruction of single crystal at high temperature [3, 11, 13]. Fig. 5.3.30 shows a temperature dependence of contact angle of liquid Al on alumina single crystal. It is clear from Fig. 5.3.30 that the contact angle depends not only on a surface orientation but also a surface reconstruction C(1 × 1)plane→C($\sqrt{31} \times \sqrt{31}$). And in this system, the contact angle decreases with an increase in temperature. In general, the contact angle decreases with an increase in temperature, but when a surface structure changes with temperature, the contact angle increases with an increase in temperature [13, 14, 16].

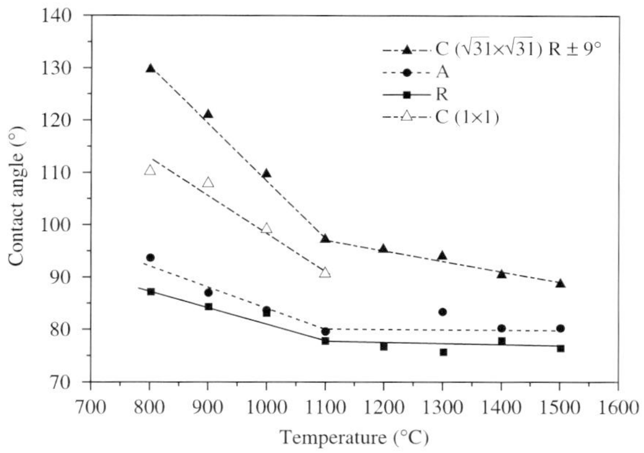

Figure 5.3.30
Temperature dependence of liquid Al on single crystal of alumina.

References

[1] R.N. Wenzel: *J. Phys. Colloid Chem.*, **53**, 1466–1470 (1949).
[2] P. Shen, H. Fujii and K. Nogi: *Mater. Trans.*, **45** (9), 2857–2863 (2004).
[3] P. Shen, H. Fujii and K. Nogi: *J. Mater. Process. Technol.*, **155–156**, 1256–1260 (2004).
[4] P. Shen, H. Fujii, T. Matsumoto and K. Nogi: *J. Am. Ceram. Soc.*, **87** (11), 2151–2159 (2004).
[5] P. Shen, H. Fujii, T. Matsumoto and K. Nogi: *Ceramic Transactions*, **146**, (2004).
[6] P. Shen, H. Fujii and K. Nogi: *Adv. Mater. Process. Technol.*, 1416–1419 (2003).
[7] P. Shen, H. Fujii, T. Matsumoto and K. Nogi: *Metall. Mater. Trans. A*, **35A**, 583–588 (2004).
[8] P. Shen, H. Fujii, T. Matsumoto and K. Nogi: *Acta Mater.*, **52**, 887–898 (2004).
[9] P. Shen, H. Fujii, T. Matsumoto and K. Nogi: *Transaction of JWRI*, **32** (2), 313–318 (2003).
[10] P. Shen, H. Fujii, T. Matsumoto and K. Nogi: *Scripta Mater.*, **49**, 563–569 (2003).
[11] P. Shen, H. Fujii, T. Matsumoto and K. Nogi: *Acta. Mater.*, **51**, 4897–4906 (2003).
[12] P. Shen, H. Fujii, T. Matsumoto and K. Nogi: *Scripta Mater.*, **49**, 563–569 (2003).
[13] P. Shen, H. Fujii, T. Matsumoto and K. Nogi: *Scripta Mater.*, **48**, 779–784 (2003).
[14] P. Shen, H. Fujii, T. Matsumoto and K. Nogi: *Transitions JWRI*, **32** (1), 155–158 (2003).
[15] K. Nogi, M. Tsujimoto, K. Ogino and N. Iwamoto: *Acta Metall.*, **40**, 1045–1050 (1992).
[16] K. Nogi, M. Nishikawa, H. Fujii and S. Hara: *Acta Mater.*, **46** (7), 2305–2311 (1998).
[17] K. Nogi, Y. Okada and K. Ogino: *Mat. Trans., JIM*, **35** (3), 156–160 (1994).
[18] B.B. Pate: *Surf. Sci.*, **165**, 83–142 (1986).
[19] K. Nogi: *Materia*, **35** (5), 523–525 (1996).
[20] P. Shen, H. Fujii and K. Nogi: *Scripta Mater.*, **52**, 1259–1263 (2005).
[21] P. Shen, H. Fujii and K. Nogi: *J. Mater. Res.*, **20**(4), Apr (2005).
[22] P. Shen, H. Fujii, T. Matsumoto and K. Nogi, *J. Am. Ceram. Soc.*, **88** (4), 912–917 (2005).
[23] B.B. Pate, P.M. Tefan, C. Binns, P.J. Jupiter, M.L. Sheck, I. Lindau and W.E. Spicer: *J. Vac. Sci. Technol., A*, **19**, 349–354 (1981).
[24] V. Hamza, G.D. Kubiak and R.H. Stulen: *Surf. Sci.*, **206**, L833–L844 (1988).
[25] S.V. Pepper: *J. Vac. Sci. Technol.*, **20**, 643–649, (1982).
[26] J.E. McDonald and J.G. Eberhart: *Transition. Met. Soc. AIME*, **40**, 1045 (1965).
[27] K, Nogi, K. Ikeda and K. Ogino: *Mater. Trans. JIM*, **54** (12), 1401–1407 (1990).

5.4 Nanopore characterization

A macroscopic body composed of nanoparticles naturally possesses microscopic void space. Some kind of nanoparticles may possess nanopores within themselves. Such pore spaces can bring functional characteristics such as adsorptive capacity, selective permeation, dielectric properties, while they would stand as a deficit as in the case for ceramic bodies. With increasing demand in functionality, recent powder-based manufacturing tends to use finer particles of submicron down to nanometer range. As a result the characterization of capillarity in the nanometer range is getting more and more important. Since the size of the void space is several to ten times smaller than the particle diameter, the characterization method for single-nanometers up to tens of nanometers should be applied for nanoparticle-based porous bodies.

In general the nitrogen adsorption method is applied for this range of pores. A nitrogen adsorption isotherm at 77 K, or the relation between adsorbed amount of nitrogen per unit mass of the solid and the relative pressure, which equals to (equilibrium pressure p)/(saturated vapor pressure p_s), is measured and analyzed with various methods described in this chapter to determine the distribution of pore volume against pore size. Measurement is done either by the volumetric method which detects pressure variation of nitrogen gas introduced in an adsorption system with constant volume, or the gravimetric method which measures weight variation of a sample contacting with the gas of given relative pressure. The great majority of commercially available automated apparatuses are based on the former principle.

In the following sections, various methods for analyzing nitrogen isotherms to determine pore size distribution (PSD) will be described with its emphasis on nanometer range. Some other experimental methods are briefly explained in the last part of this section.

5.4.1 Type of nitrogen isotherms and pore characteristics implied

The International Union of Pure and Applied Chemistry (IUPAC) established classification of pore size into the following three categories.

Macropore: Greater than 50 nm.
Mesopore: 2–50 nm.
Micropore: Smaller than 2 nm.

Brunauer et al. classified many adsorption isotherms into five categories as shown in Fig. 5.4.1, which is termed as the BDDT (Brunauer, Deming, Deming and Teller) classification. The following understanding can be made with the attention to the pore sizes and intrinsic interaction strength of solid surfaces.

- Type I: Microporous or monolayer adsorption. Strong interaction.
- Type II: Macroporous or nonporous. Strong interaction.
- Type III: Macroporous or nonporous. Weak interaction.
- Type IV: Mesoporous. Strong interaction.
- Type V: Mesoporous. Weak interaction.

Nitrogen molecules receive relatively strong interaction from most of the solids, compared with that between them. Types III and V are thus rarely seen in the case of nitrogen adsorption. Excluding here the Type II for macroporous or nonporous bodies, the remaining types of I and IV would be the isotherms to be analyzed for PSD. The so-called micropore filling phenomenon is the principle for analyzing the former type, and the capillary condensation phenomenon is for the latter.

5.4.2 Micropore filling phenomenon and PSD analysis

5.4.2.1 Micropore filling

Within the space of narrow pores in the range of 1 or 2 nm, the interaction potential energies of opposite surfaces will overlap significantly, resulting in a strongly attractive field as shown in Fig. 5.4.2 for slit-pore case. Adsorbate molecules are attracted and trapped into the space by this strong field to form up a lump of molecules with the entropy state similar to that of liquid, which differs significantly from monolayer adsorption phenomenon of covering solid surface.

5.4.2.2 Pore size estimation by average potential field

If the chemical composition of the solid wall and the geometry (shape and size) of the space is given, one would be able to model the potential field within the pore space. Horvath and Kawazoe [1] firstly employed this principle to propose a model for micropore-size estimation for slit-shaped pores, while Saito and Foley [2] later applied it for cylindrical pores.

Detailed equations should be referred to the literature, and only the basic idea of the two methods is explained here. The modeled potential energy profile within the pore space is integral averaged, which is to be equated with the free energy needed to make vapor with pressure p condense into liquid state, $kT\ln(p/p_s)$. Thus they obtained relations between the relative pressure and the pore sizes.

There have been many examples of applying this method to molecular sieving activated carbons with success. However, there are some of possible criticisms: Adsorption at a relative pressure is assumed to occur only in smaller pores than that corresponding to the relative pressure, which would be too much simplification; the potential energy between adsorptive molecules may not adequately be taken into account. The method then

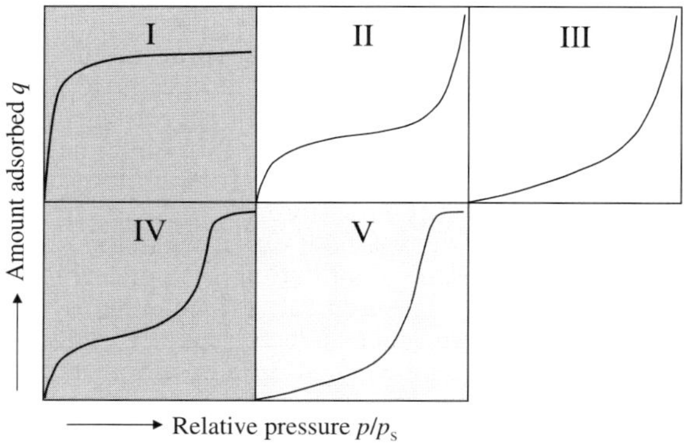

Figure 5.4.1
Type of adsorption isotherms (BDDT classification).

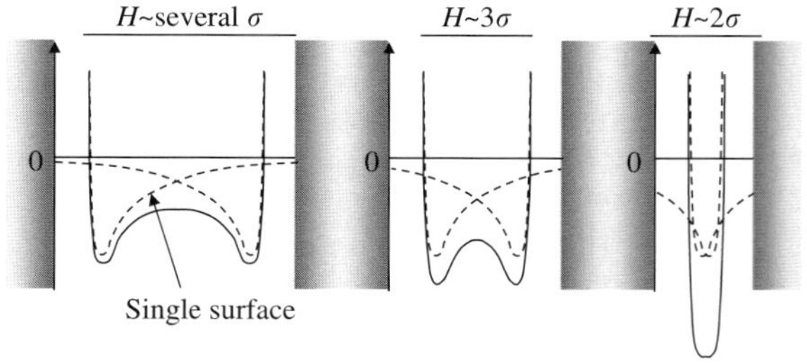

Figure 5.4.2
Potential energy profile in slit pores with width H. Overlap of potential for each surface enhances the attraction within the pore space.

may not be suitable for strict PSD analysis especially for pores with larger size or wider distribution.

5.4.2.3 Pore size estimation based on standard isotherms

Standard isotherm is the one measured on a non-porous solid with the same chemical composition as that of porous material in question, which in general exhibit Type II of BDDT classification. Converting the amount adsorbed of the standard isotherm into the statistical thickness of adsorbed film t (adsorption volume per unit surface area), and comparing it with the isotherm of the porous body, one can obtain various information of PSD, which is briefly explained in the following. Details are given elsewhere [3, 4].

The so-called t-plot can be drawn by plotting adsorption volume W for porous body against the statistical thickness t of the standard isotherm, all of which have common values of relative pressures. For macroporous solid, e.g., the volume W should simply be proportional to the thickness t up to high-pressure range, resulting in a linear t-plot whose slope gives the surface area. For micropores, on the other hand, a linear part stands only below a relative pressure at which the pore space is filled up with the adsorbed film. Above this pressure, the adsorption in the pore space will not increase any more, and possible increase is only a small one from adsorption at outer surfaces. An example of resultant t-plot is schematically shown in Fig. 5.4.3a. Note that the linear part in the low-pressure range appears only if the overlap of potential energy is not significant to exhibit similar adsorption thickness in pores to that for single surface. From a clear elbow point, one then would be able to estimate pore width because the t value at the point should be equal to half of the pore width. However, since this principle neglects the effect of potential overlap, which becomes more significant for smaller pores, the accuracy becomes low for smaller pores. Another point to be noted here is that the t-plot for mesoporous body exhibits upward shift form linearity in higher pressure range, which results from occurrence of capillary condensation.

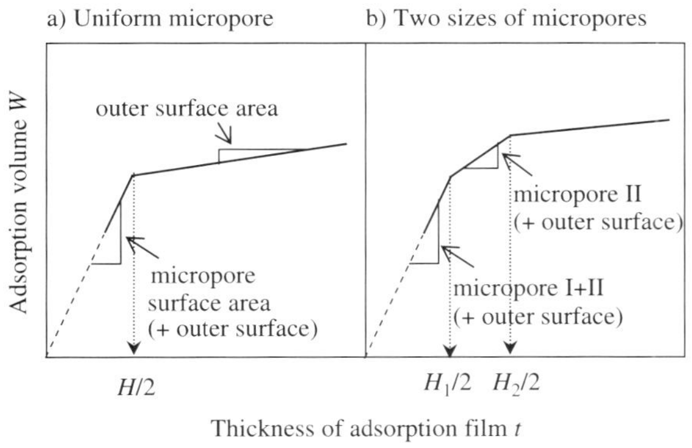

Figure 5.4.3
Basic idea of the t-plot method. (a) A micropore will be filled up when t becomes half of pore width (if potential overlap is not significant), above which only slight increase is given by outer surface. (b) Two kinds of pores will give two of linear part other than that from outer surface, slopes of which give surface areas and volumes of the two pores.

In many cases the t-plot would not exhibit the ideal behavior as shown in Fig. 5.4.3a, but show a gradually bending curve from lower linear part to higher-pressure part. Now, supposing two discrete sizes of micropores, the t-plot should be like Fig. 5.4.3b, because two of elbow points should appear. Then one must be able to obtain two of the micropore sizes from this plot. First slope means total surface area, while the slope of the second linear part gives surface area of larger pores (+outer surface). Extending this concept to continuous distribution in micropore size, one can calculate PSD from continuously curved part of the t-plot by drawing tangential lines at certain points in the curve, which is termed as MP (Micro Pore) method. Many of automated nitrogen adsorption apparatus are equipped with this method, which may be accelerating many users to apply this method. One should note, however, that this method is neglecting the overlap of potential energies in the pore space, which must produce certain extent of the errors.

5.4.3 Capillary condensation phenomenon and PSD analysis

5.4.3.1 The Kelvin equation

The capillary condensation phenomenon itself can be understood with macroscopic and classical thermodynamics. In nanoscale, especially that with so-called single-nanolength, however, they fail to express the phenomena because of hindered characteristics of liquid surface and for instance effects by potential energy exerted from pore walls.

Suppose that liquid nitrogen exists in a cylindrical pore in equilibrium with vapor of pressure p as shown in Fig. 5.4.4. The solid surface is covered with the adsorption film with the thickness t, and the condensed phase stands in the interior of the film, whose contact angle becomes zero. The Young–Laplace equation describes the pressure difference Δp brought by the pulling force of surface tension γ, which is given by equation (5.4.1), in the case of perfect wetting.

$$\Delta p \equiv p - p' = \frac{2\gamma}{r} \tag{5.4.1}$$

For a slit pore the pressure difference is given by $\Delta p = 2\gamma/H$, which quite resembles to each other, but note that r is the radius of the pore while H stands for the full width of the slit space.

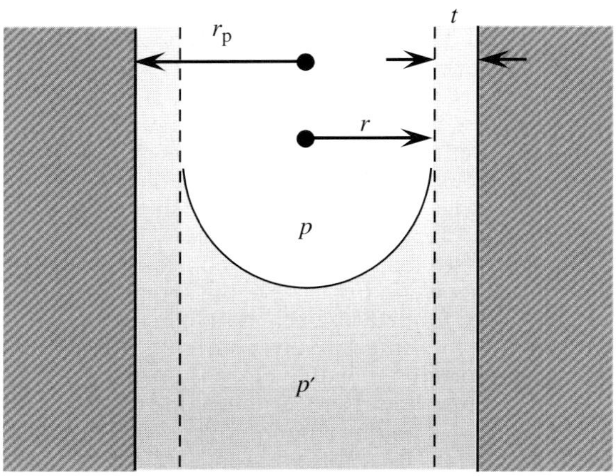

Figure 5.4.4
Capillary condensed phase in a cylindrical pore, coexisting with surface adsorbed film. Pore radius r_p is given by the sum of film thickness t and radius of the condensed phase r.

The pressure reduction of the liquid phase results in a lower free energy than a normal liquid with flat surface by $v\Delta p$, (v being molar volume of liquid). Then the equilibrium vapor pressure of pore liquid p must be smaller than the saturated vapor pressure p_s. Conversely, the vapor with a pressure smaller than p_s can condensate if it goes into a pore, which is termed as the capillary condensation phenomenon.

The equilibrium pressure p must satisfy equation (5.4.2) since the difference of free energy from p_s, or $RT\ln(p/p_s)$, equals $v\Delta p$.

$$\ln\left(\frac{p}{p_s}\right) = -\frac{2v\gamma}{rRT} \qquad (5.4.2)$$

This is the so-called Kelvin equation, which was derived originally by Lord Kelvin for vapor pressure of small droplet without the negative sign.

5.4.3.2 Calculation scheme

As shown in Fig. 5.4.4, the radius for cylindrical pore r_p is then given by the sum of the film thickness t and the radius of the condensed phase r.

$$r_p = r + t = \frac{2v\gamma}{RT\ln(p_s/p)} + t \qquad (5.5.3)$$

The width of slit pore, on the other hand, is

$$H = 2r + 2t = \frac{2v\gamma}{RT\ln(p_s/p)} + 2t \qquad (5.4.4)$$

For two adjacent points on a nitrogen isotherm, a representative pore size can be calculated by either of the above two equations. The apparent difference of adsorption volume between the two points is to be corrected by subtracting the variation of t in pores larger than the representative pore size, which should give the pore volume for the pore size. Iteration of this process gives the PSD.

Many of detailed calculation schemes were proposed by, e.g., Cranston and Inkley [5], Dollimore and Heal [6], and Barell et al. [7], at least one of which is installed in a commercial apparatus as a PSD software. Major difference between these methods is the data or equation for adsorption thickness t, and the schemes of calculation themselves do not produce significant difference if common data for t are used. Therefore one need not worry which method is installed in one's commercial apparatus, but should pay more attention to the t-data or standard isotherms in the software; how precise and how abundant for various materials they are.

5.4.3.3 Measurable range of pore size

Many of automated adsorption apparatuses declare the upper limit of the pore size to be ca. 100 nm, which corresponds to the relative pressure of ca. 0.98. In general, however, the accuracy of the adsorption measurement decreases as the equilibrium pressure approaches to the saturated vapor pressure, because the liquid nitrogen temperature would fluctuate following the variation in atmospheric pressure. For safety one would better understand the upper limit of reliability being ca. 0.95 in relative pressure, corresponding to the pore size of ca. 40 nm: Measurement for larger pores should follow the mercury porosimetry.

As for the lower limit, pore size analysis based on the capillary condensation would lose its basis if the size goes down below ca. 2 nm, because the condensation phenomenon itself (or the first order phase transition) does not occur in such small space. The micropore (<2 nm) should be analyzed by the methods based on the micropore filling explained in the preceding section.

Further rigorous analysis is presently under development by many researchers. Most of the approaches are based on the statistical thermodynamics method such as the molecular simulation and the density functional theory, which produce so-called local adsorption isotherms for series of various sizes of micropore. An experimental nitrogen isotherm is expressed as a convolution integral of the local adsorption isotherms and the pore size distribution function, the latter of which thus can be determined through minimization of the error between measured and predicted isotherms. Note that, however, this kind of technique needs further development. Sometimes calculated pore-size distribution suffers from an artifact of nonexistent pores, which may result from surface heterogeneity: One has to pay much attention on this kind of artifacts especially if the resultant distribution has bimodal or multimodal distribution.

5.4.3.4 Underestimation of pore size in single-nanometer range

The spread of commercial nitrogen adsorption apparatus has made PSD measurement quite easy. The obtained PSDs, however, should not be blindly accepted. The estimation of pore size based on the Kelvin model as given by equations (5.4.3) and (5.4.4) works well for pores larger than 10 nm. It has, however, been pointed out from late 1980s that the model underestimates the so-called single-nanometer range of pores [8, 9]. However, no method with simplicity and convenience was available in even 1990s, which forced people to use the Kelvin model though knowing its inaccuracy.

A condensation model with simple concept and easy calculation has recently been proposed [10, 11], which is explained briefly in the following. The point is that the attractive potential energy from pore walls and the stronger surface tension of curved interface will enhance the condensation in nanoscale pores. The basic equation is:

$$RT \ln \frac{p}{p_s} = -\frac{2v\gamma(\rho)}{\rho(r)} + \Delta\phi(r), \qquad (5.4.5)$$

in which the free energy for condensation is compensated not only by the Young–Laplace effect with local curvature dependent surface tension $\gamma(\rho)$, but also by the relative strength of pore wall's attractive energy compared with the liquid $\Delta\phi(r)$. The latter effect can be determined from standard isotherm data that are usually included in the automated adsorption apparatuses. Further details are given by the original papers or a review [12].

The degree of the underestimation by the Kelvin model stays almost constant regardless of the pore size, but varies depending on the pore-wall potential energy. Some examples of difference between Kelvin-based prediction and the true pore size are: ca. 1 nm or slightly greater for carbon materials, ca. 1 nm or less for silica materials, and 0.5–0.7 nm for ordered mesoporous silicates (FSM-16, MCM-41). Thus one should understand that, if the BJH (Barell, Joyner and Halenda) [7] method gives peak pore size to be 3 nm for a silica gel, the true size is about 4 nm or slightly less than 4 nm.

5.4.3.5 Hysteresis

If an adsorption isotherm goes with capillary condensation, the hysteresis between adsorption and desorption processes results in most cases. The classification of the hysteresis into four types is given by IUPAC as shown in Fig. 5.4.5. There has been a long discussion on which branch to be used for pore-size determination, or which branch represents true equilibrium process, but it is still now quite difficult to obtain a general conclusion. Limiting the topic into the structure made up by aggregated particles or sintered porous body, however, the following understanding can hold.

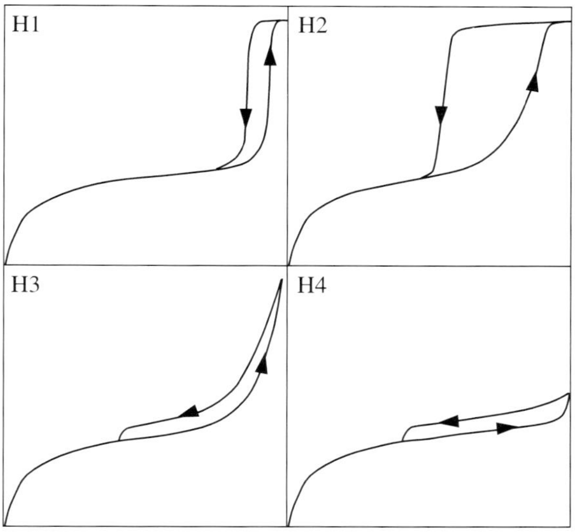

Figure 5.4.5
Classification of hysteresis given by IUPAC.

The characteristics of this kind of structure would be that there must exist particles' contacting point at the end of the pore space. Then the contacting points and its vicinity are able to provide nuclei for the condensation, and an energy barrier as may be the case for MCM-41 would not stand upon condensation process. Therefore the condensation process for this kind of materials follows an equilibrium path. Another feature is that pores are connected through narrower spaces between particles, or the pore network is formed with connecting "necks". In this case, the desorption process itself is in equilibrium, but the so-called ink-bottle effect gives hysteresis in which the evaporation of condensate in a pore space is not possible until the pressure goes down to a value corresponding to the size of the neck. The evaporated volume at this hindered process, then, does not mean the pore volume of the pore size corresponding to this pressure. This process exhibits H2-type of the hysteresis, and the adsorption branch should be used to calculate the size distribution of the pore space. Examples of materials include silica gels and porous polymer gels. Many providers of such materials use *desorption* branches for showing porous characteristics to users because it gives sharp distribution. One should note, however, that the peak in the distribution simply gives the neck size and the real distribution would be broader in most cases.

Another important topic related with hysteresis is the end-closure point of the desorption branch.

As seen in clay materials and activated carbons, hysteresis loop of type H3 and H4 often closes at the relative pressure of 0.40–0.45. As given in detail by literature [13, 14], this closure of hysteresis will result from the spinodal evaporation of condensed liquid when exceeding its tensile limit, which is determined not by the pore size but merely by the nature of liquid. This phenomenon should be considered when one uses the desorption branch for characterization.

5.4.4 Other methods of interest

5.4.4.1 Mercury porosimetry (mercury intrusion method)

An evacuated sample is immersed in mercury, and pressurization of the system gives mercury's intrusion into the pore space. The detection and analysis of the intruded volume against applied pressure gives the pore size distribution since the pore volume directly equals that of the intruded mercury. Many automated apparatuses are commercially available. The method is suitable for macropore or mesopores above tens of nanometers.

Many of automated porosimetry apparatuses declare the lower limit of the pore size to be 3–4 nm, which corresponds to the applied pressure of ca. 4,000 atm. At this level of high pressure, the porous framework might be deformed or some other influence may occur. It is also pointed out that the surface of mercury in a single-nanometer range may be different from surface of bulk liquid, which may results in a hindered contact angle. Much attention, then, should be paid on the reliability of the data in this range, for which the gas adsorption method has far superior accuracy and reliability. The upper limit of the mercury porosimetry would be around several hundred of microns: The detection of cracks or supermacro pores, which is difficult to be measured by gas adsorption, is precisely done by the intrusion method.

The extrusion process by decreasing the pressure generally gives different path from that for the intrusion process, or hysteresis. Further, it would be almost always the case that a certain amount of mercury remains in pores even after complete release of the pressure. The extrusion process, therefore, is not suitable for analysis, and the intrusion branch is used in general. The intrusion of non-wetting process corresponds to the desorption branch in gas adsorption, and then one should notice that the analysis gives the neck size for aggregated or sintered bodies.

5.4.4.2 Bubble-point method

This technique detects perforating pores while the gas adsorption method and mercury porosimetry cannot distinguish those from dead-end pores. Because of this feature the method is often applied to filters, membranes, cloths, or those porous materials whose permeation properties are of importance. Depending on the wettability the porous material is immersed either in freon or water. Pressurized air or nitrogen is then introduced to one side of the material. At a pressure that corresponds to the maximum size of the perforating pore, the gas starts to permeate. Other than the detection of maximum pore, the size distribution can be estimated applying higher pressures, because smaller pores start to open with increased pressure. Detectable pore size is usually above several tens of nm with freon, or a few hundred nm for water.

5.4.4.3 Thermoporometry (freezing point depression method)

Based upon the Gibbs–Thomson equation, which suggests that the freezing point depression in a pore from the freezing point in bulk phase is inversely proportional to the pore radius, the size distribution is estimated from calorimetric measurement. It is sometimes the case that the pore structure when wetted varies after drying because of capillary suction pressure or de-swelling of the base material. The gas adsorption or mercury porosimetry cannot characterize such porous materials because both method need evacuation before measurements. This method may be used to overcome the above problem: the measurement goes as wetted. However, not much has been clarified for the freezing behavior in confined space. Recent study, e.g., has clarified that the freezing point may even get higher than the bulk, depending on the physico-chemical nature of pore walls. The method is thus especially controversial if single-nanometer range is concerned, and it would be better not to be relied on for smaller size of nanopores.

5.4.4.4 Small angle X-ray scattering (SAXS)

The X-ray scattering with angles smaller than 10° can probe porous characteristics in single-nanometer range. Since X-ray can detect not only open pores but also closed (or isolated) pores, measurements for low-permittivity materials are typical examples of application. One has to be, however, careful upon interpretation of the data, because the resulting space distribution or correlation length does not necessarily mean the scale of pores but has resulted from the electron density distribution. Further, while an ordered material such as MCM-41 would give quite clear signals showing its lattice size or the periodicity of the regular pores, not much sensitivity can be expected for materials with disordered or random nature, which needs a rather complicated analysis for gradually decreasing scattering intensity. From the Guinier plot, e.g., one would be able to obtain the averaged radius of scattering body, but it may not necessarily be related with the pore size itself. One should understand that SAXS has only low quantitative precision when applied to nanopore characterization.

References

[1] G. Horvath, K. Kawazoe: *J. Chem. Eng. Jpn.*, **16**, 470–475 (1983).
[2] A. Saito, H.C. Foley: *AIChE J.*, **37**, 429–436 (1991).
[3] S. Kondo, T. Ishikawa and I. Abe: *Kyuchaku no Kagaku (Science of Adsorption)*, Maruzen Publishing, Tokyo, pp. 31–83 (1991).
[4] M. Senoo, ed.: *Koroido Kagaku IV – Koroido Kagaku Jikkenhou (Colloid Chemistry – Experimental Method for Colloid Chemistry)*, edited by Japan Society of Chemistry, Tokyo Kagaku Dojin, Tokyo, pp. 289–296 (1996).
[5] R.W. Cranston, F.A. Inkley: *Adv. Catal.*, **9**, 143–154 (1957).
[6] D. Dollimore, G.R. Heal: *J. Appl. Chem.*, **14**, 109–114 (1964).
[7] E.P. Barell, L.G. Joyner and P.P. Halenda: *J. Am. Chem. Soc.*, **73**, 373–380 (1951).
[8] R. Evans, U.M.B. Marconi and P. Tarazona: *J. Chem. Phys.*, **84**, 2376–2399 (1986).
[9] M. Miyahara, T. Yoshioka and M. Okazaki: *J. Chem. Phys.*, **106**, 8124–8134 (1997).
[10] M. Miyahara, T. Yoshioka and M. Okazaki: *J. Chem. Eng. Jpn*, **30**, 274–284 (1997).
[11] M. Miyahara, H. Kanda, T. Yoshioka and M. Okazaki: *Langmuir*, **16**, 4293–4299 (2000).
[12] M. Miyahara: *Shokubai (Catalyst)*, **41**, 15 (1999).
[13] O. Kadlec: *Carbon*, **27**, 141–155 (1989).
[14] P. Ravikovitch, A.V. Neimark: *Langmuir*, **18**, 1550–1560 (2002).

5.5 Grain boundaries and interfaces

The properties of polycrystalline material are different from that of single crystalline material, due to the grain boundaries and their non-periodic arrangements of atoms, as schematically shown in Fig. 5.5.1. For example, $BaTiO_3$ is one of the commercially available polycrystalline materials, which uses the grain boundaries as the origin of positive temperature coefficient of resistivity (PTCR) effect. It is very difficult to determine if the macroscopic properties of materials are dependent on the presence of grain boundaries or not, since structures and compositions of grain boundaries have rarely been characterized at the atomic scale.

Therefore, the TEM plays important roles for characterization of grain boundaries and assists the development of new polycrystalline materials.

5.5.1 The role of TEM

Electron microscopes were developed due to the restrictions of light microscopes limited by the properties of light to 500 × or 1,000 × magnification and

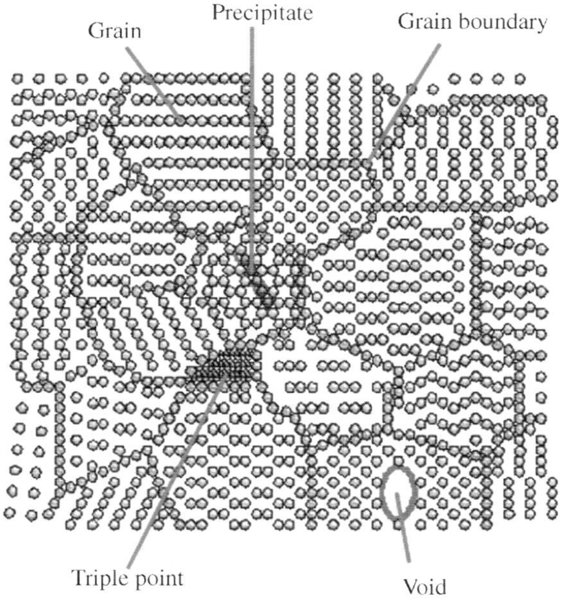

Figure 5.5.1
Schematic diagram of polycrystalline material.

a resolution of about 0.2 μm. In the early 1930s, this theoretical limit had been reached practically and there was a desire to see more details of materials, which required higher magnifications, such as more than 10,000 × magnification as shown in Fig. 5.5.2. TEM is similar to the light transmission microscope except that a focused beam of electrons is used instead of light to "see through" the specimen.

TEM is an equipment to let the incident electron beam to transmit a thin specimen at high-acceleration voltage, 80–3,000 kV, which results in generating signals caused by the interaction between the specimen and incident electrons. Structures, compositions and chemical bondings of the specimen can be determined from these signal as shown in Fig. 5.5.3.

Recently, the spatial resolution of TEM reached less than 10^{-1} nm with electron probe size less than 5×10^{-1} nm, so that structural and compositional analysis at atomic scale can be carried out easily.

TEM is not only a microscope but also a diffractometer. For example, elastically scattered electron can be selected using objective aperture to obtain dark-field image. Furthermore, electron energy-loss spectrum (EELS) is generated by inelastic scattering between the incident electrons and the specimen, which can be used as the qualitative and quantitative analysis of elements and types of chemical bondings.

In general, there are three types of transmitted electrons observed by TEM; they are unscattered

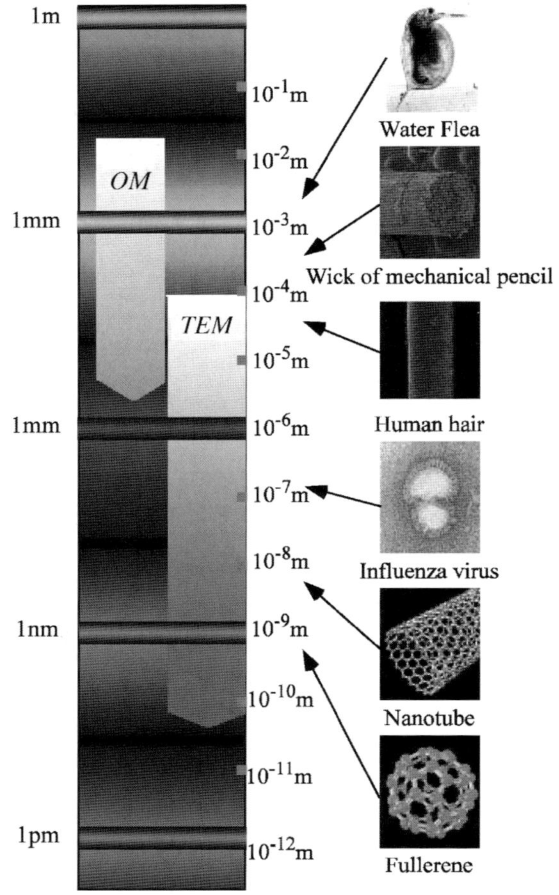

Figure 5.5.2
Spatial resolution for observing materials.

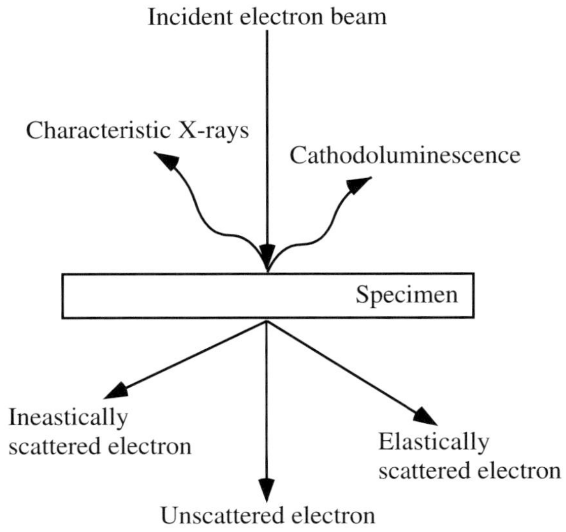

Figure 5.5.3
Interaction between the incident electron beam and the specimen, in the case of TEM.

electrons, elastically scattered electrons and inelastically scattered electrons.

1. Unscattered electrons are caused by incident electrons transmitted through the thin specimen without any interaction occurring inside the specimen. Since the amount of unscattered electrons is inversely proportional to the specimen thickness, thicker areas of the specimen have fewer unscattered electrons and appear darker.
2. Elastically scattered electrons are caused by the incident electrons that are scattered by atoms in the specimen without losing energy. These elastically scattered electrons are transmitted through the remaining portions of the specimen, and follow the Bragg's diffraction law. All incident electrons have the same energy and enter the specimen normal to its surface. All electrons that are scattered by the same atomic spacing will be scattered by the same angle. These "the same angle" scattered electrons are gathered by lens and form a pattern of spots; each spot corresponding to a specific atomic spacing. This diffracted pattern yields information about the orientation, atomic arrangements and phases present in the region of interest.
3. Inelastically scattered electrons are caused by the incident electrons that interact with atoms in specimen with loosing their energy. These inelastically scatted electrons provide two types of information. One is the inelastic loss of energy by the incident electrons, characteristic of the elements. These energies are unique to each bonding state of each element and thus can be used to extract both compositional and chemical bonding information of the specimen. Another one is the formation of bands with alternating light and dark lines, known as Kikuchi bands. These bands are also formed by inelastic scattering interactions related to the atomic spacing in the specimen.

5.5.1.1 Images and diffraction contrast

The bright-field imaging technique is the most common method for the imaging method using TEM.

The bright-field means the formation of images only with the transmitted electron beam. To switch to diffraction mode, chosen by the objective aperture. The size of the objective aperture should be small enough to remove all diffracted electron beams caused by the specimen, as schematically shown in Fig. 5.5.4.

On the other hand, dark-field images are formed if one or more diffracted electron beams are chosen by the objective aperture, as schematically shown in Fig. 5.5.5. In this case, the transmitted electron beam should be blocked. The advantage of the dark-field imaging method is its high-diffraction contrast. The dark-field imaging technique is usually used for observing grain size distributions and dislocations, as shown in Fig. 5.5.6.

Mass-thickness contrast is caused by the specimen thickness and the density, as shown in Fig. 5.5.7. The interaction of electrons with heavy atoms is stronger than with light atoms. If the thickness is homogeneous, areas in which heavy atoms or high density are concentrated appear with darker contrast than with light atoms or low density (mass contrast). Of course, more electrons are scattered in thick areas than in thin areas; thus, thick areas appear dark (thickness contrast). However, a thick area with light elements might have the same contrast as a thinner area with heavier atoms.

5.5.1.2 High-resolution TEM (HRTEM)

Using all of the diffracted and transmitted beams by a large objective aperture, instead of using a small objective aperture to select a particular diffracted electron beam or transmitted electron beam, all beams interfere each other so that a phase contrast image is formed. The image is formed by the interference of the diffracted beams with the transmitted beam. If the TEM has a sufficiently high point resolution and a suitable oriented specimen along a zone axis, then high-resolution TEM (HRTEM) images can be obtained. In many cases, the atomic structure of a specimen can directly be investigated by HRTEM (Fig. 5.5.8).

5.5.1.3 Scanning TEM (STEM)

When a focused electron beam is scanned across a region of specimen, the signals generated by the interaction between the incident electron beam and the specimen varies according to specimen characteristics such as composition and structure, orientation, and topography. When the electron beam is scanned across the specimen in a rectangular raster, the change

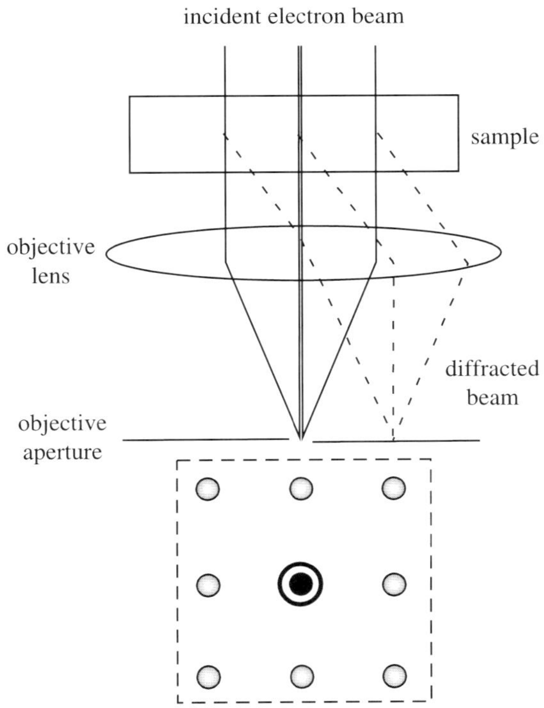

Figure 5.5.4
Locating the objective aperture to achieve bright field image.

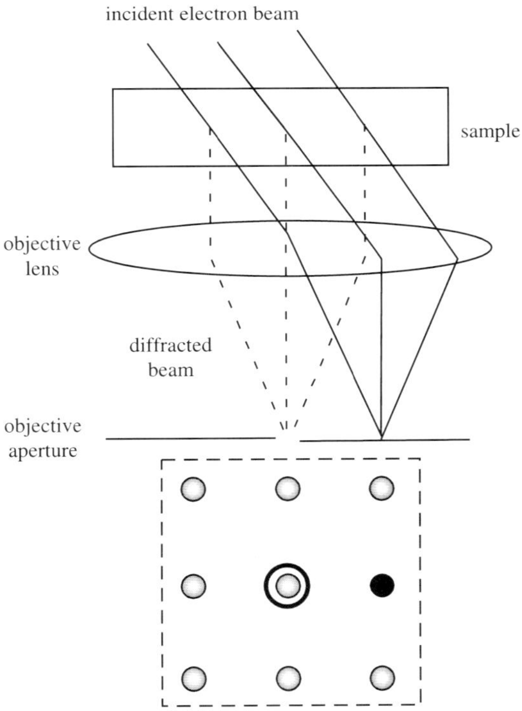

Figure 5.5.5
Tilting the entering beam to achieve dark-field imaging.

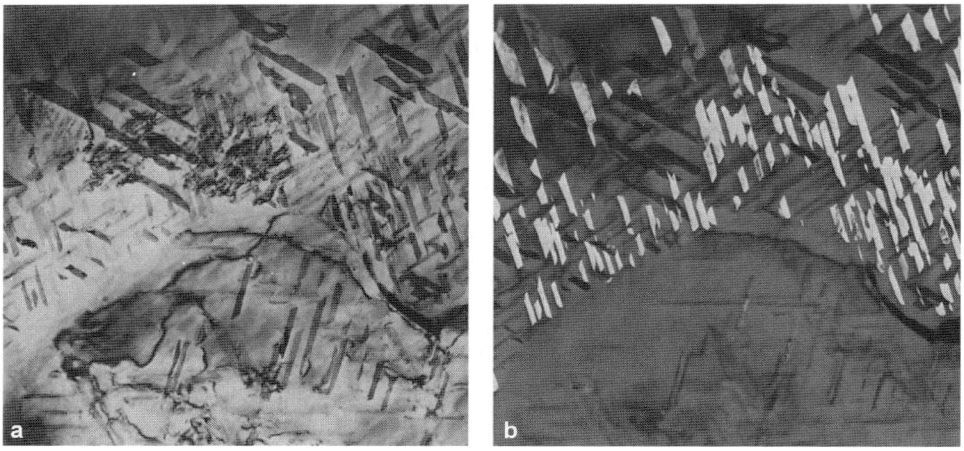

Figure 5.5.6
The comparison of bright-field image and dark-field image of aluminum alloy.

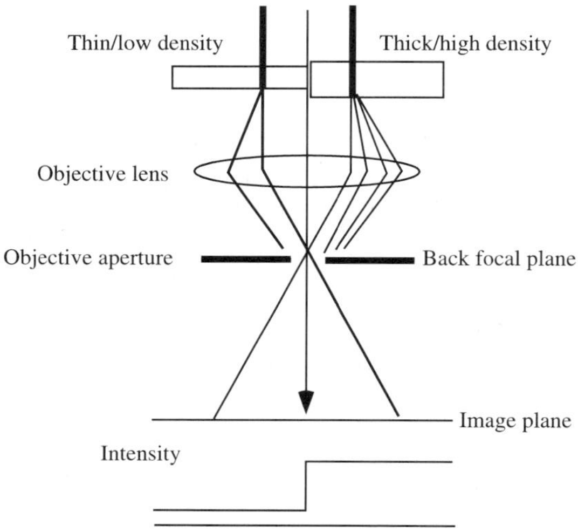

Figure 5.5.7
Mass-thickness contrast.

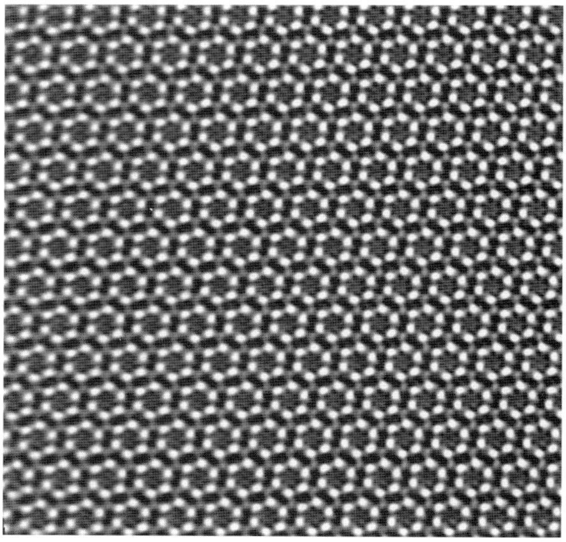

Figure 5.5.8
High-resolution TEM image of beta Si_3N_4 seen from c-axis.

in signals, detected through various STEM detectors, allows one to build up a 2D image of the specimen. Furthermore, using a high-angle annular dark-field detector, Z-contrast imaging can be carried out. By using a Scanning TEM (STEM) detector with a large inner radius, electrons are collected which are not Bragg scattered. As such Z-contrast images show little or no Bragg scattered events, their intensity is approximately proportional to Z^2. Z-contrast imaging using a high-angle annular dark-field (HAADF) detector is particularly useful for investigating high-Z materials within a low-Z material. A HAADF detector collects electrons that undergo high-angle scattering, and the signal is approximately proportional to Z^2.

5.5.2 Analytical TEM (AEM)

5.5.2.1 Energy dispersive X-ray spectroscopy (EDS)

EDS is a method for qualitative and quantitative elemental analysis using X-rays, which have unique energies from each element. The characteristic X-ray spectrum is caused by the de-energization of the atom after an inelastically scattered electron is produced (*Process 1* in Fig. 5.5.9). A lower-shell electron is emitted from the atom during process 1 and there is a vacancy at the lower shell. A higher-shell electron can

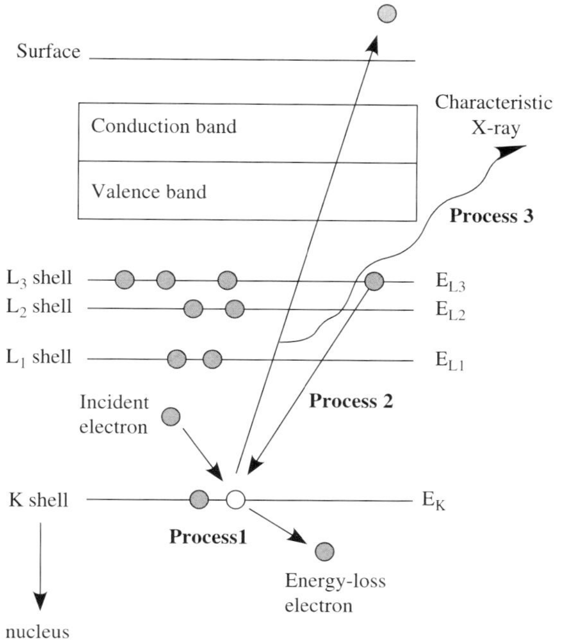

Figure 5.5.9
Interaction between the incident electron and the sample.

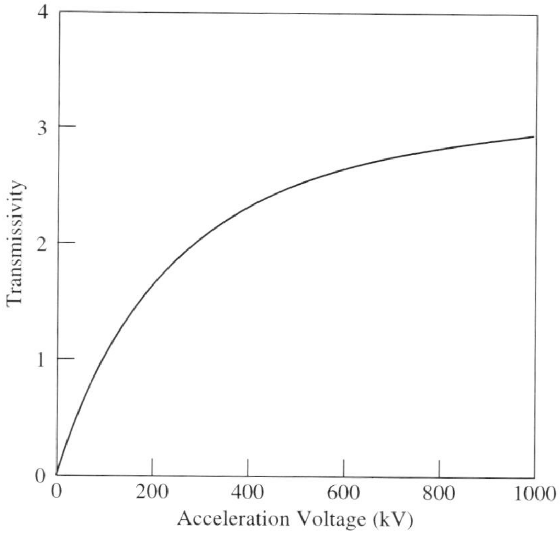

Figure 5.5.10
Transmissivity of electrons at different acceleration voltages, with comparison to at 100 kV.

"fall" into the lower shell and fill the vacancy (*Process 2* in Fig. 5.5.9). As the electron "falls", it emits X-rays to balance the total energy of the atom (*Process 3* in Fig. 5.5.9).

In the case of thinner specimen or higher acceleration voltages, the transmissibility (Fig. 5.5.10) of electron becomes larger.

Furthermore, EDS can be combined with STEM system, in which an electron beam is scanned across a defined line or region of the specimen, so that a line-profile or an elemental-map can be obtained. (Fig. 5.5.11). Within the range of sample thicknesses studied and with the mass absorption coefficients for the characteristic X-rays, absorption corrections for transforming intensity ratios to concentration ratios are unnecessary. The standard Cliff–Lorimer (1975) relationship is given by

$$\frac{C_A}{C_B} = k_{AB}\frac{I_A}{I_B} \tag{5.5.1}$$

where C_A and C_B are the weight fractions of elements A and B respectively, and I_A and I_B are the characteristic X-ray intensities of elements A and B, respectively, above background, and k_{AB} is the Cliff–Lorimer factor which is independent of composition and also independent of thickness.

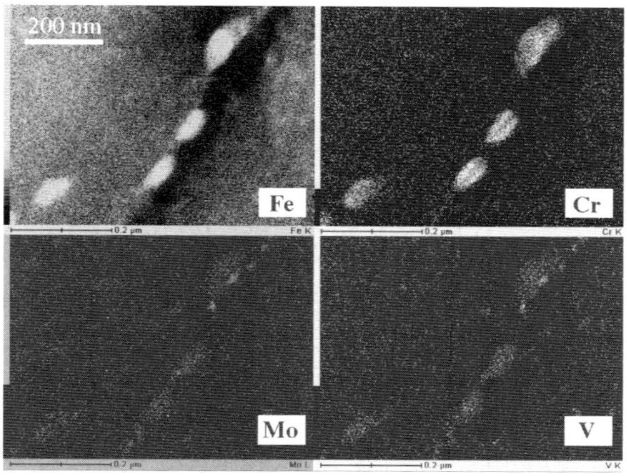

Figure 5.5.11
EDS elemental map of high-Cr Steel.

Furthermore, in the case of thin specimen, the characteristic X-ray intensity is given by equation (5.5.2),

$$I_A \propto N_0\left(\frac{Q\omega a}{A}\right)_A \rho t \tag{5.5.2}$$

I_A is the characteristic X-ray intensity of element A, N_0 the Avogadro number, Q the ionization cross section, ω the fluorescent yield, a an intensity ratio, A the atomic weight, ρ the density, and t the specimen thickness.

In the case of using a small electron probe for acquiring EDS signals, it usually causes the irradiation damage on the specimen during the acquisition.

5.5.2.2 Electron energy-loss spectroscopy (EELS)

Collisions of incident electrons with the specimen can be classified into three types of scatterings, unscattered, elastically or inelastically scattered, as illustrated in Fig. 5.5.3. If incident electrons suffer collisions resulting in energy transfer, the energy transferred to the specimen must be measured for achieving information of the specimen. In practice, electron trajectories are bent by magnetic prisms; the trajectories of electrons which suffered inelastic collisions are bent more than those which interacted with the specimen elastically. In other words the electrons are dispersed according to their amount of energy loss. A photodiode array can be used to detect the different trajectories of the electrons emerging from the spectrometer.

The inelastic scattering event caused between the incident electron and the specimen can be classified into two types. One is the interaction with conduction-band electrons leading to plasmon oscillations in the material. The information on those oscillations appears in the low-loss part of a spectrum and known as the plasmon peak. The other is the interaction of fast electrons with core-level electrons resulting in higher energy losses, known as the core-loss peak. The spectral features corresponding to the core-level excitations have often an appearance similar to what can be seen in Fig. 5.5.12. The shape of such a core-loss edge, after subtraction of non-characteristic background, can provide information on chemical bonding state of the atoms involved during the interaction, known as the near-edge fine structure. Although point analysis by parallel-detection EEL spectrometer was mainstream in past, energy-filtering TEM (EF-TEM) became the standard method recently with advances of methods and equipments.

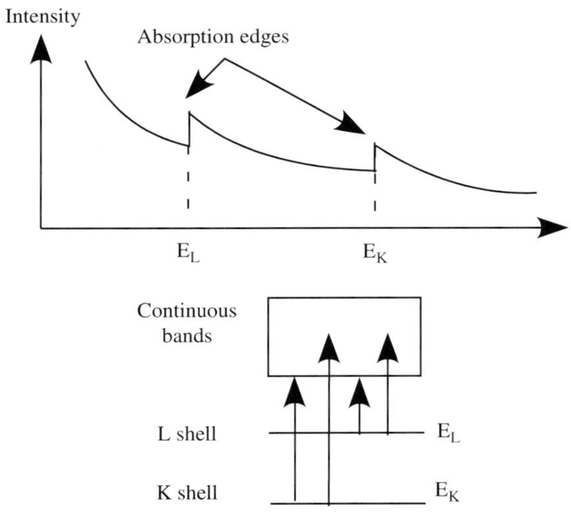

Figure 5.5.12
The generation of ELNES in EELS.

Comparison of EELS and EDXS

EELS	EDXS
High detection efficiency for low *Z* elements	High detection efficiency for high *Z* elements
Elemental, chemical, and dielectric information	Elemental information only
Energy resolution 0.3–2 eV results in far fewer overlaps; fine structures can be analyzed	Energy resolution > 100 eV causes frequent overlaps
Very efficient and higher sensitivity to most elements ≥ very efficient mapping technique	Inefficient signal generation, collection and detection ≥ inefficient X-ray mapping
Fast technique; but complex processing required	Slow technique; only simple processing required

For example, diamond, graphite, carbon nanotube, fullerene, and amorphous carbon are allotropes of carbon, and it is impossible to differentiate them by EDS. However, it is relatively easy to classify them by comparison of electron energy-loss near-edge structure (ELNES), which provides information of chemical bonding states, as shown in Fig. 5.5.14. The localized differences in structures can be judged from the comparison of these spectra.

In the case of carbon-related materials, C–K edge represents the transition from the 1s electronic core state to the unoccupied states above the Fermi level. For graphite, the shape of this edge is shown in Fig. 5.5.14 and corresponds to a material with 100% ($sp^2 + 2p_z$) bonding. It is composed of two major peaks, a first one around 285 eV, which is related to the transition from the 1s to the unoccupied π^* states, and a second one around 292 eV, which is related to the transition from the 1s to the unoccupied σ^* states. The 100% sp^3 bonding of diamond yields an ELNES spectrum as shown in Fig. 5.5.13 with only one major σ^* peak around 292 eV. By applying this prior knowledge, the spectra from graphite and diamond can be used as reference cases for other carbon-related materials, although it is unclear at present whether the σ^* and π^* labeling is still valid when dealing with amorphous material since the orbital symmetry related to s- and p-bonding in crystallographic materials is lost in amorphous materials.

Furthermore, EELS signal can be applied to quantify elements from the region of interest as EDS as equation (5.5.3). The number of atoms, N_A, of the element analyzed can be quantified by equation (5.5.3) using EELS signal [1].

$$N_A = \frac{I_A(\beta, \Delta E, E_0)}{I_t} \times \frac{1}{t\sigma_A(\beta, \Delta E, E_0)} \quad (5.5.3)$$

I_t is the total incident electron current, I_A the number of electron having excited *k*th inner shell from element *A*, *t* the specimen thickness, β the collection angle, ΔE the energy slit width, E_0 the energy of entering electron, and $\sigma(t, \Delta E, E_0)$ an ionization cross-section for the *k*th shell of element *A*. Furthermore, the ratio N_A/N_B can be obtained from equation (5.5.4).

$$\frac{N_A}{N_B} = \frac{I_A(\beta, \Delta E, E_0)}{I_B(\beta, \Delta E, E_0)} \times \frac{\sigma_B(\beta, \Delta E, E_0)}{\sigma_A(\beta, \Delta E, E_0)} \quad (5.5.4)$$

Predicting theoretical reference and identifying ELNES-structure correlation, ELNES is related to the local chemical environment of the excited atom and is often called "fingerprint" to indicate chemical and bonding information of materials. From the comparison between acquired spectra and available reference spectra, one can deduce the chemical bonding information of the specimen. For example, a fine structure in carbon K-edge can be used to identify the sp^2 or sp^3 bonding nature in carbon materials by comparison with available reference experimental spectra, as shown in Fig. 5.5.13. However, the standard experimental reference is not always available, so there are no experimental spectra that can be taken as a reliable reference. In other cases, there is no existing reference spectrum for a first synthesized material, which did not exist or was unknown before. For these situations, theoretical simulation can predict the main features of ELNES and provides a theoretical reference for experimental characterization [2, 3].

There are basically two ways to obtain EELS signal from grain boundaries and interfaces, either using a converged electron beam irradiated exactly at the interface or using a relatively broad beam. The former case provides EELS signal with higher intensity though it easily cause irradiation damage at the region of interest. On the other hand, the interface-specific components of the ELNES can be extracted by employing the spatial difference method [4, 5].

5.5.2.3 EF-TEM method

In the case of CTEM (Conventional TEM), the electrons used for imaging are selected only via the aperture diaphragm according to their scattering angles. Therefore, only electrons with a large scattering angle contribute to the contrast generation. The energy of the electrons and the difference in their energies remain unaccounted for, despite the fact that the bandwidth of the energy differences caused by the chromatic

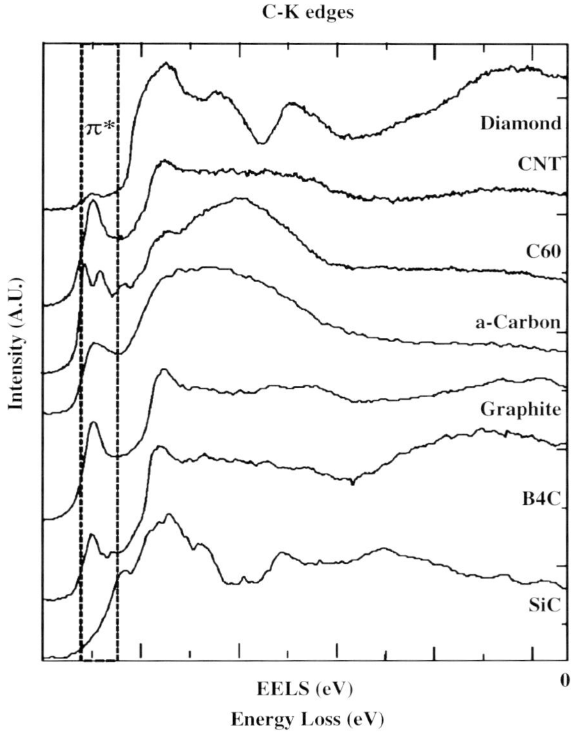

Figure 5.5.13
C–K edge ELNES of various carbon-related materials.

aberration of the objective lens has considerable influence on contrast and resolution. In the case of EF-TEM, the contrast is optimized by filtering out the contrast-reducing electrons from the spectrum of transmitted electrons, or by only allowing electrons containing specific energy information to be used for imaging.

For EF-TEM, the microscope is equipped with in-column type energy filter or with post-column type energy filter. The post-column type energy filter is mounted underneath the projection chamber, as shown in Fig. 5.5.14, and the in-column filter itself is a part of the microscope column, as shown in Fig. 5.5.15. The electron beam passes through a magnet, which separates the electrons according to their energy into an energy spectrum. The lenses restore the energy-filtered image at the entrance aperture so that it can be viewed on a TV or recorded on a slow-scan camera. Post-column energy-filters are manufactured by Gatan (GIF) and can be installed as an attachment on the exit surface of the electron-optical column of almost any TEM, as shown Fig. 5.5.14. Alternatively dedicated EF-TEM microscope with an in-column filter, the so-called "omega filter", was developed by Zeiss and later by JEOL, as shown in Fig. 5.5.15 and its example of high-Cr steel in Fig. 5.5.16.

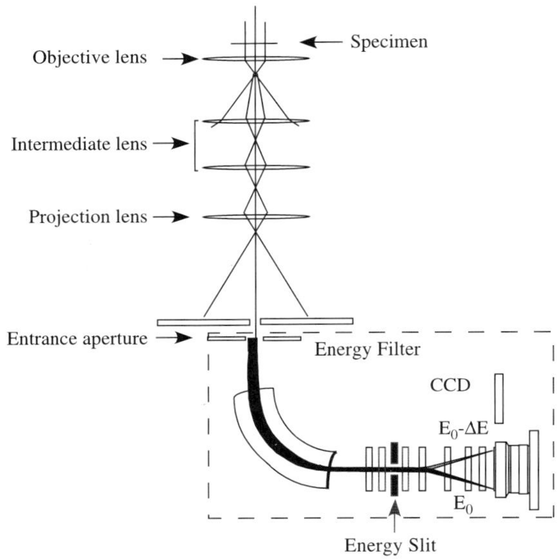

Figure 5.5.14
Post-column type energy filter.

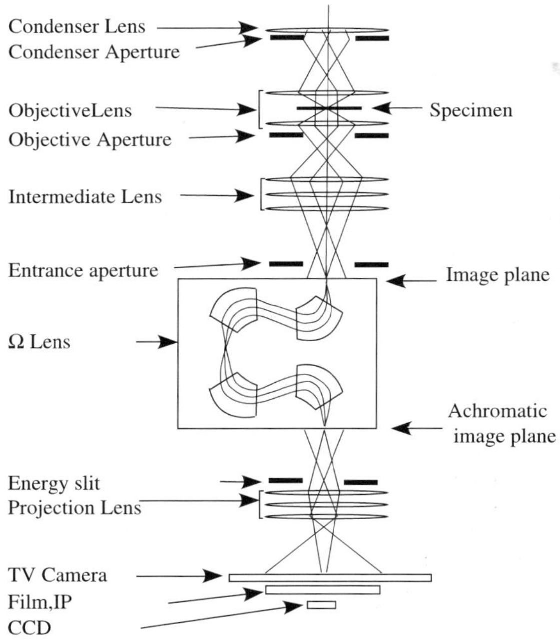

Figure 5.5.15
Ω-type in-column energy-filter.

Both types of energy-filters are devices that can form an image with electrons of only a small energy range. The inelastic scattering process in EELS is strongly localized, hence, it is used to obtain analytical information from a small specimen area. By using any spectral feature in an EEL spectrum, the change of this feature's intensity can be monitored simultaneously with high spatial resolution over a broad area of the specimen. Zero-loss filtering is done by simply selecting the electrons over a narrow energy window that includes only the zero-loss peak. The contribution of inelastically scattered electrons to an image or diffraction pattern leads to a blurring due to chromatic aberration. Zero-loss filtering removes the contribution of all inelastically scattered electrons to both images and diffraction patterns. This is more serious for thicker specimens; hence, zero-loss filtering will improve the contrast and the resolution of such images considerably.

If an energy window at a somewhat higher energy-loss is selected, the plasmon resonance peak that is proportional to the number density of valence electrons can be mapped. An efficient way to increase contrast significantly is to find an energy range where certain features can be clearly seen.

The most extensive analytical use of energy-filters is for core-loss imaging and elemental mapping. The ability of an energy filter to show a 2D distribution of a specific element, integrated over the thickness of the thin foil specimen, makes it a powerful tool for analytical studies.

5.5.3 Three-dimensional electron tomography (3D-ET)

Electron tomography (ET) consists of obtaining a 3D reconstruction of an object from a series of projection images. Data collection is accomplished by tilting the specimen in the electron beam to produce, as a tilt series. Like other imaging techniques, TEM provides a translucent view of the specimen where the details from different depths are superimposed in a 2D projection. A computational operation called "back-projection" is then used to create a 3D object from the tilt-series images. The back-projection algorithms project back the mass of the specimen into the reconstruction volume. When this process is repeated for a series of projection images from different angles, back projected rays intersect and are reinforced at the points where mass is found in the original structure. The rays pass through different amounts of specimen mass producing a different image for each angle.

The resolution and quality of the reconstruction depends primarily upon two factors. First, the resolution of tilt angle between successive images has a dramatic effect upon reconstruction quality. The smaller the tilt angle, the better is the reconstruction. Second, the total tilt range determines the amount of 3D data "seen" over all tilt angles. The larger the total tilt angle, the more data are seen in the reconstruction. Generally, images are acquired at 1°−2°of tilt angle between successive images over a total tilt range of about ±60°−70°. A CCD camera is generally used for image acquisition. Radiation damage sets a limit

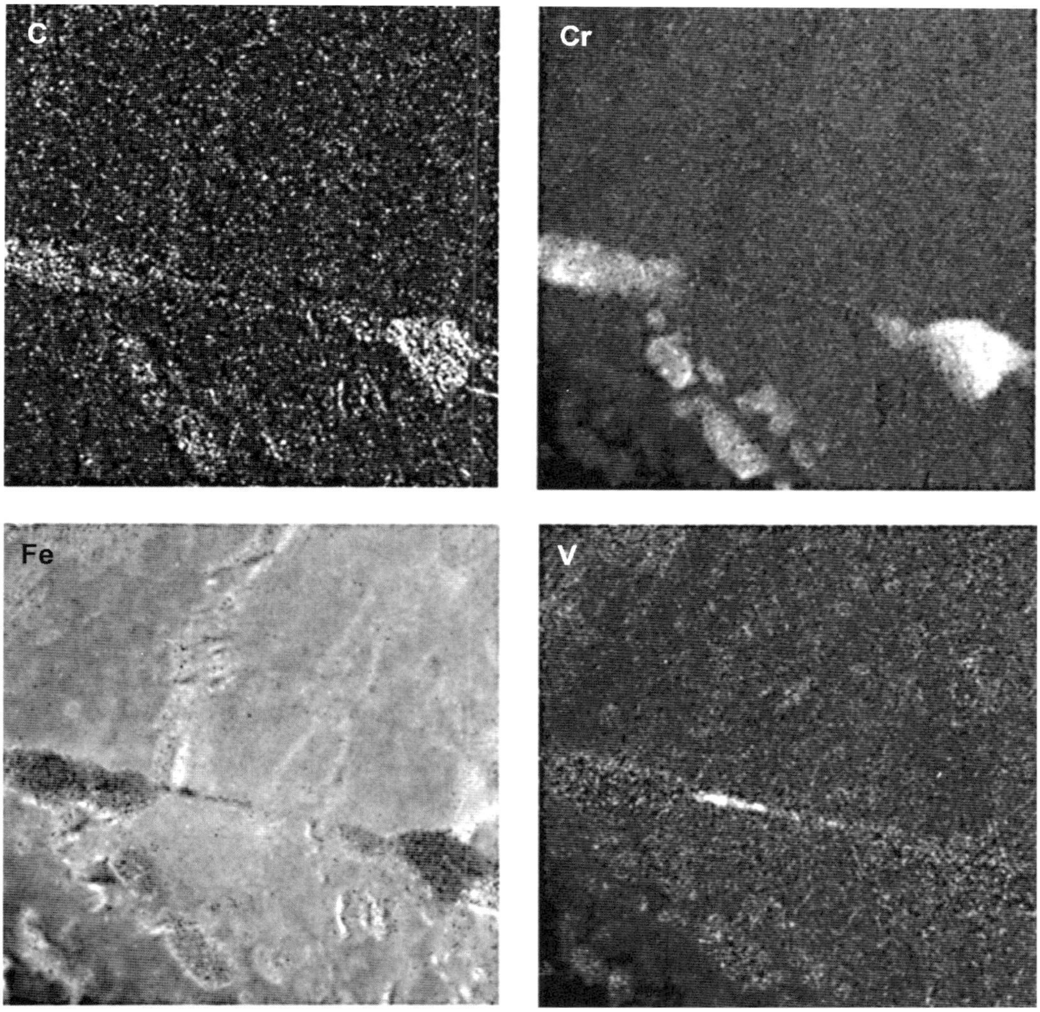

Figure 5.5.16
Core-loss images of high-Cr steel (elemental maps of C, Cr, Fe, and V).

to the effective resolution in frozen-hydrated samples (the samples are heated and melt in the radiation beam over time). ET requires a fully automated and fully digitized TEM with an accurate tilt stage and a specially designed high-tilt specimen holder. In addition, it is necessary to consider that the increase of thickness, i.e., the path length of the electron beam through the specimen is a factor of 2 at 60° and approximately a factor of 3 at 70°.

Furthermore, in the case of crystalline materials, diffraction contrast appears at particular angle, which usually degrades the quality of the reconstructed volume. To overcome such degradation, atomic number, Z-contrast, 3D-ET is applied as a method for determining the 3D structure from (poly-)crystalline materials [6], [7]. Furthermore, Z-contrast imaging is useful in the study of (poly-)crystalline materials because of the reduction of coherent diffraction contrast (Figure 5.5.17).

Abbreviations	
TEM	Transmission electron microscopy
C-TEM	Conventional TEM
EF-TEM	Energy-filtering TEM
EELS	Electron energy-loss spectroscopy
ELNES	Energy-loss near edge structure
EDS	Energy dispersive X-ray spectroscopy
HAADF	High-angle annular dark field
3D-ET	Three-dimensional electron tomography

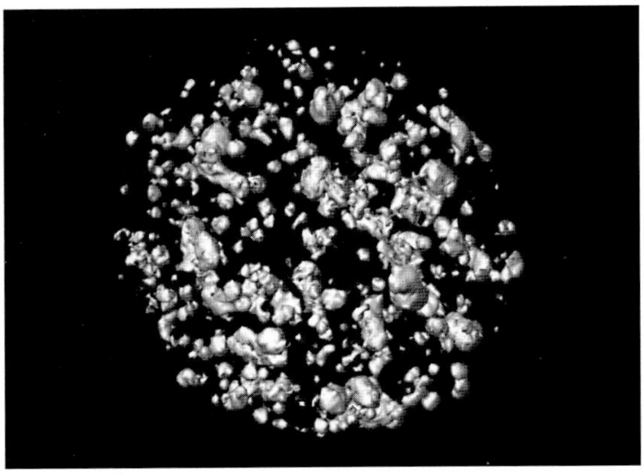

Figure 5.5.17
Reconstructed image of TiO_2–Ag composites, showing the distribution of Ag particles [7].

References

[1] R.F. Egerton: *Electron Energy-Loss Spectroscopy in the Electron Microscope*, Plenum Press, New York (1996).
[2] D.D. Vvedensky, D.K. Saldin and J.B. Pendry: *Comput. Phys. Commun.*, **40**, 421–440 (1986).
[3] J.C. Fuggle, J.E. Inglesfield: *Unoccupied Electronic States*, Springer-Verlag, Berlin (1992).
[4] J. Bruley: *Microsc. Micronanal. Microstruct.*, **4**, 23–39 (1993).
[5] J.M. Titchmarsh: *Micron*, **30**, 159–171 (1999).
[6] P.A. Midgley, M. Weyland: *Ultramicroscopy*, **96**, 413–451 (2003).
[7] K. Kaneko, W.J. Moon, K. Inoke, Z. Horita, S. Ohara, T. Adschiri, H. Abe and M. Naito, *Mater. Sci. Eng. A*, **403**, 30–36 (2005).

5.6 Evaluation methods for oxide heterostructures

Two recent technologies for the evaluation of oxide heterostructures will be introduced in this section. First a method to measure the in-plane grain size of ZnO epitaxial layer [1], which is expected to be applied for an ultra-violet LED, by the X-ray reciprocal space mapping method [2] will be explained. Secondly, the nanoprobe measuring method [3] applied for evaluation of local electronic structure of the artificial superlattice consisting of perovskite oxides will be briefly described. In the epitaxial growth of the system with large lattice mismatch like ZnO thin film on the c-plane sapphire substrate, the thin film grows by coalescence of crystal grains of a size from a few tens to hundreds of nm [4]. Fig. 5.6.1 shows the growth mechanism schematically. The crystal nuclei generated on the substrate surface at the initial stage grow along the surface keeping the crystal orientation against the substrate by the domain matching epitaxy [5] with the coinciding common multiplier of lattice constant of the substrate and the thin film. As the crystal lattices are not always connected coherently on the coalescence of crystal grains, incoherent grain boundaries occur.

The incompleteness of crystallinity in the real space of nanostructure having grown in this way and expansion of diffraction peaks in the reciprocal space are explained in Fig. 5.6.2. Considering the crystallographic tilting and the in-plane average size of the grains (average distance between the coherent grain boundaries) as a measure of the incompleteness of crystalline, they give diffused pattern to all the diffraction spots. Focusing on a plane in the reciprocal space, the crystallographic tilting gives the diffraction peak circularly diffused vertical to the reciprocal space vector and the finite grain size gives diffused spread in the direction parallel to the plane, of which width is inversely proportional to grain size. As seen from Fig. 5.6.2a, these spread directions overlap on the diffraction of the symmetric plane but meet at an angle with the asymmetric faces. Therefore it is possible to measure the width of the crystallographic tilting of crystal grains and the grain size independently by measuring the reciprocal space mapping of an asymmetric plane and dividing the spread width of diffraction spots in the individual directions.

The evaluation of the structure of nanocrystalline film by this method is a very simple and direct method without using the etching method or TEM observation. Besides, it can evaluate the grain size as small as 10 nm, which is out of the range of X-ray topography. Actually, this method is reported to be useful for the investigation of grain boundary scattering of conduction electrons in the ZnO nanocrystalline films [6].

Fig. 5.6.3 shows the dependence of electron mobility on the grain size in ZnO thin film grown on the c-plane sapphire substrate under various preparation conditions. It is seen that the electron mobility decreases with the decreasing size because the double Schottky barrier formed due to deep level electrons trapped in the grain boundaries acts as major scattering centers electronic conduction. After annealing, the mobility becomes independent of the grain size because the effect of Schottky barrier becomes weak in terms of both the larger grain size and higher electron density. The relationship between the grain size and electric properties is relavant to the operating characteristics of transparent transistors using ZnO polycrystalline film as a channel [7].

An interesting phenomenon has been found concerning the incoherent grain boundary in ZnO thin film that it works as a cavity of laser emission [4]. The reciprocal space mapping with high resolution optics

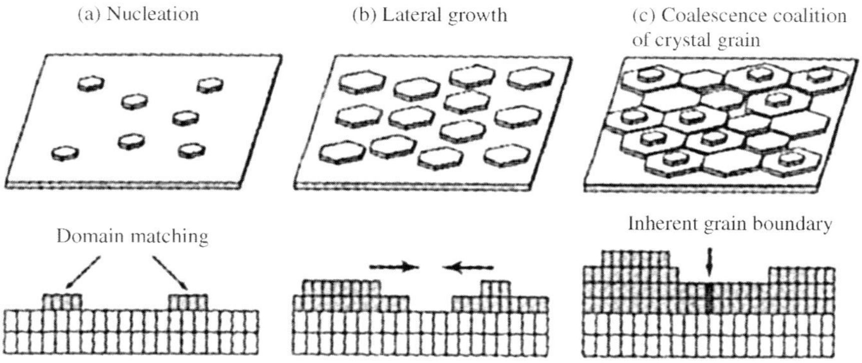

Figure 5.6.1
Domain matching and formation of incoherent grain boundary.

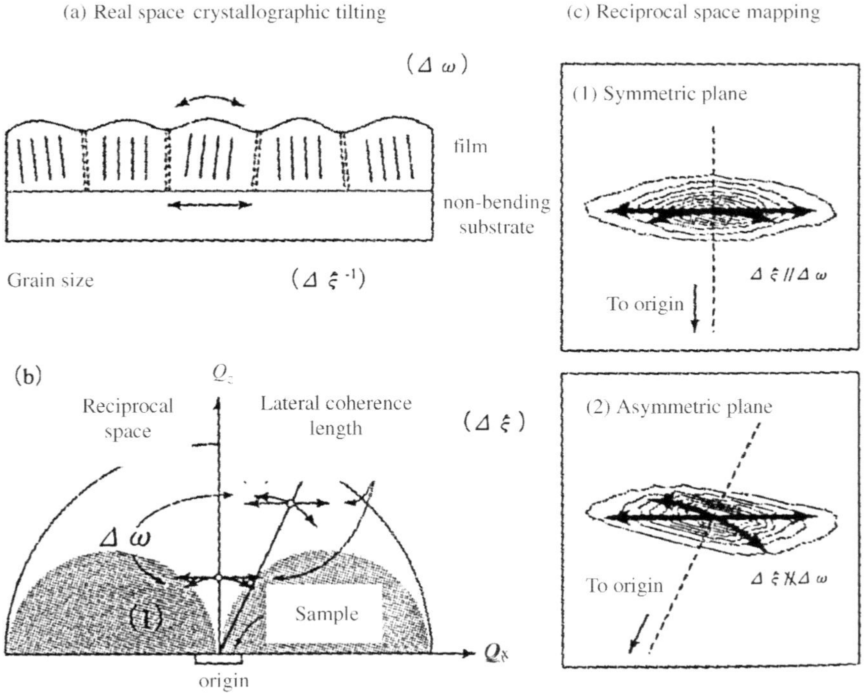

Figure 5.6.2
X-ray reciprocal space mapping of oriented nanocrystalline film.

can presently be conducted in a few minutes with using even a laboratory apparatus thanks to the advancement of a high-intensity X-ray source and a multi-dimensional detector [8].

Let us describe another topics. A number of studies have recently been carried out intensively to make use of diverse properties of perovskite oxide toward developing new functional devices. Electronic reconstruction at the surface or heterointerface often influences the operating characteristics of the devices. For example, there is a problem that the tunneling magnetic resistance (TMR) device composed of manganese oxide does not show the TMR ratio expected from the magnetic properties of the bulk material.

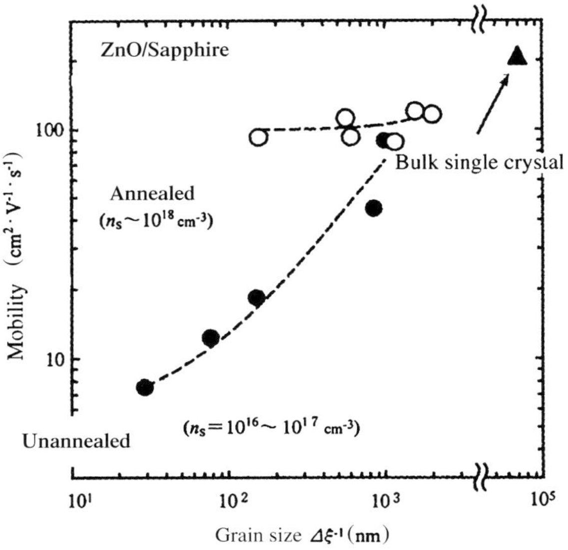

Figure 5.6.3
Dependence of electron mobility of ZnO thin films on the grain size.

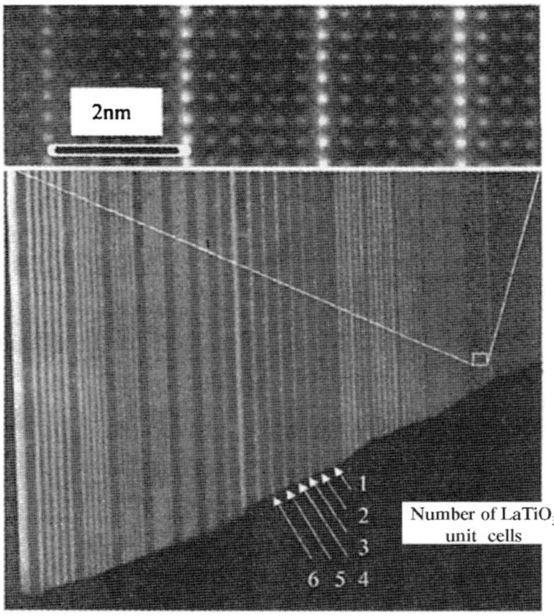

Figure 5.6.4
Annular dark field (ADF) STEM image of cross section of $SrTiO_3$/$LaTiO_3$ superlattice.

This is considered to be attributable to that because the charge transfer occurs between the $La_{1-x}Sr_xMnO_3$ ferromagnetic layer and the $SrTiO_3$ barrier layer, the electron spin polarization near the grain boundary is weakened compared to the bulk materials [9].

Therefore, the development of nanoprobing technique, which is capable of simultaneous measurement of local atomic structure and electronic states is recently getting more important. In the following is introduced an example of measurement of electron density distribution of a model interface between $SrTiO_3$ as a band insulator and $LaTiO_3$ as a Mott insulator using scanning transmission electron microscopy (STEM) and electron energy loss spectroscopy (EELS) [10].

Fig. 5.6.4 shows a STEM image of cross section of $SrTiO_3$/$LaTiO_3$ superlattice film prepared for controlling the thickness of each layer precisely as integral multiples of lattice constant by the pulsed laser deposition. The incident electrons focused down to 0.07 nm in diameter were scattered at high angles on the backside of the sample, which was thinned to a thickness of 30 nm by scanning of an accelerated ion beam, and corrected by a circular detector (annular dark field mode). It differs from the observation with high resolution conventional TEM in the point that the atomic column images with high contrast corresponding to the atomic number are obtained [11].

Because information of excitation of the core-level electron is obtained by the energy analysis of electron beam transmitting through the sample, the composition and the electronic state of each atomic column

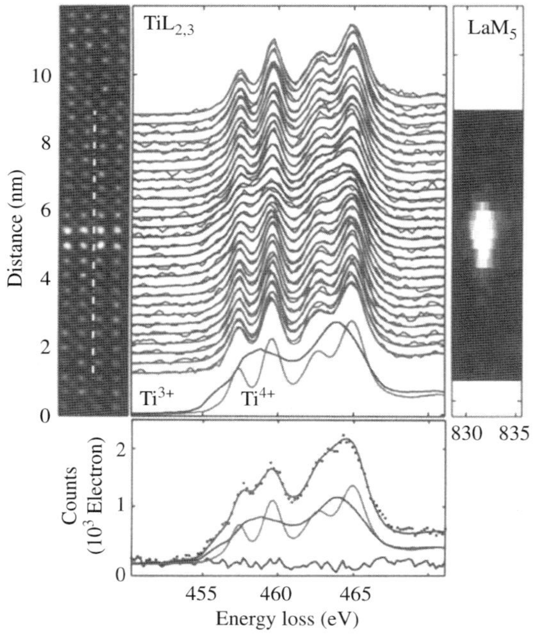

Figure 5.6.5
EELS spectra for La and Ti simultaneously recorded across a 2-unit-cell $LaTiO_3$ layer in $SrTiO_3$.

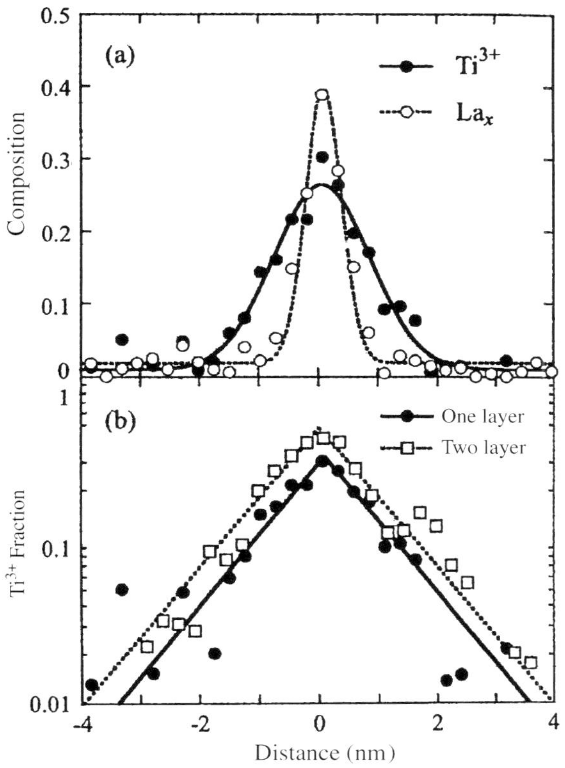

Figure 5.6.6
Spatial distribution of the Ti^{3+} signal in the vicinity of the one and two $LaTiO_3$ layers.

can be investigated. Fig. 5.6.5 shows the EELS spectrum of L edge in Ti column (in the middle) and M edge in La column (on the right) detected over the two-layer $LaTiO_3$ lattice indicated as the brightest spot in the left picture. As shown at the bottom of the figure, the spectrum of Ti column can be fitted by linear combination of the spectra of Ti^{3+} and Ti^{4+} obtained independently from pure $SrTiO_3$ and $LaTiO_3$, respectively. The average valence of Ti for each column can be determined from the weighting factor of Ti for each spectrum.

Since the excess electrons generated by replacing La^{3+} for Sr^{2+} occupies 3d orbit of Ti, the profile of the Ti^{3+} component corresponds directly to the electronic density distribution. The Ti^{3+} fraction of $LaTiO_3$ one and two layer lattice is plotted with the La^{3+} composition in Fig. 5.6.6(a). From the fact that the Ti^{3+} peak is broader than the La^{3+} peak, it is seen that the distribution of the extra electrons of $LaTiO_3$ lattice tend to be extended into the $SrTiO_3$ lattice. From the semi-log plot of the Ti^{3+} fraction, the screening length of the charge transfer is estimated to be about 1 nm [12].

Considering such charge transfer, it is attempted to form the electronic channel of high mobility on the polarity discontinuity heterointerface between the band insulators of perovskite oxide [13]. Furthermore, by use of this nanoprobing technique, the quantitative determination of oxygen vacancy in the artificially prepared in $SrTiO_3$ thin film or in the natural grain boundary and the structural evaluation on the atomic level are conducted [14, 15]. The nanoprobing technique using STEM can directly observe the electronic state at the buried interface with the atomic resolution and will be more and more important for the research of this field in the future.

References

[1] A. Tsukazaki, A. Ohtomo and M. Kawasaki: *Oyo Buturi*, **74**, 1359–1364 (2005) (in Japanese).
[2] P.F. Fewster: *Rep. Prog. Phys.*, **59**, 1339–1407 (1996).
[3] D.A. Mueller, T. Sorsch, S. Moccio, F.H. Baumann, K. Evans-Lutterodt and G. Timp: *Nature*, **339**, 758–761 (1999).
[4] M. Kawasaki, A. Ohtomo: *Solid State Phys.*, **33**, 59–64 (1998) (in Japanese).
[5] J. Narayan, P. Tiwari, X. Chen, J. Singh, R. Chowdhury and T. Zheleva: *Appl. Phys. Lett.*, **61**, 1290–1292 (1992).
[6] A. Ohtomo, H. Kimura, K. Saito, T. Makino, Y. Sefawa, H. Koinuma and M. Kawasaki: *J. Cryst. Growth*, **214/215**, 284–288 (2000).
[7] A. Ohtomo, M. Kawasaki and H. Ohno: *Solid State Phys.*, **40**, 407–414 (2005) (in Japanese).
[8] K. Saito, T. Kurosawa, S. Ueki and H. Funakubo: *Shinku*, **49**, 42–18 (2006) (in Japanese).
[9] H. Yamada, Y. Ogawa, Y. Ishii, H. Sato, M. Kawasaki, H. Akoh and Y. Tokura: *Science*, **305**, 646–648 (2004).
[10] A. Ohtomo: *Butsuri*, **58**, 425–429 (2003) (in Japanese).
[11] K. Saito, S. Tsurekawa and I. Tanaka: *Materia Japan*, **37**, 938–944 (1998) (in Japanese).
[12] A. Ohtomo, H.Y. Hwang: *Parity*, **19**, 18–25 (2004) (in Japanese).
[13] A. Ohtomo, H.Y. Hwang: *Oyo Buturi*, **73**, 605–609 (2004) (in Japanese).
[14] D.A. Mueller, N. Nakagawa, A. Ohtomo, J.L. Grazul and H.Y. Hwang: *Nature*, **430**, 657–661 (2004).
[15] M. Kim, G. Duscher, N.D. Browning, K. Sohlberg, S.T. Pantelides and S.J. Pennycook: *Phys. Rev. Lett.*, **86**, 4056–4059 (2001).

CHAPTER 6

EVALUATION METHODS FOR PROPERTIES OF NANOSTRUCTURED BODY

CONTRIBUTORS

Takehisa Fukui (6.1)
Junichi Tatami (6.2.1)
Shuji Sakaguchi (6.2.2)
Fumihiro Wakai (6.2.3)
Tetsuya Senda (6.2.4)
Takashi Akatsu (6.2.5)
Tetsuya Baba (6.3)
Yuji Noguchi (6.4.1)
Masaru Miyayama (6.4.2)
Atsushi Yamamoto (6.4.3)
Susumu Yonezawa (6.5.1)
Tomoichiro Okamoto (6.5.2)
Motohide Matsuda (6.5.2)
Masanobu Awano (6.5.3)
Tsutomu Katamoto (6.6)
Kenji Toda (6.7.1)
Soshu Kirihara (6.7.2)
Akihiko Suda (6.8)
Kiyoshi Nogi (6.9)

6.1 Functionality of nanostructures and their characteristic evaluation

Nanoparticles are defined in "Nanoparticle Technology" (The Nikkan Kogyo Shimbun, Ltd., 2003) as "ultimate fine particles with sizes in the nanometer region, formed with their respective chemical characteristics maintained, whose characteristics differ significantly from those in the form of bulk." The vast increase in specific surface area by scaling down fine particles to nanoscale elevates their original physical and chemical characteristics as materials or particles extensively. For instance, such nanosizing enhances the reactivity of each particle, enabling the low-temperature sintering of ceramics and offering remarkable improvement in functions such as catalytic activity. This nanosizing can also generate the effects of unprecedented peculiar electromagnetic and optical characteristics; for instance, nanosizing iron particles to the size of a single magnetic domain, the smallest unit capable of producing magnetism, allows them to change from a soft magnetic material that is magnetized according to a magnetic field to a hard magnetic material (permanent magnet). Such characteristics and effects are the reason that nanoparticles are called "the ultimate fine particles". However, nanosizing is prone to produce a physically and chemically unstable state due to the tremendously increased surface energy. Nanoparticles are, therefore, difficult to exist independently and undergo the agglomeration of particles, grain growth by sintering and other changes of state by oxidation and other reactions. Such poor stability, dispersibility and handling ability have hampered the wide application of nanoparticles as ultimate fine particles.

As a means to cope with such problems, the application of nanoscale structures is being developed. The functions inherent in nanoparticles can be maintained while lowering their reactivity by fabricating nanoparticles into nanostructures. Thus, stable materials, fully utilizing the functions of nanoparticles, can be created by fabricating composite and bulk materials from nanoparticles, depending on their intended use.

This chapter deals with such nanostructured materials and their various functions and the characteristic evaluation by function. As an introduction, this section summarizes the nature of nanostructures, relevant examples of the relations between nanostructures and their characteristics and the functionality and the evaluation of the characteristics of nanostructures.

6.1.1 What are nanostructures?

Nanostructures play an important role in the actual applications of nanoparticles. Typical images of nanostructures, such as composite nanoparticles, dense and porous materials, are illustrated in Fig. 6.1.1, where nanostructures composed of two types of nanoparticles and micron-size host particles are exemplified. Nanostructures can mainly be divided into particle structure and bulk structure, all of which contain nanoscale particles.

Composite nanoparticles can be roughly classified into structures comprising nanoscale particles and the nanostructures fabricated through the presence of host particles. Composite nanoparticle structures, such as catalysts, have nanoparticles fixed onto the surface of host particles as a carrier, and their single nanoparticles generate high catalytic activity. They are also often attached with catalytic promoter particles together with catalyst particles, enabling the creation of multi-functional catalysts. In fields other than catalysts, there are some types of toners having such a composite nanoparticle structure, where toner resin particles about 10 μm in particle size are coated with silica particles to improve fluidity and magnetic particles for transfer printing. The above-mentioned structures of partial coating, such as for catalysts, aim to stabilize the functions of nanoparticles. Meanwhile, the purpose of full coating structures is to regulate the high surface activity of nanoparticles or control the surface characteristics of the micron-size particles, which is called "surface modification". This surface modification can be utilized to improve the dispersibility and handling ability of nanoparticles as well as enhance the powder characteristics of micron-size particles such as in terms of fluidity. These structures are also utilized for the preparation of drugs to protect or compound active ingredients and to provide the slow-release of ingredients. Furthermore, it is possible to improve the applicability or handling ability of nanoparticles while maintaining their inherent functions by fabricating nanoparticle composites. Forming a collective structure of nanoparticles is an effective way to lower their surface activity, for which internal-dispersion and surface-coating structures are employed. In addition to the powder form, the nanoparticle composite structure is often utilized for materials and members in the form of bulk, where the functions of nanoparticles are performed inside the bulk form. Four representative types of bulk-form structures are shown in Fig. 6.1.1.

As dense structures, those fabricated from nanoparticles and those in which nanoparticles are

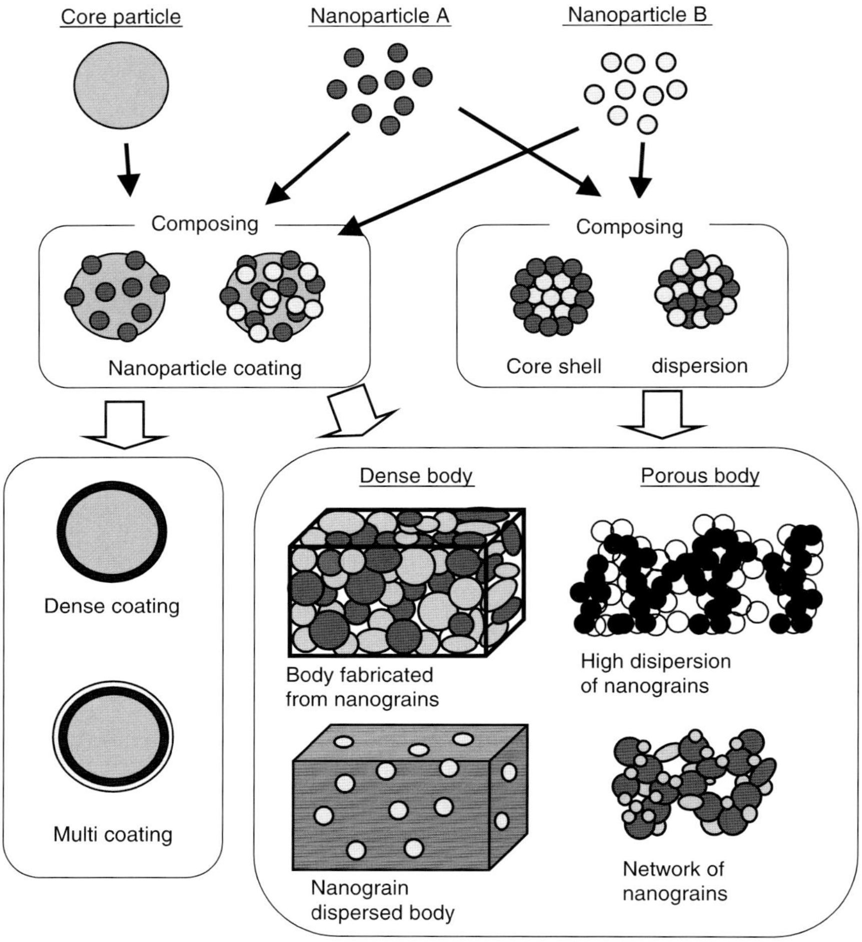

Figure 6.1.1
Nanostructures fabricated from several types of nanoparticles.

dispersed can be cited as representative examples. The dense-structure fabricated from nanoparticles can be obtained by low-temperature sintering and has very microscopic tissues because of the good sinterability of nanoparticles. Such microscopic tissues are expected to serve to enhance the performances of electronic parts as electric characteristics are said to be dependent on the sizes of constituent particles. It is also reported that the dense bulk forms, in which nanoparticles are dispersed, contribute to improving the strength of structural materials and others [1].

Typical examples for nanoporous structures are the high dispersion of nanoparticles and the networks they construct. Nanoporous materials have applications for many uses such as electrodes, filters, separation membranes, thermal insulations, etc. The functions of all these materials are strongly affected by the microstructures being limited in both bulk and porous parts.

6.1.2 Examples showing how the functions of nanostructures are performed

The detailed functions of nanostructures will be described by function later. In this section, an electrode of a solid oxide fuel cell (SOFC), the structure of which is controlled using surface-coating type composite particles and the inner-dispersion type composite particles, as shown in the composite fabrication of nanoparticles in Fig. 6.1.1, is taken as an example to explain the relation between nanostructures and their functional performance.

Electrode reaction, the conduction of electrons and ions and the diffusion of reaction gases are progressing simultaneously, and multiple functions are performed in the electrode of a fuel cell, the structure of which requires strength, heat resistance and chemical stability. The relation between the functions of SOFC electrodes and porous nanostructures is shown in Table 6.1.1, and an example of the structure of Ni-YSZ (Yttria Stabilized Zirconia) anode of SOFC in Fig. 6.1.2. The functions necessary for an are greatly dependent upon its structure, and the electrode functions are improved by control of the same. In the case of Ni-YSZ anodes, the following are required for the structure:

1. increase of the reactive interface for both Ni and YSZ, due to its microscopic, homogeneous and high-dispersion structure, since the electrode reaction occurs in the interface of Ni and YSZ;
2. networks of Ni and YSZ for the fabrication of good electron and ion paths;
3. a proper porous structure for the diffusion of fuel and reaction gases; and
4. association of constituent particles and thermostable structure for strength and thermostability.

Without any of the functions, the high performance of the electrode would be impossible. It is, therefore, essential to control the structure for a striking improvement in power-generation capacity.

The nanostructure of the SOFC electrode is controlled by employing composite nanoparticles as raw materials, which contributes to achieving a high functional performance. For the control of the structure, composite nanoparticles are utilized. Fig. 6.1.3 illustrates the microscopic structure of an Ni-YSZ anode, fabricated by the screen printing and baking method, using NiO–YSZ composite particles synthesized by spray pyrolysis (see Fig. 1.7.6, [2]) as raw materials. NiO–YSZ composite particles are those of the core–shell type with NiO particles coated with microscopic YSZ particles. The structure of electrodes

Table 6.1.1
Relation between the functionality of SOFC electrodes and nanostructures.

Function	Nanostructure necessary for the function
Electrode reaction characteristics	Constituent particles of the electrode should be nanoscale, uniform and highly dispersed
Electron and ion conductivity	Networks of constituent particles of the electrode
Diffusion of reaction gases	Proper pore size, pore size distribution and porosity
Strength, heat and chemical stability	Binding force of electrode constituent particles and proper microstructure

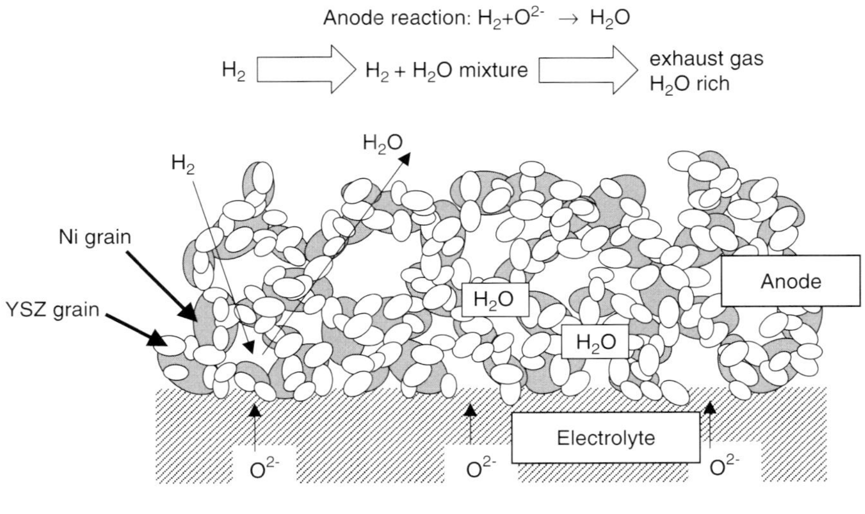

Figure 6.1.2
Morphology control of Ni-YSZ Y_2O_3 stabilized ZrO_2 cermet anode.

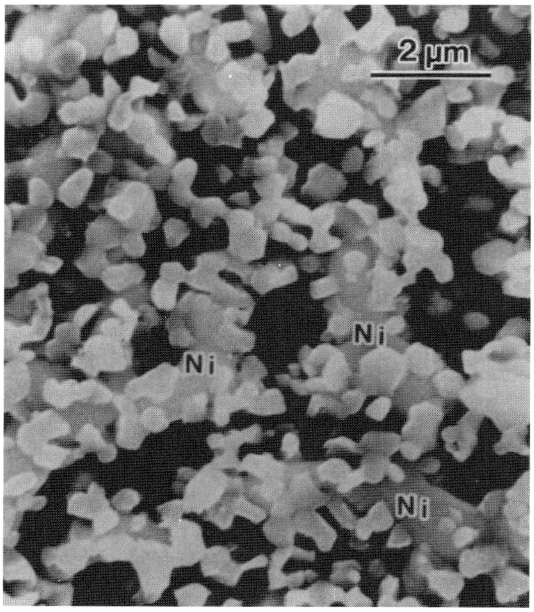

Figure 6.1.3
Nanoporous structure (Ni-YSZ cermet anode) [2].

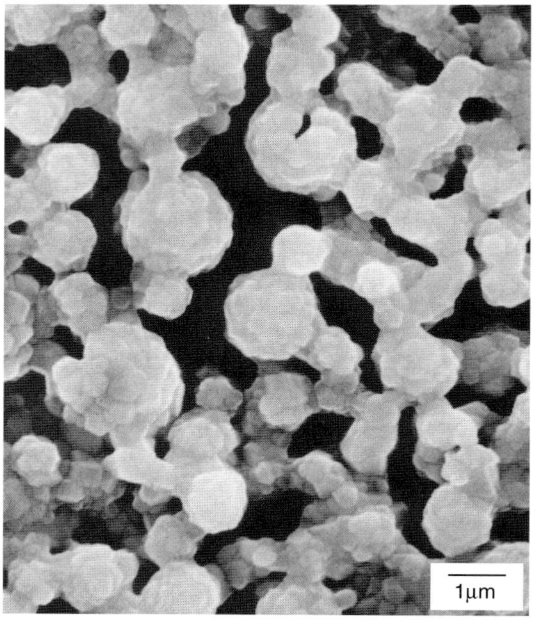

Figure 6.1.4
Nanoporous structure (LSM–YSZ cathode) [5].

constructed from such composite nanoparticles constitutes a porous structure where Ni networks are partially surrounded by microscopic YSZs. Surrounding Ni with microscopic YSZ particles not only allows the reactive interface to expand but facilitates the grain growth control of Nickel particles during operations at high temperatures (1,000°C). As a result of long-term evaluation of power-generation performance, the performance of the electrode reaction was demonstrably improved and stability maintained [3].

Next, the structure control and functions of SOFC cathode composed of La(Sr)MnO_3(LSM)–YSZ composite particles (see Fig. 1.7.5, [4]), synthesized by spray pyrolysis in the same manner, are exemplified as follows: LSM–YSZ composite particles are those of the internal-dispersion type with the microscopic LSM and YSZ particles highly dispersed in particles with a size of about 1 μm. Fabrication of the composite particles enables LSM or YSZ particles to be prevented from agglomerating each other, and ensures dispersibility among them. Fig. 6.1.4 illustrates the microscopic structure of LSM–YSZ cathode fabricated from these composite particles as raw materials using the screen printing and baking method [5]. It can be realized from the figure that it is a porous structure, comprising nanoparticles that maintain the structure of composite particles wherein LSM and YSZ microparticles are highly dispersed. Fabrication of the nanostructure of constituent particles and proper porous structure are realized, via which the performance of air electrodes is improved [5].

6.1.3 Functionality and characteristic evaluation

The preceding sections described how remarkable improvements in physical and chemical characteristics and the performance of peculiar functions may be realized by fabricating microscopic and nanoscale structures, as well as the maintenance of existing functions and the creation of new functions by fabricating nanostructures in the application of nanoparticles. Moreover, their functionality is expected to produce many applications ranging over almost all fields in the coming years, with the advancement of technology. The functional characteristics that are expected to be implemented in the coming future can be roughly cited as mechanical characteristics, thermal characteristics, electrical characteristics, electrochemical characteristics, electromagnetic characteristics, optical characteristics, catalytic characteristics and gas permeability and separation characteristics. Those functional characteristics are classified and shown in Table 6.1.2, where the respective expected effects, performances and applications are summarized and listed. The nanostructures specified to the respective characteristics and their characteristic evaluation are described in detail in the subsequent sections. Note that the functional characteristics are arranged in the order of subsequent appearance preceded by the section numbers.

Table 6.1.2
Nanostructures and functions.

Functionality	Expected effects and performances	Applications
Mechanical characteristics Strength, fracture toughness and fatigue characteristics Elasticity and hardness characteristics Creep and superplasticity characteristics Tripology characteristics Nanoindentation	• Improvement of mechanical strength such as strength, fracture toughness, etc. by fabricating composite nanostructures • Performance of superplasticity by fabricating nanostructures (making defects smaller, suppressing crack growth and grain boundary control through the control of nanostructure)	High-strength, heat-resistant and anti-corrosive parts such as bearings, crushing media, turbine blades, etc.
Thermal characteristics Thermal diffusivity, heat conductivity and interfacial heat resistance Specific heat Coefficient of thermal expansion	• Improvement of thermal characteristics, such as thermal conductivity, thermal insulation, etc. by fabricating nanostructures (control of phonon conductivity and thermal conductivity by fabricating nanostructures)	High thermal-conductivity parts such as substrates, thermal insulation, high-temperature parts, etc.
Electric characteristics Dielectric characteristics Electrical conductivity characteristics Thermoelectric characteristics	• Improvement of dielectric characteristics such as dielectric and piezoelectric properties and conductivity characteristics by fabricating nanostructures (fabrication of nanostructure of constituent particles and improvement of crystallinity)	Memory, capacitors, piezoelectric substances, electrical conductors, resistive elements, ionic conductors, etc.
Electrochemical characteristics Electrode reaction characteristics Sensor characteristics Electrochemical reactivity	• Improvement of composite functions by the control of nanostructure (miniaturization, networking, and appropriate porous structure of constituent particles etc.)	Secondary cells, fuel cells, gas centers, etc.
Electromagnetic characteristics	• Improvement of coercivity and fabrication of permanent magnetism (obtaining a single magnetic-domain size by the miniaturization of constituent particles)	Magnets, storage media, read heads, etc.
Optical characteristics Light transmission Photonic characteristics	• Improvement of emission intensity, color change and photonic expression by fabricating nanostructures (change of band gaps by miniaturization and control of wave guides by regular patterns)	Luminescent devices, display devices, paints, cosmetics, etc.
Catalytic characteristics	• Improvement of catalytic performances, lowering of reaction temperatures, etc. by fabricating nanostructures (improvement of reactivity by fabricating microscopic structures, stabilization of particles by fabricating nanostructures)	Environmental cleanup, chemical reactions, treatment of car exhaust gases, etc.
Gas permeability and separation characteristics	• Improvement of gas and liquid separation performance by fabricating nanostructures (control of porous structure)	Solid–gas separation membranes, solid–liquid separation membrane, environmental filters, etc.

References

[1] A. Sawaguchi, T. Toda and K. Niihara: *J. Am. Ceram. Soc.*, **74**, 1142–1144 (1991).

[2] T. Fukui, S. Ohara and K. Mukai: *Electrochem. Solid-State Lett.*, **1**, 120–122 (1998).

[3] T. Fukui, S. Ohara, M. Naito and K. Nogi: *J. Eur. Ceram. Soc.*, **23**, 2963–2967 (2003).

[4] T. Fukui, T. Obuchi, Y. Ikuhara, S. Ohara and K. Kodera: *J. Am. Ceram. Soc.*, **80**, 261–263 (1997).

[5] T. Fukui, S. Ohara, M. Naito and K. Nogi: *J. Nanopart. Res.*, **3**, 171–174 (2001).

6.2 Mechanical properties

6.2.1 Strength, fracture toughness and fatigue behavior

6.2.1.1 Strength

Strength is one of the most important parameters to design the structural components. In recent study, strength evaluation is needed in functional material and/or MEMS in order to improve their reliability. The strength is evaluated by bending or tensile test.

In many cases, strength measurement of metallic materials is carried out by tensile test because of machinability. Fig. 6.2.1 shows the relationship between strength and grain size of metallic materials. In the case of grain size over micrometer, strength increases with decrease in grain size according to Hall–Petch relationship. Moreover, the metallic material has excellent ductility and shows work hardening. Further grain refinement results in restriction of work hardening. As a result, strength decreases with decrease in grain size (inverse Hall–Petch relationship). In ceramics, tensile test is carried out to compare with metallic materials. Use of the specimen with ϕ 6 mm in diameter and 30 mm in gauge length is recommended in tensile test. Radius of curvature at the shoulder of gauge should be over 30 mm. The specimen should be held by using universal joint to restrict bending and twisting mode.

Since ceramics are generally hard to be machined compared with metallic materials, speaking, it is difficult to make tensile specimens with complex shape and high accuracy. Therefore, bending test is usually applied for strength measurement, which is regulated in JIS R1601. Fig. 6.2.2 shows the relationship between the strength of SiC nanoparticle dispersed Al_2O_3 and the volume fraction of SiC nanoparticles [1]. It is shown that the strength improves by adding a small amount of nanoparticles. Furthermore, it is reported that the strength also improves by annealing [1]. As shown in Fig. 6.2.3, strength variation decreases in nanocomposites [1].

6.2.1.2 Fracture toughness

According to fracture mechanics, strength is a function of fracture toughness K_{IC} and flaw size. Therefore, the fracture toughness is an important parameter to understand strength and reliability.

In metallic material, plane strain fracture toughness measurement is regulated in JIS G 0564 and ISO 12737. Many kinds of fracture toughness tests of advanced ceramics have also been proposed. Indentation fracture (IF) method and single edge precracked beam (SEPB) method is standardized as the fracture toughness test in JIS R1607. In ISO, SEPB method ISO15732:2003, surface crack in flexure (CSF) method ISO18756:2003 and chevron notched beam (CNB) method ISO/DIS24379:2003 have been established. Fig. 6.2.4 shows the relationship between the strength and fracture toughness and the SiC content [1]. Both strength and fracture toughness increase with an increase in SiC content.

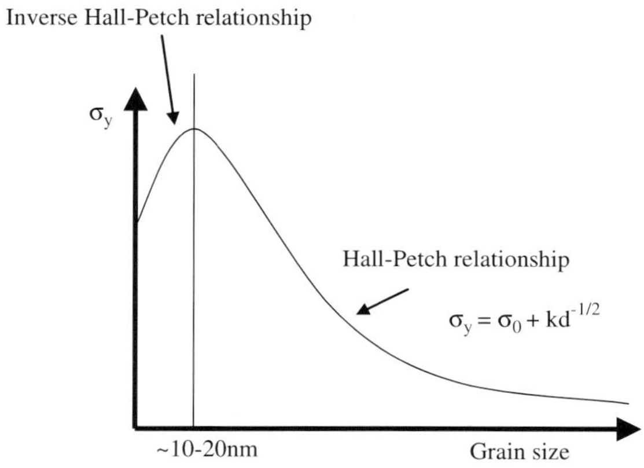

Figure 6.2.1
Grain size dependence of strength of metallic materials.

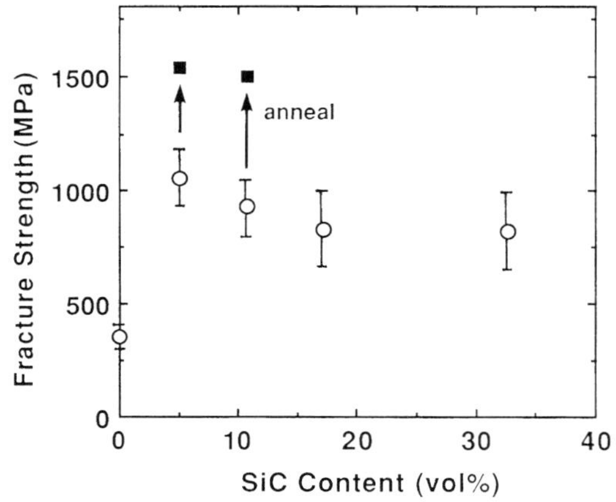

Figure 6.2.2
Effect of SiC content on the strength of Al_2O_3–SiC nanocomposites.

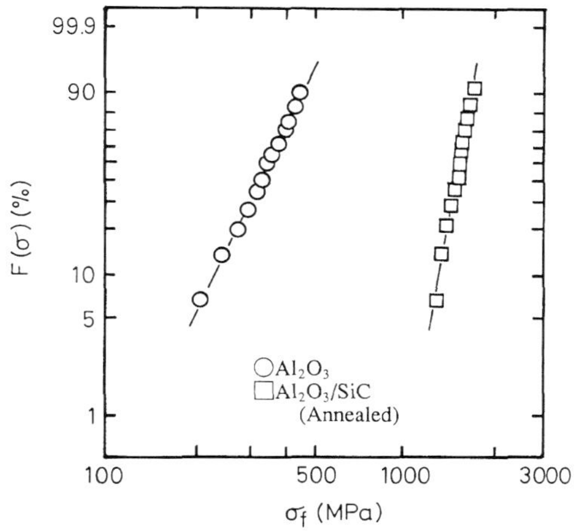

Figure 6.2.3
Weibull plot of the strength of Al_2O_3 and Al_2O_3–SiC nanocomposites.

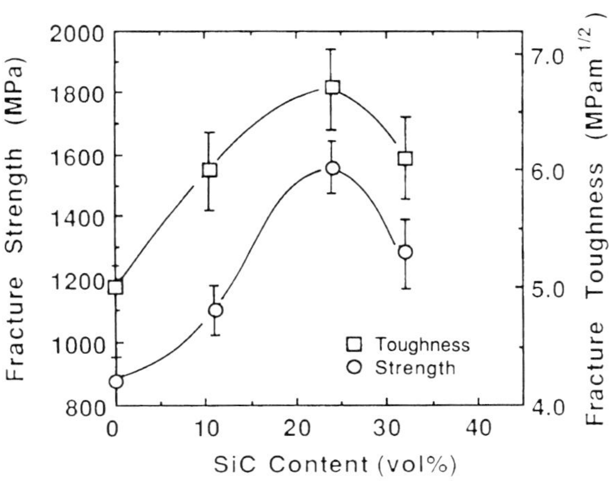

Figure 6.2.4
Influence of SiC content on the strength and the fracture toughness of Si_3N_4–SiC nanocomposites.

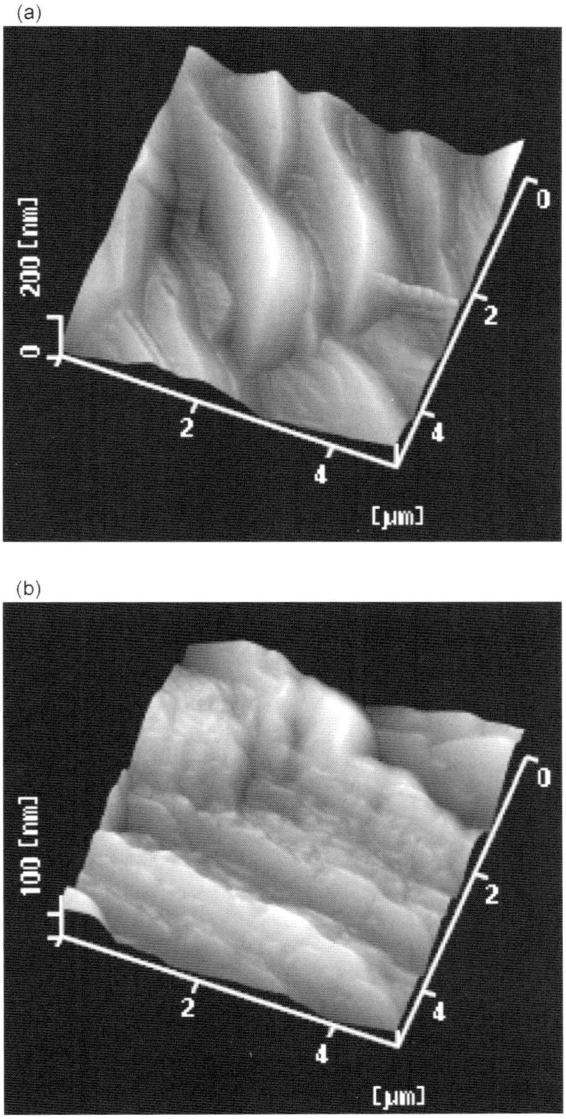

Figure 6.2.5
Scanning probe microscope images of fracture surface before and after corrosion test of Ca-α SiAlON nanoceramics. (a) Before corrosion test. (b) After corrosion test in 80°C H_2SO_4 aq.

Fractography is important to evaluate fracture origin and/or fracture process. Methodology of fractography should be referred to the other specialized book. Optical microscope and scanning electron microscope are generally used for fractography. Fractography of scanning probe microscope in nanoscale (nanofractography) has also been carried out in recent study. Fig. 6.2.5 shows fracture surface of Ca-α SiAlON nanoceramics before and after corrosion test observed by scanning probe microscope [2]. Although there was no glass phase in grain boundary because it was fabricated using 30 nm of Ca-α SiAlON nanoparticles without sintering aids, corrosion advanced along grain boundary to change fracture mode from transgranular to intergranular fracture.

6.2.1.3 Fatigue behavior

Fatigue behavior of nanostructured materials is evaluated in the same way as strength measurement by applying cyclic loading. Fig. 6.2.6 shows fatigue behavior of polypropylene nanocomposites [3].

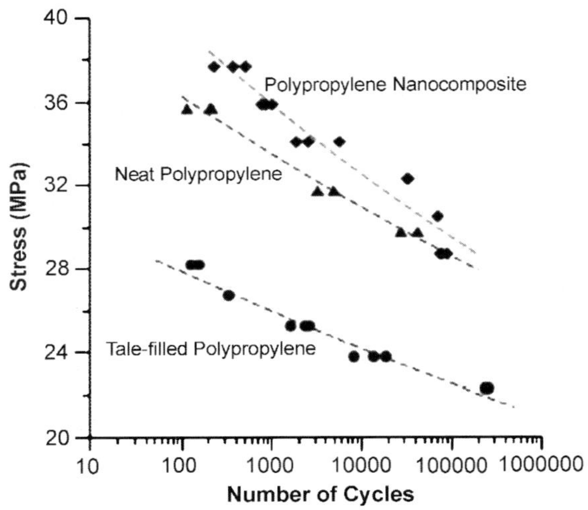

Figure 6.2.6
Fatigue behavior of polypropylene nanocomposites.

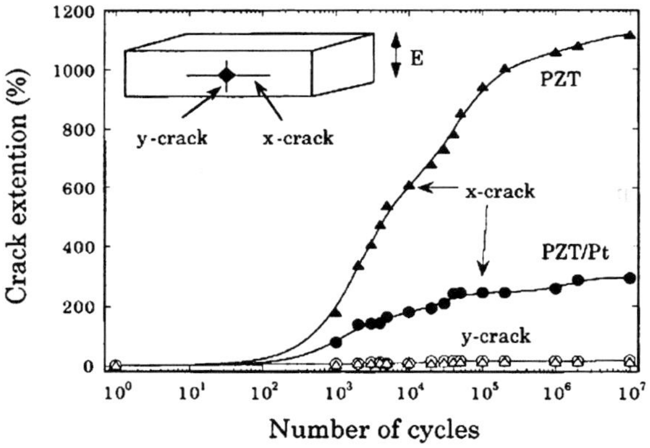

Figure 6.2.7
Fatigue crack propagation behavior of PZT–Pt nanocomposites.

The nanocomposite has 13 and 50% of higher fatigue strength than neat polymer and talc filled polymer, respectively. Furthermore, it is reported that the fatigue crack growth by AC voltage was restricted in piezoelectric nanoceramic composites (Fig. 6.2.7).

References

[1] K. Niihara: *J. Ceram. Soc. Jpn.*, **99**, 974–982 (1991).
[2] S. Kamiwaki, M. Iguchi, M. Hotta, J. Tatami, K. Komeya, T. Meguro, M. Omori and T. Hirai: Corrosion resistance of Ca-α SiAlON nano ceramics: Proceedings of 2nd MP3 176–183 (2004).
[3] Y. Zhou, V. Rangari, H. Mahfuz, S. Jeelani and P.K. Mallick: *Mater. Sci. Eng. A*, **402**, 109–117 (2005).
[4] H.J. Hwang, K.-I. Tajima, M. Sando, M. Toriyama and N. Niihara: *J. Ceram. Soc. Jpn.*, **108**, 339–344 (2000).

6.2.2 Elastic constants: hardness

For discussing elastic constants and hardness on nanomaterials, instrumented hardness tester (or called 'nano-indenter') is an important equipment. It measures the relation between the indenting depth and the applied force on the specimen, and the elastic constants and the hardness can be calculated [1–4]. The detail of this method is described in Section 6.2.5. In this section, the outline of elastic constants and hardness are explained.

Elastic constants are the parameters expressing the relation between the stress and the strain on the materials within the stress range that the materials exhibit elastic behavior. They are very fundamental parameters for discussing the mechanical properties of the materials, and they are used for numerical analysis of the components. In isotropic material, we have two independent elastic parameters. Usually, those are given as Young's modulus and Poisson's ratio. Fig. 6.2.8 illustrates the meanings of these two parameters. Young's modulus is the proportional constant for the strain of the material which is in the same direction of the stress. Poisson's ratio expresses the strain perpendicular to the stress. In metallic materials, 0.3 is the typical value for Poisson's ratio, but in materials which contain some complex microstructure, such as polycrystalline materials, they often show smaller values of Poisson's ratio, compared to the metallic materials [5].

For the measurement of elastic constants, several kinds of techniques exist, such as measuring the strain directly with applied static stress, measuring the sound resonance frequency on a certain shape of the specimen. For nanomaterials, calculation from the ultrasonic sound velocity can be the most common technique.

Young's modulus and Poisson's ratio can be calculated with the following equations, using the longitudinal sound velocity, V_L and the transversal sound velocity, V_T.

$$E = \rho \frac{3V_L^2 V_T^2 - 4V_T^4}{V_L^2 - V_T^2} \tag{6.2.1}$$

$$\nu = \frac{1}{2} \frac{V_L^2 - 2V_T^2}{V_L^2 - V_T^2} \tag{6.2.2}$$

where E is the Young's modulus, ν the Poisson's ratio and ρ the density. Fig. 6.2.9 shows the typical equipments for the measurement of the sound velocity [6].

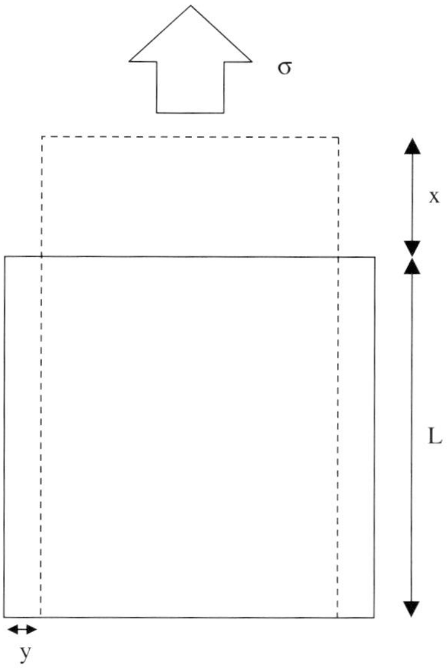

s: tensile stress
L: original length
x: strain parallel to stress
y: strain perpendicular to stress

E (Young's modulus) = s / (x / L)
n (Poisson's ratio) = y / x

Figure 6.2.8
Schematic figure on Young's modulus, E, and Poisson's ratio, v.

With the measurement of surface sound velocity, it can be applied to the measurement of elastic constants on thin films. It is reported that laser is used for generating and measuring the surface wave, and the sound velocity on thin films can be measured [7].

In nanomaterials, the elastic constants on the materials with two phases can be discussed. For the Young's modulus of two phase materials, two models are reported for giving upper and lower limit of the value. The two phase with Young's modulus of E_1 and E_2, with the volume fraction of V_1 and V_2, the upper limit of the Young's modulus of the mixed material, E_U is

$$E_U = V_1E_1 + V_2E_2 \tag{6.2.3}$$

and the lower limit, E_L becomes

$$E_L = \frac{E_1E_2}{V_1E_2 + V_2E_1} \tag{6.2.4}$$

For discussing the mechanical vibration, dumping of the vibration is another point of interest in the materials. Internal friction is a material parameter relating to elastic constants [8, 9]. Internal friction, Q^{-1}, expresses the energy loss on the vibration, or the delay of the vibration, and it is called loss tangent (tan δ), too. Internal friction can be determined with the applied vibration energy, W, and the lost energy ΔW with one cycle of vibration.

$$Q^{-1} = \frac{\Delta W}{2\pi W} \tag{6.2.5}$$

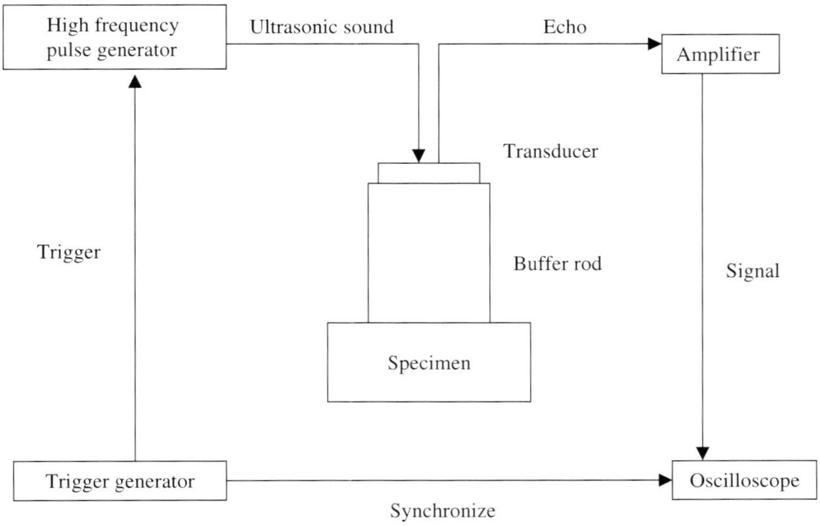

Figure 6.2.9
Example of the measurement system for ultrasonic sound velocity.

In nanomaterials, internal friction measurement can be required for expecting to minimize the vibration of the materials or components with adding secondary nanophase. For internal friction measurement, we have some techniques such as measuring the dumping of the vibration or the sharpness of the resonance of the vibration of the materials. Each technique can be applied at a certain vibration frequency, so, we can choose the technique considering the frequency for the applications.

Hardness is the parameter expressing the value for deformation of the materials with concentrated force on the surface. Vickers hardness test [10,11] is one of the most common technique for the measurement of hardness, which uses the diamond indenter, and measures the indent size after applying force on the surface.

On Vickers hardness test, diamond with pyramid shape is used for forming indents, and the tip of the indenter is put and the force is applied on the surface of the materials. The following equation is used for calculating the Vickers hardness value.

$$H_V = \frac{1.8544P}{d^2} \quad (6.2.6)$$

where, H_V is Vickers hardness, P the applied force and d the diagonal length of the formed indents (see Fig. 6.2.10). The average value of the diagonal lengths is applied as d, as we have two different diagonal lengths on one indent. For the unit expressing the Vickers hardness, kgf mm^{-2} (force by kgf and area by mm^2) was used, but recently we can see the expression with SI units, hardness in GPa.

In the measurement of brittle materials, cracks can be generated from the corner of the indents. For precise measurement, we must keep enough distance on every indent for avoiding the influence of the cracks. Fig. 6.2.10 shows the guideline for the distance of the indents, recommended in the testing on ceramic materials.

For the measurement on thin films, it is anticipated that the properties of the substrate may affect the hardness measurement results. It is said that if the film thickness is more than three times of the indent depth, the hardness can be measured with no influence by the substrate. In the case of Vickers hardness, the indentation depth, h, and the area of the indent, S, has the following relation. The indentation depth can be calculated from the result of the indent area measurement.

$$S = 24.5h^2 \quad (6.2.7)$$

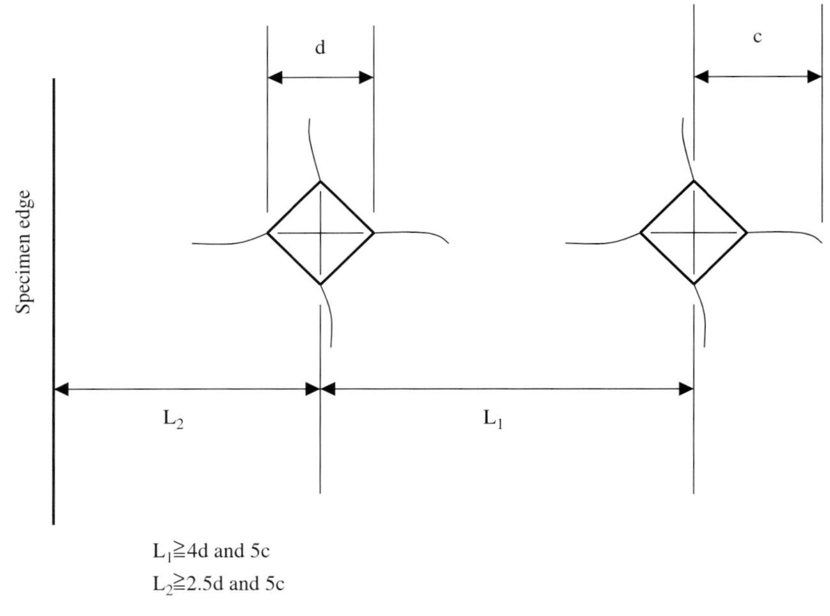

Figure 6.2.10
Indentation on Vickers hardness test and the recommended distance of indents.

Hardness test is effective for the evaluation of mechanical properties on coatings and printed circuits. It can be applied on the characterization on nanomaterials, too.

References

[1] W.C. Oliver, G.M. Pharr: *J. Mater. Res.*, **7**, 1564 (1992).
`[2] M. Sakai, S. Shimizu: *Ceramic Trans.*, **133**, 105 (2002).
[3] M. Sakai, S. Shimizu: *J. Mater. Res.*, **14**, 1471 (1999).
[4] M. Sakai: *J. Mater. Res.*, **14**, 3630 (1999).
[5] S. Sakaguchi, N. Murayama, Y. Kodama and F. Wakai: *J. Mater. Sci. Lett.*, **10**, 282 (1991).
[6] JIS R 1602: 1995, *Testing methods for elastic modulus of fine ceramics.*
[7] K. Yamanaka, H. Cho and Y. Tsukahara: *Appl. Phys. Lett.*, **76**, 2797 (2000).
[8] JIS K 7244: 1998, *Plastics – Determination of dynamic mechanical properties.*
[9] JIS R 1642: 2002, *The test method for internal friction test of fine ceramics at elevated temperatures.*
[10] JIS Z 2244: 2003, *Vickers hardness test – test method.*
[11] JIS R 1610: 2003, *Test methods for hardness of fine ceramics.*

6.2.3 Creep/superplasticity

To use nanomaterials in structural applications, design engineers must have knowledge of the mechanical response of the material. Deformation is the change in shape due to an applied force. While elastic deformation is reversible, plastic deformation is not reversible. Creep refers to the process in which a material continues to strain with time upon application of stress. As the grain size in metals and ceramics moves into the nanoscale, an increasingly large proportion of atoms in the solids are found on the grain boundaries. In those nanocrystalline materials, the diffusional creep occurs at an intermediate temperature range ($0.5\ T_m > T > 0.4T_m$, T_m: Melting temperature), because the strain rate is enhanced by grain boundary diffusion much more than that of the conventional microcrystalline materials [1–3]. Creep is a property that is related to the mechanical strength of nanocrystalline materials.

On the other hand, superplasticity refers to the ability of polycrystalline solids to exhibit exceptionally large elongations in tension at elevated temperatures [4]. It is a property commonly found in many metals, alloys, intermetallics and ceramics when the grain size is very small: less than several micrometers for metals and less than one micrometer for ceramics. Further grain refinement leads to high-strain rate superplasticity or low-temperature superplasticity. Such behavior is of interest because the ability to achieve large strains makes superplastic forming an attractive option for the manufacture of complex shaped components. Nanocrystalline materials are often fabricated from nanoparticles, for example, by hot pressing, hot isostatic pressing, sinter forging and spark plasma sintering (SPS). The creep and superplasticity are also related to those stress assisted densification processes [5].

Creep and superplasticity behavior of nanomaterials have been characterized in creep tests [6] and superplasticity tests at elevated temperatures.

6.2.3.1 Creep test

In a creep test a constant load is applied to a tensile specimen maintained at a constant temperature. Strain ε is then measured over a period of time t. A typical creep curve is shown in Fig. 6.2.11, and the slope of the curve is the strain rate $\dot{\varepsilon}$. The creep curve is usually divided into three regions. Primary creep is a period of decreasing creep rate. The strain rate eventually reaches a minimum and becomes almost constant. This is known as secondary creep or steady creep. The "creep strain rate" is typically the rate in the secondary stage, or the minimum creep rate. Tertiary creep occurs when there is a reduction in cross sectional area due to necking or effective reduction in area due to internal cavity formation.

Japanese Industrial Standards for testing creep have been established for bulk materials in tension, compression and bending.

JIS Z 2271: Method of creep and creep rupture for metallic materials.

JIS R 1631: Method for tensile creep of fine ceramics.

JIS R 1612: Method for bending creep of high-performance ceramics.

JIS R 2658: Testing method for compressive creep of refractory bricks.

Similar testing methods can be applied to nanocrystalline materials as well. When the available specimen

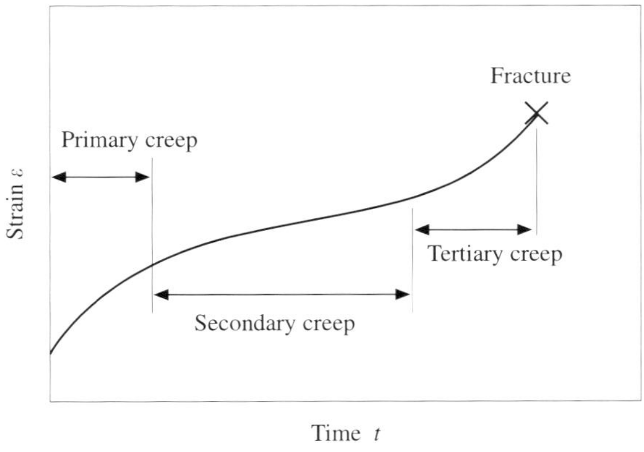

Figure 6.2.11
Typical creep curve.

size of the nanocrystalline material is very small, special testing which does not comply with the agreed standards is often performed to acquire data.

(a) Tensile creep test

For analysis of creep and creep rupture a test technique is required which uses a specimen with a uniform, stationary stress distribution in its gauge length. This is achieved in a tensile creep test most satisfactorily. In Fig. 6.2.12 a tensile specimen with a circular cross section is presented as an example. The gauge length L of the specimen is measured by: (1) mechanical strain extensometer or (2) optical strain extensometer. Engineering stress or nominal stress is defined by

$$\sigma = \frac{P}{A_0} \tag{6.2.8}$$

where A_0 is the initial cross-sectional area prior to the application of the load P. Engineering strain or nominal strain is represented by

$$\varepsilon = \frac{(L - L_0)}{L_0} \tag{6.2.9}$$

where L_0 is the initial gauge length.

One of the major concerns in tensile creep apparatus is specimen alignment to minimize bending moments that are introduced by the tensile load train. This requires the flexibility in the pull rods, that is a flexible coupling, and a grip that transmits the load through the geometric center of the specimen.

(b) Compressive creep test

The Uniaxial compression test is a useful procedure for measuring creep of a small nanocrystalline specimen that has either cylindrical or square cross section. In order to prevent buckling, specimens with length/diameter ratio $1 < L/d < 2$ are used usually. Barreling occurs when $L/d < 2$ and friction is present at the contact surface. Homogeneous deformation can be achieved by using smooth surface without friction.

(c) Bending creep test

A method of four-point bending creep for high-performance ceramics (JIS R 1612) is schematically shown in Fig. 6.2.13. The nominal stress on the surface of the specimen is calculated by using the simple beam theory

$$\sigma = \frac{3P(L-l)}{2wt^2} \tag{6.2.10}$$

where P is the applied load, L the outer span length, l the inner span length, w the specimen width, and t the specimen thickness. The nominal strain is calculated from the specimen deflection δ

$$\varepsilon(\%) = \frac{4t\delta}{l^2} \times 100 \tag{6.2.11}$$

This method is relatively easy to measure small strain, but the simple beam theory is applicable only under following assumptions: (a) Creep behavior in tension and compression is identical. (b) Friction at contact points is negligible. (c) Creep behavior is Newtonian. When the creep is non-Newtonian, the analysis of deformation is complicated due to the stress gradient inside the specimen.

6.2.3.2 Superplasticity test

Tension and compression tests are performed at a constant displacement rate by using a universal test machine in order to determine superplastic forming properties. Instead of engineering stress and engineering strain, true stress σ_t − true strain ε_t curves are obtained for large deformations

$$\varepsilon_t = \ln\left(\frac{L}{L_0}\right) \tag{6.2.12}$$

$$\sigma_t = \frac{P}{A_0}\frac{L}{L_0} \tag{6.2.13}$$

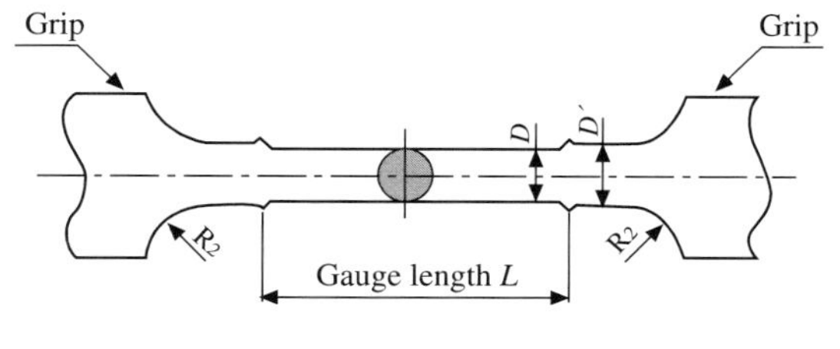

Figure 6.2.12
Tensile creep specimen (JIS R 1631).

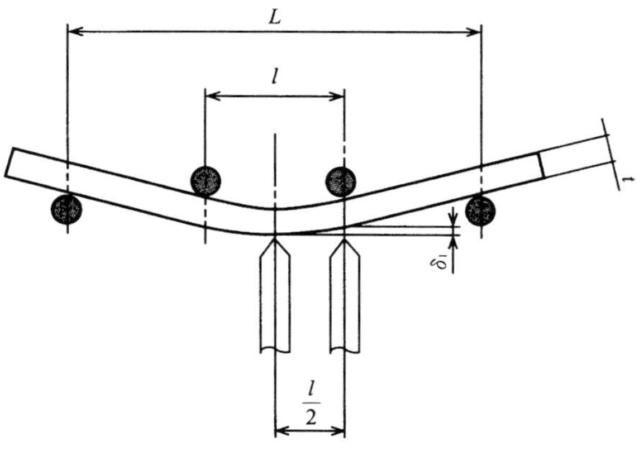

Figure 6.2.13
Four-point bending creep test method (JIS R 1612).

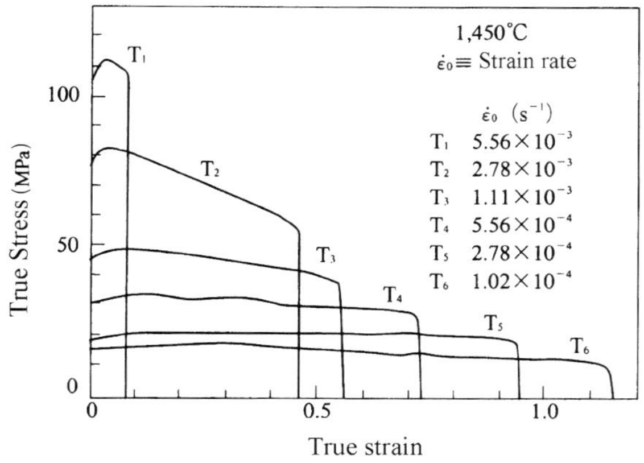

Figure 6.2.14
Stress–strain curves in superplasticity of zirconia/alumina composite [7].

A typical example of stress–strain curve is shown in Fig. 6.2.14. Important properties for superplastic forming are maximum elongation, flow stress and strain rate sensitivity index (m-value). Japanese Industrial Standards for testing superplasticity have been established for metals.

JIS H 7007: Glossary of terms used in metallic superplastic materials.
JIS H 7501: Method for evaluation of tensile properties of metallic superplastic materials.
JIS H 7502: Method for evaluation of compressive properties of metallic superplastic materials.
JIS H 7503: Method for measurement of cavity volume fraction of superplastically deformed metallic materials.

The superplastic elongation or elongation to fracture [%] is defined by multiplying 100 to the nominal strain of equation (6.2.9). Superplasticity is defined as an ability of polycrystalline material to exhibit large elongations more than several hundreds percent in the glossary (JIS H 7007). It is more than 300% for metals and about 100% for ceramics usually.

The relationship between strain rate and stress at elongation of 10% is shown in Fig. 6.2.15 as an example. The slope of the curve is stress exponent n, and the strain rate sensitivity index is the inverse of the stress exponent, $m = 1/n$. The stability of tensile deformation, that is a uniform elongation of metals without necking, is dependent on strain rate sensitivity index $m > 0.3$. Generally the strain rate sensitivity index varies with stress.

Cavities or voids may be formed during superplastic deformation. Since the mechanical performance of the superplastically formed products drops with

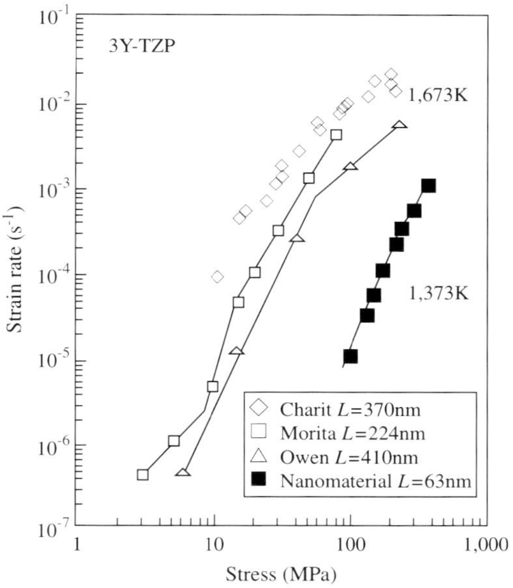

Figure 6.2.15
Relationship between stress and strain rate in nanocrystalline zirconia [8].

cavity volume fraction C_v, it is necessary to evaluate it for quality assurance

$$C_v = \frac{\rho - \rho_0}{\rho_0} \times 100[\%] \qquad (6.2.14)$$

where ρ is the density and ρ_0 the initial density.

6.2.3.3 Creep and superplasticity in nanocrystalline materials

The strain rate is often expressed by the following semi-empirical equation [9]

$$\dot{\varepsilon} = \frac{AGb}{k_B T}\left(\frac{b}{d}\right)^p \left(\frac{\sigma - \sigma_0}{G}\right)^n D \qquad (6.2.15)$$

where $\dot{\varepsilon}$ is the strain rate, b the Burgers vector, G the shear modulus, σ the stress, σ_0 the threshold stress, n the stress exponent, d the grain size, p the grain size exponent, $D = D_0 \exp(-Q/k_B T)$ the diffusion coefficient, Q the activation energy, T the temperature and k the Boltzmann's constant.

At high stresses, dislocation creep is controlled by vacancy diffusion for dislocation climb. The stress exponent is $n > 3$, and the grain size exponent is $p = 0$. At lower stress level, diffusional creep takes place. While atoms diffuse through the lattice in Nabarro–Herring creep ($n = 1$ and $p = 2$), atoms diffuse along grain boundaries in Coble creep ($n = 1$ and $p = 3$). The diffusional creep is enhanced with decrease in the grain size, and then, is responsible for the softening phenomena of nanocrystalline materials in the intermediate temperature range [1–3]. Fig. 6.2.15 shows that nanocrystalline material can be deformed at lower temperatures.

Ceramics, e.g., oxides, nitrides and carbides, are hard, strong and stiff materials. They are brittle, and lack the ductility of metals at ambient temperatures. Since the discovery of superplasticity of Y_2O_3-stabilized tetragonal ZrO_2 polycrystals (Y-TZP) [10], the research on superplasticity has further expanded to various oxides and non-oxides [11]. A remarkably high elongation of 2,510% has been achieved at high-strain rate in ZrO_2-based composites [12].

The deformation of polycrystalline solid is schematically shown in Fig. 6.2.16 [13]. The grains are elongated in classical models of diffusional creep in Fig. 6.2.16a. On the other hand, fine grains retain approximately equiaxed shape even after the extremely large elongations (>1,000%) in superplasticity. The essential mechanism of superplasticity resembles to that in the flow of soap froth as shown in Fig. 6.2.16b. While the topological change of grain is not considered in diffusional creep, the motion of grains involves grain switching, which means the rearrangement of grains by grain boundary sliding. It is considered that the boundary "mantle" of a grain behaves differently from the central "core" of the grain as shown in Fig. 6.2.16c.

References

[1] E. Arzt: *Acta Mater.*, **46**, 5611–5626 (1998).

[2] F.A. Mohamed, Y. Li: *Mater. Sci. Eng. A*, **298**, 1–15 (2001).

[3] I.A. Ovi'dko: *Int. J. Mater. Rev.*, **50**, 65–82 (2005).

[4] T.G. Nieh, J. Wadsworth and O.D. Sherby: *Superplasticity in Metals and Ceramics*, Cambridge University Press, Cambridge (1997).

[5] Z.J. Shen, H. Peng and M. Nygren: *Adv. Mater.*, **15**, 1006–1009 (2003).

[6] B.F. Dyson, R.D. Lohr and R. Morrel (Eds.): *Mechanical Testing of Engineering Ceramics at High Ttemperatures*, Elsevier Applied Science, London (1989).

[7] F. Wakai, H. Kato: *Adv. Ceram. Mater.*, **3**, 71–76 (1988).

[8] M. Yoshida, et al.: *J. Am. Ceram. Soc.*, **85**, 2834–2836 (2002).

[9] M. Jiménez-Melendo, et al.: *J. Am. Ceram. Soc.*, **81**, 2761–2776 (1998).

[10] F. Wakai, S. Sakaguchi and Y. Matsuno: *Adv. Ceram. Mater.*, **1**, 259–263 (1986).

[11] F. Wakai, Y. Kodama, S. Sakaguchi, N. Murayama, K. Izaki and K. Niihara: *Nature*, **344**, 421–423 (1990).

[12] B.N. Kim, K. Hiraga, K. Morita and Y. Sakka : *Nature*, **413**, 288–291 (2001).

[13] F. Wakai, Y. Shinoda, S. Ishihara and A. Domínguez-Rodríguez: *Acta Mater.*, **50**, 1177–1186 (2002).

6.2.4 Tribological properties

In the relative motion of two solid bodies in contact, a force acts in the direction interfering the motion as shown in Fig. 6.2.17 and the region beneath the contact point is highly stressed. The force is called friction and the material removal process by microfracture occurring near the surface is called

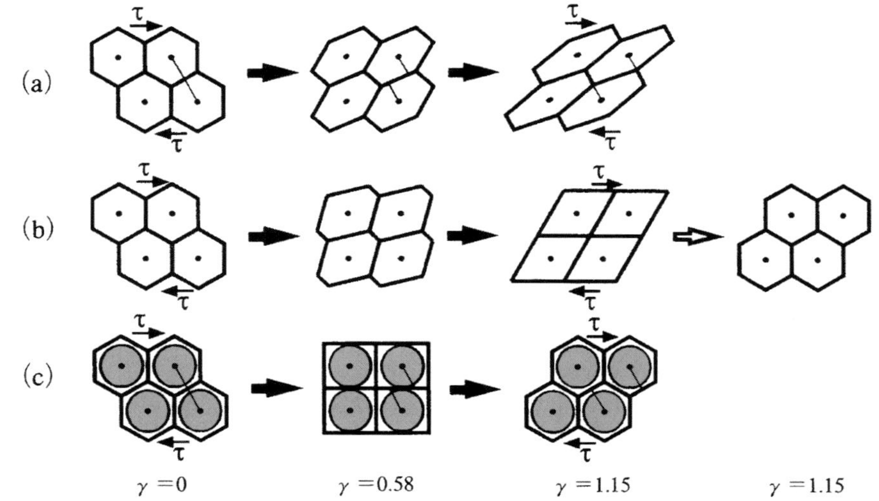

Figure 6.2.16
Shear deformation in regular array of grains [13]. (a) Diffusional creep. (b) Soap froth model. (c) Core–mantle model. Shear strain γ is induced by shear stress τ. Marker line shifts due to the grain boundary sliding.

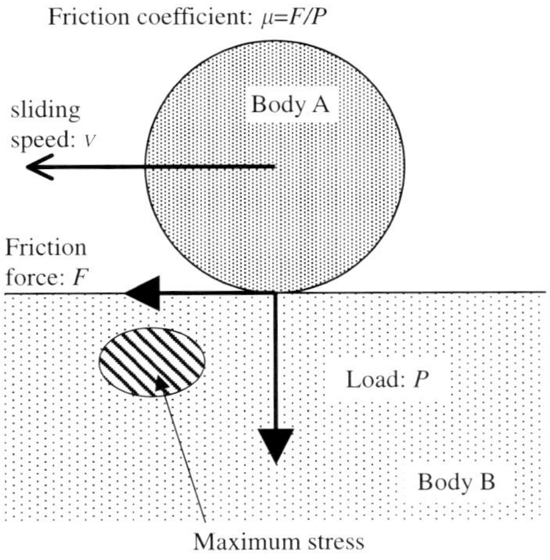

Figure 6.2.17
Stress field around the surface of body B when body A slides on body B under a load of P at a speed of v.

wear. Wear often affects the life of machinery or equipment, for example, only a wear loss of a few milligrams may cause a serious trouble of a large machine of several tons. It is, therefore, critical to suppress the wear in engineering design.

Wear, in general, depends on a number of factors including environment, temperature and materials as well as the sliding speed and the applied load. In particular, material selection for components in the relative motion is critical to the wear control. In the frictional surface, true contact area is extremely smaller than nominal contact area due to the microscopic roughness or asperity, which is inevitably present on the surface. It results in a highly stressed field forming around the true contact point, suggesting that metallurgical bonding may occur under a localized high pressure during the sliding motion. Adhesive wear, which is often observed in friction of metal contact, occurs by microfracture around the metallurgical bond, generating debris fragments. An effective method to reduce this type of wear is, therefore, to select material pairs of less attractive combination for bonding.

Ceramics, on the other hand, do not readily adhere and are hard, and therefore are expected to exhibit a good performance as sliding components. However, the wear properties of ceramics are actually not so excellent as expected when sliding against similar materials. It is mainly due to brittle microfracture occurring beneath the contact point which is caused by resultant stress of contact pressure and frictional shear force. Collision of asperities and microfracture at the contact area causes a severe wear. In order to avoid such wear, it is recommendable to choose materials with high threshold stress for microcracking and to apply some lubrication to reduce the frictional stress.

When a hard material slides on a relatively soft material, it often happens that material removal from the hard material surface occurs by cutting or ploughing by fragment of hard material embedded on the soft material surface. To reduce this type of abrasive wear, it is effective to increase the hardness of the soft material for minimizing the invasive depth of the fragments to the soft materials surface by using a harder material for the mating component.

For improving the wear-resistant performance, there are some advantages in using composite materials as described below.

1. Ceramic particles, dispersed in adhesive metals, retard formation of metallurgical joints in the contact surface.
2. Addition of the second-phase particles enhances deformation resistance and suppresses the plastic flow as well as the microfracture in the surface. As an additional effect, the second-phase particles increase the heat resistance to possible temperature increase due to frictional heating in a local area around the contact point.
3. When adding second-phase particles with lubricating effects, the frictional force can be reduced and the surface stress lowered.

As described above, additions of the second-phase particles often enhance the wear resistance, however, it is also known that an excessive addition degrades the mechanical properties as well as the wear resistance. It suggests that the second-phase particles have an optimized amount of addition. Microstructure with a fine grain size of a submicrometer level generally exhibits better wear resistance. To improve the mechanical properties further as well as wear resistance, it needs nanoparticles as the second-phase additives. If the nanosized second-phase particles enhance the uniformity of dispersion, they also affect the adhesion resistance and lubrication effects.

Several experiments have been reported on the second-phase effects of nanoparticles. Addition of alumina particles of 50 nm size by 1.11 vol% to a magnesium alloy improves the wear resistance of the alloy up to 1.8 times that of pure magnesium in sliding wear tests against a tool steel [1]. It has been simply interpreted as the effect of reinforcement and hardness increase by the addition, however, suppression of the wear by delamination in the surface area has also been observed.

It has been reported that alumina containing carbon nanotubes (CNT) exhibits a low friction coefficient and a low wear loss [2]. As shown in Fig. 6.2.18, the friction coefficient decreases with an increase of CNT addition, while the wear loss shows a minimum value as low as a half of that for pure alumina, but increases by further addition. The decrease in the friction coefficient is attributed to the lubrication effects by CNT, which has the same structure as graphite, a well-known solid lubricant. Initial addition of CNT improves the mechanical properties such as hardness, however further addition may cause poor cohesion between alumina matrix and CNT particles, resulting in exhibition of the minimum value of the wear loss by the CNT addition.

Addition of SiC particles suppresses the abrasive wear of alumina. It has been reported that this is because addition of nanoparticles refines the microstructure, decreasing the size of pulled-out particles [3]. In an alumina with addition of SiC whisker by 10–20 vol%, the wear resistance is obviously improved particularly in case of sliding along the direction perpendicular to whickers at elevated temperatures above 673 K [4]. It has been concluded that reaction products of SiC forming a soft layer reduces surface stress, resulting in the better wear resistance at high temperatures.

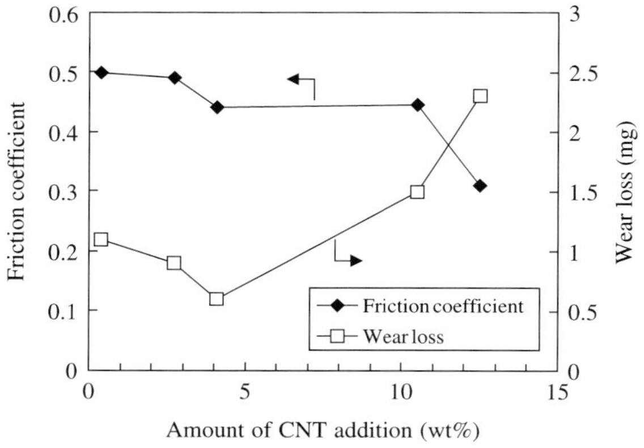

Figure 6.2.18
Friction coefficient and wear loss of hot-pressed alumina with addition of carbon nanotubes (CNT) [2].

Fig. 6.2.19 shows a mechano-chemical grinding (MCG) process for preparation of a silicon nitride base composite using a ball mill in nitrogen environment. The composite, where Ti compound and self-lubricating particles are dispersed in the silicon nitride matrix, shows a low friction coefficient of 0.2 level and a specific wear loss (wear volume divided by load and sliding distance in a unit of $m^2 N^{-1}$) nearly two order lower in magnitude than conventional silicon nitride ceramics [5].

Improvement of wear resistance by adding ceramics to plastic materials has also been investigated [6]. It has been reported that addition of alumina nanoparticles to poly(ethylene) terephthalate (PET) leads to a friction coefficient comparable to, or slightly higher than PET without addition for a sliding against a steel. The specific wear loss exhibits a half value of the PET with addition of 2 wt%, however, is slightly higher in case of addition of 10 wt%. The improvement of the wear resistance with a small addition of alumina particles is attributed to coherent transfer film formation and decrease in crystallinity by adding the nanoparticles. Further addition may accelerate abrasive wear by aggregates of alumina particles contained in wear debris.

Supplying nanoparticles to sliding interface has also been studied by dispersing nanoparticles in lubrication oil [7]. When synthetic base oil (polyalfaolefin) with addition of molybdenum base nanowire ($Mo_6S_3I_6$) is supplied to the friction surface, the friction coefficient drastically decreases under boundary lubrication conditions where solid contact occurs by depletion of lubrication oil from the contact interface. Since analysis of the friction surface has revealed the presence of MoS_2, it has been concluded that nanowire supplied on the friction surface transforms into solid-lubricating substance on the surface. In this case, dispersed particles in the lubrication oil exhibit solid-lubricating effects by adhering to the friction surface where lubrication oil is poorly supplied.

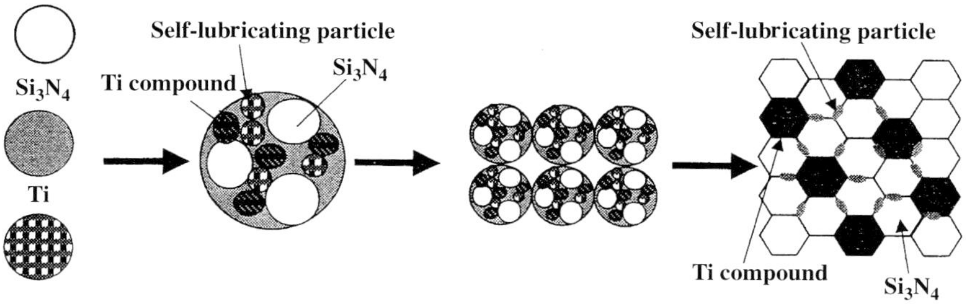

Figure 6.2.19
Preparation process and microstructure of silicon nitride with dispersion of self-lubricating particles [5].

References

[1] C.Y.H. Lim, D.K. Leo, J.J.S. Ang and M. Gupta: *Wear*, **259**, 620–625 (2005).

[2] J.-W. An, D.-H. You and D.-S. Lim: *Wear*, **255**, 677–681 (2005).

[3] J.L. Oritiz-Merino, R.I. Todd: *Acta Matarialia*, **53**, 3345–3357 (2005).

[4] D.-S. Lim, D.-S. Park, B.-D. Han, T.-S. Kan and H. Jang: *Wear*, **251**, 1452–1458 (2001).

[5] M. Yoshimura, *Bull. Jpn. Ceram. Soc.*, **39**, 177–180 (2004).

[6] P. Bhimaraj, D.L. Burris, J. Action, W.G. Sawyer, C.G. Toney, R.W. Siegel and L.S. Schadler: *Wear*, **258**, 1437–1443 (2005).

[7] L. Joly-Pottuz, F. Dassenoy, J.M. Martin, D. Vrbanic, A. Mrzel, D. Mihailovic, W. Vogel and G. Montagnac: *Tribiol. Lett.*, **18**, 385–393 (2005).

6.2.5 Nanoindentation

What is the most suitable mechanical test for characterizing materials with nanostructure? Tensile, compressive and bending tests as well as a test based on the fracture mechanics, in which the size of a main crack in a specimen is measured, are usually utilized to evaluate the mechanical properties of conventional materials. Also, tests to obtain hardness, friction- and wear-resistance are often adopted. Although these mechanical tests can be applied for the nanostructured materials, correlation between the mechanical properties and the nanostructure is hardly obtained through the tests. Nanoindentation is a unique mechanical test that can evaluate local mechanical properties related with the nanostructure of materials. Details of nanoindentation technique are introduced in this section.

Indentation technique has been conventionally utilized to evaluate the hardness of materials, in which a geometrically defined stylus made of a hard material such as diamond is indented onto the surface of a specimen. Meyer hardness is conventionally evaluated by dividing indentation load by the projected area of a residual impression on the surface. In nanoindentation defined as depth sensing indentation, mechanical properties are extracted from indentation loading–unloading curves obtained from continuous correlation between indentation load and the penetration depth of the stylus from the surface of a specimen. In comparison with the conventional indentation, the advantages of nanoindentation technique are listed as follows:

1. Measurement of the size of a residual impression on a surface is not required.
2. Preparation of a specimen dimensionally and geometrically defined well is not required.
3. Mechanical properties related with texture or microstructure are locally evaluated.
4. Highly concentrated stress field under an indentation brings plastic deformation at room temperature even for brittle materials such as ceramics.

The third advantage listed above realizes to evaluate mechanical properties related with the nanostructure of materials, because the indentations can be made just at the place where the mechanical properties related with texture or microstructure are to be obtained. The second one shows another advantage of nanoindentation that realizes the evaluation of the mechanical properties of thin films and MEMS devices. Moreover, advantage to evaluate mechanical properties with high accuracy is given due to the first characteristic of nanoindentation listed above, because experimental errors in conventional indentation technique are usually recognized in the measurement of the size of an extremely small residual impression. A sharp tip of a conical or pyramidal indenter is required to obtain load–unload curves in nanoindentation, because the effect of the roundness of an indenter tip on the curves cannot be ignored for very shallow indentations. For this reason, a Berkovich type indenter is more often chosen in nanoindentation technique in comparison with a Vickers indenter.

Meyer hardness H_M and reduced modulus E_r are driven from indentation loading and unloading curves respectively, as follows [1];

$$H_M = \frac{P_{max}}{gh_c^2} \tag{6.2.16}$$

$$E_r = \frac{S}{2}\sqrt{\frac{\pi}{gh_c^2}} \tag{6.2.17}$$

where P_{max} is the maximum indentation load and, g is the geometrical factor of an indenter defined by use of contact depth h_c and projective contact area A_c in the case of point-sharp indenters (Fig. 6.2.20) as follows;

$$g \equiv \frac{A_c}{h_c^{\,2}} \tag{6.2.18}$$

The Vickers and Berkovich type indenters utilized conventionally are geometrically designed to have a g-value of 24.5. In addition, S is the slope of an unloading curve at the maximum load (Fig. 6.2.21). According to the contact mechanics, the h_c-value is given as follows [2] (Fig. 6.2.20);

$$h_c = h_{max} - \varepsilon\frac{P_{max}}{S} \tag{6.2.19}$$

where ε is a constant, which is set to be 0.75 for the Vickers and Berkovich indenters [1]. Reduced Modulus

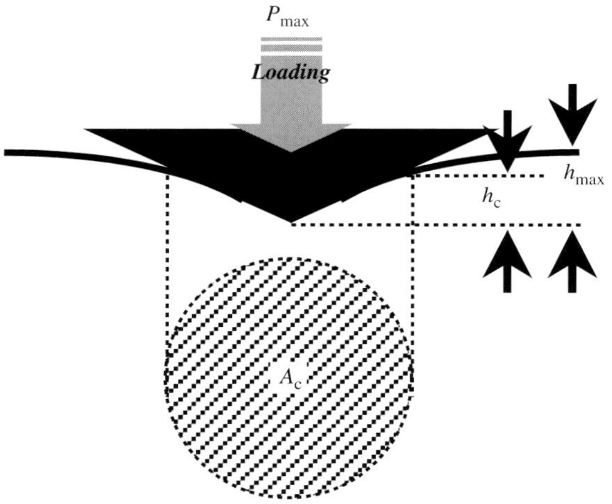

Figure 6.2.20
Surface deformation during indentation at the maximum load.

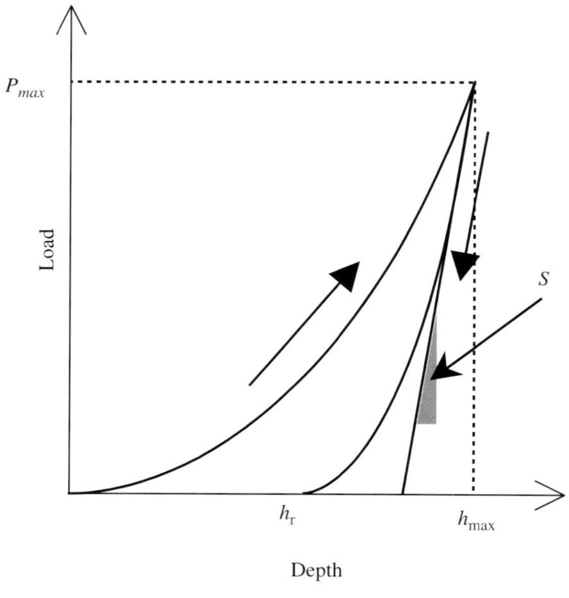

Figure 6.2.21
Loading–unloading curves.

E_r has a relationship with the Young's modulus E and Poisson's ratio of a specimen as follows;

$$\frac{1}{E_r} = \frac{1-\nu^2}{E} + \frac{1-\nu_i^2}{E_i} \tag{6.2.20}$$

where E_i and ν_i are the Young's modulus and Poisson's ratio of an indenter respectively.

In this section, conventional analysis for indentation loading and unloading curves to evaluate the Meyer hardness and reduced modulus of a specimen is introduced. According to the recent researches on nanoindentation techniques, there are several problems in the conventional analysis to evaluate mechanical properties accurately. The effect of irreversible surface deformation on indentation loading and unloading curves should be taken into account [3,4]. Furthermore, inverse analysis of the loading and unloading curves [5,6] and analysis of indentation on thin films on a substrate have progressively been studied.

References

[1] W.C. Oliver, G.M. Pharr: *J. Mater. Res.*, **7**, 1564–1583 (1992).
[2] I.N. Sneddon: *Int. J. Eng. Sci.*, **3**, 47–57 (1965).
[3] G.M. Pharr, A. Bolshakov: *J. Mater. Res.*, **17**, 2660–2671 (2002).
[4] M. Sakai, T. Akatsu and S. Numata: *Acta Mater.*, **52**, 2359–2364 (2004).
[5] Y.-T. Cheng, C.-M. Cheng: *Int. J. Solids Struct.*, **36**, 1231–1243 (1999).
[6] Y.-T. Cheng, Z. Li: *J. Mater. Res.*, **15**, 2830–2835 (2000).

6.3 Thermophysical properties

Reliable thermophysical property values of thin films are important to develop advanced industrial technologies such as highly integrated electric devices, optical disks, magneto-optical disks and thermoelectric devices. In order to meet these requirements, the National Metrology Institute of Japan of the National Institute of Advanced Industrial Science and Technology (NMIJ/AIST) has developed thermoreflectance methods by picosecond pulse heating and nanosecond pulse heating under the same geometrical configuration to the laser flash method which is the standard method to measure thermal diffusivity of bulk materials [1–4]. These light pulse heating methods observe one-dimensional heat diffusion across well-defined length of the specimen thickness. Since the geometry is very simple, thermal diffusivity can be determined reliably with uncertainty evaluation based on Guide to the Expression of Uncertainty in Measurement (GUM) [5–7].

6.3.1 Thermophysical properties related to transfer and storage of heat

Thermal conductivity λ, thermal diffusivity α, specific heat capacity c and thermal effusivity b, which are thermophysical properties related to transfer and storage of heat, are defined as follows.

6.3.1.1 Thermal conductivity

The efficiency of heat transfer by conduction is represented by thermal conductivity, which is defined by the ratio of density of heat flow transmitted through the material to the temperature gradient in the material. The SI unit of thermal conductivity is $\mathrm{W m^{-1} K^{-1}}$.

6.3.1.2 Specific heat capacity

The amount of heat required to raise unit temperature for unit mass of the material by unit temperature. The SI unit of specific heat capacity is $\mathrm{J kg^{-1} K^{-1}}$.

6.3.1.3 Thermal diffusivity

When temperature of material is raised temporarily by local heating for a short time, the heat spreads with time and the temperature will be eventually uniform over the material if it is adiabatic to the environment. Such a time-dependent heat transfer is represented by thermal diffusivity. The SI unit of thermal diffusivity is $\mathrm{m^2 s^{-1}}$. Thermal diffusivity α is related to thermal conductivity λ, specific heat capacity c, density ρ by the following equation.

$$\alpha = \frac{\lambda}{(c\rho)} \tag{6.3.1}$$

If direct measurement of thermal conductivity is difficult, thermal conductivity can be calculated from thermal diffusivity, specific heat capacity and density based on the above equation.

6.3.1.4 Thermal effusivity

The capability to absorb heat when surface of material is heated is the thermal effusivity. The temperature rise of the heated face is smaller for larger thermal effusive material like iron. On the contrary, that is larger for smaller thermal effusive material like cotton. The SI unit of a thermal effusivity is $\mathrm{J/(m^2 s^{1/2} K)}$ and is related to thermal conductivity λ, specific heat capacity c and density ρ by the following equation.

$$b = \sqrt{\lambda c \rho} \tag{6.3.2}$$

Generally, the transient heat transfer can be represented by thermal diffusivity and thermal effusivity. If arbitrary two quantity selected from thermal conductivity, volume heat capacity (=specific heat capacity × density), thermal diffusivity or thermal effusivity are fixed for a specified material, the other two quantities are also determined.

6.3.1.5 Laser flash method

When thermal diffusivity is measured with the laser flash method, front face of a planar specimen kept at constant temperature is pulsewise heated uniformly as shown in Fig. 6.3.1. Heat diffuses one-dimensionally from the heated face to the opposite face, and temperature of the whole specimen finally becomes

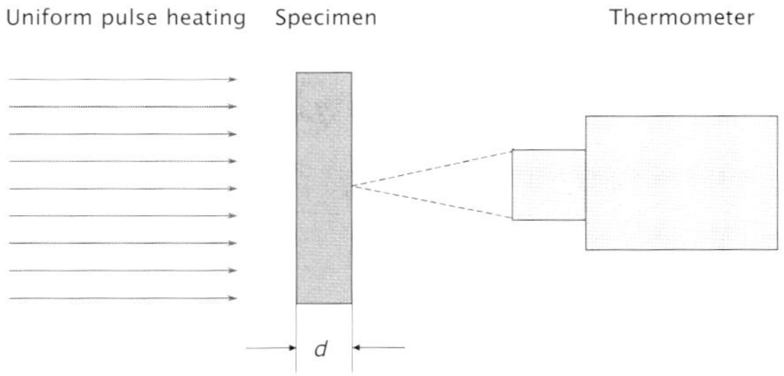

Figure 6.3.1
Principle of the laser flash method.

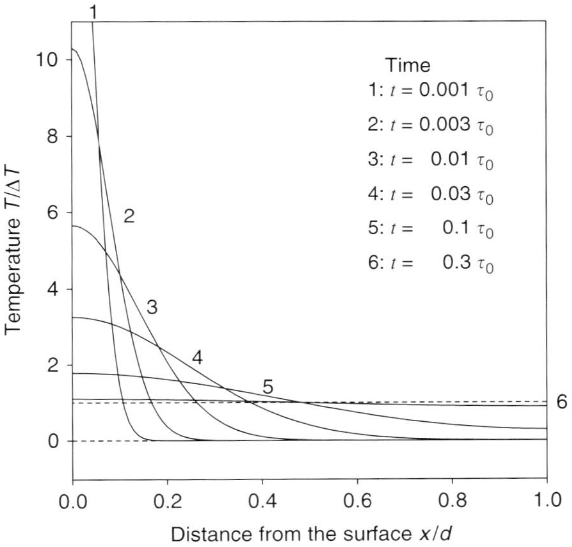

Figure 6.3.2
Change of temperature distribution inside a planar specimen of thickness d after pulsewise heating uniformly on the front face with a parameter of nondimensional time (=Fourier number, normalized by the characteristic time, τ_0).

uniform. Fig. 6.3.2 shows change of temperature distribution inside a planar specimen of thickness d after pulsewise heated uniformly on the front face with a parameter of nondimensional time (=Fourier number, normalized by the characteristic time τ_0). Fig. 6.3.3 shows temperature changes at the heated face and the opposite face of the specimen as functions of the nondimensional time [1, 5].

Since temperature of the heated face decreases quickly for materials of large thermal diffusivity and slowly for materials of small thermal diffusivity, the thermal diffusivity can be calculated from

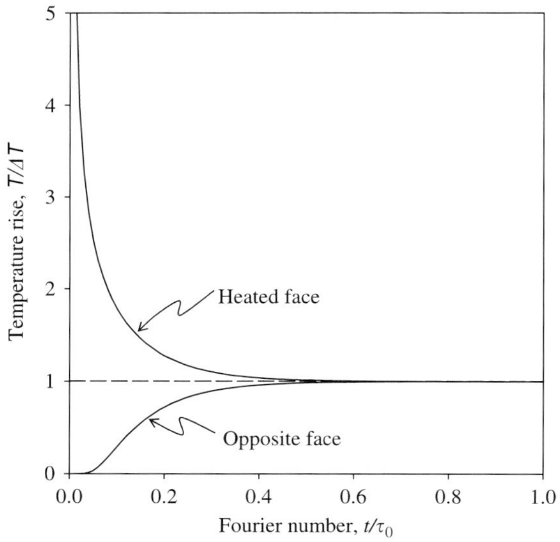

Figure 6.3.3
Temperature changes at the front face and rear face of a planar specimen as a function of the nondimensional time after uniform pulse heating.

the temperature change of the heated face. And since the temperature of the rear face opposite to the heated face increases quickly for materials of large thermal diffusivity and slowly for materials of small thermal diffusivity, the thermal diffusivity can also be calculated from the temperature change of the rear face. Because speed of a temperature change of specimen rear face is in proportion to a thermal diffusivity and is in inverse proportion to the square of thickness of specimen, a thermal diffusivity is calculated by thickness of specimen and heat diffusion time [1,5].

The following conditions are assumed for the ideal measurement:

1. Duration of the laser pulse is negligibly short compared with the heat diffusion time.
2. Specimen is adiabatic to the environment.
3. Specimen front face is heated uniformly.
4. Temperature change of specimen rear face is measured precisely.
5. Specimen is dense, uniform and opaque.
6. Change of thermal diffusivity by temperature rise of specimen after pulse heating is negligibly small.

Under the assumptions mentioned above, when front face of a planar specimen of thermal diffusivity α, specific heat capacity c, density ρ, thickness d is heated pulsewise with uniform energy density; temperature

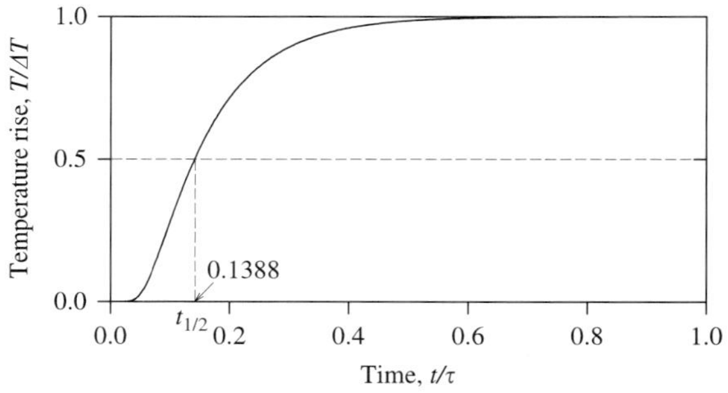

Figure 6.3.4
Half time method to calculate thermal diffusivity from the temperature change of specimen rear face after pulse heating.

change of the specimen rear face is expressed by the following equation:

$$T(t) = \Delta T\left[1 + 2\sum_{n=1}^{\infty}(-1)^n \exp\left(-(\pi n)^2 \frac{t}{\tau_0}\right)\right] \qquad (6.3.3)$$

where $\Delta T = Q/C$, Q the total energy absorbed by the specimen, C the heat capacity of the specimen, τ_0 the characteristic time of heat diffusion across the specimen. Graph of equation (6.3.3) is shown in Fig. 6.3.4.

When 0.1388 times of the characteristic time τ_0 passed after pulse heating, specimen rear face temperature reached half of the maximum temperature rise. The characteristic time is determined by fitting a theoretical curve to the rear face transient temperature curve, and the thermal diffusivity is calculated. Conventional standard data analysis algorithm is the half time method where the time reaching half of the maximum temperature rise value is determined from the measured curve, as shown in Fig. 6.3.4. The thermal diffusivity value is calculated from the following equation:

$$\alpha = \frac{0.1388 d^2}{t_{1/2}} \qquad (6.3.4)$$

Nd-YAG laser, Nd glass laser and ruby laser are commonly used for pulse heating light source. Pulse duration of light emission is from 100 μs to 1 ms, and energy of one pulse is 1–10 J.

6.3.2 Front-face heating/front-face detection picosecond thermoreflectance method

It was not easy by conventional measurement technique to determine thermal conductivity and thermal diffusivity in the thickness direction of a thin film of a thickness less than 1 μm. In order to solve this problem, "A picosecond thermoreflectance method" to observe change of the temperature of thin film front face by heat diffusion to the inside was developed [8]. The optical reflection intensity of the temperature probe pulse is detected by photodiode. Since reflectivity of material surface changes as a function of temperature, a change of specimen front-face temperature is observed by a change of reflected light amplitude. This temperature measurement method with a temperature change of such a reflectivity is called as thermoreflectance method. NMIJ/AIST of Japan developed a thin film thermal diffusivity measurement system based on a picosecond thermoreflectance method as shown in Fig. 6.3.5. The laser beam emitted from a mode-locked titanium sapphire laser is divided into transmitted beam and reflected beam by a quartz plate and about 90% is used for pulse heating and about 10% is used for temperature detection by thermoreflectance method. Area of a diameter of several 10 μm of thin film face is heated by the picosecond laser beam and the same position is irradiated by the probe beam. Then, the history of surface temperature is observed by the thermoreflectance method.

Since light travels 0.3 mm in 1 ps. By adjusting distance to specimen after it was divided, the time difference that a heating light and a probe light arrive at a specimen front face can be controlled. Response time of thermoreflectance method is incomparably faster than thermocouples, resistance temperature sensor or radiation thermometers. According to the pump probe method, ultra fast thermometry is possible only

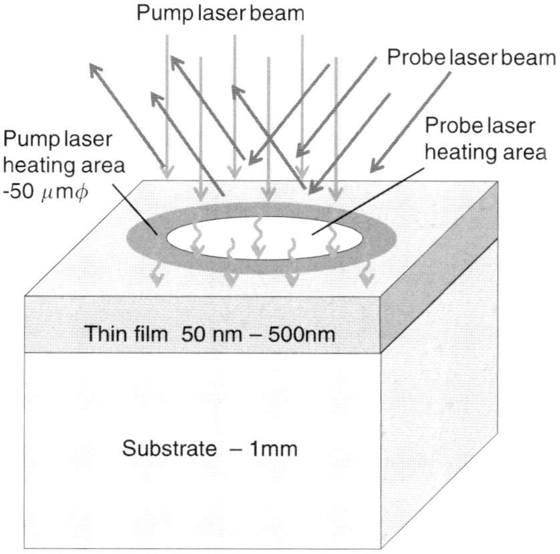

Figure 6.3.5
Specimen and heating/temperature measurement geometry of front-face heating/front-face detection picosecond thermoreflectance method.

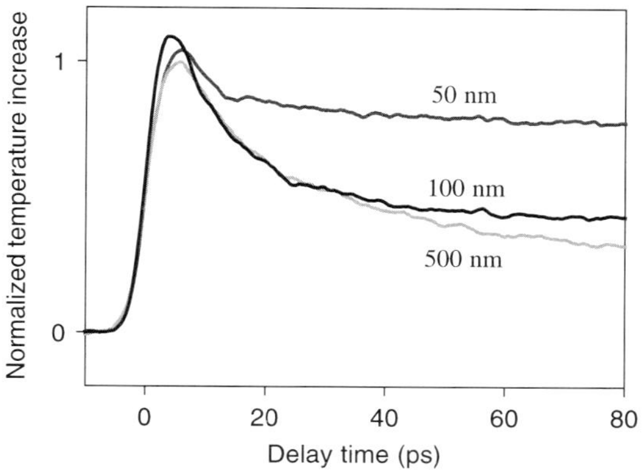

Figure 6.3.6
Thermoreflectance signals of three different thicknesses of aluminum thin films synthesized on glass substrates.

limited by time duration of the light pulses. On the other hand, it is a weak point that sensibility of temperature detection is low.

Fig. 6.3.6 shows the results of front-face temperature change by a picosecond thermoreflectance method about three aluminum thin films of different thickness synthesized on glass substrates [9]. For the specimen of 500 nm thickness, the heat has not arrived at the substrate within 120 ps after pulse heating and the surface temperature changes corresponding to internal heat diffusion of the aluminum thin film. On the other hand, when the film thickness is 100 nm, the temperature decrease after pulse heating is suppressed and deviates from a temperature change of film thickness 500 nm in 30 ps as shown in Fig. 6.3.6. Because the thermal effusivity of glass substrate is much smaller than thermal effusivity of aluminum thin film, the heat effusion into the substrate is suppressed when the heat arrives at the interface between the thin film and the substrate. For the specimen of 50 nm thickness, since the heat effuses to the substrate just after pulse heating, the temperature change only for the thin film inside cannot be observed as shown in Fig. 6.3.6.

6.3.3 Picosecond thermoreflectance method by rear face heating/front-face detection

According to the picosecond thermoreflectance method by front-face heating/front-face detection stated in the above section, a thermal diffusivity is calculated by a cooling rate after picosecond pulse heating which changes depending on a thermal diffusivity of the thin film and the penetration depth of the laser beams. Since the cooling rate of the observed temperature history changes the condition of thin film surface sensitively, it has been difficult to obtain reproducible and reliable thermal diffusivity value of the thin film by this method [10].

AIST/NMIJ of JAPAN has developed ultra fast laser flash methods which are evolution of the conventional laser flash method and the picosecond laser flash method [3,4]. As shown in Fig. 6.3.7, the ultra fast laser flash method corresponds to the picosecond thermoreflectance measurement in geometry of rear face heating/surface or front-face heating to detect a temperature change of the thin film surface where metallic thin film deposited on transparent substrate was heated from the transparent substrate side facing the heated region. Fig. 6.3.8 shows the block diagram of a measurement system. This geometry is essentially equivalent to the laser flash method which is the standard measurement method to measure thermal diffusivity of bulk material. Thermal diffusivity of the film can be calculated with small uncertainty from the thickness of a thin film and the heat diffusion time across the thin film.

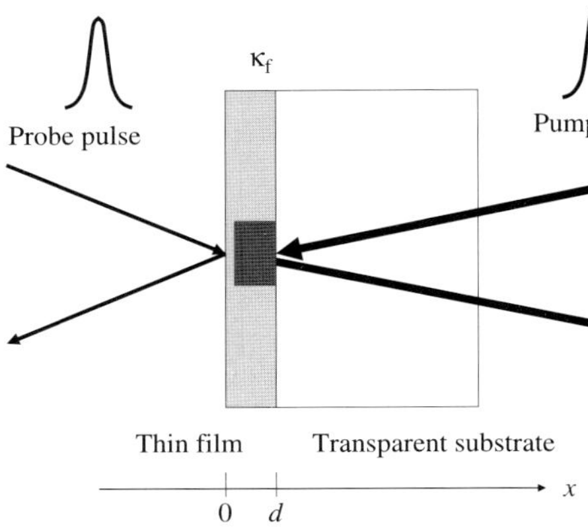

Figure 6.3.7
Principle of rear face heating/front-face detection picosecond thermoreflectance method.

The temperature probe beam is focused to a spot diameter of about 50 μm at the specimen front face just opposite to the heating light focus position. Reflected light intensity of the temperature probe light is in proportion to a change of specimen front-face temperature, and change of the reflected light intensity is detected by photodiode. Temperature coefficient of reflectivity of normal metal such as aluminum is as small as 10^{-5}/K. Since temperature rise of the specimen front face after picosecond pulse heating is smaller than a few degrees, sinusoidal component corresponding to a temperature change is small in comparison with offset level. Such a small signal can be measured by lock-in detection at modulation frequency of heating light by an acoustic–optic modulator.

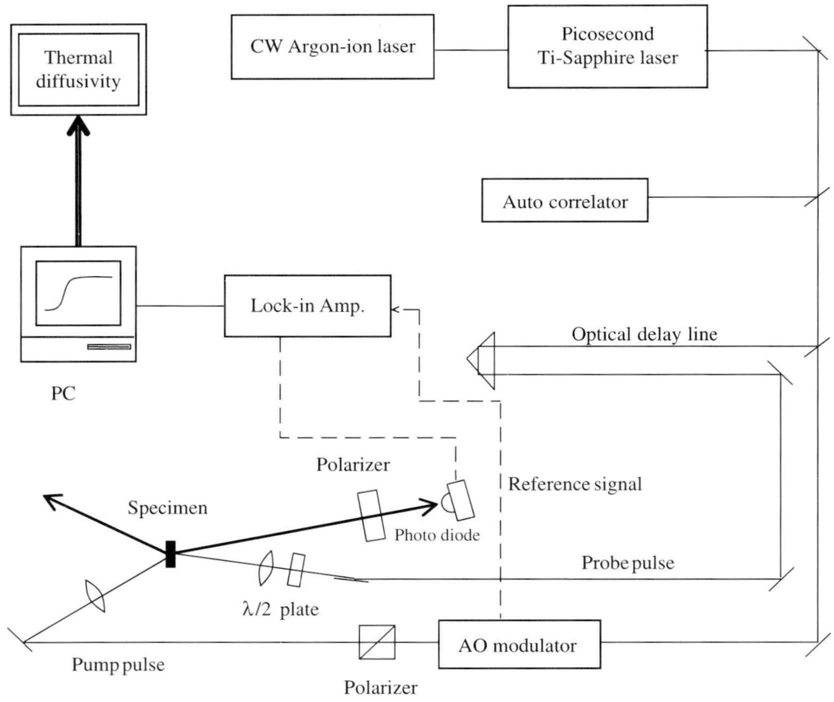

Figure 6.3.8
Block diagram of the rear face heating/front-face detection picosecond thermoreflectance system.

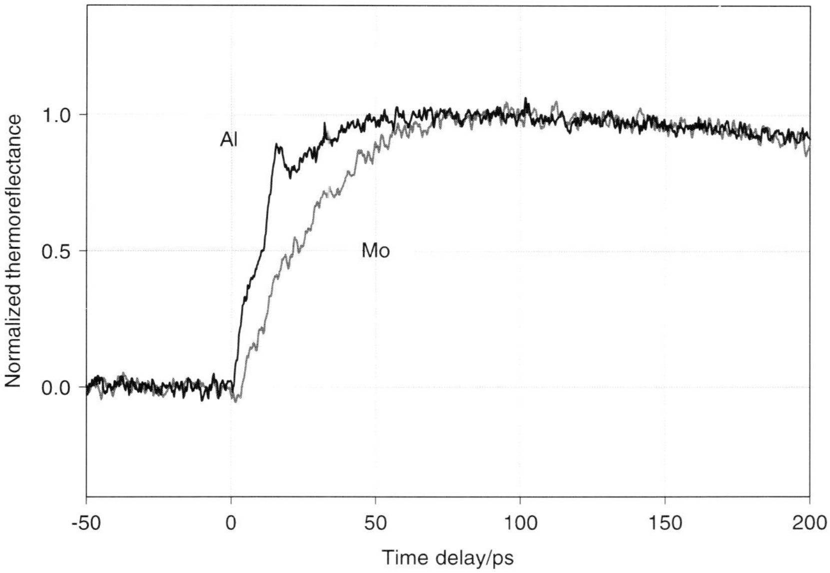

Figure 6.3.9
Thermoreflectance signals of aluminum thin film and molybdenum thin film of 100 nm thickness by rear face heating/front-face detection picosecond thermoreflectance method.

Two types of metallic thin films, an aluminum single-layered thin film of 100 nm thickness and a molybdenum single-layered thin film of 100 nm, both of which were synthesized on a Pyrex glass substrate by magnetron DC sputtering method, where the film thicknesses are of nominal values. Fig. 6.3.9 shows the temperature history curves measured by the ultra fast laser flash method.

Since the shapes of observed thermoreflectance signals are similar to those observed by the laser flash method for bulk specimen, the heat energy transport of these metallic thin films of a thickness around 100 nm at room temperature in a time scale of order of several 100 ps can be expressed by the classic heat diffusion equation.

6.3.4 Nanosecond thermoreflectance method

Initially, the electrical delay method was developed to expand the observation time of the picosecond thermoreflectance method longer for measurements of thicker films [11]. Since the pulse duration can be longer than picosecond and the repetition period of pulses is flexible, nanosecond pulse laser can be used for the ultra fast laser flash method [12]. Fig. 6.3.10 shows the block diagram of the nanosecond thermoreflectance method. The pulse duration was changed from 2 ps to 2 ns and the pulse repetition period was changed from 76 MHz (13.2 pump specific speed interval) to 50 kHz (20 μs interval). This system can measure thermal diffusivity of thin films with thickness up to several micrometers. Pulse duration is 2 ns and pulse interval is 20 μs with intensity modulation of 1 kHz by a modulator (AOM).

Irradiate it from the upper part of Fig. 6.3.10 to substrate side of specimen set in specimen measurement unit. As shown in Fig. 6.3.10, the specimen is irradiated from bottom and the reflected light is detected by a differential detector from the top.

A molybdenum thin film with thickness of 3 μm synthesized on a glass substrate by sputtering deposition was measured as shown in Fig. 6.3.11.

6.3.5 Thin film thermophysical property reference material and traceability

The high-speed laser flash method in the extended definition mentioned above can be generally used and is popular after commercial instruments are available and traceability to the national standard and/or the international standard is established. The national standard of the laser flash method was established in 2002 by NMIJ/AIST. Uncertainty of thermal diffusivity standard by the laser flash method was evaluated based on "Guide to the Expression of Uncertainty in Measurement, GUM" [7] and a quality system corresponding to ISO 17025 was constructed [2]. The standard value of thermal diffusivity of the graphite was determined with uncertainty evaluation based on GUM and supplied as a reference material in 2005.

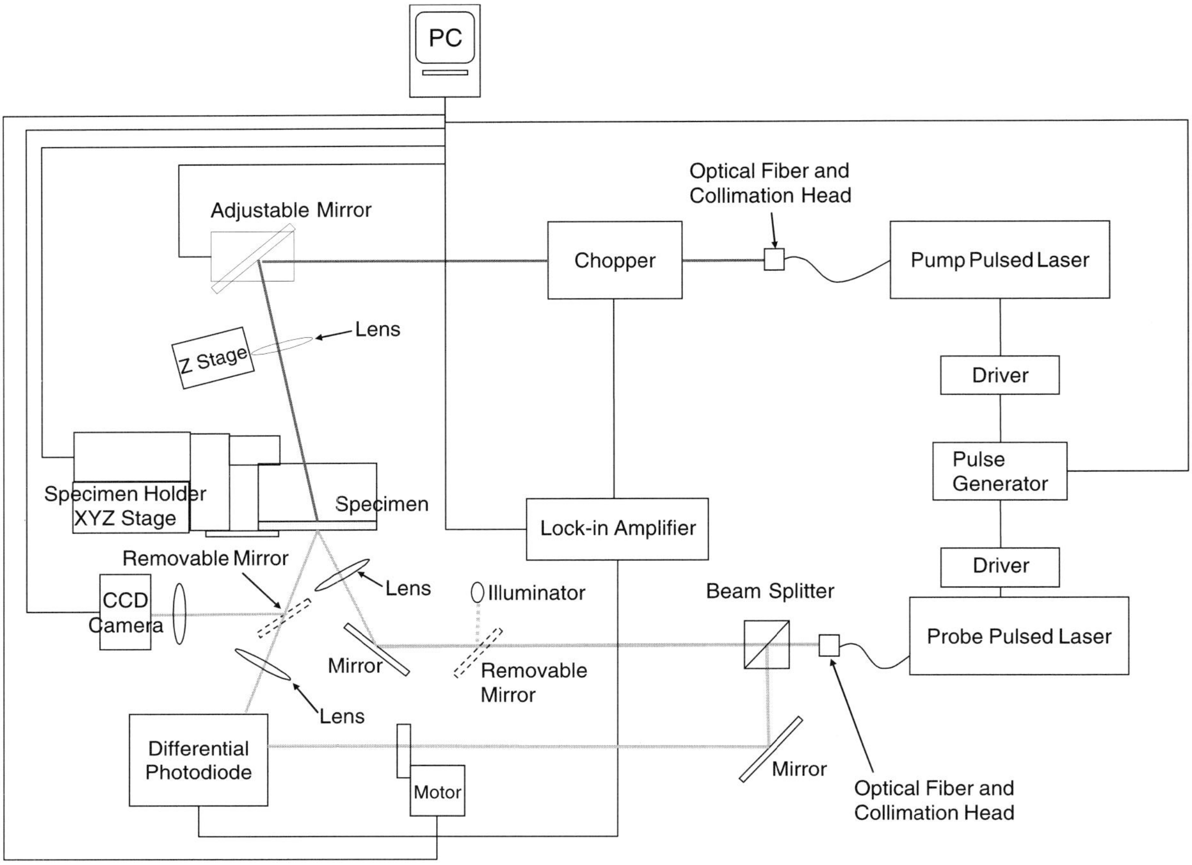

Figure 6.3.10
Block diagram of the nanosecond thermoreflectance system.

NMIJ/AIST is also developing the standard of thin film thermophysical property by the high-speed laser flash method under the same scheme as shown in Table 6.3.1. The national standard for picosecond thermoreflectance method was established in 2005. The national standard and thin film reference material are to be developed and supplied in 2007 for nanosecond thermoreflectance method.

6.3.6 Summary

Thermoreflectance methods by picosecond and nanosecond light pulse heating have been developed under the same geometrical configuration as the laser flash method which is the standard method to measure thermal diffusivity of bulk materials. Thermal diffusivity values across thin films were measured with small uncertainty. These high-speed laser flash methods observe heat diffusion time across well-defined length of the specimen thickness under

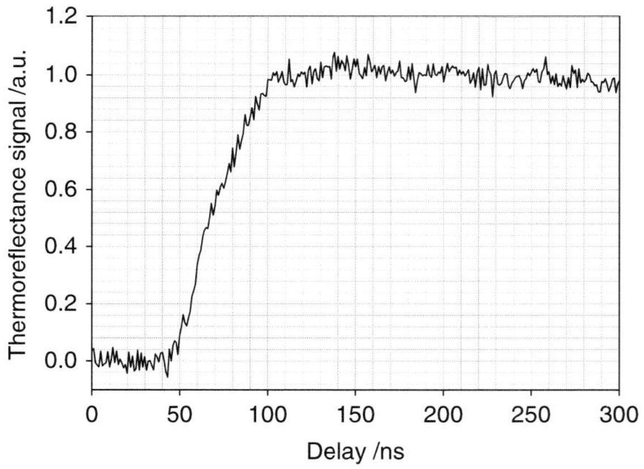

Figure 6.3.11
Thermoreflectance signal of molybdenum thin film of 3 μm thickness sputtered on a glass substrate measured with the nanosecond thermoreflectance system.

Table 6.3.1
Program to develop thermal diffusivity standards by light pulse heating methods planned by NMIJ/AIST.

Light pulse heating methods	Observable heat diffusion time	Planned year to start service	
		Metrological standard	Reference material
Laser flash method			
Conventional	1 ms 10 s	FY2002	FY2005
Application to thin plates and coatings	1 μs 10 ms	Under consideration	Under consideration
Nanosecond thermoreflectance method	10 ns 100 μs	FY2007	FY2007
Picosecond thermoreflectance method			
Electrical delay	100 ps 10 ns	FY2005	Under consideration
Optical delay	10 ps 1 ns		

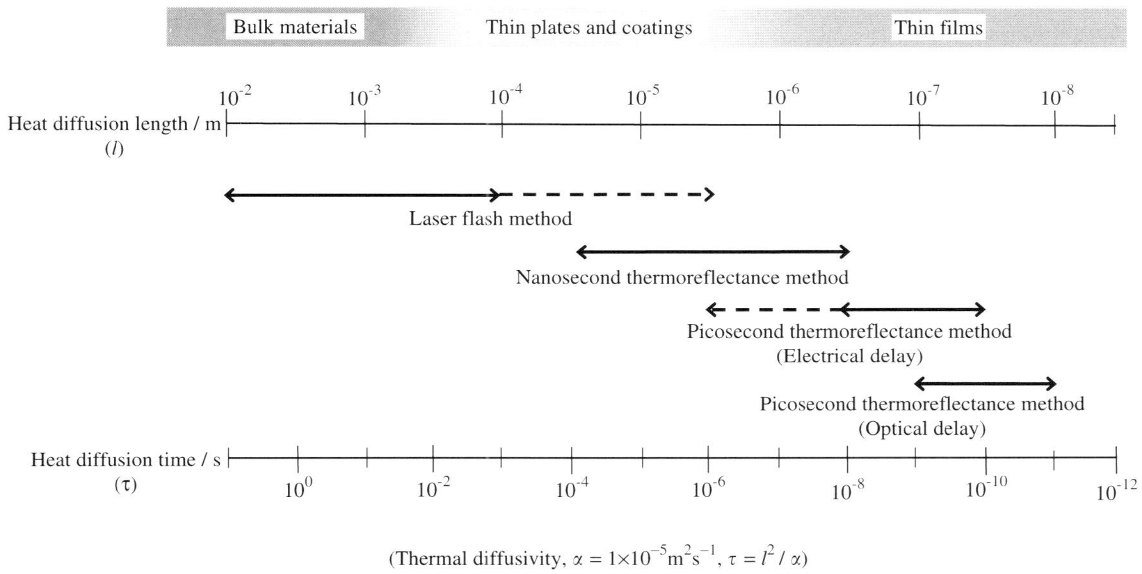

Figure 6.3.12
Observable heat diffusion time and observable thickness of the specimen covered by four types of light pulse heating methods when the thermal diffusivity is 10^{-5} m^2 s^{-1}.

one-dimensional heat flow as shown in Fig. 6.3.12. Since the geometry is very simple, thermal diffusivity can be determined reliably with uncertainty evaluation based on GUM [7]. NMIJ/AIST has been engaged in research for establishment of national standard, international standard for thermal diffusivity measurements of thin films.

Using these light pulse heating methods, thermal diffusivity of each layer of multilayered thin films and boundary thermal resistance between the layers can be determined from the observed transient temperature curves based on the response function method [5,6,13,14].

Boundary thermal resistance between thin films and thermal diffusivity of each layer of multilayer films constituting next generation light heating phase change record disc is measured by the picosecond thermoreflectance method and the nanosecond thermoreflectance method. Reliability of thermal design is improved drastically by heat-transfer simulation using those thermophysical property data [15,16]. High-speed laser flash methods are also applied to thermophysical property measurement of transparent conductive films used for a flat panel display [17] and hard coating films [18].

References

[1] T. Baba, A. Ono: Improvement of the laser flash method to reduce uncertainty in thermal diffusivity measurements, *Meas. Sci. Technol.*, **12**, 2046–2057 (2001).

[2] M. Akoshima, T. Baba: Thermal diffusivity measurements of candidate reference materials by the laser flash method, *Int. J. Thermophys.*, **26**, 151–163 (2004).
[3] N. Taketoshi, T. Baba and A. Ono: Observation of heat diffusion across submicrometer metal thin films using a picosecond thermoreflectance technique, *Jpn. J. Appl. Phys.*, **38**, L1268–L1271 (1999).
[4] N. Taketoshi, T. Baba and A. Ono: Development of a thermal diffusivity measurement system for metal thin films using a picosecond thermoreflectance technique, *Meas. Sci. Technol.* **12**, 2064–2073 (2001).
[5] T. Baba: *Thermophysical property measurement by light pulse heating*, Progress in Heat Transfer, New Series Vol. 3, Yokendo, Tokyo (2000).
[6] T. Baba: General needs on nanoscale thermal metrology and the Japanese program on this subject, Proceedings of Therminic Workshop 2004 Sophia Antipolis France.
[7] BIPM, IEC, IFCC, ISO, IUPAP, and OIML, Guide to the Expression of Uncertainty in Measurement (ISO, 1995).
[8] C.A. Paddock, G.L. Eesley: Transient thermoreflectance from thin metal films, *J. Appl. Phys.*, **60** 285–290 (1986).
[9] N. Taketoshi, T. Baba and A. Ono: Picosecond thermoreflectance measurements of thermal diffusion in film/substrate two-layer systems, *Therm Cond.*, **24**, 289–302 (1999).
[10] N. Taketoshi, T. Baba and A. Ono: Development of a thermal diffusivity measurement system with a picosecond thermoreflectance technique, *High Temp. High Press.*, **29** 59–66 (1996).
[11] N. Taketoshi, T. Baba and A. Ono: Electric delay technique in a picosecond thermoreflectance method for thermophysical property measurements of thin films, *Rev. Sci. Instrum.*, **76** 1–8 (2006).
[12] T. Baba, N. Taketoshi, K. Hatori, K. Shinzato, T. Yagi, Y. Sato and Y. Shigesato: Development of a High Speed Laser Flash System to Measure Thermophysical Property of Thin Films – Nanosecond Optical Pulse Heating Thermoreflectance Method, Proc. 25th Jpn. Symp. Thermophys. Prop., 2004, Nagano, pp. 240–242.
[13] T. Baba: Introduction of a response function method for analysis of functionally gradient materials, *Jpn. J. Thermophys. Prop.*, **7**, 14–19 (1993).
[14] T. Baba, N. Taketoshi: Analysis of thermal diffusion in multi-layer thin films by a response function method, Eurotherm 57 “Microscale Heat Transfer” (Edizioni Ets, 1998), pp. 285–292.
[15] K. Ichihara, K. Todori, T. Nakai, K. Yusu, S. Ashida, S. Tatsuta, N. Taketoshi and T. Baba: Proceedings of First International Symposium on Standard Materials and Metrology for Nanotechnology (SMAM-1) (2004), Tokyo.
[16] S. Ashida: Measurement of thermal properties of nano-thin films by pico-second thermo-reflectance method and its application to the design of next-generation DVD, Proceedings of the 25th Japanese Symposium on Thermophysical Properties (2004), Nagano, pp. 65–67.
[17] K. Tamano, T. Yagi, Y. Sato, Y. Shigesato, N. Taketoshi and T. Baba: Proceedings of the 25th Japanese Symposium on Thermophysical Properties (2004), Nagano, pp. 246–248.
[18] T. Ohtsuka, T. Yagi, N. Taketoshi, T. Baba, A. Miyamura, Y. Sato and Y. Shigesato: Thermal diffusivity measurement of TiN_x thin films using a thermoreflectance technique, Proceedings of the 25th Japanese Symposium on Thermophysical Properties, 2006, Kyoto, pp. 167–169.

6.4 Electric properties

6.4.1 Dielectric properties

Recent miniaturization and weight saving of the electronic devices requested by society have been accelerated, and the internal constitution of the devices is of the order of several-to-tens nanometer scale. Among the electronic materials, ferroelectrics have been used in a wide variety of applications such as multilayer ceramics condensers and piezoelectric devices [1]. The properties of the ferroelectrics depend strongly not only on microstructure composed of grains and grain boundaries but also on the domain structures formed in the grains and their dynamics induced by applying an electric field [2,3]. The domains in ferroelectrics are defined by the volume in which spontaneous polarization (P_s) is aligned in three dimensions, and the size of the ferroelectric domains are in the range from several nanometers to micrometers [4]. The dielectric properties such as dielectric permittivity and tangent loss are related to the physical properties determined by the domain structure and the mobility of the domain walls as well as ionic polarization. Additionally, a hysteresis behavior observed for the polarization as a function of electric field is a nonlinear phenomenon induced by the switching of the ferroelectric domains. Establishing ferroelectric devices in the order of nanoscale and controlling the device performance require an advanced technology that enables us to estimate local dielectric and ferroelectric properties.

In this chapter, the ferroelectric domains and recent characterization technology of the local ferroelectric properties are briefly explained. At first, the general information on dielectric permittivity and piezoelectric constant in perovskite-type ferroelectrics are provided, and then the domain structures of lead titanate ($PbTiO_3$) crystals are shown to be visualized on the nanometer scale by piezoresponse force microscope (PFM) which belongs to the family of scanning force microscope (SPM) [5]. Here, the theory of dielectrics/ferroelectrics and the methods for observing fundamental physical properties cannot be described due to limitations of space. The details have been published in the great book [6–8] and the reviews [9,10].

6.4.1.1 Crystal structure of perovskite-type ferroelectrics

Fig. 6.4.1 shows the crystal structure of $BaTiO_3$ with perovskite structure. In the crystal structure, divalent Ba (Ba^{2+}) occupies the A site, and tetravalent Ti (Ti^{4+}) is positioned at the B site. Above the Curie temperature (T_C of 135°C for $BaTiO_3$), ferroelectrics have a center of symmetry and do not exhibit a ferroelectric behavior. This state is categorized as paraelectric and Ti ions are located at the center of the BO_6 octahedra. During cooling, ferroelectric phase transition occurs at the T_C and the crystals show a ferroelectric behavior. The significant structural changes induced by the phase transition are the elongation along the *c* axis and shrinkage along the *a* axis of the unit cell and the off-center displacement of the Ti ions along the *c* axis. In the ferroelectric crystals with low-symmetry tetragonal, the relative displacement between the centers of "+" and "−" charges generates dipole moment in the unit cell, and the ferroelectric crystals possess electric charge spontaneously.

6.4.1.2 Ferroelectric polarization hysteresis and the applications of ferroelectrics

In the ferroelectric hysteresis loop drawn in Fig. 6.4.2, let us consider the polarization change of the crystal with the state "1" (negatively charged) when positive electric field is applied to it. As soon as the field exceeds coercive field (E_c), the polarization state changes drastically to the positive one. Even if the electric field is turned off, the charged state is retained with the state "0" and the crystals have spontaneous positive charge. In the crystal, Ti ions are displaced downward to the "1" state, while the upward displacement of the Ti ions are seen in the "0" state. Namely, the ferroelectrics can have positive or negative charge without applying an electric field. The polarization value with zero electric field is defined as remnant polarization (P_r). In the ferroelectric crystals, the sign of the P_r can be controlled by electric field. The perovskite ferroelectrics such as $BaTiO_3$ belong to the displacive-type ferroelectric in which ferroelectricity originates from the displacement of the constituent ions from the corresponding positions in the parent cubic structure.

Ferroelectrics exhibit piezoelectricity consequently from the crystallographic point of view. When stress is applied to piezoelectrics, the crystals generate electric dipole proportional to the stress value, which is called piezoelectric effect. In the ignition devices, spark discharge by a high voltage of the piezoelectrics caused by hitting the piezoelectrics is often used. In contrast, applying a voltage enables us to obtain a thickness change (Δl) of the piezoelectrics which is proportional to electric field (inverse piezoelectric effect). The inverse piezoelectric effect has found a wide variety of applications such as actuators etc., in which ferroelectric $Pb(Zr,Ti)O_3$ is exclusively used. Recently, $(K,Na)NbO_3$ [11,12] has attracted a great deal of attention as a candidate ferroelectric for Pd-free piezoelectric materials.

6.4.1.3 Dielectric permittivity and dielectric loss

In insulating materials under electric field $\boldsymbol{E}$ (V m^{-1}), charged species respond to $\boldsymbol{E}$ and dipole moments are

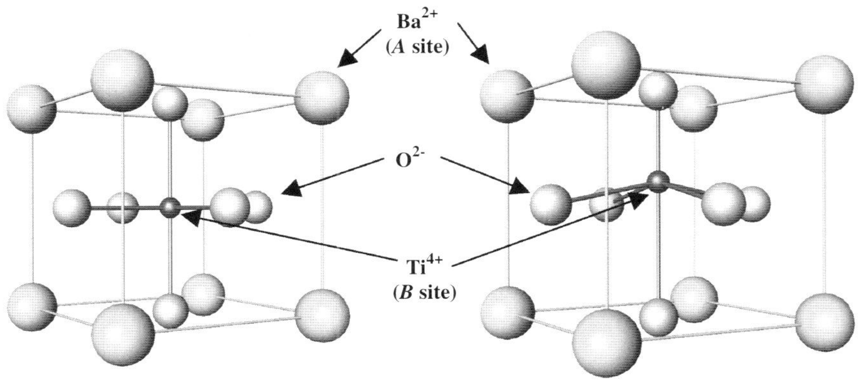

(a) Paraelectric Phase (Cubic, $Pm\bar{3}m$)

(b) Ferroelectric Phase (Tetragonal, *P4mm*)

Figure 6.4.1
Crystal structures of barium titanate ($BaTiO_3$) with perovskite structure.

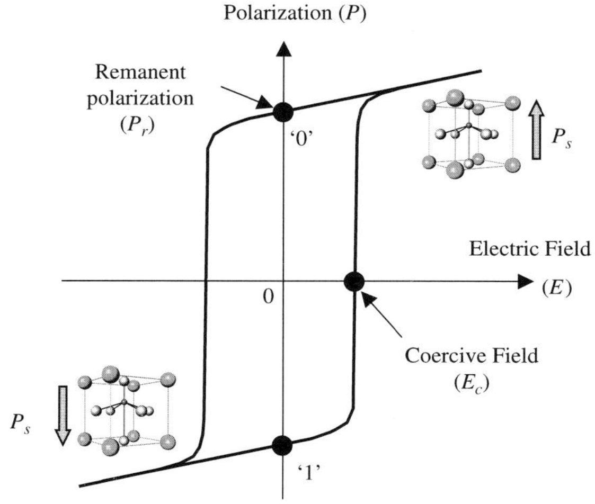

Figure 6.4.2
Polarization hysteresis loop observed for ferroelectric materials.

induced. This phenomenon is called "dielectric polarization" and polarized materials are classified into dielectrics. The charged species of point defects, ions and electrons are displaced from the neutral position by $\boldsymbol{E}$, and electric polarization $\boldsymbol{P}$ (C cm^{-2}) appears. Dielectric displacement $\boldsymbol{D}$ (C m^{-2}) is expressed by:

$$\boldsymbol{D}_i = \varepsilon_0 \cdot \varepsilon_{ij} \cdot \boldsymbol{E}_j = \varepsilon_0 \cdot \boldsymbol{E}_j + \boldsymbol{P}_i \qquad (6.4.1)$$

Where, ε_0 (scalar) is the dielectric constant of vacuum and ε_{ij} (tensor) denotes dielectric permittivity. In this equation, higher-order terms related to nonlinear dielectric behavior are ignored. Here, i and j are the directions of 1, 2, or 3, and ε_{ij} is expressed as:

$$\varepsilon_{ij} = \begin{pmatrix} \varepsilon_{11} & \varepsilon_{12} & \varepsilon_{13} \\ \varepsilon_{21} & \varepsilon_{22} & \varepsilon_{23} \\ \varepsilon_{31} & \varepsilon_{32} & \varepsilon_{33} \end{pmatrix} \qquad (6.4.2)$$

ε_{ij} is defined as the coefficient between $\boldsymbol{E}$ along the j direction and electric charge generated in the plane normal to the i direction. The crystal orientation of 1, 2 and 3 is based on the notation of the principal axis, which is different from the normal one defined for individual crystallographic symmetry. Equation (6.4.2) is simplified for the perovskite-type ferroelectrics and the case of tetragonal $BaTiO_3$ (*P4mm*) is explained here.

Paraelectric $BaTiO_3$ above the T_C belongs to cubic $Pm\bar{3}m$ and its ε_{ij} is described as

$$\varepsilon_{ij} = \begin{pmatrix} \varepsilon_{11} & 0 & 0 \\ 0 & \varepsilon_{11} & 0 \\ 0 & 0 & \varepsilon_{11} \end{pmatrix} \qquad (6.4.3)$$

where only ε_{11} is an individual parameter in the cubit phase. Below the T_C, P_s appears along the c axis, and ferroelectric $BaTiO_3$ is transformed to tetragonal *P4mm* symmetry. In the tetragonal structure, ε_{ij} is expressed by two parameters of ε_{11} normal to the P_s (along the a axis) and ε_{33} parallel to the P_s (along the c axis) as follows:

$$\varepsilon_{ij} = \begin{pmatrix} \varepsilon_{11} & 0 & 0 \\ 0 & \varepsilon_{11} & 0 \\ 0 & 0 & \varepsilon_{33} \end{pmatrix} \qquad (6.4.4)$$

The dielectric properties in a frequency range from kHz to GHz of $BaTiO_3$ are known to be governed by ionic polarization (polarization originating from the displacements of the constituent ions). Recently, the dynamics of polar nanoregions (ferroelectric nanodomains) that are classified into orientational polarization have been revealed to play an essential role in dielectric properties for perovskite ferroelectrics [10].

The electric properties of dielectrics can be explained in view of the equivalent circuit shown in Fig. 6.4.3a. Let us consider the dielectrics with a capacitance of C to which an ac voltage of $V = V_0 e^{j\omega t}$ with an angular frequency of $\omega = 2\pi f$ (f is frequency) is applied. Electrical current I passing through the dielectrics is divided into two components: one is displacement current I_C ($= j\omega CV$) with a phase difference of 90° against V, and the other is in-phase current with V due to conductance G (resistance $R = 1/G$). As can be depicted in Fig. 6.4.3b, total current I has a phase difference of δ with respect to the ideal case that energy loss is zero, i.e., $I_R = 0$ and $I = I_C$.

The energy loss of dielectrics is usually discussed in terms of dielectric dissipation factor expressed as

$$\tan\delta = \frac{|I_R|}{|I_C|} = \frac{G}{\omega C} \qquad (6.4.5)$$

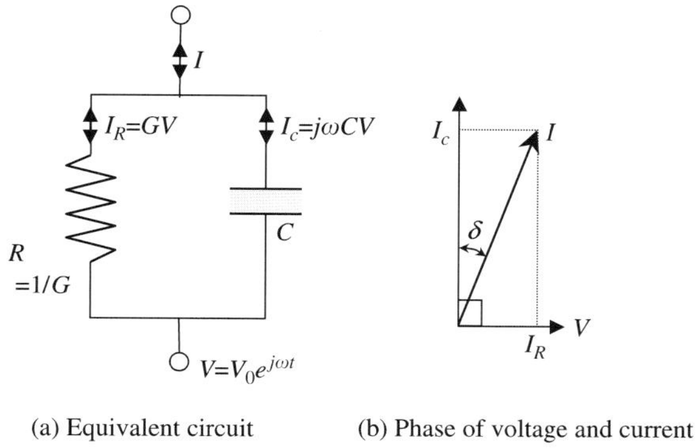

Figure 6.4.3
(a) Equivalent circuit and (b) phase of voltage and current of condensers.

An increase in I_R leads to a higher tan δ. The dielectrics with a relatively high resistance exhibit a tan δ less than 0.05. A tan δ higher than 0.5 interferes with precise measurements of I_C, i.e., ε. A tan δ higher than 1 which has often observed for high-temperature dielectric properties is due to a higher electrical conductivity (higher I_R). Except for ferroelectric materials, the main factor of R is leakage current. The dielectric loss of ferroelectrics is influenced by the movement of domain walls as well as by leakage current, as described later. A peak in the temperature dependence of tan δ near T_C is ascribed to domain wall motion.

6.4.1.4 Piezoelectric constant

In elastic bodies, an application of stress T (N m^{-2}) generates a strain S (dimensionless). In the case of piezoelectrics with a non-centrosymmetric crystal structure, T induces not only S but also polarization (P). Using piezoelectric constant d (C N^{-1}) and elastic compliance s (m^2 N^{-1}), we express the piezoelectricity (d type) as

$$D = \varepsilon^T E + dT \tag{6.4.6}$$

$$S = dE + s^E T$$

where ε^T denotes dielectric constant ($\varepsilon_0\varepsilon_{ij}$) with a constant T. Similarly, s^E shows s with a constant E. Since S and T are second-rank tensors, the coefficient s between them becomes fourth-rank tensor. The coefficient d between T and first-rank tensor D is a third-rank tensor. Here, only d is focused because the signal of PFM is directly related to d.

Considering the crystal symmetry ($P4mm$) of tetragonal $BaTiO_3$ and $PbTiO_3$, we can simplify d as follows:

$$d = \begin{bmatrix} 0 & 0 & 0 & 0 & d_{15} & 0 \\ 0 & 0 & 0 & d_{15} & 0 & 0 \\ d_{31} & d_{31} & d_{33} & 0 & 0 & 0 \end{bmatrix} \tag{6.4.7}$$

where d_{31}, d_{33} and d_{15} are the individual parameters representing piezoelectric response. For piezoelectric ceramics, the poling direction is set to be the "3" direction and the properties along the "1" and "2" axes are the same. Thus, the dielectric permittivity ε and piezoelectric d constant are described by equations (6.4.4) and (6.4.7), respectively. If $T = 0$, equation (6.4.6) is simplified as $S_1 = d_{31}E_3$, $S_2 = d_{31}E_3$, $S_3 = d_{33}E_3$, $S_4 = d_{15}E_2$, $S_5 = d_{15}E_1$. For tetragonal $BaTiO_3$ crystals the values of $d_{31} = 35$ pC N^{-1}, $d_{33} = 86$ pC N^{-1} and d_{15}=590 pC N^{-1} have been reported [13]. When an $\boldsymbol{E}$ is applied along the direction of P_s (the "3" axis), the crystals elongate along the "3" direction by S_3 (the proportionality factor is d_{33}) while the crystals shrink along the "1" and "2" axes by S_1 ($= S_2$, the proportionality factor is d_{31}, and d_{31} is a negative constant). The application of $\boldsymbol{E}$ normal to the P_s direction induces thickness-share piezoelectric vibration, leading to a share strain S_4 and S_5 ($S_4 = S_5$, the proportionality factor is d_{15}).

6.4.1.5 Domain structure and domain probing

In general, two kinds of domain walls are observed for tetragonal ferroelectrics: one is 180° domain wall across which the direction of P_s is different by 180° (antiparallel), and the other is a 90° domain wall across which the P_s vector rotates by 90°. The minimization of strain energy results in the 90° domain structure, while the minimization of electrostatic energy requires the 180° domain structure. The domain walls in the 90° and 180° domain structures have usually charge-free configurations. The domain in which the P_s vector is parallel to the direction of observation is called "c domain", while the domain with a P_s vector normal to it is "a domain". For PFM observations, the c domains are visualized by the out-of-plane mode, while the a domains are revealed by the in-plane mode.

Here, let us consider the piezoelectric vibration induced by applying an E. For longitudinal vibration, the domains with a P_s parallel to E elongate in proportion to d_{33}, and the domains with a P_s antiparallel to E shrink. When the direction of E is away from the P_s vector, E induces thickness-share vibration in proportion to d_{15} as well as longitudinal one. For the thickness-share vibration, the slip direction of the crystal surface depends on the direction of P_s.

PFM can visualize the P_s vector from the piezoresponse signal detected by the tip of the cantilever of SPM. Fig. 6.4.4 shows the schematic representation of the PFM measurement system used in this study. An ac voltage is applied between the conductive cantilever and bottom electrode and induces piezoelectric vibration just beneath the tip of the cantilever. The cantilever acts as a detector of the piezoelectric vibration as a result of the deflection or torsion of the cantilever as well as the top electrode. In theory, local dielectric constant such as d_{33} and d_{15} can be determined by PFM investigations. When E is parallel to P_s, the cantilever deflects vertically, i.e., parallel to E, due to piezoelectric longitudinal vibration (the out-of-plane mode in Fig. 6.4.4a). For the domains in which the direction of P_s is parallel to the crystal surface, i.e., E normal to P_s, thickness-share vibration induces the torsion of the cantilever (the in-plane mode in Fig. 6.4.4b). The combined observations of the out-of-plane and in-plane PFM images enable us to obtain a three-dimensional domain structure near the crystal surface [14].

6.4.1.6 Domain structure of $PbTiO_3$ crystals observed by PFM

Fig. 6.4.5 represents the example of the domain structure of $PbTiO_3$ crystals observed by PFM [5]. In the measurements, an E is applied along the a or c axis. Comparative investigations of the amplitude (A)

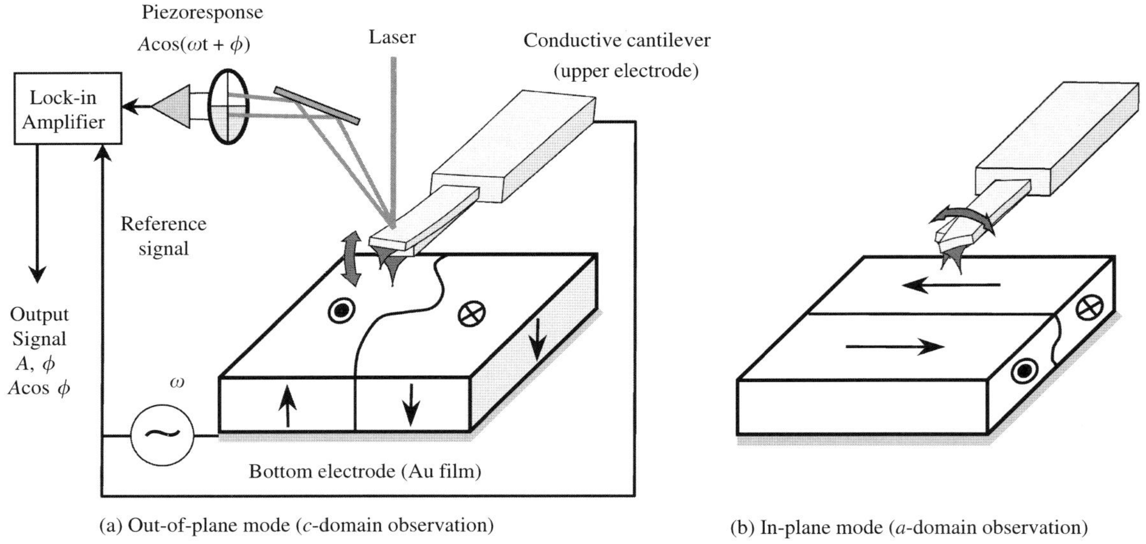

Figure 6.4.4
Schematic view of piezoresponse force microscope: (a) out-of-plane P_s for c-domain observations, (b) in-plane P_s for a-domain observations. Conductive cantilever acts as an upper electrode. Piezoresponse is formed from the piezoelectric displacement of the tip of the cantilever such as (a) deflection and (b) torsion.

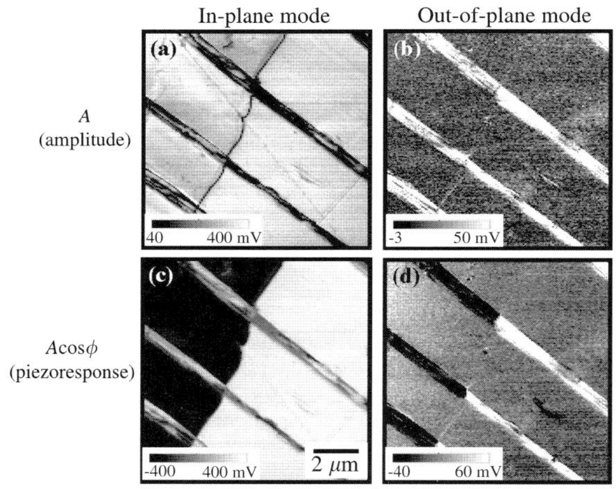

Figure 6.4.5
Piezoresponse ($A\cos\phi$) images of $PbTiO_3$ surfaces. The upper figures are the amplitude (A) images of piezoresponse and the lower ones depict the piezoresponse ($A\cos\phi$) images. The left and right images are observed in the in-plane and out-of-plane modes, respectively. The white area has a higher value of A or $A\cos\phi$.

images of the out-of-plane and in-plane observations reveal that the area with a high A in the out-of-plane mode (white region) has a low A in the in-plane mode. In contrast, the domains with a high A in the in-plane mode indicate a small A in the out-of-plane mode. The images of $A\cos\phi$ clearly show the inverse of the $A\cos\phi$ signals for the right- and left-areas due to a difference in ϕ by 180°.

Fig. 6.4.6 depicts the schematic domain structure determined by the PFM investigation. The irregular boundary between the right- and left areas is 180° domain wall. The 180° domain wall is formed irregularly in the c–c domains, while the 180° domain wall constitutes a flat surface in the a–a domains. Although the 180° domain wall is curved in the c–c domains, 180° domain wall still is charge free. The curved 180° domain wall is observed even though the energy of domain wall is minimized for the flat surface. This result shows that the formation energy of the 180° domain wall is relatively small compared with isotropic domain wall energy and elastic energy. For the a–a domains, a charge-free plane is established only when the domain wall is parallel to the P_s vector. In the formation process of domain structures, a higher-energy domain wall with an electric charge rapidly changes to the lower-energy facetted domain wall observed in Fig. 6.4.6, and the plate-like 180° domain wall is immediately established in the a–a domains.

The tensors of dielectric permittivity and piezoelectric constant are briefly explained for tetragonal ($P4mm$) ferroelectrics such as $BaTiO_3$ with perovskite structure. The principle of PFM for observing three-dimensional domain structure is described, and the nanometer-scale domain structure of tetragonal $PbTiO_3$ crystal is shown. The domain structure of ferroelectrics can be visualized

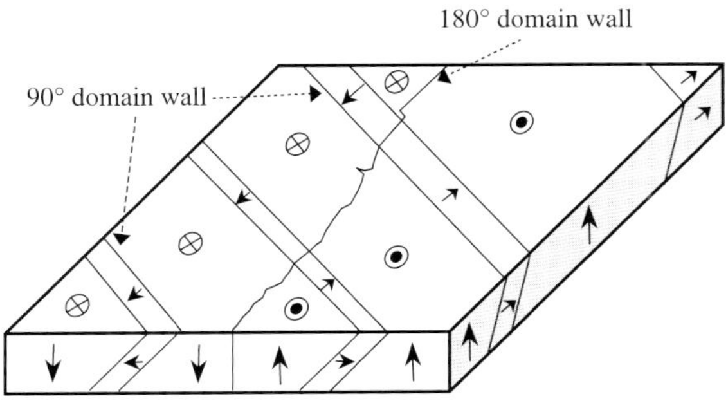

Figure 6.4.6
Schematic domain structure near the surface of the $PbTiO_3$ crystal. 90° domain walls form a flat plane due to its lowest strain energy, while 180° domain walls have a curved and complex face because of the arbitrary configuration of char-free surface.

also for tunneling acoustic microscope [15], Kelvin-force microscope [16] and nonlinear dielectric microscope [17]. Among these techniques, PFM has an advantage in view of a convenient method for revealing the direction of P_s. Recently, contact-resonance PFM with an ultrahigh sensitivity has been developed by Okino et al. [18] to visualize polar nanoregions in relaxor ferroelectrics. Nonlinear dielectric microscope [17] is expected to provide a local characterization technique with a high special resolution. The technological innovation of SPM is expected to provide a novel characterization tool for investigating local physical properties of nanostructured materials.

References

[1] M. Okuyama: *Denki Gakkai Gakujyutsu Ronbunshi E*, **121**, 537–541 (2003), in Japanese.
[2] M. Iwata, T. Ishibashi: *Kino Zairyo*, **12**, 5–11 (2002), in Japanese.
[3] S. Wada, T. Tsurumi: *Kino Zairyo*, **12**, 53–65 (2002), in Japanese.
[4] M. Takashige, S. Hamasaki: *Kotai Butsuri*, **415**, 681–688 (2000), in Japanese.
[5] Y. Noguchi, M. Miyayama: *Hyomen Kagaku*, **26**, 208–214 (2004), in Japanese.
[6] T. Nakamura: Ferroelectricity involved in structural phase transitions, *Syokabo Tokyo* (1988), in Japanese.
[7] T. Sakudo: Solid state physics; lattice dynamics and dielectric phase transitions, *Syokabo Tokyo* (1993), in Japanese.
[8] K. Uchino: *Ferroelectric Devices*, Morikita Publishing (Tokyo, Japan) (1986), in Japanese.
[9] T. Takenaka: *Ceramics*, **39**, 749–754 (2004), in Japanese.
[10] T. Tsurumi, S. Wada: *Ceramics*, **40**, 354–368 (2005), in Japanese.
[11] Y. Saito, H. Takao, T. Tani, T. Nonoyama, K. Takatori, T. Homma, T. Nagaya and M. Nakamura: *Nature*, **432**, 84–86 (2004).
[12] Y.P. Guo, K. Kakimoto and H. Ohsato: *Appl. Phys. Lett.*, **85**, 4121–4123 (2004).
[13] Landolt-Bornstein, Vol. 11, New Series Group 3, Springer-Verlag Heidelberg, New York (1979).
[14] H. Okino, T. Ida, H. Ebihara, H. Yamada, K. Matsushige and T. Yamamoto: *Jpn. J. Appl. Phys.*, **40**, 5828–5832 (2001).
[15] K. Takata, K. Kushida, K. Torii and H. Miki: *Jpn. J. Appl. Phys.*, **33**, 3193–3197 (1994).
[16] T. Yamamoto, K. Kawano, M. Saito and S. Omika: *Jpn. J. Appl. Phys.*, **36**, 6145–6149 (1997).
[17] Y. Cho, A. Kirihara and T. Saeki: *Rev. Sci. Instrum.*, **67**, 2297–2303 (1996).
[18] H. Okino, J. Sakamoto and T. Yamamoto: *Jpn. J. Appl. Phys.*, **42**, 6209–6213 (2003).

6.4.2 Electrical conduction properties

6.4.2.1 Electrical conduction in nanostructured materials

The electrical conduction properties at particle interfaces and particle surfaces, in addition to particle interior properties, greatly contribute to the overall electrical conduction properties of materials composed of nanosized particles. Molecules of oxygen and water and hydroxide ions generally adsorb onto surfaces of inorganic materials in air, which affect the electrical conduction properties of such materials. Even in dense polycrystals prepared by high-temperature heating, the grain interior and grain surface differ in electrical properties due to defects formed by reactions with gas-phase molecules at high temperatures. Thus, when evaluating electrical conduction properties, such effects should be considered.

A structural schema of porous sintered bodies of an n-type semiconductor is shown in Fig. 6.4.7. Oxygen molecules adsorbed onto the surface in air ionize and become negatively charged by the transfer of electrons from the semiconductor. That is, they act as electron acceptors. As a result, an electron depletion layer with a thickness in the range of approximately 5–50 nm is formed on the surface, and thus the surface becomes highly resistive. Fig. 6.4.8 shows calculation results showing the relationship between effective electron density and surface acceptor density, corresponding to the amount of adsorbed oxygen, for ZnO particles of various sizes [1].

A high effective electron density is maintained regardless of the amount of adsorbed oxygen for large particles. In contrast, a low effective electron density is maintained for very small particles because

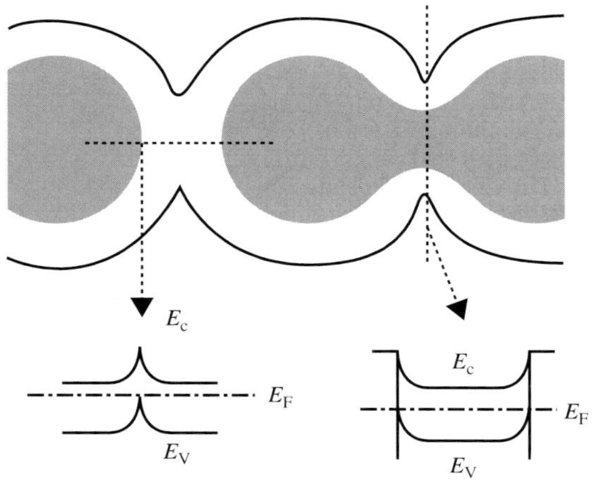

Figure 6.4.7
Structural schema of porous sintered bodies and their band structures. Insulating electron depletion layer is formed on grain surface (white region). E_F: Fermi level, E_C: energy level at the bottom of conduction band, E_V: energy level at the top of valence band.

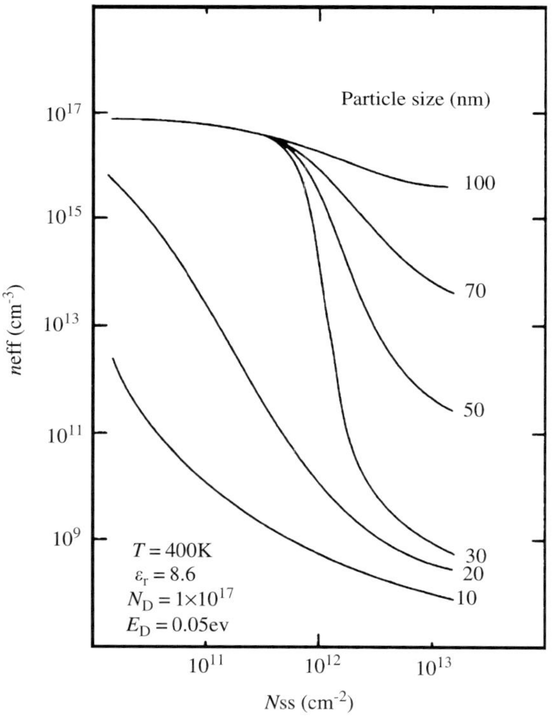

Figure 6.4.8
Relationship between effective electron density n_{eff} and surface acceptor density N_{ss} for ZnO particles of various sizes [1]. T: temperature, ε_r: dielectric permittivity N_D: donor density, E_D: donor level.

electron depletion layers cover the entire region of these particles. For particles with intermediate sizes, effective electron density changes markedly depending on the amount of adsorbed oxygen. For electrical conduction through a porous structure, as shown in Fig. 6.4.7, electric current flows by crossing electron depletion layers or through paths narrowed by electron depletion layers. Accordingly, highly resistive electron depletion layers on the surface contribute largely to the overall resistance, and electrical conductivity changes markedly depending on particle size, particle configuration and surface acceptor density. When adsorbed water layers exist on the surface of insulating particles, electric current flows mainly through these layers. Also, in this case, highly conductive water layers on the surface contribute largely to the overall resistance.

6.4.2.2 Direct current (DC) measurement

By DC measurement, the overall conduction properties including the contributions of interfaces and surfaces are measured as described above. Fig. 6.4.9 shows general measurement methods for resistivity ρ [Ω m], which is the inverse of conductivity σ [S m^{-1}] and their features [2]. It also shows the measurement methods for sheet resistance ρ_s [$\Omega/\square$] ($\rho_s = \rho/d$) used for the evaluation of thin films and surfaces. The two-probe and four-probe methods are used for high-resistance and low-resistance samples, respectively. The Van der Pauw method can be applied to the measurement of film samples of any shape. In the Van der Pauw method, electrodes are provided at any of the four positions A, B, C and D on the edge of a film sample, the voltage V_{CD} between electrodes C and D is measured when an electric current I_{AB} is made to flow between electrodes A and B, and the resistance $R_{AB,CD}$ is determined using V_{CD}/I_{AB}. The resistance $R_{BC,DA}$ is determined in the same way, and the resistivity of the sample can be obtained from the two resistances. f is the coefficient related to the sample configuration and electrode position, and $f = 1$ when $R_{AB,CD} = R_{BC,DA}$. In most cases, measurement is conducted for samples with the shape shown in Fig. 6.4.9f.

When leakage current is made to flow in the voltage measurement circuit, a conductivity larger than the true value is estimated. Accordingly, it is necessary for all measurement methods to use a voltmeter with a high internal impedance. It is also necessary to maintain the atmosphere (e.g., water vapor pressure and oxygen partial pressure) unchanged and to conduct measurement in a dry atmosphere, especially for high-resistance samples. The following treatments are effective in improving accuracy: (1) two measurements are conducted in opposite current directions and the average is adopted to avoid the effect of thermoelectric power and (2) an electric shield is applied to avoid the effect of the electric field from the outside.

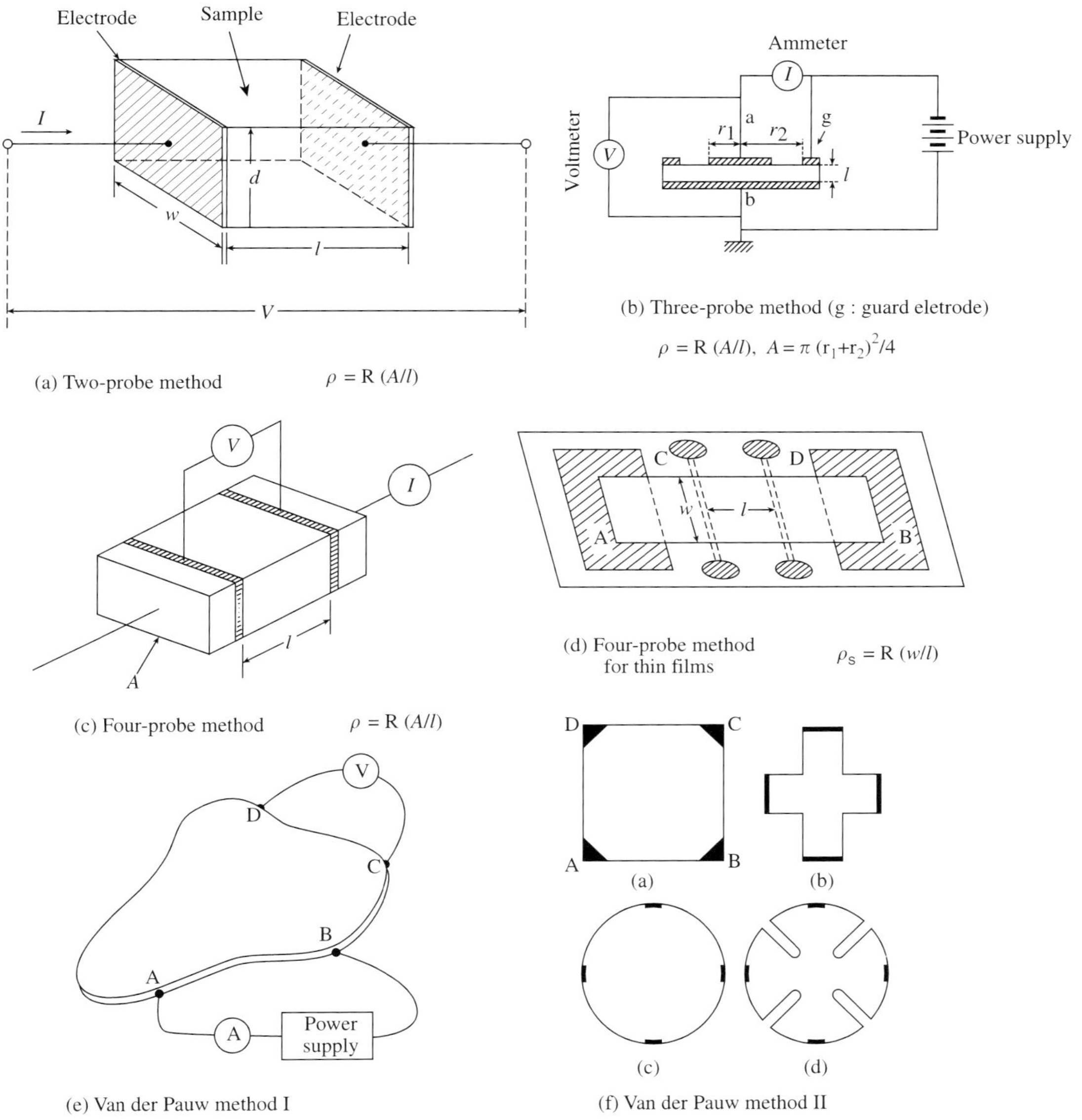

Figure 6.4.9
Measurement methods for resistivity (ρ) and sheet resistance (ρ_s). R: electrical resistance between electrodes (voltage V/current I), A: electrode area ($=wd$), l: distance between electrodes, w and d: width and length of an electrode.

6.4.2.3 Associate current measurement

The contributions of different resistances, such as resistances at the grain interior and grain boundary, can be separated by complex impedance measurement. In the case of a parallel circuit with the component resistance R and the component capacitance C, the complex impedance Z ($Z = Z'-jZ''$) is expressed by

$$Z^{-1} = R^{-1} + j\omega C, \tag{6.4.8}$$

where j and ω are the imaginary number unit and angle frequency, respectively. The real part Z' and the imaginary part Z'' are $Z' = R/(1+\omega^2R^2C^2)$ and $Z'' = \omega R^2C/(1+\omega^2R^2C^2)$, respectively; thus, the relationship between Z' and Z'' can be expressed by

$$\left(Z' - \frac{R}{2}\right)^2 + Z''^2 = \left(\frac{R}{2}\right)^2 \tag{6.4.9}$$

for a circle with a radius of $R/2$ and a center position of $(R/2, 0)$ [3]. An equivalent circuit for ceramic materials is often represented as a complex circuit in which parallel circuits are connected in series. An impedance plot for the above equivalent circuit is shown in Fig. 6.4.10, and the component resistances of the grain interior and grain boundary can be estimated from the diameter of individual semicircles. However, markedly different time constants τ ($\tau=1/\omega^* = RC$; ω^* is the angle frequency at the top of a semicircle) of individual semicircles are necessary for the separation of semicircles.

Actual composite materials have various component resistances from constituent grains, interfaces and electrodes, and various equivalent circuits are assumed. Recently, a method of estimating component resistances and capacitances has been developed and is now widely used. This method involves comparing a measured impedance plot with a calculated one assuming appropriate circuit configuration and component resistances/capacitances. An example of results by this method for an inorganic-polymer composite is shown in Fig. 6.4.11 [4]. The composite was prepared by mixing a WO_3 precursor solution with an epoxy-based prepolymer, hardening the mixture into a thick film and then precipitating oxide hydrate (WO_3 plate-like particles of approximately 300 nm size) in the composite by acid treatment. The subscripts 0, 1, 2 and 3 in the equivalent circuit indicate inorganic precipitates, interfaces (between inorganic particles and particle/matrix), a polymer matrix and electrodes, respectively. Individual component values can be obtained by selecting appropriate component values so as to obtain agreement between measured and calculated impedance plots. The software applicable to each measurement system has already been commercialized for such analysis [5].

For a clear separation of semicircles in the impedance plots, it is necessary that the component resistances in an equivalent circuit have the same values within one order of magnitude. When the component resistances are considerably different but the component capacitances are the same, the use of a complex modulus plot is effective. The complex modulus M is the inverse of complex permittivity ε and is expressed by $M = M' + jM'' = \varepsilon^{-1} = (\varepsilon' - j\varepsilon'')^{-1}$ [6]. An example is shown in Fig. 6.4.12. It is found that a separation of component circuits can be realized in the modulus plot even for an equivalent circuit in which the separation of such components is difficult in an impedance plot. A study to determine an equivalent circuit for a layer-structured ferroelectric crystal has been reported [7].

6.4.2.4 Nanoscale evaluation using microprobes

Recently, there has been rapid progress in the technology of scanning tunneling microscopes (STMs) and atomic force microscopes (AFMs), which enable the determination of the electronic states and topography of material surfaces through the use of microprobes. The conduction properties of a local area of thin films and crystal surfaces can be examined using such microprobes.

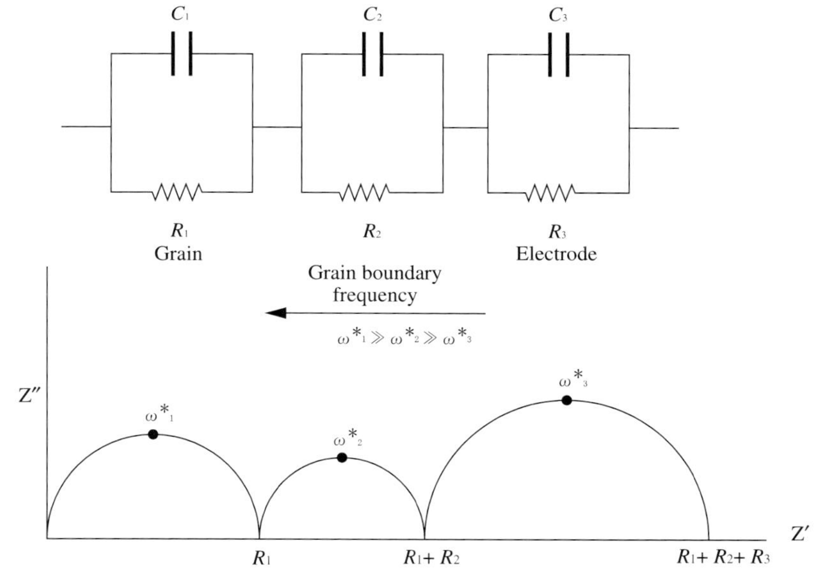

Figure 6.4.10
Equivalent circuit for ceramic materials and impedance plot. R: component resistance, C: component capacitance, ω^*: specific angle frequency. Subscripts 1, 2, 3 indicate grain interior, grain boundary and electrode, respectively.

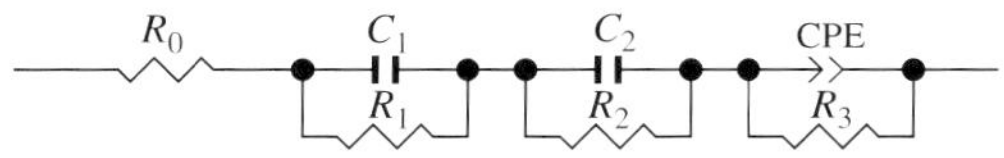

Subscript	Resistance R(Ω)	Capacitance C(F)
0	1.51×10^{2}	—
1	1.11×10^{2}	1.02×10^{-10}
2	4.65×10^{3}	2.3×10^{-5}
3	6.41×10^{14}	3.34×10^{-4}(CPE–T) 5.02×10^{-1}(CPE–P)

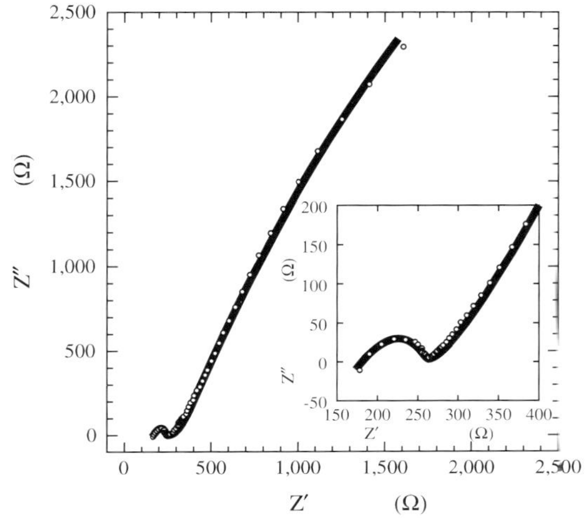

Figure 6.4.11
Equivalent circuit and impedance plot for $WO_3{\cdot}2H_2O$ (5.5 vol%) epoxy resin composite [4]. Circles and solid lines indicate measured values and calculated values, respectively. CPE: constant phase element.

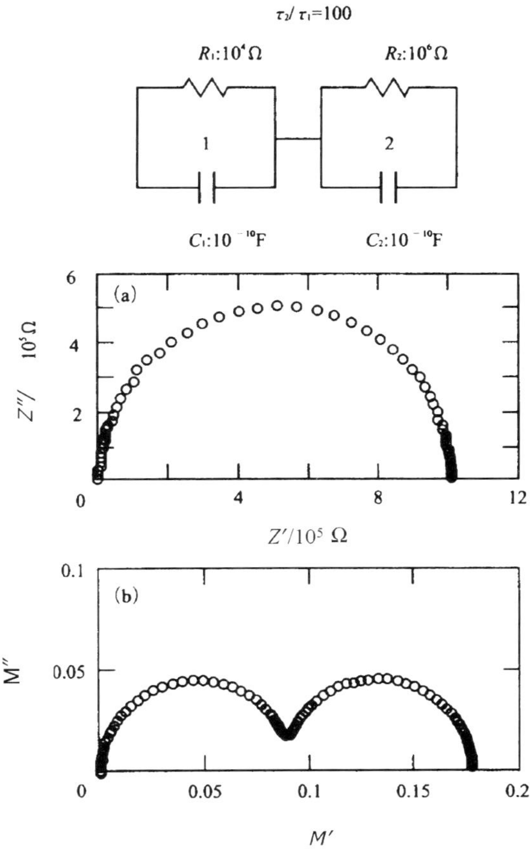

Figure 6.4.12
Simulated impedance plot (a) and modulus plot (b) for an equivalent circuit with largely different component resistances R and the same component capacitance C. τ is the time constant. The modulus is represented as the inverse of the complex permittivity.

The mapping of conduction regions in lead titanate zirconate (PZT) thin films on a Si/Pt substrate has been reported, in which electrical conduction was measured in the vertical direction of the films by attaching and scanning a conductive cantilever [8]. A conductive area of approximately 100 nm width was distributed along grain boundaries. The results showed that the grain boundaries work as current paths, degrading the insulating/ferroelectric properties of the PZT thin films.

Two microscopic four-point probe methods of examining the electrical conductivity of silicon crystal surfaces, have been developed (Fig. 6.4.13) [9]. One uses linearly arranged monolithic four-point probes fabricated on one tip, whose probe spacing is fixed around several micrometers. The other one uses four probes on four independently driven STM tips, whose probe spacing can be changed from 1 μm to 1 mm. In the latter method, a square arrangement of the probes is possible [10]; thus, this method can be regarded as a modified, infinitesimal Van der Pauw method. When the probe spacing is very small, only current flowing near the surface is detected. Accordingly, the method can accurately determine

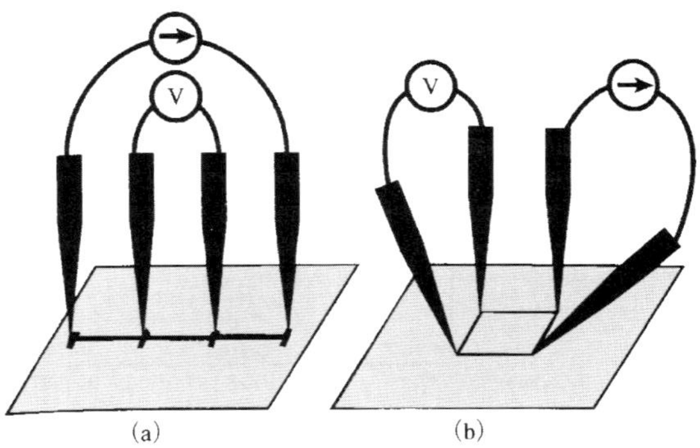

Figure 6.4.13
Schematic drawings of the four-point probe method in (a) linear and (b) square arrangements.

surface properties. In addition, the square arrangement of probes enables the evaluation of the anisotropy of surface conduction [10].

In the microprobe method, conductivity measurements for one particle and infinitesimal parts of composites will be possible with the control of probe spacing and the subsequent improvement in resolving power. Therefore, the microprobe method will be further developed as a promising evaluation method for electrical conduction properties of nanostructured materials.

References

[1] D.J. Leary, J.O. Barnes and A.G. Jordan: *J. Electrochem. Soc.*, **129**, 1382–1386 (1982).
[2] Jikken Kagaku Koza 9, *Electricity and Magnetism*, The Chemical Society of Japan (Ed.), Maruzen, 161 (1992).
[3] T. Kawada: *Chapter 8 in Kotai Dennki Kagaku*, Kodansha Scientific, Tokyo, 173–197 (2001).
[4] Y. Tanaka, M. Miyayama, M. Hibino and T. Kudo: *Solid State Ionics*, **171**, 33–39 (2004).
[5] For example: http://www.scribner.com/
[6] I.M. Hodge, M.D. Ingram and A.R. West: *J. Electroanal. Chem.*, **74**, 125–143 (1976).
[7] S.K. Kim, M. Miyayama and H. Yanagida: *J. Ceram. Soc. Jpn.*, **103**, 315–318 (1995).
[8] H. Fujisawa, M. Shimizu, T. Horiuchi, T. Shiosaki and K. Matsushige: *Appl. Phys. Lett.*, **71**, 416–418 (1997).
[9] T. Hasegawa, I. Shiraki, F. Tanabe, R. Horiba, T. Kanagawa, T. Tanikawa, I. Matsuda, C.L. Petersen, T.M. Hanssen, P. Boggild and F. Grey: *J. Surf. Sci. Jpn.*, **23**, 740–752 (2002).
[10] T. Kanagawa, R. Hobara, I. Matsuda, T. Tanikawa, A. Natori and S. Hasegawa: *Phys. Rev. Lett.*, **91**, 036805-1-4 (2003).

6.4.3 Thermoelectric properties

Thermoelectric conversion is a direct energy conversion between thermal and electrical energies. It is based on Seebeck effect and Peltier effect in the electrical conductors. Most of degenerated semiconductors and special metals with high carrier density used for this purpose are called "thermoelectric materials." The energy conversion performance of the material is defined by electrical and thermal transport coefficients, more exactly, electrical conductivity σ, Seebeck coefficient S and thermal conductivity κ.

Let us consider a closed circuit consisting of serial connections of p- and n-type thermoelectric materials and under an electrical load as shown in Fig. 6.4.14a. Applying forced temperature gradient $T_H–T_L$ along the sample length and assuming perfect thermal insulation, the thermal to electrical energy conversion efficiency is expressed as follows [1].

$$\eta = \frac{P_{OUT}}{Q_{IN}} = \frac{T_H - T_L}{T_H}\frac{\sqrt{1+ZT_{ave}}-1}{\sqrt{1+ZT_{ave}}+T_L/T_H} \quad (6.4.10)$$

where $T_{ave} = (T_H + T_L)/2$ and $Z = \sigma S^2/\kappa$

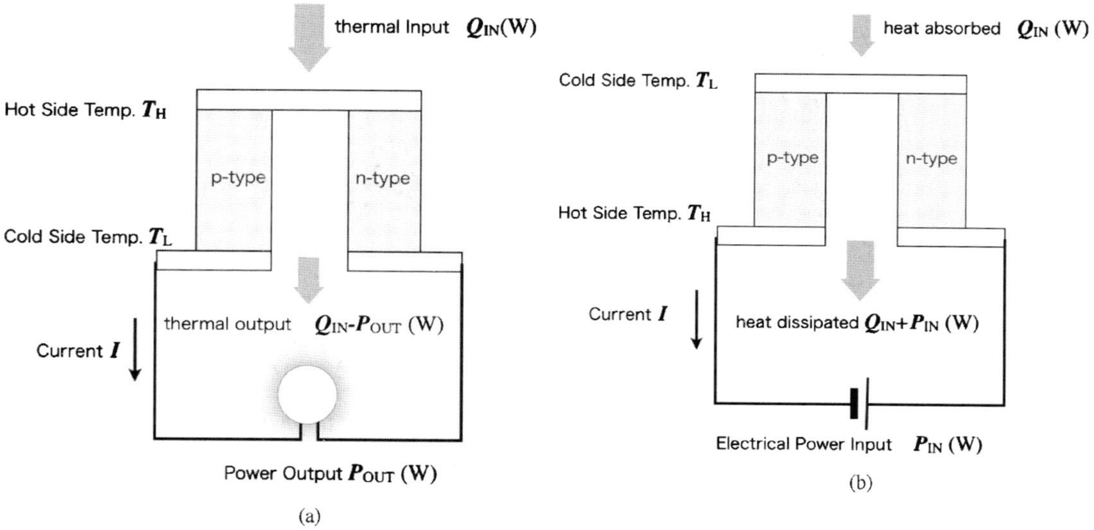

Figure 6.4.14
Schematics of thermoelectric power generation (a) and Peltier cooling (b).

Z is called "Thermoelectric Figure of Merit (FOM)" and is an important performance measure of thermoelectric materials. It has a unit of inverse temperature (K^{-1}).

Since the thermoelectric effects are reversible phenomena, the electrical-to-thermal energy conversion has a similar configuration as shown in Fig. 6.4.14b. In this case, the thermoelectric materials work as a solid-state heat pump operated by an external electric current source. The maximum heat transfer from the low temperature to high temperature side at specific power consumption is called as "Coefficient of Performance (COP)," and can be expressed as follows [1].

$$\phi = \frac{Q_{\mathrm{IN}}}{P_{\mathrm{IN}}} = \frac{T_{\mathrm{L}}}{T_{\mathrm{H}} - T_{\mathrm{L}}} \frac{\sqrt{1 + ZT_{\mathrm{ave}}} - T_{\mathrm{H}}/T_{\mathrm{L}}}{\sqrt{1 + ZT_{\mathrm{ave}}} + 1} \qquad (6.4.11)$$

As seen in equations (6.4.10) and (6.4.11), an evaluation of Z (or dimensionless value ZT) is of importance in thermoelectric energy conversion applications, which require the measurement of σ, S and κ of a sample at a specific range of temperatures. For examples, Z value for $(Bi_2Te_3)_{1-x}(Sb_2Te_3)_x$ (x=0.7–0.9), typical thermoelectric material used in Peltier cooling module, is known around 0.003 K^{-1} at room temperature; on the other hand, $Si_{80}Ge_{20}$ alloy has a Z value around 0.0008 K^{-1} at 1,000 K. Fig. 6.4.15 shows thermoelectric properties of polycrystalline $Si_{95}Ge_5$ alloy measured from 300 to 1,300 K [2].

6.4.3.1 Measurement method for thermoelectric materials

Practically, the thermoelectric properties σ, S and κ, should be evaluated with a same sample to determine the Z value, because we cannot expect uniformity among samples. The details to measure the electrical conductivity and thermal properties are mentioned in the Sections 6.4.2 and 6.3.1, respectively. References on the measurement techniques for thermoelectric materials are available at Japanese Industrial Standard (JIS), such as JIS-R-1650-1 (Seebeck coefficient), 1650-2 (Electrical resistivity), 1650-3 (thermal conductivity).

In this section, the principles of measuring important physical properties of thermoelectric materials, such as Seebeck coefficient and Figure of Merit, are outlined.

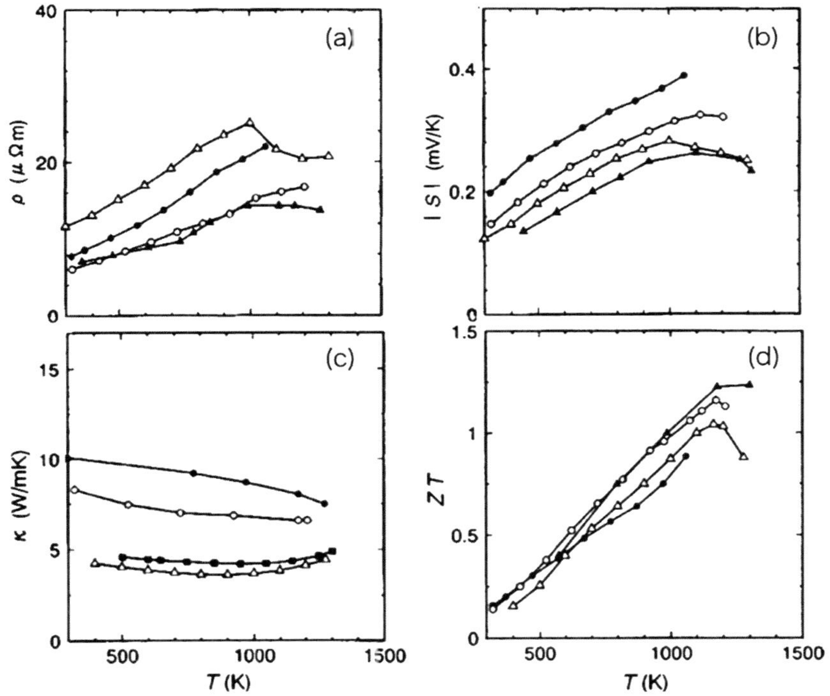

Figure 6.4.15
Thermoelectric properties of SiGe alloy [2]. (a) The electrical resistivity (inverse electrical conductivity σ), (b) the Seebeck coefficient S, (c) thermal conductivity κ and (d) the thermoelectric Figure of Merit Z. Open circles, filled circles, filled square and filled triangle correspond to P-doped $Si_{95}Ge_5$, P/GaP doped $Si_{95}Ge_5$, $Si_{80}Ge_{20}$, respectively.

(1) Measurements for bulk thermoelectric materials (including bulk materials with micro-/nanostructures)

Seebeck coefficient is defined as the ratio of electromotive force V_{emf} to temperature difference ΔT and can be expressed as below.

$$S = \frac{V_{emf}}{\Delta T} = \frac{V_{emf}}{T_H - T_L} \qquad (6.4.12)$$

$\Delta T \rightarrow 0$ is an ideal situation, but a typical value of ΔT is 2–7 K to minimize the measuring errors. However, high ΔT value is not recommended, since Seebeck coefficient is sensitive to temperature for semiconducting materials. Also, V_{emf} is a sum of thermoelectric voltage created in the sample and the electrical wire; it is necessary to know the Seebeck coefficient of the wire material. Typical wires used in the Seebeck coefficient measurement are constantan, chromel and platinum.

Four possible configurations for the measurement are shown in Fig. 6.4.16a to d. Fig. 6.4.16a is an example using metal electrodes. With this method, the possibility of having reaction between the sample and the metal electrode at elevated temperatures is of concern. Fig. 6.4.16b is taking one lead of the thermocouple wire to measure electromotive force V_{emf}. In this case, a spring-loaded mechanical contact, using electrically conducting paste or spot-welding directly onto the sample, is applied to attach the wire of thermocouple to the sample. It should watch for the offset voltage, which appears at $\Delta T = 0$ and can cause inconsistency of the measurement. A bad contact produces larger offset voltage and results in larger error in the measurement of Seebeck coefficient.

For low-temperature measurements, the use of conductive pastes or adhesive is most popular to attach the lead wire as shown in Fig. 6.4.16d. Because there is little chance of having chemical reaction between the sample and the paste at the low temperature, reliable contacts can be made by the pastes. In Fig. 6.4.16d the fine copper pieces attached to the sample are used as pick-up-leads for both temperature and voltage.

Harman method [3] is a unique method to determine thermoelectric Figure of Merit, Z, directly using the Peltier effect of test samples. Applying external current source to the sample adiabatically, the temperature gradient grows along the sample length due to Peltier effect as shown in Fig. 6.4.17. The electrical resistance measured by the direct current, $R_{Peltier}$, is therefore influenced by the thermoelectric voltage inside the sample. This $R_{Peltier}$ is always larger than the true resistance R_0 measured with the alternative current. The thermoelectric Figure of Merit Z can be determined by using the ratio $R_{Peltier}/R_0$ as follows.

$$Z = \frac{1}{T}\left(\frac{R_{Peltier}}{R_0} - 1\right) \qquad (6.4.13)$$

The electrical conductivity σ can be calculated from the sample dimension and R_0; and, Seebeck coefficient is simply determined by the equation (6.4.12). Thus, the thermal conductivity κ can be calculated as $\kappa = \sigma S^2/Z$.

Since the transporting phonon can be scattered by the microstructure (e.g., crystal grain interface and defects) of samples, this reduces thermal conductivity and enhances thermoelectric Figure of Merit. It is expected to have high thermoelectric Figure of Merit for the nano-/microtextured materials. For example, $Si_{80}Ge_{20}$ alloy with an addition of BN or Si_3N_4 fine particles (diameter of the particles: 4–20 nm) is reported as having 30% lower thermal conductivity κ and 15% increase in the thermoelectric Figure of

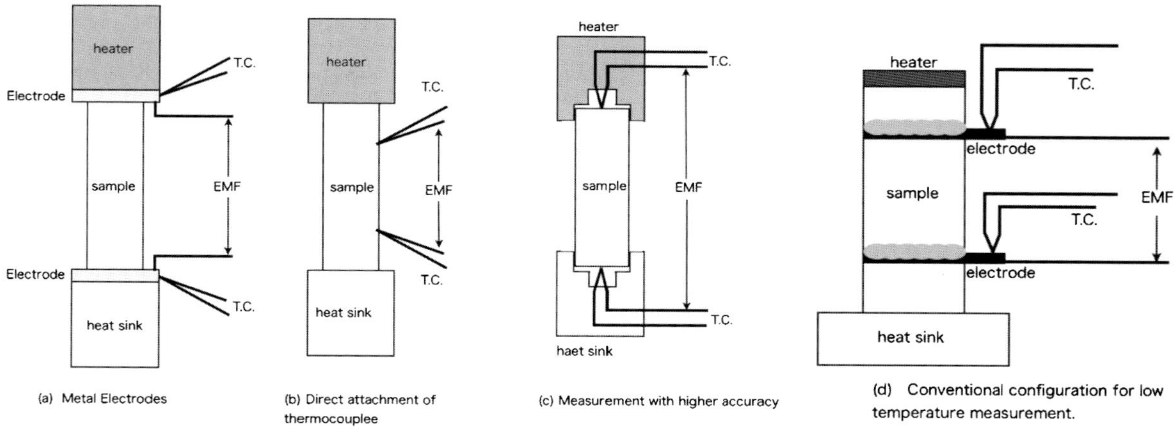

Figure 6.4.16
Configurations for the measurement of Seebeck coefficient.

Merit Z [4]. In this case, the fine particles appear to disperse uniformly inside the crystal grain, which is considered being the key contributor to the enhancement of materials performance. Kishimoto et al. also reported the improvement of Seebeck coefficient due to the reduction of crystal grain size in the PbTe bulk sample prepared by the SPS process [5]. They interpreted that it was caused by the potential barrier existing at the interface between crystal grains. $AgPb_{18}SbTe_{20}$ with nanometer-sized additives and fine lamellae structures can self-organize during the solidification process and possess thermal conductivity as low as $1\,W(mK)^{-1}$. This leads to significant increase in Figure of Merit ZT ($ZT = 2.2$ at 800 K [6]). It is a good example to demonstrate the effect of nano-/micro-structures on the bulk thermoelectric materials.

(2) Measurements of thin film thermoelectric materials with nanostructure

It is well known that transport properties of electrons and phonons are strongly influenced by their dimensions. Therefore, artificially designed and fabricated nanostructures such as quantum dots and super-lattices are interesting research topics for thermoelectric materials.

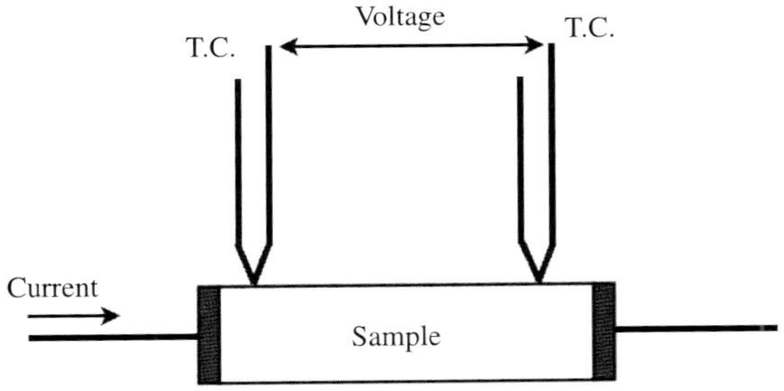

Figure 6.4.17
Harman method.

In general, the thin film sample is grown on the unusual substrate materials and we need to consider their influences in the measurement of thermoelectric properties. It is usually difficult to create large temperature difference perpendicular to the thin film with a thickness of several micrometers and below. Therefore, it is quite often to apply non-steady-state measurement and device performance measurement.

Venkatasubramanian et al. reported that the enhanced Figure of Merit ZT ($ZT = 2.4$ at 300 K) could be obtained from the Bi_2Te_3/Sb_2Te_3 super-lattice prepared by MOVPE method [7,8]. The electrical conductivity perpendicular to the film $\sigma_\perp$ is determined by TLM method using several samples with different thickness ranging from 1 to 5 μm. The thermal conductivity $\kappa_\perp$ determined by 3ω method [9] is extremely low at $0.22\,W(mK)^{-1}$ compared to $0.49\,W\,(mK)^{-1}$ of conventional bulk materials having the same composition. They used modified Harman method to determine ZT and observed steady-state cooling effect at the maximum temperature difference $\Delta T = 32$K under optimum current injection. The ZT value was much higher than that of plain $(Bi_2Te_3)(Sb_2Te_3)$ alloy film. Therefore, they concluded that it was a strong evidence demonstrating the advantage of using nanotextured thin film thermoelectric materials (Fig. 6.4.18).

Harman et al. reported the thermoelectric properties of $PbTe/PbTe_{0.02}Se_{0.98}$ quantum dot super-lattice prepared by MBE method [10]. The thickness of the film grown on the BaF_2 substrate was 104 μm and this could be a self-supporting film even after removing the BaF_2 substrate. The Peltier cooling demonstration with the p–n unicouple consisting of the self-supporting $PbTe/PbTe_{0.02}Se_{0.98}$ sample and a fine gold wire showed a maximum temperature drop of $\Delta T = 44$K. Because the Figure of Merit (ZT) of gold wire was almost zero, the observed large cooling capability was attributed to the extraordinary high ZT (in their estimation $ZT = 1.6$ at 300 K) of $PbTe/PbTe_{0.02}Se_{0.98}$ quantum dot super-lattice.

To further investigate and understand the thermoelectric transport properties of materials with nanostructures,

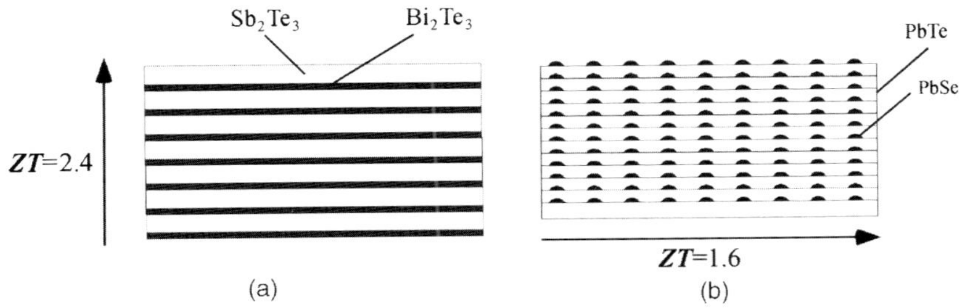

Figure 6.4.18
Thermoelectric thin films with super-lattice structure. (a) Bi_2Te_3/Sb_2Te_3 super-lattice with the period of 6 nm and total thickness of 5.4 μm [8]. (b) PbTe/PbSe quantum dot super-lattice with the period of 13 nm and total thickness of 104 μm [10].

novel technique such as Scanning Seebeck Coefficient Microprobe is being developed and applied to analyze the interface of semiconductor [11].

References

[1] Y. Suge Ed.: *Thermoelectric Semiconductors (Netsuden Handoutai)*, Maki Shoten (1966).

[2] O. Yamashita: *J. Appl. Phys.*, **89**, 6241–6246 (2001).

[3] T.C. Harman, J.H. Cahn and M.J. Logan: *J. Appl. Phys.*, **30**, 1351–1359 (1959).

[4] J.W. Vandersande and J-P Fleurial: Proceedings of the 11th International Conference on Thermoelectrics, 21–23 (1992).

[5] K. Kishimoto, T. Koyanagi: *J. Appl. Phys.*, **92**, 2544–2549 (2002).

[6] K.F. Hsu, S. Loo, F. Guo, W. Chen, J.S. Dyck, C. Uher, T. Hogan, E.K. Polychroniadis, M.G. Kanatzidis: *Science*, **303**, 818–821 (2004).

[7] R. Venkatasubramanian, E. Siivola, T. Colpitts, B. O'Quinn: Proceedings of the 18th International Conference on Thermoelectrics, 100–103 (1999).

[8] R. Venkatasubramanian, E. Siivola, T. Colpitts, B. O'Quinn: *Nature*, **413**, 597–602 (2001).

[9] D.G. Cahill: *Rev. Sci. Inst.*, **61**, 802–808 (1990).

[10] T.C. Harman, P.J. Taylor, M.P. Walsh, B.E. LaForge: *Science*, **297**, 2229–2232 (2002).

[11] H.K. Lyeo, A.A. Khajetoorians, L. Shi, K.P. Pipe, R.J. Ram, A. Shakouri, C.K. Shih: *Science*, **303**, 816–818 (2004).

6.5 Electrochemical properties

6.5.1 Electrode reaction

There has been considerable research concerning the interface between the electrode and the electrolyte where the electrochemical reaction takes place. For example, the electrochemical reaction has been basically studied by using gold and platinum plates as the inert electrode. The concept of the electric double layer (EDL) is one of the important results from those theoretical studies. However, to date, there has been no successfully conducted direct observation of the electrode surface on such a fine scale, because the EDL thickness is less than several microns and the fact that the situation changes dynamically. Therefore, attempts have been made to investigate the electrochemical reactions by monitoring the current response at a certain voltage. Further, there is the case where the electrode itself is one of the reactants, whereby it undergoes a redox reaction in the battery and plating system. Because it is attractive to study these electrochemical systems from a practical perspective, it is important to analyze the electrochemical process, including the mass transfer across the electrode surface, which must be affected by the nanoscale structure of the electrode materials. In this section, the electrochemical method used to evaluate the properties of the electrodes for batteries with a nanostructure will be mentioned, with reference to literature concerning lithium ion and nickel hydrogen batteries.

6.5.1.1 Charge and discharge test

The charge/discharge test is one of the simplest methods used to evaluate the electrochemical properties of the electrode for the battery. It must be noted that the electrode for the battery is usually the mixture. For example, the cathode of a lithium ion battery consists of active material, like $LiCoO_2$, $LiMn_2O_4$ and so on, acetylene black as an electric conductor and PVDF as a binder. This mixture is pressed onto the titanium or SUS mesh or painted onto the Al foil as a current collector to prepare the practical cathode. The electrochemical response of such an electrode must be affected by many factors, each of which is difficult to distinguish separately in terms of effect. Even the mechanical stress in the electrode can be reflected in the electrochemical response. Therefore, the charge/discharge data must be discussed very carefully, even including consideration of the cell design. Two electrode cells and three electrode cells respectively provide different information. The two electrode cell is constructed similarly to the practical cell and facilitates simple estimation of the total charge/discharge behavior. The simplest two electrode cell construction involves two electrodes, the anode and cathode, immersed in an electrolyte solution in a beaker. Since it is difficult to establish the same arrangement and distance between the electrodes with this type of cell, causing low reproducibility in the data, the simple beaker cell is infrequently used. Generally, the use of the coin type cell is preferred as a two electrode cell. The electrolyte solution permeated separator is held by the electrodes in the coin cell. In this case, it must be noted that the pressure applied to the electrodes must be uniform, otherwise there may be variable reproducibility of the measurements or leakage of the electrolyte solution. With these points in mind, a two electrode test cell for lab use is commercially provided, with an example shown in Fig. 6.5.1. The three electrode cell, meanwhile, includes a reference electrode in addition to the working and counter electrodes [1, 2]. In this case, the working and counter electrodes correspond to the cathode and anode in a two electrode cell, while the potential of the working electrode is controlled by using that of the reference electrode. The electrochemical response of the working electrode can be measured by using the three electrode cell, the electric circuit of which is shown in Fig. 6.5.2. The current which flows between the working and reference electrodes is so small that the reference electrode can maintain a constant potential during the measurement.

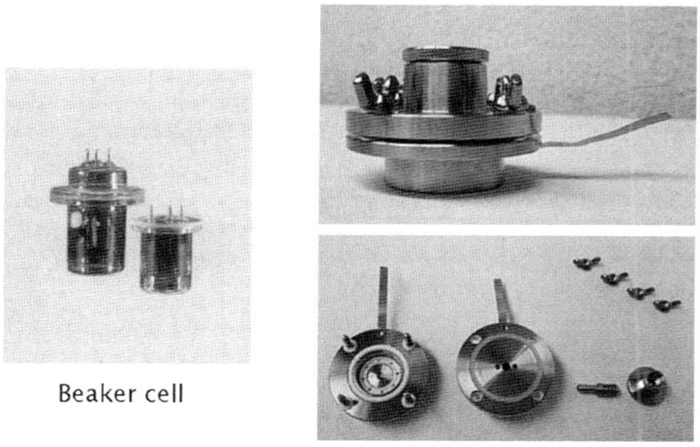

Figure 6.5.1
The experimental cell for the electrochemical measurement concerning the electrode process.

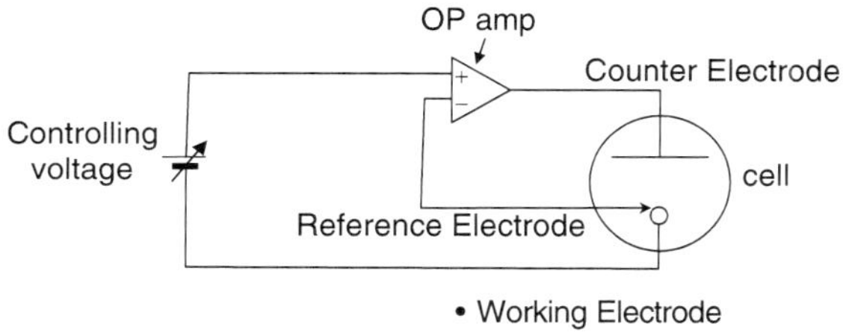

Figure 6.5.2
An electric circuit with a three electrode cell for electrochemical measurement.

Conversely, the counter electrode used is so large that the electrochemical process on it never limits the total process. The potential difference between the working and counter electrodes is relatively unimportant in this case. Lithium metal with a sufficiently large area can be used to investigate the electrochemical properties of the anode for a lithium ion battery. Fig. 6.5.3 shows the cells which are practically used to evaluate the electrochemical properties of the electrodes for nickel–hydrogen and lithium ion batteries. When using electrolyte with high resistivity, like the nonaqueous solution for the electrolyte solution of a lithium ion battery, the arrangement of the electrodes, especially that of the reference electrode, can affect the results. In the case of a lithium ion battery, the water must be removed from the system. Therefore the cell is constructed in a globe box filled with inert gas, such as Ar and N_2. A simple sealed test cell for a lithium ion battery, which only requires an inert gas atmosphere during the cell construction, has been developed as shown in Fig. 6.5.3.

As an example of the charge/discharge tests for material with a nanostructure, the results for $LiNiO_2$ fine particles prepared by the mist-dry method were shown here. These fine powder materials, prepared by the mist-dry method, have a large surface area and uniform spherical shape, which may improve the cycling efficiency, high rate performance and so on. Fig. 6.5.4 shows the scanning electron micrographs (SEM) of $LiNiO_2$ prepared by the mist-dry method [3]. The average diameter of the particle was 0.8 μm, with relatively narrow dispersion. It was found that the particle was secondly agglomerated particle by SEM observation in high magnification. This fact was reflected in the large specific surface area, measured using the BET method of 10 $m^2\ g^{-1}$, compared to that of $LiNiO_2$ prepared by the conventional method of several $m^2\ g^{-1}$. The results of the charge/discharge cycle test of $LiNiO_2$ at 0.1, 0.3 and 0.5 mA cm^{-2} respectively were shown in Fig. 6.5.5. After 10 cycles, constant charge/discharge could be carried out for more than 50 cycles for the $LiNiO_2$ fine particle sample, prepared here by the mist-dry method. Recently it has been reported that $LiMn_2O_4$ fine particles can also be prepared using the mist-dry method. $LiMn_2O_4$ fine particles were prepared with varying size and shape and their large specific area was found to be potentially capable of helping improve the properties at a high rate of charge/discharge [4,5].

Comparing the discharge capacities is the simple and convenient method used to discuss the electrochemical properties of various samples, while a Peukert equation can also be used to discuss the rate dependence of the charge/discharge capacity:

$$i^x \times t = \text{const} \qquad (6.5.1)$$

where i, t and x correspond to the charge/discharge current density, charge/discharge time and the factor which reflects the effect of the charge/discharge rate on the capacity respectively. Although this equation (6.5.1) is empirical, it is often used to quantitatively express the effect of the reaction rate at the electrode on the charge/discharge capacity. The following equation (6.5.2) can be derived from equation (6.5.1) by using discharge capacity C, as follows:

$$-(x-1) \times \log i = \log C + \text{const} \qquad (6.5.2)$$

Thus, the effect of the electrode reaction on the discharge capacity can be quantitatively expressed by the gradient of the log i *vs.* log C plot. The Peukert plots for polyaniline, prepared in various organic solvents such as propylene carbonate, γ-butylolactone and acetonitrile by electropolymerization, were shown in Fig. 6.5.6. The nanostructure of the polyaniline was influenced by the kind of solvent used for electropolymerization. A fine fibril structure was observed for polyaniline prepared in acetonitrile solution, which reflects the larger x value. This also means polyaniline with a fine porous structure has an advantage as an electrode material for frequent use.

The experimental techniques of a charge/discharge test at a constant current, as described above, are so basic that they can be applied to other methods, such

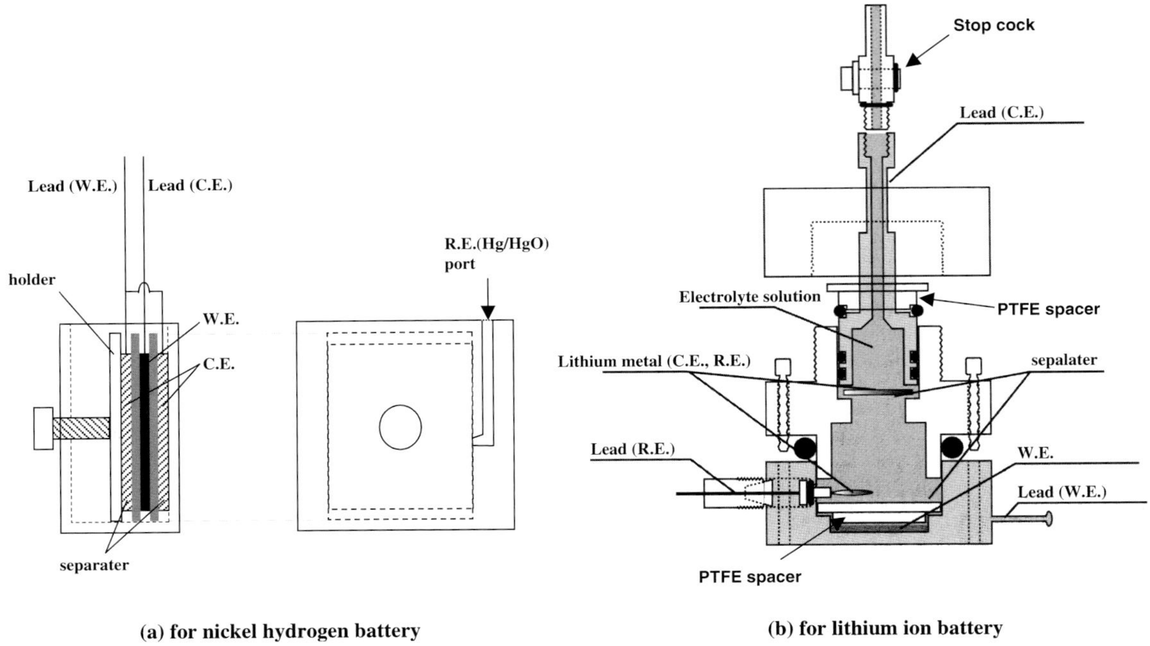

Figure 6.5.3
Construction of the three electrode cell.

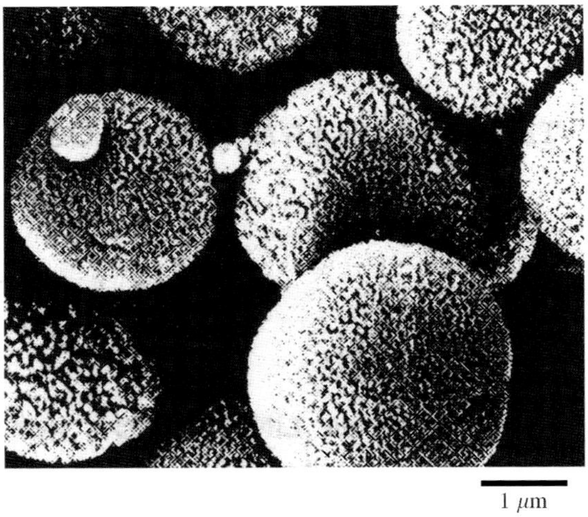

Figure 6.5.4
SEM photograph of a $LiNiO_2$ particle, prepared by the mist dry method.

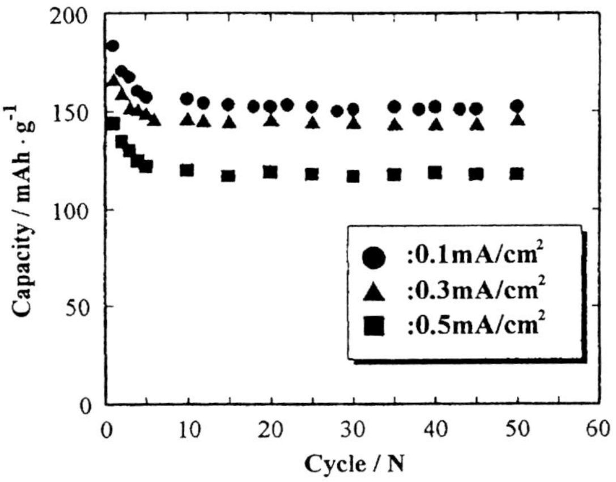

Figure 6.5.5
Charge/discharge cycling ability of $LiNiO_2$, prepared by the mist dry method.

as charge/discharge at constant potential or under a constant load or a combination of both.

General electrochemistry techniques can also be applied to estimate the properties of the electrode. For example, the $i-V$ curve and Tarfel equation are used to evaluate the performance of the electrode in a PEFC (polymer electrolyte fuel cell). Based on this kind of discussion, the porosity of the electrode was one of the important factors which affected the performance of PEFC [7].

6.5.1.2 Cyclic voltammetry

The discharge curve is drawn with the plot of E vs. t where E and t are the electrode potential and discharge

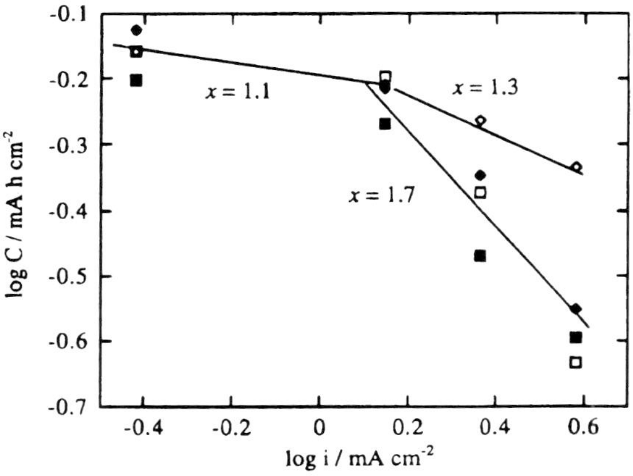

Figure 6.5.6
Log I vs. log C plots of polyaniline prepared in various organic solvents (◆, propylene carbonate, ◆, γ -butylolactone, ◇, acetonitrile, □, acidic aqueous solution, $i^x \times t$ = const. t is the discharge time).

time respectively. The value t can be substituted by the amount of electricity C with the relation of $C = i \times t$, where i means the current density. The differentiation of the discharge curve (E vs. C plots) by time gives dE/dt vs. i, where $dC/dt = i$. dE/dt has the dimension of Vs^{-1} and indicates the electrode potential change over time. In the case of cyclic voltammetry, the electrode potential is swept linearly against time (dE/dt = constant) and the current i is recorded with the electrode potential. Easily understandable is the fact that virtually the same information can be obtained from cyclic voltammetry and the differentiation form of the charge/discharge curves. Therefore cyclic voltammetry is the technique whereby the change in the electrochemical reaction rate can be conveniently detected. A three electrode cell is used for cyclic voltammetry because the electrode potential must be measured as accurately as possible, meaning a two electrode cell would be unsuitable for this experiment. The sweep rate is set at a rate ranging from several mV to several hundred mV per second in the case of aqueous electrolyte solutions and several mV per minute in the case of nonaqueous solutions and the current is recorded vs. the electrode potential. The peak in the voltammogram corresponds to the flat region in the charge/discharge curves. In the case of the secondary battery system, at least one pair of the redox peaks appears in the voltammogram. The potential difference between the redox peaks includes information concerning the internal resistance and the reversibility of the electrochemical process. Evidently, the current corresponds to the rate of the electrochemical process and the integration of the peak corresponds to the capacity for each process. However, the peak in the cyclic voltammogram of the electrode for the secondary battery system normally has a broad shape because it is the mixture of several materials as described above. With this in mind, it was difficult to accurately separate information concerning the properties of the active material from that concerning the total electrode. Dokko et. al have developed a technique to measure the cyclic voltammogram of a single particle and subsequently, the electrochemical properties of a single particle have come to be measured and discussed [8]. The cantilever of the scanning probe microscope comes into contact with the particle and functions as the working electrode. Sharp peaks were observed in the case of the voltammogram for a single particle rather than that for the whole electrode, making it possible to evaluate the electrochemical properties of the active material very accurately. This fact must promote research into active materials having a nanostructure.

6.5.1.3 AC impedance method

Detailed investigation of the internal resistance in the electrode can be carried out using the AC impedance method. Using a three electrode cell, the electrode potential of the working electrode is perturbed by several mV from around a certain electrode potential in sine waves and the response of the current is monitored. Alternatively, the current is perturbed and the response of the electrode potential is recorded, as in the case of the fuel cell. The phase difference and the ratio of the voltage and current (namely the resistivity) are obtained for each frequency. The data are often transformed to the set of the frequency (ω), the impedance in the real part Z_{real} and that in the imaginary part Z_{img}. Generally speaking, a potentio/galvanostat and frequency response analyzer (FRA) are required for this measurement, although a lock-in amp can be used instead of FRA. However, in view of the S/N and the wide frequency range, there has been a recent tendency to choose the FRA. The data set (ω, Z_{real}, Z_{img}) is usually plotted in a Cole–Cole plot, as shown in Fig. 6.5.7, where R_{sol} is the resistivity due to the polarization of the electrode, which is the

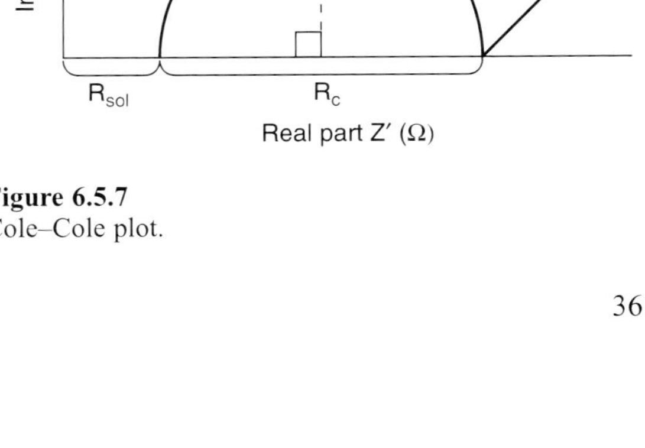

Figure 6.5.7
Cole–Cole plot.

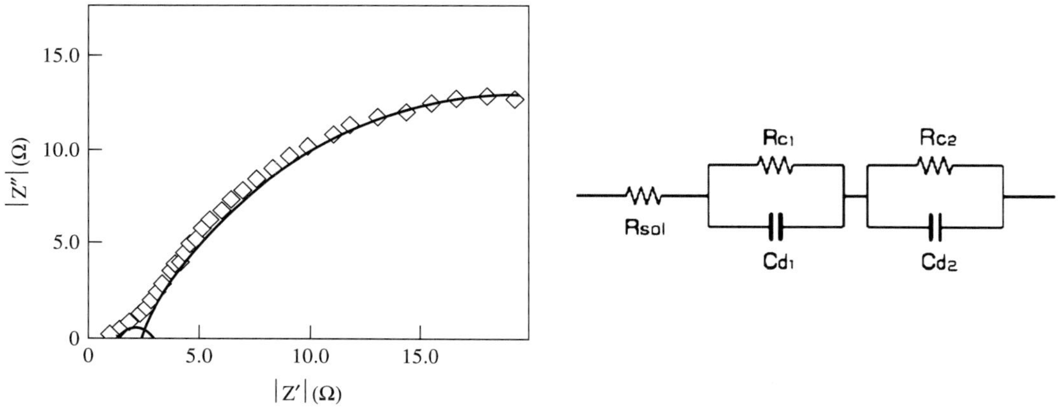

Figure 6.5.8
Cole–Cole plot for the fuel cell system and the equivalent circuit.

sum of the DC resistivity of the electrolyte solution, the separator and the current collector. In the case of a typical electrochemical reaction, it is believed that R_c and C_d indicate the reaction resistivity and the EDL capacity respectively. Some reports exist concerning a secondary battery system, in which these factors have been analyzed in detail. The impedance spectra for the fuel cell system were shown in Fig. 6.5.8, although those of the electrode for the battery are often so complicated that accurate analysis is difficult. An equivalent circuit, as shown in Fig. 6.5.8, must be found to discuss the resistivities and capacities within the system. Many trials, circuit model construction > simulation > fitting to the experimental data > reconstructing the model > . . . are required before obtaining the result. Even in the case of the fuel cell shown here as an example, there are few results involving the analysis of R_c, meaning no model sufficiently accurate to understand the fuel cell system has yet been constructed. It is, however, a convenient method by which the internal resistance of the electrochemical system can be roughly separated into several components, such as the resistances due to the electrolyte, due to the reaction and the electric double layer capacitance.

Reference

[1] A.J. Bard, L.R. Faulkner: *Electrochemical Method*, Wiley, New York, USA, p. 500 (1980).

[2] John O'M. Bockris, Amulya K.N. Reddy: *Modern Electrochemistry 2*, Plenum Publishing Co. New York, USA, p. 632 (1977).

[3] T. Ogihara, N. Ogata, K. Yoshida, S. Yonezawa, M. Takashima and N. Mizutani: *Denki Kagaku*, **66**, 1202–1205 (1998).

[4] K. Matsuda, I. Taniguchi: *Kagakukougakuronbunnsyu*, **29**, 232–237 (2003).

[5] T. Ogihara, T. Yanagawa, N. Ogata, K. Yoshida, Y. Mizuno, S. Yonezawa, M. Takashima, N. Nagata and K. Ogawa: *Denki Kagaku*, **61**, 1339–1341 (1993).

[6] S. Yonezawa, K. Kanamura and Z. Takehara: *J. Electrochem. Soc.*, **142**, 3309–3313 (1995).

[7] Y. Takasu, N. Ohashi, X.-G. Zhang, Y. Murakami, H. Minagawa, S. Sato and K. Yahikozawa: *Electrochim. Acta*, **41**, 1595 (1996).

[8] K. Dokko, M. Mohamed, M. Umeda and I. Uchida, *J. Electrochem. Soc.*, **150**(4), A425–A429 (2003).

6.5.2 Characteristics of sensors

Although the clear definition to the term "sensor" is not found, it is generally used for a device which can provide information required for a system. The existing various sensors can be classified broadly into two types according to the difference in properties to measure. Physical sensors mainly measure the change of a physical quantity such as force, heat, magnetism and light. Chemical sensors primarily detect the change of a chemical quantity. In addition, there are biosensors using biological material. Because of the principle of the detection based on a chemical phenomenon, the biosensors are usually classified as the chemical sensors.

The sensors are required to have various functions for each application. In most cases the following characteristics are needed.

6.5.2.1 High sensitivity

In order to avoid the influence of noise, it is desirable that the sensitivity is high as much as possible. The sensitivity of the sensors differs in expression in every application and is usually estimated by comparison of the signal intensities before and after detection. Moreover, proportional relationship between the output signal and the quantity of the measured property is favorable in the range of the measurement.

6.5.2.2 Quick response

The change of the property should be detected as quickly as possible. Also the recovery to an initial state after the detection should be prompt.

6.5.2.3 Selectivity

The sensor should respond only to the property to measure, not being affected by the other coexisting properties.

6.5.2.4 Durability

It is necessary for the sensor to have high physical and chemical stability in the actual environment. Moreover, it is important to endure the repetition of the detections.

These characteristics are influenced by various factors such as the material and the structure of the sensor. Here, the effects and features of the nanostructural control on the functions of sensors are described, surveying the characteristics of a chemical sensor with nanostructure.

(1) Characteristics of oxide semiconductor gas sensor with nanostructure

An oxide semiconductor gas sensor is a practical device since its structure is simple and the sensitivity is high. The gas detection is based on the electrical resistance change attributed to contact between the sensor and the gas to detect. Fig. 6.5.9 shows the schematics of the surface of n-type oxide semiconductor particles at 200–400°C in air. Chemisorbed oxygen species, such as O_2^-, O^-, and O^{2-}, exist on the particle surface, and a space charge layer is formed near the grain boundary. When oxygen molecules come in contact with the surface of particles, the surface levels which act as acceptors are generated at energy levels lower than a Fermi level. The electrons inside particles are trapped into the acceptors, and then the space charge layer arises.

If reduction gas molecules appear on the surface, the chemisorbed oxygen species will react with them, the trapped electrons will be released, potential barrier will lower and then the electrical resistance of the sensor will decrease. On the other hand, the resistance will increase for oxidation gas.

The sensitivity of gas detection is calculated as follows:

$$\text{Sensitivity} = \frac{\text{Resistance in air}}{\text{Resistance in gas to detect}}$$

The sensing characteristic of the oxide semiconductor can be improved by reducing the particle size to nanosize. Fig. 6.5.10 shows the influence of particle size on the sensitivity of gas detection for SnO_2

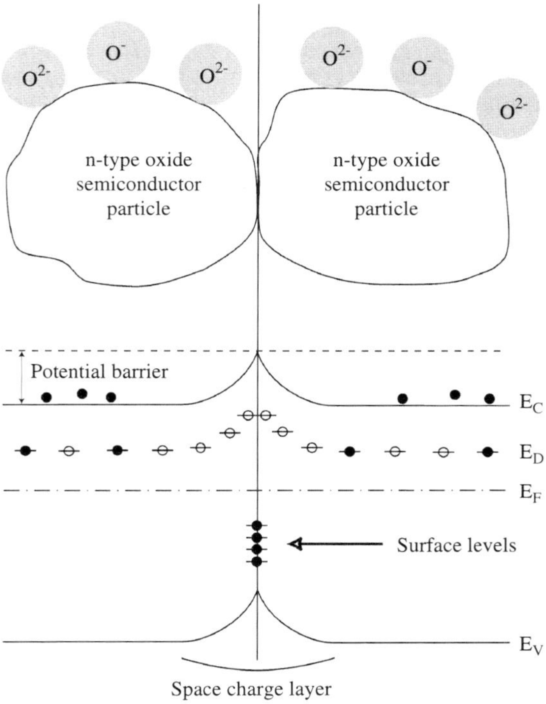

Figure 6.5.9
Schematic of the surface state of n-type oxide semiconductor particles placed at 200–400°C in air.

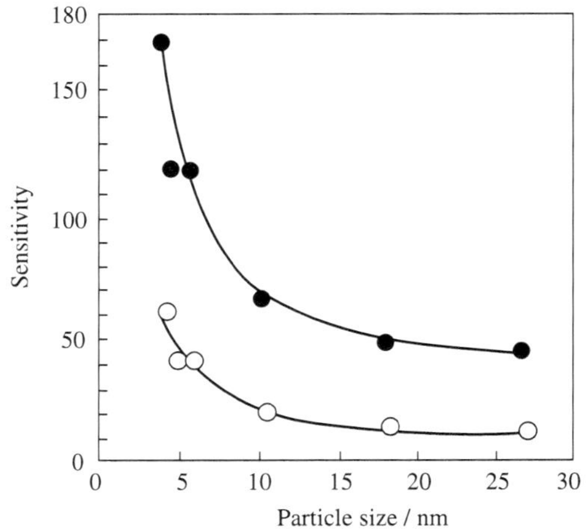

Figure 6.5.10
Relationship between particle size and sensitivity of SnO_2 ceramics at 300°C [1]. ●: 800 ppm H_2; ○: 800 ppm CO.

which is a typical oxide semiconductor gas sensor material [1]. The measurement was performed at 300°C in air containing 800 ppm H_2 or CO. The sensitivity increases significantly with decreasing particle size lower than 10 nm. It has been reported that the thickness of the space charge layer of SnO_2 is approximately 3 nm [2]. From these results, the sensitivity of gas detection is considered to be improved dramatically when the space charge layer occupies almost all the inside of each particle. Moreover, it has been indicated that not only the sensitivity but also the response improves by decreasing the particle size of CuO-doped SnO_2 to a few nanometers [3]. Thus, the gas detection is based on the chemical reaction on the surface of oxide semiconductor particles, and the characteristics can be remarkably improved by controlling the particles size in the nanometer scale.

In order to increase contact frequency between the surface of the particle and the gas to detect, the oxide semiconductor gas sensors are commonly made to have porous microstructure. For this reason, it is very important in the improvement of the sensing characteristics to control the porosity of sensor structure in the nano/meso-size. Recently, ordered mesoporous ceramics, such as SnO_2 and TiO_2 which have stable structure even at high operating temperatures in the range of 200–400°C, has been synthesized and their gas-sensing characteristics have been evaluated. Fig. 6.5.11 describes the sensitivity to 500 ppm H_2 for ordered mesoporous SnO_2 ceramics having a large specific surface area of 305 m^2 g^{-1} [4]. Compared with the ceramics having a specific surface area of 8.4 m^2 g^{-1}, the advantage of the sensor having the mesoporous structure is clear. Incidentally, since the SnO_2 ceramics used for the measurement had large secondary particles, it is questionable whether mesoporous structure contributed to the enhancement of gas detection effectively. Therefore, it is expected that the sensitivity of the oxide semiconductor gas sensor with ordered mesoporous structure increases further with decreasing size of the secondary particles.

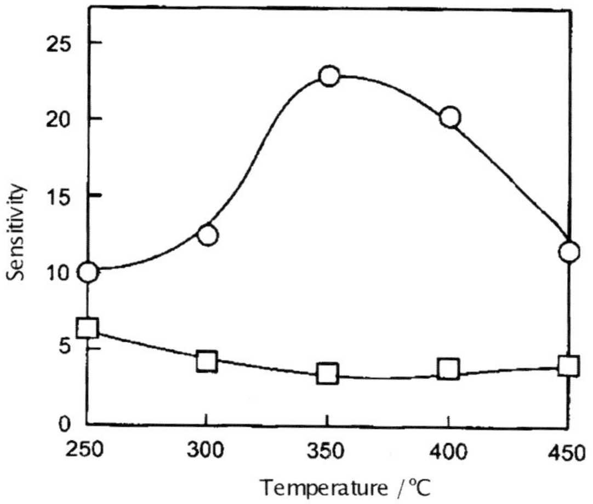

Figure 6.5.11
The sensitivity to 500 ppm H_2 of SnO_2 with various specific surface areas [4]. ○: 305 m^2 g^{-1}; □: 8.4 $m^2 \cdot g^{-1}$.

(2) Characteristic of nanosized metal sensors using surface plasmon resonance

It is known that a surface plasmon resonance which is an excitation of plasmon by light occurs on the surface of metal, and light with a specific wavelength is absorbed. For this reason, the metal nanoparticles used for coloring of the stained glass in church and ornament glass exhibit a vivid color for a long time. The chemical sensors using the surface plasmon resonance attract attention in recent years.

Since plasmon which is the quantum of plasma oscillations is vibration of electron density, it normally does not interact with light which is a transverse wave. However, for example, if a nanosized thin metal deposited on a prism will be irradiated with p-polarized light through the prism under the conditions of total internal reflection, an exponentially decaying wave called an evanescent wave will be generated on the metal surface, and the excitation of plasmon will take place.

The penetration depth of the evanescent wave is comparable with a wavelength of light, and the wave number k_e is given by the following formula:

$$k_e = \left(\frac{\omega}{c}\right) n_p \sin\theta \qquad (6.5.3)$$

where ω is the angular frequency of light, c the speed of light in vacuum, n_p the refractive index of the prism and θ an incident angle of light. The surface plasmon resonance arises when the incident light is absorbed under a coupling condition that k_e is equal to the wave number k_{sp} of the surface plasma oscillation:

$$k_{sp} = \left(\frac{\omega}{c}\right) \sqrt{\frac{\varepsilon_m \varepsilon_s}{\varepsilon_m + \varepsilon_s}} \qquad (6.5.4)$$

where ε_m and ε_s are dielectric constants of the metal and thin layer on the metal surface, respectively. If a laser is used as a light source, the coupling condition will depend on the incident angle θ. Therefore, if θ is changed, reflected light will be absorbed at a certain incident angle.

Thus the coupling condition for the surface plasmon resonance is remarkably sensitive to change of the dielectric constant near the metal surface. Therefore, as shown in Fig. 6.5.12, a sensor can be composed of a prism and a metal thin-film supporting receptors which selectively bind with specific molecule to detect. Schematics of signal changes of the sensor using the surface plasmon resonance are shown in Fig. 6.5.13. If the supported receptor binds with the molecule, the dielectric constant near the metal surface will change, and the absorption angle will shift. The detection of the specific molecule is performed based on the incident angle dependence of the reflected intensity. Moreover, if the incident angle θ_1 is fixed near the absorption angle, real-time measurement will be realized as shown in Fig. 6.5.13b.

The metal nanoparticles as well as a metal thin-film are applicable to the sensor using surface plasmon resonance. If the nanoparticles smaller than the wavelength of light are irradiated, the excitation of plasmon named as a localized surface plasmon resonance will occur resulting in absorption and scattering of the light of a specific wavelength. Therefore, the detection of the nanoparticle sensor is based on the change of transmission and/or scattering of light. Since the sensing system needs no optical system with a prism, the sensor has an advantage that the whole sensing system can be made compact compared with the thin-film sensor. Theoretically, a single metal nanoparticle can be used as a detection medium. For example, a sensor can be made by attaching the nanoparticle on the end of an optical fiber.

Fig. 6.5.14 shows the plasmon scattering from a single gold nanoparticle in various solvents measured through a optical fiber [5]. The nanoparticle with 50–100 nm in diameter was attached on the tip of an axicon microlens fabricated on the end of an optical fiber. The position of the peak changes in the various solvent, implying that the nanoparticle is sensitive to the deference of the solvent. Since the peak change is attributed to the change of the dielectric constant near the metal surface, it is expected that the single nanoparticle can be used for chemical and biological sensors of a nanometer scale.

The chemical sensor using surface plasmon resonance is indispensable in the field of the present biochemistry or biotechnology because of the following reasons:

1. The real-time measurement for a small quantity of specimen is possible.
2. Labeling substances, such as a fluorescent material, are not required.
3. The manufacturing cost of a sensor is low.
4. Disposable use is possible.

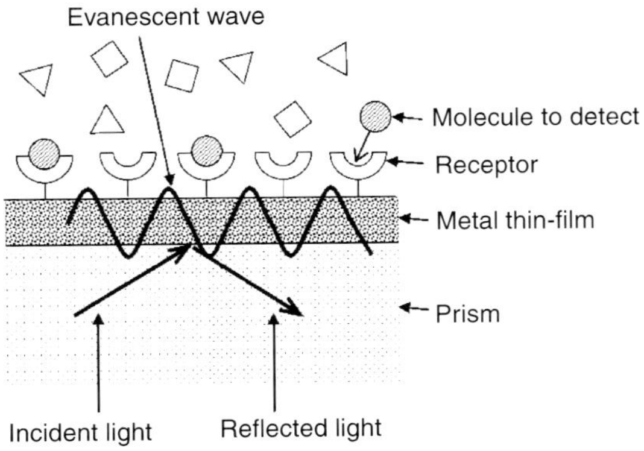

Figure 6.5.12
Schematic diagram of surface plasmon resonance sensor.

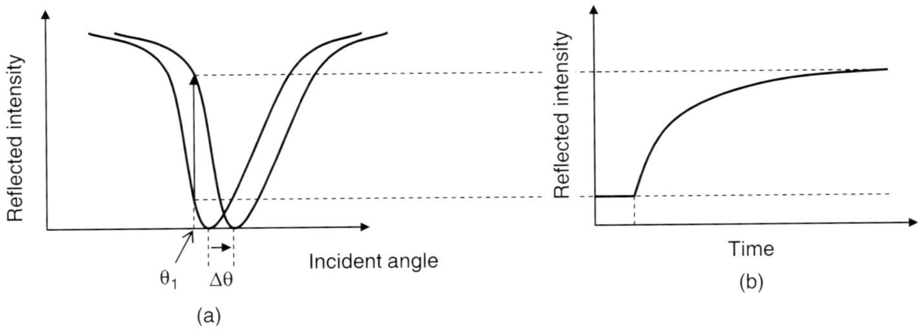

Figure 6.5.13
Schematic of detection characteristics of surface plasmon resonance sensor. (a) Change in the absorption angle that the surface plasmon resonance takes place. (b) Change in reflected light intensity with time.

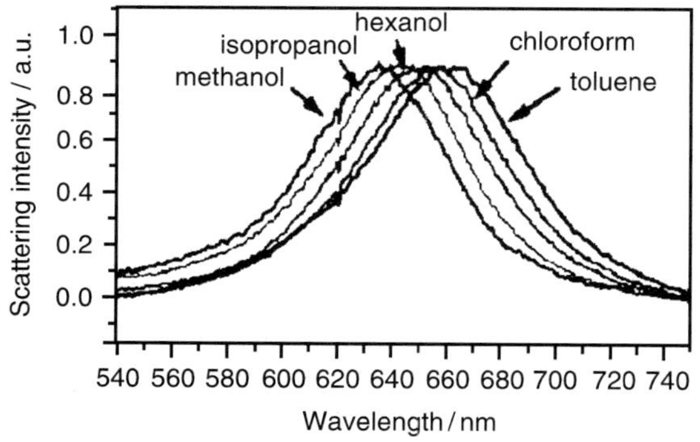

Figure 6.5.14
Plasmon scattering from a single gold nanoparticle in various solvents measured through optical fiber [5].

It is expected that the sensor using surface plasmon resonance will be applied to the new field where the conventional sensors were not able to be used.

References

[1] C. Xu, J. Tamaki, N. Miura and N. Yamazoe: *Sens. Actuat. B*, **3**, 147–155 (1991).
[2] H. Ogawa, M. Nishikawa and A. Abe: *J. Appl. Phys.*, **53**, 4448–4455 (1982).
[3] G. Zhang and M. Liu: *Sens. Actuat. B*, **69**, 144–152 (2000).
[4] Y. Shimizu, T. Hyodo and M. Egashira: *J. Eur. Ceram. Soc.*, **24**, 1389–1398 (2004).
[5] S. Eah, H. Jaeger, N. Scherer, G. Wiederrecht and X. Lin: *Appl. Phys. Lett.*, **86**, 031902 1–3 (2005).

6.5.3 Electrochemical reactivity

6.5.3.1 Introduction

Electrochemical reactivity is a phenomenon that transports electrons through electrodes with a chemical reaction [1]. For example, Solid Oxide Fuel Cell is considered as one of electrochemical energy converters. Since essential electrochemical reaction proceeds at the nanometer scale in the electrodes, nanoparticles take an important role in the enhancement of the electrochemical reaction of electrochemical cells. Another well-known phenomenon concerning nanoscale reaction occurs at the "space charge" layer. The concept is called "Nano-ionics" [2], newly proposed by Professor Y. Maier. It explained the remarkable increase in the ionic conductivity by polarization [3], which is caused by the nanoscale alternation of thin films of different types of ionic conductors such as fluorides and oxides. Nanoscale electrochemical reaction is expected to enhance the reaction by networking the nanoparticles. Furthermore, there are many possibilities to design and fabricate the interfaces of matrix and nanoparticles for hyper-ionic conductivity.

In this chapter, electrochemical properties are described considering the contribution of nanoparticles, especially from the standpoint of view of chemical reaction with material transportation and energy conversion.

6.5.3.2 Electrochemical reaction influenced by nanoparticles

SOFC is the most well-known example of the devices using the electrochemical reaction. The reaction occurs in the electrodes with the electrolyte. Electrochemical reaction is defined as a reaction that transports electrons at both electrodes in an electrochemical cell (Fig. 6.5.15). In the case of SOFCs, the concentration difference of oxygen between fuel (hydrogen, hydrocarbon etc.) and air or oxygen promotes the diffusion of oxygen ion through the electrolyte (oxygen ionic conductor) at an elevated temperature, which allows enough mobility of the ions.

The reaction in SOFCs is the conversion of chemical to electrical energy. Its revered reaction is the electro-decomposition of water. Most important factors influencing the electrochemical reaction are the mobility of ion and the micro-structural control for the enhancement of chemical or electrochemical reaction.

Fig. 6.5.16 shows a schematic drawing of SOFC cathode and gas molecules passing through the surfaces of solid particles are ionized by the electrochemical reaction at the Three Phase Boundary (TPB) site, which is usually distributed at the interface of nanoparticle and surrounding space (gas phase).

Another typical example of electrochemical reactions occurring at the interfaces is the photo-catalytic reaction, which causes effective separation of electrons and holes [4].

Controlling the electrical and ionic conductivities and the microstructure of electrodes to improve the electrochemical reactivity is the main target of the research and development in the SOFC applications. Fig. 6.5.17 shows the ionic conductivities of various electrolyte materials as a function of cell working temperature. Yttrium stabilized zirconia has been a most popular electrolyte material because of its relatively high ionic conductivity with good mechanical property and phase stability at high working temperatures. Recently, scandium-doped zirconia, Lanthanum garate (LSGM) and ceria based doped oxides are expected to be new electrolyte candidates for low-temperature applications. Electrochemical reaction proceeds at the interfaces between the electrodes and electrolyte. The electrolyte transports the reaction media (ions). That is why thinner and even electrolyte films are preferred for high performance cells to produce more electrons. On the other hand, a thick electrode with fine particles is required for having high reaction volume, good ionic connectivity and electrical conducting grains. Therefore, controlling the structure of

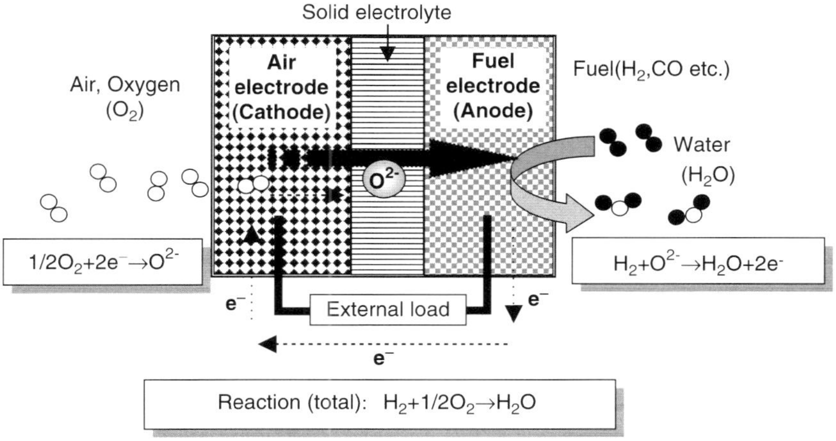

Figure 6.5.15
Schematics of the electrochemical reaction (e.g., solid oxide fuel cell).

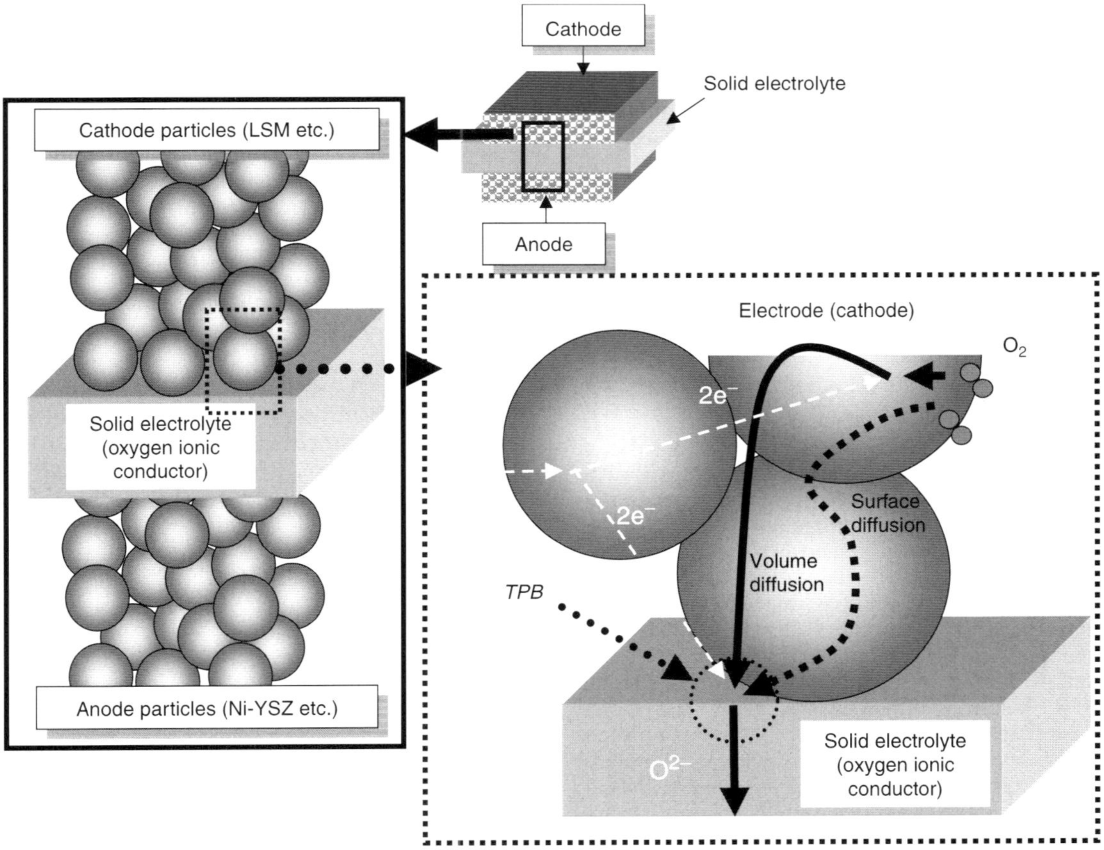

Figure 6.5.16
Schematics of the electrochemical reaction that converts gas molecules to ions at the TPB in the cathode with nanoparticles and nanospace.

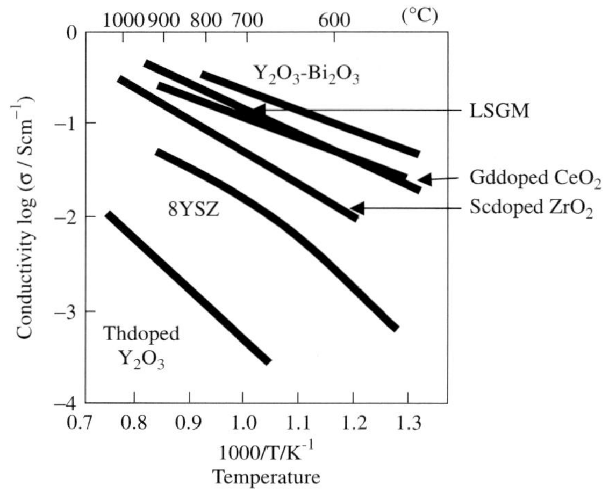

Figure 6.5.17
Oxygen ionic conductivity of various materials as a function of cell working temperatures.

various components in the electrochemical cells in the range of nano, micro and macro scale is well being studied in the processing technology development such as the fabrication of nanocomposite powders.

The concept of "percolation" in the electrodes is very important to improve the performance of electrochemical reaction. It is to maintain network structure for both ionic and electrical conducting phases such as in the SOFCs. This network structure includes the pore size distribution from the surface of electrode to the permeable area where the gas molecules are converted to ions.

6.5.3.3 Control of the electrochemical reaction and its application on the development of SOFCs

Reaction at the TPB in SOFC is the most significant example of applying nanoparticles in the field of electrochemistry. The reaction at TPB to ionize gas molecules and its reversed reaction are not only affected by nanoparticles themselves but also by the nanostructures they form and their surroundings. In this chapter, the nanoscale reaction zone in SOFC is described in detail from the view point of electrochemistry.

Since fuel cells have high efficiencies in power-generation, they can drastically reduce the emission of CO_2, which is considered causing the global warming. Various types of fuel cells such as polymer electrolyte fuel cells (PEFCs), molten carbonate fuel cells (MCFCs), phosphoric acid fuel cells (PAFCs) and SOFCs, have been developed. Among them, PEFCs have gained great attention for home and vehicle applications so far. On the other hand, SOFCs have great potential because they have higher power-generation efficiency, and utilize ceramics technologies.

Recent trends on the SOFC R&D focus on: (1) the improvement of reactivity of gas molecules and ions at electrodes (2) the enhancement of ionic conductivity of the electrolyte. In the former case, nanoparticles and their surrounding nanomicron scale area play an important role in their electrochemical reactivity. As mentioned previously, the enhancement of electrochemical reaction at the TPB area is most important in the development of SOFC. Recently, the introduction of nanotechnology with an attempt to apply nanoparticles to construct nanostructure becomes the focus of SOFC R&D.

Because of the restriction of the electrochemical reaction, electrodes should have good ionic and electrical conductivities to expedite the reaction. Therefore, the electrode materials should exhibit a high catalytic activity for the desired electrochemical reactions. Nanoparticles are expected to have high reactivity as catalysts. Over-potential in the fuel cell reflects the loss of reactivity, which can be caused by various reasons when converting gas molecules to ions. Fig. 6.5.18 indicates the voltage drops at each cell component as a function of cell operating temperature. Suppressing over-potential in the SOFC can significantly enhance its electrochemical reactivity with proper ionic conductivity. The over-potential is induced by the loss of reactivity between gas molecules and ions at the electrodes, which is mainly due to the polarization losses at the cathode. This is the reason of applying nanoparticles to make electrodes with better performance controlling factors, such as gas diffusion, distribution of nanoreaction zone (TPB) and ionic or electrical conductivity, to reduce the over-potential.

Figs. 6.5.15 and 6.5.16 show the schematic reaction of a SOFC cell, which indicates a cathode and an anode in the cell for oxidation and reduction reaction respectively. Reaction of a gas molecule to an ion is reversible in principle. In the cathode, oxygen molecule is absorbed on the surface of an electrode particle and migrated or transported by the surface

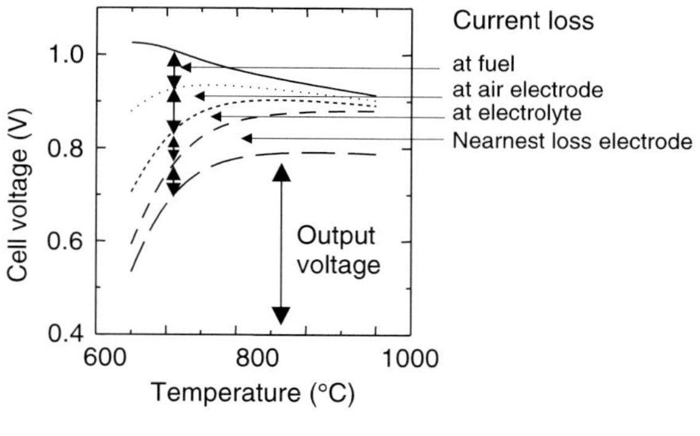

Figure 6.5.18
Factors influencing electrical losses of a SOFC in relation to cell working voltage and temperature.

diffusion or volume diffusion mechanism, respectively. Oxygen molecule turns into oxygen ion with the supplied electrons after arriving at the TPB and is transported under a concentration gradation of oxygen between the cathode and anode thorough the electrolyte. In the anode, oxygen ion turns back to oxygen molecule or water or carbon dioxide depending on the kind of fuel supplied to the anode. The released electrons cause output current of the SOFC by connecting both electrodes thorough an outside load. The differences in the ionization energy of the electrodes derived from their composition and microstructure affect the reactivity between gas molecules and ions [5–8].

The state-of-the-art anode, Ni-YSZ cermet, fine particles are used to produce a large number of active TPB sites. The electrochemical active zone consists of percolated matrix having ionic and electrical conductors with the following advantages: (1) high porous surface area with excellent electrochemical conductivity, (2) high electronic conductivity, (3) high stability through prolonged electrochemical reaction time, (4) good match with electrolyte in thermal expansion properties, (5) long lasting catalytic properties for hydrogen–proton converting reaction and (6) good sustainability as a material for its supply sources and cost effectiveness for massive production [6].

Novel applicable fabrication technology for the electrodes has been developed including screen printing [7], plasma coating [8] and electrostatic deposition method to layer the electrode materials on the electrolyte [9]. After the reduction of coated Ni-YSZ film, metallic Ni particles were precipitated to form a skeleton structure in the porous composite anode with YSZ particles.

6.5.3.4 Structural control of the electrode supported thin film electrolyte

An important factor for improving the SOFC properties (especially for the power density) is the control of the nanomicron structure in the macro electrode construction, that is the hyperstructural control of the electrode to support the thin electrolyte. To reduce the polarization loss at the electrolyte, electrode supported thin film electrolyte is believed to be most effective. The entire electrode should be porous for good gas permeability to maintain high reactivity, while dense enough at the coated thin-film electrolyte interface to have high ionic conductivity. Ordinary layering procedure is difficult to accomplish all the requirements. A novel processing technology, "colloidal processing", was developed to fabricate the unidirectional pore shape (cylindrical) and size distribution through the electrode with nano to micron scale structure at both of its surfaces. This enabled simultaneously achieving high permeability and high reactivity of the electrode (Fig. 6.5.19), and resulted in the increase of ionic conductivity of the fuel cell [10].

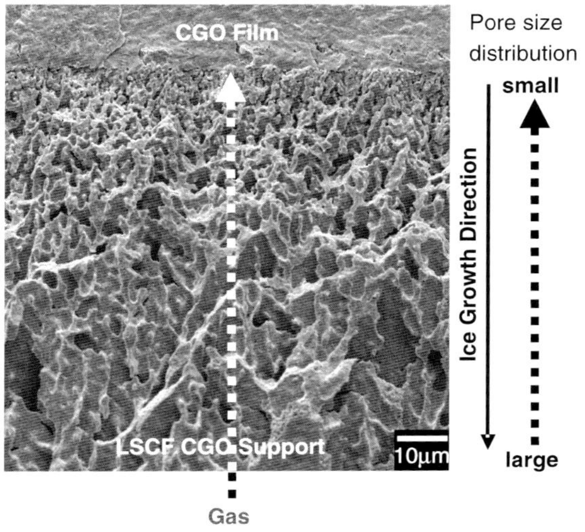

Figure 6.5.19
An example of controlling the internal structure of SOFC in the range of nano, micro and macro scale, showing uniaxial pore shape and pore size distribution vertical to the cell surface.

Ceria-based composite electrode is one of the typical examples for using nanoparticle-derived microstructure to enhance the properties of SOFCs. The ceria composite is well known for its high catalytic performance in the exhaust gas purification. Ceria has a high oxygen ionic conductivity at relatively lower cell working temperatures because of its nanoscale crystal structure. There have been many attempts to apply ceria compounds to make the cells with the chemical processing method such as spin coating. It was reported that SOFCs could be successully fabricated to perform at the working temperature of 500°C [11].

Recently new approaches to support the improvement of SOFCs are the use of "real time", "in situ" and "nano scale" analyses, such as the electrochemical STM [12], for the electrochemical cells. In addition, galvanometric measurement in the nanoscale range will further be developed because of its advantages in providing the essential analysis for the electrochemical phenomenon.

As described above, nanoparticles play an important role in the electrochemical reaction. Further applications of the nanoparticles are expected to provide solutions to various energy and environmental problems.

References

[1] B.C.H. Steele, A. Heinzel: *Nature*, **414**, 345 (2002).

[2] O. Nakamura, J.B. Goodenough: *Solid State Ionics*, **7**, 119–124 (1982).

[3] N. Sata, K. Eberman, K. Ebert and J. Maier: *Nature*, **408**, 21/28 Dec. (2000).

[4] M. Adachi, Y. Murata, J. Takao and J. Jiu: *J. Am. Chem. Soc.*, **126**, 14943 (2004).
[5] T.E. Ivers, A. Webee and D. Herbstritt: *J. Eur. Ceram. Soc.*, **21**, 1805–1811 (2001).
[6] A. Atokinson, S. Barnett, R.J. Gorte, J.T.S. Irvine, A.J. Mcevoy, M. Mogensen, S.C. Singhal, J. Vohs: *Nature Materials,* 3, January, 17–27 (2004).
[7] H. Yokogawa: *Ceramics*, **36**(7), 472–476 (2001).
[8] T. Yamaguchi, M. Matsumoto and H. Matsubara: *Ceramics*, **39**(4), 281–285 (2004).
[9] M. Aizawa: *Ceramics*, **36**(7), 493–495 (2001).
[10] J.-W. Moon, H. Hwang and M. Awano: *J. Ceram. Soc. Jpn.*, **110**(5), 479–484 (2002).
[11] S. Wang, M. Awano and K. Maeda: *J. Ceram. Soc. Jpn.*, **110**(8), 703–709 (2002).
[12] T. Yamada, N. Batina and K. Itaya: *J. Phys. Chem.*, **335**, 204 (1995).

6.6 Magnetic properties

The areas of research on magnetic nanomaterials are – magnetic fluids, magnetic granular structure thin films, magnetic multi-layer films, magnetic recording media, magnetic metallic glasses, magnetic amorphous alloys, permanent magnet materials, medical magnetic materials, etc.

While outstanding progress has been made in recent years in each of these areas, some of the areas that utilize the behavior of nanoparticles are – magnetic fluids and coated magnetic recording media, drug-delivery systems in the field of medicine, etc. Magnetic fluids utilize the super paramagnetism and dispersion stability in liquids of nanoparticles, and in general, they use materials with low crystalline magnetic anisotropy (such as magnetite or iron-based alloys).

On the other hand, in a magnetic recording medium, since it is desirable that there are more magnetic particles per unit area, fine particles that also have a large magnetic coercivity are required. In the case of medical magnetic materials, because they are used inside the human body, biological safety and not blocking the blood vessels, etc., become important requirements and hence magnetite, etc., with particle sizes of less than 50 nm are being used in their research.

6.6.1 Super paramagnetism

When the size of magnetic particles is made small, in a region of sizes less than a certain critical diameter, since the total magnetic energy becomes small when there is no magnetic wall within the particle, it becomes a single magnetic domain particle. In addition, if the particle size is made small, it exhibits super paramagnetism because of thermal fluctuations of the magnetic moment of the particle.

In the magnetic nanomaterials used in magnetic recording media or materials for magnets, it is desirable that they have a large magnetic anisotropy so that they do not become super paramagnetic. The origin of magnetic anisotropy of magnetic nanomaterials is not only the crystalline magnetic anisotropy of bulk materials, but also the shape of the particles and the specific magnetic structures at the surface of the particles are considered to be important factors. Kawamura et al. [1] have calculated the magnitudes of the coercive force of cylindrical shaped nanoparticles using a core shell model with the crystalline magnetic anisotropy constant K_1 of bulk material and the specific magnetic anisotropy (surface magnetic anisotropy) edge K_1, and compared them with actually measured values.

From the results of these calculations and experiments, it became clear that, for the same size, the intrinsic coercivity of particles increases gradually as the value of the surface magnetic anisotropy edge K_1 increases, and also, that the coercivity becomes higher as the particle size becomes small in the case of single particles [1].

Therefore, in the design of the coercivity of the magnetic nanoparticles, it is necessary to consider the crystalline magnetic anisotropy that is intrinsic to the material, the shape of the particles (the magnetic anisotropy due to the shape) and the specific surface magnetic anisotropy.

While it is common to carry out the evaluation of the magnetic characteristics of super paramagnetic materials using a Vibrating Sample Magnetometer (VSM) or a Superconducting Quantum Interference Device (SQUID), in recent years it has been found that even X-ray Magnetic Circular Dichroism (XMCD) measurements using high-energy radiation X-rays are effective.

Since XMCD measurement can evaluate the magnetic polarization at which a specific element is placed from the difference between the left-polarized light and the right-polarized light due to the X-ray absorption edge of the specific element, there is the advantage that it is possible to measure the process of magnetization of that specific element without being affected by impurities.

6.6.2 Material-specific discussion

6.6.2.1 Nanospinel ferrite

Spinel ferrite is a material that has been used from the past. Even today its research is being done actively as a magnetic fluid, drug delivery material and as a magnetic recording material. Tanaka et al. [2] have prepared, using the hydrothermal method, nanoparticles of zinc ferrite which are a typical normal spinel ferrite, and have evaluated the magnetic characteristics.

Although bulk zinc ferrite is an antiferromagnetic material having a Neel temperature around 10 K, it has been reported to have a magnetization value of 11 emu g^{-1} at room temperature when the crystallite size is made small up to 4 nm with magnetic field of 1 T (tesla). Further, it has become clear from measurements of susceptibility that these particles have a magnetic transition point at a temperature in the range of several tens of K (kelvin). The reason that zinc ferrites, which have a small magnetization value at bulk sizes, have a large magnetization value at room temperature is considered to be that zinc ferrite nanoparticles exhibit super paramagnetism [2].

Further, Yamamoto et al. [3] have synthesized Co–Ni–Mn ferrite particles having a high coercivity using the chemical coprecipitation method. It has been reported that Cobalt ferrites have a large crystalline magnetic anisotropy constant K_1 compared to other spinel ferrites, and is hard to become super paramagnetic even when very fine particles are made. However it has been reported that these Co–Ni–Mn ferrites with a chemical equation of $[(CoO)_{0.5}\ (NiO)_{0.4}(MnO)_{0.1}{}^{-}1.125(Fe_2O_3)]$ exhibit unique magnetic characteristics; coercivity of 7.12 kOe, saturation magnetization of 44.2 emu g^{-1}, average particle size of 43 nm, crystalline magnetic anisotropy constant K_1 of $+1.36 \times 10^6$, K_2 of -11.3×10^6 and anisotropic magnetic field HA of 44.7 kOe.

Although the origin of such high coercivity is not clear, it has been reported, from measurements of the torque curve of magnetic sheets, that a large anisotropic magnetic field is present. If it is possible to reduce the particle size still further, it is considered that there is the possibility of use as a coated type magnetic material for high-density magnetic recording.

6.6.2.2 Pd nanoparticles

In the 4d transition metals Ru, Rh and Pd, it has been predicted from theoretical calculations [4,5] that the band structure changes when there is a reduction in the coordination number or a change in the symmetrical property due to low-dimensionalization, and there is a possibility of showing ferromagnetism. Shinohara et al. [6] obtained Pd nanoparticles (of a polyhedron shape) with less impurities using a sample preparation method in an Ar gas environment. From the measurement of magnetic field dependency of magnetization of these particles with an average diameter of 11 nm it was found that the magnetization at low magnetic fields increased suddenly, and after that it increased gradually in proportion to the magnetic field linearly. Further, from the hysterisis curve it became clear that they had a coercivity of about 40 Oe and a residual magnetization of 0.1 emu g^{-1}.

On the other hand, from the result of measuring the temperature dependence of magnetization it became clear that the saturation magnetization component is not lost up to 400 K. Further, there is complex relationship between the saturation magnetization component and the size of the Pd nanoparticles, and it became clear that ferromagnetic component appears in the size range of 6–14.4 nm.

Shinohara et al. [6] inferred that the reason for the Pd nanoparticles to have a ferromagnetic component is that the (100) surface of the polyhedron nanocrystalline particle is ferromagnetic. This also does not contradict the results of the theoretical calculations. Further, it is suggested [6] that there is a possibility that the magnetic anisotropic energy of these nanoparticles has about the same magnitude as that of 3d transition metals.

6.6.2.3 FePt nanoparticles

While FePt is a ferromagnetic material known from the past, recently it is being studied [7] actively as a material having the possibility of achieving a recording density of the order of 1 Tb in^{-2} because it has a large crystalline magnetic anisotropy. The $L1_0$-FePt phase which is an ordered alloy has a multilayer structure of Fe atoms and Pt atoms in the C-axis direction, and its crystalline magnetic anisotropic energy *Ku* reaches [8] 7×10^7 erg cm^{-3}.

The super paramagnetism limiting diameter obtained from this *Ku* and the thermal fluctuation energy *kT* is about 4 nm and is one of the smallest among the magnetic materials known at present. In addition, it is possible to obtain nanoparticles of these FePt particles relatively easily using the polyol process, etc., and these particles are also superior in resistance to oxidation and resistance to chemicals.

However, since FePt nanoparticles that have been obtained using the synthesizing methods reported so far are made of unordered phases, they require some form of heating process. However, since fusion between FePt nanoparticles takes place due to the heating process, this has become a problem from the point of view of dispersibility of nanoparticles and their orderly arrangement.

In order to solve this problem, doping of a third element to FePt particles or fusion prevention treatment on the surface of FePt particles, etc. are being investigated. The size of $L1_0$-FePt particles obtained so far is 3–10 nm, and their magnetic characteristics are a coercivity of 1,000–10,000 Oe, and a saturation magnetization of about 50 emu g^{-1}. Further, research is being made actively not only on wet type chemical processes but also on FePt thin film growth using a vapor phase method [9,10].

There is the possibility of very big technical advancement in materials science and technology because many people from various fields are involved in this materials research. In order to realize coating type magnetic recording materials it is considered necessary to give sufficient attention to global resources and economical matters.

6.6.2.4 Metal nitrides

Many types of iron nitrides, which are penetration type nitride materials, are present such as Fe_3N, Fe_4N, $Fe_{16}N_2$, etc., and ever since Kim and Takahashi et al. [11] found that thin film $Fe_{16}N_2$ has a saturation magnetization exceeding that of α-Fe, it has attracted the attention of a large number of researchers. Also, a lot of research has been done on the synthesis of bulk material and thin films of $Fe_{16}N_2$.

Nagatomi et al. [12] succeeded in synthesizing bulk material of $Fe_{16}N_2$ by a process of nitriding α-Fe thin film or nanoparticles of α-Fe under an ammonium gas atmosphere at a low temperature (of 200°C or less). Further, Hattori et al. [13] obtained an almost single phase $Fe_{16}N_2$ by a process of nitriding, ultrafine particles (10–50 nm) of γ-Fe_2O_3 under an ammonium gas atmosphere. The saturation magnetization of the obtained powder (with a specific surface area of about 20 m^2 g^{-1}) was about 200 emu g^{-1}, which was about 20% smaller than the predicted theoretical value (257 emu g^{-1}).

The reason for the saturation magnetization being lower than the predicted value is considered to be the disturbance in the particle surface due to reduction in the particle size. On the other hand, the coercivity of this powder is about 1,200 to 2,300 Oe and suggests a large crystalline magnetic anisotropy [13]. Further, Sasaki et al. [14] obtained single phase $Fe_{16}N_2$ with a size of 17 nm using a precursor of 20 nm size magnetite particles covered with an yttrium and silicon compound. The saturation magnetization of this powder was 89 emu g^{-1} and its coercivity was 2,900 Oe.

Although the saturation magnetization is much smaller than the theoretical value, it has reached a practical level as a high-density magnetic recording material [14]. $Fe_{16}N_2$ nanoparticles do not have the large size or large magnetic anisotropy of FePt particles, but they can be considered to be a very useful material considering the global resources and the manufacturing cost.

Although the history of development of magnetic nanomaterials is old, need for its development has not dwindled due to the development of new functions or applications. If it is possible to control in the nanosize levels the magnetic particle size and the state of aggregation, it is considered that new applications can be developed. To explain the specific characteristics of nanosized magnetic materials, it is necessary to carry out further accurate characterization of the surfaces of particles and the boundaries between them and even theoretical developments in this field are eagerly awaited.

References

[1] S. Kawamura, K. Haneda: *J. Jpn. Soc. Powder Powder Metall.*, **51**(9), 703–707 (2004).

[2] K. Tanaka, M. Katsuta, S. Nakashima and K. Fujita: *J. Jpn. Soc. Powder Powder Metall.*, **52**(4), 221–227 (2004).

[3] H. Yamamoto, H. Nishio and N. Yoshida: *Magn. Soc. Jpn.*, **29**(3), 269–273 (2005).

[4] S. Blugel: *Phys. Rev. Lett.*, **68**, 851 (1992).

[5] B.V. Reddy, S.N. Khanna and B.I. Dunlap: *Phys. Rev. Lett.*, **3323**, 70 (1993).

[6] T. Shinohara, T. Sato: *Kotai Butsuri (Solid Stage Physics)*, **40**(8), 535–544 (2005).

[7] L. Graf, A. Kussmann: *Phisik. Z.*, **36**, 544–551 (1935).

[8] S. Sun, C.B. Murray, D. Weller, L. Folks and A. Moser: *Science*, **287**(5460), 1989–1992 (2000).

[9] Y. Takahashi, K. Hono: *Magn. Soc. Jpn.*, **29**(2), 72–79 (2005).

[10] T. Shima, K. Takanashi, Y.K. Takahasi and K. Hono: *Appl. Phys. Lett.*, **85**, 2571–2573 (2004).

[11] T.K. Kim, M. Takahasi, *Appl. Phys. Lett.*, **20**, 492–494 (1972).

[12] A. Nagatomi, S. Kikkawa, T. Hinomura, S. Nasu and F. Kanamaru: *J. Jpn. Soc. Powder Powder Metall.*, **46**(2), 151–155 (1999).

[13] T. Hattori, N. Kamiya and Y. Kato: *Magn. Soc. Jpn.*, **25**, 927–930 (2001).

[14] Y. Sasaki, N. Usuki, K. Matsuo and M. Kishimoto: *INTERMAG*, **GR02**, 198 (2005).

[15] T. Inoue, K. Nakiri, H. Mitsuhashi, M. Fukumoto and T. Doi: *Y. INTERMAG*, **GR03**, 862 (2005).

6.7 Optical properties

6.7.1 Transparency of nanoparticle

Free electrons in a conductor (metal) move in a collective motion called plasma oscillation. An electromagnetic wave with a frequency that is higher than a plasma frequency can penetrate plasma, whereas that with a lower frequency is absorbed or reflected. In order for a conductor to be transparent to visible light, but absorb or reflect near-infrared rays, its plasma frequency must be 800–1,000 nm, near the border separating the visible light region and the near infrared region. Because conductive metal oxides like indium tin oxide (ITO) have a plasma frequency in the near infrared region, they are transparent to visible light, but are opaque to near infrared light. Common metals with high free electron density have a high plasma frequency. Therefore, they are usually opaque to both near-infrared rays and visible light. On the other hand, insulators such as ceramics with no free electron are normally transparent to both.

In order for a substance to be transparent to visible light, it is necessary that it absorb, reflect or scatter light only slightly. When light penetrates a substance, some of the incident light is reflected on its surface; the remainder passes into its bulk. Reflection is a phenomenon by which photons rebound from the surface of a nanoparticle with basically no wavelength change. Unreflected light is absorbed or scattered within a substance. A fraction of the absorbed light is converted to heat or is reemitted as light (electromagnetic waves) of a different wavelength (photoluminescence). Scattering takes place based on the inhomogeneity of a substance.

The pattern of scattering depends on the light wavelength and the particle size to be measured. A large particle, with micrometer to millimeter order diameter, scatters light forward at a comparatively narrow angle. Mie scattering prevails in particulates with a particle size that is comparable to the wavelength of visible light [1]. In the Mie scattering region, as shown in Fig. 6.7.1, light is also backscattered. The smaller the particle size becomes, the higher is the ratio of backscattering to forward scattering. For particulates of 200 nm or less, the intensity distribution of forward scattered light changes only slightly, but a great change appears in the intensity distribution of backscattered light. Mie addressed the diffraction of a plane monochromatic wave with a uniform sphere with an arbitrary diameter in a homogeneous medium, using electromagnetics to obtain an exact solution for the scattering pattern in 1908. Numerical solutions calculated by many researchers are described in the work of Van de Hulst [2]. This principle has been applied also to the laser diffraction particle-size measurement technique of particulates that are commonly used today.

Rayleigh scattering prevails for even smaller particles of nanometer range [3]. Because incident light penetrates almost uniformly throughout the entire particle, light is scattered symmetrically to form a pattern like a cocoon, as shown in Fig. 6.7.2. The light intensity of Rayleigh scattering I_θ is expressed by the following equation:

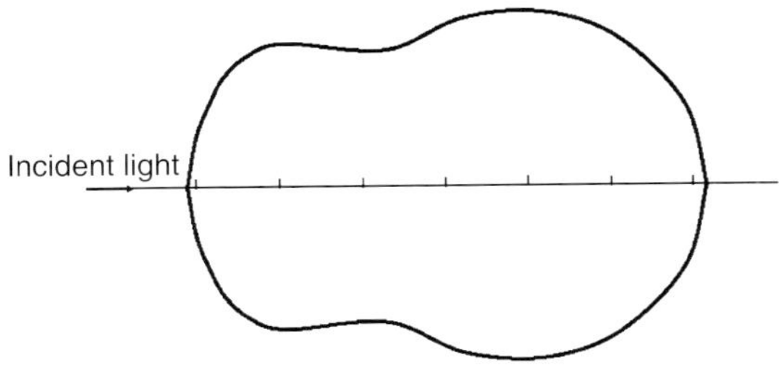

Figure 6.7.1
Mie scattering of fine particles.

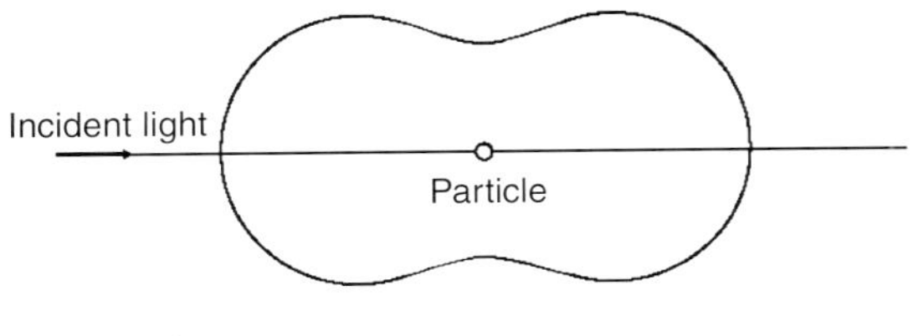

Figure 6.7.2
Rayleigh scattering of nanoparticles.

$$\frac{I_\theta}{I_0} = \frac{\pi^4 d^6}{8R^2\lambda^4}\left(\frac{m^2-1}{m^2+1}\right)(1-\cos^2\theta) \qquad (6.7.1)$$

where I_0 is the intensity of the incident light, R the distance, m the refractive index, λ the wavelength of incident light, and d the particle diameter. This equation indicates that light with a shorter wavelength is scattered more vigorously. Moreover, as Rayleigh scattering decreases in proportion to the sixth power of particle size, nanoparticles scatter light less vigorously and become transparent. That is, if an isolated nanoparticle exists, it will appear almost transparent to visible light. However in fact, it is not common to use a sole nanoparticle in a perfectly isolated state as an optical material. One practically important embodiment is to disperse nanoparticles in a transparent matrix. Its examples include sunscreens and foundations with dispersed nanoscale titanium oxide pigment, and LEDs with nanoscale phopshor materials dispersed in resins [4]. Future work will enhance the functionality of nanoparticles, such as absorption and luminescence, while retaining transparency to visible light.

Such transparent films containing nanoparticles can be evaluated using the total light transmission and haze value. Total light transmission is the transmittance of light in a transparent matrix, and the experimental procedure on visible light and UV rays are regulated. For a highly transparent material, an integrating sphere transmissometer is used. The incident light quantity and the total light quantity that pass through a specimen are measured, and the ratio is computed and expressed as a percentage, for visible radiation and UV rays, respectively. The haze value represents the degree of opacity, and is given as the scattered light transmittance divided by total light transmission in percentage. Empirically speaking, one would notice misting under conditions such as backlighting when the haze value is greater than about 2%. Therefore, to maintain the transparency of a material, particles to be dispersed therein must be nanoparticles that scatter light only slightly. The particle size should be about 100 nm or less when ceramic particles of a refractive index of about 2 are dispersed. On the other hand, too much particle size reduction increases the particle amount required to assure a desired shielding property, in spite of high transparency. Some numerical computation results have been reported for shielding of UV rays, a practically important issue [5–7]. The result of Stamatakis et al., which is most commonly used, demonstrates that the optimal particle diameters of titanium oxide for the protection of UV irradiation are 50 nm for 300 nm UV rays and 120 nm for 400 nm UV rays, respectively. However, because it is difficult to exclude the effect of dispersion of nanoparticles completely, only a few studies have actually addressed particle size dependence on the shielding of UV rays. Sakamoto et al. conducted experiments using titanium oxide particles with various primary particle diameters. They reported the effect of the diameter on the shielding ability against

UV rays as compared to theoretical calculations [8]. Their result indicates an actual shielding ability that is close to the computational result of Stamatakis et al. They have also reported that the shielding ability of titanium oxide against UV rays is attributable to the combination of scattering power and absorbing power, and that absorbing and scattering powers are dominant for UV rays of short and long wavelengths, respectively. That is, uniform particle dispersion and optical absorption based on the band structure must be examined in control of transmittance and shielding ability, as well as a simple primary particle diameter. Regarding absorption, photochromic powder, whose reflectivity to visible light decreases under strong light and recovers to the initial level so that the original powder color is restored in darkness, has been put to practical use for cosmetics [9,10]. In addition, transparency is affected by the particle shape. Flat and tabular particulates reflect light to a constant direction to produce a glossy appearance, whereas spherical particles scatter light into many directions to produce a matte appearance. The shape of a transparent matrix is also an important factor. A diffusion film for LCD backlights implements light diffusion and collection performance by means of scattering by internal particles and scattering and collection by the outline of the film.

Optical properties of powder particles as a cluster are also important. Practical embodiments include fluorescent materials applied to display panels or luminescent lamps. Unlike the case of nanoparticles in transparent matrices like solution or thin films, light scattering cannot be ignored in the measurement of reflection and absorption spectra of powder materials. Measurements of optical properties require us to observe diffuse reflection light, which penetrates inside powder particles, repeats transmission and reflection and comes out on the surface again. Because diffuse reflected light has penetrated into powder particles repeatedly, weak absorption bands tend to be emphasized. Therefore, the measured result must be comparable with the transmission spectrum and to undergo quantitative treatment with the Kubelka–Munk function by Kubelka and Munk [11–13]. Light from a powder layer comprises a specular reflection light component that has been directly reflected on the surface, and a direct light component that has passed through the gap of powder particles. It is necessary to remove these to the greatest degree possible. For smaller particle sizes, the absorbed component by the powder layer is smaller, as the relative reflecting surface increases. That is, light cannot penetrate deeply into the powder layer. Moreover, particles with a diameter smaller than the wavelength of light would engender wavelength dependence in scattered light. An integrating sphere is effective for measurement of reflection and absorption spectra to collect accurate data. The configuration of an integrating sphere is shown in Fig. 6.7.3. Powder with uniform diffuse reflection properties in a wide

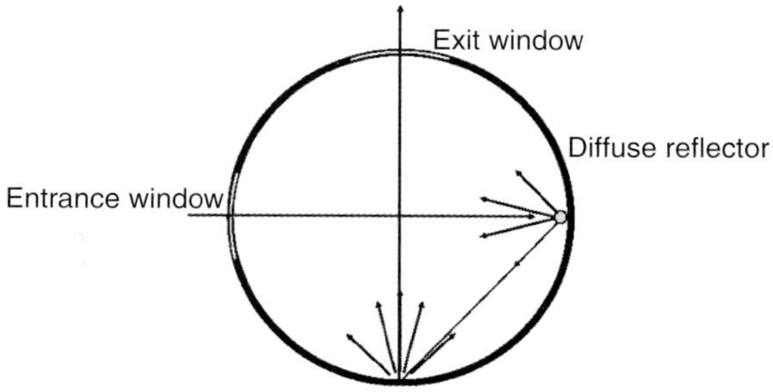

Figure 6.7.3
The configuration of an integrating sphere.

wavelength range, such as magnesium oxide or barium sulfate, is applied to the inner surface of an integrating sphere. This transforms the incident light into an almost perfect diffusion state. A slight residual specular reflection light component can be removed by an arrangement such that an entrance window and an exit window are positioned normal to each other. If the sample powder is fluorescent, the fluorescent component should be removed using a cut-off filter for the excitation wavelength region, or by synchronizing the detection side using a monochromatic light source.

Light emitted from one particle interacts again with another particle when light penetrates into a powder layer. Newly diffracted or scattered light occurs subsequently; this phenomenon will be repeated further. This process cannot be solved exactly mathematically. Therefore, a certain approximation is necessary to describe it. Typical analysis methods include:

1. the Scuster–Kubelka–Munk method, which regards a powder layer as a continuum [11–13],
2. the Johnson method, which assumes a stack of thin films [14,15] and
3. the Monte Carlo method, which computes scattered light stochastically [16–18].

Various nanoparticles used in pigments and fluorescent materials have various absorption and luminescent properties and shape. Almost no available reports describe whether the above-mentioned procedures are applicable to such nanoparticles. This will be a subject for our future study.

References

[1] G. Mie: *Ann. Phys.*, **25**, 377 (1908).
[2] H.C. Van De Hulst, H.C. Hulst: *Light Scattering by Small Particles (Structure of Matter Series.)*, Dover Publications, New York (1982).
[3] J.W.S. Rayleigh: *The Collected Optics Papers of Lord Rayleigh*, Optical Society of America, Washington (1994).

[4] T. Isobe: *Mater. Integr.*, **5**, 7 (2005).
[5] P. Stamatakis, B.R. Palmer, G.C. Salzman, C.F. Bohren and T.B. Allen: *J. Coat. Technol.*, **62**, 95 (1990).
[6] C.F. Bohren: *Am. J. Phys.*, **55**, 524 (1987).
[7] P.S. Mudgett and L.W. Richards: *Appl. Opt.*, **10**, 1485 (1971).
[8] M. Sakamoto, H. Okuda, H. Futamata, A. Sakai and M. Iida: *J. Jpn. Soc. Colour Mater.*, **68**, 203 (1995).
[9] K. Ohno, S. Kumagaya, T. Tanaka, T. Saito and F. Suzuki: *J. Soc. Cosmetic Chem.*, **27**, 314 (1993).
[10] K. Ogawa, N. Sakurai, S. Fuse, K. Ohno and S. Kumagaya: *J. Soc. Cosmetic Chem.*, **34**, 387 (2000).
[11] P. Kubelka, F. Munk: *Z. Techn. Phys.*, **12**, 593 (1931).
[12] P. Kubelka: *J. Opt. Soc. Am.*, **38**, 448 (1948).
[13] H.C. Hamaker: *Philips Res. Rep.*, **2**, 55 (1947).
[14] P.D. Johnson: *J. Opt. Soc. Am.*, **42**, 978 (1952).
[15] J.L. Ouweltjes: *Elektrizitasverwert.*, **11**, 12 (1958).
[16] T.S. Soules: *Electrochem. Soc. Extended Abstr.*, **74-1**, 311 (1974).
[17] K. Urabe: *Jpn. J. Appl. Phys.*, **19**, 885 (1980).
[18] K. Urabe: *Jpn. J. Appl. Phys.*, **20**, L28 (1981).

6.7.2 Photonic crystal

Photonic crystals composed of dielectric lattices form band gaps for electromagnetic waves [1,2]. These artificial crystals can totally reflect light or microwave at a wavelength comparable to the lattice spacings by Bragg deflection as shown in Fig. 6.7.4. Two different standing waves oscillating in the air and dielectric matrix form higher and lower frequency bands in the first and second Brillouin zones, respectively. The band gap width can be controlled by varying structure, filling ratio, and dielectric constant of the lattice. Structural modifications by introducing defects or varying the lattice spacing can control the propagation of light or microwaves. The band diagram of the photonic crystal along symmetry lines in the Brillouin zone is drawn theoretically. The Maxwell's equations (6.7.2) and (6.7.3) can be solved by means of the plane wave propagation method, where ω and c denote frequency and light velocity, respectively. Electronic and magnetic field $E\omega$ (r) and $H\omega$ (r) are described with the following plane wave equations (6.7.4) and (6.7.5), respectively. The periodic arrangement of dielectric constant $\varepsilon(r)$ can be obtained from the crystal structure [3]. G and k are reciprocal vector and wave vector, respectively.

$$\nabla \times \left[\frac{1}{\varepsilon(r)} \nabla \times H\omega(r)\right] = \left(\frac{\omega}{c}\right)^2 H\omega(r) \qquad (6.7.2)$$

$$\frac{1}{\varepsilon(r)} \nabla \times \left[\nabla \times E\omega(r)\right] = \left(\frac{\omega}{c}\right)^2 E\omega(r) \qquad (6.7.3)$$

$$\mathbf{H}_{k,n}(r) = \sum_{G} \mathbf{H}_{k,n}(\mathbf{G}) e^{i(k+G)r} \qquad (6.7.4)$$

$$\mathbf{E}_{k,n}(r) = \sum_{G} \mathbf{E}_{k,n}(\mathbf{G}) e^{i(k+G)r} \qquad (6.7.5)$$

$$\frac{1}{\varepsilon(r)} = \sum_{G} \frac{1}{\varepsilon(G)} e^{iGr} \qquad (6.7.6)$$

Fig. 6.7.5 shows expected applications of photonic crystal for light and electromagnetic wave control in various wavelength ranges. Air guides formed in a photonic crystal with nanometer order size will be used as the light wave circuit in the perfect reflective

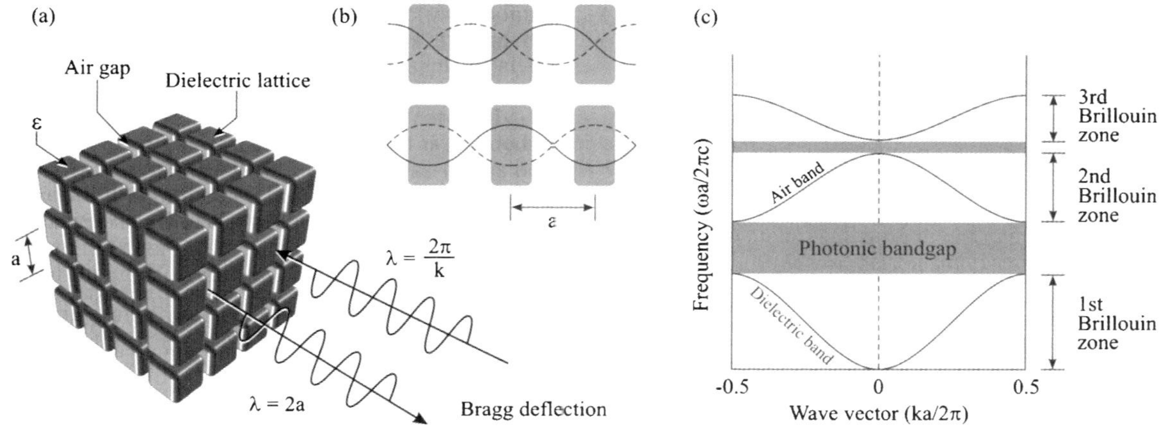

Figure 6.7.4
Schematic illustrations of photonic band gap formation. (a) Bragg deflection of electromagnetic waves in a photonic crystal, (b) standing waves in periodic arrangements of dielectric materials, (c) forbidden gaps in an electromagnetic band diagram.

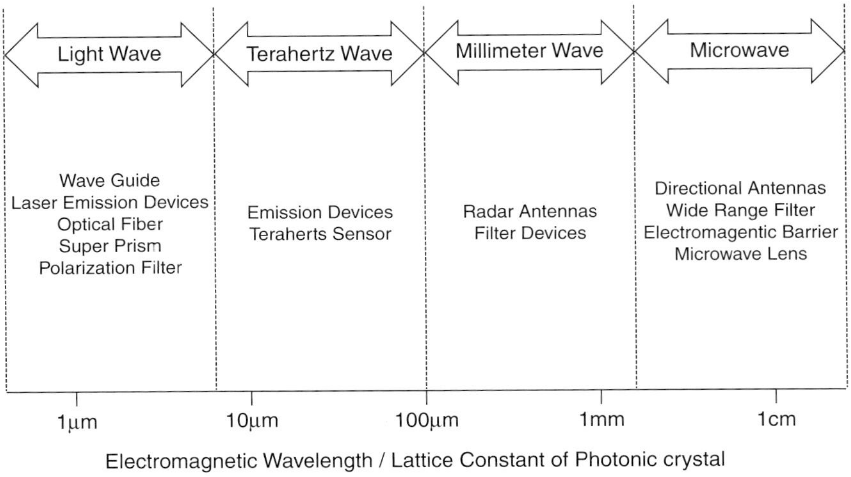

Figure 6.7.5
Expected applications of photonic crystal in various electromagnetic wavelengths.

structure. When a light emitting diode is placed in an air cavity formed in a photonic crystal, an efficient laser emission can be enhanced due to the high coherent resonance in the micro cavity. While millimeter order periodic structures can control microwaves effectively, directional antennas and filters composed of photonic crystals can be applied to millimeter wave radar devices for intelligent traffic system (ITS) and wireless communication system. The perfect reflection of millimeter wave by photonic crystal will be useful for barriers to prevent wave interference. Terahertz waves with micrometer order wavelength are expected to apply for various types of sensors to detect gun powders, drugs, bacteria in foods, microcracks in electric devices, cancer cells in human skin and other physical, chemical and living events. The micrometer order photonic crystals can be applied for the terahertz wave cavities, filters and antennas.

Fig. 6.7.6 shows typical photonic crystal structures. A woodpile structure (a) with simple structure of stacked rods can form the perfect photonic band gap. Photonic crystals composed of GaAs or InP were fabricated by using semiconductor process techniques. A light wave circuit (b) in the periodic structure of arranged AlGaAs pins is processed by using electron beam lithography and etching techniques. A layered structure (c) composed of Si and SiO_2 with the different dielectric constants realize light wave polarization and super prism effects. These layers are stacked by using self-organized growing in alternate spattering and etching. An inverse opal structure (d) is composed of air spheres with FCC structure in TiO_2, Si, Ge or CdS matrix. First, polystyrene spheres are arranged by using self-organization in colloidal solutions. Then, the slurry of these dielectric media is infiltrated into the periodic structure and sintered.

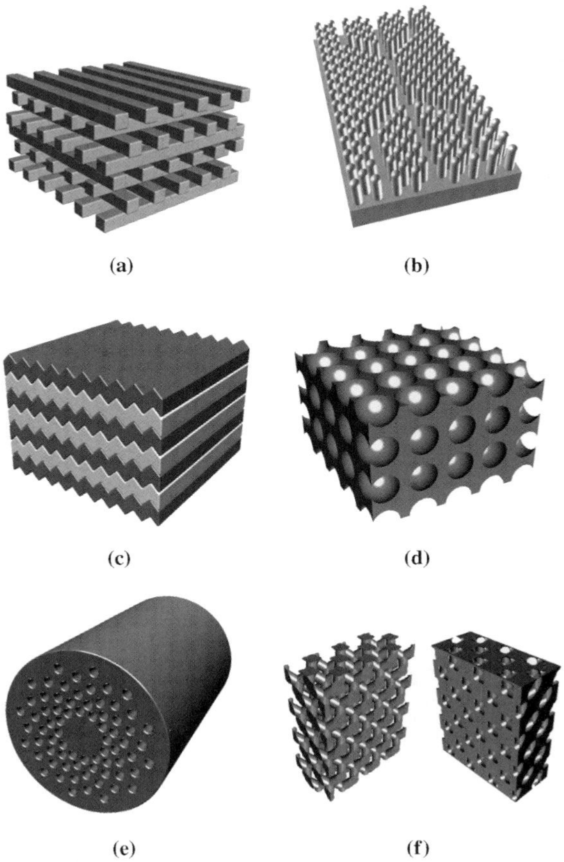

Figure 6.7.6
Typical structures of photonic crystals.

The optical fiber (e) with photonic crystal structure can guide light efficiently along the central core. Silica fibers and glass capillaries were bundled by wire drawing at high temperature. Diamond type photonic crystals (f) composed of TiO_2, SiO_2 or Al_2O_3 can be fabricated by using stereolithograpy and successive sintering process. The wider perfect band gap is obtained in microwave and terahertz wave frequency ranges.

References

[1] K. Otaka: *Phys. Rev. B*, **19**, 5057–5067 (1979).
[2] E. Yablonovitch: *Phys. Rev. Lett.*, **58**, 2059–2062 (1987).
[3] K.H. Ho, C.T. Chan and C.M. Soukoulis: *Phys. Rev. Lett.*, **65**, 3152–3155 (1990).

6.8 Catalytic property

This chapter explains catalytic property of nanoparticle. The evaluation of catalytic properties is classified mainly into following three kinds. They are physical property, chemical property and testing method of catalytic activity. The physical properties are, for example, particle morphology or microstructure; chemical properties are for example density or quality of active site. Each evaluation technique is explained as follows.

Crystal structure, crystal morphology and pore size distribution are enumerated as physical properties. The crystal structure is evaluated by powder X-ray diffraction method, and crystal morphology is analyzed by HR-TEM (High Resolution Transmission Electron Microscopy). However, they are not touched in this chapter because there are many established textbooks. In the physical properties, pore size distribution especially affects diffusivity of reactant, heat resistance and reaction selectivity. For the effective catalytic reaction, reactant has to come at active site promptly and reaction product has to eliminate from there promptly, too. For the evaluation of the reaction efficiency, diffusivity of gas molecule must be measured, however, no effective method has been established to evaluate diffusivity of them. Therefore, the evaluation of the pore size distribution of a catalyst is necessary, because the pores would act as the route of the diffusion of reactant and product.

Pore size distribution can be measured by mercury porosimetry in the range of several hundred micrometers to 0.003 μm [1]. The procedure of mercury porosimetry is as follows. Air in the pores of the sample is evacuated by vacuum pump. Then, mercury is injected into a sample cell. Then pressure is added little by little on the injected mercury, and simultaneously the volume of mercury which is intruded in the pores is measured. The higher pressure corresponds to the smaller pore size. The highest available pressure is about 400 Mpa, which corresponds to about 0.003 μm of the diameter of the pore. The relation between the pressure and the volume of intruded mercury can be converted into that of pore size and pore volume. Over 90° of the contact angle between a sample and mercury is necessary as a precondition. Pore size distribution of a material cannot be measured if the precondition would not be satisfied. Porosimetry without mercury is possible if the combination of a porous sample and liquid pressure media satisfy the precondition in which the contact angle of the pressure media on the sample is larger than 90°. However, no porosimetry technique is achieved through other liquid pressure media than mercury for practical use. On the other hand, for the case of some porous materials which have a mechanically weak framework, porous material itself is consolidated by the crumpling up of its framework under the pressure at which mercury is intruded into its pores. In that case, measured data shows only the information of the consolidation of the sample. Pore distribution of the sample cannot be measured. When data show no distinctive peak and show intrusion in increments in a wide pressure range, there is a possibility of consolidation of the sample. In such a case, pore distribution measurement must be done not only in the direction of pressure increase but also in the direction of pressure decrease and must be done for two or more cycles. If consolidation happens, the tendency of incremental intrusion does not happen after second cycle measurements. If the data shows real intrusion, the tendency of incremental intrusion is observed even after second cycle measurements reproducibly. Tubular type pore-morphology model in which one end is open and another end is closed is used as standard in mercury porosimetry. When hysteresis behavior exists in the pressure increase side and the pressure decrease side, even though, the measurement is normally done, the sample would have ink-bottle-type pore structure in which inner-size of the pores are larger than their entrance. Specific surface area of the sample also can be estimated in mercury porosimetry by using the tubular type pore-morphology model.

In the case to measure pore size distribution in the size range smaller than 0.003 μm, or the case to measure that of the material which has too weak framework to be measured by mercury porosimetry, gas adsorption method is available. The pore size is estimated by hysteresis behavior in the adsorption isotherm of adsorption and desorption of gas molecule. Pore size and pore volume are calculated with the Kelvin equation based on volume of the gas molecules which is condensed by the capillary condensation effect [2]. Physical adsorption method of nitrogen molecule at the boiling point of liquid nitrogen is being used generally. BJH (Brrett, Joyner, Halenda) method is one of the analysis methods using the capillary condensation theory. There are some analysis methods which have a difference in estimating the thickness of adsorbed gas layer on the surface of pores in the calculation for capillary condensation. The analysis methods based on the capillary condensation theory can be applied mainly for the mesoscopic pore size range (2–100 nm). The one-side opened tubular pore morphology model is also often used in the gas adsorption method. The different kinds of analysis method must be used for under the size range of mesoscopic level because the capillary condensation phenomenon does not happen there. Many analysis methods have been proposed for under mesoscopic pore size range, for example, MP method, HK (Horvath–Kawazoe) method, SF (Saito–Foley) method, DA (Dubinia–Astakhov) method, etc. The analysis methods affect the result of the pore size especially smaller than mesoscopic level. In the gas adsorption method specific surface area can be calculated from the relationship between the amount of adsorbed molecules and its partial pressure by using BET (Brunauer, Emmet and Teller's) equation and occupation area of a nitrogen molecule [3]. The bubble point method is used for measurement of maximum pore size of continuous pores in porous films. In the bubble point method, a liquid is impregnated into the sample and to measure the pressure when a bubble is observed at the beginning [4]. Mean pore size of continuous pores measured by a half-dry method [4], and pore size distribution of continuous pores can be estimated from the gas flow rate and differential pressure. As described in the Section 4.4.1, pores in catalyst or catalyst support consist of clearance between crystallites or secondary particles; therefore pore size distribution strongly depends on their aggregation and agglomeration structure. Pore size distribution affects heat resistance. As a suitable example of reaction selectivity by controlling pore size distribution of a catalyst, catalytic removal reaction of sulfur, vanadium and asphaltene are explained. The most appropriate pore sizes for them are 10, 18 and 25 nm respectively [5].

The active site is an adsorption point with the surface of the catalyst where the chemical adsorption is first caused. The surface of the metal particles and their neighborhood on their catalyst support correspond to active site in metal-supported catalyst. Acidic site has the role of active site in solid acid catalyst for example zeolite or composite oxide catalyst, and basic site has that role in solid base catalyst. For estimating the amount of active site in noble metal supported catalyst, total exposure area of noble metal particles is measured by CO adsorption method. The surface of noble metal can adsorb CO molecule selectively. CO-pulse adsorption method was standardized by Catalyst Society of Japan [6]. It is explained as follows. Noble metal in a catalyst is reduced at 400°C for 15 min in hydrogen atmosphere at the beginning. After cooling to room temperature, CO is injected into the catalyst like a pulse of a prescribed amount, and simultaneously, the quantity of the CO which is not adsorbed is measured. The amount of CO adsorption can reach the saturation after several CO-pulses . The amount of adsorbed CO is calculated from the difference of the amount of injected and exhausted CO. Total surface area is obtained from adsorbed amount of CO and occupation area of an adsorbed CO molecule. Dispersion ratio and average particle size of a noble metal can be estimated from the total surface area. Dispersion ratio means exposed ratio of atoms of a noble in the total amount of its loading. 100% of dispersion corresponds to that where all of the atoms are exposed at the surface. The shape of a metallic particle is assumed to be a cube of the same size, and average noble metal particle size is estimated.

There are three kinds of measurement methods for surface acidity and basicity, they are titration method, adsorption heat method and thermal desorption method [6]. A thermal desorption method of ammonia for measurement of density and strength of acidity is exemplified here. A sample cell of a catalyst is evacuated at 500°C for 1 h by a rotary pump to eliminate adsorbed water from a catalyst surface at the beginning. Then 100 Torr (13.3 kPa) of ammonia gas is injected in the sample cell at 100°C. After 30 min of exposure to ammonia, residue of ammonia vapor in the sample cell is eliminated by helium gas flow. Finally, the sample cell is heated in the heater whose temperature is increased by 10 K per min, and desorbed ammonia is measured at each temperature. The amount of ammonia gas which is desorbed at relatively lower temperature corresponds to the amount of acid site of relatively weak acidity, and high temperature desorption of that corresponds to relatively strong acidity [7]. As described for Au loaded catalyst in Section 4.4.1, surface acidity increases catalytic activity of loaded catalyst in some cases and decreases it in another case.

In addition, IR (infrared spectroscopy) is used to know the state of the adsorbed molecule. IR is used in a detector for qualitative and quantitative analysis in catalyst activity test described in the next paragraph as well. Moreover, there are many analysis methods used, for example, ESR, XPS, NMR, STM, AFM, EELS, LEED, AES and XAFS [8]. As an example of a particular observation of adsorbed molecules, there is an in situ observation of dehydrogenation of individual *trans*-2-butene molecules adsorbed on the Pd (110) surface. In elastically tunneled electrons from the tip of a STM proceed the reaction and the *trans*-2-butene molecules changed into 1,3-butadiene molecule under the STM observation (Fig. 6.8.1) [9].

Next, reactors for catalyst activity test are explained. They are roughly classified into three types, pulse reactor, closed reactor and flow reactor. In the pulse reactor, reactant is injected as a pulse into the reactor. Merits of the pulse reactor are high sensitivity for minute amounts of reactant or product, and its device has a simple configuration. On the other hand, the reaction in pulse reactor is unsteady state,

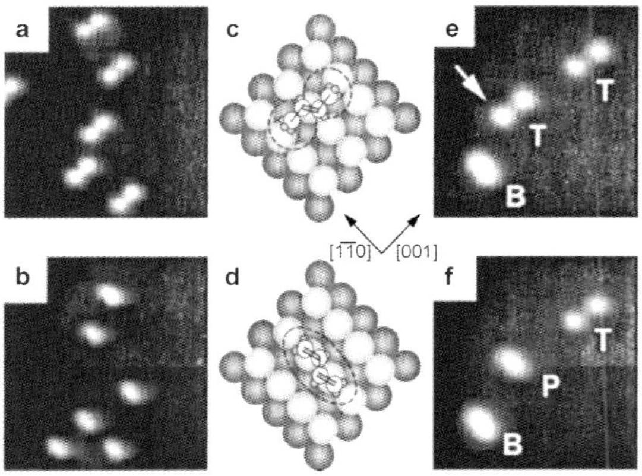

Figure 6.8.1
STM image of (a) *trans*-2-butene (C_4H_8) and (b) 1,3-buta-diene (C_4H_6) molecules (observation area: 45 × 45 Å^2) Schematic representation of the adsorption site of (c) *trans*-2-butene (C_4H_8) and (d) 1,3-butadiene (C_4H_6) molecule on Pd (110), respectively. (e) before and (f) after dosing tunneling electrons on a target molecule of *trans*-2-butene marked with arrow in (e) (observation area: 20 × 20 Å^2). T: *trans*-2-butene (C_4H_8) molecule, P: reaction product, B: 1,3-butadiene (C_4H_6) molecule.

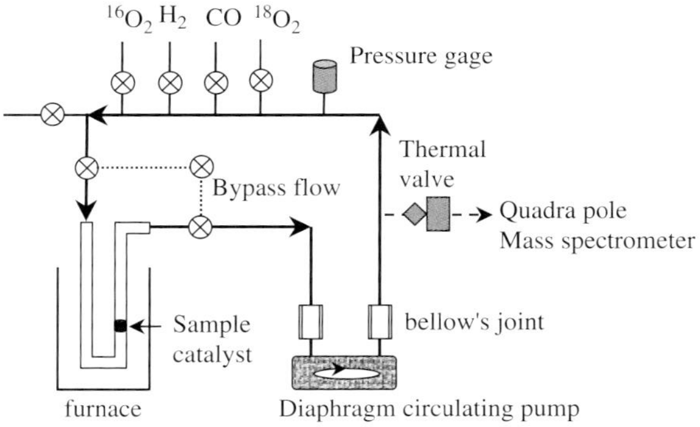

Figure 6.8.2
Schematic diagram of closed loop circulating $^{16}O/^{18}O$ isotopic exchange reaction apparatus.

so, this type is not suitable for reaction rate analysis. CO-pulse adsorption method explained above is a typical example to use the pulse reactor. Closed reactors are divided into stand-still type and recirculated type. A recirculated reactor is suitable for the slower rate reaction than convection rate or molecular movement [7]. As an example of catalyst activity test using a closed recirculated reactor (Fig. 6.8.2), a characterization of dynamic oxygen mobility in Pt/CeO_2–ZrO_2 catalysts by $^{18}O_2/^{16}O_2$ isotopic exchange reaction is shown [10]. 30 mg of a catalyst is set in the catalyst bed, and $^{18}O_2$ is recirculated in the system. Production rate of ($^{18}O^{16}O$) molecule was measured with mass spectrometer, and diffusion coefficients were estimated from the measurement. The reaction rate is relatively slow in the example, because the $^{18}O_2/^{16}O_2$ isotopic exchange reaction is controlled by volume diffusion of oxygen in Pt/CeO_2–ZrO_2 catalysts. A closed recirculation type reactor is suitable for such a characterization.

The flow reactor consists of a system in which reactants are supplied into catalyst bed continuously, and reaction products also come out continuously. The flow reactor is suitable for chemical reaction having a large reaction rate. When you have to test catalytic activity, a condition of reaction control is necessary. Catalytic activity is not examined under diffusion control condition of reactant or reaction product [7]. Generally, catalytic activity test is done in the flow reactor in which the temperature of catalyst bed is gradually increased. Reactants flow through the catalyst bed under the test. The reaction beginning temperature shows the catalytic activity of the catalyst. The lower the better. Under gradually decreasing temperature of catalyst bed the flow reactor is also available for catalytic activity test. The temperature at which the reaction cannot continue is examined. In the case of temperature-increasing direction in flow reactor, a reaction-beginning temperature is higher than that in the temperature-decreasing direction. Because of a temporary poisoning of active site by adsorption of reactant, reaction-beginning temperature in temperature-increasing direction is higher than that in the temperature-decreasing direction. High catalytic activity at lower temperature is necessary for automotive exhaust catalyst, low sensitivity for the poisoning by reactants is important in temperature increasing direction test. Bowker et al. [11] reported an observation of Pd (110) surface under CO oxidation reaction using fast XPS using synchrotron radiation. According to the report surface of the Pd was mainly covered by CO in lower temperature than activation. Adsorbed molecules are changed at the moment of the activation of the catalyst. Just after the moment, oxygen atom became the main adsorbing species (Fig. 6.8.3). They also reported that CO oxidation happened on an ultra clean surface of Pd in the reactor from the moment when CO gas was injected at 315 K, and the transitional CO oxidation was suspended after 100 s. At the moment when the CO oxidation suspended, the surface of Pd was covered by CO. The suspension of CO oxidation reaction at 315 K was understood as CO poisoning in which CO molecule adsorbed strongly on the surface of Pd, and oxygen could not be supplied to the surface of Pd. By stepwise increase of the reaction temperature, the occupation percentage of oxygen atom on the Pd surface gradually increases. The rate of the CO oxidation reaction showed steep peak at about 360 K and then decrease at higher temperature with increasing temperature. The reaction rate decrease in the high

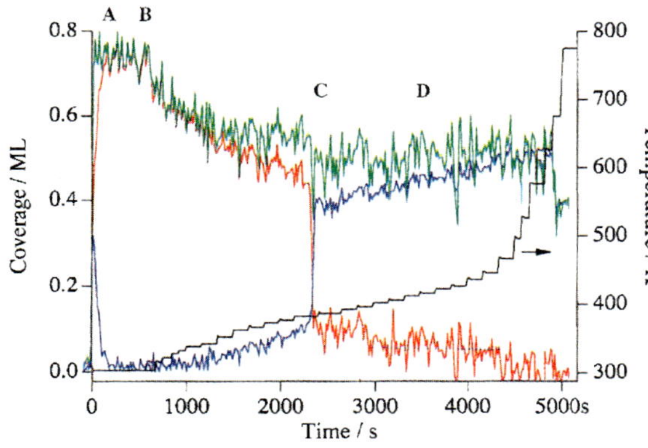

Figure 6.8.3
Coverage of CO (a) (red) and O (a) (blue) as well as their sum (green) versus time on Pd (110) plane. The black curve represents the temperature. The coverages have been calculated by integrating the spectra over the binding energy region 534.4–531.2 eV (CO (a)), respectively 531.2–528.9 eV (O (a)), using the background subtracted spectra, and by comparison to reference spectra of c (2×4)-O (0.5 ML), respectively (2×1)-CO (1 ML).

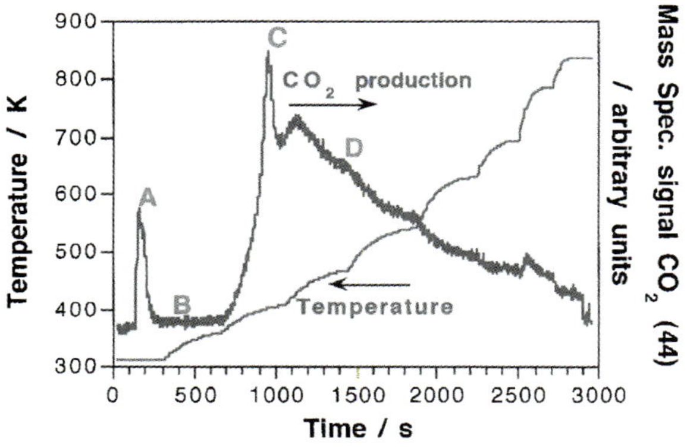

Figure 6.8.4
Production of CO_2 and stepwise change in temperature for a mixed dose of CO and O_2 onto clean Pd (110) with respect to time (s). The dose is started just prior to point A, where a transient reaction to produce CO_2 occurs, which rapidly ceased. From point B, the temperature is increased in a stepwise manner. Point C is the light-off and results in a sharp reaction rate maximum. At point D and for all higher temperatures, the reaction rate decreases.

temperature side was thought to be due to the poisoning of oxygen atom (Fig. 6.8.4).

The evaluation techniques of the catalyst characteristics are outlined above. Some of them are shown with specific examples.

References

[1] M. Oya, M. Takahashi, Y. Iwata, K. Jono, T. Hotta, H. Yamamoto, K. Washio, A. Suda, Y. Matuo, K. Tanaka and M. Morimoto: *Am. Ceram. Soc. Bull.*, **81**, 52–56 (2002).

[2] J.M. Haynes: *Mater. Struct.*, **6**(3), 209–213 (1973).

[3] JIS R 1626 Measuring methods for the specific surface area of fine ceramic powders by gas adsorption using the BET method (1996).

[4] JIS K 3832, ASTM F316-86 "Pore Size Characteristics of Membrane Filters by Bubble Point and Mean Flow Pore Test" (1990).

[5] S. Inouel, S. Asaoka and M. Nakamura: *Catal. Surv. Jpn.*, **2**(1), 87–97 (1998).

[6] The Committee of Reference Catalyst of the Catalysis Society of Japan, Shokubai (Catalyst) 31(5) 317 (1989) (in Japanese).

[7] M. Niwa, N. Katada, M. Sawa and Y. Murakami: *J. Phys. Chem.*, **99**, 8812–8816 (1995).

[8] E. Kikuchi, K. Segawa, A. Tada, Y. Imizu and H. Hattori: Atarasii shokubai kagaku, Sankyosyuppan (in Japanese) (1997).

[9] Y. Kim, T. Komeda and M. Kawai: *Phys. Rev. Lett.*, **89**, 126104 (2002).

[10] F. Dong, A. Suda, T. Tanabe, Y. Nagai, H. Sobukawa, H. Shinjoh, M. Sugiura, C. Descorme and D. Duprez: *Catal. Today*, **93–95**, 827–832 (2004).

[11] M. Bowker, I.Z. Jones, R.A. Bennett, F. Esch, A. Baraldi, S. Lizzit and G. Comelli: *Catal. Lett.*, **51** (3–4), 187–190 (1998).

6.9 Properties of gas permeation and separation membranes

The gas permeation and separation membranes can be classified into porous and non-porous membranes. The gas separation membrane has energy-saving advantages, since it is not accompanied with phase change during operation. Therefore, a number of researches have been conducted for the separation of oxygen from air, helium from natural gas, hydrogen from various gas mixtures and so on after World War II.

Although polymers are typically used as the membrane material, the research and development of alloy membranes, such as palladium–silver alloy and metal–glass amorphous alloys, for hydrogen permeation applications are very active due to their preferred mechanical strength and durability. However, the mechanism of hydrogen permeation through the polymer membrane [1], palladium–silver alloy membrane [2] and amorphous alloy membrane are very different from each other.

The polymer membrane includes the porous polymer membrane with a pore size from 5 to 100 nm and the non-porous polymer membrane with a polymer chain gap from 0.3 to 1 nm induced by thermal vibration. The effective size of the gas molecules for permeation and separation is extremely small, e.g., 0.234 nm for the smallest hydrogen, 0.323 and 0.363 nm for carbon dioxide and nitrogen respectively. Therefore, as shown in Fig. 6.9.1, gas molecules permeate not only through the porous material but also the non-porous membrane. However, the porous membrane has high gas permeation velocity but low gas separation capability, because all gas molecules can go through these pores.

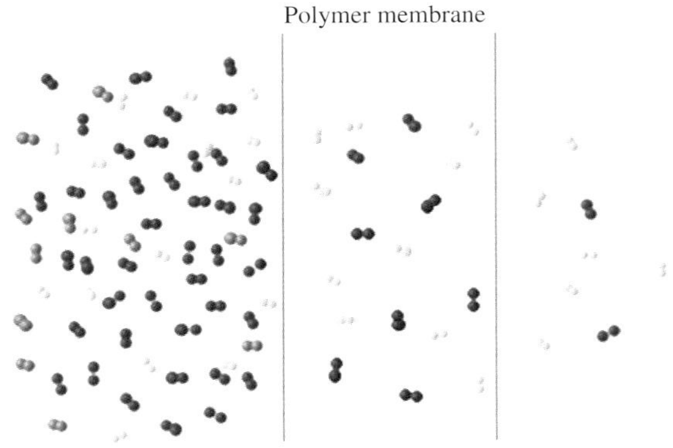

Figure 6.9.1
Gas-permeation mechanism of polymer membrane.

Membranes are used to separate oxygen and nitrogen from the air for medical and combustion fuel applications. The volume flow rate of oxygen enriched air required for the medicals purposes (40% O_2) is 4–8 L per min, while it needs air containing 30% O_2 for combustion applications as much as we can.

The relationship between the oxygen permeability coefficient and the separation coefficient of nitrogen for various kinds of polymers is indicated in Fig. 6.9.2 [3]. The polymer membrane is selected according to each individual application. In general, the membranes with high separation coefficient, presented by PO_2/PN_2, and high permeability coefficient are desirable. However, the thermal resistance of polymer needs to be first considered if the gas to be separated is at the elevated temperatures, because the polymer membranes are limited to maximum operating temperature at about 150°C. Fig. 6.9.2 shows the effect of polymer morphology and structural deformation etc. on the separation and permeability coefficients [3].

The industrial applications using polymer membranes to separate hydrogen, oxygen or carbon dioxide are listed in Table 6.9.1 [4].

To overcome the temperature limitation on polymer membranes, a porous ceramic membrane for hydrogen separation has been investigated [5]. The separation and permeability coefficient obtained so far is 500 and 10^{-8} respectively, while the target value is over 1,000 for separation coefficient and 10^{-7} for permeability coefficient [6]. The ceramic membrane generally has a three-layer structure consisting of the active separation layer made by the chemical deposition or sol–gel

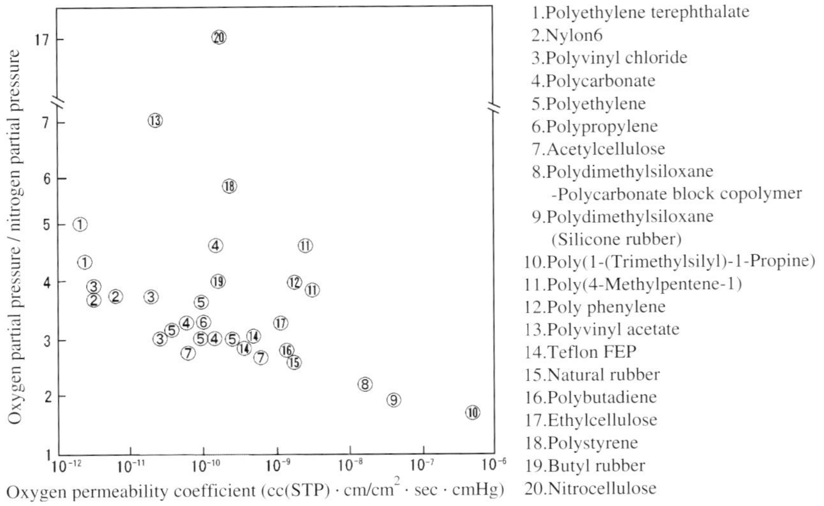

Figure 6.9.2
The relationship between the oxygen permeability coefficient and the separation coefficient of nitrogen with various kind of polymers.

Table 6.9.1
Gas-separation efficiencies of various polymer membranes.

10^{10}, P cm^3-cm cm^{-2} s^{-1}cmHg	Polysulfone	Acetylcellulose	Polyimide	Poly (4-methylpentene-1)	Silicon rubber
H_2	13	12	9	136	550
CO_2	6	6	–	93	270
O_2	1	1	0.5	32	501
α_{ij}					
H_2/CO	40	40	76	–	–
H_2/CH_4	60	60	~200	–	0.7
H_2/N_2	72	70	~200	17	2
CO_2/CH_4	25	30	–	–	0.3
O_2/N_2	6	5.5		4	2

Note: Temperature 25–30, constant pressure.

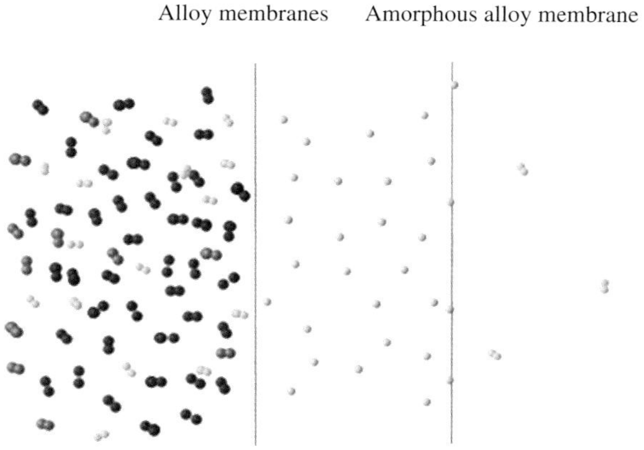

Figure 6.9.3
The mechanism of gas permeation and separation of alloy and amorphous alloy membranes.

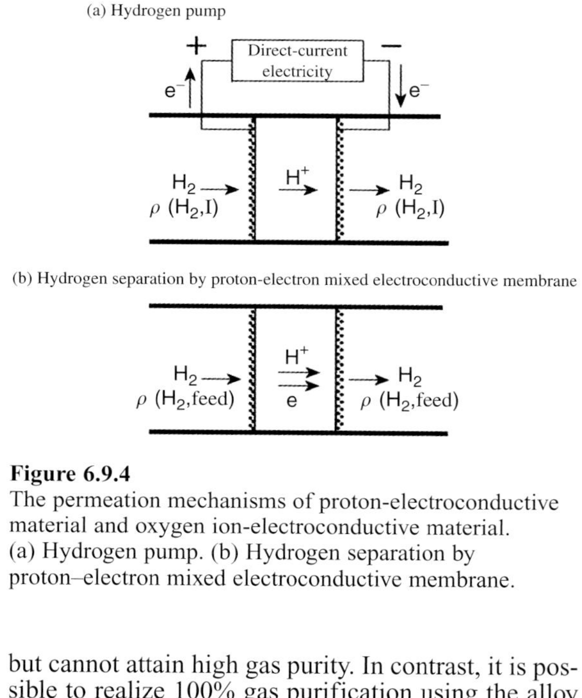

Figure 6.9.4
The permeation mechanisms of proton-electroconductive material and oxygen ion-electroconductive material.
(a) Hydrogen pump. (b) Hydrogen separation by proton–electron mixed electroconductive membrane.

method, the intermediate layer and the porous supporting plate made by the sol–gel method. By controlling the pore sizes in these structural layers at about 0.3 nm, about 4 nm and 80–150 nm respectively, high separation and permeability coefficients have been obtained with the ceramic membranes.

On the other hand, the mechanism of permeation and separation of alloy membrane and amorphous one is completely different from that of the polymer membrane, as shown schematically in Fig. 6.9.3. At the surface of the alloy and amorphous membranes, the hydrogen molecules are decomposed by the catalytic effect to generate hydrogen atoms. The hydrogen atoms are then absorbed through the membrane surface and diffuse towards its opposite side and finally leave the membrane as hydrogen molecules.

Due to such differences in the permeation and separation mechanism, the polymer membranes are able to enrich the gas molecules at a relatively high speed, but cannot attain high gas purity. In contrast, it is possible to realize 100% gas purification using the alloy and amorphous membranes, although it is only achievable for some gases.

Besides the polymer, alloy and amorphous membranes, the research and development of separation membranes for hydrogen or oxygen using proton-electroconductive oxides or oxygen ion-electro conductive oxide has also been actively carried out. These separation mechanisms include using hydrogen pump to transfer and separate hydrogen with the electric current as shown in Fig. 6.9.4a and separating hydrogen by spontaneous permeation based on the difference in the partial pressure of hydrogen as shown in Fig. 6.9.4b. The latter is regarded as a hydrogen permeation membrane [7]. However, no report has been

Table 6.9.2
Properties of permeation and separation membranes.

Type of membrane	Permeability	Dissolubility	Heat resistance	Environment resistance	General versatility
Polymer	◎	△	△	○	◎
Alloy	△	◎	◎	△	△
Amorphous alloy	△	◎	○	○	△
Ceramic	◎	△	◎	◎	◎

Note: ◎; Very good; ○: good;△: poor.

published concerning the use of electroconductive oxide as a hydrogen pump.

The oxygen-electroconductive zirconia can also be used as an oxygen pump or the oxygen permeation membrane to separate oxygen by spontaneous permeation due to the difference in the partial pressure of oxygen.

The properties of permeation and separation membranes including those under the development are summarized in Table 6.9.2, in which the environment resistance means the stability of the material against sulfur and the like; and, the versatility denotes its flexibility to separate various kinds of gases.

References

[1] T. Nakagawa: *Polym. Digest*, **38**(10), 2–11 (1989).
[2] J. Shu, B.P.A. Grandjean, A. Van Neste and S. Kaliaguine: *Can. J. Chem. Eng.*, **69**, 1036–1060 (1991).
[3] T. Nakagawa: *Development and Application of Functional Membranes for Separate*, CMC Publishing CD, Ltd., Tokyo, Japan, pp. 10–31 (2001).
[4] A.K. Fritzshe, R.S. Narayan: *CEER*, **19**(205), 19 (1987).
[5] Y. Iwamoto: *Mater. Jpn.*, **44**, 220–225 (2005).
[6] Y. Iwamoto, K. Sato, T. Kato, T. Inada and Y. Kubo: *J. Eur. Ceram. Soc.*, **25**, 257–264 (2005).
[7] H. Matsumoto: *Mater. Jpn.*, **44**, 226–232 (2005).

FUNDAMENTALS

CHAPTER 7

ENVIRONMENTAL AND SAFETY ISSUES WITH NANOPARTICLES

CONTRIBUTORS

Hisao Makino (7.1), (7.2.2~7.2.3), (7.3.1)
Hitoshi Emi (7.2.1), (7.4.1~7.4.2)
Akimasa Yamaguchi (7.2.2~7.2.3)
Eiji Iritani (7.2.4), (7.4.3)
Norikazu Namiki (7.2.5), (7.2.6)
Toshihiko Myojo (7.3.2)
Kenji Yamamoto (7.3.3)

7.1 Introduction

Since nanoparticles have superior surface activity and can be applied to the production of particles with various functions, they are extremely important for the future development of sophisticated material technologies. On the other hand, this superior activity of nanoparticles is a cause of trouble from the perspective of safety, and does not always have a positive influence on the environment. Attention must also be paid to impact on health. Nevertheless, all technologies have negative aspects, and overcoming these kinds of problems, we will be able to utilize the superior characteristics of nanoparticles for practical purposes. To achieve this goal, it is necessary to fully understand the influence of nanoparticles on the environment and the relevant safety issues.

This chapter evaluates the relationship between nanoparticles and the environment, and also describes the trouble caused by nanoparticles as well as the safety issues.

The relationship between nanoparticles and the environment will be clarified from the viewpoint of what kind of influence nanoparticles generated either artificially or naturally have on the environment. The influence on the indoor environment, where nanoparticles are produced, will also be clarified.

The safety of nanoparticles will be clearly described from the perspective of the trouble caused by the superior surface activity of nanoparticles; the effect of the compositional characteristics of nanoparticles, and also the influence on health. A method for assessing the influence of nanoparticles using quantum dots is also explained. In the final section, methods for removing nanoparticles from gas and liquid are described as technology to control the influence of nanoparticles on the environment.

7.2 Nanoparticles and environment

7.2.1 Nanoparticles in atmospheric environment

In our atmospheric environment, particles ranging from several nanometers to several tenth micron orders are suspended. They are emitted into the atmosphere at the rate of 2.5 billion tons every year. Emission sources are classified as either natural or artificial.

Natural particles occupy 60% of total particles, consisting mainly of salt particles (~1 billion ton) from the sea and soil particles (~0.5 billion ton) from the land. On the other hand, the latter particles are brought about by human activities. Although occupying only 16% of the total emitted particles, their size is mostly of submicron order and because they contain hazardous chemical components such as nitrates, sulfates, hydrocarbons, heavy metals, etc. in high concentration, their effects on the ecosystem are serious.

Fig. 7.2.1 shows an overview of the size and concentration ranges of various aerosol particles. As it can be seen, the number concentration of atmospheric aerosol which we inhale every day ranges from several thousand particles per cm^3 in clean area to several hundred thousands in dusty areas, and the size range lies between 10 nm and several tens of micrometer.

Fig. 7.2.2 shows mass-based size distribution of atmospheric aerosol particles. Since the size distribution in the nanosize range appears only when the sources of particle generation exist, the size distribution is usually bimodal with peaks in the size range of a few to 10 micron and submicron. The former peak consists of naturally generated coarse particles such as soil dust, sea salt spray, and so on. On the contrary, the latter contains plenty of artificially generated particles, some of which grow from molecules (in most cases vapor state) exhausted by human activities through chemical reaction, condensation, and coagulation. Particle growth rarely leads to particles larger than 1 μm unless high concentration of vapors or particulate matters which cause the above-mentioned growth mechanisms exist in the atmosphere.

As it can be seen from the differences in the particle generation process, fine particles generated from molecules or nanoparticles are much more complicated in their chemical component than the coarse particles, and sometimes have serious adverse health effects. Such fine particles are called PM2.5, which is defined for particles less than 2.5 μm including nanosized particles. Recent epidemiologic investigation reports that the concentration of PM2.5 showed a positive correlation to the mortality due to pulmonary diseases [1].

Various research techniques are used in order to understand the process of particle growth and to trace back to the source of pollution. An example is shown in Fig. 7.2.3 where a characteristic function of sulfur dioxide is shown taking into account all possible factors related to particle growth. Where f is the characteristic function that expresses particle size, particle concentration, particle composition, and so on [2, 3].

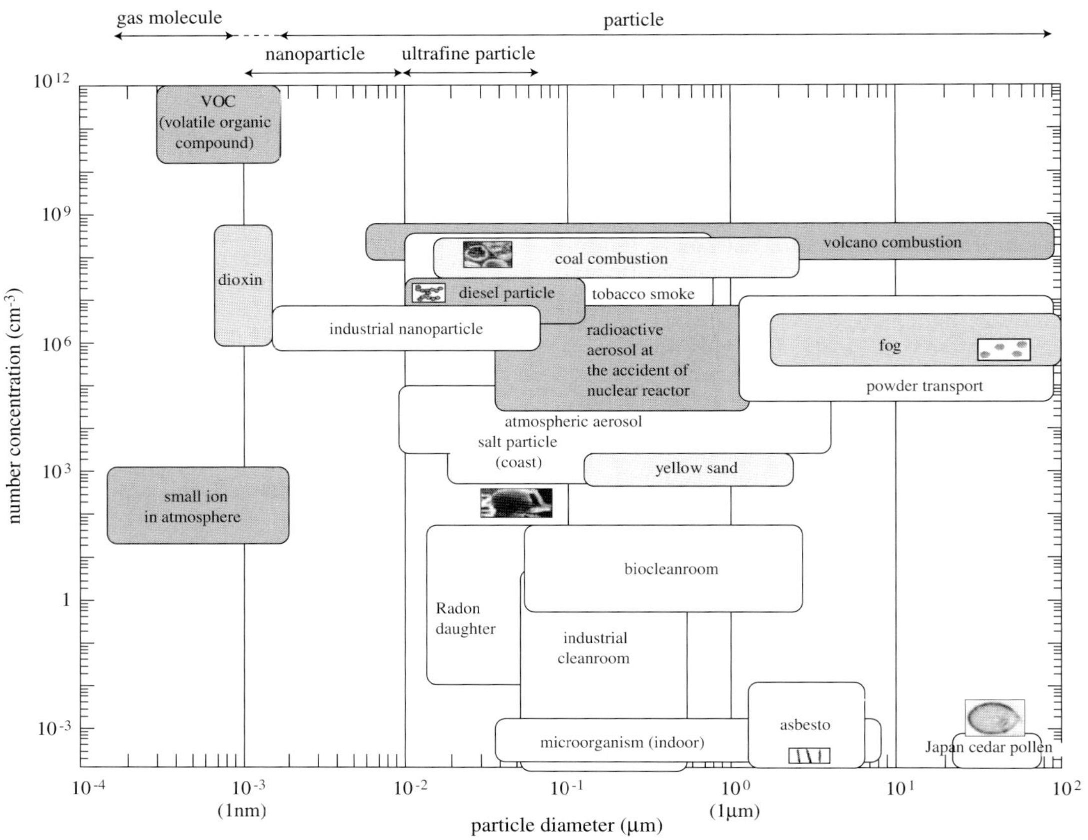

Figure 7.2.1
Particle size and concentration of various aerosols.

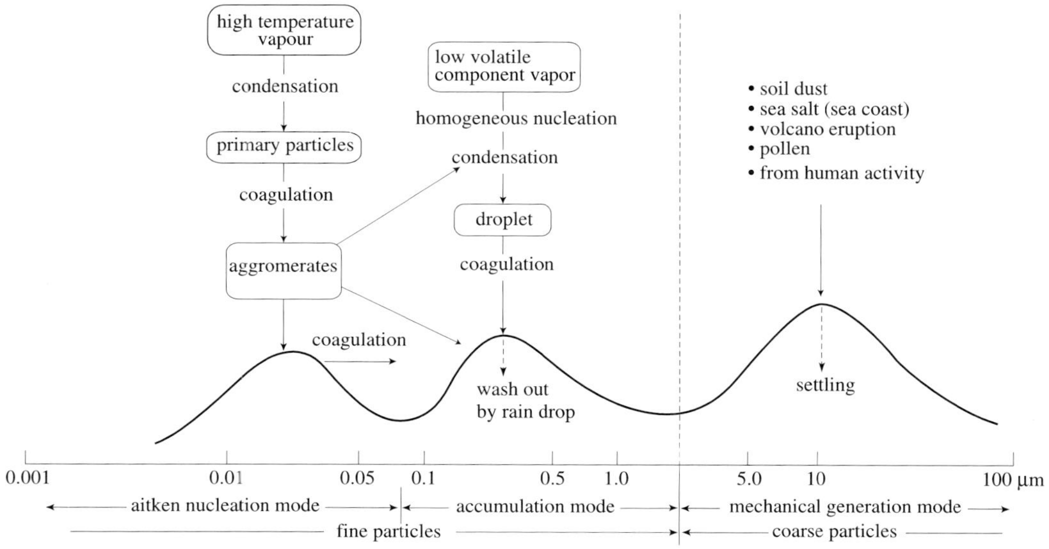

Figure 7.2.2
Mass-based size distribution of atmospheric aerosols.

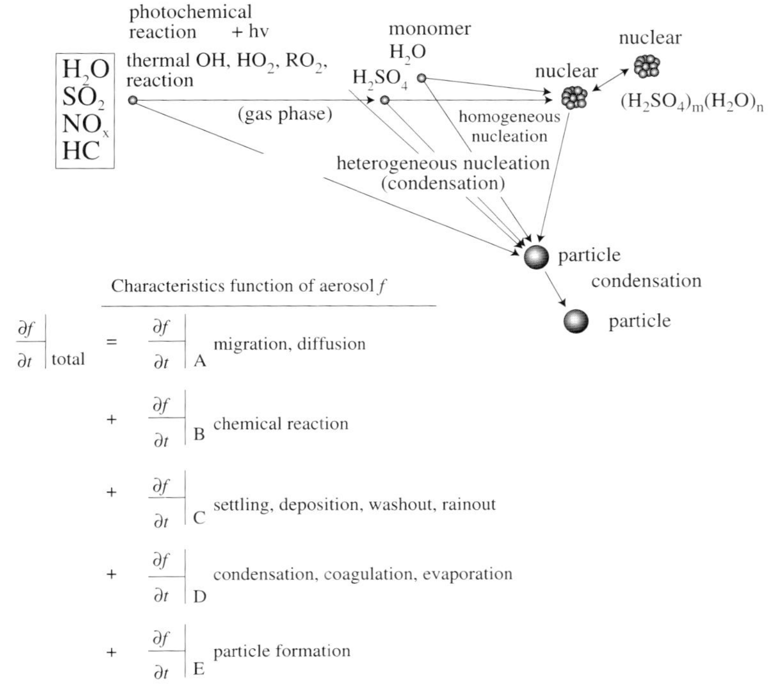

Figure 7.2.3
Generation mechanisms of atmospheric nanoparticles and characteristics function.

References

[1] C.A. Pope, J.M. Thun, M.M. Namboodiri, W.D. Dockery, S.J. Evans, E.F. Speizer and W.C. Heath: *Am. J. Respir. Crit. Care Med.*, **151**, 669–674 (1995).

[2] M. Kasahara: *J. Jpn. Soc. Air Pollut.*, **29**(6), A100 (1994).

[3] M. Kasahara: *J. Jpn. Soc. Air Pollut.*, **25**(2), 115 (1990).

7.2.2 Ground water environments and nanoparticles

Particulate materials in water are present in the form of colloids. These colloid particles are classified into inorganic colloids. Examples of the former are oxides of aluminum, silicon and other substances, and typical examples of the latter are substances such as humic acid and fulvic acid. While the structure and molecular weight of particles vary depending on the area of water, it is known that what are usually present in water are comparatively small colloids (particles smaller than 500 nm). The number concentration of colloid particles in ground water, or a typical water area environment, ranges from 10^{11} to 10^{20} (number/m^3) and varies significantly depending on the geochemical conditions of the aquifer.

It is known that in moving water, colloid particles sometimes act as a medium in conjunction with water and in some cases move faster than water.

Fig. 7.2.4 [1] is a conceptual figure showing colloid movement in soil strata. A is a heterogeneous layer composed of macropores formed in rotten roots close to the ground surface. When rain falls, colloids move through macropores saturated with water. B is a saturated

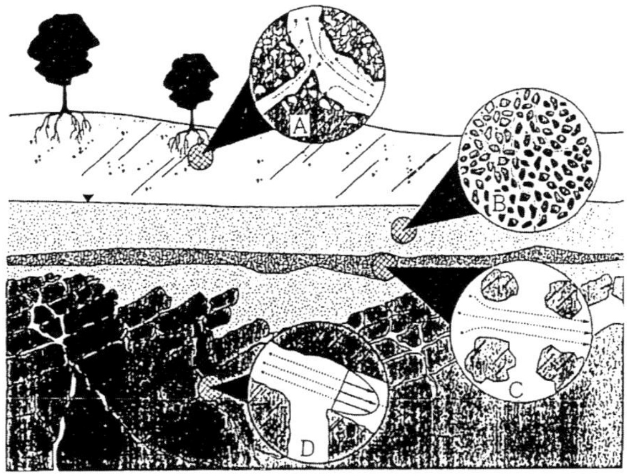

Figure 7.2.4
Concept of colloid movements in soil strata.

homogeneous porous layer such as a sand layer, and most of the colloid particles are trapped. The mechanism of partical trap in this layer is explained by the sand filtration theory. C and D are a gravel layer and a rock bed, respectively, and both have high water permeability with large gaps and cracks. Particles can also pass through easily.

Safety and movement characteristics of colloid particles have a significant influence on the movement of materials such as ionized molecules in aquatic environments. Since fine particles such as nanoparticles in particular are highly stable as colloid particles, it will be very important in the future to understand their influence. At the same time, these characteristics are considered to have a high potential to be developed for further application of nanoparticles.

Reference

[1] S. Nagasaki: *J. Jpn. Soc. Irrigat., Drain. Reclam. Eng.*, **66**(12), 1261–1269 (1998).

7.2.3 Nanoparticles in exhaust gases

In most cases, nanoparticles in exhaust gases are studied from the viewpoint of the influence of total particulate matters on the environment. The term "nanoparticles" is used only in a few cases, "fine particles" is usually used for investigation. Since nanoparticles are part of fine particles, this section will be described from this perspective.

Major sources of combustion exhaust gases are stationary large-scale combustors and diesel engines for stationary and portable use.

For stationary combustors, fuels such as coal, oil, and gas are used. Lighter fuels have a lower rate of particulate emission, but have a higher fine particle content including nanoparticles. Fig. 7.2.5 [1] shows the frequency distribution in combustion of coal and heavy oil. Fig. 7.2.5a and b are the distributions on a number and mass basis, respectively. As these figures clearly show, the total weight of particles of a size of 1 μm or smaller is extremely low, while their total number is, on the contrary, very large. It is clear that,

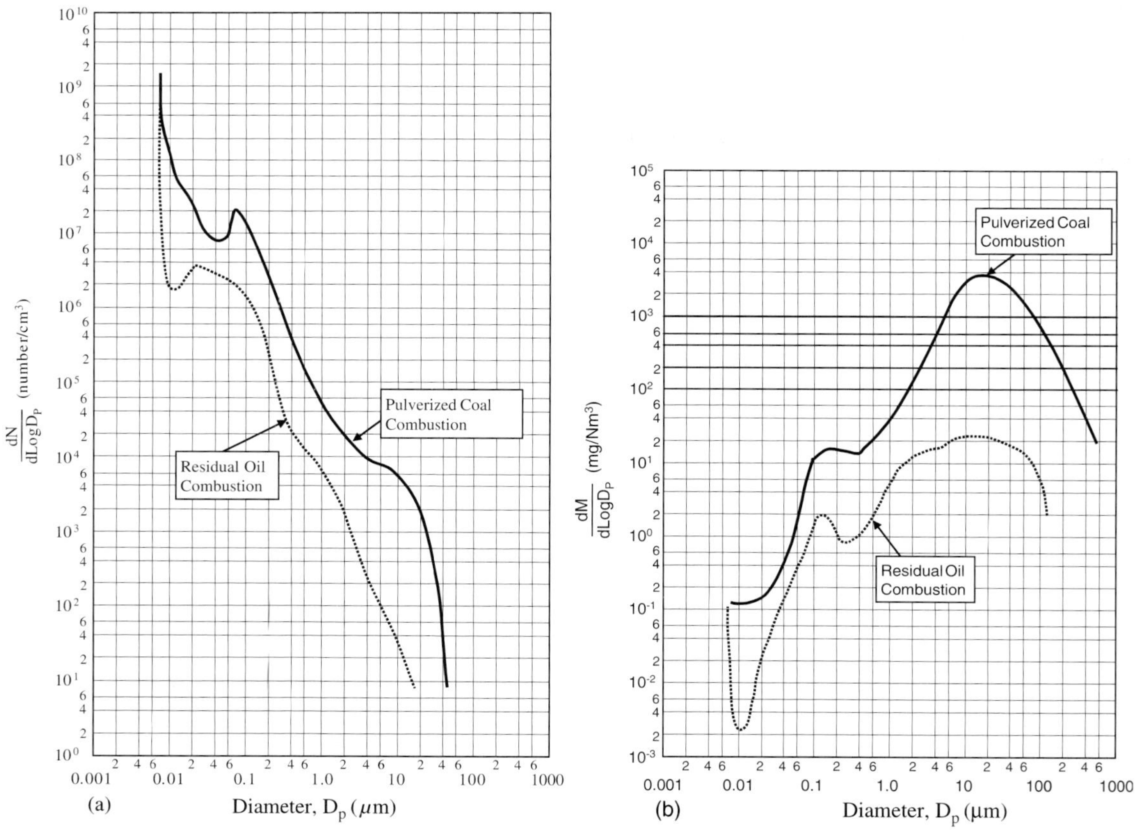

Figure7.2.5
(a) Particle size distribution on number basis at combustor outlet. (b) Particle size distribution on mass basis at combustor outlet.

while the total quantity of particulate material is far larger in coal combustion than in oil combustion, the difference is less when it comes to particles 1 μm or less in diameter, including nanoparticles.

Most of the particles contained in pulverized coal combustion exhaust gases are considered to be formed as particulate materials directly from ash content, which is originally contained in coal and also includes some unburned carbon. Particularly, almost all large-size particles are considered to be this type of particle. On the other hand, fine particles include two types. One type is formed in the process by which low boiling point metal contained in coal ash is evaporated and vaporized in a high-temperature combustion field and then becomes particles in the exhaust gas cooling process. The other includes carbon particles formed in the gas phase, or so-called soot, which is generated due to the delay in oxygen supply for combustion of evaporated volatile matter in the initial stage.

Fig. 7.2.6 [2] shows the relationship between the trace metal content in coal ash and the particle diameter. Aluminum with a high boiling point has a constant concentration regardless of the particle diameter. However it is obvious that in the case of metals with a lower boiling point, the smaller the particle diameter, the larger the content. With regard to particles with sizes 1 μm or smaller in the nano domain, it has been clarified that the generated amount is increased rapidly by reducing combustion air supply or by weakening the oxidation atmosphere in the volatile matter combustion area, for example, when air supply from a burner is reduced in two-stage combustion. This also demonstrates the significant contribution of carbon particles formed in the gas phase.

Also in the case of ash from heavy oil combustion, there are large particles of a carbon residue type generated from sprayed liquid particles and particles formed in the gas phase as well. As in the case of coal, trace metal contained in heavy oil with a low boiling point is concentrated into fine particles and discharged. Also in the case of the combustion of liquefied natural gas, carbon particles formed in the gas phase are generated, albeit in trace amounts.

In contrast, only in the case of diesel engines, fuel is injected into the high-temperature and high-pressure atmosphere produced by compressing only air to induce spontaneous ignition, and combustion continues with a heterogeneous mixture of fuel and air in the combustion chamber. Therefore, particulate materials mainly consisting of unburnt carbon are generated due to incomplete combustion.

Fig. 7.2.7 [3] shows changes in the diameter of particles according to changes in the diesel engine load. It is obvious that the overall concentration of particles increases with the increase in the load rate of the engine. According to observations using SEM, fine particles in diesel engine exhaust gases have also been found to comprise fine primary particles of a size several tens of nanometers, and coarse particles with carbon hydride condensed on the surface of secondary aggregates of primary particles.

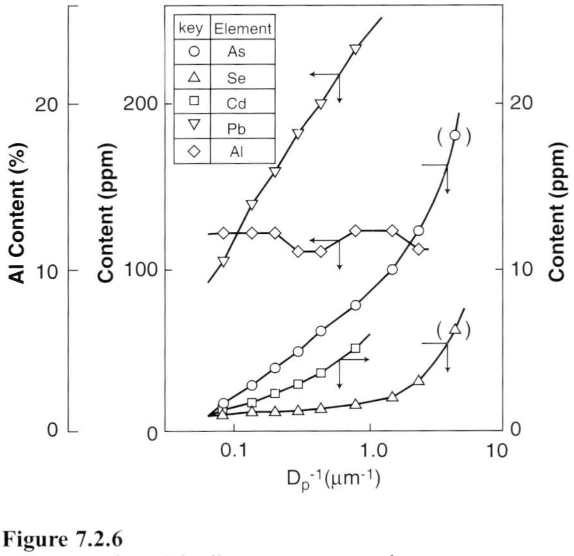

Figure 7.2.6
Influence of particle diameter on trace element contents.

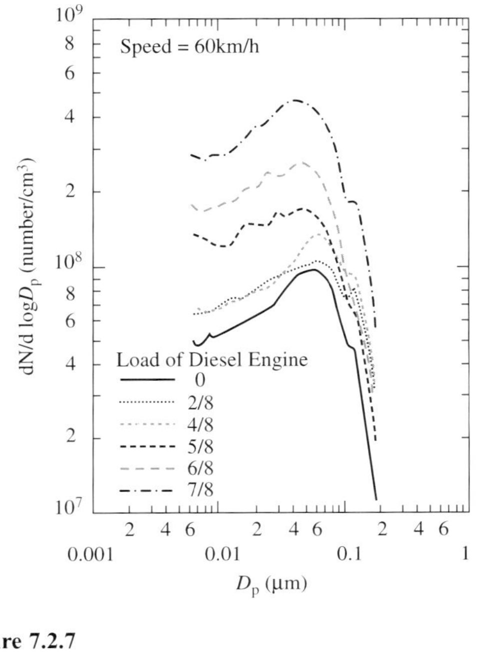

Figure 7.2.7
Particle size distribution measured by DMA.

References

[1] H. Makino: *Denchuukenhoukoku*, 281040 (1982).

[2] H. Makino: *J. Aerosol Res., Jpn.*, **4**(3) (1989).

[3] K. Saito: *J. Soc. Powder Technol., Jpn.*, **38**(7), 493–502 (2001).

7.2.4 Nanoparticles in wastewater

The volume of industrial and domestic wastewater is increasing significantly year by year with the change in the lifestyle based on mass consumption and mass disposal brought about by the dramatic development of the economy and industry. Effective advanced wastewater treatment is required because wastewater contains a variety of constituents such as particles, organic materials, and emulsion depending on the resource.

Inorganic nanoparticles are not generally stabilized in the liquid because they form aggregates of some sort more or less. For wastewater reclamation and reuse, these nanoparticles can be removed from the liquid by the advanced treatment processes such as membrane filtration following biological treatment processes. Organic materials such as macromolecules are regarded as soft nanoparticles judging from their sizes, in contrast with hard inorganic particles.

7.2.4.1 Inorganic wastewater

Chemical mechanical polishing (CMP) is one of the fastest growing processes in semiconductor industry, and it has become an integral part of the state-of-the-art fabrication line for the multilayer wiring board of large-scale integrated circuit (LSI). Besides the semiconductor devices, CMP is widely applied to the magnetic head, the magnetic memory, and the imaging devices. The process is primarily used for polishing the device side of a semiconductor wafer through the mechanical downforce of slurry abrasive in combination with chemical oxidation of wafer surface. In general, colloidal silica is used as abrasive slurry to planarize the oxide wafer surface. Particles in slurry are highly charged to avoid aggregations between particles or between particles and wafer surfaces. During the process, large volumes of ultrapure water are consumed to clean the surface of the wafer, which generates large quantity of CMP wastewater typically having high solid content resulting from slurry abrasive particles of SiO_2, Al_2O_3, or CeO_2, depending on the nature of the CMP application. The quantity of CMP wastewater generated is expected to increase proportionally with the growing needs of the CMP processes. As a result, the treatment and reuse of CMP wastewater has become increasingly necessary. The CMP wastewater has been generally treated with the conventional chemical coagulation–sedimentation process, producing large quantity of sludge. Currently, a membrane filtration process coupled with chemical pretreatment is used to separate the nanoscale particles from the CMP wastewater to reclaim the water [1, 2].

Wastewater of nanosized metal colloid, which is hard to be removed by coagulation–sedimentation process, is discharged in such diverse fields as metalworking factory, electronic components factory, and pigment-manufacturing factory [3]. It is reported that various trace elements of heavy metal are contained in wastewater discharged from a pulp production plant [4]. A spent emulsion, which contains nanosized copper colloid, is discharged from plants manufacturing copper cables for electrical industry, and the treatment for purification of effluents is examined by the integrated membrane system based on ultrafiltration (UF) and nanofiltration (NF) [5]. A glass company generates the wastewater containing fine clay and glass particles from the grinding process of glass surfaces during production of CRT glass used for TVs and monitors. Separation of fine clay and glass particles by microfiltration (MF)/UF is examined in order to treat glass industry wastewater for reuse in the manufacturing process [6].

7.2.4.2 Organic wastewater

The colored substances are free from regulatory constraint of water quality so far because they are not considered hazardous substances. However, water color is being recently used as a standard for the judgment of the purity in water because the removal of color becomes important for wastewater reclamation and reuse.

Dye works are scattered across the country as the industry with local tradition. The dyehouse effluent is discharged in large quantity, and it has extremely complex composition because it contains not only dye but also dyeing aid and finishing agent. In general, dye cannot be removed by standard biological treatment because of its low environmental biodegradability. Dye wastewater is treated by coagulation–sedimentation and activated sludge processes, and nanoparticles produced in the course of the treatment are released into the environment [7].

The color is often imparted by organic substances, predominantly humic substances. Aquatic humic substances including humic and fulvic acids are a term referring to a broad class of naturally occurring mixture of organic compounds, ubiquitous in surface waters, ground waters, and soil pore waters. They are a complex mixture of heterogeneous organic materials in terms of elemental composition, chemical functionality, and molecular size distribution since humic substances can be derived from any organic materials, including plant and animal debris, microfauna, biowaste, pesticide, and others. The molecular weights of humic acids range from several thousands to several tens of thousands daltons, and those of fulvic acids range from several tens of thousands to several hundred thousands daltons. Because of this versatility, humic substances are known to significantly affect the behavior of some pollutants in natural environments, such as trace metal speciation and toxicity, solubilization and adsorption of hydrophobic organic compounds, and disinfection by-product formation [8].

Melanoidins are natural polymeric compounds of dark brown color, and they are closely related to humic substances. They are produced by a set of consecutive and parallel nonenzymatic reactions taking place between amino compounds and carbohydrates during a Maillard reaction [9]. They are contained in the molasses wastewater from alcohol distillery, sugar processing and refinery industry, and glutamate processing industry. Such wastewater containing melanoidins has frequently caused a coloration problem of water environment, and thus the suitable decolorization treatment is required in many fermentation and sugar industries using molasses. Treatments by flocculation, ozonation, and electrolysis are promising in color removal [10].

Food-processing wastewater usually contains a variety of organic materials in varying degree of concentration. In cheese-making in the dairy products industry, only ~10% of the initial milk volume becomes product, cheese, and the other 90% becomes by-product, liquid cheese whey. Since cheese whey is a protein- and lactose-rich by-product of cheese production, its cost-effective utilization is becoming increasingly important. Recent developments in membrane technology have provided exciting new opportunities for large-scale protein and lactose fractionation in whey treatment [11]. In textile industry, typically it takes over 100 L of water to process just 1 kg of textile material. Not only the washing water must be treated to recover important by-products such as lanolin, but bleaching and dyeing chemicals must also be removed before discharge back to the rivers [12].

Surfactants are a primary constituent of the detergent used in the household routinely, and also they are widely used in industry and agriculture because they have several functions such as washing, emulsification, and dispersion. The surfactants are usually present in the solution in the form of the micelle, and large amounts of surfactant wastewater are discharged in the rivers [13]. Pesticides whose molecular weight ranged from 200 to 400 Da (~1 nm) have been used in great quantities not only for agricultural use but also in golf links and resort. Therefore, the wastewater and effluent treatments have become an important issue, and pesticide separation by NF membranes is found to be very efficient [14].

The potential reclamation of high-quality water produced by the advanced treatment of the secondary effluent of the municipal sewage has come a long way in recent years. The sewage contains various components such as virus [15], pharmaceutical substances [16], and endocrine disrupting compounds derived from zoonotic excretory substances [17]. The advanced treatment of such chemical contaminants at low level becomes increasingly important.

As mentioned above, the removal of nanoparticles contained in wastewater is stringently required to recycle the reclaimed wastewater in a wide variety of industries such as chemical industry, textile industry, pulp and papermaking industry, food-processing industry, dairy products industry, and pharmaceutical industry. Also for domestic wastewater, the reuse of the reclaimed wastewater for nonpotable purposes is becoming more and more important, and this is expected to raise awareness of the behaviors of nanoparticles contained in wastewater in order to upgrade the water treatment processes.

References

[1] J.R. Pan, C. Huang, W. Jiang and C. Chen: *Desalination*, **179**, 31–40 (2005).

[2] H. Umezawa, M. Iseki, D. Takaoka, M. Tsuihiji and T. Kasahara: *Sanyo Tech. Rev.*, **35**(2), 22–30 (2003).

[3] M. Moriya: *J. Jpn. Soc. Water Environ.*, **22**, 346–351 (1999).

[4] L. Skipperud, B. Salbu and E. Hagebø: *Sci. Total Environ.*, **217**, 251–256 (1998).

[5] K. Karakulski, A.W. Morawski: *Desalination*, **149**, 163–167 (2002).

[6] S.K. Kang, K.H. Choo: *J. Membr. Sci.*, **223**, 89–103 (2003).

[7] K. Higashi: *J. Jpn. Soc. Water Environ.*, **20**, 210–214 (1997).

[8] N. Shinozuka: *J. Jpn. Soc. Water Environ.*, **18**, 261–265 (1995).

[9] S. Homma: *Denpun Kagaku*, **38**, 73–79 (1991).

[10] A. Nagano, C. Nakamoto and M. Suzuki: *J. Jpn. Soc. Water Environ.*, **22**, 498–504 (1999).

[11] A. Rektor, G. Vatai: *Desalination*, **162**, 279–286 (2004).

[12] G. Rideal: *Filtr. Sep.*, **42**(7), 30–33 (2005).

[13] K. Yoshimura: *J. Jpn. Soc. Water Environ.*, **16**, 294–301 (1993).

[14] Y. Kiso, H.-D. Li and T. Kitao: *J. Jpn. Soc. Water Environ.*, **19**, 648–656 (1996).

[15] S.S. Madaeni, A.G. Fane and G.S. Grohmann: *J. Membr. Sci.*, **102**, 65–75 (1995).

[16] T. Urase, K. Sato: *J. Jpn. Soc. Water Environ.*, **28**, 657–662 (2005).

[17] S. Kim, Y. Suzuki: *J. Jpn. Soc. Water Environ.*, **25**, 349–354 (2002).

7.2.5 Indoor environments and nanoparticles

In recent urbanized lifestyles people tend to spend more time in enclosed buildings or residences than outdoors. Therefore, it is of great importance to characterize indoor particles and correlate between indoor and outdoor ones from the viewpoint of evaluating the influence of indoor air quality (IAQ) on human health.

As shown in Table 7.2.1, indoor nanoparticles originate from the several sources such as products of chemical reactions, nonvolatile residues (NVRs) of liquid droplets, printers/photocopiers, combustion, bioaerosols, and infiltration of outdoor air.

Table 7.2.1
Main generation sources of indoor nanoparticles.

Chemical reaction by ozone	
	Terpenes (α-pinene, limonene) – ozone reaction, formation of irritant products
Nonvolatile residues (NVRs) of droplets	
	Atomization-type air humidifier and cleaner
Printers	
	Satellite particles from ink-jet printer, toner-originated matters from laser printer
Combustion	
	Cooking by gas stoves, cigarette smoking, incense smoking
Bioaerosols	
	Virus
Outdoor sources	
	Penetration of outdoor nanoparticles through windows and other house openings

7.2.5.1 Secondary particle formation by gas phase ozonolysis

For particle formation resulting from chemical reaction via ozone, the reaction of terpenes is very common in indoor environments as well as atmospheric ones. Terpenes are emitted from fragrance-containing vegetable oils such as pine oil and citrus oil, and wooden materials including woody furniture [1].

Meanwhile, sources of emission of ozone are air cleaners, air-conditioners, laser printers using corona discharge, and infiltration of outdoor air. Terpenes are generic terms of unsaturated organic compounds that are composed of isoprene as unit (e.g. α-pinene and limonene). These compounds used for household applications readily react with ozone because they have one or more double bonds. It has been proposed that, as shown in Fig. 7.2.8, the reaction mainly proceeds to form less volatile pinonic acid via pinonaldehyde of intermediate [2].

Furthermore, the acid-catalyzed reaction allows the products to convert into higher molecular weight compounds by the polymerization via carbonyl groups in the aldehydes and the aldol condensation [3].

It has been reported that the resultant generated particles have a size distribution with a peak diameter of about 30 nm, and that the products by terpenes ozonolysis irritate human airways [4].

7.2.5.2 Nonvolatile residue (NVR) of liquid droplets

Nanoparticles are also generated from air humidifiers or negative air–ion generators in which water is atomized. In general, humidifiers are mainly categorized into vaporization type and atomization type [5].

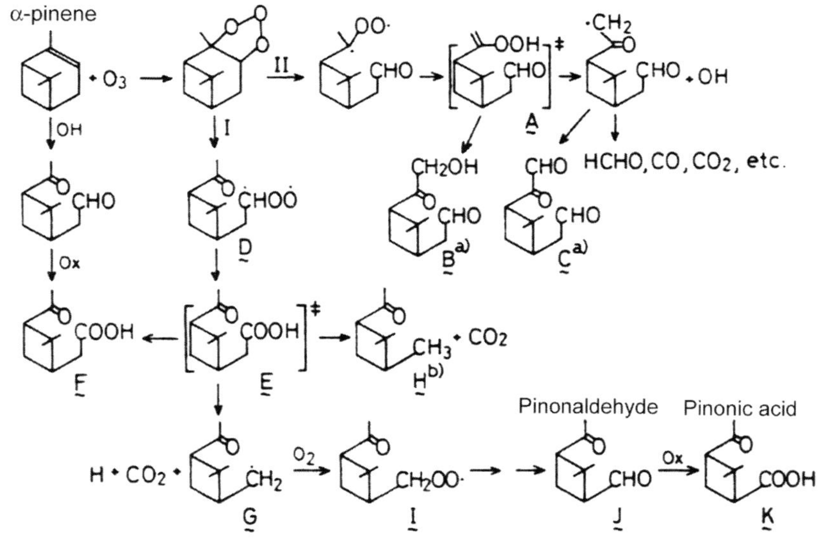

Figure 7.2.8
Reaction path of α-pinene ozonolysis.

The former does not entrain impurities in water when it is fed into indoor spaces of interest. Meanwhile, the latter has the drawback that NVRs are suspended in spaces to be humidified by feeding water via spraying and sonication. The NVRs in tap water include colloidal particles and soluble fraction such as silicates, sulfates, carbonates, and chlorides. The size of NVR particles, D_{p_r} can be estimated from the following equation:

$$D_{p\_r} = D_{p\_m}\sqrt[3]{\frac{C\rho_m}{\rho_r}} \qquad (7.2.1)$$

where D_{p_m} is the droplet size, C the mass fraction of NVR in the droplets, ρ_m the droplet density, and ρ_r the NVR particle density, respectively. Assuming that 2 µm-sized droplets ($\rho_m = 1{,}000$ kg/m^3, $\rho_r = 2{,}500$ kg/m^3) are formed by an ultrasonic nebulizer, and the mass fraction of NVR in city water is 100 ppm (10^{-4}), the NVR particle size, D_{p_r} is estimated to be 68 nm.

Recently, a wide variety of negative ion generators using the Lenard effect, corona discharge, UV/photoelectron emission and electrospray have been commercialized and attention has been focused on features such as air purification and physiological activation [6]. Among them there are the ion generators that atomize water based on the Lenard effect and electrospray form the NVR particles as by-product in addition to ion products if the supplied water contains nonvolatile impurities.

Fig. 7.2.9 shows an example of electrical mobility distribution for ions generated by the electrospray method (positive in this case) [7]. This method atomizes liquid fed to a tip of a capillary electrode to form fine droplets with large amounts of charge by applying high voltage between the tip and the downstream counter electrode. When water in the generated droplets evaporates and their surface charge density attains the charge limit called "Rayleigh limit", this phenomenon induces their self-fragmentation followed by the formation of a high concentration of cluster ions. In this figure, the high-mobility peak on the right side corresponds to the cluster ions. These ions are 2 nm in size assuming that they are singly charged. Meanwhile, another peak on the figure results from NVR in water and then its height increases with the increase in the fraction of tap water in the fed liquid. The electrical-mobility-equivalent size of NVR particles measured by differential mobility analyzer (DMA) – condensation nucleus counter (CNC) method ranges from 10 to 20 nm and their concentration is on the order of 10^4 particles/cm^3. Comparing the forementioned electrical mobility distribution with the particle size one, the NVR nanoparticles are estimated to hold about 100 charges.

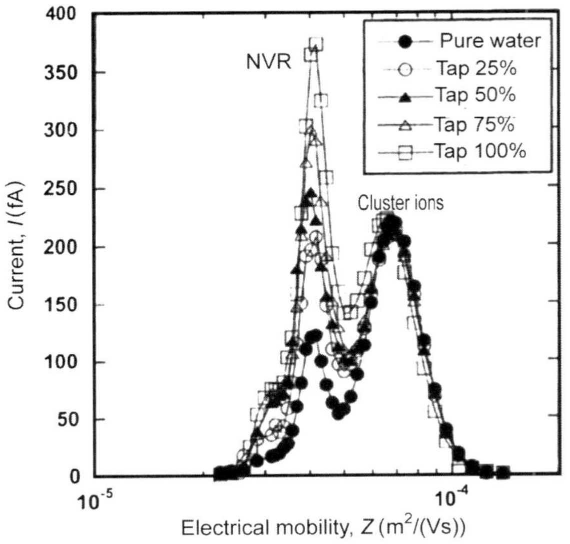

Figure 7.2.9
Electrical mobility distribution of cluster ions and nonvolatile residue (NVR) particles generated by the electrospray method.

7.2.5.3 Laser printers and photocopiers

Accompanying the recent proliferation of computers, the use of inkjet printers and electrophotographic machines such as laser printer and photocopier is becoming common in homes as well as offices. It has been reported that these devices emit various sorts of pollutants. The eco-friendliness-oriented standards such as Blue Angle Standard [8] regulate the maximum permissible limits of benzene, styrene, total volatile organic compounds (TVOC), ozone and particles. As the regulation of particles is based on the emitted mass per hour, mainly the relatively coarser particles such as toner and dust adhering to paper have been targeted. However, some reports have revealed that nanoparticles are emitted from inkjet printers or laser printers [9]. Fig. 7.2.10 depicts the size distribution of particles emitted from a laser printer measured by a scanning mobility particle sizer (SMPS). As seen in the figure, nanoparticles with a peak diameter of around 30 nm are generated in printing mode, whereas the emission in the case of feeding paper without printing is about one third of the normal printing mode. Furthermore, these particles were dried by passing them through a diffusion dryer because they are thought to originate from the nucleation of water vapor emitted from papers in the fixation process. As a result, it was found that most particles formed in the paper feed mode evaporated and then vanished, while particles in the printing mode contained nonvolatile components as well as water. From these results it is anticipated that the particles are derived from styrene remaining in the toner even though their composition has still not been identified.

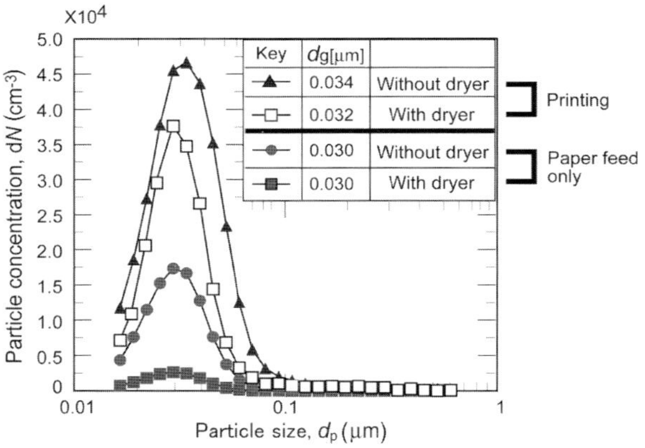

Figure 7.2.10
Size distribution of nanoparticles generated from a laser printer.

Meanwhile, it is thought that during ink discharge ink-jet printers emit not only the main ink droplets but also their satellites (about 1 μm) to result in nanometer-sized NVRs during printing.

7.2.5.4 Combustion

One of the most significant source of indoor nanoparticles relevant to combustion is cooking such as frying and sautéing [10]. Some reports said that over 90% of the particles by number were in the ultrafine fraction range during cooking with bimodal peaks at 60 and 10 nm, attaining the number concentration on the order of 10^5 particles/m^3 and the emission rate of 10^{14} particles/h. Owing to lifestyles in Asian countries, cigarette smoke, incense, and mosquito coils also contribute to indoor nanoparticle levels [11]. It was reported that especially in the Indian subcontinent the combustion of biofuels such as straw and dried cattle manure used for cooking could have a significant impact on climate change in the South Asian region [12].

Sidestream cigarette smoke also contains nanoparticles, having a concentration distribution with the main peak between 0.1 and 0.2 μm [13]. In addition, it was found that nanoparticles a peak size of 30 nm formed by the nucleation of vapor fraction in filtered sidestream smoke immediately after burning when the dilution of smoke by air was insufficient [14]. Attention should be paid to air cleaners when a high concentration of cigarette smoke has to be treated by the cleaners using a single unit air filter.

7.2.5.5 Bioaerosols

Airborne virus particles or virions are typically in the 20–400 nm size range, and are a good example of nanoparticle bioaerosols. Smaller viruses typically contain one subunit, which consists of an outer protein capsid, internal nucleic acid (e.g. DNA and RNA), and other internal proteins. Corona virus that causes SARS and influenza are good examples of them. The viruses most often are transmitted through direct contact with an infected person, such as by shaking hands, hugging or kissing, while sometimes it is spread by nasal droplets. However, it is still unknown how these virus particles behave in the case of airborne infection. Recently, the studies that attempt to elucidate the behavior have been progressing [15].

References

[1] C.J. Weschler, H.C. Shields: *Atmos. Environ.*, **33**, 2301 (1999).

[2] S. Hatakeyama: *J. Aerosol Res., Jpn.*, **6**, 106 (1991) (in Japanese).

[3] Y. Iinuma, O. Böge, T. Gnauk and H. Herrmann: *Aerosol Environ.*, **38**, 761 (2004).

[4] P. Wolkoff, P.A. Clausen, C.K. Wilkins and G.D. Nielsen: *Indoor Air*, **10**, 82 (2000).

[5] T. Ito, K. Nishimura: *J. Soc. Heat. Air-Cond. Sanit. Eng. Jpn.*, **76**, 817 (2002) (in Japanese).

[6] Y. Tanimura: *J. Aerosol Res., Jpn.*, **18**, 20 (2003) (in Japanese).

[7] B. Jan, I.W. Lenggoro, M. Choi and K. Okuyama: *Anal. Sci.*, **19**, 843 (2003).

[8] RAL German Institute for Quality Assurance and Certification: *Basic Criteria for the Award of the Environmental Label* (Printer RAL-UZ 85), p. 35 (2004).

[9] N. Namiki, Y. Otani, H. Emi, N. Kagi and S. Fujii: *Proc. Air Cleaning Contam. Control*, 118 (2003) (in Japanese).

[10] L.A. Wallace, S.J. Emmerich and C. Howard-Reed: *Environ. Sci. Technol.*, **38**, 2304 (2004).

[11] C.-S. Li, F.-T. Jenq and W.-H. Lin: *J. Aerosol Sci.*, **23**, S547 (1992).

[12] C. Venkataraman, G. Habib, A. Eiguren-Fernandez, A.H. Miguel and S.K. Friedlander: *Science*, **307**, 1454 (2005).

[13] K. Katayama, S. Kitao, M. Shimada and K. Okuyama: *J. Aerosol Res., Jpn.*, **19**, 50 (2004) (in Japanese).

[14] Y. Otani, N. Namiki: *Annu. Res. Rep. Smoking Res. Found.*, 795 (2004) (in Japanese).

[15] C J. Hogan Jr., M.-H. Lee and P. Biswas: *Aerosol Sci. Technol.*, **38**, 475 (2004).

7.2.6 Industrial processes and nanoparticles

This section describes the sources of nanoparticle generation in industrial processes by categorizing them into specific processes where a cleanroom is used and other general ones.

7.2.6.1 General industrial processes

The sources of emission of unwanted nanoparticles in general workplaces are categorized as fumes from hot processes (e.g., smelting, refining, and welding) and from (incomplete) combustion processes. Favorable conditions required for the generation of nanoparticles are found in workplaces where there is (1) presence of vaporizable material, (2) sufficiently high temperature to produce enough vapor, followed by condensation to form an independent aerosol, and (3) rapid cooling and a large temperature gradient.

There have been so many studies on occupational exposure to fine particles in the field of public health. In general, high spikes of nanoparticle concentration are observed during active operations, followed by a gradual decay after the operation, primarily because of coagulation, evaporation, dilution, and/or deposition. The fraction of the total number of nanoparticles generally decreases, whereas that of the number of submicrometer particles increases with time and distance from the point of emission. In order to accurately estimate exposure, the effects of spatial and temporal changes will need to be evaluated. Therefore, it is important to identify the time required for the concentration to decline to the normal or background levels.

As an example of reports on grinding processes, Fig. 7.2.11 shows the case where a steel substrate was ground upon using a high-speed grinder [1]. From the figure the distribution of concentration of generated particles has a distinct bimodality, one with the finer peak at around 10 nm and the coarser one at around 1 μm. The former results from within the grinder motor and the volatilization or combustion of amenable ground substrate and/or grinding materials, the latter from the mechanical abrasion and attrition. However, the resultant total concentration on the order of 10^5 particles/cm^3 is not so high.

7.2.6.2 Industrial processes with cleanrooms

Cleanrooms and associated controlled environments (e.g., in the case of an ISO Class 3 cleanroom, the maximum permissible airborne particle concentration is less than 10^3 particles/m^3 for particles with the size of 0.1 μm or larger, while the airborne particle concentration in ordinary indoor environments is on the order of 10^9 particles/m^3 or higher) are usually adopted to avoid particle contamination in industrial processes where precision products such as engineered nanoparticles, semiconductors, and other electronic or optical devices are fabricated because the deposition of particles onto product surfaces causes their yield reduction and quality deterioration. The emission sources in cleanroom environments are tabulated in Table 7.2.2. Since some of the listed emission sources emit trace amounts of nanoparticles, these nanoparticles are not regarded as particulate contaminant but as chemical or molecular one. In this section, these nanometer-sized solid substances formed on solid surfaces by chemical reaction are also included.

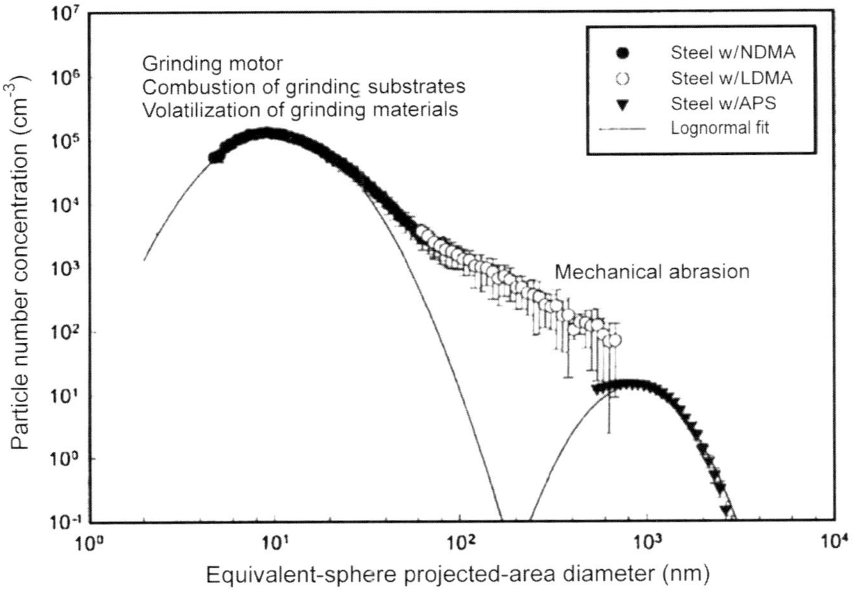

Figure 7.2.11
Size distribution of nanoparticles generated when a steel plate was ground with a high-speed grinder.

Table 7.2.2
Main sources of emission nanoparticles in cleanroom environments.

Exhaled air of human
Corona-discharge ionizer (e.g. gas-to-particle conversion of low-molecular-weight cyclosiloxane)
Boron-containing particles from borosilicate glass fibers of HEPA filter
Haze by chemical reaction on solid surfaces (precipitation of ammonium salt and silica)
Watermark on wafer surfaces at drying
Leakage from thin film and nanoparticles processing equipment

(1) Air exhaled by humans

Emissions from human bodies are a minor contribution in ordinary indoor situations because airborne particle concentration in such places is quite high, whereas the emission cannot be seen as negligible in cleanroom environments. The major human emissions are thought to be atmospheric dust deposited on clothes and skin fragments, and most of these particles are submicrometer in size. Meanwhile, particles in exhaled air are composed of fine liquid droplets from spittle (99.5% of water), and then evaporate to form nanoparticles of NVR.

In Fig. 7.2.12 an example of size distributions of particles in exhaled air before and after smoking is shown [2]. When measuring particles in exhaled air, the air was introduced into a measuring device after drying them by passing them through a diffusion dryer. The size distribution of particles in exhaled air before smoking ($N_{db} = 0$) has a bimodality, one with the peak size of 0.2 μm, and the other of 20 nm or smaller. The former peak comes from atmospheric aerosols as it decreases with the increase in number of deep breaths in a clean booth (N_{db}). The latter originates from NVR particles of spittle droplets.

Incidentally, since smoking induces the rapid increase in number concentration of particles 0.1 μm or larger by 10^4 times or more and for nanoparticles by about double, special attention should be paid to the management of personnel's clothes such as face mask when they enter a cleanroom after smoking.

(2) Emission from ionizers

Ionizers are commonly used in cleanrooms to eliminate electrostatic charge on substrates for precision electronic devices. The most popular ionizer is a corona-discharge type.

Corona-discharge type ionizers are categorized into the following three groups; AC, DC and pulsed-DC types. The issues of emission of contaminants such as ozone, NO_x and particles have been pointed out [3]. These issues are also applicable to air cleaners using a corona discharger. Among these problems is that the particle emission has a potential for particle contamination onto product surfaces and eventually decline in product yield.

The particle emission, which has been studied since the 1990s, is caused by foreign particle deposition onto electrodes, electrode erosion, and gas-to-particle conversion. The issue of electrode erosion can be solved by the improvement of electrode materials, whereas for the issue of gas-to-particle conversion, the airborne molecular contamination (AMC) control to be ionized has to be made. It was reported that

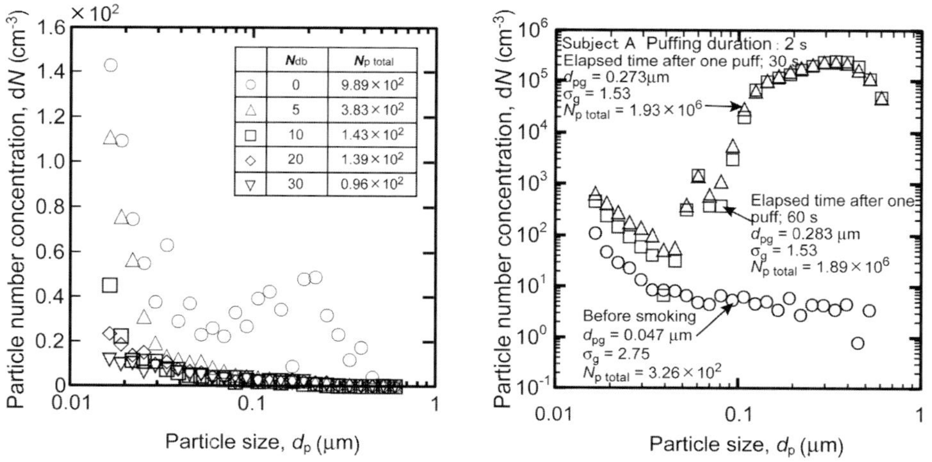

(a) Change in particle size distribution with the number of deep breathings by subject A

(b) Change in particle size distribution with the elapsed time after one puff of smoke by subject A

Figure 7.2.12
Size distribution of particles in exhaled air for smoking subjects.

silicon-containing compounds that precipitated on the electrodes result from the gas-to-particle conversion of low-molecular-weight cyclosiloxane (LMCS) from silicone sealant via corona discharge [4].

(3) Boron-containing particles from HEPA filter with borosilicate glass fibers

The use of HEPA and ULPA filters made of borosilicate glass fibers prevails in the most cleanrooms. It has been known that owing to chemical reaction in equation (7.2.2) BF_3 vapor is formed from glass fiber filters by passing HF gas leaking from wet cleaning equipment through the filters. Boron, which is a dopant element for semiconductors, has been thought to be a contaminant that might cause failure in semiconductor devices if it comes from the surroundings. In addition, it has been revealed that trace amounts of boron in the form of boric acid (H_3BO_4) are also formed from the fibers via the reaction with moisture in the surrounding air (equation (7.2.3)).

Fig. 7.2.13 depicts the change in volatilized boron mass from various filters in terms of airborne boron concentration. Especially, at the initial stage just after the initiation of ventilation, the volatilized boron mass increases with increase in relative humidity [5]. Boric acid, which is solid at room temperature, is surmised to form in the particulate form. However, its existence was identified only by chemical analysis because it is present only in trace amounts.

$$B_2O_3 + 6HF \rightarrow 2BF_3 + 3H_2O \qquad (7.2.2)$$

$$B_2O_3 + 3H_2O \rightarrow 2H_3BO_3 \qquad (7.2.3)$$

(4) Haze on solid surfaces by chemical reaction

Haze might form on glass surfaces of lenses and mirrors for optical instruments if they are exposed for a long time to cleanroom environments where AMCs are not controlled properly. The haze is more likely to bring about the insufficient light delivery onto a surface to be exposed in photolithographic processes. One of the reasons is that ammonium sulfate ($(NH_4)_2SO_4$) is formed, which then precipitates on the glass surfaces via chemical reaction of sulfur dioxide (SO_2) with ammonia or amines.

For another reason, hexamethyldisilazane (HMDS) used as additive in resist coating or LMCS from silicone sealant is adsorbed, and then decomposed to form silica precipitates on glass surfaces by photochemical reaction during laser irradiation, followed by the unwanted decline in laser penetration [6]. As another example a report said that tiny projections, which are also known as "haze", with a size of 0.2 μm or smaller were formed on silicon wafer surfaces owing to the adsorption of organosilicate compounds in thin film formation processes with CVD. It is similarly caused by the precipitation of SiO_2 [7].

(5) Watermarks on solid surfaces during drying

When a silicon wafer surface is cleaned with deionized water and then dried in air, a watermark is formed on it via the mechanisms demonstrated in Fig. 7.2.14. Oxygen in air is dissolved and diffused into water droplets or adsorbed water on a wafer surface, followed by the formation of silicate compounds via silanol reaction. The watermark on a wafer

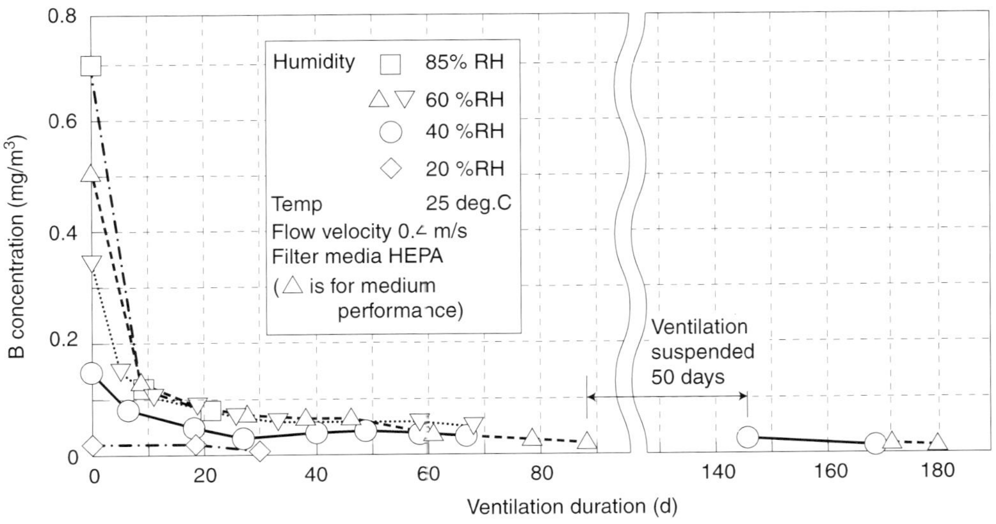

Figure 7.2.13
Change in concentration of boron emitted from various filters at different relative humidities.

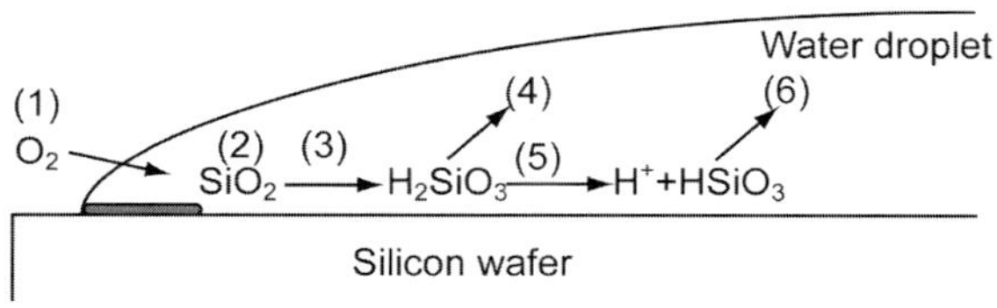

Figure 7.2.14
Mechanism of watermark formation on a silicon wafer surface.

surface is detected in the form of nanometer-sized particles by an electron microscope [8].

(6) Leakage from nanoparticle production processes

In regard to the risks to processing equipments by nanoparticle leakage from production processes, the VDI report in Germany [9] has been described in detail. The production of engineered nanoparticles can be generally categorized into two approaches. One is a "top-down" approach that is initiated with a bulk material and then breaks it into finer pieces using some form of energy such as etching, ball milling, sputtering, and laser ablation. The other approach is to synthesize materials from the atomic or molecular level by growth and assembly to form the desired nanoparticles. Processes included in this "bottom-up" category are sol-gel, chemical vapor deposition, flame synthesis laser pyrolysis, and so on.

Most of these processes are performed in a closed reaction chamber installed in a cleanroom or associated controlled environment. Human exposure to these engineered particles does not take place during synthesis unless there is an unexpected system failure (e.g. rupture of a seal). Human exposure is more likely to occur after the manufacturing when opening the reaction chamber, drying the products, or in the post-process handling of the products.

The release of nanoparticles during production chamber cleaning operations is another critical point. Cleaning typically involves using water or some solvent. Brushes, sponges, or tissues used in the cleaning will carry nanoparticles into the waste stream. Disposal of the waste and wastewater may become a source of nanoparticle release into the environment.

Further, conditioning of nanoparticles such as compression, coating, and composition to form final products may also result in the release to the environment and resultant exposure although very few studies have been carried out on this subject. Recent studies [10] to evaluate the aerosol discharge during the handling of carbon nanotubes showed that the generation of nanoparticles occurred under vacuum to remove spilled nanotube materials or vigorous mechanical agitation. However, they reported that the concentrations were very low. In addition, measurements of nanoparticle levels during final packaging of carbon black, which is a typical engineered nanoparticle material, showed that there was no increase in nearby air [11].

References

[1] A.T. Zimmer, A.D. Maynard: *Ann. Occup. Hyg.*, **46**, 663 (2002).

[2] Y. Otani, N. Namiki: *Annu. Res. Rep. Smoking Res. Found.*, 747 (2005) (in Japanese).

[3] M. Suzuki, Y. Yamaji, *J. Jpn. Air Cleaning Assoc.*, **32**, 218 (1989) (in Japanese).

[4] N. Namiki, Y. Otani, H. Emi and S. Fujii: *J. Inst. Environ. Sci.*, **39**, 26 (1996).

[5] K. Kato, A. Tanaka, A. Saiki and J. Hirata: *Proc. Air Cleaning Contam. Control*, **5** (1995) (in Japanese).

[6] Y. Ushio, T. Nakamura, S. Shimizu, T. Oshino, K. Matsuda and T. Arai: *Proc. Air Cleaning Contam. Control*, 335 (1998) (in Japanese).

[7] A. Saiki, S. Ro and T. Fujimoto: *Chemical Contamination in Semiconductor Processing Environments and its Countermeasures*, Realize, Inc., Tokyo, 426 (1997) (in Japanese).

[8] The Japan Society of Industrial Machinery Manufactures: Report on Behavior Control of Individual Sort of Contaminants – *1993 Report on Introduction of Advanced Technologies to Environmental Equipment Industry*, 171 (1994) (in Japanese).

[9] W. Luther: *Industrial Application of Nanomaterials – Chance and Risks*. Future Technologies, Division of VDI Technologiezentrum, Düsseldorf, Germany, p. 112 (2004).

[10] A.D. Maynard, P.A. Baron, M. Foley, A.A. Shvedova, E.R. Kisin and V. Castranova: *J. Toxicol. Environ. Health A*, **67**, 87 (2004).

[11] T.A.J. Kuhlbusch, S. Neumann and H. Fissan: *J. Occup. Environ. Hyg.*, **1**, 660 (2004).

7.3 Safety of nanoparticles

7.3.1 Problems caused by nanoparticles

Study on the safety of nanoparticles has started only recently, and no sufficiently systemized results have been obtained. What should be noted in particular is that the possibility of radial troubles caused by particulate matters are considered to increase by the decrease of particle diameter in nanoparticles.

One typical example is the problem of dust explosion, caused by the high surface reactivity of fine particles. In other words, since nanoparticles are extremely fine particles, dust explosion is more likely to occur. Explosion is more likely to occur because fine particles are different in their composition, in that low boiling point metal can be easily condensed, as described in Section 7.2.

However, of particular note here is that, since all particles do not necessarily exist independently in the form of a single particle, the possibility of dust explosion does not simply increase as particles become finer.

Fine particles with sizes of 1 μm or smaller such as nanoparticles have an extremely high agglomeration propensity and secondary particles can be easily generated. Therefore, in some cases they conversely behave like large particles. These are the points to be taken into consideration when studying the problems caused by nanoparticles.

As shown in Fig. 7.3.1 [1], the effect of the particle diameter on dust explosion tends to be that the smaller the particle diameter of the dust, the lower the minimum explosion concentration. In other words, explosion can be induced under conditions of lower concentration of particles in air as the particle diameter becomes smaller. Due to the difficulty of conducting experiments to suspend particles with the same size in a uniform concentration, this result was obtained from particles far larger than nanoparticles; however, it has been clarified qualitatively that the smaller the particle diameter, the higher the possibility of dust explosion.

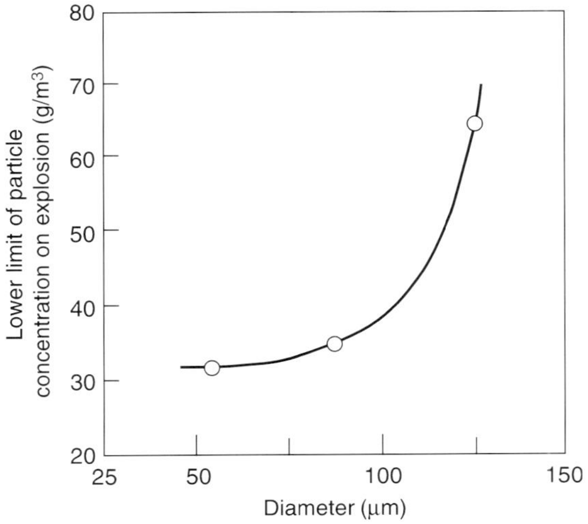

Figure 7.3.1
Influence of particle diameter on lower limit of particle concentration for explosion.

From the perspective of composition, the possibility of explosion increase if materials that react easily with oxygen at low temperature are condensed into particles of small diameter. Therefore, with regard to the effect of particle diameter on dust explosion, more careful attention needs to be paid in the case of combined materials than in the case of uniform materials, as assumed in Fig. 7.3.1. As described before, however, since nanoparticles are considered to exist often as agglomerates, it is necessary from the perspective of the particle diameter to take into consideration the diameter of not only primary particles but also of particles after agglomeration. To address the safety of nanoparticles, it will be important in the future to elucidate their behavior in detail including these factors.

Reference

[1] H. Enomoto: Funjin-Bakuhatsu–Kikensei hyouka to Boushi taisaku (Dust explosion–Estimation of danger and control policy), Ohmsha, Ltd., Tokyo, 17 (1991).

7.3.2 Health effects on nanoparticles

The terms ‘nanoparticles’ and ‘nanomaterial’ have been used for particles of which one representative dimension, for example, diameter of particles on cross-sectional diameter of fibers has at least 100 nm or less. Some people hold that the majority of such fine particles are exhaled without depositing in the respirator tract, and that therefore the particles may not cause pulmonary diseases. However, the properties of nanoparticles are known to be different from the bulk material they are derived from. In cases where the biological effects of bulk materials have been reported, nanosized particles of these materials may be expected to have stronger dose response for the health effects. Every effort must be made to clarify the uncertainty on the risks of these nanomaterials [1]. At the present time, there is no regulation or standard for assessing the biological effects of nanomaterials, and therefore there is a paucity of toxicological data concerning nanomaterials. Much more systematic and strategic studies are needed to enable risk assessments for human health [2–6].

As regards risk assessment and risk management of nanomaterials, the characterization and identification of anticipated risks should be first determined for chemical substances or foods. Conventional assessment methods are applicable for water-soluble particles. For insoluble nanoparticles, the assessment of potential health hazards should be made based on their properties or toxicity and dose–response relationship. The risk is a product of hazard and exposure; even if a nanoparticle has a hazard, the risk is lower when the possibility of exposure to the nanoparticle is small [2].

7.3.2.1 Exposure routes and uptake of nanoparticles

(1) Exposure routes for nanoparticles

Nanoparticles can either be deliberately introduced into the body for medical purposes (drug delivery systems) or absorbed involuntarily from the environment (inhalation of nanoparticle-containing dust in the air). A distinction should also be drawn between nanoparticles manufactured for industrial application and those unintentionally generated and released in the environment, such as welding fumes or diesel exhaust particles (DEP). In the fields of environmental science and toxicology, numerous studies on the potential health hazards caused by ultrafine particles have been conducted. Practically, there are several definitions of nanoparticles or ultrafine particles, however, findings regarding biological effects of the ultrafine particles are useful as a starting point for estimating the effects of nanoparticles on human health.

Human and animals contact with nanoparticles through various routes: nanoparticles can be inhaled in the air, swallowed in the water, ingested in food, and absorbed via the skin in cosmetics. For successful risk assessment, it is important to determine how nanomaterials or nanoparticles are used, such as composites, surface coating, or powders. Coatings or powders have the potential to release a part of their nanomaterials into the environment. Workers who come into contact with nanomaterials have the possibility of exposure to nanoparticles at the workplace. Consumers of products using nanotechnology can also be exposed to them. Attention needs to be paid to the environments and ecosystems in which nanoparticles and nanomaterials are released. Nanoparticles in the products may change their size, quantity, and composition during their life-cycle of manufacturing, use, transportation, and disposal.

(2) Respiratory uptake of nanoparticles

Inhalation is the main route of exposure to nanoparticles. Particles inhaled with the air through the mouth and nose pass through the throat (nasopharynx and oropharynx) and tracheobronchial tree before reaching the alveolar region where oxygen moves from the alveoli to the blood and carbon dioxide moves from the blood to the alveoli. How deeply particles can penetrate and where they become deposited on each respiratory airway such as the nasal cavity, tracheobronchial tree, and the alveoli depend on their size under the various deposition mechanisms: inertial impaction, gravitational sedimentation and diffusion, etc. The respiratory airway includes the anterior nasal passage, posterior nasal passage, pharynx, larynx, trachea, main bronchi, bronchi, bronchioles, terminal bronchioles, alveolar duct, and alveoli, as shown in Fig. 7.3.2. In the human lungs, the trachea divides asymmetrically into the right main bronchus that enters the right lung where it divides into three lobes, that is, an upper, a middle, and a lower, and the left main bronchus that enters the left lung where it divides into two lobes, that is, an upper and a lower. The trachea divides into two branches, dividing progressively to the terminal alveolus.

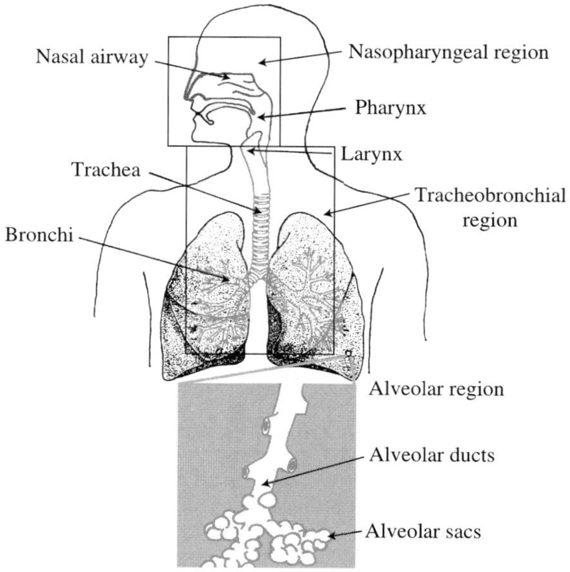

Figure 7.3.2
Anatomical overview of human respiratory tract showing major structures.

To quantitatively assess the pulmonary particle deposition needs human lung morphology models, respiratory physiology based models of the entire lung airway system and aerosol deposition models based on many experimental findings. In 1994, the ICRP Task Group on Lung Dynamic (ICRP: the International Commission of Radiological Protection) published their revised lung model [7]. The deposition, clearance, and translocation of particles in each of the compartments were described. While the model has been widely used in the nuclear field, it is applicable to conventional aerosols as well as radioactive aerosol. In the nuclear field, aerosols including radon progeny that used to be nanoparticles have been studied. Fig. 7.3.3 shows the deposition fractions of inhaled particles per adult nasal respiration of $1.2\,m^3/h$ in each region re-calculated for the nasopharynx, tracheobronchial, and the pulmonary (alveolar) region based on the model. Inhaled aerosol particles deposit on different regions depending on their size; for example, nanoparticles larger than 10 nm deposit mostly in the alveoli and those less than 10 nm deposit in the nasal cavity. How deeply particles penetrate into the lung depends on their size. Nanoparticles can reach pulmonary region in the lung and deposit more intensively and this, therefore, has become one of the reasons for concern about the effects of nanoparticles on human health. However,

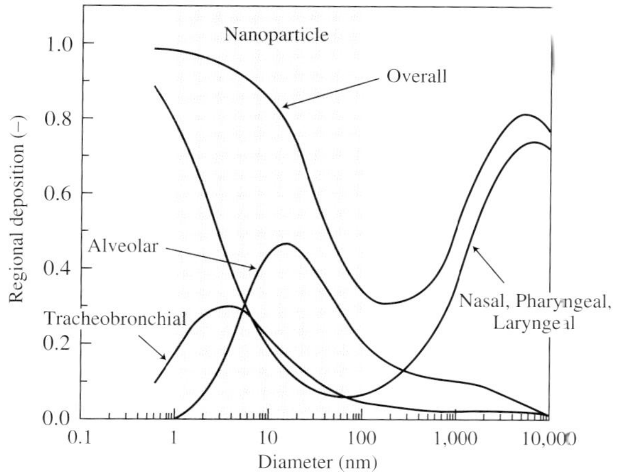

Figure 7.3.3
Predicted fractional deposition of inhaled particles in the nasopharyngeal, tracheobronchial and alveolar region of human respiratory tract during nose breathing based on ICRP Publication 66 [7] (Healthy adult and breathing rate is 1.2 m^3/h).

deposition of nanoparticle chain-like agglomeration and fibrous particles such as carbon nanotubes cannot be estimated by this model.

Most inhaled nanoparticles are deposited on the surface of the respiratory tract. Generally, if insoluble particles are deposited in the ciliated airspaces which are lined with a mucous layer, they are transported to the digestive tract with the mucous flow by mucociliary movements. Particles deposited on nonciliated bronchioles and alveoli are phagocytosed by macrophages, which is a kind of white blood cell. As a result, their residence time is longer, however, they are usually transported to the ciliated upper part of the respiratory tract. Removal of deposited particles described above is called clearance. When the amount of deposited particles is below a certain value, no health effect is produced. In relation to a macrophage response to particles, crystalline silica is hazardous, whereas titanium dioxide is not. Pneumoconiosis is a well-known lung disease that is caused by exposure to dust particles of several micrometers in diameter. Silicosis is a typical form of pneumoconiosis resulting from exposure to crystalline silica dust and characterized by a progressive fibrosis of the lungs. The macrophage-mediated clearance (phagocytes) was effective for micron and submicron particles. It has been reported that only 20% of deposited nanoparticles were removed by the clearance mechanism [8]. It has been suggested that the remaining nanoparticles may pass through the alveolar walls, penetrate into the blood or lymphatic circulation, and be transported to other organs. Many studies have shown that the smaller the particle size the greater the mobility or they pass easily through the alveolar wall and enter into the bloodstream. It is further presumed that the mechanism of health effects of nanoparticles on cardiovascular system other than the respiratory tract is similar to that in the airborne ultrafine particles from DEP.

7.3.2.2 Biological effects of nanoparticles

(1) Biological effects of particulate matters

In Japan, two reference values, the 'administrative control levels' (ACL) and the 'occupational exposure limits' (OELs), are used for the regulation of hazardous chemicals as well as dust (particle matters). OELs are recommended and revised every year by the Japan Society for Occupational Health. OEL (OEL-Mean) for mean concentration of a chemical substance is defined as the reference value to the mean exposure concentration at or below which adverse health effects caused by the substance do not appear in most workers working for 8 h a day, 40 h a week under a moderate workload [9]. The 'threshold limit value' (TLV) has the same definition (but may not be the same values as OEL of the same substance) provided by the American Conference of Governmental Industrial Hygienists (ACGIH).

ACL is an index to determine the control class to judge the propriety of the working environment control based on the results for working environment measurement, which have been implemented for the unit work area in accordance with the Working Environment Measurement Standard. The results of working environment measurement are evaluated by classifying the working environments concerned into three control classes (Control Class I, II, and III). These classes are used as the standards to classify the level of the working environment concerned. Among those subject to working environment measurement, these standards apply to workplace where dust, lead, organic solvent, and specified chemical substances are used. Article 65 of the Industrial Safety and Health Law stipulates that certain workplaces in which harmful substances are involved or harmful work operations are performed shall be the subject to working environment measurement. A 50% cut-off size of the dust particle is set at 4 μm for both standards and is far larger than nanosized particles. OEL or ACL [10] for particulate matters are usually based on mass concentration, that is, milligram per cubic meter. Therefore, if the particulate matters have broad size distribution, the contribution of nanoparticles is not large to in terms of mass of particles.

Evidence from a number of toxicological studies on insoluble particles indicates that the primary determinant of the health effect of particles depends on the surface area of particles deposited [11, 12]. On the basis of the results from a number of in vitro studies of insoluble nanoparticles, a hypothetical cellular interaction has been proposed [13]: inflammation and oxidative stress can be mediated by several primary pathways: (1) the particle surface causes oxidative stress resulting in increased intracellular calcium and gene activation; (2) transition metals released from particles result in oxidative stress, increased intracellular calcium, and gene activation; (3) cell surface receptors are activated by transition metals released from particles, resulting in subsequent gene activation; or (4) intracellular distribution of insoluble nanoparticles to mitochondria generates oxidative stress.

In the workplace, the concentration of nanoparticles may be at a high level and most of the nanoparticles become agglomerates, while nanoparticles will form single nanoparticle at low levels in the general environment. It is a matter of debate whether agglomerates of nanoparticles react as a larger particle or a single nanoparticle in the lung or other organs.

If insoluble particles are retained in the lung for a longer time without enough clearance mechanisms, they can cause pulmonary inflammation or pneumoconiosis. It is of interest that nanoparticles deposited in the lung can move into the blood vessel through alveolar epithelium and they can damage vessels or produce blood clots [14, 15]. In a recent study, nanoparticles deposited in the nose may move directly to the brain via the olfactory bulb [16].

(2) Biological effects of fullerene

The biological effects of fullerene have being investigated intensively. In rats dosed orally with radioisotope-labeled C_{60} fullerenes, most were excreted in the feces and some were found in the urine. A small amount of them can be absorbed via the gastrointestinal tract. In contrast, in the same study, 90% of the same labeled fullerenes administered intravenously were retained after 1 week, with most found in the liver [17].

LD50s (acute toxicity) by intraperitoneal injection in mice and rats were 1.2 and 0.6 g/kg, respectively. The dose of 2.5 g/kg orally in rats did not result in death. The reproductive translocation of fullerenes was also observed in mice. Fullerenes have shown mutagenic activity in Ames tests. Fullerenes have shown no skin irritation or allergic reactions [18].

On the other hand, fullerenes are being tested for possible medical use. Fullerenes are basically hydrophobic but water-soluble derivatives have been synthesized to be used as drugs or its carrier. The derivative can be anticipated as drugs, for example, anti-AIDS drug. It has been stated that the toxicity of fullerenes changes due to slight structural changes including chemical modification [18].

(3) Biological effects of carbon nanotubes

Carbon nanotubes are chemically stable and are similar in form and size to asbestos; these characteristics have given rise to concern that carbon nanotubes may have the potential to cause pulmonary diseases such as lung cancer and mesothelioma similar to asbestos. A few data are available concerning the biological effects of carbon nanotubes. The biological effects of carbon nanotubes are being researched. Epithelioid granulomas and interstitial inflammation are induced in mice and rats following exposure to single-walled carbon nanotubes [18–20]. Untreated carbon nanotubes contain the nanoparticles of transition metals such as iron and nickel, which are used as catalysts in forming carbon nanotubes. These nickel-containing carbon nanotubes have been reported to be toxic [19].

The concentration of airborne asbestos fibers is expressed as a number concentration, that is, fibres per cubic centimeter or fiber per liter. When fiber concentrations are determined by phase contrast light microscopy, the fibers with a diameter of less than 3 μm, a length longer than 5 μm, and a length-to-diameter ratio (aspect ratio) greater than 3 are counted [21]. Asbestos fibers having nanosized diameter were often observed in analyses of environmental samples using electron microscopy. International Agency for Research on Cancer (IARC) rated asbestos as a known human carcinogen (group 1) [22] and the concentration of chrysotile asbestos is expressed as a risk level of 0.15 fiber/cm^3 [9]. Health effects of vitreous fibers and other asbestos substitutes have been assessed to determine their OELs or their carcinogenicity in humans. The health effects of carbon nanotubes are being intensively investigated now.

(4) Biological effects of carbon black

The OEL for carbon black respirable dust is 1 mg/m^3 and these for activated charcoal and graphite are 0.5 mg/m^3 in each [9]. In the Ref. [23], while rats and mice inhaled carbon black with a particle diameter of ~30 nm at a concentration of 5–13 mg/m^3 did not produce any specific changes, particles (agglomerate of small particles) of ~450 nm at a concentration of 2–6 mg/m^3 produced early pulmonary changes.

(5) Biological effects of metal oxides

Micronsized titanium dioxide particles are thought to have almost no toxicity and often used as a negative control substance. The OEL for titanium dioxide is 1 mg/m^3 for respirable fraction [9]. However, the results of a series of studies by Oberdörster et al. [3, 11, 24, 25] on submicron- and nanosized titanium dioxide suggested that as size decreases, inflammatory effects are intensified, and normally nontoxic substances may assume hazardous characteristics. Fig. 7.3.4 shows a part of the results by Oberdörster et al. in which rats and mice were exposed to anatase titanium dioxide particles [25]. Their results have been frequently cited in the discussion of whether the health effects of fine particles should be based on its mass or its surface area. In Fig. 7.3.4, percentages of neutrophils in lung lavage of rats are shown as indicators of inflammation after intratracheal instillation

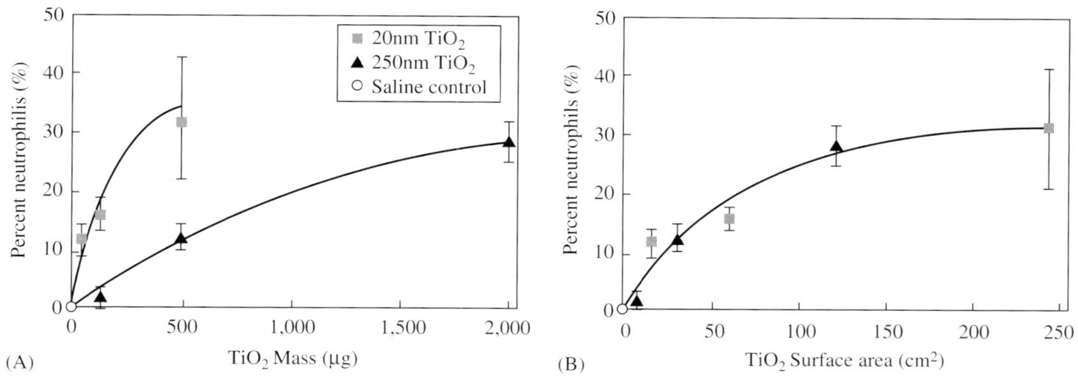

Figure 7.3.4
Percentage of neutrophils in lung lavage of rats as indicators of inflammation 24 h after intratracheal instillation of different mass doses of 20 and 250−nm TiO_2 particles in rats [3]. (A) The dose responses of two samples are different from each other as mass. (B) The same response relationship is shown as particle surface area.

of different mass doses of 20 and 250 nm TiO_2 particles. The steeper dose response of nanosized TiO_2 particles is observed than for submicron TiO_2 particles when the dose is expressed as mass (Fig. 7.3.4A). If the same dose–response relationship as in Fig.7.3.4A is indicated as particle surface area (Fig. 7.3.4B), the particle surface area seems to be a more appropriate dosimetric for comparing effects of different-sized particles of the same chemical structure.

Zinc oxide is a white powder and used in pigments. The OEL is 1 mg/m^3 for its respirable fraction [9], and the value for zinc oxide fume which causes metal fume fever is under consideration.

Nanoparticles of transition metals and rare earth elements and their oxides will be used widely. Since many of these metals and their oxides have biological effects, particular attention should be given to them.

Nickel compounds are rated as a human carcinogen (group 1) by IARC. In particular, nickel oxide is particularly insoluble among the nickel compounds and remains longer in the lung. Nanosized nickel oxide particles have greater toxicity to the lung than larger particles [26, 27]. Pulmonary inflammatory responses induced by nanosized cobalt particles have been reported [28–30]. Biological effects of nanosized particles of other transient metals such as iron and manganese have received attention [3].

Rare earth elements are a general term of 17 chemical elements consisting of scandium (Sc), yttrium (Y), and a lanthanide series of 15 elements from lanthanum (La) to lutetium (Lu). These elements have been used in magnetic alloy, fluorescent and hydrogen storage alloy. Particularly cerium oxide nanoparticles are frequently used as a fuel additive and are incorporated in cosmetics formulation. The potential biological and environmental effects of these elements have not sufficiently been investigated. It has been demonstrated that LD50s for these elements in oral and intravenous administration are in a range from several dozens to several thousands milligrams per kilograms, indicating that none of these elements has high toxicity. In the results of studies in which the biopersistence and the distribution of rare earth compounds in the body were investigated, for example, the compounds deposited in bones and teeth, and organs including lung, liver, spleen and kidney following intratracheal, oral, intravenous, and intraperitoneal administration. Although the compounds deposited predominantly in the liver other than bones, it has also been reported that the distribution of the compounds in the lung and spleen increased when the dose was increased [31].

The demand for indium compounds has been sharply increasing. The compounds have been used in the materials for transparent electrodes for flat panel displays. In Japan, the cases of pulmonary interstitial pneumonia and pulmonary fibrosis have been reported in workers engaged in cutting and grinding of sintered indium–tin oxide (ITO) and potentially having inhaled the dusts released from ITO [32, 33]. The biological effects of indium arsenic compounds and indium phosphorus compounds also have been investigated. The ACGIH has proposed a value of 0.1 mg/m^3 for their TLV, and the value has been applied tentatively in Japan.

References

[1] The Royal Society, *Nanoscience and Nanotechnologies:* ISBN 0 85403 604 0, http://www.nanotec.org.uk/finalReport.htm (2004).

[2] M. Takemura: *ChemoBio Integr. Manage.*, **1**, 57–73 (2005) (in Japanese).

[3] G. Oberdörster, E. Oberdörster and J. Oberdörster: Nano-toxicology: *Environ. Health Perspect.*, **113**, 823–840 (2005).
[4] P. Biswas, C.Y. Wu: *J. Air Waste Manage. Assoc.*, **55**, 708–746 (2005).
[5] A.D. Maynard, E.D. Kuempel: *J. Nanoparticle Res.*, **7**, 587–614 (2005).
[6] K. Thomas, P. Sayre: *Toxicol. Sci.*, **87**, 316–321 (2005).
[7] International Commission on Radiological Protection. Publication 66: *Ann. ICRP* **24**, 1–300 (1994).
[8] J. Ferin, G. Oberdörster, S.C. Soderholm and R. Gelein: *J. Aerosol Med.*, **4**(1), 57–68 (1991).
[9] Recommendation of Occupational Exposure Limits: *J. Occup. Health*, **48**(4), 296 (2005).
[10] Ministry of Health, Labor and Welfare, Japan. Notification 368, Working Environment Evaluation Standards (2004).
[11] G. Oberdörster, J. Ferin, R. Gelein, S.C. Soderholm, J. Finkelstein: *Environ. Health Perspect.*, **97**, 193–197 (1992).
[12] K. Donaldson, X.Y. Li, W. MacNee: *J. Aerosol Sci.*, **29**, 553–560 (1998).
[13] K. Donaldson, C.-L. Tran: *Inhal. Toxicol.*, **14**, 5–27 (2002).
[14] A. Nemmar, M.F. Hoylaerts, P.H.M. Hoet, D. Dinsdale, T. Smith and H. Xu, J. Vermylen, B. Nemery: *Am. J. Respir. Crit. Care Med.*, **166**, 998–1004 (2002).
[15] A. Nemmar, M.F. Hoylaerts, P.H.M. Hoet, J. Vermylen and B. Nemery: *Toxicol. Appl. Pharmacol.*, **186**, 38–45 (2003).
[16] G. Oberdörster Z. Sharp, V. Atudorei, A. Elder, R. Gelein and W. Kreyling, C. Cox: *Inhal. Toxicol.*, **16**(6/7), 437–445 (2004).
[17] S. Yamago, H. Tokuyama, E. Nakamura, K. Kikuchi, S. Kananishi, K. Sueki, H. Nakahara, S. Enomoto, F. Ambe: *Chem. Biol.*, **2**, 385–389 (1995).
[18] Y. Ishihara, R. Sakata and Y. Fujita: *J. Aerosol Res.*, **20**, 193–199 (2005) (in Japanese).
[19] C.W. Lam, J.T. James, R. McCluskey, R.L. Hunter: *Toxicol. Sci.*, **77**, 126–134 (2004).
[20] D.B. Warheit, B.R. Laurence, K.L. Reed, D.H. Roach, G.A.M. Reynolds, T.R. Webb: *Toxicol. Sci.*, **77**, 117–125 (2004).
[21] WHO: Environmental Health Series 4. World Health Organization, Copenhagen (1985).
[22] IARC (International Agency for Research on Cancer): *IARC Monogr. Eval. Carcinog. Risks Hum.*, **81**, 1–418 (2002).
[23] J.H.E. Arts, S.M. Spoor, H. Muijser: *Inhal. Toxicol.*, **12**, 261–266 (2000).
[24] G. Oberdörster, J. Ferin, J. Finkelstein, P. Wade and N. Corson: *J. Aerosol Sci.*, **21**, 384–387 (1990).
[25] G. Oberdörster: *Philos. Trans. R. Soc. Lond. A*, **358**, 2719–2740 (2000).
[26] A. Ogami, M. Hirohashi, Y. Nagafuchi, K. Kuroda: *J. Aerosol Res.*, **20**, 200–206 (2005) (in Japanese).
[27] IARC (International Agency for Research on Cancer): Summaries & Evaluations – Nickel and Nickel Compounds, Lyon, France, Vol. 49 (1990).
[28] Q. Zhang, Y. Kusaka: *Inhal. Toxicol.*, **12**, 267–273 (2000).
[29] Q. Zhang, Y. Kusaka and K. Donaldson: *J. Occup. Health*, **42**, 179–184 (2000).
[30] Q. Zhang, Y. Kusaka, X. Zhu, K. Sato, Y. Mo, T. Klutz and K. Donaldson: *J. Occup. Health,* **45**, 23–30 (2003).
[31] M. Takaya, T. Toya, A. Takata, N. Otaki, K. Yoshida and N. Kohyama: *J. Aerosol Res.*, **20**, 207–212 (2005) (in Japanese).
[32] A. Tanaka: *J. Aerosol Res.*, **20**, 213–218 (2005) (in Japanese).
[33] A. Tanaka, M. Hirata, M. Omura, N. Inoue, T. Ueno, T. Homma, K. Sekizawa: *J. Occup. Health*, **44**, 99–102. (2002)

7.3.3 Safety assessment for the nanoparticles

We have been experiencing amazing progress of the technology on the processing for the nanometer-sized materials. The applications cover even biomedical engineering in addition to information technology, material, environmental science, and energy production [1–5]. As the result, many kinds of new materials have been designed, fabricated, and discarded. From now on, this movement will be accelerated and even more new functional materials will be distributed in the world. Here, we should not forget the safety of those materials in the process of the production, usage, and discard. Without this safety assessment, we will go into the same problems as that of asbestos just we are facing now. First of all, we have to conduct experiments to reveal the minimal concentration for emergence of the toxicity. In other words, we have to fix the standard value for the threshold concentration for each material first of all. If we do not fix it, we should not use the material at any concentration, which means any engineering process could not be carried out. The applications in the various fields have started all over the world, and the safety assessment is urgently needed. In this article, we introduce the methods of the safety assessment of the semiconductor nanoparticles and describe the safety and the threshold depending on the surface treatment.

The production process and the surface treatment play one of the most important roles for the safety of the nanoparticles. Cd and Se semiconductor nanoparticles coated with ZnS are one of the most widely used for the strong intensity of the fluorescent activity. Cd oxide and Se compounds once dispersed in the tri-octhil-phosphine oxide (TOPO) heated up to 300°C generate nanoparticles by self-assembly. Then ZnS enhances the stability of the structure and raises up the fluorescent intensity than without the coated one.

Nanoparticles, thus manufactured, as materials for a novel memory in the field of intelligence technology (IT), and as super-micro devices for laser in the field of optics, have been developed all over the world [1–5]. These nanoparticles cannot be dissolved into water, but dissolved into organic solvents like toluene. Therefore, for biological and medical applications, various technologies for surface-conjugations to make them hydrophilic [6, 7] have been developed. For example, nanoparticles covered with TOPO are hydrophobic because an alkyl group on them is hydrophobic. Therefore, a technology for replacing this alkyl group with hydrophilic carbonic acid (making the whole particles soluble in water) has been developed [6].

Nanoparticles, thus surface-treated, can be dissolved into water, like sodium salt or potassium salt. With this method, various kinds of materials have been surface-conjugated.

7.3.3.1 The MTT assay method and thresholds for cyto-toxicity

The MTT assay method is a way to evaluate the hazard assessment of nanoparticles, in which the activation metabolism in a mitochondrium in a cell is measured and the influence of nanoparticles on the prolification of the cell is qualitified. The MTT is a kind of tetrazolium, whose molecular formula is $C_{18}H_{16}BrN_5S$. Taken into a cell, it is decomposed by a dehydrogenase enzyme in a mitochondrium into a pigment called 'hormazan'. The measurement of the fluorescence intensity of the pigment shows the prolification of the cell [8–10].

Fig. 7.3.5. shows the hazard assessment of Vero cells and kidney cells of the African Green Monkey against CdSe/ZnS nanoparticles. The horizontal axis shows the concentrations of the nanoparticles and the vertical axis the fluorescence intensity of the hormazan at 460 nm, that is, the metabolism of the cells. The figure indicates

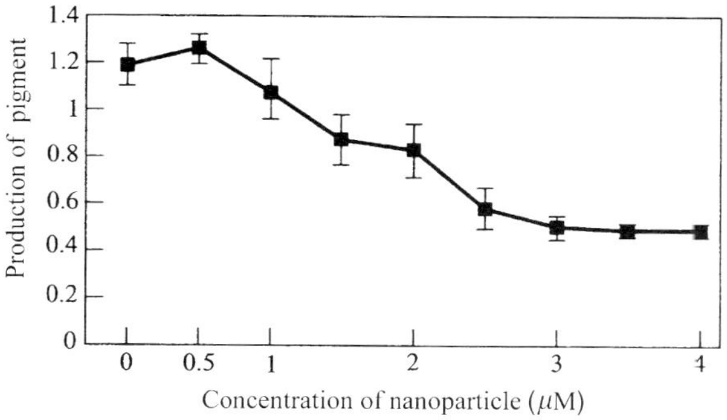

Figure 7.3.5
Cyto-toxicity of the quantum dots by the MTT assay.

that cyto-toxicity is not observed for 10^5 cells when the concentration is less than 0.5 μM. This result means that this concentration is the threshold of the cell toxicity. Likewise, cyto-toxicity is not observed for the HeLa cells and for the human primary cells. Further, in order to find out how the sizes of nanoparticles influence the cyto-toxicity, the cyto-toxicities were evaluated with three kinds of quantum dots; one whose fluorescence wavelength is 640 nm, red, one with 570 nm, yellow, and the other with 520 nm, green. The following results were obtained. The largest quantum dots whose fluorescence wavelength is 640 nm show a tendency to give cyto-toxicity. Cyto-toxicity is observed at concentrations more than 1 μM [11].

7.3.3.2 Measurement of cyto-toxicity by the flow cytometry

Another method to evaluate cyto-toxicity is the flow cytometry [12]. The MTT assay alone cannot tell whether the toxicity observed is lethal to the cells or just restrains the prolification of them. In the flow cytometry, the nuclei of dead cells are dyed with propidium iodide (PI) after the nanoparticles are taken in and the ratios of the dead cells are measured. Fig. 7.3.6 shows the lethal cyto-toxicity of MUA conjugated nanoparticles (520 nm, green) against Vero cells. The vertical axis indicates the numbers of the cells, and the horizontal axes show the fluorescence intensities and the cyto-toxicities. These experiments also show that dead cells cannot be observed at concentrations less than 0.5 μM even though nanoparticles are taken in, as was shown in the MTT assay. However, at concentrations more than 2 μM, the nanoparticles taken in cause damage to more than the half of the cells. That is, the cyto-toxicity of MUA quantum dots against cells is lethal [11].

7.3.3.3 Relations between surface-conjugations of nanoparticles and their safety

Nanoparticles have been surface-conjugated for applications for various uses.

Some surface-conjugations cause more grave cyto-toxicity than others. Therefore, relations between surface-conjugations and their safety for cells have to be considered. In order to find out the relations, the safety evaluations of nanoparticles surface-conjugated with two materials were made; one is with MUA (quantum dots-COOH) and the other is with glycerol (quantum dots-OH), and their purified and unpurified particles. Fig. 7.3.7 shows that the purification reduces the cyto-toxicity for the quantum dots-OH, and that the toxicity remains the same after the purification for the quantum dots-COOH. MUA itself, a material with which particles are conjugated, has cyto-toxicity. This experiment shows that toxicity against cells is connected not only with particles themselves but also with kinds of surface-conjugations and degrees of purification.

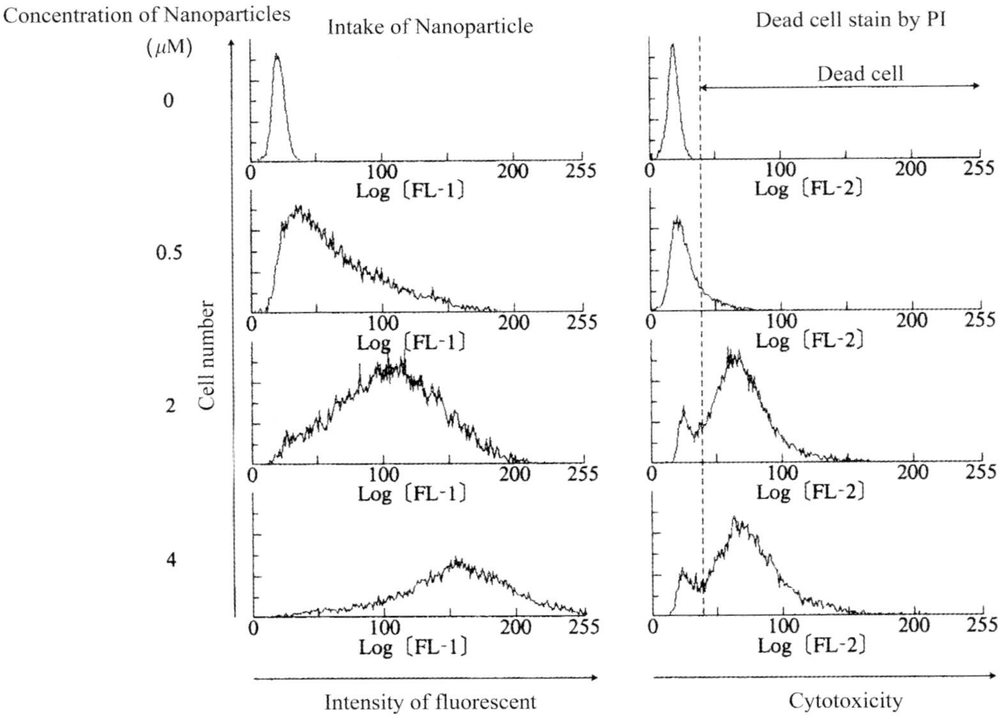

Figure 7.3.6
The detection of cyto-toxicity with flow cytometry.

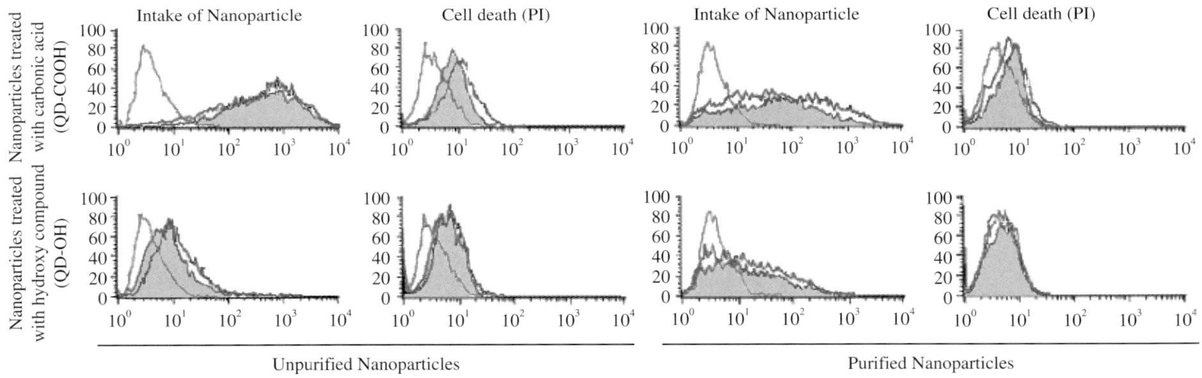

Figure 7.3.7
Differences of surface-conjugations and their cyto-toxicities.

7.3.3.4 Quantitative evaluation of damaged DNA; the comet assay

It is shown in the above experiments that the toxicity of nanoparticles is lethal to cells. Next, we found out by the Comet Assay method whether the toxicity is derived from damaged DNA. The method is a way to evaluate quantitative damage of DNA by electrophoresis. The fragmented DNA seeps out of their cells by treating cells, whose DNA has been fragmented, with agarose-gel to break their cell membrane, and then electrophoresing them. It looks like the "tales of comets". Cells, whose DNA has not been fragmented, have their nuclei keeping their spherical shape after electrophoresis, and the tales of comets cannot be observed [13–16].

Fig. 7.3.8 shows the results of the experiments with quantum dots conjugated with COOH (both purified and unpurified) to WTK-1 cells, a human lymphoblast mutation strain [17]. The vertical axis shows the lengths of the tails, and the horizontal axis the concentrations of the quantum dots. The concentration of the nanoparticles is 2 μM.

The results are as follows: the unpurified quantum dots with COOH caused damage to DNA. On the other hand, the electrophoresis with the purified quantum dots does not show DNA damage. This is probably because the DNA damage has been repaired during the longer cultivation time.

Fig. 7.3.9 shows the results of DNA damage. The nanoparticles have been purified over and over again in order to get rid of impurities derived from the process of the surface-conjugation. The concentrations are from 0.5 to 2 μM for 10^5 cells. The cultivation time after the addition of the nanoparticles is 24 h. Hydrogen peroxide (aqua) is used as the positive control. The results show; DNA damage is not observed up to 2 μM. The same experiment was

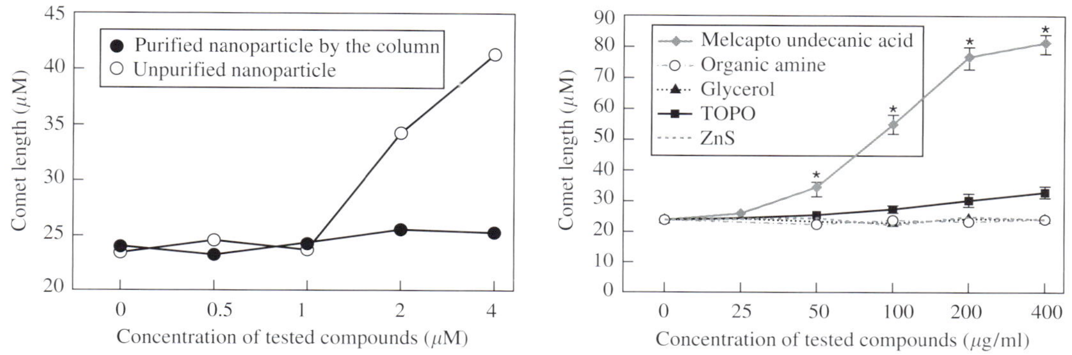

Figure 7.3.8
Cyto-toxicity of nanoparticles by the Comet Assay.

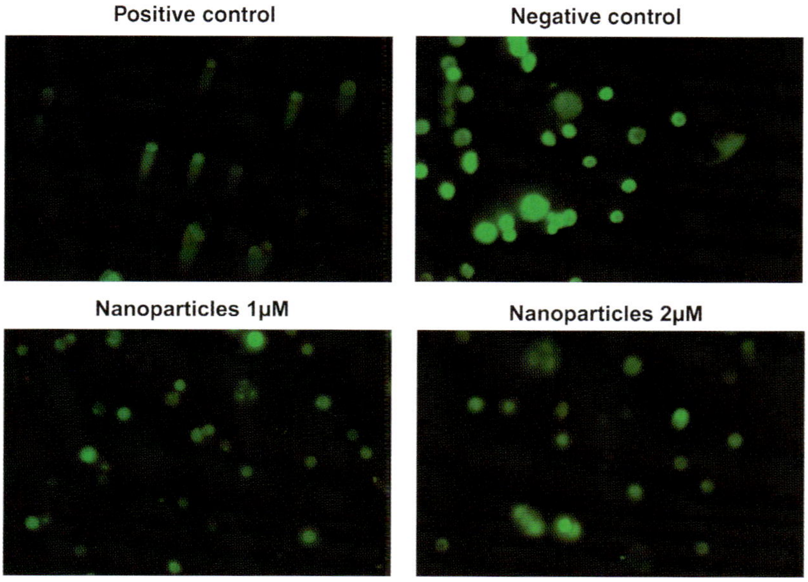

Figure 7.3.9
Microscopic study for the Comet Asssay and the estimation of the cyto-toxicity.

conducted with quantum dots conjugated with MUA, and quantum dots-NH_2 with an amino group. DNA damage was not observed in either case. The results indicate that as far as particles themselves do not break down, the cyto-toxicity of quantum dots is derived from the chemical properties of the materials covering the quantum dots.

The safety evaluation of nanoparticles has not been conducted sufficiently. As indicated above, the procedure for surface-conjugation could apply not only to Cd/Se nanoparticles but also to other nanoparticles. Today, various kinds of techniques for surface-conjugations have been available in academic papers, proceedings, and on the Internet. Some of them are widely known and others are patented. Those techniques are all shared among the human race.

Even at this moment, the human beings are making breakthroughs in various fields and developing different kinds of technologies. Sharing these technologies will lead to still more speedy developments of yet more advanced technologies. To achieve it, these technologies should be so structured that different fields, for example, bioimaging and biotechnology, structured on their own, can be linked to each other. Such structured knowledge will play an essential part in merging different fields.

References

[1] S. Coe, W.K. Woo, M. Bawendi and V. Bulovic: *Nature*, **420**, 800–803 (2002).

[2] T.C. Harman, P.J. Taylor, M.P. Walsh and B.E. LaForge: *Science*, **297**, 2229–2232 (2002).

[3] C. Santori, D. Fattal, J. Vuckovic, G.S. Solomon and Y. Yamamoto: *Nature*, **419**, 594–597 (2002).

[4] X. Li, Y. Wu, D.Y. Steel, D. Gammon, T.H. Stievater, D.S. Katzer, D. Park, C. Piermarocchi and L.J. Sham: *Science*, **301**, 809–811 (2003).

[5] A. Zrenner, E. Beham, S. Stufler, F. Findeis, M. Bichler and G. Abstreiter: *Nature*, **418**, 612–614 (2002).

[6] W.C. Chan, S. Nie: *Science*, **281**, 2016–2018 (1998).

[7] K. Hanaki, A. Momo, T. Oku, A. Komoto, S. Maenosono, Y. Yamaguchi and K. Yamamoto: *Biochem. Biophy. Res. Commun.*, **302**, 496–501 (2003).

[8] M. Ishiyama, Y. Miyazono, K. Sasamoto, Y. Ohkura and K. Ueno: *Talanta*, **44**, 1299 (1997).

[9] T. Mosman: Rapid colorimetric assay for cellular growth and survival: *J. Immunol. Method.*, **65**, 55–63 (1983).

[10] H. Tominaga, M. Ishiyama, F. Ohseto, K. Sasamoto, T. Hamamoto, K. Suzuki and M. Watanabe: *Anal. Commun.*, **36**, 47 (1999).

[11] A. Shiohara, A. Hoshino, K. Hanaki, K. Suzuki and K. Yamamoto: *Microbiol. Immunol.*, **48**, 669–675 (2004).

[12] A.N. Shatrova, N.D. Aksenov, A.I. Poletaev, V.V. Zenin: *Tsitologiia*, **45**(1), 59–68 (2003).

[13] H. Hoffmann, G. Speit: *Mutat. Res.* **581**(1–2), 105–114 (2005).

[14] S. Lemiere, C. Cossu-Leguille, A.M. Charissou and P. Vasseur: *Biomarkers*, **10**(1), 41–57 (2005).

[15] F. Mattiolo, A. Martelli, C. Garbero, M. Gosmar, V. Manfredi, F.P. Mattiolo, G. Torre and G. Brambilla: *Toxicol. Appl. Pharmacol.*, **203**(2), 99–105 (2005).

[16] F. Mouchet, L. Gauthier, C. Mailhes, M.J. Jourdain, V. Ferrier and A. Devaux: *J. Toxicol. Environ. Health A*, **68**(10), 811–832 (2005).

[17] A. Hoshino, K. Fujioka, T. Oku, M. Suga, Y. F. Sasaki, T. Ohta, M. Yasuhara, K. Suzuki and K. Yamamoto: *Nano Lett.*, **4**(11), 2163–2169 (2004).

7.4 Removal of nanoparticles

7.4.1 Principle of particle removal

In order to prevent nanoparticles release from a system so as to maintain environmental safety, the removal technique of nanoparticles must be established. In this section, separation techniques of particles from exhausted or suspended gas and liquid are described focusing on particles less than 100 nm.

Generally, as shown in Fig. 7.4.1, all particle separators for a dispersed system employ either one of three basic forms of particle separation. On the left hand side of the figure lie the separation methods in which particles are collected only by force field (electrostatic force, centrifugal force, gravity force, etc.), and the representative separator is electrostatic precipitator (ESP). If some obstacles (collectors) are placed into the particle laden stream, particle separation is facilitated because particles are collected on obstacles with a smaller deviation from the fluid flow by the force exerting on the particles compared to the case without obstacles. Typical collectors of this form are air filter, deep bed filter, etc. On the right hand side of the figure lie separators that collect particles utilizing only sieving effect of obstacles without any force field. In this case, geometrical size of channel between the obstacles must be smaller than that of particles. Membrane filter, fabric filter, etc. belong to this group.

When we apply the above collection forms to nanoparticles, the major collection mechanisms are Brownian diffusion and electrostatic force for particles in gas, while sieving effect and interception/adhesion forces for those in liquid.

7.4.2 Removal of nanoparticles suspended in gas

As mentioned above, most airborne particles are collected by separators utilizing various kinds of forces such as gravity, centrifugal force, electrostatic forces, inertia, Brownian diffusion force, and so on.

Elementary cause	Force field	Force field and obstacle	Obstacle
Form	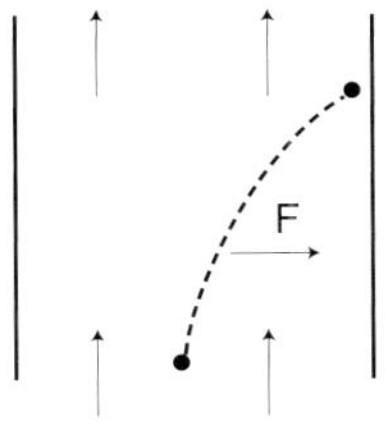	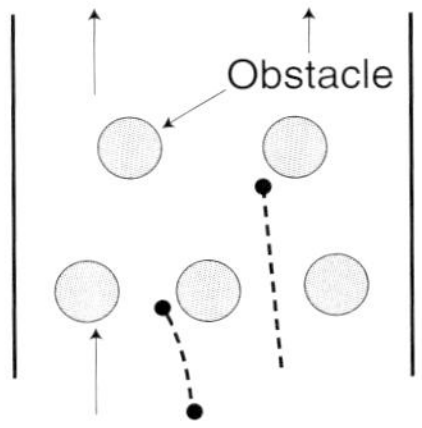	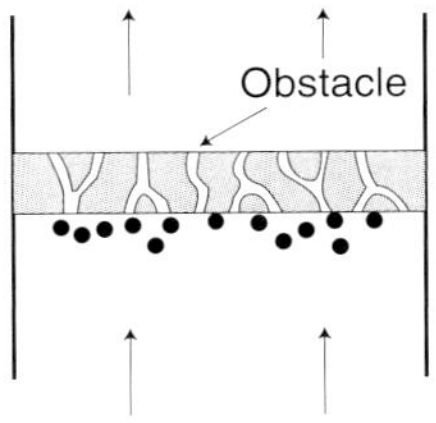
Collection efficiency	low	middle	high
Pressure drop	low	middle	high
Critical factor for performance	• Deposition velocity	• Collision efficiency • Pressure drop	• Pressure drop
Separator	• Thickner • ESP • Cyclone	• Venturi scrubber • Fibrous filter • Granular bed	• Filter press • Bag filter • Membrane filter

Figure 7.4.1
Basic forms of particle separation.

Therefore, the migration velocities or displacement of a particle per second due to the individual forces gives the basis for the comparison of removal efficiencies due to each force. In Fig. 7.4.2, migration velocities of particles due to various forces are depicted against particle diameter at normal temperature and pressure for particle density of 1 g/cm^3 [1]. As seen from the figure, the velocities due to gravity, centrifugal force, and inertia monotonically decrease with decreasing particle diameter, suggesting that the removal of nanoparticles with these forces is difficult. On the contrary, the velocities due to Brownian diffusion and electrostatic forces increase with decreasing particle diameter for particles less than 100 nm. This suggests that Brownian diffusion and electrostatic forces are most effective in collecting nanoparticles.

Fig. 7.4.3 summarizes typical conventional dust collectors. Among them, the effective collectors for nanoparticles are ESP and fabric/air filter. However, for the case of ESP, which relies on only electrostatic force, nanoparticles (<10 nm) fail to carry even one electron resulting in low collection efficiency. In this case electrically charged filters are effective because we can expect the combined effects of electrostatic forces and Brownian diffusion.

Among charged filters, so-called electret filter, which consists of permanently polarized fibers, is the most favorable filter because of its charge stability. Particle penetration data of electret filter are plotted against particle diameter in Fig. 7.4.4 and compared with that of uncharged filter with the same structure. For the three combinations of charged states of fiber and particle there exist the most penetrating particle

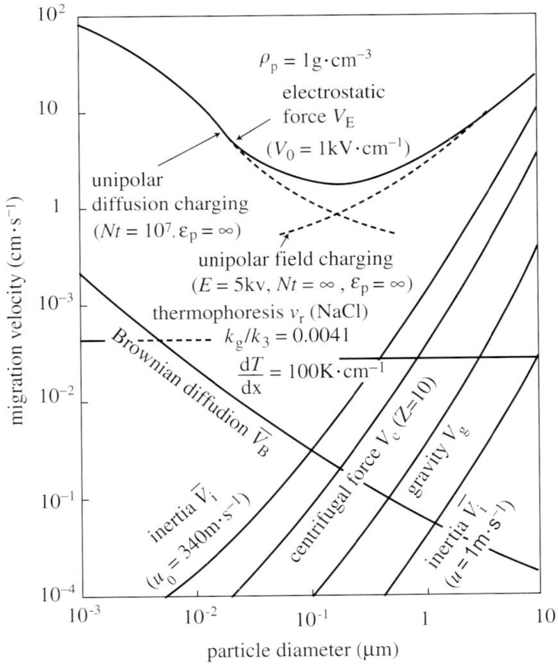

Figure 7.4.2
Migration velocity of airborne particles under force fields.

diameters. For the uncharged fiber, collection efficiency of uncharged particle has a minimum at 100 nm and increases with decreasing particle diameter, showing that 100 nm is the most penetrating

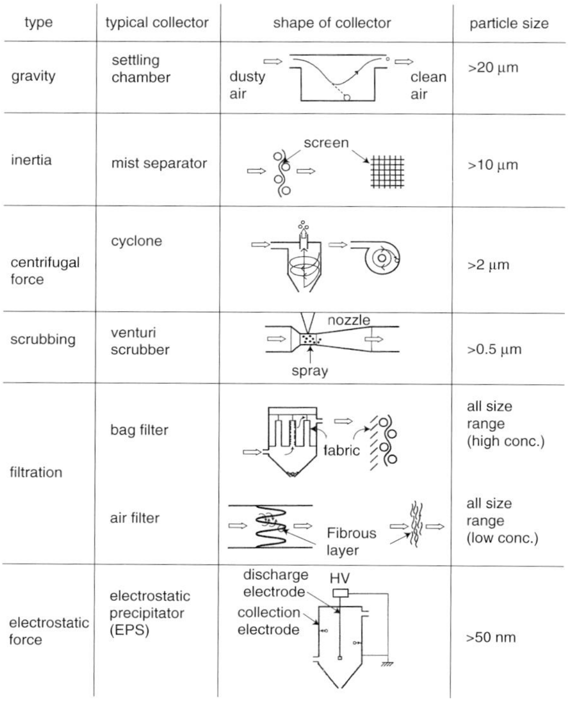

Figure 7.4.3
Classification of particle separators.

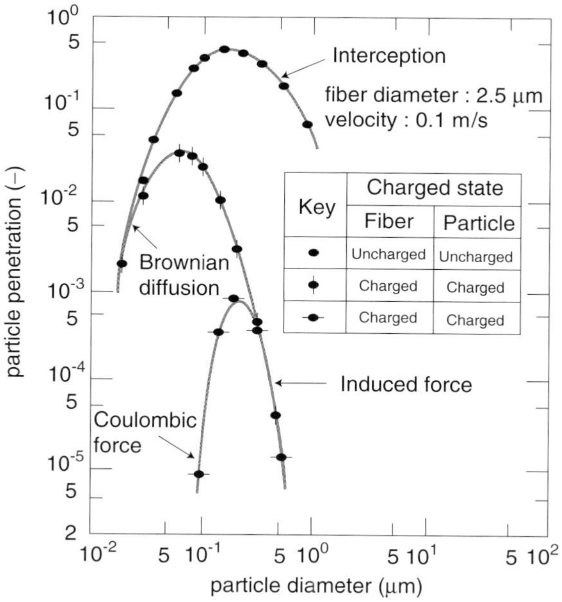

Figure 7.4.4
Particle penetration of electrically charged filter.

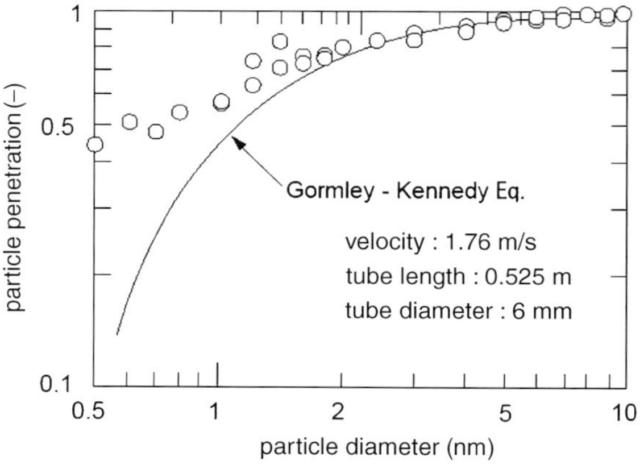

Figure 7.4.5
Penetration of airborne nanoparticles through a circular tube.

particle size. For the charged fiber, particle collection efficiency is very high even for uncharged particle, and the efficiency for charged nanoparticles is extremely high because of strong Coulombic force between fiber and particle. The experimental data plotted in Fig. 7.4.4 are qualitatively in good agreement with the theoretical prediction following the particle size dependency on particle migration velocity (shown in Fig. 7.4.2).

However, as particle size becomes smaller and comparable with the size of a molecule, particles may rebound on a collector surface, and the adhesion probability of particles drops, resulting in a decrease in collection efficiency. Fig. 7.4.5 is an example of experimental data that confirm the particle rebound [2]. The figure shows the penetration of nanoparticles through a grounded circular tube. The solid curve is the theoretical line derived by assuming that particles are deposited from a laminar flow in a tube by Brownian diffusion. It is evident that experimental penetration deviates from the theoretical line for particles less than 2 nm. This means that molecular behavior begins to appear when the particle size becomes as small as 2 nm, and as a result, the collection efficiency is reduced.

It should be noted that considerable amounts of nanosized particles are contained in diesel exhaust particles (DEP), possibly penetrating through the honeycomb type (tubular channel) diesel particulate filters (DPF).

References

[1] K. Takahashi: Ouyou Earozorugaku, Youkendou, p. 144, (1984).

[2] Y. Otani, H. Emi, S.J. Cho and N. Namiki: *Adv. Powder Technol.*, **6**(4), 271–281 (1995).

7.4.3 Removal of nanoparticles in liquid

There are two types of methods that differ in the way the nanoparticles in liquid are collected. The first group, called membrane filtration, constrains the particles by a membrane, and the liquid is allowed to flow freely through the membrane. In the second group of ultracentrifugation, the liquid is constrained in a rotating vessel, and the particles move freely within the liquid by an external field of acceleration caused by ultracentrifugal field. These methods have been quite extensively used in separation of macromolecules and molecules from liquid, and they are recently becoming important also in separation of nanoparticles from liquid.

7.4.3.1 Fouling mechanism in membrane filtration

In pressure-driven membrane filtration processes, the pressure gradient across the membrane would force solvent and smaller species through the pores of the membrane, while the larger molecules/particles would be retained. Membrane filtration processes are usually classified into three general categories according to the size of separating components, as shown in Fig. 7.4.6. Microfiltration (MF) is designed to retain suspended particles in the range of 50 nm–5 μm. Ultrafiltration (UF), on the other hand, retains macromolecules or nanoparticles in the range of 5–50 nm (nominal molecular weight cut-off (NMWCO) ranging from 5,000 to 5,000,000 Da). Nanofiltration (NF) is a relatively new process that uses charged membranes, and it covers molecular sizes ranging from 0.1 to 5 nm (NMWCO ranging from 200 to 1,000). It is useful in that it can separate dissociated forms of a compound from the undissociated form.

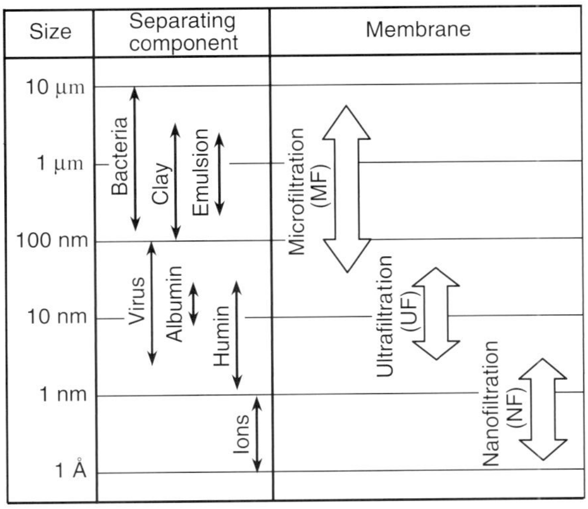

Figure 7.4.6
Useful ranges of various membrane filtration processes.

One of the major factors limiting the use of membrane filtration is membrane fouling, resulting in a dramatic decline in flux with time of operation. To account for the membrane fouling, the resistance-in-series model is frequently employed. The resistance model becomes

$$u_1 = \frac{dv}{d\theta} = \frac{p}{\mu R_t} = \frac{p}{\mu(R_{bm} + R_c + R_{cp})} \tag{7.4.1}$$

where u_1 is the permeate flux, v the filtrate volume per unit membrane area, θ the filtration time, p the applied transmembrane pressure, μ the viscosity of the permeate, R_t the total resistance, R_{bm} the resistance of the membrane per se plus the clogging of the membrane pores, R_c the resistance of the filter cake, and R_{cp} the resistance of the concentration polarization layer. Significance of each resistance in membrane filtration is as follows.

The membrane, even in the absence of any suspended particle, has a natural flow resistance. During membrane filtration, particles become attached to the pore channel of the membrane thereby reducing the flow channel dimension, or pores become blocked off completely. The last two effects lead to resistances that are due to adsorption and pore blocking. The blocking filtration model introduced by Hermans and Bredée [1], and Grace [2] is most commonly used as an interpretation of such phenomena.

The clogging of the membrane pores is strongly influenced by such solution environment as pH and the ionic strength. The permeate flux of bovine serum albumin (BSA) (pI 5.1, molecular weight 67,000, Stokes–Einstein diameter 3.55 nm) solution by permeable MF membrane (nominal pore size 0.05 μm) is lowest at around the isoelectric point [3]. As the BSA molecule carries no net charge at the isoelectric point, the molecule is in its most compact state at that point. The BSA molecules deposit themselves rather densely onto the pore walls of the membrane to form a compact configuration with a smaller lateral electrical interaction between the molecules. As a result of this, the flow resistance increases markedly at around the isoelectric point.

In dead-end membrane filtration, which has a feed and permeate stream, each with the same mass flow rate, the resistance of the filter cake plays a major part in the filtration resistance. Therefore, the cake filtration theory can be applied, and thus the permeate flux u_1 is described by

$$u_1 = \frac{p(1 - ms)}{\mu\alpha_{av}\rho s(v + v_m)} \tag{7.4.2}$$

where m is the ratio of wet to dry cake masses, s the mass fraction of solids in the solution, α_{av} the average specific filtration resistance, ρ the density of the permeate, and v_m the fictitious filtrate volume per unit membrane area, equivalent to the flow resistance of the membrane [4].

For fine particle suspensions, colloidal forces which arise from interaction between the suspended particles control the nature of the filter cake. The average specific filtration resistance α_{av} and the average porosity ε_{av} of the filter cake are strongly affected by the solution properties, including pH and electrolyte strength. For instance, in MF of suspensions of the titanium dioxide (pI 8.1, the original mean specific surface area size 470 nm), α_{av} goes through a minimum, and ε_{av} is much larger near the isoelectric point [5], as shown in Fig. 7.4.7. The titanium dioxide particles are destabilized around the isoelectric point where the van der Waals attraction is more dominant. Consequently, the particle tends to come together, that is, to flocculate, and the very porous flocs are then formed. Thus, it is speculated that the filter cake formed from such porous flocs has often loose and wet structures. On the other hand, the filter cake becomes compact and dry when the particle carries the charge. Since the most loose filter cake forms around the isoelectric pH, the filter cake is most permeable.

It is interesting to note that the results in protein UF had a distinctly different behavior. In protein UF of BSA solution, the filter cake is in its most compact state around the isoelectric point [6], as shown in Fig. 7.4.8. Since the BSA molecules are hydrophilic colloids, their stability in the solution would appear to be influenced not only by the presence of a surface charge on the protein but also by hydration of the surface layers of the protein. The BSA molecules, because of hydrated layers surrounding them, are not destabilized by such consideration as depression of the electrical double layer. Thus, the BSA molecules have water bound to them even around the isoelectric point. The hydrophilic BSA molecules maintain a dispersed state in the solution due to hydration of the surface layers of the protein even around the isoelectric point. When a BSA molecule acquires a charge, the filter cake becomes loose and wet due to electrostatic repulsion between the charged BSA molecules. This contrasts to the compact filter cake around the isoelectric point. The average specific filtration resistance α_{av} has a definite maximum around the isoelectric point since a compact filter cake provides a large hydraulic flow resistance.

Most membrane filtration processes are operated in the cross-flow mode, in which the feed is moved tangentially to the membrane surface so that the filter cake is continuously sheared off. During membrane filtration, particles in the feed are brought to the upstream surface of the membrane by convective transport, and this results in a higher local concentration of the rejected particles at the membrane surface as compared to the bulk solution which is referred to as concentration polarization [7].

The particle concentration in the solution adjacent to the membrane varies from the value at the membrane surface, C_m, to that in the bulk feed solution, C_b, over a distance equal to the concentration boundary layer thickness δ. The resulting concentration gradient causes the particles to be transported back into the bulk solution due to diffusional effects. At steady state, the rate of convective transport of particle toward the membrane is balanced by the rate of particle transport through the membrane plus the rate of

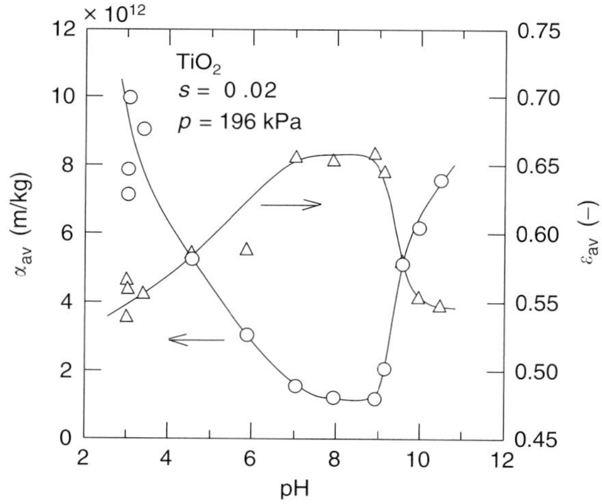

Figure 7.4.7
Effect of pH on average specific filtration resistance and average porosity of filter cake formed in microfiltration of titanium dioxide suspensions.

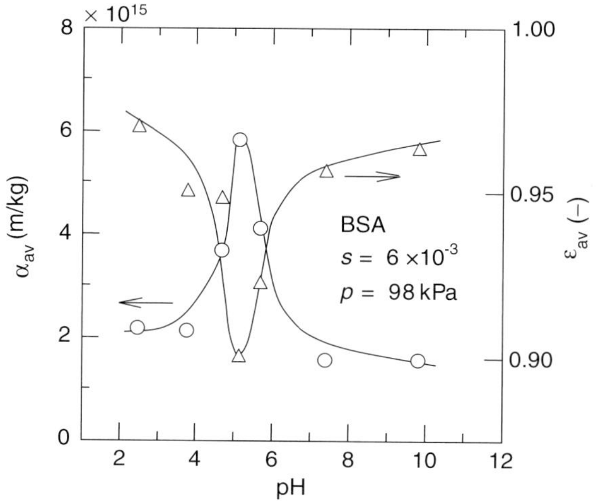

Figure 7.4.8
Effect of pH on average specific filtration resistance and average porosity of filter cake formed in ultrafiltration of BSA solutions.

the diffusive back transport of particle. Thus, the permeate flux u_1 is given by

$$u_1 = k \ln\left(\frac{C_m - C_p}{C_b - C_p}\right) \qquad (7.4.3)$$

where C_p is the particle concentration in the permeate, k $(=D/\delta)$ the mass transfer coefficient, and D the diffusion coefficient.

The osmotic pressure model assumes that the deviation from pure water flux occurs solely due to the osmotic pressure difference across the membrane, and thus the permeate flux u_1 is given by

$$u_1 = \frac{p - \{\pi(C_m) - \pi(C_p)\}}{\mu R_m} \qquad (7.4.4)$$

where π is the osmotic pressure, which is a function of the concentration. Equation (7.4.4) means that the effective driving force across the clean membrane is $p - \{\pi(C_m) - \pi(C_p)\}$. Replacing $p - \{\pi(C_m) - \pi(C_p)\}$ by the hydraulic pressure at the membrane surface, p_m, equation (7.4.4) reduces to the cake filtration equation.

7.4.3.2 Techniques for controlling membrane fouling

To minimize the effects of cake buildup and concentration polarization, membrane filtration is usually conducted using the cross-flow geometry in which the feed flow is parallel to the membrane and perpendicular to the filtrate flow [8]. However, especially in MF the energy requirements associated with pumping the feed (plus any recirculation flow) along the membrane surface are typically very high. Thus, there have been some innovations in recent years with cakeless filtration.

The rotating disk module in which the membrane disk is stationary is suited for large-scale operation [9]. It is possible to enhance the permeate flux by using the vibrating modules [10, 11]. In the rotating cylinder device with the membrane on the inner rotating cylinder, counter-rotating Taylor vortices within the annular gap are available [12, 13]. Dean vortices that twist and spiral in the direction of flow inside a highly curved channel, similar to vortices in rotary modules can result in enhanced flux [14]. These vortices, or flow instabilities, induce turbulence into the system.

Periodic removal of the formed filter cake is also effective for enhancing the permeate flux. Recently, several methods have been investigated: back washing using the filtrate or air pressurization [15], periodic rotation of the cylindrical membrane [16], pulsatile flow [17], high-frequency transmembrane pressure pulsing with a frequency around 0.1–1 Hz [18].

Dead-end upward filtration, where the filtrate flow is in the opposite direction to gravity, and dead-end inclined filtration, where the membrane is inclined, can reduce the cake formation onto the membrane in UF of nanoparticulate suspension and protein solutions. In upward UF of silica sol (mean diameter 6.2 nm) and BSA solution, a sustained permeate flux is achieved, as shown in Fig. 7.4.9 [19, 20], because the filter cake overlying the membrane is exfoliated continuously under the gravitational force acting on the particles comprising the filter cake.

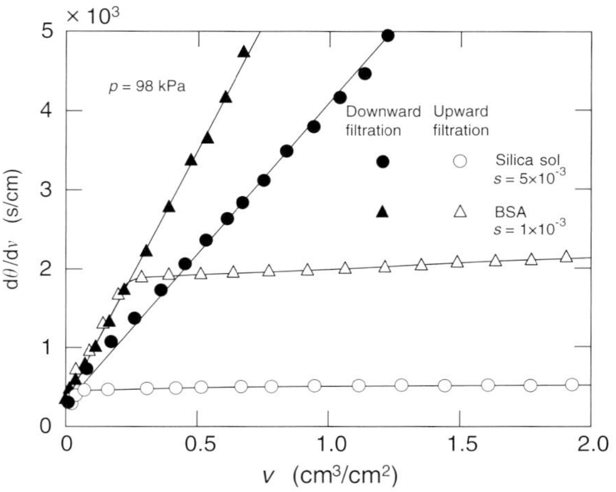

Figure 7.4.9
Permeate flux in dead-end upward ultrafiltration of silica sol and BSA solution.

7.4.3.3 Hybrid operation with membrane filtration

Another approach for enhancing the permeate flux is to employ external force fields. Electrofiltration, in which an applied electric field is used to drive charged particles away from the membrane surface, has been developed. In electrofiltration, the accumulation of the particles on the membrane surface is limited by the imposed electrophoretic force. In addition, the permeate flux through the filter cake is dramatically enhanced due to electroosmosis as a secondary electrokinetic phenomenon. This method can be applied to a broad combination ranging from MF of particulate suspension such as bentonite [21] to UF of protein solution. Fig. 7.4.10 shows the reciprocal permeate flux ($d\theta/dv$) versus the permeate volume per unit membrane area, v, for various values of the strength of the DC electric field, E [22]. The steady permeate flux increases noticeably with the magnitude of the imposed field strength. Also, a higher electric field strength causes the permeate flux to equilibrate more rapidly.

A method has been developed for removing humic substances by hybrid UF combined with both flocculation and adsorption treatments, as shown in Fig. 7.4.11 [23]. Flocculation by use of polyaluminum chloride (PACl) is particularly effective for the removal of humic acids, which constitute the relatively high molecular weight fractions of humic substances, whereas adsorption by use of powdered activated carbon (PAC) is able to remove fulvic acids of relatively low molecular weight effectively, which cannot be fully flocculated by PACl. Hybrid UF in combination with flocculation and adsorption

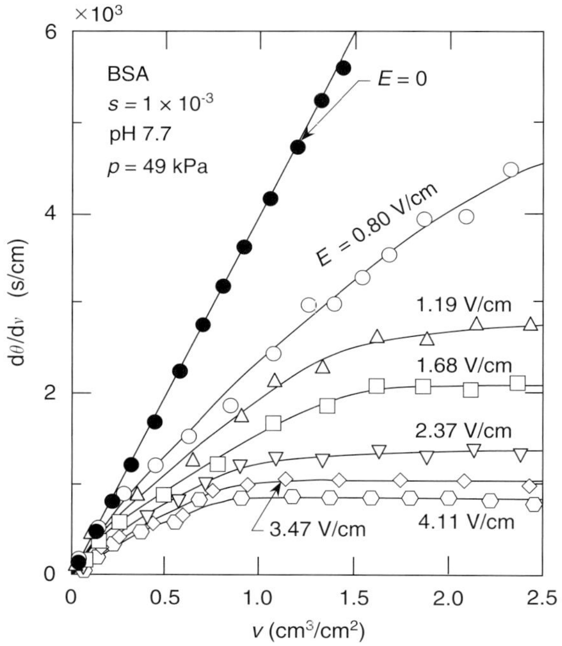

Figure 7.4.10
Permeate flux in electro-ultrafiltration of BSA solution.

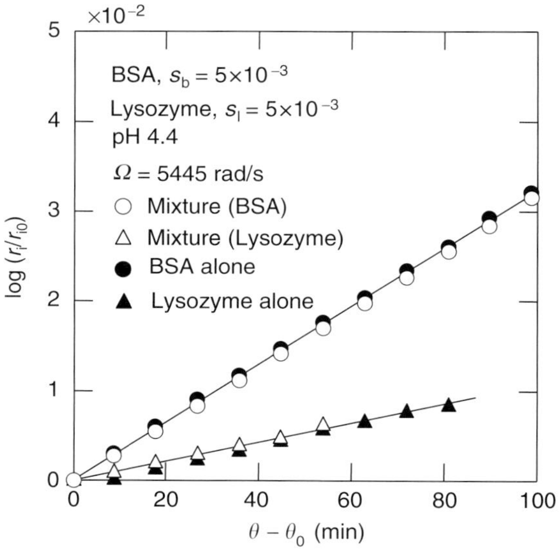

Figure 7.4.12
Behaviors of ultracentrifugal sedimentation of binary protein mixtures at pH 4.4.

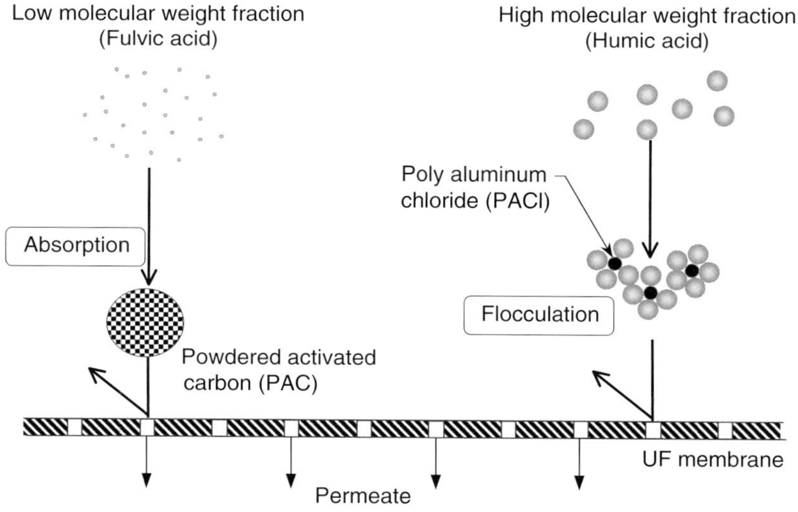

Figure 7.4.11
Mechanism of hybrid ultrafiltration of humic substances combined with both flocculation and adsorption treatments.

treatments exhibits high permeate quality because the flocs and PAC are easily retained by the UF membrane.

7.4.3.4 Ultracentrifugal sedimentation

In ultracentrifugal sedimentation, ultracentrifugal force field of several tens of thousands of revolutions per minute is applied to a rotor. In recent years, ultracentrifugal sedimentation is employed for concentrating dilute protein solutions and for separating proteins and other large biological molecules from low-molecular-weight solutes or from much larger particles. Fig. 7.4.12 shows the results for ultracentrifugal sedimentation of an aqueous solution of the mixtures of BSA and egg white lysozyme (pI 11.0, MW 14,300) measured using Schlieren optics in an analytical ultracentrifuge [24]. The angular acceleration Ω of the rotor is 5,445 rad/s. The symbol r_i and r_{i0} in the figure represent the distances from the center of rotation to the sedimentation boundary at time θ and θ_0, respectively. The electrical nature of macromolecules plays a significant role in determining the sedimentation behavior in ultracentrifugation of binary protein mixtures. In the pH range where both protein molecules were electropositive, the molecules sediment independently due to the electrostatic repulsive force acting between BSA and lysozyme molecules.

References

[1] P.H. Hermans, H.L. Bredée: *J. Soc. Chem. Ind.*, **55T**, 1–4 (1936).

[2] H.P. Grace: *AIChE J.*, **2**, 307–336 (1956).

[3] E. Iritani, Y. Mukai, Y. Tanaka and T. Murase: *J. Membr. Sci.*, **103**, 181–191 (1995).

[4] B.F. Ruth, G.H. Montillon and R.E. Montanna: *Ind. Eng. Chem.*, **25**, 153–161 (1933).

[5] E. Iritani, Y. Toyoda and T. Murase: *J. Chem. Eng. Jpn.*, **30**, 614–619 (1997).

[6] E. Iritani, S. Nakatsuka, S. Aoki and T. Murase: *J. Chem. Eng. Jpn.*, **24**, 177–183 (1991).

[7] S. Kimura, S. Sourirajan: *AIChE J.*, **13**, 497–503 (1967).

[8] J. Murkes, C.G. Carlsson: Crossflow filtration: Theory and practice, Wiley, NY (1988).

[9] T. Toda: in *Encyclopedia of Fluid Mechanics: Slurry Flow Technology*, N.P. Cheremisinoff (Ed.), Gulf Publishing, Houston, TX, p. 1149 (1985).

[10] B. Culkin, A.D. Armando: *Filtr. Sep.*, **29**, 376–378 (1992).

[11] T. Murase, E. Iritani, P. Chidpong, K. Yagishita, K. Yoshida, T. Sugiyama and M. Shirato: *Kagaku Kogaku Ronbunshu*, **14**, 135–140 (1988).

[12] W. Tobler: *Filtr. Sep.*, **19**, 329–332 (1982).

[13] T. Murase, E. Iritani, P. Chidpong, K. Kano, K. Atsumi and M. Shirato: *Kagaku Kogaku Ronbunshu*, **15**, 630–637 (1989).

[14] K.-Y. Chung, R. Bates and G. Belfort: *J. Membr. Sci.*, **81**, 139–150 (1993).

[15] V. Gekas, B. Hallström: *Desalination*, **77**, 195–218 (1990).

[16] T. Murase, E. Iritani, P. Chidpong and K. Kano: *Kagaku Kogaku Ronbunshu*, **15**, 1179–1186 (1989).

[17] V.G.J. Rodgers, R.E. Sparks: *J. Membr. Sci.*, **68**, 149–168 (1992).

[18] S.M. Finnigan, J.A. Howell: *Trans. IChemE, Part A (Chem. Eng. Res. Des.)*, **70**, 527–536 (1992).

[19] E. Iritani, T. Watanabe and T. Murase: *Kagaku Kogaku Ronbunshu*, **17**, 206–209 (1991).

[20] E. Iritani, T. Watanabe and T. Murase: *J. Membr. Sci.*, **69**, 87–97 (1992).

[21] H. Yukawa, K. Kobayashi, Y. Tsukui, S. Yamano and M. Iwata: *J. Chem. Eng. Jpn.*, **9**, 396–401 (1976).

[22] E. Iritani, K. Ohashi and T. Murase: *J. Chem. Eng. Jpn.*, **25**, 383–388 (1992).

[23] E. Iritani, Y. Mukai, N. Katagiri and T. Hirano: *Kagaku Kogaku Ronbunshu*, **30**, 353–359 (2004).

[24] E. Iritani, S. Akatsuka and T. Murase: *Kagaku Kogaku Ronbunshu*, **23**, 224–229 (1997).

APPLICATIONS

APPLICATIONS

CONTRIBUTORS

APPLICATION 1

1 DISPERSION OF FINE SILICA PARTICLES USING ALKOXYSILANE AND INDUSTRIALIZATION

1. Sol–gel hybrid

Development of composite materials that have the properties of rigid light plastics and hard heat-resistant ceramics is a key objective of R&D in nanotechnology together with the preparation of nanocomposites by the layer intercalation method and nanofiller dispersion method [1–3].

A recent technology called the sol–gel hybrid method enables the mixing of plastics and ceramics in the nanosized range. The sol–gel method is used to produce a metal oxide thin layer by hardening metal alkoxide represented by alkoxysilane [TMOS (tetramethoxysilane) or TEOS (tetraethoxysilane)] as shown in Fig. 1.1. When this sol–gel hardening reaction takes place in a selected melted polymer or polymer solution, the growth of the siloxane bond by the sol–gel hardening is hindered by the polymer and consequently the hardened composite material with the nanosized silica particles dispersed in the polymer is obtained, which is called a sol–gel hybrid. Principally, it utilizes the interaction between the silanol group (Si-OH) generated in the process of sol–gel hardening and the hydrogen bond in the polymer [1, 2]. The composite material made by the sol–gel method is completely transparent and does not look like a mixture of different kinds of materials. Therefore the terminology "hybrid" is applied for this material rather than "composite".

2. Molecular design

The finer the ceramic particles dispersed in the matrix polymer in the composite materials, the greater the interface between both the materials inversely proportional to the particle size and the more significant its effects on the properties of the composite materials. For this reason, we examined the hybrid by sol–gel method for developing a molecular hybrid dispersed with silica particles smaller than polymer chain length.

For the sol–gel hybrid, the kinds of applicable polymers are restricted to those soluble to alcohol or water having strong interaction with the generated silica. In the forming process of the sol–gel hybrid, the sol–gel hardening reaction takes place at competitive speed considering the evaporation of the solvent and the hardening of the polymer, the relative speed of which is affected by environmental changes in the kind of solvent. The hardening conditions and thickness of hybrid material etc. determines the quality of the hybrid materials, namely the dispersion state of silica. For this reason, it is difficult to utilize the sol–gel hybrid for various industrial applications with constant quality [1, 4]. Therefore we aimed at developing a hybrid method by which anyone can easily obtain good dispersion of silica for diverse sorts of polymers.

Plastic materials become harder and were brittle but gain in heat resistance when ceramic materials like glass are mixed into them. By mixing these different kinds of materials, the advantages and disadvantages of each material can be reflected in the properties of the composite material. Our research target was not the preparation of composite material of polymer and silica having arithmetically averaged properties but the creation of new material with the advantages of both the materials in consideration of individual applications.

The site selective molecular hybrid method developed was a hybrid method, which overcomes each of the above-mentioned problems of conventional composite materials. As shown in Fig. 1.2, by a site selective molecular hybrid method, an alkoxysilane compound is polymerized to produce polyalkoxysiloxane with a functional group like a glycidyl group. Using this functional group, polyalkoxysiloxane is introduced to a certain position of a polymer, producing an alkoxysilane-denatured polymer [5].

At the site of polyalkoxysiloxane, a sol–gel hardening reaction takes place like the monomer to generate silica. By covalent bonding of the alkoxysiloxane with the polymer by their prior reaction, the hybrid technology can be applied to various kinds of polymers and realize good stable dispersion of silica, which is barely affected by environmental parameters such as the film thickness and the hardening conditions.

The most important advantage of the site selective molecular hybrid method developed here, is that the structure can be designed to suit the applications (Fig. 1.3). In a similar manner used by polymer

Hydrolysis

$$Si(OR)_4 + H_2O \longrightarrow (RO)_3SiOH + ROH$$

Condensation

$$\equiv SiOH + HOSi \equiv \longrightarrow \equiv SiOSi \equiv + H_2O$$

Figure 1.1
Sol–gel curing reaction.

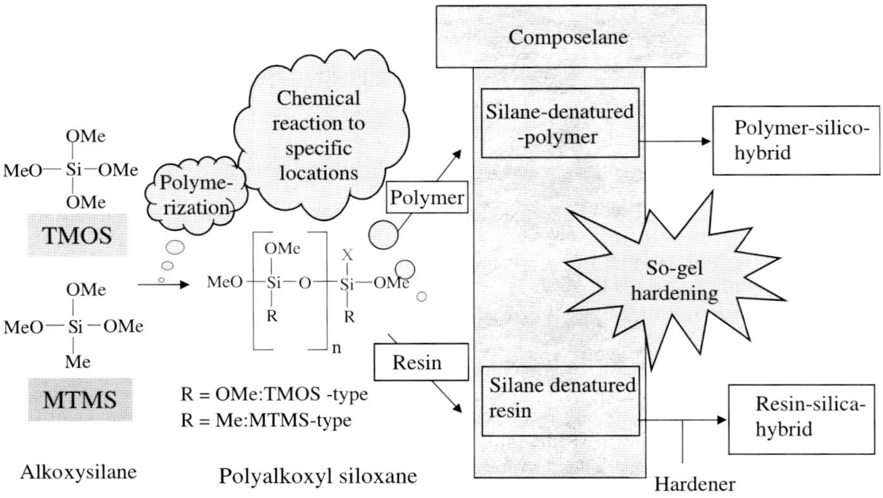

Figure 1.2
Process of location selective molecular hybrid method.

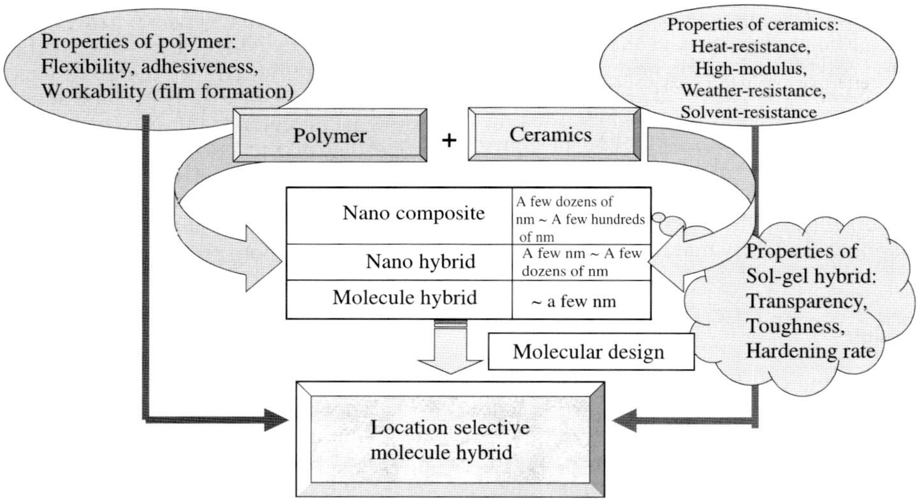

Figure 1.3
Advantages of location selective molecular hybrid.

scientists to control the bulk properties by designing the kind and arrangement of monomers, the position of silica is designed by the hybrid method regarding the generated silica as a kind of monomer that has bulk properties meeting the customers' requirements for each application.

Silica, sufficiently smaller than the polymer chain length can produce hybrid materials having the advantages of both materials by modifying the specified site of the polymer effectively and minimizing the influence on the other sites.

The polymer–silica hybrid material commercialized by Arakawa Chemical Industries, Ltd. for industrial use called "Compoceran" is a product developed using this site selective molecular hybrid method [5].

3. Unmeltable plastics: epoxy resin hybrid

The properties of polymer materials are heavily deteriorated when the temperature exceeds *Tg* (glass transition temperature) and is softened. In addition, there is increasing need for heat-resistant and flame-retarded epoxy resins in advanced fields such as the electronic device industry as halogenated epoxy resin may become unusable because of environmental issues.

To improve the heat resistance of an epoxy resin hybrid, silica was introduced at the heat-sensitive site of epoxy resin by hardening the alkoxysilane oligomer as shown in Fig. 1.4. This process realized the disappearance of *Tg* of the hardened material (Fig. 1.5), increase in the decomposition temperature and reduction of the rate of thermal expansion [6].

Additionally, owing to the advantages of the low dielectric constant and good adhesiveness with inorganic substrates of silica, this hybrid material is utilized as a base material for various industrial applications. As for the electrical insulation material related with the printed board, its high heat-resistance and good electrical insulating property are highly evaluated and used for resist ink, electric resistant resin between the build-up board layers and reinforcing materials.

For semiconductors, this hybrid material is being used as a heat-resisting component for anisotropic electroconductive film (ACF). Concerning the coating material for non-adhesive materials, the hybrid material is utilized as undercoating paint for repairing Japanese roof tiles and as a hard coating material for the polycarbonate used for car ports because of its high adhesiveness and anti-yellowing performance.

Application of the hybrid material to the anchoring agent for hot-dip galvanization has the great environmental benefit of avoiding the waste fluid produced in

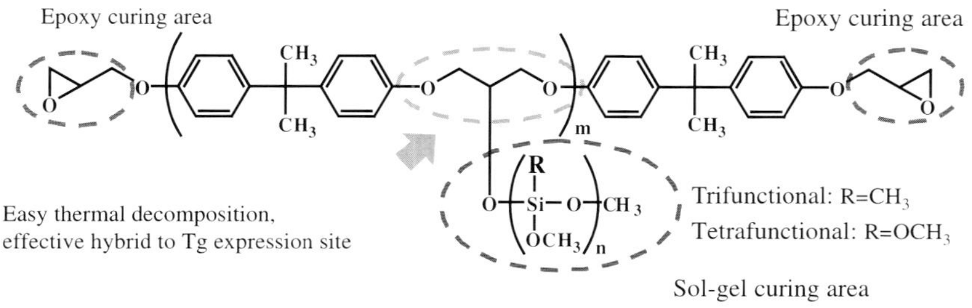

Figure 1.4
Chemical structure of silane-modified epoxy resin (Composelane E).

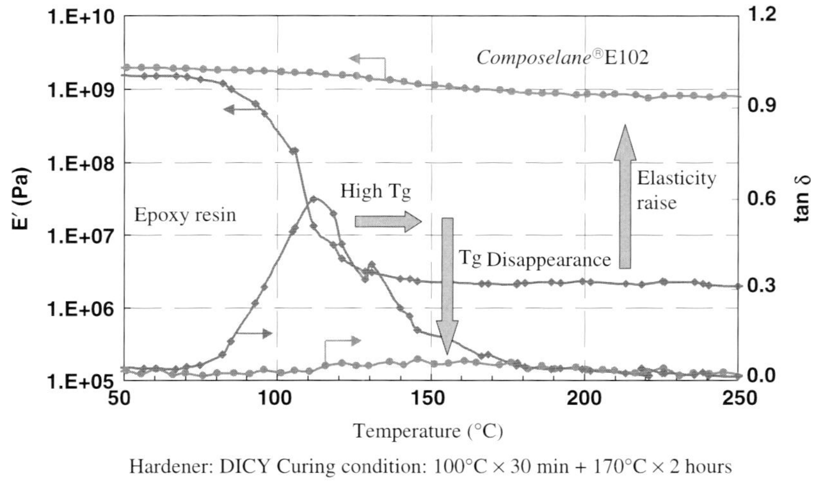

Figure 1.5
Heat-resistance of epoxy resin–silica hybrid (DMA).

chromate treatment. In the field of liquid crystal display, consumption of the hybrid material has increased as the sealing material due to its high heat-resistance and adhesiveness onto glass substrate and is being investigated for use as the protection coating for color filters.

Furthermore, three-phase hybrid consisting of epoxy resin, silica hybrid and liquid polymer phases has also been developed by this method and commercialized as adhesives for flexible printed boards with high heat-resistance, adhesiveness and flexibility.

The water absorption, electric property and mechanical strength are controlled by adjusting tetramethoxysilane and alkyltrimethoxysilane as raw silane compound. In addition, bisphenol and novolac for epoxy resin are also in the product line to meet diverse customer requirements.

4. Tough resin: hybrid of the phenol resin system

When the novolac phenol resin used as a hardening agent for epoxy resin is combined with silica, it develops mechanical toughness as well as disappearance of *Tg* as the epoxy system by the introduction of silica, although it is not superior to the epoxy system in terms of the low rate of thermal expansion and the decomposition temperature [7].

These phenol-system hybrids are commercialized as the material for loudspeakers, taking advantage of its heat-resistance, flame retardancy and mechanical strength.

5. Soft silica hybrid: hybrid of the urethane system

The specified orientation of the silica site is essential for making a hybrid of silica making use of the rubber elasticity of urethane. Urethane is a typical rubber material having a structure with solid hard segment domains (HS) dispersed in the liquid phase of soft segment domains (SS).

As shown in Fig. 1.6, it was contrived to form the hybrid domain with silica generated only in the HS phase in the urethane-silica hybrid. The silica contributes to give the heat-resistance to the HS domain and does not affect its flexibility by the existence of the SS phase.

We previously demonstrated the structure of the molecular hybrid from the photoluminescence spectrum of colorimetric substrate on the silica surface and reported the interaction between the domains using SALS [8, 9]. This material has an expanded range of uses as a coating material for flexible boards, fiberscopes, elastic paints and sealants.

6. Cheap engineering plastics in place for imide: hybrid of the amideimide system

Amideimide is hardly used in the advanced electronic industries because of its low flexibility, high dielectric constant and high water absorption, in spite of its low monomer cost of less than a fifth compared with imide. Although the ductility of the material cannot be increased by the addition of silica, water absorption and dielectric constant similar to those of polyimide were obtained whilst maintaining the original flexibility of polyamideimide limiting the influence of silica to the bridge of molecules by introducing polymethoxysiloxane at only the edge of the molecules as shown in Fig. 1.7 [10]. A hybrid of the amideimide system is being commercialized as a material for super heat-resistant enamel wire, sliding parts and belts for printers.

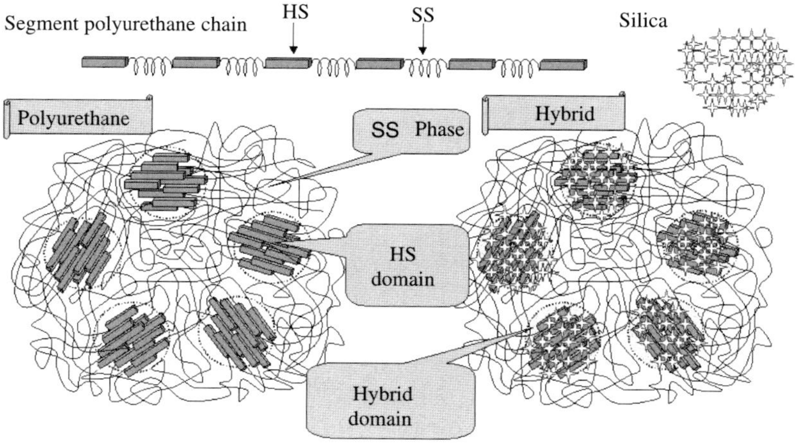

Figure 1.6
Model of urethane–silica hybrid.

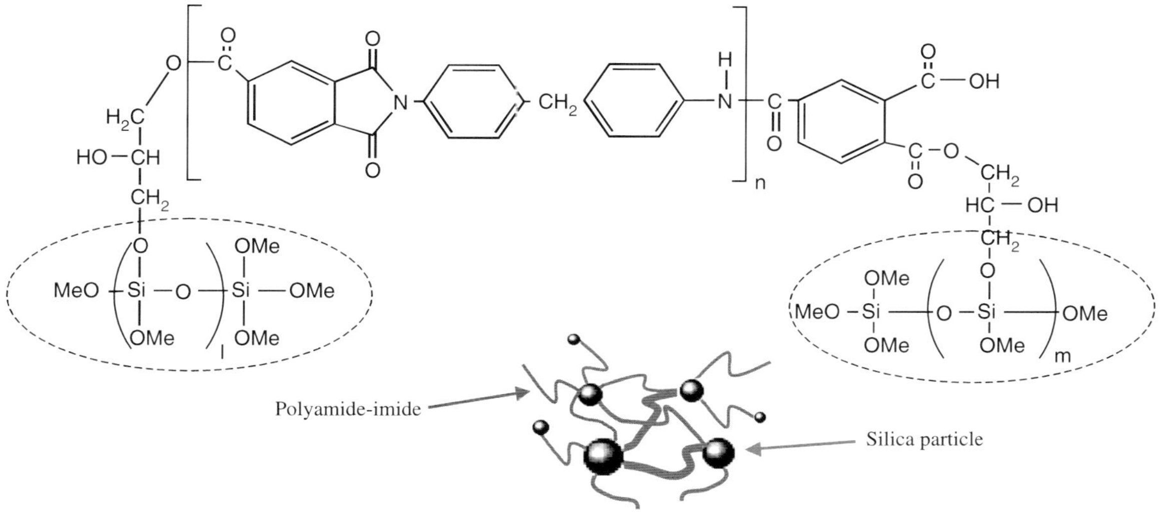

Figure 1.7
Chemical structure of silane-denatured amideimide.

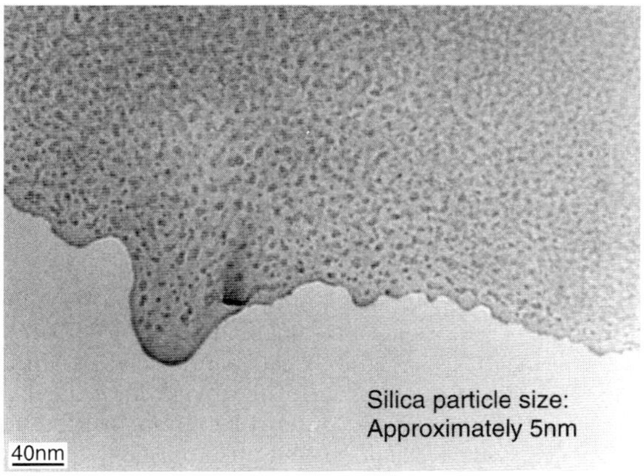

Figure 1.8
TEM picture of imide–silica hybrid.

7. Imide useful for electroless plating: hybrid of the imide system

The imide film is used as a heat-resistant and insulation material in various fields but is insufficiently adhesive onto electroconductive materials such as metal. However, when amic acid with alkoxysilane-oligomer is hardened to imide on metal film, a printed board with dispersed nanosized silica as seen in Figs. 1–8 can be obtained. It has recently been applied for industrial use as a double-layer flexible circuit board with high adhesiveness.

In imide–silica hybrid film, the silica is located as a crosslinker between imide chains and realizes wet plating with simple preprocessing without the roughening procedure (Fig. 1.9) [11]. By this wet plating method, layered products with a flat interface between the plastic film and the metal can be obtained, which is a useful technology for meeting the requirements

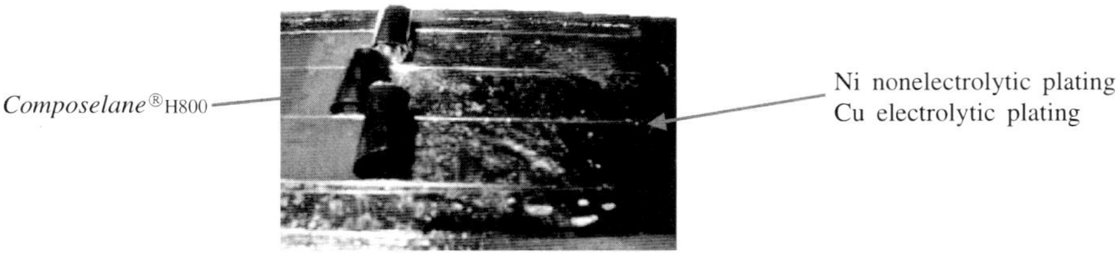

Figure 1.9
Imide–silica hybrid film after plating.

for advanced printed boards with thin conductive layers and flat interfaces.

References

[1] B.C. Novak: *Adv. Mater.*, **5** (6), 422 (1993).
[2] Y. Chujo, T. Saegusa: *Adv. Ploym. Sci.*, **100**, 11 (1992).
[3] G.L. Wilkes, B. Orler and H.H. Haung: *Polym. Prep.*, **26** (2), 300 (1985).
[4] S. Yamazaki: Polyurethane–Silica hybrid, Technical Report of National Institute for Material Science, Japan, **4**, 41 (1996).
[5] Arakawa Chemical Industries Ltd., Patent, WO01-05862, EP1123944, CN1318077T, TW483907, US6506868.
[6] H. Goda: *J. Jpn. Soc. Colour Mater.*, **77** (2), 69 (2004).
[7] H. Goda: *Polym. Prepr. Jpn.*, **50**, 2688 (2001).
[8] H. Goda, C.W. Frank: *Chem. Mater.*, **13** (7), 2783 (2001).
[9] H. Goda: *Kobunshi Ronbunshu*, **59**, 596 (2002).
[10] H. Goda, M. Mesaki: *Polym. Prepr. Jpn.*, **51**, 2245 (2002).
[11] H. Goda, T. Fujiwara: *Expected Mater. Future*, **3**, 34 (2003).

APPLICATION 2

2 GENERATION OF METAL NANOPARTICLES USING REACTIVE PLASMA ARC EVAPORATION

1. Summary of the reactive plasma arc evaporation method

(1) Features of the reactive plasma arc evaporation method

The reactive plasma arc evaporation method has been developed by M. Uda et al., of the National Research Institute for Metals, in the late of 1970's, and which is classified as one of plasma heating methods. The plasma heating methods are technique to obtain nanoparticles by metal plasma evaporation.

A gas evaporation method which has been developed previously must be conducted under several hundreds to thousands Pa of argon (Ar) or helium (He) atmosphere for nanoparticle synthesis [1].

On the other hand, the reactive plasma arc evaporation method can be under atmospheric pressure of diatomic molecular gas such as nitrogen (N_2) and hydrogen (H_2). These gases greatly influenced on the nanoparticle generation mechanism.

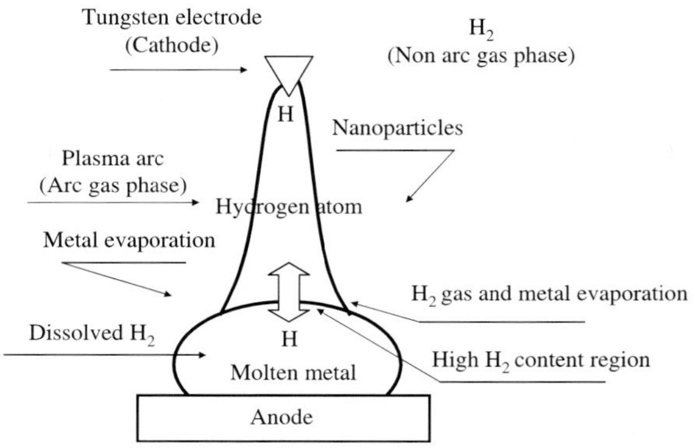

Figure 2.1
Schematic diagram of the nanoparticles generation mechanism under H_2 gas atmosphere.

(2) Nanoparticle generation mechanism

Fig. 2.1 shows a schematic diagram of the nanoparticle generation mechanism under H_2 gas atmosphere [2]. According to M. Uda et al., the mechanism is as follows: diatomic molecular gases dissociate to single atoms in the arc plasma, and then these atoms dissolve in the molten metal.

The transfer of these atoms to supersaturated area in the molten metal caused by convection flow. The atoms rebond and be discharged. At the same time, lots of excited metals evaporate from the surface of the molten metal. The metal evaporation condenses and nanoparticles are obtained [3].

H_2 gas can catalyze during metal nanoparticles generation from molten metal under H_2 arc plasma.

(3) Outline of the metal nanoparticle generation device

Fig. 2.2 shows the outline drawing of the metal nanoparticle generation device with the reactive plasma arc evaporation method.

Bulk as a starting material is placed on the water-cooled copper hearth, the bulk is irradiated with the

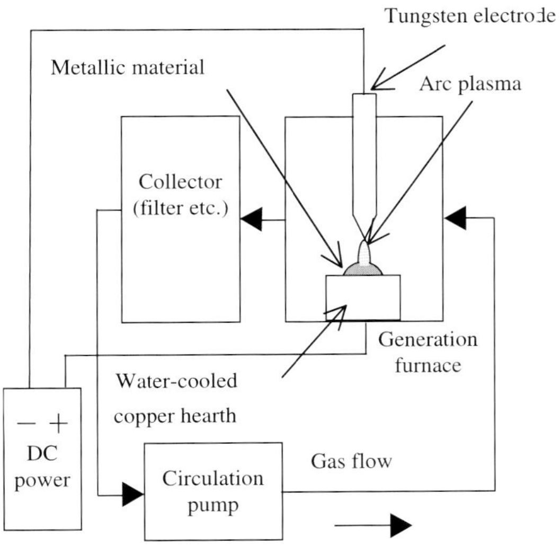

Figure 2.2
The outline drawing of metal nanoparticles generation device.

Table 2.1
The example of manufacturing nanoparticles.

Material	Atmosphere gas kind	Obtained particles
Pure metal	Hydrogen	Fe, Cu, Ni, Co, Cr, Ti, In, ...
Alloy	Hydrogen	Fe–Ni, Fe–Co, Ag–Cu, Ag–Pd, ...
Metal or nitride	Nitrogen	TiN, ZrN, AlN, AlN+Al, ...
Metal or carbide	Hydrogen	SiC, TiC, ...
Boronide	Hydrogen	LaB_6
Metal or oxide	Oxygen	WO_3, MoO_3, Nb_2O_5, ...

plasma, and the metal is melted. The nanoparticles are collected on the filter (or cyclone) by circulating a carrier gas (a mixed gas of a diatomic molecular gas and an inert gas) [4].

The arc plasma becomes unstable in a diatomic molecular gas, and hence an inert gas such as argon is added to it.

2. Nanoparticles by the reactive plasma arc evaporation method

This method is applicable to all metals. As shown in Table 2.1, an alloy, a nitride and an oxide can be obtained by control of atmospheric gas.

Nanoparticle composition varies by degree of affinity between the atmospheric gas and the metal. When Zr, Ti, Al, and Si are used as the bulk under N_2 atmosphere, three kinds of nanoparticles will be generated as follows: 1. nitride (Zr, Ti), 2. mixture of metal and nitride (Al), 3. metal (Si) according to an order of the affinities Zr > Ti > Al > Si [5]. In the case of alloy nanoparticles, for instance, the chemical composition of Fe-Ni particles is nearly identical with bulk composition. The composition of Cu in Fe-Cu particles and Si in Fe-Si remarkably deviated positively and negatively from the bulk composition, respectively [6].

The bulk is irradiated with alternative current arc plasma generated between two tungsten electrodes [7].

3. The nanoparticles generation rate, characteristic, and shape

(1) The nanoparticles generation rate

Fig. 2.3 shows the relationship between Fe nanoparticles generation rate and H_2 gas ratio in mixed gas atmosphere by this method [8]. The nanoparticles generation rate increased proportionally to the gas ratio of up to 50%. When above 50%, the generation rate reached the saturation point.

The generation rate increases with H_2 gas ratio, although the grade of the rate varies according to the metal species. In addition, the kind of atmospheric gas affects on the rate.

(2) The nanoparticles characteristics

Table 2.2 shows relationship between specific surface area of Fe nanoparticles and atmospheric gas ratio investigated by Uda et al. [9]. The specific surface area increased with increasing H_2 gas concentration. In other words, particle size becomes smaller as H_2 gas concentration. Nanoparticles characteristics can be controlled by atmospheric gas ratio and gas pressure.

Particle size obtained by this method is about 20 nm at the minimum. However, particle size of metal having lower melting point becomes larger (that is 100 nm and above) because of longer condensation time. Factors for which the particle size can be controlled are the structure of the device, the cooling condition, the gas flow velocity, the atmospheric gas pressure, the partial pressure of the mixed gas, the heat input characteristics of the arc plasma (current and voltage), and the metal species. Smaller particle size and sharpened distribution can be obtained with rapid quenching.

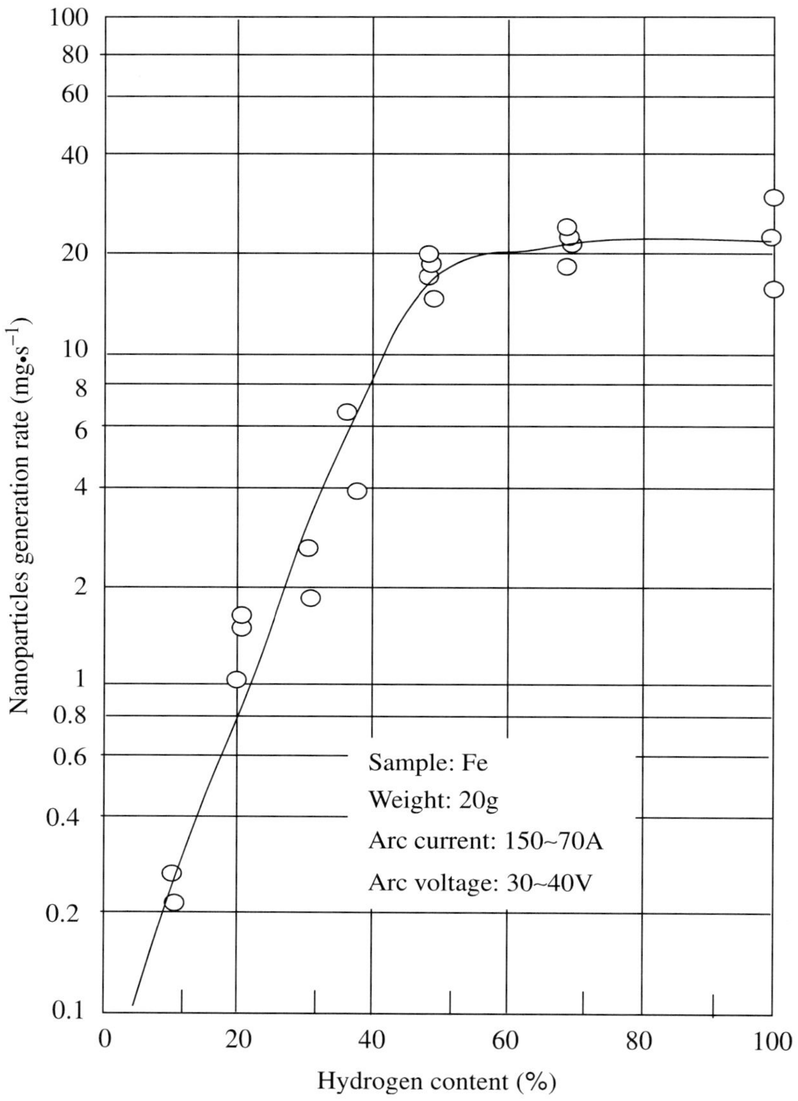

Figure 2.3
Relation between the iron nanoparticles generation rate and the hydrogen content.

Table 2.2
Specific surface area of iron nanoparticles.

Atmosphere gas	Specific surface area ($m^2\ g^{-1}$)	
	<60 kPa	>60 kPa
100%H_2	25–30	15–20
50%H_2–Ar	30–25	10–16
30%H_2–Ar	–	8–15

(3) Transmission electron microscope observation

Fig. 2.4 and Fig. 2.5 show TEM observations of Fe and Cr nanoparticles by this method, respectively [10]. Both each particles form chain-like network. In Cr nanoparticles, development of cubic crystal growth was observed.

4. Application of the nanoparticle

The reactive plasma arc evaporation method can produce larger amount of metal nanoparticles with high quality, therefore it is suitable for industrial manufacturing.

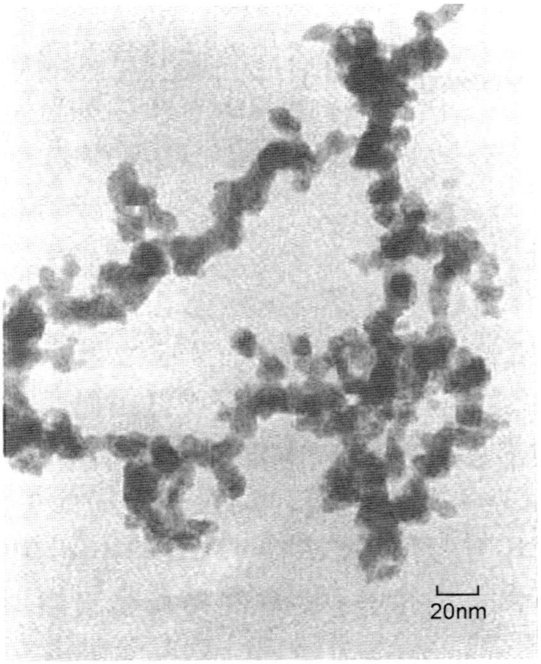

Figure 2.4
Iron nanoparticles TEM image.

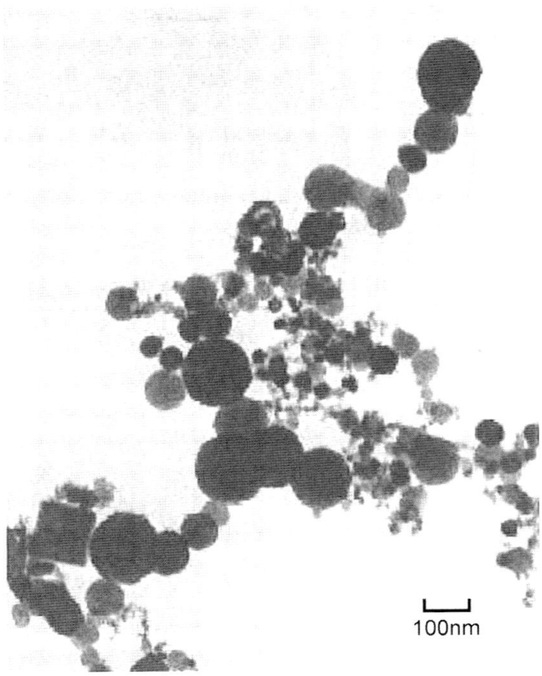

Figure 2.5
Chrome nanoparticles TEM image.

However, the produced nanoparticles are easily oxidized when exposed to an air. Widely applications such as electrical, magnetical, and optical can be accomplished by building properly production line.

References

[1] The Chemical Society of Japan (Ed.): Ultrafine Particle Science and Application, JSSP, **28** (1985).

[2] M. Uda: Research Report, Nisshin Steel Co., Ltd., **61**, 90–99 (1989).

[3] M. Ozawa, S. Ohno, H. Okuyama, M. Uda and E. Ozawa: Report of NRIM, **10**, 157–172 (1989).

[4] H. Shimoda, S. Shiozaki: The Society of Powder Technology, Japan, the 40th of round-table discussion texts, 53–54 (2005).

[5] M. Uda, S. Ohno, E. Ozawa, H. Okuyama and Y. Sakka: Report of NRIM, **7**, 59–73 (1986).

[6] S. Ohno, M. Uda: *J. Jpn. Inst. Metals*, **53**(9), 946–952 (1989).

[7] M. Uda, S. Ohno: *J. Surf. Sci. Soc. Jpn.*, **5**(4), 412–420 (1984).

[8] M. Uda: *J. Jpn. Inst. Metals*, **22**(5), 412–420 (1983).

[9] H. Okuyama, S. Ohno, K. Honma, M. Ozawa and M. Uda: *J. High Temp. Soc. Jpn.*, **15**(4), 143–150 (1989).

[10] N. Nakajima, J. Noma and H. Shimoda: POWTEX OSAKA 2005, Kurimoto Ltd. Exhibition Panel.

APPLICATION 3

3 SENSING BASED ON LOCALIZED SURFACE PLASMON RESONANCE IN METALLIC NANOPARTICLES

1. Localized surface plasmon

Gold nanospheres dispersed in a solution are ruby red, since green light interacts with electrons in the nanospheres and is adsorbed. This phenomenon is called localized surface plasmon resonance (LPR). Localized surface plasmon resonance occurs not only in metallic nanoparticles but also in a sharp metallic tip and at a rough metallic surface. These LPR phenomena are applied to enhancement of near-field scanning microscopy, Raman scattering and fluorescence spectroscopy. LPR has the following features:

(1) Confinement of light within a nanometer region.
(2) The sensitive resonance condition to the dielectric constant of the ambient medium.
(3) A large electric field produced at the LPR condition.

Recently, nano- and micro-photonic devices and sensors have been developed, based on LPR. This article deals with small and potential sensing devices by use of metallic nanoparticles.

The optical properties of nanoparticles are described in textbooks in detail [1, 2]. Here we consider the optical response of a nanosphere described in Fig. 3.1a. Suppose the dielectric constant of the nanosphere $\varepsilon_a(\lambda)$ and that of the ambient medium $\varepsilon_m(\lambda)$ at a wavelength λ. If the size of the nanosphere is much smaller than the wavelength of light, the polarizability of the nanosphere can be described as follows [3].

$$\alpha(\lambda) = 4\pi\varepsilon_m(\lambda)r^3 \frac{\varepsilon_a(\lambda) - \varepsilon_m(\lambda)}{\varepsilon_a(\lambda) + 2\varepsilon_m(\lambda)} \tag{3.1}$$

where r stands for radius of the nanosphere. The polarizability $\alpha(\lambda)$ is maximum with the minimum absolute value of the denominator. When the nanosphere is covered with a dielectric thin film with a thickness d and dielectric constant $\varepsilon_d(\lambda)$, equation (3.1) is rewritten as

$$\alpha(\lambda) = 4\pi\varepsilon_m(\lambda)(r+d)^3 \frac{\varepsilon_d(\lambda)\varepsilon_A(\lambda) - \varepsilon_m(\lambda)\varepsilon_B(\lambda)}{\varepsilon_d(\lambda)\varepsilon_A(\lambda) + 2\varepsilon_m(\lambda)\varepsilon_B(\lambda)} \tag{3.2}$$

where the parameters are described as

$$\varepsilon_A(\lambda) = \varepsilon_a(\lambda)(3 - 2P) + 2\varepsilon_d(\lambda)P \tag{3.3}$$

$$\varepsilon_B(\lambda) = \varepsilon_a(\lambda)P + \varepsilon_d(\lambda)(3 - P) \tag{3.4}$$

$$P = 1 - \left(\frac{r}{r+d}\right)^3 \tag{3.5}$$

Fig. 3.1c shows scattering efficiency of gold nanospheres of 40 nm in diameter covered with a dielectric thin film ($\varepsilon_d(\lambda)$ =2.25), in which the thicknesses of the dielectric film are 2, 5 and 10 nm. The thickness of the thin film and its optical properties can be probed by measuring scattering spectra or absorption spectra, which is determined by $\alpha(\lambda)$.

2. Two sensing method using plasmon

There are two kinds of bio and chemical sensors that utilize LPR in metallic nanoparticles. One is the sensor that probes affinity between molecules

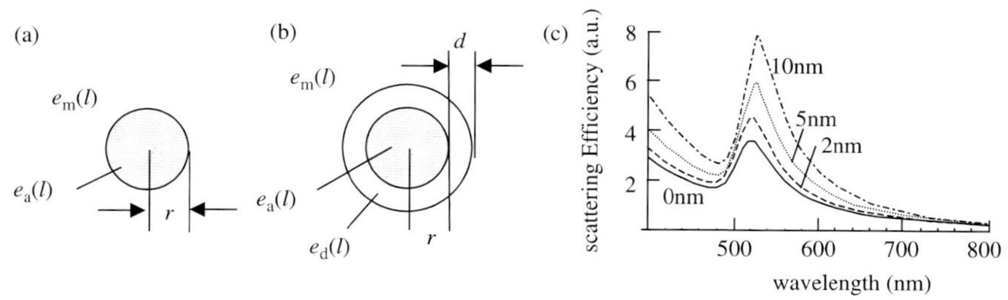

Figure 3.1
(a) Optical geometry of a nanoparticle. (b) Optical geometry of nanoparticle covered with a dielectric film. (c) Calculated scattering efficiency of gold nanoparticles with various thickness of dielectric films.

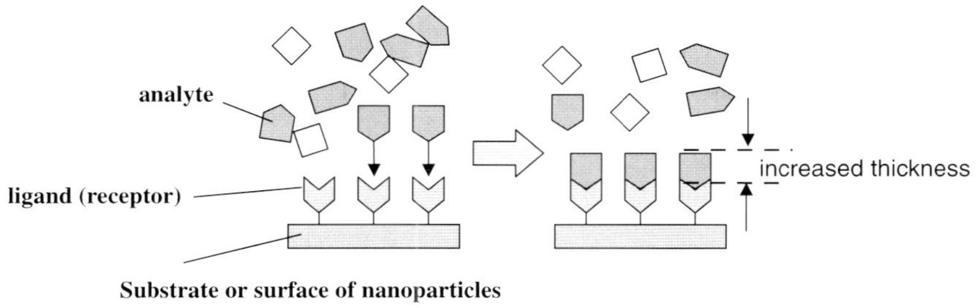

Figure 3.2
Principle of affinity biosensor. When analyte molecules are bound to receptors, the average thickness of surface layer increases.

(affinity biosensor) and the other is a Raman sensor that provides a vibration spectrum. The former is based on the feature of LPR that the resonance condition is sensitive to the dielectric constants of the ambient medium, and the latter is based on the phenomenon that a large electric field is produced at the resonance condition. The large electric field provides us with a huge enhancement of the Raman signal. This is called surface enhanced Raman scattering (SERS). Since Raman signal is usually weak, it should be enhanced to probe the signal from very thin films or nanometer structures. Both sensors have different characters as described below.

The affinity biosensor has been widely used in these days. Its principle is illustrated in Fig. 3.2. The receptor molecules, which have affinity to the analyte molecules such as proteins and DNA, are immobilized to a surface of a substrate or nanoparticles. Gold is used as a substrate or nanoparticle materials because it is inert. When a sample solution involves analyte molecules, they react with the receptor molecules immobilized on the surface, resulting in an increase in a thickness of the surface dielectric layer. Although the size of DNA and proteins are in the range of 1–10 nm, the average thickness increase is about in the order of 0.01–0.1 nm, because the molecules do not form a densely packed structure. Therefore highly sensitive method that can probe such a small increase in thickness is needed. The method widely used is labeling with fluorescent choromophore. It is a very sensitive method, but labeling process is required. Label-free methods, surface plasmon resonance (SPR) [4–6] in attenuated total reflection (ATR) geometry and quartz crystal microbalance (QCM) [7], have been developed. They are good probes that can monitor the small increase in thickness of the surface layer. However it is difficult to make the probes with a size of micrometers.

On the other hand, LPR sensors using metallic nanoparticles have the following remarkable features. The LPR sensor probes are small and the optical setup is simple. In principle, the sensor can be made with one nanoparticle. A schematic picture of the optical fiber

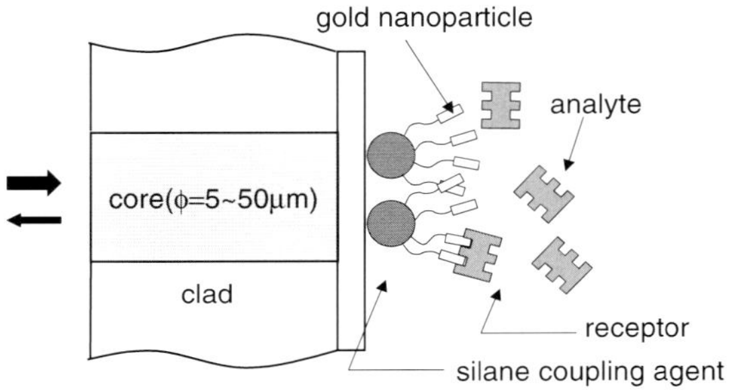

Figure 3.3
Schematic picture of the optical fiber sensor [8]. Gold nanoparticles with receptor molecules are immobilized at an endface of an optical fiber. When sample solution involves analyte molecules, they bound to the receptor, and change in the scattering efficiency occurs. As a result, the return light intensity and spectrum are changed.

LPR sensor that the authors have developed is shown in Fig. 3.3 [8]. The return light from the endface of the optical fiber is measured for biosensing. The gold nanoparticles are immobilized at the endface with a surface coverage of 10–30%. The return light involves both components of reflected light at the surface and scattered light by the particles. Since the diameter of the optical fiber is small, ~100 μm (the core diameter is 5–50 μm), we can probe proteins in a solution of 50 nL using a special sample holder, whereas a current SPR system requires a sample volume of ~50 μL. This feature has a great advantage because the amount of biological molecules is usually little.

For exhaustive investigation of proteins and DNA, the sensing array is needed to perform multichannel sensing. DNA chips have been widely used in these days, but labeling process is required. As a label-free detection, the SPR sensing array using ATR optics has been investigated. LPR in metallic nanoparticles can also be used for label-free sensing array. Since this system is simple, it is possible to make a high-density sensing array chip.

Attention has also been paid to the SERS sensor in recent years. The sensor provides information of vibration and rotation in a molecule, so that details in molecular structures are available. The SERS spectroscopy was found in the 1970s, and many studies have revealed that the origin of the large enhancement is attributed to two reasons [9]: one is the main reason that a large electric field is produced in a rough silver surface and the other is an additional reason that enlarged molecular vibration upon adsorption of molecules on a silver surface. Recently a huge enhancement of the SERS signal has been found in a system of aggregated silver nanoparticles, and Raman spectroscopy from a single molecule has been reported [10–12]. Substrates that provide a large SERS signal have been developed [13]. These efforts enable us to perform Raman spectroscopy without using special techniques. Also portable Raman systems have been reported [14, 15].

In summery, LPR in metallic nanoparticles provides many potential applications for optical sensors. The fields are wide, such as chemistry, biology, medical, genome engineering, materials science, and electronics.

References

[1] C.F. Bohren, D.R. Huffman: *Absorption and Scattering of Light by Small Particles*, Wiley, New York, pp. 335–380 (1983).
[2] M. Fukui, M. Ohtsu: Hikarinanotekunoroginokiso, Ohmsha Ltd., pp. 69–105 (2003) (in Japanese).
[3] A.E. Neeves, M.H. Birnboim: *J. Opt. Soc. Am. B*, **6**, 787–796 (1989).
[4] J. Homola, S.S. Yee and G. Gauglitz: *Sens. Actuators B*, **54**, 3–15 (1999).
[5] B. Liedberg, K. Johansen: Affinity Biosensing Based on Surface Plasmon Resonance Detection, in: K.R. Rogers, A. Mulchandani (eds.). *Affinity Biosensors Techniques and Protocols*, Humana Press, Totowa, NJ, pp. 31–54 (1998).
[6] K. Kajikawa: *Protein Nucleic Acid Enzyme*, **49**, 1772–1776 (2004) (in Japanese).
[7] Y. Okahata, H. Furusawa: *Protein Nucleic Acid Enzyme*, **49**, 1754–1758 (2004) (in Japanese).
[8] K. Mitsui, Y. Handa and K. Kajikawa: *Appl. Phys. Lett.*, **85**, 4231–4233 (2004).
[9] K. Kneipp, H. Kneipp, I. Itzkan, R.R. Dasari and M.S. Field: *Chem. Rev.*, **99**, 2957–2975 (1999).
[10] S. Nie, S.R. Emory: *Science*, **275**, 1102–1106 (1997).
[11] H. Yu, E.J. Bjereld, M. Kaell and L. Borjesson: *Phys. Rev. Lett.*, **83**, 4357–4360 (1999).
[12] Y. Maruyama, M. Ishikawa and M. Futamata: *J. Phys. Chem. B*, **108**, 673–678 (2004).
[13] N. Hashimoto, N. Ishikawa and A. Nakajima: *Chem. Phys. Lett.*, **413**, 78–83 (2005).
[14] A.J. Sommer, S.A. Stewart: *Appl. Spectrosc.*, **53**, 483–488 (1999).
[15] M.A. Young, D.A. Stuart, O. Lyandres, M.R. Glucksberg and R.P. Van Duyne: *Can. J. Chem.*, **82**, 1435–1441 (2004).

APPLICATION 4

4 MICROELECTRONICS PACKAGING BY METAL NANOPARTICLE PASTES

1. Conductive paste technique and metal nanoparticle paste

In the field of microelectronics packaging, conductive paste technique is usually used for the formation of various electronic components such as conductive circuits, electrodes, resistors, and dielectrics. Conventional technique of conductive pastes is mainly thick film pastes composed of metal powders with several micrometers to submicrometer size and organic compounds. Using conductive pastes, the electronic circuit patterns are figured on a substrate by screen printing or dispensing method, and so on, and then dried and fired to decompose the pastes and to remove organic additives such as dispersants and coherence agents. As a result, metal particles are melted and fused to afford conductive metallic thin film. However, the firing temperature of the conventional conductive pastes is over 550°C, limiting the usage of the pastes to the glass and ceramics substrates.

On the other hand, much attention has been paid to metal nanoparticle pastes including metal nanoparticles with several nanometer to dozens of nanometer size as a new technology available for down-sizing and flexibility of the electronic components. Decreasing the diameter of metal particle, the surface energy increases to become unstable compared with the bulk metal. Usually, metal nanoparticles prepared by various methods are capped by organic protecting groups to control the growth of the particles and the particle size [1, 2]. Fig. 4.1 shows the transmission electron microscope (TEM) photograph of gold nanoparticles prepared by gas evaporation

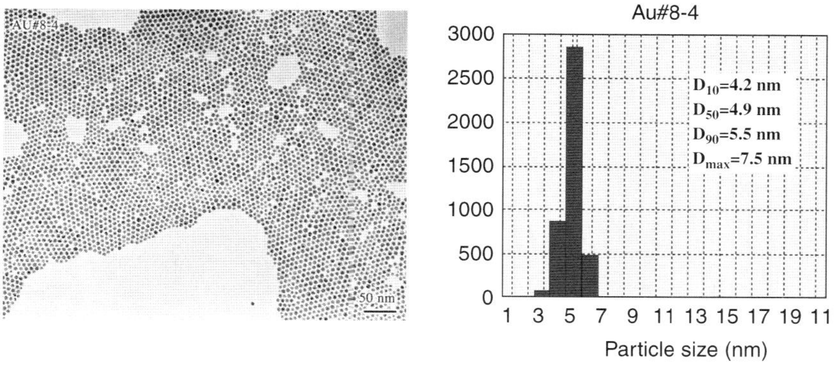

Figure 4.1
Independently dispersed gold nanoparticles and its particle size distribution (Photograph by Ulvac Corporate Center).

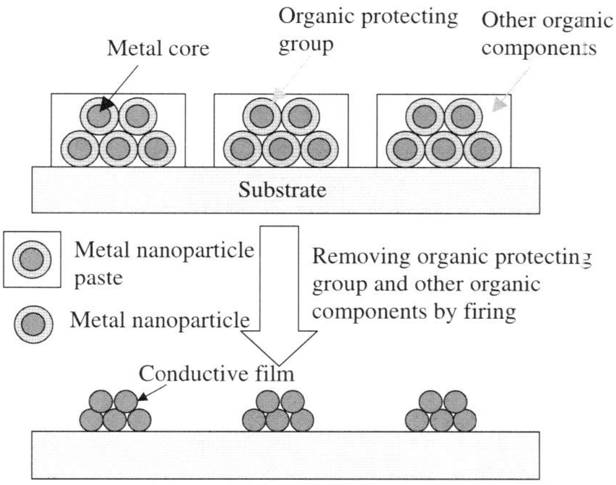

Figure 4.2
Electronic circuit pattern formation by the use of metal nanoparticle paste.

method. The monodispersed gold nanoparticles with an average diameter less than 10 nm show good dispersion for making conductive pastes. Fig. 4.2 shows the concept of the circuit pattern formation by metal nanoparticle pastes. As similar to the conventional conductive pastes, the heating process causes the decomposition of the metal nanoparticle pastes to remove the organic protecting groups around the metal core of the particles and the other organic components of the pastes, forming the metallic conductive thin film.

However, the metal nanoparticles show the low temperature decomposition property to fuse each other and form metallic thin film less than 300°C due to the quantum size effect. As a result, metal nanoparticle pastes can be applied to prepare electronic circuit pattern on the plastic substrates as well as on the glass and ceramic substrates [2–5]. In addition, fine electronic circuit pattern with the line and space less than 50 μm can be formed by screen printing. It means that the limitation by the conventional thick film pastes can be solved by the use of metal nanoparticle pastes. It is also expected to prepare much more fine circuit pattern with 10 μm line around the LSI. On the other hand, taking the advantage of superior dispersion property of nanoparticles, they have been applied to inkjet printing technique to form much more fine patterning [4]. Furthermore, it has been developed to apply metal nanoparticles as the joining materials. Thus, metal nanoparticles have just become the developing materials in the field of microelectronics packaging.

2. Low temperature firing and fine electronic circuit pattern formation by screen printing

A variety of metal nanoparticle pastes, for example, silver nanoparticle pastes, have been developed in order to prepare fine electronic circuit pattern by screen printing [1–5]. The molecular design of the nanoparticles and the control of the particle size, metal content, and the viscosity of the pastes can realize the fine pitch electronic circuit pattern with good resistivity compatible with that of bulk metal using no lithography technique.

Monodispersed gold and silver nanoparticles with average diameter of 3–7 nm [1] have been prepared by gas evaporation method as shown in Fig. 4.1. These metal nanoparticles are used for metal nanoparticle pastes dispersed in organic solvent (nanopaste) [3, 4]. Nano Pastes include the supplementary materials, which are activated by heating to remove the protecting groups of the nanoparticles in the firing process. Conductivity has appeared by the fusion of the metal cores after low temperature firing at 200°C. The conductive thin film with the thickness of 0.1–10 μm and the resistivity of 3–50 μΩ cm has been prepared on glass, silicon wafer, and organic substrate such as polyimido film by the firing at 180–250°C for 20–60 min. The appeared resistivity is consistent with that of bulk silver (1.6 μΩcm). Fig. 4.3 shows the example of fine electronic circuit pattern formation on polyimido film by screen printing.

A variety of silver nanoparticles with an average diameter of 8.8–32.7 nm and silver content over 90 wt% prepared by the controlled thermolysis of metal complexes [2] can be applied to silver nanoparticle pastes without the use of supplementary materials. This type of silver nanoparticle paste affords the conductive thin film by removing the organic protecting group and fusion of the metallic cores in the firing process less than 300°C. Fine electronic pattern of line and space 20 μm/20 μm with the film thickness of 5.6–30 μm and the resistivity of 5.8–40 μΩcm can be prepared on polyimido film by screen printing and firing at 300°C for 30 min. In addition, a new low-temperature firing type of silver nanoparticle paste [5] has been developed to prepare electronic circuit pattern with about 100 μΩcm on PET film by firing at 100°C for 60 min as shown in Fig. 4.4a.

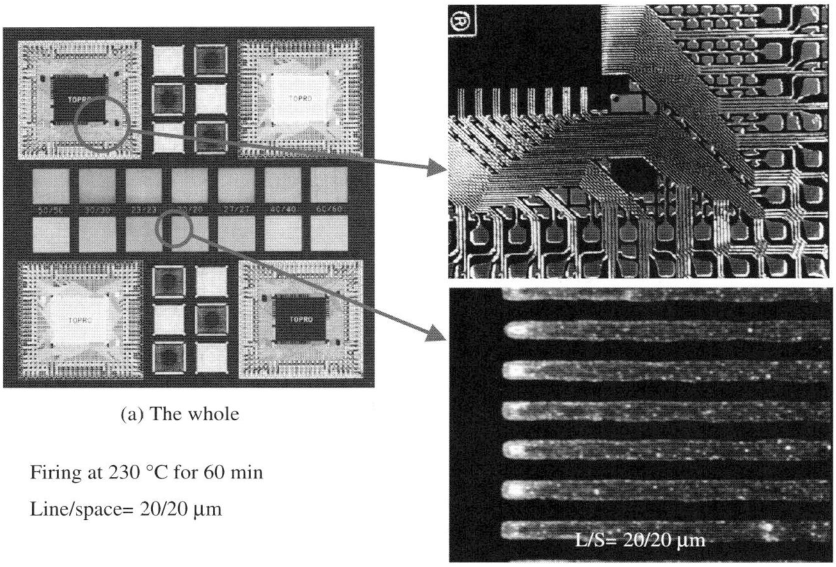

(a) The whole

(b) Magnification

Figure 4.3
Screen printing example by silver nanoparticle paste (Photograph by Harima Chemicals, Inc.)

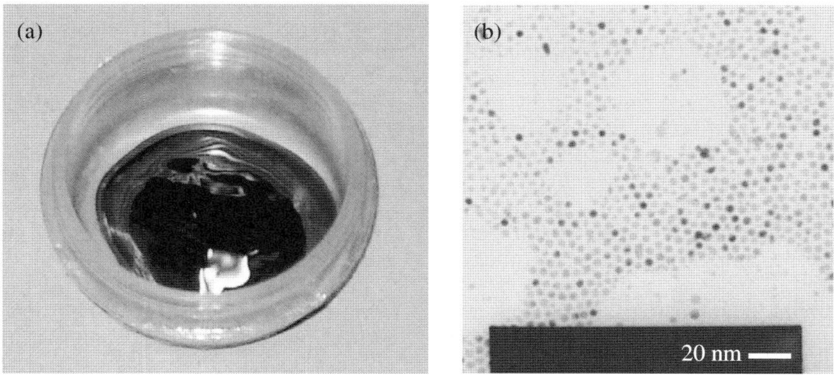

Figure 4.4
(a) Silver nanoparticle paste for the PET substrate and (b) TEM photograph of silver–palladium alloy nanoparticles for the alloy nanoparticle paste (Ag85/Pd15) (Photograph by Daiken Chemical Co., Ltd.)

There may be a risk of electromigration of silver in the case of electronic devices using silver conductors. In order to solve the problem, a new type of silver–palladium alloy nanoparticle paste with desired Ag–Pd ratio (Fig. 4.4b) has been prepared [5]. Silver–palladium alloy nanoparticle paste with Ag85–Pd15 ratio affords the conductive film of the thickness of 4 μm and the resistivity of 12 μΩcm, showing the anti-electromigration property.

Monodispersed silver nanoparticles prepared by gas evaporation method are used for a hybrid type silver paste, where silver nanoparticles are mixed with micron size silver powder (average diameter 3–5 μm) [1, 6]. After firing the hybrid silver paste, the small size of silver particles are filled in the vacancy between the large silver particles to form tight conductive film to reinforce the electric contact and reduce the resistivity. Such hybrid paste shows a good performance to form a reliable fired film compared to the conventional silver paste.

Silver nanoparticle pastes with high viscosity and thixotropy for the screen printing are also used for the patterning by the dispensing method to realize the 18 μm line width. In addition, the patterning by the dispensing technique can repair the disconnection part of the circuit. For example, metal nanoparticle pastes are injected on the disconnection part of the PDP electrodes by the dispensing method, and then the part is recovered to be conductive through the heat treatment by the laser irradiation [3].

3. Direct formation of the electronic circuit pattern by inkjet printing

The combination of the high performance of dispersion of the metal nanoparticles with inkjet printing can realize a direct formation of the superfine patterning with the line width of several micrometers, which is impossible by screen-printing technique. In the case of inkjet printing, the minimum quantity of liquid, 2 pl, coming from the printer head corresponds to a dot with a diameter of 15–16 μm. The connection of the dots forms the line width directly figured by inkjet printing.

Instead of the pastes for screen printing, the metal nanoparticle ink for inkjet printing shows the low viscosity with dozens of mPa·s, thus the ink should have the high performance of dispersion avoiding to clog the nozzle of printer head. Concerning this point, the ink of monodispersed metal nanoparticles [3, 4] shows a good performance for the inkjet printing. Thus, the fine electronic circuit pattern with 20 μm line width is directly figured as shown in Fig. 4.5.

The fired film thickness of the inkjet printed circuit pattern is about 1–2 μm, but the control of the paste viscosity and the ejection voltage of the printer head, the surface treatment of the substrates, the heat treatment conditions, and the printing technique can easily

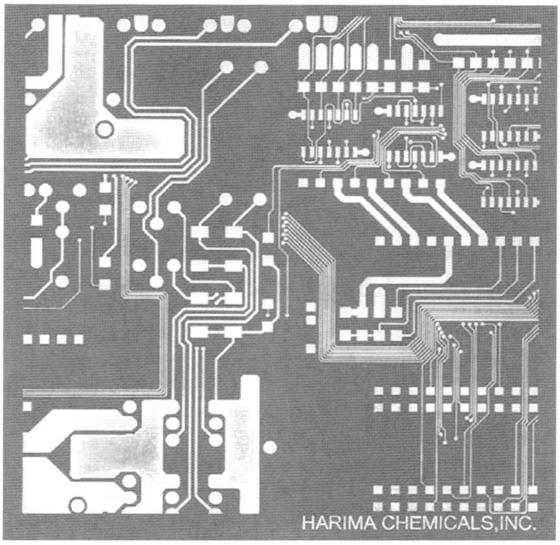

Figure 4.5
Inkjet printing example by silver nanopaste (Photograph by Harima Chemicals, Inc.).

prepare the electronic patterns as the CAD pattern prepared by the computer. Furthermore, recent super-inkjet printing technology [7] using the femto liter (fl) drop of monodispersed silver nanoparticle ink can realize the fine patterns with the line and space 3.6 μm/1.4 μm. Further development of this technology may be expected as a new wiring tool for the superfine packaging.

The inkjet printing using silver nanoparticle paste has been developing as a printable electronics technique and various samples are prepared for the estimation. For example, the electrode (line width 50 μm, fired thickness 2 μm, resistivity 2 μΩcm) of the PDP front panel (the diagonal size 10 inches) was manufactured on trial [8].

Besides the two-dimensional circuit pattern formation, the 20 layers board with alternative conductor and the insulator layers can be manufactured by the use of silver nanoparticle paste and the insulator paste, respectively, where the line width 50 μm, minimum space 110 μm, and thickness 4 μm are patterned on each layer [9]. Silver nanoparticle pastes in the sheets are fired at 150–200°C to afford the conductivity. Fine pattern formation technique has been also developed for making the low temperature co-firing ceramics (LTCC) multi-layer substrates [10]. A set of sheets with the inkjet printed pattern using silver nanoparticle paste are packaged and co-fired all at once. Furthermore, the multi-layering technique using the inkjet printing may be applied to the multi-layer substrates embedded with IC and capacitors, the flexible substrates embedded with the super high density system module and the electronic components, and the electronic circuit pattern formation on the surface of housing.

4. Application as the joining materials

The approach of silver nanoparticle pastes to the joining technique [11] has been proposed as the new type of lead-free solder materials in the field of microelectronics packaging. Silver nanoparticle pastes are used for the copper–copper, silver–silver, and gold–gold joinings under the condition of 300°C for 300 s and the pressure at 5 MPa, resulting in the joining strength of 40 MPa. Silver and the joining materials are joined through metallic binding by the firing process of the silver nanoparticle pastes at 300°C. The joining strength by silver nanoparticle pastes becomes stronger than that of the copper–copper joining by lead-rich high temperature solders.

The technology of metal nanoparticle paste is now going to be developed. Both screen printing and inkjet printing show the possibility of metal nanoparticle pastes as the new materials for fine pitch circuit pattern and multi-layers formation and also for the system in package (SiP) [3] preparing the components such as capacitor, resistance, and external terminal all at once.

References

[1] M. Oda: *J. Jpn. Inst. Electron. Packaging*, **5**(6), 523–528 (2002).
[2] M. Nakamoto: *Mater. Stage*, **3**(11), 35–40 (2004).
[3] Y. Matsuba: *J. Japn. Inst. Electron. Packaging*, **6**(2), 130–135 (2003).
[4] H. Saito, M. Ueda and Y. Matsuba: Proceedings of microelectronics symposium, p. 189–192 (2004).
[5] M. Nakamoto, M. Yamamoto, Y. Kashiwagi, H. Kakiuchi, T. Tsujimoto and Y. Yoshida: Proceedings of microelectronics symposium, p. 241–244 (2005).
[6] Y. Ohsako: Abstract of chemical engineering, 34th meeting, K122 (2001).
[7] K. Murata: *Mater. Stage*, **2**(8), 23–26 (2002).
[8] M. Furusawa: SID international symposium digest of technical papers, p. 753 (2002).
[9] K. Mizugaki, K. Wada: Proceedings of microelectronics symposium, p. 249–252 (2005).
[10] K. Koiwai, T. Sakuma, H. Kawamura, S. Yamaguchi, K. Wada, K. Sakurada and T. Kobayashi: Proceedings of microelectronics symposium, p. 245–248 (2005).
[11] E. Ide, S. Angata, A. Hirose and K.F. Kobayashi: *Acta Mater.*, **53**, 2385–2393 (2005).

APPLICATION 5

5 A DYE-SENSITIZED SOLAR CELL UTILIZING METAL NANOPARTICLE

In nanoparticles of metals such as silver, the "Surface plasmon" is excited, which involves free electrons on the surface of the nanoparticle collectively vibrating with irradiation of light of a specific wavelength. It is known that interaction with the surface plasmon causes phenomena, such as enhancing the Raman scattering intensity of a semiconductor.

In this section, the following are explained: results successfully achieved utilizing this phenomenon to enhance the light absorbance of a dye used for dye-sensitized solar cells; the effect of enhancement in a TiO_2 porous thin film, indispensable for enhancing the efficiency of solar cells; and the potential for solar cells which utilize this enhancement effect.

1. What is a dye-sensitized solar cell?

A dye-sensitized solar cell is a wet solar cell, composed of two conductive glasses, filled between with an interim electrolyte solution, one of the conductive glasses having a titania (TiO_2) film carrying a dye, the TiO_2 film sintering on the conductive glass and including a porous structure some dozens of nanometers thick, and the other conductive glass serving as an opposing electrode. A photo of the titania porous film of a dye-sensitized solar cell, taken by an electron microscope, is shown in Fig. 5.1, while a conceptual diagram of the dye-sensitized solar cell (DSC) is shown in Fig. 5.2.

The ease of production and relatively high conversion efficiency (reportedly about 11% [1]) have raised expectations of the potential for economical and highly efficient solar cells. However, even in the case of a ruthenium dye, which is known to have excellent performance as a sensitized dye, since it is adsorbed only on the surface of the TiO_2 porous film, the light absorption coefficient of a light absorption layer is lower than that of semiconductors such as silicon, which consequently hampers efforts to enhance efficiency. Therefore, if the light absorption coefficient of the TiO_2 film (light absorption layer) carrying a dye could be improved, enhanced efficiency of the DSC would be possible.

Figure 5.1
SEM Photo of titania porous film of dye-sensitized solar cell.

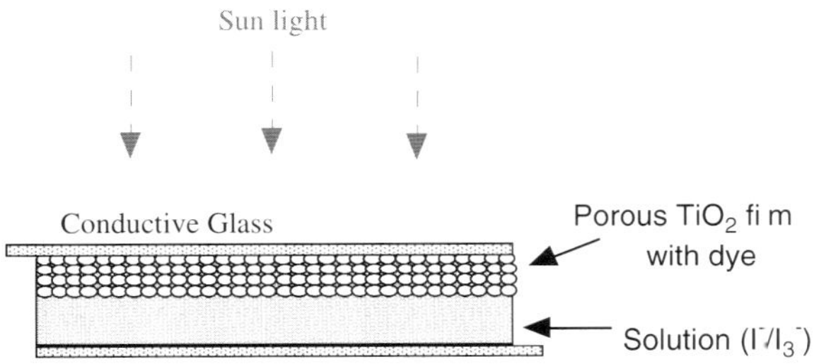

Figure 5.2
Structure of dye-sensitized solar cell.

2. Enhancement of the absorption coefficient of the ruthenium dye, with the silver nanoparticle produced via vacuum evaporation on the quartz substrate

As mentioned above, if the absorption coefficient of the dye itself can be enhanced, the enhancement of the conversion efficiency of DSC will be expected. It is known that a ray of light, incident on the surface of a metal nanoparticle, causes a local electric field enhancement effect, due to the surface plasmon vibration, resulting in an increased intensity of Raman scattering and the absorption coefficient of a semiconductor or dye, which reside adjacent to the incident position.

If the absorbance of the dye via the local electric field enhancement effect is enhanced, caused by a metal nanoparticle, and if this enhanced result is utilized to produce a dye-sensitized solar cell, enhanced efficiency will be expected. However, the magnitude of the absorbance enhancement is heavily dependent on the combination of a metal nanoparticle, semiconductor, and dye.

Therefore, we have set ourselves the challenge of enhancing the absorption coefficient of the ruthenium dye (*cis*-$(NCS)_2$bis(2,2′-bipyridyl-4,4′-dicarboxylate) ruthenium(II) dye) with a silver nanoparticle, the ruthenium dye reportedly high efficiency by about 10% in a DSC.

Fig. 5.3 exhibits the molecular structure of the ruthenium dye (*cis*-$(NCS)_2$bis(2,2′-bipyridyl-4,4′-dicarboxylate) ruthenium(II) dye) and the light absorption spectrum. The ruthenium dye used has absorptions at 400 and 535 nm, due to metal–ligand charge transfer. Due to these absorptions, the dye-sensitized solar cell generates a carrier.

Fig. 5.4 exhibits the absorption spectra of four kinds of silver/dye films of variable dye concentrations. This shows that the absorption at around 550 nm increases as the dye concentration increases.

Furthermore, although the concentration of the dye film exhibited in Fig. 5.3 was higher than that of every film in Fig. 5.4, the light absorbance of the dye film at around 550 nm still remained at a maximum of about 0.04, while that of the silver/dye films peaked at about 0.38. This proves that the dye significantly increases its light absorbance with the silver nanoparticle. Separating the absorption spectra of silver/dye films

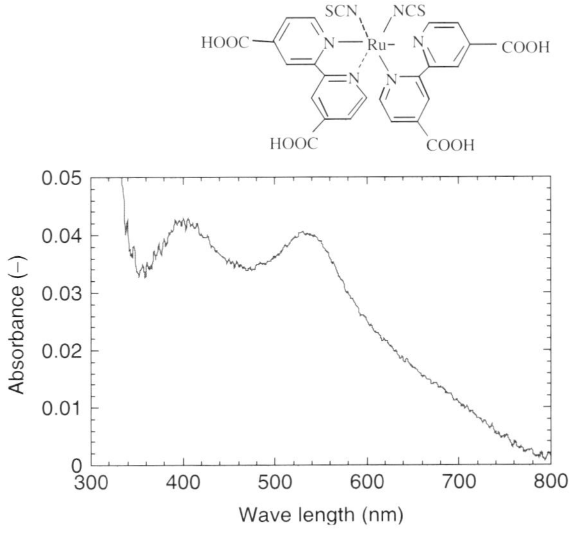

Figure 5.3
Light absorption spectrum of ruthenium dye (*cis*-$(NCS)_2$bis(2,2′ -bipyridyl-4,4′-dicarboxylate) ruthenium(II) dye) (ruthenium dye concentration: 8×10^{-9} mol/cm^2) [2].

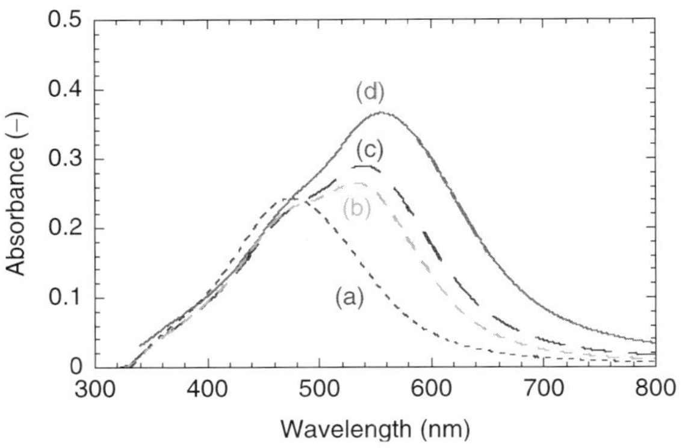

Figure 5.4
Light absorption spectra of silver/ruthenium dye composite films [2]. (a) 9×10^{-11} mol/cm^2, (b) 3×10^{-10} mol/cm^2, (c) 5×10^{-10} mol/cm^2, (d) 2×10^{-9} mol/cm^2.

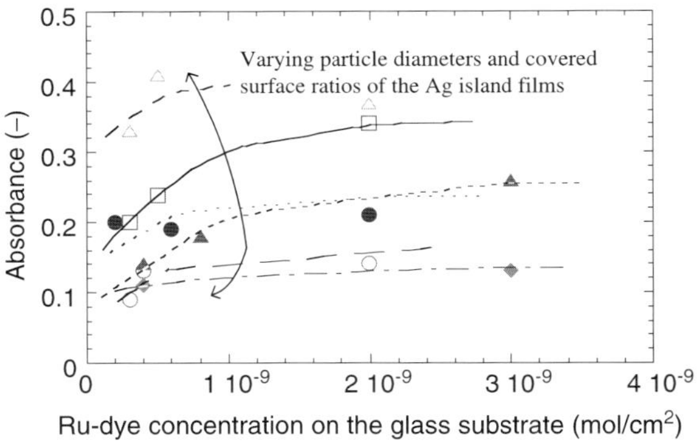

Figure 5.5
Light absorbance of Band B versus dye concentration under a constant silver nanoparticle concentration [2].

into two Lorenz-distribution type absorption peaks (hereinafter, referred to as Band A: 470–500 nm, and Band B: 520–580 nm) proved that the absorption coefficient of Band B was enhanced up to 149 times.

This enhancement ratio is far in excess of the reported ratios and reveals the expectation of applications other than for solar cells. The silver nanoparticle has converted the dye film from virtual translucency, due to the low absorption coefficient, to exhibiting deep violet color. This proves that the absorption coefficient is highly enhanced as visible with the naked eye.

The light absorbance and peak shift of Band B increased as the dye concentration increased (refer to Fig. 5.5), and then plateaued due to saturation. The enhancement ratio of the absorption coefficient, however, decreased as the concentration increased (refer to Fig. 5.6). Concerning the relation between the enhancement ratio and the surface area of silver, under the same dye concentration, the enhancement ratio increased as the surface area increased (refer to Fig. 5.7).

The following are understood based on the above facts: the effect of enhancing the light absorbance of the dye with the silver nanoparticle is heavily dependent on the distance between the silver and dye; and, since a dye directly adsorbed to the silver or residing in the extreme vicinity of the same, is particularly strongly effected, the light absorbance of the silver/dye film is mostly determined by that of the dye residing in the immediate vicinity of the silver. Therefore, it is also understood that controlling the distance between a silver nanoparticle and dye in a nanometer order is critical for the application to solar cells.

3. Enhancement of the absorption coefficient of silver nanoparticle-ruthenium dye within porous TiO_2

The results mentioned in Chapter 2 are based on silver nanoparticles, produced on a quartz substrate via a vacuum evaporation technique. In order to achieve a practical enhancement of efficiency, however, within a space having an order of dozens of nanometers, which are formed in porous TiO_2 used for dye-sensitized solar cells, silver nanoparticles and dye must interact with each other to enhance the light absorption coefficient.

For these reasons, we set ourselves the challenge to enhance the absorption coefficient of the light absorption layer of a dye-sensitized solar cell by carrying a silver nanoparticle in the nanopores of TiO_2, by using a colloidal solution of silver nanoparticles.

(1) Experimental methods

(a) Light absorption characteristics of ruthenium dye/silver nanoparticle (produced from colloidal solution) on a quartz substrate

To measure the light absorption spectrum of ruthenium dye/silver nanoparticle, polymer-modified silver nanoparticle was used. After the colloidal solution was dropped onto a quartz substrate and dried, the polymer-modified silver nanoparticles were heated to control the light absorption spectrum of the surface plasmon, and then an ethanol solution of the ruthenium dye was also dropped thereon to measure the light absorption characteristics. For comparison, ruthenium dye solution at the same concentration was dropped on quartz substrate to be measured in the same manner and the result was compared.

(b) Light absorption characteristics of ruthenium dye/silver nanoparticle (produced from colloidal solution) carried in the nanoporous film of titania

TiO_2 porous film was produced by forming a thin film on a quartz substrate, by using a TiO_2 paste with

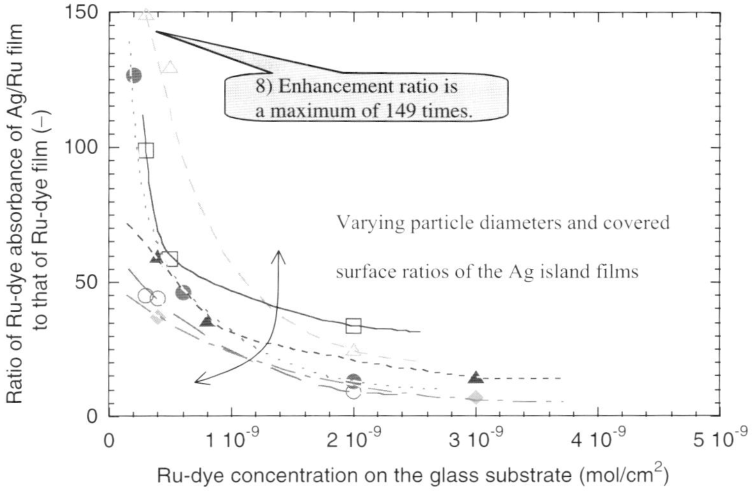

Figure 5.6
Light absorption enhancement ratio versus the dye concentration of Band B under a constant silver nanoparticle concentration [2].

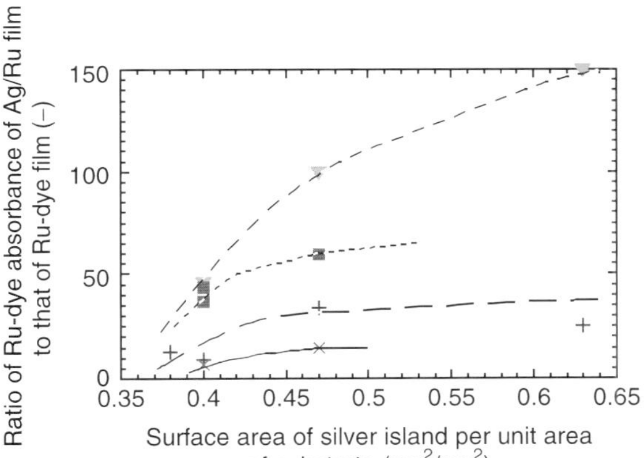

Figure 5.7
Surface area of the silver nanoparticle versus the absorption enhancement ratio of Band B [2], dye surface concentrations (×) of 3×10^{-9} mol/cm^2, (+) 2×10^{-9} mol/cm^2, (□) 4×10^{-10} mol/cm^2, and (▼) 3×10^{-10} mol/cm^2.

a spin coating method and then sintering at 450°C. The TiO_2 porous film was dipped in a solution of the polymer-modified silver colloid, dispersed in ethanol, to form a TiO_2 film carrying silver nanoparticles, and then subjected to heat treatment. By adjusting the heat treatment conditions such as time, the plasmon absorption spectrum of the silver were changed. Moreover, the ruthenium dye solution was dropped onto the silver/TiO_2 film and dried to measure the light absorption spectrum; and furthermore, compared with the sample dropped onto the TiO_2 film.

(2) Results and discussion

(a) Enhancement of the absorption coefficient of the ruthenium dye with silver nanoparticles (produced from colloidal solution) on the quartz substrate

Fig. 5.8 exhibits the light absorption spectra of the ruthenium dye/silver nanoparticle composite film, produced from the polymer-modified silver colloid solution. As well as in the case of the composite film, using the silver nanoparticles produced via a vacuum evaporation technique, enhancement of the light absorbance was observed in the case of the silver nanoparticle composite film produced from colloidal solution, relative to the case of the ruthenium dye only. The absorption peaks derived from the ruthenium dye and silver were separated by fitting the obtained spectrum data for the Lorenz-distribution function. Consequently, the light absorbance of the ruthenium dye was enhanced with the silver nanoparticle, a maximum of about 27 times.

(b) Enhancement of the absorption coefficient of the ruthenium dye with silver nanoparticles. (produced from colloidal solution) in the TiO_2 porous film

Fig. 5.9 exhibits the differential spectra of the light absorption spectra, before and after the dropping, where the ruthenium dye solution was dropped on each of the silver/TiO_2 films (A, B, and C) and the TiO_2 film. On the silver/TiO_2 films (A, B, and C),

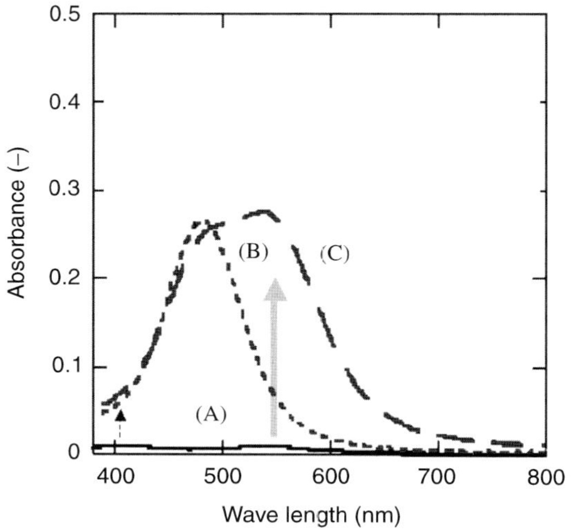

Figure 5.8
Light absorption spectra of heat-treated polymer-base modified silver nanoparticles, ruthenium dye, and ruthenium dye/silver nanoparticle composite film: (A) ruthenium dye (ruthenium surface concentration: 1×10^{-9} mol/cm^2); (B) silver nanoparticle; (C) ruthenium/silver nanoparticle composite film (ruthenium surface concentration: 1×10^{-9} mol/cm^2)

absorption peaks caused by the enhanced absorption of the dye at 540 nm were observed at around 513 nm, thus we succeeded in enhancing the light absorbance at around 513 nm a maximum of 4 times, or 14 times if the light absorbance of the silver nanoparticle is included.

Explained are the results whereby the light absorption of the ruthenium dye was enhanced with the silver nanoparticle produced on the quartz substrate via the vacuum vaporization technique or the colloidal solution for the purpose of enhancing the efficiency of the dye-sensitized solar cell, and that the light absorption of the dye in the TiO_2 porous film used for the dye-sensitized solar cell was also enhanced depending on the same effects.

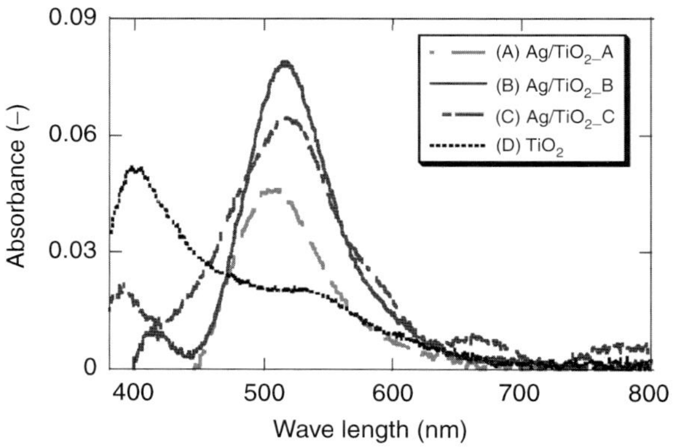

Figure 5.9
Differential spectra of the light absorption characteristics, before and after the dropping of the ruthenium dye solution in ruthenium dye/silver nanoparticle/TiO_2 composite film and ruthenium dye/TiO_2 composite film (ruthenium surface concentration: 8×10^{-10} mol/cm^2).

Based on these results, we will aim to further enhance the efficiency of "A dye-sensitized solar cell utilizing a metal nanoparticle" employing a silver nanoparticle/dye/titania hybrid film. It would be appreciated if you would refer to our findings in our reports to be issued.

References

[1] Nd. K. Nazeeruddin, R. Humphry-Baker, P. Liska and M. Gratzel: Investigation of sensitizer adsorption and the influence of protons on current and voltage of dye-sensitized nanocrystalline TiO_2 solar cell, *J. Phys. Chem. B*, **107**(34), 8981–8987 (2003).

[2] M. Ihara, K. Tanaka, K. Sakaki, I. Honma and K. Yamada: Enhancement of the absorption coefficient of *cis*-$(NCS)_2$bis(2,2′-bipyridy1-4,4′-dicarboxylate) ruthenium(II) dye in dye-sensitized solar cells by a silver island film, *J. Phys. Chem. B*, **101** (26), 5153–5157 (1997).

APPLICATION 6

6 DESIGN OF NANOPARTICLES FOR ORAL DELIVERY OF PEPTIDE DRUGS

Bioavailability of peptide-based pharmaceuticals is extremely low in general because these drugs can be easily decomposed by proteolytic enzymes existing in the body after their administration. Such a tendency becomes significant particularly if the pharmaceutical preparations are administered orally. Due to this nature, most peptide and protein pharmaceuticals are delivered by invasive routes such as subcutaneous injections, often leading to the poor patient compliance. Alternatively, oral drug delivery offers painless (non-invasive), self-administrable, inexpensive and convenient usage, and thus can be expected as a patient-friendly dosing system. Considerable research efforts have been thus invested in the development of oral drug delivery systems. Above all, use of particulate carriers, especially 'nanoparticulate' carriers has attracted much attention in the pharmaceutical field since the late 1970s in advance of recent progress on nanotechnologies [1–4].

1. Particulate design and functions

Principal functions required for nanoparticles as a carrier of oral peptide delivery are summarized in Fig. 6.1. Two major functions for the successful oral peptide delivery are discussed below.

(1) Drug loading

Loading of peptide drugs into carriers should be carried out so as to make the loading efficiency and the content high enough to ensure the therapeutic efficacy while it should not affect the chemical stability of peptide drugs (Fig. 6.1). The loading procedures can be classified into two categories, i.e., peptides can be loaded during (in situ loading) or after the preparation of carriers (pro-loading). The in situ loading often provides high drug content, while attention must be paid to a possible denaturation of peptide drugs if preparation conditions of the carriers require the use of organic solvents (or even aqueous solvents with large alteration of pH), high temperatures and/or high share forces. In contrast, the pro-loading is usually accomplished by simple soaking of the carrier into an aqueous peptide solution, so that it may have an ability to stably load the drug under a relatively mild condition. Nevertheless, the loading efficiency and/or the drug content may be limited to a certain extent if simple equilibrium partitioning without specific interaction between the drug and the carrier takes place.

(2) Behaviors of nanoparticles in the gastrointestinal tract

Orally administered nanoparticulate carriers should have an ability to protect the loaded peptides from the digestive enzymes, such as pepsin, in the stomach with acidic conditions while they do not release the loaded peptides (Fig. 6.1). Followed by passing through the stomach, the nanocarriers reach the small intestine that is a major absorption site for peptide drugs. Here, two major biological barriers for the drug absorption exist, i.e., the brush border membrane and mucus layer. Mucus is a fully hydrated viscoelastic gel (water occupies almost 95% of the contents under nonpathological conditions) overlying epithelial cell surfaces as a continuous gel blanket [5–7]. Its neo-genesis continuously takes place with relatively rapid turnover rate [8]. The thickness of human mucus varies depending on the anatomical location with a range of 1–400 μm (the average thickness is around 200 μm) [9]. Mucus consists of non-mucin components and mucin. The non-mucin components are IgA antibodies, enzymes, surfactants and free lipids. Mucin is defined as the stainable component of mucus. The major macromolecular component of mucin is a group of high-molecular weight glycoproteins; the polydisperse glycoprotein has a molecular weight in excess of 2×10^6 kDa, but can range from 2×10^5 to 15×10^6 kDa. Most mucus glycoproteins carry a net negative charge due to the presence of sialic acids and ester sulfates at the terminus of some sugar units. The brash border membrane is composed of epithelial monolayer of absorptive enterocytes that are conjugational through tight junctions. In addition, solitary lymphoid nodules are widespread along the entire intestine as oval aggregates in the antimesometrial gut wall, the so-called Peyer's patches [3]. The epithelial part of Peyer's patches is composed of highly specialized enterocytes, the membranous microfold (M) cells, having a function to collect various antigens from the gastrointestinal surface.

In order to overcome these physiological and morphological barriers against peptide delivery, two possible approaches have been considered. One is to provide muco-penetrative and/or muco-adhesive properties to drug carriers for prolonging their residence time at the absorption site. The particles with smaller size are likely to be diffusive in the mucus layer much higher, provided that diffusion of the particles in the mucus layer obeys the Stokes–Einstein equation. Compared with the microparticles, therefore, the nanoparticles can penetrate into the deeper zone of the mucus layer [10–12]. This is one of major reasons why nanoparticulate drug carriers have been attracted much attention for the purpose of oral peptide delivery. Modification of nanoparticle surface with muco-adhesive components may lead to further prolongation of the residence time at the absorption site. Drug release from the nanoparticles at the epithelial cell lining followed by penetration of the nanoparticles into the depths of the mucus layer can give rise to high drug concentration gradient across the intestinal membrane, possibly leading to the increased drug absorption by passive diffusion. Alternative approach is to directly traverse the nanoparticles themselves to the blood stream through the intestinal membrane. This approach makes it possible to protect peptide drugs from proteolytic degradation by digestive enzymes such as trypsin and chemotripsin existing in the intestinal mucus layer. The nanoparticulate carriers used for this purpose must be constructed with biocompatible and/or biodegradable materials since they are absorbed into the body. As transport pathways through the intestinal membrane, three possible routes have been suggested: (1) uptake by the Peyer's patches, (2) gap between epithelial absorptive enterocytes (paracellular route), (3) uptake by absorptive enterocytes (transcellular route) [3]. There are

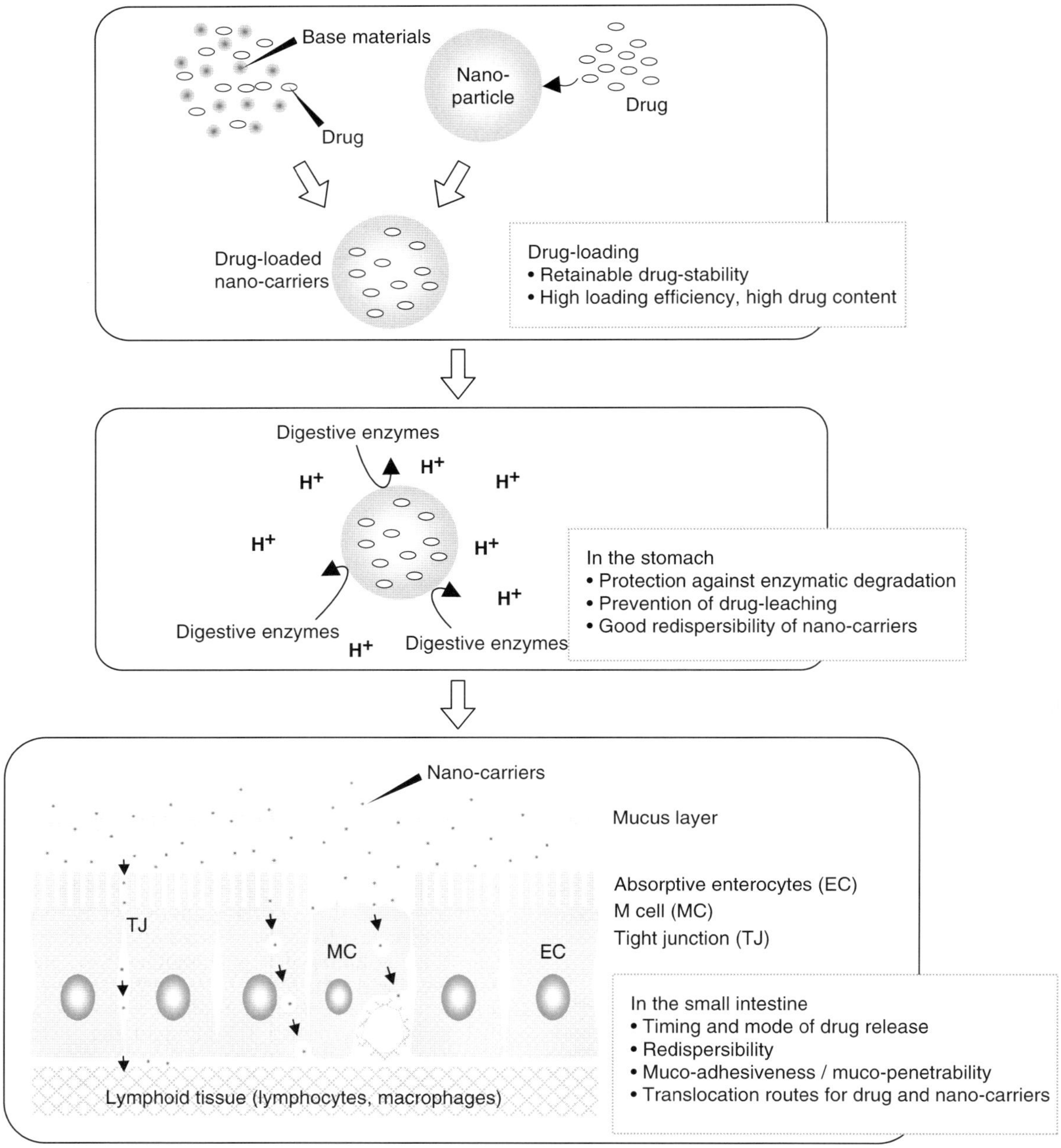

Figure 6.1
Principal functions required for nanoparticles as a carrier of oral peptide delivery.

several literatures investigating relation between particle size and the magnitude of particle uptake by enterocytes [13–15]. These studies suggest that the amounts of particles absorbed through the intestinal membrane tend to increase with decrease of the particle size. However, the observed extent of absorption, excluding certain peculiar cases, is shown to be only few percent of the dose. All the three major transport routes may contribute the intestinal absorption of the nanoparticles independently and/or mutually. Among these, the paracellular spaces sealed by thigh junctions, however, occupy less than 1% of the mucosal surface area. In addition, the pore diameter of these junctions was reported to be smaller than 10 Å [2]. Therefore, significant paracellular transport of particles is an unlikely event. Although certain treatments to enhance the opening of the tight junctions are thus necessary, their relevance for the in vivo situation is yet agnostic [2]. The transport mechanism of nanoparticles through the intestinal membrane is still phenomenalistic and factors governing the transport of nanoparticles are also unclear. Thus further systematic studies are necessary to clarify such translocation events.

2. Case studies

Because of its protective ability of peptide drugs from harsh gastrointestinal environments as well as its possible translocation in the intestinal membrane, the nanoparticulate carriers for oral peptide delivery are still being studied extensively. Excellent reviews in relation to this topic are readily available [1–4]. Historical research trends on typical nanoparticle-based formulations can be found in these reviews. Therefore, the present chapter will mainly survey recent reports published in the last decade (Table 6.1).

(1) Liposomes

Liposome is an artificial vesicle consisting of an aqueous core enclosed in one or more phospholipids bilayers. Since its discovery by Bangham, it has been one of the most extensively studied classes as DDS carriers for a number of drugs including peptides. Because its main component is phospholipids (lecithin) similar to biomembrane, the liposome possesses an excellent biocompatibility/biodegradability. This nature has significant advantage over other types of nanoparticulate drug carriers because toxicological issues can be less concerned even if the liposome is transported to the systemic circulation as a consequence of its intestinal absorption. Another feature of the liposome is its relatively large capacity of the inner aqueous phase. Since most of peptide- and protein-based drugs are water-soluble, this structural feature allows us to incorporate and retain these drugs in the liposomes stably and efficiently. For instance, in the cases of loading of insulin and calcitonin into the liposomes, high loading efficiencies around 80% were successfully achieved, depending on the liposomal formulations and preparation conditions [15].

Instability of the liposomes in the gastrointestinal tract has been concerned in effective oral peptide delivery, but it depends on the lipid composition. Use of a membrane stabilizing agent (cholesterol) [16] and phospholipids with a high gel–liquid crystalline transition temperature, and/or surface modification of liposome with polyethylene glycol [17] were shown to be effective to enhance the stability of liposome against gastric juice, bile and digestive enzymes. Surface modification of liposome is also proposed to enhance its residence time at the absorption site of the peptide drugs. For example, surface-coating of liposomes with muco-adhesive polymers such as chitosan and poly(acrylic acid) was reported to be effective to increase the intestinal calcitonin absorption due to the enhanced adhesion of liposomes to the intestinal mucus layer [18]. The use of lectins (proteins that recognize and bind to sugar complexes attached to most cell surface proteins and many lipids in cell membranes) as surface-modifying agents was also shown to promote the oral absorption of insulin because of the adhesive interaction between the lectin-modified liposomes and the intestinal absorptive cells [15].

(2) Polymeric nanoparticles

Nanoparticles constituted with polymeric materials can be usually fabricated by polymerizing monomeric materials as sources of the polymers by a variety of polymerization techniques, or nanoprecipitating the established polymers by certain methods followed by dissolving them in solvents (Table 6.1).

(a) Polyacrylates

Alkylcyanoacrylates and alkyl(meth)acrylates have been often used as monomers to prepare nanoparticles through ionic polymerization for the former and radical one for the latter. In particular, poly(alkylcyanoacrylates) have been extensively studied as the promising candidate materials of nanoparticulate drug carriers since 1980s because they have shown to have biodegradability. Among a number of researches regarding poly(alkylcyanoacrylates) nanoparticles, one of the most interesting results can be found in the report of Damge's group [19]. They prepared insulin-containing poly(isobutylcyanoacrylate) nanocapsules (the mean diameter was 220 nm) by the interfacial polymerization technique. It is worthwhile to note that oral administration of this insulin-loaded nanocapsules to diabetic rats showed surprisingly prolonged hypoglycemic effect lasted up to 20 days after the administration [19]. More recently, poly(isobutylcyanoacrylate) nanocapsules double-coated with Tween 80 and PEG20000 was shown to effective for brain targeting of hexapeptide (dalargin) via oral administration [20].

While most poly(alkyl(meth)acrylates) are practically non-biodegradable, a variety of (meth)acrylic monomers with different chemical properties are available and thus the combination of two or more different types of acrylic monomers enables us to freely design multi-functional nanoparticles. A representative example can be seen in nanospheres synthesized by radical polymerization of methacrylic acid (MAA) and poly(ethylene glycol) mono methacrylate (PEGMA) in the presence of a cross-linking agent (hereafter abbreviated as p(MAA-g-EG)) [21–27]. The p(MAA-g-EG) nanospheres are hydrogel nanoparticles exhibiting pH-responsive swelling through the formation of interpolymer complexes as a

Table 6.1
Characteristics of representative nanoparticulate carriers for oral peptide delivery.

Carrier type and base materials	Preparation method	Particle size	Surface-modifying agents or additives	Peptide drug	References
Liposomes					
Phosphatidylethanol based lipids	Thin-layer hydration/ ultrasonication	50–250 nm	None	Porcine insulin	Kisel et al. [47]
DSPC/DCP/Chol	Thin-layer hydration/ ultrasonication	470, 660, 4100 nm	Chitosan	Elcatonin	Takeuchi et al. [12]
Soyalecithin/Chol	Reverse evaporation	190 nm	WGA, TL, or UEA1-*N*-Glut-PE conjugates	Insulin	Zhang et al. [15]
DPPC/DCP or SA	Thin-layer hydration	N/A	Chitosan or Carbopol	Elcatonin	Takeuchi et al. [18]
DPPC/Chol/SA	Thin-layer hydration	348, 453, 479 nm	Cetyl-mucin, DSPE-PEG	Bovine insulin	Iwanaga et al. [17]
DPPC/Chol/DOPE-PEG	Thin-layer hydration	306 nm	None	rhEGF	Li et al. [48]
Poly(alkyl(meth)acrylates)					
P(IBCN)	Interfacial emulsion polymerization	220 nm	Miglyol core	Procine insulin	Damge et al. [19]
P(IBCN)	In situ ionic polymerization	85 nm	Pluronic acid	Human insulin	Mesiha et al. [49]
P(IBCN)	Ionic polymerization	100 nm	Tween-80 and PEG20000	Dalargin	Das and Lin [20]
P(MAA-g-EG)	Dispersion polymerization	300 nm	Pluronic F68	Bovine insulin	Donini et al. [22]
P(MAA-g-EG)	Dispersion polymerization	200–1200 nm	None	Salmon calcitonin	Lugo et al. [26]
P(AA-g-EG)	Dispersion polymerization	230–340 nm	Trehalose	Bovine insulin	Foss et al. [50]
P(NIPAAm)/P(MAA-g-EG)	Dispersion polymerization	170–350 nm	None	Vancomycin, bovine insulin	Ichikawa et al. [28, 29]
P(St)/hydrophilic polymer chains	Dispersion polymerization	400–1250 nm	None	Salmon calcitonin	Sakuma et al. [30]
Polyesters/polyanhydrides					
PLGA	Emulsion-solvent diffusion	250 nm	None	TRH	Kawashima et al. [32]
PLGA	Emulsion-solvent diffusion	240 nm	Chitosan, poly(acrylic acid)	Elcatonin	Kawashima et al. [33]
PLGA	Phase inversion	> 1 μm (80%)	FAO/Fe3O4	Insulin	Carino et al. [31]
PLGA	Solvent diffusion	171–315 nm	None	Salmon calcitonin	Yoo and Park [34]
PCL	Multiple emulsion evaporation	350 nm	Polycationic polymer (Eudragit RS)	Human insulin	Attivi et al. [35]

Dextran-PCL copolymer	Multiple emulsion evaporation	200 nm	None	BSA, B moLL, LC	Rodrigues et al. [37]
DEAPA-co-PVA-g-PLLA	Spontaneous self-assembling	200–500 nm	None	Recombinant human insulin	Simon et al. [38]
P(MVE/MA)	Solvent replacement	299 nm	PEG2000	N/A	Yoncheva et al. [36]
Polysaccharides					
Chitosan	Iontropic gelation	290 nm	Tripolyphosphate	N/A	Behrens et al. [39]
Chitosan	Iontropic gelation	418 nm, 504 nm	Tripolyphosphate	Insulin	Ma and Lim [44]
Chitosan	Iontropic gelation	269–688 nm	Tripolyphosphate	Porcine insulin	Ma et al. [40]
Chitosan	Precipitation/coacervation	643 nm	Sodium alginate	Ovalbumin	Borges et al. [45]
Chitosan/poly(gamma-glutamic acid)	Iontropic gelation	218 nm	None	N/A	Lin et al. [42]
Chitosan/CMKG	Polyelectrolyte complexation	330–900 nm	None	BSA	Du et al. [51]
Pectin	Iontropic gelation	589–1043 nm	Sodium chloride	Porcine insulin	Cheng and Lim [41]
Others					
Porous silicon dioxide	Liquid filling	N/A	Labrasol	Erythropoietin	Venkatesan et al. [46]
Carbon nanotube	Liquid filling	N/A	Labrasol, Gelucire, HCO-60, Casein	Erythropoietin	Venkatesan et al. [46]
Carbon nanohorn	Liquid filling	N/A	Labrasol	Erythropoietin	Venkatesan et al. [46]
Fullerene	Liquid filling	N/A	Labrasol	Erythropoietin	Venkatesan et al. [46]
Tripalmitin (lipid nanoparticle)	Double emulsion-solvent emulsification	538 nm	Chitosan	Salmon calcitonin	Fuentes et al. [52]
PC/NaTDC (Proliposome)	Sovent dropping evaporation	741 nm	Sorbitol	Salmon calcitonin	Song et al. [53]

Abbreviations: BmoLL, *Bauhinia monandra* lectin; BSA, bovine serum albumin; CMKG, carboxymethyl konjac glucomannan; Chol, cholesterol; DCP, dicetyl phosphate; DEAPA-co-PVA-g-PLLA, poly[(vinyl-3-(diethylamino)-propylcarbamate-co-(vinyl acetate)-co-(vinyl alcohol)]-graft-poly(L-lactic acid); DOPE-PEG, dioleoylphosphatidylethanolamine-poly(ethylene glycol) 2000; DPPC, L-alpha-dipalmitoylphosphatidylcholine; DSPC, L-alpha-distearoylphosphatidylcholine; DSPE-PEG, distearoylphosphatidylethanolamine-poly(ethylene glycol) 2000; FAO, fumaric acid oligomer; LC, *Lens culinaris* lectin; NaTDC, sodium taurodeoxycholate; *N*-Glut-PE, *N*-glutaryl-phosphatidylethanolamine; P(AA-g-EG), poly(acrylic acid-g-ethylene glycol); PC, L-alpha-phosphatidylcholine; PCL, poly(epsilon-caprolactone); PEG, poly(ethylene glycol); P(IBCN), poly(isobutylcyanoacrylate); PLGA, DL-lactide/glycolide copolymer; P(MAA-g-EG), poly(methacrylic acid-g-ethylene glycol); P(MVE/MA), poly(methyl vinyl ether-co-maleic anhydride); P(NIPAAm), poly(*N*-isopropylacrylamide); P(St), poly(styrene); rhEGF, recombinant human epidermal growth factor; SA, stearylamine; TL, tomato lectin; TRH, thyrotropin releasing hormone; UEA1, *Ulex europaeus* agglutinin 1; WGA, wheat germ agglutinin.

result of hydrogen bonding between etheric groups of the graft PEG chains and the proton of the carboxylic groups of the PMAA network. At low pH values corresponding to a gastric juice, the p(MAA-g-EG) nanospheres remain shrunken, while they swell to a high degree in neutral or basic media. Thus, the loaded peptides can be protected by the collapsed hydrogels from the acidic environment of the stomach, but released as the nanospheres pass from the stomach to the upper small intestine because the drastic pH changes causes the hydrogel swell [21–24]. Additionally, the p(MAA-g-EG) nanospheres were molecularly designed to contain PEG tethered chains promoting mucosal adhesion as well as a PMAA backbone with carboxylic pendant groups which can act as calcium binders leading to inactivation of certain calcium-dependent luminal enzymes [25] and epithelial tight-junction opening effect [26, 27]. Such multi-functions of the p(MAA-g-EG) nanospheres were aimed at realizing the enhanced intestinal absorption of peptide drugs such as insulin and calcitonin. In addition to such molecular design of poly(alkyl(meth)acrylates), composite particle structures of nanoparticles are achievable by elaborating polymerization techniques. As such, a novel composite nanoparticle consisting of a thermo-sensitive poly(*N*-ispropylacrylamide) (p(NIPAAm)) core having a layer of pH-sensitive p(MAA-g-EG) was prepared by the photo-initiated dispersion polymerization method [28, 29]. This core-shell nanoparticle revealed specific functions including the pro-loading of peptides at an aqueous environment with low temperature, the suppressed release of the loaded peptide in the acidic media and the sustained release of the loaded peptide in neutral pH [28, 29]. As another types of poly(alkyl(meth)acrylate) nanoparticles, hydrophobic polystyrene core particles having various hydrophilic polymer chains were proposed [30]. Comprehensive studies were carried out to evaluate the in vitro and in vivo performances of the proposed nanoparticles as an oral delivery device of calcitonin [30]. Through these studies, introduction of p(NIPAAm) chains to the polystyrene core was found to be particularly effective to obtain the calcition absorption-enhancing effect after the oral administration of the formulation to rats.

(b) Polyesters

Biodegradable polyesters, in particular poly(lactic acid), poly(lactic acid-co-glycolic acid) (PLGA) [31–34] and poly(caprolactone) (PCL) [35] have been widely used as a constitute of nanoparticles. The preparation of nanoparticles using these polyesters is accomplished with the methods based on solvent diffusion via an oil-in-water type emulsion. The use of an aqueous medium as a dispersing phase usually leads to low loading efficiency because most peptide drugs are water soluble and thus leached out from the polymeric phase readily. In order to improve the loading efficiency, therefore, several methods including the use of non-aqueous medium [32, 33] and the utilization of ionic complex formation between peptide and fatty acids [34] have been proposed so far. According to these methods, the drug content became two to ten times higher than that of the conventional methods, but the absolute drug content still remained to be few percent. Similar to the case of liposomes, there are several literatures indicating the effectiveness of non-covalently coating of PLGA nanoparticles with muco-adhesive polymers, e.g., poly(acrylic acid) and chitosan, for the improved intestinal absorption of peptide drugs. As other approaches, attempts were made to covalently attach the PEG chains to the PLGA or acid-anhydride polymers for providing muco-adhesive property [36], to introduce hydrophilic dextran to PCL nanoparticle with core–corona structure for further surface-modification with lectins [37], and to prepare self-assembled nanoparticles through the ion-pair formation between insulin and .the newly synthesized amphiphilic polyesters [38].

(c) Polysaccharides

In addition to synthetic polymers, natural polymers such as polysaccharides have also been widely investigated as a polymeric source of nanoparticulate carriers for drug delivery. Most polysaccharides are more or less soluble in aqueous media. Thus, polysaccharide-based nanoparticles can be prepared by a precipitation through ion-pair formation with certain additives (e.g., tripolyphosphate for chitosan [39, 40], calcium ion for pectin [41]) the so-called iontropic gelation. Addition of counter ionic polymers to polysaccharides is another useful way to make polysaccharide nanoparticles through the polyelectrolyte-based ionic complexation [42, 43]. Among polysaccharides, chitosan-based nanoparticles has been shown to be endocytosed by in vitro cultured cells [39, 40, 44] and also interactive with intestinal absorptive enterocytes and/or the Peyer's patches in vivo [39, 40]. Due to these properties, the chitosan-based nanoparticles were also utilized as a delivery system of protein (ovalbumin) for the purpose of oral vaccination [45].

(3) Others

Attempts have been made to utilize several nanoporous absorbents such as carbon-based nanomaterials (carbon nanotubes, carbon nanohorns, fullerene, etc.) and porous silicon dioxide (Sylysia 550) as carriers for oral peptide delivery [46]. In order to improve the oral absorption of erythropoietin (EPO), liquid filled nanoparticulate systems (LFNPs) were prepared by simply mixing these absorbents and EPO along with surfactants and enzyme inhibitors. The in vivo animal studies revealed that the intestinal absorption of EPO after the intra-jejunum administration of EPO-containing LFNPs was 5–20 times enhanced in comparison to that of the EPO solution. Among the series of LFNPs, the most significant absorption-enhancing effect was found in the LFNPs using carbon nanotubes; the bioavailability of EPO was 11.5%.

References

[1] E. Allemann, J.C. Leroux and R. Gurny: *Adv. Drug Del. Rev.*, **34**, 171–189 (1998).

[2] F. Delie: *Adv. Drug Del. Rev.*, **34**, 221–233 (1998).

[3] T. Jung, W. Kamm, A. Breitenbach, E. Kaiserling, J.X. Xiao and T. Kissel: *Eur. J. Pharm. Biopharm.*, **50**, 147–160 (2000).

[4] J. Kreuter: *Int. J. Pharm.*, **331**, 1–10 (2007).

[5] C. Marriott, N.P. Gregory: Mucus physiology and pathology in *Bioadhesive Drug Delivery Systems*, Eds. V. Lenaerts, R. Gurny, CRC Press, Inc., Florida, p. 1 (1990).

[6] X. Yang, J.R. Robinson: Bioadhesion in Mucosal Drug Delivery in *Biorelated Polymers and Gels*, Ed. T. Okano, Academic Press, U.S.A., pp. 135–192 (1998).

[7] R. Bansil, E. Stanley and T. LaMont: *Annu. Rev. Physiol.*, **57**, 635–657 (1995).

[8] C.M. Lehr, F.G.J. Poelma, H.E. Junginger and J.J. Tukker: *Int. J. Pharm.*, **70**, 235–240 (1991).

[9] S. Kerss, A. Allen and A. Garner: *Clin. Sci.*, **63**, 187–195 (1982).

[10] D.A. Norris, P.J. Sinko: *J. Appl. Polym. Sci.*, **63**, 1481–1492 (1997).

[11] N.N. Sanders, S.C.D. Smedt and J. Demeester: *J. Pharm. Sci.*, **89**, 835–849 (2000).

[12] H. Takeuchi, Y. Matsui, H. Sugihara, H. Yamamoto and Y. Kawashima: *Int. J. Pharm.*, **303**, 160–170 (2005).

[13] P.U. Jani, G.W. Halbert, J. Langridge and A.T. Florence: *J. Pharm. Pharmacol.*, **42**, 821–826 (1990).

[14] P.G. Jenkins, K.A. Howard, N.W. Blackhall, N.W. Thomas, S.S. Davis and D.T. O'Hagan: *J. Control. Release*, **29**, 339–350 (1994).

[15] N. Zhang, Q.N. Ping, G.H. Huang and W.F. Xu: *Int. J. Pharm.*, **294**, 247–259 (2005).

[16] Y. Aramaki, H. Tomizawa, T. Hara, K. Yachi, H. Kikuchi and S. Tsuchiya: *Pharm. Res.*, **10**, 1228–1231 (2003).

[17] K. Iwanaga, S. Ono, K. Norioka, K. Morimoto, M. Kakemi, S. Yamashita, M. Nango and N. Oku: *Int. J. Pharm.*, **157**, 73–80 (1997).

[18] H. Takeuchi, Y. Matsui, H. Yamamoto and Y. Kawashima: *J. Control. Release*, **86**, 235–242 (2003).

[19] C. Damge, C. Michel, M. Aprahamian and P. Couvreur: *Diabetes*, **37**, 246–251 (1988).

[20] D. Das, S. Lin: *J. Pharm. Sci.*, **94**, 1343–1353 (2005).

[21] H. Ichikawa, N.A. Peppas: in *New Trend in Polymers for Oral and Parental Administration*, Eds. G. Barratt, D. Duchêne, F. Fattal, J.Y. Legendre, De Sante, Paris, pp. 261–264 (2001).

[22] C. Donini, D.N. Robinson, P. Colombo, F. Giordano and N.A. Peppas: *Int. J. Pharm.*, **245**, 83–91 (2002).

[23] M.T. Lugo, N.A. Peppas: *J. Nanopart. Res.*, **4**, 73–81 (2002).

[24] D.N. Robinson, N.A. Peppas: *Macromolecules*, **35**, 3668–3674 (2002).

[25] F. Madsen, N.A. Peppas: *Biomaterials*, **20**, 1701–1708 (1999).

[26] M.T. Lugo, M. Garcia, R. Record and N.A. Peppas: *Biotechnol. Prog.*, **18**, 612–616 (2002).

[27] H. Ichikawa, N.A. Peppas: *J. Biomed. Mater. Res.*, **67A**, 609–617 (2003).

[28] H. Ichikawa, Y. Yamasaki and Y. Fukumori: in *New Trend in Polymers for Oral and Parental Administration*, Eds. G. Barratt, D. Duchêne, F. Fattal, J.Y. Legendre, De Sante, Paris, pp. 257–260 (2001).

[29] S. Haruna, H. Ichikawa and Y. Fukumori: Proceedings of the 2nd Asian particle technology symposium 2003, Malaysia, 277–284 (2003)

[30] S. Sakuma, M. Hayashi and M. Akashi: *Adv. Drug Del. Rev.*, **47**, 21–37 (2001).

[31] G.P. Carino, J.S. Jacob and E. Mathiowitz: *J. Control. Release*, **65**, 261–269 (2000).

[32] Y. Kawashima, H. Yamamoto, H. Takeuchi, T. Hino and T. Niwa: *Eur. J. Pharm. Biopharm.*, **45**, 48 (1998).

[33] Y. Kawashima, H. Yamamoto, H. Takeuchi and Y. Kuno: *Pharm. Dev. Tech.*, **5**, 77–85 (2000).

[34] H.S. Yoo, T.G. Park: *J. Pharm. Sci.*, **93**, 488–495 (2004).

[35] D. Attivi, P. Wehrle, N. Ubrich, C. Damge, M. Hoffman and P. Maincent, *Drug Dev. Ind. Pharm.*, **31**, 179–189 (2005).

[36] K. Yoncheva, E. Lizarraga and J.M. Irache: *Eur. J. Pharm. Sci.*, **24**, 411–419 (2005).

[37] J.S. Rodrigues, N.S. Santos-Magalhaes, L.C.B.B. Coelho, P. Couvreur, G. Ponchel and R. Gref: *J. Control. Release*, **92**, 103–112 (2003).

[38] M. Simon, M. Wittmar, U. Bakowsky and T. Kissel: *Bioconjug. Chem.*, **15**, 841–849 (2004).

[39] I. Behrens, A.I.V. Pena, M.J. Alonso and T. Kissel: *Pharm. Res.*, **19**, 1185–1193 (2002).

[40] Z. Ma, T.M. Lim and L.Y. Lim: *Int. J. Pharm.*, **293**, 271–280 (2005).

[41] K. Cheng, L.Y. Lim: *Drug Dev. Ind. Pharm.*, **30**, 359–367 (2004).

[42] Y.H. Lin, C.K. Chung, C.T. Chen, H.F. Liang, S.C. Chen and H.W. Sung: *Biomacromolecules*, **6**, 1104–1112 (2005).

[43] J. Du, R. Sun, S. Zhang, L.F. Zhang, C.D. Xiong and Y.X. Peng: *Biopolymers*, **78**, 1–8 (2005).

[44] Z. Ma, L.Y. Lim: *Pharm. Res.*, **20**, 1812–1819 (2003).

[45] O. Borges, G. Borchard, J.C. Verhoef, A. de Sousa and H.E. Junginger: *Int. J. Pharm.*, **299**, 155–166 (2005).

[46] N. Venkatesan, J. Yoshimitsu, Y. Ito, N. Shibata and K. Takada: *Biomaterials*, **26**, 7154–7163 (2005).

[47] M.A. Kisel, L.N. Kulik, I.S. Tsybovsky, A.P. Vlasov, M.S. Vorob'yov, E.A. Kholodova and Z.V. Zabarovskaya: *Int. J. Pharm.*, **216**, 105–114 (2001).

[48] H. Li, J.H. Song, J.S. Park and K. Han: *Int. J. Pharm.*, **258**, 11–19 (2003).
[49] M.S. Mesiha, M.B. Sidhom and B. Fasipe: *Int. J. Pharm.*, **288**, 289–293 (2005).
[50] A.C. Foss, T. Goto, M. Morishita and N.A. Peppas: *Eur. J. Pharm. Biopharm.*, **57**, 163–169 (2004).
[51] J. Du, S. Zhang, R. Sun, L.F. Zhang, C.D. Xiong and Y.X. Peng: *J. Biomed. Mater. Res.*, **72B**, 299–304 (2005).
[52] M.G. Fuentes, D. Torres and M.J. Alonso: *Int. J. Pharm.*, **296**, 122–132 (2005).
[53] K.H. Song, S.J. Chung and C.K. Shim: *J. Control. Release*, **106**, 298–308 (2005).

7

APPLICATION 7

FORMATION OF THICK ELECTRONIC CERAMIC FILMS WITH BONDING TECHNIQUE OF CRYSTALLINE FINE PARTICLES AND THEIR APPLICATIONS

In connection with the developments of information technology (IT), thick electronic ceramic films have become important constitutive materials for various IT devices and systems. For instance, good piezoelectric or ferromagnetic thick films having thickness from several microns to several hundreds microns are required in MEMS micro-actuator devices, optical devices and high-frequency devices. It is well known that good electronic ceramic films whose thickness is approximately less than 1 μm can be formed by means of physical or chemical methods including sputtering and CVD. On the other hand, ceramic bulks with good electrical or magnetic properties are obtained by sintering.

For forming the electronic ceramic films with several microns to several hundreds microns thickness located between thin films and bulk materials, the above fabrication methods become less useful unfavorably. To overcome this situation, Akedo et al. [1] recently showed that aerosol deposition (AD) of crystalline fine particles onto substrate is an effective method to form the thick electronic ceramic films. Fundamental properties and applications of AD method have been studied as NEDO project (Nano-level electronic ceramics low temperature formation and integration technology), in which the AD-formed magnetic garnet, PZT and PLZT thick films have also been investigated by the authors for applying them in optical devices.

In this article, the AD formation of thick electronic ceramic films and their applications in optical devices are described mostly based on their results.

1. Aerosol deposition method (ADM)

Fig. 7.1 shows the fundamental setup for obtaining thick ceramic films by aerosol deposition method (ADM). Crystalline fine particles whose diameter is approximately submicron are introduced in a powder vessel so as to obtain their aerosol with an inert gas such as nitrogen. The aerosol is then led into a vacuum chamber kept at a pressure of several Pa, and is sprayed directly onto a substrate from the slit-shaped nozzle as high-speed jet. The film formation mechanism is now called as room-temperature shock-compaction phenomenon, where thick and high-density films are considered to be formed by breaking crystalline fine particles via impact and their local activation through mechano-chemical reaction. In fact, the film deposition rate of ADM is very fast in comparison with those of other existing techniques, and the density of thus obtained film reaches 90% of bulk materials.

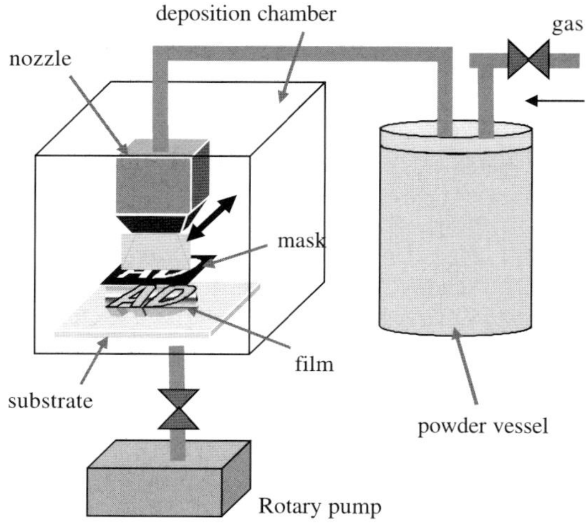

Figure 7.1
Fundamental set-up for aerosol deposition (AD).

2. Formation of thick electronic ceramic films with ADM

(1) Thick magnetic garnet films

Magnetic garnet materials such as yttrium iron garnet ($Y_3Fe_5O_{12}$; YIG), known as transparent ferromagnetic materials, are useful and key magneto-optical (MO) media in optical isolator devices for optical communication and spatial light modulators for holographic data storage.

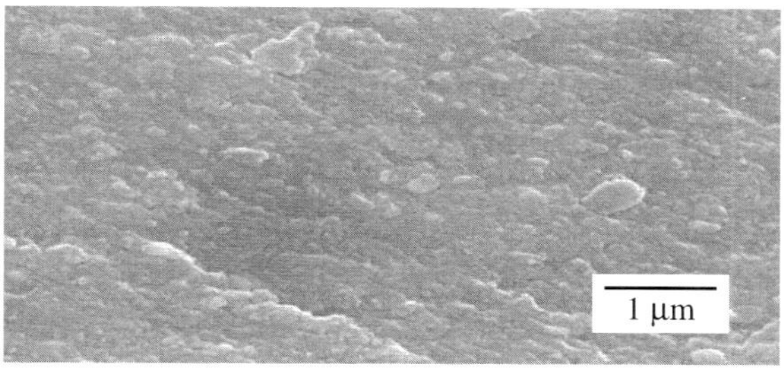

Figure 7.2
Cross-sectional image of the AD-formed thick Bi:YIG film.

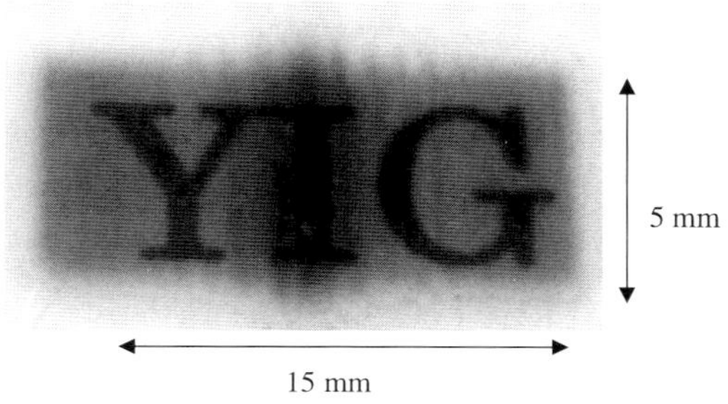

Figure 7.3
Surface photograph of the AD-formed thick Bi:YIG film. The word "YIG" which was placed behind the film is seen.

To obtain thick magnetic garnet films, Bi-substituted YIG ($Bi_{0.5}Y_{2.5}Fe_5O_{12}$; Bi:YIG) films were formed by ADM with crystalline Bi:YIG fine particles whose average diameter was approximately 500 nm. As shown in Fig. 7.2, ADM enabled us to form thick Bi:YIG film with considerably high density. The as-deposited film was transparent as seen in Fig. 7.3. In this case, the film deposition rate was more than 1 μm/min, which is very fast and satisfactory for practical applications, although the crystallographic structure of thus obtained films was found to be deteriorated from that of the starting Bi:YIG fine particles, mainly due to the introduction of strains through the impact of fine particles. Favorably, such strains could be released and crystallographic structure was recovered after post annealing. As a result, the post-annealed Bi:YIG film showed magneto-optical responses similar to those of sputtered films (Fig. 7.4).

(2) Multiferroic ceramic thick films

ADM enables us to form thick ceramic films at a relatively low temperature. This leads to a natural consequence that the technique will be useful for obtaining composite multiferroic ceramic films, mixture of ferromagnetic and ferroelectric materials. To obtain such a composite film, difference in formation temperatures of the ferromagnetic and ferroelectric

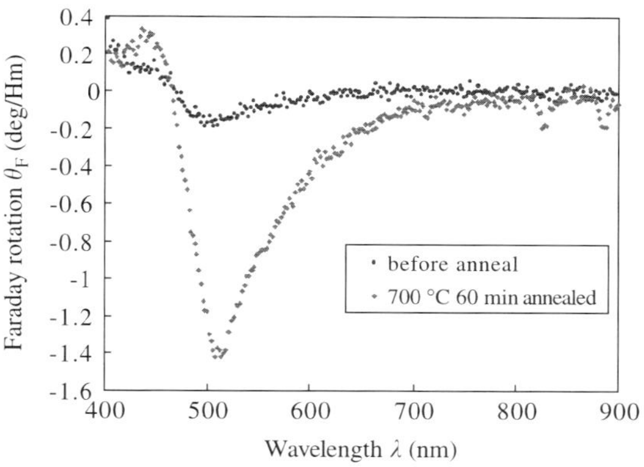

Figure 7.4
Wavelength spectra of magneto-optical Faraday rotation angles of the AD-formed thick Bi:YIG films before and after the post annealing.

phases was always a problem, because the difference easily led to the formation of undesired phases and the resultant film normally showed poor ferromagnetic and ferroelectric properties.

ADM inherently has no such problem because of the use of crystalline particles whose crystallographic structures are the same as those of the objective films. In this respect, the authors fabricated a Bi:YIG/PZT composite thick film with ADM. Fig. 7.5 shows the X-ray diffraction chart of thus obtained films, indicating clear coexistence of Bi:YIG and PZT perovskite phases. The film was identified as ferromagnetic through VSM measurements where the saturation magnetization of film was similar to that of Bi:YIG single film. At the moment, ferroelectric properties of film have been studied in detail. If the ferromagnetism and ferroelectricity are both maintained at room temperature, strong coupling between the magnetization and electric polarization is also expected. This kind of thick multiferroic films are very attractive because the magnetization (electric polarization) of the film can be controlled by the application of electric (magnetic) field.

3. Applications of AD ceramic films

(1) Magneto-optic spatial light modulators

A novel type of magneto-optic spatial light modulators (MOSLM) are now under investigation by utilizing thick AD PZT films. SLMs are optical micro-devices for modulating amplitude or phase of traveling optical waves, and are now widely used in projectors, for instance. Recently, associated with the developments of holographic memory, SLMs with faster operation speed are required for ensuring the high data transfer speed of the memory. To meet the requirements, the authors have been developing the MOSLM, where optical wave is

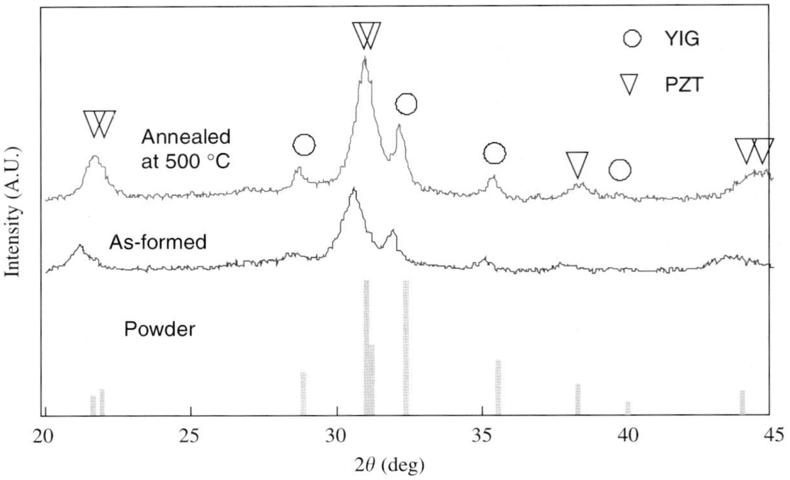

Figure 7.5
XRD patterns of Bi:YIG, starting powder, as-formed and after annealing.

modulated via MO effect corresponding to the direction of magnetization. The switching of magnetization takes place within several nanoseconds, and hence the MOSLM is essentially very fast.

Normally, magnetization direction is controlled by a magnetic field produced by currents. This is, however, always associated with a heat problem, and alternative method for controlling the direction of magnetization in MOSLM is needed instead of the current drive. To solve this problem, piezoelectric film was employed for applying stresses to designated magnetic pixels. Because of the inverse magnetostriction, the applied stresses are equivalent to effective magnetic field, which attributes to the control of magnetization. Based upon this consideration, the authors constructed a MOSLM with the AD PZT film, as shown in Fig. 7.6. The device was tested by applying a voltage to the PZT film for controlling the direction of magnetization (Fig. 7.7). In fact, magnetizations in the voltage-applied pixels switched and the contrast of these pixels was clearly changed. To our knowledge, this is the first MOSLM that is driven by piezoelectricity or voltage.

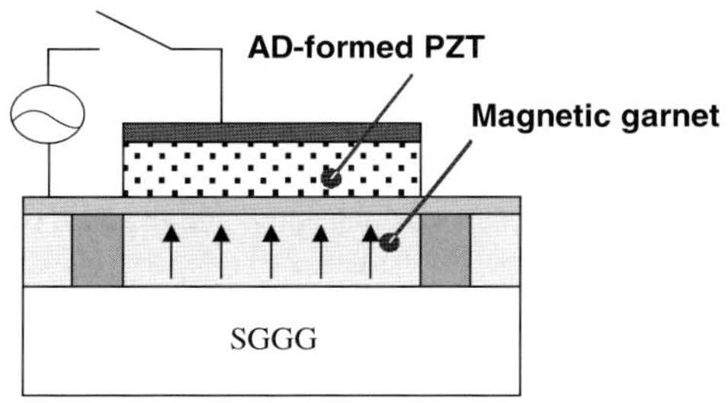

Figure 7.6
Fundamental configuration of MOSLM in which the magnetization in pixel is switched by applying a voltage to the AD-formed PZT film.

(2) PLZT multimode optical waveguides

To realize recent network design such as "fiber to the home (FTTH)", low-price and small optical switches are desired for consumer use. To meet this requirement, a novel optical cross-connect system is now under development by combining holographic optical filter and optical speckle pattern modulation. In this system, optical waves are controlled in multimode optical waveguides possessing the electro-optic (EO) effect, and AD-formed PLZT thick films were used for such waveguides. As shown in Fig. 7.8, good PLZT thick films showing low optical absorption at 1.5 μm wavelength of light were obtained by controlling the preparation conditions during the aerosol deposition. Fig. 7.9 is a cross-sectional image of the optical waveguide composed of thus obtained PLZT film. As seen in the figure, dense and thick PLZT optical waveguide has been formed. By utilizing this waveguide, the optical cross-connect system is now being developed.

The work was supported in part by NEDO nanolevel electronic ceramics low temperature formation and integration project and MEXT novel technology development project. The author expresses his sincere thanks to Hosokawa Micron Corp. which supplied the Bi:YIG fine particles for the project.

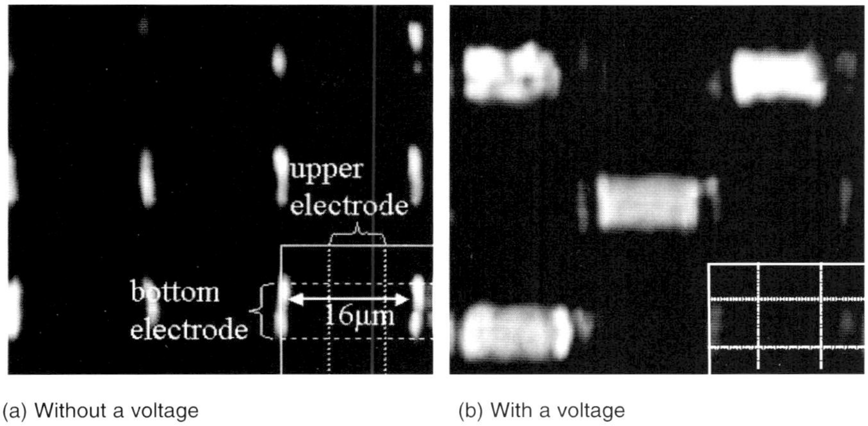

Figure 7.7
Pixel drive by applying a voltage to the AD-formed PZT film.

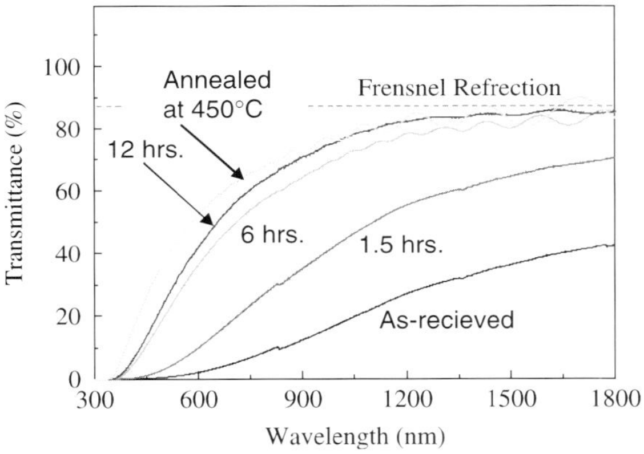

Figure 7.8
Wavelength spectra of transmissivity of AD-formed PLZT films.

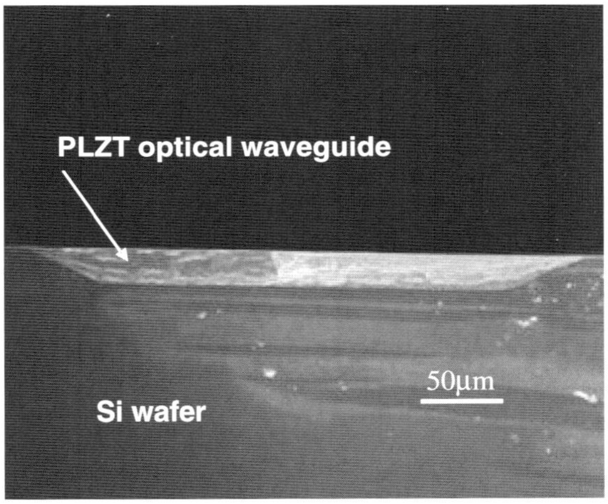

Figure 7.9
Cross-sectional image of the optical waveguide with AD-formed PLZT film.

Reference

[1] J. Akedo, M. Lebedev, *J. Crystal Growth*, **235**, 415–420 (2002).

APPLICATION 8

8 DEVELOPMENT AND MULTI-FUNCTIONALIZATION OF HIGH-FUNCTIONAL SEPARATION MEMBRANES

A number of studies have been carried out on nanostructured separation membranes, and gas and liquid separation technologies have been developed. These can be classified as shown in Fig. 8.1 [1] according to principles and scopes of application.

1. Gas separation

Gas separation membranes can be classified into non-porous membranes like high molecular membranes, and porous membranes like inorganic membranes.

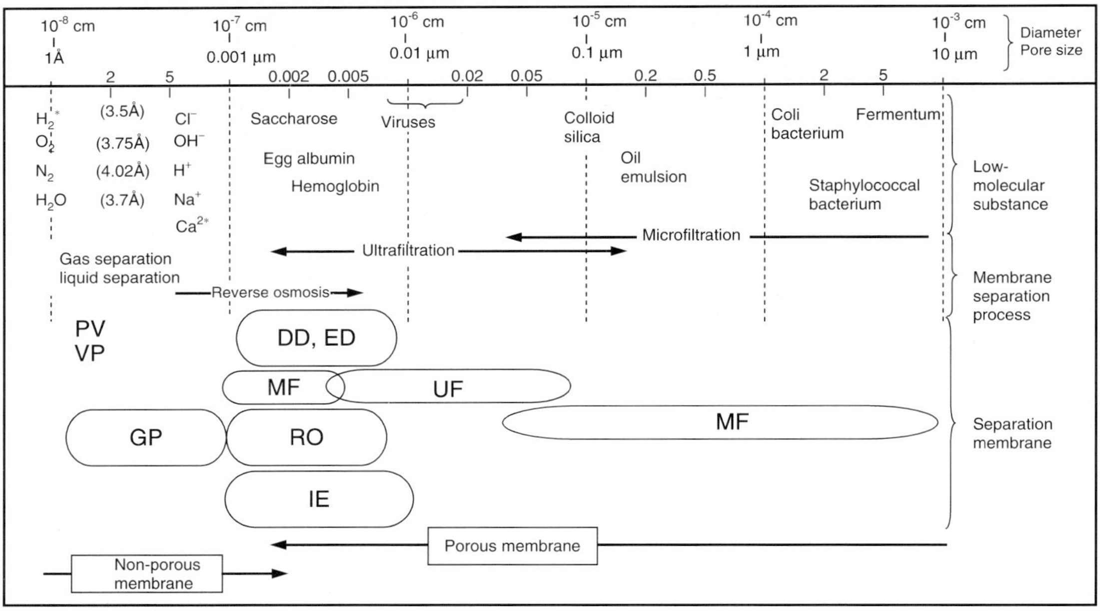

GP; gas permeation
VP; vapor permeation
ED; electrodialysis
NF; nanofiltration
MF; microfiltration
PV; pervaporation
DD; dialysis
RO; reverse osmosis
UF; ultrafiltrationtion

Figure 8.1
Principles and scopes of application.

Among high molecular membranes, hydrogen separation membranes are drawing attention as CO_2 separation membranes, and Research Institute of Innovative Technology for the Earth (RITE) [2] is developing a hollow-filament type polyimide membrane module. Inorganic separation membranes can be further classified into non-porous dense membranes such as Pd membranes, fine porous membranes utilizing diffusion velocity or molecular sieves such as γ-alumina and zeolite-silica, and facilitated transport liquid membranes using carriers.

As high-functional membranes with modified porous membranes, which are superior to conventional separation membranes, Iwamoto et al. [3–5] have developed highly selective hydrogen separation membranes using Ni nanoparticles. As shown in Fig. 8.2, Ni nanoparticle-dispersed silica membranes were synthesized on alumina porous substrates. The diameter of pores on the top surface was controlled to 3–5 nm by the formation method on the spot using a chemical solution deposition.

The hydrogen permeability at 500°C using the membranes was found to be of the order of 10^{-7} (mol/m^2 s Pa) and it was confirmed that this value was

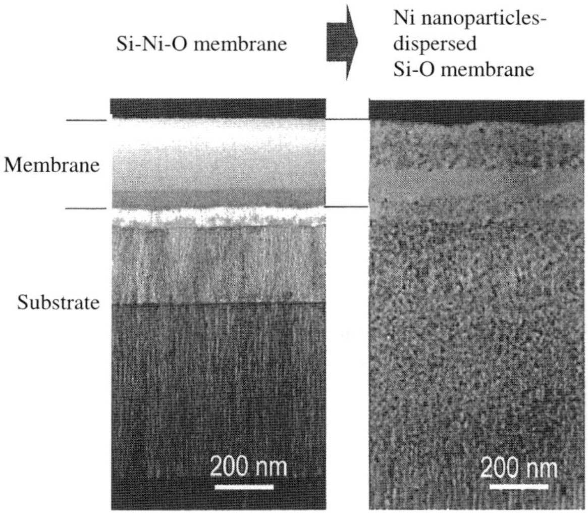

Figure 8.2
Structure of highly selective hydrogen separation membrane.

about five times higher than even an unachieved helium permeability when a conventional molecular sieving function was used (Fig. 8.3).

Translucent porous glass with fine pore diameters controlled to the nanometer scale is used for separation membranes and catalysts, due to its characteristics.

Kuraoka et al. [6] produced inorganic–organic hybrid solid electrolytic membranes by introducing organic silicon compounds with conductive functional groups to the inside of pores and the surface of porous glass (Fig. 8.4). As a result, development work is now being carried out to produce membranes expected to provide new solid electrolytic membranes. The new membranes have the heat resistant, anti-organic solvent and anti-swelling properties of glass medium. In addition to these properties, the new membranes have high conductivity thanks to conductive functional groups, and also being usable under severe conditions where conventional solid electrolytic membranes cannot be used.

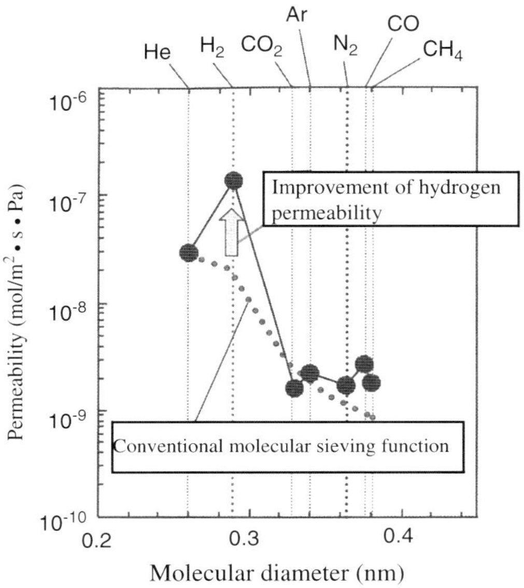

Figure 8.3
Characteristics of highly selective hydrogen separation membrane.

A remarkable increase in the contact area of photocatalyst materials and decomposed materials can be achieved by loading and foaming materials with photocatalystic functions in the pores of porous glass membranes without hindering translucency (Fig. 8.5). The membrane permeation system also facilitates very efficient decomposition. Studies are being carried out on this type of high-functional membranes with extremely high photo-oxidation functions [7].

Various studies have also been conducted on gas separation membranes using inorganic–organic nanohybrid membranes [8–10]. Gomes et al. [8] produced nano-hybrid membranes by sol–gel copolymerization by adding various organic alkoxysilane and tetraethoxysilane to a tetrahydrofuran solution of PTMSP (Poly (1-trimethylsilyl-1-propyne)) and evaluated the permeability and selectivity of the membranes. As a result, they found that the conversion rate of silica in the membranes was high and that both the permeability of butane and the butane/methane selectivity were simultaneously high when the diameter of particles was 20–40 nm.

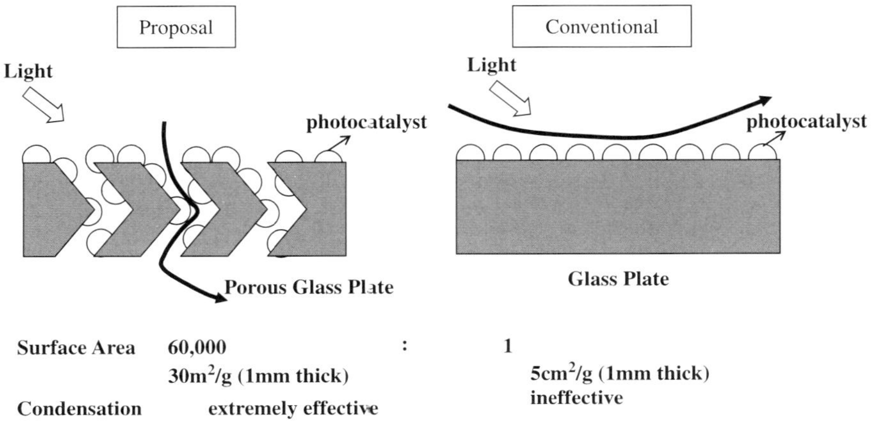

Figure 8.4
Characteristics of porous glass plate supported by photocatalyst.

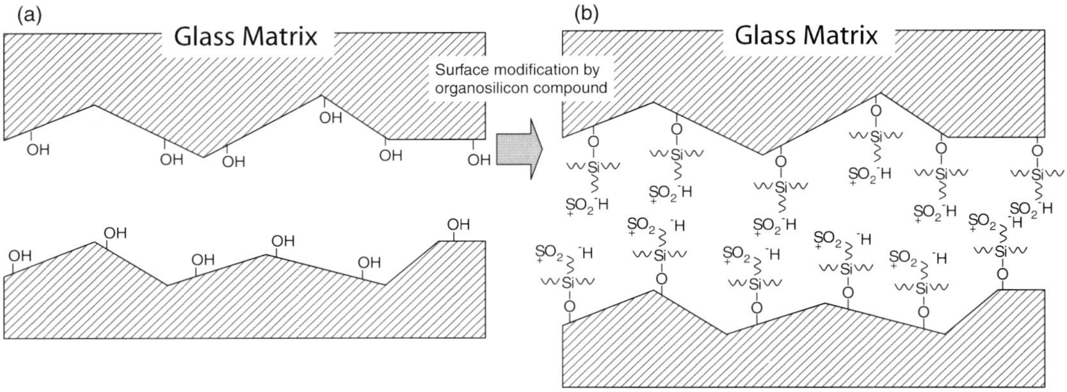

Inorganic-organic hybrid solid electrolytic membrances are produced by introducing organic silicon compounds with conductive functional groups to the inside of pores and the surface of porous glass.

Figure 8.5
Surface modification in pores of porous glass.

Table 8.1
Characteristics and application of NF membrane.

Characteristics	Sodium chloride rejection: 10–90% (low rejection, middle rejection) Surface charging of membrane: charged species (+,−), charge population Sodium chloride rejection monovalent ion < multiply-charged ion Organics fraction: molecular weight cutoff – 200–1,000 Permeability coefficient of purified water: high permeability coefficient, low pressure operation		
Application	Water purifying	Drainage treatment	Food and beverage
	Softening Color removal Removal of natural organics Removal of precursor of disinfection byproduct Removal of harmful inorganic ion (arsenic ion, sulfate ion) Removal of synthetic chemicals (agrichemicals, interfacial active agent)	BOD removal COD removal Nitrogen elimination Phosphorus elimination Dye removal Color removal Partial dechlorination Recovery of valuable resource	Demineralization Condensation Fractionation Development of flavor Decolorization

2. Liquid separation

We introduce the characteristics of nanofiltration (NF) membranes and studies on NF membranes. NF membranes belong to a new domain among separation membranes that include reverse osmosis (RO) membranes, NF membranes, ultrafiltration (UF) membranes and microfiltration (MF) membranes.

Both RO membranes and NF membranes are driven by pressure difference. RO membranes are used for separating dissolved substances with molecular weight of several hundreds, while NF membranes are used for separating dissolved substances with molecular weight of several thousands. In addition, NF membranes are different in characteristics from RO membranes, as shown in Table 8.1 [11].

Table 8.2
Various NF hybrid processes.

Hybrid System	Municipal	Process	Wastewater	Food and dairy
NF + AC (of FAC)	Pesticide removal micropollutants		Micropollutants dioxin dye removal	
NF + IEX	Nitrate removal		Nitrate removal	
MF/UF + NF	Satellite® high-quality water from main	Tailor-made process water	Water mining	Fermentation products concentration and purification
(MF/UF) +NF + RO	Seawater desalination	Tailor-made process water high recovery system		Fractionation and concentration
RO + NF	High recovery seawater desalination	High recovery system dual-purpose process water	Dumpsite leachate treatment	Multistage RO combined system (MRC system)
(Bioreactor, chemical oxidation) + (UF) + NF	Highly polluted surface water treatment		Advanced wastewater treatment	
NF + ED (EDR)		High recovery system	Caustic recovery pulp bleaching effluent	
Complexation + NF			Heavy metal removal	Coffee concentration
NF + PV	VOC removal		VOC removal	High concentration

One of the significant characteristics of NF membranes is that they have an electric charge. Therefore, high differential characteristics are required, such as the separation of salt and organic substances and the separation of monatomic ions and diatomic ions.

With regard to NF membranes, Maeda et al. [12] took a broad survey of studies and summarized that high-functional performance could be achieved by hybridization (Table 8.2). Inoue et al. [13] developed new membranes (DKL-3000X, DKL-4000X) with the same performance as conventional membranes even at an ultra-low pressure of 0.3 MPa. These membranes were found to exhibit a high removal rate of humic acid, or the precursor of trihalomethane.

Large-scale NF membrane plants are already in operation in the USA and Europe. NF membranes seem highly promising as membranes for practical use in the future and can be used for sophisticated water treatment, capable of removing the precursor of trihalomethane, agrichemicals, environmental hormones, etc., from tap water.

References

[1] K. Matsumoto: *Yuzanotameno Jituyou Maku Bunnri Gijutsu (Practical Membrane Separation Technology for Users)*, The Nikkan Kogyo Shimbun Ltd., Tokyo, Japan, p. 9 (1996).
[2] RITE: http://www.rite.or.jp/.
[3] Y. Iwamoto: Proceedings of meeting for reading research papers in Japan Fine Ceramics Center, pp. 18–19 (2004).
[4] K. Inada, Y. Iwamoto and Y. Hirayama: Proceedings of meeting for reading research papers in Japan Fine Ceramics Center, p. 28 (2004).
[5] T. Kato, K. Inada, Y. Iwamoto, T. Hirayama and Y. Ikuhara: Proceedings of annual meeting of the ceramic society of Japan, p. 323 (2004).
[6] K. Kuraoka, T. Kikukawa and T. Yazawa: *Chem. Ind.*, **57**(1), 41–44 (2004).
[7] T. Yazawa, N. Kubo: Proceeding of the 22nd annual meeting of the membrane society of Japan, p. 46 (2000).
[8] D. Gomes, S.P. Nunes and K.V. Peinemann: *J. Memb. Sci.*, **246**, 13–25 (2005).
[9] S.P. Nunes, K.V. Peinemann, K. Ohlrogge, A. Alpers, M. Keller and A.T.N. Pires: *J. Memb. Sci.*, **157**, 219–226 (1999).
[10] P.A. Sermon and F.P. Getton: *J. Sol-Gel. Sci. Technol.*, **33**(1) (2005).
[11] K. Matsumoto: *J. High Pressure Gas Safety Inst. Jpn*, **39**, 8 (2002).
[12] K. Maeda: *Membrane*, **23**(5), 235–244 (1998).
[13] G. Inoue, M. Murakami and Y. Fusaoka: *Membrane*, **26**(5), 231–233 (2001).

APPLICATION 9

9 DEVELOPMENT OF POLYMER-CLAY NANOCOMPOSITES BY DISPERSION OF PARTICLES INTO POLYMER MATERIALS

Synthetic resin and rubber are widely used for diverse articles from automobile parts to daily commodities. It seems to be considerably attributable to the improvement in strength, rigidity, and heat resistance of synthetic resin and rubbers by composing glass fibers or fillers represented by carbon in addition to the development of various synthetic resins and rubbers suitable for individual applications. With much attention focused on the function of intercalation of montmorillonite, a kind of clay, the uniform dispersion of its basic unit layer called "silicate layer" in synthetic resin or rubber has been investigated.

The crystalline structure of montmorillonite consists of laminated silicate layers made up of a tetrahedral silica layer, octahedral alumina layer, and another tetrahedral silica layer. These silicate layers are sheet-like with a thickness of 1 nm and a side length of 100 nm. If this silicate layer could be dispersed in synthetic resins, it could be expected to work literally as "molecular-size filler."

In this section, nylon 6–clay hybrid (nanocomposite) and polypropylene–clay hybrid, in which the silicate layer could be successfully dispersed uniformly in synthetic resins, are introduced [1] mentioning their recent applications to rubbers and alloys.

1. Nylon 6–clay hybrid

(1) Polymerization after intercalation of monomer

First the silicate layer was dispersed by polymerization of ε caprolactam as a monomer of nylon 6 between the silicate layers of montmorillonite. After ion exchange with ammonium salt of 12-amino dodecanoic acid ($H_2N(CH_2)_{11}COOH$) and mixing with melted ε caprolactam (melting point: 70°C), the ring-opening polymerization of ε caprolactam took place from the starting points of 12-amino dodecanoic acid at 250°C. As the polymerization proceeds, the layer clearance increases up to over 10 nm and the Nylon 6–clay hybrid (NCH) with the silicate layers of montmorillonite dispersed in nylon 6 [2].

(2) Properties of NCH

(a) Physical properties

Table 9.1 compares the properties of NCH with those of nylon 6. The NCH has 1.5 times higher tensile strength and about double the elasticity of nylon 6 and its heat distortion temperature is 152°C, an increase of 80°C compared with nylon 6 with the addition of 4.2% montmorillonite.

(b) Gas barrierability

Gas barrierability is one of the new functions obtained by combining the silica layer of montmorillonite as molecule-sized filler. On the formation of NCH film, the sheet-like silicate layers are oriented parallel to the film surface, which was observed by SEM of the film cross section cut by oxygen plasma etching (Fig. 9.1). As the molecules such as water and oxygen penetrate circumventing the silicate layer, the gas barrierability of NCH is higher than that of nylon 6 compared at the same thickness. For example, the diffusion coefficient of water reduced down to a half by adding only 2% montmorillonite. Furthermore, since the increase of barrierability is attributable not to chemical interaction but to the circumventing effect by the silicate layer, this function is expected regardless of the kind of gas or liquid.

(c) Intercalation of nylon between the silicate layers

Besides nylon 6, there are several kinds of nylon resins like nylon 66, nylon 610, nylon 12, nylon 11,

Table 9.1
Properties of NCH.

Characteristics	Unit	NCH2*	NCH5*	Nylon 6
Clay content	wt%	1.6	4	0
	vol%	0.6	1.6	0
Tensile strength	MPa	76.4	97.2	68.6
Temperature of thermal deformation	°C	118	152	65
Gas barrier (Permeability of hydrogen)	$\times 10^{-11}$ $cm^3 \cdot cm \cdot cm^{-2} \cdot s \cdot cmHg^{-1}$	1.79	1.28	2.57
Impact intensity (Charpy, without notch)	kJ/m^2	102	52.5	>150

Note: The figures 2 and 5 before * in the table denote the amount of feed of organificated clay on the synthesis.

Figure 9.1
Cross-sectional SEM picture of NCH film (after oxygen plasma etching for 30 s): (a) 0.5% clay, (b) 1.85% clay, (c) 5.0% clay.

and nylon 46. The synthesis of NCH was realized by polymerizing monomer between the layers of montmorillonite, but this method was difficult to apply to other kinds of nylon resins. However, the compounding method based on the optimization of organic treatment of montmorillonite and screw shape of biaxial extruder has made it possible to produce the NCH [3].

NCH made by the conventional polymerization method and that made by the compounding method show similar tensile strength and softening points. The gas barrierability of nylon 12, which is commonly used as a tube material, was found to be improved by dispersing clay in it like the NCH.

2. Synthesis and properties of polypropylene–clay hybrid

After the success with NCH, an examination was started of polypropylene consumed in large quantities in the automobile industry. The montmorillonite after ion exchange with octadecyl ammonium was added to polypropylene by melting and kneading. It does not attain complete dispersion by only reducing the polarity of montmorillonite because the polypropylene is nonpolar. It was tried with polarized polypropylene denaturalized by maleic anhydride and then it was found that the possibility of intercalation is determined by the amount of the polar radical [4].

The creation of polypropylene–clay hybrid (PPCH) was realized by dispersing the intercalation compound made from the clay and denaturalized polypropylene obtained on the basis of this investigation into polypropylene [5]. It was also made clear that the PPCH can be produced by mixing, melting and kneading of polypropylene, denaturalized polypropylene, and organized clay without prior preparation of intercalation compound [6].

Furthermore, on the basis of the knowledge obtained regarding polypropylene–clay hybrid, hybridization of polystyrene and clay was successfully carried out with polyethylene by making clay hybrid and using denaturalized polystyrene as a dispersant [7].

3. Synthesis and properties of EPDM–clay hybrid

In case of EPDM representing the rubber of an olefin system, the rubber molecules were found to be intercalated into the clay layers using denaturalized EPDM as polypropylene. However, as for the rubber material, the intercalation was realized even if non-denaturalized EPDM was used since the rubber molecules were polarized during the vulcanization processing [8].

From the fact that the intercalation of EPDM takes place in such different ways depending upon the kind of vulcanization accelerator, it is considered to bond to EPDM radically and give the polarity to cause the intercalation between the clay layers.

4. Morphology control by polymers with clay

Here we introduce the morphology control observed with SEBS–clay hybrid (SEBS: hydrogenated styrene/ butadiene/ styrene block polymer). A nanocomposite was prepared by melting and kneading at 200°C of SEBS pellets and organized montmorillonite. A tri-block polymer consisting of 30% styrene and 70% hydrogenated butadiene was used as SEBS.

Fig. 9.2 shows a TEM photograph of SEBS–clay hybrid. The black fine line indicates the cross section of clay with a thickness of 1 nm, and the gray-colored part corresponds to the polystyrene.

In comparison to the random microphase-separated structure of polystyrene and hydrogenated butadiene in case of the original SEBS, the regularly controlled microphase-separated structure along the clay layer was observed in the clay hybrid [9]. It suggests the potentiality of the control of microphase construction of polymers by the fine dispersion of nanosized

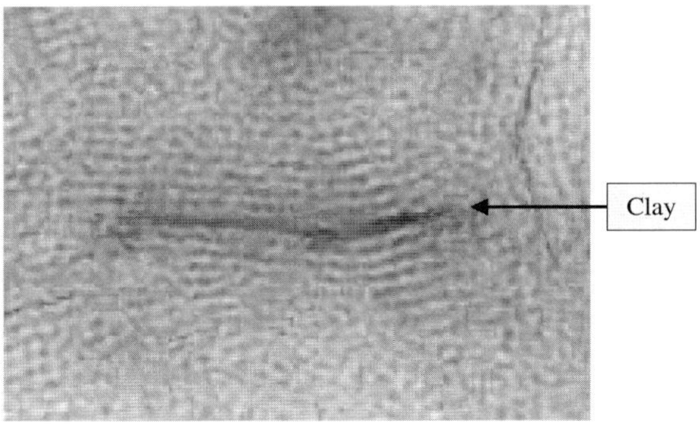

Figure 9.2
TEM picture of SEBS–clay hybrid (ruthenium oxide staining).

particles beyond the clay and it is expected that its application will expand in the future.

References

[1] A. Usuki, N. Hasegawa and M. Kato: *Adv. Polym. Sci.*, **179**, 135–195 (2005).
[2] A. Usuki, A. Okada: *Koubunsi*, **43**, 360 (1994).
[3] M. Kato, H. Okamoto, N. Hasegawa, A. Usuki and N. Sato: Proceedings of 6th Japan international SAMPE symposium, Vol. 2, p. 693 (1999).
[4] M. Kato, A. Usuki and A. Okada: *J. Appl. Polym. Sci.*, **66**, 1781 (1997).
[5] N. Hasegawa, M. Kawasumi, M. Kato, A. Usuki and A. Okada: *J. Appl. Polym. Sci.*, **67**, 87 (1998).
[6] M. Kawasumi, N. Hasegawa, M. Kato, A. Usuki and A. Okada: *Macromolecules*, **30**, 6333 (1997).
[7] N. Hasegawa, H. Okamoto, M. Kawasumi and A. Usuki: *J. Appl. Polym. Sci.*, **74**, 3359 (1999).
[8] A. Usuki, A. Tsukigase and M. Kato: *Polymer*, **43**, 2185 (2002).
[9] N. Hasegawa, A. Usuki: *Polym. Bull.*, **51**, 77 (2003).

APPLICATION 10

10 DEVELOPMENT OF NOVEL FERROELECTRIC MATERIALS

Nonvolatile memories using ferroelectric function have attracted a great deal of world-wide interest for next-generation memory devices because of the low-energy consumption, high-speed read/write, and potentially high density. The widespread applications of the ferroelectric memories require ferroelectric thin films with a larger remanent polarization (P_r), high endurance against repeated switching of bistable polarization states, and low-temperature synthesis as low as 650°C [1–4]. Up to now, two kinds of ferroelectric oxides have been recognized as candidate for the memory materials: one is lead zirconate titanate (PZT) with perovskite structure, and the other is bismuth layer-structured ferroelectrics (BLSFs, see Fig. 10.1) with layered perovskite structure such as $SrBi_2Ta_2O_9$ (SBT) and $Bi_4Ti_3O_{12}$ (BiT) [3, 4].

Ferroelectric PZT shows a superior polarization property such as a relatively larger P_r of 20–40 μC/cm^2, whereas PZT films with Pt electrodes suffer from fatigue problems, and then expensive Ir or IrO_x electrodes are required to be used in memory devices [1, 3]. Furthermore, PZT-based materials contain a large amount of toxic lead, and the use of lead element in these electronic devices would be restricted in the near future by the RoHS directives. Thus, lead-free materials with sufficient ferroelectric properties are required in order to protect the environment and the ecosystem.

Layered ferroelectric SBT and BiT have been widely studied as candidate materials for ferroelectric memories because of their high fatigue endurance [4]. SBT, however, shows a relatively low P_r of 7–10 μC/cm^2, leading to a difficulty in establishing high-density ferroelectric memories using SBT. The substitution of La into BiT enables us to obtain ferroelectric films at a relatively low deposition temperature of 650°C with a larger P_r (20 μC/cm^2), but this P_r value is still lower than that of PZT films. Thus, novel Pb-free ferroelectrics with a larger P_r are expected for next-generation high-density ferroelectric memories.

In this chapter, it is described that novel ferroelectrics composed of two kinds of BLSFs, so-called "superlattice-structured ferroelectrics" show superior polarization properties [5–10]. The novel ferroelectrics with a large P_r of 50 μC/cm^2 compared with PZT are promising not only for memory materials but also for Pb-free piezoelectric ones.

1. Crystal structure of bismuth layer-structured ferroelectrics (BLSFs)

In the crystal structure of BLSFs, perovskite blocks ($A_{m-1}B_mO_{3m+1}$) are sandwiched between Bi_2O_2 layers, and the perovskite blocks are composed of m layers of BO_6 octahedra with A-site cations (see Fig. 10.1). The Bi_2O_2 layers play a significant role not only in the polarization properties but also in the high durability of ferroelectric capacitor. Furthermore, spontaneous polarization (P_s) and insulating properties strongly

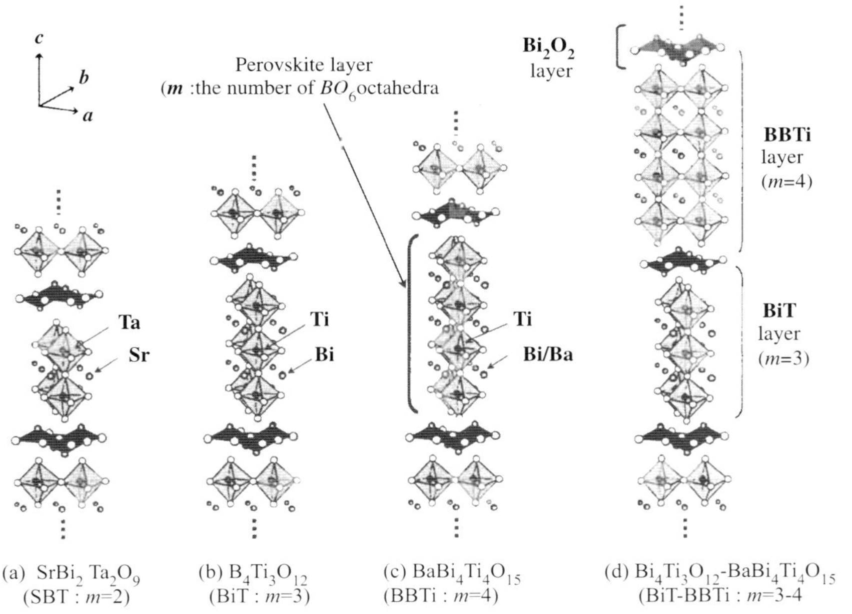

Figure 10.1
Crystal structure of bismuth layer-structured ferroelectrics (BLSFs). BLSFs are composed of the alternate stacking of Bi_2O_2 layers and perovskite layers along the *c* axis. The number of BO_6 octahedra in one perovskite layer is defined as m.

depend on m of the perovskite blocks [11–13]. The Bi_2O_2 layers act as insulating paraelectric layers and control the electronic response (electrical conductivity, band gap, etc.), while the ferroelectricity arises mainly in the perovskite blocks. The crystal structure of BLSFs is briefly characterized by m, and the dielectric and ferroelectric anisotropy strongly depends on the value of m.

Recently, superlattice-structured BLSFs, discovered by Kikuchi et al. [5, 6], have received a renewed interest as a promising candidate for ferroelectric materials with a relatively large P_r [7–10]. For superlattice-structured BiT–SBTi, two kinds of perovskite blocks of BiT (m=3) and $SrBi_4Ti_4O_{15}$ (SBTi: m=4) are in turn sandwiched between Bi_2O_2 layers. A lattice mismatch between the two perovskite blocks and their different chemical character induce the large lattice distortion in the Bi_2O_2 layers, leading to a quite distinct type of the ferroelectric displacement of the Bi in the Bi_2O_2 layers along the ***a*** axis [7]. This Bi displacement is suggested to contribute partly to a larger P_r observed for the BiT–SBTi ceramics than the constituent BiT and SBTi [7]. While most investigations on the superlattice-structured BLSFs to data have been performed on bulk ceramics to elucidate fundamental properties, their dielectric and ferroelectric tensor is not understood, mainly because of the lack of the single crystals.

Here, the properties of the single crystals of superlattice-structured BiT–$BaBi_4TiO_{15}$ (m=3–4) are reviewed. BiT–BBTi has an ideal chemical formula of $BaBi_8Ti_7O_{27}$, and the m=3 (BiT) and m=4 ($BaBi_4TiO_{15}$:BBTi) layers are alternately stacked in the structure along the ***c*** axis. The measurements of the single crystals have demonstrated that the superlattice-structured BiT–BBTi exhibited a P_s of 52 μC/cm^2 [9, 10], which was larger that that of BiT (46 μC/cm^2) and BBTi (16 μC/cm^2).

2. Crystal growth and experimental procedure

First, BiT, BBTi, and BiT–BBTi powders were prepared by a solid-state reaction. The mixture of $BaCO_3$ (99.99%), Bi_2O_3 (99.99%), and TiO_2 (99.99%) with its stoichiometric composition was calcined at 1,050°C for 5 h several times with intermediate grinding. The X-ray powder diffraction data agreed well with the pattern calculated by the Rietveld analysis. Time-of-flight neutron powder diffraction was collected using the Vega diffractometer at KENS, and structural parameters were refined using the program RIETAN-TN.

Single crystals of BiT–BBTi were grown by a self-flux method using the BiT–BBTi powder. The BiT–BBTi powder was mixed with Bi_2O_3 as a flux and put into a Pt crucible. The material was heated at 1,200°C for 5 h, and then cooled to room temperature at 200°C/h. The single crystals obtained were plate-like thin sheets, with 0.1–0.4 mm thickness and 3×3 mm^2

planar dimensions. After heat treatment at 950°C for 24 h in air, annealing under high-pressure oxygen at 35 MPa was performed at 750°C for 12 h to reduce oxygen vacancies in the crystals. The composition of the crystals determined by inductively coupled plasma (ICP) emission spectrometry was $Ba_{0.75}Bi_{8.32}Ti_7O_{27}$, which is a Ba-deficient and Bi-excess composition compared with the ideal one ($BaBi_8Ti_7O_{27}$).

For electrical measurements, the crystals were cut so as to apply electric field along the ***a*(*b*)** directions according to the crystallographic axis determined by the 90°-domain structure. The small crystals of $1\times0.2\times0.25$ mm^3 (Au electrodes were sputtered on 1×0.2 mm^2 faces) were used for polarization measurements. Single crystals of BiT and BBTi were also grown separately, and their properties were compared with those of the BiT–BBTi crystals.

3. Layered structure, dielectric and leakage current properties of BiT–BBTi crystals

Fig. 10.2a shows the θ - 2θ XRD pattern of BiT–BBTi single crystals. Apparent 00*l* reflections originating from the superlattice structure were observed. Fig. 10.2b shows the transmission electron micrograph of a BiT–BBTi crystal. Since the crystals were easily damaged by electron beam irradiation, this photograph was taken within a few seconds. Fig. 10.2c shows a micrograph of the region enclosed

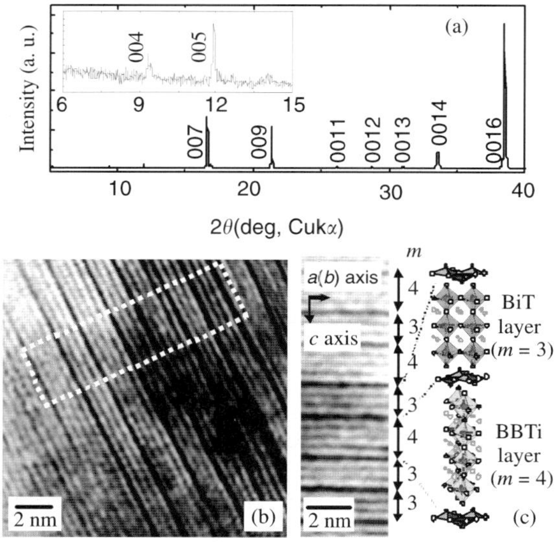

Figure 10.2
Results of the structural analysis of BiT–BBTi single crystals: (a) θ - 2θ XRD pattern showing 00*l* reflections from the crystals surface; (b) transmission electron micrograph in the ***a*(*b*)**–***c*** plane; and (c) the schematic crystal structure.

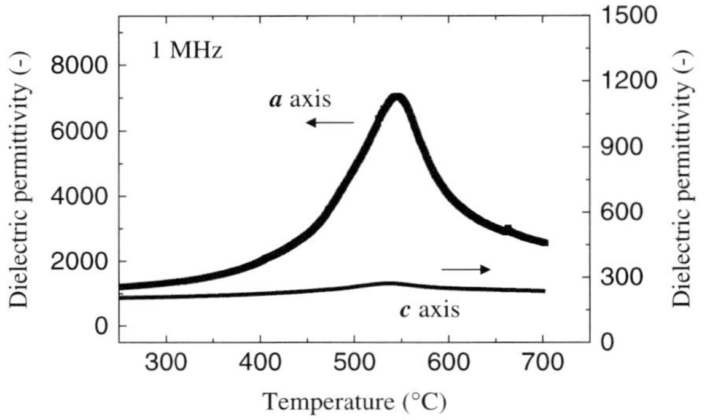

Figure 10.3
Temperature dependence of dielectric permittivity of BiT–BBTi single crystals (1 MHz).

by the dashed line in Fig. 10.2b. The thickness of the constituent layers was approximately 1.5 and 2.0 nm, and these values are in good agreement with half the lattice parameters *c* of BiT and BBTi, respectively. These results clearly indicate that the alternative superlattice stacking of $m=3$ and $m=4$ layers are maintained not only in nanometer scale but also in a wider-ranging area. In addition to the alternate stacking, a stacking fault along the ***c*** axis was observed (m=-3-4-3-4-3-3-4-).

Fig. 10.3 shows the temperature dependence of dielectric permittivity of BiT–BBTi crystals along the ***a*(*b*)** and ***c*** axes at a frequency of 1 MHz. The dielectric permittivity along the ***a*** axis was approximately 1,000 at 25°C, while that along the ***c*** axis was 195. The Curie temperature (T_C) estimated from the permittivity peak was 540°C, and this T_C was intermediate between those of BiT (T_C=675°C) and BBTi (T_C=410°C) [7–10]. Compared with BiT–BBTi ($BaBi_8Ti_7O_{27}$) ceramics with T_C of 500°C, the BiT–BBTi crystals showed a 40°C higher T_C. The higher T_C observed for Ba-deficient BiT–BBTi crystals is suggested to originate from the Bi substitution [14] at the perovskite *A* site.

The leakage currents along the ***a*(*b*)** axis as a function of electric field at 25°C (Fig. 10.4) show that the BiT crystals exhibited a poor value of $\sim10^{-7}$ A/cm^2 above 25 kV/cm, while the BBTi crystals showed a lower current of $\sim10^{-9}$ A/cm^2. Note that the leakage current of the BiT–BBTi crystals remained a sufficiently low value, which was one-fifth that of the BiT crystals. A simple equivalent circuit of parallel conduction of individual BiT and BBTi layers cannot explain this behavior.

4. Giant polarization in BiT–BBTi crystals

Fig. 10.5 indicates the polarization hysteresis loops of single crystals along the ***a*(*b*)** axis at 25°C. The P_s

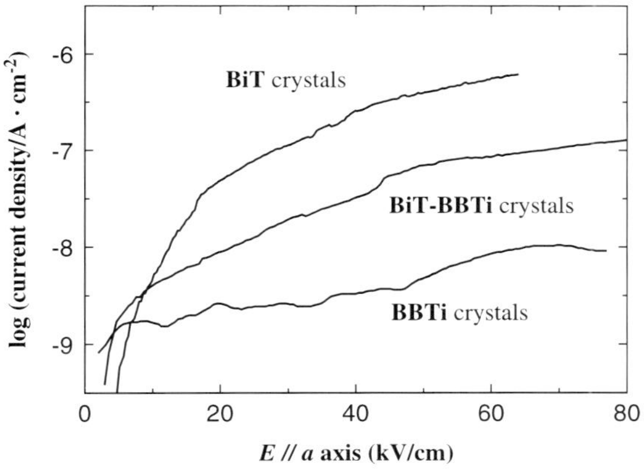

Figure 10.4
Leakage current density as a function of electric field applied along the ***a*** axis (25°C).

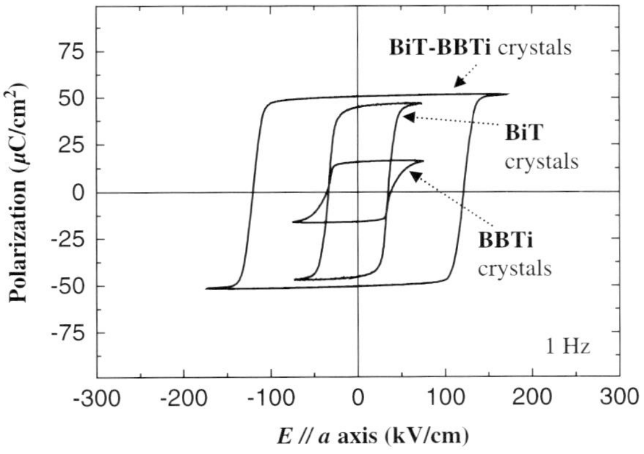

Figure 10.5
Polarization hysteresis loops along the ***a*** axis (25°C).

and the coercive field (E_c) of BiT–BBTi were 52 μC/cm^2 and 120 kV/cm, respectively. Note that this P_s value is larger than those of BiT (46 μC/cm^2) and BBTi (16 μC/cm^2) crystals. It has been widely known that BiT has the largest P_s among the BLSFs, and that the P_s of BBTi is much smaller [13]. If the intrinsic ferroelectric distortions in BiT and BBTi were maintained in the superlattice structure, the P_s of BiT–BBTi would be the average of those for BiT and BBTi (approximately 30 μC/cm^2). However, the P_s observed for the BiT–BBTi crystals is much larger than the average P_s. The enhanced P_s of the BiT–BBTi crystals implies that the lattice strain induced by the alternate stacking of $m=3$ and $m=4$ layers with different cell size promotes ferroelectric distortion.

One of the origins of the larger P_s of the BiT–BBTi crystals is the Bi substitution at the Ba site. For strontium bismuth tantalate with Sr-deficient and Bi-excess composition, the Bi substitution at the Sr site occurs, and the charge difference between Bi^{3+} and Sr^{2+} is compensated through the formation of Sr vacancies [14]. This enhances P_s from 19 μC/cm^2 ($SrBi_2Ta_2O_9$) to 29 μC/cm^2 ($Sr_{0.81}Bi_{2.13}Ta_2O_9$), and raises T_C from 295 to 410°C [14]. It has been reported for BBTi ceramics with random orientation that the Bi substitution at the Ba site enhances P_r from 14 μC/cm^2 ($BaBi_4Ti_4O_{15}$) to 18 μC/cm^2 ($Ba_{0.9}Bi_{4.07}Ti_4O_{15}$), and leads to an increase in T_C from 415 to 440°C. The compositional analysis of the BiT–BBTi crystals shows the Ba-deficient and Bi-excess composition ($Ba_{0.75}Bi_{8.32}Ti_7O_{27}$), indicating that excess Bi is substituted at the Ba site. Indeed, higher T_C was observed for the BiT–BBTi crystals (540°C) than for the stoichiometric ceramics (490°C). This is direct evidence of the Bi substitution at the Ba site. However, the effects of the Bi substitution alone cannot explain the large P_s of the BiT–BBTi crystals.

It is considered that the lattice distortion induced by the alternate stacking of $m=3$ and $m=4$ layers plays a dominant role in enhancing P_s of BiT–BBTi. Since BBTi contains Ba ions with large ionic radius at the *A* site, parameters *a* and *b* are larger than those of BiT. The lattice mismatch between the BiT and the BBTi reaches 0.42% along the ***a*** axis and 0.88% along the ***b*** axis. The parameters *a* and *b* of BiT–BBTi were the averages of those of values for the BiT and the BBTi. These results imply that tensile and compressive stresses built up in the $m=3$ and $m=4$ layers of the superlattice structure, respectively. These stresses are suggested to be concentrated on the Bi_2O_2 layers that interleave the two kinds of perovskite block. This crystallographic environment induces a local symmetry breaking of the Bi_2O_2 layers. All Bi ions of the Bi_2O_2 layers in BiT and BBTi are identical from the crystallographic point of view. In contrast, the Bi ions of the Bi_2O_2 layers in the BiT–BBTi are divided into two cations: one is connected to the perovskite blocks of the $m=3$ layer, and the other is adjacent to that of the $m=4$ layer. The symmetry breaking leads to an unusual ferroelectric displacement of the Bi ions of the Bi_2O_2 layers in the BiT–BBTi. It has been reported that the Bi ions of the Bi_2O_2 layers in BiT–$SrBi_4Ti_4O_{15}$ are displaced along the ***a*** axis (the polar direction) by 2% of the parameter *a* from the corresponding positions of the high-temperature tetragonal structure. Similar displacements of the Bi ions are expected to the BiT–BBTi, which enhance P_s. In addition to the lattice distortions of the Bi_2O_2 layers, the alternate stacking of the $m=3$ and $m=4$ layers is considered to promote the ferroelectric distortions in the perovskite blocks in the BiT–BBTi.

Superlattice-structured BiT–BBTi single crystals were grown by a self-flux method, and the structure was confirmed by X-ray diffraction and transmission electron microscope analyses. Dielectric measurements showed that T_C of the BiT–BBTi was 540°C, which is 40°C higher than that of stoichiometric BiT–BBTi ceramics. The P_s of the BiT–BBTi crystals along the ***a***(***b***) axis was 52 μC/cm^2, which is larger than those of BiT and BBTi crystals. It is suggested that the Bi substitution at the Ba site and the local symmetry breaking of the Bi_2O_2 layers are partially responsible for the larger P_s of BiT–BBTi.

References

[1] Y. Arimori, T. Eshita: *Oyo Butsuri*, **69**, 1080–1084 (2000) in Japanese.
[2] H. Ishiwara: *FED Journal*, **11**, 52–66 (2000) in Japanese.
[3] M. Okuyama: *Denki Gakkai Gakujyutsu Ronbunshi E*, **121**, 537–541 (2003) in Japanese.
[4] Y. Fujisaki: *Kino Zairyo*, **23**, 22–30 (2003) in Japanese.
[5] T. Kikuchi: *J. Less-Common Met.*, **48**, 319–323 (1976).
[6] T. Kikuchi, A. Watanabe and K. Uchida: *Mater. Res. Bull.*, **12**, 299–304 (1977).
[7] Y. Noguchi, M. Miyayama and T. Kudo: *App. Phys. Lett.*, **77**, 3639–3641 (2000).
[8] Y. Goshima, Y. Noguchi and Masaru Miyayama: *App. Phys. Lett.*, **81**, 2226–2228 (2002).
[9] T. Kobayashi, Y. Noguchi and Masaru Miyayama: *Jpn. J. Appl. Phys.*, **43**, 6653–6657 (2004).
[10] T. Kobayashi, Y. Noguchi and Masaru Miyayama: *App. Phys. Lett.*, **86**, 012907 (2005).
[11] T. Takenaka: *Choonpa TECHNO*, **13** (8), 2–12 (2001) in Japanese.
[12] Y. Noguchi, M. Miyayama: *Kino Zairyo*, **21** (9), 31–36 (2001) in Japanese.
[13] H. Irie, M. Miyayama and T. Kudo: *Jpn. J. Appl. Phys.*, **40**, 239–243 (2001).
[14] Y. Noguchi, M. Miyayama and T. Kudo: *Phys. Rev. B*, **63**, 214102 (2001).

APPLICATION 11

11 DEVELOPMENT OF NEW PHOSPHORS

1. History of development of nanophosphor

There are many famous working hypotheses as legends in various research fields. If famous one is chosen in the phosphor research field, it is that "luminescence efficiency of fine(nano)particle phosphor is low". When the particle size of phosphor material was reduced, the luminescence efficiency decreased remarkably. The nanophosphors have large surface area with many defects than that of bulk phosphor [1]. This phenomenon appears especially in submicron-size phosphor. Therefore, bulk phosphor with several microns diameter has been usually used in many applications such as displays and lamps.

The synthesis method of a nanophosphor is classified into chemical synthesis methods such as a solution method and physical synthesis methods such as mechanical grinding. In the chemical synthesis method, the nanophosphor obtained at low temperature has a low crystallinity and high temperature annealing leads to the grain growth. In the physical method, the luminescence efficiency decreases by the formation of the surface defects with the mechanical contact. On the other hand, many attempts to obtain a high-luminance nanophosphor were carried out by new synthesis method and the surface modification. In 1994, Bhargava et al. reported that the luminescence efficiency of surface-modified ZnS:Mn^{2+} nanophosphor increased with decrease of particle size [2]. The high-efficiency ZnS:Mn^{2+} nanophosphor has the following two characteristics. One is new surface modification by methacrylic acid. As indicated above, the surface of the nanophosphor has many defects and non-radiative relaxation is dominant at the phosphor surface. The other is new hypothesis that there is strong coupling between ZnS s–p electron and Mn^{2+} d electron in the nanophosphor by quantum effect. Bhargava claimed that the hypothesis was supported by observation of shorter luminescence lifetime (3.7 ns and 20.5 ns) of nanophosphor than the bulk one (1.8 ms). However, it is considered that the short lifetimes of the ZnS:Mn^{2+} nanophosphor are due to the structural defects at present [3–6].

Apart from the quantum effect, the former surface modification technique by the organic materials became a useful processing method for the nanophosphors. Isobe et al. reported that emission intensity of ZnS:Mn^{2+} nanophosphor increased with the surface modification of carboxylic acid [7–9]. This is not only a capping effect of surface defects but also efficient energy transfer process from the carboxylic acid to the Mn^{2+} ion.

2. Properties of rare earth nanophosphor

However, these nanophosphors have a semiconducting character even if they contain the localized emission ions such as Mn^{2+}. The mainstreams of current phosphor are the oxide phosphors including the rare earth ions as the emission center because of high luminous efficiency. When rare earth phosphor is activated by small-energy blue and near-UV light, light absorption and emission occur around the emission ions. The semiconducting character of oxide host lattice was not reflected at all by the photoluminescence. There was the 4f orbital of rare earths located near an atomic nucleus and comparatively covered by an outside 5s orbital. Therefore, the electron of 4f orbital is hard to take a quantum effect for the nanoparticles. Recent results of rare earths doped nanophosphors are reviewed below.

In 2000, Park et al. reported synthesis of $Y_3Al_5O_{12}$:Tb nanosize phosphor (25–45 nm) from nitric acid solution by sol–gel method [10]. This sample with heat treatment until 800°C was amorphous and the excitation spectrum was different from that of the sample with high-temperature heating. From decay time and excitation spectrum, they claimed that the nanophosphor sample was useful for plasma display. Hasse et al. synthesized nanosize YVO_4 and $LaPO_4$ by the hydrothermal technique. The sample was colloidal solution and separated by centrifugation. The YVO_4 nanophosphor has the same crystal structure as its bulk sample. On the other hand, monazite structure is observed for the nano $LaPO_4$. The particle size depends on pH value of the solution and other reaction parameters. In alkaline solution, the sample was fine particles from 10 to 50 nm. In acid solution, morphology of the sample was nanofiber [11, 12].

In 2001, Konrad synthesized 10 nm sized cubic yttria phosphor by chemical vapor technique [13]. The crystallite size increased to 20 and 50 nm upon heating at 900 and 1,100°C, respectively, in air for 5 h. Depending on the particle size, a broadening of absorption edge and blue shift of photoluminescence spectra were found. The change of spectra was explained by the change of a configurational coordinate diagram.

In 2002, Lu et al. reported synthesis of small-size LED phosphor $Y_3Al_5O_{12}:Ce^{3+}$ by sol–gel method [14]. Raw material aqueous solution reacts with urea in the presence of polyvinyl alcohol at 150°C. Gelling was achieved at 250°C. Although the heat treatment at 800–1,100°C yielded the nanosize phosphor, the emission intensity was remarkably low compared with the conventional bulk sample synthesized at 1,450°C.

In 2003, Pan et al. reported the synthesis of a red $CaTiO_3:Pr^{3+}$ nanophosphor from polymer precursor [15]. The nanophosphor powder (10 nm) was obtained by heating at 600°C for 5 h. Higher temperature and longer heating time increased particle size of the phosphors and photoluminescence intensity. The emission spectrum is comparable with the conventional phosphor synthesized by the solid-state reaction. Wang et al. synthesized Y_2O_3:Tb nanophosphor by the combustion synthesis [16]. The particle size estimated by XRD pattern was 35–70 nm. It is an interesting feature that the emission intensity is increased by irradiating the 250 nm UV light. It is considered by ESR measurement that the behavior is due to the passivation of the dangling bond on the surface. Tissue et al. reported on the vapor-phase synthesis of Y_2O_3:Eu nanophosphor [17]. By CO_2 laser heating under 10 and 400 Torr nitrogen, 5 and 12 nm nanophosphor samples were obtained, respectively. The particle size grew up to about twice by annealing for a long time at 800°C. The crystal structure of 5 nm particle changed from mixed phase to cubic single phase and 13 nm particle did not change the monoclinic structure. He et al. reported synthesis of Y_2O_3:Eu nanophosphor by wet chemical synthesis [18]. Chloride solution of yttrium and europium was mixed with butanol, which acted as a surfactant. The nanophosphor was obtained by adding sodium carbonate to the solution and heat treatment at 800°C. When butanol was added, the sample of a particle size was smaller than that of the non-surfactant sample.

In 2004, Chander et al. synthesized the nanocrystal of a long-persistence phosphor $SrAl_2O_4$:Eu, Dy by modified combustion method [19]. The nitrate solution of starting materials was mixed with urea and boric acid heated from 400 to 600°C. The nanophosphor sample of 50 nm or less was obtained after combustion reaction. However, the decay time is shorter than that of the sample synthesized by a conventional solid-phase reaction. Although Peng et al. also have synthesized the 25 nm phosphor by the combustion synthesis method [20], the afterglow time was still shorter that of the conventional ones.

3. Development trend of new nanophosphor

Summarizing these, it has been observed that "luminescence efficiency of fine(nano)particle phosphor is low". Does the nanophosphor have practical use? White LED application is one of the promising candidates. Rayleigh scattering decreases in proportion to the sixth power of particle size. Therefore, the nanoparticles dispersed in resins scatter light less vigorously and become transparent. Rare earth complex phosphors are dispersed in the plastics [21]. Fukui et al. reported nanocluster phosphors containing the rare earth ions [22]. Isobe et al. reported the transparent phosphor sheet of nano ZnS:Mn and $Y_3Al_5O_{12}$:Ce dispersed in the resins [23].

On the other hand, there are some trials to improve the luminescence efficiency of the nanophosphors by precise process control. Nishisu et al. reported the

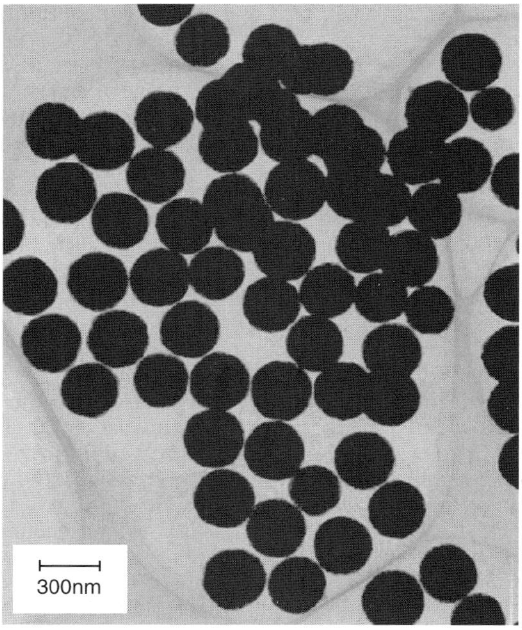

Figure 11.1
TEM photograph of the spherical Y_2O_3:Eu phosphor heated at 650°C [24].

Figure 11.2
SEM photograph of high-luminance rare earth oxycarbonate phosphor [26].

synthesis of 300–400 nm Y_2O_3:Eu spherical phosphor by uniform coprecipitation method [24]. The luminance of the nanophosphor under 147 nm excitation is comparable to that of the conventional sample synthesized at high temperature. Fig. 11.1 shows a TEM photograph of the Y_2O_3:Eu phosphor heated at 650°C. The spherical phosphor could maintain almost a monodispersed state. Kakihana et al. reported synthesis of high-luminance 200–300 nm Y_2O_2S:Eu phosphor by complex homogeneous precipitation method [25]. Masui et al. synthesized new layered $Gd_2O_2CO_3$:xTb^{3+} green phosphor in 0.476 Li_2CO_3–0.270 Na_2CO_3–0.254 K_2CO_3 flux. This nanophosphor shows higher luminance than that of commercial lamp phosphor [26]. Fig. 11.2 showed an SEM photograph of rare earth oxycarbonate phosphor. In these processing studies, the choice of the precursor and optimization of heating condition to prevent the aggregation are most important.

References

[1] T. Welker: *Ext. Abst. Electrochem. Meeting*, **91–92**, 973 (1991).

[2] R.N. Bhagava, D. Gallagher, X. Hong and A. Narmikko: *Phys. Rev. Lett.*, **72**, 416 (1994).

[3] J. Yu, H. Liu, Y. Wang and W. Jia: *J. Lumin.*, **79**, 191 (1998).

[4] A.A. Bol, A. Meijerink: *Phys. Rev.*, **B58**, R15997 (1998).

[5] N. Murase, R. Jagannathan, Y. Kanematsu, M. Watanabe, A. Kurita, K. Hirata, T. Yazawa and T. Kushida: *J. Phys. Chem.*, **B103**, 754 (1999).

[6] M. Tanaka, Y. Masumoto: *Chem. Phys. Lett.*, **324**, 249 (2000).

[7] T. Isobe: *Funct. Mater.*, **19**, 17 (1999).

[8] M. Konishi, T. Isobe and M. Sennna: *J. Limin.*, **93**, 1 (2001).

[9] T. Isobe: *J. Surf. Sci. Soc. Jpn*, **22**, 315 (2001).

[10] C.H. Park, S.J. Park, B.U. Yu, H.S. Bae, C.H. Kim, C.H. Pyun and H.G. Yan: *J. Mater. Sci. Lett.*, **19**, 335 (2000).

[11] M. Hasse, K. Riwotzki, H. Meyssamy and A. Kornowski: *J. Alloys Compd.*, **303–304**, 191 (2000).

[12] H. Meyssamy, K. Riwotzki, A. kornnowki, S. Naused and M. Hasse: *Adv. Mater.*, **11**, 840 (1999).

[13] A. Conrad, U. Herr, R. Tidecks, F. Kummer and Samwer: *J. Appl. Phys.*, **90**, 3516 (2001).

[14] C.H. Lu, J. Jagannathan: *Appl. Phys. Lett.*, **80**, 3608 (2002).

[15] Y. Pan, Q. Su, H. Xu, T. Chen, W. Ge, C. Yang and M. Wu: *J. Solid State Chem.*, **23**, 4300 (2003).

[16] J. Wang, H. Song, B. Sun, X. Ren, B. Chen and W. Xu: *Chem. Phys. Lett.*, **379**, 507 (2003).

[17] B.M. Tissue, H.B. Yuan: *J. Solid State Chem.*, **171**, 12 (2003).

[18] C. He, Y. Guan, L. Yao, W. Cai, X. Li and Z. Yao: *Mater. Res. Bull.*, **38**, 973 (2003).
[19] H. Chander, D. Haranath, V. Shanker and P. Sharma: *J. Cryst. Growth*, **271**, 307 (2004).
[20] T. Peng., H. Yang, X. Pu, B. Hu, Z. Jiang and C. Yan: *Mater. Lett.*, **58**, 352 (2004).
[21] S. Yanagida, Y. Wada and Y. Hasegawa: *Phosphor Research Society of the 293rd Meeting Technical Digest*, 15 (2002).
[22] T. Fukui, H. Mataki: *Kidorui*, **46**, 212 (2005).
[23] T. Isobe: *Mater. Stage*, **5**(6), 7 (2005).
[24] Y. Nishisu, M. Kobayashi: Report of the millennium project (Environment, No. 12317; Research and development of phosphors for a mercury-free lamp) by Science and Technology Agent (2006).
[25] M. Kakihana, Y. Kawahara, T. Ichihara and N. Kijima: *Phosphor Research Society of the 304th Meeting Technical Digest*, 25 (2004).
[26] T. Masui, Y. Mayama, K. Koyabu and N. Imanaka: *Chem. Lett.*, **34**, 1236 (2005).

APPLICATION 12

12 ZEOLITE MEMBRANE

Zeolites are crystalline hydrated aluminosilicates with small pores comparable to molecule in frameworks based on extensive three-dimensional networks of oxygen ions. The zeolites have been routinely used as adsorbents, ion exchangers and catalysts. In these applications, the zeolites are usually used in powder and pellet forms. Recently, much interest has been paid to membranes of the zeolites, because the unique structures and physicochemically properties of zeolites make the zeolite membranes attractive for a wide variety of practical applications including gas and vapor separations, membrane reactors, chemically selective sensors and optoelectronic devices. In this section, synthesis and separation properties of zeolite membranes are described in brief.

1. Characteristics

Many types of membranes which have been used for separation of various mixtures are usually composed of polymers. Although most of the polymeric membranes show sufficient separation properties, the applications have been limited to the environment around room temperature, except for a few heat-resistant polymeric membranes such as polyamide. In addition, the polymeric membranes often face a serious problem with non-aqueous mixtures that their separation properties are degraded, due to swelling of membranes. On the other hand, compared with the polymeric membranes, organic ones promise high thermal and chemical stability as well as high strength. However, their separation properties are, in many cases, insufficient for practical applications.

The zeolites known as "molecular sieves" can exclude the entrance of molecules larger than the pores and allow only the passage of smaller ones. With the zeolites as membranes, it is therefore expected to achieve molecular-selective separation based on size and shape in a continuous manner. Besides, the zeolites have some inherent properties: hydrophilic or organophilic properties, acid or basic properties and selective adsorption of polar molecules due to a local electrostatic field within the pores. From combining their unique physical and chemical properties with the dimensionality, zeolite membranes are very attractive for separation of various mixtures containing non-adsorbing molecules, or different organic compounds, or mixtures of gases/vapors or water/organics.

2. Synthesis

In considering application of the zeolite membranes for separation technology, pinhole-free microstructure and orientation of zeolite crystals are needed to achieve true "molecular sieves" with high permeance. In addition, the membranes should be as thin as possible to maximize the throughput.

A variety of methods have been reported for the synthesis of zeolite membranes. Most of these fabrications reported can be basically classified into three categories: (1) in situ crystallization method, (2) secondary growth method and (3) dry-gel conversion method. The three methods are here introduced in brief.

(1) In situ crystallization method

In this synthesis, membranes are directly synthesized on appropriate porous substrates in synthetic solutions containing constituents of zeolites under hydrothermal conditions. Fig. 12.1 shows schematically an experimental flow for the synthesis of zeolite membranes by the in situ crystallization method. Some of mechanisms of growing zeolite membranes on the surface of substrate under hydrothermal conditions have been proposed. For example, zeolite crystals formed in the synthetic solution are continuously deposited on the substrate, and intercrystal voids are

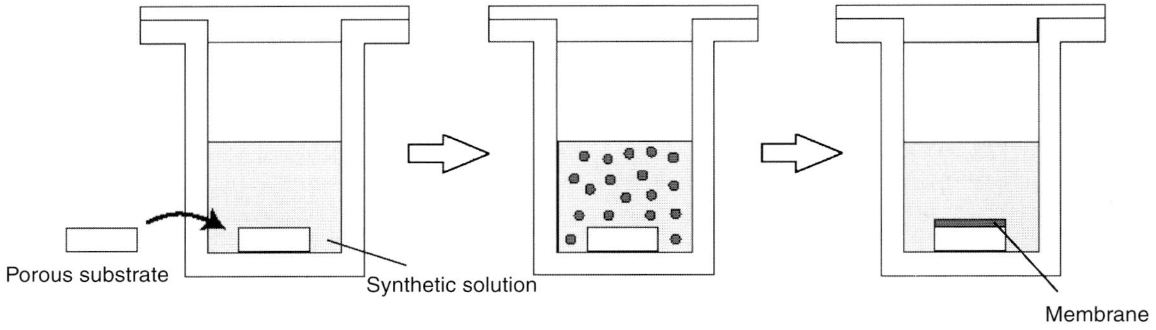

Figure 12.1
Schematic experimental flow for the synthesis of zeolite membranes by the in situ crystallization method.

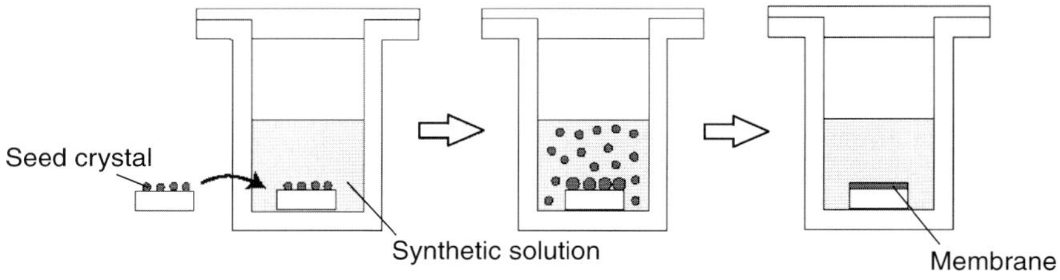

Figure 12.2
Schematic experimental flow for the synthesis of zeolite membranes by the secondary growth method.

filled with fine crystals formed newly through nucleation. According to another mechanism proposed, the synthesis of zeolite membranes is based on combination of heterogeneous nucleation in a gel adsorbed on substrate with intergrowth of the zeolite crystals formed. However, clear explanation has not yet been given for the synthesis of zeolite membranes by the in situ crystallization method.

Although the in situ crystallization method is a practical process because of its simplicity, it is difficult to control the nucleation and crystal growth. In most cases, the zeolite membranes synthesized by the in situ crystallization method are randomly oriented. Wang et al. [1] have recently succeeded in synthesizing *b*-axis oriented MFI zeolite membranes with thickness less than 1 μm. They exactly controlled process parameters including chemical composition of synthetic solution, hydrothermal temperature and time. Seike et al. [2] have synthesized dense and *b*-axis oriented 170 nm-thick MFI zeolite membranes in a unique in situ crystallization process including slow dissolution of source material [2]. Dense and *b*-axis oriented MFI zeolite membranes have been also synthesized by Zhang et al. [3] in a different process. Thus, remarkable technological progress has been made in the recent works.

(2) Secondary growth method

As illustrated schematically in Fig. 12.2, the secondary growth method involves the growth of a layer of seed crystals deposited in advance on a substrate, followed by transformation of the precursor layer composed of seed crystals into a dense form under hydrothermal conditions. It is possible in the secondary growth method to synthesize oriented membranes by pre-assembly of well-aligned seed crystals on the substrate. A report by Lai et al. [4] is cited here as a reference.

This method has potential advantages in terms of reproducibility as compared with the in situ crystallization method. However, this method often produces thick membranes unless thin precursor layers are formed.

(3) Dry-gel conversion method

This method is based on deposition of a gel layer containing constituents of zeolites on a substrate followed by treatment with a vapor of some structure-directing agents. The structure-directing agents are included in the synthetic solution or gel. In the latter case, the conversion of precursor gel layer to highly crystallized zeolite membrane is assisted with steam. The conversion process has been considered to proceed with nucleation and crystal growth of zeolite crystals via osmosis of the structure-directing agents and water

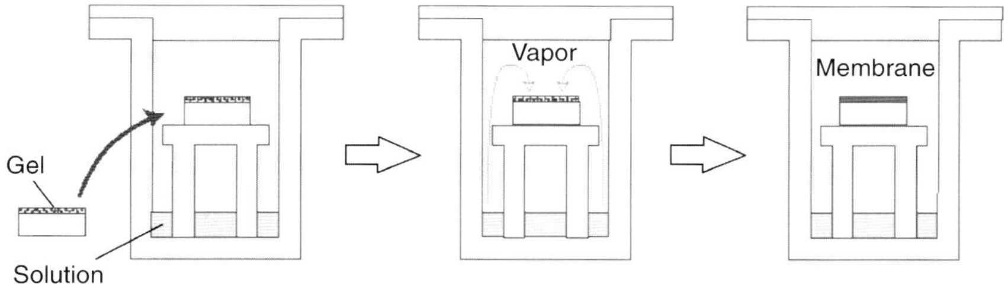

Figure 12.3
Schematic experimental flow for the synthesis of zeolite membranes by the dry-gel conversion method.

into the gel layer with reaction time. A representative experimental set-up is schematically illustrated in Fig. 12.3 for the dry-gel conversion method.

The synthesis of ultra-thin zeolite membranes with both dense body and orientation is still difficult, although dense membranes without orientation or oriented membranes without pinhole-free microstructure have been synthesized until now. However, since recent progress in microstructure-controlling technologies including nanotechnology is dramatic, it is expected that synthesis of pinhole-free and oriented ultra-thin zeolite membranes with thickness in a nanoscale is achieved in no distant future.

3. Separation properties of zeolite membranes

Zeolite membranes have the potential for continuous separation of gas, vapor and non-aqueous mixtures. Various attempts have been reported. For example, hydrophilic A-type zeolite membranes with excellent separation properties for alcohol dehydration and solvent dewatering have been applied to production of bio-ethanol and dehydration of waste isopropyl alcohol. The dehydration of isopropyl alcohol in pervaporation process using the A-type zeolite membranes is the first commercial application of zeolite membranes.

On the other hand, gas separation using zeolite membranes has been limited for practical applications. One of the reasons comes from the difficulty in syntheses of zeolite membranes. Recently, selective separation of *p*-xylene from xylene isomers has been reported on MFI zeolite membranes, and evaluation of MFI membranes for xylene separation has been made from the industrial side.

As mentioned above, various types of separation can be expected using zeolite membranes, because of the unique physical and chemical properties of zeolites. In this part, separation of CO_2 from a binary mixture of CO_2 and N_2 on Y-type zeolite membrane is reported. The selective separation of CO_2 results from a difference in affinity of CO_2 and N_2 to pore walls of the Y-type zeolite crystal.

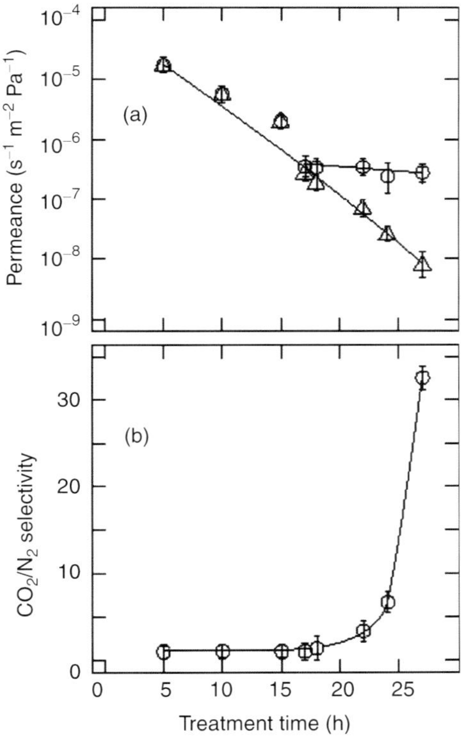

Figure 12.4
Permeance changes found for Y-type zeolite membranes in the binary mixture of CO_2 and N_2 as a function of hydrothermal reaction time for the secondary growth of precursors formed electrophoretically (a) permeances of CO_2 (O) and N_2 (Δ) and (b) the CO_2/N_2 selectivities.

Fig. 12.4 shows the permeances of N_2 and CO_2 and the CO_2/N_2 selectivities in the binary mixture for Y-type zeolite membranes. Membranes with thickness of about 35 μm synthesized by secondary growth of electrophoretically fabricated Y-type zeolite precursor films were used [5]. The values on the abscissa in Fig. 12.4 indicate the hydrothermal treatment times for the secondary growth. The permeances were measured at room temperature with the Wicke-Kallenbach set-up with no pressure drop across the membranes.

As is seen in Fig. 12.4, the permeances of both N_2 and CO_2 decreased with elongation of reaction time. This was due to a progress in densification of membranes. When the membranes were hydrothermally treated with a period longer than 16 h, a difference was observed in the permeances between CO_2 and N_2. The difference brought about the separation of CO_2 from the binary mixture. The CO_2/N_2 selectivity rapidly increased with the treatment time, and reached a value of about 32 at the treatment time of 27 h. The kinetic diameters of CO_2 and N_2 are 0.33 and 0.36 nm, respectively, and these values are much smaller than the pore mouth opening of the Y-type zeolite crystals which is about 0.74 nm. Since both molecules of CO_2 and N_2 can intrude into the pores of Y-type zeolite crystals, size-selective separation is not expected for the separation of CO_2 on the Y-type zeolite membranes.

The Y-type zeolite crystal has a local electrostatic field within the pores with charge-balancing cations, interacting strongly with polar molecules. Because the quadrupole moment of CO_2 molecule is about four times that of N_2, preferential adsorption of CO_2 molecules on the Y-type zeolite crystals is expected. Since a chemical potential gradient in the pores generally drives surface diffusion of molecules, it is considered that the permeance is accelerated for CO_2 molecule. As another mechanism, it is hypothesized that the preferentially adsorbed CO_2 prevents permeation of less adsorptive N_2 along the pore surface. However, since the diffusion of molecules in zeolite crystals is much more complex compared with the adsorption, a clear explanation has not yet made for the CO_2 selective separation mechanism on the Y-type zeolite membranes. Kusakabe et al. [6] have found the effect of cation exchange of the charge-balancing cations in the Y-type zeolite crystals on the CO_2/N_2 separation, and succeeded in obtaining high selectivity of about 100 on cation-exchanged Y-type zeolite membranes [7]. Since CO_2/N_2 separation is of significance for protection of the environment, further improvement is desired.

References

[1] Z. Wang, Y. Yan: *Chem. Mater.*, **13**, (2001).

[2] T. Seike, M. Matsuda and M. Miyake: *J. Am. Ceram. Soc.*, **87**, 1585 (2004).

[3] F.-Z. Zhang, M. Fuji and M. Takahashi: *Chem. Mater.*, **17**, 1167 (2005).

[4] Z. Lai, G. Bonilla, I. Diaz, J.G. Nery, K. Sujaoti, M.A. Amat, E. Kokkoli, O. Terasaki, R.W. Thompson, M. Tsapatsis and D.G. Vlachos: *Science*, **300**, 456 (2003).

[5] T. Seike, M. Matsuda and M. Miyake: *J. Mater. Chem.*, **12**, 366 (2002).

[6] K. Kusakabe, T. Kuroda and S. Morooka: *J. Memb. Sci.*, **148**, 13 (1998).

[7] K. Kusakabe, T. Kuroda, A. Murata and S. Morooka: *Ind. Eng. Chem. Res.*, **36**, 649 (1997).

APPLICATION 13

13 ENHANCEMENT OF THE PERFORMANCE OF INSULATING MATERIALS

Some solid particles, several tens of microns in size, have been conventionally added to organic insulating materials; mainly for the purpose of cost reduction by using more reasonable particles as a bulking filler [1].

On the other hand, it has long been known that the electrical performance of insulating materials may be improved by adding nanoparticles to the organic insulating materials, since some reports were presented at the IEEE CEIDP (Conference of Electrical Insulation and Dielectric Phenomena) in 2002. Subsequently, a large amount of measurement data of various properties have been reported at international conferences, including the one organized by the Institute of Electrical Engineers of Japan [2] and so on. Typical examples of reported data from these events will be introduced here after the properties have been classified.

1. Withstand voltage characteristics

The space charge distribution was measured using two kinds of epoxy resins with different sizes (1.5 μm and 38 nm) of TiO_2, subjected to direct voltage applied on the test pieces [3]. Following the 10 wt% particle addition to the resin, the electric field in front of the cathode was as high as 40 kV/mm with a distribution of complicated pattern, as shown in Fig. 13.1, when micron-sized particles were used.

On the other hand, with the nanosized particles used as filler, the electric field in front of the cathode was only 2 kV/mm and its distribution was relatively uncomplicated. This is considered attributable to the fact that where nanoparticles are added, the transportation ability and charge storage are low, owing to the high concentration of shallow charge trapping.

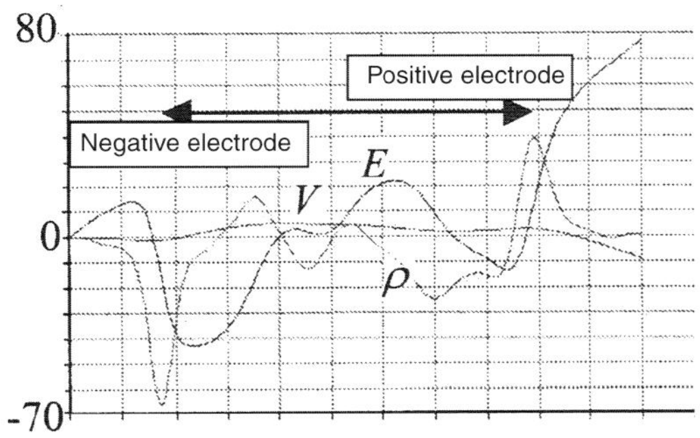

(a) In case of micron particles

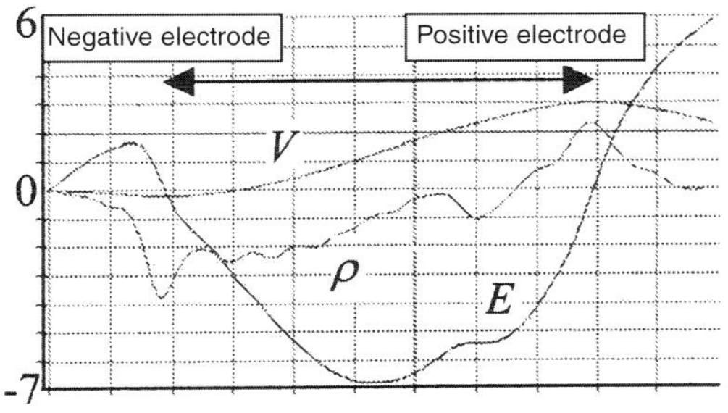

(b) In case of nanoparticles

Figure 13.1
Distributions of voltage, electric charge and field of epoxy resin dispersed with TiO_2 particles [3]. (V: Electric voltage [kV], ρ: Electric charge[C/m^3], E: Electric field [kV/mm]).

Following the withstand voltage test, conducted on the above-mentioned sample under an alternating voltage of 60 Hz, the lifetime was extended about 100 times using nanoparticles compared to the case where micron particles were used at a voltage of 400 kV/mm, as seen in Fig. 13.2 [4]. This is regarded as attributable to the reduction of internal charge, which causes insulating materials to deteriorate at the interface between the resin and particles.

On the other hand, there is a report where Al_2O_3 particles 10 nm in size were added to epoxy resin [5]. Following the withstand voltage tests, conducted on the material under the alternating voltage of 60 Hz, the time required to reach the insulation breakdown was found to nearly double with the addition of 5 wt% Al_2O_3 compared with the epoxy resin, as shown in Fig. 13.3. This was explained by the progress of the electric trees being hindered by the nanoparticles.

In another report [6], the properties were measured using two kinds of ZnO, of particle sizes 1 μm and 65 nm, which were added to epoxy resin. The withstand voltage test under an alternating voltage of 50 Hz showed that the time to the insulation breakdown increased with the increasing addition of ZnO and that it doubled when nanosized particles were used, compared to the case using micron-sized particles, as seen in Fig. 13.4. This phenomenon was reportedly attributable to the existence of submicron-sized void-like defects near the nanoparticles, which enlarged the damage process zone at the tip of the electric tree and hindered the tree progress.

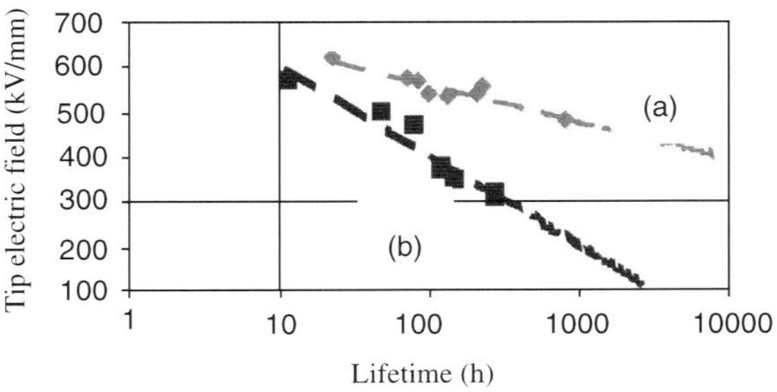

Figure 13.2
Withstand voltage characteristics of epoxy resin dispersed with TiO_2 particles [4]: (a) Nanoparticles, (b) micron particles.

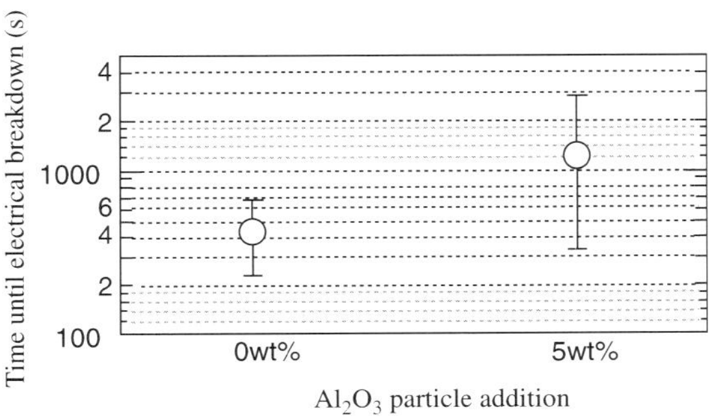

Figure 13.3
Time to reach the electrical breakdown of epoxy resin with Al_2O_3 particle addition [5].

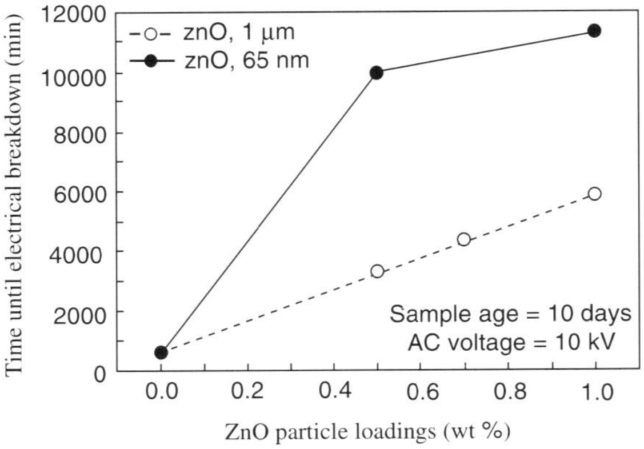

Figure 13.4
Time to reach the electrical breakdown of epoxy resin with ZnO particle addition [6].

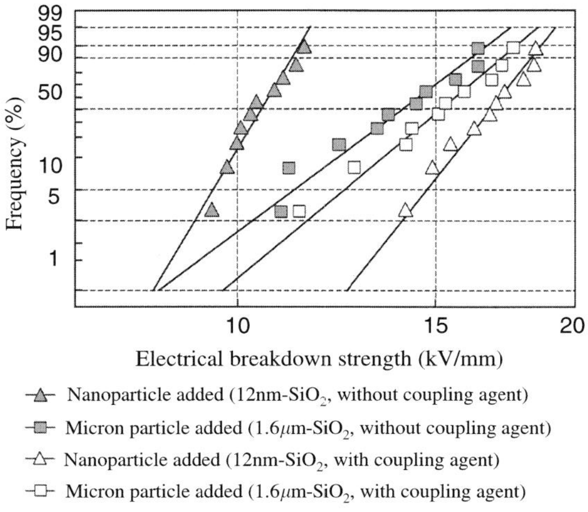

Figure 13.5
Withstand voltage of epoxy resin [7].

There is another report, in which two kinds of SiO_2 particles, of sizes 1.6 µm and 12 nm, were used and the effect of the coupling agent used for the particle surface modification was compared [7]. As a result of the withstand voltage test under an alternating voltage of 50 Hz, only modest difference in the effect in the case of micron-sized particles was apparent, as seen in Fig. 13.5.

Conversely however, the electrical breakdown strength was increased by 50% using the surface treated nanoparticles and became larger than that with the micron-sized particles. This is considered attributable to the fact that the electric tree progress is hindered, due to the reduced inter-particle distance when the nanoparticles are applied, and that the weak points at the interface between the resin and particles can be decreased by treating the nanoparticle surface.

2. Tracking-resistance and erosion-resistance characteristics

There is a report [8] comparing the properties of silicon rubber with the addition of SiO_2 particles of differing sizes, namely 5 µm and 12 nm. The erosion volume in the test sample was then measured under 2–2.5 kV AC.

Consequently, it was found to be reduced with the increasing addition of both micron- and nanosized particles, as shown in Fig. 13.6. The addition of just 10% of nanoparticles resulted in an erosion volume similar to that obtained by the 50% addition of micron-sized particles, while that reportedly caused by the decomposition of silicon rubber was hindered by the silica layer that formed when the nanoparticles were added.

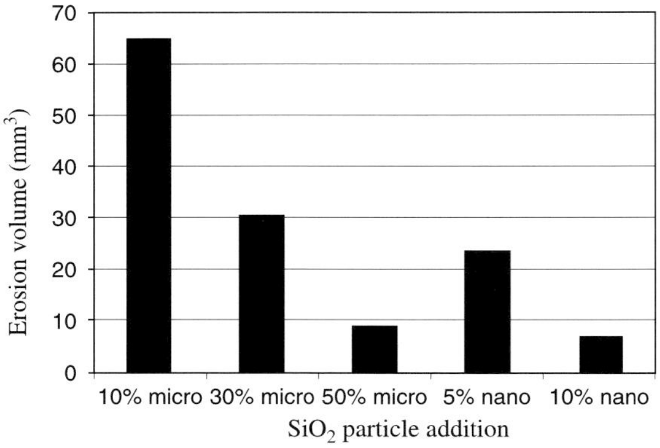

Figure 13.6
Erosion volume of silicon rubber [8].

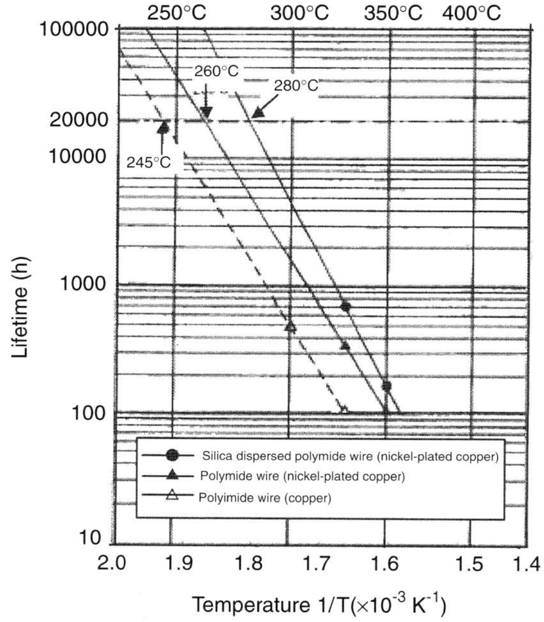

Figure 13.7
Heat-resistance index [11].

In addition, the strong adhesion of nanoparticles to the silicon rubber is also considered to be one of the reasons hampering the decomposition of the silicon rubber [9].

It is also reported that the addition of 5 wt% SiO_2 to the silicon rubber increased the tracking resistance [10]. It was explained that oxygen and flammable gas hardly penetrated into the silicon rubber because the nanoparticles occupied the free volume space inside it.

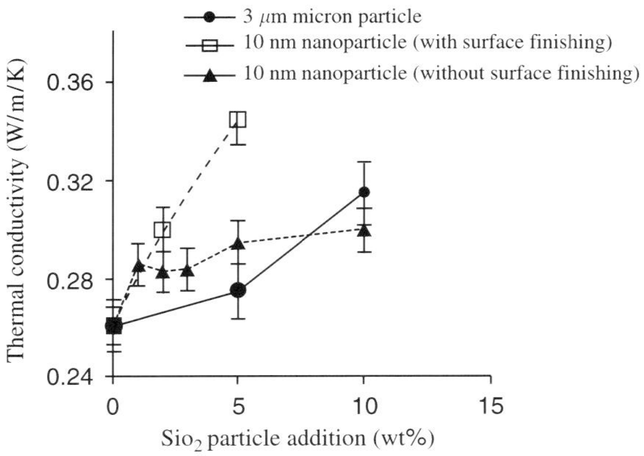

Figure 13.8
Thermal conductivity of polyimide resin [12].

3. Thermal characteristics

The heat-resistance index of a sample of polyimide resin, dispersed with SiO_2 particles several tens of nanometers in size, was measured [11]. As shown in Fig. 13.7, the heat-resistance index was 280°C at a lifetime of 20,000 h, which was some 35°C higher than that (245°C) of conventional polyimide resin. This was reportedly due to the suppression of the decomposition of polyimide resin due to the addition of the nanoparticles.

The thermal conductivity was measured using samples of polyimide resin with dispersed SiO_2 particles of two different sizes, namely 3 μm and 10 nm [12]. The values were similar when compared in the range of addition from 5 to 10 wt%, as seen in Fig. 13.8, but the thermal conductivity increased by 30% using a nanosized coupling agent. This is considered attributable to the improved interaction between the particles and polymer chains due to the particle surface treatment.

References

[1] H. Kakiuchi: *New Epoxy Resin*, Shokodo, Tokyo, p. 293 (1985).
[2] T. Tanaka, M. Iwata and M. Kozako: *IEEJ Electrical Insulation News in Asia*, Vol.11, 18–20 (2004).
[3] J.K. Nelson, J.C. Fothergill, L.A. Dissado and W. Peasgood: 2002 Annual Report of Conference on Electrical Insulation and Dielectric Phenomena, pp. 295–298 (2002).
[4] J.K. Nelson, Y. Hu: 2004 International Conference on Solid Dielectrics, pp. 832–835 (2004).
[5] M. Kozako, S. Yamano, R. Kido, Y. Oki, M. Koto, S. Okabe and T. Tanaka: 2004 International Symposium on Electrical Insulating Materials, pp. 231–234 (2005).
[6] H.Z. Ding, B.R. Varlow: 2004 Annual Report of Conference on Electrical Insulation and Dielectric Phenomena, pp. 332–335 (2004).
[7] T. Imai, F. Sawa, T. Ozaki, T. Shimizu, R. Kido, M. Kozako and T. Tanaka: 2004 International Symposium on Electrical Insulating Materials, pp. 239–242 (2005).
[8] A.H. El-Hag, S.H. Jayaram and E.A. Cherney: 2004 Annual Report of Conference on Electrical Insulation and Dielectric Phenomena, pp. 385–388 (2004).
[9] A.H. El-Hag, L.C. Simon, S.H. Jayaram and E.A. Cherney: 2004 Annual Report of Conference on Electrical Insulation and Dielectric Phenomena, pp. 688–691 (2004).
[10] C. Dengke, W. Xishan, L. Lei and Y. Jianhui: 2004 International Conference. on Solid Dielectrics, pp. 800–803 (2004).
[11] K. Asano, K. Suzuki, K. Soma and K. Murouchi: 2004 Annual Meeting Record IEEJ, 2-S5-6 (2004).
[12] P.C. Irwin, Y. Cao, A. Bansal and L.S. Schadler: 2003 Annual Report of Conference on Electrical Insulation and Dielectric Phenomena, pp. 120–123 (2003).

APPLICATION 14

14 BARIUM TITANATE NANOPARTICLES SYNTHESIZED UNDER SUB AND SUPERCRITICAL WATER CONDITIONS

Memory materials with a large capacity are of great demand with the development of IT industry. A cell area of a dynamic random access memory (DRAM), which is one of the leading mainstreams of memory, is determined by its capacitor. A high dielectric insulating surface layer is attracting much attention instead of a conventional silicon oxide film. Barium titanate ($BaTiO_3$) is one of candidates of a high dielectric insulator. A dielectric constant of cubic $BaTiO_3$ is 100 times larger than that of SiO_2 or V_2O_5. When $BaTiO_3$ is used as an insulating surface layer material of a capacitor, the thickness of layer can be several 10 nm and thus it is expected to make an extremely fine capacitor [1, 2].

The high-dielectric insulator using $BaTiO_3$ was investigated by sol-gel routed film processing [5]. However, since a precursor of $BaTiO_3$ should be heated at about 800°C to crystallize, substrates and transistors were damaged by this heat treatment.

Tetragonal $BaTiO_3$ particles are obtained by solid-state reaction, but the size of these particles is in order of several 10 μm [3]. In hydrothermal synthesis, the crystalline $BaTiO_3$ particle is produced by hydrothermal treatment at 100–150°C from barium hydroxide and titanium dioxide with high alkali concentration [4, 5]. Although particles several 10 nm in diameter could be produced, some hydroxyl groups remained inside or on the surface of the particle and thus the crystalline degree of the hydroxide is not necessarily high [6]. This resulted from low reaction temperature and high alkali concentration. In this way, nanosized $BaTiO_3$ particles with high crystalline degree could not be obtained by any methods.

Tetragonal $BaTiO_3$ nanoparticles have been synthesized using supercritical water [7]. The features of hydrothermal reaction under supercritical conditions are ability to control ionic reaction and metal oxide solubility in water; these are determining factors of particle formation on hydrothermal synthesis, by changing temperature and pressure. Especially, when hydrothermal crystallization takes place near the critical point of water, nucleation occurs at a high degree of supersaturation and then nanoscaled particles can be produced. This resulted from an enhancement of ionic reaction rate and an extremely low solubility of metal oxide (product) in water due to decrease of dielectric constant of water [8]. Here, selective production of tetragonal $BaTiO_3$ nanoparticles by hydrothermal synthesis in supercritical water using a flow reactor is described.

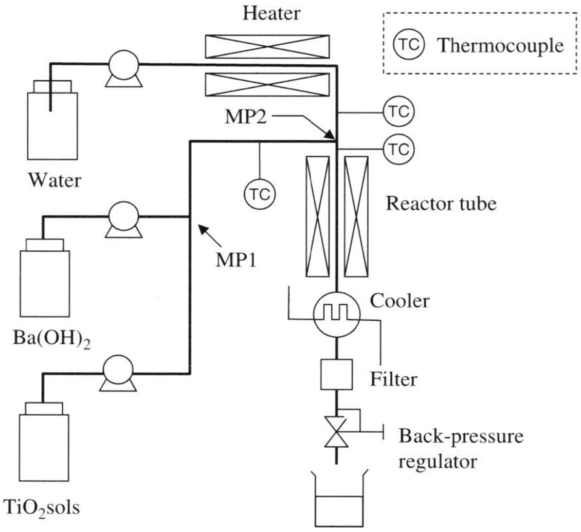

Figure 14.1
Continuous equipment to produce barium titanate nanoparticles.

1. Experiment for producing tetragonal $BaTiO_3$ nanoparticles by supercritical hydrothermal synthesis

Titanium oxide sol was used as a source of titanium. Barium hydroxide was used as a source of barium. In the cases to investigate the effect of Ba/Ti molar ratio of starting solution and pH, sodium hydroxide was used as pH controller.

Fig. 14.1 shows a schematic diagram of a flow reactor system for producing $BaTiO_3$ particles. The system consists of three high-pressure pumps, two electric furnaces, a particle collector, a cooler and a back-pressure regulator. The titanium oxide sol and barium hydroxide aqueous solution were fed by separate pumps. Two streams are mixed at the mixing tee (MP1) and then fed to the reactor. On the other hand, distilled water is fed by another pump and then heated to supercritical water by an electric furnace. The mixed stream is mixed at the entrance of reactor (mixer) with supercritical water and then is heated to the reaction temperature rapidly. Hydrothermal reaction occurs and metal oxide particle are formed. The reactants are cooled and then the products are captured in the inline filter (pore size: 500 nm). The reactant is recovered in a vessel followed by depressurization. Reaction temperatures were 300–420°C, and pressures were 20–40 MPa. The collected particles were dried in an oven at 60°C for 24 h. Crystal structure and phase were determined by an X-ray diffraction method and Raman spectroscopy. Morphologies and size of particles were observed by a transmission electron microscopy.

2. Selective production of tetragonal $BaTiO_3$

In order to determine the appropriate solution condition for producing $BaTiO_3$, we conducted several preparatory experiments at 400°C, 30 MPa. The Ba/Ti molar ratio was changed in the range of 1–10 under various NaOH concentrations from 0.02 to 0.2 M. As a result, $BaTiO_3$ was obtained from barium-rich (Ba/Ti > 1.2) and basic conditions (NaOH > 0.1 M). So, the Ba/Ti molar ratio of the starting solution was fixed to 1.2 in this study [9].

We investigated the effects of the reaction temperature on the crystal phase by changing from 300 to 420°C at a constant pressure of 30 MPa. Fig. 14.2 shows XRD profiles of particles obtained at various reaction temperatures. According to XRD analyses, the product at 300°C was the mixture of the residual TiO_2 and $BaTiO_3$. The samples obtained at 350°C and above were single-phase $BaTiO_3$. A decision on the polymorphism of $BaTiO_3$, cubic or tetragonal, was taken based on the existence of several doublets in

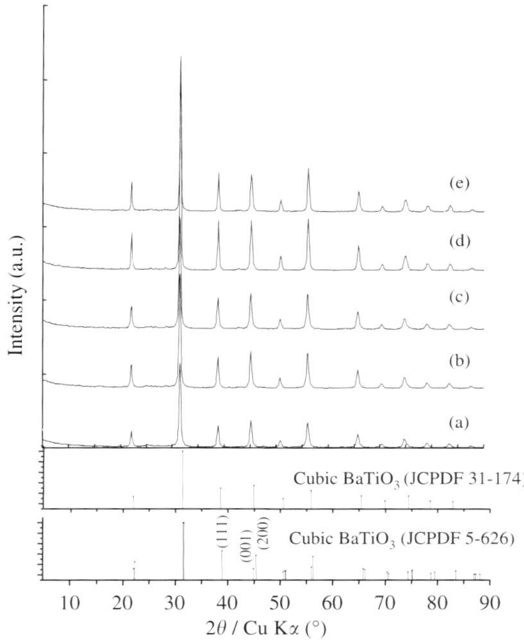

Figure 14.2
XRD profiles of barium titanate particles obtained at various temperatures of: (a) 300°C, (b) 350°C, (c) 380°C, (d) 400°C, (e) 420°C. Pressure is 30 MPa in all cases.

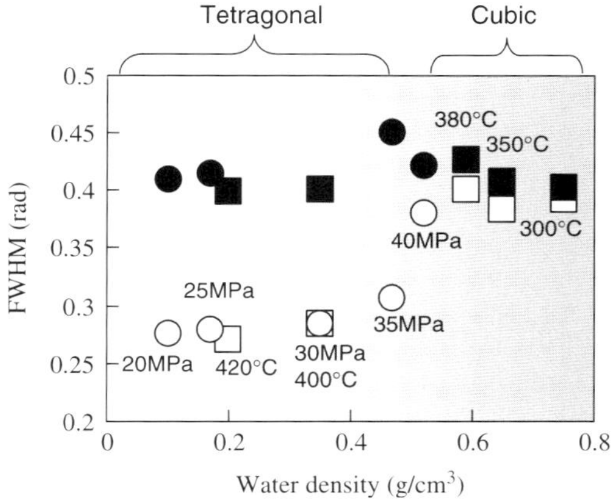

Figure 14.3
Water density dependencies on the peak height ratio, *R* and FHWM values of specific two peaks of the products. Symbols: Rectangles – FWHM200 (opened at 300–420°C and 30 MPa; filled at 400°C and 20–40 MPa); circle – FWHM111 (opened, filled – same as above); square – R (opened, filled – same as above).

the X-ray diffraction peaks, which are specific to the tetragonal $BaTiO_3$. In case of the fine particles several 10 nm in diameter, it is difficult to detect the doublet peaks due to the peak broadening. So we evaluated the full widths at half maximum (FWHM) of the selected two peaks, which corresponded to a singlet (1 1 1) and doublet (2 0 0, 0 0 2) of the tetragonal $BaTiO_3$. These FWHMs were named as $FWHM_{111}$ and $FWHM_{200}$, respectively. When the $FWHM_{200}$ was greater than the $FWHM_{111}$, the product was regarded as a tetragonal $BaTiO_3$. As a result of the phase transformation from cubic to tetragonal, the $FWHM_{200}$ was thought to become greater than the $FWHM_{111}$, since the doublet (200, 102) had overlapped. In addition, we also evaluated the peak height ratio ($R = H_{111}/H_{200}$) of each peak. By means of the standard peak pattern, the *R* value of cubic BT was smaller than 1, while that of tetragonal BT was larger than 1.

So, the FWHMs of each peak of the products are plotted in Fig. 14.3 for water density, which are represented by the opened symbols. At 380°C and below, the *R* values were smaller than 1, but the they became larger than 1 over 400°C. The FWHMs of the two peaks of the products were nearly the same at 380°C and below; however, the $FWHM_{200}$ remained constant around 0.4 rad but the $FWHM_{111}$ decreased to 0.28 rad at 400°C and above. We also performed analyses of crystal phase of products by the Raman spectroscopy. Tetragonal BT reveals a specific RAMAN shift around 300 cm^{-1}. The samples obtained at both 400 and 420°C had the obvious peak around 300 cm^{-1}. These results of XRD, TEM and RAMAN reveal that the products obtained under supercritical conditions were tetragonal BT.

Next, we evaluated the effects of pressure on the crystal phase at 400°C, whose results are represented in Fig. 14.3 by the filled symbols. The *R* value was greater than 1 and the $FWHM_{200}$ was greater than the $FWHM_{111}$ at the low pressure of 20–35 MPa, which corresponded to the water density of 0.1–0.47 g/cm^3. However at 40 MPa (0.524 g/cm^3), the *R* value was less than 1 and the $FWHM_{111}$ was nearly equal to $FWHM_{200}$. This means that the particle obtained at 40 MPa were close to cubic BT.

Fig. 14.4 shows TEM images of the products at 300 and 420°C. The particles obtained at 300°C were spheres around 30 nm in diameter. The particles obtained at 420°C were cubic particles around 50–100 nm in diameter. Since the cubic particles did not represent anisotropic crystal growth, it was found that the difference between these FWHMs was not due to crystal growth directed in (1 1 1) phase.

Fig. 14.5 shows TEM images and particle size distribution of $BaTiO_3$ obtained at 400°C, 30 MPa and 2 s of reaction time. As shown in the figure, cube-shaped $BaTiO_3$ particles with average particle size of 32 nm were obtained [10].

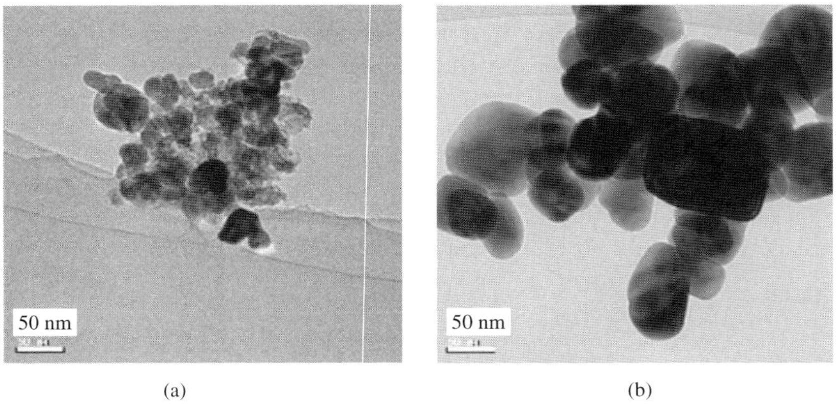

Figure14.4
TEM photographs of $BaTiO_3$ nanoparticles at: (a) 300°C, 30 MPa; (b) 420°C, 30 MPa.

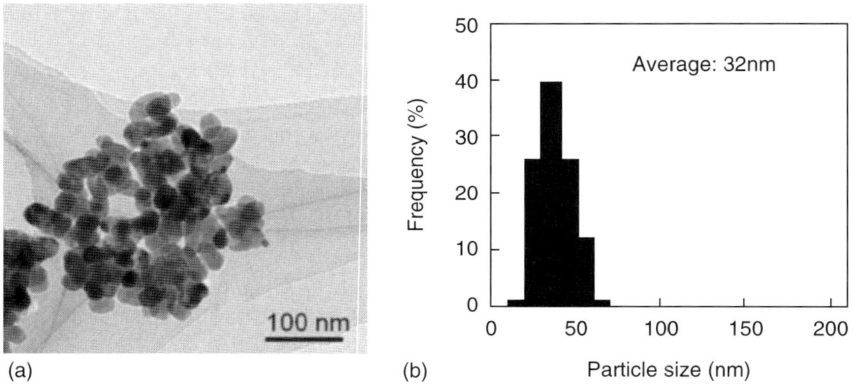

Figure 14.5
(a) TEM photograph and (b) particle size distribution of barium titanate nanoparticles obtained at 400°C, 30 MPa, 2 s of reaction time.

Polymorphism of $BaTiO_3$ depended not only on reaction temperature but also on reaction pressure (water density). As mentioned above, one reason for the formation of cubic BT during hydrothermal synthesis is said to be the residual OH ions in the crystal structure. Water density under supercritical conditions is smaller than that under subcritical conditions. In addition, since the diffusivity of supercritical water is higher than that of subcritical water, dehydration could proceed rapidly and the residual OH ions in the lattice were subsequently reduced. We think that this is the reason why tetragonal BT could be obtained under supercritical conditions.

Our work on $BaTiO_3$ is summarized as follows. Tetragonal $BaTiO_3$ nanoparticles can be obtained from titanium oxide sols and barium hydroxide aqueous solution by supercritical hydrothermal synthesis. Polymorphism of $BaTiO_3$ particles depended on water density, and selective production of tetragonal $BaTiO_3$ particles was achieved under supercritical conditions.

References

[1] J. Zeng, C. Lin, J. Li and K. Li: *Mater. Lett.*, **38**, 112–115 (1999).

[2] V.M. Fuenzalida, J.G. Lisoni, N.I. Morimoto and J.C. Acquadro: *Appl. Surf. Sci.*, **108**, 385–389 (1997).

[3] E. Brzozowski, M.S. Castro: *J. Eur. Ceram. Soc.*, **20**, 2347–2355 (2000).
[4] R. Asiaie, W. Zhu, S.K. Akbar and P.K. Dutta: *Chem. Mater.* **8**, 226–234 (1996).
[5] S.W. Lu, B.I. Lee, Z.L. Wang and W.D. Samuels: *J. Cryst. Growth*, **219**, 269–276 (2000).
[6] I.J. Clark, T. Takeuchi, N. Ohtori and D.C. Sinclair: *J. Mater. Chem.*, **9**, 83–91 (1999).
[7] Y. Hakuta, H. Ura and H. Hayashi: Japanese Patent No. 289737A (2005).
[8] T. Adschiri, K. Sue, Y. Hakuta and K. Arai: *J. Nanopart. Res.*, **3**, 227–232 (2001).
[9] Y. Hakuta, H. Ura, H. Hayashi and K. Arai: *Mater. Lett.*, **59**, 1387–1390 (2005).
[10] Y. Hakuta, H. Ura, H. Hayashi and K. Arai: *Ind. Eng. Chem. Res.*, **44**, 840–846 (2005).

APPLICATION 15

15 CERAMIC FILTER FOR TRAPPING DIESEL PARTICLES

Exhaust gas discharged from diesel engine vehicles, plant facilities, etc. has become a major concern as a cause of air pollution, health disturbances, and offensive odors. Gas emissions, especially those from diesel engine vehicles, are drawing close attention for their PM (Particulate Matter: suspended particulate matter defined in Japan as particles with a grain diameter of 10 μm or less). The PM in gas emissions can be easily seen as black fumes. PM is now established to be carcinogenic, and may act in ways similar to environmental hormones [1]. This redoubles the need to reduce the discharge of gas emissions and nitrogen oxide. There are various ways to reduce the harmful substances discharged from diesel engine vehicles. One is the common rail fuel injection system. Through recent technical innovations, gas emissions from diesel engine vehicles can now be purified to some extent. Better purification has recently enabled the design of high-performance diesel engine vehicles. Thanks to this improvement, in combination with the excellent fuel efficiency and durability of diesel engines, diesel engine vehicles now make up about 50% of the total sales of passenger cars in Europe. With the demand for diesel engine vehicles increasing, further purification of gas emissions is extremely important. Automakers will have to make diesel engine vehicles more environment-friendly while maintaining their high performance. Fig. 15.1 shows the regulatory trends for the control of gas emissions from diesel engine vehicles around the world. This control value for the discharge of PM from diesel engines is equal to or stricter than that from gasoline engines, and it continues to grow stricter still. The elimination of air pollution by the DPF (Diesel Particulate Filter) is a very effective means to achieve the control value. The following sections describe the basic properties and characteristics of a ceramic filter for trapping diesel particles with a porous silicon carbide obtained by a recrystallization method. Details are given on several basic study items applied in the testing of the filter medium for practical use: the trapping efficiency, pressure loss, strength, and heat resistance etc.

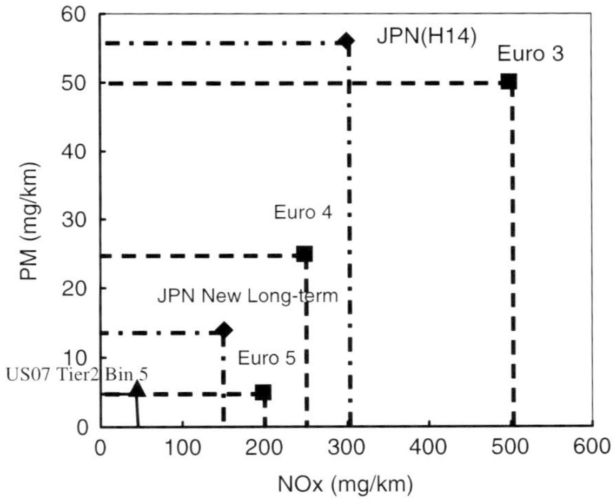

Figure 15.1
Trends in exhaust gas regulation in each country.

1. Production of PM

The production of PM is thought to involve three processes: the PAH (Polycyclic Aromatic Hydrocarbon) production process, a process in which PAH grows into crystallites of soot, and a process in which PAH aggregates into soot (see Fig. 15.2) [2]. Because diffusion combustion is the main process in diesel combustion, nothing can be done to eliminate the formation of local areas with excessive concentration. The unwanted outcome is

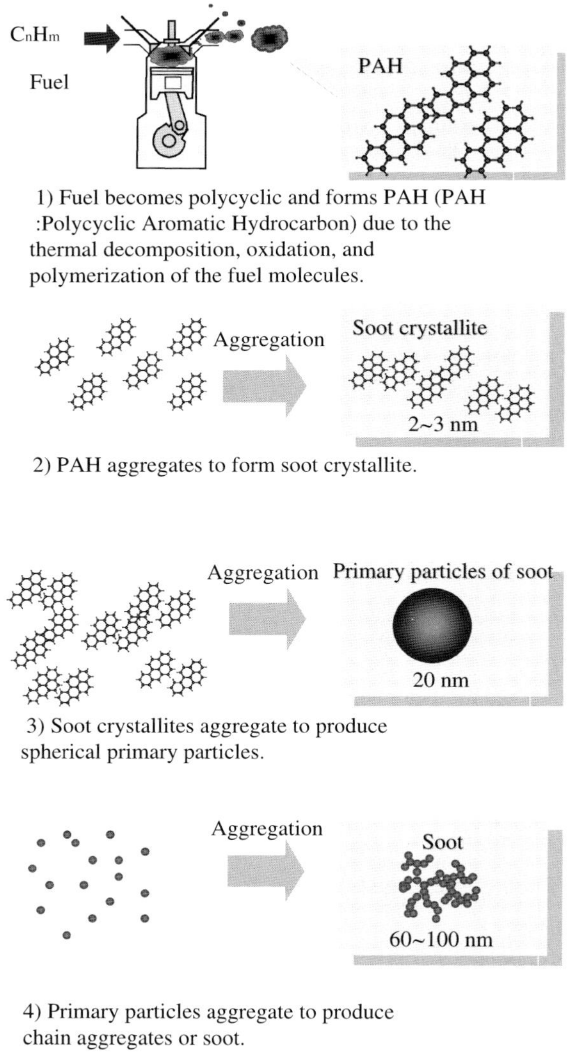

Figure 15.2
Process of soot production.

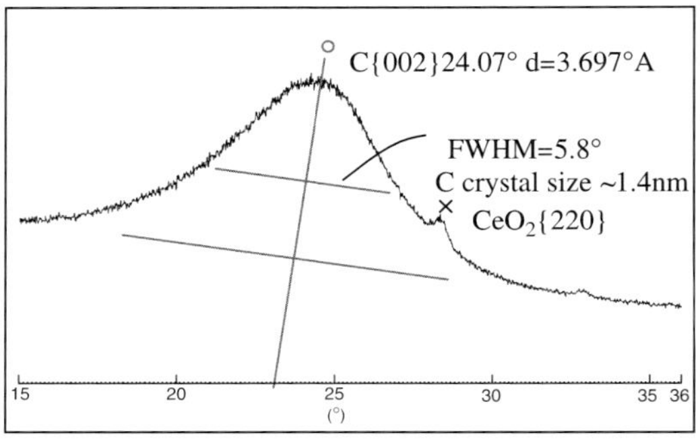

Figure 15.3
Result of X-ray diffraction measurement.

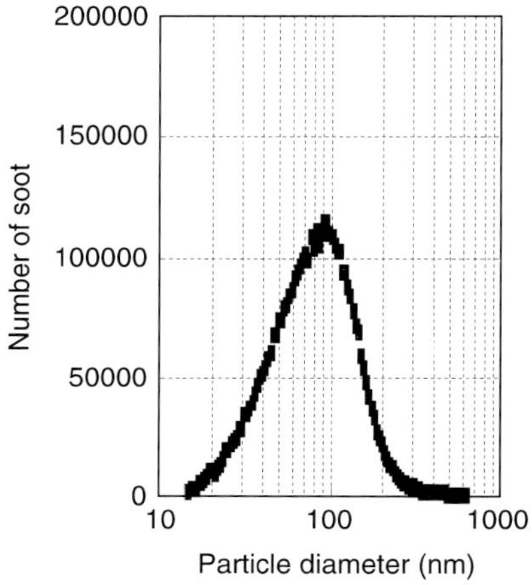

Figure.15.4
Size and number of particles discharged from an engine.

the production and discharge of soot. Fig. 15.3 shows the result of an X-ray diffraction analysis performed to determine the size of the soot crystallites discharged from diesel engine vehicles. The size of the crystallites determined from the half-value width was on the order of 1.4 nm. This result agrees with the result of a TEM image observation. The soot discharged from diesel engine vehicles is a product of aggregation of crystallites of about 1.4 nm in size. Fig. 15.4 shows the testing results on the number of soot particles discharged and the particle diameters. PM measurement methods can be broadly classified into the gravimetric method and particle diameter method. A gravimetric procedure called the tunneling method is now the only method accepted for certification tests. The measurement result shown here, however, was obtained using SMPS (Scanning Mobility Particle Sizer, a measurement solution now under study for future adoption), a method capable of counting the number of particles of given diameters. It can be seen from this result that PM discharged from diesel engine vehicles is made up mostly of particles on the order of 100 nm in diameter.

2. Trapping of PM

Fig. 15.5 shows the results of measurement of trapping characteristics performed by SMPS at an average pore diameter of 12 μm, a porosity of 40%, and a 60% DPF.

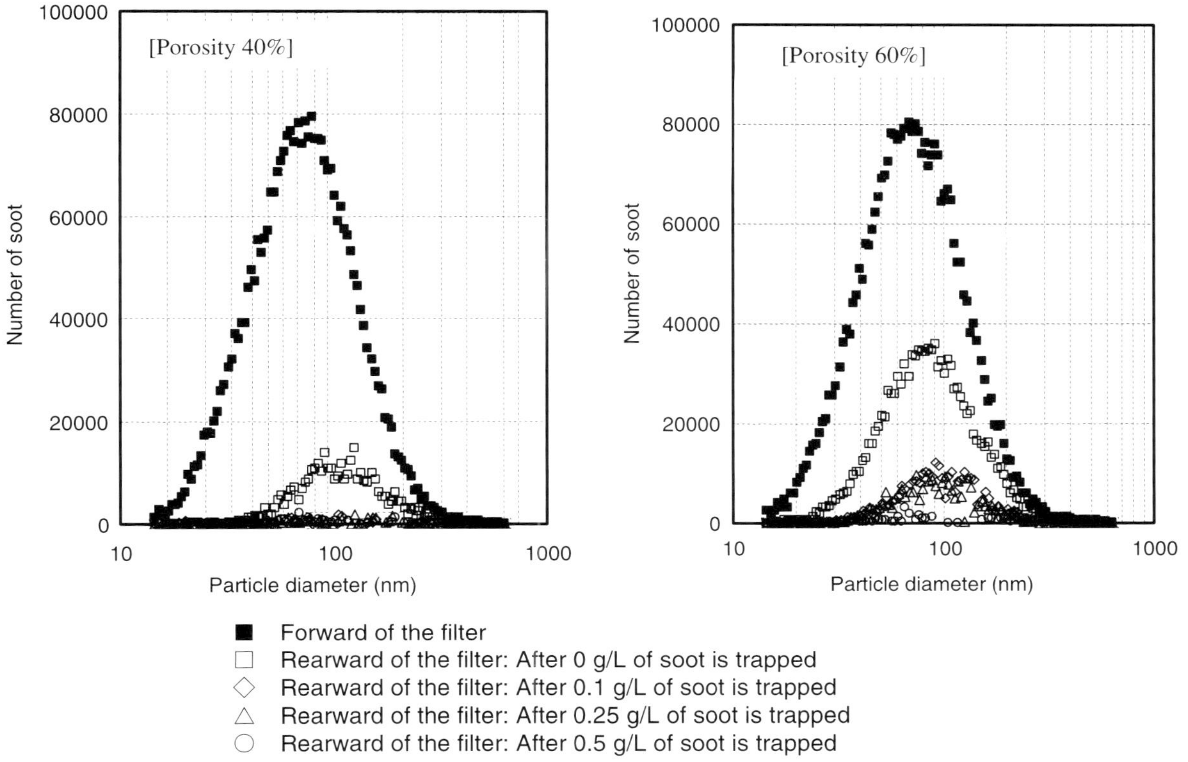

Figure 15.5
Characteristics of trapping efficiency by DPF, and their change as time goes by.

As this figure shows, the trapping efficiency is worst in the area near 100 nm where the largest amount of PM particles is discharged. In general, three forms of trapping act on filters [3–5]: (1) diffusion, (2) interception, and (3) impaction.

(1) shown above is Brownian diffusion, a phenomenon which grows in strength as the particle diameters get smaller. (2) and (3), on the other hand, become stronger as the particle diameters get larger. It can be also seen from above that the trapping efficiency is the worst for medium-sized particles with a diameter of about 100 nm. This is consistent with the earlier result shown in Fig. 15.5, in which the same tendency was conspicuous in a high-porosity filter. However, the trapping efficiency for particles with diameters of several tens nanometers or less is high from the initial stage one. This seems to be a characteristic of DPF. Further, in the case of a 40% porosity, the trapping efficiency after the trapping of 0.1 g/l of PM (time of use: a few minutes or less) reaches almost 100%. The PM initially trapped is thought to improve the trapping efficiency by forming another filter layer.

3. Pressure loss

When pressure loss is considered, it is necessary to study pressure loss by dividing it as shown below:

(1) Initial condition.
(2) When PM is trapped.
(3) After a filter is regenerated (after the ash piles up).
(4) After a catalytic coating is applied (for the catalyzed DPF).

The method to reduce the pressure loss differs from case to case. It is easily speculated that the pressure loss can be effectively reduced by increasing the pore diameter, making the porosity higher, and thinning the filter wall. But if these methods are implemented at the same time, the trapping efficiency will noticeably decline. Therefore, it is necessary to select the most effective method for a given filter to ensure that the trapping efficiency is never impaired. In the initial condition, for example, a reduced wall thickness will provide the highest effect. Later, when the PM is deposited, the pressure loss is reduced more effectively by increasing the filter cell density and increasing the filter surface area in order to thin the deposit layer. Thus, it is important to skillfully combine the methods most effective for each case and eventually to design a system that performs optimally and reduces the loss of fuel for regeneration without imposing a burden on the engine.

4. Features of porous silicon carbide

Silicon carbide is a covalently bound substance. The strength of covalent bonding of a substance increases with an increase in the melting temperature and decomposition temperature of the substance. For silicon carbide, the decomposition temperature is as high as 2,400°C or even 2,500°C. It can also be seen from this that silicon carbide is a material that shows very strong covalent bonding and is also excellent in chemical stability. As a result of these properties, silicon carbide also provides high hardness and excellent resistance to heat, corrosion, oxidation, and wear. Silicon carbide also has a high modulus of elasticity, and with the light weights of Si and C, the constituent atoms of silicon carbide, and the small difference in atomic weights, the lattice vibrations tend to become harmonic vibrations. Hence, silicon carbide provides a high coefficient of thermal conductivity among the ceramics [6]. If pores exist inside a ceramic, the mechanical properties of the ceramic are generally thought to deteriorate. In addition, ceramics at low temperature tend to fracture with brittleness, and ceramics with little or no plastic deformation tend to fracture rapidly under critical stress. This can be easily understood by contemplating the fracture phenomenon in the context of the conditional equation for brittle fracture proposed by Griffith [7] in 1920 based on the assumption that microcracks exist inside brittle materials and by contemplating the crack growth from the viewpoint of energy balance. From this approach, it is seen that the strength will be determined by the size of the cracks inside the material. The strengths of ceramics that manifest this type of behavior are frequently treated in the Weibull distribution presented in 1950 [8] based on the hypothesis (weakest link hypothesis) that the chain strength is determined not by the average strength of the individual rings running in a line as a chain, but by the weakest ring [9–11]. This is because the strength of a dense ceramic is governed by the size of the defects present in the material independently of the average characteristics of the material. The strength is determined, for example, by cracks or portions with air gaps occurring during formation. This means that ceramics can fracture at strengths much lower than the strengths at which cracking is theoretically predicted. It is therefore extremely important, in the manufacture of ceramics, to remove the causes of defects from the processes. Ceramics manufacturers have made considerable efforts to this end. We next come to the question of structures with only pores, in other words, porous matter in which the defects can be considered to be distributed more or less evenly. As in the case of dense matter, it is essential, with porous matter, to eliminate the processes that produce critical defects at every stage of manufacture, from material procurement up to commercialization of the final product. The theoretical strength of porous matter is lower than that of dense matter as a matter of course. It must be asked, however, whether small variation in strength can be safely predicted if the defects are thought to have been distributed evenly from the outset. Thinking this way, this point becomes advantageous, as porous matter is easier to work with than dense matter in the design stage. Moreover, recrystallized porous silicon carbide has a very small shrinkage of 1% or less in the burning process. This makes it much easier to control the dimensions during manufacture.

5. Functions and characteristics of DPF

The function of DPF, as its name indicates, is to purify gas emissions by trapping PM through filtration. In this case, the DPF must not allow the PM deposited on the filter through trapping to impede the flow of gas emissions, as this can prevent an engine from running. Further, deposited PM is removed by methods such as combustion [12]. Of the removal methods, the regeneration method by combustion works by igniting a part of the trapped PM forcibly in various ways and dispersing combustion to ensure that all of the PM is burned. This method regenerates the filters. In this case, the temperature of DPF reaches high values of 900°C or above. A rapid increase in temperature occurs during the regeneration of DPF and a temperature gradient develops during propagation of heat from the ignition area. Thermal shock resistance, heat resistance, and low reactiveness at high temperatures are therefore required. The honeycomb-type recrystallized silicon carbide DPF has a higher rate of trapping compared with that of other types, such as foam type and fiber type. It is also capable of trapping PM on the nano-order with low pressure loss. The requirements for this PM trapping efficiency and low pressure loss are fulfilled by a pore structure obtainable by virtue of the properties of recrystallized silicon carbide, namely, because silicon carbide is difficult to sinter and has little or no sintering shrinkage. Because of these properties, the pore structure can be controlled by the particle diameter of the coarse grain and particle size distribution, or by adding a pore-making agent. Further, recrystallized silicon carbide is poorly reactive during regeneration up to the temperatures of 900°C or above, and thus it fully meets the characteristics requirements for a DPF. A high coefficient of thermal conductivity can also be listed as a feature. This high coefficient of thermal conductivity is a very important characteristic for DPF. When trapped PM is burned, the heat is efficiently diffused radially, and from the upstream side of the gas to the downstream side. Because of this characteristic, the production of local high-temperature portions can be avoided and maximum temperature inside the DPF can be reduced [13] even if PM is burned inside a filter. This also improves the durability. Another problem is deterioration in the Pt catalyst due to the local temperature increase in the catalyzed DPF. The activity is reduced due to Pt sintering. Fortunately, this also can be controlled. The drawbacks of silicon carbide are its large coefficient of thermal expansion [3] (4.3×10^{-6}/°C) and its tendency to crack due to thermal stress. These problems have been solved by developing a technique in which DPF is divided into small segments and then combined into a required size [14].

6. Future of filters for trapping diesel particles

This section has described filters that use porous silicon carbide to trap diesel particles. As a result of the development based on the characteristics of silicon carbide, we have been able to trap nanoparticles and provide the market with filters with a trapping efficiency of almost 100%. In the future we would like to research porous silicon carbide with the target of further enhancing the superiority of diesel engine vehicles. If it succeeds, it will be possible to develop exhaust gas systems for diesel engine vehicles, which help to purify and improve the global environment.

References

[1] N. Kajiwara (Ed.): in *Technology for Removing Fine Particles Contained in Gas Emissions Discharged from Diesel Engine Vehicles*, CMC Inc., Tokyo p. 23 (2001).

[2] S. Kubo: Text for the Society of Automotive Engineers of Japan's Symposium, No. 01-04 20044070 (2004).

[3] A. G. Konstandopoulos: SAE Paper No. 2000-01-1016 (2000).

[4] H. Emi: *Res. Aerosol*, **4**(4), 246–255 (1989).

[5] T. Kusuda: *Ceramics*, **23**(8), 723–726 (1988).

[6] S. Soumiya, Y. Inomata (Eds.): in *Basics, Application and Product Introduction on Silicon Carbide Ceramics*, Junkudo Co. Ltd., Kobe (1988).

[7] A. A. Griffith: The theory of rupture, in C.B. Biezeno, J. M. Burgers (Eds.) *Proceedings of the First International Congress on Applied Mechanics*, J. Waltman, Delft, p. 55 (1924).

[8] W. Weibull: *J. App. Mech.*, Vol. 51, Sept., 293–297 (1951).

[9] Edited by the subcommittee on basic engineering courses under the editorial committee of The Ceramic Society of Japan, Mechanical properties of ceramics, The Ceramic Society of Japan, Tokyo, p. 21 (1979).

[10] R.W. Davidge: *Strength and Fracture of Ceramics*, Translated by H. Suzuki and T. Iseki, Kyoritsu Shuppan Co. Ltd., Tokyo, p. 34 (1982).

[11] T. Nishida, K. Yasuda (Eds.), *Evaluation of Mechanical Properties of Ceramics*, p. 63, The Nikkan Kogyo Shimbun Ltd., Tokyo (1986).

[12] Edited by the 124th committee on high-temperature ceramic materials, Japan Society for the Promotion of Science, New SiC-based ceramic materials – Recent development, Junkudo Co. Ltd., Kobe, p. 239 (2001).

[13] K. Ohno, K. Shimato, N. Taoka, H. Santae, T. Ninomiya, T. Komori and O. Salvat: SAE paper 2000-01-0185 (2000).

[14] A. Itoh, K. Shimato, T. Komori, H. Okazoe, T. Yamada, K. Niimura and Y. Watanabe: SAE paper No. 930360 (1993).

APPLICATION 16

16 NANOPARTICLE FORMATION OF DNA (GLOBULE TRANSFORMATION)

Structure of DNA (coiled structure) in solution can be transformed to globule structure by raising concentrations of condensing reagents such as combination of low-molecular-weight cation and polyethylene glycol (PEG) or multivalent cation (e.g. spermidine). The globule structure is highly condensed and it is induced by dehydration by PEG and suppression of repulsive force between strands by screening negative charge of DNA backbone. This phase transition depends on concentrations of both PEG and cations, and both coiled and globule structure are found in the transition region (Fig. 16.1) [1]. This phase transition is reversible; therefore, decreasing the concentration of condensing reagents induces reversal phase transition to the coiled structure. Structure of globule DNA depends on species and concentrations of condensing reagents. Typical condensing reagents, combination of PEG and cation, induce a highly condensed toroidal structure, which is several ten nanometers in diameter (Fig. 16.2).

This phase transition is induced immediately under sufficiently high concentration of condensing

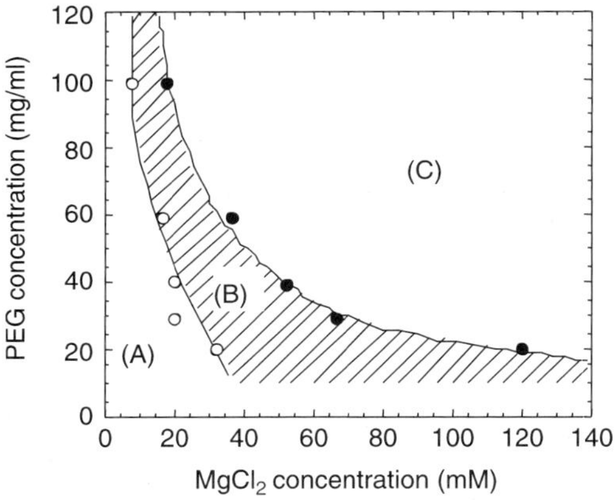

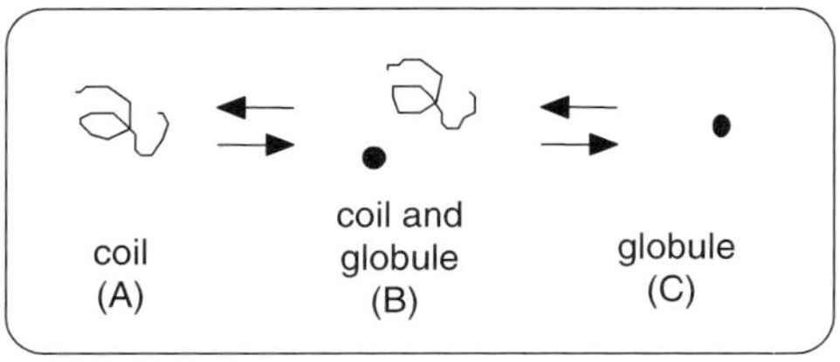

Figure 16.1
Dependence of DNA structure on the concentration of PEG and $MgCl_2$. In PEG/$MgCl_2$, the state of the DNA is represented by: (A) coiled; (B) coiled/globule coexistence; (C) globule.

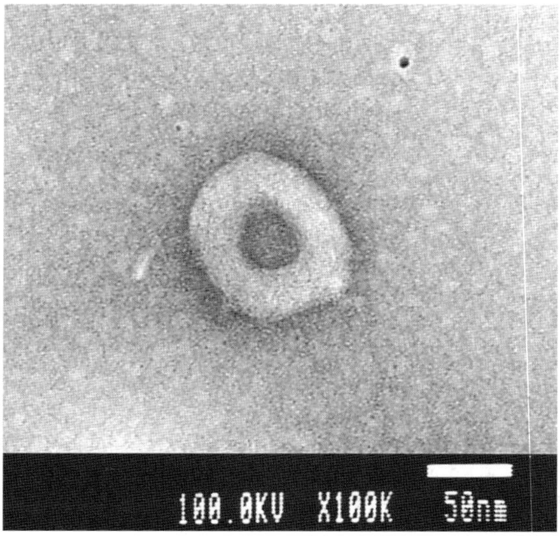

Figure 16.2
Image of globule DNA captured by a transmission electron microscope (provided by Dr. Y. M. Urano, Toyohashi Univ. of Tech.).

reagents [2]. Fig. 16.3 shows sequential photographs of the globule transition in 15% PEG20000 solution by aluminum ion supplied by an electrode reaction. When DC voltage is applied to electrodes, the aluminum ion is generated at an aluminum anode, and migrates toward the cathode (left direction in the figures) by electrophoresis. When the aluminum front (dotted lines in Fig. 16.3) migrates across an area of DNA molecules, globule transition is induced immediately due to exceeding of a critical concentration of the aluminum ion because the solution contains PEG. As the result of the globule transition, blurred spots of coiled DNA change to condensed blight spots as shown in Fig. 16.3.

1. Tolerance of DNA nanoparticles against mechanical stress

Since the globule structure is highly condensed, effect of mechanical stress can be suppressed, and this property permits handling of giant DNA in solution as demonstrated previously [3]. In this experiment, first of all, yeast chromosomal DNA embedded in agarose gel plugs was prepared. Some plugs were treated with condensing reagents of combination of PEG and NaCl for globule transition. After this treatment, the samples were stained with a fluorescent dye and observed with a fluorescent microscope. To evaluate the effect of mechanical stress, the gel plugs on coverslips were melted by increasing the temperature and then immediately solidified by cooling. Effects of the solution flow on the shape of DNA molecules were evaluated because melting of agarose plug generates shear stress upon DNA molecules accompanying the flow in solution. Fig. 16.4 shows the shape of the DNA molecules treated as described previously. Fig. 16.4 demonstrates that globule transition condensed DNA molecules significantly, and the condensed structure remained even after exposure of mechanical stress. On the other hand, coiled DNA molecules were stretched by flow. This strongly suggests that breakdown due to shear stress of flow can be suppressed by globule transition.

This property was also examined by pulsed field gel electrophoresis. Agarose gel plugs containing coiled DNA and globule DNA were melted and exposed to shear stress by vortexing solution at a different speed. Those samples were re-solidified and reverted to coiled structure by soaking gel plugs in electrophoresis buffer, and analyzed by pulsed field gel electrophoresis. As shown in the electrophoregram of Fig. 16.5, solution mixing produced many short fragments and DNA remained after mixing in the case of coiled DNA no longer. On the other hand, long DNA still remained after mixing in the case of globule DNA. This result suggests that fragmentation of long DNA can be strongly suppressed by globule transition, and globule transition

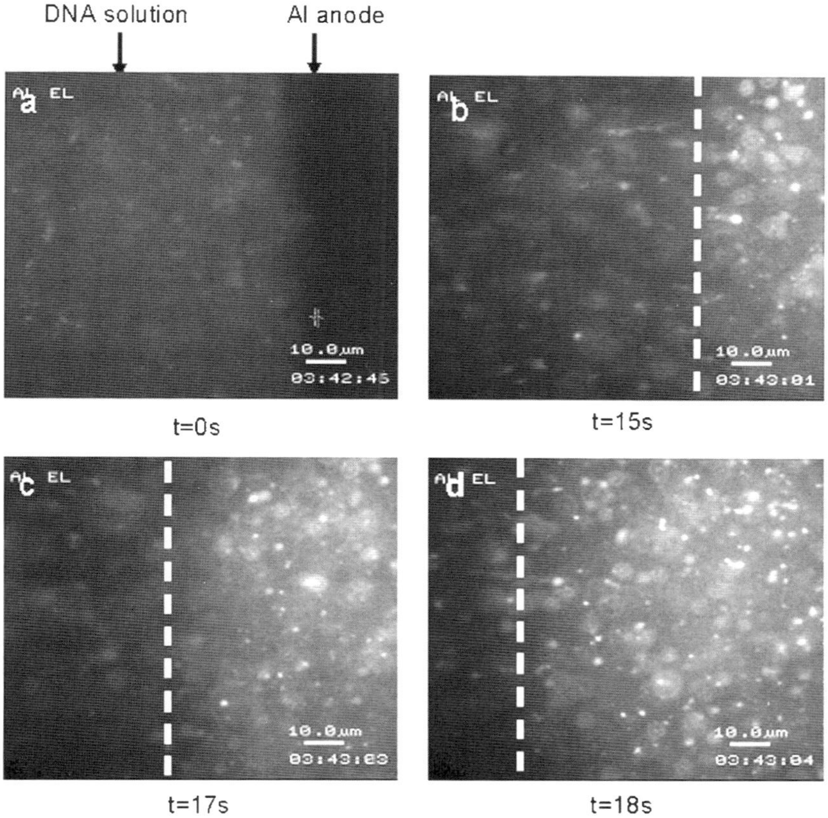

Figure 16.3
Coil–globule transition of DNA molecules by electrolysis of Al electrodes. Phase transition of T4 dc DNA due to electrolysis was fluorescently observed. (a) DNA molecules around the solution–electrode interface before voltage application. (b–d) Time-lapse photographs of the coil–globule transition of the DNA molecules. Dotted line in each photograph indicates the aluminum ion front. Scale bars represent 10 μm.

is a quite effective method to manipulate giant DNA without breakdown.

2. Micromanipulation of DNA nanoparticles

To analyze DNA molecules or to apply DNA molecules as templates for addressing nanoparticles, a target of single DNA molecules should be manipulated and transported to intended positions for suitable analysis or templates. Even though electric force is quite effective to manipulate small particles such as DNA molecules, charge of globule DNA is almost canceled; therefore, Coulomb force cannot be applied for manipulation. Gradient force is proportional to square of electric field gradient and is much smaller than Coulomb force; however, this force acts on electrically neutral particles such as globule DNA. The gradient force is also generated by AC electric field because the gradient force directs toward higher electric field regardless of its direction. Application of AC electric field is quite effective to suppress electrolysis on electrodes, which sometimes interferes manipulation due to bubble production.

Light, electromagnetic wave of extremely high frequency, also generates gradient force. This method, called as laser tweezers, permits non-contact micromanipulation of particles; however, tapping force for coiled DNA is quite small, and coiled DNA cannot be manipulated by this method. So far, laser manipulation for coiled DNA was achieved by conjugation of coiled DNA and a microbead that was several microns in diameter. The conjugate of DNA–microbead can be manipulated by laser tweezers or a needle magnet. However, the conjugation method for long DNA is not

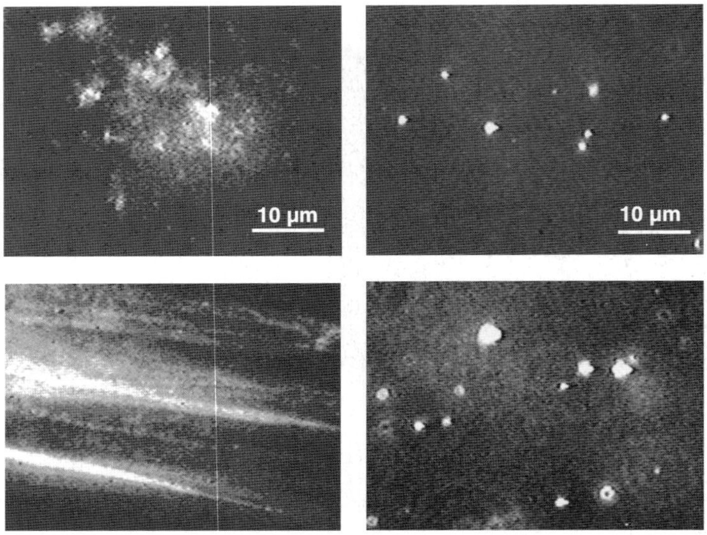

Figure 16.4
Photographs of coiled and globule DNA before and after exposure of shear stress. Fragmentation of DNA molecules was evaluated by observing shape of coiled and globule DNA before and after exposure of shear stress; (a) and (b) show coiled and globule yeast chromosomal DNA molecules, respectively, prepared in gel plugs. These molecules were exposed to shear stress by melting the gel plugs; (c) and (d) show coiled and globule DNA, respectively, exposed to shear stress.

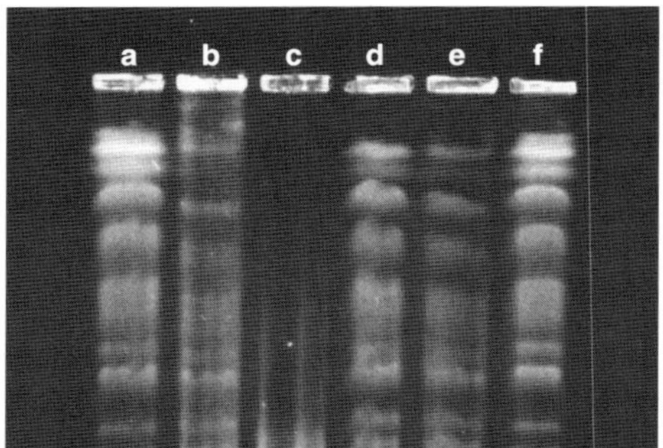

Figure 16.5
Pulsed field gel electrophoresis to analyze effects of shear stress. Effects of shear stress on coiled and globule DNA were analyzed by pulsed field gel electrophoresis. Gel plugs containing coiled or globule DNA were vortexed. Lanes (a–c) and (d–f) correspond to coiled and globule DNA, respectively. Samples in lane (b) and (d) were vortexed at setting 3 for 10 s. Samples in lane (c) and (e) were vortexed at setting 6, representing considerably higher stress than setting 3. No vortexing was treated with sample (a) and (f).

easy due to low efficiency of conjugation reaction and fragility of DNA molecules. Because of difficulty in conjugation, new manipulation methods without conjugation procedure have been desired. Low trapping force of DNA is ascribed to low density of materials and low difference of refractive index. As described previously, globule transformation induces highly condensed state, and it increases difference in refractive index between DNA molecule and surrounding, and also increases trapping force. Laser trapping of globule DNA was already demonstrated [4]. T4 phage DNA was transformed to globule state by 60 mg/ml PEG, 50 mM $MgCl_2$, followed by staining with a fluorescent dye, DAPI (4′, 6-diamidino-2-phenylindole). The transformed DNA was optically trapped by infrared laser of 180 mW in optical power. Fig. 16.6 shows laser manipulation of the T4 phage DNA molecule. The white arrow in this figure denotes a focal point of the laser beam. When a microscope stage was moved, free DNA moved accompanied by stage moving leftward (Fig. 16.6a–c) and upward (Fig. 16.6c–e). During this stage movement, the trapped DNA was retained at the focal point.

Breakdown of large DNA molecules can be suppressed by globule transition. The transition contributes to not only breakdown suppression but also manipulation of DNA without modification. These manipulation methods facilitate addressing of nanoparticles by using DNA template.

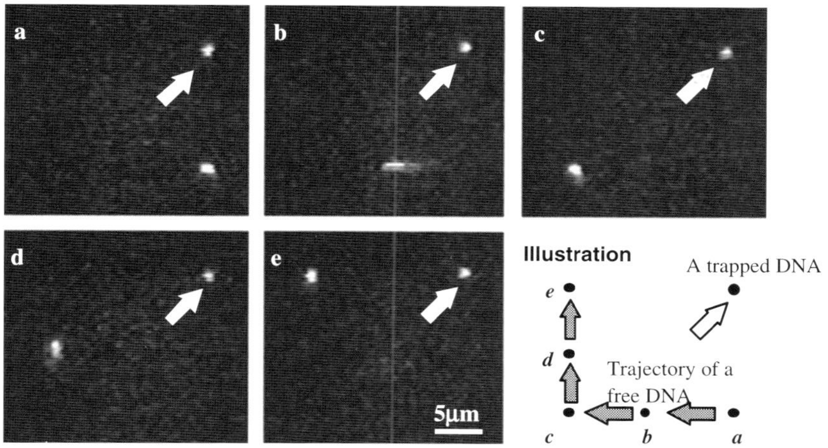

Figure 16.6
Sequential photographs of trapped and free T4 phage DNA molecules in the globule state at $t = 0$ (a), 0.24 s (b), 0.64 s (c), 0.87 s (d), and 1.14 s (e). Scale bar 5 μm. The position of the focal point is indicated by white arrows. At $t = 0$, the microscope stage started to move leftward until $t = 0.64$ s and the free DNA molecules moved with the stage. During the motion, the optically trapped DNA molecule remained stationary. (f) The trajectories of the free and the trapped DNA molecules are illustrated schematically.

References

[1] K. Yoshikawa, Y. Matsuzawa: *J. Am. Chem. Soc.*, **118**, 929–930 (1996).

[2] J. Komatsu, M. Nakano, H. Kurita, K. Takashima, S. Katsura and A. Mizuno: Electrophoresis, **26**, 4296–4303 (2005).

[3] S. Katsura, A. Yamaguchi, K. Hirano, Y. Matsuzawa and A. Mizuno: *Electrophoresis*, **21**, 171–175 (2000).

[4] S. Katsura, K. Hirano, Y. Matsuzawa, K. Yoshikawa and A. Mizuno: *Nucleic Acids Res.*, **26**, 4943–4945 (1998).

APPLICATION 17

17 ADDRESSING OF NANOPARTICLES BY USING DNA MOLECULES

DNA serves as the stores of genetic information, which are expressed as a sequence of four bases – adenine (A) thymine (T), guanine (G) and cytosine (C). The structure of DNA is a double helix form of 2 nm in diameter as shown in Fig. 17.1.

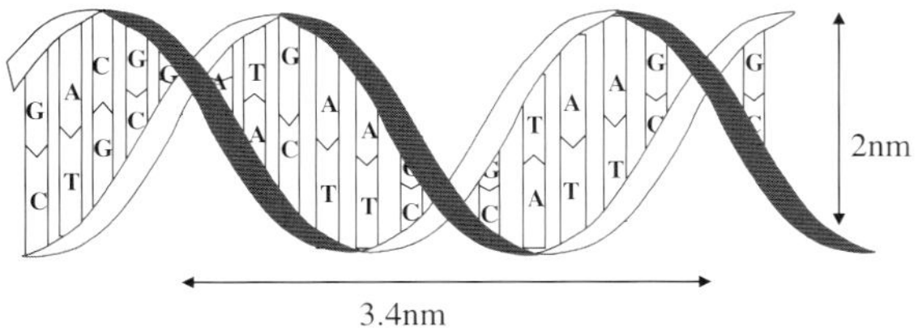

Figure 17.1
Tertiary structure of DNA.

Specific hydrogen bonds are formed between bases of each strand (Fig. 17.2). 'A' specifically links only 'T', and 'G' specifically links only 'C'. The hydrogen bonds can be destabilized by alkaline treatment or heat treatment, and double-stranded DNA (dsDNA) is converted into single-stranded DNA (ssDNA). When denatured ssDNA is restored in normal condition, specific hydrogen bonds are formed between two complementary DNA strands. This formation of the hydrogen bonds is called 'hybridization'. Hybridization has been widely applied for detection of DNA molecules having same sequence by labeling small DNA of known sequence (probe) by radioisotope or fluorescent dye.

Recently it was proposed that hybridization technique can be applied for addressing nanoparticles. The energy band is affected by quantum effects

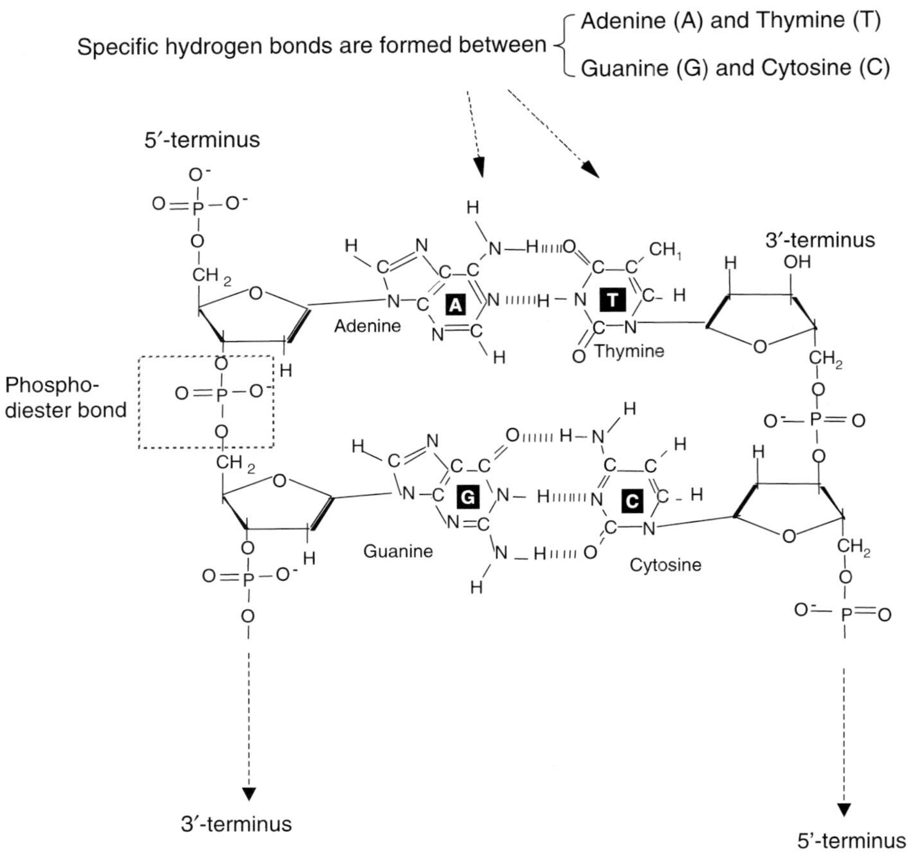

Figure 17.2
Sequence-specific hybridization of DNA molecules.

of particle boundary because dimensions of nanoparticles are extremely small. The possibility was pointed out that addressing of nanoparticles creates new optical and/or electrical devices. Application of DNA hybridization is expected to permit spontaneous addressing of nanoparticles.

1. Stretching of DNA molecules

Addressing pattern of nanoparticles can be designed by choosing sequences of DNA and oligonucleotides as shown in Fig. 17.3. For this purpose, DNA molecules should be artificially stretched; otherwise DNA shrinks because entropy of DNA molecules increases by bringing DNA terminals close together.

Several stretching methods have already been proposed as shown in Fig. 17.4. Fig. 17.4a illustrates the molecular combing method, which is based on a meniscus movement. DNA in air is stretched because DNA molecules behave to raise length in water phase due to high hydrophilic property. Yokota et al. [1] stretched DNA molecules using this method followed by restriction digestion, and succeeded in determination of the restriction map by observing DNA molecules treated with the restriction enzyme. Flow in solution also stretches DNA molecules of which one terminal is anchored on the substrate surface (Fig. 17.4b).

Fig. 17.4c illustrates a method based on electric field application. Since DNA is negatively charged, DC electric field stretches DNA molecules of which one terminal is anchored on the substrate surface. However, DC current promotes electrode reaction especially under high ionic condition, and therefore stretching by AC electric field has also been developed. The force to stretch DNA by the AC electric field is lower than that by the DC

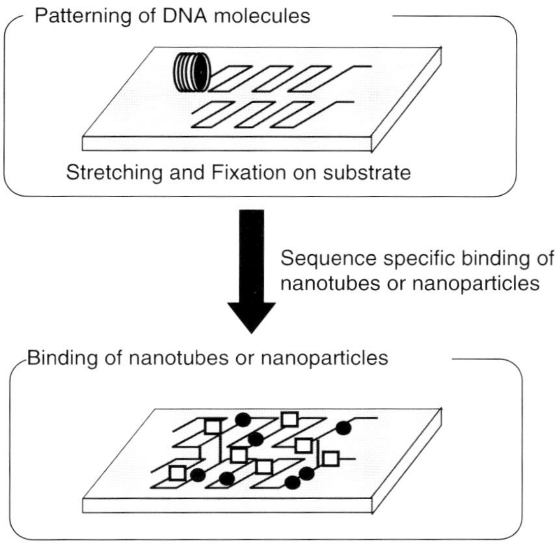

Figure 17.3
Functional structure based on nanoparticles aligned by DNA hybridization.

electric field, because the force by the AC electric field arises by polarization of DNA molecules. Therefore, gap between electrodes is shortened and electric field strength is intensified for AC stretching operation [2].

As another method, a stretching method based on movement of interface between ice and solution was also proposed [3]. DNA molecules tend to be excluded into the solution area. Therefore, if the terminus of DNA molecules is anchored on substrates or in ice crystal during unidirection freezing of solution with DNA, the remaining parts of those DNA are stretched (Fig. 17.4d).

2. Addressing of nanoparticles

Holiday junction arrays were designed and applied for addressing of nanoparticles [4, 5]. When one strand of DNA is replaced with conjugate DNA–gold particles, the gold particles are aligned along the holiday junction arrays.

Further development of addressing method can be expected if positions and shapes of DNA molecules can be manipulated on the substrate. This addressing

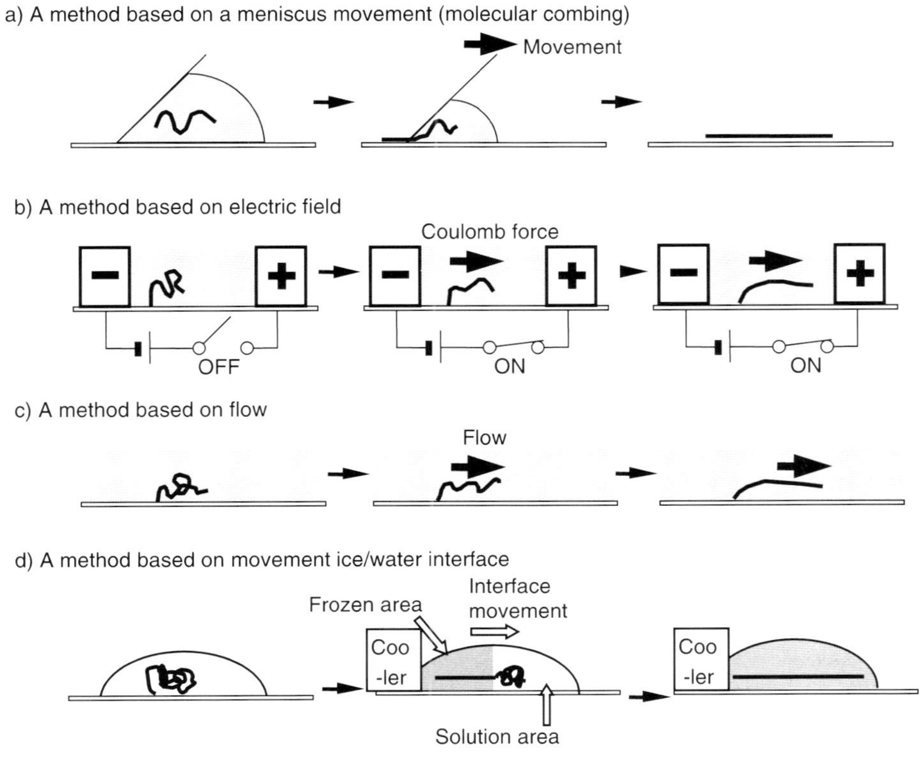

Figure 17.4
Various methods for DNA stretching.

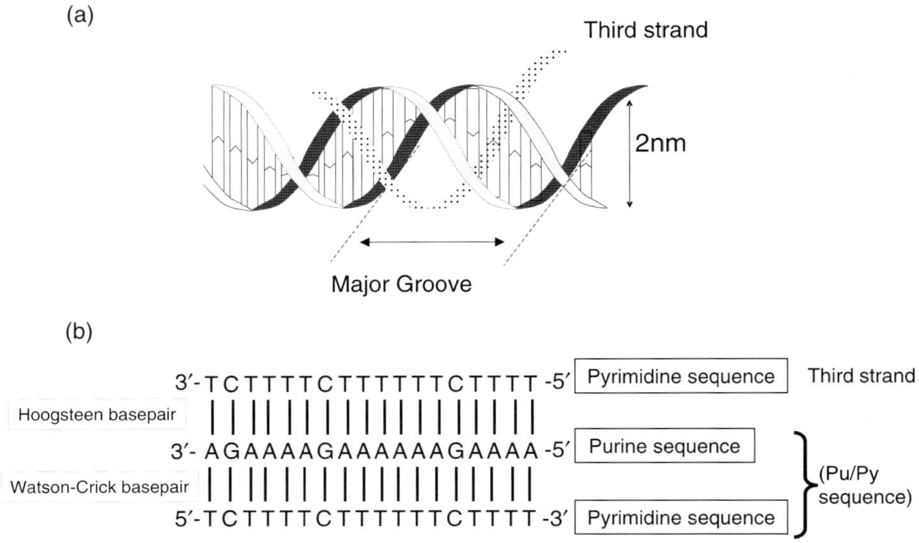

Figure 17.5
Structure of triple DNA helix. (a) Illustration of triple DNA helix. Third strand intrudes into major groove of the double-stranded DNA. (b) Purine (A or G) or pyrimidine (T or C) tract is essential for triple helix formation. The third strand is antiparallel to purine tract of the double-stranded DNA.

method is expected to permit addressing of many kinds of nanoparticles as designed. However, it is difficult to hybridize template DNA on the substrate and conjugate probe DNA–nanoparticles, because this hybridization requires handling of single-stranded template DNA molecules, which are so fragile that they can be easily broken down by shear stress.

Even though template DNA can be denatured by the alkaline condition or heat treatment after anchoring on the substrate, renaturation with the original strands may occur immediately after removal of denaturation condition.

To solve this problem, hybridization of triple-helix structure has been proposed. Triple-helix structure is formed in purine (A or G) or pyrimidine (T or C) tract. The third DNA strand intrudes into the major groove of the double-stranded DNA. Since the hybridization with the third strand is formed by the Hoogsteen base pair, the new hybridization occurs without breakdown of ordinary Watson–Crick base pair (Fig. 17.5). This property permits hybridization of conjugate probe DNA–nanoparticles with double-stranded DNA [6] and addressing of nanoparticles without denaturation of template DNA.

Addressing methods of nanoparticles, based on manipulation of DNA and sequence-specific hybridization, are expected to permit spontaneous assembly of large-scale functional structure of nanoparticles, and can be applied in several industrial fields.

References

[1] H. Yokota, F. Johnson, H. Lu, M. Robinson, A.M. Belu, M.D. Garrison, B.D. Ratner, B.J. Trask and D.M. Miller: *Nucleic Acids Res.*, **25**, 1064–1070 (1997).

[2] H. Kabata, O. Kurosawa, I. Arai, M. Washizu, S.A. Margarson, R.E. Glass and N. Shimamoto: *Science*, **262**, 1561–1563 (1993).

[3] J. Komatsu, M. Nakano, H. Kurita, K. Takashima, S. Katsura and A. Mizuno: *J. Biomol. Struct. Dyn.*, **22**, 331–338 (2004).

[4] C. Mao, W. Sun and N.C. Seeman: *J. Am. Chem. Soc.*, **121**, 5437–5443 (1999).

[5] Y. Hatakeyama, M. Minami, M. Umetsu, S. Ohara, S. Takami and T. Adschiri: *Jpn. J. Polym. Sci. Tech.*,**61**, 617–622 (2004).

[6] Y. Morii, J. Komatsu, K. Sei, K. Takashima, S. Katsura and A. Mizuno: 70th Conference of Society of Chemical Engineers, Japan, B309 (2005).

APPLICATION 18

18 DEVELOPMENT OF HIGH-PERFORMANCE ELECTROCHEMICAL REACTORS

1. Environmental purification by electrochemical reactors

The electrochemical reaction is a directional conversion of materials and energy. The highly efficient electrochemical reaction is primarily the oxidation and reduction occurring at electrodes. Many applications with the ideal hydrogen energy take a long time to realize. For example, fuel cell cars are expected to be with a zero-emission system, but the substitution of fuel cell cars for gasoline and diesel engine cars will not occur until 2030s or later. Therefore, technology development for purifying the exhaust gas from cars has to be continued for at least 20–30 years more. Electrochemical reactors are expected to be applied for environmental purifications as well as for energy generation.

In this chapter, the development of electrochemical reactors for NO_x decomposition and a novel reactor combining the electrochemical and thermoelectric devices are described [1–6].

2. NO_x decomposition in the exhaust gas with electrochemical reactors

The reduction of NO_x emission from vehicles has become one of the most important research targets for the environmental protection. Air pollution from nitrogen oxides (NO_x) in the combustion waste causes serious environmental problems, especially in the urban areas. Without the co-existence of oxygen, a successful decomposition of NO gas into oxygen and nitrogen was demonstrated over 30 years ago using the "three-way catalyst (TWC)" and other effective catalysts. However, in the presence of oxygen (such as under the lean-burn conditions), TWC does not provide sufficient decomposition to NO_x. The widely used catalytic decomposition process for NO_x requires additional 2–3% of fuel as reducing reagent, and thus increases the amount of fuel used by the engine. A high-efficient diesel engine is expected to save energy even when the exhaust gas contains the partial pressure of oxygen up to 10%. This is why the development of high-performance catalytic systems is based on the high P_{O_2} conditions. Environmental solutions must work hand in hand with energy-saving technologies.

From this point of view, the reduction of NO_x emissions can be achieved not only by catalytic NO_x decomposition but also by electrochemical decomposition. Without coexisting oxygen, a successful decomposition of NO gas into oxygen and nitrogen in a primitive electrochemical cell was first demonstrated over 25 years ago as shown in Fig. 18.1. Recently, many attempts to improve the properties of electrochemical cells operating in the presence of excess oxygen have been carried out using various catalysts as the cathode materials. Unfortunately, for

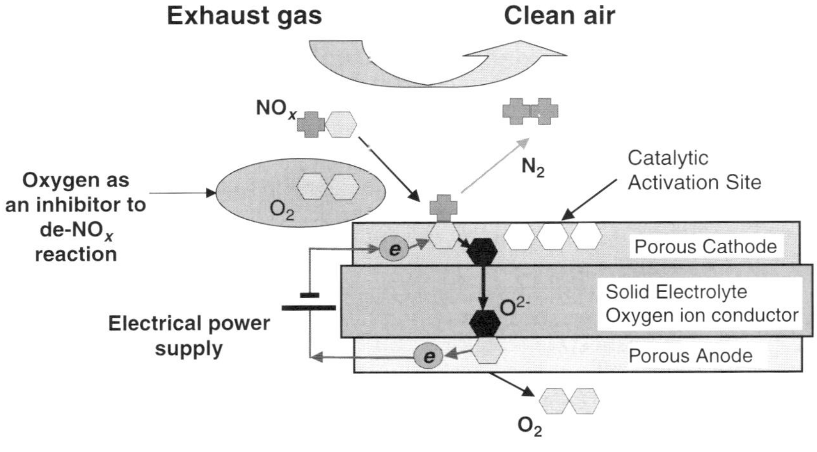

Figure 18.1
Schematics of the exhaust gas purification by an electrochemical cell.

all of the known cells to date, the co-existing oxygen is decomposed in preference to the NO_x gas; and thus, the electrical power requirements for NO_x decomposition stay at an extremely high level. In these electrochemical cells, their cathodes were used both as a current collector and a catalyst for NO_x decomposition.

Upon applying a voltage on these cells, NO gas is directly reduced at the three-phase boundary (cathode–YSZ–gas) forming gaseous N_2 and oxygen ions. The oxygen ions are transported through the solid-state electrolyte from cathode to anode and gaseous O_2 is evolved at the anode. At the same time, the excess O_2 in the combustion exhaust gas is adsorbed and decomposed at the three-phase boundary (TPB) in preference to the NO gas. As a result, the additional current associated with the oxygen ions produced due to this unwanted reaction far exceeds the current associated with the desired reaction. The supply of coexisting oxygen molecules should be reduced in order to decrease the electrical energy consumption to a level of practical use.

To ensure effective NO_x decomposition in the presence of excess oxygen, a new type of electrochemical cell with a functional multi-layer electrode was developed. This new electrochemical cell could be represented by the asymmetric arrangement of the general symmetric type SOFC cells, which consisting of an electrolyte sandwiched by electrodes. The external voltage applied between the cathode and the anode led to the polarization of the YSZ disk and generated a high concentration of oxygen vacancies in the near cathode region. Due to the concentration gradient of the oxygen ions, the diffusion of oxygen ions from the electro-catalytic electrode to the YSZ disk took place. To have effective cell operation, the cathode should be a good electric and oxygen ionic current conductor, having high electric and ionic conductivities along the cathode plane and from the electro-catalytic electrode to the YSZ solid electrolyte.

In addition to improving the cell configuration in the macro and microscale, applying nanoscale control at the reaction sites is expected to be effective for enhancing the selective absorption and decomposition of NO_x molecules over the predominant oxygen molecules. In addition, the accessibility of gas molecules to restricted spaces and electrochemically active surfaces of ceramics is also important in improving the reduction and oxidation reactions. Thus, nanostructural electro-catalytic electrode was investigated in relation to the de-NO_x characteristics of the cell. Combining the NO_x reduction accelerator with the oxygen acceptor and the nanoscale reaction spaces can promote very efficient NO_x decomposition, even under high P_{O_2} conditions. This is expected to be applied to treat the exhaust gas from diesel engines.

Reaction mechanism is described as follows. The electrochemical reactors for selective NO_x decomposition can be represented by the following asymmetric reactor arrangement, i.e. functional multi-layer electrode-cathode–YSZ-anode. The YSZ disk is used as an oxygen ionic conducting substrate for manufacturing the electrochemical reactor. The functional multilayer electrode on the cathode, in particular the nanoporous NiO–YSZ electro-catalytic electrode, plays most important role on the de-NO_x properties. The nanostructure artificially formed at the NiO/YSZ interface of the electro-catalytic electrode is thought to be a distinguishing feature of this type of reactors. In this reactor, the external voltage is applied between the cathode and anode; and, the electro-catalytic electrode on the cathode is free from the voltage drop. This geometry is very important to the de-NO_x operation, which depends not only on the ionic conductivity of the circuit as in the SOFC, but also on the diffusion process of the reactants. This cell configuration contributes to the decrease in the cell working temperature from 800 to 500°C, which is close to the expected temperature range for the purification of exhaust gas from diesel engines.

Applying an external electric field leads to the polarization of the YSZ solid electrolyte and to the generation of a high concentration of oxygen vacancies inside the three-dimensional network of YSZ particles in the NiO–YSZ electro-catalytic electrode. This causes the reduction of NiO grains into Ni at the area near its surface (Fig. 18.2). As a result, the formation of nanosized nickel grains takes place near the YSZ/NiO interface region. When the external voltage is switched off, the penetration of NO_x and O_2 gases through the nanopores between the NiO and YSZ grains leads to the oxidation of the newly formed nanosized Ni grains and transforms them to NiO grains. As shown in the Fig. 18.2, Ni grains with diameter of 10–50 nm are located around the untransformed central part of primary grains. Also, there are pores located both in the interfacial vicinity and in front of the newly formed nano Ni grains. This phenomenon is an electrochemical reduction reaction occurring at the YSZ/NiO interfaces caused by the induced electrical field in the cell when applying voltage across the electrodes.

Nanoparticles were formed by an electrochemical reduction reaction at the interfaces of ionic conducting phase and electrical conducting phase. An applied voltage caused the reduction reaction and re-crystallization of NiO to Ni. This resulted in a decrease in the volume. Voltage induced generation and diffusion of the oxygen vacancy accelerated the phase transformation and Ni nanoparticles were distributed at the interfaces. The strength of applied electrical field affected the thickness of the nanoparticle layer at the interface. The thickness of nanoparticle layer increased with an increase in the applied voltage. However, applying excess voltage could result in extra removal of oxygen and caused generation of micron size pores in the nanoparticle layer.

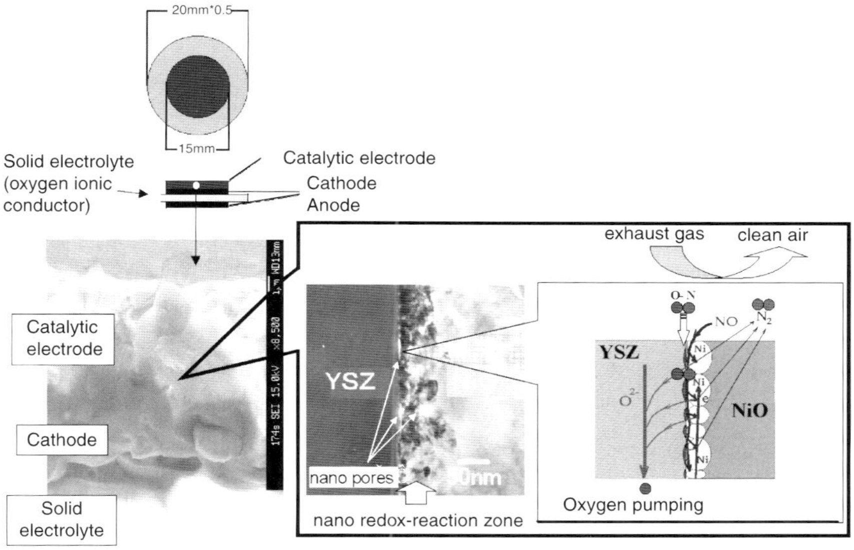

Figure 18.2
TEM image of the interface between NiO and YSZ; and, the reaction model of selective NO_x decomposition.

In this case, selective decomposition of NO_x molecules was drastically suppressed due to the existence of large pores. The formation of the micron size pores by the electrochemical reaction was irreversible. Therefore, increasing the efficiency of the selective decomposition of NO_x requires an increase in the thickness of the interface layer with nanoparticles and nanopores. This can be achieved by applying low voltage for a prolonged period of time.

The unique structure of the YSZ/NiO interface is the existence of the 10–50 nm Ni grains reversibly produced during the electrochemical reaction and cell operation. It is well known that the adsorption and decomposition of NO_x gas molecules occur in preference to oxygen gas molecules on Ni grain surfaces. In addition, it should be noted that rough surfaces with nanosized Ni grains are much more active on breaking NO_x chemical bonds than smooth, flat surfaces. Based on the above results, the following reaction mechanism was proposed for NO_x decomposition on the nanosized Ni grains produced during the reactor operation.

$$NO + Ni \rightarrow Ni - NO \qquad (18.1)$$

$$2Ni - NO \rightarrow 2NiO + N_2 \qquad (18.2)$$

NO gas molecules are first chemisorbed on Ni. And then, the chemisorbed NO decomposes to form N_2 and oxidizes Ni to NiO. The NO_x gas molecules in the nanospaces have extremely high possibility to interact with Ni and oxygen vacancy in YSZ comparing to the oxygen molecules through the redox reaction.

Oxygen ionic current passes through the network of YSZ particles surrounding the Ni grains; removes the produced oxygen species; and, allows the reaction to reoccur. The regeneration of Ni grains from the reduction of NiO takes place at the NiO/YSZ interface by the following reaction

$$NiO + Y_O(ZrO_2) + 2e^- \rightarrow Ni + O^{2-}(YSZ) \qquad (18.3)$$

The reduction of NiO grains to Ni grains and the oxidation of Ni grains to NiO take place continuously during the reactor operation. As a result, the catalytic activity for NO_x decomposition is independent of the operating time.

As shown in Fig. 18.3, the efficiency of NO_x decomposition by the newly developed electrochemical reactor far exceeds the efficiency of the conventional electrochemical cells. Previous tests applying electrochemical cells to decompose NO_x were unsuccessful in suppressing the supplied electrical power due to the effect of co-existing oxygen molecules. Precise nanostructural control on the electrocatalytic layer enabled us to improve the cell properties and reduced its energy consumption to half of those using presently employed catalysts (as calculated from the engine power and fuel loss equivalent to the electrical energy).

To commercialize the electrochemical reactors, there are still many problems to overcome. For example, stacking up cells to obtain sufficient

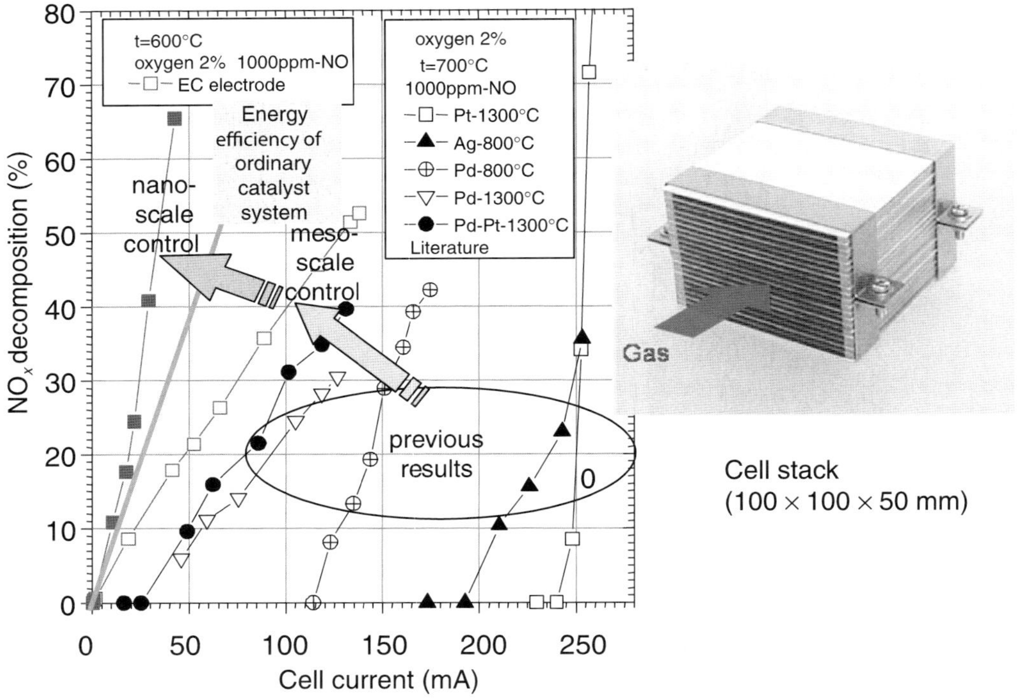

Figure 18.3
Improve de-NO_x efficiency of electrochemical cell by controlling its nanoscale structure in comparison to previous results.

reaction area and improving the cell properties with actual gas compositions and flow rates are the issues remained to be evaluated and resolved. Techniques such as stack assembly and module construction will allow this "nanotechnology" to be employed in the commercial applications to solve the environmental problems. The stack fabrication for large cells used as the gas-purifying module in vehicles has been done. Fig. 18.3 shows a large cell stack in the size of 10 cm square. Uniform nanostructure developed by the electrochemical cell treatment enabled us to operate the reactor with sufficiently low electrical power consumption and long operating durability.

The newly developed electrochemical reactor with nanostructured multi-layer electro-catalytic electrode can be used effectively for NO_x decomposition in the presence of high oxygen concentrations. It works even at the low concentration of NO_x (300–500 ppm) with high concentration of oxygen (10%) in the exhaust gas.

Other problems to be solved before commercialization include the cell durability when operating with actual exhaust gases containing H_2O, CO, CH, SO_x, lifetime, and overall cell fabrication costs. The NO_x decomposition properties of the cell before and after introducing the humidity at the high temperatures indicated that good cell performance could be achieved even with 10% humidity at high working temperatures. The cell had good durability to the co-existence of CO and CH; but, showed a slight degradation when co-existing high concentrations of SO_x.

3. Development of the electrochemical reactors for simultaneous purification of NO_x/PM

(1) Electrochemical reactors for purifying NO_x/PM

Taking the advantage of having "oxidation/reduction" taking place simultaneously at both electrodes of the electrochemical cell, further electrode improvements were made to enhance oxidation activity based on the radical oxygen on the anode side and NO_x reduction on the cathode side. As a result, the decomposition of particulate matter (PM: soot-like mass) in exhaust gas became possible. Unlike the DPF (a filter commonly used today), the new technology (as shown in Fig. 18.4) has substantial potential as an energy-saving purification method to remove NO_x and PM simultaneously, independent of the engine operating conditions and capable of being activated on demand.

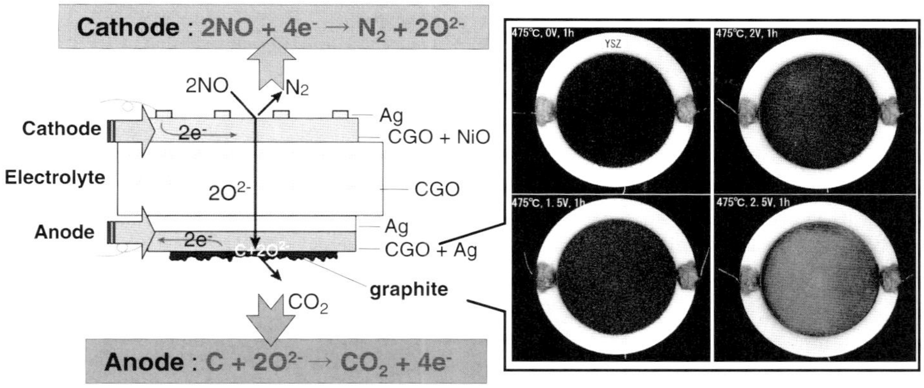

Figure 18.4
Schematics and photographs of anode for simultaneous removal of NO_x/PM in an electrochemical cell.

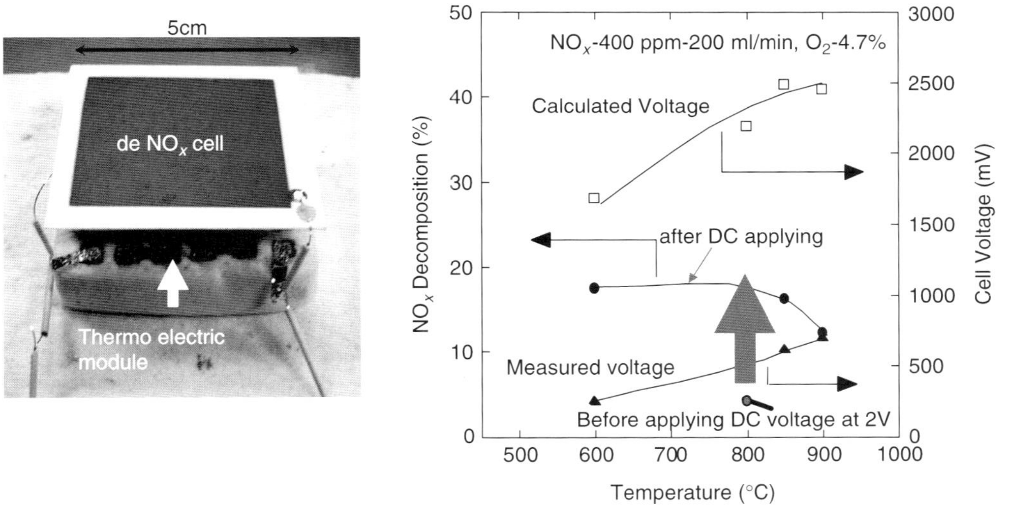

Figure 18.5
Demonstration of de-NO_x using the converted energy from the exhaust gas heat without any external electrical power.

(2) Interactive ceramic reactor made of the de-NO_x electrochemical cell with a power module of thermoelectric energy conversion ceramics

The electrochemical cells for de-NO_x reaction have properties well suited for the de-NO_x/PM purification; particularly, their electrical power consumption is very low for the application. With an attempt to develop an autonomous continuous purification system, the concept of operating the device for environment purification without energy consumption was finally materialized through the improvement and assembly of the ceramic materials. Fig. 18.5 shows an interactive ceramic reactor made of the de-NO_x electrochemical cell with a power module of thermoelectric energy conversion ceramics. Without any external electrical power, 20% of NO_x decomposition was achieved continuously using the electrical current produced from the thermoelectric power module, which performed energy conversion from the waste heat in the hot exhaust gas. This experiment indicates that a complete purification of exhaust gas is possible by stacking up a number of thermoelectric power modules.

Promoting energy efficiency in the transportation sector is crucial to achieve the goal of CO_2 emission reduction under Kyoto Protocol. A new electro-catalytic reactor applying nanostructural control of an electrical conductor (NiO) and an ionic conductor (YSZ) in the electro-catalytic electrode for ultimate NO_x decomposition was proposed. Electrochemically induced nano-sized Ni grains surrounded by the nanopores at the NiO/YSZ interface of the electro-catalytic electrode have resulted in a remarkable improvement on the de-NO_x efficiency. Oxygen ionic current passed through the network of YSZ particles surrounding the Ni grains and removed the produced oxygen species repeatedly. Gas molecules trapped in the nanospaces at the interfaces accelerated the selective reaction of NO_x molecules in presence of excess amount of oxygen. The electrical power required for NO_x decomposition was reduced to an applicable level under this situation. Furthermore, the concept of enhancing reaction efficiency of gas molecule in the nanospaces is applicable to many other applications of electrochemical cells such as SOFC and chemical syntheses.

References

[1] K. Hamamoto, T. Hiramatsu, O. Shiono, S. Katayama, Y. Fujishiro, S. Bredikhin and M. Awano: *J. Ceram. Soc. Jpn.*, **112**, S1071–S1074 (2004).

[2] Y. Fujishiro, K. Hamamoto and M. Awano: *J. Mater. Sci., Mater. Electron.*, **15**, 769–773 (2004).

[3] M. Awano, S. Bredikhin, A. Aronin, G. Abrosimova, S. Katayama and T. Hiramatsu: *Solid State Ionics*, **175**, 605–608 (2004).

[4] M. Awano, Y. Fujishiro, K. Hamamoto, S. Katayama and S. Bredikhin: *Int. J. Appl. Ceram. Technol.*, **1**, 277–286 (2004).

[5] M. Awano, Y. Fujishiro, H-J. Hwang, S. Bredikhin, K. Matsuda, K. Maeda, T. Kanai, M. Miyata: Japanese Patent No. 3626971 (2004).

[6] M. Awano, Y. Fujishiro, H-J. Hwang, S. Bredikhin, K. Matsuda, K. Maeda, T. Kanai, M. Miyata: US Patent No. US6818107B2 (2004).

APPLICATION 19

19 DENDRIMERS AND THEIR APPLICATION TO ORGANIC ELECTRONICS DEVICES

1. Synthesis and structure of dendrimer

In macromolecule science, various kinds of linear chain polymers have been reported. The molecule shape of these polymers has a statistical distribution due to their molecular weight distribution and structural flexibility. In contrast, the protein, a single structural polymer, produces advanced functions. The fact indicates the importance to precisely control a multi-dimensional structure.

Dendrimers have recently been drawing much attention as a nano-material, because of their precisely controlled branch structures (Fig. 19.1). Tomalia et al. synthesized a dendrimer with a regular structure by the Divergent method, precisely constructing the branch units from a single core [1]. The synthetic method provides the dendrimer with a single molecule weight, the shape of which becomes spherical with the increasing generation (polymerization degree) [2].

Fréchet et al. proposed Convergent method – a synthetic method to obtain a dendrimer by reversely synthesizing dendron stepwise from the terminal points, and then combining a number of dendrons with a core at the last stage [3]. This method prevents branches from defecting in the synthesis process and makes it easier to introduce functional groups into the predictable three-dimensional structure of dendrimers [4]. Using the divergent and convergent methods, we can design novel advanced structures and functions easily.

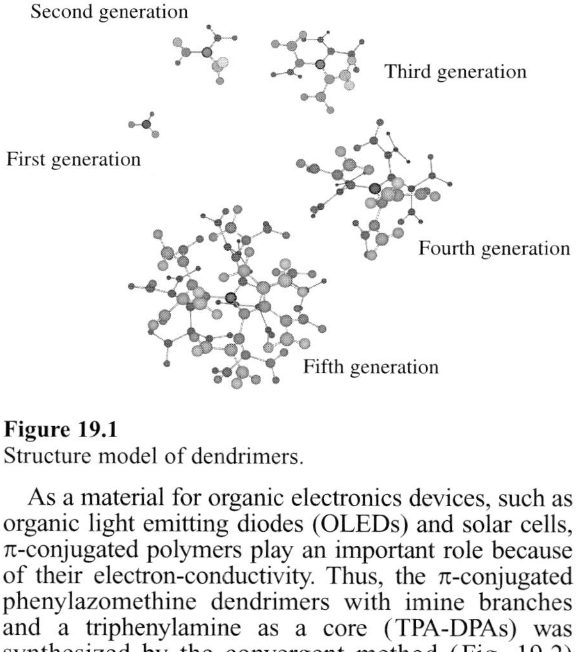

Figure 19.1
Structure model of dendrimers.

As a material for organic electronics devices, such as organic light emitting diodes (OLEDs) and solar cells, π-conjugated polymers play an important role because of their electron-conductivity. Thus, the π-conjugated phenylazomethine dendrimers with imine branches and a triphenylamine as a core (TPA-DPAs) was synthesized by the convergent method (Fig. 19.2) [5, 6]. Their structure, consisting of a π-conjugated framework, gives comparable high thermal-stability with engineering plastics; a 10% thermo gravimetrical

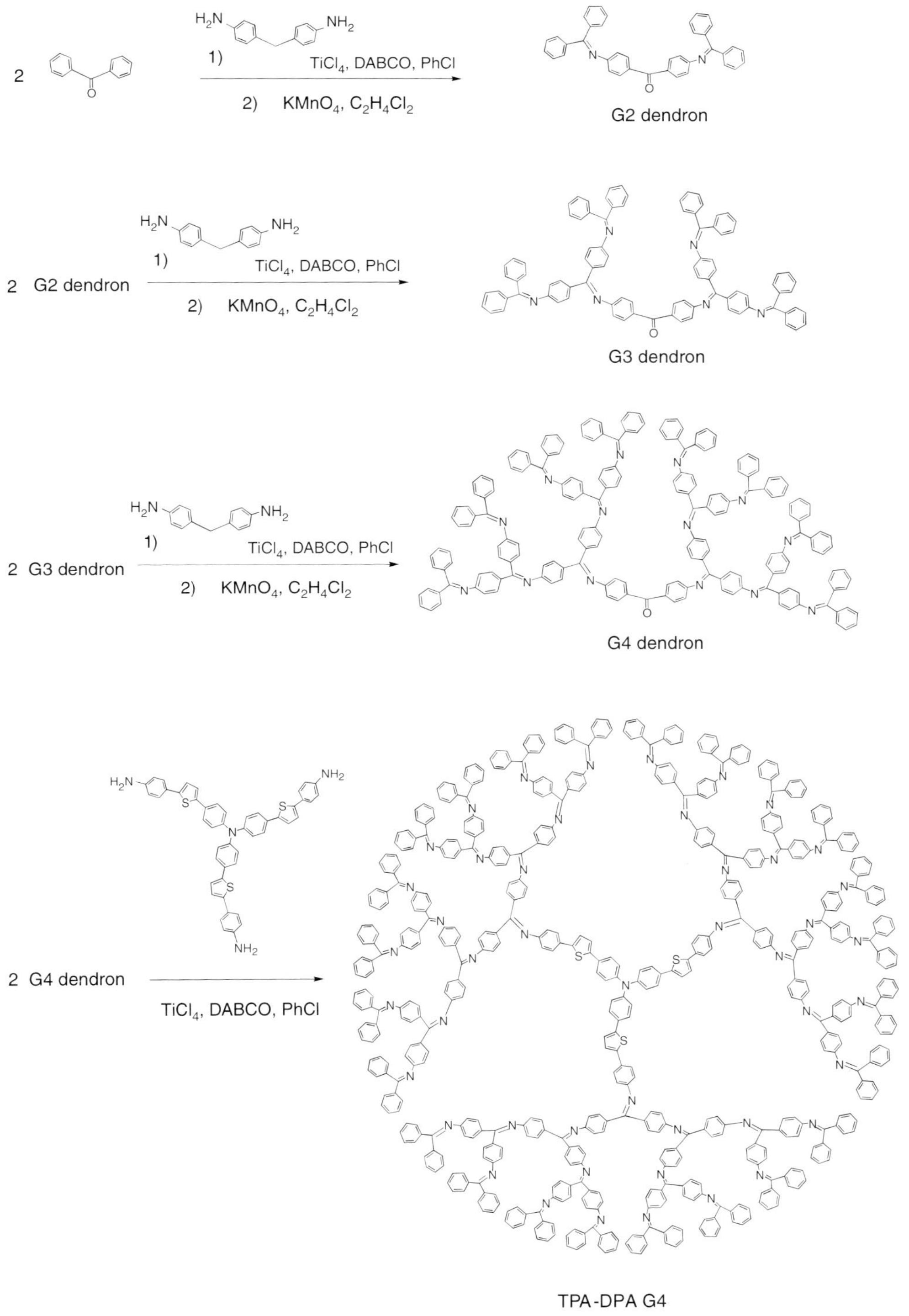

Figure 19.2
Synthesis scheme of TPA-DPA G4.

loss temperature is over 500ºC. Their rigid spherical structure forms a supramolecular assembly, the close packing structure on the substrate by only cast [7, 8]. Furthermore, the triphenylamine core is well known as a hole transfer material and a electron donor.

2. Metal assembling property of dendrimer

Maciejewski proposed an idea to use the highly branched molecules as a container holding the finer particles in the pores [9]. The host–guest chemistry using dendrimers has developed remarkably and established an important position in supramolecular science.

Tomalia et al. analyzed the metal assembly of Cu(II) ions of a polyamide amine (PAMAM) dendrimer in water, using ESR [10]. A Cu(II) ion combines with the four nitrogen atoms of the amine and amino groups in the dendrimer. Thus, the terminal amino groups of the fourth generation dendrimer can hold maximum 15.5 equivalent Cu(II) ions. The reduction of the included metals produces the metal clusters, held stably in the dendrimer [11]. The Au clusters, made by this method, show strong fluorescence. It is the first time as a metal cluster that an ultra fast attenuation of the fluorescence confirmed [12]. Additionally, Crooks et al. applied the included metal clusters to the catalysts [13].

In contrast with PAMAM dendrimers, the complexation of metal ions with TPA-DPAs precisely proceeds in radial step-wise manner. It means that the complexation to the outer layers never occurs without filling the inner layers. The metal ion complexes with a imine site at a ratio of 1:1, resulting in assembling of 45 equivalent $SnCl_2$ into TPA-DPA G4 in the organic solvent (Fig. 19.3) [6]. Such stepwise complexation occurs because of the electron density gradient; the electron donation from the imines concentrates along the inside dendritic framework, leading to higher imine basicity in the inner layers [14]. The metal assembly reduces the electric resistance of the dendrimer, which improves device performances.

3. Application to electronic devices

The good solubility and easiness in preparing a homogeneous film are also absolutely necessary for a material used in organic devices; dendrimers satisfy the requirement. Additionally, dendrimers isolate a chromophore by the dendritic shell.

Yokoyama et al. applied the dendrimers, fixed rhodamine on the core, to the organic optical electronics like the optical amplification luminescence and solid laser [15]. The confinement of dyes within dendrimers suppresses the energy transfer between the dyes, especially the self quenching in high-concentration solution or solid state [16]. The systems like this, including rhodamine, coumalin or DCM dyes, are also reported as luminescent materials in high concentration [17, 18].

These methods are also applied to a light emitting layer in OLEDs. The blue light-emitting dye or polymer in solid state is difficult, because of the red shift in luminescent wavelength and the self-quenching. Müllen and List et al. obtained the high efficient blue OLEDs to synthesize the novel polyfluorenes, shielded by the bulky dendron side chain, as blue light-emitting materials [19].

Thompson and Fréchet et al. tried the color tuning by mixing luminescent dendrimer dye for the light emitting single layer. The simultaneous luminescence of the mixing fluorescent dyes is generally prevented from the energy transfer between the dyes. However, the electroluminescence from the two enraptured dyes, coumalin 343 and pentathiophene, within the high generation dendrimers is observed by suppressing the energy transfer [20].

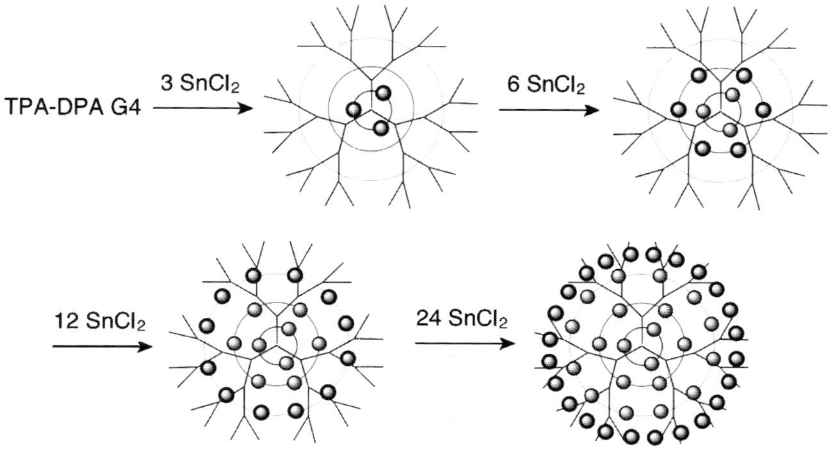

Figure 19.3
Assembling behavior of $SnCl_2$ toward TPA-DPA G4.

Recently, phosphorescent dyes have drawn much attention as light-emitting materials because of their high luminescence efficiency. In contrast with fluorescent dyes, the phosphorescent quantum efficiency can reach nearly 100%, because both the single and triplet excited states contribute luminescence. However the film of polymers doped with a phosphorescent dyes like indium complex does not yield high luminescent OLEDs because of their inhomogeneous film and/or dye's aggregation. On the contrary, Samuel et al. synthesized dendrimers with an indium complex, introducing an alkyl group at the terminal of a π-conjugated dendron for high affinity to the solvent. The casting film from the mixture of this dendrimer, hole transfer material (CBP) and electron transfer material (TPBI), works as the high efficiency monolayer OLEDs [21].

On the other hand, TPA-DPAs are applied to a hole-transporting layer [6, 7]. Though a triphenylamine (TPA) has high crystallinity, the TPA-DPAs are easy to prepare a homogeneous film by cast. The OLEDs, using TPA-DPAs as a hole transfer layer and Alq as an emitting layer, decrease not in the brightness but in the electric current density as the generation advances, resulting in the improvement of the luminescent efficiency. The phenomena are attributed to the concentration of the hole and electron density in the emitting layer; the large dendrimer shells suppress not the hole transfer to the emitting layer but the electron pass from the emitting layer. Additionally, the assembly of $SnCl_2$ into TPA-DPA improves the brightness and efficiency because of the reduction in the electric resistance.

In the energy diagram of the OLEDs, the energy level of the π-conjugated orbital matches with that of the hole transfer; it does not exist in the energy level for the electron pass. Thus, the hole transfer, assisted with the π-conjugated orbital, predominates over the electron pass. This effect is magnified, as the radius of the dendrimer increases with the advancing generation. Such a vector control of electron transfer using TPA-DPAs also functions as a kind of charge separating layer in an organic solar cell.

The dye-sensitized solar cell (DSSC), developed by Grätzal et al., can separate the charge efficiently by the ideal photo-induced primary reaction [22]. The back electron transfer from the titanium oxide to the electrolyte, however, restricts the open-circuit voltage and the energy conversion efficiency at the interface. This problem is resolved by TPA-DPAs, cast on the dye absorbing titanium oxide film, which assists the hole transfer (the reduction of the dye, oxidized by the primary reaction) and suppress the back electron transfer from the titanium oxide (Fig. 19.4). Furthermore, the accepter in this back electron transfer, I_3^-, complexes with the imines at the branch structure of the dendrimer at a ratio of 1:2, and then produces I^- working as an electron donor to the oxidized dye.

Because of the above-mentioned synergy effect, the open-circuit voltage improves with the generational growth of the dendrimer. Moreover, the complexion with $SnCl_2$ into the dendrimer enhances the fill factor because of the reduction in the resistance, leading to the improvement of the energy conversion efficiency. In total, the energy conversion efficiency is improved by 34% by using the TPA-DPA G5 complex with $SnCl_2$ [23].

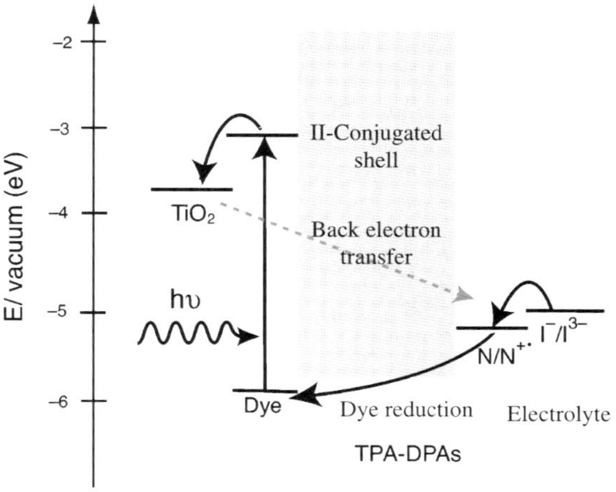

Figure 19.4
Energy diagram of the dye sensitization solar cell using TPA-DPA.

References

[1] D.A. Tomalia, H. Baker, J. Dewald, M. Hall, G. Kallos, S. Martin, J. Roeck, J. Ryder and P. Smith: *Polym. J.*, **12**, 117–132 (1985).

[2] D.A. Tomalia: *Prog. Polym. Sci.*, **30**, 294–324 (2005).

[3] C.J. Hawker, J.M.J. Fréchet: *J. Am. Chem. Soc.*, **112**, 7638–7647 (1990).

[4] S.M. Grayson, J.M.J. Fréchet: *Chem. Rev.*, **101**, 3819–3867 (2001).

[5] K. Takanashi, H. Chiba, M. Higuchi and K. Yamamoto: *Org. Lett.*, **6**, 1709–1712 (2004).

[6] N. Satoh, J.-S. Cho, M. Higuchi and K, Yamamoto: *J. Am. Chem. Soc.*, **125**, 8104–8105 (2003).

[7] M. Higuchi, S. Shiki, K. Ariga and K. Yamamoto: *J. Am. Chem. Soc.*, **123**, 4414–4420 (2001).

[8] N. Satoh, J.-S. Cho, M. Higuchi and K. Yamamoto: *J. Photopol. Sci. Technol.*, **18**, 55–58 (2005).

[9] M. Maciejewski: *J. Macrom. Sci. Chem.*, **17A**, 689–703 (1982).

[10] M.F. Ottaviani, F. Montalti, N.J. Turro and D.A. Tomalia: *J. Phys. Chem. B*, **101**, 158–166 (1997).

[11] L. Balogh, D.A. Tomalia: *J. Am. Chem. Soc.*, **120**, 7355–7356 (1998).

[12] O. Varnavski, R.G. Ispasoiu, L. Balogh, D. Tomalia and T. Goodson III: *J. Chem. Phys.*, **114**, 1962–1965 (2001).

[13] R.W.J. Scott, O.M. Wilson and R.M. Crooks: *J. Phys. Chem. B*, **109**, 692–704 (2005).

[14] K. Yamamoto, M. Higuchi, S. Shiki, M. Tsuruta and H. Chiba: *Nature*, **415**, 509–511 (2002).
[15] S. Yokoyama, T. Nakahama, H. Miki and S. Mashiko: *Thin Solid Films*, **438–439**, 452–456 (2003).
[16] S. Yokoyama, A. Otomo, T. Nakahama and S. Mashiko: *Thin Solid Films*, **393**, 124–128 (2001).
[17] S. Yokoyama, A. Otomo and S. Mashiko: *Appl. Phys. Lett.*, **80**, 7–9 (2002).
[18] N. Satoh, T. Watanabe, Y. Iketaki, T. Omatsu, M. Fujii and K. Yamamoto: *Polym. Adv. Technol.*, **15**, 159–163 (2005).
[19] A. Pogntsch, F.P. Wenzl, E.J.W. List, G. Leising, A.C. Grimsdare and K. Muellen: *Adv. Mater.*, **14**, 1061–1064 (2002).
[20] P. Furuta, J. Brooks, M.E. Thompson and J.M.J. Fréchet: *J. Am. Chem. Soc.*, **125**, 13165–13172 (2003).
[21] T.D. Anthopoulos, J.P.J. Markham, E.B. Namdas and I.D.W. Samuel: *Appl. Phys. Lett.*, **82**, 4824–4826 (2003).
[22] A. Hagfeldt, M. Grätzel: *Chem. Rev.*, **95**, 49–68 (1995).
[23] N. Satoh, T. Nakashima and K. Yamamoto: *J. Am. Chem. Soc.*, **127**, 13030–13038 (2005).

APPLICATION 20

20 ELECTRICAL CONDUCTIVE CNT DISPERSED Si_3N_4 CERAMICS

Si_3N_4 ceramics is one of the typical engineering ceramics with high strength, hardness, fracture toughness, corrosion resistance and wear resistance. Although electrical insulation is also characteristic of Si_3N_4 ceramics, electrical conductivity is needed depending on the application. Recently, carbon nanotubes (CNTs) with high thermal conductivity, electrical conductivity and mechanical properties have been developed. CNT/metal or polymer composites have been developed to improve some properties by adding the small amount of CNTs [1, 2] as well as CNT dispersed Al_2O_3 ceramics is also reported [3, 4, 5]. Although it should be possible to improve electrical conductivity of Si_3N_4 ceramics by using CNTs, there is no literature on the dense CNT dispersed Si_3N_4 ceramics. The Y_2O_3–Al_2O_3–TiO_2–AlN [6] as the additive for lower temperature densification of Si_3N_4 has been studied. In this section, CNT dispersed Si_3N_4 ceramics developed by using the above sintering aid will be introduced [7].

Fig. 20.1 shows the electrical conductivity of CNT-dispersed Si_3N_4 ceramics fabricated by gas-pressure sintering technique. Although Si_3N_4 ceramics without CNTs are an insulator, their electrical conductivity appeared suddenly in the sample with 1.8 wt% CNT addition, having a value of 2.8 S/m. It decreased with an increasing amount of CNTs and disappeared at 3.6 wt% CNT addition. It then began increasing over 4.2 wt% CNT addition. The SEM photographs in Fig. 20.2 confirmed that β-Si_3N_4 grains in Si_3N_4 ceramics containing up to 3 wt% CNTs elongated to as large as those in the sample without CNTs, and that CNTs remained in the grain boundary of Si_3N_4 ceramics. CNTs should form an electrical conduction path in Si_3N_4 ceramics, and it was observed that the largest amount of CNTs remained in the sample with

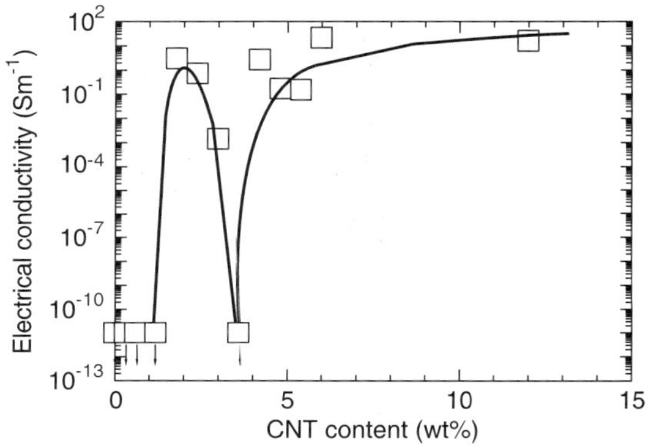

Figure 20.1
Electrical conductivity of CNT dispersed Si_3N_4 ceramics.

Figure 20.2
Microstructure of CNT dispersed Si_3N_4 ceramics (CNT: 1.8 wt%, 1,800 °C, 2 h, 0.9 $MPaN_2$).

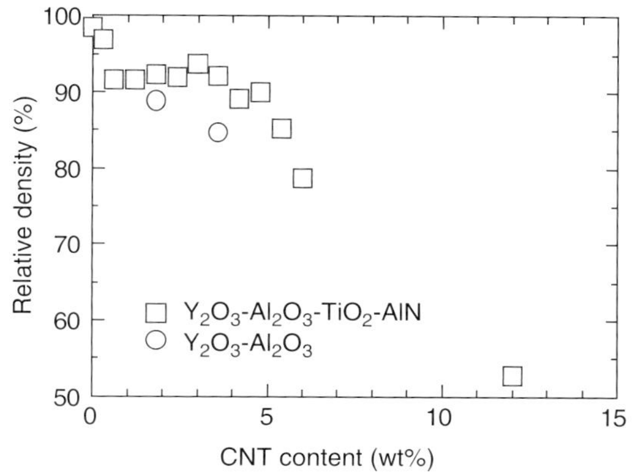

Figure 20.3
Relative density of CNT-dispersed Si_3N_4 ceramics.

1.8 wt% CNTs. The residual CNTs decreased as more CNTs were added, and CNTs were not found in the sample with over 4.2 wt% CNTs.

Fig. 20.3 depicts the relative density of Si_3N_4 ceramics fabricated using Y_2O_3–Al_2O_3–TiO_2–AlN by the gas-pressure sintering technique. The relative density exceeded 92% in the sample with 1.8 wt% CNTs, though it decreased slightly with an increasing amount of CNTs due to the inhibition of densification by CNTs. Thus, electroconductive and dense Si_3N_4 ceramics by addition of CNTs can be realized.

XRD analysis confirmed that the sample with up to 1.8 wt% CNT exhibited the same XRD profile as the sample without CNT, namely only β-Si_3N_4 and TiN were identified. SiC was found to be included in the Si_3N_4 ceramics with over 2.4 wt% CNT addition. When a larger amount of CNT was added, α-sialon was also found in the sample, which means that there was a reaction between CNTs and liquid phase at high temperature because α-sialon is expected to be generated from the liquid at high N concentration [8]. However, when Y_2O_3 and Al_2O_3 were used as sintering aids, the density was lowered (Fig. 20.3), SiC was formed and the residual CNTs could not be observed in the Si_3N_4 ceramics. SiC should be formed in the following reaction between CNTs and SiO_2, which is the main component of the grain boundary glassy phase and/or Si_3N_4 itself.

$$SiO_2(s,l) + C(s) \rightarrow SiO(g) + CO(g)$$

$$SiO(g) + 2C(s) \rightarrow SiC(s) + CO(g)$$

$$2SiO(g) + 3C(s) \rightarrow 2SiC(s) + CO_2(g)$$

$$Si_3N_4(s) + 3C(s) \rightarrow 3SiC(s) + 2N_2(g)$$

These chemical reactions must follow gas formation and depend on the partial pressure of SiO and N_2. For instance, SiC is in a stable phase at 1,600°C under 1.0 MPa N_2 [9]. If the sintered body has a lot of open pores, the above reactions advance by discharging the gas out of the sample like previous study [10, 11]. However, if the pores become closed during sintering, the gas formation reactions should be restrained and CNTs remain in the sample. In the system of Si_3N_4–Y_2O_3–Al_2O_3–TiO_2–AlN, CNTs remained in the sample in spite of their characteristics since it was densified at lower temperature to form closed pores before the reaction between CNTs and SiO_2 or Si_3N_4.

Hot isostatic pressing treatment was carried out in order to evaluate mechanical properties of the electroconductive Si_3N_4 ceramics. Furthermore, the ceramics was fabricated by hotpressing. As a result, more CNTs remained in the hotpressed samples. Table 20.1 shows the mechanical properties of CNT dispersed Si_3N_4 ceramics. The CNT-dispersed Si_3N_4 ceramics made by the authors has higher strength (722 MPa produced by gas-pressure sintering and hot isostatic pressing, and 1,173 MPa produced by hotpressing) than previous ceramics due to the excellent sinterability of the Si_3N_4–Y_2O_3–Al_2O_3–TiO_2–AlN system. Fracture toughness of CNT-dispersed Si_3N_4 ceramics, fabricated by gas-pressure sintering followed by hot isostatic pressing and hotpressing, is 6.6 and 6.5 MPa $m^{1/2}$, and hardness is 14.3 and 14.8 GPa, respectively. These are nearly the same values as the sample without CNTs.

Table 20.1
Mechanical properties of CNT dispersed Si_3N_4 ceramics.

	1.8 wt% CNTs GPS-HIP	1.8 wt% CNTs HP	No CNTs GPS-HIP
Relative density (%)	96.4	98.3	98.1
Electrical conductivity (Sm^{-1})	30	79	$<10^{-5}$
Bending strength (MPa)	721	904	950
Fracture toughness (MPa $m^{1/2}$)	6.6	6.5	6.5
Vicker's hardness (GPa)	14.3	14.8	15.2

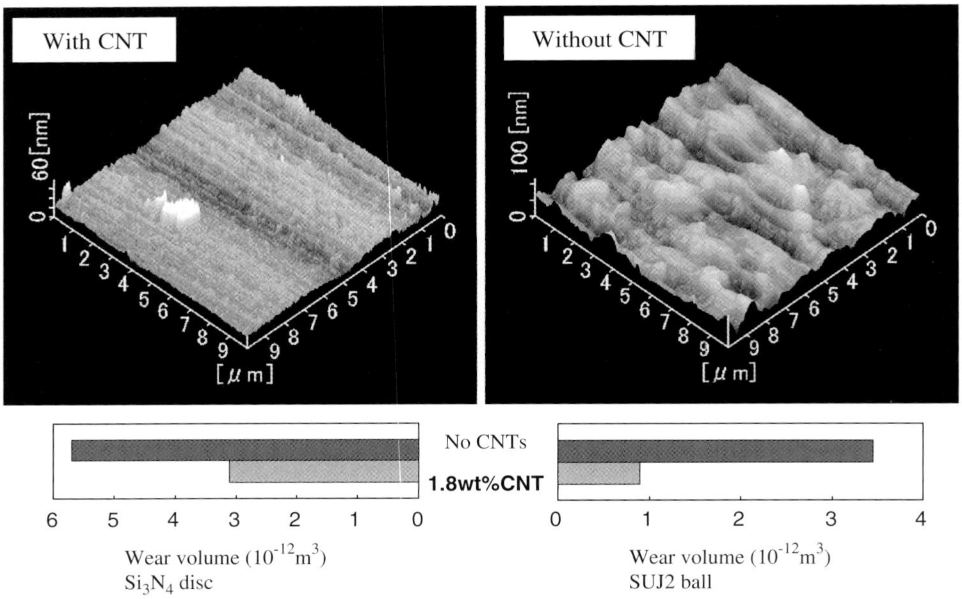

Figure 20.4
Wear property of CNT dispersed Si_3N_4 ceramics.

Consequently, electrical conductivity and superior mechanical properties can be achieved simultaneously by adding Y_2O_3–Al_2O_3–TiO_2–AlN to Si_3N_4. The electroconductivity of the HIPped sample was 30 S/m. Therefore, electrical property improved without degradation of mechanical properties. Hotpressed sample had higher electrical and mechanical properties than HIPped sample. These electroconductive Si_3N_4 ceramics can be applied in extremely severe mechanical and environmental conditions such as ceramic bearings for high-speed, high-temperature, high-corrosion and high-vacuum environments, and as novel functional materials with high durability and electrical conductivity, which result from achievement of densification at lower temperature. Furthermore, as a result of ball-on-disk test for wear properties, it was confirmed that wear volume of not only Si_3N_4 itself but also SUJ2 mating metallic material decreased (Fig. 20.4).

Si_3N_4 ceramics was applied to bearing materials with longer lifetime and higher reliability to replace metallic materials. The developed electroconductive Si_3N_4 ceramics with high strength, toughness and wear resistance can restrict degradation of system lifetime due to dust adhesion. For this reason, they are expected to be applied as bearing materials for the electronic components and so on for which the conventional one cannot be used.

References

[1] C.L. Xu, B.Q. Wei, R.Z. Ma, J. Liang, X.K. Ma and D.H. Wu: *Carbon*, **37**(5), 855–858 (1999).
[2] J. Sandler, M.S.P. Shaffer, T. Prasse, W. Bauhofer, K. Schulte and A.H. Windle: *Polymer*, **40**(21), 5967–5971(1999).
[3] G.-D. Zhan, J.D. Kuntz, J. Wan and A.K. Mukherjee: *Nat. Mater.*, **2**, 38–42 (2003).
[4] A. Peigney, C. Laurent, E. Flahaut and A. Rousset: *Ceram. Int.*, **26**, 677–683 (2000).
[5] Ch. Laurent, A. Peigney, O. Dumortier and A. Rousset: *J. Eur. Ceram. Soc.*, **18**, 2005–2013 (1998).
[6] J. Tatami, M. Toyama, K. Noguchi, K. Komeya, T. Meguro and M. Komatsu: *Key Eng. Mater.*, **247**, 83–86 (2003).
[7] J. Tatami, T. Katashima, K. Komeya, T. Meguro and T. Wakihara: *J. Am. Ceram. Soc.*, **88**, 2889–2893 (2005).
[8] S. Hampshire and K.H. Jack: in: *Progress in Nitrogen Ceramics*, F.L. Riley (Ed.), Martinus Nijhoff, Boston, p. 263 (1983).
[9] H. Wada., M.J. Wang and T.Y. Tien: *J. Am. Ceram. Soc.*, **71**, 837–840 (1988).
[10] Cs. Balázsi, F.S. Cinar, O. Addemir, F. Wéber and P. Arató: *J. Eur. Ceram. Soc.*, **24** (12), 3287–3294 (2004).
[11] Cs. Balázsi, Z. Kónya, F. Wéber, L.P. Biró and P. Arató: *Mater. Sci. Eng. C*, **23**, 1133–1137 (2003).

APPLICATION 21

21 DEVELOPMENT OF FUNCTIONAL SKINCARE COSMETICS USING BIODEGRADABLE PLGA NANOSPHERES

Cosmetics incorporating "nanotechnology" and claiming the inclusion of "nano-size ingredients" and "nanoparticles" have appeared considerably on the consumer market over the past few years. Many manufacturers of the skincare products (e.g. creams, lotions, beauty essences, etc.) use ultra high-pressure emulsification and other techniques to nano size their active ingredients down to tens of nanometers, which are smaller than the gaps between the skin cells in the corneum layer (70 nm), with an attempt to improve their skin permeability and absorbability. Many liquid liposomes and solid particles, such as the pro-vitamin A loaded calcium carbonate nanoparticles [1] and the pro-vitamin C loaded PLGA (poly lactide-co-glycolide) nanospheres are recently utilized as nanocosmetics carriers to enhance the performance of cosmetic products. In this section, we would focus on the pro-vitamin C loaded PLGA nanospheres technology and use it as an example of solid type carrier for creating functional skincare cosmetics.

1. Nanocosmetics that whiten skin and eliminate wrinkles

(1) The pharmacological effect of vitamins on the skin

Many people are bothered by the blemishes and wrinkles on the skin. Vitamins C, E, and A are believed to be effective in mitigating the blemish and wrinkle problems. Vitamin C suppresses the blemishes because it limits the activity of tyrosinase, which promotes melanin production. Furthermore, it increases collagen formation to reduce wrinkles, and prevents cell oxidation by eliminating active oxygen. As to vitamin E and A, they also act as antioxidant and collagen promoter, respectively. Therefore, many whitening and anti-aging cosmetic products apply various vitamin complexes and their derivatives. But, not many products offer sufficient tangible results, because the corneum layer (the skin's outermost layer) acts as a barrier that considerably reduces the absorption rate of the vitamins and other active ingredients. In addition, the vitamins are decomposed easily under ordinary conditions. To be effective, the vitamins have to be functionally refined to enhance their stability and to increase their cutaneous permeability.

(2) Nanocosmetics

Many liquid-type nanocosmetics carriers, such as liposomes, are structurally unstable. Specifically, when passing through the skin, they adhere to the inside walls of the skin cells causing the collapse of phospholipid association bodies and the leak of their encapsulated ingredients. As a result, their ability to transport active ingredients to deep skin is not likely good. Meanwhile, the use of macromolecular (PLGA) nanosphere, a shape-stable nanoparticle carrier, is being developed.

We encapsulated three vitamin derivatives in PLGA nanospheres, including vitamin C (ascorbyl tetraisoplamitate), vitamin E (tocopherol acetate) and vitamin A (retinyl palmitate); and marketed a new type of functional skin care product, which requires mixing of these nanospheres with beauty essence at the time of use [2]. These nanospheres have an average size of 200 nm, which was found to be the largest size that permeates not only skin pores (200–300 nm) but also the horny layer in the cutaneous permeability experiment as described below. The job of the PLGA nanospheres is to permeate the melanocytes in the stratum basale of the epidermis and the fibroblasts in the dermis; and then, gradually release their vitamin derivatives over 48 h due to hydrolysis by the skin moisture. The released vitamin derivatives are converted to pure vitamins by esterase in the body and show pharmacological effect as illustrated in Fig. 21.1 near the target area.

For skin whitening and anti-wrinkle/sagging applications, it is important to have a technology which can nano size the vitamin derivatives, amino acids, and other active ingredients, so that they can reach the target areas in a stable form, and sustain the pharmacological effect for a long time. In a sense, the idea of nanoparticle design for drug delivery systems in cosmetics applications is important. Presented below are the evaluation of the cutaneous permeability of PLGA nanospheres and their effects on the whitening and anti-wrinkle of skin [3, 4].

2. Evaluation of the cutaneous permeability of PLGA nanospheres and their functional effect [3, 4]

(1) Evaluation of the cutaneous permeability of PLGA nanospheres

The cutaneous permeability of PLGA nanospheres was evaluated by using the modified Bronaugh diffusion chamber developed by Miwa et al. with human skin biopsies [3, 5]. Unlike those methods using artificial skin or animal skin, this method allows test fluid to penetrate the human skin biopsy by natural diffusion, and therefore produces data very close to those obtained in the clinical study.

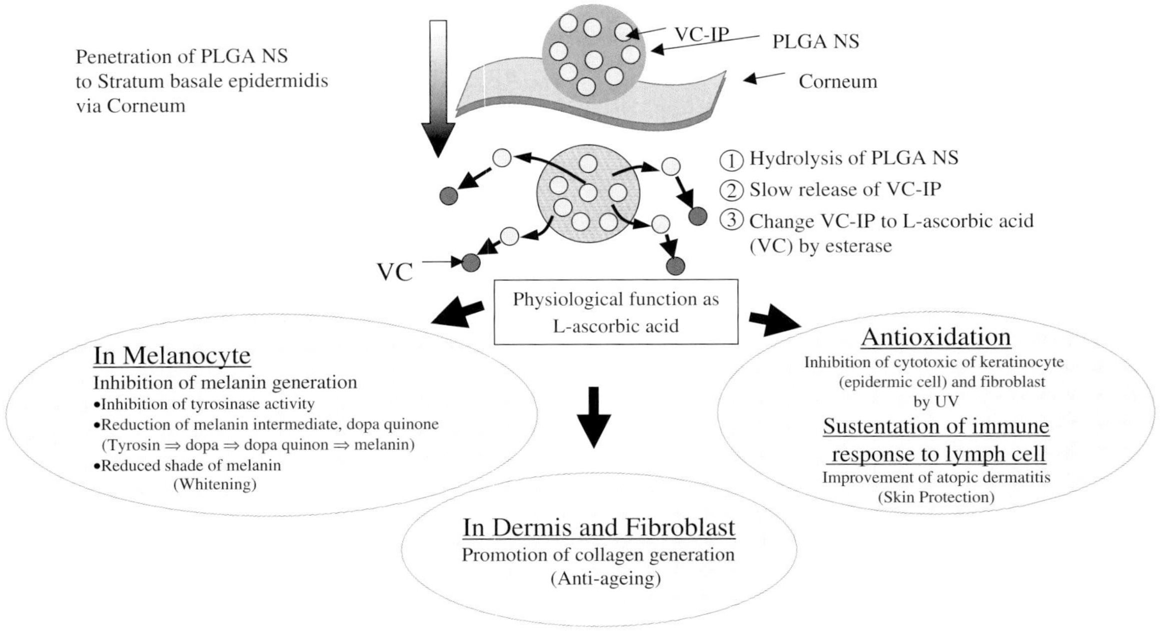

Figure 21.1
Vitamin C delivered to deep skin by VC-IP encapsulated PLGA nanospheres and its pharmacological effects.

Fig. 21.2 shows the SEM photographs of skin samples taken over several hours after applying 0.2% aqueous dispersion of coumarin-6 loaded PLGA nanospheres (0.0001 wt% coumarin-6 concentration, $D_{50} = 215$ nm) in comparison to those applying coumarin-6 dispersion (10 wt% coumarin-6 concentration, $D_{50} = 4.63$ μm, with 15 wt% of the particles in the sub-micrometer range). Coumarin-6 is a fluorescent substance and serves as the labeling agent in the experiment. The coumarin-6 loaded PLGA nanospheres permeated the skin far better and sustained much longer in the tissue than the coumarin-6 solution. The coumarin-6 loaded nanospheres could reach the dermis by passing through the skin pores. In addition, the concentration of coumarin-6 in the PLGA nanospheres suspension was 1/100,000th of that of coumarin-6 solution used in the experiment. The intense image of the highly diluted coumarin-6 loaded PLGA nanospheres permeating through the skin further demonstrated the effectiveness of drug loaded PLGA nanospheres for transdermal drug delivery applications. Also, although not shown here, a separate skin sample evaluation showed that the coumarin-6 loaded PLGA nanospheres could effectively permeate the skin in places without skin pores.

This effect is attributed to two factors. First, some of the PLGA nanospheres were small enough to directly slip through the gaps between the skin cells in the corneum layer; and secondly, the interfacial energy of the dispersion and skin was weakened, and the nanospheres acquired enough thigmotaxis in the dispersion to permeate the corneum layer.

(2) Effectiveness of the VC-IP encapsulated PLGA nanospheres

Vitamin C can prevent oxidation of biological materials in the skin caused by the UV irradiation, promote the formation of collagen, and suppress the occurrence of melanin pigment. It is known for its effectiveness in preventing spots, wrinkles and aging of skin, and is often used in the skincare products. But, vitamin C is very unstable and can easily be decomposed under ordinary conditions. In addition, it has very poor skin permeability due to its hydrophilic characteristics. In this study, taking the advantage of their superb skin permeability, PLGA nanospheres were used as the carriers to deliver vitamin C to the epidermis and dermis. The vitamin C used in the experiment was an oil-soluble vitamin C derivative, "ascorbyl tetraisoplamitate (VC-IP)". The esterase in the skin transformed VC-IP to reduced vitamin C (ascorbic acid), known as an antioxidant.

After the skin samples were treated for a certain period of time, the epidermis and dermis were separated, and the contained amounts of vitamin C in each skin layer were measured with HPLC and coulometric ECD. Applying the conventional VC-IP aqueous dispersion (emulsion), only small amounts of the

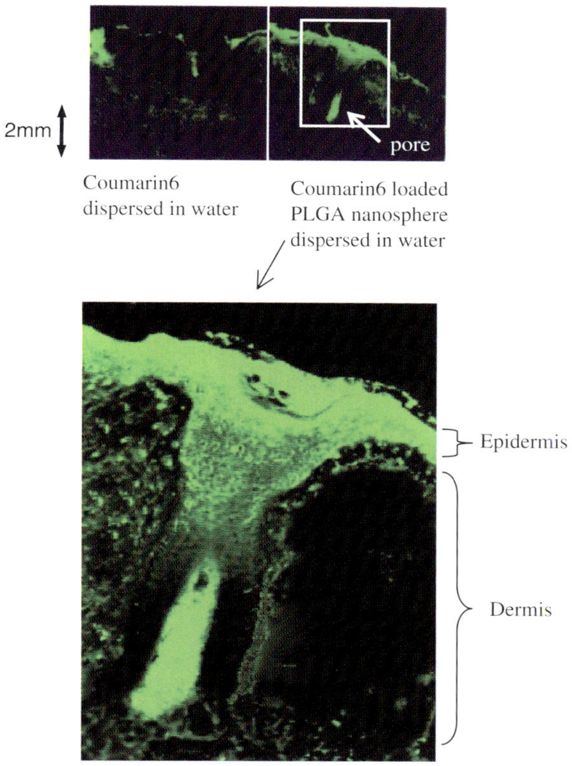

(Evaluation : Prefectural University of Hiroshima/Miwa lab.)

Figure 21.2
Permeability comparison between micro dispersion and PLGA NS dispersion in water.

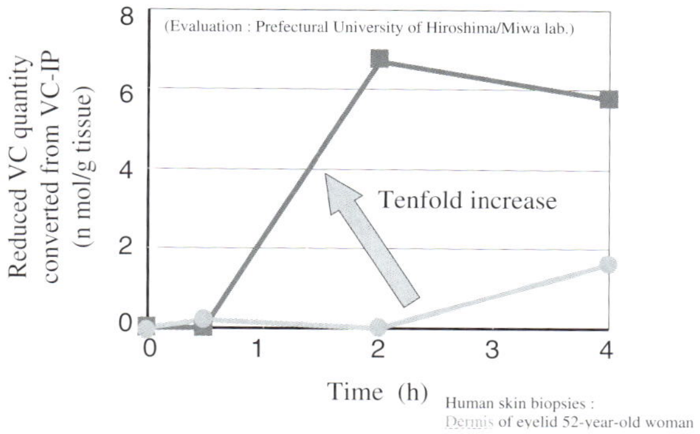

Figure 21.3
Evaluation of Vitamin C delivery efficiency to dermis by VC-IP encapsulated PLGA nanospheres.

vitamin C were detected, and none would exist 18 h after the treatment. But, when applying VC-IP encapsulated PLGA nanospheres, the vitamin C could be detected even 48 h after the treatment. Fig. 21.3 shows the changes over time in the amount of reduced Vitamin C at the dermis after applying the aqueous dispersion containing 7.69 wt% of VC-IP encapsulated PLGA nanospheres (D_{50} = 215 nm) on the skin samples in comparison to that of applying 1 wt% VC-IP oil–water emulsion (D_{50} = 276 nm). Both of them had a VC-IP concentration of 1.0 wt%. However, VC-IP encapsulated in the PLGA nanospheres was more effective than applied as it is. With the VC-IP encapsulated PLGA nanospheres, immediate effect could be observed 2 h after the treatment; and, after 4 h its cumulative amount of vitamin C could reach 10 times of that applying VC-IP emulsion. This quantitatively showed the benefits of using PLGA nanospheres to deliver vitamin C for skincare applications.

If VC-IP is applied alone, the agglutination of its molecules would be hard to pass through the gaps between skin cells, but the above effect is achieved by encapsulating VC-IP in the PLGA nanospheres, allowing them to pass through the horny layer and be delivered stably through the skin. The VC-IP is gradually released due to hydrolysis of PLGA and the metabolism to pure vitamin C by esterase in the epidermis and dermis. The immediate effect, however, is attributed to the VC-IP on nanosphere surfaces. These two effects enable PLGA nanospheres to be useful nanoparticle carriers that can deliver pro-vitamin C to hard-to-reach skin area and provide both immediate and sustained cosmeceutical effects.

(3) Defense against UV by VC-IP encapsulated PLGA nanospheres

Recently, owing to the depletion of ozone layer, a larger amount of UV radiation (UV-A, 320–400 nm; UV-B, 280–320 nm) reaches the Earth. UV-A can penetrate the skin to dermis and cause damages to collagen and elastin from long-term exposure. Although UV-B does not penetrate to the dermis, it can cause inflammation (sunburn) such as erythema and edema in a short time. Daily exposure may induce skin aging and skin cancer. For verifying the effect of vitamin C delivered to the skin tissue by PLGA nanospheres, we evaluated DNA double-strand breakages in the cell nuclei. This experiment demonstrated the effectiveness of applying VC-IP encapsulated PLGA nanospheres to control the DNA damage by UV irradiation. The emulsion containing VC-IP encapsulated PLGA nanosphere was applied to human skin samples (from the back of the ear of a healthy 47-year-old woman), which were then given accelerated UV irradiation for several tens of seconds to provide a cumulative amount equivalent to that a person could receive in a day (UV-A, 200 J/cm^2; UV-B, 2 J/cm^2). The skin samples were evaluated with the TUNEL method 24 h later. The TUNEL method selectively added the green fluorescein isothiocyanate to DNA strand broken in the skin samples. As shown in Fig. 21.4, the fluoresced

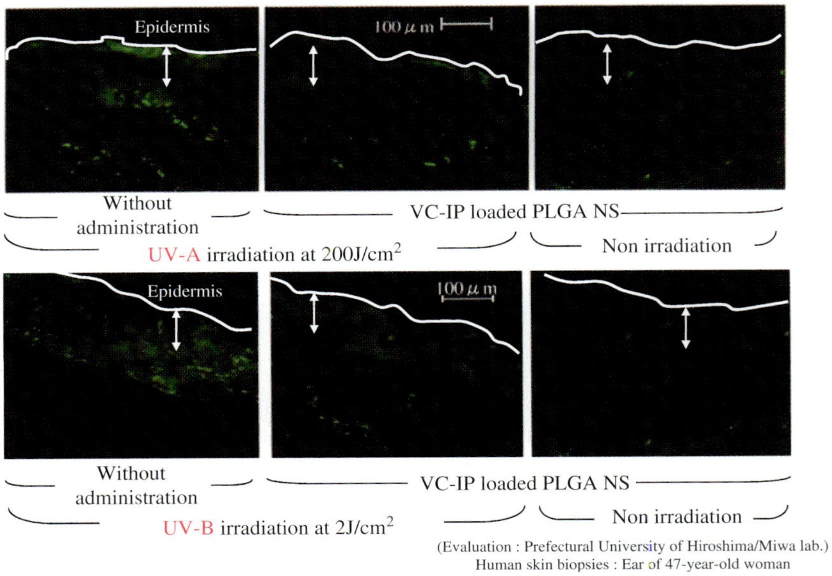

Figure 21.4
Effect of VC-IP loaded PLGA nanospheres on DNA double strand under UV irradiation.

areas were the damaged DNA strands. The skin samples treated with the VC-IP encapsulated PLGA nanospheres closely resembled the un-irradiated skin samples. This demonstrated that the VC-IP encapsulated PLGA nanospheres have effectively reduced the DNA breakage due to the antioxidation effect of the reduced vitamin C, efficiently delivered and protected by the PLGA nanospheres.

3. Functional cosmetics using PLGA nanospheres

As stated previously, because we were able to verify the benefits of applying vitamin encapsulated PLGA nanospheres, we developed skincare products manifesting skin whitening and anti-aging capabilities as shown in Figs. 21.5 and 21.6 (Hosokawa Micron Cosmetics and NanoCrysphere Prime Serum (functional beauty essence)). PLGA nanospheres in these products encapsulate, in addition to VC-IP, vitamin E derivative (tocopherol acetate) and vitamin A derivative (retinyl palmitate). They are mixed with beauty essence at the time of use. Applying our proprietary manufacturing process, the encapsulated PLGA nanospheres are compounded with water-soluble vitamin C derivatives and sugar alcohol to form nanocomposite powder with improved storage stability and redispersibility in beauty essence. The efficacy of these products were evaluated and described below.

(1) Whitening effect of functional cosmetics [4]

We used human HMV-II melanin-producing cells to evaluate the effect of functional beauty essence (NanoCrysphere) on the melanin production. These human melanin-producing cells as shown in Fig. 21.5a can produce more melanin if treated with theophylline (a caffeine-like substance found in tea leaves). After being treated with theophylline, the cells turned to have brownish-black color as shown in Fig. 21.5b, which confirmed the production of melanin. To explore its effect on the melanin production, the functional beauty essence (0.1 μl/ml) was applied to the cells 4 h before the theophylline treatment. Fig. 21.5c shows that the functional beauty essence suppressed the melanin production that could have increased due to the theophylline treatment. This result was attributed to the sustained release of vitamins encapsulated PLGA nanospheres in the beauty essence.

(2) Anti-aging effect of functional cosmetics

It takes more than several decades to develop wrinkles on the skin. Therefore, it is not easy to evaluate the efficacy of wrinkle prevention agents in a short time for new product development. However, using the high-speed artificial wrinkle formation system developed by Miwa et al., accelerated test results can be obtained.

The wrinkle formation system intermittently irradiates human skin biopsies with a critical dose of UV rays while keeping the skin cells alive. By this way, artificial wrinkles can be formed in a short time to evaluate the efficacy of wrinkle prevention agents. Fig. 21.6 showed skin samples and their wrinkle-line histograms. The skin samples were prepared by evenly dividing a human skin biopsy (from the ear of a 55-year-old woman), cultured in a modified Bronaugh

(a) Non-administration of Theophyllin (before pseudo-sunburn)

(b) Theophyllin only (3 days later) (After pseudo-sunburn by UV radiation)

(c) NanoCrysphere + Theophyllin (3 days later)

(Evaluation: Prefectural University of Hiroshima/Miwa lab.)

Figure 21.5
SEM photos (1000×) of melanocyte derived from human cells (HMV-II) with and without Teophylin and nanospheres treatment.

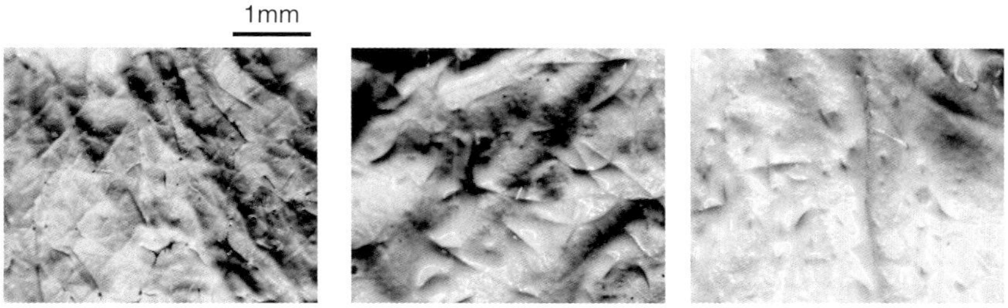

(a) Before UV-A irradiation (control)

(b) After UV-A irradiation: wrinkle level +++

(c) After UV-A irradiation (administration of Nanocrysphere TM): wrinkle level +-

(Evaluation: Prefectural University of Hiroshima/Miwa lab.)

Figure 21.6
Antiaging effect of NanoCrysphere™ on human skin biopsies before and after UV-A irradiation. UV-A: 3 J/cm^2, 2 times/min, 2 times/day. Concentration of NanoCrysphere: 200 μl. Human skin biopsies: ear skin from 55-year-old woman.

diffusion chamber, and then exposed to the UV-A irradiation at 3 J/cm^2 twice a day for five consecutive days.

After receiving the UV-A irradiation for some time, large wrinkles shown as deep shadows on the skin (Fig. 21.6a) and small wrinkles shown as narrow but deep-cut indentation (Fig. 21.6b) could be observed. The wrinkle-line histograms were also irregular and rough, as furrows and hills. However, if the functional beauty essence (0.1% NanoCrysphere in saline, 200 μl) was applied on the skin sample before UV-A irradiation, there were hardly any large or small wrinkles formed on the skin sample (Fig. 21.6c), and the wrinkle-line histogram was comparatively regular with fine grooves (texture), indicating that the wrinkle formation was remarkably suppressed. This was likely because the encapsulated nanospheres significantly penetrated through the skin and helped to form vitamin derivatives in the dermis, which eliminated the wrinkle causing UV-A-derived active oxygen.

PLGA nanospheres can encapsulate low skin permeability drugs and deliver them to deep skin (dermis). We applied this basic technology to skincare products and demonstrated their skin whitening and anti-aging efficacy. The PLGA nanosphere can be a useful transdermal drug delivery carrier.

References

[1] Yomiuri Shinbun Tyokan, pp. 2, 11 January, 2005 Oyobi Yomiuri Shinbun Yukan, pp. 3, 1 March, "Kyokushou kapuseru shiwa tori keshouhin".

[2] Hosokawa Micron Cosmetics, Nanocrysphere on: http://www.nanocrysphere.com./lineup/index.html
[3] H. Tsujimoto, K. Hara, C.C. Huang, T. Yokoyama, H. Yamamaoto, H. Takeuchi, Y. Kawashima, K. Akagi and N. Miwa: Percutaneous absorption study of biodegradable PLGA nano-spheres via human skin biopsies, *J. Soc. Powder Technol. Jpn.*, **41** (12), 867–875 (2004).
[4] H. Tsujimoto: Development of functional skin and scalp care cosmetics using biodegradable PLGA nanospheres, *Drug Deliv. Syst.*, **21**(4), 405–416 (2006).
[5] N. Miwa: *Rejuvenation and Skin Protection, and Biotechonology* (N. Miwa, Ed.), CMC Syutsupan, Tokyo, Japan (2003).

APPLICATION 22

22 DEVELOPMENT OF PHOTONIC CRYSTALS BASED ON NANOPARTICLE ASSEMBLY

Photonic crystals are new types of optical materials, in which small dielectric cells with the size of optical wavelength (from several 100 nm to several 10 μm) are periodically arranged in two- or three-dimensions [1, 2]. They are expected to contribute to the drastic improvement of the performances of lasers and to the miniaturization of integrated optical circuits. While self-assembly techniques are extensively developed to construct crystalline structures of nanoparticles, a very different approach is introduced here: a method to build up crystals by accurately arranging individual nanoparticles by use of a micromanipulator under the observation of a scanning electron microscope (SEM).

1. Nanoparticle assembly technique [3]

There are several techniques for assembling individual small particles: such as micromanipulation under the observation of an optical microscope, laser trapping, and manipulation using a scanning probe microscope [4]. Among them the micromanipulation under SEM observation is the most suitable for three-dimensional assembly of optical-wavelength-sized objects.

Fig. 22.1a illustrates an actual nanoparticle assembly system. The system comprises of a field-emission-type SEM coupled with a piezoelectric micromanipulator [5]. An operator maneuvers the manipulator with joysticks while observing a real-time SEM image with a magnification from 5,000 to several tens of thousands. Micro objects are stuck to a tip of a probe (needle) and then transferred.

The manipulator is composed of a sample stage and a probe-positioning unit. The probe tip can be positioned at an arbitrary point in a 15 mm cube on the sample stage with an accuracy of 10 nm by means of a unique coarse/fine positioning mechanism. The probe is a glass needle made by drawing a heated glass cylinder and then coating of Au. The adhesion forces acting on the object originate from electrostatic forces or van der Waals forces [6, 7]. The magnitude of the adhesion forces strongly depends on the material combination, the curvature of the probe tip, and the electron beam condition (particularly, the acceleration voltage). The object can be reversibly picked up or deposited just by proper movement of the probe in most cases [8]. However, if necessary, a polymer film is coated on the

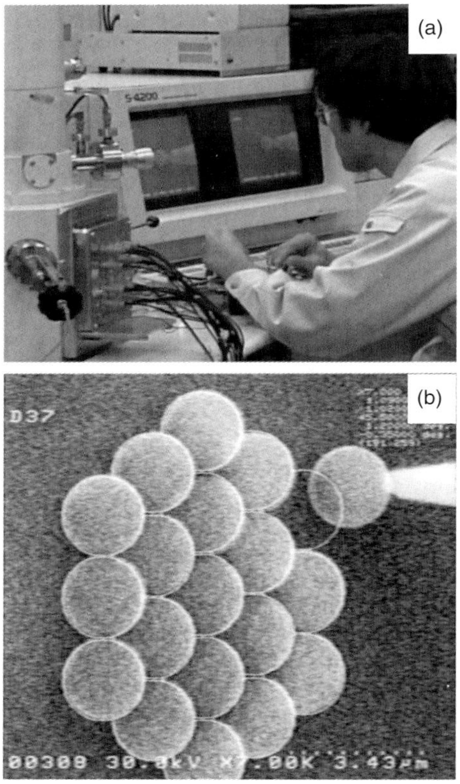

Figure 22.1
Nanoparticle assembly.

substrate to increase the adhesion of the object to the substrate, or a voltage of several 10 V is applied to the probe to enhance the adhesion to the probe. Although the application of the voltage to the probe is very effective in terms of the adhesion control, it is not necessarily useful because the electric fields seriously deteriorate the SEM image.

It has been confirmed that micro objects with diverse shapes, such as spheres, whiskers, and plates made of various materials, such as metals, polymers, ceramics, and semiconductors, can be manipulated. The smallest object manipulated to date is a metallic sphere with a diameter of 70 nm. Typical materials used for constructing photonic crystals are silica and polystyrene microspheres with a diameter of several 100 nm to several microns.

These spheres might be too large to be called nanoparticles. However, such diameters are inevitably determined based on the wavelength of the light. Prior to the assembly, the microspheres are suspended in an appropriate solvent, dropped and then dried, so that individual isolated spheres are dispersed over the substrate. The spheres do not need to have electric conductivity. This is because an electron beam, which is accelerated to several kilovolts, penetrates objects if its size is below several microns. Therefore, even insulating spheres do not exhibit remarkable charging which would prevent the SEM observation.

Polystyrene and silica spheres commercially available as monodispersed microspheres typically have a standard deviation in diameter of about 1%. However, this dispersion is not at all sufficient for fabricating a defect-free lattice. Therefore, the size and the shape of each sphere are measured on the SEM image and only appropriate spheres are used for the assembly. Although the permissible level of the deviation depends on the types of the crystals, e.g., a limit of ±0.5% is typically required. Moreover, in order to arrange the spheres with sufficient accuracy, a computer-aided-design (CAD) image is superimposed on the SEM monitor, and the spheres are placed so that the actual SEM image agrees with the CAD image. A computer-generated drawing can be seen around the upper right-hand corner in Fig. 22.1b.

The substrate needs to be electroconductive or electron-beam transmissive for the SEM observation. Glass substrates on which an electroconductive indium tin oxide film is deposited or elecron-beam-transmissive SiN membranes with a thickness of several 100 nm are usually used. If the acceleration voltage is carefully selected, an insulating substrate can also be used as it is without any conductive coating.

It takes a few minutes to arrange a sphere, on the average. Picking up, transferring, and deposition of a sphere need only a few seconds. Most of the time is spent in searching usable spheres, in adjusting to the exact position, and in confirming the accuracy of the arrangement.

2. Fabrication of photonic crystals by nanoparticle assembly technique

Most of the two-dimensional photonic crystals are fabricated by semiconductor processes [2]. In contrast, both the semiconductor [9, 10] and the self-assembly techniques [11] are employed to realize three-dimensional photonic crystals. Each technique has both merits and demerits, and thus there is no decisive method yet for the fabrication of three-dimensional crystals. In such status quo, nanoparticle assembly technique under SEM observation, being able to produce completely controlled precise three-dimensional structures, is important for trial fabrication of new crystals for verifying theoretical predictions, for investigation of underlying physics based on systematic experiments, and for exploring a novel phenomenon.

Here is introduced a representative example of the application of the nanoparticle assembly to a trial fabrication of complicated structures which cannot be realized by other methods.

One of the most important features of photonic crystals is the presence of a photonic bandgap (PBG). Semiconductor crystals such as Si have a special energy range called a bandgap, within which the electrons cannot have corresponding energy. A photonic crystal can have a similar energy (frequency) range, which should be regarded as a bandgap for photons (PBG). In a photonic crystal, the light with a frequency in the PBG cannot exist. An incident light from the outside is completely reflected, and a light emission from the inside is prohibited. For realizing high-performance lasers or ultra-small integrated optical circuits, PBG is indispensable.

However, crystal structures that exhibit PBGs are quite limited. The close packed lattices obtained by self-assembly of microspheres cannot possess any PBGs. Although we can realize PBGs by infiltrating the open sites with another material to invert the structure, we need a material with a very high refractive index. Since the PBG of such inverted structures is formed at a very high frequency domain, the gap is easily smeared out by the imperfectness of the lattices. By contrast, if the microspheres are stacked into a diamond lattice, a robust PBG should open at a low frequency region even for a low index material according to a theoretical calculation [12]. Nonetheless, any method to realize such a lattice has not been known.

A diamond lattice has a small volume fraction (34%) and is mechanically fragile. Therefore, direct fabrication is difficult even if the nanoparticle assembly is employed. However, a diamond lattice has an important feature that superposition of two lattices makes a body-centered-cubic (bcc) lattice (68%). The bcc lattice has a sufficient mechanical stability and can be easily assembled. Therefore, as Fig. 22.2 illustrates, if a bcc lattice is once produced by alternative staking of two kinds of spheres with different chemical compositions, and then

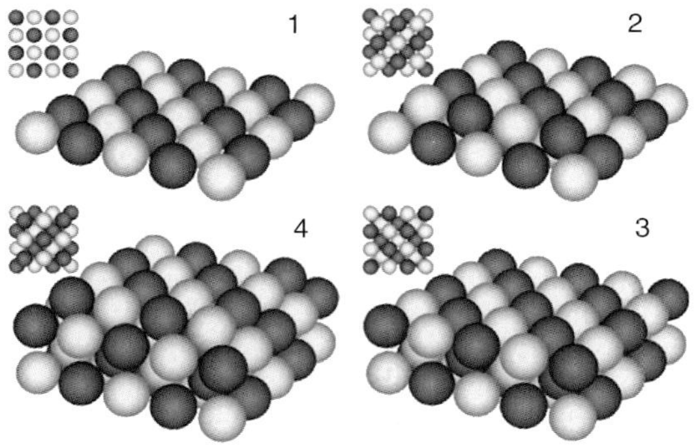

Figure 22.2
Stacking procedure of the bcc lattice for fabricating a diamond-type photonic crystal [14].

Figure 22.3
Photonic crystal with a diamond lattice.

one of the spheres is removed, it would be possible to obtain a diamond lattice made of microspheres [13, 14].

Fig. 22.3a shows a bcc lattice made of silica and polystyrene spheres with a diameter of 1.18 μm [15]. When this lattice is exposed to oxygen plasma, only the polystyrene spheres will be decomposed. Fig. 22.3b demonstrates the first photonic crystal with a diamond lattice made of silica spheres left after the plasma process. The disappearance of alternative spheres is easily recognized at the lowest row.

The optical properties of systems made of microspheres are extensively studied only for both extreme situations of isolated single spheres and periodical array of infinite spheres. In contrast, the intermediate systems between these two extremes have hardly been investigated. Since the nanoparticle assembly technique enables systematic experiments by varying the number of the arranged spheres, it has been utilized for the investigation of the intermediate domains [16, 17].

Furthermore, stacking of plates is possible as well [18]. Though the manipulation is currently carried out by skilled operators, automatic systems of the manipulation processes are being developed. Fully automatic execution from the searching of spheres to the precise arrangement of them has been reported [19].

The fundamental principle of the nanoparticle assembly technique is the adhesion phenomenon between the nanoparticles and other particles, the probe or the substrate. However, the adhesional interaction of the particles in a vacuum under the irradiation of the electron beam has hardly been investigated; this is one of the new important frontiers in the particle technology.

References

[1] J.D. Joannopoulos, R.D. Meade and J.N. Winn: *Photonic Crystals -Molding the Flow of Light*, Princeton University Press, Princeton (1995).

[2] S. Noda, T. Baba: *Roadmap on Photonic Crystals*, Kluwer Academic Publication, Dordrecht (2002).

[3] H.T. Miyazaki: *Ceramics Jpn.*, **39**, 931–934 (2004) (in Japanese).

[4] NIMS Particle Assembly Research Group: *Particle Assembly Technologies*, Kogyo Chosakai Publishing Co. Ltd., Tokyo (2001) (in Japanese).

[5] H. Morishita, Y. Hatamura: Proceedings of IEEE/RSJ international conference of intelligent robots and systems, Yokohama, pp. 1717–1721 (1993).

[6] H. Miyazaki, T. Sato: *Adv. Robot.*, **11**, 169–185 (1997).

[7] H.T. Miyazaki, Y. Tomizawa, S. Saito, T. Sato and N. Shinya: *J. Appl. Phys.*, **88**, 3330–3340 (2000).

[8] S. Saito, H.T. Miyazaki, T. Sato and K. Takahashi: *J. Appl. Phys.*, **92**, 5140–5149 (2002).

[9] S. Noda, K. Tomoda, N. Yamamoto and A. Chutinan: *Science*, **289**, 604–606 (2000).

[10] S.Y. Lin, J.G. Fleming, D.L. Hetherington, B.K. Smith, R. Biswas, K.M. Ho, M.M. Sigalas, W. Zubrzycki, S.R. Kurtz and J. Bur: *Nature*, **394**, 251–253 (1998).

[11] Y. Xia, B. Gates, Y. Yin and T. Lu: *Adv. Mater.*, **12**, 693–713 (2000).

[12] K.M. Ho, C.T. Chan and C.M. Soukoulis: *Phys. Rev. Lett.*, **65**, 3152–3155 (1990).
[13] F. Garcia-Santamaria, C. Lopez, F. Meseguer, F. Lopez-Tejeira, J. Sanchez-Dehesa and H.T. Miyazaki: *Appl. Phys. Lett.*, **79**, 2309–2311 (2001).
[14] H.T. Miyazaki: *J. Soc. Inorg. Mater. Jpn.*, **11**, 474–480 (2004).
[15] F. Garcia-Santamaria, H.T. Miyazaki, A. Urquia, M. Ibisate, M. Belmonte, N. Shinya, F. Meseguer and C. Lopez: *Adv. Mater.*, **14**, 1144–1147 (2002).
[16] H.T. Miyazaki, H. Miyazaki, K. Ohtaka and T. Sato: *J. Appl. Phys.*, **87**, 7152–7158 (2000).
[17] H.T. Miyazaki, H. Miyazaki, N. Shinya and K. Miyano: *Appl. Phys. Lett.*, **83**, 3662–3664 (2003).
[18] K. Aoki, H.T. Miyazaki, H. Hirayama, K. Inoshita, T. Baba, K. Sakoda, N. Shinya and Y. Aoyagi: *Nat. Mater.*, **2**, 117–121 (2003).
[19] T. Kasaya, H.T. Miyazaki, S. Saito, K. Koyano, T. Yamaura and T. Sato: *Rev. Sci. Instrum.*, **75**, 2033–2042 (2004).

APPLICATION 23

23 LIQUID-CRYSTALLINE INORGANIC NANO AND FINE PARTICLES

1. Organic liquid crystals and lyotropic liquid-crystalline inorganic fine particles

Liquid crystal (LC) is one of the most representative functional materials, which shows both fluidity and crystallinity. Material showing liquid crystalline property induced by the changes of the temperatures is called as thermotropic LC. On the other hand, matter exhibiting such property by the changes of the concentration of the solutions is named lyotropic LC. The discovery of the LC goes back to the report about a cloudy liquid state of cholesteryl benzoate, observed by heating, by Reinitzer in 1888. This is the first example of the thermotropic LC and the pioneering work about LCs. Further detailed descriptions about the histories and the general characters of LCs are well summarized in the literatures [1–4].

In general, LCs used as devices of LC displays are organic thermotropic LCs with low molecular weight that consist of calamitic shapes, and exhibit electric responsively and optical anisotropy. In this regard, extensive efforts have been paid for the development of novel types of organic LCs applicable to high-performance electro-optic devices, and extremely various types of organic LCs have been reported so far [5]. Recent remarkable progress in organic synthetic chemistry might also contribute such rapid development in the fields of organic LCs.

As other examples of LCs except organic LC systems, lyotropic liquid-crystalline dispersions of inorganic fine particles are included, which show optical anisotropy under fluidity and electric and/or magnetic fields. Such colloidal dispersions have previously been called as *tactosols*. The pioneering study, concerning about iron oxyhydroxides sols with optical anisotropy under magnetic field, has been carried out by Majorana in 1902 [6]. Researches about V_2O_5 sols showing optical anisotropy under fluidity and electric field, reported by Freundlich et al. in 1915 and 1916, also provide important knowledge at the dawn of this fields [7, 8]. As mentioned above, studies on both organic LCs and LC dispersions of inorganic particles have been begun almost at the same time. However, in spite of the remarkable progress of the studies about organic LCs, only a few examples of the LC dispersions of inorganic particles were reported until 1990s. Such difference might be due to the following reasons: (i) the knowledge concerning precise syntheses of inorganic fine particles had lagged behind compared with the synthetic organic chemistry; (ii) there had been no chance of discovery for industrial applications of LC dispersions of inorganic particles compared with organic LCs. However, researches on LC dispersions of inorganic particles that showed lyotropic LC behavior came to begin to be reported rapidly for these several years. Refer to the introduction, etc. for details concerning these lyotropic LC dispersions of inorganic particles [9–13]. In addition, interesting researches have been reported in recent years such as lyotropic liquid-crystallinity of organometallic clusters [14, 15] and self-assembling behavior of gold nanorods [16]. Lyotropic LC dispersions except inorganic systems have also been observed, for example, dispersions of self-assembled rod-like viruses [17], monodispersed polypeptides [18], carbon nanotubes [19, 20], fullerene derivatives [21], and cellulose fine particles [22] show lyotropic LC states, spontaneously. As a new concept to form lyotropic lamellar structure, hybridization of virus and zinc oxides nanoparticle has recently been demonstrated [23].

Here, let us consider the different points of organic LCs and LC dispersions of inorganic particles.

First of all, the digit of lattice constants of LC structures such as lamellar and columnar is different. That is, even if it is large, the lattice constants of organic LCs are about 5 nm at the largest. This originates in the size of the organic compounds or the molecular self-assembled structures that forms the LC phases. The virtual limits of the molecular sizes of the organic compounds, obtained by organic synthesis except polymer and the oligomer, are several nanometers. Furthermore, it is difficult to form the organic LC phases that consist of the lattice parameters of more than 10 nm even if molecular self-assembly is utilized. On the other hand, the lattice parameters of lyotropic LC dispersions obtained so far are at least 10 nm or more. It is due to the sizes of the inorganic particles to form the LC dispersions. In order to obtain LC order in the LC dispersions, utilization of inorganic particles with shape anisotropy such as spindle-, rod-, sheet-, and disk-like structures with high aspect ratios is essential. In general, such shape induction has been achieved by kinetically controlled particle growth. The particle growth mechanism means that the resulting particle sizes towards the long axis reach at least more than 10 nm or larger sizes. Recent remarkable progress in synthetic inorganic fine particles and nanoparticles is making it possible to obtain various types of monodispersed nanoparticles with shape anisotropies. In near future, we can expect that we will have opportunities to meet new horizons of novel functional materials using size- and shape-controlled inorganic nanoparticles as key components.

On the other hand, organic–inorganic hybrid materials have attracted a great deal of interest especially in the fields of materials science and nanotechnology, because it is not rare to find unexpected novel properties with such materials. Here, it is described about the "organic–inorganic hybrid LCs". The authors have first succeeded in the induction of thermotropic liquid crystallinity into monodispersed inorganic fine particles with shape anisotropy [24, 25]. Their studies, applied the "Gel–Sol method" which was invented by Sugimoto and Muramatsu in 1991 for the synthesis of monodispersed fine particles in large quantities at low cost [26]. The invention was based on an idea that a highly viscous gels such as metal hydroxide gel network may be useful as a matrix for holding the nuclei and growing particles of the final product to protect them from coagulation even in the presence of a high concentration of electrolyte, and also as a reservoir of metal ion and/or hydroxide ion to release the precursor monomer in highly concentrated systems. The particles with high monodispersity, shape uniformity, and variety of the resulting shapes, obtained by this method, are quite suitable for the development of "organic–inorganic hybrid LCs".

2. Development of organic–inorganic hybrid liquid crystals

(1) Hybridization of calamitic liquid-crystalline amines with monodispersed TiO_2 nanoparticles [24]

By the utilization of the "Gel–Sol method", size and shape control of anatase TiO_2 nanoparticles are readily achieved, and highly monodispersed TiO_2 nanoparticles with ellipsoidal, acicular, spherical, and cubic shapes are obtained in large quantities [27, 28]. The authors focused on the anisotropic shapes of the TiO_2 particles for the induction of thermotropic liquid crystallinity into the particles by the organic–inorganic hybridization. For the nanolevel hybridization between organic molecules and inorganic nanoparticles, introduction of interaction moiety between them might be one of the most useful ways. Here, if electric and magnetic fields and light responsive functional molecules as organic molecules are used for the hybridization, addition of such functions into TiO_2 nanoparticles might also be expected. Based on this idea, it was focused on organic LCs as electro-optical materials, and designed organic LCs having adsorption moieties on the surfaces of TiO_2 nanoparticles. The key for the anisotropic shapes of TiO_2 nanoparticles obtained by the "Gel–Sol method" is specific adsorption of organic small molecules and/or anions used as shape controllers. For example, the acicular shape of TiO_2 nanoparticles is induced by specific adsorption of ammonia or primary amines to the *c*-axis of tetragonal anatase TiO_2 particles during their growth. The strong adsorption of amines gives us a hint that we would obtain novel organic–inorganic hybrid LCs would be obtained by the adsorption of amino-substituted mesogenic organic LC molecules to the surfaces of the anisotropic TiO_2 nanoparticles.

We designed and synthesized amino-substituted **L1** and **L2**, as well as **L3**, which is same as **L1** but without an amino group, as shown in Fig. 23.1. The phase transition temperatures and LC phases of **L1–L3** were determined by differential scanning calorimetry (DSC) and polarized optical microscopy (POM) with a hot stage. Both fluorophenyl-substituted **L1** and cyanobiphenyl-substituted **L2** show thermotropic nematic (N) phases. On the other hand, TiO_2 particles with various shapes, in the transmission electron micrographs of Fig. 23.2, were prepared by the "Gel–Sol method" and thoroughly washed with 2 M HNO_3, 0.01 M NaOH, and distilled water by centrifugation to completely remove the shape controllers. The rather polydispersed and irregular-shaped particles **T4** were prepared by mixing uniform TiO_2 particles of different shapes. All TiO_2 powders were confirmed to be anatase-type TiO_2 by X-ray diffraction analysis, and their surfaces were all those of pure anatase titania as revealed by X-ray photoelectron spectroscopy.

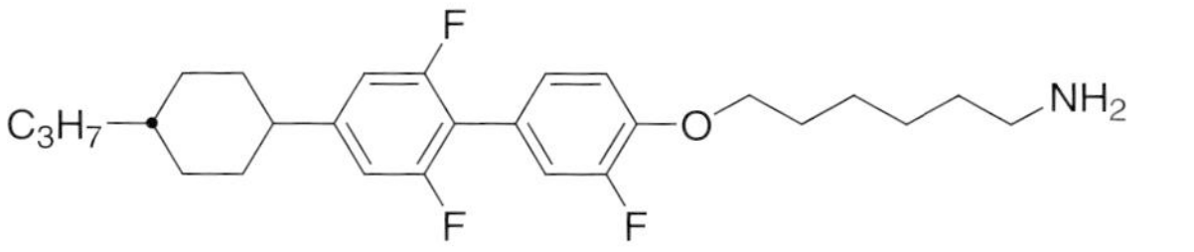

L1: Cr 35 S_X 73 N 101 Iso

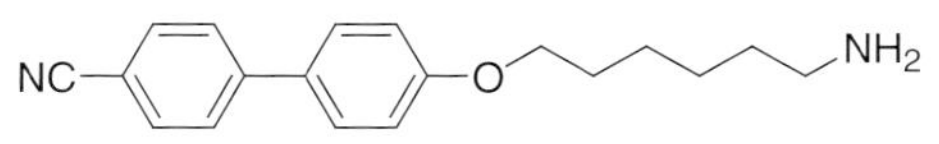

L2: Cr 72 N 92 Iso

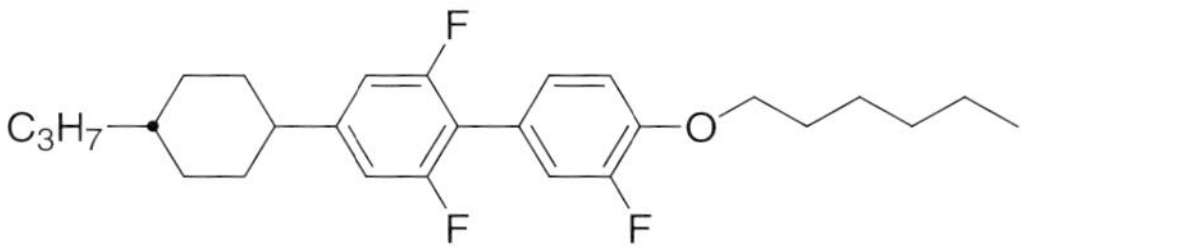

L3: Cr 58 S_X 64 N 100 Iso

Figure 23.1
Structures of thermotropic LCs **L1–L3** for hybridization with TiO_2 particles. Cr: crystal; S_X: higher order smectic; N: nematic; Iso: isotropic phases. The figures in between are transition temperatures in °C.

Hybridization of **L1**/**T1**,**2**,**3**,**4** and **L2**,**3**/**T1** was carried out as follows. Twenty milligrams of **L** and the same weight of **T** were mixed together and dispersed in 2 ml of methanol by ultrasonication for 30 min. Then, the solvent was removed by evaporation at 50°C and subsequent suction under a reduced pressure. Fig. 23.3 exhibits POM images of a hybrid **L1** and **T1** (**L1**/**T1**) at different temperatures. The hybrid melted at 48°C, and a mesophase with strong birefringence was formed as shown in Fig 23.3a. Phase transition of the mesophase occurred at 73°C as revealed from DSC, and a schlieren texture as a characteristic of an N phase was observed up to 113°C (Fig.23.3b). On reaching 113°C, the schlieren texture disappeared and an optically isotropic mesophase was seen up to ca. 130°C (Fig. 23.3c), at which a schlieren texture in a fluidized state was formed again (Fig. 23.3d). The mesophase was readily transformed to a uniaxially aligned monodomain by shearing as shown in Fig.23.3e. The birefringence was observed up to 250°C. Interestingly, such liquid crystallinity owing to the hybridization was not observed at all with hybrids **L1**/**T2**,**3**,**4**. Thus, the aspect ratio of the TiO_2 particles and their uniformity in morphology are decisive factors for the formation of the LCs of the hybrids. None of the mesomorphic phenomena was seen either with hybrid **L2**/**T1** until **L2** of the **L2**/**T1** hybrid was decomposed (ca. 240°C). Hence, in this case, the original liquid crystallinity of **L2** was lost by the hybridization with **T1**. Surface situations of fluorophenyl- covered **L1**/**T1** and cyanobiphenyl-covered **L2**/**T1** become lipophilic

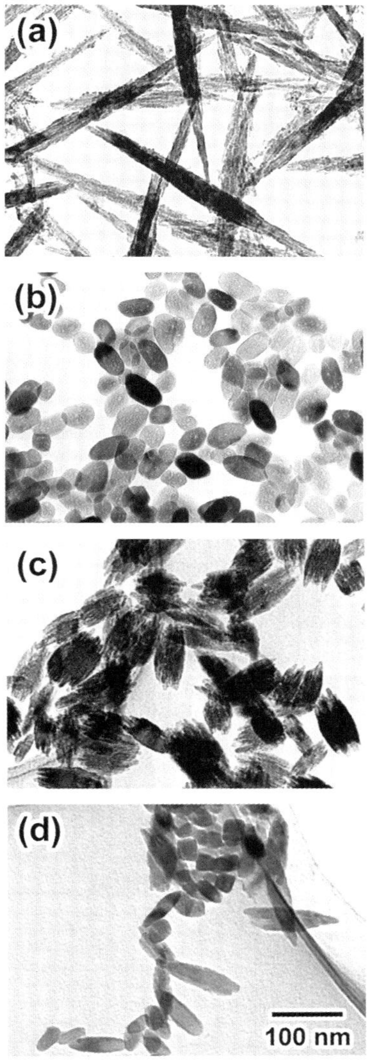

Figure 23.2
TEM images of TiO_2 particles **T1–T3** prepared by the "Gel–Sol method" and irregular shaped **T4** obtained by mixing of different particles. (a) **T1**; (b) **T2**; (c) **T3**; (d) **T4**.

and hydrophilic states, respectively. The fluidity, essential for the induction of the liquid crystallinity, might be brought by the lipophilization of the surfaces of the TiO_2 nanoparticles. Furthermore, since the phase separation was observed only with the hybrid **L3**/**T1** through POM, amino group of **L1** was likely to play an essential role for the formation of the hybrid LC of **L1** with the anisotropic TiO_2 particles. The definite interaction between the amino group of

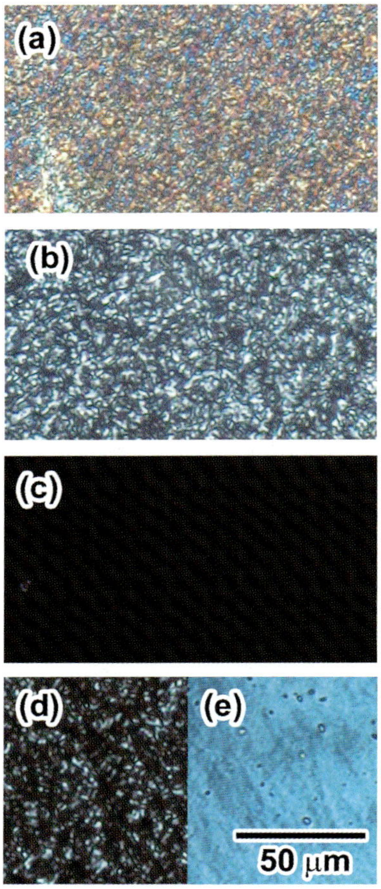

Figure 23.3
Polarized photomicrographs of thermotropic hybrid LC phases of hybrid **L1**/**T1** on the heating. (a) 70°C; (b) 100°C; (c) 120°C; (d) 140°C, polydomain; (e) 140°C, sheared monodomain.

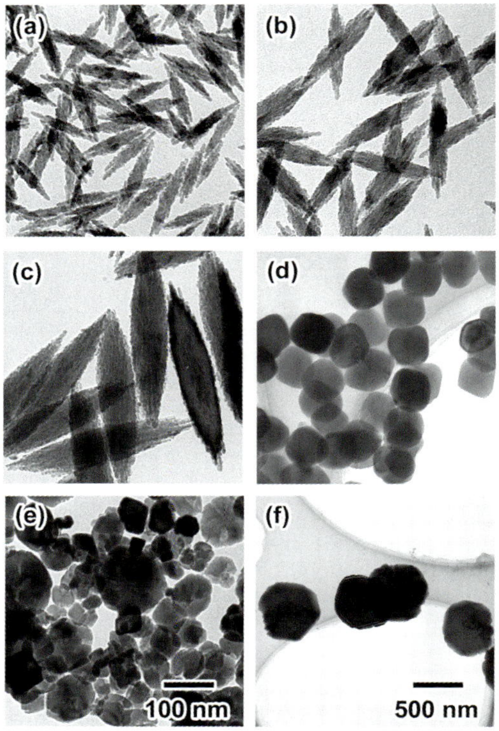

Figure 23.4
TEM images of α-Fe_2O_3 particles with various shapes **H1–H6**. The scale bar in **H5** is common for **H1–H4**.

L1 and the surfaces of **T1** was confirmed by temperature variable infrared spectroscopy. The mixing ratio of **L1** and **T1** also influences the formation of hybrid LC phases. The **L1**/**T1** = 1/2 hybrid exhibits none of liquid crystallinity on heating, and only shows a solid state by the coagulation of **T1**. In contrast, the **L1**/**T1** = 2/1 hybrid forms thermotropic LC state, and the phase transition behavior on heating is as follows: a mesomorphic phase (50–72°C); an N phase (72–121°C); an isotropic phase (121–150°C); a hybrid N phase (150–250°C). Further increase of the **L1** component in the **L1**/**T1** hybrids results in the partial phase separation. The structural characterization of the hybrids is described in the literature [24].

(2) Hybridization of calamitic liquid-crystalline phosphates with monodispersed α-Fe_2O_3 fine particles [25]

By using the "Gel–Sol method", the authors have prepared monodispersed α-Fe_2O_3 particles with various shapes and sizes by the addition of shape controllers and seed particles [26, 29–31]. For example, in the presence of PO_4^{3-} ions, monodispersed spindle-type α-Fe_2O_3 particles have been readily obtained [30–32]. The morphological control was derived from the specific adsorption of the PO_4^{3-} ions on the crystal planes parallel to the *c*-axis, which retarded the particle growth toward perpendicular to the *c*-axis. Such strong adsorption of PO_4^{3-} ions gives us a possibility to obtain novel types of organic–inorganic hybrid LCs [25] by adsorption of organic LC molecules with a phosphate moiety to the surfaces of α-Fe_2O_3 particles with different shapes. Based on this idea, α-Fe_2O_3 particles **H1–H6** with different sizes and morphologies were prepared (Fig. 23.4). Seeding technique [29, 30] enabled us the control of particle mean size of spindle-type **H1–H3** without changing of the aspect ratio. Monodispersed cuboidal **H4**, polydispersed **H5**, and hexagonal platelet **H6**, prepared by using the specific adsorption of OH^- ions *perpendicular* to the *c*-axis

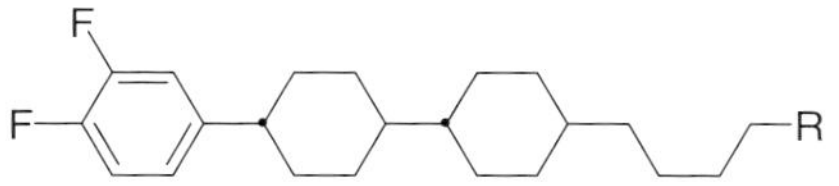

PL1: R = OPO_3H_2;
G-50 M_1 77 M_2 129 S_A 193 N 194 Iso
PL2: R = H;
Cr 41 S_B 45 N 121 Iso

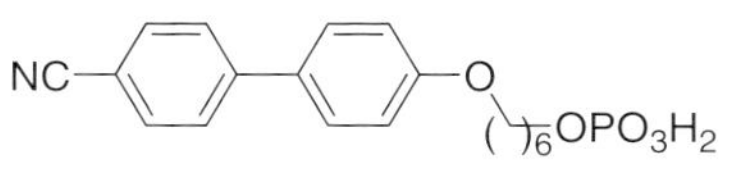

PL3: G -4 M 57 S_A 146 Iso

Figure 23.5
Structures of thermotropic LCs **PL1–PL3** for hybridization with α-Fe_2O_3 particles. G: glass; Cr: crystal; M: mesomorphic; S_A: smectic A; N: nematic; Iso: isotropic phases. The figures in between are transition temperatures in °C.

[29, 30], were used in our study. All particles were thoroughly washed with 1 M NH_3 and water by centrifugation to completely remove impurities and freeze dried. The resulting powders **H1–H6** were confirmed as α-Fe_2O_3 by X-ray diffraction analysis, and no difference in their surface conditions was observed by X-ray photoelectron spectroscopy.

Also, we designed and synthesized **PL1** and **PL3** with a phosphate moiety as shown in Fig. 23.5. Phosphate group-free **PL2** was also prepared. The phase transition temperatures and LC phases of **PL1–PL3** were determined by POM with a hot stage, DSC, and X-ray diffraction. The monodispersed spindle-type **H1–H3** are expected to adsorb the phosphate moiety of **PL1**, because they were obtained with PO_4^{3-} ions as a shape controller [32, 33].

Hybridization of **PL1** and **H1–H6** or **PL2–PL3** and **H2** was carried out as follows. Twenty milligrams of **PL** and the same weight or 40 mg of **H** (**PL/H** = 1/1 or 1/2) were mixed together and dispersed in 2 ml of methanol and 2 ml of $CHCl_3$ by ultrasonication for 30 min followed by removal of the solvents at 60°C under Ar flowing. Fig. 23.6 exhibits optical microscopic images of 1/2 hybrids of **PL1/H2**, **PL1/H5**, and **PL1/H6** at 90°C in the presence of a cross polarizer (a) and its absence (b). For the **PL1/H2** hybrid, a marbled texture as a characteristic of an N phase with strong birefringence and fluidity was observed. The mesomorphic phase was readily transformed to a uniaxially aligned monodomain by shearing (Fig. 23.6i-a inset). DSC measurement on the second heating scan revealed that the hybrid showed the glass–mesophase transition at −35°C. After an endothermic peak at 148°C, the hybrid showed a highly fluidized state with strong birefringence observed by POM. The birefringence was kept up to 250°C. Similar mesomorphic behavior was also seen for 1/2 hybrids of **PL1/H1**,**H3** and 1/1 hybrids of **PL1/H1**–**H3**. However, the further increase of the **PL1** component resulted in the partial phase separation of **PL1** with **H1**–**H3**. On the other hand, 1/2 as well as 1/1 hybrids of **PL1/H5** also formed a uniform fluidized material without phase segregation of **PL1** with **H5** (Fig. 23.6ii-b), however no birefringence was found and formed an optically isotropic state. The **PL1/H4** = 1/2 and 1/1 hybrids, consisted from monodispersed cuboidal α-Fe_2O_3 particles, also showed an optically isotropic state with fluidity. Thus, the aspect ratio of the α-Fe_2O_3 particles

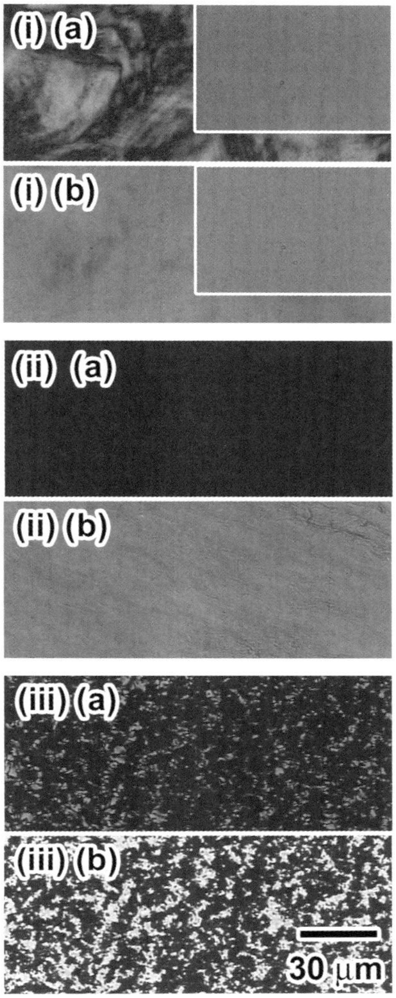

Figure 23.6
Optical photomicrographs of thermotropic hybrid LC phases of 1/2 hybrids (i) **PL1/H2**, (ii) **PL1/H5**, and (iii) **PL1/H6** at 90°C (a) with a polarizer; (i)-(a) and (b) insets: sheared monodomain.

with different shapes was a decisive factor for the formation of the nematic LCs of the hybrids. Interestingly, birefringence owing to **PL1** single component and coagulated **H6** is seen for a 1/2 mixture of **PL1/H6** as shown in Fig. 23.6iii. It means phase segregation between **PL1** and **H6** without hybridization. The platelet **H6** particles have *c*-plane as a principal one, formed by the specific adsorption of OH^- on the basal plane [29, 30]. It shows the factor for the hybridization might be brought by the specific adsorption of the phosphate group of **PL1** on a plane parallel to *c*-axis. None of the mesomorphic phenomena was seen with **PL3/H2** hybrids and it was decomposed at ca. 260°C. Hence, in this case, the original liquid crystallinity of **PL3** was lost by the hybridization with **H2**. Furthermore, since the phase separation was observed only with **PL2/H2** hybrids through POM, the phosphate group of **PL1** played an essential role for the formation of the hybrid LC of **PL1**. The definite interaction between the phosphate group of **PL1** and the surfaces of **H1–H5** was confirmed by temperature variable infrared spectroscopy, and no free PO_3H_2 vibration was observed by the hybridization. The mesomorphic phase structures of the **PL1/H1–H5** hybrids were examined by small-angle X-ray scattering measurements. The profiles of the **PL1/H2** = 1/2 hybrid at 90 and 170°C suggested the existence of periodic particle interactions at intervals of 49.7 and 46.5 nm, respectively, corresponding to the total width of an **H2** covered with the **PL1** molecules toward the short axis of the spindle-type particles. Therefore, the mesomorphic phase seems to have a nematic-like one-dimensional order in the direction of the long axis of the spindle-type particles. Such a periodic scattering was not observed for the single components of **H1–H6** and the **PL1/H5** hybrids. On the other hand, peaks of 45.6, 30.4, and 22.8 nm at 170°C were observed for the hybrid of **PL1/H4** = 1/1, which were corresponding to d_{200}, d_{300}, and d_{400} planes of a superlattice structure with a simple cubic LC structure. The lattice parameter could be assigned as a = 91.2 nm, which was consistent with the particle mean sizes of **H4** covered with the **PL1** molecules by the adsorption on their surfaces. The schematic illustration for the formation of hybrid LC phases is shown in Fig. 23.7.

3. Summary and prospect

Recent remarkable progress in synthetic chemistry concerning about monodispersed inorganic nanoparticles and fine particles with various shapes such as the "Gel–Sol method" enabled us to obtain size- and shape-controlled TiO_2 and α-Fe_2O_3 particles in large quantities. By using these particles, we have succeeded in the development of the "organic–inorganic hybrid LCs" as a novel functional nanohybrid material. To date, LC behaviors are known for some inorganic particles with anisotropic shapes in highly dilute dispersions, as mentioned above. In such systems, the liquid-crystalline performance is governed by the balance of the repulsive and attractive forces among the anisotropic particles themselves. On the other hand, surface modification of rare-metal nanospheres by thiol derivatives of LC compounds has been examined [34–36]. However, the optical anisotropy due to the particle alignment was not observed. Hence, to the best of our knowledge, our observations are the first results on new optically anisotropic LC phases with a novel organic–inorganic hybrid of a liquid-crystalline compound and monodispersed inorganic particles. In view of the almost infinite possibilities in the combination of organic and inorganic matters, such hybrid materials may be of unfathomable potentials as one of the most advanced functional materials in the future. For example, monodispersed α-Fe_2O_3 particles used in our study are readily converted into Fe_3O_4 and α-Fe_2O_3 as magnetic materials, precisely keeping their morphology without sintering [31]. The nanolevel hybridization of

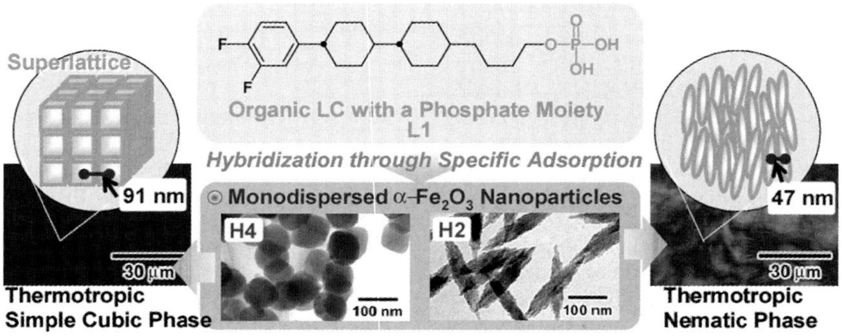

Figure 23.7
A schematic illustration for the formation of organic–inorganic hybrid LC phases.

inorganic nanoparticles with fair magnetic properties and LCs with good electric response will make them one of the most advanced functional materials with excellent multi-responsive performance such as active devices.

References

[1] Ekisho Binran, Ekisho Binran Henshuu Iinkai (Eds.), Maruzen, Tokyo (2000).
[2] Ekisho no Kagaku, Japan Chemical Society (Eds.), Japan Scientific Societies Press, Tokyo (1994).
[3] Ekisho Zairyo, N. Kusabayashi (Ed.), Kodan-shya Scientific, Tokyo (1991).
[4] P.J. Collings, M. Hird: *Introduction to Liquid Crystals*, Taylor & Francis, London (1997).
[5] http://liqcryst.chemie.uni-hamburg.de/en/lolas.php.
[6] Q. Majorana: *Phys. Z.*, **4**, 145 (1902).
[7] H. Diesselhorst, H. Freundlich: *Phys. Z.*, **16**, 419 (1915).
[8] H. Freundlich: *Z. Elektrochem.*, **22**, 27 (1916).
[9] K. Kanie, T. Sugimoto: in *A New Phase on Organic/Inorganic Nanocomposite Materials*, NTS, pp. 175–188, Tokyo (2004).
[10] P. Davidson, P. Batail, J.C.P. Gabriel, J. Livage, C. Sanchez and C. Bourgaux: *Prog. Polym. Sci.*, **22**, 913 (1997).
[11] A.S. Sonin: *Colloid J.*, **60**, 129 (1998).
[12] A.S. Sonin: *J. Mater. Chem.*, **8**, 2557 (1998).
[13] J.-C.P. Gabriel, P. Davidson: *Adv. Mater.*, **12**, 9 (2000).
[14] J. Sayettat, L.M. Bull, J.-C.P. Gabriel, S. Jobic, F. Camerel, A.-M. Marie, M. Fourmigué, P. Batail, R. Brec and R.-L. Inglebert: *Angew. Chem. Int. Ed.*, **37**, 1711 (1998).
[15] F. Camerel, M. Antonietti and C.F.J. Faul: *Chem. Eur. J.*, **9**, 2160 (2003).
[16] N.R. Jana: *Angew. Chem. Int. Ed.*, **43**, 1536 (2004).
[17] Z. Dogic, S. Fraden: *Langmuir*, **16**, 7820 (2000).
[18] S.M. Yu, V.P. Conticello, G. Zhang, C. Kayser, M.J. Fournier, T.L. Mason and D.A. Tirrell: *Nature*, **389**, 167 (1997).
[19] V.A. Davis, L.M. Ericson, A. Nicholas, G. Parra-Vasquez, H. Fan, Y. Wang, V. Prieto, J.A. Longoria, S. Ramesh, R.K. Saini, C. Kittrell, W.E. Billups, W.W. Adams, R.H. Hauge, R.E. Smalley and M. Pasquali: *Macromolecules*, **37**, 154 (2004).
[20] W. Song, I.A. Kinloch and A.H. Windle: *Science*, **302**, 1363 (2003).
[21] M. Sawamura, K. Kawai, Y. Matsuo, K. Kanie, T. Kato and E. Nakamura: *Nature*, **419**, 702 (2002).
[22] J. Araki, M. Wada, S. Kuga and T. Okano: *Langmuir*, **16**, 2413 (2000).
[23] S.-W. Lee, C. Mao, C.E. Flynn and A.M. Belcher: *Science*, **296**, 892 (2002).
[24] K. Kanie, T. Sugimoto: *J. Am. Chem. Soc.*, **125**, 10518 (2003).
[25] K. Kanie, A. Muramatsu: *J. Am. Chem. Soc.*, **127**, 11578 (2005).
[26] T. Sugimoto: *Monodispersed Particles,* Elsevier, Amsterdam (2001).
[27] T. Sugimoto, X. Zhou and A. Muramatsu: *J. Colloid Interface Sci.*, **259**, 53 (2003).
[28] K. Kanie, T. Sugimoto: *Chem. Commun.*, 2004(14) 1584 (2004).
[29] T. Sugimoto, M.M. Khan, A. Muramatsu and H. Itoh: *J. Colloids Surf. A*, **79**, 233 (1993).
[30] T. Sugimoto, Y. Wang: *J. Colloid Interface Sci.*, **207**, 137 (1998).
[31] H. Itoh, T. Sugimoto: *J. Colloid Interface Sci.*, **265**, 283 (2003).
[32] T. Sugimoto, A. Muramatsu, *J. Colloid Interface Sci.*, **184**, 626 (1996).
[33] M. Ozaki, S. Kratohvil and E. Matijevic: *J. Colloid Interface Sci.*, **102**, 146 (1984).
[34] N. Kanayama, O. Tsutsumi, A. Kanazawa and T. Ikeda: *Chem. Commun.*, (24) 2640 (2001).
[35] H. Yoshikawa, K. Maeda, Y. Shiraishi, J. Xu, H. Shiraki, N. Toshima and S. Kobayashi: *Jpn. J. Appl. Phys.*, **41**, L1315 (2002).
[36] L. Cseh, G.H. Mehl: *J. Am. Chem. Soc.*, **128**, 13376 (2006).

APPLICATION 24

24 CLOSELY PACKED COLLOIDAL CRYSTAL ASSEMBLED WITH NANOPARTICLES AND ITS APPLICATION FOR SMART MATERIALS WITH TUNABLE STRUCTURAL COLOR

Opal, colloidal crystal, is three-dimensionally and closely packed monodispersed silica particles of a few 100 nm. Opals have been used in gemstone for a long time because they show iridescent palette described as the play of rainbow color. Sanders et al. revealed the mystery of the play of rainbow color in 1960s. They explained that the mechanism of the iridescent color in opal is caused by Bragg diffraction of visible light from the periodic array of silica particles, i.e. a kind of structural color [1]. Afterward, artificial opals were in industrial production and sold as synthetic jewelry in commercial market.

Recently colloidal crystals have been attracting much attention due to their novel properties and their applications, such as photonic/optical materials, photonic crystals and novel optical devices. In this chapter, the author would like to describe the works on structural color of colloidal crystal films and their potential applications for sensing materials.

1. Closely packed colloidal crystal films

The author developed a new coating method for colloidal crystal film with uniform structural color on a solid substrate [2]. The solid surface was firstly modified to hydrophilic by oxygen plasma cleaning. The surface of the substrate was covered with colloidal suspension. Then the colloidal suspension on the substrate was covered with the silicone oil. The water in the suspension was evaporated through the covering silicon oil layer. In the conventional method, a ring shaped particle deposition was formed during the evaporating process. However, in the new method, the covering silicon oil inhibits capillary flow in the suspension. Consequently, the method enables the formation of a colloidal crystal film with uniform and homogeneous structural color as shown in Fig. 24.1A. Optical microscope image Fig. 24.1B shows a colloidal crystal film made of 100 μm sized domains assembled with nanometer sized colloidal particles. Fig. 24.1C shows an SEM image of closely packed colloidal particles in a single domain of colloidal crystal film. Fig. 24.1D shows a cross-section SEM image of the colloidal crystal film. In the film, colloidal particles are cubic-closely packed (CCP) structure and (1 1 1) planes are parallel to the substrate. The CCP (1 1 1) planes cover almost a 4 inch silicon wafer except the

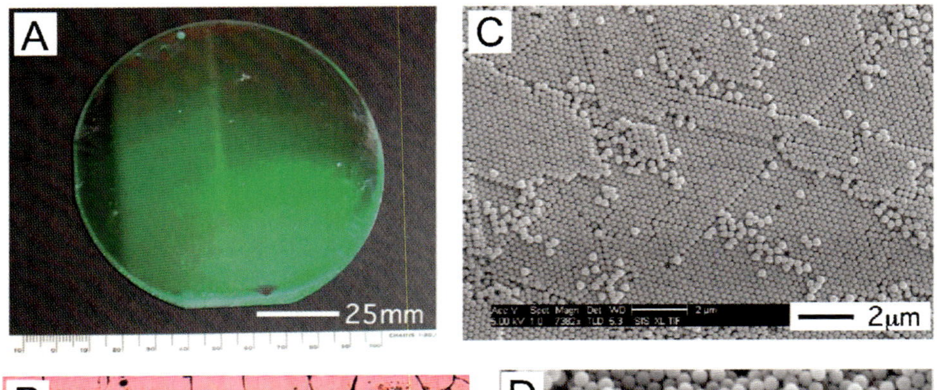

Figure 24.1
Closely packed colloidal crystal film on a silicone wafer having a 4-inch diameter. (A) CCD camera image, showing green structural color; (B) transmission optical microscope image, pink color due to complementary color of green; (C) scanning microscope image, point and line defects in colloidal crystal; and (D) cross sectional image of cubic close packed structure and (1 1 1) plane oriented parallel with a substrate. (Copyright by Materials Science Society of Japan [3]).

edge surrounding the substrate. The colloidal crystal film indicates the mono structural color of green without the play of color phenomena.

Recently colloidal crystals have gathered much attention on the application of photonic crystals, PhC [4]. Colloidal crystal is a low cost process to form a three-dimensional microstructure. However, 3D-PhC requires a perfect and large-scale single colloidal crystal without domains, point and line defects. There are still many problems remaining for 3D-PhC using colloidal crystal film as shown in Fig. 24.1. The most important task is to fabricate a single and uniform colloidal crystal film for cm scale. The colloidal crystal film is limited in availability to 1D-PhC and structural color only using CCP (1 1 1) planes.

2. Structural color of colloidal crystal and its tuning mechanism

Fig. 24.2A shows that colloidal crystal selectively diffract light according to Bragg's equation combined with Snell' law. One can recognize the diffraction light as structural color within its peak position located in the visible wavelength 400–750 nm.

$$\lambda = 2D\sqrt{(n_{\text{eff}}^2 - \sin^2\theta)} \tag{24.1}$$

where D is interplaner spacing of CCP (1 1 1) planes, n_{eff} the average refractive index and θ the incident angle.

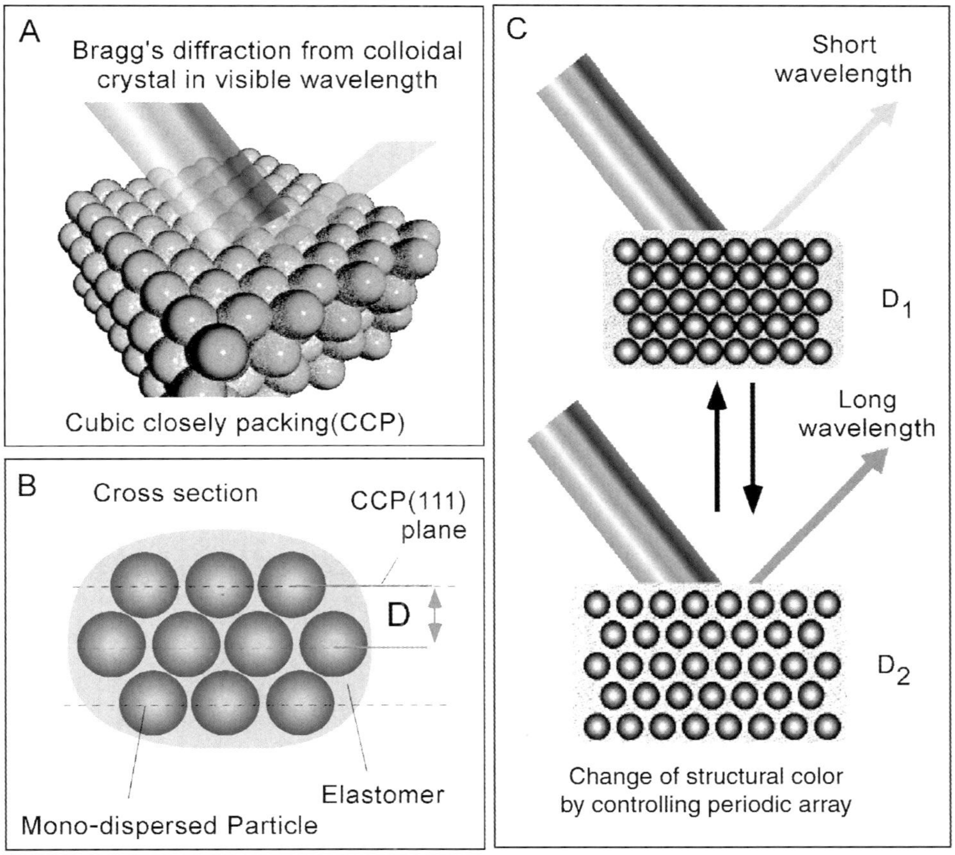

Figure 24.2
Colloidal crystal composite assembly of cubic closely packed (CCP) colloidal particles. The composite enables tuning structural color. (A) The structural color caused by Bragg's diffraction from array of CCP planes. (B) Composite material assembled with monodispersed colloidal particles and among them infilling with elastomer. The structural color is determined by the interspacing of CCP (1 1 1) planes. (C) The wavelength shifts to a longer wavelength by expanding the interspacing. In contrast, the wavelength shifts to a shorter wavelength by shrinking the interspacing. The structural color is reversible due to tunable interspacing between D_1 and D_2.

The structural color can be changed by tuning interplaner spacing, D, or changing the average refractive index, n_{eff}. In earlier work, structural color was changed by n_{eff}. Stein et al. reported colorful structural color of PhC pigments infilling with solvents or polymer having different reflex index [5]. Recent works were changing structural color by control of D. In this chapter, the author describes the changing structural color by expanding or shrinking the interplaner spacing in the 10 nm scale under constant incident angle [6].

Fig. 24.2B shows an illustration of composite colloidal crystal with submicron periodic array of particles. The colloidal crystals were embedded with elastomer for tuning the interplaner spacing of the colloidal particles. In the composite, monodispersed polystyrene (PS) particles are separated with the polydimethylsilicon (PDMS) elastomer. Fig. 24.2C shows the change of color of the composite by expanding D_2 or shrinking D_1, the interplaner spacing. Bragg's diffraction peak is moving to short wavelength at D_1 and to long wavelength at D_2. The difference of the mechanical property between the PS colloidal particle and PDMS elastomer cause the tuning of the interplaner spacing of CCP (1 1 1) planes and the structural color of the composite.

3. Tunable structural color by swelling with liquid

Fig. 24.3 shows the composite made of arrayed 202 nm PS colloidal particles and PDMS elastomer infilling the voids. Photo image Fig. 24.3A shows the composite film which indicates different structural colors in and out of the iso-propanol, IPA. The composite film on a glass substrate indicates green in a dry condition (Fig. 24.3A-i). In contrast, the film indicates red in a wet condition (Fig. 24.3A-ii). This causes swelling PDMS elastomer with IPA solvent. The PDMS absorbs IPA and its volume is expanded. Fig. 24.3B shows an SEM image of the composite film. The volume of PDMS elastomer, 24%, is smaller than that of cubic-closely packed PS particles, 76%. However, expanding the PDMS elastomer in a swollen condition changed the interplaner spacing of CCP (1 1 1) planes. Fig. 24.3C shows the reflection spectroscopy of the composite films corresponding to dry (i) and

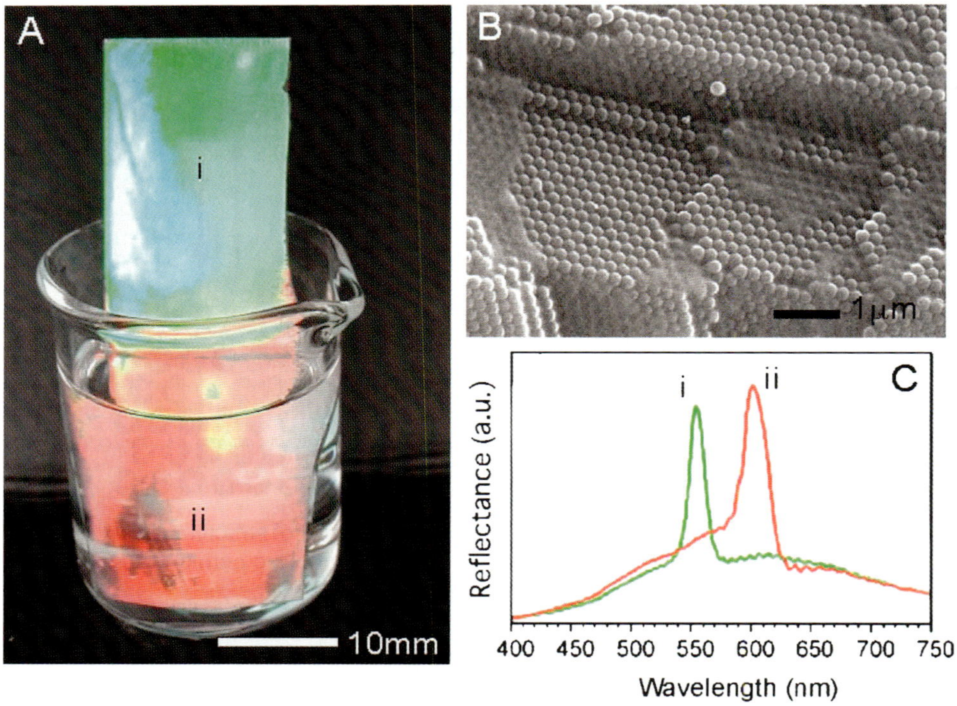

Figure 24.3
Change of structural color of PDMS elastomer by swelling with an iso-propanol liquid: (A) A specimen put into the liquid; (B) an SEM image of the specimen; and (C) comparison of two spectrums between dry and wet conditions.

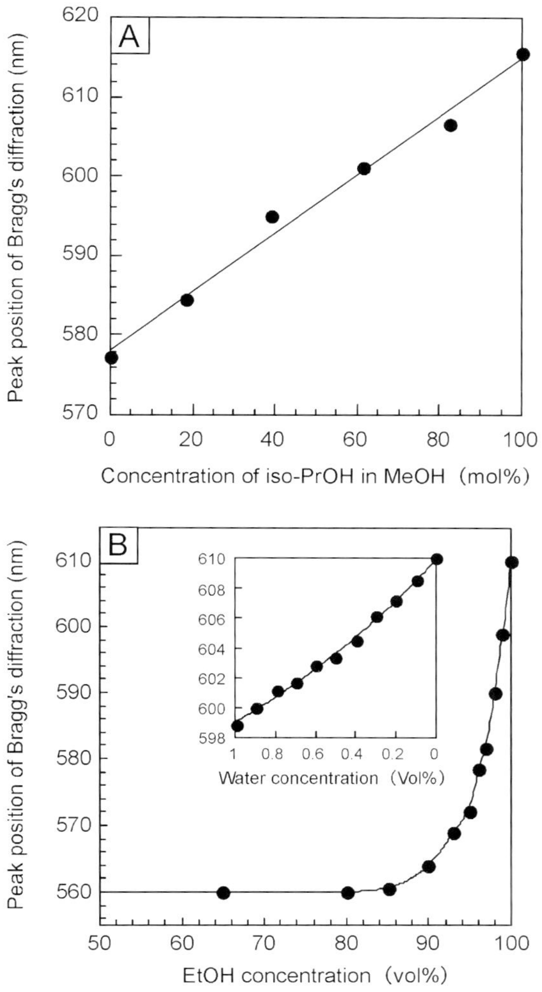

Figure 24.4
Sensing for concentration of liquids using swelling phenomena. (A) Methanol and iso-propanol solvents and (B) ethanol and water solvents, having low amount of water sensing.

wet (ii) areas. Bragg's peak shifted to high wavelength about 50 nm using IPA solvent. The substrate is pulled out from the beaker and then the composite film was completely dried, the structural color of the wet area is recovered to the original dry condition indicating green. The composite film reversibly and repeatably changes structural color by swelling and drying with IPA solvent. One can identify a solvent by chromogenic phenomenon, such as litmus test paper without using special equipment.

Furthermore, the structural color varies according to the solvents due to swelling ability of PDMS elastomer. Fig. 24.4 shows quantitative analysis of the swelling phenomenon by measuring Bragg's diffraction peaks. Fig. 24.4A shows the peak shifting on mixed solvents, methanol, MeOH and iso-propanol, iso-PrOH. The peak position (nm) is proportional to the solvent concentration (mol%). Fig. 24.4B shows a hydrophobic PDMS elastomer that does not swell in hydrophilic condition in a water–ethanol system. In this system, there is little peak shift up to about 80 vol% of ethanol concentration. Above this concentration, the peak quickly increases and its change reaches about 50 nm. The inset graph shows an enlargement of above the concentration of 99 vol% ethanol. The graph enables the estimation of a small amount of water in ethanol less than 1 vol% and the peak position shifts about 12 nm. In this work, a commercially available fiber spectrometer, resolution of 0.1 nm detects a small amount of water in ethanol less than 0.1 vol% level.

4. Tunable structural color by applying mechanical stress

A change of periodic array enables changing structural color of the composite material by mechanical deformation. Fig. 24.5 shows a composite film that was coated on a fluoride rubber sheet with tunable structural color. The color of the rubber sheet reversibly changes by elongation. Fig. 24.5A shows a red structural color sheet in its initial condition before stretching was applied. The rubber sheet was stretched in the horizontal direction. Fig. 24.5B shows the elongated sheet change structural color from red, a circle area, to green, an ellipsoid area. The structural color recovered its initial condition of red color after releasing the applied mechanical stress. The change of color in elastic deformation was measured by peak shift of Bragg's diffraction. Fig. 24.5C shows spectroscopy at different strain conditions. The reflectance peak shifts toward low wavelength as a function of elongation of the rubber sheet. In addition, the peak intensity also decreased. Fig. 24.5D shows the relationship between the strain ($\Delta L/L$) and peak position (nm). The peak position decreases as a function of elongation. This relationship can be applied to strain gages. As a further potential application, the composite film coated on a plastic sheet can display local strain distribution before and after plastic deformation of specimen substrate.

5. Summary and outlook

In this chapter, the author described the colloidal crystal composite having tunable structural color and its sensing applications. Furthermore, an idea

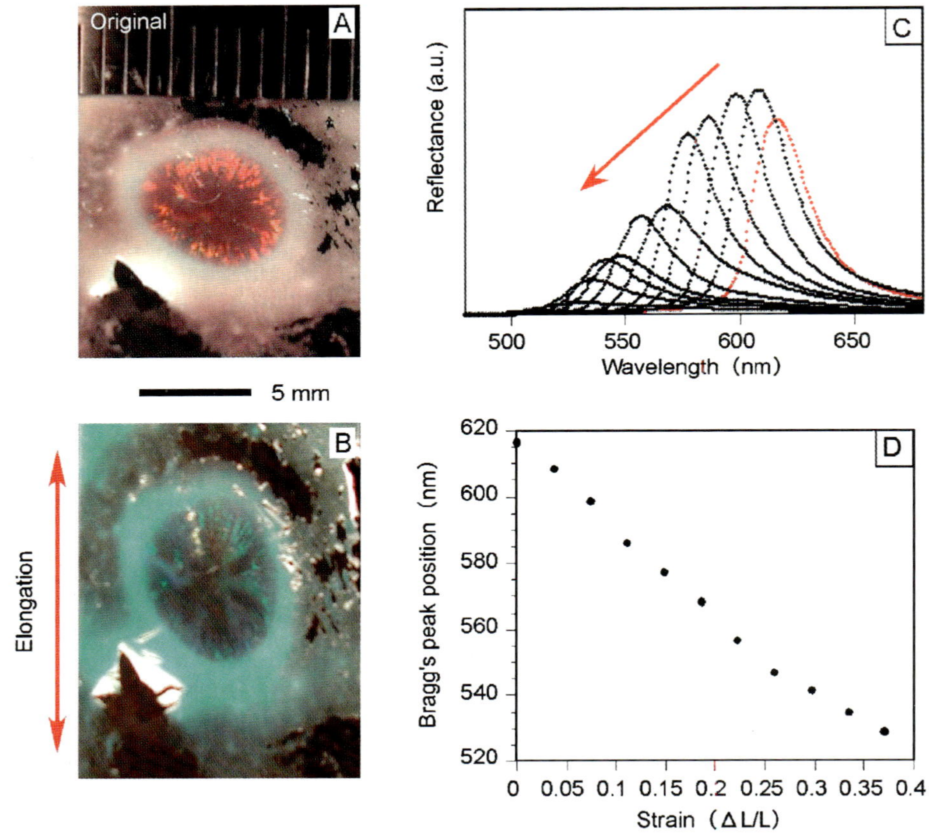

Figure 24.5
Structural color changed by mechanical deformation: (A) Initial state, red color; (B) compressing periodic lattice by tensile force, change to green color; (C) Blue shifting Bragg's diffraction peak as evolution strain; and (D): the relation between strain and peak position.

of rewritable and reusable display media were proposed such as "P-ink" and "Photonic Paper" based on tunable structural color [7]. The author believes that structural color of colloidal crystal will be applied to new devices and new materials in wider fields in the future. In addition, a technique of self- assembly of nanosized particles enables the formation of an ordered structure in the range from nano- to micro-meter size without using expensive equipment. The function of colloidal crystal is anticipated not only as photonic function, but also as a wider variety of functions, such as porous and magnetic materials.

References

[1] J.V. Sanders: *Acta Crystallogr.*, **A24**, 427 (1968).

[2] H. Fudouzi: *J. Colloid Interface Sci.*, **275**, 277 (2004).

[3] H. Fudouzi: *Mater. Sci. Technol.*, **41** (4), 194 (2004).

[4] Y. Xia (Ed.): *Adv. Mater.*, **13**, 369 (2001); C. Lopez: *Adv. Mater.*, **15**, 1679 (2003); J.D. Joannopoulos, R.D. Meade and J.N. Winn: *Photonic Crystals: Molding the Flow of Light*, Princeton University Press, Princeton, NJ (1995); A. Polman, P. Wiltzius: *MRS Bull.*, **26**, 608 (2001); Y.A. Vlasov, X.Z. Bo, J.C. Sturm and D.J. Norris: *Nature*, **414**, 289 (2001); A. Arsenault, S.B. Fournier-Bidoz, B. Hatton, H. Miguez, N. Tetrault, E. Vekris, S. Wong, S.M. Yang, V. Kitaev and G.A. Ozin: *J. Mater. Chem.*, **14**, 781 (2004).

[5] A. Stein, R.C. Schroden: *Curr. Opin. Solid State Mater. Sci.* **5**, 553 (2001).

[6] H. Fudouzi, Y. Xia: Langmuir, 19, 9653 (2003); H. Fudouzi, N. Shinya and Y. Xia: Trans. MRS Jpn., 29, 865 (2004).

[7] A.C. Arsenault, H. Miguez, V. Kitaev, G.A. Ozin and I. Manners, Adv. Mater., 15, 503 (2003); H. Fudouzi, Y. Xia, Adv. Mater., 15, 892 (2003).

APPLICATION 25

25 DEVELOPMENT OF NEW COSMETICS BASED ON NANOPARTICLES

The application of nanoparticles in the field of cosmetics goes back to the blending of ultrafine particles of titanium dioxide in sunscreen in the 1980s. Using titanium dioxide ultrafine particles of about 30 nm size was the first step in sunscreen having a high ultraviolet ray protection capacity as well as high transparency. Even at present, while 10–30 nm titanium dioxide or zinc oxide particles are being used, such ultrafine particles of sizes less than 100 nm will be referred to here as nanoparticles.

Apart from the cosmetics in which the nanoparticles themselves are dispersed in this manner, compound particles in which nanoparticles are arranged in the base particles thereby realizing various optical characteristics are being used for makeup. In an emulsified system, by making the emulsified particles to have nanosizes, new values are given by changing the tactile feel and transparency.

The application of nanoparticles in cosmetics that include emulsified particles and the application of compound particles in which form control is done using nanoparticles are described here.

1. Use of nanoparticles

(1) Nanoparticles in emulsion

Emulsification and solubilization technologies are used in skin care cosmetics such as skin lotions, skin milks, and creams. For example, by changing the sequence of mixing or the pressure in the process of mixing water and oil, it is possible to control the size of the emulsified particles from the usual 1–10 μm to several tens of nanometers.

Among the methods of preparing ultrafine emulsions are the surface chemical methods and the mechanical methods. In the surface chemical methods, although the dispersion methods such as the phase transfer emulsification method [1], the HLB temperature emulsification method [2], the D phase emulsification method [3], the amino acid gel emulsification method [4], etc. are common, sometimes the aggregation methods too are used. As an example of an aggregation method, Nakajima et al. [5] have reported that it is possible to prepare ultrafine emulsions of particles with sizes of less than 50 nm by initially solubilizing a system of surfactant, oil, and water, and then cooling it quickly to room temperature thereby transferring to a dual phase region in the phase diagram.

Further, an emulsifying machine having a large shear force is required in order to obtain an emulsion with small particle diameters. The mechanical forces that can be used for emulsification are shear force, shock force, and cavitation force. The mechanical forces other than these are frictional force, compression force, squeezing force, etc. and this kind of forces can be obtained in crushing equipment such as mills, etc. In general, the mechanical forces required for emulsification and for crushing are different.

The apparatuses used for emulsification are: (1) low speed stirring machine; (2) high speed shear stirring machine; (3) high pressure homogenizer; (4) ultrasonic emulsifying machine; and (5) static mixing machine; etc.

Even in a formulation that includes the same oil as a white opaque cream, by changing the emulsified particles into nanoparticles by applying pressure as shown in Fig. 25.1, the emulsion becomes transparent and fluid like water. When this is applied to skin, there is a feeling of smoothness and high penetration.

(2) Nanoparticles in dispersed systems

A typical application of nanoparticles is cosmetics for sunscreen. In sunscreen, nanoparticles of titanium dioxide or zinc oxide are mixed in addition to organic ultraviolet ray absorbing materials. The causes that control the ultraviolet ray protection by nanoparticles are the absorption effect and the scattering effect.

The electronic structure of a crystalline compound such as titanium dioxide can be represented by model of "a valence band filled with electrons and a conduction band where no electrons are present but there are empty orbits" as is shown in Fig. 25.2 [6]. The electrons are not allowed to have energy values in between these bands. This band is called the forbidden band, and its width is called the band gap.

When light is incident on a solid, only the lights with energies greater than the band gap are absorbed and electrons in the valence band are excited to the conduction band. The band gaps of titanium dioxide and zinc oxide are about 3 eV, which correspond to the energy of light with a wavelength of about 410 nm. Since the electrons in titanium dioxide and zinc oxide absorb lights of wavelengths shorter than this and become excited, they have the effect of absorbing ultraviolet light.

The particle diameters of fine particles have a large effect on light scattering and concealing power. Table 25.1 shows a comparison of light blocking effect of particles dividing the particle diameters into three regions taking the wavelength of light as the basis [7]. When the particle diameter is extremely large compared to the wavelength of light, in proportion to the

	(a)	(b)	(c)
Emulsified particle diameter	1~10 μm	120 nm	30 nm(0.03 μm)
Viscosity	1020 mPa · s	18 mPa · s	10 mPa · s
Preparation condition	Homogenizing mixer	High pressure homogenizer	High pressure homogenizer

(a) (b) (c)

Figure 25.1
Size of emulsified particles in and external appearance of emulsions.

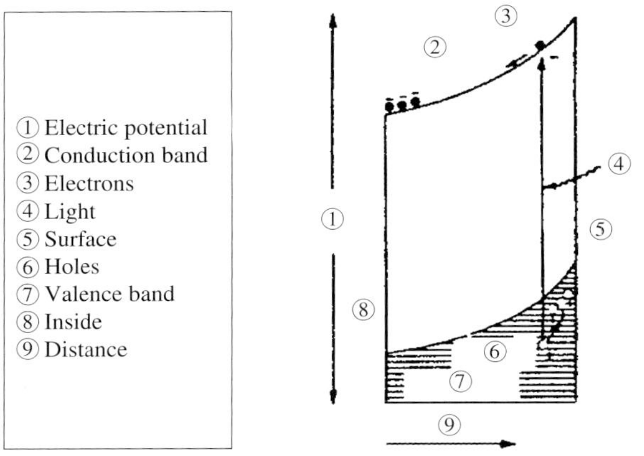

Figure 25.2
Band structure of n-type semiconductor TiO_2 and generation of electron–hole pairs due to optical excitation.

cross-sectional area of the particles, the light blocking area increases as the particle diameter decreases. When the particle diameter decreases further, the geometrical optics region is exceeded and the Mie region is entered in which scattering of light takes place. In the Mie region in which the particle diameter is of the same level as the wavelength of light, several experimental equations have been proposed for the particle diameter at which the optical scattering becomes a maximum.

When the particle diameter is extremely small compared to the wavelength of light, the Rayleigh region is entered in which the optical scattering coefficient decreases and the transparency increases. Particles that have a high transparency and still can prevent ultraviolet light are desirable for use in cosmetics. Stamatakis et al. [8] have carried out computer calculations on the dependence of optical scattering and ultraviolet ray absorption on the particle diameter of titanium dioxide and zinc oxide based on the Mie theory.

A concrete example of application is given here. While flower petal shaped zinc oxide has taken that shape as shown in Fig. 25.3 because of weakening of the binding between nanoparticles of a size of 30–50 nm, this easily gets disturbed when applied to skin, and spread evenly on the skin after becoming round nanoparticles. Turbidity is removed while the ability to protect against ultraviolet light is not diminished.

Dispersion is very important when using nanoparticles effectively in cosmetics, and various dispersion methods are being applied. In addition, since the surface area increases and the catalytic activity of the surface becomes strong when the particles become nanoparticles, deactivation becomes necessary.

(3) Functional nanocoatings

Catalytic activity is present on the surface of particles such as acid, base, oxidization, reduction, etc., and since there are cases [9] when they deteriorate the perfumes or oils that are present together, and it is very important to inactivate the catalytic activity on the surface when nanoparticles with a large specific surface area are applied in cosmetics.

A functional nanocoating utilizing the chemical vapor deposition (CVD), which is used in the formation of thin films of semiconductors, is ideally suitable for surface treatment of nanoparticles. The functional nanocoating is that of making cyclic silicone called tetra-methyl-cyclo-tetra-siloxane come into contact with the surface of the particles in the vapor phase thereby forming on the surface a mesh shaped silicone film with a thickness of 1 nm or less [10].

Table 25.1
Light reflection and scattering effect [7] of particles.

Particle diameter	Ratio of particle diameter (D) to wavelength (λ) of light	Light reflection and scattering		Efficiency of light scattering
		Region	Scattering, concealing equation	
Large	$D \gg \lambda$	Geometrical optics region	$A = \dfrac{3M}{2\rho D}$ (25.1) [A: optical shut off area; M: mass of pigment; ρ: density of pigment]	Small
Medium	$D \doteqdot \lambda$	Mie scattering region	$D_{opt} = \dfrac{0.9}{n_B \pi} \times \dfrac{m^2+2}{m^2-1}$ (25.2) $D_{opt} = \dfrac{\lambda}{2.1(n_p - n_B)}$ (25.3) $D_{opt} = \dfrac{\lambda}{1.414 n_B \pi} \times \dfrac{m^2+2}{m^2-1}$ (25.4) $D_{opt} = \dfrac{2}{\pi} \times \dfrac{\lambda}{(n_p - n_B)}$ (25.5) [D_{opt}: particle diameter at maximum light scattering; $m = \dfrac{n_p}{n_B}$; n_p: refelctive index of pigment; n_B: refelctive index of colour spreading plastic]	Big
Small	$D \ll \lambda$	Rayleigh scattering region	$S = \left(\dfrac{m^2-1}{m^2+1}\right)^2 \times \dfrac{4\lambda^2\alpha^6}{3\pi}$ (25.6) [S: optical scattering coefficient; $\alpha = \pi D/\lambda$; $m = \dfrac{n_p}{n_B}$]	Small

Because of this nanocoating, the catalytic activity of the nanoparticles is inactivated thereby preventing the deterioration of the fragrance, oils and agents present together with the nanoparticles in cosmetics. Si–H groups are present in this networked silicone, and it is possible here to introduce unsaturated compounds in a hydrosilylation chemical reaction [11]. Fig. 25.4 shows a schematic diagram of functional nanocoating.

Using this method, it is possible to give the desired surface property to the core particles. For example, it is possible to control the dispersion by introducing hydrophilic or hydrophobic groups, or to introduce alkyl groups thereby enhancing the dispersion in oil of the pigment and preparing lipsticks with strong hues. Further, by introducing glycerin residue, dual-purpose foundations having moisture retentivity that can be used both with and without water [12] have been developed.

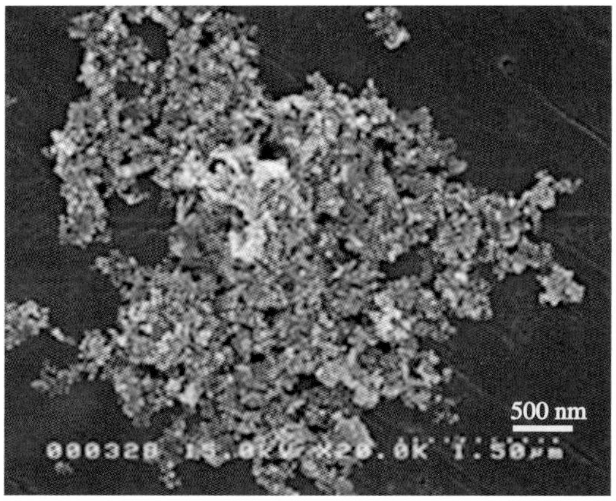

Figure 25.3
An electron microscope photograph of flower petal shaped zinc oxide.

2. Use as compound particles

(1) Color correction

Structural colors that give out color due to interference of light such as the colors in the wings of a butterfly are being used actively. It has been known that blue moles or red moles such as nevus of Ota or vascular neoplasm disappear giving a natural appearance by the use of titanium dioxide coated mica having interference color. However, since titanium dioxide coated mica has gloss, it gives an unnatural finish when used in a foundation. In order to improve its shine, a compound particle was developed with the surface of titanium dioxide coated mica covered with spherical particle of PMMA (Fig. 25.5). While correcting the reflectivity of red color using its red interference color, since the gloss of titanium dioxide coated mica is controlled by the diffused reflection of spherical PMMA particles, it was possible to realize a fine makeup skin texture having transparency as well.

An example of skin color correction using fluorescence is reduced form of zinc oxide. The reflectivity about 500 nm is increased due to the ultraviolet light present in sunlight or fluorescent light thereby attempting to correct the reflectivity of light in the blue to green region [13]. There are items in which the reflectivity on the long wavelength side is increased using the fluorescence of carthamin.

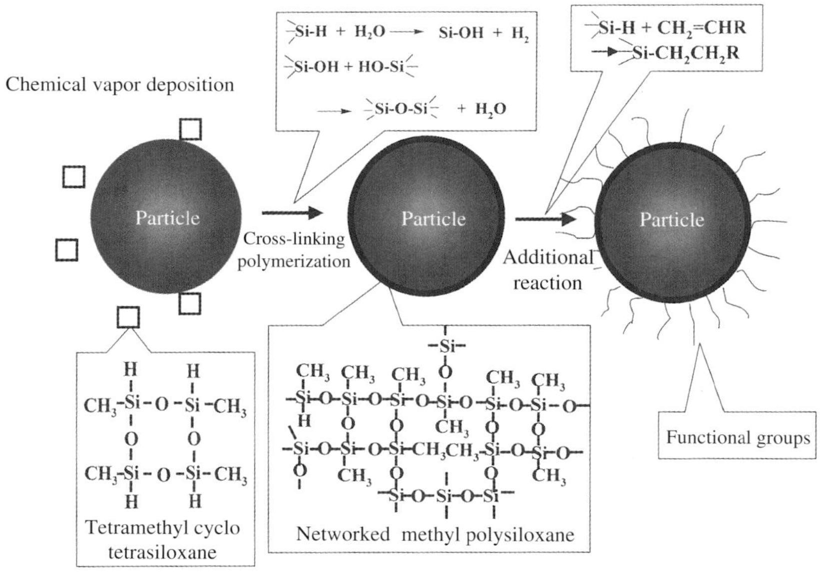

Figure 25.4
Schematic diagram of functional nanocoating.

Figure 25.5
Electron microscope photograph of PMMA coated particle.

(2) Adapting to the environmental light

While a skin with makeup applied to it looks whitish and has a floating appearance in sunlight outdoors, but appears as a blurred yellow surface indoors in fluorescent light. In order to solve this problem, a photochromic powder was developed using 25 nm titanium dioxide doped with iron [14]. These skin colored nanoparticles change to golden brown color when exposed to ultraviolet light, and have not only the effect of preventing whitishness outdoors, but also have the effect of protecting against ultraviolet light.

By further developing this type of photochromic nature, color-rendering type pearl material has been developed [15] that is not only bright but also has changing color phase due to the effect of interference with titanium dioxide coated mica. For example, by coating iron doped titanium dioxide having photochromic action on the surface of titanium dioxide coated mica having red interference color, the material will appear white when no ultraviolet light is incident on it, and when ultraviolet light is incident on it, the outer layer changes to black and the interference color becomes clear. Therefore, the particle changes from white to red due to ultraviolet light. This color rendering type red pearl material prevents the blurred yellowish appearance under a fluorescent light with slight yellowishness by generating a red interference color.

On the other hand, under a strong light source such as sunlight, the overall reflectivity decreases and exhibits superior color rendering that prevents whitishness. Because of this, it is possible to realize a beautiful makeup skin in any light environment.

(3) Shape correction

It is possible to use the defocusing effect by light scattering of spherical particles in order to correct asperity such as skin wrinkles or skin pores. Recently, by coating spherical-shaped particles on plate particles, the powdery appearance is reduced while increasing the adherence to skin and increasing the defocusing effect. Even the PMMA nanoparticle coated red interference color pearl material mentioned above is an example of this application, and in this case the two technologies of color correction by interference colors and shape correction by scattering have been combined.

A functional optical particle with a fine feather-like coating has also been developed by growing fiber shaped crystals of zinc oxide in the nanolevel on the surface of titanium dioxide coated mica (Fig. 25.6). The diffused reflection characteristic is excellent due to the fibers of zinc oxide, and a finish with a very good natural translucent color is obtained. In addition, zinc oxide has the effect of solidifying sebum cutaneum, and has the feature that the makeup does not come off with elapsing of time.

Even functional optical particles have been developed that have the effect of hiding sagging skin. For this, spherical moss shaped barium sulfate crystals of 1.0–1.5 μm size are grown on the surface of red interference color titanium dioxide coated mica (Fig. 25.7). Since light is incident from above on the forehead and cheeks of the face, while the moss shaped crystals reflect light in a diffused manner, the titanium dioxide coated mica reflects light directly and generates a red color close to the skin color thereby rendering a natural skin color.

On the other hand, since the light is incident at an inclined angle on the shadow areas such as sagging parts, etc. of the skin, the moss shaped crystals reflect in a diffused manner almost the entire incident light thereby making the sagging parts of the skin not prominent. In a sensory test conducted with women in the age group of 40–50 years, the sags were decreased resulting in a lifted-up appearance of the face, and even in computer simulations, a result of the apparent age being 10 years younger than the actual age was obtained.

(4) Skin care particles [16]

While the major purpose of cosmetics is skin care, the so-called "skin care particles" have been developed that improve skin problems using the dermatology for this purpose. This particle, which cares for the skin by efficiently adsorbing and deactivating an enzyme called "urokinase" that induces dry skin, has, as is shown in Fig. 25.8, a coating of very thin silica of 2–3 nm on zinc oxide of about 30 nm size. This skin care particle has a much higher skin care effect compared to only zinc oxide or only silica independently. The reason for this is considered to be that the amount of urokinase that is pulled and adsorbed by electrical charge has been increased by the silica coating on the zinc oxide, which itself has deactivating effect on urokinase.

Figure 25.6
Electron microscope photograph of zinc oxide fiber coated pearl material.

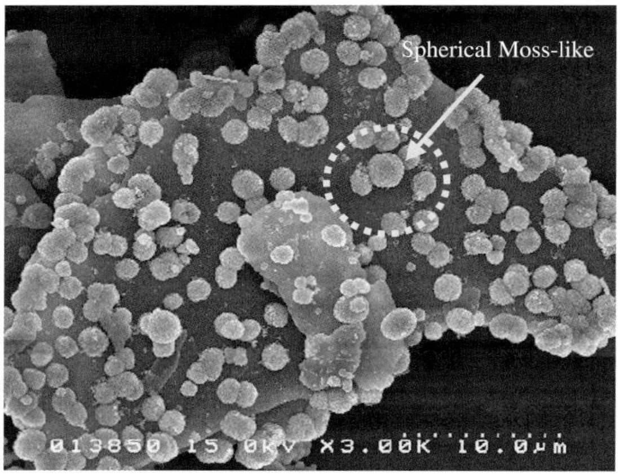

Figure 25.7
Electron microscope photograph of spherical moss shaped barium sulfate coated pearl material.

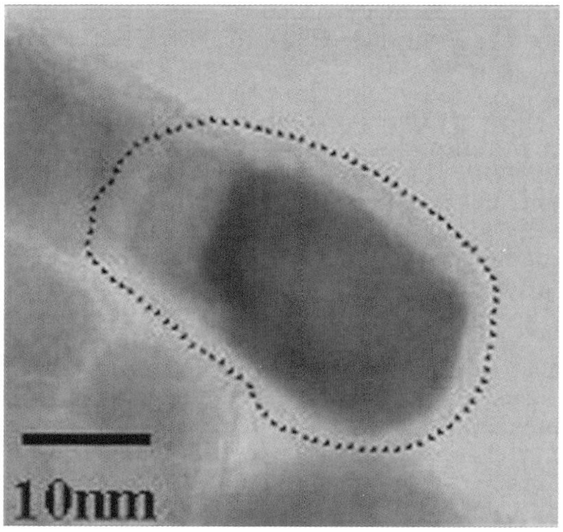

Figure 25.8
Electron microscope photograph of a skincare particle.

3. Future development

We have introduced the application of nanoparticles in cosmetics. In the field of emulsions, by making the emulsified particles with nanosizes, it is possible to change the external appearance and the usability. In the field of powders, use of zinc oxide or titanium dioxide for sunscreen and also nanoparticles having biological and dermatological value such as skin care particles have appeared, and these indicate the future direction of cosmetics.

In addition, the nanoparticles are not only added individually, but also their optical characteristics are designed while carrying out their form control by compounding with large particles, and then applying them to products for makeup. Although not discussed this time, it has been reported that fullerene has anti-oxidation effect, and it is considered that such new materials too have a big future in the skincare area. Further, while nanoclusters such as gold or silver exhibit vibrant red or yellow coloration, even such plasmon coloration will be used in make-ups in the future.

References

[1] H. Sagitani: *J. Am. Chem. Soc.*, **58**, 738 (1981).

[2] K. Shinoda, H. Saito: *J. Colloid Interface Sci.*, **26**, 70 (1968).

[3] H. Sagitani, Y. Hirai, K. Nabeta and M. Nagai: *J. Jpn. Oil Chem. Soc.*, **35**, 102 (1986).

[4] Y. Kumano, S. Nakamura, S. Tahara and S. Ohta: *J. Soc. Cosmet. Chem.*, **28**, 285 (1977).

[5] H. Nakajima, T. Tomomasa and M. Kouchi: *J. Soc. Cosmet. Chem. Jpn.*, **23**, 288 (1990).

[6] M. Fujii, T. Kawai and N. Kawai: *Oyo Buturi*, **53**, 916 (1984).

[7] M. Kiyono: *Titanium Oxide*, Gihodo Shuppan, Tokyo, p. 131 (1991).

[8] P. Stamatakis, B.R. Palmer, G.C. Salzman, C.F. Bohren and T.B. Allen: *J. Coat. Technol.*, **62**, 95 (1990).

[9] H. Fukui, R. Namba, M. Tanaka, M. Nakano and S. Fukushima: *J. Soc. Cosmet. Chem.*, **38**, 385 (1987).

[10] H. Fukui, T. Ogawa, M. Nakano, M. Yamaguchi and Y. Kanda: in *Controlled Interphases in Composite Materials*, H. Ishida (Ed.), Elsevier Science Publishing Co. Inc., New York, p. 469 (1990).

[11] F.C. Whitemore, P.W. Pietrusza and L.H. Sommer: *J. Am. Chem. Soc.*, **69**, 2108 (1947).

[12] Japan Patent Number 1635593.

[13] K. Ohno: *Fragrance J.*, **22**, 11 (1994).

[14] K. Ohno: *J. Soc. Cosmet. Chem. Jpn.*, **27**, 314 (1993).

[15] K. Ogawa: Proceedings of 43rd SCCJ conference (1998).

[16] E. Kawai, Y. Kohno, K. Ogawa, K. Sakuma, N. Yoshikawa and D. Aso: *IFSCC Mag.*, **5**, 269 (2002).

APPLICATION 26

26 DISPERSION CONTROL OF Al_2O_3 NANOPARTICLES IN ETHANOL

It is quite difficult to handle a nanoparticle powder, because of aggregation due to the high surface interaction between nanoparticles, which have a large specific surface area and high surface energy. Therefore, by using a dense suspension, the nanoparticle powder makes easier to disperse up to primary particles and to handle and facilitate transport, and then that can produce its original performance. To obtain a homogeneous suspension with a high concentration and adequate fluidity, the aggregation and dispersion behavior of particles is usually controlled by using a polymer dispersant. However, there are some problems to produce a nanoparticle suspension using a polymer dispersant [1]. For example:

1. It is difficult to disperse a nanoparticle in a dense suspension with a high concentration, because the distance between particles will be less than several nanometers and the particles will readily aggregate at the distance.
2. As for a nanoparticle, the steric repulsion promotes a dominant dispersibility and stability in suspensions, because the surface potential between nanoparticles is at least over 100 mV in order to disperse nanoparticles.
3. The molecular size of a polymer dispersant with approximately several nanometers affects the dispersibility of nanoparticles in dense suspensions.

In this chapter, the action mechanism and selection of the polymer dispersant on Al_2O_3 nanoparticles ethanol suspension is clarified, based on the surface interaction using nanocolloidal probe AFM [2, 3].

1. Effect of molecular weight of PEI on nanoparticle suspension viscosity

When using 7, 30, 100 and 370 nm scaled Al_2O_3 nanoparticles, the aggregation and dispersion behavior of the Al_2O_3 nanoparticles in dense ethanol suspensions was investigated. Six kinds of branched polyethyleneimines (PEIs) (Epomin, Nippon Syokubai Co., Ltd., Japan) with a molecular weight (MW) ranging from 300 to 70,000 were used in this study as polymer dispersants which is effective in the dispersion of ceramic particles in ethanol [4, 5]. Fig. 26.1 shows the representative molecular structure of the branched PEI.

To clarify the relationship between the suspension viscosity of each nanoparticle and the MW of branched PEIs, the apparent viscosity at the same shear rate (300 1/s) of suspensions as a function of the MW of branched PEIs is plotted in Fig. 26.2. The suspension with the Al_2O_3 nanoparticles of 100 and

370 nm displayed the minimum viscosity at the optimum MW of 10,000. When the Al_2O_3 nanoparticles of 7 and 30 nm were used, the optimum MW of PEIs for the minimum viscosity was determined to be 1,200. When using the PEI with MW of 70,000, the suspension viscosities remarkably increased. To examine the dispersion and aggregation behavior of nanoparticles, the packing condition of freeze-dried

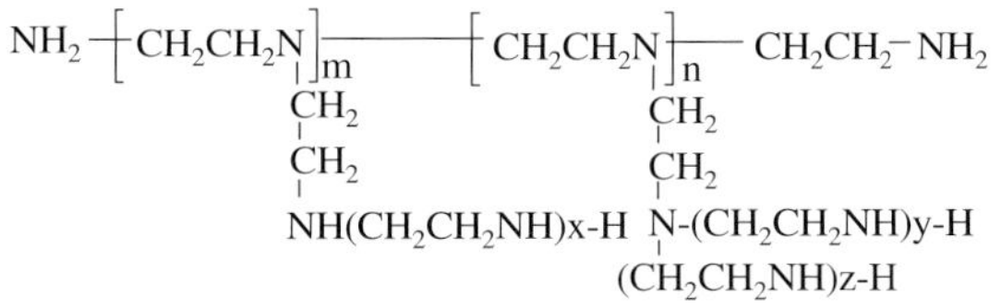

Figure 26.1
The molecular structure of branched PEI.

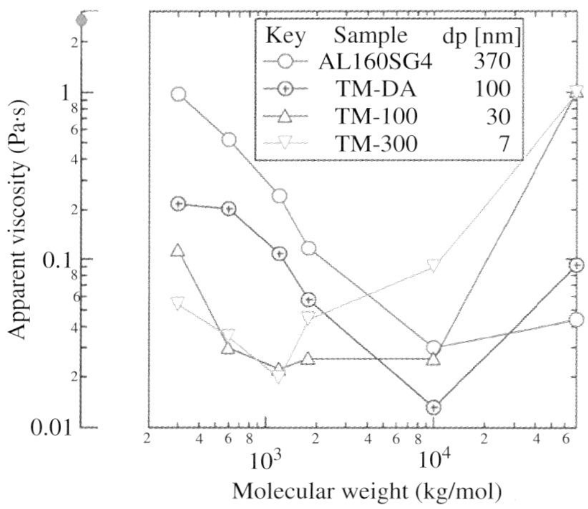

Figure 26.2
The apparent viscosity at the same shear rate (300 1/s) of the ethanol suspensions containing the Al_2O_3 nanoparticles with diameters ranging from 7 to 350 nm as a function of the MW of the branched PEIs.

suspensions were directory observed through FE-SEM. Fig. 26.3 shows the FE-SEM micrographs of the suspensions with 30 nm scaled θ-Al_2O_3 particles as a function of MW of PEIs. The suspension prepared with PEIs of MW 300 and 70,000 had large flaws and pores, and the aggregate structure with 400–500 nm scaled aggregates were observed. When using the PEI of MW 1,200, the aggregates decreased to less than 100 nm and the large flaws and pores were not observed. These results show that the optimum MW of a PEI to obtain a low suspension viscosity decreased as the particle size of Al_2O_3 nanoparticle decreased.

2. Relationship between molecular size of PEI and suspension viscosity

The molecular size of a polymer dispersant with approximately several nanometer affects the dispersibility of nanoparticles in dense suspensions. Fig. 26.4 shows the molecular size distribution of the branched PEIs in ethanol by dynamic light scattering. The mean molecular sizes of the PEIs of MWs 300, 600, 1,200, 1,800, 10,000 and 70,000 were 0.5, 0.6, 0.7, 1.7, 3.1 and 4.2 nm, respectively. The molecular size increased with an increase in MW, and the molecular size distribution of the PEI also became wider. To clarify the relationship between the molecular size and particle size, the molecular sizes of the PEIs for suspensions under an apparent viscosity of 0.04 Pa s in Fig. 26.2 are plotted in Fig. 26.5. As the particle size of Al_2O_3 nanoparticles decreased, the optimum molecular size of the PEI to obtain a low suspension viscosity decreased. The optimum molecular size of PEIs for the minimum viscosity of the suspension with the 7 nm scaled Al_2O_3 nanoparticle was determined to be about 7 nm.

3. Surface interaction between Al_2O_3 nanoparticles using nanocolloidal probe AFM

Using the branched PEIs with different MWs, the microscopic surface interaction between 7 nm scaled

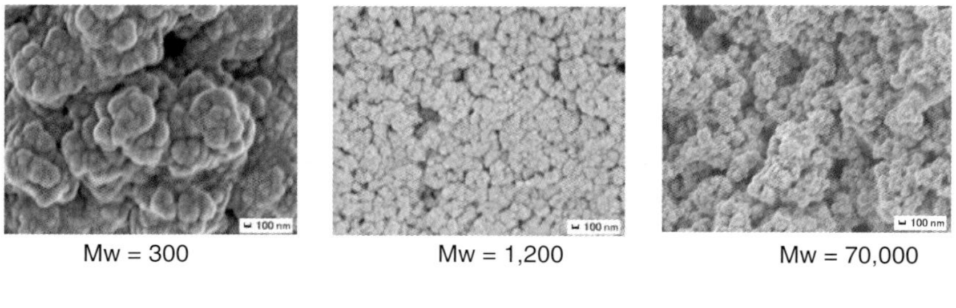

Figure 26.3
FE-SEM observations of freeze-dry suspensions with the 30 nm scaled θ-Al_2O_3 particle when using the branched PEIs with MW of 300, 1,200 and 70,000.

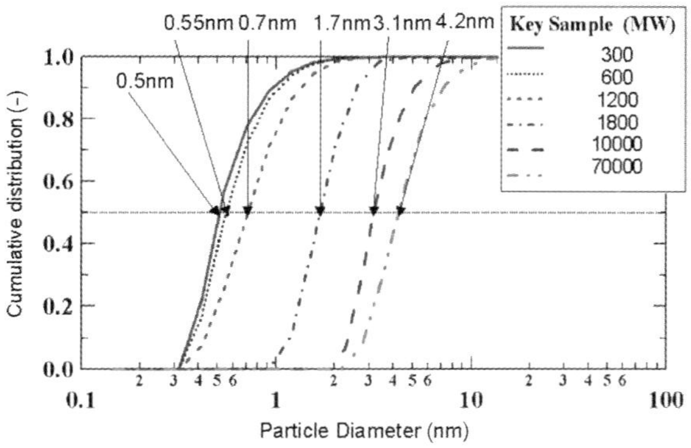

Figure 26.4
The molecular size distribution of the branched PEIs in ethanol by dynamic light scattering method.

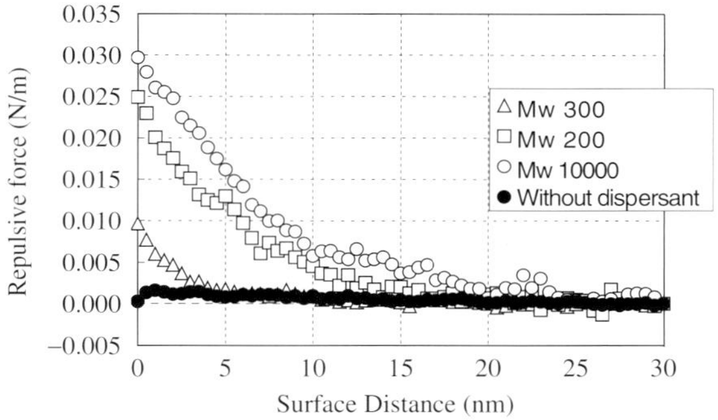

Figure 26.6
The repulsive force curve between the γ-Al_2O_3 (7 nm) nanocolloidal probe and the substrate in ethanol with the branched PEIs and without a dispersant.

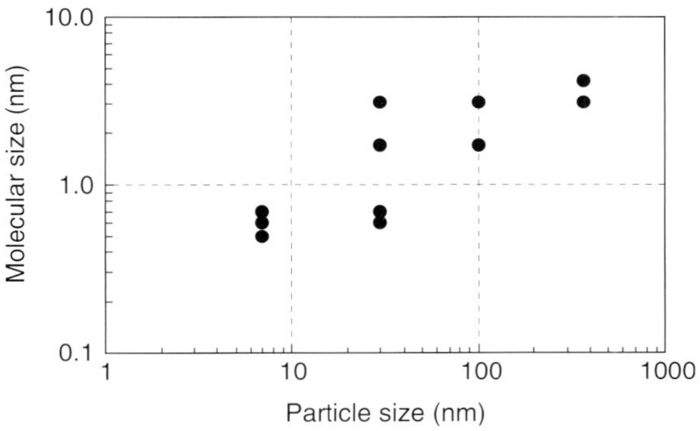

Figure 26.5
The relationship between the Al_2O_3 nanoparticle size and the molecular size of PEIs for the suspensions below an apparent viscosity of 0.04 Pa · s.

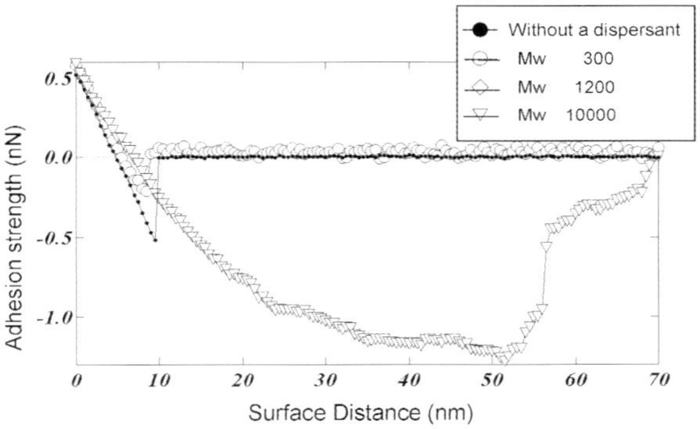

Figure 26.7
The attractive force curve between the γ-Al_2O_3 (7 nm) nanocolloidal probe and the substrate in ethanol with the branched PEI and without a dispersant.

Al_2O_3 nanoparticles was investigated by using a nanocolloidal probe AFM [6]. The repulsive and attractive force curves between the Al_2O_3 nanocolloid probe and the Al_2O_3 substrate in ethanol solutions with the PEIs are shown in Figs. 26.6 and 26.7. The repulsive force increased with an increase in MW of the PEIs. The repulsive force of the PEI with MW of 10,000 displayed the extra large value ($\sim$0.030 N/m). The adhesion strengths based on van der Waals attractive force in ethanol containing the PEI of MW 300 and without a dispersant were observed. The adhesion strength of the PEI with MW of 1,200 did not appear. Moreover, when the PEI with MW of 10,000 was used, the larger adhesion force which was over twice of that without a dispersant, appeared from a long distance of 60 nm between the nanoparticle and the substrate. The attractive force of the nanoparticle significantly depended on the MW of the PEIs, compared to that of the sub-micron scaled Al_2O_3 particle.

4. Action mechanism of polymer dispersant on Al_2O_3 nanoparticle suspension

Based on the results, the action mechanism of the polymer dispersant on the dispersion and aggregation behavior of the nanoparticle in ethanol suspension is shown in Fig. 26.8. The nanoparticles in ethanol suspension without a dispersant aggregate by

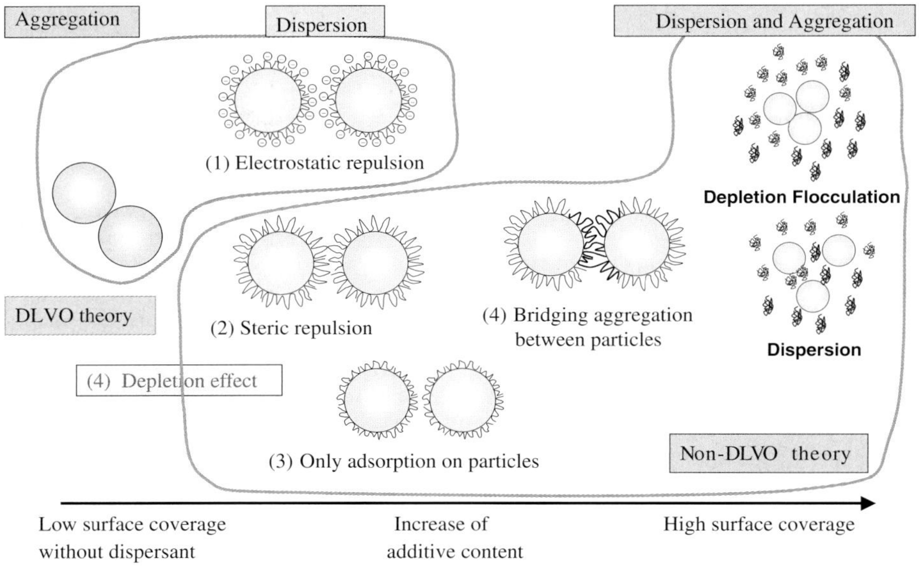

Figure 26.8
The conceptual scheme of the action mechanism of polymer dispersant in dense suspension with increasing additive content.

van der Waals type attractive force, and its viscosity increases. When the optimum PEI with MW of 1,200 and 0.7 nm in diameter was used for the minimum viscosity of a suspension with 7 nm scaled Al_2O_3 particle, the steric repulsion force displayed a relatively large value, and an adhesive force was not observed. The steric repulsion will facilitate the dispersion of the nanoparticles in the suspension and decrease the suspension viscosity, as shown in Fig. 26.8 (2). When using the PEI with low MW of 300, the adhesive force displayed a lower value than that without a dispersant, and the repulsive force was observed smaller than the attractive force. This is why the PEI with MW of 300 sufficiently did not reduce the suspension viscosity. On the contrary, the adhesion force of the PEI with MW of 10,000 displayed was quite a large value at the distance between 10 and 40 nm, where corresponded with almost the double of the maximum value of molecular size distribution (2–20 nm) of the PEI. The re-bridging aggregation rather than the steric repulsion effect between the 7 nm scaled Al_2O_3 particles by the PEI increases the suspension viscosity, as shown in Fig. 26.8 (3).

Using the Al_2O_3 particles with the mean particle size ranging from 7 to 370 nm, investigation was made on the action mechanism of the branched PEI with different MWs for the aggregation and dispersion behavior of the Al_2O_3 particles in ethanol suspension. As the particle diameter of nanoparticle decreases, the optimum MW and molecular size of the PEI also decreased for the ethanol suspension with good dispersibility of particle and high fluidity. By using the nanocolloidal probe AFM, the micro surface interaction; repulsive and attractive forces, between nanoparticles, was also analyzed. In general, the aggregation and dispersion behavior of particles in a liquid phase is facilitated by polymers with the MW ranging from the thousands to the millions. However, in the case of the nanoparticles, their behavior appeared at the low MW ranging from the hundreds to 10,000.

References

[1] H. Kamiya, M. Naito, M.I. Baraton (Ed.), Functionalization and Surface Treatment of nanoparticle, American Scientific Publishers, California (2003).

[2] T. Kakui, *The Micromeritics*, **48**, 46–55 (2004).

[3] T. Kakui, T. Miyauchi and H. Kamiya: ACS, Annual meeting, Indianapolis USA, April 22–24 (2004).

[4] T. Kakui, H. Kamiya: *Ceram. Tran.*, **146**, 43–50 (2004).

[5] T. Kakui, T. Miyauchi and H. Kamiya, *J. Euro. Ceram. Soc.*, **25**, 655–661 (2005).

[6] H. Kamiya, S. Matsui and T. Kakui, *Ceram. Trans.*, **152**, 83–92 (2003).

APPLICATION 27

27 DEVELOPMENT OF THE THERMORESPONSIVE MAGNETIC NANOPARTICLE AND ITS DEPLOYMENT IN THE BIOTECHNOLOGY FIELD

1. Magnetic nanoparticle material

Among granular materials, magnetic microparticle material has been widely used in the biotechnology field [1, 2]. A further increase in its use is also expected in a variety of fields, such as bioseparation, various assays, diagnosis methods, and drug delivery systems (DDS). However, the size of the micro-magnetic particles conventionally used is around several micrometers [3].

Reducing the particles to nanosize will dramatically increase the adsorption amount available for separation because of their increased surface area for mutual interactions, resulting in a remarkable improvement in analytical sensitivity. This size reduction facilitates rapid and highly sensitive diagnosis, and the nanomaterial has become useful to establish high-throughput screening systems (HTS), typically like those for proteome analysis, for various types and kinds of proteins. However, making the particle diameter smaller than several hundred nanometers will weaken their magnetism, which hampers separation using magnets.

To solve this dilemma, we have developed stimuli-responsive magnetic nanoparticles, on whose surface polymers responding to external stimuli, including temperature, light, electric field, and pH, are immobilized. For example, Fig. 27.1 shows thermoresponsive magnetic nanoparticles where magnetic nanopolymers are coated with thermoresponsive polymers. In these particles, coagulation and distribution conditions vary, responding to changes in stimuli through the hydration–dehydration of polymers or varied interaction between polymers according to the temperature change (details to be described later).

Accordingly, coagulating magnetic nanoparticles by temperature change will enable rapid collection by magnets. In other words, these particles will become innovative material that can be rapidly separated magnetically, even when nanosized [4–6].

2. What is a thermoresponsive polymer?

Thermoresponsive polymers, which vary their physical properties reversibly and discontinuously when exposed to slight temperature change, respond to such general stimulus as temperature changes. Accordingly, many studies on their application, including actuators, separating agents, and DDS agents, are underway,

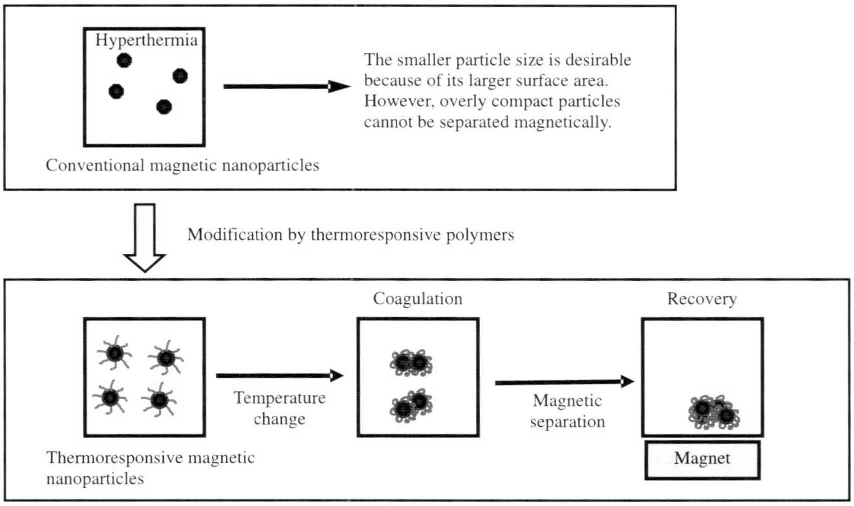

Figure 27.1
Thermoresponsive magnetic nanoparticles.

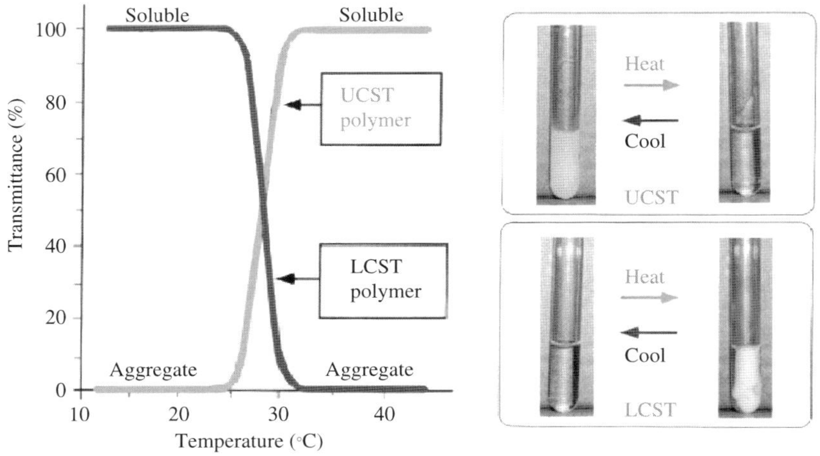

Figure 27.2
Thermoresponsivity of LCST and UCST polymers.

some of which have even reached the stage of practical use already. There has been information for some years about thermoresponsive polymers that show lower critical solution temperature (LCST). These polymers become insoluble in water at a higher temperature than LCST, and dissolve at a lower temperature (Fig. 27.2).

A typical example is poly(*N*-isopropyl acrylamide) (NIPAM) [7], which is easily obtained by radical polymerization of NIPAM, and whose LCST is 32°C. On the other hand, the polymers that show upper critical solution temperature (UCST) dissolve in water at a higher temperature than UCST, and become insoluble at a lower temperature (Fig. 27.2).

Polymers that show UCST in buffer solution have been long anticipated, because they can be used in the separation and purification of thermolabile compounds such as proteins, due to their insolubility at a lower temperature (to be described later). We have recently succeeded in developing such thermoresponsive polymers to maintain UCST, even in buffer solution [6, 8]. This is a polymer mainly composed of the copolymer of nonionic *N*-acryloyl glycine amide (NAGAm) and a biotin derivative (*N*-methacryloyl-*N'*-biotinyl propylene diamine; MBPDA). The polymer is supposed to become insoluble at a low temperature by forming hydrogen bonds between polymer chains and to dissolve at a high temperature due to weakening of the hydrogen bonds. By changing the ratios of monomers in the copolymer, UCST polymers of various transition temperatures can be prepared.

3. Thermoresponsive magnetic nanoparticles

By coating magnetic nanoparticles with the recently developed polymer materials, showing a variety of LCST and UCST, the preparation of magnetic nanoparticles, whose coagulation and dispersion can be controlled by heat, has become possible (Therma-Max®). The critical point is establishing techniques to construct the nanocomposite by forming chemical bonds strong enough to prevent their desorption between the magnetic nanoparticle surface and the polymer materials. In addition, during the preparation process, since no coagulation of the magnetic nanoparticles should occur, special care is to be taken.

As an example of Therma-Max, the preparation process of a thermoresponsive nanoparticle having UCST from NAGAm and a biotin derivative MBPDA is specifically described. To form chemical bonds between the polymers and magnetic nanoparticles, polyalcohol-coated (in this case dextrin) magnetic nanoparticles are used.

Initially, NAGAm and MBPDA are synthesized in single-step reactions shown in the schemes in Fig. 27.3. Biotin is a ligand that forms chemical bonds with avidin (a glycoprotein tetramer with molecular weight of 66,000 and having 4 binding sites with biotin) with a specific and extremely strong affinity ($K_a = 1.3 \times 10^{-15}$) and also a compound with a strong tendency to form hydrogen bonds. The polymers made by copolymerization between the NAGAm and MBPDA (copolymer of NAGAm/MBPDA) show UCST in water at a low temperature by forming hydrogen bonds.

Thermoresponsive magnetic nanoparticles having UCST in water were prepared by immobilizing the NAGAm/MBPDA copolymer on magnetite nanoparticles coated with dextran with molecular weight of several thousands (average particle diameter – around 70 nm). Since the NAGAm/MBPDA copolymers

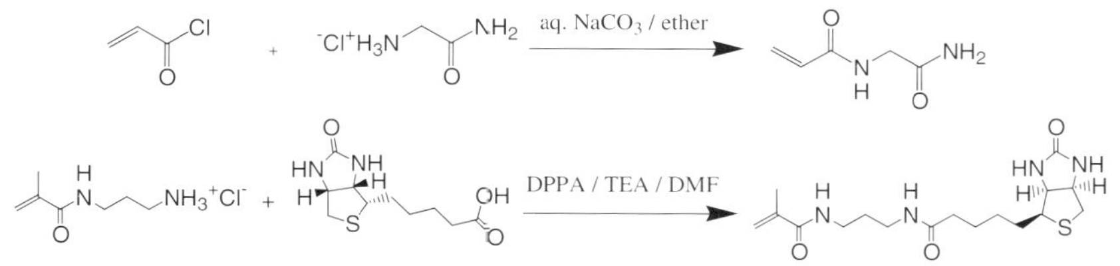

Figure 27.3
Synthesis schemes of *N*-acryloyl glycine amide and a biotin derivative.

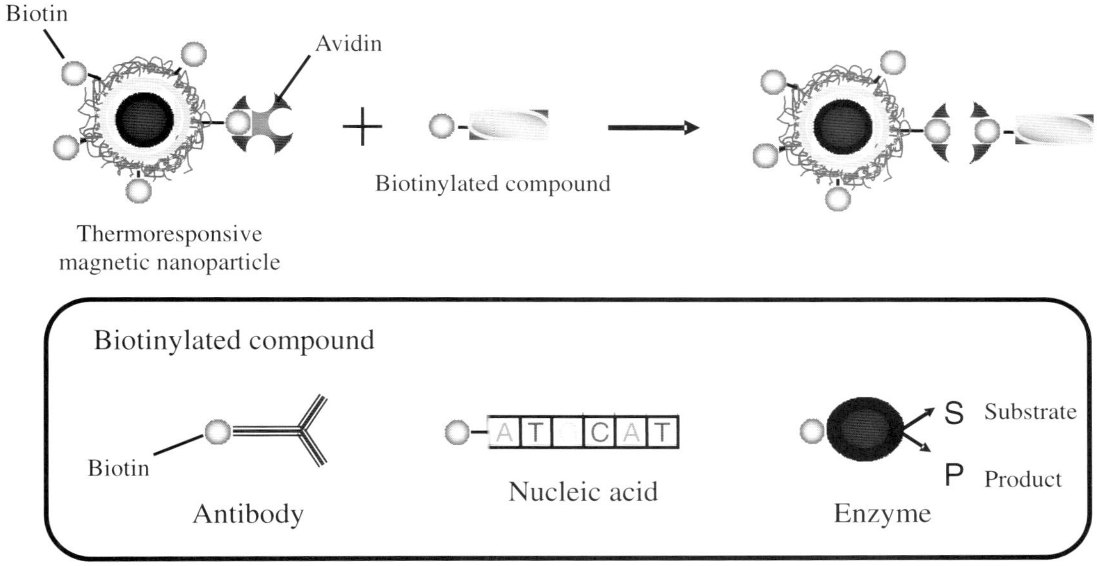

Figure 27.4
Immobilization of a biomolecule to a thermoresponsive magnetic nanoparticle using the biotin–avidin interaction.

contain biotins, these thermoresponsive magnetic nanoparticles can specifically form bonds with avidins and a variety of biological molecules, such as biotinylated proteins and DNAs, through the biotins on the particle surface (Fig. 27.4).

The thermoresponsive magnetic nanoparticles do not settle spontaneously at room temperature and cannot be separated magnetically, even by using a fairly strong magnet. However, when placed in an ice bath, they instantly coagulate and can be easily separated magnetically (Fig. 27.5). Moreover, the coagulated particles can be made to revert to a completely dispersed state by raising the temperature. As just described, the thermoresponsive magnetic nanomolecules can be stored as a completely uniform solution, but can rapidly coagulate to become magnetically separable by applying heat, rendering them a suitable material for a variety of automatic rapid analyses.

The following sections introduce specific applications of Therma-Max to the biotechnology field.

4. Application examples of the thermoresponsive magnetic nanoparticles to the biotechnology field

(1) Application to the diagnosis field

Due to its easy separation from the sample after connecting antigens, the magnetic nanoparticle coupled with antibodies is a diagnosis agent of some potential. Fig. 27.6 shows a typical example of an antigen detection method via a sandwich method using the antibody-coupled magnetic nanoparticles.

In case specific antigens exist in the sample, these will be adsorbed by the antibody-coupled magnetic nanoparticles. Following the adsorption, the labeled antibodies (secondary antibodies) are added and the

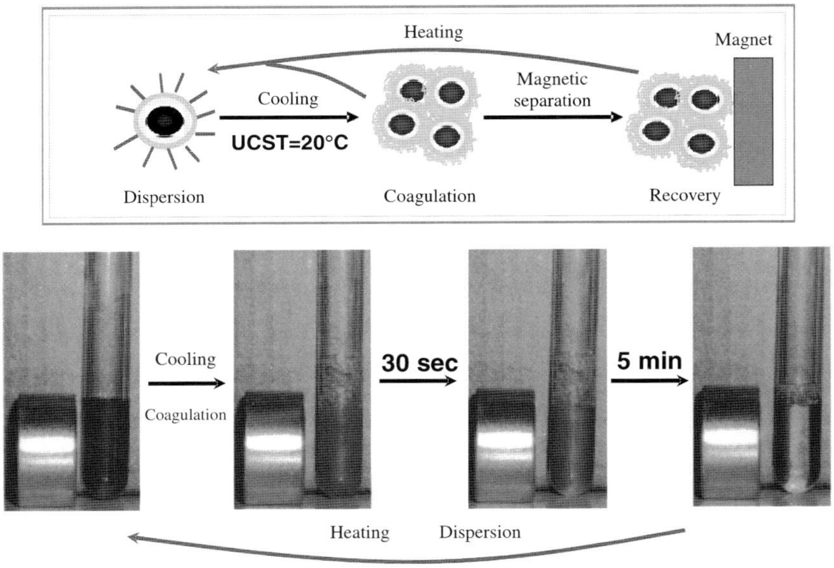

Figure 27.5
Magnetic separation of thermoresponsive magnetic nanoparticle.

labeled magnetic nanoparticles are separated and recovered from foreign substances and excessively labeled antibodies (called B/F separation).

Finally, quantitative analysis of the antigen can be conducted by determining the intensity of the labels using a spectroscopic method. Conducting a diagnosis more rapidly and with higher sensitivity critically requires the antigens to be coupled with magnetic nanoparticles via a method involving high binding specificity, and separating/condensing them quite rapidly. Therma-Max, because of its very large surface area, low nonspecific adsorption (due to having a surface less hydrophilic than those of polyethylenes), and capability to conduct rapid separation shows significant potential for various analyses.

(2) Application for the separation of proteins and cells

In cases of the separation of biological molecules such as proteins, nanosized particles are essential to increase the adsorption amount per particle quantity. By using nanoparticles, adsorption is expected to reach its equilibrium relatively swiftly, because adsorption occurs on their surface, and the adsorption operation can be completed within about 5 min. Therma-Max adsorbed about 0.5 mg of avidin per 1 mg, thus verifying its significant potential to increase the adsorption amount [5].

Also in Fig. 27.7, the results of the study on the bond formation of avidin to Therma-Max and the possibility of its detachment after repeated washing are shown. The bonding between them is proved to be stable, because the quantity of avidins connected to Therma-Max remains constant, even after repeated washing. These results indicate that Therma-Max and avidins have formed stable bonds. Since the bondings between avidins and biotins are extremely strong, those between the polymers and magnetic nanoparticles are also certainly strong.

In addition, as shown in Fig. 27.8, microparticles are being developed for a wide variety of affinity separation by coupling, on the particle surface, the molecules (referred to as ligand) capable of forming a specific bond with the target molecule. Such particles developed as above can be utilized for the purification of a variety of recombinant proteins. In other words, the thermoresponsive magnetic nanoparticles shown in Fig. 27.8 can be utilized as general-purpose separating agents by producing, in a integrated manner, the target protein and the affinity tags (protein or peptides that selectively form a bond with a specific ligand), including GST (glutathione s-transferase), avidin, and 6-residue histidine (which form an affinity bond with glutathione, biotin, and metal chelate, respectively).

Thermoresponsive magnetic nanoparticles are also quite effective in separating large targets such as biological cells [5, 6]. For example, *Escherichia coli* cells are successfully separated using Therma-Max, on whose surface the biotinylated antibodies against *E. coli* are attached. The number of *E. coli* cells in the supernatant solution and those attached to the magnetic particles, both after magnetic separation, is estimated, showing clearly that magnetic separation efficiency using Therma-Max is much higher than

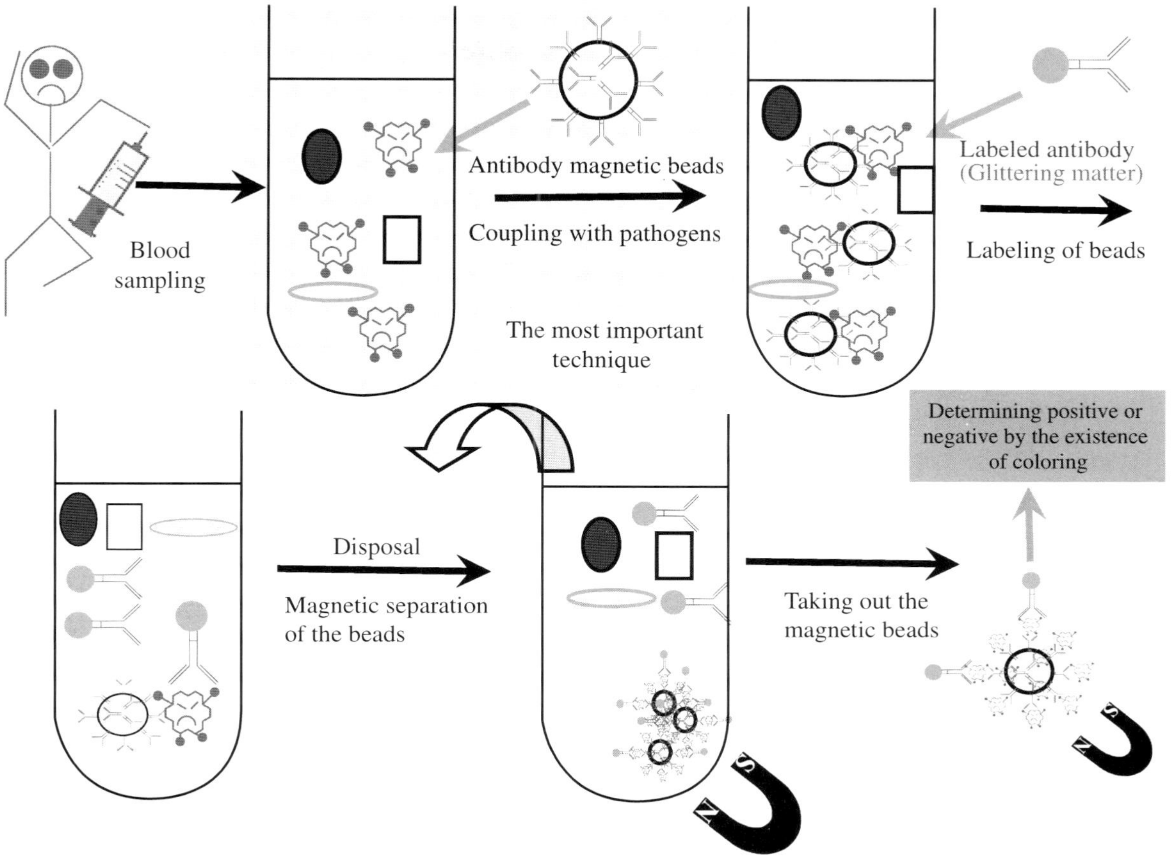

Figure 27.6
Immunodiagnosis using antibody-immobilized magnetic nanoparticles.

that using micron-size magnetic particles. In addition, its effectiveness in the separation of larger cells, such as yeast, animal, and plant cells, has also been proved [6]. As shown above, nanosize magnetic particles are also considerably effective in cell separation and assays.

(3) Application to transcriptome/proteome analyses

In April 2003, the International Joint Study Team of Japan, the USA, and Europe announced that the mapping of the human genomes (total human genetic codes) had almost been completed, and declared the total number of human genes to be around 30,000. However, which part of which genome takes charge of which function remains unknown and intense competition to solve this problem, as a major issue of the post-genome era, has since been conducted worldwide. Accordingly, exhaustive analyses of developed mRNA and proteins (transcriptome and proteome analysis) are cited as an important technique in the post-genome era.

In transcriptome analysis, thermoresponsive magnetic nanoparticles are effective in isolating sparsely developed and unstable molecules such as mRNAs. We have successfully conducted a rapid analysis of mRNA using Therma-Max. Specifically Therma-Max, on whose surface oligo (dT) 25 is immobilized, is prepared through avidins, and then the Therma-Max is used to isolate mRNA from the cell extract. The isolated mRNAs are immobilized on the surface of the magnetic nanoparticles, following which RT-PCR amplification is made possible (because the particles are perfectly dispersed), which enables rapid amplification.

Moreover, the thermoresponsive magnetic nanoparticles are expected to be extremely effective in the analysis of protein interactions, which also represents an important procedure of proteome analysis. A representative method is called the pull-down method

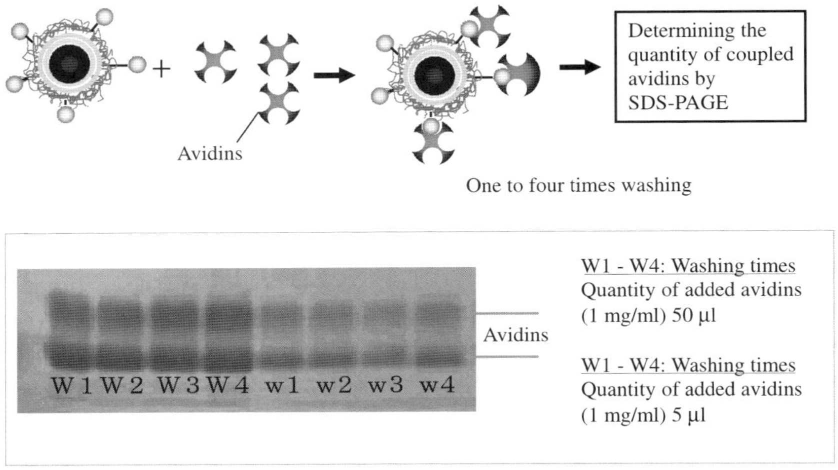

Figure 27.7
Stability of the bond between avidin and Therma-Max®.

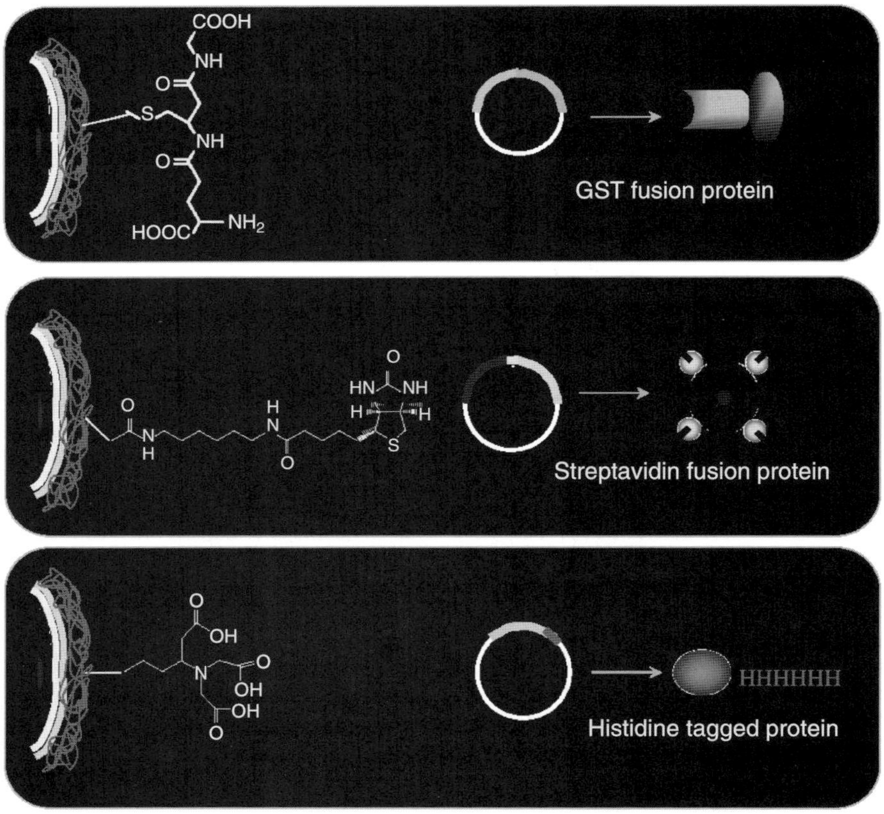

Figure 27.8
Thermoresponsive magnetic nanoparticles and targeted proteins for various affinity separations.

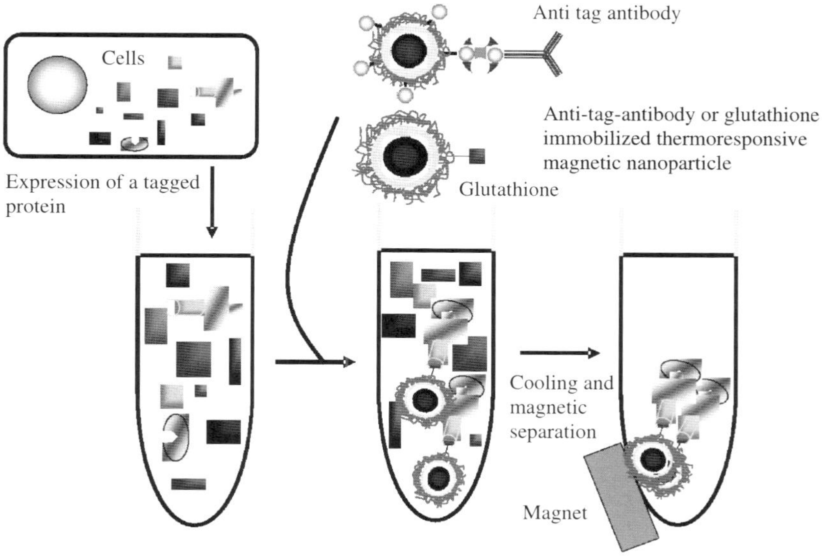

Figure 27.9
Pull-down assay to analyze the interactions between biomolecules.

(Fig. 27.9), which is simply described as follows. The protein fused with such affinity tags, as shown in Fig. 27.8, is expressed in the cells and then separated using the corresponding affinity magnetic nanoparticles. The molecules in the cells that have interactions with the expressed fusion proteins are then isolated at the same time in a coupled state, so they can be identified one by one.

During the process of accumulation of such basic research results, the relations between transcriptome/proteome and pathological conditions are being elucidated and utilized in a wide variety of diagnoses, where the magnetic nanoparticles are surely expected to play a major role.

(4) Application to medical field

The effectiveness of the application of magnetic nanoparticles to the medical field, such as their use as contrast agents [9] in MRI scans and in cancer hyperthermia [10], has been proved and further development is highly anticipated, as in the application to DDS. Containing some drugs in the magnetic nanoparticles will enable simultaneous detection (by MRI and others) and drug delivery. At the same time, guiding the particles to the target area using magnets becomes possible, depending on the circumstances. When microparticles are injected into a living body for such applications, the size of the particles becomes critically important.

Usually, granular foreign matter taken into a body from outside is expelled by the biodefense system, involving seizure by macrophages in the liver. To avoid this seizure by microphages and to send the nanoparticles to the targeted organ, it is critical that the particles be nanosized. In addition, particles smaller than 100 nm in diameter are relatively effective, because they are not associated with any risk of accumulation in the spleen, can easily reach the cells of the targeted organ by penetrating through blood vessel walls, and can easily access the cells.

Furthermore, if a ligand that forms a specific bond with the cells of the targeted organ can be coupled to the microparticle surface, the introduction of drugs and genes with high specificity will become possible. Such drugs and genes are expected to be deliverable to the specific point where ordinary drugs are not, avoiding the biodefense systems by covering the drugs or genes with nanosized and surface-property-controlled microparticles that are coupled with the target-tissue-specific ligands.

5. Future perspective

Fig. 27.10 illustrates the expanding application field of the thermoresponsive magnetic nanoparticle (Therma-Max) and clearly shows that it is a very important nanomaterial with an extremely wide scope of applications. Therma-Max will contribute significantly to various analysis methods (which may be a keyword of the 21st century), including SNPs (single nucleotide polymorphisms) analysis/proteome analysis (which can represent challenges of the post-genome era) and ultramicroanalysis of environmental hormones.

Therma-Max, a thermoresponsive magnetic nanoparticle, has extremely high dispersibility in water and high

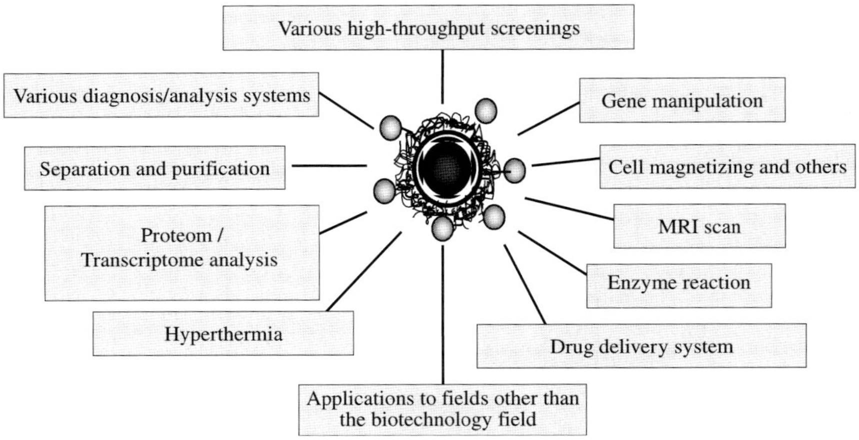

Figure 27.10
Application fields of thermoresponsive magnetic nanoparticles.

molecule-recognition ability, resulting in easy recovery by magnets because of its rapid coagulation, in response to slight temperature change. Accordingly, replacing the magnetic microparticles used in the conventional analyses by Therma-Max will greatly increase sensitivity and reduce the measuring time (for high-throughput measurement). Moreover, further development in medical diagnosis fields and DDS areas is highly anticipated.

References

[1] A. Kondo: *J. Inst. Electrostatics Jpn.*, **23**, 16–22 (1999).
[2] A. Kondo: *Chem. Eng.*, 80–86 (1994).
[3] T. Lea, F. Vertdal, K. Nustad, S. Funderud, A. Berge, T. Ellingsen, R. Shmid, P. Stenstad and J. Ugelstad: *J. Mol. Recognit.*, **1**, 9–18 (1988).
[4] A. Kondo, H. Kamura and K. Higashitani: *Appl. Microbiol. Biotechnol.*, **41**, 99–105 (1994).
[5] H. Furukawa, R. Shimojo, N. Ohnishi, H. Fukuda and A. Kondo: *Appl. Microbiol. Biotechnol.*, **62**, 478–483 (2003).
[6] N. Ohnishi, H. Furukawa, H. Hata, J.M. Wang, C.I. An, E. Fukusaki, K. Kataoka, K. Ueno and A. Kondo: *Nanobiotechonology*, **2**, 43–49 (2006).
[7] M. Heskins, J.E. Guillet: *J. Macromol. Sci., Chem.*, **A2**, 1441–1445 (1968).
[8] N. Onishi, H. Furukawa, K. Kataoka and K. Ueno: *Polym. Prepr. Jpn.*, **47**, 2359 (1998).
[9] S. Hirohashi, R. Hirohashi, H. Uchida, K. Kachi, K. Otomo, G. Uchiyama, M. Niitsu and Y. Itai: *J. Jpn. Radiol. Soc.*, **54**, 776–783 (1994).
[10] M. Mitsumori, M. Hiraoka, T. Shibata, Y. Okuno, S. Masunaga, M. Koishi, K. Okajima, Y. Nagata, Y. Nishimura and M. Abe: *Int. J. Hyperthermia*, **10**, 785–793 (1994).

APPLICATION 28

28 DEVELOPMENT OF FUEL CELLS

The demand for energy has been expanding rapidly as it cannot be separated from the human life, which is upgrading continuously. The majority of the energy depends on the fossil fuel and the energy is used naturally as electricity, power and heat in daily life. Japan is not rich in fossil resource such as oil, but has made great efforts to develop the new energy and energy conservation technologies from the experience of energy crisis. As a result, Japan has become a country where the energy is used with highest efficiency in the world. However, it is clear that fossil fuel will dry up in the future, and further development of both new energy and energy conservation technologies is indispensable. The fuel cell has been paid much attention as a new energy technology, and active research and development have been advanced all over the world.

In the current power generating system such as thermal power plant, the fuel is first made to heat, the turbine is rotated by the heat energy, and this energy is converted into electricity. The power generation efficiency of such method is normally less than 40% according to the restriction at Carnot cycle. On the other hand, because the fuel cells do not face the restriction at Carnot cycle and can directly convert the energy of fuel into the electricity, highly effective power generation is possible. Fig. 28.1 shows the power generation principle of the fuel cells. The fuel cell has a three-layered structure that consists of the electrolyte with two electrodes, similar to a general battery. However, the fuel cell differs from the battery and includes no chemical material for power generation in the electrode. The fuel cell continues power generation during the supply of fuel and oxidant (air) from outside. The power generation principle of the fuel cell will be explained using Fig. 28.1. Hydrogen as a fuel is fed into the anode, and air is fed into the cathode. Oxide ion (O_{2-}) that ionizes in the cathode moves in the electrolyte to the anode side. The oxide ion reacts electrochemically with hydrogen to form water in the anode, and chemical energy of the fuel is converted into electricity at that time.

Among the several types of fuel cells that are expected as next power generation systems, solid oxide fuel cells (SOFCs) can achieve highest efficiency. The development of SOFC has accelerated, and the fine powders are chiefly used as raw materials for manufacturing SOFCs. The nanoparticle technology, as well as the conventional powder technology, is applied for SOFC development to improve its generation performance and reliability. This chapter introduces a method to make high-performance SOFC that applies the nanosized powder processing and the nanostructure control of SOFC components.

1. Development task of fuel cells

The fuel cell is classified into some kinds after the electrolyte used as follows: PAFC (phosphoric acid fuel cells), PEFC (polymer electrolyte fuel cells), MCFC (molten carbonate fuel cells) and SOFCs. In PAFC and PEFC, proton (H^+) moves in the electrolyte. Also, in MCFC, the carbonate ion moves and in SOFC, the oxygen ion moves in the electrolyte. Ion conductivity (mobility) in the electrolyte depends on the temperature, and the operating temperatures of PEFC, PAFC, MCFC and SOFC are 80, 200, 650 and 1000°C, respectively. PAFC and PEFC are classified into the low-temperature type, and MCFC and SOFC are classified into the high-temperature type fuel cell. The high power-generation efficiency is more possible for the high-temperature type fuel cell than for the low-temperature type. The high-temperature type fuel cell combined with the gas turbine is provisionally calculated to achieve the power generation efficiency that exceeds 50%. However, as for the high-temperature type fuel cell, high operation temperature leads to some problems such as heat resistance, durability and reliability of the material and system.

An applicable material for SOFC has been limited to the ceramic because of high operational temperature such as about 1,000°C, and few hundred kilowatts SOFC has been demonstrated in the world. As for the national project in Japan, development that aimed at high power-generation efficiency and massive power generation had been advanced until recently. However, it has currently not been connected with immediate practical use because of low durability and reliability of the material and system. Under such a situation, development to decrease the operating temperature of SOFC has been activated aiming at lowering the cost and improving the reliability. The SECA (solid state energy conversion alliance) project in United States is a representative one for the development of the

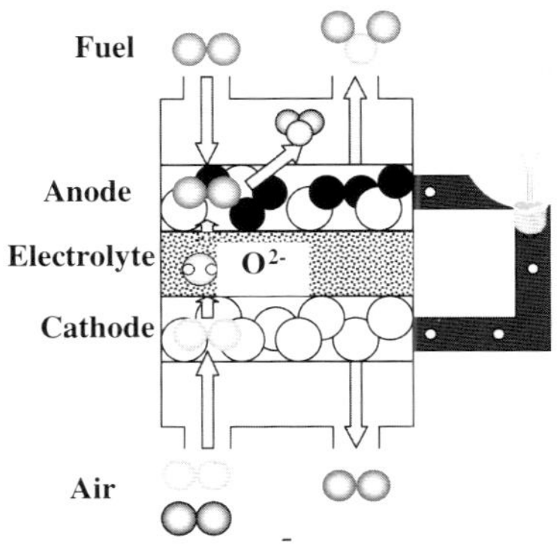

Solid Oxide Fuel Cell (SOFC) Component materials

Electrolyte: YSZ (Y_2O_3 stabilized ZrO_2)

Anode: Ni-YSZ cermet

Cathode: Perovskite oxide ($La(Sr)MnO_3$, $La(Sr)Fe(Co)O_3$ etc)

Figure 28.1
Principle of fuel cell generation.

lower-temperature operating SOFC, and such developments are the present trends in Europe and Japan. These developments have aimed to decrease the SOFC operating temperature to 800°C or less. Further developments aiming at the operation temperature of 600°C or less have also begun. The lower operating temperature expands the applicability of the material for SOFC and enables application of metallic materials such as stainless steel and also a rapid start and stop.

2. Development of high-performance solid oxide fuel cells using nanoparticle technology

Open circuit voltage (OCV) of the fuel cell depends on some parameters such as kinds of fuels and oxidants as well as operation conditions (temperature, pressure and concentration of fuel and oxidant). Theoretical OCV (E) can be calculated from the Gibbs free energy of standard formation (ΔG_0) and concentration of fuel and oxidant, when hydrogen is used as a fuel and air as an oxidant, by the following equation:

In the case of reaction $H_2 + \frac{1}{2}O_2 \rightarrow H_2O$:

$$E = E_0 + \left(\frac{RT}{nF}\right)\ln\left(\frac{PH_2O}{PH_2 \cdot PO_2^{1/2}}\right) \qquad (28.1)$$

$$\Delta G_0 = -nFE_0$$

The terminal voltage (CCV: closed circuit voltage) of the fuel cell descends by the internal resistance (IR) when electricity is generated. Fig. 28.2 shows the relationship between the current density and terminal voltage as well as the voltage loss during power generation. The voltage loss consists of the IR loss of cell components, polarization loss of both electrodes and the Nernst loss. The Nernst loss means OCV decrease by the concentration change of fuel and air during power generation (see equation (28.1)). In Fig. 28.2, IR loss, polarization loss and Nernst loss are subtracted from the OCV, and the terminal voltage of the cell is calculated. Therefore, the power density of the cell increases with a decrease in voltage loss, and how much IR loss and polarization loss are reduced leads to the improvement of the power generation efficiency of fuel cells. In SOFC, the development of the high ion-conductive material as an electrolyte and making a thin-film electrolyte have been tried to decrease the IR loss. Moreover, the development of high-activity and conductive electrode materials and their microstructure control have been attempted to decrease the polarization loss.

For standard SOFC component materials (see Fig. 28.1), the lower temperature (700°C) operation of SOFC has been achieved by development of an anode-supported thin-electrolyte cell with nanopowder processing and by microstructure control of electrodes with nanocomposite powder 1–4). Fig. 28.3 [1] shows the anode-supported cell developed by nanopowder processing. Fig. 28.4 [2] shows the power generation performance of the cell (current density versus terminal voltage and power-density characterizations). The trial fabrication of the cell up to a diameter of 12 cm has been succeeded, and the high power density was achieved by the anode-supported cell. Fig. 28.5 [2] shows a Scanning Electron Microscope (SEM) image of the cross section of the developed cell and schematic diagram of the cell structure. It is confirmed in Fig. 28.5 that a thin-Y_2O_3 stabilized ZrO_2 (YSZ)

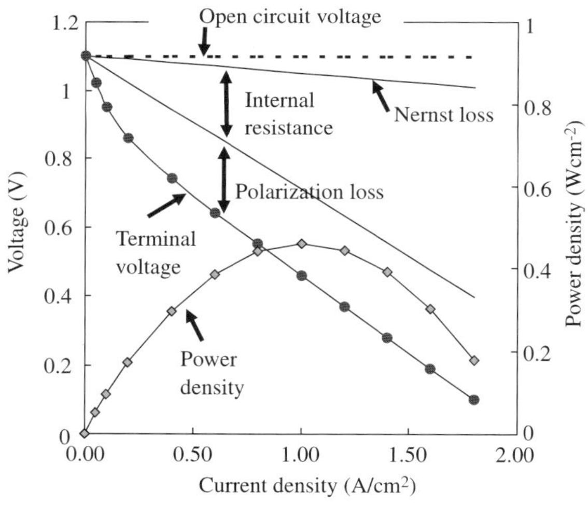

Figure 28.2
Relationship between V–I characterization and voltage loss.

Figure 28.3
Anode-supported cell [1].

electrolyte of 10 μm in thickness has been obtained in the developed cell. This thin-film electrolyte depends on the cell structure, and the anode plays the role of the radical material and maintains the structure. Such a structure is called the anode-supported cell structure. In this nanopowder processing, multitape casting is used to fabricate such anode-supported thin electrolyte. Optimizing the characteristics of the raw powders (particle-size and distribution), the slurry properties (powder density, viscosity and kinds of binder) and casting conditions are indispensable to succeed in trial fabrication.

The decrease of polarization loss was also successfully realized by microstructure control of the electrodes in addition to making of the thin-film electrolyte. Fig. 28.6 [3] shows a SEM image of the cross section of Ni–YSZ cermet anode developed. Fig. 28.7 [4] shows results of Ni and O elemental mapping of the cermet anode. In Fig. 28.7, O mapping means the existence of YSZ. It is confirmed from Fig. 28.6 to have achieved the porous network structure with fine grains. In addition, it can be confirmed from the mapping results in Fig. 28.7 that the network structure in which the YSZ

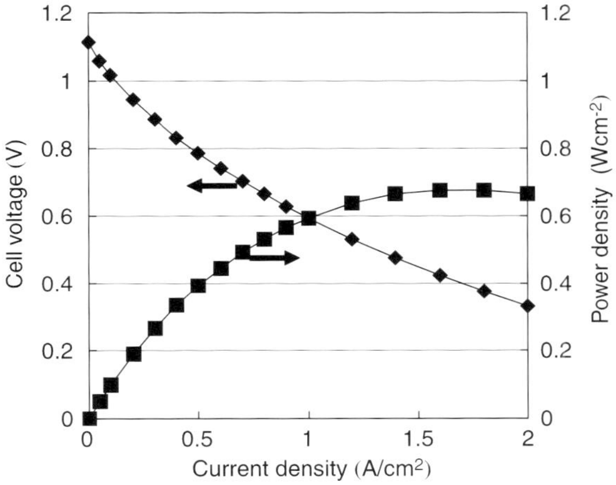

Figure 28.4
I–V and I–P characterizations of an anode-supported cell [2].

Figure 28.6
SEM image of Ni–YSZ cermet anode [3].

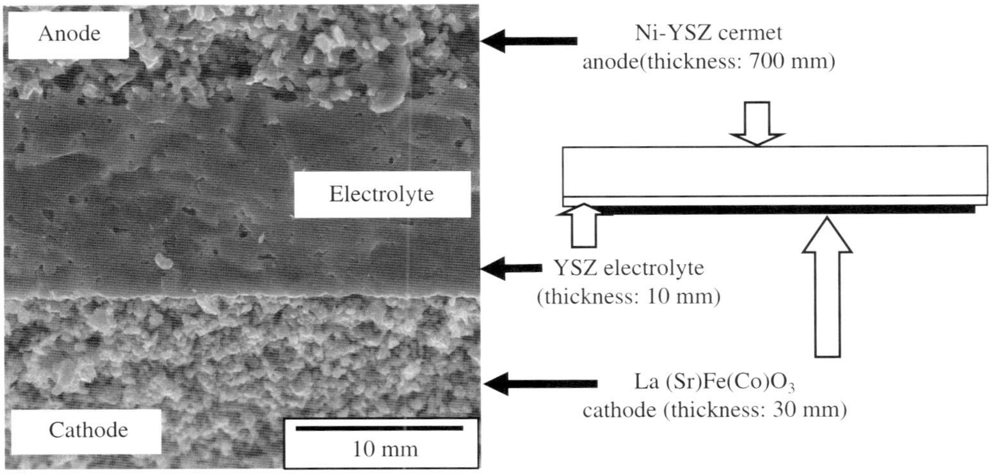

Figure 28.5
SEM image of the cross section of anode-supported cell [2].

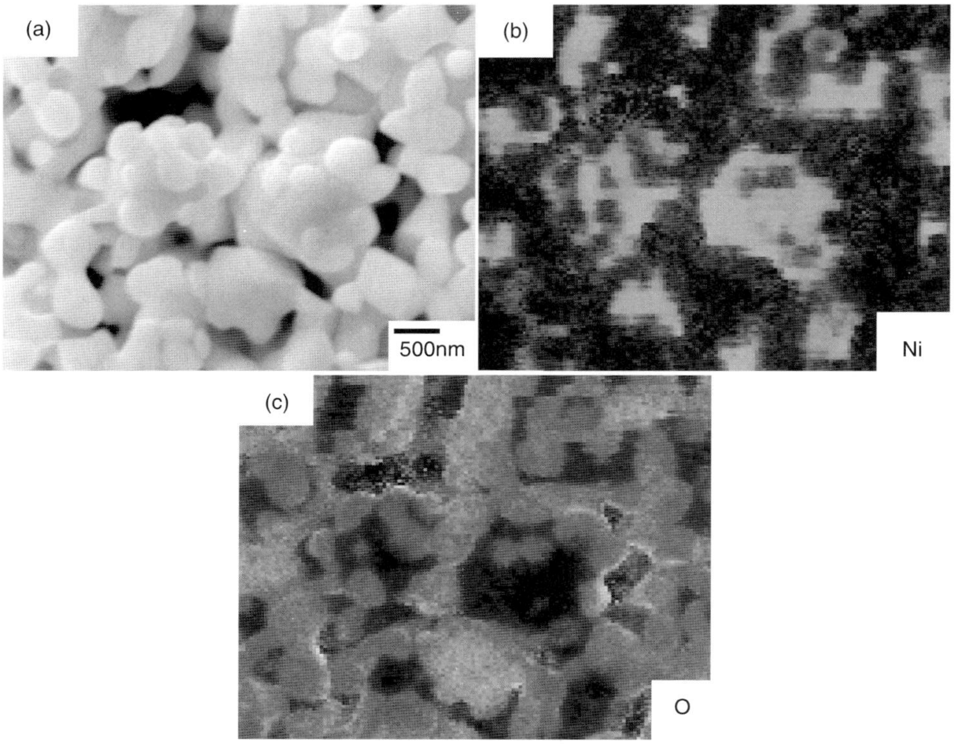

Figure 28.7
Morphology of Ni–YSZ cermet anode. (a) SEM image, (b) Ni mapping, (c) O mapping [4].

nanoparticles-embedded Ni grains are connected has been achieved. Control of porous structure and grain network and expansion of electrochemical reaction area (TPB: triple phase boundary) are indispensable for the decrease in polarization loss. The NiO–YSZ composite powder has been used to control such anode morphology. Fig. 28.8 [5] shows the microstructure of the NiO–YSZ composite particles made by the advanced mechanochemical technique. This technique is defined as a novel method to create chemical bonds between particles in dry process without any binder. It is confirmed from Fig. 28.8 that YSZ nanoparticles of 100 nm or less partially cover the surface of NiO particles. The average particle size of the NiO–YSZ composite powder was about 0.5 μm or less and it had narrow particle distribution [3, 4]. In conventional fabrication process of Ni–YSZ cermet anode, fabrication of fine Ni grain is difficult because NiO grain is sintered more easily compared with YSZ grain. In the fabrication process with NiO–YSZ composite powder, composite structure partially covered with YSZ particles will lead to fine Ni grains and good network in the cermet anode. Narrow particle-size distribution will also lead to the homogeneous porous structure.

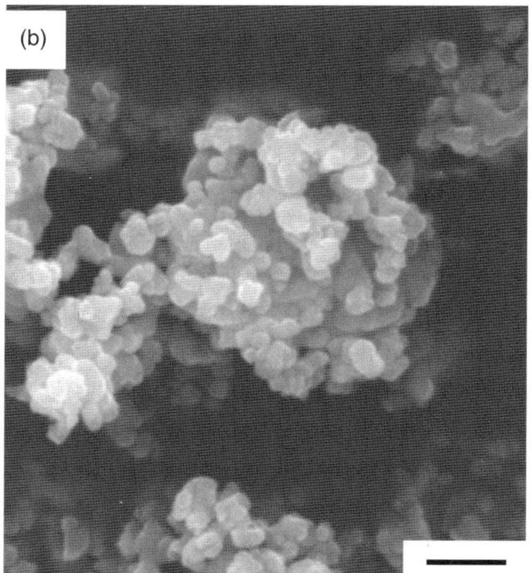

Figure 28.8
SEM image of NiO–YSZ composite particles [3].

References

[1] T. Fukui: *J. Fuel Cell Technology*, **5**, 41–46 (2006).
[2] K. Murata, T. Fukui, H. Abe, M. Naito and K. Nogi: *J. Power Sources*, **145/2**, 257–226 (2005).
[3] T. Fukui, K. Murata, S. Ohara, H. Abe, M. Naito and K. Nogi: *J. Power Sources*, **125**, 17–21 (2004).
[4] T. Fukui, K. Murata, C.C. Huang, M. Naito, H. Abe and K. Nogi: Proceedings of 28th International Conference on Advanced Ceramics and Composites, Vol. 25 (3 and 4), American Ceramic Society, Ohio, 263–267 (2004).

APPLICATION 29

29 DELIVERY TO THE BRAIN

The central nervous system (CNS) diseases have presently become one of the five most serious diseases. Representative typical CNS diseases are Alzheimer's disease, brain tumor, Parkinson's disease, white matter dystrophy symptom, Creutzfeldt–Jakob disease, meningitis, encephalitis and so on. With development of new medicine for CNS diseases, the percentage of dropout during clinical trial phases is rather high in comparison with other drugs. This tendency relates with a unique structure of the blood–brain barrier (BBB) playing as a barrier for absorption of drugs.

Goldman et al. proved the function of BBB with the injection of trypan blue solution into rats. In their study, the body of the rat was colored blue except CNS after i.v. injection, whereas after intranervous injection, CNS was colored blue but the color of the body did not change. Based on this finding, numerous researchers focused on studying the peculiar structure of blood vessels. Endothelial cells of the brain capillary joined more tightly with deep tight junction compared with endothelial cells of other tissues. Therefore, material transports via paracellular route are restricted. Generally, permeability of substance into the brain is proportional to its partition coefficient. However, nutrients, e.g. glucose, amino acid and so on, are preferably permeable, whereas cellular toxic drugs, e.g. anticancer drugs, are less permeable although they are highly lipophilic. These phenomena are related with the efflux of absorbed drug by a transporter such as P-glycoprotein in the epithelial cell membranes. According to this drug-transport restriction mechanism of BBB, drug delivery to brain and CNS becomes quite difficult. In this section, the application of biodegradable polymeric nanoparticles used for drug transport to brain targeting will be explained.

1. Surface modification to improve the nanoparticle distribution in the brain

1.1 Brain targeting by active targeting

Active targeting with body distribution controlled and drug carrier behavior by the modification of surface properties with various polymers, surfactants, reagents and so on have been widely studied. Kreuter et al. reported that drug distribution into the brain could be increased by using polysorbate-coated nanoparticles of submicron size [1]. Fig. 29.1 shows doxorubicin concentration in the rat brain–time profiles after i.v. injection via tail vein. In the administration groups of doxorubicin solution, doxorubicin solution including polysorbate 80 and unmodified doxorubicin-loaded nanoparticles, no increase in doxorubicin concentration could be detected in the brain, whereas the doxorubicin concentration in brain of the polysorbate 80-coated nanoparticles increased significantly for 4 h after injection. Furthermore, survival period of glioblastoma 101/8 bearing rats was significantly prolonged with the injection of polysorbate 80-coated nanoparticles with loading doxorubicin.

The mechanism of drug distribution in the brain increased by using polysorbate 80-coated nanoparticles has not been completely elucidated yet. At present, a possible mechanism in Fig. 29.2 is speculated [2]. First, APO-E proteins adsorb on the polysorbate 80-coated nanoparticles after i.v. injection. Then, complexes of adsorbed APO-E protein and polysorbate 80 on the nanoparticle surface are recognized by low-density lipoprotein (LDL) receptor on the epithelial cells of BBB as LDL and are taken into epithelial cells. After uptake into the cells, enclosed drug in the nanoparticles is released with degradation of the polymer. Other possibility for the increase of drug concentration in brain is that polysorbate 80 might increase the permeability of drug through the paracellular route by opening tight junction and inhibit the drug efflux of P-glycoprotein.

Up to now, drugs that have been successfully delivered into the brain using polysorbate 80 coated nanoparticles include the hexapeptide dalagin, the dipeptide kyotrophin, loperamide, tubocuranine, the NMDA receptor antagonist MRZ 2/576, and doxorubicin. Several drug carriers for targeting to the receptors on the epithelial cell membrane of the BBB, e.g. nanoparticles modified with thiamine [3], have been reported.

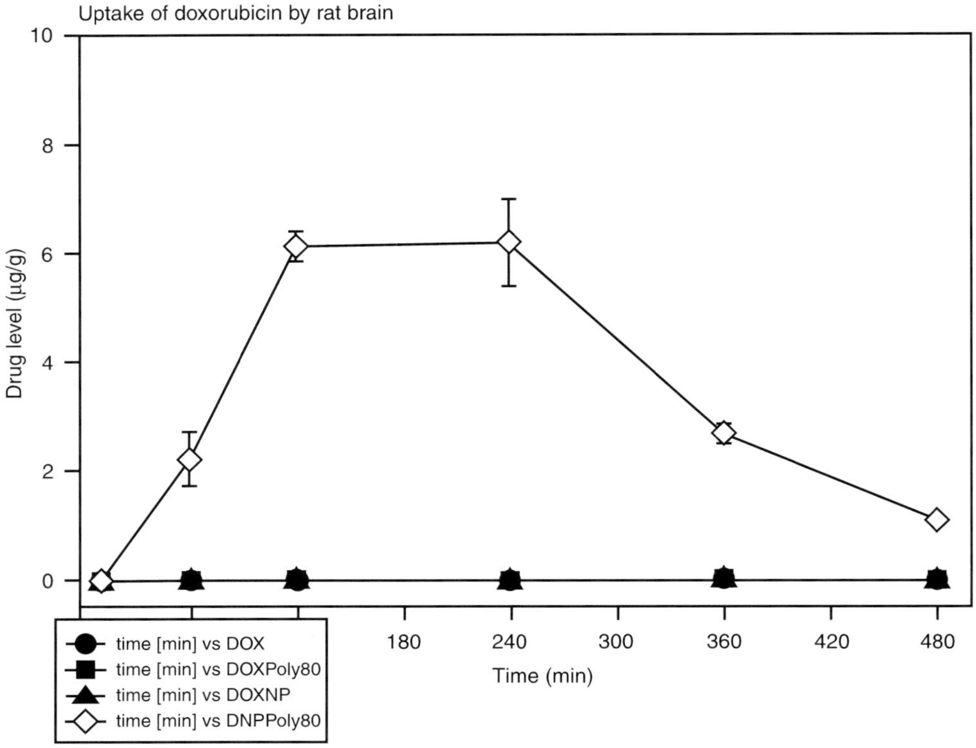

Figure 29.1
Uptake of doxorubicin by rat brain. Doxorubicin brain concentration after intravenous injection of one of the following preparations: ●, Doxorubicin [5 mg/kg] solution in saline; ■, doxorubicin [5 mg/kg] solution in saline plus 1% polysorbate 80; ▲, doxorubicin [5 mg/kg] bound to poly (butyl cyanoacrylate) nanoparticles; ◇, doxorubicin [5 mg/kg] bound to poly (butyl cyanoacrylate) nanoparticles plus 1% polysorbate 80.

1.2. Brain targeting by passive targeting

The particles smaller than 5 µm in diameter are taken into the reticuloendothelial system (RES), such as liver, spleen, lung, etc., after injection and the particles are eliminated rapidly from the blood circulation. An immunoglobulin (IgG or IgM) and a complement protein (C3 fragment) are related in their uptake by RES. First, these blood proteins adsorb on the particulate surface and then particles are taken up by macrophages possessing opsonin receptor and so on. Adsorption of such proteins on the particulate surface increases with increased hydrophobicity and positive charge. Numerous papers have described the avoidance of particles taken in by increasing hydrophilicity, e.g., modifying particulate surface with surfactants or hydrophilic polymers.

Neovascular developments in tumor tissue have a loose structure compared with blood vessels in normal tissue. Therefore, the submicronized particles (less than 100 nm) can penetrate into tumor tissue through vascular wall. This phenomenon is called "enhanced permeability and retention effect" (EPR effect). According to this EPR effect, nanoparticles can prolong circulation in blood, resulting in distribution preferable in the brain tumor.

Couvreur et al. reported that distribution in the brain could be improved by using such long-circulating nanoparticles modified by polyethylene glycol (PEG) [4]. Furthermore in this report, PEGylated nanoparticles could penetrate the brain tissue more deeply compared with those modified with poloxamine 908 and polysorbate 80. Especially, PEGylated nanoparticles showed higher brain distribution percentage in the inflammatory site and invasion of macrophages was also found. Therefore, distribution of PEGylated nanoparticles might be explained with passive targeting and reduction in uptake by macrophages at the inflammatory site.

2. Effect of administration route on the brain distribution

For the chemotherapy against the brain tumor, therapeutic effect against the brain tumor can be improved

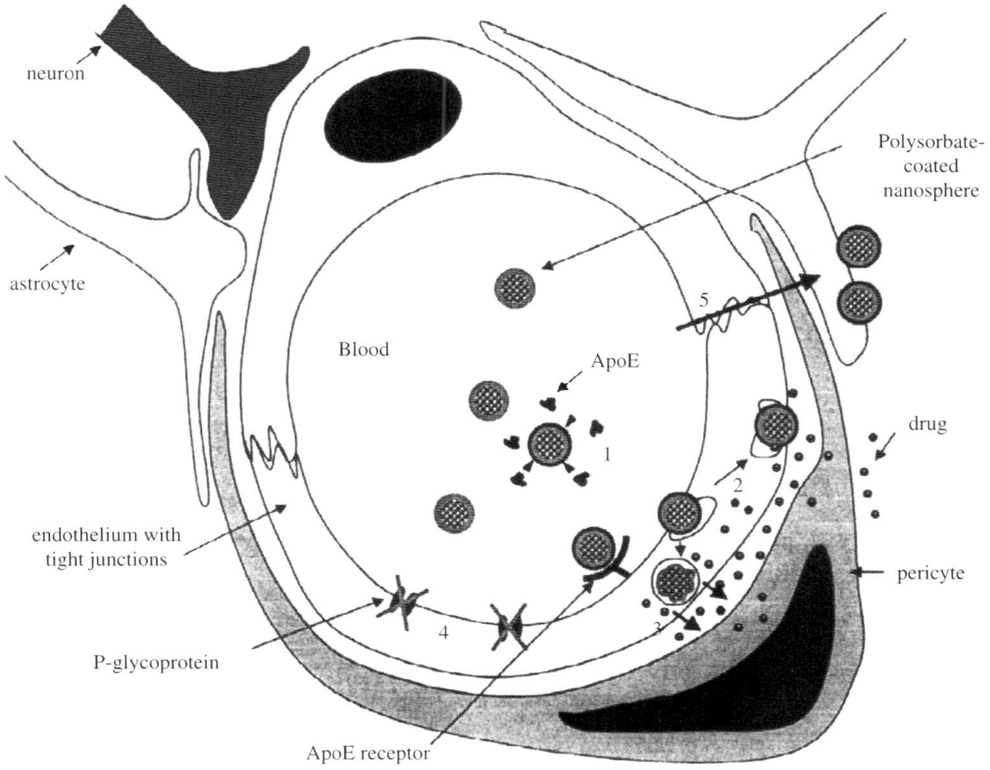

Figure 29.2
Hypothetical mechanism of drug delivery to the brain by means of polysorbate 80-coated polyalkylcyanoacrylate PACA nanospheres.

by injection of methotrexate via the carotid artery. Conventionally, the drug delivery for brain targeting has been carried out by injecting it into the vein. The authors evaluated the effect of the administration route of the surface-modified nanoparticle with polysorbate 80, poloxamer 188 and chitosan on brain distribution percentage and blood concentration (Fig. 29.3).

All types of nanoparticles after injection via the carotid artery showed higher brain-distribution percentage than that after the jugular vein injection. Elimination rate of uncoated nanoparticles was the fastest among the nanoparticles evaluated. Remaining percentage of uncoated nanoparticles in the brain reduced to 25% at 15 min after injection. At that time, the brain distribution percentage of the uncoated nanoparticles after the carotid artery injection was almost corresponding to that after the jugular vein injection. Furthermore, the elimination rate of uncoated nanoparticles in the blood was the fastest among the evaluated nanoparticles. According to these results, uncoated nanoparticles might not interact with the brain endothelial cell and were not taken from the blood stream.

The interaction between poloxamer 188-coated nanoparticles and brain endothelial cells might be weakened by this rapid elimination. On the other hand, polysorbate 80-coated nanoparticles after injection into the carotid artery and into the jugular vein showed higher brain distribution percentage compared with other modified nanoparticles. These results consist with the data obtained by Kreuter's group. Furthermore, polysorbate 80-coated nanoparticles had prolonged circulation in blood compared with other nanoparticles. This long circulating effect might be one of the factors for increasing the brain distribution by the avoidance of uptake by RES.

Although the rapid clearance from blood circulation occurred, chitosan-coated nanoparticles prolonged their retention in brain. Furthermore, the concentrations of nanoparticles in brain and their retention behavior were not changed by the change in injection dose. Chitosan-coated nanoparticles prolonged their retention on the brain blood capillary by their adherence to endothelial cell with electrostatic interaction, whereas unabsorbed particles were eliminated rapidly from blood circulation by the uptake in RES.

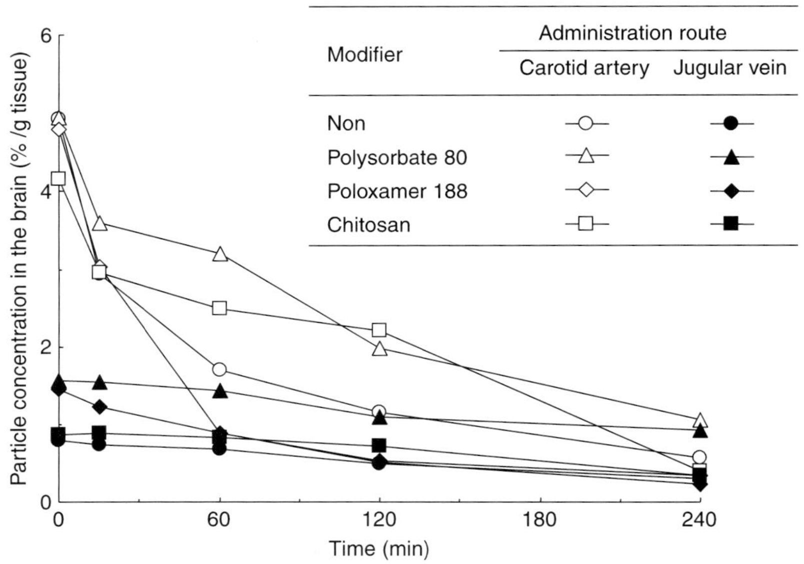

Figure 29.3
Effect of administration route on the brain distribution of surface-modified poly(DL-lactic-*co*-glycolic acid) nanoparticles.

3. Perspective of brain targeting with nanoparticles

Investigations on brain function are progressing extensively from physiological, organizational and cytological viewpoints. Therapy against the brain and CNS diseases is becoming more and more important. By elucidating the function of brain and the mechanism of brain diseases clearly, the drug delivery system with nanoparticles can be designed more precisely to cure currently incurable diseases of brain and CNS.

References

[1] J. Kreuter: *Adv. Drug Deliv. Rev.*, **47**, 65–81 (2001).
[2] C. Vautier, C. Dubernet, E. Fattal, H. Pinto-Alphandary, P. Couvreur: *Adv. Drug Deliv. Rev.*, **55**, 519–548 (2003).
[3] P.R. Lockman, M. Oyewumi, J. Koziara, K. Roder, R. Mumper, D. Allen: *J. Control Release* **93**, 271–282 (2003).
[4] P. Calvo, B. Gouritin, H. Chacun, D. Desmaele, J. D'Angelo, JP. Noel, D. Georgein, E. Fattal, JP. Andreux, P. Couvreur: *Pharm. Res.*, **18**, 1157–1166 (2001).

APPLICATION 30

30 NOZZLE-FREE INKJET TECHNOLOGY

Inkjet technology was developed as a new printing method and due to the rapid proliferation of the computer, it has been widely used in low-cost, small-sized printing units such as printers or copying machines for domestic use. The inkjet printing method is classified into two categories. One method uses a piezoelectric device that gets deformed when a voltage is applied to it. The other is the thermal or the bubble jet method that utilizes the generation of air bubbles due to heating. Both methods require very fine nozzles. The ink is pushed out from the nozzle for the printing due to the deformation of the piezoelectric device in the piezoelectric method, and due to the pressure of an air bubble created inside the ink by heating in the thermal or the bubble-jet method. Therefore, the ink needs to have properties suitable for the nozzle such as high fluidity, low viscosity, and appropriate drying property, and a cleaning system is always provided for clearing clogged ink. In addition, this inkjet technology is not merely used in printing machines but is also being used for the manufacture of devices such as multilayer printed circuit boards, displays, or DNA chips, etc., and for accurate injection of very small quantities of materials for quantitative analysis. In these new applications, unlike the dye-based inks used conventionally, the inks containing organic materials as well as those dispersed with organic, inorganic, and metallic fine particles (nanoparticles) are used. In the latter case, it is used as the liquid raw materials for device manufacture. In order to widen the applicability of inkjet technology from simple printing to new applications, improvements in the technology and formulation of the target inks become very important. However, in order to prepare the samples of a device, inclusion of organic materials or dispersion of particles is necessary, which causes problems such as reduction in the fluidity or increase in the viscosity of the ink (the solution of the raw materials) as well as wearing out or clogging of the nozzles. Therefore, in order to widen the applicability of inkjet technology it is necessary to widen its technical applicability to solutions of different raw materials. For the purpose of solving such problems caused by nozzles, nozzle-free inkjet technology and a system using it were jointly developed by the Joining and Welding Research Institute of Osaka University and Hosokawa Powder Technology Research Institute. In this chapter, an introduction is given about the principle of nozzle-free inkjet technology, an outline of the development system, and some samples of their application.

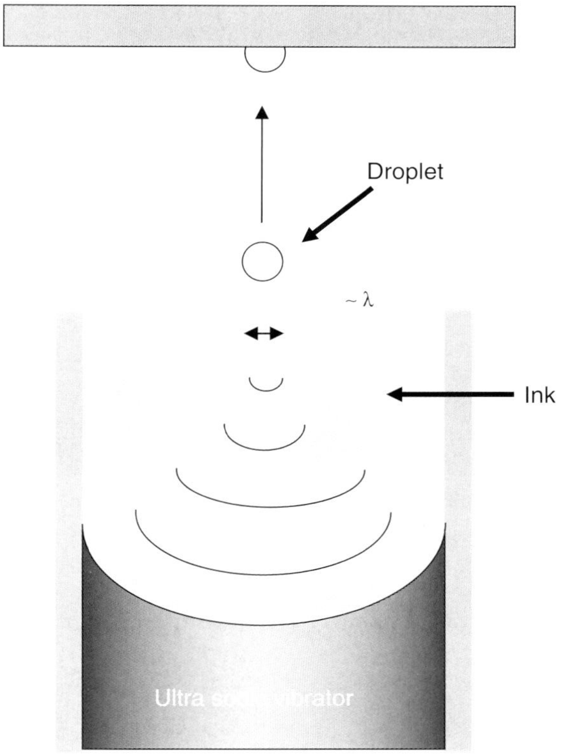

Figure 30.1
Schematic diagram of the basic principle of nozzle-free inkjet technology.

1. Principle of nozzle-free inkjet technology and outline of developed system

The basic principle of nozzle-free inkjet technology is shown in Fig. 30.1. Using an ultrasonic transducer, ultrasonic waves are converged onto the surface of a raw material solution (ink) in the container. The basic principle of inkjet printing is that due to the converged ultrasonic waves (converged acoustic wave), very small droplets of the liquid are generated from the surface of the liquid solution, and the inkjet printing is carried out by emitting and forming these droplets onto the substrate. The state of emitting liquid droplets is shown in Fig. 30.2. Initially, by the radiation pressure of the converging acoustic waves, the surface of the liquid solution rises in the form of a mountain. Due to the balance between the surface tension of the liquid solution and the radiation force, this rise gradually becomes higher and a droplet is formed from it. Finally, the radiation force becomes greater than the surface tension and the droplet is emitted. The droplet that is emitted or sprayed is then impinged on a substrate and the prescribed printing or formation is completed. The biggest feature of this nozzle-free inkjet printing method is that it is possible to carry out printing similar to the conventional piezoelectric device or the thermal method without using nozzles in this manner. Further, when selecting a suspension of particles as the liquid solution, it is also possible to form particle structures on the substrate by removing the solvent by drying, and the range of selection of the raw material solution becomes wider since no nozzle is used.

An outline of the droplet spraying system of the nozzle-free inkjet method using acoustic wave convergence as its principle is shown in Fig. 30.3. By the use of computer control, a tone burst wave is produced from an arbitrary waveform generator, amplified by a power amplifier, and is applied as the electrical signal to the ultrasonic transducer. A frequency from 10 to 100 MHz is being used as the burst frequency for this ultrasonic wave generation and the ultrasonic transducer used is the one that has an acoustic lens and is used in sonar detectors. The relationship between the acoustic wave frequency and the spray droplet diameter in water is shown in Fig. 30.4. The droplet diameter decreases with increasing acoustic wave frequency. Since this relationship has a proportionality of almost power of $-2/3$, it is possible to control the droplet diameter using the acoustic frequency. In addition, the dependency of this droplet diameter on the frequency matches with other experimental results thereby substantiating the reliability of

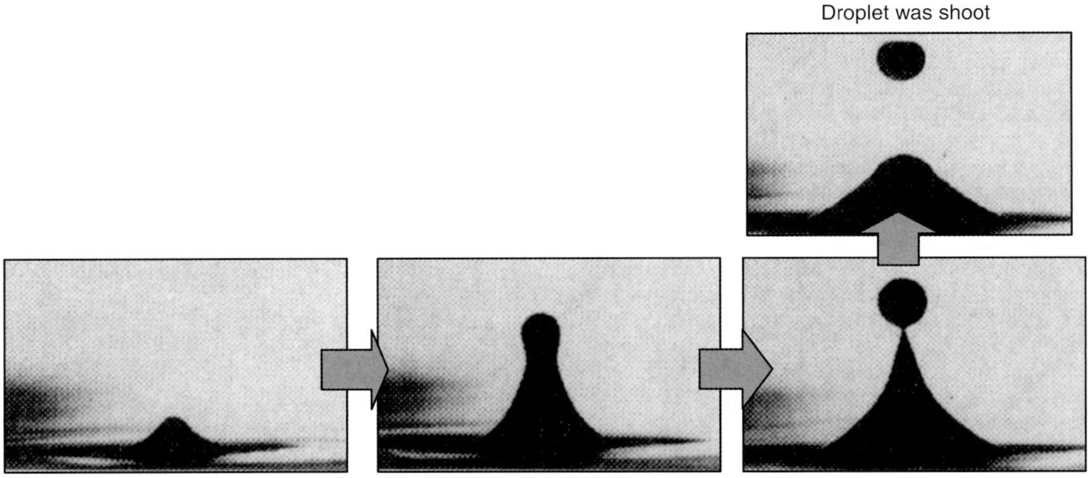

Figure 30.2
Droplet shooting process by the nozzle-free technique.

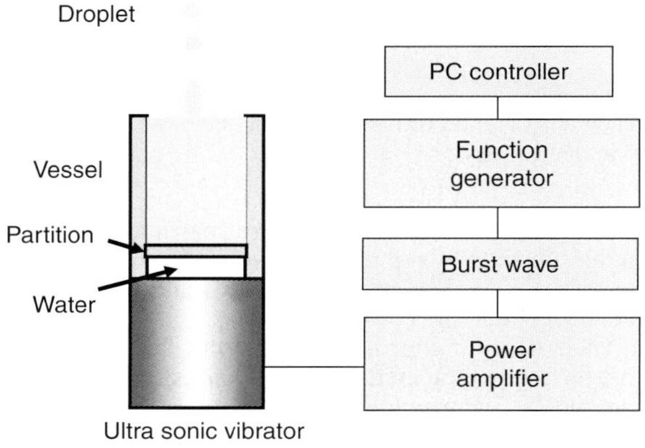

Figure 30.3
Schematic diagram of the developed system.

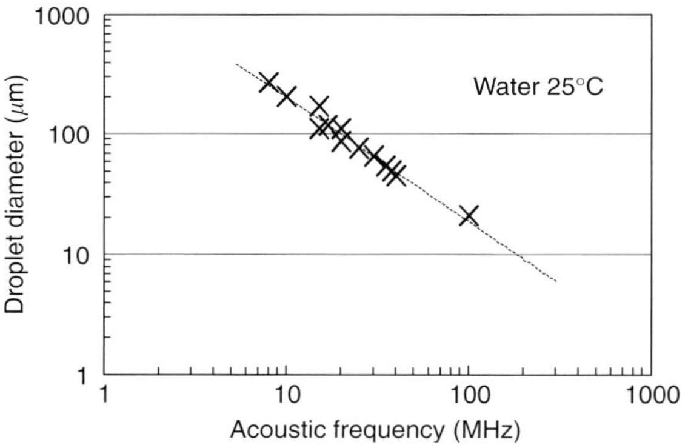

Figure 30.4
Relation between acoustic wave and droplet diameter.

these experimental results [1]. Further, in order to protect the ultrasonic transducer from the corrosive liquid such as a strong acid, a polyethylene container is used for holding the liquid solution of raw material, a partition wall is provided between the liquid solution container and the ultrasonic transducer, and water is inserted in between as the medium for transmitting the acoustic waves. This structure is being adopted for replacement of liquid solution reagent, or for continuous spraying of different types of liquid solutions. Since different types of materials are arranged in the direction of propagation of the ultrasonic waves, there will be reflections of the acoustic wave due to mismatching of the acoustic wave impedance. Because the reflection of the ultrasonic waves is as less as 1% at the interface with water, polyethylene is used as the material for the container. In addition, by making the thickness of the bottom of the container less than about 300 μm, a transmittivity of more than 90% has been achieved thereby aiming to maintain the efficiency of the droplet spray. An outline of the development system is shown in Fig. 30.5 [2]. The plate to be

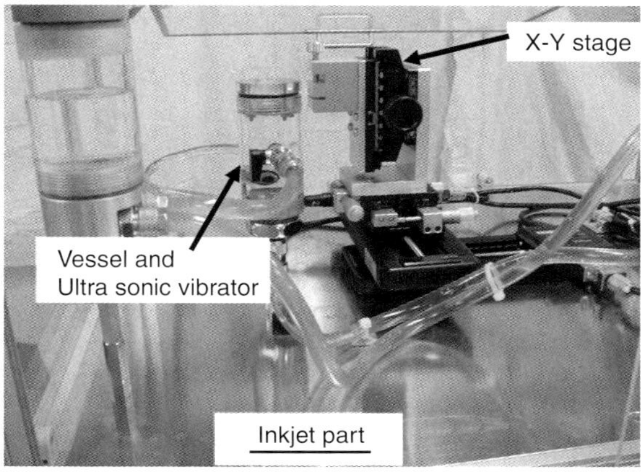

Figure 30.5
Overview of the developed system.

printed is fixed on the side of a moving type X–Y stage, and the droplets emitted from the liquid solution container can be impinged at any required position, thereby making it possible to form patterns of the particles.

2. Formation of slurry using nozzle-free inkjet technology

The biggest feature of nozzle-free inkjet technology is that it is possible to carry out printing and formation using suspensions of particles (slurries) as the raw material. Compared to the nozzle-type printing methods such as the piezoelectric device or the thermal type, it has the advantage to be able to spray the droplets of the liquid raw material with a high viscosity, and even when a relatively dense slurry is used, there is no problem of the nozzle getting clogged. However, for spraying slurries, it is necessary to increase the output power of the ultrasonic transducer that generates the acoustic waves. The ultrasonic waves in a slurry container are subject to attenuation due to the viscosity between the solvent medium and the particles, heat generation by the particles, vibration of the particle structure, etc., and a power greater than such attenuation will be necessary to spray the droplets by the converged acoustic waves. In addition, in order to operate over long periods in a stable manner, it is necessary to cool the ultrasonic transducer and maintain it at a constant temperature. In the developed system, in order to realize the slurry spraying, improvements in the apparatus such as cooling of the ultrasonic transducer, thinning of the slurry container, etc., have been made (see Fig. 30.6). The result of

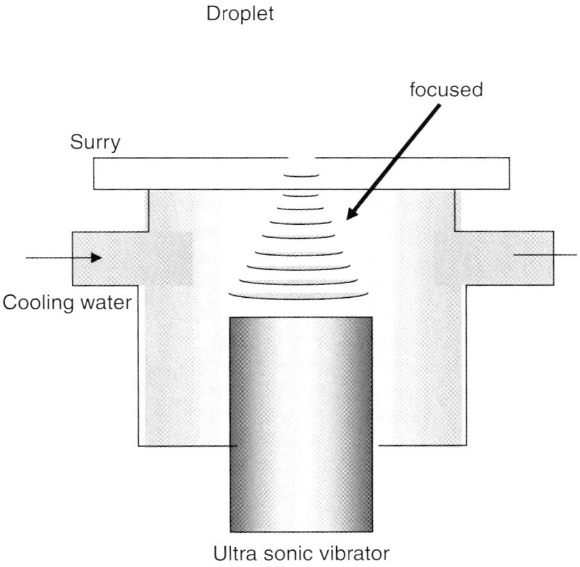

Figure 30.6
Application of the nozzle-free inkjet technology to slurry droplet formation.

forming state of slurry droplet by the nozzle-free inkjet method with polystyrene (PS) beads of about 1 μm size is shown in Fig. 30.7. A uniform film formation on the surface of the substrate has been confirmed, thereby

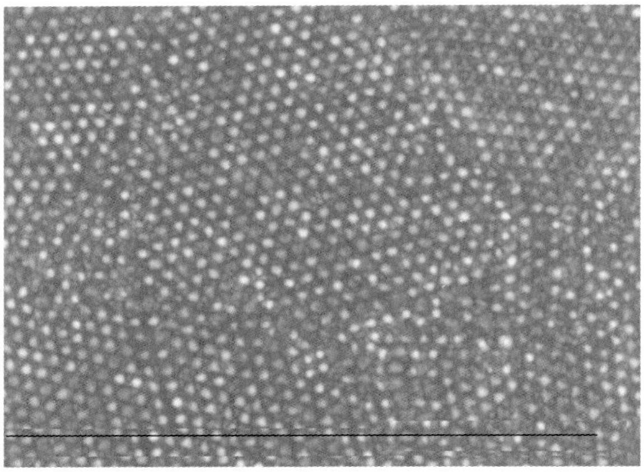

Figure 30.7
Forming state of slurry droplets (7 vol.% PS slurry).

experimentally proving the applicability of this method to the slurry spray formation.

The outline of the newly developed nozzle-free inkjet technology has been introduced. This technology has a potential to extend the areas of printing and formation to the use of materials such as suspensions, which has been difficult by the conventional inkjet technology. The extension of this nozzle-free inkjet technology to new applications is looked forward to in the near future.

References

[1] S.A. Elrod, B. Hadimioglu, B.T. Khuri-Yakub, E. Richley, C.F. Quate, N.N. Mansour and T.S. Lundgren: *J. Appl. Phys.*, **65**, 3441–3447 (1989).

[2] Technical information: *The Micromeritics*, published by Hosokawa Powder Technology Research Institute, **48**, 91 (2004).

APPLICATION 31

31 DEVELOPMENT OF EXHAUST CATALYST

Three-way catalysts that are developed by using new technology for nanolevel powders are explained in this section.

1. Supported metal catalyst

The temperature of an automobile exhaust reaches as high as 1,000°C under a high-load driving condition. Therefore, exhaust catalyst must endure such a high temperature and must keep up its catalytic activity for cleaning up automobile exhaust throughout a car's life. For the purpose of keeping up the high catalytic activity of a fresh catalyst, the particles of platinum group element (mainly Pt, Rh and Pd), which are loaded in the catalyst, must be kept at single-nanometer level. Because these noble metal particles play a role of active sites of the catalyst, the smaller size of the noble metal particle means that a larger number of active sites exist in the catalyst. The noble metals are scarce and expensive resources; hence, more effective utilization of them is strongly needed. For keeping up such a high dispersion state of the noble metals, they must be loaded on a catalyst support, which has a high specific surface area. Their high dispersion state enables the components of exhaust gas, hydrocarbons (HC), carbon monoxide (CO) and nitrogen oxides (NO_x), to collide more frequently and efficiently with active sites in a high-velocity stream of exhaust gas.

On the other hand, the three-way catalyst must clean up HC and CO by oxidation and must clean up NO_x by reduction at the same time. Simultaneous cleaning by the reduction and oxidation can be achieved in an operating window of A/F around the stoichiometric composition of reductive components and oxidative composition in the three-way catalyst (Fig. 31.1). For keeping up the stoichiometric composition of an exhaust gas, the three-way catalyst is used in the three-way catalyst system, which is composed of not only a catalyst itself but also electronically controlled fuel injector, air flow meter, oxygen sensors and computer to control their working. An oxygen sensor set just upstream of a catalyst sends a signal of oxygen concentration in an exhaust gas to the electronically controlled fuel injector for controlling A/F (air to fuel ratio) at the stoichiometric composition. However, there is a small delay between the signal of oxygen sensor and the response of controlling A/F; an exhaust gas composition fluctuates from fuel rich to fuel lean especially in the transient condition, acceleration or deceleration. The A/F deviance from the stoichiometric composition makes the conversions of HC, CO and NO_x worse. For solving this problem, catalyst itself has to have a function of buffer of oxygen to keep the stoichiometric composition in the transient condition in which the mechanical and electronic control cannot follow. Oxygen-storage materials in a three-way catalyst

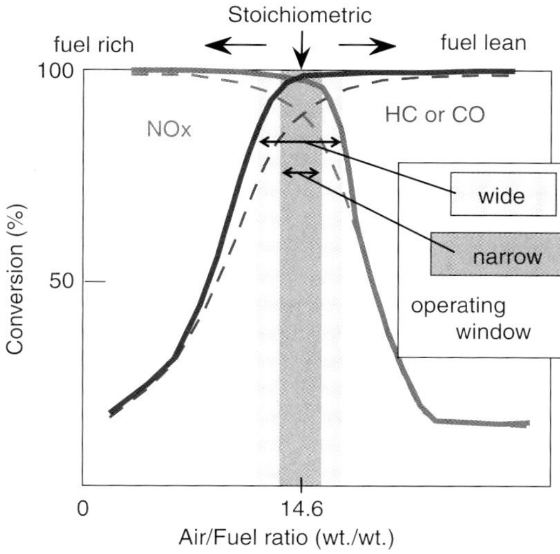

Figure 31.1
The relation between air/fuel ratio and conversion efficiency of a three-way catalyst.

play an important role to give a function of oxygen buffer to three-way catalyst.

2. Oxygen storage capacity (OSC) of catalyst

The availability of oxygen storage material was found in 1972 for automobile exhaust catalyst [1–4] The term, "oxygen-storage capacity: OSC" was coined by Gandhi et al. in 1976 [5], and several base metal oxides were shown to function as oxygen storage materials [5]. The three-way catalyst system was put to practical use in 1977 [6]. In the three-way catalysts of early years, BaO-doped CeO_2 [7, 8] or La_2O_3-doped CeO_2 [6, 9–11] was used as oxygen-storage materials. The effect of the additives to CeO_2 is to introduce oxygen vacancy by dissolving lower valence cation into CeO_2 lattice and to accelerate volume diffusion of oxygen ion. The function of oxygen storage capacity of CeO_2 was reported in detail by Yao et al. [12]. CeO_2 shows reversible OSC according to the following chemical equation.

$$CeO_2 \rightarrow CeO_{2(1-X)} + XO_2 \quad (31.1)$$

in fuel-rich region, and

$$CeO_{2(1-X)} + XO_2 \rightarrow CeO_2 \quad (31.2)$$

in fuel-lean region.

CeO_2 has cubic fluorite structure. Ce ions in the crystal structure are tetravalent in fuel-lean atmosphere. In fuel-rich atmosphere, partial Ce ions are trivalent. However, molar fraction of trivalent Ce ions is only a small amount in the total Ce ions. The molar fractions of trivalent Ce ions of CeO_2, $X/2$, at 500°C in fuel-rich atmosphere, after durability test at 900 and 1200°C were 0.01 and 0.002, respectively, in the study by present authors [13]. The reduction in the value of $X/2$ at 500°C is ascribed by the fact that only the oxygen ions on the surface can contribute OSC [14]. Therefore, oxygen-storage materials of early years have a weak point that OSC decreases with the decreasing specific surface area of CeO_2 after such a high-temperature durability test. BaO and La_2O_3 had an effect in preventing sintering and maintaining specific area of CeO_2; however, it was not enough [10].

3. Improvement of OSC of catalyst

Two types of new technological developments are introduced concerning the nanolevel particles achieved in the three-way catalyst system in recent years as follows. The first one is the improvement of OSC, and the second aimed at the improvement of heat resistance. The first one is a technology to dissolve ZrO_2 into CeO_2 and improves OSC remarkably. The phenomenon that ZrO_2 dissolving into CeO_2 increases its OSC several times was discovered by Ozawa et al. [15], and it was put into practical use in 1989. The CeO_2–ZrO_2 was made by impregnation of zirconium salt in CeO_2 or coprecipitation. There was a solubility limit at around 20 mol% of ZrO_2 dissolving into CeO_2 at that time. The first technology introduced here concerns the dissolution of ZrO_2 into CeO_2 at any composition beyond solubility limit. According to the phase diagram of CeO_2–ZrO_2 binary system [16], they can make uniform solid solution over 1500°C. Therefore, if they are heated to a very high temperature and then quenched abruptly, you could make an entire solid solution [17]. However, you could not make nanosized particles that have a high specific surface area of several tens of square meters per gram and is adequate for catalytic use. There are two types of preparation methods proposed for the entire solid solution of CeO_2 and ZrO_2. They are classified into break-down method and build-up method. The break-down methods use mechanical milling [18, 19], and the build-up methods use precipitating methods from liquid phase [20, 21]. The former has been understood as a mechanical alloying mechanism. Fig. 31.2 shows the appearance of the X-ray diffraction patterns. Each peak of CeO_2 shifts to the higher angle side by the progress of the milling process. It was reported that mixing ZrO_2 from zirconia milling media was dissolved into CeO_2 lattice. The build-up process from a liquid phase has higher manufacturing efficiency. To increase dissolving concentration of ZrO_2 in CeO_2 from 20 to 50 mol% makes its OSC three times larger (Fig. 31.3) [21, 22]. This increase of OSC, which is caused by dissolution

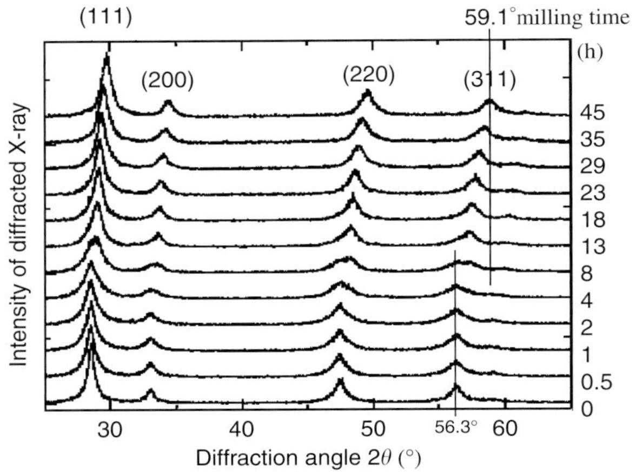

Figure 31.2
X-ray diffraction angle of milled CeO_2 powders. (Each peak shifted to the higher angle with progressing milling.)

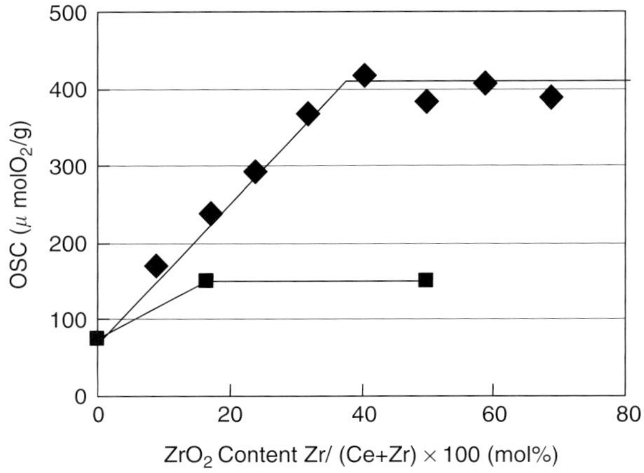

Figure 31.3
The relation between ZrO_2 content and OSC of CeO_2–ZrO_2 solid solution. (◆): By mechanical alloying method by using high-energy milling. (■): By solid-phase reaction of CeO_2 and impregnated ZrO_2.

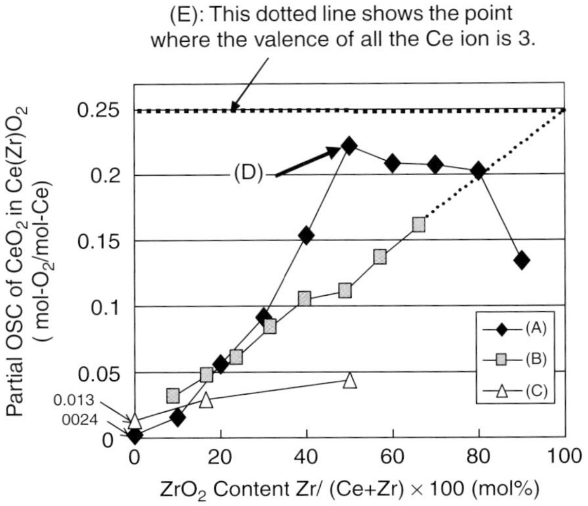

Figure 31.4
The relation between ZrO_2 content and partial OSC of CeO_2–ZrO_2 solid solution: (A) high-temperature reduction at 1200°C in CO (B) mechanical alloying of CeO_2–ZrO_2 (C) solid-phase reaction of CeO_2–ZrO_2 (CeO_2 and impregnated ZrO_2 heated at 700 °C for 5 h in air) (D) κ phase $Ce_2Zr_2O_8$, dotted line (E) dotted line corresponds to theoretical value.

of ZrO_2, came from increasing the amount of oxygen not only on the surface but also inside the CeO_2 lattice [14]. This technology was put into practical use in 1997.

Adding some explanation about CeO_2–ZrO_2 binary phase, the existence of pyrochlore phase (ϕ phase $Ce_2Zr_2O_7$) has been known in Ce_2O_3–ZrO_2 in reductive atmosphere for many years [23]. Yao et al. found metastable $Ce_2Zr_2O_8$ phase in CeO_2–ZrO_2 system [24]. The metastable $Ce_2Zr_2O_8$ phase was named ϕ' phase because the phase was made by the oxidation of ϕ phase ($Ce_2Zr_2O_7$) in early years, and the phase was renamed as κ phase $Ce_2Zr_2O_8$ [25]. The κ phase ($Ce_2Zr_2O_8$) has extremely large OSC. Its maximum OSC reaches almost theoretical limit at the composition of 50 mol% of ZrO_2. In the composition, almost all cerium ions included in the material can become trivalent. Fig. 31.4 shows a comparison of OSC data of (A) κ phase $Ce_2Zr_2O_8$, (B) mechanically alloyed CeO_2–ZrO_2 and (C) solid-phase reaction of CeO_2–ZrO_2 (zirconium salt was impregnated onto CeO_2 powder and then neutralized and heated at 700°C for 5 h in air), in which OSC are shown by using the amount of O_2 per mol of Ce (partial OSC). The pyrochlore phase has an ordered arrangement of cation, and the ordered arrangement is kept in the κ phase $Ce_2Zr_2O_8$ (Fig. 31.5). The crystal structure of the κ phase $Ce_2Zr_2O_8$ was analyzed by Kisimoto et al. [26]. Many analytical researches of the structure are being carried out aggressively by many researchers. At present, intermediate oxidation phase, $Ce_2Zr_2O_{7.5}$, has been found by Sasaki et al. The new phase, $Ce_2Zr_2O_{7.5}$, was named as β phase [27]. These analytical researches are valuable for understanding the mechanism of high OSC of the solid solutions and compounds in CeO_2–ZrO_2 binary system, and also would provide a guideline for the development of next-generation OSC material.

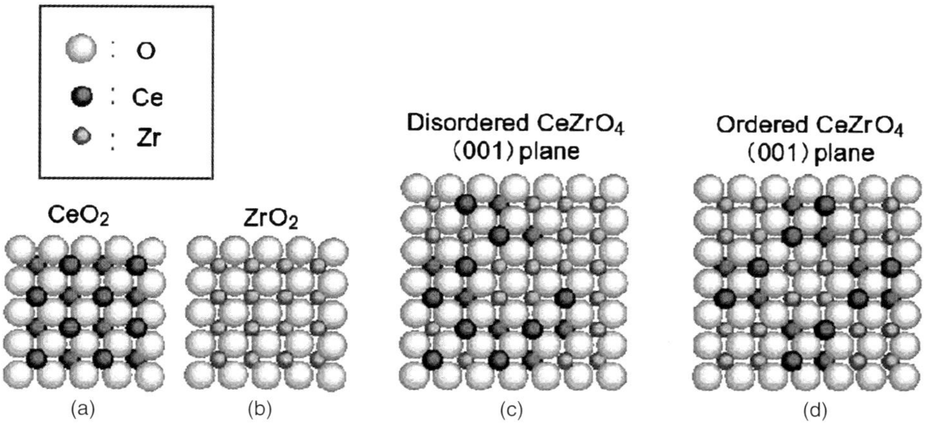

Figure 31.5
Crystal model of (0 0 1) plane of ordered or disordered cation structure. (a) Cerium and oxygen ions in CeO_2 (b) Zirconium and oxygen ions in cubic ZrO_2 (c) CeO_2–ZrO_2 solid solution (disordered cations) (d) $Ce_2Zr_2O_8$ double oxide (ordered cations).

4. Improvement of thermal resistance of catalyst

As mentioned above, the second technological development is the improvement of heat resistance of the OSC material. This technology consists of making a nanolevel mixture of CeO_2–ZrO_2 solid solution particles and alumina particles. CeO_2–ZrO_2 solid solution particles are dispersed among alumina particles, which have better heat resistance than those of CeO_2–ZrO_2 solid solution and are kept in smaller size at a higher temperature than the material composed only of CeO_2–ZrO_2 solid solution. The concept of this technique was named as "diffusion barrier concept", because alumina particles play a role of preventing diffusion of Ce and Zr ions [28]. Fig. 31.6 shows a schematic diagram of diffusion barrier concept in Al_2O_3 and CeO_2–ZrO_2 solid solution system. Moreover, it is confirmed that alumina and the CeO_2–ZrO_2 solid solution hardly react mutually during the durability test of Al_2O_3 and CeO_2–ZrO_2 solid solution nanolevel dispersion system in the real engine exhaust.

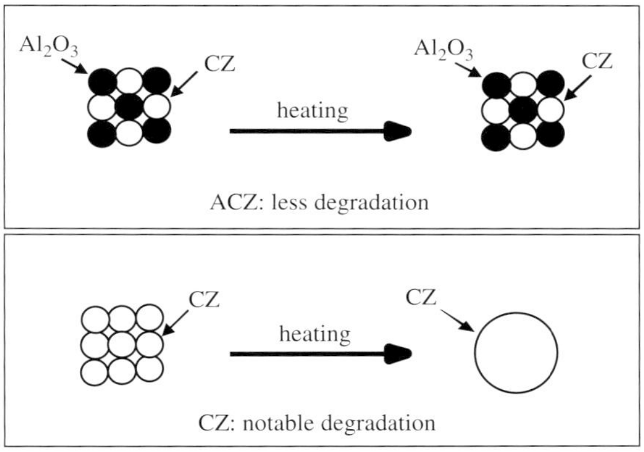

Figure 31.6
Diffusion barrier concept.

Progress about partial OSC in each of these four kinds of OSC materials, {CeO_2 (pure-CeO_2)}, {CeO_2 20 mol%; ZrO_2 (CZ8020)}, {CeO_2 50 mol%; ZrO_2 (CZ5050)} and {Al_2O_3–(CeO_2–ZrO_2) (ACZ)} is shown in Fig. 31.7. These values were measured after durability test at 1000°C for 5 h with 1 wt.% of platinum loading. Advancements achieved in each generation are demonstrated clearly. Crystallite sizes of platinum and CeO_2–ZrO_2 solid solution were calculated by Scherrer's equation from full width at half maximum in X-ray diffraction patterns. Platinum sizes on CZ5050 and ACZ were 23.7 and 19.6 nm, and those of CeO_2–ZrO_2 solid solution were 17.2 and 8.7 nm, respectively, in the full-size catalyst after durability test at 950°C in the real engine exhaust. These data obviously show that diffusion barrier of alumina works effectively to prevent sintering of CeO_2–ZrO_2 solid solution in ACZ. Furthermore, platinum also prevented sintering effectively in ACZ. Platinum is known to sinter more easily in oxidation atmosphere [29], so catalyst using ACZ seems to keep stoichiometric condition better. The description above proved the effectiveness of the diffusion barrier concept. Fig. 31.8 shows an SEM micrograph of back-scattered electron image of ACZ after 1000°C for 10-h durability test [30]. About 10 nm size of dispersed CeO_2–ZrO_2 solid solution can be observed (white part) in alumina (dark part).

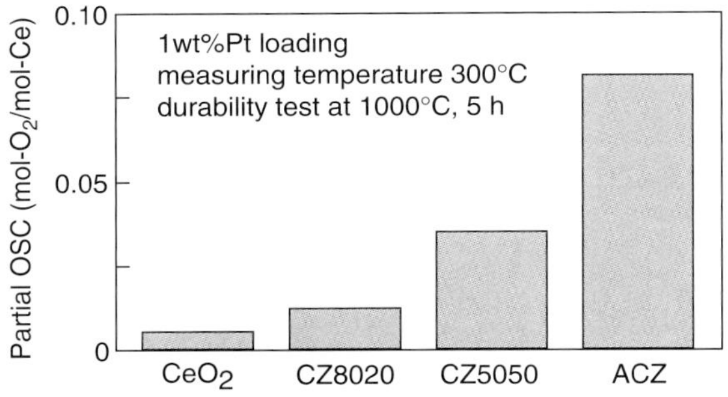

Figure 31.7
Progress of partial OSC in each generation of CeO_2–ZrO_2 solid solution partial OSC: Molar amount of O_2 per 1 mol of Ce.

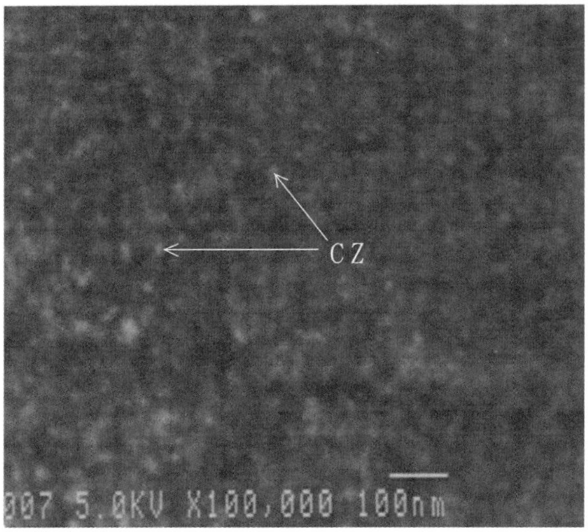

Figure 31.8
Back-scattered electron image of ACZ, using scanning electron microscope, after 10-h durability test in air at 1,000°C.

Applications of concerning nanoparticle technology in some developments of exhaust catalysts have been described above.

References

[1] Japanese Patent No. 865370 (1977).
[2] Japanese Patent No. 935113 (1978).
[3] Japanese Patent No. 958489 (1979).
[4] Japanese Patent No. 1033368 (1981).
[5] H.S. Gandhi, A.G. Piken, M. Shelef and R.G. Delesh: SAE Paper 760201 (1976).
[6] M. Ohashi: *Shokubai*, **29**, 598–604 (1987).
[7] M. Funabiki, T. Yamada: SAE Paper 881684 (1988).
[8] M. Funabiki, T. Yamada: *Shokubai*, **31** (1989) 566.
[9] Japanese Patent No. 1330556 (1986).
[10] N. Miyoshi, S. Matsumoto, M. Ozawa and M. Kimura: SAE Paper 891970 (1989).
[11] S. Matsumoto, N. Miyoshi, T. Kanazawa, M. Kimura and M. Ozawa: *Catalytic Science and Technology*, Vol. 1, (eds. S. Yoshida, T.N. Tabezawa and T. Ono), (1991), 335, Kodansha-VCH, Tokyo, Weinheim, NY.
[12] H.C. Yao and Y.F. Yu Yao: *J. Catal.*, **86** (1984) 254.
[13] A. Suda, Y. Ukyou, H. Sobukawa and M. Sugiura: *J. Ceram. Soc. Jpn.*, **110** (2002) 126–130.
[14] A. Suda, K. Yamamura, Y. Ukyo, T. Sasaki, H. Sobukawa, T. Tanabe, Y. Nagai, M. Sugiura: *J. Ceram. Soc. Jpn*, **112**, 581–585 (2004).
[15] M. Ozawa, M. Kimura and A. Isogai: *J. Alloys Compd.*, **193** (1993) 73.
[16] M. Yashima, M. Yoshimura: *Materia. Jpn.*, **34**, 448–454 (1995).
[17] P. Fornasiero, R. Di Monte, Rao G. Ranga, J. Kaspar, S. Meriani, A. Trovarelli and M. Graziani: *J. Catal.*, **151**, 168–177 (1995).
[18] A. Trovarelli, F. Zamar, J. Llorca, C. Leitenburg, G. Dolcetti and J.T. Kiss: *J. Catal.*, **169**, 490–502 (1997).
[19] A. Suda, T. Kandori, N. Terao, Y. Ukyo, H. Sobukawa and M. Sugiura: *J. Mater. Sci. Lett.*, **17**, 89–90 (1998).
[20] Japanese Patent No. 3238316 (1995).
[21] A. Suda, H. Sobukawa, T. Suzuki, T. Kandori, Y. Ukyo and M. Sugiura: *J. Ceram. Soc. Jpn.*, **109**, 177–180 (2001).
[22] M. Sugiura, M. Ozawa, A. Suda, T. Suzuki and T. Kanazawa: *Bull. Chem. Soc. Jpn.*, **78** (2005) 752–767.
[23] J.J. Casey, L. Katz and W.C. Orr: JCPDS-CARD 08-0221, *J. Am. Chem. Soc.*, **77**, 2187 (1955).
[24] S. Yao, H. Morikawa, N. Izu and K. Okuda: *J. Jpn. Inst. Metals*, **59–12**, 1237 (1995).
[25] S. Otsuka-Yao-Matsuo, T. Omata, N. Izu and H. Kishimoto: *J. Solid State Chem.*, **138**, 47–54 (1998).
[26] H. Kishimoto, T. Omata, S. Otsuka-Yao-Matsuo, K. Ueda, H. Hosono and H. Kawazoe: *J Alloys Compd.*, **312**, 94–103 (2000).
[27] T. Sasaki, Y. Ukyo, A. Suda, M. Sugiura, K. Kuroda, S. Arai and H. Saka: *J. Ceram. Soc. Jpn.*, **111**, 382–385 (2003).
[28] T. Kanazawa, J. Suzuki, T. Takada, T. Suzuki, A. Morikawa, A. Suda, H. Sobukawa and M. Sugiura, SAE Paper, 01-0811 (2003).
[29] P. Wynblatt, N.A. Gjostein: *Prog. Solid. State. Chem.*, **9**, 21 (1975).
[30] T. Kanazawa, J. Suzuki, T. Takada, T. Suzuki, A. Morikawa, A. Suda, H. Sobukawa and M. Sugiura: *Stud. Surf. Sci. Catal.*, **145** (2002) 415.

APPLICATION 32

32 DEVELOPMENT OF OPTICAL MEMORY USING SEMICONDUCTOR NANOPARTICLES

While nanotechnology has been drawing a lot of attention in recent years, the "quantum size effect" is one of the properties unique to "nanosizes" that do not appear in either the bulk form or in the atomic and molecular levels. In this phenomenon when the size of a material becomes smaller than the effective Bohr radius (about 10 nm or less), the band levels become discrete and the band gap becomes larger. In other words, since it is possible to control the band gap of the material by controlling the size, it is possible to expect applications as a material with characteristics not present conventionally.

Although all materials such as metals, magnetic materials, and semiconductors are researched as types of nanoparticles, here, we consider CdSe, which is a II–VI compound semiconductor. Large volume synthesis of CdSe nanoparticles became easy after the 1990s when a method of synthesis called the "hot soap method" was developed [1, 2] in which organic metal compounds are injected into an organic phosphorus compound at a high temperature.

The features of these CdSe nanoparticles are, as is shown in Fig. 32.1a, that the light emission wavelength can be controlled easily in the visible range, and that a high quantum efficiency of 85% can be realized [2]. In order to establish this method of synthesis, research on CdSe nanoparticles was carried out widely, and a large number of characteristics peculiar to nanoparticles were reported [3–5].

1. Fluorescence characteristics of semiconductor nanoparticles

These nanoparticles are modified on the particle surface with an organic material in order to prevent aggregation (Fig. 32.1b). Further, it is also possible to synthesize core-shell type of nanoparticles (Fig. 32.1c) in which a different type of semiconductor

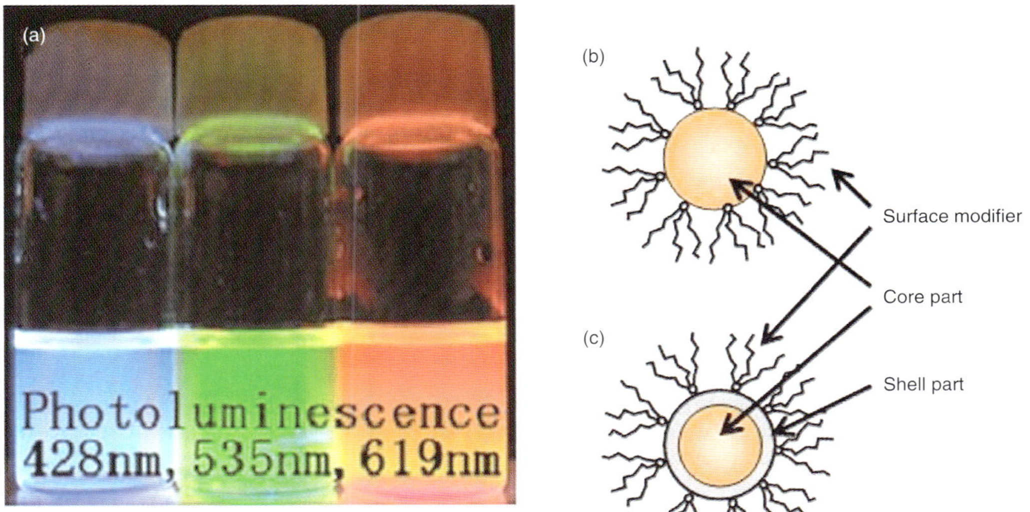

Particle diameter Small Medium Large

Figure 32.1
(a) Fluorescence of solutions of CdSe nanoparticles with different particle diameters. (b) Core-type nanoparticle. (c) Core-shell-type nanoparticle.

(such as ZnS, ZnSe, CdS, etc.) is coated around the core part of the nanoparticles.

Some other features of nanoparticles are that it is possible to generate a wide range of optical characteristics by changing the material of the shell part, and that it is possible to control over a wide range the hydrophobicity, the hydrophilicity, and pH durability, etc., by changing the surface-active agent. It is also possible to modify the functional moiety that reacts to a specific molecule or DNA, and because of the characteristic that color quenching is less likely to occur compared to the fluorescent dyes being used at present, these nanoparticles are expected to be applied in the medical field as biological labeling agents [6–8]. Further, since the specific surface area of nanoparticles having diameters of 10 nm or less is large, the optical characteristics are extremely sensitive to the surrounding environment. Application to sensors using this characteristic has also been studied [9, 10]. However, how the boundary surface of nano sizes affects the optical characteristics is not yet fully explained.

2. Optical memory effect of semiconductor nanoparticle thin films

The present authors have reported [11–13] a phenomenon that, when CdSe nanoparticles are coated on a substrate thereby preparing a thin film, and excitation light is irradiated on the thin film continuously, the intensity of fluorescence light from the thin film increases up to a specific value.

If the thin film whose fluorescence intensity has increased is stored in a dark place, and after a specific period of time, if it is irradiated again with excitation light, the fluorescence intensity will be almost the same value as the fluorescence intensity before storing in the dark place. In other words, this phenomenon can be said to be that "the excitation light irradiation time interval is being recorded as the fluorescence intensity".

The possibilities of multiple-value recording using the continuous increase in the fluorescence intensity or of recording in smaller areas using near-field light have already been reported [12], and it is expected to apply these phenomena to optical memories and to realize higher recording capacity than conventional methods. In addition, since these samples are prepared by a coating process, there is also another big advantage that it is possible to prepare large-area films on flexible substrates.

In this paper, we report that the recording and read-out operations of information in the form of fluorescence intensity were actually carried out, that these operations could be made with a single-wavelength light, and that reversible changes of the recorded value are possible [11].

3. Methods of preparing and evaluating CdSe thin films

CdSe/ZnS core/shell-type nanoparticles with a particle diameter of about 4 nm were coated on a glass substrate using the spin coating method. The nanoparticle thin films were observed under a nitrogen gas atmosphere using a confocal laser scanning microscope. First, a 60 μm^2 area was observed with an irradiation intensity of 0.6 nW (read-out operation). At this time, the observed fluorescence intensity is uniform (Fig. 32.2a). Next, an excitation light of 11, 220, or 1,370 nW was emitted onto a 7.4 μm^2 area at the center (writing operation). After that, the observation area is returned to 60 μm^2, and this area was observed again with an irradiation intensity of 0.6 nW (Fig. 32.2b). At this time, it was observed that the fluorescence intensity of the central part in which the writing operation was made had increased. The read-out operation at 0.6 nW and the writing operation using a strong excitation light were repeated, and the changes in the value of the fluorescence intensity at the center during the read-out operation were observed. In other words, the fluorescence intensity at the central part in which the writing operation was carried out was taken as I, and the fluorescence intensity at the surrounding part in which no writing operation was carried out was taken as I_0, and the ratio I/I_0 was obtained as an index of the fluorescence intensity. The wavelength of the excitation light used in all the operations was 488 nm.

4. Dependency of intensity of fluorescence on the excitation light intensity

The dependency on the excitation light intensity of the behavior of I/I_0 is shown in Fig. 32.3. The horizontal axis represents the total energy radiated during the writing operation, and the vertical axis represents the value of I/I_0 during the read-out operation. When an excitation light of an intensity of 11 nW is used during the writing operation, the fluorescence intensity I increases successively as the

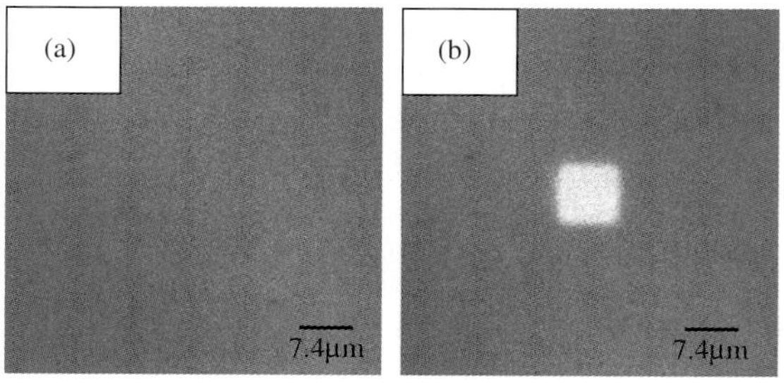

Figure 32.2
(a) Fluorescence image before writing operation.
(b) Fluorescence image after writing operation.

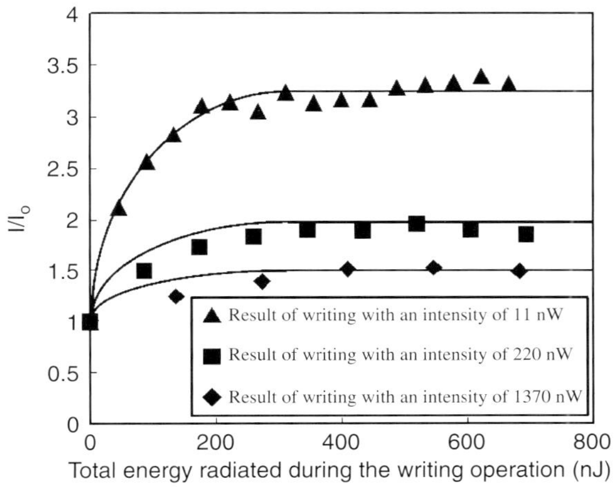

Figure 32.3
Multiple-value recording of fluorescence intensity due to writing operation using different excitation light intensities.

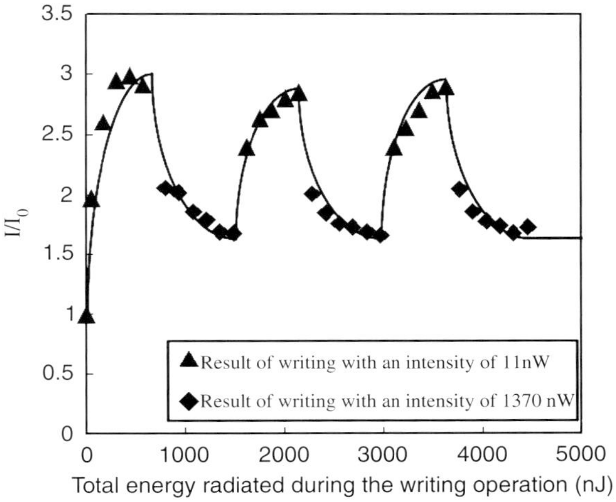

Figure 32.4
Rewriting of recorded value by alternatingly irradiating with an excitation light having an intensity of 11 nW and an excitation light having an intensity of 1,370 nW.

writing operation is repeated and gets saturated at an I/I_0 value of about 3.0.

On the other hand, when an excitation light having an intensity of 1,370 nW is used, although the fluorescence intensity increases in a similar manner, it reaches saturated value I/I_0 of about 1.5. This attained an intermediate value of I/I_0 of about 1.8 when excited with light having an intermediate intensity of 220 nW. From the above results, it is shown that multiple-value recording of fluorescence intensity is possible by changing the excitation light intensity.

Next, Fig. 32.4 shows the result of writing again using an excitation light having an intensity of 1,370 nW in the same area after recording a fluorescent intensity of I/I_0 of about 3.0 in an area of about 7.4 μm^2 using an excitation light having an intensity of 11 nW. The fluorescence intensity that increased up to an I/I_0 value of about 3.0 due to the writing operation using an excitation light having an intensity of 11 nW, decreased to an I/I_0 value of about 1.5 due to the writing operation using an excitation light having an intensity of 1,370 nW. Further, thereafter, when a writing operation was carried out in the same area with an excitation light intensity of 11 nW, the fluorescence intensity increased again up to an I/I_0 value of about 3.0. This fluorescence intensity of an I/I_0 value of about 3.0 due to an excitation light having an intensity of 11 nW and the fluorescence intensity of an I/I_0 value of about 1.5 due to an excitation light having an intensity of 1,370 nW can be recorded repeatedly (Fig. 32.4).

From the above result, it is shown that the fluorescence intensity that is recorded once can be changed to a newly recorded value by irradiating with an excitation light having a different intensity, that is, that it is possible to rewrite the recorded value.

5. Future topics

The two features of multiple-value recording and rewriting were described above in an optical memory device using semiconductor nanoparticles. The problem that has to be solved in the future is that of storing the recorded value for a long time. In the present stage, it has only been reported that the value can be stored for a period of only about a few months, and it is necessary to realize a storage period on the order of several tens of years in order to realize practical products. At the same time, it has also been reported that the optical characteristics of semiconductor nanoparticles are strongly dependent on the surrounding environment.

In view of this, as a method of simultaneously achieving lengthening of the recording storage period and shutting out the effect of the surrounding environment, it is necessary to study measures such as providing a protective layer on top of the coated thin film of nanoparticles.

Further, in this device, improvement in the S/N ratio of the fluorescence intensity is directly related to achieving multiple levels in multiple-value recording. Since there are factors that are causing deterioration in the S/N ratio, such as the processing problem that the nanoparticles are not coated uniformly, and the problems in the apparatus, such as vibrations during the writing and read-out operations, the recording capacity is expected to be increased greatly by solving these problems.

Finally, the mechanism of this phenomenon that the fluorescence intensity increases with irradiation with excitation light has not yet been fully clarified. By explaining this mechanism, it is expected not only that it will be possible to control more accurately the fluorescence intensity, but also that it can be extended to nanoparticles of other materials.

References

[1] C.B. Murray, D.J. Norris and M.G. Bawendi: *J. Am. Chem. Soc.*, **115**, 8706 (1993).
[2] L. Qu, X.G. Peng: *J. Am. Chem. Soc.*, **124**, 2049 (2002).
[3] S.A. Empedocles, M.G. Bawendi: *Science*, **278**, 2114 (1997).
[4] W.C.W. Chan, S.M. Nie: *Science*, **281**, 2016 (1998).
[5] B.R. Fisher, H.J. Eisler, N.E. Scott and M.G. Bawendi: *J. Chem. Phys.*, **108**, 143 (2004).
[6] W.C.W. Chan, S.M. Nie: *Science*, **281**, 2016 (1998).
[7] X. Michalet, F. Pinaud, T.D. Lacoste, M. Dahan, M.P. Bruchez, A.P. Alivisatos and S. Weiss: *Single Mol.*, **2**, 261 (2001).
[8] A.P. Alivisatos: *Nat. Biotechnol.*, **22**, 47 (2003).
[9] D.Y. Godovsky: *Adv. Polym. Sci.*, **153**, 163 (2000).
[10] A.Y. Nazzal, L. Qu, X. Peng and M. Xiao: *Nano Lett.*, **3**, 819 (2003).
[11] N. Eiha, S. Maenosono, K. Hanaki, K. Yamamoto and Y. Yamaguchi: *Jpn. J. Appl. Phys.*, **42**, L310 (2003).
[12] S. Maenosono, E. Ozaki, K. Yoshie and Y. Yamaguchi: *J. Appl. Phys.*, **40**, L638 (2001).
[13] J. Kimura, S. Maenosono and Y. Yamaguchi: *Nanotechnology*, **14**, 69 (2003).
[14] T. Uematsu, J. Kimura and Y. Yamaguchi: *Nanotechnology*, **15**, 822 (2004).

APPLICATION 33

33 DEVELOPMENT OF BRIGHT PHOSPHORS USING GLASSES INCORPORATING SEMICONDUCTOR NANOPARTICLES

Recently, a demand for bright phosphors is increasing because displays and lightings with high luminance, ultrafine resolution and low-energy consumption are required. In this field ceramics phosphors doped with rare earth ions and transition metal ions have been mainly researched and developed so far, because those ceramics phosphors are highly stable. However, short-wavelength light is required for excitation, and the control of the emission wavelength is not easy. In addition, the photoluminescence (PL) of such ceramics phosphors has long decay time (ca. 1 ms) because the transition has a forbidden character. This causes the saturation of PL brightness when the excitation light intensity is increased.

Under such a situation, much attention is directed to nanoparticles of group II–VI semiconductors such as CdTe and ZnSe because they emit bright PL whose wavelength depends on the particle size. These semiconductor nanoparticles have a direct transition bandgap and short PL decay time of ca. 10 ns that is shorter than those of rare earth ions and transition metal ions by ca. 5 orders of magnitude. The fast PL decay leads to high brightness that does not saturate easily when the excitation light intensity is increased. Another feature of these semiconductor nanoparticles is that the emission wavelength and emission efficiency are almost independent of excitation wavelengths. The authors have carried out research and development of glass phosphors showing various fluorescent colors with improved long-term stability by incorporating semiconductor nanoparticles. This study was supported in part by the Nanotechnology Glass Project (2001–2006) related to the Nanomaterial Process Technology of the Nanotechnology Program, sponsored by the New Energy and Industrial Technology Development Organization (NEDO), Japan [1, 2].

Semiconductor nanoparticles also find applications to fluorescent markers for biological tissues and materials from a living body [3]. The position and the movement of virus and various materials bound to the semiconductor nanoparticles can be monitored by luminescence. So far organic fluorescent dyes have been widely used in this field. However, the wavelength difference between the excitation light and emission light is not very large, and therefore different excitation wavelengths are needed for obtaining PL from different dyes. Another disadvantage of organic fluorescent dyes is their low stability that causes deterioration and luminescence quenching in a short time of irradiation. On the other hand, semiconductor nanoparticles have much slower deterioration rate compared with organic dyes, and the long-term stability can be further improved by incorporating them into glass matrices. Furthermore, two or more kinds of semiconductor nanoparticles with different emission wavelengths can be used at the same time, using a single-excitation light wavelength.

In the following are introduced the syntheses of highly photoluminescent CdTe and ZnSe nanoparticles and the preparation of glass phosphors incorporating those nanoparticles with three different forms.

1. Syntheses of highly photoluminescent semiconductor nanoparticles by an aqueous solution method

There is a general rule that the bandgap (Eg) of group II–VI semiconductors becomes narrower as the constituent atoms become heavier. In the case of nanoparticles with diameters of ca. 2–10 nm, the Eg is increased due to the quantum size effect compared with the bulk semiconductor, and it leads to various fluorescent colors reflecting small differences in the particle size.

When the semiconductor nanoparticles are coated with sulfur- and phosphor-containing surfactants, the PL efficiency improves. The uncoated nanoparticles have many surface defects, which cause nonradiative deactivation after excitation. Also, such nanoparticles easily agglomerate, and the agglomeration causes imperfect chemical bonds between nanoparticles. The coating with surfactants decreases the number of surface defects on the nanoparticles and suppresses the agglomeration of the nanoparticles. Highly photoluminescent nanoparticles can be synthesized in organic solutions [4] and aqueous solutions [5]. However, the organic solution route uses reactions of explosive materials at high temperatures that need complex, expensive apparatus and devices, and the synthesized nanoparticles tend to lose their PL quickly when contacting with water.

On the other hand, aqueous solution route enables the syntheses of highly photoluminescent nanoparticles under mild conditions, and the nanoparticles obtained are stable in aqueous solution at least for about several days at room temperature. First, clusters of a group II–VI semiconductor are generated by reacting the ion of a group VI element with an alkaline solution of a group II element in the presence of a surfactant under inactive atmosphere at room temperature. Next, nanoparticles are grown by refluxing the cluster solution. When reflux time is prolonged, the particle size increases and PL wavelength shifts to longer wavelengths. The CdTe nanoparticles can be synthesized by using cadmium perchlorate, hydrogen telluride gas, and a surfactant such as thioglycolic acid (TGA). The diluted aqueous solutions of the CdTe nanoparticles having diameters of 3–4 nm are almost colorless and transparent under visible light. They emit PL of green to red colors depending on the particle size under ultraviolet light (Fig. 33.1). The authors optimized the conditions of synthesis of CdTe nanoparticles by reducing the amount of TGA, etc., and it increased

Figure 33.1
Aqueous solutions of CdTe nanoparticles having various diameters between 3 and 4 nm. (Above) Colorless and transparent appearance under visible light. (Below) Photoluminescence under ultraviolet light (wavelength: 365 nm).

the PL efficiencies more than 65% [6,7]. The ZnSe nanoparticles can be synthesized by using zinc perchlorate, hydrogen selenide gas, and a surfactant such as TGA and thioglycerol (TG). A diluted aqueous solution of the ZnSe nanoparticles with a diameter of ca. 3 nm is almost colorless and transparent under visible light, and emits blue PL under ultraviolet light [8,9]. When the ZnSe nanoparticle core is coated with ZnS shell, the PL efficiency improves up to ca. 40% [10,11].

It is known that the PL intensity of an aqueous solution of semiconductor nanoparticles changes with pH [12]. Zeta potential measurement of the TGA-coated CdTe nanoparticles has shown that in the acidic region, TGA loses the charge reflecting its acid dissociation constant (pK_a). Therefore the agglomeration of noncharged nanoparticles is promoted, which gives rise to the decrease of PL intensity at low pH [13].

2. Preparation of glass phosphors incorporating semiconductor nanoparticles by a sol–gel method

Previously, Mulvaney et al. tried to incorporate CdSe nanoparticles prepared by an organic solution method in glass matrix by a sol–gel method [14]. However, their method does not seem to be appropriate to obtain solidified glass phosphors with high concentration of nanoparticles because the hydrophobic CdSe nanoparticles used are not very compatible with the sol–gel process. By contrast, the authors have successfully prepared well-solidified glass phosphors having three forms, i.e., (a) bulk, (b) small bead, and (c) thin film, by a sol–gel method using water-dispersible semiconductor nanoparticles.

(1) Preparation of bulk glass phosphors

Sol–gel derived bulk glass phosphors incorporating semiconductor nanoparticles can be prepared by hydrolysis of alkoxysilane or alkoxide in solutions mixed with aqueous solutions of semiconductor nanoparticles [6,15,16]. When using alkoxysilane such as 3-aminopropyltrimethoxysilane (APS) having an amino group that adsorbs TGA on the surface of nanoparticles, the agglomeration of the nanoparticles was suppressed and led to bright glass phosphors. It was possible to incorporate CdTe nanoparticles up to the concentration of ca. 2×10^{-4} mol/l in the bulk glass. The obtained bulk glass phosphors showed PL spectra similar to those of the aqueous solutions of nanoparticles, and exhibited PL efficiencies higher than 40%. These glass phosphors retained PL efficiencies of higher than 30% even after 6 months of preparation, demonstrating superior long-term stability [6,16]. In a similar manner, blue-emitting bulk glass phosphors incorporating ZnSe nanoparticles can be prepared. It is also possible to fabricate bulk glass phosphors having various shapes and PL colors under atmospheric pressure and at room temperature, by applying the sol solutions to a patterned substrate (Fig. 33.2).

(2) Preparation of small glass bead phosphors

Small glass bead phosphors incorporating semiconductor nanoparticles can be prepared by sol–gel reaction in the small space inside inverse micelles [17,18]. The agglomeration of nanoparticles is suppressed by involving water-dispersible semiconductor nanoparticles in the individual inverse micelle. On the other hand, a few methods have been reported to prepare small glass bead phosphors without using inverse micelles. Bawendi et al. prepared photoluminescent small glass beads by forming thin silica layer containing semiconductor nanoparticles by a sol–gel method onto the surface of a silica bead without containing nanoparticles [19]. Alivisatos et al. reported a complex chemical synthetic method to deposit mono- and multilayers of silica onto the surface of a single semiconductor nanoparticle [20]. Compared with those methods, inverse micelle-based methods have advantages such as good controllability of glass bead size and nanoparticle concentration in a glass bead. The following two preparation methods have been developed by the authors using inverse micelles.

In the first preparation method, an inverse micelle of an ionic surfactant Aerosol OT (AOT) (sodium bis(2-ethylhexyl) sulfosuccinate) is formed in a hydrophobic organic solvent isooctane. Then aqueous solution of CdTe nanoparticles is mixed to generate inverse micelles that involve small droplets of the nanoparticle solution. Afterwards, an alkoxide tetraethoxysilane (TEOS) is added, and the solution is stirred for 1–3 days in order to complete the hydrolysis of TEOS. As a result, small glass bead phosphors are obtained [17]. The small glass beads emit PL both in the isooctane solution and in the powder form (Fig. 33.3). The PL and absorption spectra of the nanoparticles were almost unchanged before and after the incorporation into small glass beads. However, electron microscopic observation revealed that the nanoparticles with diameters of ca. 3 nm were fixed in the vicinity of the outer surface of the glass beads whose diameters were around 25 nm. It was speculated that the nanoparticles were pushed out of the silica network structure that developed slowly during the sol–gel reaction. The PL efficiencies of thus obtained small glass beads were ca. 10% at most. Such relatively low PL efficiencies were thought to be due to possible deterioration of nanoparticles during the long stirring in preparation.

In order to obtain brighter glass beads, the authors have developed the second inverse-micelle-based preparation method with shorter reaction time. In this method, an aqueous solution containing

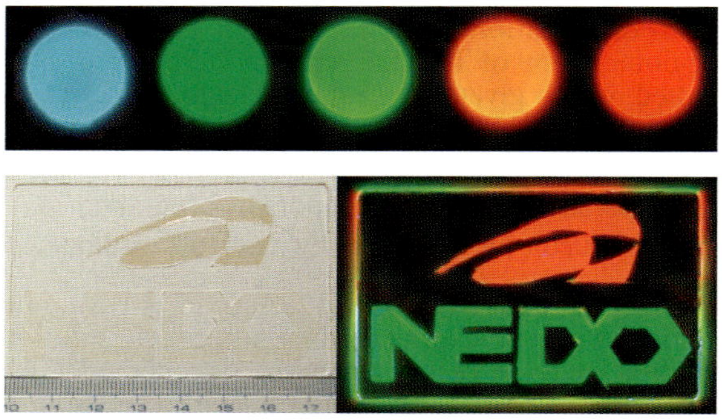

Figure 33.2
(Above) Photoluminescence of bulk glass phosphors incorporating CdTe and ZnSe nanoparticles. (Below) Multi-color photoluminescence of prototype indicator using bulk glass phosphors incorporating CdTe and ZnSe nanoparticles (Left: under visible light. Right: under ultraviolet light (wavelength: 365 nm)).

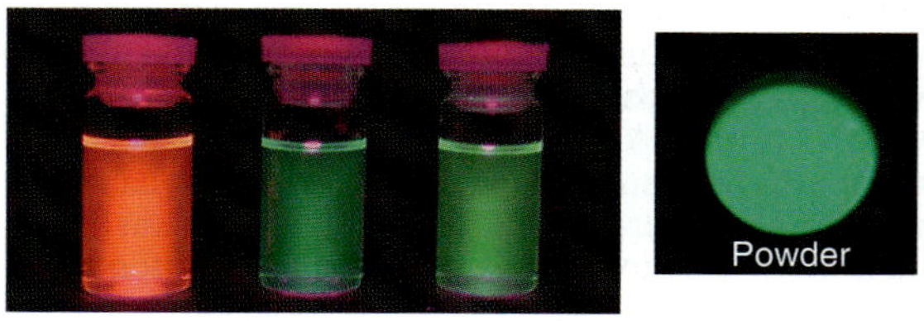

Figure 33.3
Photoluminescence of small glass beads incorporating CdTe nanoparticles. (Left: Isooctane solution of the glass beads. Right: Powder of the glass beads.)

semiconductor nanoparticles and partially hydrolyzed TEOS (solution (1)) and a cyclohexane solution containing a nonionic surfactant Igepal CO-520 (polyoxyethylene(5) nonylphenyl ether) (solution (2)) are prepared. Mixing of solution (1) with solution (2) results in the formation of inverse micelles involving small droplets of solution (1) because cyclohexane is a hydrophobic organic solvent. Afterwards, hydrolysis of TEOS proceeds and completes in inverse micelles in a relatively short time (several hours), leading to the formation of highly photoluminescent small glass beads [18]. The shortened reaction time results in the incorporation of nanoparticles in the small glass beads at high concentration without deterioration. The PL efficiencies of the green- and red-emitting small glass beads incorporating CdTe nanoparticles reached 27% and 65%, respectively. These values were the same as those of the initial aqueous solutions of the nanoparticles.

(3) Preparation of glass film phosphors

Glass film phosphors incorporating highly concentrated semiconductor nanoparticles can be prepared by a layer-by-layer (LbL) self-assembly method. In this method alkoxysilanes such as APS and 3-mercaptopropyltrimethoxysilane (MPS) are used to link the layers of nanoparticles. A glass substrate coated with APS or MPS is dipped in an aqueous solution of nanoparticles and then in a toluene solution of APS or MPS. By repeating these - dip-coating processes, glass film phosphors having alternating layered structure of nanoparticles layers and glass layers are obtained [21]. The glass film phosphors prepared were found to incorporate semiconductor nanoparticles at very high concentrations around 0.01 mol/l. It was possible to prepare glass film phosphors incorporating red- and green-emitting CdTe nanoparticles and blue-emitting ZnSe nanoparticles. The PL efficiency of the red-emitting glass film phosphor was estimated to be ca. 24%.

The authors have successfully synthesized group II–VI semiconductor nanoparticles showing PL of three primary colors (RGB) with high PL efficiencies by the aqueous solution method. Bright glass phosphors having three different forms (bulk, small bead and thin film) incorporating the nanoparticles were prepared by using sol–gel method, inverse micelle method and LbL self-assembly method. The incorporation of semiconductor nanoparticles into glass matrices have improved the long-term stability of nanoparticles, and retained strong PL that does not saturate easily when the excitation light intensity is increased. Research and development of glass phosphors that do not contain toxic Cd have also been explained. These glass phosphors are expected to be applied to various luminescent devices and markers using blue and ultraviolet light-emitting diodes (LEDs) as excitation light sources that have recently come into wide use.

References

[1] M. Ando, C.L. Li and N. Murase: *Monthly Display*, **10**(8), 67–73 (2004).

[2] N. Murase, K. Hirao: *J. Soc. Powder Technol., Jpn.*, **42**(11), 790–798 (2005).

[3] M. Bruchez Jr., M. Moronne, P. Gin, S. Weiss and A.P. Alivisatos: *Science*, **281**, 2013–2016 (1998).

[4] B.O. Dabbousi, J. Rodriguez-Viejo, F.V. Mikulec, J.R. Heine, H. Mattoussi, R. Ober, K.F. Jensen and M.G. Bawendi: *J. Phys. Chem. B*, **101**, 9463–9475 (1997).

[5] A.L. Rogach, L. Katsikas, A. Kornowski, D.Su, A. Eychmüller and H. Weller: *Ber. Bunsen-Ges. Phys. Chem.*, **100**, 1772–1778 (1996).

[6] C.L. Li, N. Murase: *Langmuir*, **20**, 1–4 (2004).

[7] C.L. Li, Murase: *Chem. Lett.*, **34**, 92–93 (2005).

[8] N. Murase, M.Y. Gao, N. Gaponik, T. Yazawa and J. Feldmann: *Int. J. Mod. Phys. B*, **15**, 3881–3884 (2001).

[9] N. Murase, M.Y. Gao: *Mater. Lett.*, **58**, 3898–3902 (2004).

[10] K. Nishikawa, C.L. Li, H. Enomoto and N. Murase: Annual Meeting of the Ceramic Society of Japan, 2005 (Okayama), Proceedings, p. 98 (2E02) (2005).

[11] C.L. Li, K. Nishikawa, M. Ando, H. Enomoto and N. Murase: *Colloid Surf. A*, **294**, 33–39 (2007).

[12] M.Y. Gao, S. Kirstein, H. Möhwald, A.L. Rogach, A. Kornowski, A. Eychmüller and H. Weller: *J. Phys. Chem. B*, **102**, 8360–8363 (1998).

[13] M. Ando, C.L. Li and N. Murase: *Mater. Res. Soc. Symp. Proc.*, **789**, 123–128 (2004).

[14] S.T. Selvan, C. Bullen, M. Ashokkumar and P. Mulvaney: *Adv. Mater.*, **13**, 985–988 (2001).

[15] C.L. Li, M. Ando and N. Murase: *Phys. Status Solidi C*, **0**, 1250–1253 (2003).

[16] C.L. Li, M. Ando and N. Murase: *J. Non-Cryst. Solids*, **342**, 32–38 (2004).

[17] S.T. Selvan, C.L. Li, M. Ando and N. Murase: *Chem. Lett.*, **33**, 434–435 (2004).

[18] P. Yang, M. Ando and N. Murase: 18th Fall Meeting of the Ceramic Society of Japan (Sakai), Extended Abstracts, p. 200 (1G05) (2005).

[19] Y. Chan, J.P. Zimmer, M. Stroh, J.S. Steckel, R.K. Jain and M.G. Bawendi: *Adv. Mater.*, **16**, 2092–2097 (2004).

[20] D. Gerion, F. Pinaud, S.C. Williams, W.J. Parak, D. Zanchet, S. Weiss and A.P. Alivisatos: *J. Phys. Chem. B*, **105**, 8861–8871 (2001).

[21] P. Yang, C.L. Li and N. Murase: *Langmuir*, **21**, 8913–8917 (2005).

APPLICATION 34

34 DEVELOPMENT OF PHOTOCATALYST INSERTED INTO SURFACE OF POROUS ALUMINOSILICATE

Anatase-type TiO_2 with nanosize particles is well known as a photocatalyst [1]. The TiO_2 absorbs ultraviolet light (UV) of 390 nm or less and generates active oxygen such as hydroxyl radicals and the super anion oxide by the oxidation of water and oxygen in air. The active oxygen with a strong oxidizing activity decomposes organic compounds, and demonstrates self-cleaning, deodorizing, and an anti-bacterial effect. However, the photocatalyst has some disadvantages. For example, the active oxygen generated by TiO_2 generally decomposes organic materials such as paper, resin, and textile [2,3]. The TiO_2 cannot be processed into the organic materials because of a strong oxidizing activity. Next, its efficiency is not enough for each practical condition because of a lack of adsorption capacity. On the other hand, it is difficult for a consumer to recognize its performance. Accordingly, to solve these problems, the photocatalyst inserted into a porous aluminosilicate, which has anti-bacterial and deodorizing effects, was developed. Characteristics of this TiO_2–aluminosilicate complex will be introduced in this chapter [4, 5].

1. Structure of TiO_2–aluminosilicate complex

The mean particle size of a porous aluminosilicate (Lionite from LION Co., Japan) was about 5–8 μm. As shown in Fig. 34.1, the SEM (Scanning Electron Microscope) photograph of the surface of aluminosilicate, a highly porous surface of the particle was observed. The pore-size distribution was determined to be from 4 to 100 nm by using N_2 adsorption–desorption isotherms by the BET (Brunauer Emmitt and Teller) method (Fig. 34.2). The surface area was 160 m^2/g. Moreover, this aluminosilicate has the deodorizing effects of some gases such as ammonia, hydrogen sulfite, and acetic acid, because its composition is 8.6-SiO_2–Al_2O_3–4.8-ZnO and has acid and alkaline functional groups on the surface.

Using colloidal TiO_2 (STS-1 from Ishihara Sangyo Co., Ltd. Japan) as a photocatalyst, the TiO_2 complexes containing 1, 5 and 10 wt.% of TiO_2 were prepared from the TiO_2 and the porous aluminosilicate by using a cationic surfactant. As the cross-section TEM (Transmission Electron Microscope) photograph of a particle of the TiO_2 complex is shown in Fig. 34.3, several black points with about 10 nm were observed near the surface of aluminosilicate. The Ti element was detected only on the black points by EDS (Energy Dispersive X ray Spectrometer) analysis in Fig. 34.4. These results suggest that the TiO_2 nanoparticles of the Ti complex were inserted into the pore near the surface of aluminosilicate.

2. Photocatalysis of TiO_2–aluminosilicate complex

A 3 L closed polyvinyl fluoride bag was used in which 20 mg of sample powder and formaldehyde gas were put. The decomposition of formaldehyde by the TiO_2–aluminosilicate complex with 1-wt.%

Figure 34.1
The SEM photograph of the surface of aluminosilicate.

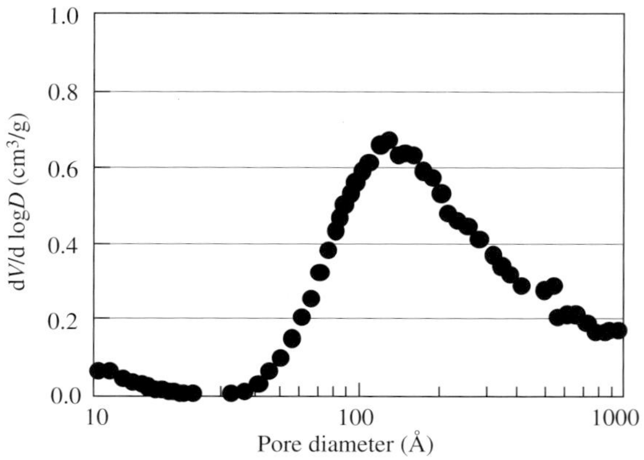

Figure 34.2
The pore size distribution of the aluminosilicate by BET method.

TiO_2 was evaluated with a gas detector under UV light (1 mW/cm^2) or in the dark. The original aluminosilicate decreased formaldehyde concentration to around 14 ppm under both conditions as shown in Fig. 34.5. The difference between without and with UV light was not observed. Accordingly, this decrease seems to be due to absorption. However, as shown in Fig. 34.6 the TiO_2 complex drastically decreased formaldehyde to 7 ppm after only 24 h under UV light. Then, the gas was injected again; the gas concentration also decreased down to 10 ppm after 24 h. The results suggest that the TiO_2 complex showed photocatalysis.

To clarify the photocatalysis of TiO_2 complex, the carbondioxide (CO_2) that was generated by decomposing formaldehyde was determined quantitatively by FT-IR analysis as shown in Fig. 34.7 [6]. The initial formaldehyde of 180 ppm was adsorbed down to almost 45 ppm on the TiO_2 complex without UV light, because CO_2 was not generated. After the irradiation of UV light began, the concentration of CO_2 increased with a decrease in that of formaldehyde.

In order to investigate the difference of the photocatalytic efficiency between the TiO_2–aluminosilicate complex and the powder (TiO_2 mixture) prepared from TiO_2 and aluminosilicate by dry blending, the decomposition of formaldehyde was examined as a function of the TiO_2 quantity. When the initial concentration of formaldehyde was fixed at 20 ppm, the formaldehyde concentrations after 2 h under UV light are shown in Fig. 34.8. The TiO_2 complexes with 1- and 5-wt.% TiO_2 efficiently decomposed formaldehyde compared to the TiO_2 mixtures. In particular, the TiO_2 complex with 5 wt.% TiO_2 decomposed formaldehyde down to 4 ppm, though the TiO_2 mixture hardly decomposed it. It seems that the reason why TiO_2 complex has high efficiency is because of good dispersibility of the TiO_2 nanoparticles on the surface of aluminosilicate. However, when the amount of TiO_2 was 10 wt.%, the formaldehyde concentration of both powders was about 5ppm. The reason can be that the 10 wt.% TiO_2 has enough photocatalysis to decompose formaldehyde under this condition.

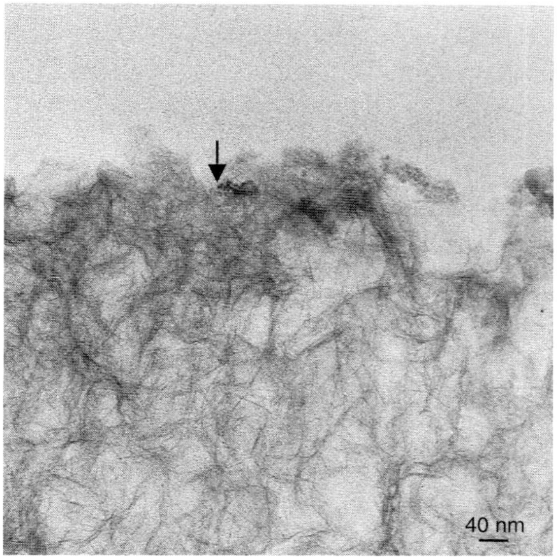

Figure 34.3
The cross-section TEM photograph of a particle of the TiO_2 complex with 10 wt.% TiO_2.

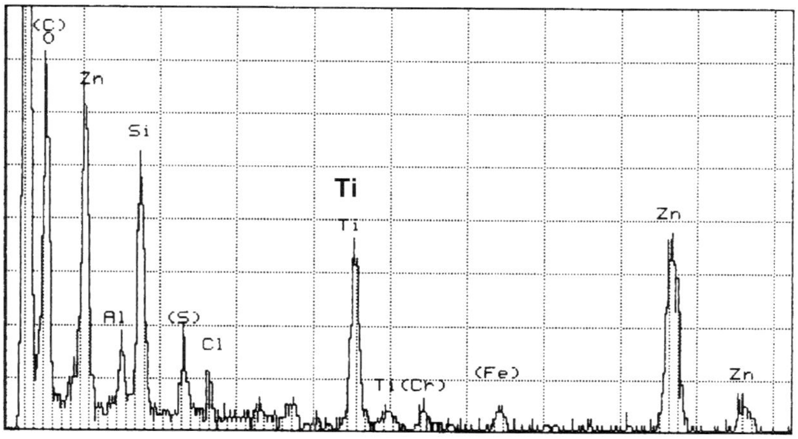

Figure 34.4
The EDS analysis of the black point on the surface of the TiO_2 complex in Fig. 34.3.

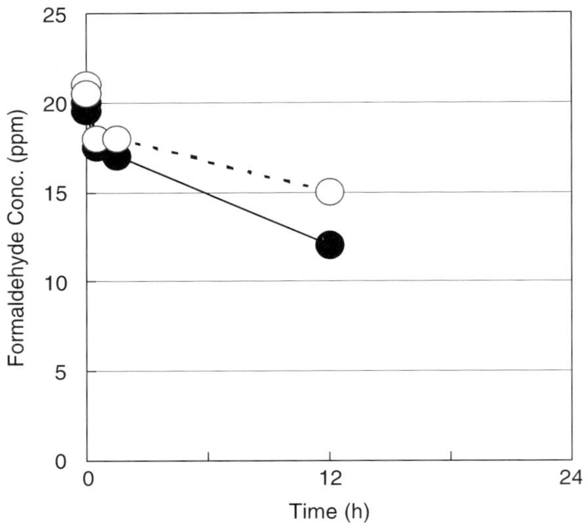

Figure 34.5
The change of the formaldehyde concentration under UV light (○) and without UV light (●) when using the original aluminosilicate.

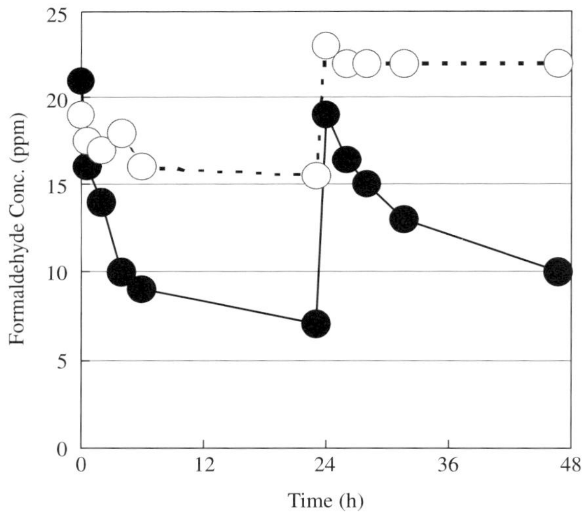

Figure 34.6
The change of the formaldehyde concentration under UV light (○) and without UV light (●) when using the TiO_2 complex.

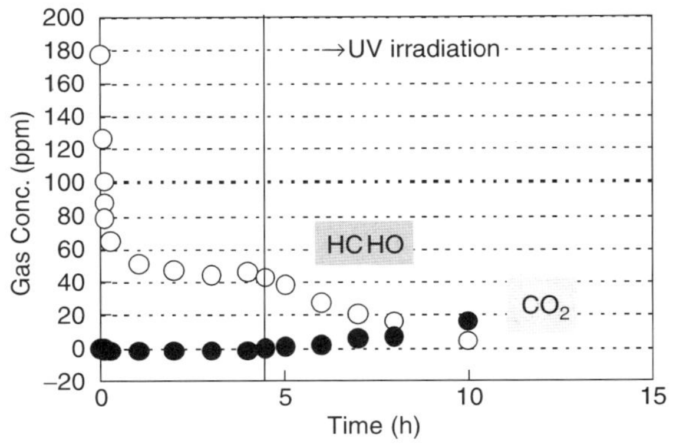

Figure 34.7
The FT-IR analysis of the concentration of formaldehyde (HCHO) and carbondioxide (CO_2).

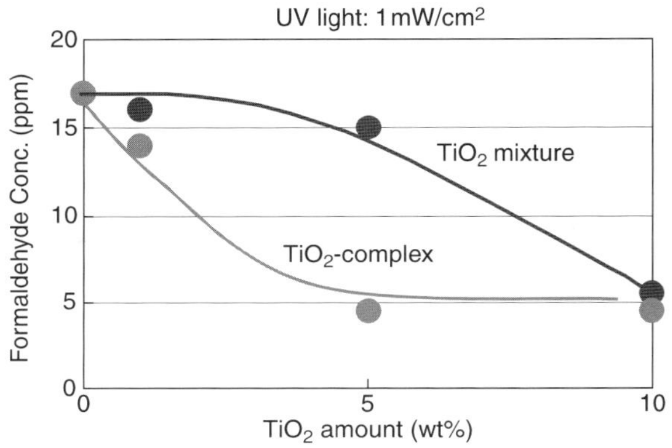

Figure 34.8
The comparison between the TiO_2 complex and the TiO_2 mixture on decomposition of formaldehyde as a function of the TiO_2 amount.

Next, the self-cleaning effect of paper containing TiO_2 complex was examined using the paper that was stained with tar of a cigarette. As shown in the photographs of Fig. 34.9, the paper with the TiO_2 complex was bleached by photocatalysis after 6 h under UV light. However, the paper containing the original aluminosilicate was not bleached. Accordingly, the TiO_2 complex has photocatalysis.

3. Photoenduarance of paper with TiO_2–aluminosilicate complex

In general, the active oxygen generated by photocatalysis of TiO_2 decomposes organic materials such as paper, resin, and textile. The decomposition of the paper containing the TiO_2 complex with 5-wt.% TiO_2 was examined under UV light compared with the paper with TiO_2 mixture. The UV light was

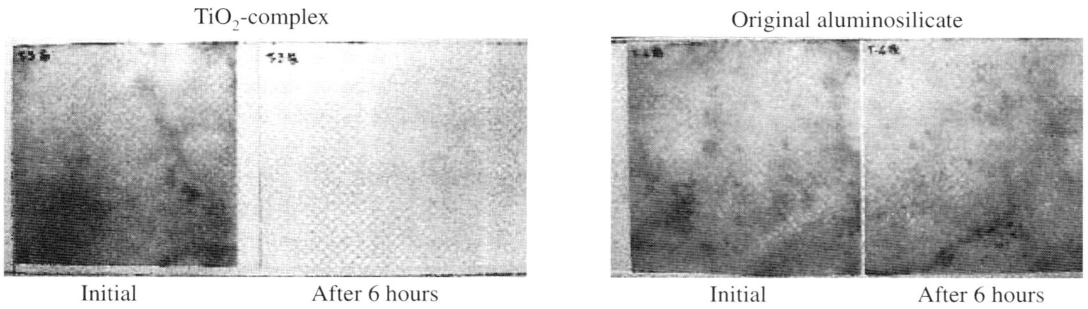

Figure 34.9
The comparison of bleaching effect of paper with the TiO_2 complex and the original aluminosilicate after 6 h under UV light.

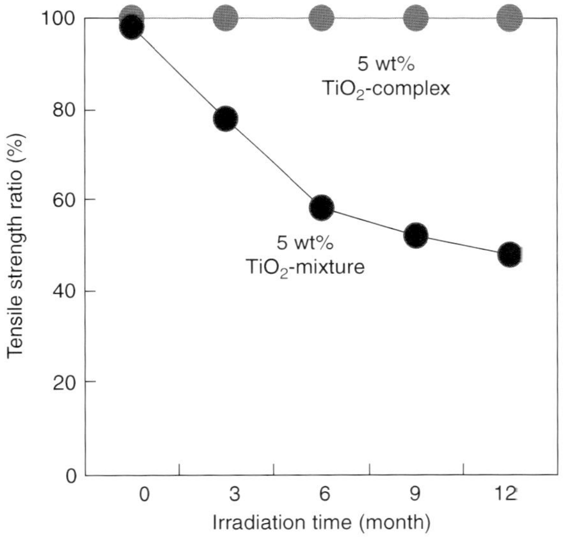

Figure 34.10
The tensile strength ratio of the paper with the TiO_2 complex and the TiO_2 mixture under UV light in a fade meter.

irradiated on the papers with 20 wt.% of these powders in a fade-meter. The irradiation for 3 weeks in the fade meter is equivalent to a year under ordinary sunlight. The tensile strength ratios of papers that are represented by the value that tensile strengths were divided in each initial tensile strength are shown in Fig. 34.10. The tensile strength of paper with the TiO_2 complex did not decrease, although that with the TiO_2 mixture decreased down to a half of its initial strength. The paper containing the TiO_2 complex displayed photoendurance on having photocatalysis. It seems that the TiO_2 complex does not decompose organic materials, because the TiO_2 photocatalyst inserted on the porous surface of aluminosilicate is not directly in contact with a fiber of the paper.

The TiO_2 as a photocatalyst and the porous aluminosilicate were combined in an aqueous solution with a cationic surfactant. The presence of TiO_2 nanoparticles, which inserted and uniformly dispersed on the porous surface of aluminosilicate, was observed by TEM–EDS analysis. The TiO_2 complex has photocatalysis, and its photocatalytic efficiency was also improved compared with that of TiO_2 mixture. On the other hand, the tensile strength of the paper with the TiO_2 complex did not decrease under UV light, although the TiO_2 mixture decreased down to a half of its initial strength. Accordingly, the TiO_2 aluminosilicate complex had photocatalysis without decomposing organic materials.

References

[1] A. Fujishima, K. Hashimoto and T. Watanabe: *TiO_2 Photocatalyst, Fundamentals and Applications*, BKC, Inc., Tokyo (1999).

[2] H. Nimura, S. Imaizumi, K. Hayashi and N. Yamashita: *J. Eng. Technol. Ind.*, **49**, 97 (2001).

[3] T. Nonami: *J. Odor Res. Eng.*, **33**, 333 (2002).

[4] M. Nikaido, S. Furuya and T. Kakui: *J. Surf. Finish. Soc. Jpn.*, **55**, 33 (2004).

[5] T. Kakui, M. Nikaido and S. Furuya: Proceedings of the 107th Annual Meeting of The American Ceramic Society on CD-ROM (2005).

[6] M. Nikaido, S. Furuya, T. Kakui and H. Kamiya: 18th Fall Meeting of the Ceramic Society of Japan, Preprints, Osaka, Japan, 376 (2005).

APPLICATION 35

35 AC OVERHEAD TRANSMISSION LINE AUDIBLE-NOISE REDUCTION MEASURES USING SURFACE IMPROVEMENT

1. Audible noise of AC overhead transmission lines

In AC overhead transmission lines, particularly in high-voltage AC overhead transmission lines of 270 kV or higher, normally in order to lower the surface electric field so that corona discharge does not take place from power lines to which a high voltage has been applied, the wires are made thick, or a number of wires are bundled together in a power line while maintaining a constant spacing between them. However, if there are any scratches on a normally smooth surface of a wire, or if there are any projections on it due to the adhesion of some foreign matter such as bird droppings, the surface electric field there becomes high thereby making it easy for corona discharge to occur. Even during rain, it is easy for corona discharge to take place since water droplets also get adhered to the surface of the power lines. Further, this corona discharge can generate audible noise in the form of screeching or booming sounds [1].

In general, during rain in AC overhead transmission lines, the audible noise is large at the time of new installation, and it has been recognized that there is a trend of the audible noise becoming small with the passing of years due to aging of the surface of the power lines. This is because the states of adhesion of water droplets on the two are different.

In actuality, the water droplets adhered on the surface of a new power line as is shown in Fig. 35.1 have more projections that become corona discharge points as compared with the water droplets adhered on the surface of an aged power line such as that shown in Fig. 35.2. This means that one method of reducing the audible noise is to improve the wetting property of the surface of the lines that determines the state of adhesion of water droplets. And instead of letting natural aging improve power lines, surface improvement of lines can be considered a possible way to reduce the corona discharge and the audible noise generated from power lines.

2. Wetting property of power lines

New power lines exhibit hydrophobicity due to the lubricating oil adhered to it during manufacture, and when exposed to nature for a long time after being mounted as power lines on steel towers, the surface has the oil removed and gets oxidized and its nature changes to hydrophilicity. While this is due to aging, an index indicating the extent of wetting property that expresses the hydrophobicity or the hydrophilicity of the surface of lines is the contact angle of water.

The contact angle is, as is shown in Fig. 35.3, the angle formed by the contact line of the liquid with respect to the solid surface as viewed from the side after a small quantity of a liquid is dropped on the surface of a solid, and the contact angle θ on a smooth surface is given by the Young's equation shown in equation (35.1) below, and is determined by the surface tension γ_s of the solid, the surface tension γ_L of the liquid, and the surface tension γ_{SL} of the

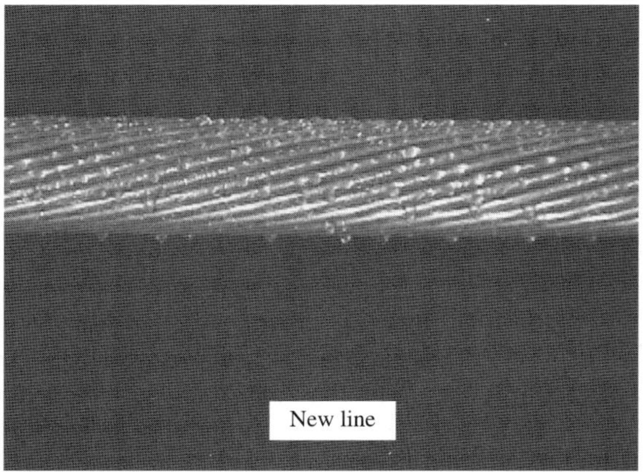

Figure 35.1
New power line.

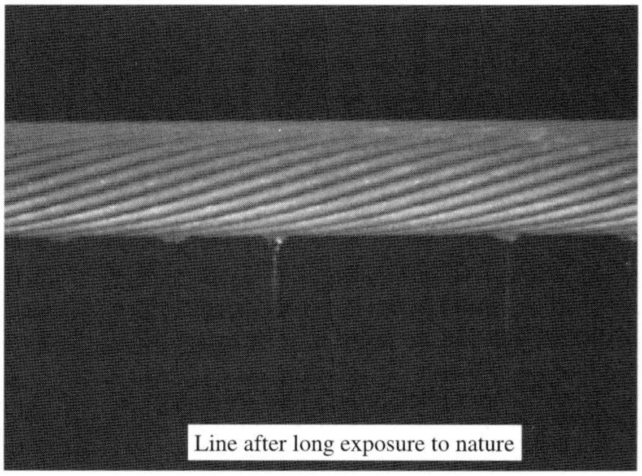

Figure 35.2
Aged power line.

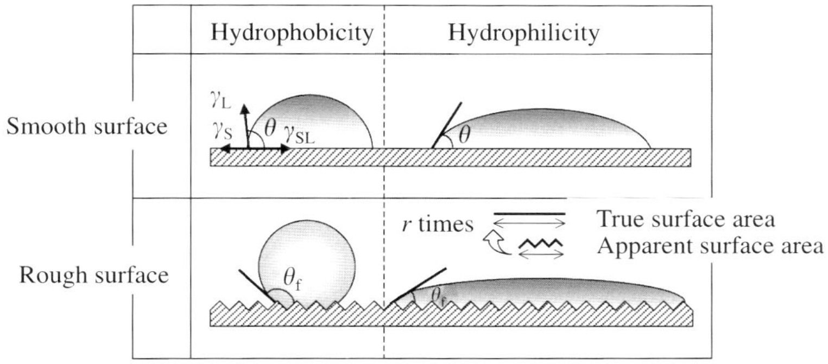

Figure 35.3
Contact angle.

solid–liquid boundary. When the surface has become rough and the surface area has become large, the apparent contact angle θ_f is given by Wenzel's equation given by equation (35.2) below, and the contact angle becomes still larger on a hydrophobic surface causing the water droplet to have a spherical shape, whereas the contact angle becomes still smaller on a hydrophilic surface causing the water droplet to spread in the lateral direction. Here, r is the surface-area magnification factor.

$$\frac{\gamma_S - \gamma_{SL}}{\gamma_L} = \cos\theta \tag{35.1}$$

$$\cos\theta_f = \frac{r(\gamma_S - \gamma_{SL})}{\gamma_L} = r\cos\theta \tag{35.2}$$

Further, in recent years, practical application of super-hydrophobic or super-hydrophilic surfaces is being made for the purpose of preventing adhesion of ice or snow or for dirt resistance in the exteriors of buildings. A super-hydrophobic surface has a remarkably large effect of repelling water, and water droplets roll off easily from such a surface. On the other hand, a super-hydrophilic surface has an extremely large effect of receptivity to water adhesion, and any water dropped on such a surface spreads very easily. As a guideline, the contact angle of super-hydrophobicity is more than 150°, and the contact angle of super-hydrophilicity is about a few degrees.

In this report, we investigate surface improvement methods that can be applied even to power line surfaces that are aluminum surfaces, prepare various power lines having a variety of wetting property characteristics from super-hydrophobicity to super-hydrophilicity, and from experiments simulating transmission lines under rain, we clarify the relationship between the wetting property of power lines and audible noise.

3. Preparation of test power lines

As is shown in Table 35.1, the test power lines used in the experiments have nine types of wetting property with the contact angle being in the range from 0.3 to 156.5°. In the case of a twisted wires structure in which aluminum alloy strands are bundled together, there is some restriction on the methods of improving the surface because the wires are very long being more than a few hundred meters in length. Therefore, in the present study, we attempted three types of methods using which it is possible to realize uniform surface improvement on the aluminum surface in open air.

To begin with, a simple method is that of coating the wires with a paint. Depending on the type of paint, it is possible to obtain different types of wetting properties, and it is also possible to easily apply this method on power lines that have already been installed on towers. However, in an environment in which the paint on the power lines is continuously exposed to strong wind, rain, and sunlight, and in which there is a very severe difference between hot and cold weathers, the life of the painted covering film is much shorter than that of the power lines, and hence is not considered to be realistic, except in special cases.

The next method of improving the surface properties is the use of blasting. Sandblasting is actually being used, and its purpose is to roughen the power line surface, which is glossy in order to blend with the environment by suppressing reflection of sun's rays. In addition, using the blasting process, depending on its conditions, it is possible to implant a part of the blasting material into the surface of the power line, and as a result, a wetting property that cannot be obtained by aluminum alone can be obtained. The blasting process can be realized easily as one of the processes in the manufacture of power lines.

Finally, thermal spraying is a method of surface improvement by spraying a material in the molten state and at a high temperature onto the surface of the power

Table 35.1
Surface properties of test power lines.

Type of power line	Manufacturing method	Contact angle	Features		
			Surface condition	Adhesion	Others
Super-hydrophobic power line	Paint coating	156.5°	Smooth	Extremely low	Deterioration with years present
Hydrophobic power line	Paint coating	107.8°	Smooth	High	Deterioration with years present
Super-hydrophilic power line	Paint coating	5.3°	Smooth	High	Deterioration with years present
New power line	–	98.6°	Smooth, glossy	–	–
Aged power line	Exposure to nature	–	Smooth, no gloss	–	–
Sandblasted power line	Blasting	74.0°	Undulating surface	–	–
Titanium oxide-blasted power line	Blasting	87.6°	Undulating surface	–	–
Alumina thermal-sprayed power line	Spraying	11.6°	Undulating surface	Extremely high	–
Titanium oxide thermal-sprayed power line	Spraying	0.3°	Undulating surface	Extremely high	Has photocatalytic function

line, and it is possible to change the surface wetting property of the line to the characteristics of the sprayed material itself. Depending on the spraying conditions, it is possible to control the coated film thickness and the structure as desired, and it is possible to obtain a coating that has a high adhesion to the surface of the power line and has a high durability. Even the thermal spraying process can be realized easily as one of the processes in the manufacture of power lines.

Further, since the measurement of the contact angle is difficult in case of a twisted wires structure, the contact angles in Table 35.1 are the ones measured by carrying out the same surface improvement on a separate flat aluminum plate.

4. Features of titanium oxide thermal-sprayed films

In recent years, titanium oxide is a material that is attracting attention due to its photocatalytic effect. Titanium oxide is excited by the ultraviolet rays in sunlight, etc. and decomposes organic material, or causes super-hydrophilicity. These functions maintain the surfaces of outdoor constructions that are exposed to sunlight clean without manual cleaning due to the multiplied effects of direct decomposition of soiling material and the cleaning effect of rain [4]. In the case of transmission lines whose cleaning cannot be done easily, even if the wetting property is improved by surface improvement, since there is the likelihood of those characteristics being lost due to soiling of the surface, it is considered effective to apply self-cleaning functions.

The conditions of the surfaces of the thermal-sprayed films of titanium oxide sprayed power lines prepared during the present study are shown in Fig. 35.4. There are pores of 1 μm diameter and cracks of 5 μm length in the surface of the sprayed titanium oxide film, and there are fine spheres of sizes from 500 nm to about 2 μm. In addition, there are undulations on the micrometer order over the entire surface, the area is very large, and hence the super-hydrophilicity is promoted.

5. Audible-noise measurement

In the evaluation of the test power line during the experiments, the audible-noise measurements were made using a small-sized water-spraying voltage-application equipment shown in Fig. 35.5. A uniform water spray simulating rain is made to fall onto the test power line below from a water-injection nozzle above. In this equipment, it is possible to set the surface electric field of the test power line to any desired value by adjusting the applied voltage.

In the position relationship between the test power line and the ground plane as in this equipment, the maximum electric field (G_{max}) on the surface of the test power line occurs at the lowermost position facing the ground plane on the surface of the line, and can be obtained by calculation. In this study, the experiments were conducted under conditions of values of G_{max} up to 12–16 kV/cm, which includes the range of values of actual use of the high-voltage AC overhead transmission lines. The measured item is the A-weighted sound pressure level stipulated in the JIS (Japanese Industrial Standard) C 1505 standards, and the measurements were made by placing a microphone at a location 2 m from the power line.

Sample results of the experiments are shown in Fig. 35 6. The horizontal axis represents the contact

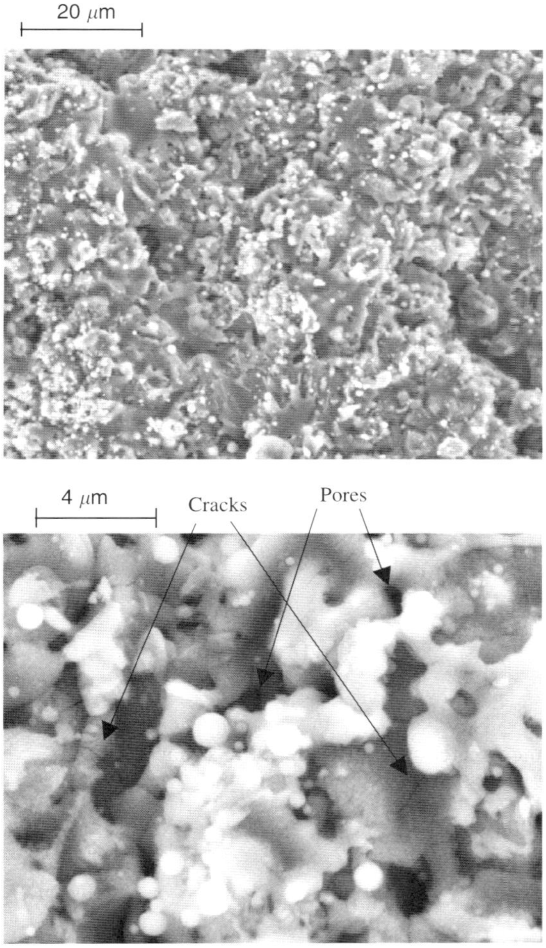

Figure 35.4
Surface of titanium oxide (thermal sprayed) film.

angle of the test power lines given in Table 35.1. The vertical axis represents the amount of audible noise that subtracts the background noise level from measured noise level, and is the value of only the corona discharge noise generated from the test power lines. Further, the curves in Fig. 35.6 are categorized into the G_{max} values of 12, 14, and 16 kV/cm.

From Fig. 35.6, it can be seen that although there is some fluctuation in the values when G_{max} is low, the audible noise becomes a minimum in the super-hydrophobic region of contact angles of more than 150°, but on the other hand, it increases in the hydrophobic region. At this time, although a number of water droplets that become the discharge points are adhered on the surface of the hydrophobic power line as in the case of a new line shown in Fig. 35.1, there are almost no water droplets adhered to the super-hydrophobic power line surface and all the water sprayed onto it has rolled off as if being repelled by the surface. Although it can be considered that an super-hydrophobic power line is the best method of reducing audible noise, since the audible noise increases as its surface deteriorates becoming merely hydrophobic, it is considered difficult to put super-hydrophobic power lines to practical use, unless the surface improvement method can maintain the characteristics over a long period of time.

Further, from Fig. 35.6, it can be seen that the amount of audible noise generated is small in the hydrophilic region with contact angles less than 90°, and also the changes are small. Further, in the super-hydrophilic region, the audible noise becomes still smaller.

At this time, water droplets are present only on the lowermost part of the surface of the power line, and the adhered water droplets become larger as the hydrophilic function is stronger and the number of droplets tends to become small. The super-hydrophilic power line also has small amount of audible noise generated, and if the surface improvement

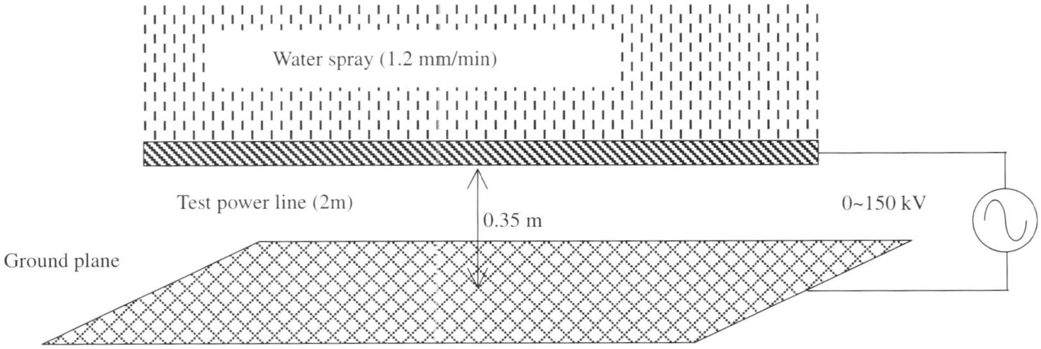

Figure 35.5
Experimental setup.

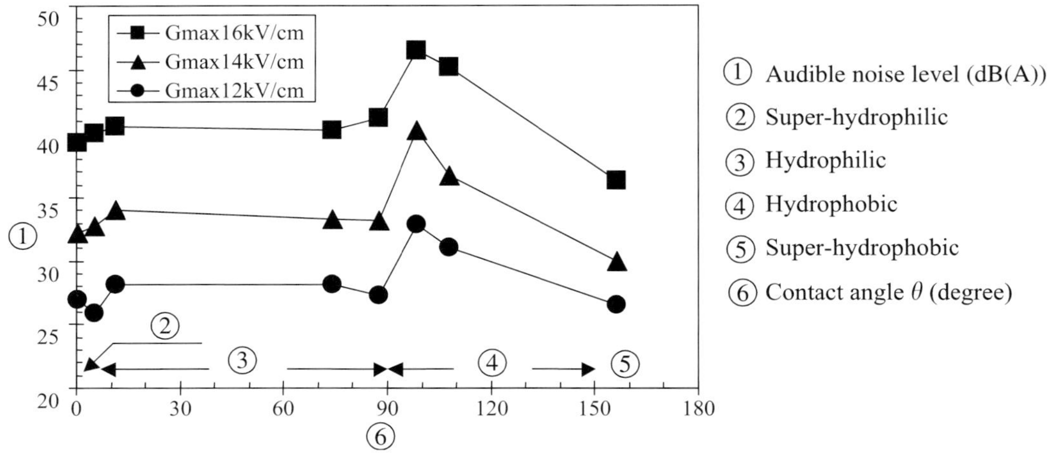

Figure 35.6
Experimental results.

method is titanium oxide thermal spraying, since even the characteristics can be maintained over a long period of time due to the photocatalytic function and a film having superior durability can be obtained, it is considered to be excellent as a method of reducing audible noise in transmission lines.

Only about two water droplets are present per meter on a titanium oxide thermal-sprayed power line, which is smaller than the aged power line shown in Fig. 35.2. At present the super-hydrophilic power lines using titanium oxide are considered to be the surface improvement method superior in corona characteristics and durability.

Various parts are sometimes affixed to AC overhead transmission lines for the purpose of suppressing the wind roar of the lines during strong winds or of suppressing snow accumulation. However, these accumulate rainwater during rainfall and often cause increase in the audible noise. We intend to suppress the audible noise by carrying out surface improvement of these fixtures other than the power lines, and to develop transmission lines of the completely environment-compatible type that are not affected by the natural environment.

References

[1] Committee to Investigate Technology Countermeasures for Audible Noise and Aeolian Noise Overhead Transmission Lines: Technology countermeasures for audible noise and aeolian noise in overhead transmission lines, Institute of Electrical Engineers of Japan Technical Report, No. 976, 5–6, (2004).

[2] K. Miyajima, K. Tanabe: Reduction of audible noise by super-water-repellent treatment of transmission line surfaces: *Trans. IEE Jpn.*, **121-B** (1), 136–137 (2001) (in Japanese).

[3] K. Tsujii: Super water- and oil-repellent surfaces resulting from fractal structure. *Hyomen (Surface)*, **35** (12), 629–639 (1997) (in Japanese).

[4] A. Fujishima, K. Hashimoto and T. Watanabe: *The Optical Clean Revolution*, CMC, Tokyo, 8–12, (1997).

APPLICATION 36

36 DEVELOPMENT OF A HIGH-PERFORMANCE SECONDARY BATTERY BY CONTROLLING THE SURFACE STRUCTURE

Since the electrochemical reaction takes place at the surface of the electrode active material, modification of the surface of the electrode active material must have a significant effect on the battery performance and it would be possible to improve the electrode performance through various kinds of surface modification. There is considerable research relating to this concept and some research concerning improvements to the anodes and cathodes of nickel hydrogen and lithium ion batteries by surface modification has been mentioned in this chapter.

1. Anode of a nickel hydrogen battery

The reactions of the nickel hydrogen battery system are as follows:

Cathode reaction: $NiOOH + H_2O \Rightarrow Ni(OH)_2 + OH^-$

Anode reaction: $MH + OH^- \Rightarrow M + H_2O + e^-$

Total reaction: $NiOOH + MH \Rightarrow Ni(OH)_2 + M$ (M is the hydrogen storage alloy)

Most major metals M, used as anode active materials for nickel hydrogen batteries, are AB_5 type alloys such as $LaNi_5$, in which the A site is occupied by rare earth trivalent ions, Nb^{3+} and Zr^{4+} etc. and the B site by nickel, cobalt and aluminum ions. $MmNi_5$ and its derivatives, such as $La_{0.8}Nb_{0.2}Ni_{2.5}Co_{2.4}Al_{0.1}$, $La_{0.8}Nb_{0.2}Zr_{0.03}Ni_{3.8}Co_{0.7}Al_{0.5}$, $MmNi_{3.65}Co_{0.75}Mn_{0.4}Al_{0.3}$, $MmNi_{2.5}Co_{0.7}Al_{0.8}$ and $Mm_{0.85}Zr_{0.15}Ni_{1.0}Al_{0.8}V_{0.2}$ are used in a practical cell where Mm represents a mixed metal, namely a mixture containing 40–50% Ce and 20–40% La, Pr and Nd, while Laves phase (an AB_2 type alloy) has also been studied as an anode active material. This alloy has the closest packing structure, with an atomic radii ratio of 1.225. The Ti–V, Ti–Cr and Zr–Mn phases have also been investigated as anode active materials, with an electrolyte solution of 7 M KOH aq. based aqueous solution. Since the alloys are less noble metals, their surfaces corrode to form a Ni rich layer on the surface [1]. Ni–PTFE composite plating on the hydrogen storage alloy is one of the trial means to control corrosion via surface treatment. Kiyokawa et al. have developed a new electroplating technique to prepare Ni–PTFE composite film on the $MmNi_5$ fine powder [2, 3]. The corrosion of the anode active material in the alkaline aqueous solution was controlled by the composite film plating while the film never inhibits the hydrogen transfer across the film. In addition, the electroplated Ni–PTFE composite film on the $MmNi_5$ powder particles had a large specific surface area, which contributed to improving the high rate charge/discharge performance and extending the battery life [4]. Fig. 36.1 shows SEM photographs of Ni–PTFE film, electroplated on $MmNi_{3.65}Co_{0.75}Al_{0.25}Mn_{0.35}$ fine particles. It was found that the composite film of 2 μm

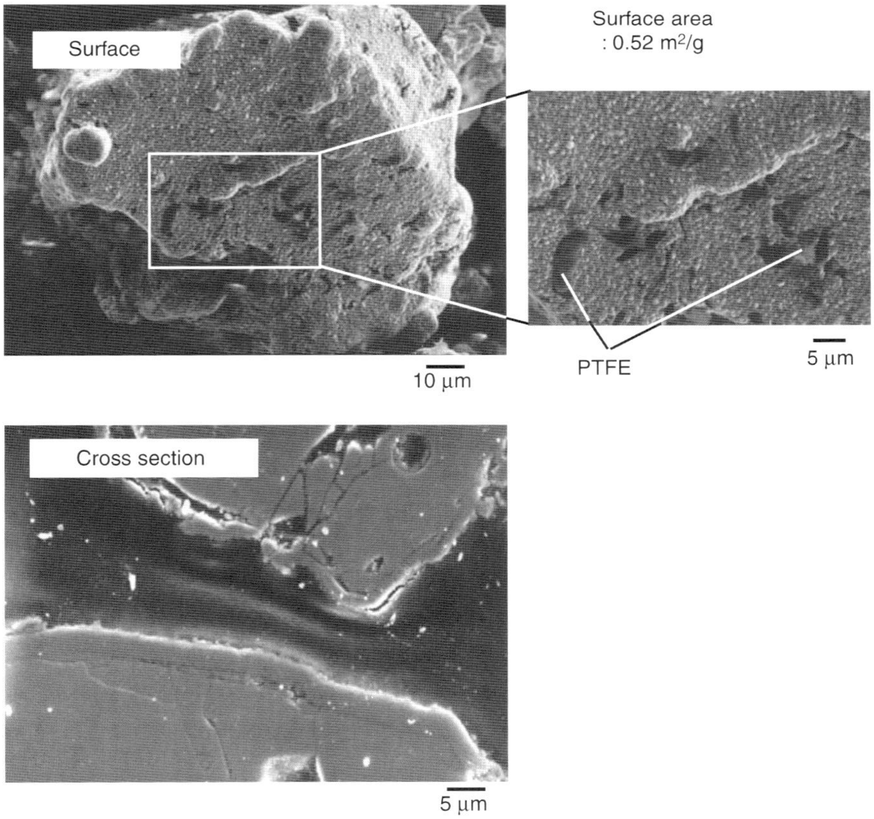

Figure 36.1
SEM photographs of the Ni–PTFE composite film coated $MmNi_{3.65}Co_{0.75}Al_{0.25}Mn_{0.35}$ powder particle.

thickness covered the particles uniformly and was relatively porous. Meanwhile, the dissolution of the manganese from the alloy was limited to less than one tenth while no particular loss of discharge capacity was observed. Consequently the Ni–PTFE composite film works as an anticorrosive film and a catalyst to promote the redox reaction of hydrogen. It appears important to design the nano structure of Ni–PTFE composite film on the anode active material to improve the electrochemical properties of hydrogen storage.

2. Cathode of the nickel hydrogen battery

As shown in the previous section, the cathode of the nickel hydrogen battery is nickel hydroxide, $Ni(OH)_2$. During the charge process, $Ni(OH)_2$ is converted into NiOOH. Both $Ni(OH)_2$ and NiOOH are oxide ceramics, with very low electric conductivities compared to the metal electrode. The electron conductivity of $Ni(OH)_2$/NiOOH was shown, summarized in relation to the valency of the nickel in Fig. 36.2. The electron conductivity decreased when the mean valency was less than 2.2, meaning the resistivity of the cathode in a nickel hydrogen battery increases throughout the discharge process. This fact affects the discharge capacity. For example, the discharge capacity of the $Ni(OH)_2$ cathode, fabricated with nickel foam having larger pores as a current collector, was relatively smaller than that having smaller pores. Oshitani et al. proposed the formation of an electron conduction path network via surface modification by using cobalt compounds as an effective means of solving the aforementioned internal resistance problem. In their method, CoO, $Co(OH)_2$ or Co powders were added to the cathode. These cobalt compounds were dissolved into the alkaline electrolyte, whereupon the cathode was covered with $Co(OH)_2$. During the oxidation process (charge process), this $Co(OH)_2$ was converted to CoOOH, which has high electric conductivity, allowing a very effective electron conduction path network to be prepared. Since CoOOH is stable within the potential range where $Ni(OH)_2$ works as a cathode, this network is retained during the charge/discharge process. A usage rate of 95% for the cathode active material was achieved by this method, as shown in Fig. 36.3 [5,6]. Meanwhile, trials to obtain the improved electron conduction network remain ongoing with various methods used.

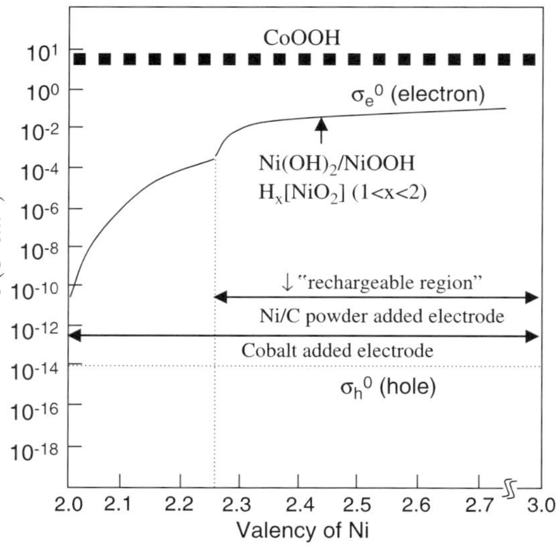

Figure 36.2
Change in the electric conductivity of $Ni(OH)_2$ throughout the charge/discharge process [6].

Furthermore, the nano structure of the fine powder particle is also important to obtain the higher performance electrode. Since the $Ni(OH)_2$ particle is the aggregate of the fine crystals, it contains many mesopores of 2–10 nm. It is possible to decrease the pore volume in the $Ni(OH)_2$ particle to less than 10% using the reaction precipitation method. By using high density $Ni(OH)_2$, spherically shaped and of uniform size, the capacity of the nickel hydrogen battery has increased (150% was achieved) [7].

3. Cathode of lithium ion battery

The reactions of the lithium ion battery system are as follows:

Cathode reaction: $Li_{1-y}MO_x + yLi^+ + ye^- \Leftrightarrow LiMO_x$

Anode reaction: $C_xLi \Leftrightarrow C_{x-y}Li + yLi^+ + ye^-$

Total reaction: $Li_{1-y}MO_x + C_xLi \Leftrightarrow LiMO_x + C_{x-y}Li$ (M: Co, Mn, Ni, etc.)

The first cathode active material for a commercial cell was $LiCoO_2$, following which many other lithium containing oxides have been studied. However, almost all of these are ceramics, with low electric conductivity, meaning conducting material like acetylene black has always been added to the cathodes [8, 9]. Surface modification, involving coating the active materials with some electric conducting material, is one of the other approaches to improve the electric conductivity. Evidently, the transfer of Li^+, in this case across the interface between the electrode and the electrolyte, need not be inhibited. It has been reported that nano thickness carbon coating using the arc discharge method has been effective in improving the electrochemical properties of $LiMn_2O_4$ [10], although the carbon coating did not contribute to improving the electric conductivity of the electrode in this case. However, the cycling ability was improved by carbon coating, which may mean that

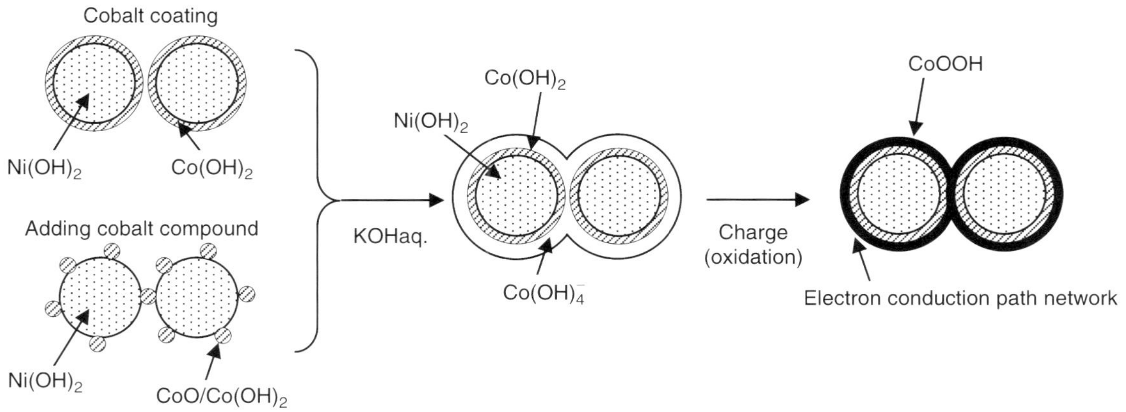

Figure 36.3
Formation of the electron conduction path network by cobalt coating on $Ni(OH)_2$.

the presence of nano thickness carbon film on the active material affects the lithium ion transfer between the electrode and the electrolyte, controlling the decomposition of the electrolyte on the surface of the active material and the affinity with the electrolyte solution. Concerning the control over the electrolyte decomposition, Endo et al. have reported that a diamond-like carbon coating can control the decomposition [11]. However, in this case, the resistivity in the cathode increased significantly. Surface fluorination can be used to improve the stability of the cathode active material surface, while preciously controlled surface fluorination can achieve some advantage [12]. Fig. 36.4 shows the effect of the surface fluorination of $LiMn_2O_4$, by using F_2 gas for its cycling ability. During surface fluorination under 1.3 kPa F_2 at room temperature, the discharge capacity increased by 5% and the loss of discharge capacity after 50 cycles was limited to just 2%, while the untreated sample exhibited a 13% loss in discharge capacity. On the other hand, excess fluorination such as less than 1.3 kPa F_2 at 200°C caused a particular decrease in discharge capacity and an increased loss of discharge capacity after 50 cycles. Since excess fluorination may result in resistive LiF film formation on the $LiMn_2O_4$ particle, it is important to introduce fluorine only on the surface and control its amount preciously. Hybrid coating with carbon and fluorine has also been proposed and investigated by Takashima et al. It has been reported that this type of surface modification may have a synergic effect of improving the electrochemical properties of the $LiMn_2O_4$ particle [10,13].

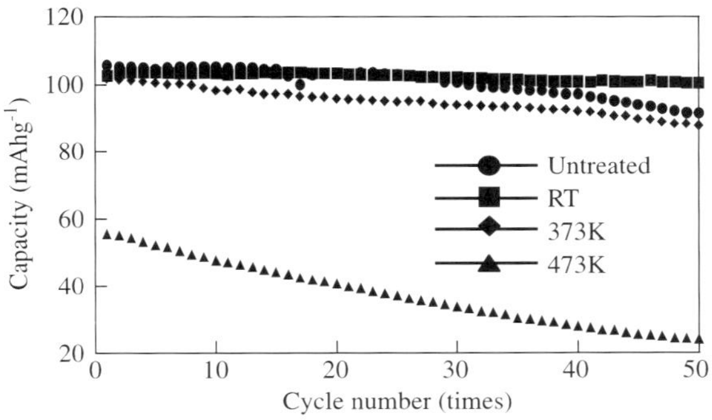

Figure 36.4
Change in the discharge capacity along the cycle number of $LiMn_2O_4$ after fluorinating the surface by F_2 gas.

4. Anode of the lithium ion battery

The anode active material used in the lithium ion battery is generally carbon material. Amorphous carbon was first applied to the practical cell, followed by investigation of other carbon materials as the anode material for the lithium ion battery to boost capacity. To date, a charge/discharge capacity of 340 mA h g^{-1} has been achieved, which is almost the theoretical capacity for the C/C_6Li system. In the case of a carbon anode of a lithium ion battery, the electrolyte is decomposed to generate a Solid Electrolyte Interface (SEI) in the first charge process, which plays an important role in maintaining stable charge/discharge cycling, so various approaches have been implemented to prepare an SEI suitable for charge/discharge cycling by surface treatment and combination of the species in the electrolyte. Surface fluorination [14–16], surface oxidation [17–19], heat treatment under high vacuum [20], metal coating [21], carbon coating [22–24] and so on have been reported as methods to modify the surface of the anode active materials. Treatment with an oxidative process may affect the morphology near the surface, while heat treatment under a high vacuum may have the effect of removing the oxide film on the

Table 36.1
Initial charge/discharge columbic efficiency of surface fluorinated natural graphite.

Sample	Initial columbic efficiency (%)		
	~7 μm	~25 μm	~40 μm
Untreated natural graphite	79.8–80.8	85.6	85.1
Surface fluorinated graphite by using F_2 gas[a]			
150–300°C	77.0–79.0	85.9–86.4	85.5–86.3
300–500°C	72.0–75.4	66.5–83.9	65.1–83.4
Fluorinated natural graphite by plasma method	80.2–85.7	85.4–85.8	84.0–85.1

[a] F_2: 30 kPa, 2 min.

surface. The oxide film is also reportedly removed by oxidative treatment [18]. When the carbon material is fluorinated at a high temperature, the edge of the graphite structure is fluorinated to generate a C–F covalent bond, the specific surface area and the pore distribution are evidently affected and the concentration of oxygen on the surface of the carbon material decreases. As the typical case of the effect of surface fluorination of the carbon material on its electrochemical properties, the initial charge/discharge columbic efficiency of surface fluorinated natural graphite was summarized in Table 36.1. While the columbic efficiency decreased during surface fluorination at 350–500°C, it remains at the same value as untreated natural graphite during surface fluorination at less than 350°C. The small amount of fluorine introduction on the surface does not increase the irreversible capacity in the first cycle. It was found that the surface structure of the natural graphite became more random after the fluorination, based on the results of Raman spectra. Since a surface with random structure may be suitable for storing the solvated lithium ion, it seems possible to facilitate SEI formation by decomposing the solvent. SEI containing fluorine, however, tends to have a higher electric resistivity and may be unsuitable for use under a larger charge/discharge rate.

References

[1] H. Tamura (Ed.): *Hydrogen Storage Alloy*, NTS Inc., Tokyo (1998).

[2] T. Ibe, Y. Matsumura, T. Ueno, K. Kiyokawa, S. Yonezawa and M. Takashima: *Nihonkaakukaioshi*, **7**, 387 (2001).

[3] H. Kiyokawa, R. Ikeda, Y. Chong, S. Yonezawa and M. Takashima: *Hyomengijyutu*, **48**, 939 (1997).

[4] Y. Osumi: Hydrogen *Storage Alloy – Its Properties and Application*, p. 144 Agne Gijyutu Center, Tokyo (1993).

[5] M. Oshitani, H. Yufu, K. Takashima, S. Tsuji and Y. Matsumaru: *J. Electrochem. Soc.*, **136**, 1590 (1989).

[6] H. Tamura (Ed.): *Recent Technology for Nickel Hydrogen Battery*, Chapter 2, NTS Inc., Tokyo (2001).

[7] Japanese Patent 2140564 and 2679274.

[8] D. Aurbach, M.D. Levi, E. Levi, H. Teller, B. Markovsky, G. Salitra, U. Heider and L. Heider: *J. Electrochem. Soc.*, **145**, 3024 (1998).

[9] Y.C. Chang, H.J. Sohn: *J. Electrochem. Soc.*, **147**, 50 (2000).

[10] M. Takashima, S. Yonezawa and M. Ozawa: *Mol. Cryst. Liq. Cryst.*, **388**, 153–159 (2002).

[11] E. Endo, T. Yasuda, A. Kita, K. Yamamura and K. Sekai: *J. Electrochem. Soc.*, **147**, 1291 (2000).

[12] S. Yonezawa, M. Yamasaki and M. Takashima: *J. Fluorine Chem.*, **125**, 1657–1661 (2005).

[13] S. Yonezawa, M. Ozawa and M. Takashima: *Tanso*, **205**, 260–262 (2002).

[14 T. Nakajima, M. Koh, R.N. Singh and M. Shimada: *Electrochim. Acta*, **44**, 2879 (1999).

[15] V. Gupta, T. Nakajima, Y. Ozawa and H. Iwata: *J. Fluorine Chem.*, **112**, 233 (2001).

[16] T. Nakajima, V. Gupta, Y. Ozawa, M. Koh, R.N. Singh, A. Tressaud and E. Durand: *J. Power Sources*, **104**, 108 (2002).

[17] J.S. Xue, J.R. Dahn: *J. Electrochem. Soc.*, **142**, 3668 (1995).

[18] M. Hara, A. Satoh, N. Tamaki and T. Osaki: *Tanso*, **165**, 261 (1994).

[19] E. Peled, C. Menachem, D. Bar-Tow and A. Melman: *J. Electrochem. Soc.*, **143**, L4 (1996).

[20] T. Takamura, M. Kikuchi: *Battery Tech.*, **7**, 29 (1995).

[21] R. Takagi, T. Okubo, K. Sekine and T. Takamura: *Denki Kagaku*, **65**, 333 (1997).

[22] M. Yoshino, H. Wang, K. Fukuda, Y. Hara and Y. Adachi: *J. Electrochem. Soc.*, **147**, 1245 (2000).

[23] H. Wang, M. Yoshino, T. Abe and Z. Ogumi: *J. Electrochem. Soc.*, **149**, A499 (2002).

[24] Y. Ohzawa, M. Mitani, T. Suzuki, V. Gupta and T. Nakajima: *J. Power Sources*, **122**, 153 (2003).

APPLICATION 37

37 PINPOINT DRUG AND GENE DELIVERY

The administration of general drugs systemically is often limited because of the adverse side effects, so improving the drug delivery system (DDS) is an important means of optimizing drug therapy. For example, the administration of some anticancer drugs has been halted due to their adverse side effects.

In the 1970s, the term DDS was used to indicate the controlled release of a drug. Nowadays, the term DDS is used to indicate the technology of delivering a drug selectively to only the diseased part [1]. Since 1980 in the United States and since 1995 in Japan, gene therapy has been developed in order to cure diseases and reduce symptoms at the genetic level. DDS technology is essential for delivering the therapeutic gene to the target tissues and cells accurately and efficiently.

Representative examples of carriers include viral vectors such as adenovirus and retrovirus, and liposomes. The viral vectors' transduction efficiency of genes is extremely high, however, there are disadvantages such as reproduction of the virus within the living body, infection accidents, and untargeted nonspecific infection [2, 3].

Although liposomes are safer, there are disadvantages such as low transduction efficiency and low target specificity. Therefore, the development of a carrier that delivers chemicals and genes in the living body safely and efficiently is being sought in a wide range of areas, such as pharmacy, medicine and engineering. With the advancement of nanotechnology, carriers with a controlled size such as microcapsules and nanocapsules are being developed. However, there are very few carriers with high target specificity on the market.

The effects and potential for the practical application of a bio-nanocapsule (BNC), which has been developed as a viral genome-free new carrier with strong infectivity and high target specificity, is introduced in the next section [4].

1. Bio-nanocapsules (BNC)

We use the term BNC to refer to nanosized hollow particles created by biotechnology. At present, a hollow capsule has been realized, having a diameter of 100 nm with the fundamental skeleton of the Hepatitis B virus surface antigen (HbsAg) produced using yeast.

BNC is not pathogenic because it does not include any gene of the Hepatitis B virus or enzyme necessary for replication. The HBsAg particle has already been used as the recombinant Hepatitis B vaccine worldwide for the last two decades, and it is considered safe in the human body.

In the HBsAg molecule, there are 3 kinds of protein, S, M and L that have C terminals in common but different N terminals. The L protein, which has M protein, to whose N terminal the pre-S1 region is added, is used as BNC [5]. As shown in Fig. 37.1, the pre-S1 region, which combines directly to liver cells, develops at the particle surface and plays the important role of introducing specifically into human liver cells.

(1) Drug delivery

BNC delivers a drug to only the target cells, which is expected to reduce the side effects on the non-targeted cells. Verification of the drug delivery was attempted with fluorescent material calcein as a model compound. Calcein was encapsulated inside BNC by electroporation and added into the cultured cells in the presence of fetal bovine serum.

Fluorescence derived from calcein, shown in Fig. 37.2, was observed in human hepatoma-derived cell line (HepG2, NuE), however, it was not observed as a contrast in the human colon cancer-derived cell line [4].

Human hepatoma-derived cell line (NuE) and human colon cancer-derived cell line (WiDr) were implanted subcutaneously in the back of nude mice, and BNC including calcein was administrated

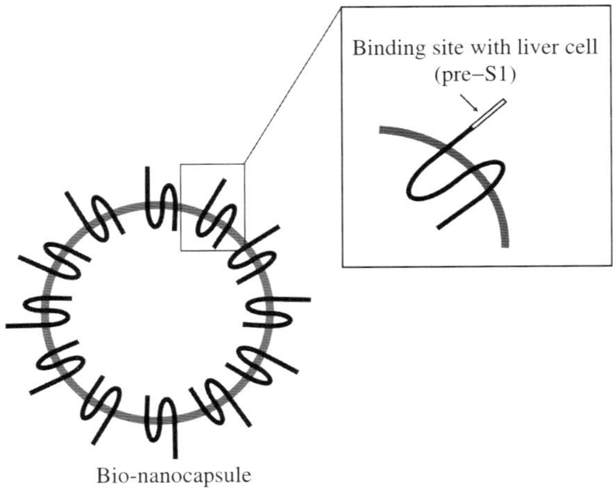

Figure 37.1
Bio-nanocapsule.

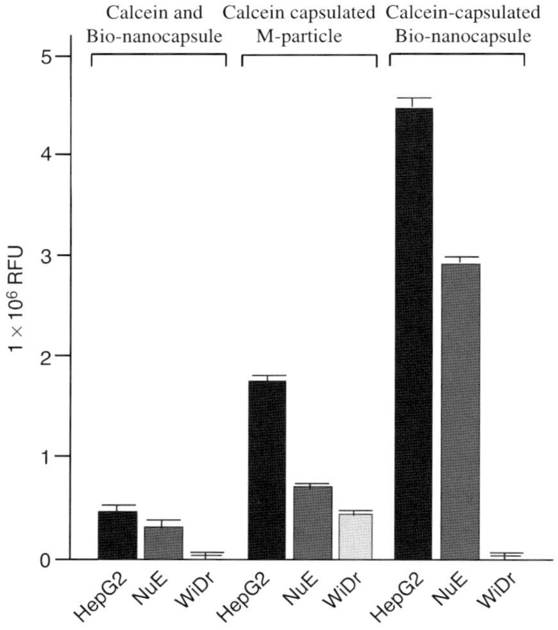

Figure 37.2
Pinpoint delivery of BNC encapsulated with calcein to the human liver cell.

intravenously from the tail vein. Fluorescence was observed in the implanted human hepatoma – derived tumor tissue but not in internal organs or human colon cancer-derived tumor tissue in the mouse. Kupffer cells in the liver usually phagocytize xenobiotics, however, BNC was not recognized as a xenobiotic.

Therefore, with low-molecular compounds, BNC has great potential for drug delivery specific to the human liver cells and tissues.

(2) Gene delivery

Gene delivery by BNC was conduced by BNC with the expression vector for *Aequorea coerulescens*-derived green fluorescence protein (GFP). BNC encapsulated with GFP expressed gene was added to the human normal liver cell and human hepatoma-deprived cell line by electroporation as described in the previous section, and fluorescence expressed by GFP was observed. However, expression of GFP was not observed in the non-hepatoma cell line. In a tumor-bearing mouse that had been implanted human hepatoma-deprived cell line, GFP expression vector was delivered only in human hepatoma-derived solid cancer. Expression of GFP was not observed in other tissues [4].

Therefore as a model of gene therapy, an experiment to verify the treatment of hemophilia B with blood clotting factor IX, FIX expression gene was conducted. Consequently, by administering BNC containing FIX expression gene only once to the tail vein intravenously, expression of FIX was detected in the blood of the mouse. This was a promising expressed level that is expected to show a therapeutic effect in patients with advanced hemophilia B.

In addition, although adenovirus vector is used in ordinary gene treatment and excluded by immunity in a few days, a few weeks of successive expression was confirmed in the administration of BNC.

These results show that BNC has great potential as a gene delivery system targeting liver cells and tissues.

(3) Retargeting of BNC

We attempted to deliver the altered BNC to tissues other than liver (Fig. 37.3). Using epidermal growth factor; EGF as a model, EGF-displaying BNC replacing the liver cell binding site within the pre-S1 region was prepared. The calcein was introduced into the cell expressing EGF receptor on its cell surface A431 with the calcein encapsulated EGF-displaying BNC but was not introduced to the liver cell that had originally been transfected [4].

By presenting various cytokines and antibodies on the surface, many substances have been successfully delivered with its specificity. Research on the single-chain peptide, which enables high-level transfection to tissue inside a living body is advancing and is likely to widen the range of BNC application range. In addition, by integrating the protein at the C terminal of the HBsAg L protein forming the BNC, the delivery of BNC containing protein in a living body succeeded.

Therefore, responding to various targeting mechanisms, BNC is expected to be the ultimate carrier to enable the pinpoint delivery of drugs, proteins and genes within a living body.

2. Potential applications of BNC

Example applications of BNC are shown below.

(1) Treatment for hepatic disease by BNC

As the original target of BNC is human liver, it is a natural flow to use BNC for the treatment of hepatic disease.

It is possible for BNC to deliver a treatment gene (tumor suppressor gene, apoptosis-inducing gene, etc.) to liver cancer, and more accurate targeting will be possible by presenting a molecule that recognizes the antigen specific to the hepatic cancer cell selectively.

Interferon is administered for the treatment of Hepatitis virus, but serious systemic side effects and the production of neutralizing antibodies are major problems. Specifically administering interferon genes or protein to liver using BNC may provide a safe treatment of Hepatitis virus.

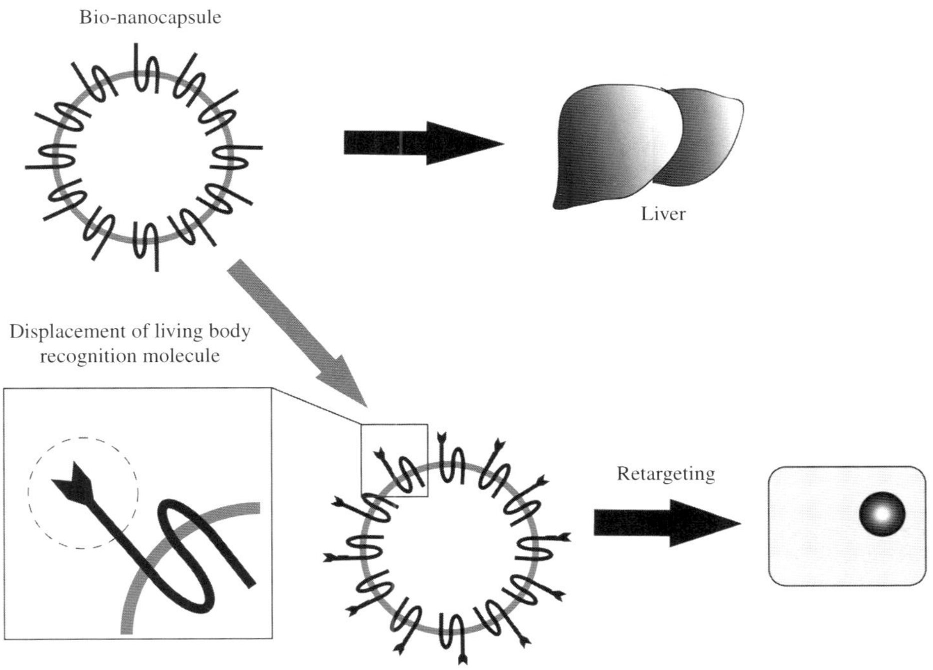

Figure 37.3
Retargeting of BNC to other cells than liver cell.

(2) Treatment by BNC including anticancer drug

In many cases of cancer treatment, an anticancer drug is used simultaneously. However, whole body administration of an anticancer drug often causes a serious side effect. The anticancer drug itself cannot distinguish cancer cells from healthy cells, but delivery of the anticancer drug to only the diseased part would minimize the side effects and maximize the medicinal effects. Certain sugar chains and antigens specific to cancer are usually present on the surface of the cancer tissue. By making altered BNC that presents the relevant molecule (antibody, etc.) that recognizes the sugar chain and antigen, pinpoint drug delivery may be realized.

(3) Gene therapy by BNC

There are a number of genetic illnesses and most of them are caused by lack of expression of a single protein, such as hemophilia and phenylketonuria. BNC is a method of gene replacement therapy by introducing the lacking protein gene into the human liver. RNA interference is a new living body level gene regulatory technique that can be classified broadly into two categories, one technique using a plasmid vector and the other technique using small interfering RNA (siRNA). In the former, as there is a risk of gene integration with the chromosomes of the patients, this complete gene therapy will require considerable research before practical application. The latter techniques show adequate bioactivity in spite of the extremely low stability in the body, and practical application may be realized sooner as there is less possibility of damaging chromosomes of the patients.

Consequently, BNC is attracting attention for raising the stability of siRNA in blood and achieving tissue specific siRNA delivery in a living body. Although the major pharmaceutical companies are focusing attention on RNA interference as a new medical means, it cannot be systemically administered. Therefore, a combination of an *in vivo* pinpoint delivery system and RNA interference is sought.

With the spread of tailor-made health care based on increases in gene information availability, BNC will be applied in gene therapy personalized for each patient, selecting target organs, target molecules and genes for the treatment.

3. Assignment

In practical application of BNC, the main target is realizing administration by lowering immunogenicity and antigenicity. In particular there is a possibility that BNC would not be effective in patients with Hepatitis B virus antibody by vaccine inoculation, although the immunization rate for vaccine is low in Japan but high in the United States and Europe. Therefore, it is necessary to complete stealth BNC by mimicking the amino acid mutated part of the Hepatitis B virus escape variant, which is hardly recognized by the human immune system.

Electroporation, the conventional method of including materials into BNC, is unsuitable for including large molecules due to the variation in reproducibility. The electroporation is carried out only by manual operation, which is a major obstacle in producing a medical drug.Therefore, efforts are being made to reduce the dosage by the improved method, (e.g., lipsome-mediated method) which will realize inclusion with much greater efficiency than electroporation. In addition, safety and the dynamic state in the human body are being studied.

4. Conclusion

There have been a number of studies on carriers for DDS such as retrovirus and adenovirus, liposome and fusogenic forms of hemagglutination virus of Japan and liposomes. However, few carriers have been commercialized after clinical testing.

To date, BNC with the following features have been developed.

1. Liver cell-binding site by HBsAg (Pre-S1 region) indicates specific binding with human liver cells.
2. As it does not include genes derived from a virus, it is not pathogenic.
3. By replacing liver cell-binding sites presented on the surface for another living organism recognition molecule, pinpoint delivery to the targeted cell is made possible.

A drug BNC that is capable of excellent pinpoint delivery is sought to become a carrier that can be used for drug and gene therapy.

The findings of BNC introduced here are results from a collaborative study of Masaharu Seno (Okayama University), Akihiko Kondo (Kobe University), Masakazu Ueda (Keio University), Katsuyuki Tanizawa (Osaka University). The practical application of this study is proceeding with Beacle, Inc. which is a university venture, established by the authors.

References

[1] H. Sezaki: *Drug Delivery System*, Hirokawa, Tokyo (1989).

[2] E. Marshall: *Sciences*, **298**, 32 (2002).

[3] J. Savulescu: *J. Med. Ethics*, **27**, 148 (2001).

[4] T. Yamada, Y. Iwasaki, H. Tada, H. Iwabuki, M.K.L. Chuah, T.V. Driessche, H. Fukuda, A. Kondo, M. Ueda, M. Seno, K. Tanizawa and S. Kuroda: *Nat. Biotechnol.*, **21**, 885–890 (2003).

[5] S. Kuroda, S. Otaka, T. Miyazaki, M. Nakao and Y. Fujisawa: *J. Biol. Chem.*, **267**, 1953–1961 (1992).

APPLICATION 38

38 EXPRESSION OF OPTICAL FUNCTION BY NANOSTRUCTURE USING FEMTOSECOND LASER PROCESSING

It is well known that laser light can be pulsed and focused to a spot of wavelength order. Ultrashort pulsed lasers have been used as powerful tools to clarify elementary processes, such as excitation-energy relaxation and both electron and proton transfer on nanosecond and picosecond time scales that occur in a micrometer-sized area. In addition, ultrashort pulsed laser can be used to make microscopic modifications to transparent materials [1–9]. The reason for using this laser is that the strength of its electric field can reach 100 TW/cm^2, which is sufficient for inducing nonlinear optical effects in materials by use of a focusing lens, when the pulse width is 100 fs and the pulse energy is 1 μJ. The photo-induced reaction is expected to occur only near the focused part of the laser beam due to multiphoton processes.

1. Space selective valence state manipulation of rare earth ions inside glasses

The authors have observed space-selective room-temperature permanent photoreduction of rare-earth ions of Sm^{3+} to Sm^{2+} in glasses by an infrared femtosecond laser. Room-temperature permanent photoreduction of Sm^{3+} to Sm^{2+} was observed in borate

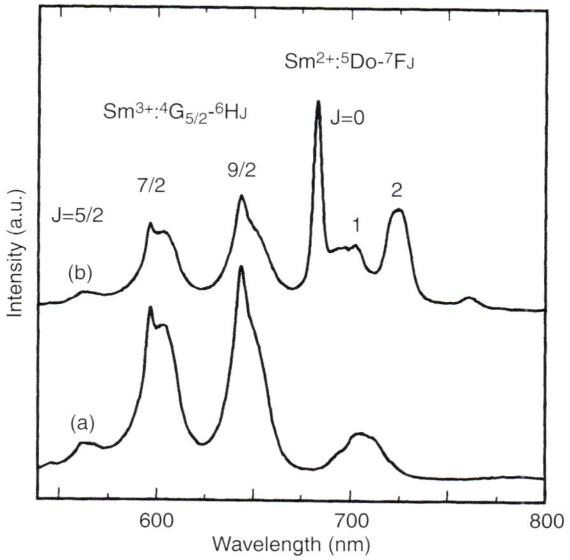

Figure 38.1
Photoluminescence spectra of the Sm^{3+} borate glass sample before (a) and after (b) laser irradiation.

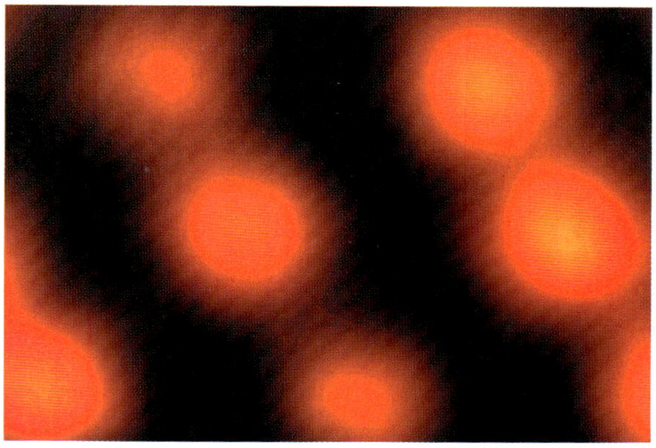

Figure 38.2
The irradiated spot of Sm^{2+} shown in red color. The diameter of one circle is about 200 nm. The black color is occupied by Sm^{3+} ions in glass samples.

and other glass samples. In Fig. 38.1 the photoluminescence spectra of the Sm^{3+}-doped glass sample before (a) and after (b) laser irradiation show that only emissions at 560, 600, 645 and 705 nm were observed in the unirradiated glass sample due to the 4f–4f transitions of Sm^{3+}. Four new peaks at 683, 700, 724 and 760 nm were observed in the photoluminescence spectrum of the laser-irradiated glass sample due to the 4f–4f transitions of Sm^{2+}. Therefore, a part of Sm^{3+} was converted to Sm^{2+} after the laser irradiation. The results demonstrated the possibility of selectively inducing a change of valence state of Sm^{3+} ions on the micrometer scale inside a glass sample by use of a focused nonresonant femtosecond pulsed laser as shown in Fig. 38.2. On the other hand, a three-dimensional optical memory has approximately 10^{13} bits/cm^3 storage density, which means that data information can be stored in the form of a change in refractive index in a spot; optical memory using a valence-state change of rare-earth ions in a spot may have the same storage density and allow one to read out data in the form of luminescence, thus providing the advantage of a high signal-to-noise ratio. Therefore, the present technique will be useful in the fabrication of three-dimensional optical memory devices with high storage density. One example by using the characters of A, B and C is shown in Fig. 38.3. These characters are written at each 10 layer. Moreover, femtosecond laser-photoreduced Sm^{3+}-doped glasses exhibited a photochemical spectral hole burning memory property. The microspot induced by the focused femtosecond laser inside a glass sample can be further used to store data information via the irradiation of laser light with different wavelengths. As a result, the data information can be read out in the form of spectral holes. Sm^{2+}-doped glasses could become an ultimate optical memory device with an ultrahigh storage density.

2. Precipitation control of gold nanoparticles inside transparent materials by a femtosecond laser

Nanoparticles exhibit a wide range of electrical and optical properties due to the quantum size effect, surface effect and conjoint effect of the nanostructures. Noble metal nanoparticles doped into materials exhibit large third-order nonlinear susceptibility and ultrafast nonlinear response. They are expected to be promising materials for ultrafast all-optical switches in the THz region. Nanoparticles need to be arranged into well-defined configurations or to be distributed space-selectively in materials in order to build integrated systems. Up to now, many studies have been carried out on the fabrication of nanoparticle-doped materials. However, there are no effective methods of preparation so that the distribution of nanoparticles is space-selectively well controlled. The authors demonstrated three-dimensional precipitation and control of nanoparticles in materials by using focused femtosecond laser irradiation and successive annealing in detail.

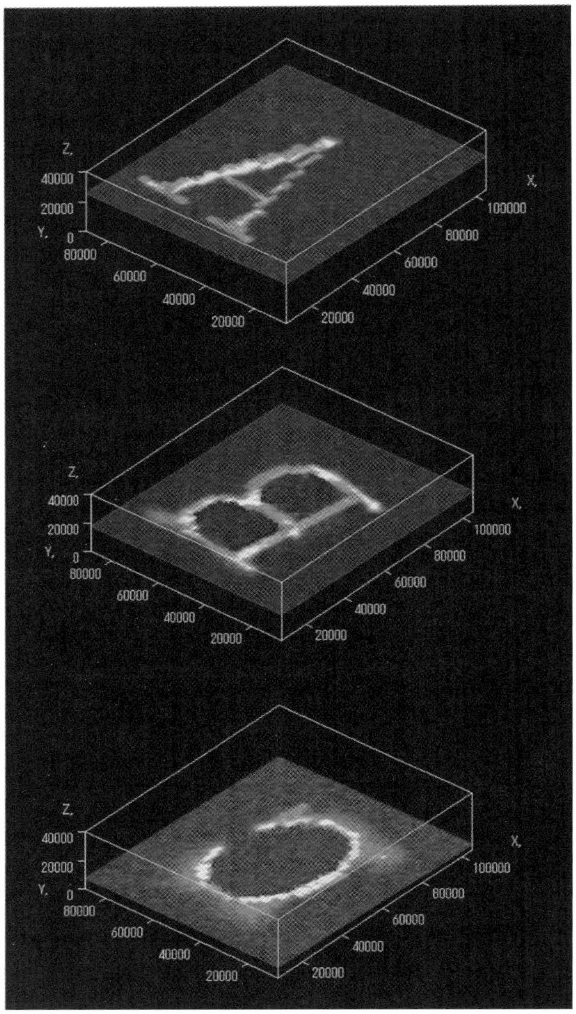

Figure 38.3
Three-dimensional optical memory using the Sm^{2+} luminescence at each 10 layer.

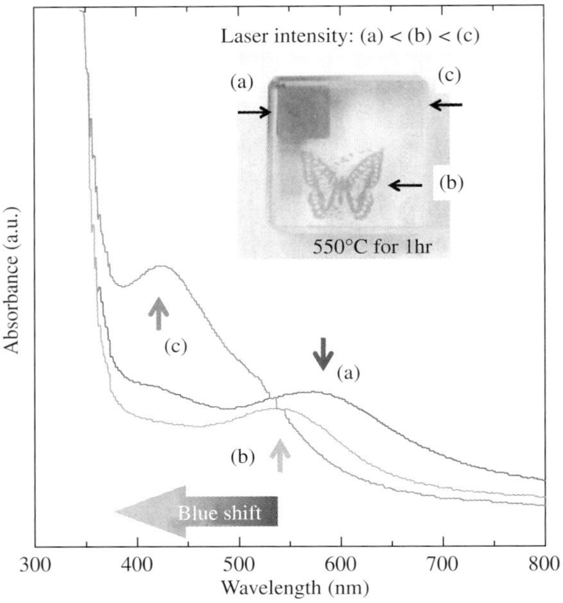

Figure 38.4
Photograph of images drawn inside the glass sample by using a femtosecond laser: gray-colored image (after femtosecond laser irradiation) and red-colored butterfly (after femtosecond laser irradiation and further heat treatment at 550°C for 1 h).

It is also demonstrated that the size distribution of nanoparticles can be controlled by the laser irradiation conditions. This method should be useful not only for practical applications such as three-dimensional optical memory and fabrication of integrative all-optical switches, but also for controlling nucleation and crystal growth processes. After irradiation by the focused infrared femtosecond laser with an average power of 300 mW and focused by a 10 × objective lens with a numerical aperture of 0.30 on each spot for 1/63 s, a gray-colored spot with a diameter of about 40 μm was formed near the focused area of the laser beam. Then, the glass sample was annealed at 550°C for 1 h. The laser-irradiated part became red caused by plasmon resonance shown in Fig. 38.4b after the heat treatment. By using these phenomena, a red-colored butterfly and then a gold-colored image were obtained by further annealing caused by the growth of Au particles inside the glass sample, as shown in Fig. 38.4c. An absorption spectra of the glass sample before and, after the femtosecond laser irradiation and successive annealing at various temperatures for 1 h show that there was an apparent increase in (c) absorbance in the wavelength region from 300 to 800 nm in the irradiated region due to hole trap centers. It was suggested that a gold ion was reduced to a gold atom by capturing an electron from non-bridging oxygen during the femtosecond laser irradiation and gold atoms aggregated to form nanoparticles after the heat treatment. Fig. 38.5 shows a TEM picture of precipitated gold nanoparticles in the femtosecond laser-irradiated part after successive annealing at 550°C for 1 h, which were

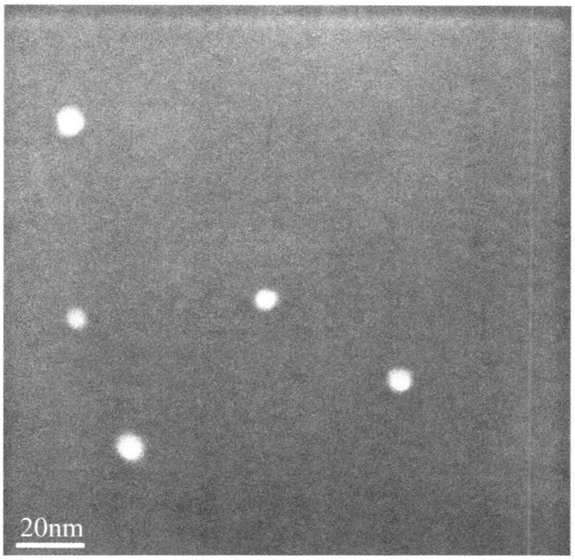

Figure 38.5
TEM image of gold nanoparticles in the femtosecond laser-irradiated glass sample after annealing at 550°C for 1 h.

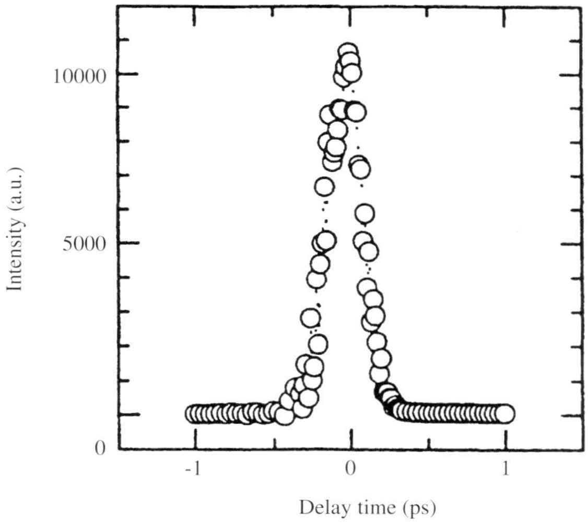

Figure 38.6
FWHM signal of nanoparticle precipitated specimen annealed at 550°C for 1 h.

found to be spherical gold nanoparticles with sizes ranging from 5 to 8 nm. A femtosecond optical Kerr shutter (OKS) experiment was performed with a 2-mm long gold nanoparticle-precipitated sample annealed at 550°C for 1 h. The full width at half maximum of the incident pulse was estimated as 500 fs at the position of the sample. The photo -energy of the pulse was set to the surface plasmon resonance peak (2.3 eV). The Kerr signal raised and decayed suddenly at around $t = 0$. The FWHM of the signal is 240 fs, and no slow decay component was observed as shown in Fig. 38.6. This is a very fast nonlinear response time for a gold nanoparticle system. The $\chi^{(3)}$ of the gold nanoparticle-precipitated part is estimated to be 0.93×10^{-11}, which is 300 times the magnitude of that for SiO_2 glass (2.8×10^{-14} esu).

3. Nanograting fabrication

Until now there has been no observation of periodic structures being generated within the bulk of a material just by a single writing laser beam and the mechanism of its appearance has not been fully understood. Here, the authors have shown the first observation of the emergence of periodic structurally changed regions of nanometer size inside silica glass after irradiation by intense femtosecond light pulses. The phenomenon is interpreted in terms of interference between the incident light field and the electric field of bulk electron plasma density wave, resulting in periodic modulation of electron plasma concentration and permanent structural changes in glass. After laser irradiation, the sample was polished to the depth of the beam waist location. The surface of the polished sample was analyzed by scanning electron microscope (JEOL, model JSM-6700F) and Auger electron spectroscopy (PHI, model SAM-680). Secondary electron (SE) images and backscattering electron (BE) images of the same surface were compared (Fig. 38.7). It is well known that the SE image reveals the surface morphology of a sample, while the BE image is sensitive to the atomic weight of the elements or the density of material constituting the observed surface. The SE images of the polished silica sample indicate that the morphology of an irradiated sample in the examined cross-section hardly changes, namely, a void does not exist. On the other hand, the BE images reveal a periodic structure of stripe-like dark regions with low density of material and of ca. 20 nm width which are aligned perpendicular to the writing laser polarization direction. Horizontal striation, slightly visible in the SE images, could be explained by a weak surface relief created in the polishing process due to the density variations in the grating.

The authors have observed various femtosecond laser-induced phenomena in glasses. It was confirmed that the femtosecond laser-induced microstructure will open new possibility in the realization of novel optical functions for glass.

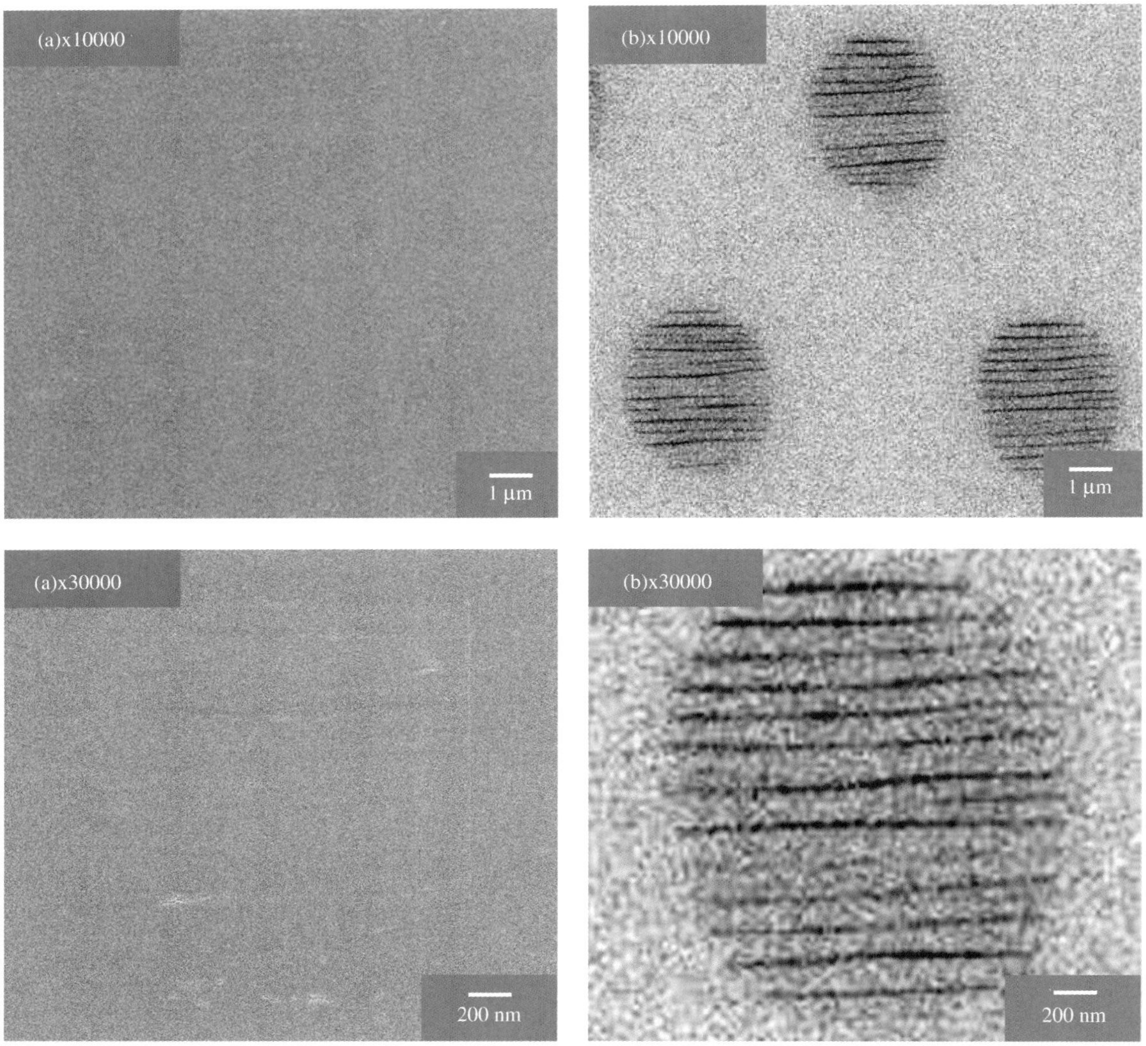

Figure 38.7
(a) Secondary electron images of silica glass surface polished close to the depth of focal spot. (b) Light "fingerprints": Backscattering electron images of the same surface. The magnification of the upper and lower images is ×10,000 and × 30,000, respectively.

References

[1] K.M. Daivis, K. Miura, N. Sugimoto and K Hirao: *Opt. Lett.*, **21**, 1729 (1996).

[2] K. Miura, J. Qiu, H. Inouye, T. Mitsuyu and K. Hirao: *Appl. Phys. Lett.*, **71**, 3329 (1997).

[3] K. Hirao, K. Miura: *J. Non-Cryst. Solids*, **239**, 91 (1998).

[4] K. Miura, J. Qie, T. Mitsuyu and K. Hirao: *Proc. SPIE*, **3618**, 141 (1999).

[5] K. Miura, J. Qiu, T. Mitsuyu and K. Hirao: *Opt. Lett.*, **25**(6), 408 (2000).

[6] K. Miura, J. Qiu, S. Fujiwara, S. Sakaguchi and K. Hirao: *Appl. Phys. Lett.*, **80**, 2263 (2002).

[7] J. Qiu, K. Miura and K. Hirao: *Proc. SPIE-The Int. Soc. Opt. Eng.*, **5061** (2003) 82.

[8] K. Hirao, J. Qiu, K. Miura, T. Nakaya and J. Si: *United Approaches Mater. Sci.*, **29**(1), 3 (2004).

[9] J. Qiu, Y. Shimotsuma, K. Miura, P.G. Kazanski and K. Hirao: *SPIE*, **5713**, 137 (2005).

APPLICATION 39

39 INSTANTANEOUS NANOFOAMING METHOD FOR FABRICATION OF CLOSED-POROSITY SILICA PARTICLE

Most previous methods for fabricating nanoporous solids have focused on formation of open nanopores. Those open-porosity nanoporous solids are advantageous in case where reactants or fluids need to have access to the inner space of those nanopores. From the viewpoint of the characterization of the nanopores, a significant portion of the information on their structure can be obtained from their adsorption behavior. Bulk properties of nanoporous solids largely depend on the amount and structure of their nanopores. As a result, the material designing is normally considered to be realized by tailoring the nanoporous structure. For example, the electrical inductivity can be varied as a function of the porosity. In principle, the porosity as a relevant structural parameter is determined by the concentration of the structure-directing additives. Those additives normally form liquid crystalline phases at higher concentration. Since these liquid crystalline phases have highly interconnected structure, the resultant nanopores tend to be open to the outside. For these reasons, nanoporous solids of the open-porosity type have been the major object of studies and a quite large number of researchers are engaged in these works. Nevertheless, there are certain cases where closed nanopores are more desirable than open ones. The authors have sought fabrication methods of closed-porosity nanoporous silica [1–3]. For example, nanoscale assemblies of tri-block copolymers (Pluronic®) can be immobilized in instantaneously solidified silica matrix by rapid drying [2]. In this case, the block copolymers are not given sufficient time to form complete liquid crystalline phase structure. As a result, the finally obtained solid retains the structural features of the dispersed phase of those molecular assemblies. Resultantly, the volume fraction of the closed pore exceeds 50% of the total nanopores. In the author's experience, this example is the superior limit of the fraction of the closed pores that can be attained by the rapid drying method. The complete enclosure of those assemblies of the block copolymer by the rapid drying method is not realistic because the drying rate cannot be infinitely enhanced. Actually, the rate of drying process is severely determined by the heat transfer. It should be noted that molecular assemblies normally tend to reach a liquid crystalline phase which works as a template for the formation of open-porosity nanoporous structure. As one of the solutions to avoid the opening of the nanopores toward the outside, the author has introduced a method for the selective formation of closed nanopores. This method is based on so-called foaming in which a dilute phase (gaseous phase) is enclosed in a closed space in a condensed phase [4]. Normally, foaming is considered as one of the phase separation phenomena which take place in a mobile condensed phase such as liquid. Under such circumstances, the size of the unit cell of the foam (bubble) easily enhances being driven by the interfacial tension. The rate of the growth of the cell is much lower when the medium for the bubble formation is solid or extremely viscous liquid. Therefore, the growth of the cells is expected to be restricted when it occurs in solid silica matrix.

The experimental procedure for preparing the foamed silica is quite simple as follows. Tetraethyl orthosilicate (TEOS) was hydrolyzed with ten folds larger molar number of water at a weakly acidic condition (pH 5) with hydrochloric acid. The hydrolysis was readily completed in an hour and gave macroscopically homogeneous transparent silica sol. The silica sol was pretreated by a heat treatment to give solid silica. In case where the hydrolysis is carried out at an acidic condition, the obtained pretreated silica has many residual hydroxyl functions. At the moment of the instantaneous heat treatment above 500°C, the residual hydroxyl functions instantly underwent the condensation to form siloxane bondings. At the same time, the silica matrix instantaneously hardened due to the increase in the density of siloxane bondings. During the instantaneous condensation, the generated water immediately vaporized and formed bubbles (balloons) being enclosed in the densely developed siloxane bondings [4]. Fig. 39.1 shows a typical transmission electron microscope (TEM) image of the foamed silica. Many spherical balloon-like cells with the diameter of 1–10 nm are seen. Their spherical shape indicates that they were formed as the balloons were blown by the inner pressure of the enclosed gas (vapor). These bubbles are quite small compared to those formed in liquids or normal polymer resins. The hardened silica matrix effectively blocked the bubbles from continuation of the growth or coalescence. In the case of foaming in normal plastic resin, the size of the bubble is quite hard to restrict below 1 μm. This is because these resins normally need to be softened by increasing the temperature up to the vicinity of their softening point. The formed bubbles tend to readily grow or undergo mutual coalescence. Therefore, the hardening which occurs simultaneously with the condensation of the residual hydroxyl functions is considered to play a crucial role in the realization of the

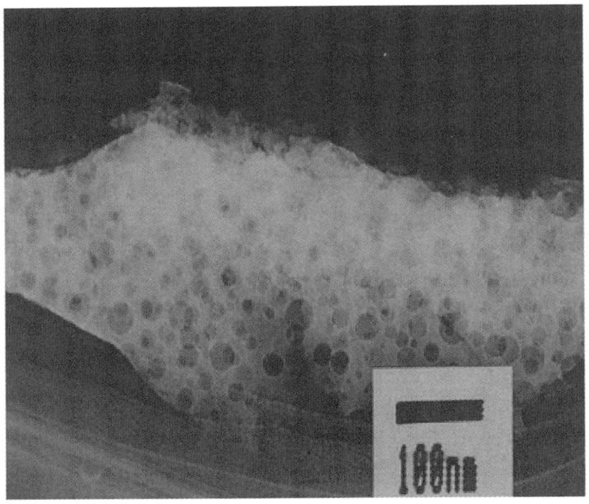

Figure 39.1
Typical TEM image of nanofoamed silica.

nanofoaming in the present case. Fig. 39.2 shows TEM images of the foamed silicas which were instantaneously heated at various temperatures ranging from 500 to 950°C. At 500°C, the shape of each cell was not completely spherical. The foaming behavior was enhanced with the temperature for the instantaneous heat treatment. Both the average size of the balloons and total pore volume are seen to increase with the temperature.

In addition to the nanoscopic size of the balloon, the high fraction of the closed pores is to be stressed as an important structural feature of the foamed silica. Fig. 39.3 shows nitrogen adsorption isotherms at 77 K of the foamed silicas obtained by instantaneous heating at various temperatures from 500 to 950°C. On the whole, the adsorbed amount of nitrogen for a unit weight of the foamed silica increased with the temperature for instantaneous heating. This tendency corresponds to the enhancement in the extent of the foaming behavior. The amount of the cell which was directly linked to the outer surface of the foamed silica particles increased with the extent of the foaming.

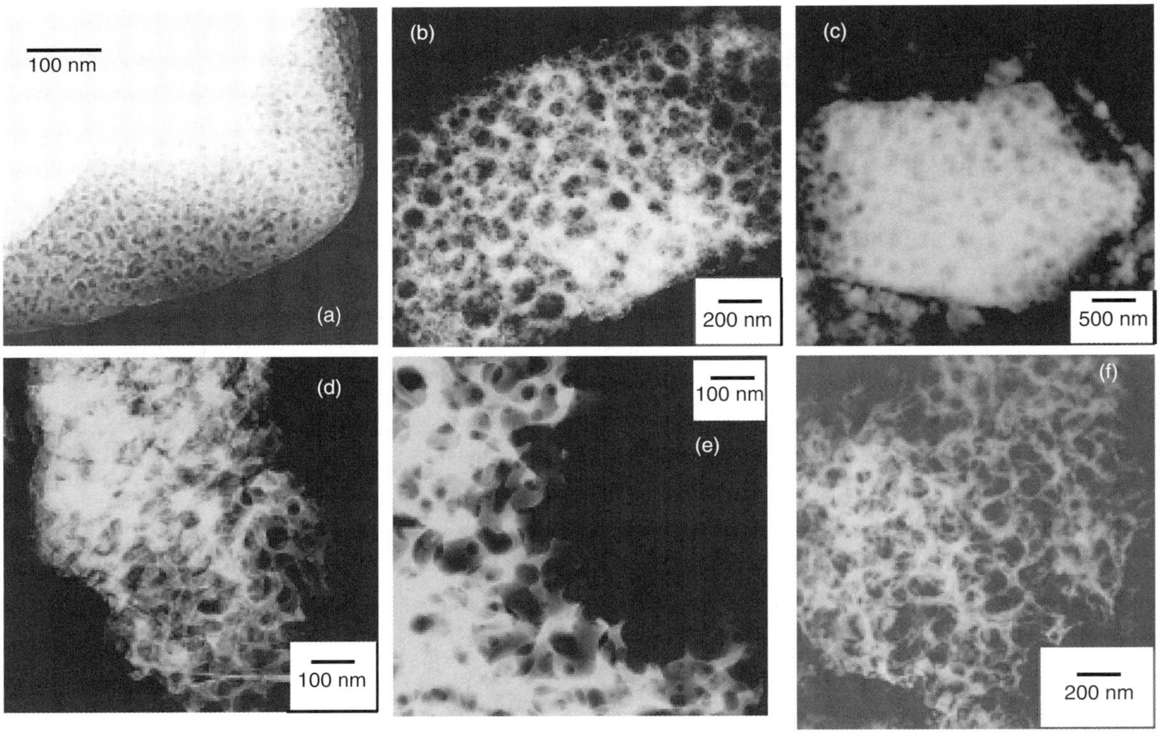

Figure 39.2
TEM images of nanofoamed silica. The foaming was carried out at various temperatures of instantaneous heat treatment as given below: (a) 500°C, (b) 600°C, (c) 700°C, (d) 800°C, (e) 900°C, (f) 950°C.

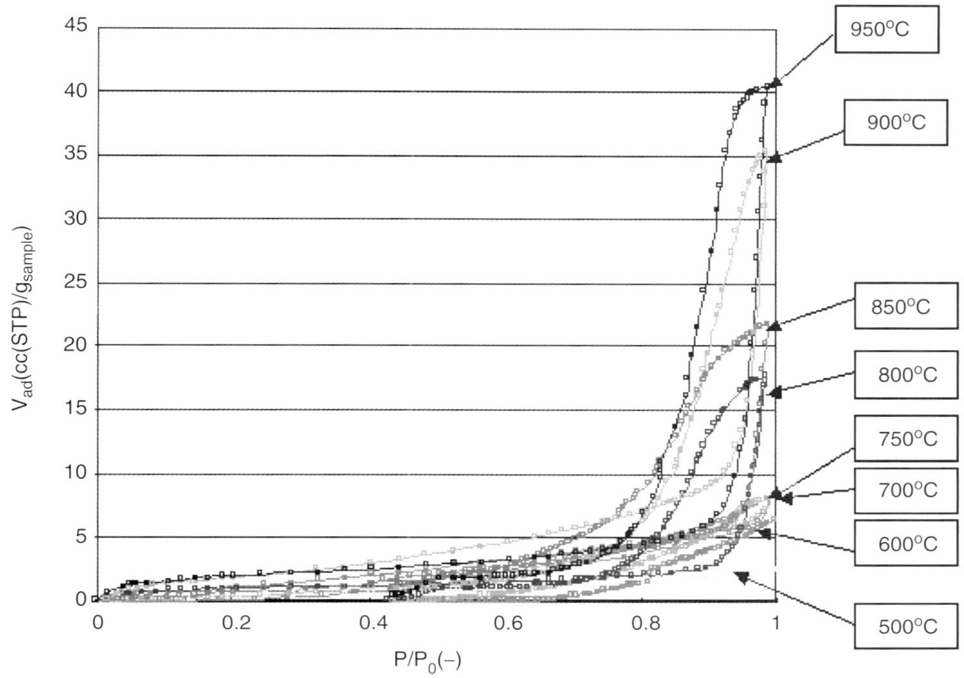

Figure 39.3
Nitrogen adsorption isotherms at 77 K obtained for nanofoamed silicas instantaneously heated at various temperatures. The temperatures for the heat treatment are shown in the right of the figure.

Nevertheless, it should be noted that the adsorbed amount of nitrogen itself was only 1/50 to 1/10 of that of normal open-porosity nanoporous silicas replicated from liquid crystalline phases. Therefore, the formed nanoballoons as seen in Fig. 39.1 mostly contributed to the closed porosity and there are almost no channels through which nitrogen molecules can go in or out of the formed silica. In the region of the relative pressure $0 \leq P/P_0 \leq 0.2$, the formed silicas had almost no adsorption. This result clearly shows that the silica matrix had no micropores. The absence of the micropores is distinctive of sol–gel derived silica prepared at weakly acidic contributions (pH 5). Contrastively, these micropores captured by the nitrogen adsorption at a very low relative pressure ($P/P_0 \approx 0$) are formed when sol–gel derived silica is prepared at basic conditions, for example, with the help of ammonia. Fig. 39.4 shows a TEM image of a silica sample which was prepared with the same procedure except that the condensation was carried out at a basic condition (pH 10) with ammonia. The sample in Fig. 39.4 was instantaneously heated at 800°C. Nevertheless, no bubble was found in the TEM image. Therefore, the silica matrix needs to be free of micropores for

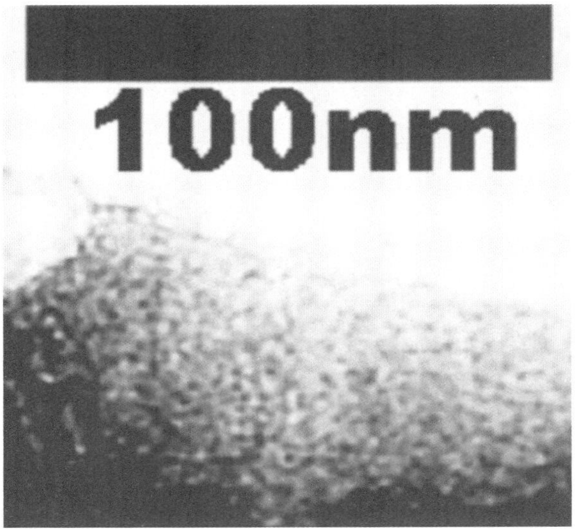

Figure 39.4
Silica samples prepared at ammonia basic condition at pH 10. Note that no bubbles were formed at this condition.

triggering the ballooning phenomena. The weakly acidic condition is required to form the nonporous silica matrix which enables the generated vapor to cause the nanoballooning on the instantaneous heat treatment for the foaming. Another possible reason for the absence of the foamed structure in Fig. 39.4 is that the condensation of hydroxyl function had been terminated prior to the instantaneous heat treatment for the foaming. This is also quite likely to be the case because the condensation of silanol is effectively catalyzed by a basic condition. In this case, the silica matrix cannot yield vapor available for the foaming on the instantaneous heat treatment.

Fig. 39.5 shows the dependencies of the closed porosity and maximum adsorbed amount of nitrogen on the temperature. The closed porosities are shown for both the cases – one where a transparent quartz crucible was used and another where a normal ceramic one was used. Here, the closed porosity can be evaluated from the apparent density assuming that the immersing fluid (gas or liquid) has no access to the inside space of the closed pores. The closed porosity was estimated using 2,200 and 0 kg m^{-3} for the neat silica matrix and the inside vacant space in the closed pore, respectively. The adjustment of the density of the immersing liquid was carried out by mixing

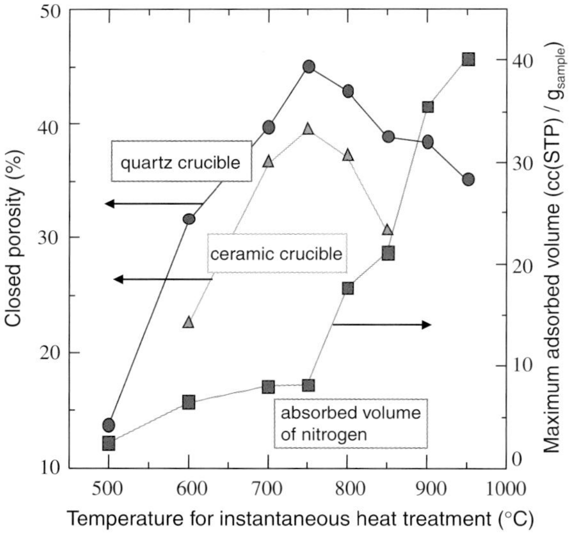

Figure 39.5
Dependencies of closed porosity and saturated adsorbed amount of nitrogen at 77 K. Data of the closed porosity are shown both for the cases where the silica samples were instantaneously heated in a transparent quartz vessel and in a ceramic one. The corresponding maximum adsorbed amount of nitrogen at 77 K is shown together on the right-hand ordinate.

bromoform and ethanol, which are completely miscible with each other. The range of the density measurement was between 800 and 2,800 kg m^{-3} in the present case. The transparent quartz crucible is obviously advantageous for the instantaneous heat transfer because radiation works more effectively particularly at a higher temperature. The experimental result in Fig. 39.5 that the closed porosity was larger when the quartz crucible was used indicates that more closed balloons were formed when the heating took place more rapidly with the help of the heat transfer by direct radiation. Furthermore, the maximum closed porosity was obtained at 750°C in both the cases described above. Above 850°C, the closed porosity clearly showed decrease. As indicated by the results in Figs. 39.2 and 39.3, more cells had open channels as the extent of foaming was enhanced. The abrupt increase in the adsorbed amount of nitrogen above 750°C is consistent with the cell-opening behavior as shown in Fig. 39.5 (right-hand ordinate). Therefore, the total amount of the closed pores increased below 750°C mainly due to the increase in the size of the mutually isolated pores. Above that temperature, the formed bubbles became partly interconnected resulting in increase in the nitrogen adsorption. Nevertheless, it should be noted that the proportion of the bubbles which converted to the open pores was still quite small. It does not seem that those bubbles are readily transformed to interconnected pores during the process of the transformation. Therefore, significant nanoscale transformation such as mutual fusion of the nanobubbles did not occur in the present system. The silica matrix behaved as a "hard" and "solid" matrix. Even in the case of 950°C, the open porosity was estimated to be approximately 5% to the volume of silica matrix from the saturated adsorbed amount of nitrogen at $P/P_0 \approx 1$. Thus, the foaming by the instantaneous heat treatment was shown to dominantly generate closed nanopores compared to open pores.

Assuming that the "pore opening" is brought about by partial fracture of the silica matrix during the ballooning process, preliminary reinforcement of the silica matrix prior to the instantaneous heating is expected to prevent the opening the bubbles during ballooning. A preliminary heat treatment of the silica matrix enhances its hardness by increasing the extent of the condensation of the residual silanol groups. Fig. 39.6 shows the dependence of the closed porosity on the temperature of the instantaneous heat treatment for multiple cases where the preliminary reinforcement of the silica matrices was carried out at various temperatures ranging from 50 to 450°C. On the whole, closed porosity clearly exhibited increase with the temperature for the preliminary reinforcement of the silica matrix. Furthermore, the peak temperature giving the maximum closed porosity shifts to the higher temperature. In the case of 450°C, closed porosity

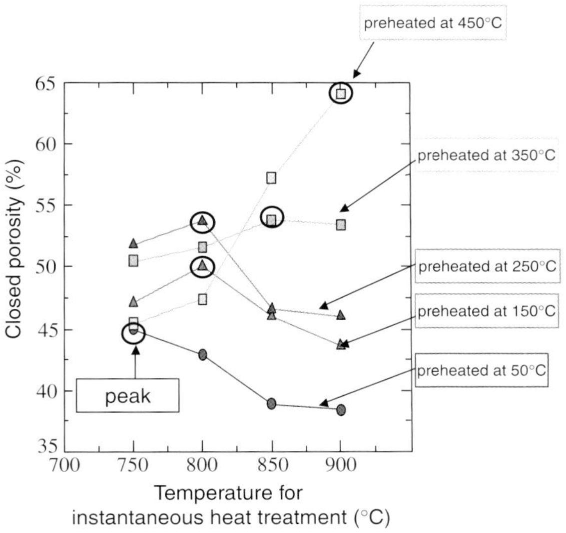

Figure 39.6
Dependencies of closed porosity on the temperature of the instantaneous heat treatment for the nanofoaming in various cases of temperatures of pretreatment for preliminary mechanical reinforcement of the silica matrix.

showed that the preliminary reinforcement in the mechanical strength of the silica matrix by preheating significantly contributed to enhancing the foaming without opening the formed nanobubbles to the outer surface. The maximum closed porosity that was actually obtained in the author's experiments was 64%, which was estimated from the apparent density of 790 kg m^{-3}. This value is almost the same as that of neat ethanol at room temperature. Here, it should be noted that the amount of the bubbles also increases with the total extent of the reinforcement by the preheating. Thus, the superiority in the mechanical strength of the silica matrix works more dominantly than the amount of the residual silanol groups which are directly responsible for generating water at the moment of the condensation.

Fig. 39.7 shows the nitrogen adsorption isotherms of the foamed silica samples before and after the instantaneous heat treatment for foaming. Fig. 39.7 reveals a drastic structural change caused by the instantaneous heat treatment. Before the instantaneous heat treatment, the silica matrix had micropores. They were captured as the prominent adsorption peaks seen at the left-hand side of Fig. 39.7 ($P/P_0 \approx 0$). These micropores were entirely eliminated by the instantaneous heat treatment. Thus, the enclosure of the generated vapor in the nanoscale bubbles was

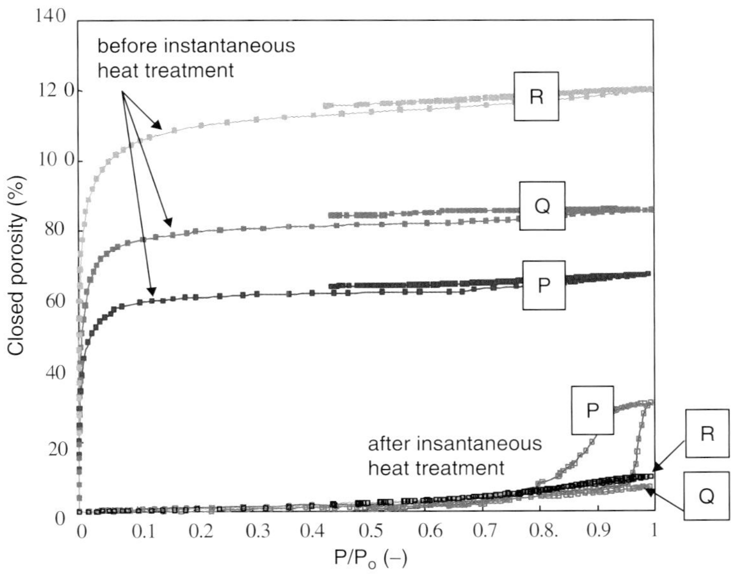

Figure 39.7
Nitrogen adsorption isotherms before and after the instantaneous heat treatment for the nanofoaming at three different conditions as given below: (a) preheated at 250°C, (b) preheated at 350°C, (c) preheated at 350°C with twice larger added amount of water for hydrolysis.

shown to be caused by the elimination of the micropores. The micropores instantaneously disappeared at the moment of the completion of the condensation of the residual silanol groups. That is, the completion of the siloxane bondings simultaneously blocked the generated water molecules from being fugitive from the silica matrix and the enclosed water molecules ubiquitously formed balloons all over the silica matrix. The preheating for the reinforcement is considered to contribute to the densification of the silica matrix. The siloxane network moderately densified by preheating would also be advantageous in preventing the generated water molecules from having way out of the silica matrix and form the nanoscale balloons.

Finally, the author estimated the number of water molecules enclosed in a single nanoscale bubble. Assuming that the final inner pressure inside the nanoscale bubble levels off at the atmospheric pressure, the occupied volume by a single gaseous molecule is roughly estimated as 4.1×10^{-26} m^3, which gives its cubic root as 3.5 nm. Based on this estimation, one nanoscopic bubble is considered to contain only a few decades of water molecules. Interestingly, even such a small number of the enclosed gaseous molecules can form entirely isotropic spherical geometry of the nanoscopic balloons.

References

[1] L. Pei, K. Kurumada, M. Tanigaki, M. Hiro and K. Susa: *J. Colloid Interface Sci.*, **284**, 222–227 (2004).

[2] L. Pei, K. Kurumada, M. Tanigaki, M. Hiro and K. Susa: *J. Mater. Sci.*, **39**, 4045–4047 (2004).

[3] L. Pei, K. Kurumada, M. Tanigaki, M. Hiro and K. Susa: *J. Mater. Sci.*, **39**, 663–665 (2004).

[4] K. Kurumada, N. Kitao, M. Tanigaki, M. Hiro and K. Susa: *Langmuir*, **20**, 4771–4773 (2004).

APPLICATION 40

40 EVALUATION AND APPLICATIONS OF DISPERSING CARBON NANOTUBE IN THE POLYMERS

1. Carbon nanotube

The carbon nanotube, abbreviated as CNT hereafter, was found by Iijima in 1991 and has drawn much attention as one of the remarkable new materials in the field of nanotechnology [1, 2]. Lately, many intensive CNT studies in very diverse applications were reported. CNT is expected to be applied in the development of various new materials.

There are two main streams in the research and development of CNT. One of them is to design and prepare a piece of CNT with high accuracy and advanced control on its orientation; and, the other is to conduct massive production of CNT. As an example of the former applications, CNT is developed to be used as the electron emission source of Field Emission Display (FED), cantilever of SEM, negative electrode material of Li battery, wiring of LSI, electronic devices and so on [3–6]. On the other hand, a large amount of CNT is required for making hydrogen storage materials and resin composite materials [7–9]. The ability to massively produce CNT plays an important role on its cost reduction for general applications. In this section, the dispersion of CNT in the molten resin, a common difficult task, is introduced.

2. Fracture model of agglomerates of carbon nanotube

Palmgren explained how to disperse carbon black agglomerates into the matrix rubber material. As shown in Fig. 40.1 [10], he proposed the following four steps [11].

1. Crush and micronize the large CNT agglomerates into smaller ones by impactions and other external forces.
2. Mix the smaller CNT agglomerates into the matrix material.
3. The mixture of CNT agglomerates and the matrix material is further micronized by shearing force to disperse and deagglomerate CNT fiber.
4. Repeat the process until the deagglomerated CNT fiber is well distributed and diffuses into the matrix material.

3. Dispersion of CNT by an extruder

To disperse CNT into the molten resin, the biaxial extruder with high kneading power is used in

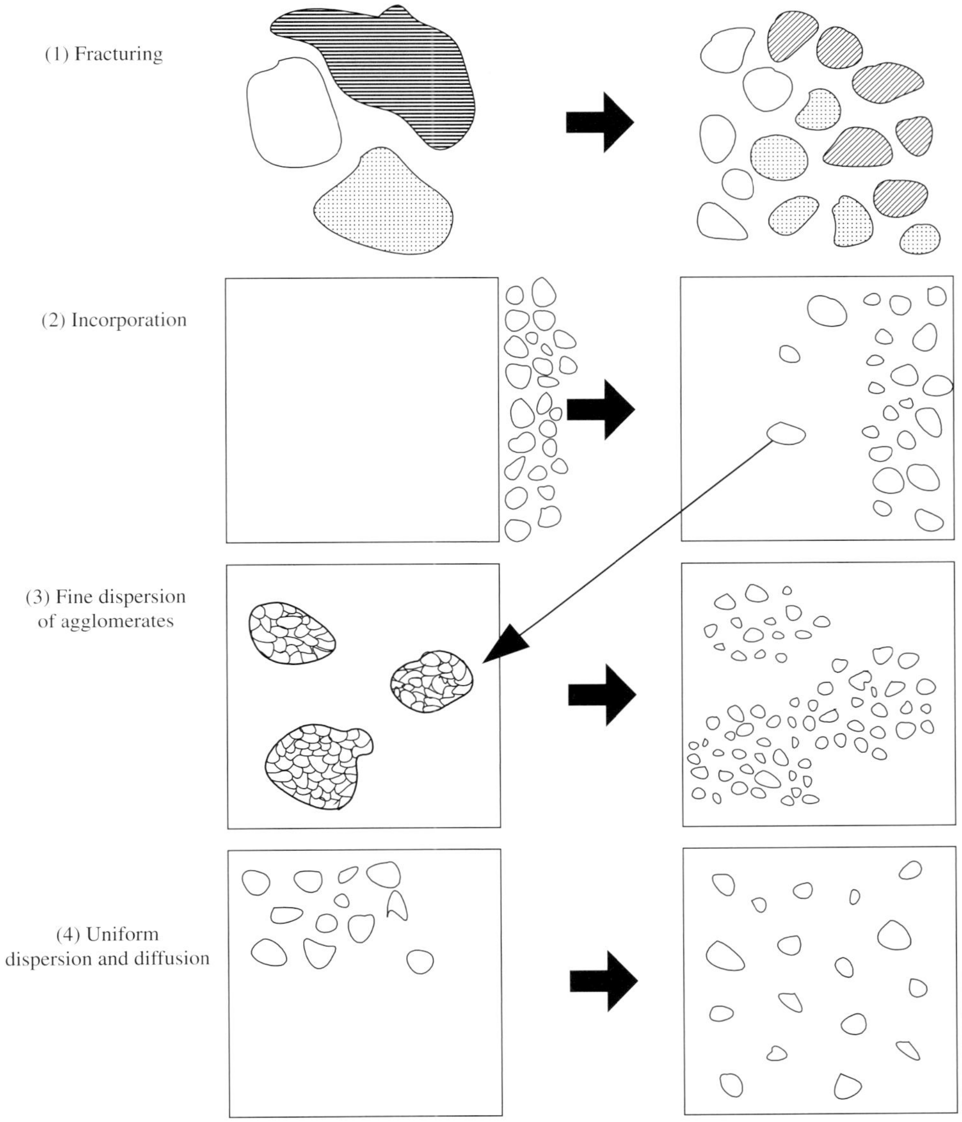

Figure 40.1
Four steps to disperse carbon black into rubber: (1) fracturing, (2) incorporation, (3) fine dispersion of agglomerates, (4) uniform dispersion and diffusion.

many cases. It is important that the CNT is dispersed to its individual fiber in order to bring out its original functionalities.

It is well known that the dispersing state of CNT is related to the operating conditions of the extruder and its screw design. However, only few reports elucidated these relationships. Takase et al. prepared CNT composite materials using a biaxial extruder and clarified the relationship between the dispersing state of CNT and the extruder processing conditions [12, 13]. From their report, the quantitative evaluation of CNT dispersion in the resin will be explained as follows.

4. Dispersion of composites and its evaluation

Although it was not necessary for nanomaterials, a number of reports discussed the dispersion of composite materials and investigated their particle morphology and the composite properties using transmission electron microscope (TEM) or scanning electron microscope (SEM). However, most of them, the pictures of TEM or SEM, were compared and evaluated visually; but the dispersion state of composites was not handled numerically. For quantitatively evaluating the dispersion state of composites, it is suggested to apply the image analysis of the microscope pictures, taken from the sliced samples of the composites, to define parameters such as *Ar* (the fraction of the area occupied by the agglomerates in the picture). This method will further be explained in the following section [12–14].

5. Relationship between the agglomerate fraction *Ar* and composite properties

It is well known that the mixing and dispersion states of fillers by extrusion depend upon the extrusion conditions such as the residence time and the mechanical construction of extruder such as the shape of screws. The relationship among the following three factors is quantitatively discussed below.

1. The effects of the mechanical construction and the operating conditions of the extruder are calculated in terms of the total shearing stress $\dot{\gamma} \cdot t$, which is the product of the shearing speed ($\dot{\gamma}$) and the residence time (t).
2. The volume resistivity (ρ_v) of the composite material is measured as a composite property.
3. The dispersion state of the composite is quantitatively evaluated as follows.

A sliced sample with a thickness of several microns is taken off from the CNT composite material. Fig. 40.2 shows a picture taken by a transmission stereomicroscope at 30-fold magnification, where the CNT agglomerates are seen as black spots. Using image analysis software, the particles having cross-section area larger than a certain value are selected and the fraction of the area of agglomerates *Ar*(%) in the picture is calculated. The *Ar* shows the degree of dispersion numerically and is defined as the dispersibility of CNT in the composite.

The screw rotation speed of the biaxial extruder *N* [rpm] and the resin flow rate from the extruder *Q* [kg/h], as noted in Fig. 40.2, are indicated as parameters of the forming condition. It can be seen that *Ar* describing the dispersion state of composite is affected by the processing conditions in the Fig. 40.2.

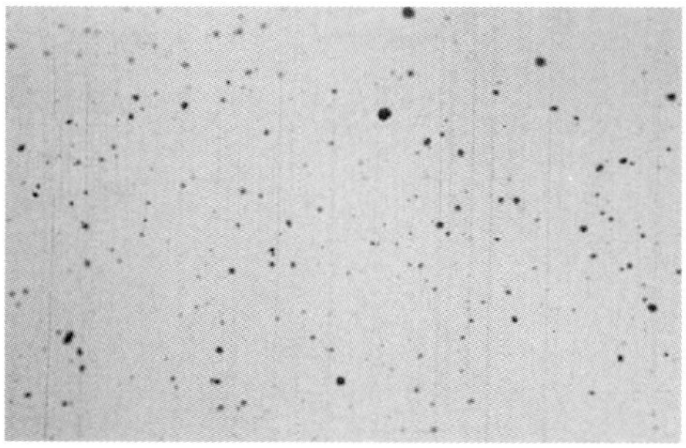

(a) N=20rpm [Q=1.7kg, Temp.=270 °C, Ar=1.59, L/D=33]

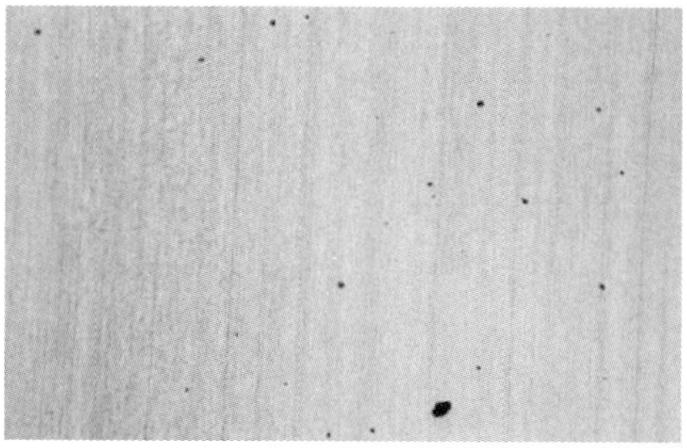

(b) N=60rpm [Q=1.7kg, Temp.=270 °C, Ar=0.33, L/D=33]

(c) N=100rpm [Q=1.7kg, Temp.=270 °C, Ar=0.18, L/D=33]

Figure 40.2
Observation of dispersion state of CNT by optical microscope.

The reason for the difference in the number and size of the agglomerates in spite of the same additive amount of CNT is that the CNT agglomerates are dispersed by the shearing force of the extruder to sizes below the detecting limit of the image analysis. The more CNT agglomerates are observed, as seen in Fig. 40-2a, the larger is the *Ar*, which reduces as the dispersion state of composite improves. It also shows that the dispersion of CNT increases with the higher screw rotation and the larger shearing speeds per unit mass of the material. Fig. 40.3 shows the TEM picture taken from the sample with good CNT dispersion (*Ar* below 0.2%) as shown in Fig. 40.2c. In the picture, a piece of CNT fiber obtained by the extruding deagglomeration and the fine CNT dispersion structure in the matrix can be observed.

An extruder with various types of screw segments and shapes is used to prepare CNT composites, so that a wide range of extruding conditions, shearing speed and the total shearing stress can be applied. In Fig. 40.4, (1) the total shearing stress $\dot{\gamma}\cdot t$, as a factor representing mechanical construction and operating conditions, and (2) the volume resistivity ρ_v, as the composite property, are plotted against (3) the agglomerate fraction *Ar*, as a parameter of the dispersion degree of filler. It is seen that as the total shearing stress increases, *Ar* reduces; namely, the dispersion is improved. In this way, the total shearing stress affects and controls the dispersion state, and by adjusting the extrusion condition the dispersion degree can be optimized.

The *Ar* value introduced here as a dispersion measurement is under the assumption that CNT is fully dispersed when the size of agglomerates is below the observation limit. The evaluation can be done with TEM picture as shown in Fig. 40.3, if it is necessary to confirm the CNT dispersion down to a individual fiber.

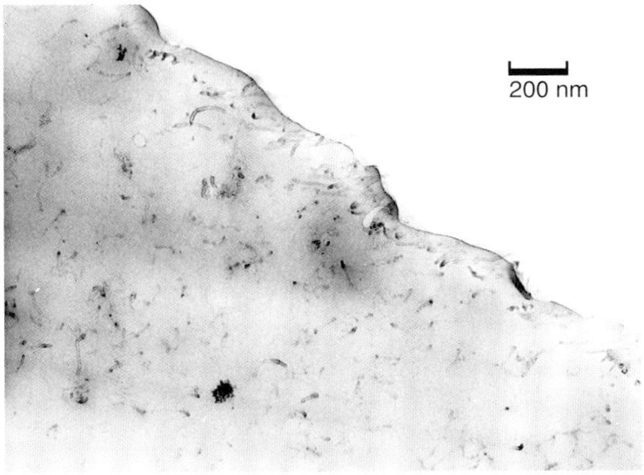

Figure 40.3
Observation of dispersion state of CNT by TEM.

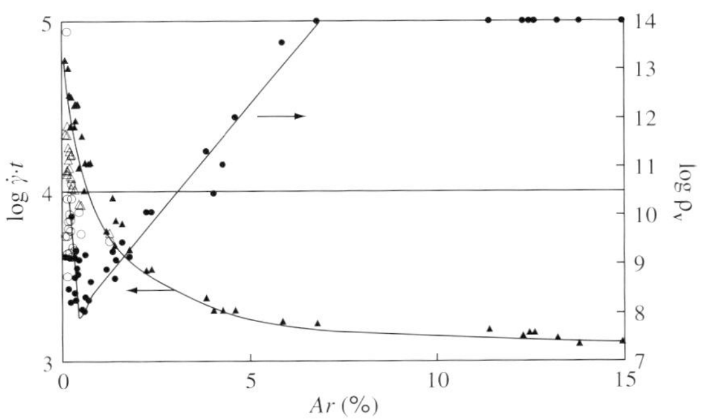

Figure 40.4
Dependences of dispersion ratio (*Ar*) on the total shear strain ($\dot{\gamma}\cdot t$) and the volume resistivity (ρ_v). ○, △ Screw profiles of high shear region, ●, ▲ Screw profiles of low-middle shear region.

6. Percolation

Percolation is the phenomenon which cannot be neglected on making electroconductive composite material. Fig. 40.5 shows the percolation model and the change of volume resistivity with the concentration of CNT and conventional electroconductive fillers. When an electroconductive filler such as carbon black is added into the resin, it has very low conductivity with a small amount of the filler. However, the electroconductivity shoots up abruptly as the addition increases. It appears that, at the low filler concentration, the filler only exists sparsely in the resin as in

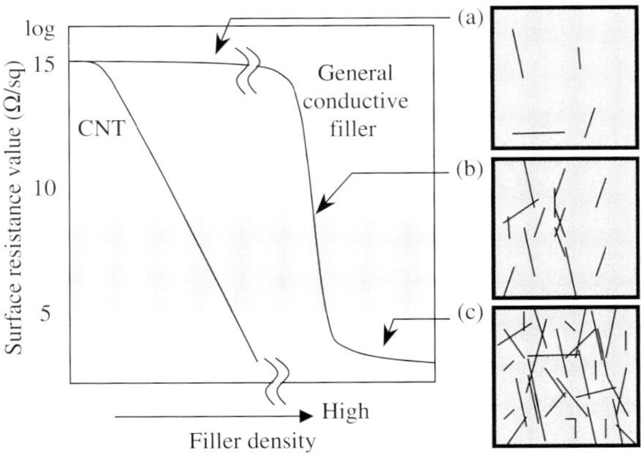

Figure 40.5
Percolation model. Changes of volume resistivity with CNT and conventional electroconductive fillers. (Conventional electroconductive fillers: carbon black, acetylene black, Kechin black, etc.).

Fig. 40.5a. As the amount of filler fibers increases in the resin, a continuous structure, called electroconductive path, is formed due to the inter-contacts as shown in Fig. 40.5b.

The electroconductivity does not increase proportionally with the filler amount but goes up suddenly. This phenomenon is called percolation and the critical filler concentration is defined as the percolation threshold.

Comparing the percolation curves of CNT and conventional electroconductive fillers, it is found that the electroconductivity of composites is more controllable with CNT than with the conventional fillers, because the percolation tends to take place at the surface resistance around 10^5–10^9 Ω/sq as required by the antistatic materials.

7. Development of CNT composite resin materials

The most potential application of CNT composites is to avoid electro-static discharge (ESD). The recent advanced semiconductors and hard disks often suffered from the problems caused by the ESD. For reducing the surface resistance down to the above-mentioned value to eliminate the electrostatic charging, conventional electroconductive carbon fillers have to overcome the following problems and are required to meet severe specifications as follows.

Problems:

- The properties of composites including their flatness degrade with an increase in high filler content.
- The surface resistance of composites fluctuates and leads to instability because of the percolation.
- The filler particles drop off from the composites because of their high concentration and particle shape. These free filler particles contaminate the clean room and adhere onto the devices causing the short circuit of wiring.

Required specifications:

- The surface resistance needs to be controllable arbitrarily in the range of $10^5 - 10^9$ Ω/sq with an accuracy of less than $10^{\pm 1}$ Ω/sq.
- There should be no surface resistance fluctuation at different locations.
- There should be no degradation of composite properties.
- The drop-off of fillers needs to be reduced.

It is possible to reduce the filler amount down to one-tenth with CNT while keeping the same surface resistance as that with conventional electroconductive fillers. It results in less degradation of the composite properties. In addition, the surface resistance is easy to control with less scattering, because its percolation curve as shown in Fig. 40.5 is not steep.

Furthermore, CNT hardly drops off from the composites due to the locking effect caused by its fibrous shape. Apart from the semiconductor industry, there are a number of applications where the electrostatic problems need to be mediated. CNT seems to have a big potential market if its price becomes reasonable.

CNT has excellent features in mechanical strength and heat conductivity in addition to the electroconductivity and is expected to be used in various areas. Its dispersion method should be selected considering the shape and properties of CNT. It is important to evaluate the dispersion state with proper parameters.

Although there are still many obstacles to overcome in the dispersion technology as well as the price and quality, CNT composite is regarded as its nearest commercial application on a mass basis.

References

[1] S. Iijima: *Nature*, **354**, 56 (1991).
[2] *Nikkei Mechanical*, **567**, 36 (2001).
[3] Y. Saito: *J. Nanosci. Nanotechnol.*, **3**, 39 (2003).
[4] S. Fan, M.C. Chapline, N.R. Franklin, T.W. Tombler, A.M. Cassell and H. Dai: *Science*, **283**, 512 (1999).
[5] Y. Nakayama: *Ultramicroscopy*, **91**, 49 (2002).
[6] S. Akita, Y. Nakayama, S. Mizooka, Y. Takano, T. Okawa, Y. Miyatake, S. Yamanaka, M. Tsuji and T. Nosaka: *Appl. Phys. Lett.*, **79**, 1691 (2001).
[7] A.C. Dillon, K.M. Jones, T.A. Bekkedahl, C.H. Kiang, D.S. Bethune and M.H. Heben: *Nature*, **386**, 377 (1997).
[8] X.B. Wu, O. Chen, J. Lin and K.L. Tan: *Int. J. Hydrogen Energy*, **25**, 261 (2000).
[9] H. Kato: *Plastics*, **52** (9), 75 (2001).
[10] H. Palmgren: *Rubber Chem. Technol.*, **48**, 462 (1975).
[11] S. Hashizume: Resin/Filler Mixing & Dispersion, 2, 50, Technical Information Institute (2000).
[12] H. Takase, Y. Mikata, S. Matsuda and A. Murakami: *Seikei-Kakou*, **14**(2), 126 (2002).
[13] H. Takase, M. Furukawa, H. Kishi and A. Murakami: *Seikei-Kakou*, **17**(1), 50 (2005).
[14] H. Takase, Y. Mikata, S. Matsuda and A. Murakami: *Seikei-Kakou*, **15**(1), 80 (2003).
[15] K. Mitsuishi, M. Kawano and N. Nagayama: *Seikei-Kakou*, **8**(3), 183 (1996).
[16] K. Terashita, K. Miyanami, N. Yabe, K. Izumida: *Zairyou*, **37**(422), 1344 (1988).
[17] ASTM-D-2663-89.

APPLICATION 41

41 SURFACE MODIFICATION OF INORGANIC NANOPARTICLES BY ORGANIC FUNCTIONAL GROUPS

Nanoparticles are considered as key materials in informational, environmental and medical technologies. Various methods are proposed to synthesize the nanoparticles of metal, metal oxide and organic materials. The merits to use nanoparticles in these fields are summarized as below:

(a) We can design new materials with hybridized functions by dispersing nanoparticles in a liquid or solid.
(b) We can reduce resources and costs required to enable various functions that emerge at surfaces.
(c) We can bind nanoparticles with biomolecules to enable diagnosis and cure with the minimal invasion.
(d) We can fabricate nanostructures with less effort by organizing different kinds of nanoparticles.
(e) We can explore new properties that arise based on quantum size effects.

These merits emerge from the two characters of nanoparticles: (1) the new phase of materials that can mix with solids or liquids to form virtually continuous phases, and (2) the minimum unit of materials that have defined structure and functions, which arise from the decreased size of nanoparticles itself. However the decreased size also results in the difficult handling. We cannot manipulate each nanoparticle and the nanoparticles tend to aggregate. This nature of nanoparticles prohibits the handling similar to that of micrometer-sized particles.

These tendencies come from the effects of surface atoms, which plays dominant role in the dispersion, aggregation and hybridization in the nanometer-sized materials. However, we can exploit this tendency to control the behavior of nanoparticles. The surface properties of nanoparticles are possibly changed by functional groups on the surface nanoparticles. Based on this idea, we propose the chemical modification of the surface of nanoparticles to realize better handling of nanoparticles. In this chapter, we discuss the attachment of organic functional groups on the surface of inorganic nanoparticles.

1. Surface-modified noble metal nanoparticles

Sulfur atom has large affinity with noble metals. This affinity is used to produce surface-modified noble metal nanoparticles. Thiol-capped noble metal nanoparticles are synthesized by reducing noble metal ions in the presence of thiols [1]. While the reduced noble metal atoms aggregate to form nanoparticles, thiol molecules attach on the surface of the nanoparticle to passivate the growing surface and minimize the chance to aggregate. The surface modification by alkanethiol ($C_nH_{2n+1}SH$) produces hydrophobic nanoparticles because SH groups bind with noble metal atoms and the nanoparticles are covered with alkyl groups. The modification can be performed with other molecules that have sulfur atoms [2]. The binding energy between noble metals and sulfur atoms is not so strong as covalent bonds. Based on this nature, we can realize surface modification with two kinds of thiols by dispersing thiol-modified noble metal nanoaparticles in a solution with a different kind of thiol. The exchange reaction between the thiol on the nanoparticles and the thiol in solution occurs to produce binary modified noble metal nanoparticles. Most noble metal nanoparticles including gold, silver, copper, palladium, platinum and nickel can be modified with thiols. Due to its easy experimental procedure, many researches have studied the surface modification of noble metal nanoparticles by thiols.

2. Organic modification of metal oxide nanoparticles

Metal oxides have various properties including electron transport, semiconducting property, ferromagnetism, giant magnetoresistance, luminescence, ferroelectric property and catalysis (Table 41.1). In addition, most oxides are stable even in air and water. Due to these practical merits, various metal oxides are used in functional devices. However, incorporation of these properties of metal oxides into nanoscale devices requires two techniques, that is, synthesis of metal oxide nanoparticles and their surface modification. So far, few studies have been performed to synthesize surface-modified metal oxide nanoparticles. We are studying the metal oxide nanoparticles whose surface is covered by organic functional groups. We have developed simultaneous synthesis and modification of metal oxide nanoparticles because nanoparticles easily aggregate together irreversibly. Fig. 41.1 shows our method to synthesize surface-modified metal oxide nanoparticles. During hydrothermal synthesis, the growth of metal oxide nanoparticles proceeds by

Table 41.1
Functions of metal oxides arranged by inputs and outputs.

Input\output	Photonic	Electronic	Magnet	Thermal	Chemical
Photonic	Luminescence Laser	Photoconductivity Photovoltaic effect	Photoinduced magnet		Photo catalysis
Electronic	Light-emitting devices Electro luminescence	Semiconductor Dielectrics	Magnetoelectric effect		
Magnetic	Faraday effect	Magnetic induction Colossal magnetic resistance		Magnetic heating	
Thermal		Piezoelectric materials	Magnetic phase transition		
Chemical		Fuel cell Chemical sensor			Catalysis

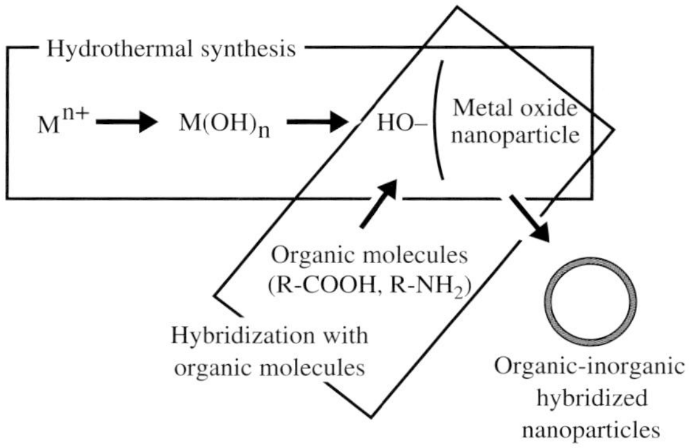

Figure 41.1
Schematic of the surface modification of metal oxide nanoparticles during their hydrothermal synthesis.

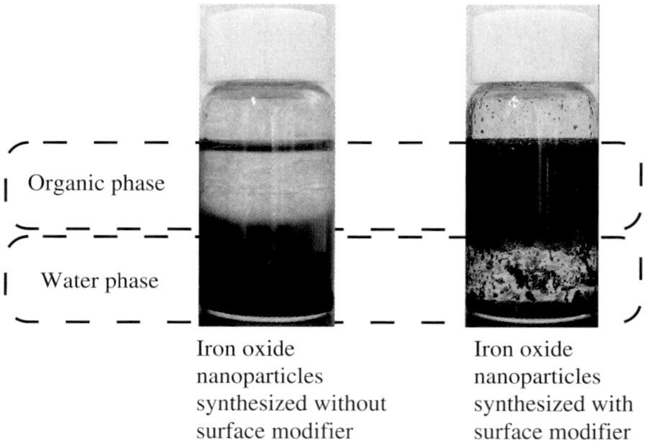

Figure 41.2
Change in dispersion of nanoparticles by surface modification.

the dehydration reaction between surface hydroxyl groups and metal hydroxides. We proposed surface modification through chemical bonding between the surface hydroxyl groups on the surface of metal oxide nanoparticles and organic reagents during the synthesis. Fig. 41.2 shows the surface-modified metal oxide nanoparticles. The hydrothermally synthesized metal oxide nanoparticles were covered with hydroxyl groups and therefore hydrophilic. On the other hand, the surface-modified metal oxide nanoparticles were covered with alkyl chains, which made the surface of metal oxide nanoparticles hydrophobic. This result confirms that the synthesis of metal oxide nanoparticles in the presence of organic reagents produces the surface-modified metal oxide nanoparticles. We also succeeded in modifying the surface of metal oxide nanoparticles with $-COOH$ or $-NH_2$ groups using the similar method. These modifications allow us to hybridize metal oxide nanoparticles with biomolecules, polymers and solid surfaces through strong chemical bonds.

3. Hybridization of inorganic nanoparticles with biomolecules

The synthesis of surface-modified inorganic nanoparticles enables the use of inorganic nanoparticles in various medical applications. The surface-modified nanoparticles can bind with biomolecules as well as inorganic materials. The hybridized nanoparticles have both inorganic properties and biological specificity. In this section, we discuss the alignment of metal nanoparticles on a ladder structure of deoxyribonucleic acid (DNA) through hybridization between nanoparticles and DNA single strands. Recently, Mao et al. [3] succeeded to build up ~10 nm rhombic lattices from 6 or 8 DNA single strands with designed sequences. They used the nature of DNA single strands to form Holliday junction that appears during the exchange of genetic information (Fig. 41.3). They also succeeded in combining the lattices together to form lattice and ladder structures of DNA. Based on this method, we tried to hybridize metal nanoparticles with the DNA lattice structure to align the metal nanoparticles. We designed the sequences of DNA single strands to prepare the similar structure as Mao et al. [3]. The designed sequence prepares a DNA ladder structure that allows a guest DNA single strand to attach on both side of the DNA ladder. We also designed a DNA single strand that can hybridize with the DNA ladder on one end and with a gold nanoparticle with the other end. By mixing gold nanoparticles, the designed single strand and the DNA ladder structure, we succeeded in aligning gold nanoparticles as shown in Fig. 41.4 [4]. In addition to this work, we are trying to hybridize inorganic nanoparticles with enzymes and antibodies to realize various inorganic-bimolecular complexes.

In this chapter, we discussed the surface modification of inorganic nanoparticles by organic functional groups. Recent development in the preparative methods of inorganic nanoparticles has opened the way to the application of the nanoparticles. The organic modification of the inorganic nanoparticles is the most suitable method to modify the surface properties including dispersion and hybridization of the nanoparticles. We believe that the control of the surface properties of the nanoparticles widens the range of application.

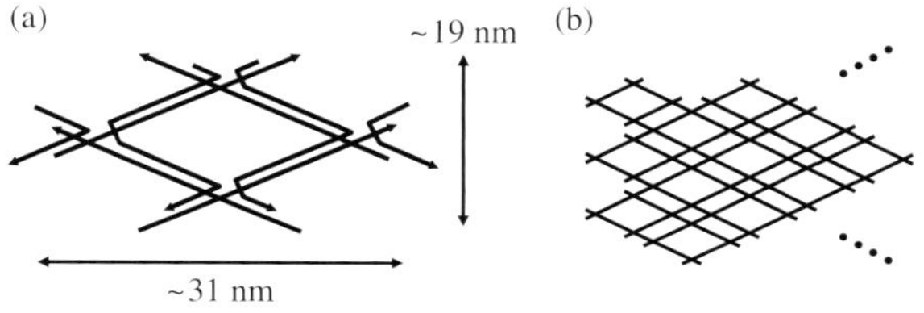

Figure 41.3
(a) Eight DNA single strands compose the rhombic structure as proposed by Seeman et al. (b) The rhombic structures attach together to from a planar structure.

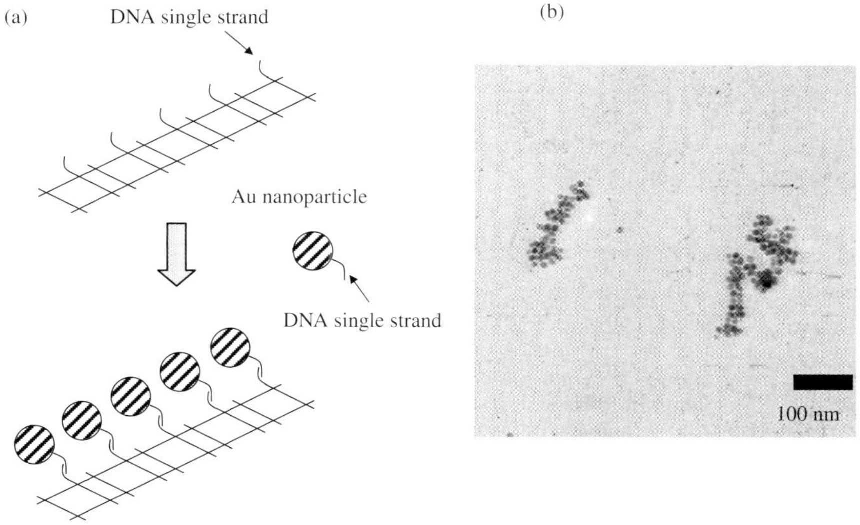

Figure 41.4
(a) Designed arrangement of Au nanoparticles using the DNA ladder structure, (b) transmission electron micrograph of prepared gold structures.

References

[1] M. Burst, M. Walker, D. Bethel, D.J. Stiffing and R. Why man: *J. Chem. Soc. Chem. Common.*, 801–802 (1994).

[2] A.C. Templeton, W.P. Wuelfing and R.W. Murray: *Acc. Chem. Res.*, **33**, 27–36 (2000).

[3] C. Mao, W. Sun and N.C. Seeman: *J. Am. Chem. Soc.*, **121**, 5437–5443 (1999).

[4] Y. Hatakeyama, M. Minami, S. Ohara, M. Umetsu, S. Takami and T. Adschiri: *Kobunshi Ronbunshu*, **61**, 617–622 (2004).

APPLICATION 42

42 FABRICATION TECHNIQUE OF ORGANIC NANOCRYSTALS AND THEIR OPTICAL PROPERTIES AND MATERIALIZATION

Organic nanocrystals are, for the definition, crystals of nanometer order of an organic compound. Organic nanocrystals are the nanomaterials that have been attracting attention in recent years, because they show some interesting characteristics for basic science, and many applications using organic nanocrystals are expected as well.

Regarding the fabrication techniques of organic nanocrystals until around 1990, an evaporation technique in inert gas was reported as the bottom-up method by Toyotama or Yase et al. [1, 2]. Evaporation techniques, however, were not applicable for preparing nanocrystals of different organic compounds, because ordinary organics are unstable at high temperatures. On the other hand, the fabrication of organic pigments nanocrystals has been performed using milling technique mainly by company researchers. The milling method is effective to obtain the nanocrystals with particle size around 50–100 nm. However it is not easy for the milling method to reduce the particle size to 50 nm below because of structural rearrangement in crystals induced by a mechano–thermal effect. In this review, the authors introduce "the reprecipitation method" [3] and the related technique to fabricate organic and polymer π-conjugated nanocrystals. And their size-dependent optical properties and application are described here.

1. The organic compounds used for nanocrystallization

The π-conjugated compounds attempted so far for the preparation of nanocrystals are summarized in Fig. 42.1.

Polydiacetylenes are π-conjugated polymer materials that can be obtained by irradiation of UV light or γ-ray or by thermal treatment through solid-state polymerization of diacetylenes [4]. The diacetylene derivative used in this paper was 1,6-di(*N*-carbazolyl)-2,4-hexadiyne (DCHD), which has the carbazolyl group at both ends. 4-Dimethylamino-N-Methyl-4-Stilbazolium Tosylate (DAST) is an ionic compound with stilbazolium structure. As the electric dipole moment of each molecule almost falls in the same direction in a DAST crystals, it is well known that the crystals have large dipole moments and show very large second-order nonlinear optical properties. TPB and perylene are organic dye compounds having strong fluorescence even in crystalline state, and attract attention as fluorescence and electroluminescence (EL) materials. Titanylphthalocyanine (TiOPc), quinaridone, C_{60} and so on were used as hardly soluble compounds. TiOPc is known to have a superior optical photoconductive property and has been really used in usual photocopiers. Quinaridone has been used very well as a high-quality red pigment. All of the compounds shown in Fig. 42.1 are of interest for their optical and electronic properties in the crystalline state.

2. Fabrication techniques of organic nanocrystals

(1) Reprecipitation method

The scheme of reprecipitation method [3], which we found for a fabrication technique of organic nanocrystals is shown in Fig. 42.2. The reprecipitation method is a pure chemical technique to prepare organic nanocrystals simply by injecting solution of the target compound into a poor solvent. As distilled water is usually used as poor solvent, water-soluble solvent such as alcohol, acetone or THF was selected as the injected solution. The characteristics of reprecipitation method are that it is simple, quick and inexpensive. In addition, nanocrystallizations using this method are applicable for many kinds of organic compounds. In case of water-soluble organic dyes, the above-described reprecipitation method cannot be applied. In such a case, the reprecipitation method was improved like the following two ways: the reprecipitation of dissolved salts into saturated aqueous solution instead of pure water, i.e., salting-out method [5], and the reprecipitation into a hydrocarbon solvent such as hexane, decalin, and so on, i.e., inverse-reprecipitation method [6]. In the reprecipitation method, the size and morphology of nanocrystals could be controlled by changing experimental condition such as combination between poor and good solvents, concentration of

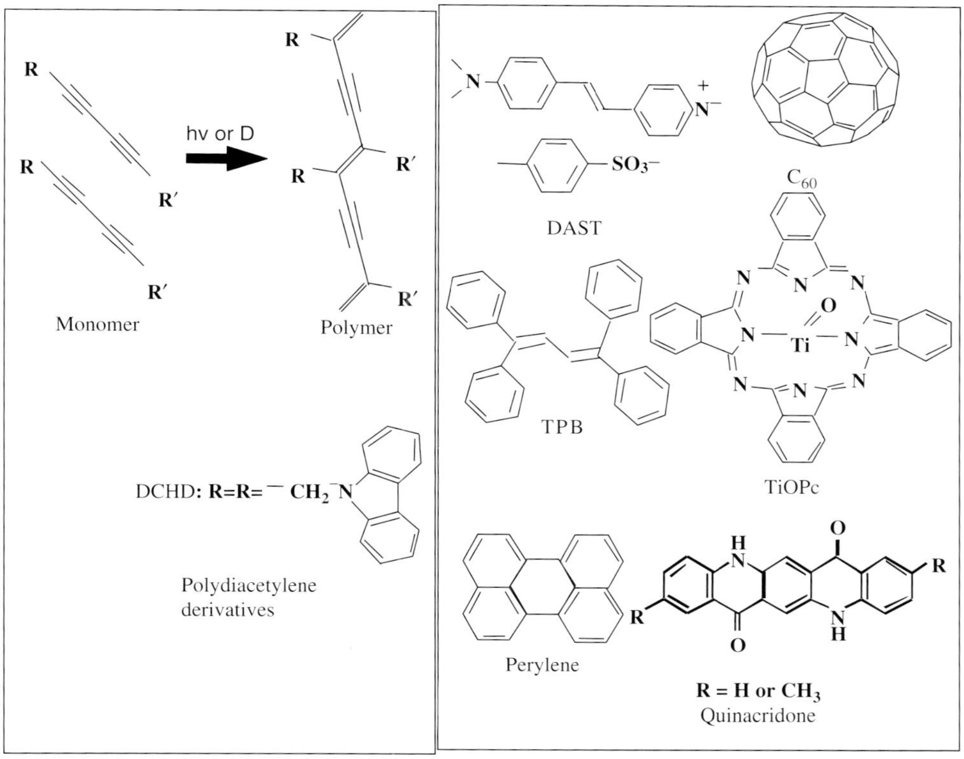

Figure 42.1
Structural formula of compounds tried for nanocrystallization.

injected solution, temperature of poor solvent and so on. And then, the nanocrystals could be finally obtained in the poor solvent in dispersed state.

One typical example of our reprecipitation method for preparing DCHD (see Fig. 42.1) nanocrystals is described as follow. A 200 μL of acetone solution (0.1–10 M) was injected into 10 mL of well-stirred distilled water at room temperature. It was found that, just after injection treatment, amorphous DCHD nanoparticles were produced in water by SEM observation. At that time, size of the amorphous particles was clearly reduced by injecting solution of DCHD with lower concentration. And then, nanocrystallization of DCHD nanoparticles while maintaining the size was done by keeping at room temperature for more than 20 min. As above, DCHD nanocrystals with minute size were finally obtained [7]. When heat treatment was given to the amorphous nanoparticles dispersed in water, crystal size of the nanocrystals tended to increase with the progress of the crystallization, and the nanofibers were finally fabricated [8, 9]. It was also recognized, in the case of diacetylene compounds, that a light milk-white dispersion of monomer nanocrystals was changed to a blue dispersion of polydiacetylene nanocrystals by irradiating UV light or γ-ray, as shown in Fig. 42.2. In summary, it was found to produce poly-DCHD nanocrystals with controlled size, and could be prepared by controlling both concentration of solution and temperature of poor solvent as shown in Fig. 42.3.

(2) Microwave irradiation method

An example that the amorphous particles were produced just after reprecipitation treatment is introduced in Chapter 3.1. In some compounds, however, the supersaturated state of keeping the cluster or monomer was formed. In this state, it took a long time to finish nanocrystallization, and it was hard to fabricate the nanocrystals with less than 100 nm in size, because the crystal grows due to the adhesion among clusters, monomer or nanocrystals even at the room temperature. In order to apply for the above conditions, the microwave irradiation method was developed [13] where microwave was irradiated to the dispersion in supersaturated state just after reprecipitation treatment.

When the usual reprecipitation method was applied for TPB or perylene (see Fig. 42.1), it was difficult to obtain fine crystals less than 1 μm in size and the

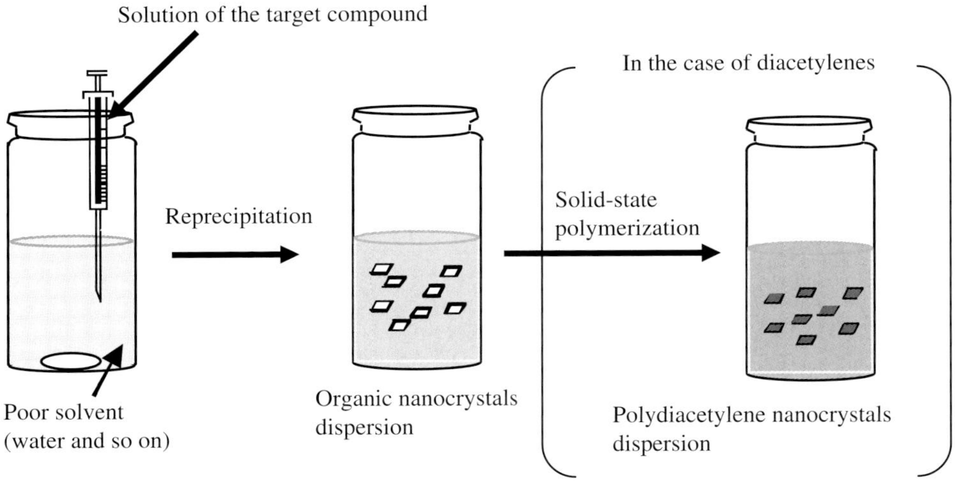

Figure 42.2
Scheme of reprecipitation method.

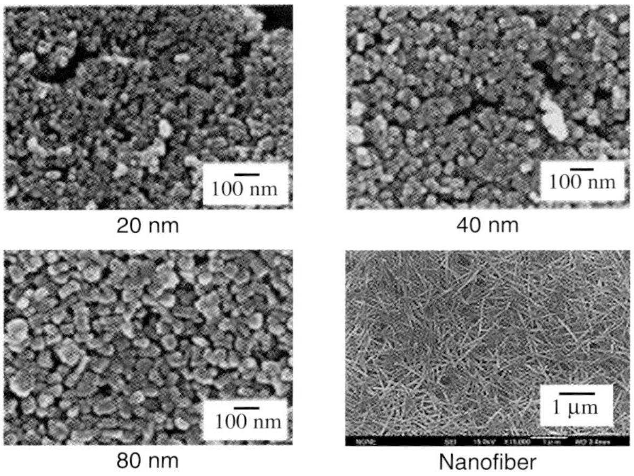

Figure 42.3
SEM photographs of size-controlled poly-DCHD nanocrystals.

final size distribution tended to spread, because the liquid obtained after injection became supersaturated [10]. On the other hand, when the microwave method was used for TPB or perylene by irradiating microwave into the supersaturated liquid in around 1 min, nanocrystallization promptly progressed. As a result, it was recognized for both compounds, that the crystal size was almost monodispersed about 100 nm in average, as shown in Fig. 42.4 [11, 12]. Here, the microwave irradiation was effective for rapid evaporation of acetone out of the system because of the heating and then the supersaturation, just after the reprecipitation treatment broke quickly. Concomitantly, the nanocrystallization quickly occurred. Finally, it was considered that the nanocrystals were obtained with a narrow size distribution.

(3) Fabrication of pigment nanocrystals by supercritical fluid crystallization method

It was also difficult to apply the reprecipitation for hardly soluble organic compounds such as pigments and so on. Generally, it is well known that the solubility of materials can be widely changed by operating the temperature and pressure of solution. Especially, high solubility is attained at high temperature and high pressure such as supercritical fluid (SCF). And then, the nanocrystallization of hardly soluble compounds was attempted by the reprecipitation

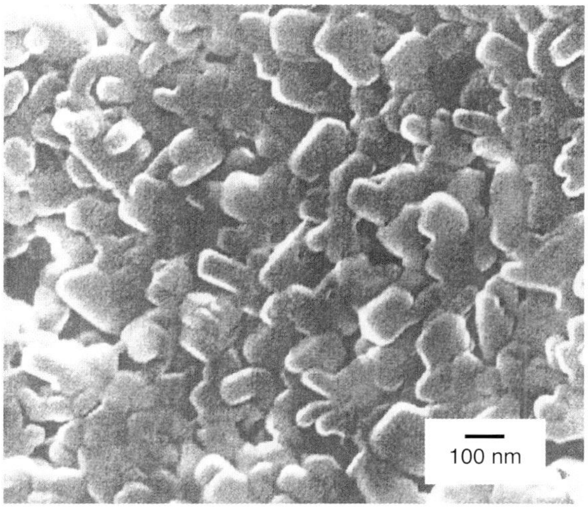

Figure 42.4
SEM photographs of TPB nanocrystals fabricated by microwave irradiation method.

from the solution of SCF as the advanced technique of conventional reprecipitation [13, 14]. That is, after the target compounds were dissolved in SCF, the nanocrystals were fabricated by the reprecipitation into poor solvent with cooling down. This technique is called SCF crystallization (SCFC) method.

As an example, the fabrication of TiOPc (see Fig. 42.1) nanocrystals using SCFC method is introduced. First, it is significant that the solubility of TiOPc to acetone at 670 K and 22 MPa was about 0.5 g/L, i.e., 0.9 mM, because the value improved more than six figures than that at room temperature. In the supercritical acetone system, crystal size and morphology of TiOPc nanocrystals could be controlled by setting suitable experimental conditions such as temperature of SCF (T_{SCF}) and the ratio of acetone in water for cooling solvent (R_A). When only acetone was used for the cooling solvent; i.e., $R_A = 1$, the size of TiOPc nanocrystals obtained was larger than 500 nm. On the other hand, as R_A became decreasing, the crystal size tended to become smaller. When T_{SCF} was set to be lower than 300 K, the nanocrystals with 30 nm in size were fabricated. Regarding the modification forms, R_A gave larger influence than T_{SCF}. When R_A was 1, the nanocrystals of stable β type were produced. And when the ratio of water to acetone for cooling solvent was increased, the nanocrystals of metastable α type appeared. In summary, size and morphology of TiOPc nanocrystals could be controlled by adjusting R_A and T_{SCF} in SCFC method [14]. Moreover, it was recognized that SCFC method was successfully applicable for other hardly soluble compounds such as quinaridone or C_{60} [15, 16].

3. Size-dependence of optical properties for organic nanocrystals

Fig. 42.5 indicates the visible absorption spectra for poly-DCHD nanocrystals having different sizes, which were prepared by the already-described reprecipitation method. It was found that the maximum absorption peak positions around 650 nm, being assigned to excitonic absorption, were shifted to high-energy region with a decreasing in size. The relationship between absorption peak positions and sizes is shown in Fig. 42.6, and the shift trends to become remarkable below 200 nm in size [8, 17]. Although, such tendency would be similar to quantum confinement effects observed in metal and/or semiconductor nanoparticles less than 10 nm in size, this experimental result seems to be due to a peculiar size effect in π-conjugated organic and polymer nanocrystals, since the shift phenomenon would appear even in the range of about more than ten times greater size.

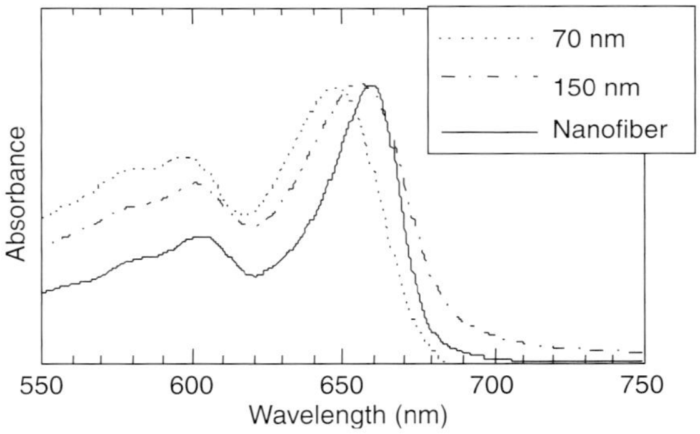

Figure 42.5
Absorption spectra of aqueous dispersion of poly-DCHD nanocrystals having different sizes.

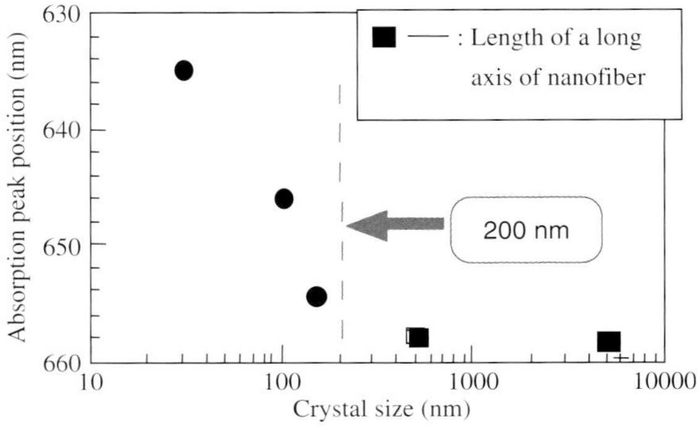

Figure 42.6
Relationship between absorption peak positions and sizes of poly-DCHD crystals.

It is speculated that the decrease in number of π-conjugated polymer chains in one nanocrystal as well as the softening of crystal lattice is responsible for the origins of the above-mentioned shift on the basis of so-called single nanocrystal spectroscopy and the visible absorption spectral changes measured at low temperature [17]. The similar behavior was also observed in organic nanocrystals of perylene, C_{60}, etc. [16, 18]. For example, the absorption peak position is 350 nm in C_{60} nanocrystals, fabricated by the conventional reprecipitation method, with ca. 270 nm in size. On the other hand, C_{60} nanocrystals (size: ca. 40 nm) provides the absorption peak around 340 nm, when produced by the supercritical reprecipitation method.

This absorption peak arises from CT-excitonic absorption band in C_{60} nanocrystals, and this phenomenon is also said to be the characteristic size effect in organic nanocrystals. A similar tendency was confirmed in the emission spectra of organic nanocrystals (for details, see references).

4. Orientation control of dispersed organic nanocrystals by external field

One of the most characteristic points in the reprecipitation method is that organic nanocrystals are dispersed stably and are freestanding in a dispersion medium. This means that the dispersion system of organic nanocrystals has both properties of liquid and crystals, that is to say, "liquid and crystals" system. If one could prepare organic nanocrystals having huge dipole moment and control their orientation by applying electric field, the optical transmittance and refractive index of the dispersion liquid would be changeable conveniently.

DAST nanocrystals (see Fig. 42.1) were employed in the demonstration. First, DAST nanocrystals (size: ca. 400 nm) was dispersed in low dielectric medium, and the electric field response was investigated by the measurement with visible absorption spectrum [5].

Fig. 42.7 exhibits the absorbance changes at maximum absorption peak (λ_{max} = 550 nm) with and without applying 300 V/m of electric field. The (a) and (b) in this figure represent the perpendicular and parallel configurations between applied electric field and probe light for measurement of absorbance, respectively. The absorbance increased or decreased by turning-on or -off of the electric field, being similar to the so-called liquid crystal system. In addition, the absorbance was inversely changed in perpendicular and parallel configurations, which suggests that the orientation of DAST nanocrystals was induced obviously by the applied electric field.

Fig. 42.8 shows the dependence of absorbance changes on applied electric field. It has become apparent that the maximum absorbance change was approximately 75% even by applying 500 V/cm of low electric field, in comparison with conventional liquid crystal system [19, 20]. The application for display is restricted to some extent, because the response time is typically of order of 200 ms. The "liquid and crystals" system, however, has a high potential of another type in novel display devices.

Recently, the oriented DAST nanocrystals induced by magnetic field were successfully fixed by using photocuring polymer as a dispersion medium [21]. This fact means that organic nanocrystals are aligned to any direction in three-dimensional space, which derives possibly the fabrications of highly functional crystal materials.

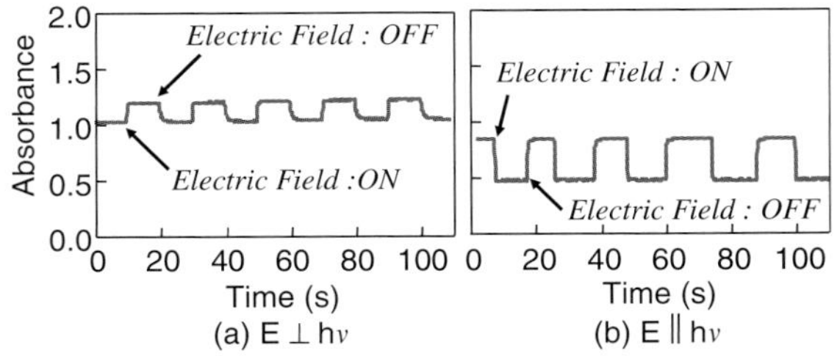

Figure 42.7
Absorbance changes at maximum absorption peak with and without applying 300 V/m of electric field. (The (a) and (b) in the figure represents the perpendicular and parallel configurations between applied electric field and probe light for measurement of absorbance, respectively.)

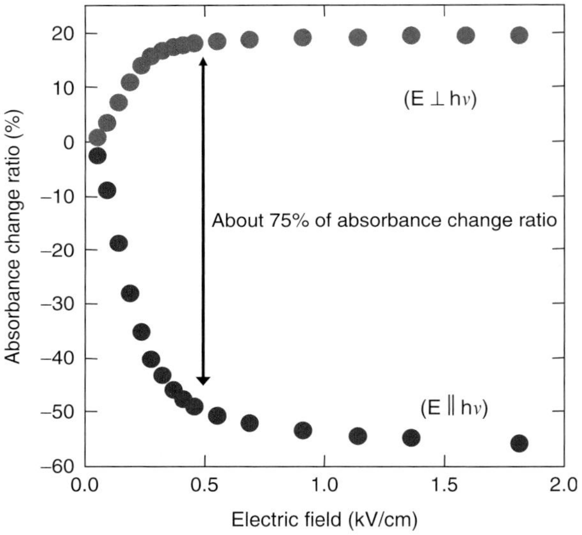

Figure 42.8
Dependence of absorbance changes of DAST nanocrystals dispersion on applied electric field.

We introduced size-controlled nanocrystals for some kinds of organic materials, which are fabricated mainly by the reprecipitation method, and the size-dependence of optical properties and "liquid and crystals" system were discussed in detail. Recently, the reprecipitation method is further being developed by utilizing chemical engineering techniques, and it would be possible stepwise to prepare mono-dispersed and mass-produced organic nanocrystals. In the near future, well-defined organic nanocrystals fabricated by the reprecipitation method are expected to be applied as practically industrial products in the fields of pigments, ink, organic photoconductor materials, medical supplies, agricultural chemicals, and so on.

References

[1] H. Toyotama: *Kinozairyo* (in Japanese), **6**, 44 (1987).
[2] K. Yase, T. Inoue, M. Okada, T. Funada and J. Hirano: *Hyomen Kagaku* (in Japanese), **8**, 434 (1987).
[3] H. Kasai, H. S. Nalwa, H. Oikawa, S. Okada, H. Matsuda, N. Minami, A. Mukoh and H. Nakanishi: *Jpn. J. Appl. Phys.*, **31**, L1132 (1992).
[4] G. Wegner: *Z. Naturforsch*, **24b**, 824 (1969).
[5] S. Fujita, H. Kasai, S. Okada, H. Oikawa, T. Oikawa, T. Fukuda, H. Matsuda, S.K. Tripathy and H. Nakanishi: *Jpn. J. Appl. Phys.*, **38**, L659 (1999).
[6] H. Kamatani, H. Kasai, S. Okada, H. Matsuda, N. Minami, A. Kakuta, K. Ono, A. Mukoh and H. Nakanishi: *Mol. Cryst. Liq. Cryst.*, **252**, 233 (1994).
[7] H. Katagi, H. Kasai, S. Okada, H. Oikawa, K. Komatsu, H. Matsuda, Z. Liu, and H. Nakanishi: *Jpn. J. Appl. Phys.*, **35**, L1364 (1996).
[8] H. Katagi, H. Kasai, S. Okada, H. Oikawa, H. Matsuda and H. Nakanishi: *J. Macromol. Sci.-Pure Appl. Chem.*, **A34**, 2013 (1997).
[9] T. Oshikiri, H. Kasai, H. Katagi, S. Okada, H. Oikawa and H. Nakanishi: *Mol. Cryst. Liq. Cryst.*, **337**, 25 (1999).
[10] H. Kasai, H. Oikawa, S. Okada and H. Nakanishi: *Bull. Chem. Soc. Jpn.*, **71**, 2597–2601 (1998).
[11] K. Baba, H. Kasai, S. Okada, H. Oikawa and H. Nakanishi: *Jpn. J. Appl. Phys.*, **39**, L1256 (2000).
[12] K. Baba, H. Kasai, S. Okada, H. Oikawa and H. Nakanishi: *Opt. Mater.*, **21**, 591 (2002).
[13] Y. Komai, H. Kasai, H. Hirakoso, Y. Hakuta, H. Katagi, S. Okada, H. Oikawa, T. Adschiri, H. Inomata, K. Arai and H. Nakanishi: *Jpn. J. Appl. Phys.*, **38**, L81 (1999).
[14] Y. Komai, H. Kasai, H. Hirakoso, Y. Hakuta, S. Okada, H. Oikawa, T. Adschiri, H. Inomata, K. Arai and H. Nakanishi: *Mol. Cryst. Liq. Cryst.*, **322**, 167 (1998).
[15] B. Li, H. Kasai, H. Oikawa, S. Okada, Arai and H. Nakanishi: *J. Nanosci. Nanotechnol.*, **3**, 365 (2003).
[16] H. Kasai, S. Okazaki, T. Hanada, S. Okada, H. Oikawa, T. Adschiri, K. Arai, K. Yase, and H. Nakanishi: *Chem. Lett.*, 1392 (2000).
[17] V.V. Volkov Victor, T. Asahi, H. Masuhara, H. Kasai, H. Oikawa and H. Nakanishi: *J. Phys. Chem. B*, **108**, 7674 (2004).
[18] H. Kasai, H. Kamatani, S. Okada, H. Oikawa, H. Matsuda and H. Nakanishi: *Jpn. J. Appl. Phys.*, **35**, L221 (1996).
[19] T. Onodera, H. Kasai, H. Oikawa, and H. Nakanishi: *Ekisho* (in Japanese), **6**, 131 (2003).
[20] T. Onodera, M. Yoshida, S. Okazoe, S. Fujita, H. Kasai, S. Okada, H. Oikawa and H. Nakanishi: *Inter. J. Nanosci.*, **1**, 737 (2002).
[21] Y. Kaneko, S. Shimada, T. Fukuda, T. Kimura, H. Yokoi, H. Matsuda, T. Onodera, H. Kasai, S. Okada, H. Oikawa, and H. Nakanishi: *Adv. Mater.*, **17**, 160 (2005).

APPLICATION 43

43 BIO-IMAGING WITH QUANTUM DOTS

1. Developments of quantum dots

Today, extremely rapid progress in developments of nanosized quantum dots is being made. Until recently, '1 μm' is one of the stumbling blocks in the manufacture and engineering technology of small particles. In the course of overcoming the difficulty, new technologies have emerged to create some new materials, such as functional materials, micro-machines and macro-reactors.

However, recent demands for stronger materials, and for new functional materials that absorb electromagnetic waves call for new manufacturing and engineering technologies. The technologies must deal with materials much smaller than those in conventional technologies. With conventional technologies, the manufacture of materials whose sizes are a little less than 1 μm (sub-micron) is possible, but completely new technologies are necessary for particles smaller than such a range.

In other fields, new ways of calculation of orbitals and the dynamic calculation have made it possible to determine structures of lower molecular compounds. Thus, prediction of their absorbent wavelength and fluorescence wavelength has become possible. All these advances have made it possible to design new medicines with new functions.

Further, with electric microscopes and nucleus microscopes, it has become possible to observe smaller particles. By crystal analysis, molecular structures not only of metals and non-organic materials, but also of proteins and nucleic acids have become apparent. With data obtained through these new technologies, calculations of higher molecular materials and metal crystals (more than 100) are now in progress in higher power computers.

However, the calculation of the energy of super-small metal crystal consisting of 500 nuclei, whose surface is conjugated with organic materials, is difficult with the present technology because the calculation is far beyond its capacity (the dimension is sometimes lowered to avoid the problem).

Thus, manufacturing technologies dealing with nanosized materials are unknown to us, and therefore, they are treasure boxes from which innovative and more valuable treasures are expected to be taken out.

The assembly of 1 digit nanosized particles, where each of them reacts to one another and as a whole they have a particular function, is called a 'supermolecule'. The 'nanotechnology' can be defined as one, which deals with the design, manufacture and application of supermolecules.

Quantum dots are extremely small metal crystals consisting of a hundred to a few hundred nuclei. The behaviors of small metal particles are very different from those of ordinary-sized particles. In the 1960s, Ryogo Kubo theoretically postulated the quantum dot effect, which shows quite peculiar behaviors. In the 1990s, a group in the Department of Metalology at Massachusetts Institute of Technology succeeded in manufacturing quantum dots and the 'Kubo theory' was verified.

At first, quantum dots were manufactured with semiconductors in metals belonging to the second family or to the sixth family, such as cadmium or serene. The group used tri-*n*-octyl phosphine oxide as surfactants, and in their micelle, quantum dots were manufactured via 'self-assemble'. The reaction is carried out at 350°C, therefore this new technology is called the 'Hot Soap Method'.

Next, after the use of the second and the sixth families, a method of manufacturing of quantum dots with metals belonging to the third and the fifth families was developed. In this method, nitrogen or phosphorus is used instead of serene or cadmium, which makes the quantum dots all the more safe. Now, with metals belonging to the fourth family, such as silicon or titan, quantum dots are manufactured.

Quantum dots have fluorescence and a particular surface conjugation makes the fluorescence stronger. The fluorescence has longer photostability. To make these properties longer and stronger, various kinds of surface-conjugating technologies have been developed. For example, Silica coating makes quantum dots more stable. ZnS coating makes the fluorescence stronger. Polymer-conjugated ones obtain a hydrophilic property.

2. Development of bio-imaging

When we try to understand something, visible information is helpful. The visible sense is the most important of all in terms of cerebral physiology. Visible information can be understood much faster than information obtained through the other sensory organs.

Therefore, when we try to understand a system with complicated information, we often transfer it to visible information for easier understanding. The visible information thus obtained is called 'image' and the process of changing complicated information into visible information is 'imaging'. Especially, when the object matter is concerned with life, the process is called 'bio-imaging'.

The bio-imaging, as a category, is thought to have existed long ago. Pictures on the walls in a cave might have been drawn with the intention of informing whether a particular food was edible or an animal was dangerous.

Thus, the human race has been progressing with bio-imaging. Along the way, inventions of new imaging instruments brought about grave social changes. One example is the invention of the microscope. A technology of polishing lenses led to the development of the microscope, which eventually led to a hypothesis that all plants consist of cells. Afterward, the hypothesis was extended to animals.

The human race came to think that getting sick was derived from the deviation of cells from the normal state, causing the malfunctioning of the body as a whole. The invention of the microscope brought about revolutionary upward changes in various fields, not only in botany, zoology and medical sciences, but also in pharmacy, agriculture and sociology.

Even in recent years, the development of the electric microscope made it possible to 'image' a whole particle of virus and to verify the hypothesis that the gene of the virus is circular. Now, various kinds of optical technologies are used to clarify a lot of life phenomena: electric microscopes, confocal laser microscopes, total reflection microscopes, soft X-ray microscopes, nuclear force microscopes and electron tunnel microscopes, to name only a few.

3. Bio-imaging and quantum dots

One of the most crucial phenomena in life is the behavior of molecules. It will bring about tremendous results to find out how all molecules, supermolecules

Minute 0 5 30

Quantum dots

FITC

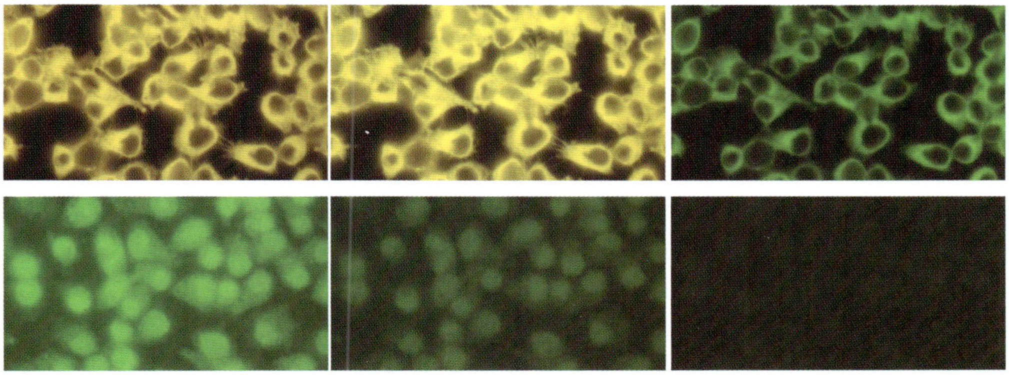

Figure 43.1
Characteristics of quantum dots: higher brightness and photostability.

and cells in the body behave. One of the objectives of the present bio-imaging is to pursue their behavior in the body. A method for it is with quantum dots. If objective molecules (including supermolecules and cells) can be marked with quantum dots, they can be traced during the course of time. Quantum dots are superior to conventional fluorescence dyes like FITC in their longer photostability.

Fig. 43.1 shows the results of experiments with Vero cells from the kidney of the African Green Monkey. FITC and quantum dots were introduced into the cells and their photostability was recorded. The photostability of FITC is weaker and tracing it for more than 5 min is difficult. On the other hand, the quantum dots were traced for more than 30 min.

These results mean that quantum dots can be traced much longer than conventional organic dyes. In terms of photostability, quantum dots are far superior. Cd/Se (belonging to the second and the sixth family) quantum dots surface-conjugated with ZnS are known to have much stronger photostability.

It is also known that quantum dots belonging to the third and the fifth family are about 30 times brighter than non-conjugated ones if their surface is conjugated with ZnS. It can be predicted that quantum dots will be much brighter if their surface is conjugated with metals belonging to the fourth family, which is much safer.

In order to make applications of quantum dots with the properties stated above, they have to be properly surface-conjugated. For example, Cd/Se quantum dots are hydrophobic because they are a cluster manufactured in high-temperatured micelle (tri-*n*-octyl phosphine oxide). In order to make them hydrophilic, the three materials in the micelle have to be replaced with hydrophilic molecules. For that purpose, various molecules, with suitable technologies, have been suggested.

The following surface-conjugate molecules have been suggested so far: organic acids, amine and amino acids (lower molecules), sugar, peptides, polyethylene glycol and chemically synthesized polymers. The distribution and behaviors of the quantum dots conjugated hydrophilically with one of the materials and can be traced and observed in the body. Further, dyed with certain quantum dots, the distribution and behaviors of cells in a particular organ and in the whole body can be observed. Here are some examples:

4. Quantum dots label for the antibody

The strong intensity of the fluorescent makes it possible to detect immunological reaction by the small amount of the antibody by using quantum dots labeling.

A small amount of the antigen can also be detected by the quantum dots-labeled antibody, even in the single molecular label. Fig. 43.2 shows the fluorescent microscopic image of the tubulin in the Vero cell. The Vero cell was first treated with anti-tubulin mouse IgG antibody and then stained with anti-mouse IgG conjugated with FITC, which is shown in Fig. 43.2, panel (b). Panel (a) shows that the same sample was stained with anti-mouse IgG conjugated with quantum dots as the secondary reaction. The microscopic image of the quantum dots labeling is quit clear and makes it possible to continue the observation for a much longer time.

5. In vivo imaging of the quantum dots stained cell: the localization in organs

The quantum dots-labeled cell can be introduced into the body and observe the localization in the organs.

Fig. 43.3 shows the result. Black mouse-lymph cell (EL4) was stained with quantum dots and introduced in the nude mouse through the tail vein. After a while, a thin-sliced sample extracted from the organs can be studied with a fluorescent microscopy to obtain the information of the localization of the labeled cell in them. In Fig. 43.3 [1], the labeled-cell image can be detected in Spleen, Lung, Kidney and Liver.

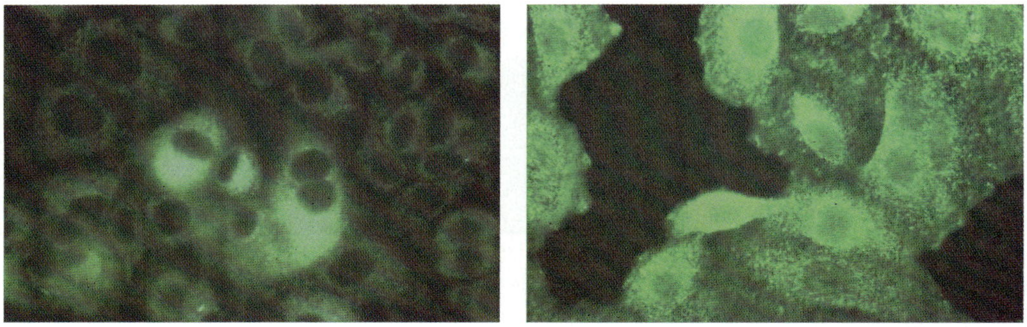

(a) Anti-tublin mouse IgG and FITC-labeled anti-mouse IgG antibody

(b) Anti-tublin mouse IgG and QD-labeled anti-mouse IgG antibody

Figure 43.2
Stain of cells with antigens tagged with quantum dots (Vero cells).

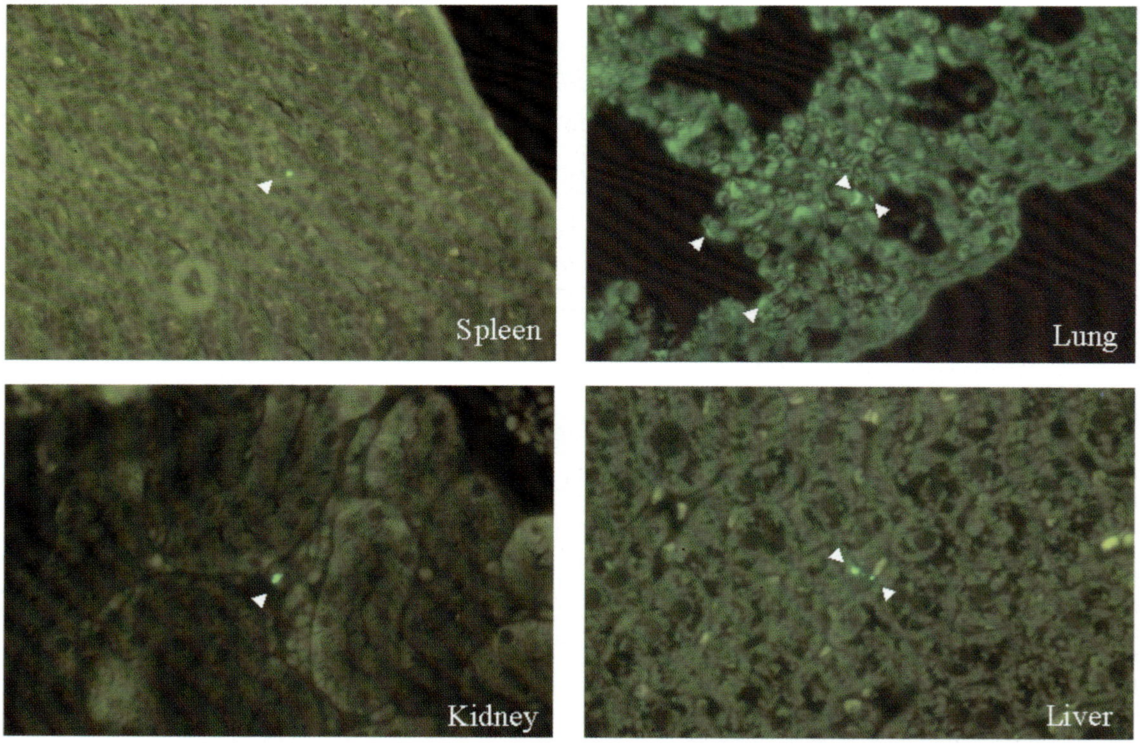

Figure 43.3
Behaviors of EL4 cells stained with quantum dots in the body of mice.

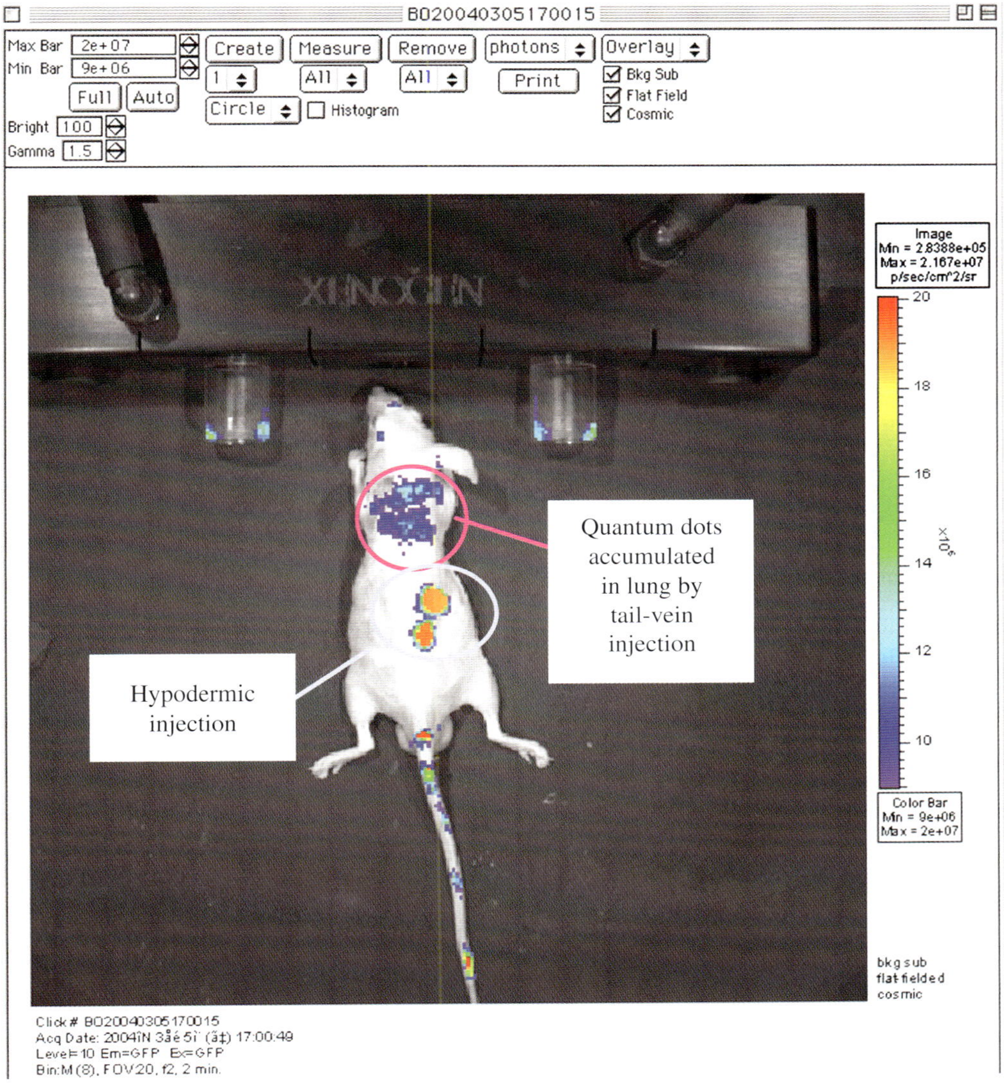

Figure 43.4
Detection of the fluorescence from the inside of the body of a mouse (courtesy of Dr. Takahiro Ochiya of the National Cancer Center Research Institute, Tokyo, Japan).

6. Observation of the localization from outside of the body

Quantum dots can be conjugated with protein, nucleic acids and other bio-molecule including the drug. We can make the excitation for the quantum dots conjugated with the molecules from outside the body and observe the emission light. Fig. 43.4 shows the localization of the water-soluble colloidal quantum dots injected in the mouse tail vein. We observe the fluorescent light from the quantum dots accumulated in the lungs. Quantum dots light up the pathway to study the mobility of the bio-molecules and the stained cell inside the body.

References

[1] A. Hoshino, K. Hanaki, K. Suzuki and K. Yamamoto: *Biochem. Biophys.l Res. Commun.*, **314**, 46–53 (2004).

APPLICATION 44

44 APPLICATION OF QUANTUM DOTS FOR BIO-MEDICAL ENGINEERING

The accuracy and the real-time process are expected for diagnosis these days. For example, the conventional diagnosis method for tuberculosis depends on the result of the culture test for the infectious agents if the microscopic study is negative for the patient's sputum. Since this culture test takes one to two months to produce results, antibiotics are prescribed for tuberculosis for the prevention of another infection based on the symptom and the X-ray study. The development of polymerase chain reaction has made the diagnosis more accurate and the results are obtained within several hours.

The development of polymerase tests expanded various medical fields such as biochemistry, immunology, cell biology, microbiology, radiology, NMR studies, mass spectrometry, ultrasound scanning, molecular diagnosis, respiratory physiology, endoscopy, catheter studies and so on. Intensive efforts for development have been made in all of these fields of study toward the same orientation, that is, less sample volume and shorter time for getting results. The environmental fields of these technologies also show amazing progress including the method for sampling of ultrasmall amount of blood.

In order to correspond to this inquiry, laboratory test is extensively developed by using optical engineering. The research and the development that have been done for this purpose are the ones using fluorescent biomarkers such as quantum dots. Quantum dots are the semiconductor nanoparticles made from materials such as CdSe and ZnS. They have strong fluorescent activity excited by laser. Though metals are widely used, other materials should be used for safety in bio-medical engineering.

Also, the field of photonics has developed rapidly, and this field has produced important useful devices such as a photo-multiple counter and a CCD camera. In the field of medical laboratory test, engineering technology has progressed significantly owing to the development of photonics. One example is the photoluminescence method for the immunological laboratory test. This method is superior to the conventional radioisotope method from the viewpoint of cost and convenience. The fluorescent probe plays an important role in the field of photonics used in biotechnology and medical engineering. Here, we focus on the quantum dots as biomarkers for their use as follows.

1. Application for laboratory test

(1) Latex beads

Fig. 44.1 shows antigen–antibody reaction with the sub-micron-sized latex beads. The surfaces of the beads are conjugated both with diphtheria-specific antibody and with the quantum dots. As the quantum dots do not show any inhibition for the antigen–antibody reaction, the beads treated with this antibody on the surface show the aggregation reaction soon after the diphtheria antigen is added in the solution. White dots on the surface of the beads after the reaction can be observed in the Fig. 44.1. Even though the amount of sample is very small, the change of the fluorescence activity makes it possible to detect the reaction.

(2) Detection of the antigen–antibody reaction by evanescent light

Evanescent light is available by the total reflection and the other method. With evanescent light, the single molecular study by using a microscope and the quantum dots as the probe is also useful for

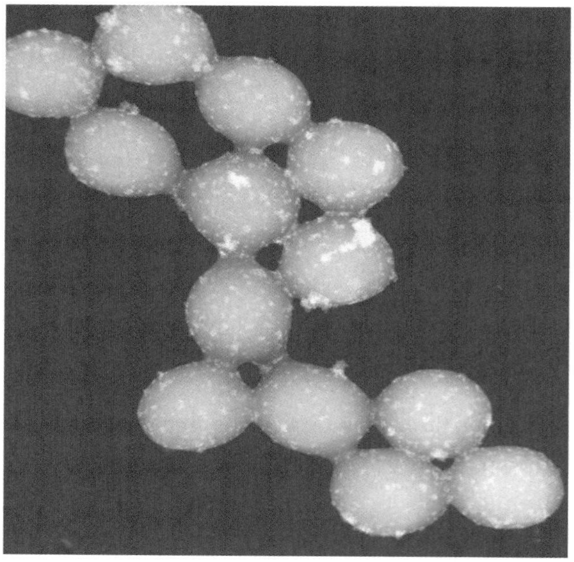

Figure 44.1
Antigen-antibody reaction with the latex beads.

the reduction of the total amount of the sample volume much more than the method described above. A multiple and simultaneous study has been attained for infectious diseases. The application, including the tumor marker of the allergy, will be expanded to other fields.

2. Diagnosis by imaging analysis

(1) Immunological stain and diagnosis

The disease-specific antibody tagged with quantum dots is useful for detection and diagnosis. Although the organic fluorescent dye is widely used for the conventional method, both the weak intensity of the fluorescent light and the photo–bleaching together make it hard to stain the target with organic dye-conjugated specific antibody.

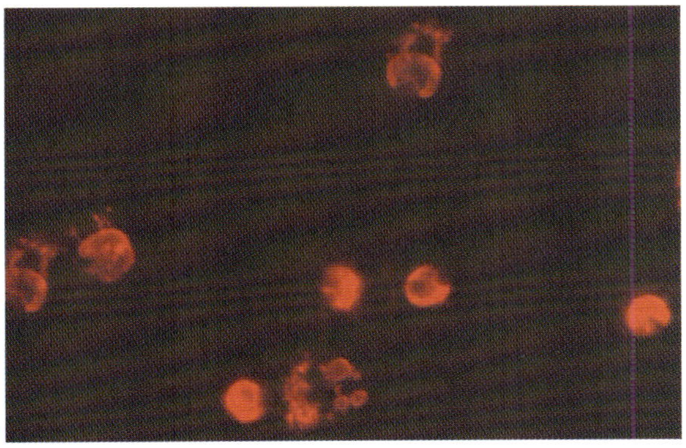

Figure 44.2
Observation of vasculitis-related molecule of neutrophil.

In a case where the number of target molecules is small, the result will be much worse. Quantum dots conjugated on antibody have made it possible to detect a small amount of signal because of the strong fluorescent activity. Fig. 44.2 shows the immunological stain of the vasculitis-related molecule with quantum dots conjugated on specific antibody. Immunological stain with quantum dots conjugated on specific antibody will be used widely for supplying useful information for diagnosis.

(2) Living cell stain and in vivo study

Quantum dots are also useful for staining a living cell without any loss of function. The stain stays for more than 10 days inside the mouse body. By the introduction of the quantum dots-stained cell into the living body, we can analyze the mobility of the specific cells inside the body. Fig. 44.3 shows quantum dots-stained strain cells, derived from African green monkey, 10 days after their introduction into the living nude mouse body. Even 10 days later, after the staining, we can recognize the fluorescent light by the stained cell. Quantum dots-conjugated pathogens and the specific tumor marker will make it possible to study the translocation and the chemotaxis.

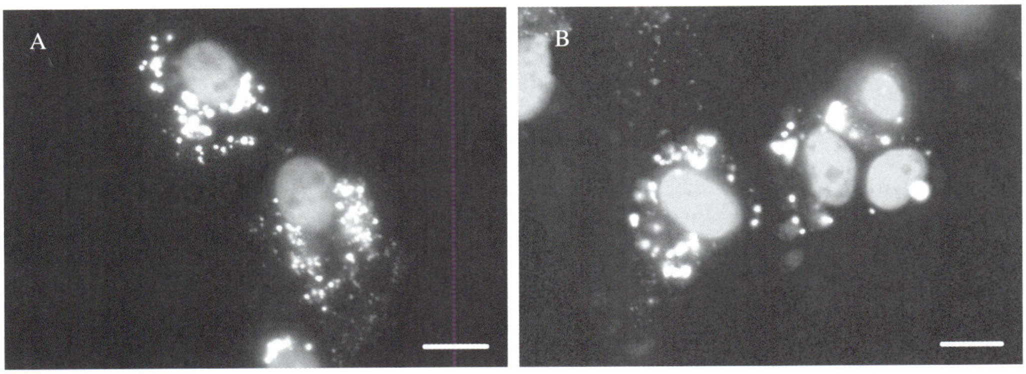

Figure 44.3
Cell stain with quantum dots in the animal body.

Subject Index

C

D

Q

R

S

T